Fourth Edition

Algebra: Introductory and Intermediate

D0022285

Richard N. Aufmann
Palomar College, California

Vernon C. Barker
Palomar College, California

Joanne S. Lockwood
New Hampshire Community Technical College

Houghton Mifflin Company
Boston New York

Vice President and Publisher: Jack Shira
Senior Sponsoring Editor: Lynn Cox
Associate Editor: Melissa Parkin
Assistant Editor: Noel Kamm
Senior Project Editor: Tamela Ambush
Editorial Assistant: Sage Anderson
Manufacturing Buyer: Florence Cadran
Executive Marketing Manager: Brenda Bravener-Greville
Senior Marketing Manager: Katherine Greig

Cover photo: © Karim Rashid

Photo Credits:
CHAPTER 1: p. 1, Gary Conner/PhotoEdit; **p. 3,** Tony Freeman/PhotoEdit; **p. 15,** Paula Bronstein/Getty Images; **p. 36,** AP/Wide World Photos; **p. 53,** Image Bank/Getty Images. **CHAPTER 2: p. 71,** Michael Newman/PhotoEdit; **p. 77,** ©Terres Du Sud/CORBIS; **p. 89,** Martin Fox/Index Stock Imagery; **p. 89,** Bill Aron/PhotoEdit; **p. 90,** ©Lawrence Manning/CORBIS; **p. 91,** Davis Barber/PhotoEdit; **p. 111,** ©Steve Prezant/CORBIS; **p. 121,** ©Renee Comet/PictureArts Corp/CORBIS; **p. 122,** Photodisc; **p. 135,** ©Richard Cummins/CORBIS; **p. 153,** Alan Oddie/PhotoEdit. **CHAPTER 3: p. 159,** ©Kevin Fleming/CORBIS; **p. 176,** ©2002 Photodisc; **p. 177,** ©CORBIS; **p. 190,** ©Topham/Syndicated Features Limited/The Image Works; **p. 192,** ©Roger Ressmeyer/CORBIS; **p. 192,** ©CORBIS; **p. 201,** ©Kevin R. Morris/CORBIS. **CHAPTER 4: p. 217,** Jeff Greenberg/PhotoEdit; **p. 223,** ©Craig Tuttle/CORBIS; **p. 229,** Ulrike Welsh, PhotoEdit; **p. 239,** Robert W. Ginn/PhotoEdit; **p. 260,** Eric Fowke/PhotoEdit; **p. 270,** ©CORBIS; **p. 271,** ©David Keaton/CORBIS; **p. 283,** The Granger Collection, New York. **CHAPTER 5: p. 295,** David R. Stoecklein/CORBIS; **p. 325,** Jose Carillo/PhotoEdit; **p. 326,** ©CORBIS; **p. 327,** Michael Newman/PhotoEdit; **p. 333,** AP/World Wide Photos; **p. 335,** Michael Newman/PhotoEdit; **p. 340,** Susan Van Etten/PhotoEdit. **CHAPTER 6: p. 343,** AP/World Wide Photos; **p. 355,** Stocktrek/CORBIS; **p. 356,** ©NASA/JPL Handout/Reuters/CORBIS; **p. 356,** John Neubauer/PhotoEdit; **p. 387,** ©Duomo/CORBIS; **p. 388,** ©Roger Ressmeyer/CORBIS; **p. 394,** Susan Van Etten/PhotoEdit. **CHAPTER 7: p. 399,** Digital Vision/Getty Images; **p. 439,** ©Pierre Ducharme/Reuters/CORBIS; **p. 439,** AP/Wide World Photos; **p. 441,** Bill Aron/PhotoEdit; **p. 446,** ©CORBIS. **CHAPTER 8: p. 451,** ©Jeff Henry/Peter Arnold, Inc.; **p. 478,** Clayton Sharrard/PhotoEdit; **p. 482,** ©CORBIS; **p. 482,** ©Alinari Archives/CORBIS; **p. 493,** Tom Carter/PhotoEdit; **p. 493,** David Young-Wolff/PhotoEdit; **p. 494,** ©Sheldan Collins/CORBIS; **p. 494,** ©Galen Rowell/CORBIS; **p. 495,** Billy E. Barnes/PhotoEdit; **p. 496,** ©Lee Cohen/CORBIS; **p. 499,** Tony Freeman/PhotoEdit; **p. 512,** ©CORBIS. **CHAPTER 9: p. 515,** Benjamin Shearn/TAXI/Getty Images; **p. 539,** ©Bettmann/CORBIS; **p. 541,** Sandor Szabo/EPA/Landov; **CHAPTER 10: p. 561,** David Young-Wolff/PhotoEdit; **p. 576,** Photex/CORBIS; **p. 595,** Bill Aron/PhotoEdit; **p. 596,** ©Reuters/CORBIS. **CHAPTER 11: p. 607,** Tim Boyle/Getty Images; **p. 608,** Lon C. Diehl/PhotoEdit; **p. 616,** ©Jim Craigmyle/CORBIS; **p. 623,** Rich Clarkson/Getty Images; **p. 624,** Vic Bider/PhotoEdit; **p. 627,** ©Nick Wheeler/CORBIS; **p. 633,** ©Jose Fuste Raga/CORBIS; **p. 637,** ©Joel W. Rogers/CORBIS; **p. 638,** Robert Brenner/PhotoEdit; **p. 649,** Michael Newman/PhotoEdit; **p. 658,** ©Kim Sayer/CORBIS; **CHAPTER 12: p. 659,** Rudi Von Briel/PhotoEdit; **p. 672,** The Granger Collection, New York; **p. 689,** Express Newspaper/Getty Images; **p. 690,** ©Richard T. Nowitz/CORBIS; **p. 690,** Courtesy of the Edgar Fahs Smith Image Collection/University of Pennsylvania Library, Philadelphia, PA 19104-6206; **p. 691,** ©Bettman/CORBIS; **p. 693,** Mark Harmel/Getty Images; **p. 694,** Myrleen Fergusun Cate//PhotoEdit; **p. 695,** Mike Johnson-www.earthwindow.com; **p. 696,** ©Roger Ressmeyer/CORBIS; **p. 699,** David Young-Wolff/PhotoEdit; **p. 704,** ©Macduff Everton/CORBIS; **p. 708,** Frank Siteman/PhotoEdit.

Copyright © 2007 by Houghton Mifflin Company. All rights reserved.

No part of this work may be reproduced or transmitted in any form or by any means, electronic or mechanical, including photocopying and recording, or by any information storage or retrieval system without the prior written permission of Houghton Mifflin Company unless such copying is expressly permitted by federal copyright law. Address inquiries to College Permissions, Houghton Mifflin Company, 222 Berkeley Street, Boston, MA 02116-3764.

Printed in the U.S.A.

Library of Congress Control Number: 2005936314

Instructor's Annotated Edition:
ISBN 13: 978-0-618-61131-7
ISBN 10: 0-618-61131-2

For orders, use student text ISBNs:
ISBN 13: 978-0-618-60953-6
ISBN 10: 0-618-60953-9

89-WC-10 09 08

Contents

Copyright © Houghton Mifflin Company. All rights reserved.

Copyright © Houghton Mifflin Company. All rights reserved.

Copyright © Houghton Mifflin Company. All rights reserved.

Copyright © Houghton Mifflin Company. All rights reserved.

Copyright © Houghton Mifflin Company. All rights reserved.

Copyright © Houghton Mifflin Company. All rights reserved.

Copyright © Houghton Mifflin Company. All rights reserved.

Preface

The fourth edition of *Algebra: Introductory and Intermediate* examines the fundamental ideas of algebra. Recognizing that the basic principles of geometry are a necessary part of mathematics, we have also included a separate chapter on geometry (Chapter 3) and have integrated geometry topics, where appropriate, throughout the text. The text has been designed not only to meet the needs of the traditional college student, but also to serve the needs of returning students whose mathematical proficiency may have declined during years away from formal education.

In this new edition of *Algebra: Introductory and Intermediate,* we have continued to integrate some of the approaches suggested by AMATYC. Each chapter opens with a photo and a reference to a mathematical application within the chapter. At the end of each section there are "Applying the Concepts" exercises, which include writing, synthesis, critical thinking, and challenge problems. At the end of each chapter there is a "Focus on Problem Solving," which introduces students to various problem-solving strategies. This is followed by "Projects and Group Activities," which can be used for cooperative-learning activities.

NEW! Changes to This Edition

We have found that students who are taught division of a polynomial by a monomial as a separate topic are subsequently more successful in factoring a monomial from a polynomial. Therefore, we have added a new objective, "To divide a polynomial by a monomial," to Section 4 of Chapter 6.

In Section 11.3, the material on composition of functions has been expanded, and students are given more opportunities to apply the concept to applications.

In the previous edition, complex numbers were presented in Section 9.3. In this edition, complex numbers have been moved to the last section of the chapter. This provides for a better flow of the material in Chapter 9 and places complex numbers immediately before Chapter 10, "Quadratic Equations," where it is used extensively.

In Section 2 of Chapter 12, the introduction to logarithms has been rewritten. Motivating students to understand the need for logarithms is developed within the context of an application. This topic is presented at a slower pace to help students better understand and apply the concept of logarithm.

The in-text examples are now highlighted by a prominent HOW TO bar. Students looking for a worked-out example can easily locate one of these problems.

As another aid for students, more annotations have been added to the Examples provided in the paired Example/You Try It boxes. This will assist students in understanding what is happening in key steps of the solution to an exercise.

Throughout the text, data problems have been updated to reflect current data and trends. Also, titles have been added to the application exercises in the exercise sets. These changes emphasize the relevance of mathematics and the variety of problems in real life that require mathematical analysis.

The Chapter Summaries have been remodeled and expanded. Students are provided with definitions, rules, and procedures, along with examples of each. An objective reference and a page reference accompany each entry. We are confident that these will be valuable aids as students review material and study for exams.

Copyright © Houghton Mifflin Company. All rights reserved.

In many chapters, the number of exercises in the Chapter Review Exercises has been increased. This will provide students with more practice on the concepts presented in the chapter.

The calculator appendix has been expanded to include instruction on more functions of the graphing calculator. Notes entitled Integrating Technology appear throughout the book and many refer the student to this appendix. Annotated illustrations of both a scientific calculator and a graphing calculator appear on the inside back cover of this text.

Feedback from users of the third edition informed us that the material on Cramer's Rule and the material on conic sections were not covered in the majority of classes. Those who did present these topics presented only abbreviated coverage. Consequently, these topics are not included in the textbook in this fourth edition, which, of course, reduces the size of the text and lowers the cost to the students. However, material on Cramer's Rule and conic sections is available online to instructors and students who use the text. Please see the Table of Contents, Sections 5.5 and 11.5.

Copyright © Houghton Mifflin Company. All rights reserved.

chapter 8 Rational Expressions

In order to monitor species that are or are becoming endangered, scientists need to determine the present population of that species. Scientists catch and tag a certain number of the animals and then release them. Later, a group of the animals from that same habitat is caught and the number tagged is counted. A proportion is used to estimate the total population size in that region, as shown in **Exercise 27 on page 482.** Tracking the tagged animals also assists scientists in learning more about the habits of that species.

OBJECTIVES

Section 8.1
A To simplify a rational expression
B To multiply rational expressions
C To divide rational expressions

Section 8.2
A To rewrite rational expressions in terms of a common denominator
B To add or subtract rational expressions

Section 8.3
A To simplify a complex fraction

Section 8.4
A To solve an equation containing fractions

Section 8.5
A To solve a proportion
B To solve application problems
C To solve problems involving similar triangles

Section 8.6
A To solve a literal equation for one of the variables

Section 8.7
A To solve work problems
B To use rational expressions to solve uniform motion problems

Section 8.8
A To solve variation problems

Need help? For online student resources, such as section quizzes, visit this textbook's website at **math.college.hmco.com/students.**

Copyright © Houghton Mifflin Company. All rights reserved.

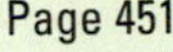

Page 451

Chapter Opening Features

NEW! Chapter Opener

New, motivating chapter opener photos and captions have been added, illustrating and referencing a specific application from the chapter.

The at the bottom of the page lets students know of additional online resources at math.college.hmco.com/students.

Objective-Specific Approach

Each chapter begins with a list of learning objectives that form the framework for a complete learning system. The objectives are woven throughout the text (i.e., Exercises, Prep Tests, Chapter Review Exercises, Chapter Tests, Cumulative Review Exercises) as well as through the print and multimedia ancillaries. This results in a seamless learning system delivered in one consistent voice.

Prep Test and Go Figure

Prep Tests occur at the beginning of each chapter and test students on previously covered concepts that are required in the coming chapter. Answers are provided in the Answer Section. Objective references are also provided if a student needs to review specific concepts.

The **Go Figure** problem that follows the *Prep Test* is a playful puzzle problem designed to engage students in problem solving.

Page 608

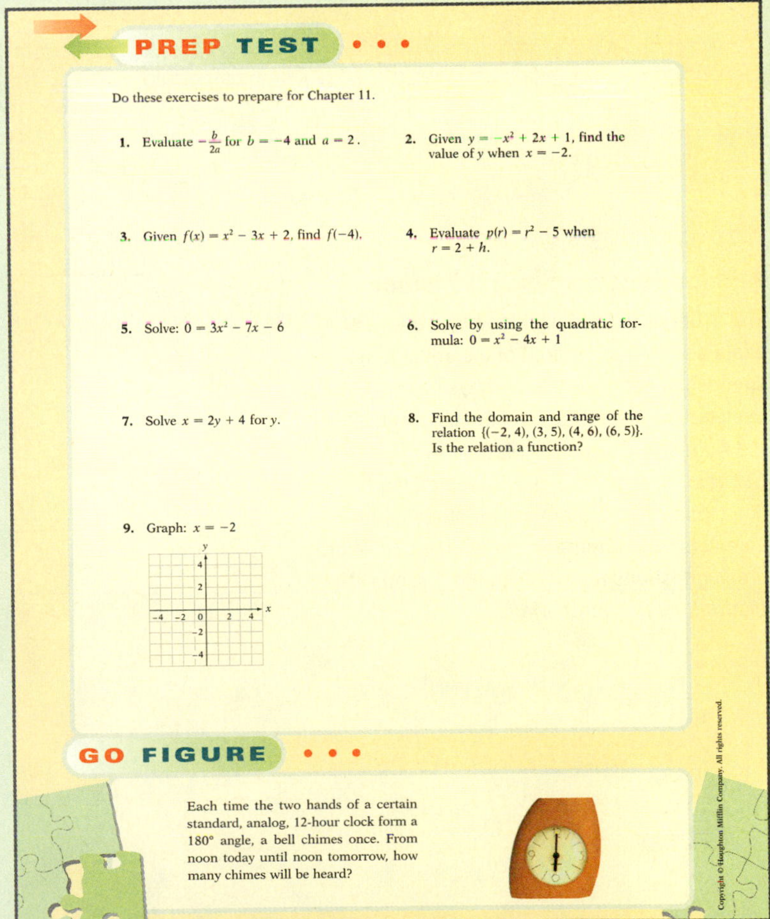

PREP TEST

Do these exercises to prepare for Chapter 11.

1. Evaluate $-\frac{b}{2a}$ for $b = -4$ and $a = 2$.
2. Given $y = -x^2 + 2x + 1$, find the value of y when $x = -2$.
3. Given $f(x) = x^2 - 3x + 2$, find $f(-4)$.
4. Evaluate $p(r) = r^2 - 5$ when $r = 2 + h$.
5. Solve: $0 = 3x^2 - 7x - 6$
6. Solve by using the quadratic formula: $0 = x^2 - 4x + 1$
7. Solve $x = 2y + 4$ for y.
8. Find the domain and range of the relation $\{(-2, 4), (3, 5), (4, 6), (6, 5)\}$. Is the relation a function?
9. Graph: $x = -2$

GO FIGURE

Each time the two hands of a certain standard, analog, 12-hour clock form a 180° angle, a bell chimes once. From noon today until noon tomorrow, how many chimes will be heard?

Copyright © Houghton Mifflin Company. All rights reserved.

Copyright © Houghton Mifflin Company. All rights reserved.

Aufmann Interactive Method (AIM)

An Interactive Approach

Algebra: Introductory and Intermediate uses an interactive style that provides a student with an opportunity to try a skill as it is presented. Each section is divided into objectives, and every objective contains one or more sets of matched-pair examples. The first example in each set is worked out; the second example, called "You Try It," is for the student to work. By solving this problem, the student actively practices concepts as they are presented in the text.

There are complete worked-out solutions to these examples in an appendix. By comparing their solution to the solution in the appendix, students obtain immediate feedback on, and reinforcement of, the concept.

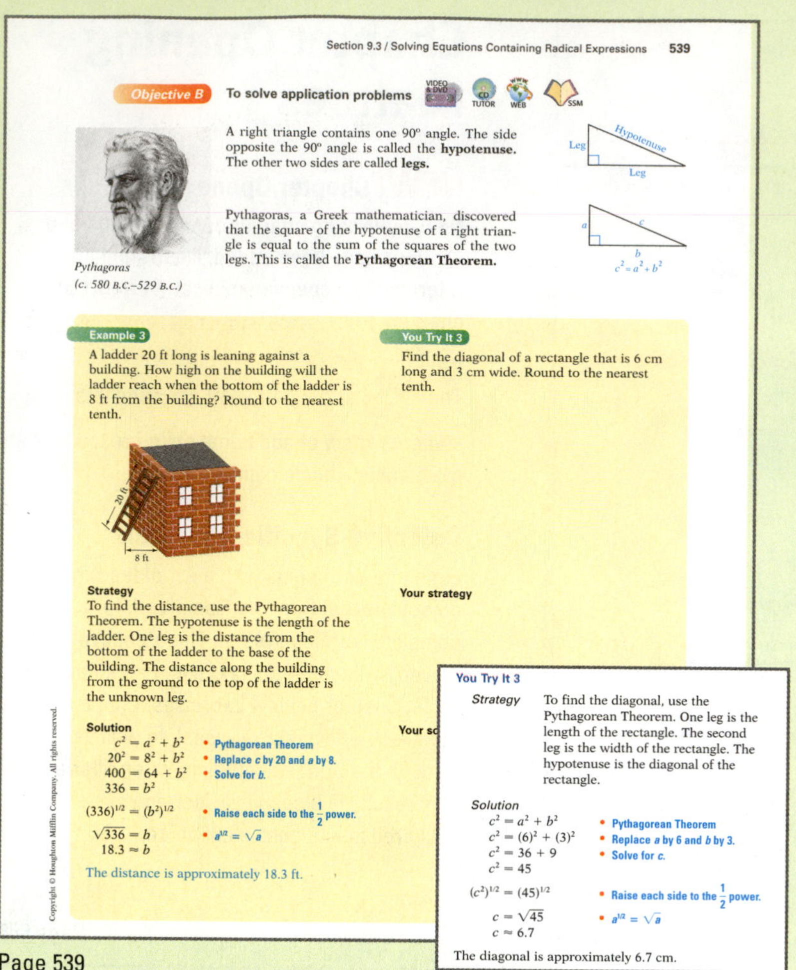

Section 9.3 / Solving Equations Containing Radical Expressions 539

Objective B To solve application problems

A right triangle contains one 90° angle. The side opposite the 90° angle is called the **hypotenuse.** The other two sides are called **legs.**

Pythagoras, a Greek mathematician, discovered that the square of the hypotenuse of a right triangle is equal to the sum of the squares of the two legs. This is called the **Pythagorean Theorem.**

Pythagoras (c. 580 B.C.–529 B.C.)

Example 3

A ladder 20 ft long is leaning against a building. How high on the building will the ladder reach when the bottom of the ladder is 8 ft from the building? Round to the nearest tenth.

Strategy

To find the distance, use the Pythagorean Theorem. The hypotenuse is the length of the ladder. One leg is the distance from the bottom of the ladder to the base of the building. The distance along the building from the ground to the top of the ladder is the unknown leg.

Solution

$c^2 = a^2 + b^2$ • Pythagorean Theorem

$20^2 = 8^2 + b^2$ • Replace c by 20 and a by 8.

$400 = 64 + b^2$ • Solve for b.

$336 = b^2$

$(336)^{1/2} = (b^2)^{1/2}$ • Raise each side to the $\frac{1}{2}$ power.

$\sqrt{336} = b$ • $a^{1/2} = \sqrt{a}$

$18.3 \approx b$

The distance is approximately 18.3 ft.

You Try It 3

Find the diagonal of a rectangle that is 6 cm long and 3 cm wide. Round to the nearest tenth.

Your strategy

You Try It 3

Strategy To find the diagonal, use the Pythagorean Theorem. One leg is the length of the rectangle. The second leg is the width of the rectangle. The hypotenuse is the diagonal of the rectangle.

Solution

$c^2 = a^2 + b^2$ • Pythagorean Theorem

$c^2 = (6)^2 + (3)^2$ • Replace a by 6 and b by 3.

$c^2 = 36 + 9$ • Solve for c.

$c^2 = 45$

$(c^2)^{1/2} = (45)^{1/2}$ • Raise each side to the $\frac{1}{2}$ power.

$c = \sqrt{45}$ • $a^{1/2} = \sqrt{a}$

$c \approx 6.7$

The diagonal is approximately 6.7 cm.

Page 539

Page S29

AIM for Success Student Preface

This student 'how to use this book' preface explains what is required of a student to be successful and how this text has been designed to foster student success, including the Aufmann Interactive Method (AIM). *AIM for Success* can be used as a lesson on the first day of class or as a project for students to complete to strengthen their study skills. There are suggestions for teaching this lesson in the *Instructor's Resource Manual.*

Page xxiii

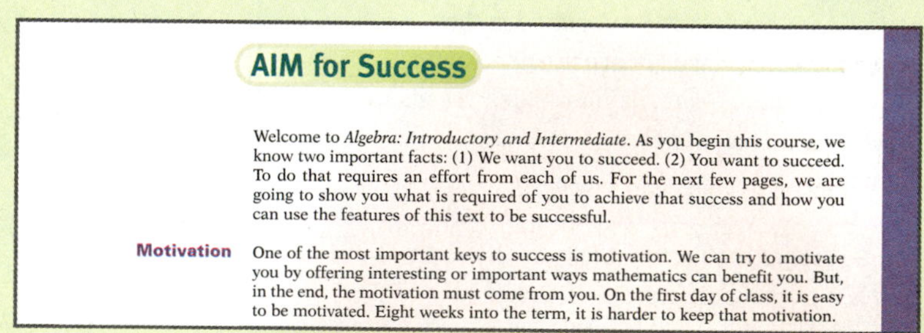

AIM for Success

Welcome to *Algebra: Introductory and Intermediate*. As you begin this course, we know two important facts: (1) We want you to succeed. (2) You want to succeed. To do that requires an effort from each of us. For the next few pages, we are going to show you what is required of you to achieve that success and how you can use the features of this text to be successful.

Motivation One of the most important keys to success is motivation. We can try to motivate you by offering interesting or important ways mathematics can benefit you. But, in the end, the motivation must come from you. On the first day of class, it is easy to be motivated. Eight weeks into the term, it is harder to keep that motivation.

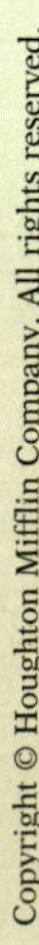
Copyright © Houghton Mifflin Company. All rights reserved.

Problem Solving

Focus on Problem Solving

At the end of each chapter is a Focus on Problem Solving feature which introduces the student to various successful problem-solving strategies. Strategies such as drawing a diagram, applying solutions to other problems, working backwards, inductive reasoning, and trial and error are some of the techniques that are demonstrated.

Focus on Problem Solving

Find a Pattern One approach to problem solving is to try to find a pattern. Karl Friedrich Gauss supposedly used this method to solve a problem that was given to his math class when he was in elementary school. As the story goes, his teacher wanted to grade some papers while the class worked on a math problem. The problem given to the class was to find the sum

$$1 + 2 + 3 + 4 + \cdots + 100$$

Gauss quickly solved the problem by seeing a pattern. Here is what he saw.

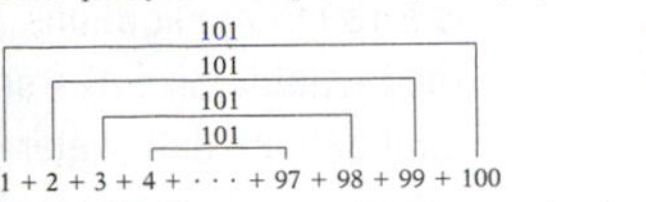

Note that

$1 + 100 = 101$
$2 + 99 = 101$
$3 + 98 = 101$
$4 + 97 = 101$

Gauss noted that there were 50 sums of 101. Therefore, the sum of the first 100 natural numbers is

$$1 + 2 + 3 + 4 + \cdots + 97 + 98 + 99 + 100 = 50(101) = 5050$$

Point of Interest

Karl Friedrich Gauss

Karl Friedrich Gauss (1777–1855) has been called the "Prince of Mathematicians" by some historians. He applied his genius to many areas of mathematics and science. A unit of magnetism, the *gauss*, is named in his honor. Some types of electronic equipment (televisions, for instance) contain a *degausser* that controls magnetic fields.

Try to solve Exercises 1 to 6 by finding a pattern.

1. Find the sum $2 + 4 + 6 + \cdots + 96 + 98 + 100$.
2. Find the sum $1 + 3 + 5 + \cdots + 97 + 99 + 101$.
3. Find another method of finding the sum $1 + 3 + 5 + \cdots + 97 + 99 + 101$ given in the preceding exercise.
4. Find the sum $\frac{1}{1 \cdot 2} + \frac{1}{2 \cdot 3} + \frac{1}{3 \cdot 4} + \cdots + \frac{1}{49 \cdot 50}$.
   Hint: $\frac{1}{1 \cdot 2} = \frac{1}{2}, \frac{1}{1 \cdot 2} + \frac{1}{2 \cdot 3} = \frac{2}{3}, \frac{1}{1 \cdot 2} + \frac{1}{2 \cdot 3} + \frac{1}{3 \cdot 4} = \frac{3}{4}$
5. A *polynomial number* is a number that can be represented by arranging that number of dots in rows to form a geometric figure such as a triangle, square, pentagon, or hexagon. For instance, the first four *triangular* numbers, 3, 6, 10, and 15, are shown below. What are the next two triangular numbers?

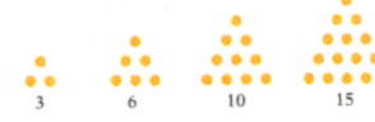

6. The following problem shows that checking a few cases does not always result in a conjecture that is true for *all* cases. Select any two points on a circle (see the drawing in the left margin) and draw a *chord*, a line connecting the points. The chord divides the circle into two regions. Now select three different points and draw chords connecting each of the three points with every other point. The chords divide the circle into four regions. Now select four points and connect each of the points with every other point. Make a conjecture as to the relationship between the number of regions and the number of points on the circle. Does your conjecture work for five points? six points?

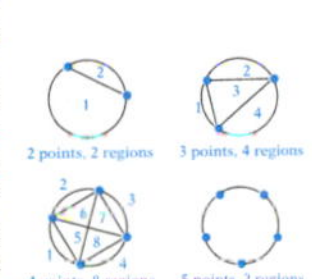

Copyright © Houghton Mifflin Company. All rights reserved.

Page 283

Page 114

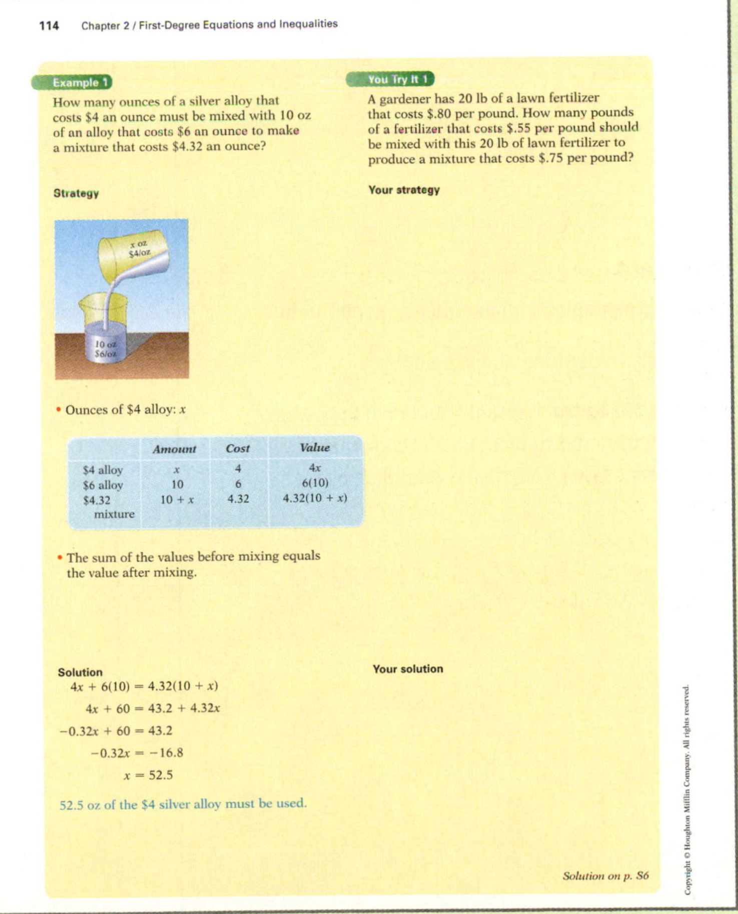

Example 1

How many ounces of a silver alloy that costs \$4 an ounce must be mixed with 10 oz of an alloy that costs \$6 an ounce to make a mixture that costs \$4.32 an ounce?

Strategy

- Ounces of \$4 alloy: x

	Amount	Cost	Value
\$4 alloy	x	4	$4x$
\$6 alloy	10	6	$6(10)$
\$4.32 mixture	$10 + x$	4.32	$4.32(10 + x)$

- The sum of the values before mixing equals the value after mixing.

Solution

$$4x + 6(10) = 4.32(10 + x)$$
$$4x + 60 = 43.2 + 4.32x$$
$$-0.32x + 60 = 43.2$$
$$-0.32x = -16.8$$
$$x = 52.5$$

52.5 oz of the \$4 silver alloy must be used.

You Try It 1

A gardener has 20 lb of a lawn fertilizer that costs \$.80 per pound. How many pounds of a fertilizer that costs \$.55 per pound should be mixed with this 20 lb of lawn fertilizer to produce a mixture that costs \$.75 per pound?

Your strategy

Your solution

Solution on p. S6

Copyright © Houghton Mifflin Company. All rights reserved.

Problem-Solving Strategies

The text features a carefully developed approach to problem solving that emphasizes the importance of *strategy* when solving problems. Students are encouraged to develop their own strategies—to draw diagrams, to write out the solution steps in words—as part of their solution to a problem. In each case, model strategies are presented as guides for students to follow as they attempt the "You Try It" problem. Having students provide strategies is a natural way to incorporate writing into the math curriculum.

Copyright © Houghton Mifflin Company. All rights reserved.

Real Data and Applications

Applications

One way to motivate an interest in mathematics is through applications. Wherever appropriate, the last objective of a section presents applications that require the student to use problem-solving strategies, along with the skills covered in that section, to solve practical problems. This carefully integrated applied approach generates student awareness of the value of algebra as a real-life tool.

Applications are taken from many disciplines including astronomy, business, carpentry, chemistry, construction, Earth science, education, manufacturing, nutrition, real estate, and telecommunications.

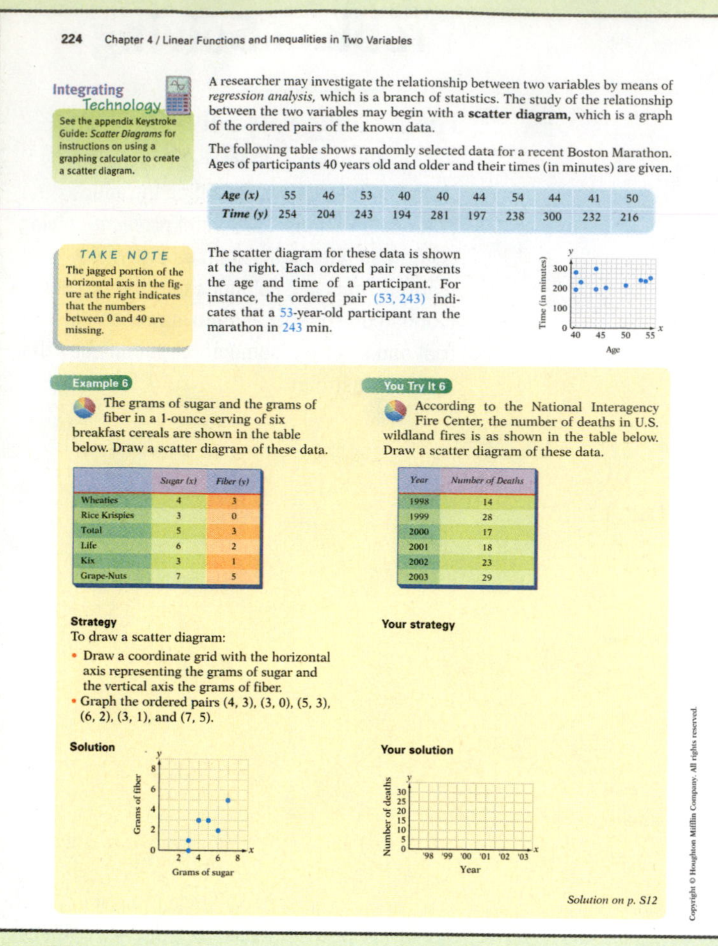

224 Chapter 4 / Linear Functions and Inequalities in Two Variables

Integrating Technology
See the appendix Keystroke Guide: *Scatter Diagrams* for instructions on using a graphing calculator to create a scatter diagram.

A researcher may investigate the relationship between two variables by means of *regression analysis,* which is a branch of statistics. The study of the relationship between the two variables may begin with a **scatter diagram,** which is a graph of the ordered pairs of the known data.

The following table shows randomly selected data for a recent Boston Marathon. Ages of participants 40 years old and older and their times (in minutes) are given.

Age (x)	55	46	53	40	40	44	54	44	41	50
Time (y)	254	204	243	194	281	197	238	300	232	216

TAKE NOTE
The jagged portion of the horizontal axis in the figure at the right indicates that the numbers between 0 and 40 are missing.

The scatter diagram for these data is shown at the right. Each ordered pair represents the age and time of a participant. For instance, the ordered pair (53, 243) indicates that a 53-year-old participant ran the marathon in 243 min.

Example 6
The grams of sugar and the grams of fiber in a 1-ounce serving of six breakfast cereals are shown in the table below. Draw a scatter diagram of these data.

	Sugar (x)	*Fiber (y)*
Wheaties	4	3
Rice Krispies	3	0
Total	5	3
Life	6	2
Kix	3	1
Grape-Nuts	7	5

Strategy
To draw a scatter diagram:
- Draw a coordinate grid with the horizontal axis representing the grams of sugar and the vertical axis the grams of fiber.
- Graph the ordered pairs (4, 3), (3, 0), (5, 3), (6, 2), (3, 1), and (7, 5).

Solution

You Try It 6
According to the National Interagency Fire Center, the number of deaths in U.S. wildland fires is as shown in the table below. Draw a scatter diagram of these data.

Year	*Number of Deaths*
1998	14
1999	28
2000	17
2001	18
2002	23
2003	29

Your strategy

Your solution

Solution on p. S12

Copyright © Houghton Mifflin Company. All rights reserved.

Page 224

Real Data

Real data examples and exercises, identified by , ask students to analyze and solve problems taken from actual situations. Students are often required to work with tables, graphs, and charts drawn from a variety of disciplines.

Page 356

356 Chapter 6 / Polynomials

117. **Astronomy** The distance from Earth to Saturn is 8.86×10^8 mi. A satellite leaves Earth traveling at a constant rate of 1×10^5 mph. How long does it take for the satellite to reach Saturn?

118. **The Federal Government** In 2004, the gross national debt was approximately 7×10^{12} dollars. How much would each American have to pay in order to pay off the debt? Use 3×10^8 as the number of citizens.

119. **Physics** The mass of an electron is 9.109×10^{-31} kg. The mass of a proton is 1.673×10^{-27} kg. How many times heavier is a proton than an electron?

120. **Geology** The mass of Earth is 5.9×10^{24} kg. The mass of the sun is 2×10^{30} kg. How many times heavier is the sun than Earth?

121. **Physics** How many meters does light travel in 8 h? The speed of light is 3×10^8 m/s.

122. **Physics** How many meters does light travel in 1 day? The speed of light is 3×10^8 m/s.

Sojourner

123. **Astronomy** It took 11 min for the commands from a computer on Earth to travel to the rover Sojourner on Mars, a distance of 119 million miles. How fast did the signals from Earth to Mars travel?

124. **Measurement** The weight of 31 million orchid seeds is 1 oz. Find the weight of one orchid seed.

Centrifuge

125. **Physics** A high-speed centrifuge makes 4×10^8 revolutions each minute. Find the time in seconds for the centrifuge to make one revolution.

APPLYING THE CONCEPTS

126. Correct the error in each of the following expressions. Explain which rule or property was used incorrectly.
a. $x^0 = 0$
b. $(x^4)^5 = x^9$
c. $x^2 \cdot x^3 = x^6$

127. Simplify. **a.** $1 + [1 + (1 + 2^{-1})^{-1}]^{-1}$
b. $2 - [2 - (2 - 2^{-1})^{-1}]^{-1}$

Copyright © Houghton Mifflin Company. All rights reserved.

Copyright © Houghton Mifflin Company. All rights reserved.

Student Pedagogy

Icons

The    

icons at each objective head remind students of the many and varied additional resources available for each objective.

Key Terms and Concepts

Key terms, in bold, emphasize important terms. The key terms are also provided in a **Glossary** at the back of the text.

Key concepts are presented in orange boxes in order to highlight these important concepts and to provide for easy reference.

Point of Interest

These margin notes contain interesting sidelights about mathematics, its history, or its application.

Take Note

These margin notes alert students to a point requiring special attention or are used to amplify the concept under discussion.

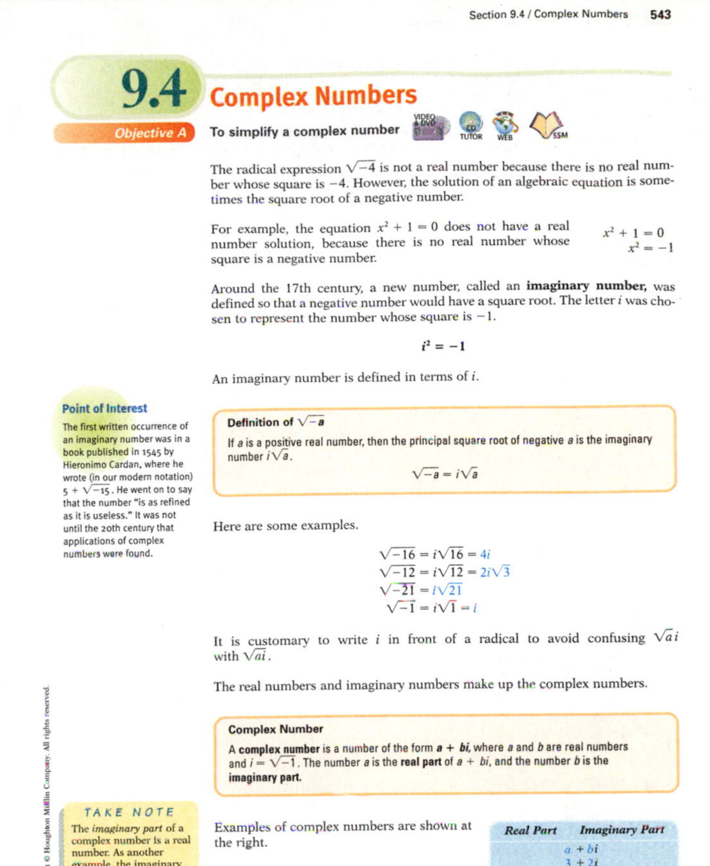

Section 9.4 / Complex Numbers 543

9.4 Complex Numbers

Objective A **To simplify a complex number**

The radical expression $\sqrt{-4}$ is not a real number because there is no real number whose square is -4. However, the solution of an algebraic equation is sometimes the square root of a negative number.

For example, the equation $x^2 + 1 = 0$ does not have a real number solution, because there is no real number whose square is a negative number.

$$x^2 + 1 = 0$$
$$x^2 = -1$$

Around the 17th century, a new number, called an **imaginary number,** was defined so that a negative number would have a square root. The letter i was chosen to represent the number whose square is -1.

$$i^2 = -1$$

An imaginary number is defined in terms of i.

Point of Interest
The first written occurrence of an imaginary number was in a book published in 1545 by Hieronimo Cardan, where he wrote (in our modern notation) $5 + \sqrt{-15}$. He went on to say that the number "is as refined as it is useless." It was not until the 20th century that applications of complex numbers were found.

Definition of $\sqrt{-a}$
If a is a positive real number, then the principal square root of negative a is the imaginary number $i\sqrt{a}$.
$$\sqrt{-a} = i\sqrt{a}$$

Here are some examples.

$$\sqrt{-16} = i\sqrt{16} = 4i$$
$$\sqrt{-12} = i\sqrt{12} = 2i\sqrt{3}$$
$$\sqrt{-21} = i\sqrt{21}$$
$$\sqrt{-1} = i\sqrt{1} = i$$

It is customary to write i in front of a radical to avoid confusing $\sqrt{a}\,i$ with $\sqrt{ai}$.

The real numbers and imaginary numbers make up the complex numbers.

Complex Number
A **complex number** is a number of the form $a + bi$, where a and b are real numbers and $i = \sqrt{-1}$. The number a is the **real part** of $a + bi$, and the number b is the **imaginary part.**

TAKE NOTE
The *imaginary part* of a complex number is a real number. As another example, the imaginary part of $6 - 8i$ is -8.

Examples of complex numbers are shown at the right.

Real Part	Imaginary Part
$a + bi$	
$3 + 2i$	
$8 - 10i$	

Copyright © Houghton Mifflin Company. All rights reserved.

Page 543

Page 544

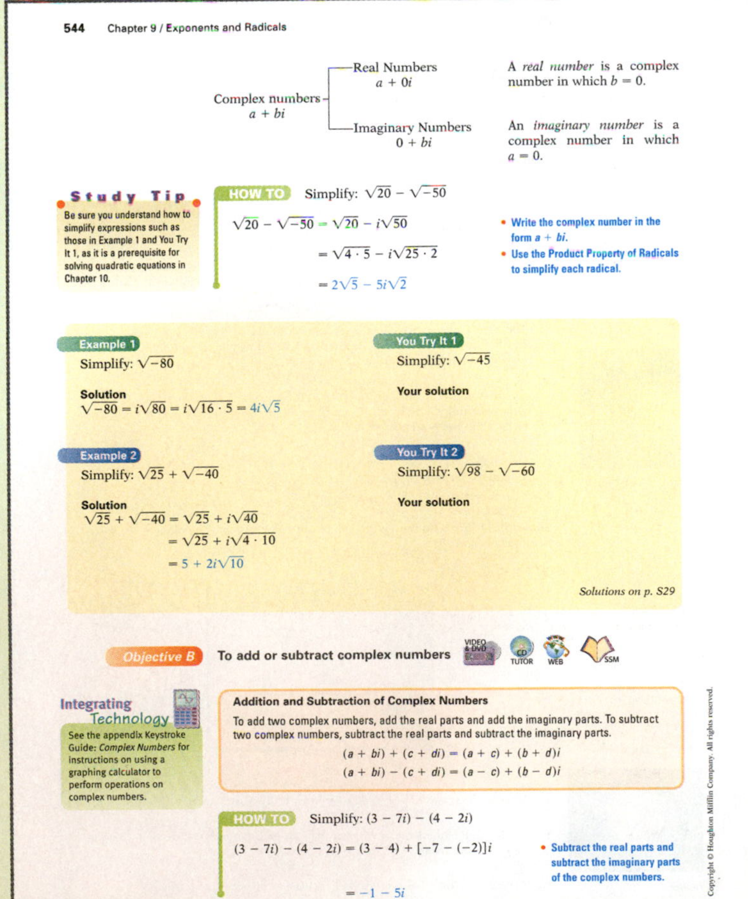

544 Chapter 9 / Exponents and Radicals

Complex numbers $a + bi$ — Real Numbers $a + 0i$; Imaginary Numbers $0 + bi$

A *real number* is a complex number in which $b = 0$.

An *imaginary number* is a complex number in which $a = 0$.

Study Tip
Be sure you understand how to simplify expressions such as those in Example 1 and You Try It 1, as it is a prerequisite for solving quadratic equations in Chapter 10.

HOW TO Simplify: $\sqrt{20} - \sqrt{-50}$

$$\sqrt{20} - \sqrt{-50} = \sqrt{20} - i\sqrt{50}$$
$$= \sqrt{4 \cdot 5} - i\sqrt{25 \cdot 2}$$
$$= 2\sqrt{5} - 5i\sqrt{2}$$

- Write the complex number in the form $a + bi$.
- Use the Product Property of Radicals to simplify each radical.

Example 1
Simplify: $\sqrt{-80}$

Solution
$\sqrt{-80} = i\sqrt{80} = i\sqrt{16 \cdot 5} = 4i\sqrt{5}$

You Try It 1
Simplify: $\sqrt{-45}$

Your solution

Example 2
Simplify: $\sqrt{25} + \sqrt{-40}$

Solution
$\sqrt{25} + \sqrt{-40} = \sqrt{25} + i\sqrt{40}$
$= \sqrt{25} + i\sqrt{4 \cdot 10}$
$= 5 + 2i\sqrt{10}$

You Try It 2
Simplify: $\sqrt{98} - \sqrt{-60}$

Your solution

Solutions on p. S29

Objective B **To add or subtract complex numbers**

Integrating Technology
See the appendix Keystroke Guide: *Complex Numbers* for instructions on using a graphing calculator to perform operations on complex numbers.

Addition and Subtraction of Complex Numbers
To add two complex numbers, add the real parts and add the imaginary parts. To subtract two complex numbers, subtract the real parts and subtract the imaginary parts.
$$(a + bi) + (c + di) = (a + c) + (b + d)i$$
$$(a + bi) - (c + di) = (a - c) + (b - d)i$$

HOW TO Simplify: $(3 - 7i) - (4 - 2i)$

$$(3 - 7i) - (4 - 2i) = (3 - 4) + [-7 - (-2)]i$$
$$= -1 - 5i$$

- Subtract the real parts and subtract the imaginary parts of the complex numbers.

Copyright © Houghton Mifflin Company. All rights reserved.

Study Tips

These margin notes remind students of study skills presented in the *AIM for Success*; some notes provide page references to the original descriptions. They also provide students with reminders of how to practice good study habits.

HOW TO Examples

HOW TO examples use annotations to explain what is happening in key steps of the complete, worked-out solutions.

Integrating Technology

These margin notes provide suggestions for using a calculator or refer the student to an appendix for more complete instructions on using a calculator.

Copyright © Houghton Mifflin Company. All rights reserved.

Exercises and Projects

Exercises

The exercise sets of *Algebra: Introductory and Intermediate* emphasize skill building, skill maintenance, and applications. Concept-based writing or developmental exercises have been integrated within the exercise sets. Icons identify appropriate writing 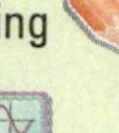, data analysis , and calculator  exercises.

Included in each exercise set are **Applying the Concepts** that present extensions of topics, require analysis, or offer challenge problems. The writing exercises ask students to explain answers, write about a topic in the section, or research and report on a related topic.

Page 682

14. $f(x) = \log_3(2 - x)$
15. $f(x) = -\log_2(x - 1)$
16. $f(x) = -\log_2(1 - x)$

APPLYING THE CONCEPTS

Use a graphing calculator to graph the functions in Exercises 17 to 22.

17. $f(x) = x - \log_2(1 - x)$
18. $f(x) = -\frac{1}{2}\log_2 x - 1$
19. $f(x) = \frac{x}{2} - 2\log_2(x + 1)$
20. $f(x) = x + \log_3(2 - x)$
21. $f(x) = x^2 - 10\ln(x - 1)$
22. $f(x) = \frac{x}{3} - 3\log_2(x + 3)$

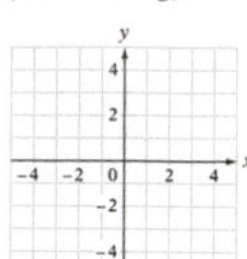
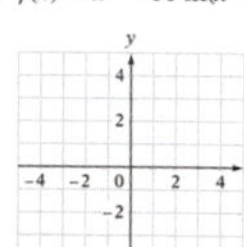
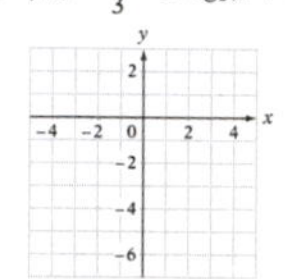

23. **Astronomy** Astronomers use the *distance modulus* of a star as a method of determining the star's distance from Earth. The formula is $M = 5\log(s - 5)$, where M is the distance modulus and s is the star's distance from Earth in parsecs. (One parsec $\approx 1.9 \times 10^{13}$ mi)
   a. Graph the equation.
   b. The point with coordinates (25.1, 2) is on the graph. Write a sentence that describes the meaning of this ordered pair.

24. **Typing** Without practice, the proficiency of a typist decreases. The equation $s = 60 - 7\ln(t + 1)$, where s is the typing speed in words per minute and t is the number of months without typing, approximates this decrease.
   a. Graph the equation.
   b. The point with coordinates (4, 49) is on the graph. Write a sentence that describes the meaning of this ordered pair.

Copyright © Houghton Mifflin Company. All rights reserved.

Projects and Group Activities

The Projects and Group Activities featured at the end of each chapter can be used as extra credit or for cooperative learning activities. The projects cover various aspects of mathematics, including the use of calculators, collecting data from the Internet, data analysis, and extended applications.

Page 388

Projects and Group Activities

Astronomical Distances and Scientific Notation

Astronomers have units of measurement that are useful for measuring vast distances in space. Two of these units are the astronomical unit and the light-year. An **astronomical unit** is the average distance between Earth and the sun. A **light-year** is the distance a ray of light travels in 1 year.

1. Light travels at a speed of 1.86×10^5 mi/s. Find the measure of 1 light-year in miles. Use a 365-day year.
2. The distance between Earth and the star Alpha Centauri is approximately 25 trillion miles. Find the distance between Earth and Alpha Centauri in light-years. Round to the nearest hundredth.
3. The Coma cluster of galaxies is approximately 2.8×10^8 light-years from Earth. Find the distance, in miles, from the Coma cluster to Earth. Write the answer in scientific notation.
4. One astronomical unit (A.U.) is 9.3×10^7 mi. The star Pollux in the constellation Gemini is 1.8228×10^{12} mi from Earth. Find the distance from Pollux to Earth in astronomical units.
5. One light-year is equal to approximately how many astronomical units? Round to the nearest thousand.

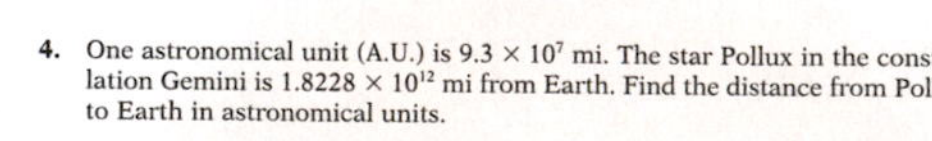

Gemini

Shown below are data on the planets in our solar system. The planets are listed in alphabetical order.

Planet	*Distance from the Sun (in kilometers)*	*Mass (in kilograms)*
Earth	1.50×10^8	5.97×10^{24}
Jupiter	7.79×10^8	1.90×10^{27}
Mars	2.28×10^8	6.42×10^{23}
Mercury	5.79×10^7	3.30×10^{23}
Neptune	4.50×10^9	1.02×10^{26}
Pluto	5.87×10^9	1.25×10^{22}
Saturn	1.43×10^9	5.68×10^{26}
Uranus	2.87×10^9	8.68×10^{25}
Venus	1.08×10^8	4.87×10^{24}

Point of Interest

In November 2001, the Hubble Space Telescope took photos of the atmosphere of a planet orbiting a star 150 light-years from Earth in the constellation Pegasus. The planet is about the size of Jupiter and orbits close to the star HD209458. It was the first discovery of an atmosphere around a planet outside our solar system.

Jupiter

6. Arrange the planets in order from closest to the sun to farthest from the sun.
7. Arrange the planets in order from the one with the greatest mass to the one with the least mass.
8. Write a rule for ordering numbers written in scientific notation.

Copyright © Houghton Mifflin Company. All rights reserved.

Copyright © Houghton Mifflin Company. All rights reserved.

End of Chapter

Chapter 2 Summary 149

Chapter 2 Summary

Key Words	Examples
An *equation* expresses the equality of two mathematical expressions. [2.1A, p. 73]	$3 + 2(4x - 5) = x + 4$ is an equation.

Page 149

Essential Rules and Procedures	Examples
Addition Property of Equations [2.1B, p. 74] The same number can be added to each side of an equation without changing the solution of the equation. If $a = b$, then $a + c = b + c$.	$x + 5 = -3$ $x + 5 - 5 = -3 - 5$ $x = -8$

Page 150

Chapter Summary

At the end of each chapter there is a Chapter Summary that includes Key Words, Essential Rules and Procedures, and an example of each. Each entry includes an objective reference and a page reference indicating where the concept is introduced. These chapter summaries provide a single point of reference as the student prepares for a test.

Chapter Review Exercises

Chapter Review Exercises are found at the end of each chapter. These exercises are selected to help the student integrate all of the topics presented in the chapter.

Page 152

152 Chapter 2 / First-Degree Equations and Inequalities

Chapter 2 Review Exercises

1. Solve: $3t - 3 + 2t = 7t - 15$
2. Solve: $3x - 7 > -2$
3. Is 3 a solution of $5x - 2 = 4x + 5$?
4. Solve: $x + 4 = -5$

Chapter Test

Each Chapter Test is designed to simulate a possible test of the material in the chapter.

Page 155

Chapter 2 Test 155

Chapter 2 Test

1. Solve: $x - 2 = -4$
2. Solve: $b + \frac{3}{4} = \frac{5}{8}$
3. Solve: $-\frac{3}{4}y = -\frac{5}{8}$
4. Solve: $3x - 5 = 7$

Cumulative Review Exercises

Cumulative Review Exercises, which appear at the end of each chapter (beginning with Chapter 2), help students maintain skills learned in previous chapters.

Page 157

Cumulative Review Exercises 157

Cumulative Review Exercises

1. Subtract: $-6 - (-20) - 8$
2. Multiply: $(-2)(-6)(-4)$
3. Subtract: $-\frac{5}{6} - \left(-\frac{7}{16}\right)$
4. Simplify: $-4^2 \cdot \left(-\frac{3}{2}\right)^3$

The answers to all Chapter Review Exercises, all Chapter Test exercises, and all Cumulative Review Exercises are given in the Answer Section. Along with the answer, there is a reference to the objective that pertains to each exercise.

Page A6

CUMULATIVE REVIEW EXERCISES

1. 6 [1.1C] **2.** −48 [1.1D] **3.** $-\frac{19}{48}$ [1.2C] **4.** 54 [1.2E] **5.** $\frac{49}{40}$ [1.3A] **6.** 6 [1.4A] **7.** $-17x$ [1.4B] **8.** $-5a - 4b$ [1.4B] **9.** $2x$ [1.4C] **10.** $36y$ [1.4C] **11.** $2x^2 + 6x - 4$ [1.4D] **12.** $-4x + 14$ [1.4D] **13.** $6x - 34$ [1.4D] **14.** $A \cap B = [-4, 0]$ [1.5A] **15.** −5 −4 −3 −2 −1 0 1 2 3 4 5 [1.5C] **16.** Yes [2.1A]

Copyright © Houghton Mifflin Company. All rights reserved.

Instructor Resources

Algebra: Introductory and Intermediate has a complete set of support materials for the instructor.

Instructor's Annotated Edition This edition contains a replica of the student text and additional items just for the instructor.

Online Instructor's Resource Manual with Solutions The *Instructor's Resource Manual with Solutions* contains worked-out solutions for all exercises in the text. It also contains suggested *Course Sequences* and a printout of the *AIM for Success* PowerPoint slide show. It is available on the *ClassPrep CD* and at the Online Teaching Center.

Online Instructor's Test Bank with Chapter Tests This resource contains a static version of the *HM Testing* files. It also contains eight ready-to-use Chapter Tests per chapter. All resources are also available on the *ClassPrep CD* and at the Online Teaching Center.

HM ClassPrep™ with HM Testing (powered by Diploma™) *HM ClassPrep* offers a combination of two class management tools including supplements and text-specific resources for the instructor. *HM Testing* offers instructors a flexible and powerful tool for test generation and test management. Now supported by the Brownstone Research Group's market-leading *Diploma* software, this new version of *HM Testing* significantly improves on functionality and ease of use by offering all the tools needed to create, author, deliver, and customize multiple types of tests—including authoring and editing algorithmic questions.

Online Teaching Center For an abundance of instructor resources, visit the free Houghton Mifflin Teaching Center on our website, **college.hmco.com/pic/aufmannAIAI4e.**

Blackboard®, WebCT®, and eCollege® Houghton Mifflin can provide you with valuable content to include in your existing Blackboard, WebCT, and eCollege systems. This text-specific content enables instructors to teach all or part of their course online. Contact your Houghton Mifflin sales rep for cartridge availability.

TeamUP Integration Services TeamUP, our integration program, offers flexible, personalized training and consultative services by phone, online, or on campus. The TeamUP Integration Team will:

- Show you how to use our products.
- Provide ideas and best practices for incorporating all elements of our text and technology program into your course.
- Customize our programs or products to achieve your teaching objectives.
- Provide technical assistance.
- Link you to faculty development opportunities.

Visit **teamup.college.hmco.com** for more information, or contact your Houghton Mifflin sales representative to schedule one of our customized programs.

NEW! Eduspace® Eduspace, powered by Blackboard, is Houghton Mifflin's online learning tool. Eduspace makes it easy for instructors to create all or part of a course online. Homework exercises, quizzes, tests, tutorials, and supplemental study materials all come ready-to-use. Instructors can choose to use the content as is or modify it, and they can even add their own. Visit **www.eduspace.com** for more information.

NEW! HM Assess for Mathematics HM Assess is a new diagnostic assessment tool from Houghton Mifflin that tests core concepts in specific courses and provides students with *individualized study paths* for self-remediation. These paths are carefully designed to offer self-study options and to appeal to a variety of learning styles with concept reviews, interactive lessons, similar examples, and additional practice exercises. Instructors can use HM Assess to quickly gauge which

Copyright © Houghton Mifflin Company. All rights reserved.

students are in jeopardy and which concepts require additional review. Both Course Assessments and Chapter Assessments are available. HM Assess is offered as part of Eduspace. Visit **hmassess.college.hmco.com** for more information.

Student Resources

Student Solutions Manual The *Student Solutions Manual* contains complete solutions to odd-numbered exercises in the text.

Math Study Skills Workbook by Paul D. Nolting This workbook is designed to reinforce skills and minimize frustration for students in any math class, lab, or study skills course. It offers a wealth of study tips and sound advice on note taking, time management, and reducing math anxiety. In addition, numerous opportunities for self-assessment enable students to track their own progress.

Eduspace® Eduspace, powered by Blackboard, is Houghton Mifflin's online learning tool. Eduspace is a text-specific, web-based learning environment offering students a combination of practice exercises, multimedia tutorials, video explanations, online algorithmic homework, and more. Specific content is available 24 hours a day to help you succeed in your course.

HM mathSpace® Tutorial CD-ROM For students who prefer the portability of a CD-ROM, this tutorial provides opportunities for self-paced review and practice with algorithmically generated exercises and step-by-step solutions.

SMARTHINKING® Houghton Mifflin's unique partnership with SMARTHINKING brings students real-time, online tutorial support when they need it most. Using state-of-the-art whiteboard technology and feedback tools, students interact and communicate with "e-structors." These specially trained tutors guide students through the learning and problem solving process without providing answers or rewriting a student's work.

SMARTHINKING offers three levels of service.*

- **Live Tutorial Help** provides real-time, one-on-one instruction.
- **Questions Any Time** allows students to e-mail questions to a tutor outside of the scheduled tutorial sessions and receive a reply, usually within 24 hours.
- **Independent Study Resources** connects students around-the-clock to additional educational resources, ranging from interactive websites to Frequently Asked Questions. Visit **smarthinking.com** for more information.

**Limits apply; terms and hours of SMARTHINKING service are subject to change.*

Houghton Mifflin Instructional Videos and DVDs Text-specific videos and DVDs, hosted by Dana Mosely, cover all sections of the text and provide a valuable resource for further instruction and review.

Online Study Center For an abundance of student resources, visit the free Houghton Mifflin Study Center on our website, **college.hmco.com/pic/aufmannAIAI4e.**

Acknowledgments

The authors would like to thank the people who have reviewed this manuscript and provided many valuable suggestions.

Donna Foster, *Piedmont Technical College*
William Graesser, *Ivy Tech State College*
Anne Haney
Tim R. McBride, *Spartanburg Technical College*
Michael McComas, *Marshall Community and Technical College*
Linda J. Murphy, *Northern Essex Community College*

Copyright © Houghton Mifflin Company. All rights reserved.

AIM for Success

Welcome to *Algebra: Introductory and Intermediate*. As you begin this course, we know two important facts: (1) We want you to succeed. (2) You want to succeed. To do that requires an effort from each of us. For the next few pages, we are going to show you what is required of you to achieve that success and how you can use the features of this text to be successful.

Motivation

One of the most important keys to success is motivation. We can try to motivate you by offering interesting or important ways mathematics can benefit you. But, in the end, the motivation must come from you. On the first day of class, it is easy to be motivated. Eight weeks into the term, it is harder to keep that motivation.

> TAKE NOTE
> Motivation alone will not lead to success. For instance, suppose a person who cannot swim is placed in a boat, taken out to the middle of a lake, and then thrown overboard. That person has a lot of motivation but there is a high likelihood the person will drown without some help. Motivation gives us the desire to learn but is not the same as learning.

To stay motivated, there must be outcomes from this course that are worth your time, money, and energy.

List some reasons you are taking this course.

____

____

____

Although we hope that one of the reasons you listed was an interest in mathematics, we know that many of you are taking this course because it is required to graduate, it is a prerequisite for a course you must take, or because it is required for your major. Although you may not agree that this course is necessary, it is! If you are motivated to graduate or complete the requirements for your major, then use that motivation to succeed in this course. Do not become distracted from your goal to complete your education!

Commitment

To be successful, you must make a commitment to succeed. This means devoting time to math so that you achieve a better understanding of the subject.

List some activities (sports, hobbies, talents such as dance, art, or music) that you enjoy and at which you would like to become better.

ACTIVITY	TIME SPENT	TIME WISHED SPENT

Thinking about these activities, put the number of hours that you spend each week practicing these activities next to the activity. Next to that number, indicate the number of hours per week you would like to spend on these activities.

Whether you listed surfing or sailing, aerobics or restoring cars, or any other activity you enjoy, note how many hours a week you spend doing it. To succeed in math, you must be willing to commit the same amount of time. Success requires some sacrifice.

The "I Can't Do Math" Syndrome

There may be things you cannot do, such as lift a two-ton boulder. You can, however, do math. It is much easier than lifting the two-ton boulder. When you first

Copyright © Houghton Mifflin Company. All rights reserved.

learned the activities you listed above, you probably could not do them well. With practice, you got better. With practice, you will be better at math. Stay focused, motivated, and committed to success.

It is difficult for us to emphasize how important it is to overcome the "I Can't Do Math" Syndrome. If you listen to interviews of very successful athletes after a particularly bad performance, you will note that they focus on the positive aspect of what they did, not the negative. Sports psychologists encourage athletes to always be positive—to have a "Can Do" attitude. Develop this attitude toward math.

Strategies for Success

Textbook Review Right now, do a 15-minute "textbook review" of this book. Here's how:

First, read the table of contents. Do it in three minutes or less. Next, look through the entire book, page by page. Move quickly. Scan titles, look at pictures, notice diagrams.

A textbook review shows you where a course is going. It gives you the big picture. That's useful because brains work best when going from the general to the specific. Getting the big picture before you start makes details easier to recall and understand later on.

Your textbook review will work even better if, as you scan, you look for ideas or topics that are interesting to you. List three facts, topics, or problems that you found interesting during your textbook review.

____

____

____

The idea behind this technique is simple: It's easier to work at learning material if you know it's going to be useful to you.

Not all the topics in this book will be "interesting" to you. But that is true of any subject. Surfers find that on some days the waves are better than others, musicians find some music more appealing than other music, computer gamers find some computer games more interesting than others, car enthusiasts find some cars more exciting than others. Some car enthusiasts would rather have a completely restored 1957 Chevrolet than a new Ferrari.

Know the Course Requirements To do your best in this course, you must know exactly what your instructor requires. Course requirements may be stated in a *syllabus*, which is a printed outline of the main topics of the course, or they may be presented orally. When they are listed in a syllabus or on other printed pages, keep them in a safe place. When they are presented orally, make sure to take complete notes. In either case, it is important that you understand them completely and follow them exactly. Be sure you know the answer to each of the following questions.

1. What is your instructor's name?
2. Where is your instructor's office?
3. At what times does your instructor hold office hours?
4. Besides the textbook, what other materials does your instructor require?
5. What is your instructor's attendance policy?
6. If you must be absent from a class meeting, what should you do before returning to class? What should you do when you return to class?

Copyright © Houghton Mifflin Company. All rights reserved.

7. What is the instructor's policy regarding collection or grading of homework assignments?
8. What options are available if you are having difficulty with an assignment? Is there a math tutoring center?
9. If there is a math lab at your school, where is it located? What hours is it open?
10. What is the instructor's policy if you miss a quiz?
11. What is the instructor's policy if you miss an exam?
12. Where can you get help when studying for an exam?

Remember: Your instructor wants to see you succeed. If you need help, ask! Do not fall behind. If you are running a race and fall behind by 100 yards, you may be able to catch up but it will require more effort than had you not fallen behind.

TAKE NOTE

Besides time management, there must be realistic ideas of how much time is available. There are very few people who can *successfully* work full-time and go to school full-time. If you work 40 hours a week, take 15 units, spend the recommended study time given at the right, and sleep 8 hours a day, you will use over 80% of the available hours in a week. That leaves less than 20% of the hours in a week for family, friends, eating, recreation, and other activities.

Time Management We know that there are demands on your time. Family, work, friends, and entertainment all compete for your time. We do not want to see you receive poor job evaluations because you are studying math. However, it is also true that we do not want to see you receive poor math test scores because you devoted too much time to work. When several competing and important tasks require your time and energy, the only way to manage the stress of being successful at both is to manage your time efficiently.

Instructors often advise students to spend twice the amount of time outside of class studying as they spend in the classroom. Time management is important if you are to accomplish this goal and succeed in school. The following activity is intended to help you structure your time more efficiently.

List the name of each course you are taking this term, the number of class hours each course meets, and the number of hours you should spend studying each subject outside of class. Then fill in a weekly schedule like the one printed below. Begin by writing in the hours spent in your classes, the hours spent at work (if you have a job), and any other commitments that are not flexible with respect to the time that you do them. Then begin to write down commitments that are more flexible, including hours spent studying. Remember to reserve time for activities such as meals and exercise. You should also schedule free time.

	Monday	Tuesday	Wednesday	Thursday	Friday	Saturday	Sunday
7–8 a.m.							
8–9 a.m.							
9–10 a.m.							
10–11 a.m.							
11–12 p.m.							
12–1 p.m.							
1–2 p.m.							
2–3 p.m.							
3–4 p.m.							
4–5 p.m.							
5–6 p.m.							
6–7 p.m.							
7–8 p.m.							
8–9 p.m.							
9–10 p.m.							
10–11 p.m.							
11–12 a.m.							

Copyright © Houghton Mifflin Company. All rights reserved.

We know that many of you must work. If that is the case, realize that working 10 hours a week at a part-time job is equivalent to taking a three-unit class. If you must work, consider letting your education progress at a slower rate to allow you to be successful at both work and school. There is no rule that says you must finish school in a certain time frame.

Schedule Study Time As we encouraged you to do by filling out the time management form on the previous page, schedule a certain time to study. You should think of this time the way you would the time for work or class—that is, reasons for missing study time should be as compelling as reasons for missing work or class. "I just didn't feel like it" is not a good reason to miss your scheduled study time.

Although this may seem like an obvious exercise, list a few reasons you might want to study.

____

____

____

Of course we have no way of knowing the reasons you listed, but from our experience one reason given quite frequently is "To pass the course." There is nothing wrong with that reason. If that is the most important reason for you to study, then use it to stay focused.

One method of keeping to a study schedule is to form a ***study group***. Look for people who are committed to learning, who pay attention in class, and who are punctual. Ask them to join your group. Choose people with similar educational goals but different methods of learning. You can gain insight from seeing the material from a new perspective. Limit groups to four or five people; larger groups are unwieldy.

There are many ways to conduct a study group. Begin with the following suggestions and see what works best for your group.

1. Test each other by asking questions. Each group member might bring two or three sample test questions to each meeting.
2. Practice teaching each other. Many of us who are teachers learned a lot about our subject when we had to explain it to someone else.
3. Compare class notes. You might ask other students about material in your notes that is difficult for you to understand.
4. Brainstorm test questions.
5. Set an agenda for each meeting. Set approximate time limits for each agenda item and determine a quitting time.

And finally, probably the most important aspect of studying is that it should be done in relatively small chunks. If you can study only three hours a week for this course (probably not enough for most people), do it in blocks of one hour on three separate days, preferably after class. Three hours of studying on a Sunday is not as productive as three hours of paced study.

Text Features That Promote Success

There are 12 chapters in this text. Each chapter is divided into sections, and each section is subdivided into learning objectives. Each learning objective is labeled with a letter from A to G.

Copyright © Houghton Mifflin Company. All rights reserved.

Preparing for a Chapter Before you begin a new chapter, you should take some time to review previously learned skills. There are two ways to do this. The first is to complete the ***Cumulative Review Exercises***, which occurs after every chapter (except Chapter 1). For instance, turn to page 341. The questions in this review are taken from the previous chapters. The answers for all these exercises can be found on page A17. Turn to page A17 now and locate the answers for the Chapter 5 Cumulative Review Exercises. After the answer to the first exercise, which is $-6\sqrt{10}$, you will see the objective reference [1.2F]. This means that this question was taken from Chapter 1, Section 2, Objective F. If you missed this question, you should return to that objective and restudy the material.

A second way of preparing for a new chapter is to complete the ***Prep Test***. This test focuses on the particular skills that will be required for the new chapter. Turn to page 296 to see a Prep Test. The answers for the Prep Test are the first set of answers in the answer section for a chapter. Turn to page A14 to see the answers for the Chapter 5 Prep Test. Note that an objective reference is given for each question. If you answer a question incorrectly, restudy the objective from which the question was taken.

Before the class meeting in which your professor begins a new section, you should read each objective statement for that section. Next, browse through the objective material, being sure to note each word in bold type. These words indicate important concepts that you must know in order to learn the material. Do not worry about trying to understand all the material. Your professor is there to assist you with that endeavor. The purpose of browsing through the material is so that your brain will be prepared to accept and organize the new information when it is presented to you.

Turn to page 3. Write down the title of the first objective in Section 1.1. Write down the words under the title of the objective that are in bold print. It is not necessary for you to understand the meaning of these words. You are in this class to learn their meaning.

__________ __________ __________ __________

__________ __________ __________ __________

__________ __________ __________ __________

__________ __________ __________ __________

__________ __________ __________ __________

Math Is Not a Spectator Sport To learn mathematics you must be an active participant. Listening and watching your professor do mathematics is not enough. Mathematics requires that you interact with the lesson you are studying. If you filled in the blanks above, you were being interactive. There are other ways this textbook has been designed to help you be an active learner.

Annotated Examples The HOW TO feature indicates an example with explanatory remarks to the right of the work. Using paper and pencil, you should work along as you go through the example.

Copyright © Houghton Mifflin Company. All rights reserved.

$3x + 2 < -4$
$3x < -6$
$\frac{3x}{3} < \frac{-6}{3}$
$x < -2$
The solution set is $\{x | x < -2\}$.

HOW TO Solve: $3x + 2 < -4$

$$3x + 2 < -4$$
$$3x < -6$$
$$\frac{3x}{3} < \frac{-6}{3}$$
$$x < -2$$

- Subtract 2 from each side of the inequality.
- Divide each side of the inequality by the coefficient 3.

The solution set is $\{x | x < -2\}$.

Page 127

When you complete the example, get a clean sheet of paper. Write down the problem and then try to complete the solution without referring to your notes or the book. When you can do that, move on to the next part of the objective.

Leaf through the book now and write down the page numbers of two other occurrences of a HOW TO example.

You Try Its One of the key instructional features of this text is the paired examples. Notice that in each example box, the example on the left is completely worked out and the "You Try It" example on the right is not. Study the worked-out example carefully by working through each step. Then work the You Try It. If you get stuck, refer to the page number at the end of the example, which directs you to the place where the You Try It is solved—a complete worked-out solution is provided. Try to use the given solution to get a hint for the step you are stuck on. Then try to complete your solution.

Example 2

Solve: $3x - 5 \le 3 - 2(3x + 1)$

Solution

$$3x - 5 \le 3 - 2(3x + 1)$$
$$3x - 5 \le 3 - 6x - 2$$
$$3x - 5 \le 1 - 6x$$
$$9x - 5 \le 1$$
$$9x \le 6$$
$$\frac{9x}{9} \le \frac{6}{9}$$
$$x \le \frac{2}{3}$$

$\left\{x \middle| x \le \frac{2}{3}\right\}$

You Try It 2

Solve: $5x - 2 \le 4 - 3(x - 2)$

Your solution

$$5x - 2 \le 4 - 3(x - 2)$$
$$5x - 2 \le 4 - 3x + 6$$
$$5x - 2 \le 10 - 3x$$
$$8x - 2 \le 10$$
$$8x \le 12$$
$$\frac{8x}{8} \le \frac{12}{8}$$
$$x \le \frac{3}{2}$$

$\left(-\infty, \frac{3}{2}\right]$

Solution on p. S7

Page 128

When you have completed your solution, check your work against the solution we provided. (Turn to page S7 to see the solution of You Try It 2.) Be aware that frequently there is more than one way to solve a problem. Your answer, however, should be the same as the given answer. If you have any question as to whether your method will "always work," check with your instructor or with someone in the math center.

Browse through the textbook and write down the page numbers where two other paired example features occur.

Remember: Be an active participant in your learning process. When you are sitting in class watching and listening to an explanation, you may think that you understand. However, until you actually try to do it, you will have no confirmation of the new knowledge or skill. Most of us have had the experience of sitting in class thinking we knew how to do something only to get home and realize that we didn't.

TAKE NOTE

There is a strong connection between reading and being a successful student in math or any other subject. If you have difficulty reading, consider taking a reading course. Reading is much like other skills. There are certain things you can learn that will make you a better reader.

Copyright © Houghton Mifflin Company. All rights reserved.

Word Problems Word problems are difficult because we must read the problem, determine the quantity we must find, think of a method to do that, and then actually solve the problem. In short, we must formulate a *strategy* to solve the problem and then devise a *solution*.

Note in the paired example below that part of every word problem is a strategy and part is a solution. The strategy is a written description of how we will solve the problem. In the corresponding You Try It, you are asked to formulate a strategy. Do not skip this step, and be sure to write it out.

Example 6

Find three consecutive positive odd integers whose sum is between 27 and 51.

Strategy
To find the three integers, write and solve a compound inequality using n to represent the first odd integer.

Solution

$$\text{Lower limit of the sum} < \text{sum} < \text{upper limit of the sum}$$
$$27 < n + (n + 2) + (n + 4) < 51$$
$$27 < 3n + 6 < 51$$
$$27 - 6 < 3n + 6 - 6 < 51 - 6$$
$$21 < 3n < 45$$
$$\frac{21}{3} < \frac{3n}{3} < \frac{45}{3}$$
$$7 < n < 15$$

The three odd integers are 9, 11, and 13; or 11, 13, and 15; or 13, 15, and 17.

You Try It 6

An average score of 80 to 89 in a history course receives a B. Luisa Montez has grades of 72, 94, 83, and 70 on four exams. Find the range of scores on the fifth exam that will give Luisa a B for the course.

Your strategy
To find the scores, write and solve an inequality. Let N be the score on the last test.

Your solution

$$80 \le \frac{72 + 94 + 83 + 70 + N}{5} \le 89$$
$$80 \le \frac{319 + N}{5} \le 89$$
$$5(80) \le 5\left(\frac{319 + N}{5}\right) \le 5(89)$$
$$400 \le 319 + N \le 445$$
$$400 - 319 \le 319 + N - 319 \le 445 - 319$$
$$81 \le N \le 126$$

The range of scores to get a B is $81 \le N \le 100$.

Solutions on p. S8

Page 130

TAKE NOTE
If a rule has more than one part, be sure to make a notation to that effect.

Rule Boxes Pay special attention to rules placed in boxes. These rules give you the reasons certain types of problems are solved the way they are. When you see a rule, try to rewrite the rule in your own words.

I can add the same number to both sides of an inequality and not change the solution set.

When solving an inequality, we use the **Addition and Multiplication Properties of Inequalities** to rewrite the inequality in the form *variable* < *constant* or in the form *variable* > *constant*.

The Addition Property of Inequalities

If $a > b$, then $a + c > b + c$.
If $a < b$, then $a + c < b + c$.

Page 125

TAKE NOTE
If you are working at home and need assistance, there is online help available at math.college.hmco.com/students, at this text's website.

Chapter Exercises When you have completed studying an objective, do the exercises in the exercise set that correspond with that objective. The exercises are labeled with the same letter as the objective. Math is a subject that needs to be learned in small sections and practiced continually in order to be mastered. Doing all of the exercises in each exercise set will help you master the problem-solving techniques necessary for success. As you work through the exercises for an objective, check your answers to the odd-numbered exercises with those in the back of the book.

Copyright © Houghton Mifflin Company. All rights reserved.

Preparing for a Test There are important features of this text that can be used to prepare for a test.

- Chapter Summary
- Chapter Review Exercises
- Chapter Test

After completing a chapter, read the Chapter Summary. (See page 285 for the Chapter 4 Summary.) This summary highlights the important topics covered in the chapter. The page number following each topic refers you to the page in the text on which you can find more information about the concept.

Following the Chapter Summary are Chapter Review Exercises (see page 288) and a Chapter Test (see page 291). Doing the review exercises is an important way of testing your understanding of the chapter. The answer to each review exercise is given at the back of the book, along with its objective reference. After checking your answers, restudy any objective from which a question you missed was taken. It may be helpful to retry some of the exercises for that objective to reinforce your problem-solving techniques.

The Chapter Test should be used to prepare for an exam. We suggest that you try the Chapter Test a few days before your actual exam. Take the test in a quiet place and try to complete the test in the same amount of time you will be allowed for your exam. When taking the Chapter Test, practice the strategies of successful test takers: (1) scan the entire test to get a feel for the questions; (2) read the directions carefully; (3) work the problems that are easiest for you first; and perhaps most importantly, (4) try to stay calm.

When you have completed the Chapter Test, check your answers. If you missed a question, review the material in that objective and rework some of the exercises from that objective. This will strengthen your ability to perform the skills in that objective.

Is it difficult to be successful? YES! Successful music groups, artists, professional athletes, chefs, and Write your major here have to work very hard to achieve their goals. They focus on their goals and ignore distractions. The things we ask you to do to achieve success take time and commitment. We are confident that if you follow our suggestions, you will succeed.

Copyright © Houghton Mifflin Company. All rights reserved.

chapter

1 Real Numbers and Variable Expressions

When you take a multiple-choice test such as a class exam, the ACT, or the SAT, there is usually a point system for scoring your answers. Correct answers receive a positive number of points, and incorrect answers receive a negative number of points. For the ACT and the SAT, you will score higher if you leave a question blank when you are unsure of the answer. An unanswered question will cause fewer points to be deducted from your score; sometimes it will not cost you any points at all. **Exercises 182 and 183 on page 15** show how professors can adjust the grading systems on multiple-choice exams to discourage students from guessing randomly.

Need help? For online student resources, such as section quizzes, visit this textbook's website at **math.college.hmco.com/students.**

OBJECTIVES

Section 1.1

- **A** To use inequality symbols with integers
- **B** To find the additive inverse and absolute value of a number
- **C** To add or subtract integers
- **D** To multiply or divide integers
- **E** To solve application problems

Section 1.2

- **A** To write a rational number as a decimal
- **B** To convert among percents, fractions, and decimals
- **C** To add or subtract rational numbers
- **D** To multiply or divide rational numbers
- **E** To evaluate exponential expressions
- **F** To simplify numerical radical expressions
- **G** To solve application problems

Section 1.3

- **A** To use the Order of Operations Agreement to simplify expressions

Section 1.4

- **A** To evaluate a variable expression
- **B** To simplify a variable expression using the Properties of Addition
- **C** To simplify a variable expression using the Properties of Multiplication
- **D** To simplify a variable expression using the Distributive Property
- **E** To translate a verbal expression into a variable expression

Section 1.5

- **A** To write a set using the roster method
- **B** To write a set using set-builder notation
- **C** To graph an inequality on the number line

Copyright © Houghton Mifflin Company. All rights reserved.

PREP TEST • • •

Do these exercises to prepare for Chapter 1.

1. What is 127.1649 rounded to the nearest hundredth?

2. Add: $49.147 + 5.96$

3. Subtract: $5004 - 487$

4. Multiply: 407×28

5. Divide: $456 \div 19$

6. What is the smallest number that both 8 and 12 divide evenly into?

7. What is the greatest number that divides evenly into both 16 and 20?

8. Without using 1, write 21 as a product of two whole numbers.

9. Represent the shaded portion of the figure as a fraction.

GO FIGURE • • •

If you multiply the first 20 natural numbers

$$(1 \cdot 2 \cdot 3 \cdot 4 \cdot 5 \cdot \ldots \cdot 17 \cdot 18 \cdot 19 \cdot 20),$$

how many zeros will be at the end of the number?

Copyright © Houghton Mifflin Company. All rights reserved.

1.1 Integers

Objective A To use inequality symbols with integers

It seems to be a human characteristic to group similar items. For instance, nutritionists classify foods according to food groups: pasta, crackers, and rice are among the foods in the bread group.

Mathematicians likewise place objects with similar properties in *sets* and use braces to surround a list of the objects in the set, which are called **elements.** The numbers that we use to count elements, such as the number of people at a baseball game or the number of horses on a ranch, have similar characteristics. These numbers are the *natural numbers.*

$$\textbf{Natural numbers} = \{1, 2, 3, 4, 5, 6, 7, 8, 9, 10, 11, \ldots\}$$

The natural numbers alone do not provide all the numbers that are useful in applications. For instance, a meteorologist needs numbers below zero and above zero.

$$\textbf{Integers} = \{\ldots, -5, -4, -3, -2, -1, 0, 1, 2, 3, 4, 5, \ldots\}$$

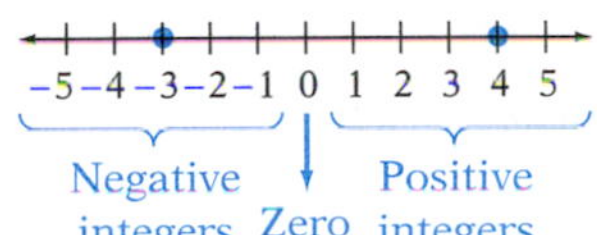

Each integer can be shown on a **number line.** The **graph of an integer** is shown by placing a heavy dot on the number line directly above the number. The graphs of -3 and 4 are shown on the number line at the left.

The integers to the left of zero are **negative integers.** The integers to the right of zero are **positive integers.** Zero is neither a positive nor a negative integer.

Point of Interest

The Alexandrian astronomer Ptolemy began using *omicron*, *O*, the first letter of the Greek word that means "nothing," as the symbol for zero in A.D. 150. It was not until the 13th century, however, that Fibonacci introduced 0 to the Western world as a placeholder so that we could distinguish, for example, 45 from 405.

Consider the sentences below.

The quarterback threw the football and the receiver caught *it*.
An accountant purchased a calculator and placed *it* in a briefcase.

In the first sentence, *it* means football; in the second sentence, *it* means calculator. In language, the word *it* can stand for many different objects. Similarly, in mathematics, a letter of the alphabet can be used to stand for a number. Such a letter is called a **variable.** Variables are used in the next definition.

Definition of Inequality Symbols

If a and b are two numbers and a is to the left of b on the number line, then a is **less than** b. This is written $a < b$.

If a and b are two numbers and a is to the right of b on the number line, then a is **greater than** b. This is written $a > b$.

There are also inequality symbols for **less than or equal to** ($\leq$) and **greater than or equal to** ($\geq$). For instance,

$$6 \leq 6 \text{ because } 6 = 6. \qquad 7 \leq 15 \text{ because } 7 < 15.$$

It is convenient to use a variable to represent, or stand for, any one of the elements of a set. For instance, the statement "x is an element of the set $\{0, 2, 4, 6\}$" means that x can be replaced by 0, 2, 4, or 6. The symbol for "is an element of" is $\in$; the symbol for "is not an element of" is $\notin$. For example,

$$2 \in \{0, 2, 4, 6\} \qquad 6 \in \{0, 2, 4, 6\} \qquad 7 \notin \{0, 2, 4, 6\}$$

Copyright © Houghton Mifflin Company. All rights reserved.

Example 1

Let $x \in \{-6, -2, 0\}$. For which values of x is the inequality $x \leq -2$ a true statement?

Solution

Replace x by each element of the set and determine whether the inequality is true.

$x \leq -2$

$-6 \leq -2$ True. $-6 < -2$

$-2 \leq -2$ True. $-2 = -2$

$0 \leq -2$ False.

The inequality is true for -6 and -2.

You Try It 1

Let $y \in \{-5, -1, 5\}$. For which values of y is the inequality $y > -1$ a true statement?

Your solution

Solution on p. S1

Objective B To find the additive inverse and absolute value of a number

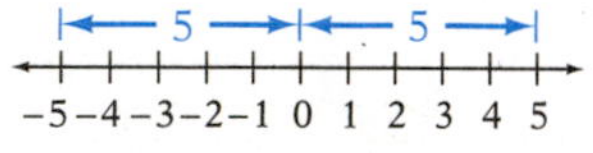

On the number line, the numbers 5 and -5 are the same distance from zero but on opposite sides of zero. The numbers 5 and -5 are called **opposites** or **additive inverses** of each other. (See the number line at the left.)

TAKE NOTE

The distance from 0 to 5 is 5; $|5| = 5$. The distance from 0 to -5 is 5; $|-5| = 5$.

The opposite (or additive inverse) of 5 is -5. The opposite of -5 is 5. The symbol for opposite is $-$.

$-(5)$ means the opposite of *positive* 5. $-(5) = -5$

$-(-5)$ means the opposite of *negative* 5. $-(-5) = 5$

The **absolute value** of a number is its distance from zero on the number line. The symbol for absolute value is two vertical bars, $|\ |$.

Point of Interest

The definition of absolute value that we have given in the box is written in what is called "rhetorical style." That is, it is written without the use of variables. This is how all mathematics was written prior to the Renaissance. During that period, from the 14th to the 16th century, the idea of expressing a variable symbolically was developed.

Absolute Value

The absolute value of a positive number is the number itself. The absolute value of zero is zero. The absolute value of a negative number is the opposite of the negative number.

$|9| = 9$
$|0| = 0$
$|-7| = 7$

HOW TO Evaluate: $-|-12|$

$-|-12| = -12$

- **The absolute value sign does not affect the negative sign in front of the absolute value sign.**

Example 2

Let $a \in \{-12, 0, 4\}$. Find the additive inverse of a and the absolute value of a for each element of the set.

Solution

Replace a by each element of the set.

$-a$	$\lvert a \rvert$
$-(-12) = 12$	$\lvert -12 \rvert = 12$
$-(0) = 0$	$\lvert 0 \rvert = 0$
$-(4) = -4$	$\lvert 4 \rvert = 4$

You Try It 2

Let $z \in \{-11, 0, 8\}$. Find the additive inverse of z and the absolute value of z for each element of the set.

Your solution

Solution on p. S1

Copyright © Houghton Mifflin Company. All rights reserved.

Objective C To add or subtract integers

A number can be represented anywhere along the number line by an arrow. A positive number is represented by an arrow pointing to the right, and a negative number is represented by an arrow pointing to the left. The size of the number is represented by the length of the arrow.

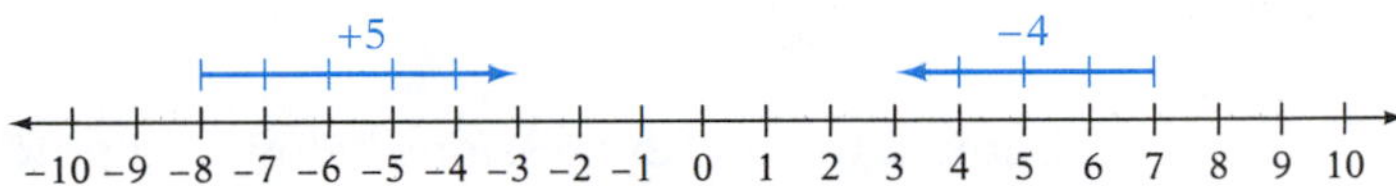

Addition of integers can be shown on the number line. To add integers, start at zero and draw an arrow representing the first number. At the tip of the first arrow, draw a second arrow representing the second number. The sum is below the tip of the second arrow.

TAKE NOTE

Each number in a sum is called an **addend.** For instance, 4 and 2 are addends in the sum $4 + 2 = 6$.

$4 + 2 = 6$

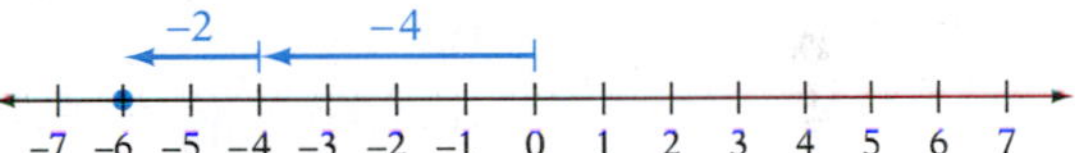

$-4 + (-2) = -6$

$-4 + 2 = -2$

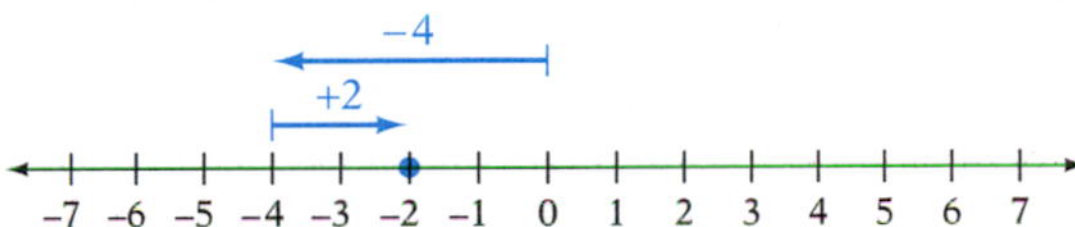

$4 + (-2) = 2$

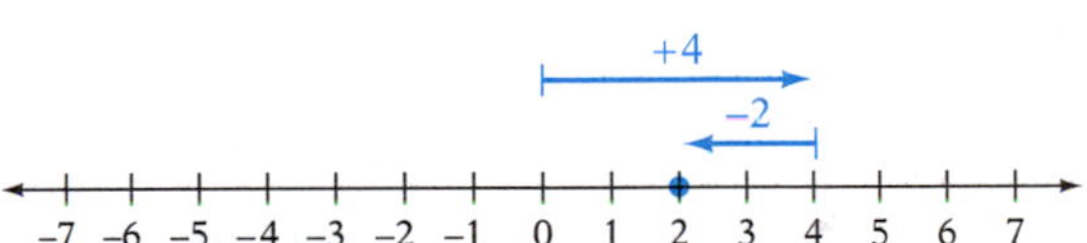

The pattern for the addition of integers shown on the number line can be summarized in the following rule.

Addition of Integers

- *Numbers with the same sign*
  To add two numbers with the same sign, add the absolute values of the numbers. Then attach the sign of the addends.
- *Numbers with different signs*
  To add two numbers with different signs, find the absolute value of each number. Then subtract the smaller of these numbers from the larger. Attach the sign of the number with the larger absolute value.

HOW TO Add: $(-9) + 8$

$|-9| = 9 \quad |8| = 8$ • **The signs are different. Find the absolute value of each number.**

$9 - 8 = 1$ • **Subtract the smaller number from the larger.**

$(-9) + 8 = -1$ • **Attach the sign of the number with the larger absolute value. Because $|-9| > |8|$, use the sign of -9.**

Copyright © Houghton Mifflin Company. All rights reserved.

HOW TO Add: $(-23) + 47 + (-18) + 5$

To add more than two numbers, add the first two numbers. Then add the sum to the third number. Continue until all the numbers are added.

$$\begin{aligned}(-23) + 47 + (-18) + 5 &= 24 + (-18) + 5\\ &= 6 + 5\\ &= 11\end{aligned}$$

Look at the two expressions below and note that each expression equals the same number.

$8 - 3 = 5$ 8 minus 3 is 5.

$8 + (-3) = 5$ 8 plus the opposite of 3 is 5.

This example suggests that to subtract two numbers, we add the opposite of the second number to the first number.

first number	−	second number	=	first number	+	the opposite of the second number	
40	−	60	=	40	+	(−60)	= −20
−40	−	60	=	−40	+	(−60)	= −100
−40	−	(−60)	=	−40	+	60	= 20
40	−	(−60)	=	40	+	60	= 100

HOW TO Subtract: $-21 - (-40)$

Change this sign to plus.

$-21 - (-40) = -21 + 40 = 19$

Change −40 to the opposite of −40.

- **Rewrite subtraction as addition of the opposite. Then add.**

HOW TO Subtract: $15 - 51$

Change this sign to plus.

$15 - 51 = 15 + (-51) = -36$

Change 51 to the opposite of 51.

- **Rewrite subtraction as addition of the opposite. Then add.**

HOW TO Subtract: $-12 - (-21) - 15$

$$\begin{aligned}-12 - (-21) - 15 &= -12 + 21 + (-15)\\ &= 9 + (-15)\\ &= -6\end{aligned}$$

- **Rewrite each subtraction as addition of the opposite. Then add.**

Copyright © Houghton Mifflin Company. All rights reserved.

Example 3 Add: $(-52) + (-39)$

Solution The signs are the same. Add the absolute values of the numbers: $52 + 39 = 91$

Attach the sign of the addends: $(-52) + (-39) = -91$

You Try It 3 Add: $100 + (-43)$

Your solution

Example 4 Add: $37 + (-52) + (-21) + (-7)$

Solution

$$\begin{aligned} &37 + (-52) + (-21) + (-7) \\ &= -15 + (-21) + (-7) \\ &= -36 + (-7) \\ &= -43 \end{aligned}$$

You Try It 4 Add: $(-51) + 42 + 17 + (-102)$

Your solution

Example 5 Subtract: $-11 - 15$

Solution

$$\begin{aligned} -11 - 15 &= -11 + (-15) \\ &= -26 \end{aligned}$$

You Try It 5 Subtract: $19 - (-32)$

Your solution

Example 6 Subtract: $-14 - 18 - (-21) - 4$

Solution

$$\begin{aligned} &-14 - 18 - (-21) - 4 \\ &= -14 + (-18) + 21 + (-4) \\ &= -32 + 21 + (-4) \\ &= -11 + (-4) \\ &= -15 \end{aligned}$$

You Try It 6 Subtract: $-9 - (-12) - 17 - 4$

Your solution

Solutions on p. S1

Objective D To multiply or divide integers

Multiplication is the repeated addition of the same number. The product 3×5 is shown on the number line below.

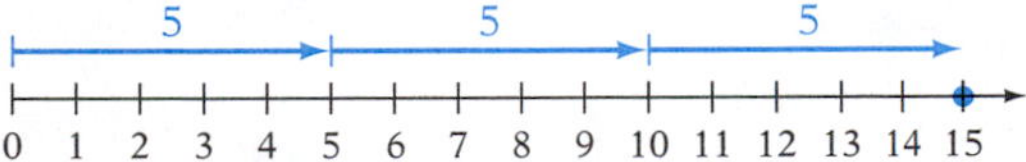

$$3 \times 5 = \underbrace{5 + 5 + 5}_{\text{5 is added 3 times.}} = 15$$

> **TAKE NOTE**
> Each number of a product is called a **factor.** For instance, 3 and 5 are factors of the product $3 \cdot 5 = 15$.

To indicate multiplication, several different symbols are used.

$3 \times 5 = 15$ $\quad 3 \cdot 5 = 15$ $\quad (3)(5) = 15$ $\quad 3(5) = 15$ $\quad (3)5 = 15$

Note that when parentheses are used and there is no arithmetic operation symbol, the operation is multiplication.

Copyright © Houghton Mifflin Company. All rights reserved.

Now consider the product of a positive and a negative number.

TAKE NOTE
$3(-5)$ is 3 times -5.

$$3(-5) = \underbrace{(-5) + (-5) + (-5)}_{-5 \text{ is added 3 times.}} = -15$$

• **Multiplication is repeated addition.**

This suggests that the product of a positive number and a negative number is negative. Here are a few more examples.

$4(-7) = -28 \qquad (-6)5 = -30 \qquad (-5) \cdot 7 = -35$

To find the product of two negative numbers, look at the pattern at the right. As -5 multiplies a sequence of decreasing integers, the products increase by 5.

The pattern can be continued by requiring that the product of two negative numbers be positive.

These numbers decrease by 1. These numbers increase by 5.

$$\begin{aligned} -5(3) &= -15 \\ -5(2) &= -10 \\ -5(1) &= -5 \\ -5(0) &= 0 \\ -5(-1) &= 5 \\ -5(-2) &= 10 \\ -5(-3) &= 15 \end{aligned}$$

Multiplication of Integers

- *Numbers with the same sign*
  To multiply two numbers with the same sign, multiply the absolute values of the numbers. The product is positive.
- *Numbers with different signs*
  To multiply two numbers with different signs, multiply the absolute values of the numbers. The product is negative.

HOW TO Multiply: $-2(5)(-7)(-4)$

$$\begin{aligned} -2(5)(-7)(-4) &= -10(-7)(-4) \\ &= 70(-4) \\ &= -280 \end{aligned}$$

• **To multiply more than two numbers, multiply the first two numbers. Then multiply the product by the third number. Continue until all the numbers are multiplied.**

For every division problem there is a related multiplication problem.

TAKE NOTE
In division, the **quotient** is the result of dividing the divisor into the dividend.

$\frac{8}{2} = 4$ because $4 \cdot 2 = 8$

Division — Related multiplication

This fact and the rules for multiplying integers can be used to illustrate the rules for dividing integers.

Note in the following examples that the quotient of two numbers with the same sign is positive.

$\frac{12}{3} = 4$ because $4 \cdot 3 = 12 \qquad \frac{-12}{-3} = 4$ because $4 \cdot (-3) = -12$

The next two examples illustrate that the quotient of two numbers with different signs is negative.

$\frac{12}{-3} = -4$ because $(-4)(-3) = 12 \qquad \frac{-12}{3} = -4$ because $(-4) \cdot 3 = -12$

Copyright © Houghton Mifflin Company. All rights reserved.

Division of Integers

- *Numbers with the same sign*
  To divide two numbers with the same sign, divide the absolute values of the numbers. The quotient is positive.
- *Numbers with different signs*
  To divide two numbers with different signs, divide the absolute values of the numbers. The quotient is negative.

HOW TO Simplify: $-\frac{-56}{7}$

$$-\frac{-56}{7} = -\left(\frac{-56}{7}\right) = -(-8) = 8$$

Note that $\frac{-12}{3} = -4$, $\frac{12}{-3} = -4$, and $-\frac{12}{3} = -4$. This suggests the following rule.

TAKE NOTE
The symbol ≠ is read "is not equal to."

If a and b are integers, and $b \neq 0$, then $\frac{-a}{b} = \frac{a}{-b} = -\frac{a}{b}$.

Properties of Zero and One in Division

- Zero divided by any number other than zero is zero.

  $\frac{0}{a} = 0$ because $0 \cdot a = 0$ For example, $\frac{0}{7} = 0$ because $0 \cdot 7 = 0$.

- Division by zero is not defined.

  To understand that division by zero is not permitted, suppose that $\frac{4}{0}$ were equal to n, where n is some number. Because each division problem has a related multiplication problem, $\frac{4}{0} = n$ means $n \cdot 0 = 4$. But $n \cdot 0 = 4$ is impossible because any number times 0 is 0. Therefore, division by 0 is not defined.

- Any number other than zero divided by itself is 1.

  $\frac{a}{a} = 1$, $a \neq 0$ For example, $\frac{-8}{-8} = 1$.

- Any number divided by one is the number.

  $\frac{a}{1} = a$ For example, $\frac{9}{1} = 9$.

Integrating Technology
Enter 4 ÷ 0 = on your calculator. You will get an error message.

TAKE NOTE
"To the Student" on page xxviii explains how best to use the boxed examples in this text, such as the one below.

Copyright © Houghton Mifflin Company. All rights reserved.

Example 7 Multiply: $(-3)4(-5)$

Solution $(-3)4(-5) = (-12)(-5) = 60$

You Try It 7 Multiply: $8(-9)10$

Your solution

Solution on p. S1

Example 8 Multiply: $12(-4)(-3)(-5)$

Solution $12(-4)(-3)(-5)$
$= (-48)(-3)(-5)$
$= 144(-5) = -720$

You Try It 8 Multiply: $(-2)3(-8)7$

Your solution

Example 9 Divide: $(-120) \div (-8)$

Solution $(-120) \div (-8) = 15$

You Try It 9 Divide: $(-135) \div (-9)$

Your solution

Example 10 Divide: $\frac{95}{-5}$

Solution $\frac{95}{-5} = -19$

You Try It 10 Divide: $\frac{-72}{4}$

Your solution

Example 11 Divide: $-\frac{-81}{3}$

Solution $-\frac{-81}{3} = -(-27) = 27$

You Try It 11 Divide: $-\frac{36}{-12}$

Your solution

Solutions on p. S1

Objective E To solve application problems

To solve an application problem, first read the problem carefully. The Strategy involves identifying the quantity to be found and planning the steps that are necessary to find that quantity. The Solution involves performing each operation stated in the Strategy and writing the answer.

Example 12

The average temperature on Mercury's sunlit side is 950°F. The average temperature on Mercury's dark side is −346°F. Find the difference between these two average temperatures.

Strategy
To find the difference, subtract the average temperature on the dark side (−346) from the average temperature on the sunlit side (950).

Solution
$950 - (-346) = 950 + 346$
$= 1296$
The difference between these average temperatures is 1296°F.

You Try It 12

The daily low temperatures (in degrees Celsius) during one week were recorded as follows: −6°, −7°, 0°, −5°, −8°, −1°, −1°. Find the average daily low temperature.

Your strategy

Your solution

Solution on p. S1

Copyright © Houghton Mifflin Company. All rights reserved.

1.1 Exercises

TAKE NOTE
"To the Student" on page xxiii discusses the exercise sets in this textbook.

Objective A To use inequality symbols with integers

Place the correct symbol, $<$ or $>$, between the two numbers.

1. 8 −6 **2.** −14 16 **3.** −12 1 **4.** 35 28 **5.** 42 19

6. −42 27 **7.** 0 −31 **8.** −17 0 **9.** 53 −46 **10.** −27 −39

Answer true or false.

11. $-13 > 0$ **12.** $-20 > 3$ **13.** $12 > -31$ **14.** $9 > 7$ **15.** $-5 > -2$

16. $-44 > -21$ **17.** $-4 > -120$ **18.** $0 > -8$ **19.** $-1 > 0$ **20.** $-10 > -88$

21. Let $x \in \{-23, -18, -8, 0\}$. For which values of x is the inequality $x < -8$ a true statement?

22. Let $w \in \{-33, -24, -10, 0\}$. For which values of w is the inequality $w < -10$ a true statement?

23. Let $a \in \{-33, -15, 21, 37\}$. For which values of a is the inequality $a > -10$ a true statement?

24. Let $v \in \{-27, -14, 14, 27\}$. For which values of v is the inequality $v > -15$ a true statement?

25. Let $n \in \{-23, -1, 0, 4, 29\}$. For which values of n is the inequality $-6 > n$ a true statement?

26. Let $m \in \{-33, -11, 0, 12, 45\}$. For which values of m is the inequality $-15 > m$ a true statement?

Objective B To find the additive inverse and absolute value of a number

Find the additive inverse.

27. 4 **28.** 8 **29.** −9 **30.** −12 **31.** −28 **32.** −36

Evaluate.

33. $-(-14)$ **34.** $-(-40)$ **35.** $-(77)$ **36.** $-(39)$ **37.** $-(0)$ **38.** $-(-13)$

39. $|-74|$ **40.** $|-96|$ **41.** $-|-82|$ **42.** $-|-53|$ **43.** $-|81|$ **44.** $-|38|$

Copyright © Houghton Mifflin Company. All rights reserved.

Place the correct symbol, < or >, between the values of the two numbers.

45. $|-83| \quad |58|$ **46.** $|22| \quad |-19|$ **47.** $|43| \quad |-52|$ **48.** $|-71| \quad |-92|$

49. $|-68| \quad |-42|$ **50.** $|12| \quad |-31|$ **51.** $|-45| \quad |-61|$ **52.** $|-28| \quad |43|$

53. Let $p \in \{-19, 0, 28\}$. Evaluate $-p$ for each element of the set.

54. Let $q \in \{-34, 0, 31\}$. Evaluate $-q$ for each element of the set.

55. Let $x \in \{-45, 0, 17\}$. Evaluate $-|x|$ for each element of the set.

56. Let $y \in \{-91, 0, 48\}$. Evaluate $-|y|$ for each element of the set.

Objective C To add or subtract integers

57. **a.** Explain the rule for adding two integers with the same sign.
b. Explain the rule for adding two integers with different signs.

58. Explain how to rewrite the subtraction $8 - (-6)$ as addition of the opposite.

Add or subtract.

59. $-3 + (-8)$ **60.** $-6 + (-9)$ **61.** $-8 + 3$ **62.** $-9 + 2$

63. $-3 + (-80)$ **64.** $-12 + (-1)$ **65.** $-23 + (-23)$ **66.** $-12 + (-12)$

67. $16 + (-16)$ **68.** $-17 + 17$ **69.** $48 + (-53)$ **70.** $19 + (-41)$

71. $-17 + (-3) + 29$ **72.** $13 + 62 + (-38)$ **73.** $-3 + (-8) + 12$ **74.** $-27 + (-42) + (-18)$

75. $16 - 8$ **76.** $12 - 3$ **77.** $7 - 14$ **78.** $6 - 9$

79. $-7 - 2$ **80.** $-9 - 4$ **81.** $7 - (-2)$ **82.** $3 - (-4)$

83. $-6 - (-3)$ **84.** $-4 - (-2)$ **85.** $6 - (-12)$ **86.** $-12 - 16$

Copyright © Houghton Mifflin Company. All rights reserved.

87. $13 + (-22) + 4 + (-5)$

88. $-14 + (-3) + 7 + (-21)$

89. $-16 + (-17) + (-18) + 10$

90. $-25 + (-31) + 24 + 19$

91. $26 + (-15) + (-11) + (-12)$

92. $-32 + 40 + (-8) + (-19)$

93. $-14 + (-15) + (-11) + 40$

94. $28 + (-19) + (-8) + (-1)$

95. $-4 - 3 - 2$

96. $4 - 5 - 12$

97. $12 - (-7) - 8$

98. $-12 - (-3) - (-15)$

99. $-19 - (-19) - 18$

100. $-8 - (-8) - 14$

101. $-17 - (-8) - (-9)$

102. $7 - 8 - (-1)$

103. $-30 - (-65) - 29 - 4$

104. $42 - (-82) - 65 - 7$

105. $-16 - 47 - 63 - 12$

106. $42 - (-30) - 65 - (-11)$

107. $-47 - (-67) - 13 - 15$

108. $-18 - 49 - (-84) - 27$

109. $-19 - 17 - (-36) - 12$

110. $48 - 19 - 29 - 51$

111. $21 - (-14) - 43 - 12$

112. $17 - (-17) - 14 - 21$

Objective D To multiply or divide integers

113. Describe the rules for multiplying two integers.

114. Name the operation in each expression. Justify your answer.
a. $8(-7)$ **b.** $8 - 7$ **c.** $8 - (-7)$ **d.** $-xy$ **e.** $x(-y)$ **f.** $-x - y$

Copyright © Houghton Mifflin Company. All rights reserved.

Multiply or divide.

115. $(14)3$ **116.** $(17)6$ **117.** $-7 \cdot 4$ **118.** $-8 \cdot 7$ **119.** $(-12)(-5)$ **120.** $(-13)(-9)$

121. $-11(23)$ **122.** $-8(21)$ **123.** $(-17)14$ **124.** $(-15)12$ **125.** $6(-19)$ **126.** $17(-13)$

127. $12 \div (-6)$ **128.** $18 \div (-3)$ **129.** $(-72) \div (-9)$ **130.** $(-64) \div (-8)$ **131.** $-42 \div 6$

132. $(-56) \div 8$ **133.** $(-144) \div 12$ **134.** $(-93) \div (-3)$ **135.** $48 \div (-8)$ **136.** $57 \div (-3)$

137. $\frac{-49}{7}$ **138.** $\frac{-45}{5}$ **139.** $\frac{-44}{-4}$ **140.** $\frac{-36}{-9}$ **141.** $\frac{98}{-7}$

142. $\frac{85}{-5}$ **143.** $-\frac{-120}{8}$ **144.** $-\frac{-72}{4}$ **145.** $-\frac{-80}{-5}$ **146.** $-\frac{-114}{-6}$

147. $0 \div (-9)$ **148.** $0 \div (-14)$ **149.** $\frac{-261}{9}$ **150.** $\frac{-128}{4}$ **151.** $9 \div 0$

152. $(-21) \div 0$ **153.** $\frac{132}{-12}$ **154.** $\frac{250}{-25}$ **155.** $\frac{0}{0}$ **156.** $\frac{-58}{0}$

157. $7(5)(-3)$ **158.** $(-3)(-2)8$ **159.** $9(-7)(-4)$ **160.** $(-2)(6)(-4)$

161. $16(-3)5$ **162.** $20(-4)3$ **163.** $-4(-3)8$ **164.** $-5(-9)6$

165. $-3(-8)(-9)$ **166.** $-7(-6)(-5)$ **167.** $(-9)7(5)$ **168.** $(-8)7(10)$

169. $7(-2)(5)(-6)$ **170.** $(-3)7(-2)8$ **171.** $-9(-4)(-8)(-10)$ **172.** $-11(-3)(-5)(-2)$

173. $7(9)(-11)4$ **174.** $-12(-4)7(-2)$ **175.** $(-14)9(-11)0$ **176.** $(-13)(15)(-19)0$

Copyright © Houghton Mifflin Company. All rights reserved.

Objective E To solve application problems

Geography The elevation, or height, of places on Earth is measured in relation to sea level, or the average level of the ocean's surface. The table below shows height above sea level as a positive number and depth below sea level as a negative number. Use the table for Exercises 177 to 179.

Continent	*Highest Elevation (in meters)*		*Lowest Elevation (in meters)*	
Africa	Mt. Kilimanjaro	5895	Qattara Depression	−133
Asia	Mt. Everest	8848	Dead Sea	−400
Europe	Mt. Elbrus	5634	Caspian Sea	−28
America	Mt. Aconcagua	6960	Death Valley	−86

Mt. Everest

177. Find the difference in elevation between Mt. Aconcagua and Death Valley.

178. What is the difference in elevation between Mt. Kilimanjaro and the Qattara Depression?

179. For which continent shown is the difference between the highest and lowest elevations greatest?

Chemistry The table at the right shows the boiling point and melting point in degrees Celsius for three chemical elements. Use this table for Exercises 180 and 181.

Chemical Element	*Boiling Point*	*Melting Point*
Mercury	357	−39
Radon	−62	−71
Xenon	−107	−112

180. Find the difference between the boiling point and melting point of mercury.

181. Find the difference between the boiling point and melting point of xenon.

182. Testing To discourage random guessing on a multiple-choice exam, a professor assigns 5 points for a correct answer, −2 points for an incorrect answer, and 0 points for leaving the question blank. What is the score for a student who had 20 correct answers, had 13 incorrect answers, and left 7 questions blank?

183. Testing To discourage random guessing on a multiple-choice exam, a professor assigns 7 points for a correct answer, −3 points for an incorrect answer, and −1 point for leaving the question blank. What is the score for a student who had 17 correct answers, had 8 incorrect answers, and left 2 questions blank?

Copyright © Houghton Mifflin Company. All rights reserved.

The Atmosphere The table at the right shows the average temperatures at different cruising altitudes for airplanes. Use the table for Exercises 184 to 186.

Cruising Altitude	*Average Temperature*
12,000 ft	16°F
20,000 ft	−12°F
30,000 ft	−48°F
40,000 ft	−70°F
50,000 ft	−70°F

184. What is the difference between the average temperatures at 12,000 ft and at 40,000 ft?

185. What is the difference between the average temperatures at 40,000 ft and at 50,000 ft?

186. How much colder is the average temperature at 30,000 ft than at 20,000 ft?

Meteorology A meteorologist may report a wind-chill temperature. This is the equivalent temperature, including the effects of wind and temperature, that a person would feel in calm air conditions. The table below gives the wind-chill temperature for various wind speeds and temperatures. For instance, when the temperature is 5°F and the wind is blowing at 15 mph, the wind-chill temperature is −13°F. Use this table for Exercises 187 and 188.

Wind Speed (in mph)	Wind Chill Factors Thermometer Reading (in degrees Fahrenheit)														
	25	20	15	10	5	0	−5	−10	−15	−20	−25	−30	−35	−40	−45
5	19	13	7	1	−5	−11	−16	−22	−28	−34	−40	−46	−52	−57	−63
10	15	9	3	−4	−10	−16	−22	−28	−35	−41	−47	−53	−59	−66	−72
15	13	6	0	−7	−13	−19	−26	−32	−39	−45	−51	−58	−64	−71	−77
20	11	4	−2	−9	−15	−22	−29	−35	−42	−48	−55	−61	−68	−74	−81
25	9	3	−4	−11	−17	−24	−31	−37	−44	−51	−58	−64	−71	−78	−84
30	8	1	−5	−12	−19	−26	−33	−39	−46	−53	−60	−67	−73	−80	−87
35	7	0	−7	−14	−21	−27	−34	−41	−48	−55	−62	−69	−76	−82	−89
40	6	−1	−8	−15	−22	−29	−36	−43	−50	−57	−64	−71	−78	−84	−91
45	5	−2	−9	−16	−23	−30	−37	−44	−51	−58	−65	−72	−79	−86	−93

187. When the thermometer reading is −5°F, what is the difference between the wind-chill factor when the wind is blowing at 10 mph and when the wind is blowing at 25 mph?

188. When the thermometer reading is −20°F, what is the difference between the wind-chill factor when the wind is blowing at 15 mph and when the wind is blowing at 25 mph?

APPLYING THE CONCEPTS

189. If $-4x$ equals a positive integer, is x a positive or a negative integer? Explain your answer.

190. Is the difference between two integers always smaller than either of the integers? If not, give an example in which the difference between two integers is greater than either integer.

Copyright © Houghton Mifflin Company. All rights reserved.

1.2 Rational and Irrational Numbers

Objective A To write a rational number as a decimal

Point of Interest

As early as A.D. 630, the Hindu mathematician Brahmagupta wrote a fraction as one number over another separated by a space. The Arab mathematician al Hassar (around A.D. 1050) was the first to show a fraction with a horizontal bar separating the numerator and denominator.

A *rational number* is the quotient of two integers. A rational number written in the following way is commonly called a fraction. Here are some examples of rational numbers.

$$\frac{3}{4}, \quad \frac{-4}{9}, \quad \frac{15}{-4}, \quad \frac{8}{1}, \quad -\frac{5}{6}$$

Rational Numbers

A **rational number** is a number that can be written in the form $\frac{a}{b}$, where a and b are integers and $b \neq 0$.

Because an integer can be written as the quotient of the integer and 1, every integer is a rational number. For instance,

$$\frac{6}{1} = 6 \qquad \frac{-8}{1} = -8$$

Point of Interest

Simon Stevin (1548–1620) was the first to name decimal numbers. He wrote the number 2.345 as 2 0 3 1 4 2 5 3. He called the whole number part the *commencement*; the tenths digit was *prime*, the hundredths digit was *second*, the thousandths digit was *third*, and so on.

A number written in **decimal notation** is also a rational number.

three-tenths $0.3 = \frac{3}{10}$ forty-three thousandths $0.043 = \frac{43}{1000}$

A rational number written as a fraction can be rewritten in decimal notation.

HOW TO Write $\frac{5}{8}$ as a decimal.

The fraction bar can be read "÷".

$$\frac{5}{8} = 5 \div 8$$

```
   0.625   ← This is called a terminating decimal.
8)5.000
 -4 8
    20
   -16
    40
   -40
     0   ← The remainder is zero.
```

$$\frac{5}{8} = 0.625$$

Write $\frac{4}{11}$ as a decimal.

```
   0.3636...  ← This is called a repeating decimal.
11)4.0000
  -3 3
     70
    -66
      40
     -33
       70
      -66
        4   ← The remainder is never zero.
```

$$\frac{4}{11} = 0.\overline{36}$$ ← The bar over the digits 3 and 6 is used to show that these digits repeat.

Copyright © Houghton Mifflin Company. All rights reserved.

Example 1

Write $\frac{8}{11}$ as a decimal. Place a bar over the repeating digits of the decimal.

Solution $\frac{8}{11} = 8 \div 11 = 0.7272\ldots = 0.\overline{72}$

You Try It 1

Write $\frac{4}{9}$ as a decimal. Place a bar over the repeating digits of the decimal.

Your solution

Solution on p. S1

Objective B To convert among percents, fractions, and decimals

Percent means "parts of 100." Thus 27% means 27 parts of 100.

27% of the region is shaded.

In applied problems involving percent, it may be necessary to rewrite a percent as a fraction or decimal or to rewrite a fraction or decimal as a percent.

To write a percent as a fraction, remove the percent sign and multiply by $\frac{1}{100}$.

HOW TO Write 27% as a fraction.

$$27\% = 27\left(\frac{1}{100}\right) = \frac{27}{100}$$

• Remove the percent sign and multiply by $\frac{1}{100}$.

To write a percent as a decimal, remove the percent sign and multiply by 0.01.

$$33\% = 33(0.01) = 0.33$$

Move the decimal point two places to the left. Then remove the percent sign.

A fraction or decimal can be written as a percent by multiplying by 100%.

$$\frac{5}{8} = \frac{5}{8}(100\%) = \frac{500}{8}\% = 62.5\%, \text{ or } 62\frac{1}{2}\%$$

$$0.82 = 0.82(100\%) = 82\%$$

Move the decimal point two places to the right. Then write the percent sign.

Example 2

Write 130% as a fraction and as a decimal.

Solution

$$130\% = 130\left(\frac{1}{100}\right) = \frac{130}{100} = \frac{13}{10}$$

$$130\% = 130(0.01) = 1.30$$

You Try It 2

Write 125% as a fraction and as a decimal.

Your solution

Solution on p. S1

Copyright © Houghton Mifflin Company. All rights reserved.

Example 3

Write $\frac{5}{6}$ as a percent.

Solution $\frac{5}{6} = \frac{5}{6}(100\%) = \frac{500}{6}\% = 83\frac{1}{3}\%$

You Try It 3

Write $\frac{1}{3}$ as a percent.

Your solution

Example 4 Write 0.092 as a percent.

Solution $0.092 = 0.092(100\%) = 9.2\%$

You Try It 4 Write 0.043 as a percent.

Your solution

Solutions on p. S1

Objective C To add or subtract rational numbers

Fractions with the same denominator are added by adding the numerators and placing the sum over the common denominator.

Addition of Fractions

To add two fractions with the same denominator, add the numerators and place the sum over the common denominator.

$$\frac{a}{c} + \frac{b}{c} = \frac{a+b}{c}$$

To add fractions with different denominators, first rewrite the fractions as equivalent fractions with a common denominator. Then add the fractions.

The least common denominator is the **least common multiple** (LCM) of the denominators. This is the smallest number that is a multiple of each of the denominators.

TAKE NOTE

You can find the LCM by multiplying the denominators and then dividing by the *common factor* of the two denominators. In the case of 6 and 10, $6 \cdot 10 = 60$. Now divide by 2, the common factor of 6 and 10.

$60 \div 2 = 30$

HOW TO Add: $-\frac{5}{6} + \frac{3}{10}$

The LCM of 6 and 10 is 30. Rewrite the fractions as equivalent fractions with the denominator 30. Then add the fractions.

$$-\frac{5}{6} + \frac{3}{10} = -\frac{5}{6} \cdot \frac{5}{5} + \frac{3}{10} \cdot \frac{3}{3} = -\frac{25}{30} + \frac{9}{30} = \frac{-25+9}{30} = \frac{-16}{30} = -\frac{8}{15}$$

To subtract fractions with the same denominator, subtract the numerators and place the difference over the common denominator.

TAKE NOTE

The least common multiple of the denominators is frequently called the **least common denominator** (LCD).

HOW TO Subtract: $-\frac{4}{9} - \left(-\frac{7}{12}\right)$

Rewrite subtraction as addition of the opposite. The LCM of 9 and 12 is 36. Rewrite the fractions as equivalent fractions with the denominator 36.

$$-\frac{4}{9} - \left(-\frac{7}{12}\right) = -\frac{4}{9} + \frac{7}{12} = -\frac{16}{36} + \frac{21}{36} = \frac{-16+21}{36} = \frac{5}{36}$$

Copyright © Houghton Mifflin Company. All rights reserved.

To add or subtract decimals, write the numbers so that the decimal points are in a vertical line. Then proceed as in the addition or subtraction of integers. Write the decimal point in the answer directly below the decimal points in the problem.

HOW TO Add: $-114.039 + 84.76$

$|-114.039| = 114.039$
$|84.76| = 84.76$

• The signs are different. Find the absolute value of each number.

$$\begin{array}{r} 114.039 \\ -\ 84.76 \\ \hline 29.279 \end{array}$$

• Subtract the smaller of these numbers from the larger.

$-114.039 + 84.76 = -29.279$

• Attach the sign of the number with the larger absolute value. Because $|-114.039| > |84.76|$, use the sign of -114.039.

Example 5

Simplify: $-\frac{3}{4} + \frac{1}{6} - \frac{5}{8}$

Solution The LCM of 4, 6, and 8 is 24.

$$-\frac{3}{4} + \frac{1}{6} - \frac{5}{8} = -\frac{18}{24} + \frac{4}{24} - \frac{15}{24}$$
$$= \frac{-18 + 4 - 15}{24}$$
$$= \frac{-29}{24} = -\frac{29}{24}$$

You Try It 5

Simplify: $-\frac{7}{8} - \frac{5}{6} + \frac{3}{4}$

Your solution

Example 6 Subtract: $42.987 - 98.61$

Solution $42.987 - 98.61$
$= 42.987 + (-98.61)$
$= -55.623$

You Try It 6 Subtract: $16.127 - 67.91$

Your solution

Solutions on p. S1

Objective D To multiply or divide rational numbers

The product of two fractions is the product of the numerators divided by the product of the denominators.

$$\frac{a}{b} \cdot \frac{c}{d} = \frac{ac}{bd}$$

HOW TO Multiply: $\frac{3}{8} \cdot \frac{12}{17}$

$$\frac{3}{8} \cdot \frac{12}{17} = \frac{3 \cdot 12}{8 \cdot 17}$$

• Multiply the numerators. Multiply the denominators.

$$= \frac{3 \cdot \overset{1}{\cancel{2}} \cdot \overset{1}{\cancel{2}} \cdot 3}{2 \cdot \underset{1}{\cancel{2}} \cdot \underset{1}{\cancel{2}} \cdot 17}$$

• Write the prime factorization of each factor. Divide by the common factors.

$$= \frac{9}{34}$$

• Multiply the factors in the numerator and the factors in the denominator.

Copyright © Houghton Mifflin Company. All rights reserved.

TAKE NOTE
To invert the divisor means to write its reciprocal. The reciprocal of $\frac{18}{25}$ is $\frac{25}{18}$.

To divide fractions, invert the divisor. Then multiply the fractions.

HOW TO Divide: $\frac{3}{10} \div \left(-\frac{18}{25}\right)$

The signs are different. The quotient is negative.

$$\frac{3}{10} \div \left(-\frac{18}{25}\right) = -\left(\frac{3}{10} \div \frac{18}{25}\right) = -\left(\frac{3}{10} \cdot \frac{25}{18}\right) = -\left(\frac{3 \cdot 25}{10 \cdot 18}\right)$$

$$= -\left(\frac{\overset{1}{\cancel{3}} \cdot \overset{1}{\cancel{5}} \cdot 5}{2 \cdot \underset{1}{\cancel{5}} \cdot 2 \cdot \underset{1}{\cancel{3}} \cdot 3}\right) = -\frac{5}{12}$$

To multiply decimals, multiply as with integers. Write the decimal point in the product so that the number of decimal places in the product equals the sum of the numbers of decimal places in the factors.

HOW TO Multiply: $-6.89(0.00035)$

```
     6.89     2 decimal places
× 0.00035     5 decimal places     • Multiply the absolute values.
     3445
    2067
0.0024115     7 decimal places
```

$-6.89(0.00035) = -0.0024115$ • The signs are different. The product is negative.

To divide decimals, move the decimal point in the divisor to the right to make the divisor a whole number. Move the decimal point in the dividend the same number of places to the right. Place the decimal point in the quotient directly over the decimal point in the dividend. Then divide as with whole numbers.

TAKE NOTE
The symbol ≈ is used to indicate that the quotient is an approximate value that has been rounded off.

HOW TO Divide: $1.32 \div 0.27$. Round to the nearest tenth.

```
          4.88 ≈ 4.9
0.27.)1.32.00
     -1 08
        240
       -216
         240
        -216
          24
```

• Move the decimal point 2 places to the right in the divisor and then in the dividend. Place the decimal point in the quotient above the decimal point in the dividend.

Example 7

Divide: $-\frac{5}{8} \div \left(-\frac{5}{40}\right)$

Solution The quotient is positive.

$$-\frac{5}{8} \div \left(-\frac{5}{40}\right) = \frac{5}{8} \div \frac{5}{40} = \frac{5}{8} \cdot \frac{40}{5} = \frac{5 \cdot 40}{8 \cdot 5}$$

$$= \frac{\overset{1}{\cancel{5}} \cdot \overset{1}{\cancel{2}} \cdot \overset{1}{\cancel{2}} \cdot \overset{1}{\cancel{2}} \cdot 5}{\underset{1}{\cancel{2}} \cdot \underset{1}{\cancel{2}} \cdot \underset{1}{\cancel{2}} \cdot \underset{1}{\cancel{5}}} = \frac{5}{1} = 5$$

You Try It 7

Divide: $-\frac{3}{8} \div \left(-\frac{5}{12}\right)$

Your solution

Solution on p. S2

Copyright © Houghton Mifflin Company. All rights reserved.

Example 8 Multiply: $-4.29(8.2)$

Solution The product is negative.

$$\begin{array}{r} 4.29 \\ \times\ 8.2 \\ \hline 858 \\ 3432 \\ \hline 35.178 \end{array}$$

$-4.29(8.2) = -35.178$

You Try It 8 Multiply: $-5.44(3.8)$

Your solution

Solution on p. S2

Objective E To evaluate exponential expressions

Point of Interest

René Descartes (1596–1650) was the first mathematician to use exponential notation extensively as it is used today. However, for some unknown reason, he always used *xx* for x^2.

Repeated multiplication of the same factor can be written using an exponent.

$2 \cdot 2 \cdot 2 \cdot 2 \cdot 2 = 2^5$ ← exponent; 2 ← base

$a \cdot a \cdot a \cdot a = a^4$ ← exponent; a ← base

The **exponent** indicates how many times the factor, called the **base,** occurs in the multiplication. The multiplication $2 \cdot 2 \cdot 2 \cdot 2 \cdot 2$ is in **factored form.** The exponential expression 2^5 is in **exponential form.**

2^1 is read "the first power of 2" or just 2. Usually the exponent 1 is not written.

2^2 is read "the second power of 2" or "2 squared."

2^3 is read "the third power of 2" or "2 cubed."

2^4 is read "the fourth power of 2."

a^4 is read "the fourth power of a."

There is a geometric interpretation of the first three natural-number powers.

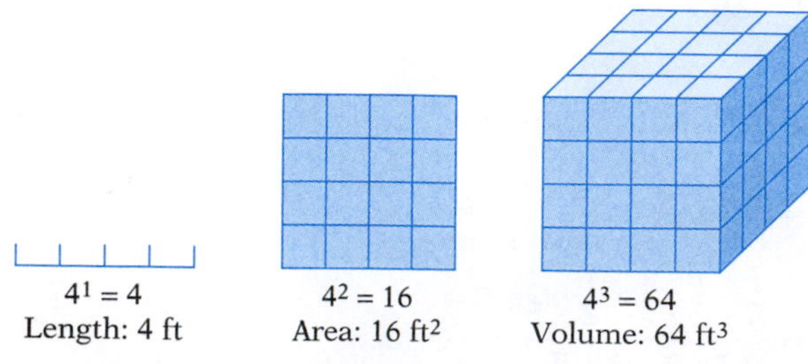

Copyright © Houghton Mifflin Company. All rights reserved.

To evaluate an exponential expression, write each factor as many times as indicated by the exponent. Then multiply.

HOW TO Evaluate $(-2)^4$.

$(-2)^4 = (-2)(-2)(-2)(-2)$ • Write (-2) as a factor 4 times.
$= 16$ • Multiply.

HOW TO Evaluate -2^4.

$-2^4 = -(2 \cdot 2 \cdot 2 \cdot 2)$ • Write 2 as a factor 4 times.
$= -16$ • Multiply.

From these last two examples, note the difference between $(-2)^4$ and -2^4.

$$(-2)^4 = 16$$
$$-2^4 = -(2^4) = -16$$

Example 9 Evaluate -5^3.

Solution $-5^3 = -(5 \cdot 5 \cdot 5) = -125$

You Try It 9 Evaluate -6^3.

Your solution

Example 10 Evaluate $(-4)^4$.

Solution $(-4)^4 = (-4)(-4)(-4)(-4)$
$= 256$

You Try It 10 Evaluate $(-3)^4$.

Your solution

Example 11 Evaluate $(-3)^2 \cdot 2^3$.

Solution $(-3)^2 \cdot 2^3 = (-3)(-3) \cdot (2)(2)(2)$
$= 9 \cdot 8 = 72$

You Try It 11 Evaluate $(3^3)(-2)^3$.

Your solution

Example 12 Evaluate $\left(-\frac{2}{3}\right)^3$.

Solution $\left(-\frac{2}{3}\right)^3 = \left(-\frac{2}{3}\right)\left(-\frac{2}{3}\right)\left(-\frac{2}{3}\right)$
$= -\frac{2 \cdot 2 \cdot 2}{3 \cdot 3 \cdot 3} = -\frac{8}{27}$

You Try It 12 Evaluate $\left(-\frac{2}{5}\right)^2$.

Your solution

Example 13 Evaluate $-4(0.7)^2$.

Solution $-4(0.7)^2 = -4(0.7)(0.7)$
$= -2.8(0.7) = -1.96$

You Try It 13 Evaluate $-3(0.3)^3$.

Your solution

Solutions on p. S2

Copyright © Houghton Mifflin Company. All rights reserved.

Objective F To simplify numerical radical expressions

A **square root** of a positive number x is a number whose square is x.

A square root of 16 is 4 because $4^2 = 16$.
A square root of 16 is -4 because $(-4)^2 = 16$.

Every positive number has two square roots, one a positive number and one a negative number. The symbol "$\sqrt{}$," called a **radical sign,** is used to indicate the positive or **principal square root** of a number. For example, $\sqrt{16} = 4$ and $\sqrt{25} = 5$. The number under the radical sign is called the **radicand.**

When the negative square root of a number is to be found, a negative sign is placed in front of the radical. For example, $-\sqrt{16} = -4$ and $-\sqrt{25} = -5$.

Square Roots of Perfect Squares

$\sqrt{1} = 1$
$\sqrt{4} = 2$
$\sqrt{9} = 3$
$\sqrt{16} = 4$
$\sqrt{25} = 5$
$\sqrt{36} = 6$
$\sqrt{49} = 7$
$\sqrt{64} = 8$
$\sqrt{81} = 9$
$\sqrt{100} = 10$
$\sqrt{121} = 11$
$\sqrt{144} = 12$

The square of an integer is a **perfect square.** 49, 81, and 144 are examples of perfect squares.

$$7^2 = 49$$
$$9^2 = 81$$
$$12^2 = 144$$

The principal square root of an integer that is a perfect square is a positive integer.

$$\sqrt{49} = 7$$
$$\sqrt{81} = 9$$
$$\sqrt{144} = 12$$

If a number is not a perfect square, its square root can only be approximated. For example, 2 and 7 are not perfect squares. The square roots of these numbers are **irrational numbers.** Their decimal representations never terminate or repeat.

$$\sqrt{2} \approx 1.4142135\ldots \qquad \sqrt{7} \approx 2.6457513\ldots$$

Recall that rational numbers are fractions such as $-\frac{6}{7}$ or $-\frac{10}{3}$, where the numerator and denominator are integers. Rational numbers are also represented by repeating decimals, such as $0.25767676\ldots$, and by terminating decimals, such as 1.73. An irrational number is neither a repeating nor a terminating decimal. For instance, $2.45445444544445\ldots$ is an irrational number.

Real Numbers

The rational numbers and the irrational numbers taken together are called the **real numbers.**

TAKE NOTE

Recall that a factor of a number divides the number evenly. For instance, 6 is a factor of 18. The perfect square 9 is also a factor of 18. 9 is a *perfect-square factor* of 18, whereas 6 is not a perfect-square factor of 18.

Radical expressions that contain radicands that are not perfect squares are frequently written in simplest form. A radical expression is in simplest form when the radicand contains no factor greater than 1 that is a perfect square. For instance, $\sqrt{50}$ is not in simplest form because 25 is a perfect-square factor of 50. The radical expression $\sqrt{15}$ is in simplest form because there are no perfect-square factors of 15 that are greater than 1.

The Product Property of Square Roots and a knowledge of perfect squares are used to simplify radicands that are not perfect squares.

Copyright © Houghton Mifflin Company. All rights reserved.

The Product Property of Square Roots

If a and b are positive real numbers, then $\sqrt{ab} = \sqrt{a} \cdot \sqrt{b}$.

TAKE NOTE

From the example at the right, $\sqrt{72} = 6\sqrt{2}$. The two expressions are different representations of the same number. Using a calculator, we find that $\sqrt{72} \approx 8.485281$ and $6\sqrt{2} \approx 8.485281$.

HOW TO Simplify: $\sqrt{72}$

$\sqrt{72} = \sqrt{36 \cdot 2}$
- Write the radicand as the product of a perfect square and a factor that does not contain a perfect square.

$= \sqrt{36}\sqrt{2}$
- Use the Product Property of Square Roots to write the expression as a product.

$= 6\sqrt{2}$
- Simplify $\sqrt{36}$.

Note that 72 must be written as the product of a perfect square and *a factor that does not contain a perfect square*. Therefore, it would not be correct to rewrite $\sqrt{72}$ as $\sqrt{9 \cdot 8}$ and simplify the expression as shown at the right. Although 9 is a perfect-square factor of 72, 8 has a perfect square factor $(8 = 4 \cdot 2)$. Therefore, $\sqrt{8}$ is not in simplest form. Remember to find the *largest* perfect-square factor of the radicand.

$$\begin{aligned}\sqrt{72} &= \sqrt{9 \cdot 8}\\ &= \sqrt{9}\sqrt{8}\\ &= 3\sqrt{8}\end{aligned}$$

Not in simplest form

HOW TO Simplify: $\sqrt{-16}$

Because the square of any real number is positive, there is no real number whose square is -16. $\sqrt{-16}$ is not a real number.

Example 14 Simplify: $3\sqrt{90}$

Solution

$$\begin{aligned}3\sqrt{90} &= 3\sqrt{9 \cdot 10}\\ &= 3\sqrt{9}\sqrt{10}\\ &= 3 \cdot 3\sqrt{10}\\ &= 9\sqrt{10}\end{aligned}$$

You Try It 14 Simplify: $-5\sqrt{32}$

Your solution

Example 15 Simplify: $\sqrt{252}$

Solution

$$\begin{aligned}\sqrt{252} &= \sqrt{36 \cdot 7}\\ &= \sqrt{36}\sqrt{7}\\ &= 6\sqrt{7}\end{aligned}$$

You Try It 15 Simplify: $\sqrt{216}$

Your solution

Solutions on p. S2

Copyright © Houghton Mifflin Company. All rights reserved.

Objective G To solve application problems

One of the applications of percent is to express a portion of a total as a percent. For instance, a recent survey of 450 mall shoppers found that 270 preferred the mall closest to their home even though it did not have as much store variety as a mall farther from home. The percent of shoppers who preferred the mall closest to home can be found by converting a fraction to a percent.

$$\frac{\text{Portion preferring mall closest to home}}{\text{Total number surveyed}} = \frac{270}{450}$$

$$= 0.60 = 60\%$$

The Congressional Budget Office projected that the total surpluses for 2001 through 2011 would be \$5.6 trillion. The number 5.6 trillion means

$$5.6 \times \underbrace{1{,}000{,}000{,}000{,}000}_{\text{1 trillion}} = 5{,}600{,}000{,}000{,}000$$

Numbers such as 5.6 trillion are used in many instances because they are easy to read and offer an approximation of the actual number.

The table below shows the net incomes, in millions of dollars, for the third quarter of 2004 for three companies. *Note:* Negative net income indicates a loss. Use this table for Example 16 and You Try It 16.

Company	*Net Income, 3rd Quarter of 2004 (in millions of dollars)*
America West	−47.1
FLYi	−82.7
Frontier Airlines	−2.1

Example 16

If throughout each quarter of 2005, America West's net income remained at the same level as in the third quarter of 2004, what would be America West's annual income for 2005?

Strategy

To determine the 2005 annual net income, multiply the net income for the third quarter of 2004 (−47.1) by the number of quarters in one year (4).

Solution

$4(-47.1) = -188.4$

The annual net income for America West for 2005 would be −\$188.4 million.

You Try It 16

If throughout each quarter of 2005, Frontier Airlines's net income remained at the same level as in the third quarter of 2004, what would be Frontier Airlines's annual net income for 2005?

Your strategy

Your solution

Solution on p. S2

Copyright © Houghton Mifflin Company. All rights reserved.

1.2 Exercises

Objective A To write a rational number as a decimal

Write as a decimal. Place a bar over the repeating digits of a repeating decimal.

1. $\frac{1}{8}$ **2.** $\frac{7}{8}$ **3.** $\frac{2}{9}$ **4.** $\frac{8}{9}$ **5.** $\frac{1}{6}$

6. $\frac{5}{6}$ **7.** $\frac{9}{16}$ **8.** $\frac{15}{16}$ **9.** $\frac{7}{12}$ **10.** $\frac{11}{12}$

11. $\frac{6}{25}$ **12.** $\frac{14}{25}$ **13.** $\frac{9}{40}$ **14.** $\frac{21}{40}$ **15.** $\frac{5}{11}$

Objective B To convert among percents, fractions, and decimals

16. 
a. Explain how to convert a fraction to a percent.
b. Explain how to convert a percent to a fraction.
c. Explain how to convert a decimal to a percent.
d. Explain how to convert a percent to a decimal.

17. Explain why multiplying a number by 100% does not change the value of the number.

Write as a fraction and as a decimal.

18. 75% **19.** 40% **20.** 64% **21.** 88% **22.** 125%

23. 160% **24.** 19% **25.** 87% **26.** 5% **27.** 450%

Write as a fraction.

28. $11\frac{1}{9}\%$ **29.** $4\frac{2}{7}\%$ **30.** $12\frac{1}{2}\%$ **31.** $37\frac{1}{2}\%$ **32.** $66\frac{2}{3}\%$

33. $\frac{1}{4}\%$ **34.** $\frac{1}{2}\%$ **35.** $6\frac{1}{4}\%$ **36.** $83\frac{1}{3}\%$ **37.** $5\frac{3}{4}\%$

Copyright © Houghton Mifflin Company. All rights reserved.

Write as a decimal.

38. 7.3% **39.** 9.1% **40.** 15.8% **41.** 16.7% **42.** 0.3%

43. 0.9% **44.** 9.9% **45.** 9.15% **46.** 121.2% **47.** 18.23%

Write as a percent.

48. 0.15 **49.** 0.37 **50.** 0.05 **51.** 0.02 **52.** 0.175

53. 0.125 **54.** 1.15 **55.** 1.36 **56.** 0.008 **57.** 0.004

58. $\frac{27}{50}$ **59.** $\frac{83}{100}$ **60.** $\frac{1}{3}$ **61.** $\frac{3}{8}$ **62.** $\frac{5}{11}$

63. $\frac{4}{9}$ **64.** $\frac{7}{8}$ **65.** $\frac{9}{20}$ **66.** $1\frac{2}{3}$ **67.** $2\frac{1}{2}$

Objective C To add or subtract rational numbers

Add or subtract.

68. $-\frac{5}{6}-\frac{5}{9}$ **69.** $-\frac{6}{13}+\frac{17}{26}$ **70.** $-\frac{7}{12}+\frac{5}{8}$ **71.** $\frac{5}{8}-\left(-\frac{3}{4}\right)$

72. $\frac{3}{5}-\frac{11}{12}$ **73.** $\frac{11}{12}-\frac{5}{6}$ **74.** $-\frac{2}{3}-\left(-\frac{11}{18}\right)$ **75.** $-\frac{5}{8}-\left(-\frac{11}{12}\right)$

76. $\frac{1}{3}+\frac{5}{6}-\frac{2}{9}$ **77.** $\frac{1}{2}-\frac{2}{3}+\frac{1}{6}$ **78.** $-\frac{5}{16}+\frac{3}{4}-\frac{7}{8}$ **79.** $\frac{1}{2}-\frac{3}{8}-\left(-\frac{1}{4}\right)$

Copyright © Houghton Mifflin Company. All rights reserved.

80. $\frac{3}{4} - \left(-\frac{7}{12}\right) - \frac{7}{8}$ **81.** $\frac{1}{3} - \frac{1}{4} - \frac{1}{5}$ **82.** $\frac{2}{3} - \frac{1}{2} + \frac{5}{6}$ **83.** $\frac{5}{16} + \frac{1}{8} - \frac{1}{2}$

84. $-13.092 + 6.9$ **85.** $2.54 - 3.6$ **86.** $5.43 + 7.925$ **87.** $-16.92 - 6.925$

88. $-3.87 + 8.546$ **89.** $6.9027 - 17.692$ **90.** $2.09 - 6.72 - 5.4$

91. $-18.39 + 4.9 - 23.7$ **92.** $19 - (-3.72) - 82.75$ **93.** $-3.07 - (-2.97) - 17.4$

94. $16.4 - (-3.09) - 7.93$ **95.** $-3.09 - 4.6 - (-27.3)$ **96.** $2.66 - (-4.66) - 8.2$

Objective D **To multiply or divide rational numbers**

Multiply or divide.

97. $\frac{1}{2}\left(-\frac{3}{4}\right)$ **98.** $-\frac{2}{9}\left(-\frac{3}{14}\right)$ **99.** $\left(-\frac{3}{8}\right)\left(-\frac{4}{15}\right)$

100. $\left(-\frac{3}{4}\right)\left(-\frac{8}{27}\right)$ **101.** $-\frac{1}{2}\left(\frac{8}{9}\right)$ **102.** $\frac{5}{12}\left(-\frac{8}{15}\right)$

103. $\frac{5}{8}\left(-\frac{7}{12}\right)\frac{16}{25}$ **104.** $\left(\frac{5}{12}\right)\left(-\frac{8}{15}\right)\left(\frac{1}{3}\right)$ **105.** $\frac{1}{2}\left(-\frac{3}{4}\right)\left(-\frac{5}{8}\right)$

106. $\frac{3}{8} \div \frac{1}{4}$ **107.** $\frac{5}{6} \div \left(-\frac{3}{4}\right)$ **108.** $-\frac{5}{12} \div \frac{15}{32}$

109. $-\frac{7}{8} \div \frac{4}{21}$ **110.** $\frac{7}{10} \div \frac{2}{5}$ **111.** $-\frac{15}{64} \div \left(-\frac{3}{40}\right)$

Copyright © Houghton Mifflin Company. All rights reserved.

112. $\frac{1}{8} \div \left(-\frac{5}{12}\right)$

113. $-\frac{4}{9} \div \left(-\frac{2}{3}\right)$

114. $-\frac{6}{11} \div \frac{4}{9}$

115. $1.2(3.47)$

116. $(-0.8)6.2$

117. $(-1.89)(-2.3)$

118. $(6.9)(-4.2)$

119. $1.06(-3.8)$

120. $-2.7(-3.5)$

121. $1.2(-0.5)(3.7)$

122. $-2.4(6.1)(0.9)$

123. $2.3(-0.6)(0.8)$

Divide. Round to the nearest hundredth.

124. $-1.27 \div (-1.7)$

125. $9.07 \div (-3.5)$

126. $0.0976 \div 0.042$

127. $-6.904 \div 1.35$

128. $-7.894 \div (-2.06)$

129. $-354.2086 \div 0.1719$

Objective E To evaluate exponential expressions

Evaluate.

130. 6^2

131. 7^4

132. -7^2

133. -4^3

134. $(-3)^2$

135. $(-2)^3$

136. $(-3)^4$

137. $(-5)^3$

138. $\left(\frac{1}{2}\right)^2$

139. $\left(-\frac{3}{4}\right)^3$

140. $(0.3)^2$

141. $(1.5)^3$

142. $\left(\frac{2}{3}\right)^2 \cdot 3^3$

143. $\left(-\frac{1}{2}\right)^3 \cdot 8$

144. $(0.3)^3 \cdot 2^3$

145. $(-2) \cdot (-2)^2$

146. $2^3 \cdot 3^3 \cdot (-4)$

147. $(-3)^3 \cdot 5^2 \cdot 10$

148. $(-7) \cdot 4^2 \cdot 3^2$

149. $(-2) \cdot 2^3 \cdot (-3)^2$

150. $\left(\frac{2}{3}\right)^2 \cdot \frac{1}{4} \cdot 3^3$

151. $\left(\frac{3}{4}\right)^2 \cdot (-4) \cdot 2^3$

152. $8^2 \cdot (-3)^5 \cdot 5$

Copyright © Houghton Mifflin Company. All rights reserved.

Objective F To simplify numerical radical expressions

Simplify.

153. $\sqrt{16}$ **154.** $\sqrt{64}$ **155.** $\sqrt{49}$ **156.** $\sqrt{144}$ **157.** $\sqrt{32}$ **158.** $\sqrt{50}$

159. $\sqrt{8}$ **160.** $\sqrt{12}$ **161.** $6\sqrt{18}$ **162.** $-3\sqrt{48}$ **163.** $5\sqrt{40}$ **164.** $2\sqrt{28}$

165. $\sqrt{15}$ **166.** $\sqrt{21}$ **167.** $\sqrt{29}$ **168.** $\sqrt{13}$ **169.** $-9\sqrt{72}$ **170.** $11\sqrt{80}$

171. $\sqrt{45}$ **172.** $\sqrt{225}$ **173.** $\sqrt{0}$ **174.** $\sqrt{210}$ **175.** $6\sqrt{128}$ **176.** $9\sqrt{288}$

Find the decimal approximation rounded to the nearest thousandth.

177. $\sqrt{240}$ **178.** $\sqrt{300}$ **179.** $\sqrt{288}$ **180.** $\sqrt{600}$ **181.** $\sqrt{256}$ **182.** $\sqrt{324}$

183. $\sqrt{275}$ **184.** $\sqrt{450}$ **185.** $\sqrt{245}$ **186.** $\sqrt{525}$ **187.** $\sqrt{352}$ **188.** $\sqrt{363}$

Objective G To solve application problems

189. **Business** The table below shows the annual net incomes for the periods ending in January 2004 and January 2003 for two companies. Figures are in millions of dollars. Profits are shown as positive numbers; losses are shown as negative numbers. Round to the nearest thousand dollars.

a. What was the average monthly net income for TiVo, Inc., for the period ending in January 2004?

b. Find the difference between the annual net income for Warnaco for the period ending January 2004 and the period ending January 2003.

Company	*Annual Net Income for Period Ending January 2004 (in millions of dollars)*	*Annual Net Income for Period Ending January 2003 (in millions of dollars)*
TiVo, Inc.	−32.018	−80.596
Warnaco Group, Inc.	2,360.423	−964.863

Copyright © Houghton Mifflin Company. All rights reserved.

190. **The Stock Market** At the close of the stock markets on December 2, 2004, the indexes were posted as shown below, along with the increase or decrease, shown as a negative number for that day. At what level were the indexes at the close of the day on December 1, 2004?

Index	*Points at Market Close*	*Decrease for the Day*
Dow Jones Industrial Average	10,585.12	-5.10
Standard & Poor's 500	1,190.33	-1.04
NASDAQ	2,143.57	+5.34

191. **Halloween Spending** In a recent year, the average consumer spent approximately $44 on Halloween merchandise. The breakdown is shown in the graph at the right (*Source:* BIGresearch for the National Retail Federation). What percent of the total is the amount spent on decorations?

192. **The Federal Budget** The table at the right shows the surplus or deficit, in billions of dollars, for the federal budget every fifth year from 1945 to 1995 and every year from 1995 to 2000 (*Source:* U.S. Office of Management and Budget). The negative sign (−) indicates a deficit.

a. Find the difference between the deficits in the years 1980 and 1985.

b. Calculate the difference between the surplus in 1960 and the deficit in 1955.

c. How many times greater was the deficit in 1985 than in 1975? Round to the nearest whole number.

d. What was the average deficit, in millions of dollars, per quarter for the year 1970?

e. Find the average surplus or deficit for the years 1995 through 2000. Round to the nearest million.

Year	*Federal Budget Surplus or Deficit*
1945	−47.533
1950	−3.119
1955	−2.993
1960	0.301
1965	−1.411
1970	−2.842
1975	−53.242
1980	−73.835
1985	−212.334
1990	−221.194
1995	−163.899
1996	−107.450
1997	−21.940
1998	69.246
1999	79.263
2000	117.305

APPLYING THE CONCEPTS

193. List the whole numbers between $\sqrt{8}$ and $\sqrt{90}$.

194. Use a calculator to determine the decimal representations of $\frac{17}{99}$, $\frac{45}{99}$, and $\frac{73}{99}$. Make a conjecture as to the decimal representation of $\frac{83}{99}$. Does your conjecture work for $\frac{33}{99}$? What about $\frac{1}{99}$?

195. Describe in your own words how to simplify a radical expression.

Copyright © Houghton Mifflin Company. All rights reserved.

1.3 The Order of Operations Agreement

Objective A

To use the Order of Operations Agreement to simplify expressions

Let's evaluate $2 + 3 \cdot 5$.

There are two arithmetic operations, addition and multiplication, in this expression. The operations could be performed in different orders.

Multiply first.	$2 + \underbrace{3 \cdot 5}$	Add first.	$\underbrace{2 + 3} \cdot 5$
Then add.	$\underbrace{2 + 15}$	Then multiply.	$\underbrace{5 \cdot 5}$
	17		25

In order to prevent there being more than one answer when simplifying a numerical expression, an Order of Operations Agreement has been established.

The Order of Operations Agreement

Step 1 Perform operations inside grouping symbols. Grouping symbols include parentheses ( ), brackets [ ], braces { }, absolute value symbols | |, and the fraction bar.

Step 2 Simplify exponential expressions.

Step 3 Do multiplication and division as they occur from left to right.

Step 4 Do addition and subtraction as they occur from left to right.

HOW TO Evaluate $12 - 24(8 - 5) \div 2^2$.

$12 - 24(8 - 5) \div 2^2 = 12 - 24(3) \div 2^2$ • Perform operations inside grouping symbols.

$= 12 - 24(3) \div 4$ • Simplify exponential expressions.

$= 12 - 72 \div 4$ • Do multiplication and division as they occur from left to right.

$= 12 - 18$

$= -6$ • Do addition and subtraction as they occur from left to right.

One or more of the above steps may not be needed to evaluate an expression. In that case, proceed to the next step in the Order of Operations Agreement.

Copyright © Houghton Mifflin Company. All rights reserved.

When an expression has grouping symbols inside grouping symbols, perform the operations inside the inner grouping symbols first.

Study Tip

The HOW TO feature gives an example with explanatory remarks. Using paper and pencil, you should work through the example. See *AIM for Success,* pages xxvii–xxviii.

HOW TO Evaluate $6 \div [4 - (6 - 8)] + 2^2$.

$6 \div [4 - (6 - 8)] + 2^2 = 6 \div [4 - (-2)] + 2^2$

- Perform operations inside grouping symbols.

$= 6 \div 6 + 2^2$

$= 6 \div 6 + 4$

- Simplify exponential expressions.

$= 1 + 4$

- Do multiplication and division as they occur from left to right.

$= 5$

- Do addition and subtraction as they occur from left to right.

Example 1

Evaluate $4 - 3[4 - 2(6 - 3)] \div 2$.

Solution

$4 - 3[4 - 2(6 - 3)] \div 2$
$= 4 - 3[4 - 2 \cdot 3] \div 2$
$= 4 - 3[4 - 6] \div 2$
$= 4 - 3[-2] \div 2$
$= 4 + 6 \div 2$
$= 4 + 3$
$= 7$

You Try It 1

Evaluate $18 - 5[8 - 2(2 - 5)] \div 10$.

Your solution

Example 2

Evaluate $27 \div (5 - 2)^2 + (-3)^2 \cdot 4$.

Solution

$27 \div (5 - 2)^2 + (-3)^2 \cdot 4$
$= 27 \div 3^2 + (-3)^2 \cdot 4$
$= 27 \div 9 + 9 \cdot 4$
$= 3 + 9 \cdot 4$
$= 3 + 36$
$= 39$

You Try It 2

Evaluate $36 \div (8 - 5)^2 - (-3)^2 \cdot 2$.

Your solution

Example 3

Evaluate $(1.75 - 1.3)^2 \div 0.025 + 6.1$.

Solution

$(1.75 - 1.3)^2 \div 0.025 + 6.1$
$= (0.45)^2 \div 0.025 + 6.1$
$= 0.2025 \div 0.025 + 6.1$
$= 8.1 + 6.1$
$= 14.2$

You Try It 3

Evaluate $(6.97 - 4.72)^2 \cdot 4.5 \div 0.05$.

Your solution

Solutions on p. S2

Copyright © Houghton Mifflin Company. All rights reserved.

1.3 Exercises

Objective A **To use the Order of Operations Agreement to simplify expressions**

1.  Why do we need an Order of Operations Agreement?

2. Describe each step in the Order of Operations Agreement.

Evaluate by using the Order of Operations Agreement.

3. $4 - 8 \div 2$

4. $2^2 \cdot 3 - 3$

5. $2(3 - 4) - (-3)^2$

6. $16 - 32 \div 2^3$

7. $24 - 18 \div 3 + 2$

8. $8 - (-3)^2 - (-2)$

9. $8 - 2(3)^2$

10. $16 - 16 \cdot 2 \div 4$

11. $12 + 16 \div 4 \cdot 2$

12. $16 - 2 \cdot 4^2$

13. $27 - 18 \div (-3^2)$

14. $4 + 12 \div 3 \cdot 2$

15. $16 + 15 \div (-5) - 2$

16. $14 - 2^2 - (4 - 7)$

17. $14 - 2^2 - |4 - 7|$

18. $10 - |5 - 8| + 2^3$

19. $3 - 2[8 - (3 - 2)]$

20. $-2^2 + 4[16 \div (3 - 5)]$

21. $6 + \dfrac{16 - 4}{2^2 + 2} - 2$

22. $24 \div \dfrac{3^2}{8 - 5} - (-5)$

23. $18 \div |9 - 2^3| + (-3)$

24. $96 \div 2[12 + (6 - 2)] - 3^2$

25. $4[16 - (7 - 1)] \div 10$

26. $18 \div 2 - 4^2 - (-3)^2$

27. $20 \div (10 - 2^3) + (-5)$

28. $16 - 3(8 - 3)^2 \div 5$

29. $4(-8) \div [2(7 - 3)^2]$

Copyright © Houghton Mifflin Company. All rights reserved.

30. $\frac{(-10) + (-2)}{6^2 - 30} \div |2 - 4|$

31. $16 - 4 \cdot \frac{3^3 - 7}{2^3 + 2} - (-2)^2$

32. $(0.2)^2 \cdot (-0.5) + 1.72$

33. $0.3(1.7 - 4.8) + (1.2)^2$

34. $(1.8)^2 - 2.52 \div 1.8$

35. $(1.65 - 1.05)^2 \div 0.4 + 0.8$

APPLYING THE CONCEPTS

36. Find two fractions between $\frac{2}{3}$ and $\frac{3}{4}$. (There is more than one answer to this question.)

37. A **magic square** is one in which the numbers in every row, column, and diagonal sum to the same number. Complete the magic square at the right.

$\frac{2}{3}$		
	$\frac{1}{6}$	$\frac{5}{6}$
		$-\frac{1}{3}$

38. For each part below, find a rational number r that satisfies the condition.
a. $r^2 < r$ **b.** $r^2 = r$ **c.** $r^2 > r$

39. **Electric Vehicles** In a survey of consumers, 9% said they would buy an electric vehicle. Approximately 43% said they would be willing to pay $1500 more for a new car if the car had an EPA rating of 80 mpg. If your car now gets 28 mpg and you drive approximately 10,000 mi per year, in how many months would your savings on gasoline pay for the increased cost of such a car? Assume the average cost for gasoline is $2.00 per gallon. Round to the nearest whole number.

40. Find three different natural numbers a, b, and c such that $\frac{1}{a} + \frac{1}{b} + \frac{1}{c}$ is a natural number.

41. The following was offered as the simplification of $6 + 2(4 - 9)$.

$$\begin{aligned} 6 + 2(4 - 9) &= 6 + 2(-5) \\ &= 8(-5) \\ &= -40 \end{aligned}$$

Is this a correct simplification? Explain your answer.

42. The following was offered as the simplification of $2 \cdot 3^3$.

$$2 \cdot 3^3 = 6^3 = 216$$

Is this is a correct simplification? Explain your answer.

Copyright © Houghton Mifflin Company. All rights reserved.

1.4 Variable Expressions

Objective A To evaluate a variable expression

Point of Interest

Historical manuscripts indicate that mathematics is at least 4000 years old. Yet it was only 400 years ago that mathematicians started using variables to stand for numbers. The idea that a letter can stand for some number was a critical turning point in mathematics.

Often we discuss a quantity without knowing its exact value—for example, the price of gold next month, the cost of a new automobile next year, or the tuition cost for next semester. Recall that a letter of the alphabet can be used to stand for a quantity that is unknown or that can change, or *vary.* Such a letter is called a variable. An expression that contains one or more variables is called a **variable expression.**

A variable expression is shown at the right. The expression can be rewritten by writing subtraction as the addition of the opposite.

$$3x^2 - 5y + 2xy - x - 7$$

$$3x^2 + (-5y) + 2xy + (-x) + (-7)$$

Note that the expression has five addends. The **terms** of a variable expression are the addends of the expression. The expression has five terms.

$$\underbrace{\overbrace{3x^2}\ \overbrace{-\ 5y}\ \overbrace{+\ 2xy}\ \overbrace{-\ x}}_{\text{variable terms}}\ \underbrace{\overbrace{-\ 7}}_{\text{constant term}}$$

(5 terms)

The terms $3x^2$, $-5y$, $2xy$, and $-x$ are **variable terms.**

The term -7 is a **constant term**, or simply a **constant.**

Each variable term is composed of a **numerical coefficient** and a **variable part** (the variable or variables and their exponents).

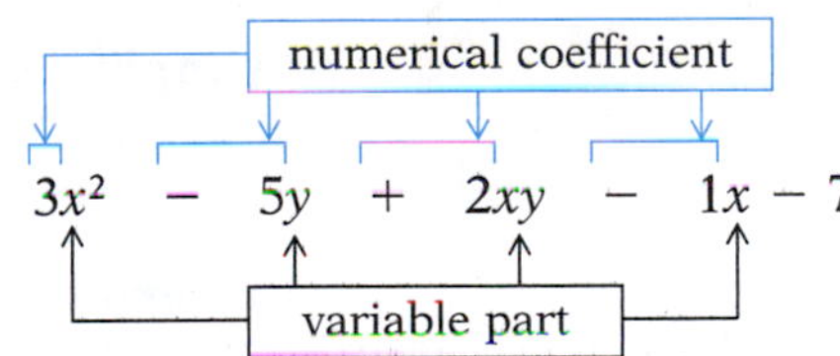

When the numerical coefficient is 1 or -1, the 1 is usually not written ($x = 1x$ and $-x = -1x$).

Replacing each variable by its value and then simplifying the resulting numerical expression is called **evaluating the variable expression.**

Integrating Technology

See the appendix Keystroke Guide for instructions on using a graphing calculator to evaluate variable expressions.

HOW TO Evaluate $ab - b^2$ when $a = 2$ and $b = -3$.

$ab - b^2$

$2(-3) - (-3)^2$ • Replace each variable in the expression by its value.

$= 2(-3) - 9$ • Use the Order of Operations Agreement to simplify the resulting numerical expression.

$= -6 - 9$

$= -15$

Copyright © Houghton Mifflin Company. All rights reserved.

Example 1

Evaluate $\frac{a^2 - b^2}{a - b}$ when $a = 3$ and $b = -4$.

Solution $\frac{a^2 - b^2}{a - b}$

$\frac{3^2 - (-4)^2}{3 - (-4)} = \frac{9 - 16}{3 - (-4)}$

$= \frac{-7}{7} = -1$

You Try It 1

Evaluate $\frac{a^2 + b^2}{a + b}$ when $a = 5$ and $b = -3$.

Your solution

Example 2 Evaluate $x^2 - 3(x - y) - z^2$ when $x = 2$, $y = -1$, and $z = 3$.

Solution $x^2 - 3(x - y) - z^2$

$2^2 - 3[2 - (-1)] - 3^2$
$= 2^2 - 3(3) - 3^2$
$= 4 - 3(3) - 9$
$= 4 - 9 - 9$
$= -5 - 9$
$= -14$

You Try It 2 Evaluate $x^3 - 2(x + y) + z^2$ when $x = 2$, $y = -4$, and $z = -3$.

Your solution

Solutions on p. S2

Objective B To simplify a variable expression using the Properties of Addition

Like terms of a variable expression are terms with the same variable part. (Because $x^2 = x \cdot x$, x^2 and x are not like terms.)

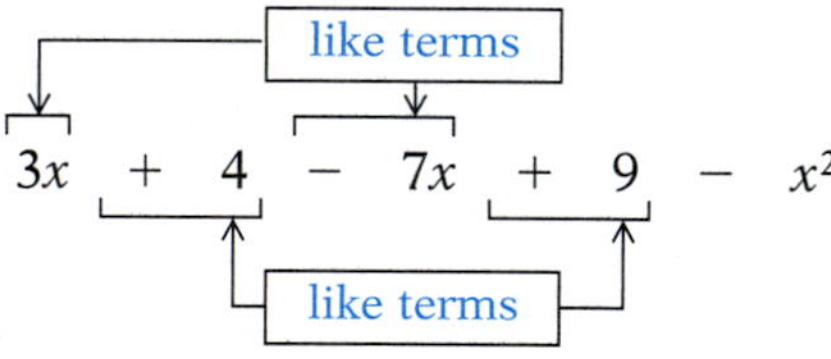

Constant terms are like terms. 4 and 9 are like terms.

To simplify a variable expression, use the Distributive Property to combine like terms by adding the numerical coefficients. The variable part remains unchanged.

Distributive Property

If a, b, and c are real numbers, then $a(b + c) = ab + ac$.

The Distributive Property can also be written as $ba + ca = (b + c)a$. This form is used to simplify a variable expression.

Copyright © Houghton Mifflin Company. All rights reserved.

HOW TO Simplify: $2x + 3x$

Use the Distributive Property to add the numerical coefficients of the like variable terms. This is called **combining like terms.**

$$2x + 3x = (2 + 3)x$$
$$= 5x$$

- Use the Distributive Property.

HOW TO Simplify: $5y - 11y$

$$5y - 11y = \boxed{(5 - 11)y}$$
$$= -6y$$

- Use the Distributive Property. This step is usually done mentally.

TAKE NOTE

Simplifying an expression means combining like terms. A constant term (5) and a variable term ($7p$) are not like terms and therefore cannot be combined.

HOW TO Simplify: $5 + 7p$

The terms 5 and $7p$ are not like terms.

The expression $5 + 7p$ is in simplest form.

In simplifying variable expressions, the following Properties of Addition are used.

The Associative Property of Addition

If a, b, and c are real numbers, then $(a + b) + c = a + (b + c)$.

When three or more like terms are added, the terms can be grouped (with parentheses, for example) in any order. The sum is the same. For example,

$$(3x + 5x) + 9x = 3x + (5x + 9x)$$
$$8x + 9x = 3x + 14x$$
$$17x = 17x$$

The Commutative Property of Addition

If a and b are real numbers, then $a + b = b + a$.

When two like terms are added, the terms can be added in either order. The sum is the same. For example,

$$2x + (-4x) = -4x + 2x$$
$$-2x = -2x$$

The Addition Property of Zero

If a is a real number, then $a + 0 = 0 + a = a$.

The sum of a term and zero is the term. For example,

$$5x + 0 = 0 + 5x = 5x$$

Copyright © Houghton Mifflin Company. All rights reserved.

The Inverse Property of Addition

If a is a real number, then $a + (-a) = (-a) + a = 0$.

The sum of a term and its opposite is zero. The opposite of a number is called its **additive inverse.**

$$7x + (-7x) = -7x + 7x = 0$$

HOW TO Simplify: $8x + 4y - 8x + y$

Use the Commutative and Associative Properties of Addition to rearrange and group like terms. Then combine like terms.

$$\begin{aligned} 8x + 4y - 8x + y &= (8x - 8x) + (4y + y) \\ &= 0 + 5y \\ &= 5y \end{aligned}$$

• This step is usually done mentally.

HOW TO Simplify: $4x^2 + 5x - 6x^2 - 2x + 1$

Use the Commutative and Associative Properties of Addition to rearrange and group like terms. Then combine like terms.

$$\begin{aligned} 4x^2 + 5x - 6x^2 - 2x + 1 &= (4x^2 - 6x^2) + (5x - 2x) + 1 \\ &= -2x^2 + 3x + 1 \end{aligned}$$

Example 3 Simplify: $3x + 4y - 10x + 7y$

Solution $3x + 4y - 10x + 7y = -7x + 11y$

You Try It 3 Simplify: $3a - 2b - 5a + 6b$

Your solution

Example 4 Simplify: $x^2 - 7 + 4x^2 - 16$

Solution $x^2 - 7 + 4x^2 - 16 = 5x^2 - 23$

You Try It 4 Simplify: $-3y^2 + 7 + 8y^2 - 14$

Your solution

Solutions on p. S2

Objective C To simplify a variable expression using the Properties of Multiplication

In simplifying variable expressions, the following Properties of Multiplication are used.

The Associative Property of Multiplication

If a, b, and c are real numbers, then $(a \cdot b) \cdot c = a \cdot (b \cdot c)$.

When three or more factors are multiplied, the factors can be grouped in any order. The product is the same. For example,

$$2(3x) = (2 \cdot 3)x = 6x$$

Copyright © Houghton Mifflin Company. All rights reserved.

Study Tip

Some students think that they can "coast" at the beginning of this course because they have been taught this material before. However, this chapter lays the foundation for the entire course. Be sure you know and understand all the concepts presented. For example, study the properties of multiplication presented in this lesson.

The Commutative Property of Multiplication

If a and b are real numbers, then $a \cdot b = b \cdot a$.

Two factors can be multiplied in either order. The product is the same. For example,

$$(2x) \cdot 3 = 3 \cdot (2x) = 6x$$

The Multiplication Property of One

If a is a real number, then $a \cdot 1 = 1 \cdot a = a$.

The product of a term and one is the term. For example,

$$(8x)(1) = (1)(8x) = 8x$$

The Inverse Property of Multiplication

If a is a real number, and a is not equal to zero, then $a \cdot \frac{1}{a} = \frac{1}{a} \cdot a = 1$.

$\frac{1}{a}$ is called the **reciprocal** of a. $\frac{1}{a}$ is also called the **multiplicative inverse** of a. The product of a number and its reciprocal is one. For example,

$$7 \cdot \frac{1}{7} = \frac{1}{7} \cdot 7 = 1$$

The multiplication properties just discussed are used to simplify variable expressions.

HOW TO Simplify: $2(-x)$

$$\begin{aligned} 2(-x) &= 2(-1 \cdot x) \\ &= [2(-1)]x \\ &= -2x \end{aligned}$$

- Use the Associative Property of Multiplication to group factors.

HOW TO Simplify: $\frac{3}{2}\left(\frac{2x}{3}\right)$

Use the Associative Property of Multiplication to group factors.

$$\begin{aligned} \frac{3}{2}\left(\frac{2x}{3}\right) &= \frac{3}{2}\left(\frac{2}{3}x\right) \\ &= \left(\frac{3}{2} \cdot \frac{2}{3}\right)x \\ &= 1 \cdot x \\ &= x \end{aligned}$$

- Note that $\frac{2x}{3} = \frac{2}{3}x$.
- The steps in the dashed box are usually done mentally.

Copyright © Houghton Mifflin Company. All rights reserved.

HOW TO Simplify: $(16x)2$

Use the Commutative and Associative Properties of Multiplication to rearrange and group factors.

$$(16x)2 = 2(16x) = (2 \cdot 16)x = 32x$$

• The steps in the dashed box are usually done mentally.

Example 5 Simplify: $-2(3x^2)$

Solution $-2(3x^2) = -6x^2$

You Try It 5 Simplify: $-5(4y^2)$

Your solution

Example 6 Simplify: $-5(-10x)$

Solution $-5(-10x) = 50x$

You Try It 6 Simplify: $-7(-2a)$

Your solution

Example 7 Simplify: $(6x)(-4)$

Solution $(6x)(-4) = -24x$

You Try It 7 Simplify: $(-5x)(-2)$

Your solution

Solutions on p. S2

Objective D To simplify a variable expression using the Distributive Property

Recall that the Distributive Property states that if a, b, and c are real numbers, then

$$a(b + c) = ab + ac$$

The Distributive Property is used to remove parentheses from a variable expression.

HOW TO Simplify: $3(2x + 7)$

$$3(2x + 7) = 3(2x) + 3(7) = 6x + 21$$

• Use the Distributive Property. Do this step mentally.

HOW TO Simplify: $-5(4x + 6)$

$$-5(4x + 6) = -5(4x) + (-5) \cdot 6 = -20x - 30$$

• Use the Distributive Property. Do this step mentally.

Copyright © Houghton Mifflin Company. All rights reserved.

HOW TO Simplify: $-(2x - 4)$

$$-(2x-4) = -1(2x-4) = -1(2x) - (-1)(4) = -2x + 4$$

- Use the Distributive Property. Do these steps mentally.

Note: When a negative sign immediately precedes the parentheses, the sign of each term inside the parentheses is changed.

HOW TO Simplify: $-\frac{1}{2}(8x - 12y)$

$$-\frac{1}{2}(8x-12y) = -\frac{1}{2}(8x) - \left(-\frac{1}{2}\right)(12y) = -4x + 6y$$

- Use the Distributive Property. Do this step mentally.

HOW TO Simplify: $4(x - y) - 2(-3x + 6y)$

$$4(x-y) - 2(-3x+6y) = 4x - 4y + 6x - 12y$$

- Use the Distributive Property twice.

$$= 10x - 16y$$

- Combine like terms.

Study Tip

One of the key instructional features of this text is the Example/You Try It pairs. Each Example is completely worked. You are to solve the You Try It problems. When you are ready, check your solution against the one in the Solutions section. The solutions for You Try It 8 and 9 below are on pages S2–S3 (see the reference at the bottom right of the You Try It box). See *AIM for Success,* page xxviii.

The Distributive Property is used when an expression inside parentheses contains more than two terms. See the example below.

HOW TO Simplify: $3(4x - 2y - z)$

$$3(4x-2y-z) = 3(4x) - 3(2y) - 3(z) = 12x - 6y - 3z$$

- Use the Distributive Property. Do this step mentally.

Example 8 Simplify: $-3(-5a + 7b)$

Solution $-3(-5a + 7b) = 15a - 21b$

You Try It 8 Simplify: $-8(-2a + 7b)$

Your solution

Example 9 Simplify: $(2x - 6)2$

Solution $(2x - 6)2 = 4x - 12$

You Try It 9 Simplify: $(3a - 1)5$

Your solution

Solutions on pp. S2–S3

Copyright © Houghton Mifflin Company. All rights reserved.

Example 10 Simplify: $3(x^2 - x - 5)$

Solution $3(x^2 - x - 5) = 3x^2 - 3x - 15$

You Try It 10 Simplify: $2(x^2 - x + 7)$

Your solution

Example 11 Simplify: $2x - 3(2x - 7y)$

Solution $2x - 3(2x - 7y) = 2x - 6x + 21y$
$= -4x + 21y$

You Try It 11 Simplify: $3y - 2(y - 7x)$

Your solution

Example 12 Simplify:
$7(x - 2y) - (-x - 2y)$

Solution $7(x - 2y) - (-x - 2y)$
$= 7x - 14y + x + 2y$
$= 8x - 12y$

You Try It 12 Simplify:
$-2(x - 2y) - (-x + 3y)$

Your solution

Example 13 Simplify:
$2x - 3[2x - 3(x + 7)]$

Solution $2x - 3[2x - 3(x + 7)]$
$= 2x - 3[2x - 3x - 21]$
$= 2x - 3[-x - 21]$
$= 2x + 3x + 63$
$= 5x + 63$

You Try It 13 Simplify:
$3y - 2[x - 4(2 - 3y)]$

Your solution

Solutions on p. S3

Objective E To translate a verbal expression into a variable expression

One of the major skills required in applied mathematics is the ability to translate a verbal expression into a variable expression. This requires recognizing the verbal phrases that translate into mathematical operations. A partial list of the verbal phrases used to indicate the different mathematical operations follows.

Addition	added to	6 added to y	$y + 6$
	more than	8 more than x	$x + 8$
	the sum of	the sum of x and z	$x + z$
	increased by	t increased by 9	$t + 9$
	the total of	the total of 5 and y	$5 + y$

Copyright © Houghton Mifflin Company. All rights reserved.

Point of Interest

The way in which expressions are symbolized has changed over time. Here is how some of the expressions shown at the right may have appeared in the early 16th century.

R p. 8 for $x + 8$. The symbol R was used for a variable to the first power. The symbol p. was used for plus.

R m. 2 for $x - 2$. The symbol R is again used for the variable. The symbol m. was used for minus.

The square of a variable was designated by Q and the cube of a variable was designated by C. The expression $x^3 + x^2$ was written **C p. Q.**

Subtraction	minus	x minus 2	$x - 2$
	less than	7 less than t	$t - 7$
	decreased by	m decreased by 3	$m - 3$
	the difference between	the difference between y and 4	$y - 4$
Multiplication	times	10 times t	$10t$
	of	one-half of x	$\frac{1}{2}x$
	the product of	the product of y and z	yz
	multiplied by	y multiplied by 11	$11y$
	twice	twice d	$2d$
Division	divided by	x divided by 12	$\frac{x}{12}$
	the quotient of	the quotient of y and z	$\frac{y}{z}$
	the ratio of	the ratio of t to 9	$\frac{t}{9}$
Power	the square of	the square of x	x^2
	the cube of	the cube of a	a^3

HOW TO Translate "14 less than the cube of x" into a variable expression.

14 *less than* the *cube* of x
- Identify the words that indicate the mathematical operations.

$x^3 - 14$
- Use the identified operations to write the variable expression.

In most applications that involve translating phrases into variable expressions, the variable to be used is not given. To translate these phrases, a variable must be assigned to an unknown quantity before the variable expression can be written.

HOW TO Translate "the sum of two consecutive integers" into a variable expression. Then simplify.

the first integer: n
- Assign a variable to one of the unknown quantities.

the next consecutive integer: $n + 1$
- Use the assigned variable to write an expression for any other unknown quantity.

$n + (n + 1)$
- Use the assigned variable to write the variable expression.

$(n + n) + 1$

$2n + 1$
- Simplify the variable expression.

Copyright © Houghton Mifflin Company. All rights reserved.

Many of the applications of mathematics require that you identify an unknown quantity, assign a variable to that quantity, and then attempt to express another unknown quantity in terms of the variable.

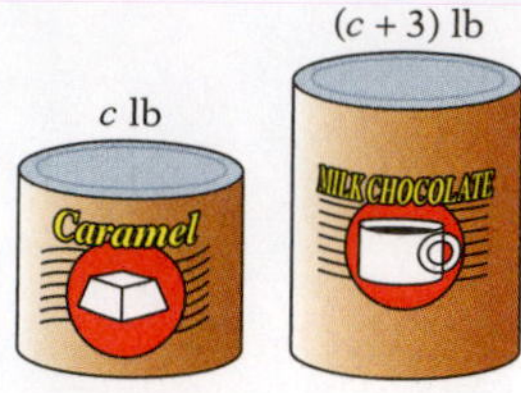

HOW TO A confectioner makes a mixture of candy that contains 3 lb more of milk chocolate than of caramel. Express the amount of milk chocolate in the mixture in terms of the amount of caramel in the mixture.

Amount of caramel in the mixture: c • Assign a variable to the amount of caramel in the mixture.

Amount of milk chocolate in the mixture: $c + 3$ • Express the amount of milk chocolate in the mixture in terms of c.

Example 14

Translate "four times the sum of half of a number and fourteen" into a variable expression. Then simplify.

Solution

the unknown number: n

half of the number: $\frac{1}{2}n$

the sum of half of the number and fourteen: $\frac{1}{2}n + 14$

$4\left(\frac{1}{2}n + 14\right)$

$2n + 56$

You Try It 14

Translate "five times the difference between a number and sixty" into a variable expression. Then simplify.

Your solution

Example 15

The length of a swimming pool is 4 ft less than two times the width. Express the length of the pool in terms of the width.

Solution

the width of the pool: w

the length is 4 ft less than two times the width: $2w - 4$

You Try It 15

The speed of a new printer is twice the speed of an older model. Express the speed of the new model in terms of the speed of the older model.

Your solution

Example 16

A banker divided \$5000 between two accounts, one paying 10% annual interest and the second paying 8% annual interest. Express the amount invested in the 10% account in terms of the amount invested in the 8% account.

Solution

the amount invested at 8%: x

the amount invested at 10%: $5000 - x$

You Try It 16

A guitar string 6 ft long was cut into two pieces. Express the length of the shorter piece in terms of the length of the longer piece.

Your solution

Solutions on p. S3

Copyright © Houghton Mifflin Company. All rights reserved.

1.4 Exercises

Objective A **To evaluate a variable expression**

Evaluate the variable expression when $a = 2$, $b = 3$, and $c = -4$.

1. $6b \div (-a)$

2. $bc \div (2a)$

3. $b^2 - 4ac$

4. $a^2 - b^2$

5. $b^2 - c^2$

6. $(a + b)^2$

7. $a^2 + b^2$

8. $2a - (c + a)^2$

9. $(b - a)^2 + 4c$

10. $b^2 - \frac{ac}{8}$

11. $\frac{5ab}{6} - 3cb$

12. $(b - 2a)^2 + bc$

Evaluate the variable expression when $a = -2$, $b = 4$, $c = -1$, and $d = 3$.

13. $\frac{b + c}{d}$

14. $\frac{d - b}{c}$

15. $\frac{2d + b}{-a}$

16. $\frac{b + 2d}{b}$

17. $\frac{b - d}{c - a}$

18. $\frac{2c - d}{-ad}$

19. $(b + d)^2 - 4a$

20. $(d - a)^2 - 3c$

21. $(d - a)^2 \div 5$

22. $3(b - a) - bc$

23. $\frac{b - 2a}{bc^2 - d}$

24. $\frac{b^2 - a}{ad + 3c}$

25. $\frac{1}{3}d^2 - \frac{3}{8}b^2$

26. $\frac{5}{8}a^4 - c^2$

27. $\frac{-4bc}{2a - b}$

28. $-\frac{3}{4}b + \frac{1}{2}(ac + bd)$

29. $-\frac{2}{3}d - \frac{1}{5}(bd - ac)$

30. $(b - a)^2 - (d - c)^2$

31. $(b + c)^2 + (a + d)^2$

32. $4ac + (2a)^2$

33. $3dc - (4c)^2$

Copyright © Houghton Mifflin Company. All rights reserved.

Objective B To simplify a variable expression using the Properties of Addition

Simplify.

34. $6x + 8x$

35. $12x + 13x$

36. $9a - 4a$

37. $12a - 3a$

38. $4y + (-10y)$

39. $8y + (-6y)$

40. $-3b - 7$

41. $-12y - 3$

42. $-12a + 17a$

43. $-3a + 12a$

44. $5ab - 7ab$

45. $9ab - 3ab$

46. $-12xy + 17xy$

47. $-15xy + 3xy$

48. $-3ab + 3ab$

49. $-7ab + 7ab$

50. $-\frac{1}{2}x - \frac{1}{3}x$

51. $-\frac{2}{5}y + \frac{3}{10}y$

52. $\frac{3}{8}x^2 - \frac{5}{12}x^2$

53. $\frac{2}{3}y^2 - \frac{4}{9}y^2$

54. $3x + 5x + 3x$

55. $8x + 5x + 7x$

56. $5a - 3a + 5a$

57. $10a - 17a + 3a$

58. $-5x^2 - 12x^2 + 3x^2$

59. $-y^2 - 8y^2 + 7y^2$

60. $7x + (-8x) + 3y$

61. $8y + (-10x) + 8x$

62. $7x - 3y + 10x$

63. $8y + 8x - 8y$

64. $3a + (-7b) - 5a + b$

65. $-5b + 7a - 7b + 12a$

66. $3x + (-8y) - 10x + 4x$

67. $3y + (-12x) - 7y + 2y$

68. $x^2 - 7x + (-5x^2) + 5x$

69. $3x^2 + 5x - 10x^2 - 10x$

Copyright © Houghton Mifflin Company. All rights reserved.

Objective C **To simplify a variable expression using the Properties of Multiplication**

Simplify.

70. $4(3x)$ **71.** $12(5x)$ **72.** $-3(7a)$ **73.** $-2(5a)$ **74.** $-2(-3y)$

75. $-5(-6y)$ **76.** $(4x)2$ **77.** $(6x)12$ **78.** $(3a)(-2)$ **79.** $(7a)(-4)$

80. $(-3b)(-4)$ **81.** $(-12b)(-9)$ **82.** $-5(3x^2)$ **83.** $-8(7x^2)$ **84.** $\frac{1}{3}(3x^2)$

85. $\frac{1}{6}(6x^2)$ **86.** $\frac{1}{5}(5a)$ **87.** $\frac{1}{8}(8x)$ **88.** $-\frac{1}{2}(-2x)$ **89.** $-\frac{1}{4}(-4a)$

90. $-\frac{1}{7}(-7n)$ **91.** $-\frac{1}{9}(-9b)$ **92.** $(3x)\left(\frac{1}{3}\right)$ **93.** $(12x)\left(\frac{1}{12}\right)$ **94.** $(-6y)\left(-\frac{1}{6}\right)$

95. $(-10n)\left(-\frac{1}{10}\right)$ **96.** $\frac{1}{3}(9x)$ **97.** $\frac{1}{7}(14x)$ **98.** $-\frac{1}{5}(10x)$ **99.** $-\frac{1}{8}(16x)$

100. $-\frac{2}{3}(12a^2)$ **101.** $-\frac{5}{8}(24a^2)$ **102.** $-\frac{1}{2}(-16y)$ **103.** $-\frac{3}{4}(-8y)$ **104.** $(16y)\left(\frac{1}{4}\right)$

105. $(33y)\left(\frac{1}{11}\right)$ **106.** $(-6x)\left(\frac{1}{3}\right)$ **107.** $(-10x)\left(\frac{1}{5}\right)$ **108.** $(-8a)\left(-\frac{3}{4}\right)$ **109.** $(21y)\left(-\frac{3}{7}\right)$

Objective D **To simplify a variable expression using the Distributive Property**

Simplify.

110. $-(x + 2)$ **111.** $-(x + 7)$ **112.** $2(4x - 3)$ **113.** $5(2x - 7)$

114. $-2(a + 7)$ **115.** $-5(a + 16)$ **116.** $-3(2y - 8)$ **117.** $-5(3y - 7)$

Copyright © Houghton Mifflin Company. All rights reserved.

118. $(5 - 3b)7$

119. $(10 - 7b)2$

120. $\frac{1}{3}(6 - 15y)$

121. $\frac{1}{2}(-8x + 4y)$

122. $3(5x^2 + 2x)$

123. $6(3x^2 + 2x)$

124. $-2(-y + 9)$

125. $-5(-2x + 7)$

126. $(-3x - 6)5$

127. $(-2x + 7)7$

128. $2(-3x^2 - 14)$

129. $5(-6x^2 - 3)$

130. $-3(2y^2 - 7)$

131. $-8(3y^2 - 12)$

132. $3(x^2 - y^2)$

133. $5(x^2 + y^2)$

134. $-\frac{2}{3}(6x - 18y)$

135. $-\frac{1}{2}(x - 4y)$

136. $-(6a^2 - 7b^2)$

137. $3(x^2 + 2x - 6)$

138. $4(x^2 - 3x + 5)$

139. $-2(y^2 - 2y + 4)$

140. $\frac{1}{2}(2x - 6y + 8)$

141. $-\frac{1}{3}(6x - 9y + 1)$

142. $4(-3a^2 - 5a + 7)$

143. $-5(-2x^2 - 3x + 7)$

144. $-3(-4x^2 + 3x - 4)$

145. $3(2x^2 + xy - 3y^2)$

146. $5(2x^2 - 4xy - y^2)$

147. $-(3a^2 + 5a - 4)$

148. $-(8b^2 - 6b + 9)$

149. $4x - 2(3x + 8)$

150. $6a - (5a + 7)$

151. $9 - 3(4y + 6)$

152. $10 - (11x - 3)$

153. $5n - (7 - 2n)$

154. $8 - (12 + 4y)$

Copyright © Houghton Mifflin Company. All rights reserved.

155. $3(x + 2) - 5(x - 7)$

156. $2(x - 4) - 4(x + 2)$

157. $12(y - 2) + 3(7 - 3y)$

158. $6(2y - 7) - (3 - 2y)$

159. $3(a - b) - (a + b)$

160. $2(a + 2b) - (a - 3b)$

161. $4[x - 2(x - 3)]$

162. $2[x + 2(x + 7)]$

163. $-2[3x + 2(4 - x)]$

164. $-5[2x + 3(5 - x)]$

165. $-3[2x - (x + 7)]$

166. $-2[3x - (5x - 2)]$

167. $2x - 3[x - (4 - x)]$

168. $-7x + 3[x - (3 - 2x)]$

169. $-5x - 2[2x - 4(x + 7)] - 6$

Objective E To translate a verbal expression into a variable expression

Translate into a variable expression. Then simplify.

170. twelve minus a number

171. a number divided by eighteen

172. two-thirds of a number

173. twenty more than a number

174. the quotient of twice a number and nine

175. ten times the difference between a number and fifty

176. eight less than the product of eleven and a number

177. the sum of five-eighths of a number and six

178. nine less than the total of a number and two

179. the difference between a number and three more than the number

Copyright © Houghton Mifflin Company. All rights reserved.

180. the quotient of seven and the total of five and a number

181. four times the sum of a number and nineteen

182. five increased by one-half of the sum of a number and three

183. the quotient of fifteen and the sum of a number and twelve

184. a number added to the difference between twice the number and four

185. the product of two-thirds and the sum of a number and seven

186. the product of five less than a number and seven

187. the difference between forty and the quotient of a number and twenty

188. the quotient of five more than twice a number and the number

189. the sum of the square of a number and twice the number

190. a number decreased by the difference between three times the number and eight

191. the sum of eight more than a number and one-third of the number

192. a number added to the product of three and the number

193. a number increased by the total of the number and nine

194. five more than the sum of a number and six

195. a number decreased by the difference between eight and the number

196. a number minus the sum of the number and ten

197. the difference between one-third of a number and five-eighths of the number

198. the sum of one-sixth of a number and four-ninths of the number

199. two more than the total of a number and five

200. the sum of a number divided by three and the number

201. twice the sum of six times a number and seven

Copyright © Houghton Mifflin Company. All rights reserved.

202. **Planets** The planet Saturn has 9 more moons than Jupiter (*Source:* NASA). Express the number of moons Saturn has in terms of the number of moons Jupiter has.

203. **The Olympics** The number of nations participating in the Olympic Games in 2004 was 1990 more than the number of nations participating in the Olympic Games in 1896 (*Source: USA Today* research). Express the number of nations participating in the Olympic Games in 2004 in terms of the number of nations participating in the Olympic Games in 1896.

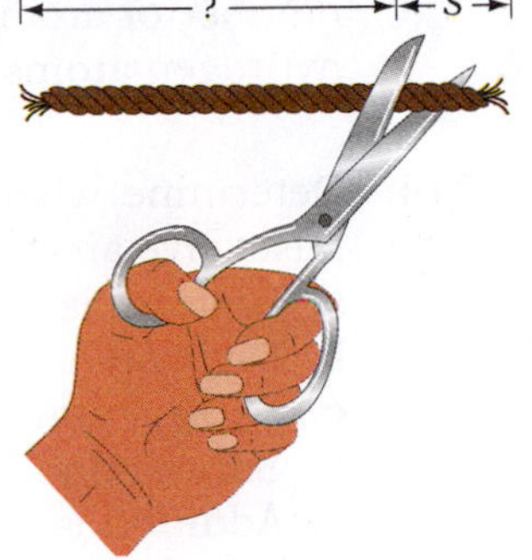

204. **Sailing** A halyard 12 ft long was cut into two pieces of different lengths. Use one variable to express the lengths of the two pieces.

205. **Natural Resources** Twenty gallons of crude oil was poured into two containers of different sizes. Use one variable to express the amount of oil poured into each container.

206. **Travel** Two cars start at the same place and travel at different rates in opposite directions. Two hours later the cars are 200 mi apart. Express the distance traveled by the faster car in terms of the distance traveled by the slower car.

207. **Agriculture** In a recent year, Alabama produced one-half the number of pounds of pecans that Texas produced that same year (*Source:* National Agricultural Statistics Service). Express the amount of pecans produced in Alabama in terms of the amount produced in Texas.

208. **Internal Revenue Service** According to the Internal Revenue Service, it takes about one-fifth as much time to fill out Schedule B (interest and dividends) as to fill out Schedule A (itemized deductions). Express the amount of time it takes to fill out Schedule B in terms of the time it takes to fill out Schedule A.

209. **Sports Equipment** The diameter of a basketball is approximately 4 times the diameter of a baseball. Express the diameter of a basketball in terms of the diameter of a baseball.

210. **World Population** According to the U.S. Bureau of the Census, the world population in the year 2050 is expected to be twice the world population in 1980. Express the world population in 2050 in terms of the world population in 1980.

Copyright © Houghton Mifflin Company. All rights reserved.

APPLYING THE CONCEPTS

211. Does every number have an additive inverse? If not, which real numbers do not have an additive inverse?

212. Does every number have a multiplicative inverse? If not, which real numbers do not have a multiplicative inverse?

213. **Chemistry** The chemical formula for glucose (sugar) is $C_6H_{12}O_6$. This formula means that there are twelve hydrogen atoms, six carbon atoms, and six oxygen atoms in each molecule of glucose. If x represents the number of atoms of oxygen in a pound of sugar, express the number of hydrogen atoms in the pound of sugar in terms of x.

H   O
 \ //
  C
  |
H – C – OH
  |
HO – C – H
  |
H – C – OH
  |
H – C – OH
  |
CH_2OH

214. Determine whether the statement is true or false. If the statement is false, give an example that illustrates that it is false.
a. Division is a commutative operation.
b. Division is an associative operation.
c. Subtraction is an associative operation.
d. Subtraction is a commutative operation.
e. Addition is a commutative operation.

215. **Metalwork** A wire whose length is given as x inches is bent into a square. Express the length of a side of the square in terms of x.

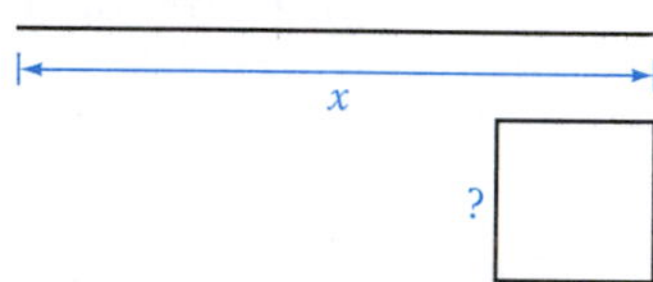

216. For each of the following, determine the first natural number x greater than 2 for which the second expression is larger than the first.

a. $x^3, 3^x$ **b.** $x^4, 4^x$ **c.** $x^5, 5^x$ **d.** $x^6, 6^x$

On the basis of your answers, make a conjecture that appears to be true about the expressions x^n and n^x, where $n = 3, 4, 5, 6, 7, \ldots$ and x is a natural number greater than 2.

217. **Pulleys** A block-and-tackle system is designed so that pulling 5 feet on one end of a rope will move a weight on the other end a distance of 3 feet. If x represents the distance the rope is pulled, express the distance the weight moves in terms of x.

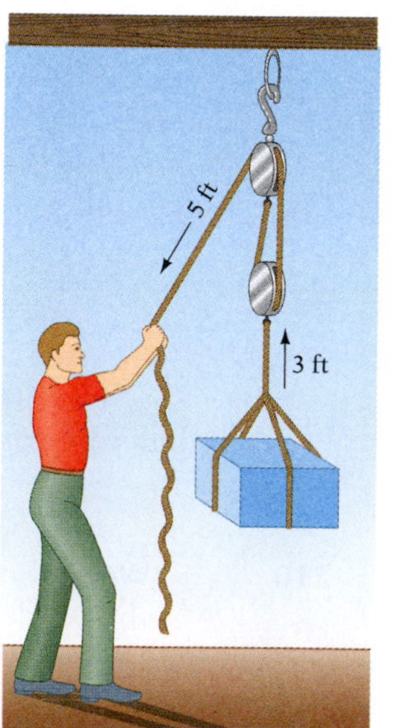

218. Give examples of two operations that occur in everyday experience that are not commutative (for example, putting on socks and then shoes).

219. Choose any number a. Evaluate the expressions $6a^2 + 2a - 10$ and $2a(3a - 4) + 10(a - 1)$. Now choose a different number and evaluate the expressions again. Repeat this two more times with different numbers. What conclusions might you draw from your evaluations?

Copyright © Houghton Mifflin Company. All rights reserved.

1.5 Sets

Objective A To write a set using the roster method

A **set** is a collection of objects, which are called the **elements** of the set. The **roster method** of writing a set encloses a list of the elements in braces.

The set of the last three letters of the alphabet is written {x, y, z}.

The set of the positive integers less than 5 is written {1, 2, 3, 4}.

HOW TO Use the roster method to write the set of integers between 0 and 10.

$A = \{1, 2, 3, 4, 5, 6, 7, 8, 9\}$

A set can be designated by a capital letter. Note that 0 and 10 are not elements of set A.

HOW TO Use the roster method to write the set of natural numbers.

$A = \{1, 2, 3, 4, \ldots\}$

• The three dots mean that the pattern of numbers continues without end.

The **empty set,** or **null set,** is the set that contains no elements. The symbol ∅ or { } is used to represent the empty set.

The set of people who have run a two-minute mile is the empty set.

The **union** of two sets, written $A \cup B$, is the set that contains the elements of A and the elements of B.

HOW TO Find $A \cup B$, given $A = \{1, 2, 3, 4\}$ and $B = \{3, 4, 5, 6\}$.

$A \cup B = \{1, 2, 3, 4, 5, 6\}$

• The union of A and B contains all the elements of A and all the elements of B. Any elements that are in both A and B are listed only once.

The **intersection** of two sets, written $A \cap B$, is the set that contains the elements that are common to both A and B.

HOW TO Find $A \cap B$, given $A = \{1, 2, 3, 4\}$ and $B = \{3, 4, 5, 6\}$.

$A \cap B = \{3, 4\}$

• The intersection of A and B contains the elements common to A and B.

Example 1

Use the roster method to write the set of the odd positive integers less than 12.

Solution

$A = \{1, 3, 5, 7, 9, 11\}$

You Try It 1

Use the roster method to write the set of the odd negative integers greater than −10.

Your solution

Solution on p. S3

Copyright © Houghton Mifflin Company. All rights reserved.

Example 2

Use the roster method to write the set of the even positive integers.

Solution

$A = \{2, 4, 6, \ldots\}$

You Try It 2

Use the roster method to write the set of the odd positive integers.

Your solution

Example 3

Find $D \cup E$, given $D = \{6, 8, 10, 12\}$ and $E = \{-8, -6, 10, 12\}$.

Solution

$D \cup E = \{-8, -6, 6, 8, 10, 12\}$

You Try It 3

Find $A \cup B$, given $A = \{-2, -1, 0, 1, 2\}$ and $B = \{0, 1, 2, 3, 4\}$.

Your solution

Example 4

Find $A \cap B$, given $A = \{5, 6, 9, 11\}$ and $B = \{5, 9, 13, 15\}$.

Solution

$A \cap B = \{5, 9\}$

You Try It 4

Find $C \cap D$, given $C = \{10, 12, 14, 16\}$ and $D = \{10, 16, 20, 26\}$.

Your solution

Example 5

Find $A \cap B$, given $A = \{1, 2, 3, 4\}$ and $B = \{8, 9, 10, 11\}$.

Solution

$A \cap B = \varnothing$

You Try It 5

Find $A \cap B$, given $A = \{-5, -4, -3, -2\}$ and $B = \{2, 3, 4, 5\}$.

Your solution

Solutions on p. S3

Objective B To write a set using set-builder notation

VIDEO & DVD   

Point of Interest

The symbol ∈ was first used in the book *Arithmeticae Principia*, published in 1889. It was the first letter of the Greek word εστι, which means "is." The symbols for union and intersection were also introduced at that time.

Another method of representing sets is called **set-builder notation.** Using set-builder notation, the set of all positive integers less than 10 is written as follows:

$\{x|x < 10, x \in \text{positive integers}\}$, which is read "the set of all x such that x is less than 10 and x is an element of the positive integers."

HOW TO Use set-builder notation to write the set of real numbers greater than 4.

$\{x|x > 4, x \in \text{real numbers}\}$

• "$x \in$ real numbers" is read "x is an element of the real numbers."

Copyright © Houghton Mifflin Company. All rights reserved.

Example 6

Use set-builder notation to write the set of negative integers greater than -100.

Solution

$\{x|x > -100, x \in \text{negative integers}\}$

You Try It 6

Use set-builder notation to write the set of positive even integers less than 59.

Your solution

Example 7

Use set-builder notation to write the set of real numbers less than 60.

Solution

$\{x|x < 60, x \in \text{real numbers}\}$

You Try It 7

Use set-builder notation to write the set of real numbers greater than -3.

Your solution

Solutions on p. S3

Objective C To graph an inequality on the number line

Point of Interest

The symbols for "is less than" and "is greater than" were introduced by Thomas Harriot around 1630. Before that, ⊏ and ⊐ were used for > and <, respectively.

An expression that contains the symbol >, <, ≥, or ≤ is called an **inequality.** An inequality expresses the relative order of two mathematical expressions. The expressions can be either numerical or variable.

$$\left.\begin{array}{l} 4 > 2 \\ 3x \le 7 \\ x^2 - 2x > y + 4 \end{array}\right\} \text{Inequalities}$$

An inequality can be graphed on the number line.

TAKE NOTE

In many cases, we assume that the real numbers are being used and omit "$x \in$ real numbers" from set-builder notation. Using this convention, $\{x|x > 1, x \in \text{real numbers}\}$ is written $\{x|x > 1\}$.

HOW TO Graph: $\{x|x > 1\}$

The graph is the real numbers greater than 1. The parenthesis at 1 indicates that 1 is not included in the graph.

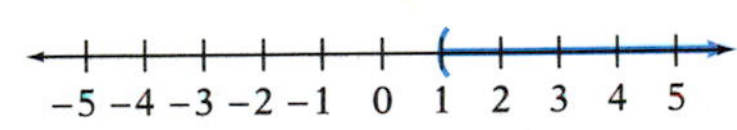

HOW TO Graph: $\{x|x \ge 1\}$

The bracket at 1 indicates that 1 is included in the graph.

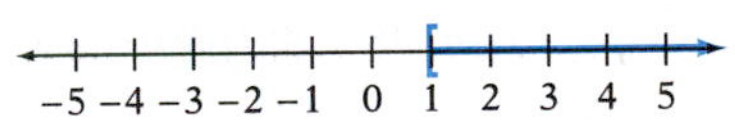

HOW TO Graph: $\{x|x < -1\}$

The numbers less than -1 are to the left of -1 on the number line.

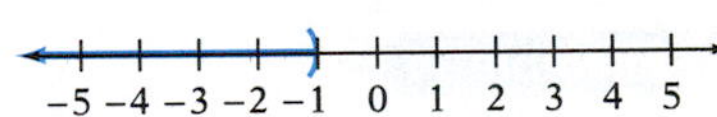

The union of two sets is the set that contains all the elements of each set.

HOW TO Graph: $\{x|x > 4\} \cup \{x|x < 1\}$

The graph is the numbers greater than 4 and the numbers less than 1.

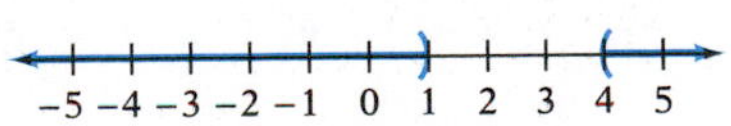

Copyright © Houghton Mifflin Company. All rights reserved.

The intersection of two sets is the set that contains the elements common to both sets.

HOW TO Graph: $\{x|x > -1\} \cap \{x|x < 2\}$

The graphs of $\{x|x > -1\}$ and $\{x|x < 2\}$ are shown at the right.

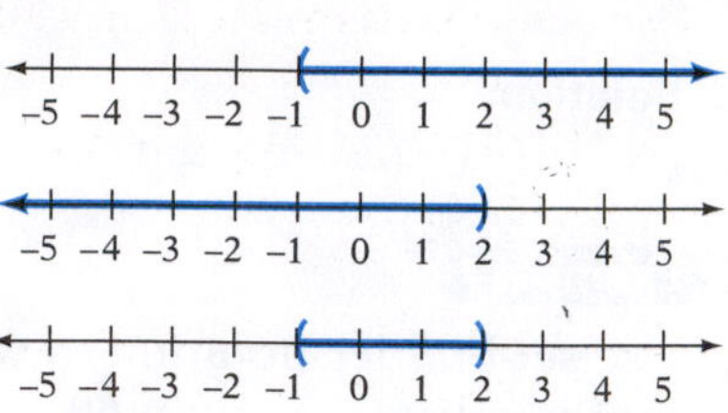

The graph of $\{x|x > -1\} \cap \{x|x < 2\}$ is the numbers between -1 and 2.

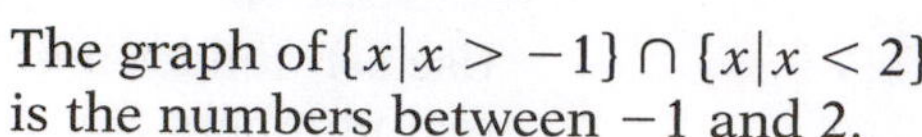

Example 8

Graph: $\{x|x < 5\}$

Solution

The graph is the numbers less than 5.

You Try It 8

Graph: $\{x|x > -2\}$

Your solution

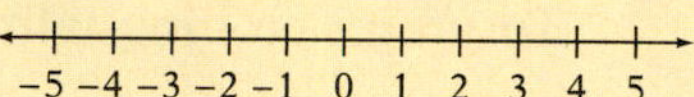

Example 9

Graph: $\{x|x > 3\} \cup \{x|x < 1\}$

Solution

The graph is the numbers greater than 3 and the numbers less than 1.

You Try It 9

Graph: $\{x|x > -1\} \cup \{x|x < -3\}$

Your solution

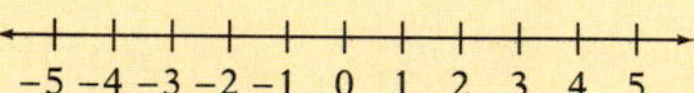

Example 10

Graph: $\{x|x > -2\} \cap \{x|x < 1\}$

Solution

The graph is the numbers between -2 and 1.

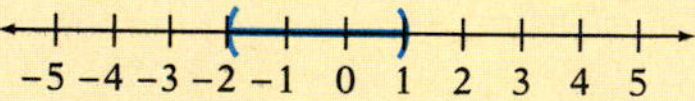

You Try It 10

Graph: $\{x|x \le 4\} \cap \{x|x \ge -4\}$

Your solution

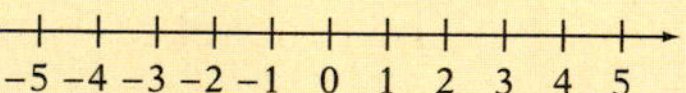

Example 11

Graph: $\{x|x \le 5\} \cup \{x|x \ge -3\}$

Solution

The graph is the real numbers.

You Try It 11

Graph: $\{x|x < 2\} \cup \{x|x \ge -2\}$

Your solution

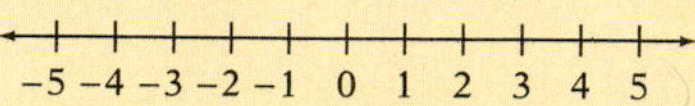

Solutions on p. S3

Copyright © Houghton Mifflin Company. All rights reserved.

1.5 Exercises

Objective A To write a set using the roster method

Use the roster method to write the set.

1. the integers between 15 and 22
2. the integers between -10 and -4
3. the odd integers between 8 and 18
4. the even integers between -11 and -1
5. the letters of the alphabet between a and d
6. the letters of the alphabet between p and v
7. Explain how to find the union of two sets.
8. Explain how to find the intersection of two sets.

Find $A \cup B$.

9. $A = \{3, 4, 5\}$ $B = \{4, 5, 6\}$
10. $A = \{-3, -2, -1\}$ $B = \{-2, -1, 0\}$
11. $A = \{-10, -9, -8\}$ $B = \{8, 9, 10\}$
12. $A = \{a, b, c\}$ $B = \{x, y, z\}$
13. $A = \{a, b, d, e\}$ $B = \{c, d, e, f\}$
14. $A = \{m, n, p, q\}$ $B = \{m, n, o\}$
15. $A = \{1, 3, 7, 9\}$ $B = \{7, 9, 11, 13\}$
16. $A = \{-3, -2, -1\}$ $B = \{-1, 1, 2\}$

Find $A \cap B$.

17. $A = \{3, 4, 5\}$ $B = \{4, 5, 6\}$
18. $A = \{-4, -3, -2\}$ $B = \{-6, -5, -4\}$
19. $A = \{-4, -3, -2\}$ $B = \{2, 3, 4\}$
20. $A = \{1, 2, 3, 4\}$ $B = \{1, 2, 3, 4\}$
21. $A = \{a, b, c, d, e\}$ $B = \{c, d, e, f, g\}$
22. $A = \{m, n, o, p\}$ $B = \{k, l, m, n\}$

Objective B To write a set using set-builder notation

Use set-builder notation to write the set.

23. the negative integers greater than -5
24. the positive integers less than 5

Copyright © Houghton Mifflin Company. All rights reserved.

25. the integers greater than 30

26. the integers less than -70

27. the even integers greater than 5

28. the odd integers less than -2

29. the real numbers greater than 8

30. the real numbers less than 57

Objective C To graph an inequality on the number line

Graph.

31. $\{x|x > 2\}$

−5 −4 −3 −2 −1 0 1 2 3 4 5

32. $\{x|x \geq -1\}$

−5 −4 −3 −2 −1 0 1 2 3 4 5

33. $\{x|x \leq 0\}$

−5 −4 −3 −2 −1 0 1 2 3 4 5

34. $\{x|x < 4\}$

−5 −4 −3 −2 −1 0 1 2 3 4 5

35. $\{x|x > -2\} \cup \{x|x < -4\}$

−5 −4 −3 −2 −1 0 1 2 3 4 5

36. $\{x|x > 4\} \cup \{x|x < -2\}$

−5 −4 −3 −2 −1 0 1 2 3 4 5

37. $\{x|x > -2\} \cap \{x|x < 4\}$

−5 −4 −3 −2 −1 0 1 2 3 4 5

38. $\{x|x > -3\} \cap \{x|x < 3\}$

−5 −4 −3 −2 −1 0 1 2 3 4 5

39. $\{x|x \geq -2\} \cup \{x|x < 4\}$

−5 −4 −3 −2 −1 0 1 2 3 4 5

40. $\{x|x > 0\} \cup \{x|x \leq 4\}$

−5 −4 −3 −2 −1 0 1 2 3 4 5

APPLYING THE CONCEPTS

41. Determine whether the statement is always true, sometimes true, or never true.
 a. Given that $a > 0$ and $b < 0$, then $ab > 0$.
 b. Given that $a < 0$, then $a^2 > 0$.
 c. Given that $a > 0$ and $b < 0$, then $a^2 > b$.

42. By trying various sets, make a conjecture as to whether the union of two sets is
 a. a commutative operation
 b. an associative operation

43. By trying various sets, make a conjecture as to whether the intersection of two sets is
 a. a commutative operation
 b. an associative operation

Copyright © Houghton Mifflin Company. All rights reserved.

Focus on Problem Solving

Inductive Reasoning

Suppose you take 9 credit hours each semester. The total number of credit hours you have taken at the end of each semester can be described in a list of numbers.

$$9, 18, 27, 36, 45, 54, 63, \ldots$$

The list of numbers that indicates the total credit hours is an ordered list of numbers, called a **sequence.** Each number in a sequence is called a **term** of the sequence. The list is ordered because the position of a number in the list indicates after which semester that number of credit hours has been taken. For example, the 7th term of the sequence is 63, and a total of 63 credit hours have been taken after the 7th semester.

Assuming the pattern continues, find the next three numbers in the pattern

$$-6, -10, -14, -18, \ldots$$

This list of numbers is a sequence. The first step in solving this problem is to observe the pattern in the list of numbers. In this case, each number in the list is 4 less than the previous number. The next three numbers are $-22, -26, -30$.

This process of discovering the pattern in a list of numbers uses inductive reasoning. **Inductive reasoning** involves making generalizations from specific examples; in other words, we reach a conclusion by making observations about particular facts or cases.

Try the following exercises. Each exercise requires inductive reasoning.

Name the next two terms in the sequence.

1. 1, 3, 5, 7, 1, 3, 5, 7, 1, . . .

2. 1, 4, 2, 5, 3, 6, 4, . . .

3. 1, 2, 4, 7, 11, 16, . . .

4. A, B, C, G, H, I, M, . . .

Draw the next shape in the sequence.

5. 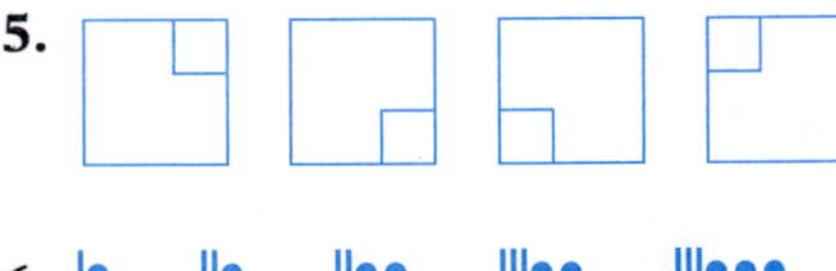

6. |• ||• ||•• |||•• |||•••

Solve.

7. Convert $\frac{1}{11}, \frac{2}{11}, \frac{3}{11}, \frac{4}{11}$, and $\frac{5}{11}$ to decimals. Then use the pattern you observe to convert $\frac{6}{11}, \frac{7}{11}$, and $\frac{9}{11}$ to decimals.

8. Convert $\frac{1}{33}, \frac{2}{33}, \frac{4}{33}, \frac{5}{33}$, and $\frac{7}{33}$ to decimals. Then use the pattern you observe to convert $\frac{8}{33}, \frac{13}{33}$, and $\frac{19}{33}$ to decimals.

Copyright © Houghton Mifflin Company. All rights reserved.

Projects and Group Activities

Calculators

Does your calculator use the Order of Operations Agreement? To find out, try this problem:

$$2 + 4 \cdot 7$$

If your answer is 30, then the calculator uses the Order of Operations Agreement. If your answer is 42, it does not use the agreement.

Even if your calculator does not use the Order of Operations Agreement, you can still correctly evaluate numerical expressions. The parentheses keys, (and), are used for this purpose.

Remember that $2 + 4 \cdot 7$ means $2 + (4 \cdot 7)$ because the multiplication must be completed before the addition. To evaluate this expression, enter the following:

Enter:	2	+	(	4	×	7	)	=
Display:	2	2	(	4	4	7	28	30

When using your calculator to evaluate numerical expressions, insert parentheses around multiplications or divisions. This has the effect of forcing the calculator to do the operations in the order *you* want rather than in the order the calculator wants.

Evaluate.

1. $3 \cdot (15 - 2 \cdot 3) - 36 \div 3$

2. $4 \cdot 2^2 - (12 + 24 \div 6) - 5$

3. $16 \div 4 \cdot 3 + (3 \cdot 4 - 5) + 2$

4. $15 \cdot 3 \div 9 + (2 \cdot 6 - 3) + 4$

Using your calculator to simplify numerical expressions sometimes requires use of the +/− key or, on some calculators, the negative key, which is frequently shown as (−). These keys change the sign of the number currently in the display. To enter −4:

- For those calculators with +/−, press 4 and then +/−.
- For those calculators with (−), press (−) and then 4.

Here are the keystrokes for evaluating the expression $3(-4) - (-5)$.

Calculators with +/− key: 3 × 4 +/− − 5 +/− =

Calculators with (−) key: 3 × (−) 4 − (−) 5 =

This example illustrates that calculators make a distinction between negative and minus. To perform the operation $3 - (-3)$, you cannot enter 3 − − 3. This would result in 0, which is not the correct answer. You must enter

3 − 3 +/− =   or   3 − (−) 3 =

Use a calculator to evaluate each of the following exercises.

5. $-16 \div 2$

6. $3(-8)$

7. $47 - (-9)$

8. $-50 - (-14)$

9. $4 - (-3)^2$

10. $-8 + (-6)^2 - 7$

Copyright © Houghton Mifflin Company. All rights reserved.

Chapter 1 Summary

Key Words	Examples
The set of *natural numbers* is $\{1, 2, 3, 4, 5, \ldots\}$. The set of *integers* is $\{\ldots, -3, -2, -1, 0, 1, 2, 3, \ldots\}$. [1.1A, p. 3]	
A number *a is less than* a number *b*, written $a < b$, if *a* is to the left of *b* on a number line. A number *a is greater than* a number *b*, written $a > b$, if *a* is to the right of *b* on a number line. The symbol $\le$ means *is less than or equal to*. The symbol $\ge$ means *is greater than or equal to*. [1.1A, p. 3]	$-5 < -3$ $9 > 0$ $3 \le 3$ $4 \le 7$ $5 \ge 5$ $-6 \ge -9$
Two numbers that are the same distance from zero on the number line but on opposite sides of zero are *opposite numbers* or *opposites*. The *additive inverse* of a number is the opposite of the number. [1.1B, p. 4]	7 and -7 are opposites. $-\frac{3}{4}$ and $\frac{3}{4}$ are opposites.
The *absolute value* of a number is its distance from 0 on the number line. [1.1B, p. 4]	$\lvert 5\rvert = 5$ $\lvert -2.3\rvert = 2.3$ $\lvert 0\rvert = 0$
A *rational number* (or fraction) is a number that can be written in the form $\frac{a}{b}$, where *a* and *b* are integers and $b \neq 0$. A rational number can be represented as a *terminating* or *repeating decimal*. [1.2A, p. 17]	$\frac{3}{8}$, $-\frac{9}{2}$, and 4 are rational numbers. 1.13 and $0.4\overline{73}$ are also rational numbers.
Percent means "parts of 100." [1.2B, p. 18]	72% means 72 of 100 equal parts.
An expression of the form a^n is in *exponential form*. The *base* is *a* and the *exponent* is *n*. [1.2E, p. 22]	5^4 is an exponential expression. The base is 5 and the exponent is 4.
A *square root* of a positive number *x* is a number whose square is *x*. The *principal square root* of a number is the positive square root. The symbol $\sqrt{}$ is called a *radical sign* and is used to indicate the principal square root of a number. The *radicand* is the number under the radical sign. [1.2F, p. 24]	$\sqrt{25} = 5$ $-\sqrt{25} = -5$
The square of an integer is a *perfect square*. If a number is not a perfect square, its square root can only be approximated. [1.2F, p. 24]	$7^2 = 49$; 49 is a perfect square.
An *irrational number* is a number that has a decimal representation that never terminates or repeats. [1.2F, p. 24]	π, $\sqrt{2}$, and 1.34334333433334 . . . are irrational numbers.
The rational numbers and the irrational numbers taken together are the *real numbers*. [1.2F, p. 24]	$\frac{3}{8}$, $-\frac{9}{2}$, 4, 1.13, $0.4\overline{73}$, π, $\sqrt{2}$, and 1.34334333433334 . . . are real numbers.

Copyright © Houghton Mifflin Company. All rights reserved.

A *variable* is a letter that is used for a quantity that is unknown or that can change. A *variable expression* is an expression that contains one or more variables. [1.4A, p. 37]	$4x + 2y - 6z$ is a variable expression. It contains the variables x, y, and z.
The *terms* of a variable expression are the addends of the expression. Each term is a *variable term* or a *constant term.* [1.4A, p. 37]	The expression $2a^2 - 3b^3 + 7$ has three terms, $2a^2$, $-3b^3$, and 7. $2a^2$ and $-3b^3$ are variable terms. 7 is a constant term.
A variable term is composed of a *numerical coefficient* and a *variable part*. [1.4A, p. 37]	For the expression $-7x^3y^2$, -7 is the coefficient and x^3y^2 is the variable part.
In a variable expression, replacing each variable by its value and then simplifying the resulting numerical expression is called *evaluating the variable expression*. [1.4A, p. 37]	To evaluate $2ab - b^2$ when $a = 3$ and $b = -2$, replace a by 3 and b by -2 and then simplify the numerical expression. $2(3)(-2) - (-2)^2 = -16$
Like terms of a variable expression are terms with the same variable part. Constant terms are like terms. [1.4B, p. 39]	For the expressions $3a^2 + 2b - 3$ and $2a^2 - 3a + 4$, $3a^2$ and $2a^2$ are like terms; -3 and 4 are like terms.
To simplify the sum of like variable terms, use the Distributive Property to add the numerical coefficients. This is called *combining like terms.* [1.4B, p. 39]	$5y + 3y = (5 + 3)y$ $= 8y$
The *multiplicative inverse* of a number is the *reciprocal of the number.* [1.4C, p. 41]	$\frac{3}{4}$ is the multiplicative inverse of $\frac{4}{3}$. $-\frac{1}{4}$ is the multiplicative inverse of -4.
A *set* is a collection of objects, which are called the *elements* of the set. The *roster method* of writing a set encloses a list of the elements in braces. The *empty set* or *null set,* written $\varnothing$, is the set that contains no elements. [1.5A, p. 55]	The set of cars that can travel faster than 1000 mph is an empty set.
The *union* of two sets, written $A \cup B$, is the set that contains the elements of A and the elements of B. [1.5A, p. 55]	Let $A = \{2, 4, 6, 8\}$ and $B = \{0, 1, 2, 3, 4\}$. Then $A \cup B = \{0, 1, 2, 3, 4, 6, 8\}$.
The *intersection* of two sets, written $A \cap B$, is the set that contains the elements that are common to both A and B. [1.5A, p. 55]	Let $A = \{2, 4, 6, 8\}$ and $B = \{0, 1, 2, 3, 4\}$. Then $A \cap B = \{2, 4\}$.
Set-builder notation uses a rule to describe the elements of a set. [1.5B, p. 56]	Using set-builder notation, the set of real numbers greater than 2 is written $\{x \mid x > 2, x \in \text{real numbers}\}$.

Essential Rules and Procedures	Examples
To add two numbers with the same sign, add the absolute values of the numbers. Then attach the sign of the addends. [1.1C, p. 5]	$7 + 15 = 22$ $-7 + (-15) = -22$

Copyright © Houghton Mifflin Company. All rights reserved.

To add two numbers with different signs, find the absolute value of each number. Subtract the smaller of the two numbers from the larger. Then attach the sign of the number with the larger absolute value. [1.1C, p. 5]

$7 + (-15) = -8$
$-7 + 15 = 8$

To subtract one number from another, add the opposite of the second number to the first number. [1.1C, p. 6]

$7 - 19 = 7 + (-19) = -12$
$-6 - (-13) = -6 + 13 = 7$

To multiply two numbers with the same sign, multiply the absolute values of the numbers. The product is positive. [1.1D, p. 8]

$7 \cdot 8 = 56$
$-7(-8) = 56$

To multiply two numbers with different signs, multiply the absolute values of the numbers. The product is negative. [1.1D, p. 8]

$-7 \cdot 8 = -56$
$7(-8) = -56$

To divide two numbers with the same sign, divide the absolute values of the numbers. The quotient is positive. [1.1D, p. 9]

$54 \div 9 = 6$
$(-54) \div (-9) = 6$

To divide two numbers with different signs, divide the absolute values of the numbers. The quotient is negative. [1.1D, p. 9]

$(-54) \div 9 = -6$
$54 \div (-9) = -6$

Properties of Zero and One in Division [1.1D, p. 9]

If $a \neq 0, \frac{0}{a} = 0.$ $\qquad \frac{0}{-5} = 0$

If $a \neq 0, \frac{a}{a} = 1.$ $\qquad \frac{-12}{-12} = 1$

$\frac{a}{1} = a$ $\qquad \frac{7}{1} = 7$

$\frac{a}{0}$ is undefined. $\qquad \frac{8}{0}$ is undefined.

To write a percent as a fraction, remove the percent sign and multiply by $\frac{1}{100}$. [1.2B, p. 18]

$$60\% = 60\left(\frac{1}{100}\right) = \frac{60}{100} = \frac{3}{5}$$

To write a percent as a decimal, remove the percent sign and multiply by 0.01. [1.2B, p. 18]

$73\% = 73(0.01) = 0.73$
$1.3\% = 1.3(0.01) = 0.013$

To write a decimal or a fraction as a percent, multiply by 100%. [1.2B, p. 18]

$0.3 = 0.3(100\%) = 30\%$
$\frac{5}{8} = \frac{5}{8}(100\%) = \frac{500}{8}\% = 62.5\%$

To add two fractions with the same denominator, add the numerators and place the sum over the common denominator. [1.2C, p. 19]

$$\frac{7}{10} + \frac{1}{10} = \frac{7+1}{10} = \frac{8}{10} = \frac{4}{5}$$

To subtract two fractions with the same denominator, subtract the numerators and place the difference over the common denominator. [1.2C, p. 19]

$$\frac{7}{10} - \frac{1}{10} = \frac{7-1}{10} = \frac{6}{10} = \frac{3}{5}$$

Copyright © Houghton Mifflin Company. All rights reserved.

To multiply two fractions, place the product of the numerators over the product of the denominators. [1.2D, p. 20]

$$-\frac{2}{3}\cdot\frac{5}{6}=-\frac{2\cdot 5}{3\cdot 6}=-\frac{10}{18}=-\frac{5}{9}$$

To divide two fractions, multiply the dividend by the reciprocal of the divisor. [1.2D, p. 21]

$$-\frac{4}{5}\div\frac{2}{3}=-\frac{4}{5}\cdot\frac{2}{3}=-\frac{2\cdot 2\cdot 3}{5\cdot 2}=-\frac{6}{5}$$

Product Property of Square Roots [1.2F, p. 25]
$\sqrt{ab}=\sqrt{a}\cdot\sqrt{b}$

$$\begin{aligned}\sqrt{50}&=\sqrt{25\cdot 2}\\&=\sqrt{25}\sqrt{2}=5\sqrt{2}\end{aligned}$$

Order of Operations Agreement [1.3A, p. 33]

Step 1 Perform operations inside grouping symbols. Grouping symbols include parentheses ( ), brackets [ ], braces { }, and the fraction bar.

Step 2 Simplify exponential expressions.

Step 3 Do multiplication and division as they occur from left to right.

Step 4 Do addition and subtraction as they occur from left to right.

$$\begin{aligned}&50\div(-5)^2+2(7-16)\\&=50\div(-5)^2+2(-9)\\&=50\div 25+2(-9)\\&=2+(-18)\\&=-16\end{aligned}$$

The Distributive Property [1.4B, p. 38]
If a, b, and c are real numbers, then $a(b+c)=ab+ac$.

$$\begin{aligned}5(4+7)&=5\cdot 4+5\cdot 7\\&=20+35=55\end{aligned}$$

The Associative Property of Addition [1.4B, p. 39]
If a, b, and c are real numbers, then $(a+b)+c=a+(b+c)$.

$-4+(2+7)=-4+9=5$
$(-4+2)+7=-2+7=5$

The Commutative Property of Addition [1.4B, p. 39]
If a and b are real numbers, then $a+b=b+a$.

$2+5=7$ and $5+2=7$

The Addition Property of Zero [1.4B, p. 39]
If a is a real number, then $a+0=0+a=a$.

$-8+0=-8$ and $0+(-8)=-8$

The Inverse Property of Addition [1.4B, p. 40]
If a is a real number, then $a+(-a)=(-a)+a=0$.

$5+(-5)=0$ and $(-5)+5=0$

The Associative Property of Multiplication [1.4C, p. 40]
If a, b, and c are real numbers, then $(ab)c=a(bc)$.

$-3\cdot(5\cdot 4)=-3(20)=-60$
$(-3\cdot 5)\cdot 4=-15\cdot 4=-60$

The Commutative Property of Multiplication [1.4C, p. 41]
If a and b are real numbers, then $ab=ba$.

$-3(7)=-21$ and $7(-3)=-21$

The Multiplication Property of One [1.4C, p. 41]
If a is a real number, then $a\cdot 1=1\cdot a=a$.

$-3(1)=-3$ and $1(-3)=-3$

The Inverse Property of Multiplication [1.4C, p. 41]
If a is a real number and a is not equal to zero, then $a\cdot\frac{1}{a}=\frac{1}{a}\cdot a=1$.

$-3\cdot -\frac{1}{3}=1$ and $-\frac{1}{3}\cdot -3=1$

Copyright © Houghton Mifflin Company. All rights reserved.

Chapter 1 Review Exercises

1. Let $x \in \{-4, 0, 11\}$. For what values of x is the inequality $x < 1$ a true statement?

2. Find the additive inverse of -4.

3. Evaluate $-|-5|$.

4. Add: $-3 + (-12) + 6 + (-4)$

5. Subtract: $16 - (-3) - 18$

6. Multiply: $(-6)(7)$

7. Divide: $-100 \div 5$

8. Write $\frac{7}{25}$ as a decimal.

9. Write 6.2% as a decimal.

10. Write $\frac{5}{8}$ as a percent.

11. Simplify: $\frac{1}{3} - \frac{1}{6} + \frac{5}{12}$

12. Subtract: $5.17 - 6.238$

13. Divide: $-\frac{18}{35} \div \frac{17}{28}$

14. Multiply: $4.32(-1.07)$

15. Evaluate $\left(-\frac{2}{3}\right)^4$.

16. Simplify: $2\sqrt{36}$

17. Simplify: $-3\sqrt{120}$

18. Evaluate $-3^2 + 4[18 + (12 - 20)]$.

Copyright © Houghton Mifflin Company. All rights reserved.

19. Evaluate $(b - a)^2 + c$ when $a = -2$, $b = 3$, and $c = 4$.

20. Simplify: $6a - 4b + 2a$

21. Simplify: $-3(-12y)$

22. Simplify: $5(2x - 7)$

23. Simplify: $-4(2x - 9) + 5(3x + 2)$

24. Simplify: $5[2 - 3(6x - 1)]$

25. Use the roster method to write the set of odd positive integers less than 8.

26. Find $A \cap B$, given $A = [1, 5, 9, 13\}$ and $B = \{1, 3, 5, 7, 9\}$.

27. Graph $\{x|x > 3\}$.

−5 −4 −3 −2 −1 0 1 2 3 4 5

28. Graph $\{x|x \le 3\} \cup \{x|x < -2\}$.

−5 −4 −3 −2 −1 0 1 2 3 4 5

29. **Testing** To discourage random guessing on a multiple-choice exam, a professor assigns 6 points for a correct answer, −4 points for an incorrect answer, and −2 points for leaving a question blank. What is the score for a student who had 21 correct answers, had 5 incorrect answers, and left 4 questions blank?

30. **Candy** The circle graph shows the amount of candy consumed by Americans during a recent year (*Source:* Candy USA). What percent of the candy consumed was chocolate? Round to the nearest tenth of a percent.

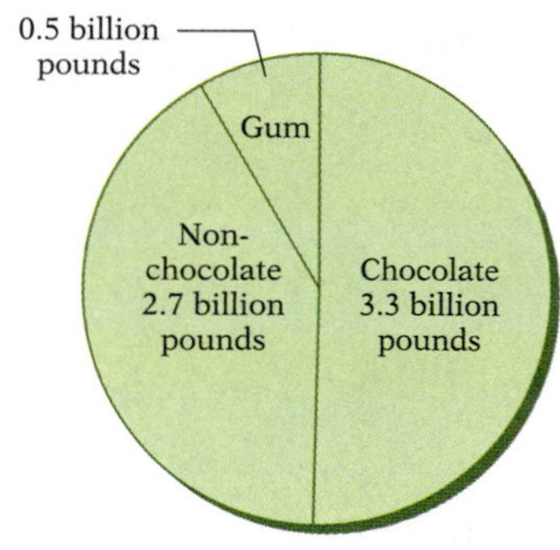

31. Translate "the difference between twice a number and one-half of the number" into a variable expression. Then simplify.

32. **Baseball Cards** A baseball card collection contains five times as many National League players' cards as American League players' cards. Express the number of National League players' cards in terms of the number of American League players' cards.

33. **Money** A club treasurer has some five-dollar bills and some ten-dollar bills. The treasurer has a total of 35 bills. Express the number of five-dollar bills in terms of the number of ten-dollar bills.

Copyright © Houghton Mifflin Company. All rights reserved.

Chapter 1 Test

1. Place the correct symbol, < or >, between the two numbers.
   $-2 \quad -40$

2. Find the opposite of -4.

3. Evaluate $-|-4|$.

4. Subtract: $16 - 30$

5. Add: $-22 + 14 + (-8)$

6. Subtract: $16 - (-30) - 42$

7. Divide: $-561 \div (-33)$

8. Write $\frac{7}{9}$ as a decimal. Place a bar over the repeating digit of the decimal.

9. Write 45% as a fraction and as a decimal.

10. Add: $-\frac{2}{5} + \frac{7}{15}$

11. Multiply: $6.02(-0.89)$

12. Divide: $\frac{5}{12} \div \left(-\frac{5}{6}\right)$

13. Evaluate $\frac{3}{4} \cdot (4)^2$.

14. Simplify: $-2\sqrt{45}$

15. Evaluate $16 \div 2[8 - 3(4 - 2)] + 1$.

16. Evaluate $b^2 - 3ab$ when $a = 3$ and $b = -2$.

17. Simplify: $3x - 5x + 7x$

18. Simplify: $\frac{1}{5}(10x)$

Copyright © Houghton Mifflin Company. All rights reserved.

19. Simplify: $-3(2x^2 - 7y^2)$

20. Simplify: $2x - 3(x - 2)$

21. Simplify: $2x + 3[4 - (3x - 7)]$

22. Use the roster method to write the set of integers between -3 and 4.

23. Use set-builder notation to write the set of real numbers less than -3.

24. Find $A \cup B$ given $A = \{1, 3, 5, 7\}$ and $B = \{2, 4, 6, 8\}$.

25. Graph $\{x \mid x < 1\}$.

−5 −4 −3 −2 −1 0 1 2 3 4 5

26. Graph $\{x \mid x \leq -3\} \cup \{x \mid x > 0\}$.

−5 −4 −3 −2 −1 0 1 2 3 4 5

27. Translate "ten times the difference between a number and 3" into a variable expression. Then simplify.

28. **Baseball** The speed of a pitcher's fastball is twice the speed of the catcher's return throw. Express the speed of the fastball in terms of the speed of the return throw.

29. **Balance of Trade** The table at the right shows the U.S. balance of trade, in billions of dollars, for the years 1980 to 2000 (*Source:* U.S. Dept. of Commerce).

a. In which years did the trade balance increase from the previous year?

b. Calculate the difference between the trade balance in 1990 and the trade balance in 2000.

c. During which two consecutive years was the difference in the trade balance greatest?

d. How many times greater was the trade balance in 1990 than in 1980? Round to the nearest whole number.

e. Calculate the average trade balance per quarter for the year 2000.

Year	*Trade Balance*
1980	−19.4
1981	−16.2
1982	−24.2
1983	−57.8
1984	−109.2
1985	−122.1
1986	−140.6
1987	−153.3
1988	−115.9
1989	−92.2
1990	−81.1
1991	−30.7
1992	−35.7
1993	−68.9
1994	−97.0
1995	−95.9
1996	−102.1
1997	−104.7
1998	−166.9
1999	−265.0
2000	−369.7

30. **Temperature** The lowest temperature recorded in North America is $-81.4°F$. The highest temperature recorded is $134.0°F$ (*Source:* National Climatic Data Center). Find the difference beween these two extremes.

Copyright © Houghton Mifflin Company. All rights reserved.

chapter

2 First-Degree Equations and Inequalities

Hourly wage, salary, and commissions are three ways to receive payment for doing work. Commissions are usually paid to salespersons and are calculated as a percent of total sales. The salesperson in this photo receives a combination of an hourly wage and commissions. The sales personnel in **Exercises 94 and 95 on page 135** receive a combination of salary and commissions. In these exercises, you will be using first-degree inequalities to determine the amount of sales needed to reach target income goals.

Need help? For online student resources, such as section quizzes, visit this textbook's website at **math.college.hmco.com/students.**

OBJECTIVES

Section 2.1

- **A** To determine whether a given number is a solution of an equation
- **B** To solve an equation of the form $x + a = b$
- **C** To solve an equation of the form $ax = b$
- **D** To solve application problems using the basic percent equation
- **E** To solve uniform motion problems

Section 2.2

- **A** To solve an equation of the form $ax + b = c$
- **B** To solve an equation of the form $ax + b = cx + d$
- **C** To solve an equation containing parentheses
- **D** To solve application problems using formulas

Section 2.3

- **A** To solve integer problems
- **B** To translate a sentence into an equation and solve

Section 2.4

- **A** To solve value mixture problems
- **B** To solve percent mixture problems
- **C** To solve uniform motion problems

Section 2.5

- **A** To solve an inequality in one variable
- **B** To solve a compound inequality
- **C** To solve application problems

Section 2.6

- **A** To solve an absolute value equation
- **B** To solve an absolute value inequality
- **C** To solve application problems

Copyright © Houghton Mifflin Company. All rights reserved.

PREP TEST

Do these exercises to prepare for Chapter 2.

1. Write $\frac{9}{100}$ as a decimal.

2. Write $\frac{3}{4}$ as a percent.

3. Evaluate $3x^2 - 4x - 1$ when $x = -4$.

4. Simplify: $R - 0.35R$

5. Simplify: $\frac{1}{2}x + \frac{2}{3}x$

6. Simplify: $6x - 3(6 - x)$

7. Simplify: $0.22(3x + 6) + x$

8. Translate into a variable expression: "The difference between 5 and twice a number."

9. A new graphics card for computer games is five times faster than a graphics card made two years ago. Express the speed of the new card in terms of the speed of the old card.

10. A board 5 ft long is cut into two pieces. If x represents the length of the longer piece, write an expression for the shorter piece in terms of x.

GO FIGURE

How can a donut be cut into eight equal pieces with three cuts of a knife?

Copyright © Houghton Mifflin Company. All rights reserved.

2.1 Introduction to Equations

Objective A

To determine whether a given number is a solution of an equation

Point of Interest

One of the most famous equations ever stated is $E = mc^2$. This equation, stated by Albert Einstein, shows that there is a relationship between mass m and energy E. As a side note, the chemical element einsteinium was named in honor of Einstein.

An **equation** expresses the equality of two mathematical expressions. The expressions can be either numerical or variable expressions.

$$\left.\begin{array}{l} 9 + 3 = 12 \\ 3x - 2 = 10 \\ y^2 + 4 = 2y - 1 \\ z = 2 \end{array}\right\} \text{Equations}$$

The equation at the right is true if the variable is replaced by 5.

$x + 8 = 13$

$5 + 8 = 13$ A true equation

The equation is false if the variable is replaced by 7.

$7 + 8 = 13$ A false equation

A **solution of an equation** is a number that, when substituted for the variable, results in a true equation. 5 is a solution of the equation $x + 8 = 13$. 7 is not a solution of the equation $x + 8 = 13$.

TAKE NOTE

The Order of Operations Agreement applies to evaluating $2(-2) + 5$ and $(-2)^2 - 3$.

HOW TO Is -2 a solution of $2x + 5 = x^2 - 3$?

$$\begin{array}{r|l} \multicolumn{2}{c}{2x + 5 = x^2 - 3} \\ \hline 2(-2) + 5 & (-2)^2 - 3 \\ -4 + 5 & 4 - 3 \\ \multicolumn{2}{c}{1 = 1} \end{array}$$

- Replace x by -2.
- Evaluate the numerical expressions.
- If the results are equal, -2 is a solution of the equation. If the results are not equal, -2 is not a solution of the equation.

Yes, -2 is a solution of the equation.

Example 1 Is -4 a solution of $5x - 2 = 6x + 2$?

Solution

$$\begin{array}{r|l} \multicolumn{2}{c}{5x - 2 = 6x + 2} \\ \hline 5(-4) - 2 & 6(-4) + 2 \\ -20 - 2 & -24 + 2 \\ \multicolumn{2}{c}{-22 = -22} \end{array}$$

Yes, -4 is a solution.

You Try It 1 Is $\frac{1}{4}$ a solution of $5 - 4x = 8x + 2$?

Your solution

Example 2 Is -4 a solution of $4 + 5x = x^2 - 2x$?

Solution

$$\begin{array}{r|l} \multicolumn{2}{c}{4 + 5x = x^2 - 2x} \\ \hline 4 + 5(-4) & (-4)^2 - 2(-4) \\ 4 + (-20) & 16 - (-8) \\ \multicolumn{2}{c}{-16 \neq 24} \end{array}$$

($\neq$ means "is not equal to")

No, -4 is not a solution.

You Try It 2 Is 5 a solution of $10x - x^2 = 3x - 10$?

Your solution

Solutions on p. S3

Copyright © Houghton Mifflin Company. All rights reserved.

Objective B To solve an equation of the form $x + a = b$

Study Tip

To learn mathematics, you must be an active participant. Listening and watching your professor do mathematics is not enough. Take notes in class, mentally think through every question your instructor asks, and try to answer it even if you are not called on to answer it verbally. Ask questions when you have them. See *AIM for Success*, page xxvii, for other ways to be an active learner.

To **solve an equation** means to find a solution of the equation. The simplest equation to solve is an equation of the form *variable* = *constant*, because the constant is the solution.

The solution of the equation $x = 5$ is 5 because $5 = 5$ is a true equation.

The solution of the equation at the right is 7 because $7 + 2 = 9$ is a true equation.

$$x + 2 = 9 \qquad 7 + 2 = 9$$

Note that if 4 is added to each side of the equation $x + 2 = 9$, the solution is still 7.

$$\begin{aligned} x + 2 &= 9 \\ x + 2 + 4 &= 9 + 4 \\ x + 6 &= 13 \qquad 7 + 6 = 13 \end{aligned}$$

If -5 is added to each side of the equation $x + 2 = 9$, the solution is still 7.

$$\begin{aligned} x + 2 &= 9 \\ x + 2 + (-5) &= 9 + (-5) \\ x - 3 &= 4 \qquad 7 - 3 = 4 \end{aligned}$$

Equations that have the same solution are **equivalent equations.** The equations $x + 2 = 9$, $x + 6 = 13$, and $x - 3 = 4$ are equivalent equations; each equation has 7 as its solution. These examples suggest that adding the same number to each side of an equation produces an equivalent equation. This is called the *Addition Property of Equations.*

Addition Property of Equations

The same number can be added to each side of an equation without changing its solution. In symbols, the equation $a = b$ has the same solution as the equation $a + c = b + c$.

In solving an equation, the goal is to rewrite the given equation in the form *variable* = *constant*. The Addition Property of Equations is used to remove a *term* from one side of the equation by adding the opposite of that term to each side of the equation.

HOW TO Solve: $x - 4 = 2$

$x - 4 = 2$ • The goal is to rewrite the equation as *variable* = *constant*.

$x - 4 + 4 = 2 + 4$ • Add 4 to each side of the equation.

$x + 0 = 6$ • Simplify.

$x = 6$ • The equation is in the form *variable* = *constant*.

Check:

$$\begin{array}{c|c} x - 4 & 2 \\ \hline 6 - 4 & 2 \\ 2 & 2 \end{array}$$

$2 = 2$ A true equation

The solution is 6.

Because subtraction is defined in terms of addition, the Addition Property of Equations also makes it possible to subtract the same number from each side of an equation without changing the solution of the equation.

Copyright © Houghton Mifflin Company. All rights reserved.

HOW TO Solve: $y + \frac{3}{4} = \frac{1}{2}$

$$y + \frac{3}{4} = \frac{1}{2}$$

- The goal is to rewrite the equation in the form *variable* = *constant*.

$$y + \frac{3}{4} - \frac{3}{4} = \frac{1}{2} - \frac{3}{4}$$

- Subtract $\frac{3}{4}$ from each side of the equation.

$$y + 0 = \frac{2}{4} - \frac{3}{4}$$

- Simplify.

$$y = -\frac{1}{4}$$

- The equation is in the form *variable* = *constant*.

The solution is $-\frac{1}{4}$. You should check this solution.

Example 3 Solve: $x + \frac{2}{5} = \frac{1}{3}$

Solution

$$x + \frac{2}{5} = \frac{1}{3}$$

$$x + \frac{2}{5} - \frac{2}{5} = \frac{1}{3} - \frac{2}{5}$$

- Subtract $\frac{2}{5}$ from each side.

$$x + 0 = \frac{5}{15} - \frac{6}{15}$$

$$x = -\frac{1}{15}$$

The solution is $-\frac{1}{15}$.

You Try It 3 Solve: $\frac{5}{6} = y - \frac{3}{8}$

Your solution

Solution on p. S3

Objective C To solve an equation of the form $ax = b$

The solution of the equation at the right is 3 because $2 \cdot 3 = 6$ is a true equation.

$$2x = 6 \qquad 2 \cdot 3 = 6$$

Note that if each side of $2x = 6$ is multiplied by 5, the solution is still 3.

$$2x = 6$$
$$5(2x) = 5 \cdot 6$$
$$10x = 30 \qquad 10 \cdot 3 = 30$$

If each side of $2x = 6$ is multiplied by -4, the solution is still 3.

$$2x = 6$$
$$(-4)(2x) = (-4)6$$
$$-8x = -24 \qquad -8 \cdot 3 = -24$$

The equations $2x = 6$, $10x = 30$, and $-8x = -24$ are equivalent equations; each equation has 3 as its solution. These examples suggest that multiplying each side of an equation by the same nonzero number produces an equivalent equation.

Multiplication Property of Equations

Each side of an equation can be multiplied by the same *nonzero* number without changing the solution of the equation. In symbols, if $c \neq 0$, then the equation $a = b$ has the same solutions as the equation $ac = bc$.

Copyright © Houghton Mifflin Company. All rights reserved.

The Multiplication Property of Equations is used to remove a coefficient by multiplying each side of the equation by the reciprocal of the coefficient.

HOW TO Solve: $\frac{3}{4}z = 9$

$\frac{3}{4}z = 9$ • The goal is to rewrite the equation in the form *variable* = *constant*.

$\frac{4}{3} \cdot \frac{3}{4}z = \frac{4}{3} \cdot 9$ • Multiply each side of the equation by $\frac{4}{3}$.

$1 \cdot z = 12$ • Simplify.

$z = 12$ • The equation is in the form *variable* = *constant*.

The solution is 12. You should check this solution.

Because division is defined in terms of multiplication, each side of an equation can be divided by the same nonzero number without changing the solution of the equation.

TAKE NOTE

Remember to check the solution.

Check: $6x = 14$

$6\left(\frac{7}{3}\right) \mid 14$

$14 = 14$

HOW TO Solve: $6x = 14$

$6x = 14$ • The goal is to rewrite the equation in the form *variable* = *constant*.

$\frac{6x}{6} = \frac{14}{6}$ • Divide each side of the equation by 6.

$x = \frac{7}{3}$ • Simplify. The equation is in the form *variable* = *constant*.

The solution is $\frac{7}{3}$.

When using the Multiplication Property of Equations, multiply each side of the equation by the reciprocal of the coefficient when the coefficient is a fraction. Divide each side of the equation by the coefficient when the coefficient is an integer or a decimal.

Example 4 Solve: $\frac{3x}{4} = -9$

Solution

$\frac{3x}{4} = -9$

$\frac{4}{3} \cdot \frac{3}{4}x = \frac{4}{3}(-9)$ • $\frac{3x}{4} = \frac{3}{4}x$

$x = -12$

The solution is −12.

You Try It 4 Solve: $-\frac{2x}{5} = 6$

Your solution

Example 5 Solve: $5x - 9x = 12$

Solution

$5x - 9x = 12$

$-4x = 12$ • Combine like terms.

$\frac{-4x}{-4} = \frac{12}{-4}$

$x = -3$

The solution is −3.

You Try It 5 Solve: $4x - 8x = 16$

Your solution

Solutions on pp. S3–S4

Copyright © Houghton Mifflin Company. All rights reserved.

Objective D To solve application problems using the basic percent equation

An equation that is used frequently in mathematics applications is the basic percent equation.

Basic Percent Equation

$$\text{Percent} \cdot \text{Base} = \text{Amount}$$
$$P \cdot B = A$$

In many application problems involving percent, the base follows the word *of*.

HOW TO 20% of what number is 30?

$P \cdot B = A$ • Use the basic percent equation.

$0.20B = 30$ • $P = 20\% = 0.20$, $A = 30$, and B is unknown.

$\frac{0.20B}{0.20} = \frac{30}{0.20}$ • Solve for B.

$B = 150$

The number is 150.

TAKE NOTE

We have written $P(80) = 70$ because that is the form of the basic percent equation. We could have written $80P = 70$. The important point is that each side of the equation is divided by 80, the coefficient of P.

HOW TO 70 is what percent of 80?

$P \cdot B = A$ • Use the basic percent equation.

$P(80) = 70$ • $B = 80$, $A = 70$, and P is unknown.

$\frac{P(80)}{80} = \frac{70}{80}$ • Solve for P.

$P = 0.875$ • The question asked for a percent. Convert the decimal to a percent.

$P = 87.5\%$

70 is 87.5% of 80.

Copyright © Houghton Mifflin Company. All rights reserved.

HOW TO The world's production of cocoa for a recent year was 2928 metric tons. Of this, 1969 metric tons came from Africa. (*Source:* World Cocoa Foundation) What percent of the world's cocoa production came from Africa? Round to the nearest tenth of a percent.

Strategy To find the percent, use the basic percent equation. $B = 2928$, $A = 1969$, P is unknown.

Solution

$$P \cdot B = A$$
$$P(2928) = 1969$$
$$P = \frac{1969}{2928} \approx 0.672$$

Approximately 67.2% of the world's cocoa production came from Africa.

The simple interest that an investment earns is given by the **simple interest equation** $I = Prt$, where I is the simple interest, P is the principal, or amount invested, r is the simple interest rate, and t is the time.

HOW TO A \$1500 investment has an annual simple interest rate of 7%. Find the simple interest earned on the investment after 18 months.

The time is given in months but the interest rate is an annual rate. Therefore, we must convert 18 months to years.

$$18 \text{ months} = \frac{18}{12} \text{ years} = 1.5 \text{ years}$$

To find the interest, solve $I = Prt$ for I.

$I = Prt$

$I = 1500(0.07)(1.5)$ • $P = 1500, r = 0.07, t = 1.5$

$I = 157.5$

The investment earned \$157.50.

Point of Interest

In the jewelry industry, the amount of gold in a piece of jewelry is measured by the *karat*. Pure gold is 24 karats. A necklace that is 18 karats is $\frac{18}{24} = 0.75 = 75\%$ gold.

The amount of a substance in a solution can be given as a percent of the total solution. For instance, if a certain fruit juice drink is advertised as containing 27% cranberry juice, then 27% of the contents of the bottle must be cranberry juice.

When solving problems involving mixtures, we use the **percent mixture equation** $Q = Ar$, where Q is the quantity of a substance in the solution, A is the amount of the solution, and r is the percent concentration of the substance.

HOW TO Part of the formula for a perfume requires that the concentration of jasmine be 1.2% of the total amount of perfume. How many ounces of jasmine are in a 2-ounce bottle of this perfume?

The amount of perfume is 2 oz. Therefore, $A = 2$. The percent concentration is 1.2%, so $r = 0.012$. To find the number of ounces of jasmine, solve $Q = Ar$ for Q.

$Q = Ar$

$Q = 2(0.012)$ • $A = 2, r = 0.012$

$Q = 0.024$

There is 0.024 ounce of jasmine in the perfume.

In most cases, you should write the percent as a decimal before solving the basic percent equation. However, some percents are more easily written as fractions. For example,

$$33\frac{1}{3}\% = \frac{1}{3} \qquad 66\frac{2}{3}\% = \frac{2}{3} \qquad 16\frac{2}{3}\% = \frac{1}{6} \qquad 83\frac{1}{3}\% = \frac{5}{6}$$

Copyright © Houghton Mifflin Company. All rights reserved.

Example 6

12 is $33\frac{1}{3}\%$ of what number?

Solution

$P \cdot B = A$ • Use the basic percent equation.

$\frac{1}{3}B = 12$ • $33\frac{1}{3}\% = \frac{1}{3}$

$3 \cdot \frac{1}{3}B = 3 \cdot 12$

$B = 36$

12 is $33\frac{1}{3}\%$ of 36.

You Try It 6

18 is $16\frac{2}{3}\%$ of what number?

Your solution

Example 7

The data in the table below show the number of households (in millions) that downloaded music files for a three-month period in a recent year. (*Source:* NPD Group)

Month	April	May	June
Downloads	14.5	12.7	10.4

For the three-month period, what percent of the files were downloaded in May? Round to the nearest percent.

Strategy

To find the percent,

- Find the total number of files downloaded for the three-month period.
- Use the basic percent equation. B is the total number of files downloaded for the three-month period; $A = 12.7$, the number of files downloaded in May; P is unknown.

Solution

$14.5 + 12.7 + 10.4 = 37.6$

$P \cdot B = A$ • Use the basic percent equation.

$P(37.6) = 12.7$ • $B = 37.6, A = 12.7$

$P = \frac{12.7}{37.6} \approx 0.34$

Approximately 34% of the files were downloaded in May.

You Try It 7

The Bowl Championship Series (BCS) received approximately \$83.3 million in revenues from various college football bowl games. Of this amount, the college representing the Pac-10 in the Rose Bowl received approximately \$3.1 million. (*Source:* BCSfootball.org) What percent of the total received by the BCS did the college representing the Pac-10 receive? Round to the nearest tenth of a percent.

Your strategy

Your solution

Solutions on p. S4

Copyright © Houghton Mifflin Company. All rights reserved.

Example 8

In April, Marshall Wardell was charged an interest fee of \$8.72 on an unpaid credit card balance of \$545. Find the annual interest rate on this credit card.

Strategy

The interest is \$8.72. Therefore, $I = 8.72$. The unpaid balance is \$545. This is the principal on which interest is calculated. Therefore, $P = 545$. The time is 1 month. Because the *annual* interest rate must be found and the time is given as 1 month, we write 1 month as $\frac{1}{12}$ year, so $t = \frac{1}{12}$. To find the interest rate, solve $I = Prt$ for r.

Solution

$$I = Prt$$

• Use the simple interest equation.

$$8.72 = 545r\left(\frac{1}{12}\right)$$

• $I = 8.72$, $P = 545$, $t = \frac{1}{12}$

$$8.72 = \frac{545}{12}r$$

$$\frac{12}{545}(8.72) = \frac{12}{545}\left(\frac{545}{12}r\right)$$

$$0.192 = r$$

The annual interest rate is 19.2%.

You Try It 8

Clarissa Adams purchased a municipal bond for \$1000 that earns an annual simple interest rate of 6.4%. How much must she deposit into a bank account that earns 8% annual simple interest so that the interest earned from each account after one year is the same?

Your strategy

(1000)(.064)(1) 64.00
64 ÷ .08 =

\$800.00

Your solution

Example 9

To make a certain color of blue, 4 oz of cyan must be contained in 1 gal of paint. What is the percent concentration of cyan in the paint?

Strategy

The cyan is given in ounces and the amount of paint is given in gallons. We must convert ounces to gallons or gallons to ounces. For this problem, we will convert gallons to ounces: 1 gal = 128 oz. Solve $Q = Ar$ for r with $Q = 4$ and $A = 128$.

Solution

$$Q = Ar$$

• Use the percent mixture equation.

$$4 = 128r$$

• $Q = 4$, $A = 128$

$$\frac{4}{128} = \frac{128r}{128}$$

$$0.03125 = r$$

The percent concentration of cyan is 3.125%.

You Try It 9

The concentration of sugar in a certain breakfast cereal is 25%. If there are 2 oz of sugar contained in the cereal in a bowl, how many ounces of cereal are in the bowl?

Your strategy

Your solution

Solutions on p. S4

Copyright © Houghton Mifflin Company. All rights reserved.

Objective E To solve uniform motion problems

TAKE NOTE

A car traveling in a *circle* at a constant speed of 45 mph is *not* in uniform motion because the direction of the car is always changing.

Any object that travels at a constant speed in a straight line is said to be in *uniform motion.* **Uniform motion** means that the speed and direction of an object do not change. For instance, a car traveling at a constant speed of 45 mph on a straight road is in uniform motion.

The solution of a uniform motion problem is based on the **uniform motion equation** $d = rt$, where d is the distance traveled, r is the rate of travel, and t is the time spent traveling. For instance, suppose a car travels at 50 mph for 3 h. Because the rate (50 mph) and time (3 h) are known, we can find the distance traveled by solving the equation $d = rt$ for d.

$$d = rt$$

$$d = 50(3) \qquad \bullet\ r = 50,\ t = 3.$$

$$d = 150$$

The car travels a distance of 150 mi.

HOW TO A jogger runs 3 mi in 45 min. What is the rate of the jogger in miles per hour?

Strategy

- Because the answer must be in miles per *hour* and the given time is in *minutes,* convert 45 min to hours.
- To find the rate of the jogger, solve the equation $d = rt$ for r.

Solution $t = 45 \text{ min} = \frac{45}{60} \text{ h} = \frac{3}{4} \text{ h}$

$$d = rt$$

$$3 = r\left(\frac{3}{4}\right) \qquad \bullet\ d = 3,\ t = \frac{3}{4}$$

$$3 = \frac{3}{4}r$$

$$\left(\frac{4}{3}\right)3 = \left(\frac{4}{3}\right)\frac{3}{4}r \qquad \bullet\ \text{Multiply each side of the equation by the reciprocal of } \frac{3}{4}.$$

$$4 = r$$

The rate of the jogger is 4 mph.

If two objects are moving in opposite directions, then the rate at which the distance between them is increasing is the sum of the speeds of the two objects. For instance, in the diagram below, two cars start from the same point and travel in opposite directions. The distance between them is changing at 70 mph.

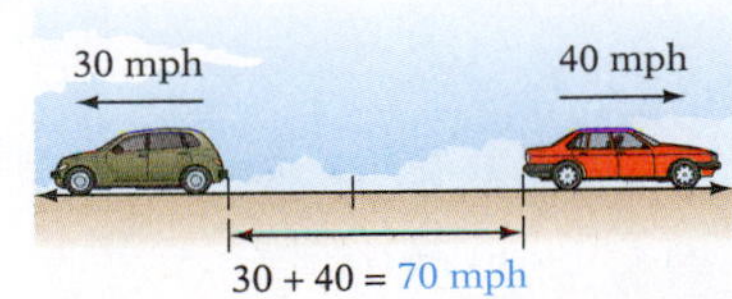

Copyright © Houghton Mifflin Company. All rights reserved.

Similarly, if two objects are moving toward each other, the distance between them is decreasing at a rate that is equal to the sum of the speeds. The rate at which the two planes at the right are approaching one another is 800 mph.

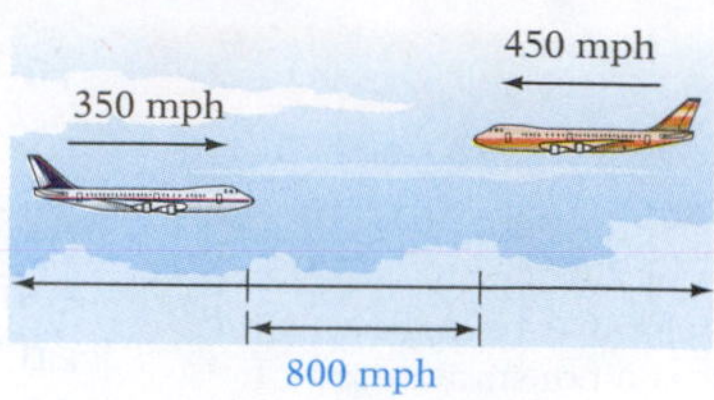

HOW TO Two cars start from the same point and move in opposite directions. The car moving west is traveling 45 mph, and the car moving east is traveling 60 mph. In how many hours will the cars be 210 mi apart?

Strategy The distance is 210 mi. Therefore, $d = 210$. The cars are moving in opposite directions, so the rate at which the distance between them is changing is the sum of the rates of each of the cars. The rate is 45 mph + 60 mph = 105 mph. Therefore, $r = 105$. To find the time, solve the equation $d = rt$ for t.

Solution

$$d = rt$$

$$210 = 105t \qquad \bullet\ d = 210,\ r = 105$$

$$\frac{210}{105} = \frac{105t}{105} \qquad \bullet\ \text{Solve for } t.$$

$$2 = t$$

In 2 h, the cars will be 210 mi apart.

If a motorboat is on a river that is flowing at a rate of 4 mph, then the boat will float down the river at a speed of 4 mph when the motor is not on. Now suppose the motor is turned on and the power adjusted so that the boat can travel 10 mph without the aid of the current. Then, if the boat is moving with the current, its effective speed is the speed of the boat using power plus the speed of the current: 10 mph + 4 mph = 14 mph. (See the figure below.)

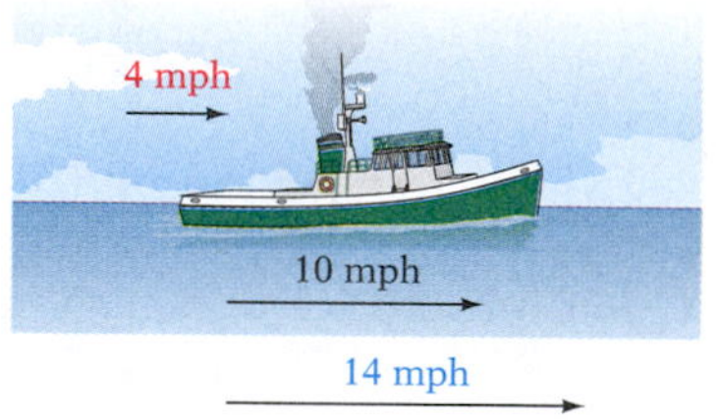

However, if the boat is moving against the current, the current slows the boat down, and the effective speed of the boat is the speed of the boat using power minus the speed of the current: 10 mph − 4 mph = 6 mph. (See the figure below.)

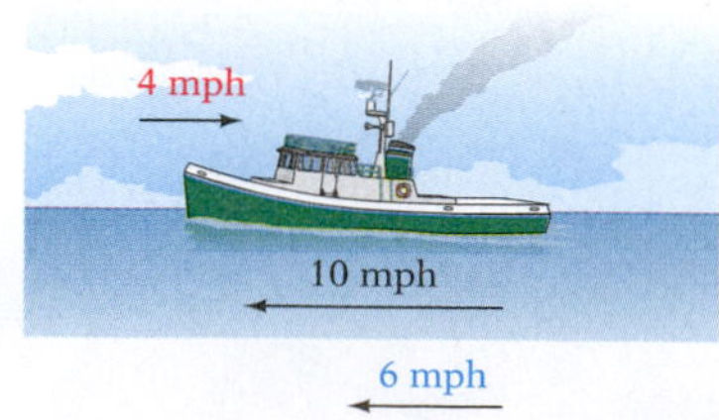

Copyright © Houghton Mifflin Company. All rights reserved.

There are other situations in which the preceding concepts may be applied.

TAKE NOTE

The term ft/s is an abbreviation for "feet per second." Similarly, cm/s is "centimeters per second" and m/s is "meters per second."

HOW TO An airline passenger is walking between two airline terminals and decides to get on a moving sidewalk that is 150 ft long. If the passenger walks at a rate of 7 ft/s and the moving sidewalk moves at a rate of 9 ft/s, how long, in seconds, will it take the passenger to walk from one end of the moving sidewalk to the other? Round to the nearest thousandth.

Strategy The distance is 150 ft. Therefore, $d = 150$. The passenger is traveling at 7 ft/s and the moving sidewalk is traveling at 9 ft/s. The rate of the passenger is the sum of the two rates, or 16 ft/s. Therefore, $r = 16$. To find the time, solve the equation $d = rt$ for t.

Solution

$$d = rt$$

$$150 = 16t \qquad \bullet\ d = 150,\ r = 16$$

$$\frac{150}{16} = \frac{16t}{16} \qquad \bullet\ \text{Solve for } t.$$

$$9.375 = t$$

It will take 9.375 s for the passenger to travel the length of the moving sidewalk.

Example 10

Two cyclists start at the same time at opposite ends of an 80-mile course. One cyclist is traveling 18 mph, and the second cyclist is traveling 14 mph. How long after they begin will they meet?

Strategy

The distance is 80 mi. Therefore, $d = 80$. The cyclists are moving toward each other, so the rate at which the distance between them is changing is the sum of the rates of each of the cyclists. The rate is 18 mph + 14 mph = 32 mph. Therefore, $r = 32$. To find the time, solve the equation $d = rt$ for t.

Solution

$$d = rt$$

$$80 = 32t \qquad \bullet\ d = 80,\ r = 32$$

$$\frac{80}{32} = \frac{32t}{32} \qquad \bullet\ \text{Solve for } t.$$

$$2.5 = t$$

The cyclists will meet in 2.5 h.

You Try It 10

A plane that can normally travel at 250 mph in calm air is flying into a headwind of 25 mph. How far can the plane fly in 3 h?

Your strategy

Your solution

Solution on p. S4

Copyright © Houghton Mifflin Company. All rights reserved.

2.1 Exercises

Objective A **To determine whether a given number is a solution of an equation**

1.  What is the difference between an equation and an expression?

2.  Explain how to determine whether a given number is a solution of an equation.

3. Is 4 a solution of $2x = 8$?

4. Is 3 a solution of $y + 4 = 7$?

5. Is -1 a solution of $2b - 1 = 3$?

6. Is -2 a solution of $3a - 4 = 10$?

7. Is 1 a solution of $4 - 2m = 3$?

8. Is 2 a solution of $7 - 3n = 2$?

9. Is 5 a solution of $2x + 5 = 3x$?

10. Is 4 a solution of $3y - 4 = 2y$?

11. Is -2 a solution of $3a + 2 = 2 - a$?

12. Is 3 a solution of $z^2 + 1 = 4 + 3z$?

13. Is 2 a solution of $2x^2 - 1 = 4x - 1$?

14. Is -1 a solution of $y^2 - 1 = 4y + 3$?

15. Is 4 a solution of $x(x + 1) = x^2 + 5$?

16. Is 3 a solution of $2a(a - 1) = 3a + 3$?

17. Is $-\frac{1}{4}$ a solution of $8t + 1 = -1$?

18. Is $\frac{1}{2}$ a solution of $4y + 1 = 3$?

19. Is $\frac{2}{5}$ a solution of $5m + 1 = 10m - 3$?

20. Is $\frac{3}{4}$ a solution of $8x - 1 = 12x + 3$?

Objective B **To solve an equation of the form $x + a = b$**

21. Can 0 ever be the solution of an equation? If so, give an example of an equation for which 0 is a solution.

22. Without solving $x + \frac{13}{15} = -\frac{21}{43}$, determine whether x is less than or greater than $-\frac{21}{43}$. Explain your answer.

For Exercises 23 to 64, solve and check.

23. $x + 5 = 7$

24. $y + 3 = 9$

25. $b - 4 = 11$

26. $z - 6 = 10$

27. $2 + a = 8$

28. $5 + x = 12$

29. $n - 5 = -2$

30. $x - 6 = -5$

31. $b + 7 = 7$

32. $y - 5 = -5$

33. $z + 9 = 2$

34. $n + 11 = 1$

35. $10 + m = 3$

36. $8 + x = 5$

37. $9 + x = -3$

38. $10 + y = -4$

Copyright © Houghton Mifflin Company. All rights reserved.

39. $2 = x + 7$

40. $-8 = n + 1$

41. $4 = m - 11$

42. $-6 = y - 5$

43. $12 = 3 + w$

44. $-9 = 5 + x$

45. $4 = -10 + b$

46. $-7 = -2 + x$

47. $m + \frac{2}{3} = -\frac{1}{3}$

48. $c + \frac{3}{4} = -\frac{1}{4}$

49. $x - \frac{1}{2} = \frac{1}{2}$

50. $x - \frac{2}{5} = \frac{3}{5}$

51. $\frac{5}{8} + y = \frac{1}{8}$

52. $\frac{4}{9} + a = -\frac{2}{9}$

53. $m + \frac{1}{2} = -\frac{1}{4}$

54. $b + \frac{1}{6} = -\frac{1}{3}$

55. $x + \frac{2}{3} = \frac{3}{4}$

56. $n + \frac{2}{5} = \frac{2}{3}$

57. $-\frac{5}{6} = x - \frac{1}{4}$

58. $-\frac{1}{4} = c - \frac{2}{3}$

59. $d + 1.3619 = 2.0148$

60. $w + 2.932 = 4.801$

61. $-0.813 + x = -1.096$

62. $-1.926 + t = -1.042$

63. $6.149 = -3.108 + z$

64. $5.237 = -2.014 + x$

Objective C To solve an equation of the form $ax = b$

65. Without solving $-\frac{15}{41}x = -\frac{23}{25}$, determine whether x is less than or greater than 0. Explain your answer.

66. Explain why multiplying each side of an equation by the reciprocal of the coefficient of the variable is the same as dividing each side of the equation by the coefficient.

For Exercises 67 to 110, solve and check.

67. $5x = -15$

68. $4y = -28$

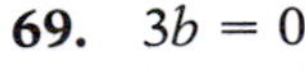

69. $3b = 0$

70. $2a = 0$

71. $-3x = 6$

72. $-5m = 20$

73. $-3x = -27$

74. $-\frac{1}{6}n = -30$

Copyright © Houghton Mifflin Company. All rights reserved.

75. $20 = \frac{1}{4}c$

76. $18 = 2t$

77. $0 = -5x$

78. $0 = -8a$

79. $49 = -7t$

80. $\frac{x}{3} = 2$

81. $\frac{x}{4} = 3$

82. $-\frac{y}{2} = 5$

83. $-\frac{b}{3} = 6$

84. $\frac{3}{4}y = 9$

85. $\frac{2}{5}x = 6$

86. $-\frac{2}{3}d = 8$

87. $-\frac{3}{5}m = 12$

88. $\frac{2n}{3} = 0$

89. $\frac{5x}{6} = 0$

90. $\frac{-3z}{8} = 9$

91. $\frac{3x}{4} = 2$

92. $\frac{3}{4}c = \frac{3}{5}$

93. $\frac{2}{9} = \frac{2}{3}y$

94. $-\frac{6}{7} = -\frac{3}{4}b$

95. $\frac{1}{5}x = -\frac{1}{10}$

96. $-\frac{2}{3}y = -\frac{8}{9}$

97. $-1 = \frac{2n}{3}$

98. $-\frac{3}{4} = \frac{a}{8}$

99. $-\frac{2}{5}m = -\frac{6}{7}$

100. $5x + 2x = 14$

101. $3n + 2n = 20$

102. $7d - 4d = 9$

103. $10y - 3y = 21$

104. $2x - 5x = 9$

105. $\frac{x}{1.46} = 3.25$

106. $\frac{z}{2.95} = -7.88$

107. $3.47a = 7.1482$

108. 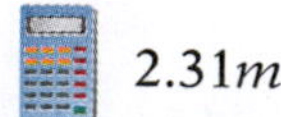 $2.31m = 2.4255$

109. 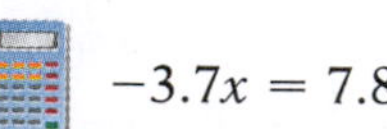 $-3.7x = 7.881$

110. $\frac{n}{2.65} = 9.08$

Copyright © Houghton Mifflin Company. All rights reserved.

Objective D To solve application problems using the basic percent equation

111. Without solving an equation, indicate whether 40% of 80 is less than, equal to, or greater than 80% of 40.

112. Without solving an equation, indicate whether $\frac{1}{4}$% of 80 is less than, equal to, or greater than 25% of 80.

113. What is 35% of 80?

114. What percent of 8 is 0.5?

115. Find 1.2% of 60.

116. 8 is what percent of 5?

117. 125% of what is 80?

118. What percent of 20 is 30?

119. 12 is what percent of 50?

120. What percent of 125 is 50?

121. Find 18% of 40.

122. What is 25% of 60?

123. 12% of what is 48?

124. 45% of what is 9?

125. What is $33\frac{1}{3}$% of 27?

126. Find $16\frac{2}{3}$% of 30.

127. What percent of 12 is 3?

128. 10 is what percent of 15?

129. 12 is what percent of 6?

130. 20 is what percent of 16?

131. $5\frac{1}{4}$% of what is 21?

132. $37\frac{1}{2}$% of what is 15?

133. Find 15.4% of 50.

134. What is 18.5% of 46?

135. 1 is 0.5% of what?

136. 3 is 1.5% of what?

137. $\frac{3}{4}$% of what is 3?

138. $\frac{1}{2}$% of what is 3?

139. What is 250% of 12?

Copyright © Houghton Mifflin Company. All rights reserved.

140. **Education** The stacked-bar graph at the right shows the number of people age 25 or older in the U.S. who have attained some type of degree beyond high school.

a. In 2002, there were approximately 182.7 million people age 25 or older. What percent of the people age 25 or older had received an associate degree or a bachelor's degree in 2002? Round to the nearest tenth of a percent.

b. In 2000, there were approximately 177.5 million people age 25 or older. Was the percent of people in 2000 with a graduate degree less than or greater than the percent of people in 2002 with a graduate degree?

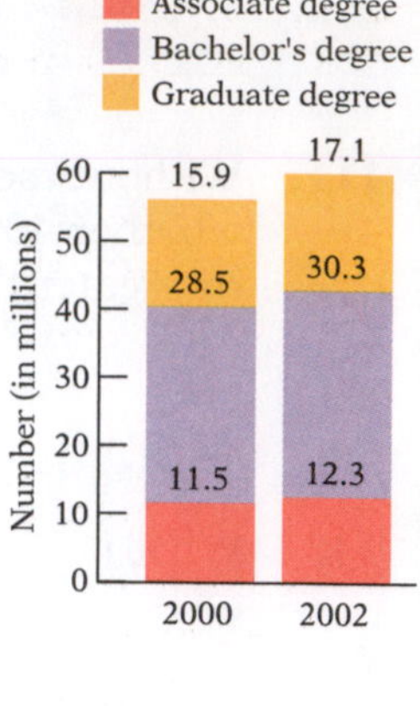

141. **Chemistry** Approximately 21% of air is oxygen. Using this estimate, determine how many liters of oxygen there are in a room containing 21,600 L of air.

142. **Record Sales** According to Nielsen SoundScan, there were approximately 680 million record albums sold in the fourth quarter of 2002. This is about 39% of the total number of record albums sold that year. How many record albums were sold in 2002? Round to the nearest million.

143. **Income** According to the U.S. Census Bureau, the median income fell 1.1% between two successive years. If the median income before the decline was $42,900, what was the median income the next year? Round to the nearest dollar.

144. **Government** To override a presidential veto, at least $66\frac{2}{3}\%$ of the Senate must vote to override the veto. There are 100 senators in the Senate. What is the minimum number of votes needed to override a veto?

145. **Sports** According to **www.superbowl.com,** approximately 138.9 million people watched Super Bowl XXXVIII. What percent of the U.S. population watched Super Bowl XXXVIII? Use a figure of 290 million for the U.S. population. Round to the nearest tenth of a percent.

146. **Advertising** Suppose 9.4 million people watch a 30-second commercial for a new cellular phone during a broadcast of the TV show *CSI*. The cost of that commercial was approximately $470,000. If the cellular phone manufacturer makes a profit of $10 on every phone sold, what percent of the people watching the commercial would have to buy one phone for the company to recover the cost of the commercial? (*Source:* Nielsen Media Research/ San Diego Union)

147. **School Enrollment** The circle graph at the right shows the percent of the U.S. population over 3 years old who are enrolled in school. To answer the question "How many people are enrolled in college or graduate school?" what additional piece of information is necessary?

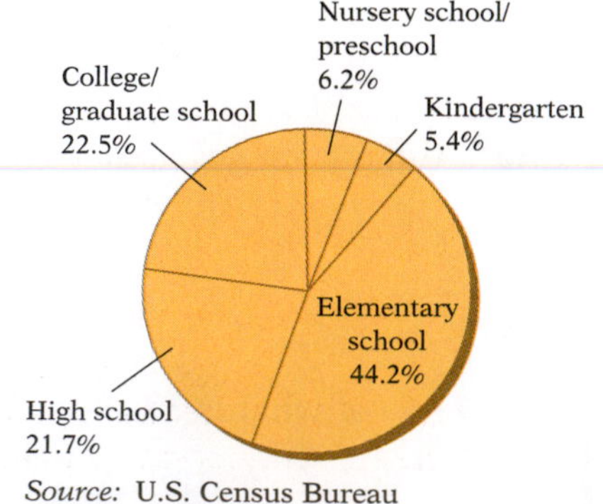

Source: U.S. Census Bureau

Copyright © Houghton Mifflin Company. All rights reserved.

148. **Investment** If Kachina Caron invested \$1200 in a simple interest account and earned \$72 in 8 months, what is the annual interest rate?

149. **Investment** How much money must Andrea invest for two years in an account that earns an annual interest rate of 8% if she wants to earn \$300 from the investment?

150. **Investment** Sal Boxer decided to divide a gift of \$3000 into two different accounts. He placed \$1000 in one account that earns an annual simple interest rate of 7.5%. The remaining money was placed in an account that earns an annual simple interest rate of 8.25%. How much interest will Sal earn from the two accounts after one year?

151. **Investment** If Americo invests \$2500 at an 8% annual simple interest rate and Octavia invests \$3000 at a 7% annual simple interest rate, which of the two will earn the greater amount of interest after one year?

152. **Investment** Makana invested \$900 in a simple interest account that had an interest rate that was 1% more than that of her friend Marlys. If Marlys earned \$51 after one year from an investment of \$850, how much did Makana earn in one year?

153. **Investment** A \$2000 investment at an annual simple interest rate of 6% earned as much interest after one year as another investment in an account that earns 8% simple interest. How much was invested at 8%?

154. **Investment** An investor placed \$1000 in an account that earns 9% annual simple interest and \$1000 in an account that earns 6% annual simple interest. If each investment is left in the account for the same period of time, is the interest rate on the combined investment less than 6%, between 6% and 9%, or greater than 9%?

155. **Metallurgy** The concentration of platinum in a necklace is 15%. If the necklace weighs 12 g, find the amount of platinum in the necklace.

156. **Dye Mixtures** A 250-milliliter solution of a fabric dye contains 5 ml of hydrogen peroxide. What is the percent concentration of the hydrogen peroxide?

157. **Fabric Mixtures** A carpet is made of a blend of wool and other fibers. If the concentration of wool in the carpet is 75% and the carpet weighs 175 lb, how much wool is in the carpet?

158. **Juice Mixtures** Apple Dan's 32-ounce apple-flavored fruit drink contains 8 oz of apple juice. A 40-ounce generic brand of an apple-flavored fruit drink contains 9 oz of apple juice. Which of the two brands has the greater concentration of apple juice?

Copyright © Houghton Mifflin Company. All rights reserved.

159. **Food Mixtures** Bakers use simple syrup in many of their recipes. Simple syrup is made by combining 500 g of sugar with 500 g of water and mixing it well until the sugar dissolves. What is the percent concentration of sugar in the simple syrup?

160. **Pharmacology** A pharmacist has 50 g of a topical cream that contains 75% glycerine. How many grams of the cream is not glycerine?

161. **Chemistry** A chemist has 100 ml of a solution that is 9% acetic acid. If the chemist adds 50 ml of pure water to this solution, what is the percent concentration of the resulting mixture?

162. **Chemistry** A 500-gram salt-and-water solution contains 50 g of salt. This mixture is left in the open air and 100 g of water evaporates from the solution. What is the percent concentration of salt in the remaining solution?

Objective E To solve uniform motion problems

163. As part of the training program for the Boston Marathon, a runner wants to build endurance by running at a rate of 9 mph for 20 min. How far will the runner travel in that time period?

164. It takes a hospital dietician 40 min to drive from home to the hospital, a distance of 20 mi. What is the dietician's average rate of speed?

165. Marcella leaves home at 9:00 A.M. and drives to school, arriving at 9:45 A.M. If the distance between home and school is 27 mi, what is Marcella's average rate of speed?

166. The Ride for Health Bicycle Club has chosen a 36-mile course for this Saturday's ride. If the riders plan on averaging 12 mph while they are riding, and they have a 1-hour lunch break planned, how long will it take them to complete the trip?

167. Palmer's average running speed is 3 kilometers per hour faster than his walking speed. If Palmer can run around a 30-kilometer course in 2 h, how many hours would it take for Palmer to walk the same course?

168. A shopping mall has a moving sidewalk that takes shoppers from the shopping area to the parking garage, a distance of 250 ft. If your normal walking rate is 5 ft/s and the moving sidewalk is traveling at 3 ft/s, how many seconds would it take for you to walk from one end of the moving sidewalk to the other end?

Copyright © Houghton Mifflin Company. All rights reserved.

169. Two joggers start at the same time from opposite ends of an 8-mile jogging trail and begin running toward each other. One jogger is running at the rate of 5 mph, and the other jogger is running at a rate of 7 mph. How long, in minutes, after they start will the two joggers meet?

170. Two cyclists start from the same point at the same time and move in opposite directions. One cyclist is traveling at 8 mph, and the other cyclist is traveling at 9 mph. After 30 min, how far apart are the two cyclists?

171. Petra and Celine can paddle their canoe at a rate of 10 mph in calm water. How long will it take them to travel 4 mi against the 2 mph current of the river?

172. At 8:00 A.M., a train leaves a station and travels at a rate of 45 mph. At 9:00 A.M., a second train leaves the same station on the same track and travels in the direction of the first train at a speed of 60 mph. At 10:00 A.M., how far apart are the two trains?

APPLYING THE CONCEPTS

173. Solve the equation $ax = b$ for x. Is the solution you have written valid for all real numbers a and b?

174. Solve. **a.** $\dfrac{3}{\frac{1}{x}} = 5$ **b.** $\dfrac{2}{\frac{1}{y}} = -2$ **c.** $\dfrac{3x + 2x}{3} = 2$

175. **a.** Make up an equation of the form $x + a = b$ that has 2 as a solution.
b. Make up an equation of the form $ax = b$ that has -1 as a solution.

176. Write out the steps for solving the equation $\frac{1}{2}x = -3$. Identify each Property of Real Numbers or Property of Equations as you use it.

177. In your own words, state the Addition Property of Equations and the Multiplication Property of Equations.

178. If a quantity increases by 100%, how many times its original value is its new value?

Copyright © Houghton Mifflin Company. All rights reserved.

2.2 General Equations

Objective A To solve an equation of the form $ax + b = c$

In solving an equation of the form $ax + b = c$, the goal is to rewrite the equation in the form *variable* = *constant*. This requires the application of both the Addition and the Multiplication Properties of Equations.

HOW TO Solve: $\frac{3}{4}x - 2 = -11$

The goal is to write the equation in the form *variable* = *constant*.

$$\frac{3}{4}x - 2 = -11$$

$$\frac{3}{4}x - 2 + 2 = -11 + 2$$ • Add 2 to each side of the equation.

$$\frac{3}{4}x = -9$$ • Simplify.

$$\frac{4}{3} \cdot \frac{3}{4}x = \frac{4}{3}(-9)$$ • Multiply each side of the equation by $\frac{4}{3}$.

$$x = -12$$ • The equation is in the form *variable* = *constant*.

The solution is -12.

TAKE NOTE

Check:

$\frac{3}{4}x - 2 = -11$	
$\frac{3}{4}(-12) - 2$	-11
$-9 - 2$	-11
$-11 = -11$	

A true equation

Here is an example of solving an equation that contains more than one fraction.

HOW TO Solve: $\frac{2}{3}x + \frac{1}{2} = \frac{3}{4}$

$$\frac{2}{3}x + \frac{1}{2} = \frac{3}{4}$$

$$\frac{2}{3}x + \frac{1}{2} - \frac{1}{2} = \frac{3}{4} - \frac{1}{2}$$ • Subtract $\frac{1}{2}$ from each side of the equation.

$$\frac{2}{3}x = \frac{1}{4}$$ • Simplify.

$$\frac{3}{2}\left(\frac{2}{3}x\right) = \frac{3}{2}\left(\frac{1}{4}\right)$$ • Multiply each side of the equation by $\frac{3}{2}$, the reciprocal of $\frac{2}{3}$.

$$x = \frac{3}{8}$$

The solution is $\frac{3}{8}$.

It may be easier to solve an equation containing two or more fractions by multiplying each side of the equation by the least common multiple (LCM) of the denominators. For the equation above, the LCM of 3, 2, and 4 is 12. The LCM has the property that 3, 2, and 4 will divide evenly into it. Therefore, if both sides of the equation are multiplied by 12, the denominators will divide evenly into 12. The result is an equation that does not contain any fractions. Multiplying each side of an equation that contains fractions by the LCM of the denominators is called **clearing denominators.** It is an alternative method, as we show in the next example, of solving an equation that contains fractions.

Copyright © Houghton Mifflin Company. All rights reserved.

HOW TO Solve: $\frac{2}{3}x + \frac{1}{2} = \frac{3}{4}$

$$\frac{2}{3}x + \frac{1}{2} = \frac{3}{4}$$

$$12\left(\frac{2}{3}x + \frac{1}{2}\right) = 12\left(\frac{3}{4}\right)$$

- Multiply each side of the equation by 12, the LCM of 3, 2, and 4.

$$12\left(\frac{2}{3}x\right) + 12\left(\frac{1}{2}\right) = 12\left(\frac{3}{4}\right)$$

- Use the Distributive Property.

$$8x + 6 = 9$$

- Simplify.

$$8x + 6 - 6 = 9 - 6$$

- Subtract 6 from each side of the equation.

$$8x = 3$$

$$\frac{8x}{8} = \frac{3}{8}$$

- Divide each side of the equation by 8.

$$x = \frac{3}{8}$$

The solution is $\frac{3}{8}$.

TAKE NOTE
Observe that after we multiply by the LCM and simplify, the equation no longer contains fractions. Also note that this is the same equation solved on the previous page.

Note that both methods give exactly the same solution. You may use either method to solve an equation containing fractions.

Example 1 Solve: $3x - 7 = -5$

Solution

$$3x - 7 = -5$$

$$3x - 7 + 7 = -5 + 7$$

- Add 7 to each side.

$$3x = 2$$

$$\frac{3x}{3} = \frac{2}{3}$$

- Divide each side by 3.

$$x = \frac{2}{3}$$

The solution is $\frac{2}{3}$.

You Try It 1 Solve: $5x + 7 = 10$

Your solution

Example 2 Solve: $5 = 9 - 2x$

Solution

$$5 = 9 - 2x$$

$$5 - 9 = 9 - 9 - 2x$$

- Subtract 9 from each side.

$$-4 = -2x$$

$$\frac{-4}{-2} = \frac{-2x}{-2}$$

- Divide each side by −2.

$$2 = x$$

The solution is 2.

You Try It 2 Solve: $2 = 11 + 3x$

Your solution

Solutions on p. S4

Copyright © Houghton Mifflin Company. All rights reserved.

Example 3 Solve: $\frac{2}{3} - \frac{x}{2} = \frac{3}{4}$

Solution

$$\frac{2}{3} - \frac{x}{2} = \frac{3}{4}$$

$$\frac{2}{3} - \frac{2}{3} - \frac{x}{2} = \frac{3}{4} - \frac{2}{3}$$

- Subtract $\frac{2}{3}$ from each side.

$$-\frac{x}{2} = \frac{1}{12}$$

$$-2\left(-\frac{x}{2}\right) = -2\left(\frac{1}{12}\right)$$

- Multiply each side by -2.

$$x = -\frac{1}{6}$$

The solution is $-\frac{1}{6}$.

You Try It 3 Solve: $\frac{5}{8} - \frac{2x}{3} = \frac{5}{4}$

Your solution

Example 4 Solve $\frac{4}{5}x - \frac{1}{2} = \frac{3}{4}$ by first clearing denominators.

Solution

The LCM of 5, 2, and 4 is 20.

$$\frac{4}{5}x - \frac{1}{2} = \frac{3}{4}$$

$$20\left(\frac{4}{5}x - \frac{1}{2}\right) = 20\left(\frac{3}{4}\right)$$

- Multiply each side by 20.

$$20\left(\frac{4}{5}x\right) - 20\left(\frac{1}{2}\right) = 20\left(\frac{3}{4}\right)$$

- Use the Distributive Property.

$$16x - 10 = 15$$

$$16x - 10 + 10 = 15 + 10$$

- Add 10 to each side.

$$16x = 25$$

$$\frac{16x}{16} = \frac{25}{16}$$

- Divide each side by 16.

$$x = \frac{25}{16}$$

The solution is $\frac{25}{16}$.

You Try It 4 Solve $\frac{2}{3}x + 3 = \frac{7}{2}$ by first clearing denominators.

Your solution

Solutions on pp. S4–S5

Copyright © Houghton Mifflin Company. All rights reserved.

Example 5

Solve: $2x + 4 - 5x = 10$

Solution

$$2x + 4 - 5x = 10$$
$$-3x + 4 = 10 \quad \text{• Combine like terms.}$$
$$-3x + 4 - 4 = 10 - 4 \quad \text{• Subtract 4 from each side of the equation.}$$
$$-3x = 6$$
$$\frac{-3x}{-3} = \frac{6}{-3} \quad \text{• Divide each side by } -3.$$
$$x = -2$$

The solution is −2.

You Try It 5

Solve: $x - 5 + 4x = 25$

Your solution

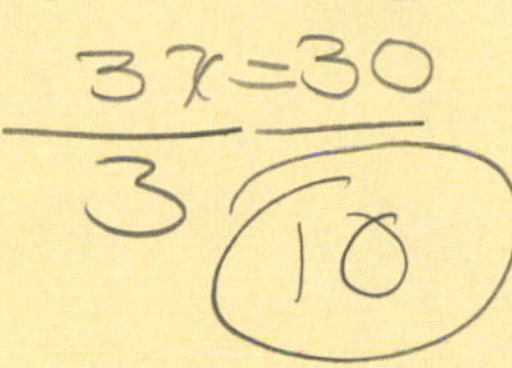

Solution on p. S5

Objective B To solve an equation of the form $ax + b = cx + d$

Study Tip

Have you considered joining a study group? Getting together regularly with other students in the class to go over material and quiz each other can be very beneficial. See *AIM for Success*, page xxvi.

In solving an equation of the form $ax + b = cx + d$, the goal is to rewrite the equation in the form *variable* = *constant*. Begin by rewriting the equation so that there is only one variable term in the equation. Then rewrite the equation so that there is only one constant term.

HOW TO Solve: $2x + 3 = 5x - 9$

$$2x + 3 = 5x - 9$$
$$2x - 5x + 3 = 5x - 5x - 9 \quad \text{• Subtract } 5x \text{ from each side of the equation.}$$
$$-3x + 3 = -9 \quad \text{• Simplify. There is only one variable term.}$$
$$-3x + 3 - 3 = -9 - 3 \quad \text{• Subtract 3 from each side of the equation.}$$
$$-3x = -12 \quad \text{• Simplify. There is only one constant term.}$$
$$\frac{-3x}{-3} = \frac{-12}{-3} \quad \text{• Divide each side of the equation by } -3.$$
$$x = 4 \quad \text{• The equation is in the form } \textit{variable} = \textit{constant}.$$

The solution is 4. You should verify this by checking this solution.

Copyright © Houghton Mifflin Company. All rights reserved.

Example 6 Solve: $4x - 5 = 8x - 7$

Solution

$$4x - 5 = 8x - 7$$

$$4x - 8x - 5 = 8x - 8x - 7$$ • Subtract $8x$ from each side.

$$-4x - 5 = -7$$

$$-4x - 5 + 5 = -7 + 5$$ • Add 5 to each side.

$$-4x = -2$$

$$\frac{-4x}{-4} = \frac{-2}{-4}$$ • Divide each side by -4.

$$x = \frac{1}{2}$$

The solution is $\frac{1}{2}$.

You Try It 6 Solve: $5x + 4 = 6 + 10x$

Your solution

Example 7 Solve: $3x + 4 - 5x = 2 - 4x$

Solution

$$3x + 4 - 5x = 2 - 4x$$

$$-2x + 4 = 2 - 4x$$ • Combine like terms.

$$-2x + 4x + 4 = 2 - 4x + 4x$$ • Add $4x$ to each side.

$$2x + 4 = 2$$

$$2x + 4 - 4 = 2 - 4$$ • Subtract 4 from each side.

$$2x = -2$$

$$\frac{2x}{2} = \frac{-2}{2}$$ • Divide each side by 2.

$$x = -1$$

The solution is -1.

You Try It 7 Solve: $5x - 10 - 3x = 6 - 4x$

Your solution

Solutions on p. S5

Copyright © Houghton Mifflin Company. All rights reserved.

Objective C **To solve an equation containing parentheses**

When an equation contains parentheses, one of the steps in solving the equation requires the use of the Distributive Property. The Distributive Property is used to remove parentheses from a variable expression.

HOW TO Solve: $4 + 5(2x - 3) = 3(4x - 1)$

$4 + 5(2x - 3) = 3(4x - 1)$

$4 + 10x - 15 = 12x - 3$ • Use the Distributive Property. Then simplify.

$10x - 11 = 12x - 3$

$10x - 12x - 11 = 12x - 12x - 3$ • Subtract $12x$ from each side of the equation.

$-2x - 11 = -3$ • Simplify.

$-2x - 11 + 11 = -3 + 11$ • Add 11 to each side of the equation.

$-2x = 8$ • Simplify.

$\frac{-2x}{-2} = \frac{8}{-2}$ • Divide each side of the equation by -2.

$x = -4$ • The equation is in the form *variable* = *constant*.

The solution is -4. You should verify this by checking this solution.

In the next example, we solve an equation with parentheses and decimals.

HOW TO Solve: $16 + 0.55x = 0.75(x + 20)$

$16 + 0.55x = 0.75(x + 20)$

$16 + 0.55x = 0.75x + 15$ • Use the Distributive Property.

$16 + 0.55x - 0.75x = 0.75x - 0.75x + 15$ • Subtract $0.75x$ from each side of the equation.

$16 - 0.20x = 15$ • Simplify.

$16 - 16 - 0.20x = 15 - 16$ • Subtract 16 from each side of the equation.

$-0.20x = -1$ • Simplify.

$\frac{-0.20x}{-0.20} = \frac{-1}{-0.20}$ • Divide each side of the equation by -0.20.

$x = 5$ • The equation is in the form *variable* = *constant*.

The solution is 5.

Copyright © Houghton Mifflin Company. All rights reserved.

Example 8

Solve: $3x - 4(2 - x) = 3(x - 2) - 4$

Solution

$3x - 4(2 - x) = 3(x - 2) - 4$

$3x - 8 + 4x = 3x - 6 - 4$ • Distributive Property

$7x - 8 = 3x - 10$

$7x - 3x - 8 = 3x - 3x - 10$ • Subtract $3x$.

$4x - 8 = -10$

$4x - 8 + 8 = -10 + 8$ • Add 8.

$4x = -2$

$\frac{4x}{4} = \frac{-2}{4}$ • Divide by 4.

$x = -\frac{1}{2}$

The solution is $-\frac{1}{2}$.

You Try It 8

Solve: $5x - 4(3 - 2x) = 2(3x - 2) + 6$

Your solution

Example 9

Solve: $3[2 - 4(2x - 1)] = 4x - 10$

Solution

$3[2 - 4(2x - 1)] = 4x - 10$

$3[2 - 8x + 4] = 4x - 10$ • Distributive Property

$3[6 - 8x] = 4x - 10$

$18 - 24x = 4x - 10$ • Distributive Property

$18 - 24x - 4x = 4x - 4x - 10$ • Subtract $4x$.

$18 - 28x = -10$

$18 - 18 - 28x = -10 - 18$ • Subtract 18.

$-28x = -28$

$\frac{-28x}{-28} = \frac{-28}{-28}$ • Divide by -28.

$x = 1$

The solution is 1.

You Try It 9

Solve: $-2[3x - 5(2x - 3)] = 3x - 8$

Your solution

Solutions on p. S5

Copyright © Houghton Mifflin Company. All rights reserved.

Objective D To solve application problems using formulas

TAKE NOTE

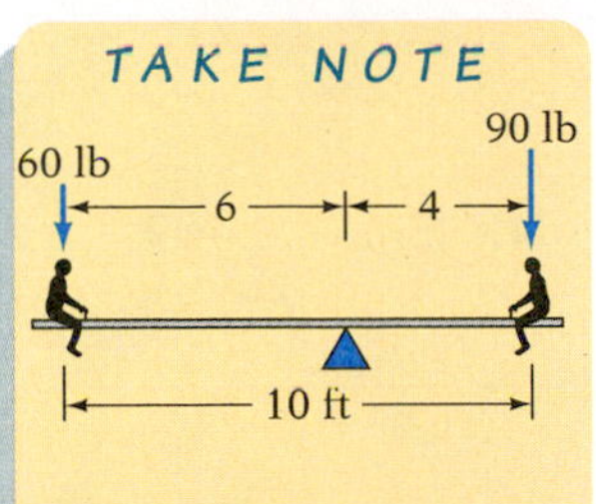

This system balances because

$$F_1x = F_2(d - x)$$
$$60(6) = 90(10 - 6)$$
$$60(6) = 90(4)$$
$$360 = 360$$

A lever system is shown at the right. It consists of a lever, or bar; a fulcrum; and two forces, F_1 and F_2. The distance d represents the length of the lever, x represents the distance from F_1 to the fulcrum, and $d - x$ represents the distance from F_2 to the fulcrum.

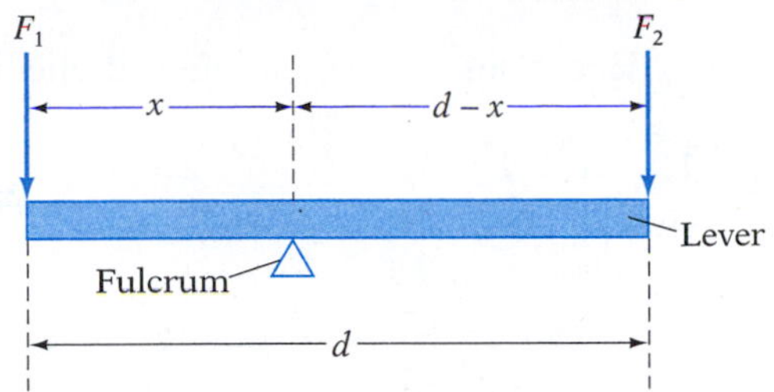

A principle of physics states that when the lever system balances, $F_1x = F_2(d - x)$.

Example 10

A lever is 15 ft long. A force of 50 lb is applied to one end of the lever, and a force of 100 lb is applied to the other end. Where is the fulcrum located when the system balances?

Strategy

Make a drawing.

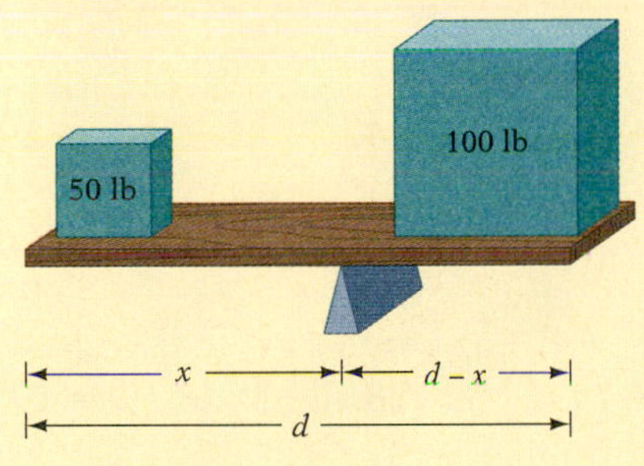

Given: $F_1 = 50$
$F_2 = 100$
$d = 15$

Unknown: x

Solution

$$F_1x = F_2(d - x)$$
$$50x = 100(15 - x)$$
$$50x = 1500 - 100x$$
$$50x + 100x = 1500 - 100x + 100x$$
$$150x = 1500$$
$$\frac{150x}{150} = \frac{1500}{150}$$
$$x = 10$$

The fulcrum is 10 ft from the 50-pound force.

You Try It 10

A lever is 25 ft long. A force of 45 lb is applied to one end of the lever, and a force of 80 lb is applied to the other end. Where is the fulcrum located when the system balances?

Your strategy

F' = 45
F² = 80
d = 25
50x = 80(d = x)

Your solution

Solution on p. S5

Copyright © Houghton Mifflin Company. All rights reserved.

2.2 Exercises

Objective A **To solve an equation of the form $ax + b = c$**

For Exercises 1 to 80, solve and check.

1. $3x + 1 = 10$
2. $4y + 3 = 11$
3. $2a - 5 = 7$
4. $5m - 6 = 9$
5. $5 = 4x + 9$
6. $2 = 5b + 12$
7. $2x - 5 = -11$
8. $3n - 7 = -19$
9. $4 - 3w = -2$
10. $5 - 6x = -13$
11. $8 - 3t = 2$
12. $12 - 5x = 7$
13. $4a - 20 = 0$
14. $3y - 9 = 0$
15. $6 + 2b = 0$
16. $10 + 5m = 0$
17. $-2x + 5 = -7$
18. $-5d + 3 = -12$
19. $-12x + 30 = -6$
20. $-13 = -11y + 9$
21. $2 = 7 - 5a$
22. $3 = 11 - 4n$
23. $-35 = -6b + 1$
24. $-8x + 3 = -29$
25. $-3m - 21 = 0$
26. $-5x - 30 = 0$
27. $-4y + 15 = 15$
28. $-3x + 19 = 19$
29. $9 - 4x = 6$
30. $3t - 2 = 0$
31. $9x - 4 = 0$
32. $7 - 8z = 0$
33. $1 - 3x = 0$
34. $9d + 10 = 7$
35. $12w + 11 = 5$
36. $6y - 5 = -7$
37. $8b - 3 = -9$
38. $5 - 6m = 2$
39. $7 - 9a = 4$
40. $9 = -12c + 5$

Copyright © Houghton Mifflin Company. All rights reserved.

41. $10 = -18x + 7$

42. $2y + \frac{1}{3} = \frac{7}{3}$

43. $4a + \frac{3}{4} = \frac{19}{4}$

44. $2n - \frac{3}{4} = \frac{13}{4}$

45. $3x - \frac{5}{6} = \frac{13}{6}$

46. $5y + \frac{3}{7} = \frac{3}{7}$

47. $9x + \frac{4}{5} = \frac{4}{5}$

48. $8 = 7d - 1$

49. $8 = 10x - 5$

50. $4 = 7 - 2w$

51. $7 = 9 - 5a$

52. $8t + 13 = 3$

53. $12x + 19 = 3$

54. $-6y + 5 = 13$

55. $-4x + 3 = 9$

56. $\frac{1}{2}a - 3 = 1$

57. $\frac{1}{3}m - 1 = 5$

58. $\frac{2}{5}y + 4 = 6$

59. $\frac{3}{4}n + 7 = 13$

60. $-\frac{2}{3}x + 1 = 7$

61. $-\frac{3}{8}b + 4 = 10$

62. $\frac{x}{4} - 6 = 1$

63. $\frac{y}{5} - 2 = 3$

64. $\frac{2x}{3} - 1 = 5$

65. $\frac{2}{3}x - \frac{5}{6} = -\frac{1}{3}$

66. $\frac{5}{4}x + \frac{2}{3} = \frac{1}{4}$

67. $\frac{1}{2} - \frac{2}{3}x = \frac{1}{4}$

68. $\frac{3}{4} - \frac{3}{5}x = \frac{19}{20}$

69. $\frac{3}{2} = \frac{5}{6} + \frac{3x}{8}$

70. $-\frac{1}{4} = \frac{5}{12} + \frac{5x}{6}$

71. $\frac{11}{27} = \frac{4}{9} - \frac{2x}{3}$

72. $\frac{37}{24} = \frac{7}{8} - \frac{5x}{6}$

73. $7 = \frac{2x}{5} + 4$

74. $5 - \frac{4c}{7} = 8$

75. $7 - \frac{5}{9}y = 9$

76. $6a + 3 + 2a = 11$

77. $5y + 9 + 2y = 23$

78. $7x - 4 - 2x = 6$

79. $11z - 3 - 7z = 9$

80. $2x - 6x + 1 = 9$

81. Solve $3x + 4y = 13$ when $y = -2$.

82. Solve $2x - 3y = 8$ when $y = 0$.

83. Solve $-4x + 3y = 9$ when $x = 0$.

84. Solve $5x - 2y = -3$ when $x = -3$.

Copyright © Houghton Mifflin Company. All rights reserved.

Objective B **To solve an equation of the form $ax + b = cx + d$**

For Exercises 85 to 111, solve and check.

85. $8x + 5 = 4x + 13$

86. $6y + 2 = y + 17$

87. $5x - 4 = 2x + 5$

88. $13b - 1 = 4b - 19$

89. $15x - 2 = 4x - 13$

90. $7a - 5 = 2a - 20$

91. $3x + 1 = 11 - 2x$

92. $n - 2 = 6 - 3n$

93. $2x - 3 = -11 - 2x$

94. $4y - 2 = -16 - 3y$

95. $2b + 3 = 5b + 12$

96. $m + 4 = 3m + 8$

97. $4y - 8 = y - 8$

98. $5a + 7 = 2a + 7$

99. $6 - 5x = 8 - 3x$

100. $10 - 4n = 16 - n$

101. $5 + 7x = 11 + 9x$

102. $3 - 2y = 15 + 4y$

103. $2x - 4 = 6x$

104. $2b - 10 = 7b$

105. $8m = 3m + 20$

106. $9y = 5y + 16$

107. $8b + 5 = 5b + 7$

108. $6y - 1 = 2y + 2$

109. $7x - 8 = x - 3$

110. $2y - 7 = -1 - 2y$

111. $2m - 1 = -6m + 5$

112. If $5x = 3x - 8$, evaluate $4x + 2$.

113. If $7x + 3 = 5x - 7$, evaluate $3x - 2$.

114. If $2 - 6a = 5 - 3a$, evaluate $4a^2 - 2a + 1$.

115. If $1 - 5c = 4 - 4c$, evaluate $3c^2 - 4c + 2$.

116. If $2y + 3 = 5 - 4y$, evaluate $6y - 7$.

117. If $3z + 1 = 1 - 5z$, evaluate $3z^2 - 7z + 8$.

Copyright © Houghton Mifflin Company. All rights reserved.

Objective C To solve an equation containing parentheses

For Exercises 118 to 138, solve and check.

118. $5x + 2(x + 1) = 23$

119. $6y + 2(2y + 3) = 16$

120. $9n - 3(2n - 1) = 15$

121. $12x - 2(4x - 6) = 28$

122. $7a - (3a - 4) = 12$

123. $9m - 4(2m - 3) = 11$

124. $5(3 - 2y) + 4y = 3$

125. $4(1 - 3x) + 7x = 9$

126. $5y - 3 = 7 + 4(y - 2)$

127. $0.22(x + 6) = 0.2x + 1.8$

128. $0.05(4 - x) + 0.1x = 0.32$

129. $0.3x + 0.3(x + 10) = 300$

130. $2a - 5 = 4(3a + 1) - 2$

131. $5 - (9 - 6x) = 2x - 2$

132. $7 - (5 - 8x) = 4x + 3$

133. $3[2 - 4(y - 1)] = 3(2y + 8)$

134. $5[2 - (2x - 4)] = 2(5 - 3x)$

135. $3a + 2[2 + 3(a - 1)] = 2(3a + 4)$

136. $5 + 3[1 + 2(2x - 3)] = 6(x + 5)$

137. $-2[4 - (3b + 2)] = 5 - 2(3b + 6)$

138. $-4[x - 2(2x - 3)] + 1 = 2x - 3$

139. If $4 - 3a = 7 - 2(2a + 5)$, evaluate $a^2 + 7a$.

140. If $9 - 5x = 12 - (6x + 7)$, evaluate $x^2 - 3x - 2$.

141. If $2z - 5 = 3(4z + 5)$, evaluate $\frac{z^2}{z - 2}$.

142. If $3n - 7 = 5(2n + 7)$, evaluate $\frac{n^2}{2n - 6}$.

Copyright © Houghton Mifflin Company. All rights reserved.

Objective D To solve application problems using formulas

Physics For Exercises 143 to 149, solve. Use the lever system equation $F_1x = F_2(d - x)$.

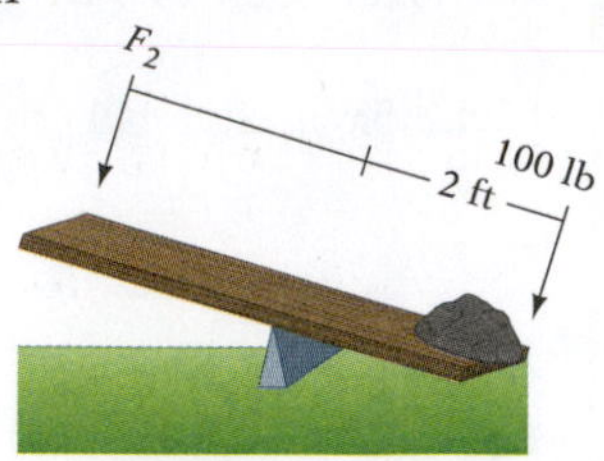

143. A lever 10 ft long is used to move a 100-pound rock. The fulcrum is placed 2 ft from the rock. What force must be applied to the other end of the lever to move the rock?

144. An adult and a child are on a seesaw 14 ft long. The adult weighs 175 lb and the child weighs 70 lb. How many feet from the child must the fulcrum be placed so that the seesaw balances?

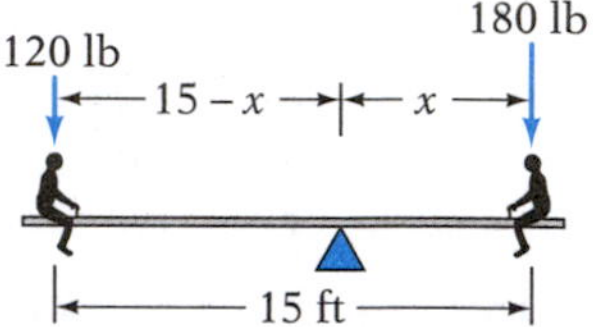

145. Two people are sitting 15 ft apart on a seesaw. One person weighs 180 lb. The second person weighs 120 lb. How far from the 180-pound person should the fulcrum be placed so that the seesaw balances?

146. Two children are sitting on a seesaw that is 12 ft long. One child weighs 60 lb. The other child weighs 90 lb. How far from the 90-pound child should the fulcrum be placed so that the seesaw balances?

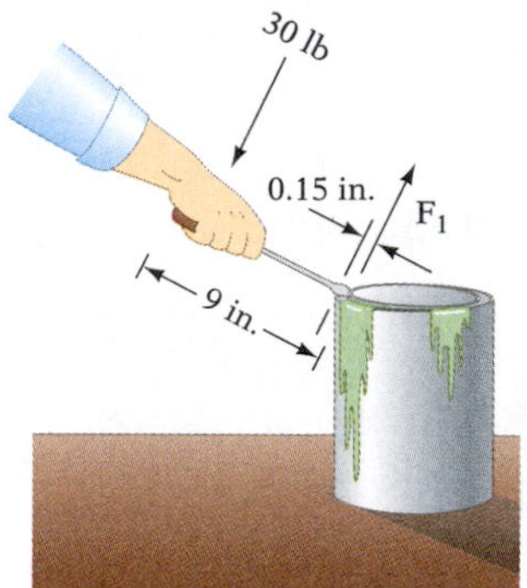

147. In preparation for a stunt, two acrobats are standing on a plank 18 ft long. One acrobat weighs 128 lb and the second acrobat weighs 160 lb. How far from the 128-pound acrobat must the fulcrum be placed so that the acrobats are balanced on the plank?

148. A screwdriver 9 in. long is used as a lever to open a can of paint. The tip of the screwdriver is placed under the lip of the can with the fulcrum 0.15 in. from the lip. A force of 30 lb is applied to the other end of the screwdriver. Find the force on the lip of the can.

149. A metal bar 8 ft long is used to move a 150-pound rock. The fulcrum is placed 1.5 ft from the rock. What minimum force must be applied to the other end of the bar to move the rock? Round to the nearest tenth.

Business To determine the break-even point, or the number of units that must be sold so that no profit or loss occurs, an economist uses the formula $Px = Cx + F$, where P is the selling price per unit, x is the number of units that must be sold to break even, C is the cost to make each unit, and F is the fixed cost. Use this equation for Exercises 150 to 155.

150. A business analyst has determined that the selling price per unit for a laser printer is \$1600. The cost to make one laser printer is \$950, and the fixed cost is \$211,250. Find the break-even point.

Copyright © Houghton Mifflin Company. All rights reserved.

151. A business analyst has determined that the selling price per unit for a gas barbecue is \$325. The cost to make one gas barbecue is \$175, and the fixed cost is \$39,000. Find the break-even point.

152. A manufacturer of thermostats determines that the cost per unit for a programmable thermostat is \$38 and that the fixed cost is \$24,400. The selling price for the thermostat is \$99. Find the break-even point.

153. A manufacturing engineer determines that the cost per unit for a desk lamp is \$12 and that the fixed cost is \$19,240. The selling price for the desk lamp is \$49. Find the break-even point.

154. A manufacturing engineer determines the cost to make one compact disc to be \$3.35 and the fixed cost to be \$6180. The selling price for each compact disc is \$8.50. Find the number of compact discs that must be sold to break even.

155. To manufacture a softball bat requires two steps. The first step is to cut a rough shape. The second step is to sand the bat to its final form. The cost to rough-shape a bat is \$.45, and the cost to sand a bat to final form is \$1.05. The total fixed cost for the two steps is \$16,500. How many softball bats must be sold at a price of \$7.00 to break even?

APPLYING THE CONCEPTS

156. Write an equation of the form $ax + b = cx + d$ that has 4 as the solution.

For Exercises 157 to 160, solve. If the equation has no solution, write "no solution."

157. $3(2x - 1) - (6x - 4) = -9$

158. $7(3x + 6) - 4(3 + 5x) = 13 + x$

159. $\frac{1}{5}(25 - 10a) + 4 = \frac{1}{3}(12a - 15) + 14$

160. $5[m + 2(3 - m)] = 3[2(4 - m) - 5]$

161. The equation $x = x + 1$ has no solution, whereas the solution of the equation $2x + 3 = 3$ is zero. Is there a difference between no solution and a solution of zero? Explain your answer.

Copyright © Houghton Mifflin Company. All rights reserved.

2.3 Translating Sentences into Equations

Objective A To solve integer problems

An equation states that two mathematical expressions are equal. Therefore, to **translate** a sentence into an equation requires recognition of the words or phrases that mean "equals." Some of these phrases are listed below.

equals
is
is equal to
amounts to
represents

} translate to =

Once the sentence is translated into an equation, the equation can be solved by rewriting the equation in the form *variable* = *constant*.

HOW TO Translate "five less than a number is thirteen" into an equation and solve.

The unknown number: n • **Assign a variable to the unknown number.**

Five less than a number	is	thirteen
$n - 5$	$=$	13

• **Find two verbal expressions for the same value.**

• **Write a mathematical expression for each verbal expression. Write the equals sign.**

$$n - 5 + 5 = 13 + 5$$ • **Solve the equation.**

$$n = 18$$

The number is 18.

TAKE NOTE
You can check the solution to a translation problem.

Check:

5 less than 18 is 13

$18 - 5 \mid 13$
$13 = 13$

Recall that the integers are the numbers {..., −4, −3, −2, −1, 0, 1, 2, 3, 4, ...}. An **even integer** is an integer that is divisible by 2. Examples of even integers are −8, 0, and 22. An **odd integer** is an integer that is not divisible by 2. Examples of odd integers are −17, 1, and 39.

Consecutive integers are integers that follow one another in order. Examples of consecutive integers are shown at the right. (Assume that the variable n represents an integer.)

11, 12, 13
−8, −7, −6
$n, n + 1, n + 2$

Examples of **consecutive even integers** are shown at the right. (Assume that the variable n represents an even integer.)

24, 26, 28
−10, −8, −6
$n, n + 2, n + 4$

Examples of **consecutive odd integers** are shown at the right. (Assume that the variable n represents an odd integer.)

19, 21, 23
−1, 1, 3
$n, n + 2, n + 4$

TAKE NOTE
consecutive even
onsecutive odd inte-
e represented
$n + 2, n + 4, ...$

Copyright © Houghton Mifflin Company. All rights reserved.

HOW TO The sum of three consecutive odd integers is forty-five. Find the integers.

Strategy

- First odd integer: n
  Second odd integer: $n + 2$
  Third odd integer: $n + 4$
  — Represent three consecutive odd integers.
- The sum of the three odd integers is 45.

Solution

$n + (n + 2) + (n + 4) = 45$ — Write an equation.

$3n + 6 = 45$ — Solve the equation.

$3n = 39$

$n = 13$ — The first odd integer is 13.

$n + 2 = 13 + 2 = 15$ — Find the second odd integer.

$n + 4 = 13 + 4 = 17$ — Find the third odd integer.

The three consecutive odd integers are 13, 15, and 17.

Example 1

The sum of two numbers is sixteen. The difference between four times the smaller number and two is two more than twice the larger number. Find the two numbers.

Solution

The smaller number: n
The larger number: $16 - n$

The difference between four times the smaller and two	is	two more than twice the larger

$$
\begin{aligned}
4n - 2 &= 2(16 - n) + 2 \\
4n - 2 &= 32 - 2n + 2 \\
4n - 2 &= 34 - 2n \\
4n + 2n - 2 &= 34 - 2n + 2n \\
6n - 2 &= 34 \\
6n - 2 + 2 &= 34 + 2 \\
6n &= 36 \\
\frac{6n}{6} &= \frac{36}{6} \\
n &= 6
\end{aligned}
$$

$16 - n = 16 - 6 = 10$

The smaller number is 6.
The larger number is 10.

You Try It 1

The sum of two numbers is twelve. The total of three times the smaller number and six amounts to seven less than the product of four and the larger number. Find the two numbers.

Your solution

Solution on p. S6

Copyright © Houghton Mifflin Company. All rights reserved.

Example 2

Find three consecutive even integers such that three times the second equals four more than the sum of the first and third.

Strategy

- First even integer: n
  Second even integer: $n + 2$
  Third even integer: $n + 4$
- Three times the second equals four more than the sum of the first and third.

Solution

$$3(n + 2) = n + (n + 4) + 4$$
$$3n + 6 = 2n + 8$$
$$3n - 2n + 6 = 2n - 2n + 8$$
$$n + 6 = 8$$
$$n = 2$$
$$n + 2 = 2 + 2 = 4$$
$$n + 4 = 2 + 4 = 6$$

The three integers are 2, 4, and 6.

You Try It 2

Find three consecutive integers whose sum is negative six.

Your strategy

Your solution

Solution on p. S6

Objective B To translate a sentence into an equation and solve

Example 3

A wallpaper hanger charges a fee of \$25 plus \$12 for each roll of wallpaper used in a room. If the total charge for hanging wallpaper is \$97, how many rolls of wallpaper were used?

Strategy

To find the number of rolls of wallpaper used, write and solve an equation using n to represent the number of rolls of wallpaper used.

Solution

\$25 plus \$12 for each roll of wallpaper	is	\$97

$$25 + 12n = 97$$
$$12n = 72$$
$$\frac{12n}{12} = \frac{72}{12}$$
$$n = 6$$

Six rolls of wallpaper were used.

You Try It 3

The fee charged by a ticketing agency for a concert is \$3.50 plus \$17.50 for each ticket purchased. If your total charge for tickets is \$161, how many tickets are you purchasing?

Your strategy

Your solution

Solution on p. S6

Copyright © Houghton Mifflin Company. All rights reserved.

Example 4

A board 20 ft long is cut into two pieces. Five times the length of the shorter piece is 2 ft more than twice the length of the longer piece. Find the length of each piece.

Strategy

Let x represent the length of the shorter piece. Then $20 - x$ represents the length of the longer piece.

Make a drawing.

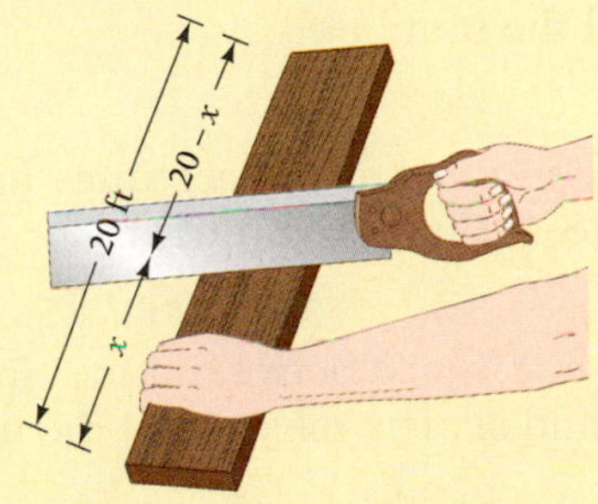

To find the lengths, write and solve an equation using x to represent the length of the shorter piece and $20 - x$ to represent the length of the longer piece.

Solution

Five times the length of the shorter piece	is	2 ft more than twice the length of the longer

$$5x = 2(20 - x) + 2$$
$$5x = 40 - 2x + 2$$
$$5x = 42 - 2x$$
$$5x + 2x = 42 - 2x + 2x$$
$$7x = 42$$
$$\frac{7x}{7} = \frac{42}{7}$$
$$x = 6$$

$20 - x = 20 - 6 = 14$

The length of the shorter piece is 6 ft.
The length of the longer piece is 14 ft.

You Try It 4

A wire 22 in. long is cut into two pieces. The length of the longer piece is 4 in. more than twice the length of the shorter piece. Find the length of each piece.

Your strategy

Your solution

Solution on p. S6

Copyright © Houghton Mifflin Company. All rights reserved.

2.3 Exercises

Objective A **To solve integer problems**

For Exercises 1 to 18, translate into an equation and solve.

1. The difference between a number and fifteen is seven. Find the number.

2. The sum of five and a number is three. Find the number.

3. The product of seven and a number is negative twenty-one. Find the number.

4. The quotient of a number and four is two. Find the number.

5. The difference between nine and a number is seven. Find the number.

6. Three-fifths of a number is negative thirty. Find the number.

7. The difference between five and twice a number is one. Find the number.

8. Four more than three times a number is thirteen. Find the number.

9. The sum of twice a number and five is fifteen. Find the number.

10. The difference between nine times a number and six is twelve. Find the number.

11. Six less than four times a number is twenty-two. Find the number.

12. Four times the sum of twice a number and three is twelve. Find the number.

13. Three times the difference between four times a number and seven is fifteen. Find the number.

14. Twice the difference between a number and twenty-five is three times the number. Find the number.

15. The sum of two numbers is twenty. Three times the smaller is equal to two times the larger. Find the two numbers.

16. The sum of two numbers is fifteen. One less than three times the smaller is equal to the larger. Find the two numbers.

17. The sum of two numbers is fourteen. The difference between two times the smaller and the larger is one. Find the two numbers.

18. The sum of two numbers is eighteen. The total of three times the smaller and twice the larger is forty-four. Find the two numbers.

19. The sum of three consecutive odd integers is fifty-one. Find the integers.

20. Find three consecutive even integers whose sum is negative eighteen.

21. Find three consecutive odd integers such that three times the middle integer is one more than the sum of the first and third.

22. Twice the smallest of three consecutive odd integers is seven more than the largest. Find the integers.

23. Find two consecutive even integers such that three times the first equals twice the second.

24. Find two consecutive even integers such that four times the first is three times the second.

25. Seven times the first of two consecutive odd integers is five times the second. Find the integers.

26. Find three consecutive even integers such that three times the middle integer is four more than the sum of the first and third.

Copyright © Houghton Mifflin Company. All rights reserved.

Objective B To translate a sentence into an equation and solve

27. **Computer Science** The processor speed of a personal computer is 3.2 gigahertz (GHz). This is three-fourths the processor speed of a newer model personal computer. Find the processor speed of the newer personal computer.

28. **Computer Science** The storage capacity of a hard-disk drive is 60 gigabytes. This is one-fourth the storage capacity of a second hard-disk drive. Find the storage capacity of the second hard-disk drive.

29. **Geometry** An isosceles triangle has two sides of equal length. The length of the third side is 1 ft less than twice the length of an equal side. Find the length of each side when the perimeter is 23 ft.

30. **Geometry** An isosceles triangle has two sides of equal length. The length of one of the equal sides is two more than 3 times the length of the third side. If the perimeter is 46 m, find the length of each side.

31. **Union Dues** A union charges monthly dues of $4.00 plus $.15 for each hour worked during the month. A union member's dues for March were $29.20. How many hours did the union member work during the month of March?

32. **Technical Support** A technical information hotline charges a customer $15.00 plus $2.00 per minute to answer questions about software. How many minutes did a customer who received a bill for $37 use this service?

33. **Construction** The total cost to paint the inside of a house was $1346. This cost included $125 for materials and $33 per hour for labor. How many hours of labor were required to paint the inside of the house?

34. **Telecommunications** The cellular phone service for a business executive is $35 per month plus $.40 per minute of phone use. In a month when the executive's cellular phone bill was $99.80, how many minutes did the executive use the phone?

35. **Computer Science** A computer screen consists of tiny dots of light called pixels. In a certain graphics mode, there are 1280 horizontal pixels. This is 768 less than twice the number of vertical pixels. Find the number of vertical pixels.

Copyright © Houghton Mifflin Company. All rights reserved.

36. **Energy** The cost of electricity in a certain city is \$.08 for each of the first 300 kWh (kilowatt-hours) and \$.13 for each kilowatt-hour over 300 kWh. Find the number of kilowatt-hours used by a family with a \$51.95 electric bill.

37. **Geometry** The distance around a rectangular path is 42 m. The length of the path is 3 m less than twice the width. Find the length and width of the path.

38. **Geometry** The fence around a rectangular vegetable garden is 64 ft. The length of the garden is 20 ft. Find the width of the garden.

39. **Carpentry** A 12-foot board is cut into two pieces. Twice the length of the shorter piece is 3 ft less than the length of the longer piece. Find the length of each piece.

40. **Sports** A 14-yard fishing line is cut into two pieces. Three times the length of the longer piece is four times the length of the shorter piece. Find the length of each piece.

41. **Education** Seven thousand dollars is divided into two scholarships. Twice the amount of the smaller scholarship is \$1000 less than the larger scholarship. What is the amount of the larger scholarship?

42. **Investing** An investment of \$10,000 is divided into two accounts, one for stocks and one for mutual funds. The value of the stock account is \$2000 less than twice the value of the mutual funds account. Find the amount in each account.

APPLYING THE CONCEPTS

43. Make up two word problems; one that requires solving the equation $6x = 123$ and one that requires solving the equation $8x + 100 = 300$.

44. A formula is an equation that relates variables in a known way. Find two examples of formulas that are used in your college major. Explain what each of the variables represents.

45. It is always important to check the answer to an application problem to be sure the answer makes sense. Consider the following problem. A 4-quart mixture of fruit juices is made from apple juice and cranberry juice. There are 6 more quarts of apple juice than of cranberry juice. Write and solve an equation for the number of quarts of each juice used. Does the answer to this question make sense? Explain.

Copyright © Houghton Mifflin Company. All rights reserved.

2.4 Mixture and Uniform Motion Problems

Objective A

To solve value mixture problems

A value mixture problem involves combining ingredients that have different prices into a single blend. For example, a coffee merchant may blend two types of coffee into a single blend, or a candy manufacturer may combine two types of candy to sell as a variety pack.

TAKE NOTE

The equation $AC = V$ is used to find the value of an ingredient. For example, the value of 4 lb of cashews costing \$6 per pound is

$$AC = V$$
$$4 \cdot \$6 = V$$
$$\$24 = V$$

The solution of a value mixture problem is based on the **value mixture equation** $AC = V$, where A is the amount of an ingredient, C is the cost per unit of the ingredient, and V is the value of the ingredient.

HOW TO A coffee merchant wants to make 6 lb of a blend of coffee costing \$5 per pound. The blend is made using a \$6-per-pound grade and a \$3-per-pound grade of coffee. How many pounds of each of these grades should be used?

Strategy for Solving a Value Mixture Problem

1. For each ingredient in the mixture, write a numerical or variable expression for the amount of the ingredient used, the unit cost of the ingredient, and the value of the amount used. For the blend, write a numerical or variable expression for the amount, the unit cost of the blend, and the value of the amount. The results can be recorded in a table.

The sum of the amounts is 6 lb.

Amount of \$3 coffee: $6 - x$
Amount of \$6 coffee: x

	Amount, A	·	Unit Cost, C	=	Value, V
\$6 grade	x	·	6	=	$6x$
\$3 grade	$6 - x$	·	3	=	$3(6 - x)$
\$5 blend	6	·	5	=	$5(6)$

TAKE NOTE

Use the information given in the problem to fill in the amount and unit cost columns of the table. Fill in the value column by multiplying the two expressions you wrote in each row. Use the expressions in the last column to write the equation.

2. Determine how the values of the ingredients are related. Use the fact that the sum of the values of all the ingredients is equal to the value of the blend.

The sum of the values of the \$6 grade and the \$3 grade is equal to the value of the \$5 blend.

$$6x + 3(6 - x) = 5(6)$$
$$6x + 18 - 3x = 30$$
$$3x + 18 = 30$$
$$3x = 12$$
$$x = 4$$

$6 - x = 6 - 4 = 2$ • **Find the amount of the \$3 grade coffee.**

The merchant must use 4 lb of the \$6 coffee and 2 lb of the \$3 coffee.

Copyright © Houghton Mifflin Company. All rights reserved.

Example 1

How many ounces of a silver alloy that costs $4 an ounce must be mixed with 10 oz of an alloy that costs $6 an ounce to make a mixture that costs $4.32 an ounce?

Strategy

- Ounces of $4 alloy: x

	Amount	*Cost*	*Value*
$4 alloy	x	4	$4x$
$6 alloy	10	6	$6(10)$
$4.32 mixture	$10 + x$	4.32	$4.32(10 + x)$

- The sum of the values before mixing equals the value after mixing.

Solution

$$4x + 6(10) = 4.32(10 + x)$$
$$4x + 60 = 43.2 + 4.32x$$
$$-0.32x + 60 = 43.2$$
$$-0.32x = -16.8$$
$$x = 52.5$$

52.5 oz of the $4 silver alloy must be used.

You Try It 1

A gardener has 20 lb of a lawn fertilizer that costs $.80 per pound. How many pounds of a fertilizer that costs $.55 per pound should be mixed with this 20 lb of lawn fertilizer to produce a mixture that costs $.75 per pound?

Your strategy

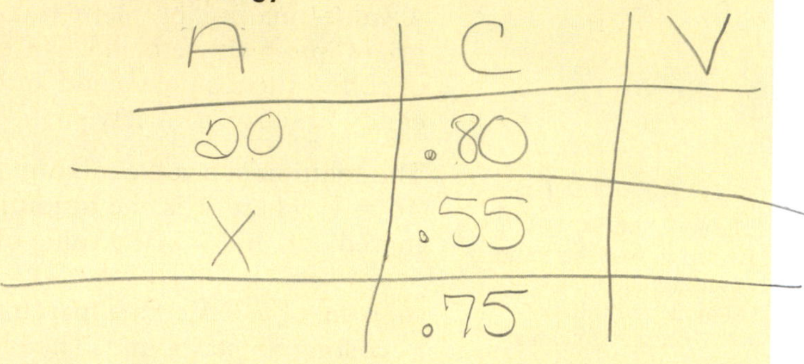

Your solution

Solution on p. S6

Copyright © Houghton Mifflin Company. All rights reserved.

Objective B To solve percent mixture problems

Recall from Section 2.1 that a percent mixture problem can be solved using the equation $Ar = Q$, where A is the amount of a solution, r is the percent concentration of a substance in the solution, and Q is the quantity of the substance in the solution.

For example, a 500-milliliter bottle is filled with a 4% solution of hydrogen peroxide.

$$Ar = Q$$
$$500(0.04) = Q$$
$$20 = Q$$

The bottle contains 20 ml of hydrogen peroxide.

HOW TO How many gallons of a 20% salt solution must be mixed with 6 gal of a 30% salt solution to make a 22% salt solution?

Strategy for Solving a Percent Mixture Problem

1. For each solution, write a numerical or variable expression for the amount of solution, the percent concentration, and the quantity of the substance in the solution. The results can be recorded in a table.

The unknown quantity of 20% solution: x

	Amount of Solution, A	·	Percent Concentration, r	=	Quantity of Substance, Q
20% solution	x	·	0.20	=	$0.20x$
30% solution	6	·	0.30	=	$0.30(6)$
22% solution	$x + 6$	·	0.22	=	$0.22(x + 6)$

TAKE NOTE
Use the information given in the problem to fill in the amount and percent columns of the table. Fill in the quantity column by multiplying the two expressions you wrote in each row. Use the expressions in the last column to write the equation.

2. Determine how the quantities of the substances in the solutions are related. Use the fact that the sum of the quantities of the substances being mixed is equal to the quantity of the substance after mixing.

The sum of the quantities of the substances in the 20% solution and the 30% solution is equal to the quantity of the substance in the 22% solution.

$$0.20x + 0.30(6) = 0.22(x + 6)$$
$$0.20x + 1.80 = 0.22x + 1.32$$
$$-0.02x + 1.80 = 1.32$$
$$-0.02x = -0.48$$
$$x = 24$$

24 gal of the 20% solution is required.

Copyright © Houghton Mifflin Company. All rights reserved.

Example 2

A chemist wishes to make 2 L of an 8% acid solution by mixing a 10% acid solution and a 5% acid solution. How many liters of each solution should the chemist use?

Strategy

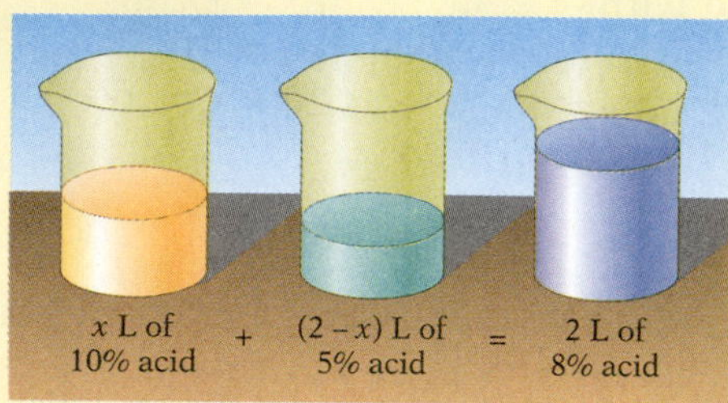

- Liters of 10% solution: x
  Liters of 5% solution: $2 - x$

	Amount	*Percent*	*Quantity*
10% solution	x	0.10	$0.10x$
5% solution	$2 - x$	0.05	$0.05(2 - x)$
8% solution	2	0.08	$0.08(2)$

- The sum of the quantities before mixing is equal to the quantity after mixing.

Solution

$$0.10x + 0.05(2 - x) = 0.08(2)$$
$$0.10x + 0.10 - 0.05x = 0.16$$
$$0.05x + 0.10 = 0.16$$
$$0.05x = 0.06$$
$$x = 1.2$$

$$2 - x = 2 - 1.2 = 0.8$$

The chemist needs 1.2 L of the 10% solution and 0.8 L of the 5% solution.

You Try It 2

A pharmacist dilutes 5 L of a 12% solution with a 6% solution. How many liters of the 6% solution are added to make an 8% solution?

Your strategy

Your solution

Solution on p. S7

Copyright © Houghton Mifflin Company. All rights reserved.

Objective C To solve uniform motion problems

Recall from Section 2.1 that an object traveling at a constant speed in a straight line is in *uniform motion.* The solution of a uniform motion problem is based on the equation $rt = d$, where r is the rate of travel, t is the time spent traveling, and d is the distance traveled.

HOW TO A car leaves a town traveling at 40 mph. Two hours later, a second car leaves the same town, on the same road, traveling at 60 mph. In how many hours will the second car pass the first car?

Strategy for Solving a Uniform Motion Problem

1. For each object, write a numerical or variable expression for the rate, time, and distance. The results can be recorded in a table.

The first car traveled 2 h longer than the second car.

Unknown time for the second car: t
Time for the first car: $t + 2$

TAKE NOTE
Use the information given in the problem to fill in the rate and time columns of the table. Find the expression in the distance column by multiplying the two expressions you wrote in each row.

	Rate, r	·	*Time, t*	=	*Distance, d*
First car	40	·	$t + 2$	=	$40(t + 2)$
Second car	60	·	t	=	$60t$

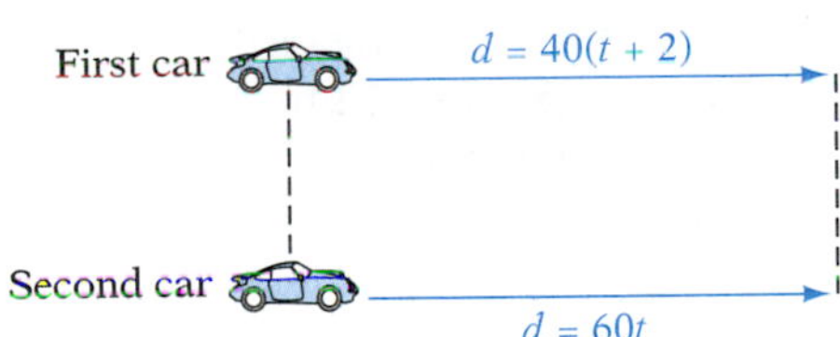

2. Determine how the distances traveled by the two objects are related. For example, the total distance traveled by both objects may be known, or it may be known that the two objects traveled the same distance.

The two cars travel the same distance.

$$40(t + 2) = 60t$$
$$40t + 80 = 60t$$
$$80 = 20t$$
$$4 = t$$

The second car will pass the first car in 4 h.

Copyright © Houghton Mifflin Company. All rights reserved.

Example 3

Two cars, one traveling 10 mph faster than the other, start at the same time from the same point and travel in opposite directions. In 3 h they are 300 mi apart. Find the rate of each car.

Strategy

- Rate of 1st car: r
  Rate of 2nd car: $r + 10$

	Rate	*Time*	*Distance*
1st car	r	3	$3r$
2nd car	$r + 10$	3	$3(r + 10)$

- The total distance traveled by the two cars is 300 mi.

Solution

$$3r + 3(r + 10) = 300$$
$$3r + 3r + 30 = 300$$
$$6r + 30 = 300$$
$$6r = 270$$
$$r = 45$$

$$r + 10 = 45 + 10 = 55$$

The first car is traveling 45 mph.
The second car is traveling 55 mph.

You Try It 3

Two trains, one traveling at twice the speed of the other, start at the same time on parallel tracks from stations that are 288 mi apart and travel toward each other. In 3 h, the trains pass each other. Find the rate of each train.

Your strategy

Your solution

Example 4

How far can the members of a bicycling club ride out into the country at a speed of 12 mph and return over the same road at 8 mph if they travel a total of 10 h?

Strategy

- Time spent riding out: t
  Time spent riding back: $10 - t$

	Rate	*Time*	*Distance*
Out	12	t	$12t$
Back	8	$10 - t$	$8(10 - t)$

- The distance out equals the distance back.

Solution

$$12t = 8(10 - t)$$
$$12t = 80 - 8t$$
$$20t = 80$$
$$t = 4 \quad \text{(The time is 4 h.)}$$

The distance out $= 12t = 12(4) = 48$ mi.

The club can ride 48 mi into the country.

You Try It 4

A pilot flew out to a parcel of land and back in 5 h. The rate out was 150 mph, and the rate returning was 100 mph. How far away was the parcel of land?

Your strategy

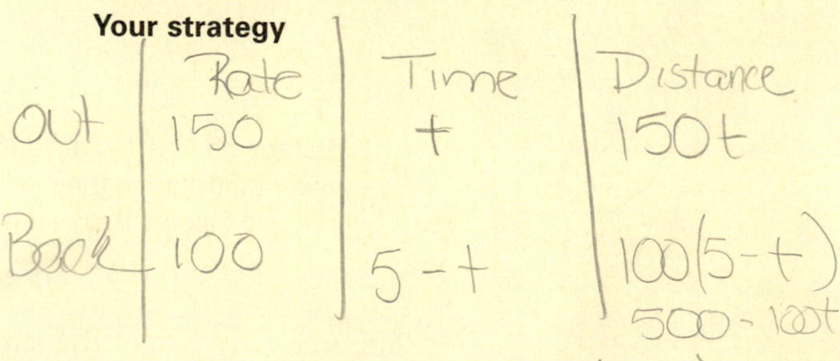

Your solution

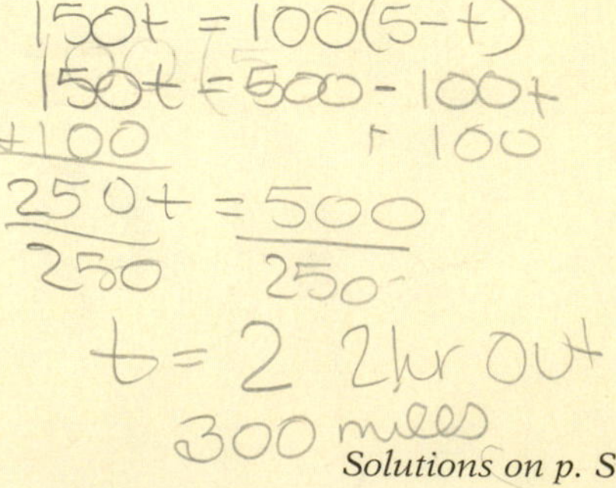

Solutions on p. S7

Copyright © Houghton Mifflin Company. All rights reserved.

2.4 Exercises

Objective A **To solve value mixture problems**

1. An herbalist has 30 oz of herbs costing $2 per ounce. How many ounces of herbs costing $1 per ounce should be mixed with the 30 oz to produce a mixture costing $1.60 per ounce?

2. The manager of a farmer's market has 500 lb of grain that costs $1.20 per pound. How many pounds of meal costing $.80 per pound should be mixed with the 500 lb of grain to produce a mixture that costs $1.05 per pound?

3. Find the cost per pound of a meatloaf mixture made from 3 lb of ground beef costing $1.99 per pound and 1 lb of ground turkey costing $1.39 per pound.

4. Find the cost per ounce of a sunscreen made from 100 oz of a lotion that costs $2.50 per ounce and 50 oz of a lotion that costs $4.00 per ounce.

5. A snack food is made by mixing 5 lb of popcorn that costs $.80 per pound with caramel that costs $2.40 per pound. How much caramel is needed to make a mixture that costs $1.40 per pound?

6. A wild birdseed mix is made by combining 100 lb of millet seed costing $.60 per pound with sunflower seeds costing $1.10 per pound. How many pounds of sunflower seeds are needed to make a mixture that costs $.70 per pound?

7. Ten cups of a restaurant's house Italian dressing is made by blending olive oil costing $1.50 per cup with vinegar that costs $.25 per cup. How many cups of each are used if the cost of the blend is $.50 per cup?

8. A high-protein diet supplement that costs $6.75 per pound is mixed with a vitamin supplement that costs $3.25 per pound. How many pounds of each should be used to make 5 lb of a mixture that costs $4.65 per pound?

9. Find the cost per ounce of a mixture of 200 oz of a cologne that costs $5.50 per ounce and 500 oz of a cologne that costs $2.00 per ounce.

10. Find the cost per pound of a trail mix made from 40 lb of raisins that cost $4.40 per pound and 100 lb of granola that costs $2.30 per pound.

Copyright © Houghton Mifflin Company. All rights reserved.

11. A 20-ounce alloy of platinum that costs $220 per ounce is mixed with an alloy that costs $400 per ounce. How many ounces of the $400 alloy should be used to make an alloy that costs $300 per ounce?

12. How many liters of a blue dye that costs $1.60 per liter must be mixed with 18 L of anil that costs $2.50 per liter to make a mixture that costs $1.90 per liter?

13. The manager of a specialty food store combined almonds that cost $4.50 per pound with walnuts that cost $2.50 per pound. How many pounds of each were used to make a 100-pound mixture that costs $3.24 per pound?

14. A goldsmith combined an alloy that cost $4.30 per ounce with an alloy that cost $1.80 per ounce. How many ounces of each were used to make a mixture of 200 oz costing $2.50 per ounce?

15. Adult tickets for a play cost $6.00 and children's tickets cost $2.50. For one performance, 370 tickets were sold. Receipts for the performance were $1723. Find the number of adult tickets sold.

16. Tickets for a piano concert sold for $4.50 for each adult. Student tickets sold for $2.00 each. The total receipts for 1720 tickets were $5980. Find the number of adult tickets sold.

17. Find the cost per pound of sugar-coated breakfast cereal made from 40 lb of sugar that costs $1.00 per pound and 120 lb of corn flakes that cost $.60 per pound.

18. Find the cost per pound of a coffee mixture made from 8 lb of coffee that costs $9.20 per pound and 12 lb of coffee that costs $5.50 per pound.

$9.20 per pound

$5.50 per pound

20 pounds

Objective B To solve percent mixture problems

19. Forty ounces of a 30% gold alloy is mixed with 60 oz of a 20% gold alloy. Find the percent concentration of the resulting gold alloy.

20. One hundred ounces of juice that is 50% tomato juice is added to 200 oz of a vegetable juice that is 25% tomato juice. What is the percent concentration of tomato juice in the resulting mixture?

Copyright © Houghton Mifflin Company. All rights reserved.

21. How many gallons of a 15% acid solution must be mixed with 5 gal of a 20% acid solution to make a 16% acid solution?

22. How many pounds of a chicken feed that is 50% corn must be mixed with 400 lb of a feed that is 80% corn to make a chicken feed that is 75% corn?

23. A rug is made by weaving 20 lb of yarn that is 50% wool with a yarn that is 25% wool. How many pounds of the yarn that is 25% wool are used if the finished rug is 35% wool?

24. Five gallons of a light green latex paint that is 20% yellow paint is combined with a darker green latex paint that is 40% yellow paint. How many gallons of the darker green paint must be used to create a green paint that is 25% yellow paint?

25. How many gallons of a plant food that is 9% nitrogen must be combined with another plant food that is 25% nitrogen to make 10 gal of a solution that is 15% nitrogen?

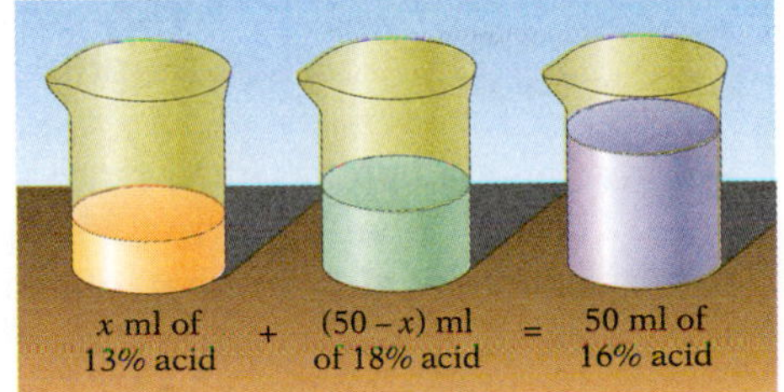

26. A chemist wants to make 50 ml of a 16% acid solution by mixing a 13% acid solution and an 18% acid solution. How many milliliters of each solution should the chemist use?

27. Five grams of sugar are added to a 45-gram serving of a breakfast cereal that is 10% sugar. What is the percent concentration of sugar in the resulting mixture?

28. A goldsmith mixes 8 oz of a 30% gold alloy with 12 oz of a 25% gold alloy. What is the percent concentration of the resulting alloy?

29. How many pounds of coffee that is 40% java beans must be mixed with 80 lb of coffee that is 30% java beans to make a coffee blend that is 32% java beans?

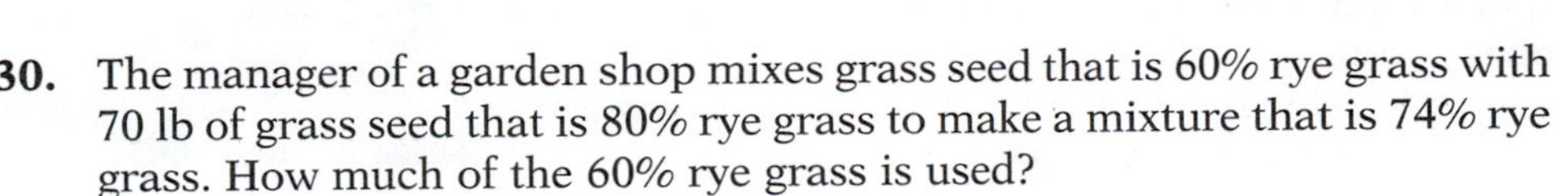

30. The manager of a garden shop mixes grass seed that is 60% rye grass with 70 lb of grass seed that is 80% rye grass to make a mixture that is 74% rye grass. How much of the 60% rye grass is used?

31. A hair dye is made by blending a 7% hydrogen peroxide solution and a 4% hydrogen peroxide solution. How many milliliters of each are used to make a 300-milliliter solution that is 5% hydrogen peroxide?

Copyright © Houghton Mifflin Company. All rights reserved.

32. A tea that is 20% jasmine is blended with a tea that is 15% jasmine. How many pounds of each tea are used to make 5 lb of tea that is 18% jasmine?

33. How many ounces of pure chocolate must be added to 150 oz of chocolate topping that is 50% chocolate to make a topping that is 75% chocolate?

34. How many ounces of pure bran flakes must be added to 50 oz of cereal that is 40% bran flakes to produce a mixture that is 50% bran flakes?

35. Thirty ounces of pure silver is added to 50 oz of a silver alloy that is 20% silver. What is the percent concentration of the resulting alloy?

36. A clothing manufacturer has some pure silk thread and some thread that is 85% silk. How many kilograms of each must be woven together to make 75 kg of cloth that is 96% silk?

Objective C **To solve uniform motion problems**

37. Two small planes start from the same point and fly in opposite directions. The first plane is flying 25 mph slower than the second plane. In 2 h, the planes are 470 mi apart. Find the rate of each plane.

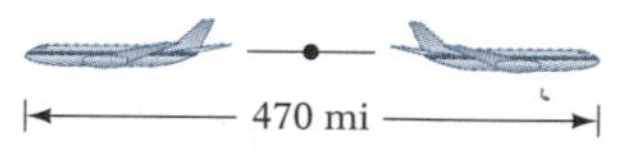

38. Two cyclists start from the same point and ride in opposite directions. One cyclist rides twice as fast as the other. In 3 h, they are 81 mi apart. Find the rate of each cyclist.

39. Two planes leave an airport at 8 A.M., one flying north at 480 km/h and the other flying south at 520 km/h. At what time will they be 3000 km apart?

40. A long-distance runner started on a course running at an average speed of 6 mph. One-half hour later, a second runner began the same course at an average speed of 7 mph. How long after the second runner starts will the second runner overtake the first runner?

41. A motorboat leaves a harbor and travels at an average speed of 9 mph toward a small island. Two hours later a cabin cruiser leaves the same harbor and travels at an average speed of 18 mph toward the same island. How many hours after the cabin cruiser leaves the harbor will it be alongside the motorboat?

Copyright © Houghton Mifflin Company. All rights reserved.

42. A 555-mile, 5-hour plane trip was flown at two speeds. For the first part of the trip, the average speed was 105 mph. For the remainder of the trip, the average speed was 115 mph. How long did the plane fly at each speed?

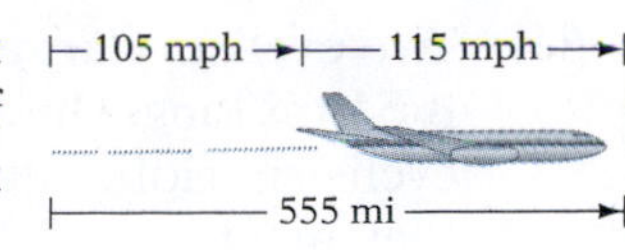

43. An executive drove from home at an average speed of 30 mph to an airport where a helicopter was waiting. The executive boarded the helicopter and flew to the corporate offices at an average speed of 60 mph. The entire distance was 150 mi. The entire trip took 3 h. Find the distance from the airport to the corporate offices.

44. After a sailboat had been on the water for 3 h, a change in the wind direction reduced the average speed of the boat by 5 mph. The entire distance sailed was 57 mi. The total time spent sailing was 6 h. How far did the sailboat travel in the first 3 h?

45. A car and a bus set out at 3 P.M. from the same point headed in the same direction. The average speed of the car is twice the average speed of the bus. In 2 h the car is 68 mi ahead of the bus. Find the rate of the car.

46. A passenger train leaves a train depot 2 h after a freight train leaves the same depot. The freight train is traveling 20 mph slower than the passenger train. Find the rate of each train if the passenger train overtakes the freight train in 3 h.

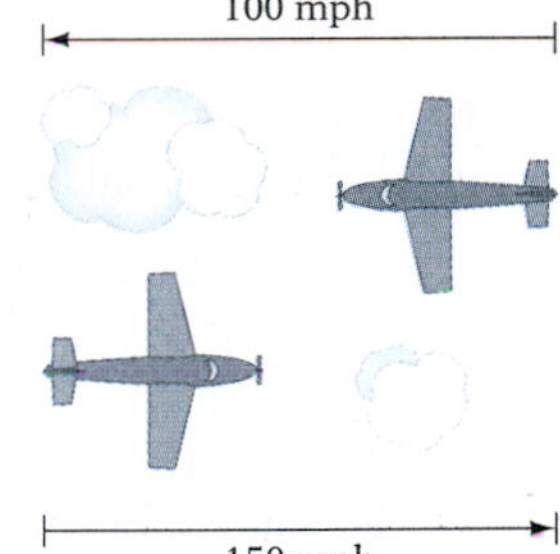

47. As part of flight training, a student pilot was required to fly to an airport and then return. The average speed on the way to the airport was 100 mph, and the average speed returning was 150 mph. Find the distance between the two airports if the total flying time was 5 h.

48. A ship traveling east at 25 mph is 10 mi from a harbor when another ship leaves the harbor traveling east at 35 mph. How long does it take the second ship to catch up to the first ship?

49. At 10 A.M., a plane leaves Boston, Massachusetts, for Seattle, Washington, a distance of 3000 mi. One hour later a plane leaves Seattle for Boston. Both planes are traveling at a speed of 500 mph. How many hours after the plane leaves Seattle will the planes pass each other?

50. At noon, a train leaves Washington, D.C., headed for Charleston, South Carolina, a distance of 500 mi. The train travels at a speed of 60 mph. At 1 P.M. a second train leaves Charleston headed for Washington, D.C., traveling at 50 mph. How long after the train leaves Charleston will the two trains pass each other?

Copyright © Houghton Mifflin Company. All rights reserved.

51. Two cyclists start at the same time from opposite ends of a course that is 51 mi long. One cyclist is riding at a rate of 16 mph, and the second cyclist is riding at a rate of 18 mph. How long after they begin will they meet?

52. A bus traveled on a level road for 2 h at an average speed that was 20 mph faster than its average speed on a winding road. The time spent on the winding road was 3 h. Find the average speed on the winding road if the total trip was 210 mi.

53. A bus traveling at a rate of 60 mph overtakes a car traveling at a rate of 45 mph. If the car had a 1-hour head start, how far from the starting point does the bus overtake the car?

54. A car traveling at 48 mph overtakes a cyclist who, riding at 12 mph, had a 3-hour head start. How far from the starting point does the car overtake the cyclist?

APPLYING THE CONCEPTS

55. **Chemistry** How many grams of pure water must be added to 50 g of pure acid to make a solution that is 40% acid?

56. **Chemistry** How many ounces of water must be evaporated from 50 oz of a 12% salt solution to produce a 15% salt solution?

57. **Automotive Technology** A radiator contains 15 gal of a 20% antifreeze solution. How many gallons must be drained from the radiator and replaced by pure antifreeze so that the radiator will contain 15 gal of a 40% antifreeze solution?

58. **Travel** At 10 A.M., two campers left their campsite by canoe and paddled downstream at an average speed of 12 mph. They then turned around and paddled back upstream at an average rate of 4 mph. The total trip took 1 h. At what time did the campers turn around downstream?

59. **Transportation** A bicyclist rides for 2 h at a speed of 10 mph and then returns at a speed of 20 mph. Find the cyclist's average speed for the trip.

60. **Travel** A car travels a 1-mile track at an average speed of 30 mph. At what average speed must the car travel the next mile so that the average speed for the 2 mi is 60 mph?

Copyright © Houghton Mifflin Company. All rights reserved.

2.5 First-Degree Inequalities

Objective A To solve an inequality in one variable

The **solution set of an inequality** is a set of numbers, each element of which, when substituted for the variable, results in a true inequality.

The inequality at the right is true if the variable is replaced by (for instance) 3, −1.98, or $\frac{2}{3}$.

$$x - 1 < 4$$
$$3 - 1 < 4$$
$$-1.98 - 1 < 4$$
$$\frac{2}{3} - 1 < 4$$

There are many values of the variable x that will make the inequality $x - 1 < 4$ true. The solution set of the inequality is any number less than 5. The solution set can be written in set-builder notation as $\{x \mid x < 5\}$.

Integrating Technology

See the appendix Keystroke Guide: *Test* for instructions on using a graphing calculator to graph the solution set of an inequality.

The graph of the solution set of $x - 1 < 4$ is shown at the right.

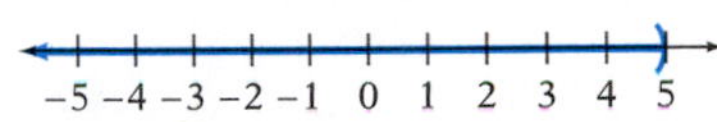

When solving an inequality, we use the **Addition and Multiplication Properties of Inequalities** to rewrite the inequality in the form *variable* < *constant* or in the form *variable* > *constant*.

The Addition Property of Inequalities

If $a > b$, then $a + c > b + c$.
If $a < b$, then $a + c < b + c$.

The Addition Property of Inequalities states that the same number can be added to each side of an inequality without changing the solution set of the inequality. This property is also true for an inequality that contains the symbol ≤ or ≥.

The Addition Property of Inequalities is used to remove a term from one side of an inequality by adding the additive inverse of that term to each side of the inequality. Because subtraction is defined in terms of addition, the same number can be subtracted from each side of an inequality without changing the solution set of the inequality.

HOW TO Solve and graph the solution set: $x + 2 \geq 4$

$$x + 2 \geq 4$$
$$x + 2 - 2 \geq 4 - 2$$
• Subtract 2 from each side of the inequality.
$$x \geq 2$$
• Simplify.

The solution set is $\{x \mid x \geq 2\}$.

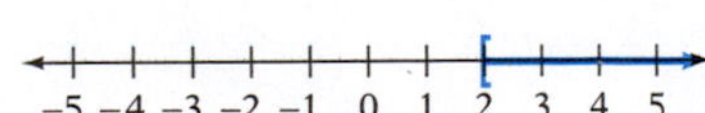

Copyright © Houghton Mifflin Company. All rights reserved.

HOW TO Solve: $3x - 4 < 2x - 1$

$$3x - 4 < 2x - 1$$
$$3x - 4 - 2x < 2x - 1 - 2x$$
• Subtract $2x$ from each side of the inequality.
$$x - 4 < -1$$
$$x - 4 + 4 < -1 + 4$$
• Add 4 to each side of the inequality.
$$x < 3$$

The solution set is $\{x \mid x < 3\}$.

The Multiplication Property of Inequalities is used to remove a coefficient from one side of an inequality by multiplying each side of the inequality by the reciprocal of the coefficient.

TAKE NOTE
$c > 0$ means c is a positive number. Note that the inequality symbols do not change.

$c < 0$ means c is a negative number. Note that the inequality symbols are reversed.

The Multiplication Property of Inequalities

Rule 1 If $a > b$ and $c > 0$, then $ac > bc$.
If $a < b$ and $c > 0$, then $ac < bc$.

Rule 2 If $a > b$ and $c < 0$, then $ac < bc$.
If $a < b$ and $c < 0$, then $ac > bc$.

Here are some examples of this property.

Rule 1		Rule 2	
$3 > 2$	$2 < 5$	$3 > 2$	$2 < 5$
$3(4) > 2(4)$	$2(4) < 5(4)$	$3(-4) < 2(-4)$	$2(-4) > 5(-4)$
$12 > 8$	$8 < 20$	$-12 < -8$	$-8 > -20$

Rule 1 states that when each side of an inequality is multiplied by a positive number, the inequality symbol remains the same. However, Rule 2 states that when each side of an inequality is multiplied by a negative number, the inequality symbol must be reversed. Because division is defined in terms of multiplication, when each side of an inequality is divided by a *positive* number, the inequality symbol remains the same. But when each side of an inequality is divided by a *negative* number, the inequality symbol must be reversed.

The Multiplication Property of Inequalities is also true for the symbols $\leq$ and $\geq$.

TAKE NOTE
Each side of the inequality is divided *by* a negative number; the inequality symbol must be reversed.

HOW TO Solve: $-3x > 9$

$$-3x > 9$$
$$\frac{-3x}{-3} < \frac{9}{-3}$$
• Divide each side of the inequality by the coefficient -3. Because -3 is a negative number, the inequality symbol must be reversed.
$$x < -3$$

The solution set is $\{x \mid x < -3\}$.

Copyright © Houghton Mifflin Company. All rights reserved.

HOW TO Solve: $3x + 2 < -4$

$$3x + 2 < -4$$

$$3x < -6$$ • Subtract 2 from each side of the inequality.

$$\frac{3x}{3} < \frac{-6}{3}$$ • Divide each side of the inequality by the coefficient 3.

$$x < -2$$

The solution set is $\{x \mid x < -2\}$.

HOW TO Solve: $2x - 9 > 4x + 5$

$$2x - 9 > 4x + 5$$

$$-2x - 9 > 5$$ • Subtract $4x$ from each side of the inequality.

$$-2x > 14$$ • Add 9 to each side of the inequality.

$$\frac{-2x}{-2} < \frac{14}{-2}$$ • Divide each side of the inequality by the coefficient -2. Reverse the inequality symbol.

$$x < -7$$

The solution set is $\{x \mid x < -7\}$.

HOW TO Solve: $5(x - 2) \geq 9x - 3(2x - 4)$

$$5(x - 2) \geq 9x - 3(2x - 4)$$

$$5x - 10 \geq 9x - 6x + 12$$ • Use the Distributive Property to remove parentheses.

$$5x - 10 \geq 3x + 12$$

$$2x - 10 \geq 12$$ • Subtract $3x$ from each side of the inequality.

$$2x \geq 22$$ • Add 10 to each side of the inequality.

$$\frac{2x}{2} \geq \frac{22}{2}$$ • Divide each side of the inequality by the coefficient 2.

$$x \geq 11$$

The solution set is $\{x \mid x \geq 11\}$.

Example 1

Solve and graph the solution set:
$x + 3 > 4x + 6$

Solution

$$x + 3 > 4x + 6$$

$$-3x + 3 > 6$$ • Subtract $4x$ from each side.

$$-3x > 3$$ • Subtract 3 from each side.

$$\frac{-3x}{-3} < \frac{3}{-3}$$ • Divide each side by -3.

$$x < -1$$

The solution set is $\{x \mid x < -1\}$.

−5 −4 −3 −2 −1 0 1 2 3 4 5

You Try It 1

Solve and graph the solution set:
$2x - 1 < 6x + 7$

Your solution

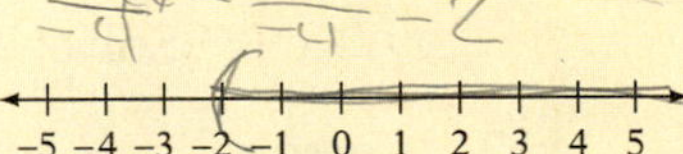

Solution on p. S7

Copyright © Houghton Mifflin Company. All rights reserved.

Example 2

Solve: $3x - 5 \le 3 - 2(3x + 1)$

Solution

$$3x - 5 \le 3 - 2(3x + 1)$$
$$3x - 5 \le 3 - 6x - 2$$
$$3x - 5 \le 1 - 6x$$
$$9x - 5 \le 1$$
$$9x \le 6$$
$$\frac{9x}{9} \le \frac{6}{9}$$
$$x \le \frac{2}{3}$$

$\left\{x \middle| x \le \frac{2}{3}\right\}$

You Try It 2

Solve: $5x - 2 \le 4 - 3(x - 2)$

Your solution

Solution on p. S7

Objective B **To solve a compound inequality**

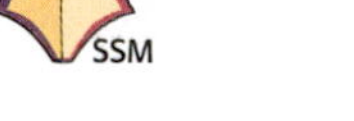

A **compound inequality** is formed by joining two inequalities with a connective word such as *and* or *or*. The inequalities at the right are compound inequalities.

$2x < 4$ and $3x - 2 > -8$

$2x + 3 > 5$ or $x + 2 < 5$

The solution set of a compound inequality with the connective word *and* is the set of those elements that appear in the solution sets of both inequalities. Therefore, it is the intersection of the solution sets of the two inequalities.

HOW TO Solve: $2x < 6$ and $3x + 2 > -4$

$2x < 6$	and	$3x + 2 > -4$	
$x < 3$		$3x > -6$	• Solve each inequality.
$\{x \mid x < 3\}$		$x > -2$	
		$\{x \mid x > -2\}$	

The solution set of a compound inequality with *and* is the intersection of the solution sets of the two inequalities.

$\{x \mid x < 3\} \cap \{x \mid x > -2\} = \{x \mid -2 < x < 3\}$

HOW TO Solve: $-3 < 2x + 1 < 5$

This inequality is equivalent to the compound inequality $-3 < 2x + 1$ and $2x + 1 < 5$.

$-3 < 2x + 1$	and	$2x + 1 < 5$	
$-4 < 2x$		$2x < 4$	• Solve each inequality.
$-2 < x$		$x < 2$	
$\{x \mid x > -2\}$		$\{x \mid x < 2\}$	

The solution set of a compound inequality with *and* is the intersection of the solution sets of the two inequalities.

$\{x \mid x > -2\} \cap \{x \mid x < 2\} = \{x \mid -2 < x < 2\}$

Copyright © Houghton Mifflin Company. All rights reserved.

There is an alternative method for solving the inequality in the last example.

HOW TO Solve: $-3 < 2x + 1 < 5$

$$-3 < 2x + 1 < 5$$
$$-3 - 1 < 2x + 1 - 1 < 5 - 1$$
$$-4 < 2x < 4$$
$$\frac{-4}{2} < \frac{2x}{2} < \frac{4}{2}$$
$$-2 < x < 2$$

- Subtract 1 from each of the three parts of the inequality.
- Divide each of the three parts of the inequality by the coefficient 2.

The solution set is $\{x \mid -2 < x < 2\}$.

The solution set of a compound inequality with the connective word *or* is the union of the solution sets of the two inequalities.

HOW TO Solve: $2x + 3 > 7$ or $4x - 1 < 3$

$2x + 3 > 7$ or $4x - 1 < 3$

$2x > 4 \qquad 4x < 4$

$x > 2 \qquad x < 1$

$\{x \mid x > 2\} \qquad \{x \mid x < 1\}$

- Solve each inequality.

Find the union of the solution sets.

$\{x \mid x > 2\} \cup \{x \mid x < 1\} = \{x \mid x > 2 \text{ or } x < 1\}$

Example 3

Solve: $1 < 3x - 5 < 4$

Solution

$$1 < 3x - 5 < 4$$
$$1 + 5 < 3x - 5 + 5 < 4 + 5$$
$$6 < 3x < 9$$
$$\frac{6}{3} < \frac{3x}{3} < \frac{9}{3}$$
$$2 < x < 3$$

$\{x \mid 2 < x < 3\}$

- Add 5 to each of the three parts.
- Divide each of the three parts by 3.

You Try It 3

Solve: $-2 \le 5x + 3 \le 13$

Your solution

Example 4

Solve: $11 - 2x > -3$ and $7 - 3x < 4$

Solution

$11 - 2x > -3$ and $7 - 3x < 4$

$-2x > -14 \qquad -3x < -3$

$x < 7 \qquad x > 1$

$\{x \mid x < 7\} \qquad \{x \mid x > 1\}$

$\{x \mid x < 7\} \cap \{x \mid x > 1\} = \{x \mid 1 < x < 7\}$

You Try It 4

Solve: $2 - 3x > 11$ or $5 + 2x > 7$

Your solution

Solutions on p. S7

Copyright © Houghton Mifflin Company. All rights reserved.

Objective C To solve application problems

Example 5

A cellular phone company advertises two pricing plans. The first is \$19.95 per month with 20 free minutes and \$.39 per minute thereafter. The second is \$23.95 per month with 20 free minutes and \$.30 per minute thereafter. How many minutes can you talk per month for the first plan to cost less than the second?

Strategy

To find the number of minutes, write and solve an inequality using N to represent the number of minutes. Then $N - 20$ is the number of minutes for which you are charged after the first free 20 min.

Solution

Cost of first plan < cost of second plan

$$\begin{aligned} 19.95 + 0.39(N - 20) &< 23.95 + 0.30(N - 20) \\ 19.95 + 0.39N - 7.8 &< 23.95 + 0.30N - 6 \\ 12.15 + 0.39N &< 17.95 + 0.30N \\ 12.15 + 0.09N &< 17.95 \\ 0.09N &< 5.8 \\ N &< 64.\overline{4} \end{aligned}$$

The first plan costs less if you talk less than 65 min.

You Try It 5

The base of a triangle is 12 in. and the height is $(x + 2)$ in. Express as an integer the maximum height of the triangle when the area is less than 50 in^2.

Your strategy

Your solution

Example 6

Find three consecutive positive odd integers whose sum is between 27 and 51.

Strategy

To find the three integers, write and solve a compound inequality using n to represent the first odd integer.

Solution

$$\text{Lower limit of the sum} < \text{sum} < \text{upper limit of the sum}$$

$$\begin{aligned} 27 &< n + (n + 2) + (n + 4) < 51 \\ 27 &< 3n + 6 < 51 \\ 27 - 6 &< 3n + 6 - 6 < 51 - 6 \\ 21 &< 3n < 45 \\ \frac{21}{3} &< \frac{3n}{3} < \frac{45}{3} \\ 7 &< n < 15 \end{aligned}$$

The three odd integers are 9, 11, and 13; or 11, 13, and 15; or 13, 15, and 17.

You Try It 6

An average score of 80 to 89 in a history course receives a B. Luisa Montez has grades of 72, 94, 83, and 70 on four exams. Find the range of scores on the fifth exam that will give Luisa a B for the course.

Your strategy

Your solution

Solutions on p. S8

Copyright © Houghton Mifflin Company. All rights reserved.

2.5 Exercises

Objective A **To solve an inequality in one variable**

1. State the Addition Property of Inequalities and give numerical examples of its use.

2. State the Multiplication Property of Inequalities and give numerical examples of its use.

3. Which numbers are solutions of the inequality $x + 7 \le -3$?
a. -17 **b.** 8 **c.** -10 **d.** 0

4. Which numbers are solutions of the inequality $2x - 1 > 5$?
a. 6 **b.** -4 **c.** 3 **d.** 5

For Exercises 5 to 31, solve. For Exercises 5 to 10, graph the solution set.

5. $x - 3 < 2$

−5 −4 −3 −2 −1 0 1 2 3 4 5

6. $x + 4 \ge 2$

−5 −4 −3 −2 −1 0 1 2 3 4 5

7. $4x \le 8$

−5 −4 −3 −2 −1 0 1 2 3 4 5

8. $6x > 12$

−5 −4 −3 −2 −1 0 1 2 3 4 5

9. $-2x > 8$

−5 −4 −3 −2 −1 0 1 2 3 4 5

10. $-3x \le -9$

−5 −4 −3 −2 −1 0 1 2 3 4 5

11. $3x - 1 > 2x + 2$

12. $5x + 2 \ge 4x - 1$

13. $2x - 1 > 7$

14. $3x + 2 < 8$

15. $5x - 2 \le 8$

16. $4x + 3 \le -1$

17. $6x + 3 > 4x - 1$

18. $7x + 4 < 2x - 6$

19. $8x + 1 \ge 2x + 13$

20. $5x - 4 < 2x + 5$

21. $4 - 3x < 10$

22. $2 - 5x > 7$

23. $7 - 2x \ge 1$

24. $3 - 5x \le 18$

25. $-3 - 4x > -11$

26. $-2 - x < 7$

27. $4x - 2 < x - 11$

28. $6x + 5 \le x - 10$

29. $x + 7 \ge 4x - 8$

30. $3x + 1 \le 7x - 15$

31. $3x + 2 \le 7x + 4$

Copyright © Houghton Mifflin Company. All rights reserved.

For Exercises 32 to 47, solve.

32. $3x - 5 \geq -2x + 5$

33. $\frac{3}{5}x - 2 < \frac{3}{10} - x$

34. $\frac{5}{6}x - \frac{1}{6} < x - 4$

35. $\frac{2}{3}x - \frac{3}{2} < \frac{7}{6} - \frac{1}{3}x$

36. $\frac{7}{12}x - \frac{3}{2} < \frac{2}{3}x + \frac{5}{6}$

37. $\frac{1}{2}x - \frac{3}{4} < \frac{7}{4}x - 2$

38. $6 - 2(x - 4) \leq 2x + 10$

39. $4(2x - 1) > 3x - 2(3x - 5)$

40. $2(1 - 3x) - 4 > 10 + 3(1 - x)$

41. $2 - 5(x + 1) \geq 3(x - 1) - 8$

42. $2 - 2(7 - 2x) < 3(3 - x)$

43. $3 + 2(x + 5) \geq x + 5(x + 1) + 1$

44. $10 - 13(2 - x) < 5(3x - 2)$

45. $3 - 4(x + 2) \leq 6 + 4(2x + 1)$

46. $3x - 2(3x - 5) \leq 2 - 5(x - 4)$

47. $12 - 2(3x - 2) \geq 5x - 2(5 - x)$

Objective B To solve a compound inequality

48. **a.** Which set operation is used when a compound inequality is combined with *or*?

b. Which set operation is used when a compound inequality is combined with *and*?

49. Explain why writing $-3 > x > 4$ does not make sense.

For Exercises 50 to 83, solve. Write the solution set in set-builder notation.

50. $3x < 6$ and $x + 2 > 1$

51. $x - 3 \leq 1$ and $2x \geq -4$

52. $x + 2 \geq 5$ or $3x \leq 3$

53. $2x < 6$ or $x - 4 > 1$

Copyright © Houghton Mifflin Company. All rights reserved.

54. $-2x > -8$ and $-3x < 6$

55. $\frac{1}{2}x > -2$ and $5x < 10$

56. $\frac{1}{3}x < -1$ or $2x > 0$

57. $\frac{2}{3}x > 4$ or $2x < -8$

58. $x + 4 \geq 5$ and $2x \geq 6$

59. $3x < -9$ and $x - 2 < 2$

60. $-5x > 10$ and $x + 1 > 6$

61. $2x - 3 > 1$ and $3x - 1 < 2$

62. $7x < 14$ and $1 - x < 4$

63. $4x + 1 < 5$ and $4x + 7 > -1$

64. $3x + 7 < 10$ or $2x - 1 > 5$

65. $6x - 2 < -14$ or $5x + 1 > 11$

66. $-5 < 3x + 4 < 16$

67. $5 < 4x - 3 < 21$

68. $0 < 2x - 6 < 4$

69. $-2 < 3x + 7 < 1$

70. $4x - 1 > 11$ or $4x - 1 \leq -11$

71. $3x - 5 > 10$ or $3x - 5 < -10$

72. $9x - 2 < 7$ and $3x - 5 > 10$

73. $8x + 2 \leq -14$ and $4x - 2 > 10$

74. $3x - 11 < 4$ or $4x + 9 \geq 1$

75. $5x + 12 \geq 2$ or $7x - 1 \leq 13$

Copyright © Houghton Mifflin Company. All rights reserved.

76. $-6 \le 5x + 14 \le 24$

77. $3 \le 7x - 14 \le 31$

78. $3 - 2x > 7$ and $5x + 2 > -18$

79. $1 - 3x < 16$ and $1 - 3x > -16$

80. $5 - 4x > 21$ or $7x - 2 > 19$

81. $6x + 5 < -1$ or $1 - 2x < 7$

82. $3 - 7x \le 31$ and $5 - 4x > 1$

83. $9 - x \ge 7$ and $9 - 2x < 3$

Objective C To solve application problems

84. **Integers** Five times the difference between a number and two is greater than the quotient of two times the number and three. Find the smallest integer that will satisfy the inequality.

85. **Integers** Two times the difference between a number and eight is less than or equal to five times the sum of the number and four. Find the smallest number that will satisfy the inequality.

86. **Geometry** The length of a rectangle is 2 ft more than four times the width. Express as an integer the maximum width of the rectangle when the perimeter is less than 34 ft.

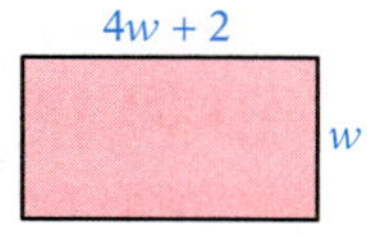

87. **Geometry** The length of a rectangle is 5 cm less than twice the width. Express as an integer the maximum width of the rectangle when the perimeter is less than 60 cm.

88. **Telecommunications** In 2003, the computer service America Online offered its customers a rate of \$23.90 per month for unlimited use or \$4.95 per month with 3 free hours plus \$2.50 for each hour thereafter. Express as an integer the maximum number of hours you can use this service per month if the second plan is to cost you less than the first.

Copyright © Houghton Mifflin Company. All rights reserved.

89. **Telecommunications** TopPage advertises local paging service for \$6.95 per month for up to 400 pages, and \$.10 per page thereafter. A competitor advertises service for \$3.95 per month for up to 400 pages and \$.15 per page thereafter. For what number of pages per month is the TopPage plan less expensive?

90. **Consumerism** Suppose PayRite Rental Cars rents compact cars for \$32 per day with unlimited mileage, and Otto Rentals offers compact cars for \$19.99 per day but charges \$.19 for each mile beyond 100 mi driven per day. You want to rent a car for one week. How many miles can you drive during the week if Otto Rentals is to be less expensive than PayRite?

91. **Consumerism** During a weekday, to call a city 40 mi away from a certain pay phone costs \$.70 for the first 3 minutes and \$.15 for each additional minute. If you use a calling card, there is a \$.35 fee and then the rates are \$.196 for the first minute and \$.126 for each additional minute. How long must a call be if it is to be cheaper to pay with coins rather than a calling card?

92. **Temperature** The temperature range for a week was between 14°F and 77°F. Find the temperature range in Celsius degrees. $F = \frac{9}{5}C + 32$

93. **Temperature** The temperature range for a week in a mountain town was between 0°C and 30°C. Find the temperature range in Fahrenheit degrees. $C = \frac{5(F - 32)}{9}$

94. **Compensation** You are a sales account executive earning \$1200 per month plus 6% commission on the amount of sales. Your goal is to earn a minimum of \$6000 per month. What amount of sales will enable you to earn \$6000 or more per month?

95. **Compensation** George Stoia earns \$1000 per month plus 5% commission on the amount of sales. George's goal is to earn a minimum of \$3200 per month. What amount of sales will enable George to earn \$3200 or more per month?

96. **Banking** Heritage National Bank offers two different checking accounts. The first charges \$3 per month and \$.50 per check after the first 10 checks. The second account charges \$8 per month with unlimited check writing. How many checks can be written per month if the first account is to be less expensive than the second account?

Copyright © Houghton Mifflin Company. All rights reserved.

97. Banking Glendale Federal Bank offers a checking account to small businesses. The charge is \$8 per month plus \$.12 per check after the first 100 checks. A competitor is offering an account for \$5 per month plus \$.15 per check after the first 100 checks. If a business chooses the first account, how many checks does the business write monthly if it is assumed that the first account will cost less than the competitor's account?

98. Education An average score of 90 or above in a history class receives an A grade. You have scores of 95, 89, and 81 on three exams. Find the range of scores on the fourth exam that will give you an A grade for the course.

99. Education An average of 70 to 79 in a mathematics class receives a C grade. A student has scores of 56, 91, 83, and 62 on four tests. Find the range of scores on the fifth test that will give the student a C for the course.

100. Integers Find four consecutive integers whose sum is between 62 and 78.

101. Integers Find three consecutive even integers whose sum is between 30 and 51.

APPLYING THE CONCEPTS

102. Let $-2 \le x \le 3$ and $a \le 2x + 1 \le b$.
 a. Find the largest possible value of a.
 b. Find the smallest possible value of b.

103. Determine whether the following statements are always true, sometimes true, or never true.
 a. If $a > b$, then $-a < -b$.
 b. If $a < b$ and $a \neq 0$, $b \neq 0$, then $\frac{1}{a} < \frac{1}{b}$.
 c. When dividing both sides of an inequality by an integer, we must reverse the inequality symbol.
 d. If $a < 1$, then $a^2 < a$.
 e. If $a < b < 0$ and $c < d < 0$, then $ac > bd$.

104. The following is offered as the solution of $2 + 3(2x - 4) < 6x + 5$.

$2 + 3(2x - 4) < 6x + 5$
$2 + 6x - 12 < 6x + 5$ • Use the Distributive Property.
$6x - 10 < 6x + 5$ • Simplify.
$6x - 6x - 10 < 6x - 6x + 5$ • Subtract $6x$ from each side.
$-10 < 5$

Because $-10 < 5$ is a true inequality, the solution set is all real numbers.

If this is correct, so state. If it is not correct, explain the incorrect step and supply the correct answer.

Copyright © Houghton Mifflin Company. All rights reserved.

2.6 Absolute Value Equations and Inequalities

Objective A

Study Tip

Before the class meeting in which your professor begins a new section, you should read each objective statement for that section. Next, browse through the objective material. The purpose of browsing through the material is to set the stage for your brain to accept and organize new information when it is presented to you. See *AIM for Success*, page xxvii.

To solve an absolute value equation

The *absolute value* of a number is its distance from zero on the number line. Distance is always a positive number or zero. Therefore, the absolute value of a number is always a positive number or zero.

The distance from 0 to 3 or from 0 to -3 is 3 units.

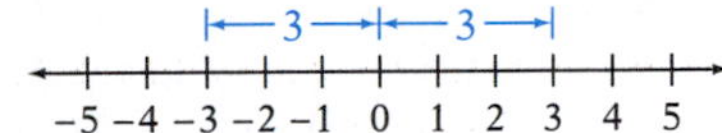

$|3| = 3$ $\qquad |-3| = 3$

Absolute value can be used to represent the distance between any two points on the number line. The **distance between two points** on the number line is the absolute value of the difference between the coordinates of the two points.

The distance between point a and point b is given by $|b - a|$.

The distance between 4 and -3 on the number line is 7 units. Note that the order in which the coordinates are subtracted does not affect the distance.

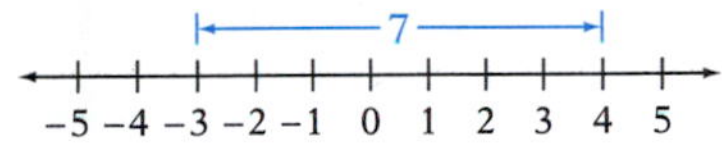

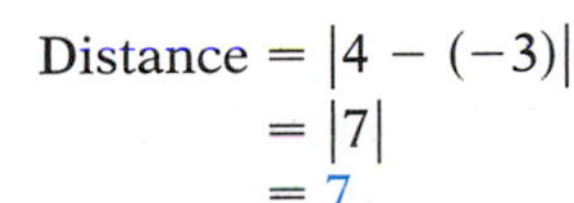

$$\text{Distance} = |-3 - 4| = |-7| = 7 \qquad \text{Distance} = |4 - (-3)| = |7| = 7$$

For any two numbers a and b, $|b - a| = |a - b|$.

An equation containing an absolute value symbol is called an **absolute value equation.** Here are three examples.

$$|x| = 3 \qquad |x + 2| = 8 \qquad |3x - 4| = 5x - 9$$

Solutions of an Absolute Value Equation

If $a \geq 0$ and $|x| = a$, then $x = a$ or $x = -a$.

For instance, given $|x| = 3$, then $x = 3$ or $x = -3$ because $|3| = 3$ and $|-3| = 3$. We can solve this equation as follows:

$|x| = 3$

$x = 3 \qquad x = -3$

• **Remove the absolute value sign from $|x|$ and let x equal 3 and the opposite of 3.**

Check:

$\|x\| = 3$		$\|x\| = 3$	
$\|3\|$	3	$\|-3\|$	3
$3 = 3$		$3 = 3$	

The solutions are 3 and -3.

Copyright © Houghton Mifflin Company. All rights reserved.

HOW TO Solve: $|x+2| = 8$

$$|x+2| = 8$$

$$x+2 = 8 \qquad x+2 = -8$$

$$x = 6 \qquad x = -10$$

- Remove the absolute value sign and rewrite as two equations.
- Solve each equation.

Check:

$$\begin{array}{c|c} |x+2| & 8 \\ \hline |6+2| & 8 \\ |8| & 8 \\ 8 = 8 & \end{array} \qquad \begin{array}{c|c} |x+2| & 8 \\ \hline |-10+2| & 8 \\ |-8| & 8 \\ 8 = 8 & \end{array}$$

The solutions are 6 and -10.

HOW TO Solve: $|5-3x| - 8 = -4$

$$|5-3x| - 8 = -4$$

$$|5-3x| = 4$$

$$5-3x = 4 \qquad 5-3x = -4$$

$$-3x = -1 \qquad -3x = -9$$

$$x = \frac{1}{3} \qquad x = 3$$

- Solve for the absolute value.
- Remove the absolute value sign and rewrite as two equations.
- Solve each equation.

Check:

$$\begin{array}{c|c} |5-3x| - 8 & -4 \\ \hline \left|5 - 3\left(\frac{1}{3}\right)\right| - 8 & -4 \\ |5-1| - 8 & -4 \\ 4 - 8 & -4 \\ -4 = -4 & \end{array} \qquad \begin{array}{c|c} |5-3x| - 8 & -4 \\ \hline |5 - 3(3)| - 8 & -4 \\ |5-9| - 8 & -4 \\ 4 - 8 & -4 \\ -4 = -4 & \end{array}$$

The solutions are $\frac{1}{3}$ and 3.

Example 1

Solve: $|2-x| = 12$

Solution

$$|2-x| = 12$$

$$2-x = 12 \qquad 2-x = -12$$

$$-x = 10 \qquad -x = -14$$

$$x = -10 \qquad x = 14$$

- Subtract 2.
- Multiply by -1.

The solutions are -10 and 14.

You Try It 1

Solve: $|2x-3| = 5$

Your solution

Example 2

Solve: $|2x| = -4$

Solution

$|2x| = -4$

There is no solution to this equation because the absolute value of a number must be nonnegative.

You Try It 2

Solve: $|x-3| = -2$

Your solution

Solutions on p. S8

Copyright © Houghton Mifflin Company. All rights reserved.

Example 3

Solve: $3 - |2x - 4| = -5$

Solution

$$3 - |2x - 4| = -5$$
$$-|2x - 4| = -8 \quad \text{• Subtract 3.}$$
$$|2x - 4| = 8 \quad \text{• Multiply by } -1.$$

$2x - 4 = 8$ | $2x - 4 = -8$
$2x = 12$ | $2x = -4$
$x = 6$ | $x = -2$

The solutions are 6 and -2.

You Try It 3

Solve: $5 - |3x + 5| = 3$

Your solution

Solution on p. S8

Objective B To solve an absolute value inequality

Recall that absolute value represents the distance between two points. For example, the solutions of the absolute value equation $|x - 1| = 3$ are the numbers whose distance from 1 is 3. Therefore, the solutions are -2 and 4. An **absolute value inequality** is an inequality that contains an absolute value symbol.

The solutions of the absolute value inequality $|x - 1| < 3$ are the numbers whose distance from 1 is less than 3. Therefore, the solutions are the numbers greater than -2 and less than 4. The solution set is $\{x \mid -2 < x < 4\}$.

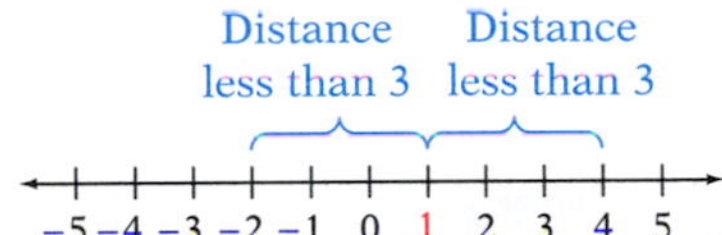

To solve an absolute value inequality of the form $|ax + b| < c$, solve the equivalent compound inequality $-c < ax + b < c$.

HOW TO Solve: $|3x - 1| < 5$

$$|3x - 1| < 5$$
$$-5 < 3x - 1 < 5 \quad \text{• Solve the equivalent compound inequality.}$$
$$-5 + 1 < 3x - 1 + 1 < 5 + 1$$
$$-4 < 3x < 6$$
$$\frac{-4}{3} < \frac{3x}{3} < \frac{6}{3}$$
$$-\frac{4}{3} < x < 2$$

The solution set is $\left\{x \mid -\frac{4}{3} < x < 2\right\}$.

The solutions of the absolute value inequality $|x + 1| > 2$ are the numbers whose distance from -1 is greater than 2. Therefore, the solutions are the numbers that are less than -3 or greater than 1. The solution set of $|x + 1| > 2$ is $\{x \mid x < -3 \text{ or } x > 1\}$.

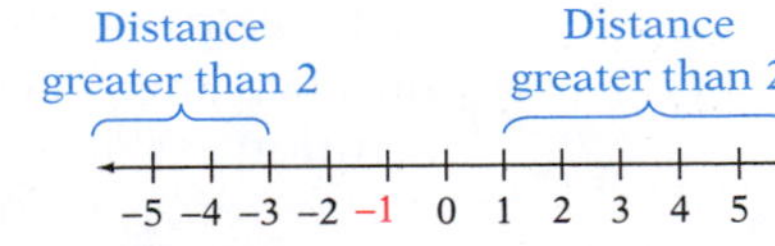

Copyright © Houghton Mifflin Company. All rights reserved.

TAKE NOTE

Carefully observe the difference between the solution method of $|ax + b| > c$ shown here and that of $|ax + b| < c$ shown on the preceding page.

To solve an absolute value inequality of the form $|ax + b| > c$, solve the equivalent compound inequality $ax + b < -c$ or $ax + b > c$.

HOW TO Solve: $|3 - 2x| > 1$

$$\begin{array}{lcl} 3 - 2x < -1 & \text{or} & 3 - 2x > 1 \\ -2x < -4 & & -2x > -2 \\ x > 2 & & x < 1 \\ \{x \mid x > 2\} & & \{x \mid x < 1\} \end{array}$$

• Solve each inequality.

The solution of a compound inequality with *or* is the union of the solution sets of the two inequalities.

$\{x \mid x > 2\} \cup \{x \mid x < 1\} = \{x \mid x > 2 \text{ or } x < 1\}$

The rules for solving these absolute value inequalities are summarized below.

Solutions of Absolute Value Inequalities

To solve an absolute value inequality of the form $|ax + b| < c$, $c > 0$, solve the equivalent compound inequality $-c < ax + b < c$.

To solve an absolute value inequality of the form $|ax + b| > c$, solve the equivalent compound inequality $ax + b < -c$ or $ax + b > c$.

Example 4 Solve: $|4x - 3| < 5$

Solution Solve the equivalent compound inequality.

$$-5 < 4x - 3 < 5$$
$$-5 + 3 < 4x - 3 + 3 < 5 + 3$$
$$-2 < 4x < 8$$
$$\frac{-2}{4} < \frac{4x}{4} < \frac{8}{4}$$
$$-\frac{1}{2} < x < 2$$
$$\left\{x \,\middle|\, -\frac{1}{2} < x < 2\right\}$$

You Try It 4 Solve: $|3x + 2| < 8$

Your solution

Example 5 Solve: $|x - 3| < 0$

Solution The absolute value of a number is greater than or equal to zero, since it measures the number's distance from zero on the number line. Therefore, the solution set of $|x - 3| < 0$ is the empty set.

You Try It 5 Solve: $|3x - 7| < 0$

Your solution

Solutions on p. S8

Copyright © Houghton Mifflin Company. All rights reserved.

Example 6 Solve: $|x + 4| > -2$

Solution The absolute value of a number is greater than or equal to zero. Therefore, the solution set of $|x + 4| > -2$ is the set of real numbers.

You Try It 6 Solve: $|2x + 7| \geq -1$

Your solution

Example 7 Solve: $|2x - 1| > 7$

Solution Solve the equivalent compound inequality.

$$\begin{aligned} 2x - 1 &< -7 \quad \text{or} \quad & 2x - 1 &> 7 \\ 2x &< -6 & 2x &> 8 \\ x &< -3 & x &> 4 \\ \{x \mid x &< -3\} & \{x \mid x &> 4\} \end{aligned}$$

$$\{x \mid x < -3\} \cup \{x \mid x > 4\} = \{x \mid x < -3 \text{ or } x > 4\}$$

You Try It 7 Solve: $|5x + 3| > 8$

Your solution

Solutions on p. S8

Objective C To solve application problems

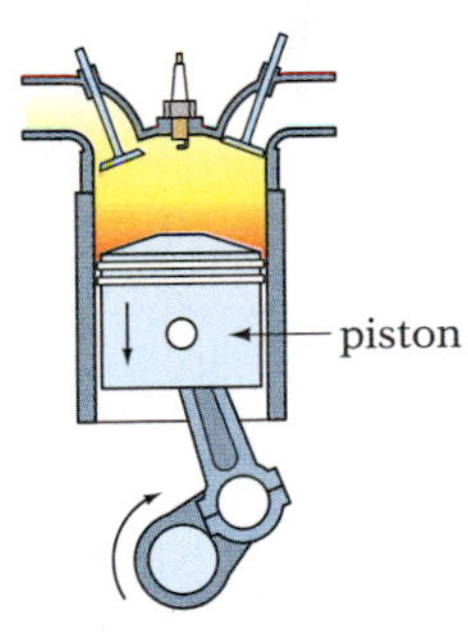

The **tolerance of a component,** or part, is the amount by which it is acceptable for the component to vary from a given measurement. For example, the diameter of a piston may vary from the given measurement of 9 cm by 0.001 cm. This is written 9 cm ± 0.001 cm and is read "9 centimeters plus or minus 0.001 centimeter." The maximum diameter, or **upper limit,** of the piston is 9 cm + 0.001 cm = 9.001 cm. The minimum diameter, or **lower limit,** is 9 cm − 0.001 cm = 8.999 cm.

The lower and upper limits of the diameter of the piston could also be found by solving the absolute value inequality $|d - 9| \leq 0.001$, where d is the diameter of the piston.

$$\begin{aligned} |d - 9| &\leq 0.001 \\ -0.001 \leq d - 9 &\leq 0.001 \\ -0.001 + 9 \leq d - 9 + 9 &\leq 0.001 + 9 \\ 8.999 \leq d &\leq 9.001 \end{aligned}$$

The lower and upper limits of the diameter of the piston are 8.999 cm and 9.001 cm.

Copyright © Houghton Mifflin Company. All rights reserved.

Example 8

The diameter of a piston for an automobile is $3\frac{5}{16}$ in. with a tolerance of $\frac{1}{64}$ in. Find the lower and upper limits of the diameter of the piston.

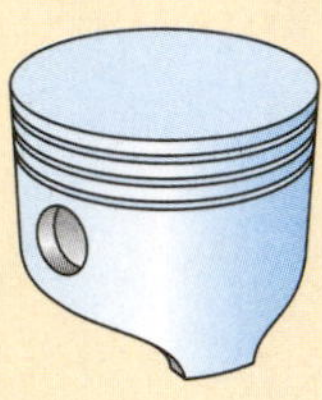

Strategy

To find the lower and upper limits of the diameter of the piston, let d represent the diameter of the piston, T the tolerance, and L the lower and upper limits of the diameter. Solve the absolute value inequality $|L - d| \le T$ for L.

Solution

$$|L - d| \le T$$

$$\left|L - 3\frac{5}{16}\right| \le \frac{1}{64} \qquad \bullet\ d = 3\frac{5}{16}$$

$$-\frac{1}{64} \le L - 3\frac{5}{16} \le \frac{1}{64}$$

$$-\frac{1}{64} + 3\frac{5}{16} \le L - 3\frac{5}{16} + 3\frac{5}{16} \le \frac{1}{64} + 3\frac{5}{16}$$

$$3\frac{19}{64} \le L \le 3\frac{21}{64}$$

The lower and upper limits of the diameter of the piston are $3\frac{19}{64}$ in. and $3\frac{21}{64}$ in.

You Try It 8

A machinist must make a bushing that has a tolerance of 0.003 in. The diameter of the bushing is 2.55 in. Find the lower and upper limits of the diameter of the bushing.

Your strategy

Your solution

Solution on p. S9

Copyright © Houghton Mifflin Company. All rights reserved.

2.6 Exercises

Objective A **To solve an absolute value equation**

1. Is 2 a solution of $|x - 8| = 6$?

2. Is -2 a solution of $|2x - 5| = 9$?

3. Is -1 a solution of $|3x - 4| = 7$?

4. Is 1 a solution of $|6x - 1| = -5$?

For Exercises 5 to 64, solve.

5. $|x| = 7$

6. $|a| = 2$

7. $|b| = 4$

8. $|c| = 12$

9. $|-y| = 6$

10. $|-t| = 3$

11. $|-a| = 7$

12. $|-x| = 3$

13. $|x| = -4$

14. $|y| = -3$

15. $|-t| = -3$

16. $|-y| = -2$

17. $|x + 2| = 3$

18. $|x + 5| = 2$

19. $|y - 5| = 3$

20. $|y - 8| = 4$

21. $|a - 2| = 0$

22. $|a + 7| = 0$

23. $|x - 2| = -4$

24. $|x + 8| = -2$

25. $|3 - 4x| = 9$

26. $|2 - 5x| = 3$

27. $|2x - 3| = 0$

28. $|5x + 5| = 0$

29. $|3x - 2| = -4$

30. $|2x + 5| = -2$

31. $|x - 2| - 2 = 3$

32. $|x - 9| - 3 = 2$

33. $|3a + 2| - 4 = 4$

34. $|2a + 9| + 4 = 5$

35. $|2 - y| + 3 = 4$

36. $|8 - y| - 3 = 1$

37. $|2x - 3| + 3 = 3$

38. $|4x - 7| - 5 = -5$

39. $|2x - 3| + 4 = -4$

40. $|3x - 2| + 1 = -1$

Copyright © Houghton Mifflin Company. All rights reserved.

41. $|6x - 5| - 2 = 4$

42. $|4b + 3| - 2 = 7$

43. $|3t + 2| + 3 = 4$

44. $|5x - 2| + 5 = 7$

45. $3 - |x - 4| = 5$

46. $2 - |x - 5| = 4$

47. $8 - |2x - 3| = 5$

48. $8 - |3x + 2| = 3$

49. $|2 - 3x| + 7 = 2$

50. $|1 - 5a| + 2 = 3$

51. $|8 - 3x| - 3 = 2$

52. $|6 - 5b| - 4 = 3$

53. $|2x - 8| + 12 = 2$

54. $|3x - 4| + 8 = 3$

55. $2 + |3x - 4| = 5$

56. $5 + |2x + 1| = 8$

57. $5 - |2x + 1| = 5$

58. $3 - |5x + 3| = 3$

59. $6 - |2x + 4| = 3$

60. $8 - |3x - 2| = 5$

61. $8 - |1 - 3x| = -1$

62. $3 - |3 - 5x| = -2$

63. $5 + |2 - x| = 3$

64. $6 + |3 - 2x| = 2$

Objective B To solve an absolute value inequality

For Exercises 65 to 94, solve.

65. $|x| > 3$

66. $|x| < 5$

67. $|x + 1| > 2$

68. $|x - 2| > 1$

69. $|x - 5| \le 1$

70. $|x - 4| \le 3$

71. $|2 - x| \ge 3$

72. $|3 - x| \ge 2$

73. $|2x + 1| < 5$

Copyright © Houghton Mifflin Company. All rights reserved.

74. $|3x - 2| < 4$

75. $|5x + 2| > 12$

76. $|7x - 1| > 13$

77. $|4x - 3| \leq -2$

78. $|5x + 1| \leq -4$

79. $|2x + 7| > -5$

80. $|3x - 1| > -4$

81. $|4 - 3x| \geq 5$

82. $|7 - 2x| > 9$

83. $|5 - 4x| \leq 13$

84. $|3 - 7x| < 17$

85. $|6 - 3x| \leq 0$

86. $|10 - 5x| \geq 0$

87. $|2 - 9x| > 20$

88. $|5x - 1| < 16$

89. $|2x - 3| + 2 < 8$

90. $|3x - 5| + 1 < 7$

91. $|2 - 5x| - 4 > -2$

92. $|4 - 2x| - 9 > -3$

93. $8 - |2x - 5| < 3$

94. $12 - |3x - 4| > 7$

Objective C To solve application problems

95. **Mechanics** The diameter of a bushing is 1.75 in. The bushing has a tolerance of 0.008 in. Find the lower and upper limits of the diameter of the bushing.

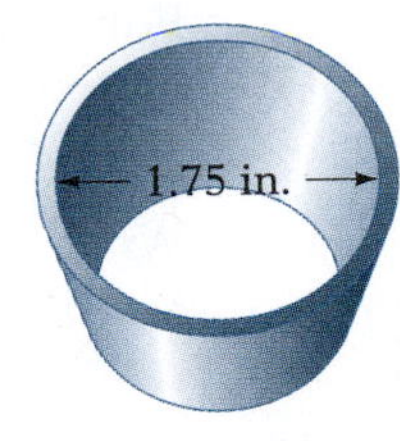

96. **Mechanics** A machinist must make a bushing that has a tolerance of 0.004 in. The diameter of the bushing is 3.48 in. Find the lower and upper limits of the diameter of the bushing.

97. **Appliances** An electric motor is designed to run on 220 volts plus or minus 25 volts. Find the lower and upper limits of voltage on which the motor will run.

Copyright © Houghton Mifflin Company. All rights reserved.

98. **Computers** A power strip is utilized on a computer to prevent the loss of programming by electrical surges. The power strip is designed to allow 110 volts plus or minus 16.5 volts. Find the lower and upper limits of voltage to the computer.

99. **Automobiles** A piston rod for an automobile is $9\frac{5}{8}$ in. long with a tolerance of $\frac{1}{32}$ in. Find the lower and upper limits of the length of the piston rod.

100. **Automobiles** A piston rod for an automobile is $9\frac{3}{8}$ in. long with a tolerance of $\frac{1}{64}$ in. Find the lower and upper limits of the length of the piston rod.

Electronics The tolerance of the resistors used in electronics is given as a percent. Use your calculator for Exercises 101 to 104.

101. Find the lower and upper limits of a 29,000-ohm resistor with a 2% tolerance.

102. Find the lower and upper limits of a 15,000-ohm resistor with a 10% tolerance.

103. Find the lower and upper limits of a 25,000-ohm resistor with a 5% tolerance.

104. Find the lower and upper limits of a 56-ohm resistor with a 5% tolerance.

APPLYING THE CONCEPTS

105. For what values of the variable is the equation true?
a. $|x + 3| = x + 3$ **b.** $|a - 4| = 4 - a$

106. Write an absolute value inequality to represent all real numbers within 5 units of 2.

107. Replace the question mark with $\leq$, $\geq$, or $=$.
a. $|x + y| \ ? \ |x| + |y|$ **b.** $|x - y| \ ? \ |x| - |y|$
c. $||x| - |y|| \ ? \ |x| - |y|$ **d.** $\left|\frac{x}{y}\right| \ ? \ \frac{|x|}{|y|}, y \neq 0$
e. $|xy| \ ? \ |x||y|$

108. Let $|x| \leq 2$ and $|3x - 2| \leq a$. Find the smallest possible value of a.

Copyright © Houghton Mifflin Company. All rights reserved.

Focus on Problem Solving

Trial-and-Error Approach to Problem Solving

The questions below require an answer of always true, sometimes true, or never true. These problems are best solved by the trial-and-error method. The trial-and-error method of arriving at a solution to a problem involves repeated tests or experiments.

For example, consider the statement

> Both sides of an equation can be divided by the same number without changing the solution of the equation.

The solution of the equation $6x = 18$ is 3. If we divide both sides of the equation by 2, the result is $3x = 9$ and the solution is still 3. So the answer "never true" has been eliminated. We still need to determine whether there is a case for which the statement is not true. Can we divide both sides of the equation by some number and get an equation for which the solution is not 3? If we divide both sides of the equation by 0, the result is $\frac{6x}{0} = \frac{18}{0}$; the solution of this equation is not 3 because the expressions on either side of the equals sign are undefined. Thus the statement is true for some numbers and not true for 0. The statement is sometimes true.

For Exercises 1 to 10, determine whether the statement is always true, sometimes true, or never true.

1. Both sides of an equation can be multiplied by the same number without changing the solution of the equation.

2. For an equation of the form $ax = b$, $a \neq 0$, multiplying both sides of the equation by the reciprocal of a will result in an equation of the form $x = constant$.

3. The Multiplication Property of Equations is used to remove a term from one side of an equation.

4. Adding -3 to each side of an equation yields the same result as subtracting 3 from each side of the equation.

5. An equation contains an equals sign.

6. The same variable term can be added to both sides of an equation without changing the solution of the equation.

7. An equation of the form $ax + b = c$ cannot be solved if a is a negative number.

8. The solution of the equation $\frac{x}{0} = 0$ is 0.

9. An even integer is a multiple of 2.

10. In solving an equation of the form $ax + b = cx + d$, subtracting cx from each side of the equation results in an equation with only one variable term in it.

Copyright © Houghton Mifflin Company. All rights reserved.

Projects and Group Activities

Water Displacement

When an object is placed in water, the object displaces an amount of water that is equal to the volume of the object.

HOW TO A sphere with a diameter of 4 in. is placed in a rectangular tank of water that is 6 in. long and 5 in. wide. How much does the water level rise? Round to the nearest hundredth.

$V = \frac{4}{3}\pi r^3$ • Use the formula for the volume of a sphere.

$V = \frac{4}{3}\pi(2^3) = \frac{32}{3}\pi$ • $r = \frac{1}{2}d = \frac{1}{2}(4) = 2$

Let x represent the amount of the rise in water level. The volume of the sphere will equal the volume displaced by the water. As shown at the left, this volume is the rectangular solid with width 5 in., length 6 in., and height x in.

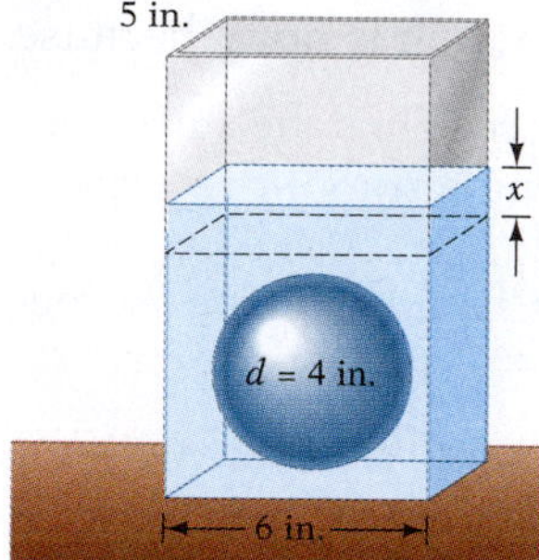

$V = LWH$ • Use the formula for the volume of a rectangular solid.

$\frac{32}{3}\pi = (6)(5)x$ • Substitute $\frac{32}{3}\pi$ for V, 5 for W, and 6 for L.

$\frac{32}{90}\pi = x$ • The exact height that the water will fill is $\frac{32}{90}\pi = \frac{16}{45}\pi$.

$1.12 \approx x$ • Use a calculator to find an approximation.

The water will rise approximately 1.12 in.

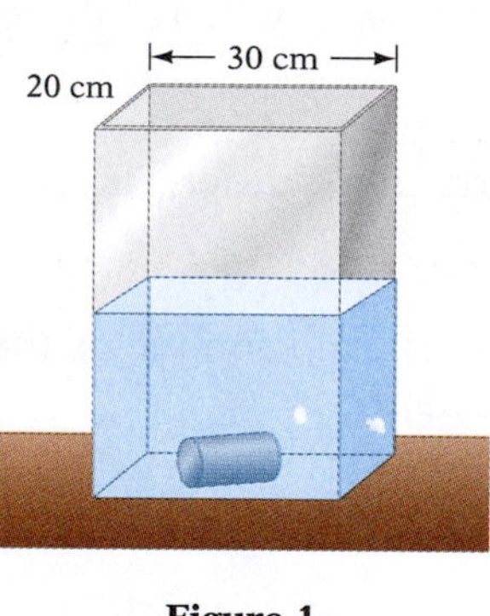

Figure 1

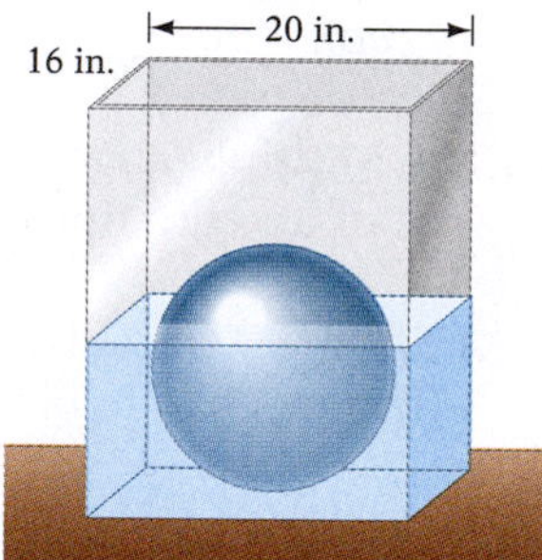

Figure 2

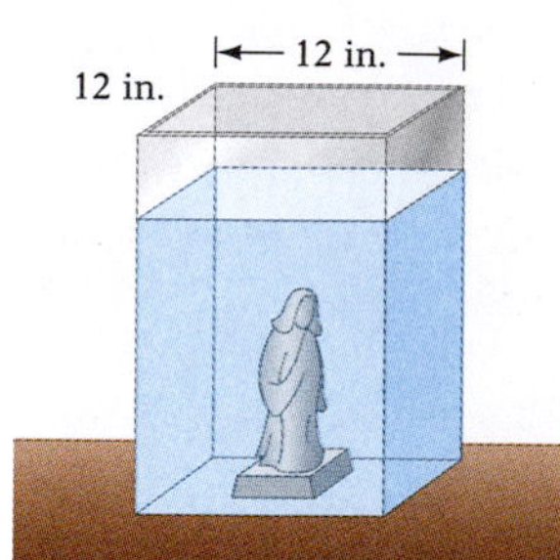

Figure 3

1. A cylinder with a 2-centimeter radius and a height of 10 cm is submerged in a tank of water that is 20 cm wide and 30 cm long (see Figure 1). How much does the water level rise? Round to the nearest hundredth.

2. A sphere with a radius of 6 in. is placed in a rectangular tank of water that is 16 in. wide and 20 in. long (see Figure 2). The sphere displaces water until two-thirds of the sphere is submerged. How much does the water level rise? Round to the nearest hundredth.

3. A chemist wants to know the density of a statue that weighs 15 lb. The statue is placed in a rectangular tank of water that is 12 in. long and 12 in. wide (see Figure 3). The water level rises 0.42 in. Find the density of the statue. Round to the nearest hundredth. (*Hint:* Density = weight ÷ volume)

Copyright © Houghton Mifflin Company. All rights reserved.

Chapter 2 Summary

Key Words	Examples
An *equation* expresses the equality of two mathematical expressions. [2.1A, p. 73]	$3 + 2(4x - 5) = x + 4$ is an equation.
A *solution of an equation* is a number that, when substituted for the variable, results in a true equation. [2.1A, p. 73]	-2 is a solution of $2 - 3x = 8$ because $2 - 3(-2) = 8$ is a true equation.
To *solve an equation* means to find a solution of the equation. The goal is to rewrite the equation in the form *variable = constant*, because the constant is the solution. [2.1B, p. 74]	The equation $x = -3$ is in the form *variable = constant*. The constant, -3, is the solution of the equation.
Equivalent equations are equations that have the same solution. [2.1B, p. 74]	$x + 3 = 7$ and $x = 4$ are equivalent equations because the solution of each equation is 4.
Multiplying each side of an equation that contains fractions by the LCM of the denominators is called *clearing denominators*. [2.2A, p. 92]	We can clear the denominators from the equation $\frac{2}{3}x - \frac{1}{2} = \frac{3}{4}$ by multiplying each side of the equation by 12.
Consecutive integers follow one another in order. [2.3A, p. 106]	5, 6, 7 are consecutive integers. $-9, -8, -7$ are consecutive integers.
The *solution set of an inequality* is a set of numbers, each element of which, when substituted in the inequality, results in a true inequality. [2.5A, p. 125]	Any number greater than 4 is a solution of the inequality $x > 4$.
A *compound inequality* is formed by joining two inequalities with a connective word such as *and* or *or*. [2.5B, p. 128]	$3x > 6$ and $2x + 5 < 7$ $2x + 1 < 3$ or $x + 2 > 4$
An *absolute value equation* is an equation that contains an absolute value symbol. [2.6A, p. 137]	$\|x - 2\| = 3$
An *absolute value inequality* is an inequality that contains an absolute value symbol. [2.6B, p. 139]	$\|x - 4\| < 5$ $\|2x - 3\| > 6$
The *tolerance* of a component or part is the amount by which it is acceptable for the component to vary from a given measurement. The maximum measurement is the *upper limit*. The minimum measurement is the *lower limit*. [2.6C, p. 141]	The diameter of a bushing is 1.5 in. with a tolerance of 0.005 in. The lower and upper limits of the diameter of the bushing are 1.5 in. ± 0.005 in.

Copyright © Houghton Mifflin Company. All rights reserved.

Essential Rules and Procedures

Examples

Addition Property of Equations [2.1B, p. 74]
The same number can be added to each side of an equation without changing the solution of the equation.

If $a = b$, then $a + c = b + c$.

$$x + 5 = -3$$
$$x + 5 - 5 = -3 - 5$$
$$x = -8$$

Multiplication Property of Equations [2.1C, p. 75]
Each side of an equation can be multiplied by the same *nonzero* number without changing the solution of the equation.

If $a = b$ and $c \neq 0$, then $ac = bc$.

$$\frac{2}{3}x = 4$$
$$\left(\frac{3}{2}\right)\left(\frac{2}{3}x\right) = \left(\frac{3}{2}\right)4$$
$$x = 6$$

Basic Percent Equation [2.1D, p. 77]
Percent · Base = Amount
$P \cdot B = A$

30% of what number is 24?

$$PB = A$$
$$0.30B = 24$$
$$\frac{0.30B}{0.30} = \frac{24}{0.30}$$
$$B = 80$$

Simple Interest Equation [2.1D, p. 78]
Interest = Principle · Rate · Time
$I = Prt$

A credit card company charges an annual interest rate of 21% on the monthly unpaid balance on a card. Find the amount of interest charged on an unpaid balance of \$232 for April.

$$I = Prt$$
$$I = 232(0.21)\left(\frac{1}{12}\right) = 4.06$$

The interest charged is \$4.06.

Consecutive Integers [2.3A, p. 106]
$n, n + 1, n + 2, \ldots$

The sum of three consecutive integers is 33.

$$n + (n + 1) + (n + 2) = 33$$

Consecutive Even or Consecutive Odd Integers [2.3A, p. 106]
$n, n + 2, n + 4, \ldots$

The sum of three consecutive odd integers is 33.

$$n + (n + 2) + (n + 4) = 33$$

Value Mixture Equation [2.4A, p. 113]
Amount · Unit Cost = Value
$AC = V$

A merchant combines coffee that costs \$6 per pound with coffee that costs \$3.20 per pound. How many pounds of each should be used to make 60 lb of a blend that costs \$4.50 per pound?

$$6x + 3.20(60 - x) = 4.50(60)$$

Copyright © Houghton Mifflin Company. All rights reserved.

Percent Mixture Problems [2.4B, p. 115]

Amount of solution · Percent of concentration = Quantity of substance

$$Ar = Q$$

A silversmith mixed 120 oz of an 80% silver alloy with 240 oz of a 30% silver alloy. Find the percent concentration of the resulting silver alloy.

$$0.80(120) + 0.30(240) = x(360)$$

Uniform Motion Equation [2.4C, p. 117]

Rate · Time = Distance

$$rt = d$$

Two planes are 1640 mi apart and are traveling toward each other. One plane is traveling 60 mph faster than the other plane. The planes meet in 2 h. Find the speed of each plane.

$$2r + 2(r + 60) = 1640$$

Addition Property of Inequalities [2.5A, p. 125]

If $a > b$, then $a + c > b + c$.
If $a < b$, then $a + c < b + c$.

$$x + 3 > -2$$
$$x + 3 - 3 > -2 - 3$$
$$x > -5$$

Multiplication Property of Inequalities [2.5A, p. 126]

Rule 1 If $a > b$ and $c > 0$, then $ac > bc$.
If $a < b$ and $c > 0$, then $ac < bc$.

$$\frac{3}{4}x > 12$$
$$\frac{4}{3}\left(\frac{3}{4}x\right) > \frac{4}{3}(12)$$
$$x > 16$$

Rule 2 If $a > b$ and $c < 0$, then $ac < bc$.
If $a < b$ and $c < 0$, then $ac > bc$.

$$-2x < 8$$
$$\frac{-2x}{-2} > \frac{8}{-2}$$
$$x > -4$$

Solutions of an Absolute Value Equation [2.6A, p. 137]

If $a \geq 0$ and $|x| = a$, then $x = a$ or $x = -a$.

$$|x - 3| = 7$$
$$x - 3 = 7 \qquad x - 3 = -7$$
$$x = 10 \qquad x = -4$$

Solutions of Absolute Value Inequalities [2.6B, p. 140]

To solve an absolute value inequality of the form $|ax + b| < c$, $c > 0$, solve the equivalent compound inequality $-c < ax + b < c$.

$$|x - 5| < 9$$
$$-9 < x - 5 < 9$$
$$-9 + 5 < x - 5 + 5 < 9 + 5$$
$$-4 < x < 14$$

To solve an absolute value inequality of the form $|ax + b| > c$, solve the equivalent compound inequality $ax + b < -c$ or $ax + b > c$.

$$|x - 5| > 9$$
$$x - 5 < -9 \quad \text{or} \quad x - 5 > 9$$
$$x < -4 \quad \text{or} \quad x > 14$$

Copyright © Houghton Mifflin Company. All rights reserved.

Chapter 2 Review Exercises

1. Solve: $3t - 3 + 2t = 7t - 15$

2. Solve: $3x - 7 > -2$

3. Is 3 a solution of $5x - 2 = 4x + 5$?

4. Solve: $x + 4 = -5$

5. Solve: $3x < 4$ and $x + 2 > -1$

6. Solve: $\frac{3}{5}x - 3 = 2x + 5$

7. Solve: $-\frac{2}{3}x = \frac{4}{9}$

8. Solve: $|x - 4| - 8 = -3$

9. Solve: $|2x - 5| < 3$

10. Solve: $\frac{2x - 3}{3} + 2 = \frac{2 - 3x}{5}$

11. Solve: $2(a - 3) = 5(4 - 3a)$

12. Solve: $5x - 2 > 8$ or $3x + 2 < -4$

13. Solve: $|4x - 5| \geq 3$

14. 30 is what percent of 12?

15. Solve: $\frac{1}{2}x - \frac{5}{8} = \frac{3}{4}x + \frac{3}{2}$

16. Solve: $6 + |3x - 3| = 2$

17. Solve: $3x - 2 > x - 4$ or $7x - 5 < 3x + 3$

18. Solve: $2x - (3 - 2x) = 4 - 3(4 - 2x)$

Copyright © Houghton Mifflin Company. All rights reserved.

19. Solve: $x + 9 = -6$

20. Solve: $\frac{2}{3} = x + \frac{3}{4}$

21. Solve: $-3x = -21$

22. Solve: $\frac{2}{3}a = \frac{4}{9}$

23. Solve: $3y - 5 = 3 - 2y$

24. Solve: $4x - 5 + x = 6x - 8$

25. Solve: $3(x - 4) = -5(6 - x)$

26. Solve: $\frac{3x - 2}{4} + 1 = \frac{2x - 3}{2}$

27. Solve: $5x - 8 < -3$

28. Solve: $2x - 9 \leq 8x + 15$

29. Solve: $\frac{2}{3}x - \frac{5}{8} \geq \frac{3}{4}x + 1$

30. Solve: $2 - 3(2x - 4) \leq 4x - 2(1 - 3x)$

31. Solve: $-5 < 4x - 1 < 7$

32. Solve: $|2x - 3| = 8$

33. Solve: $|5x + 8| = 0$

34. Solve: $|5x - 4| < -2$

35. **Uniform Motion** A ferry leaves a dock and travels to an island at an average speed of 16 mph. On the return trip, the ferry travels at an average speed of 12 mph. The total time for the trip is $2\frac{1}{3}$ h. How far is the island from the dock?

Copyright © Houghton Mifflin Company. All rights reserved.

36. **Mixtures** A grocer mixed apple juice that costs \$4.20 per gallon with 40 gal of cranberry juice that costs \$6.50 per gallon. How much apple juice was used to make cranapple juice costing \$5.20 per gallon?

37. **Compensation** A sales executive earns \$800 per month plus 4% commission on the amount of sales. The executive's goal is to earn \$3000 per month. What amount of sales will enable the executive to earn \$3000 or more per month?

38. **Integers** Translate "four less than the product of five and a number is sixteen" into an equation and solve.

39. **Mechanics** The diameter of a bushing is 2.75 in. The bushing has a tolerance of 0.003 in. Find the lower and upper limits of the diameter of the bushing.

40. **Integers** The sum of two integers is twenty. Five times the smaller integer is two more than twice the larger integer. Find the two integers.

41. **Education** An average score of 80 to 90 in a psychology class receives a B grade. A student has scores of 92, 66, 72, and 88 on four tests. Find the range of scores on the fifth test that will give the student a B for the course.

42. **Uniform Motion** Two planes are 1680 mi apart and are traveling toward each other. One plane is traveling 80 mph faster than the other plane. The planes meet in 1.75 h. Find the speed of each plane.

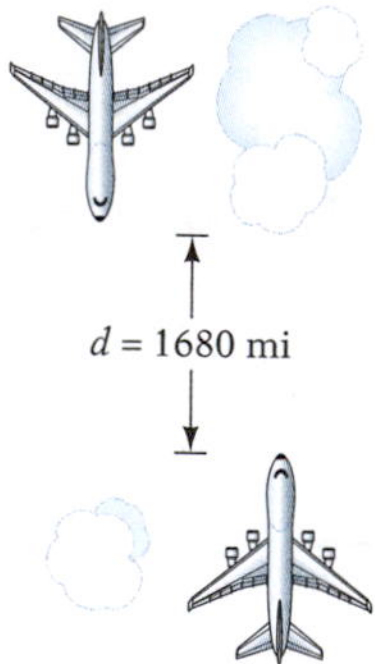

43. **Mixtures** An alloy containing 30% tin is mixed with an alloy containing 70% tin. How many pounds of each were used to make 500 lb of an alloy containing 40% tin?

44. **Automobiles** A piston rod for an automobile is $10\frac{3}{8}$ in. long with a tolerance of $\frac{1}{32}$ in. Find the lower and upper limits of the length of the piston rod.

Copyright © Houghton Mifflin Company. All rights reserved.

Chapter 2 Test

1. Solve: $x - 2 = -4$

2. Solve: $b + \frac{3}{4} = \frac{5}{8}$

3. Solve: $-\frac{3}{4}y = -\frac{5}{8}$

4. Solve: $3x - 5 = 7$

5. Solve: $\frac{3}{4}y - 2 = 6$

6. Solve: $2x - 3 - 5x = 8 + 2x - 10$

7. Solve: $2[a - (2 - 3a) - 4] = a - 5$

8. Is -2 a solution of $x^2 - 3x = 2x - 6$?

9. Solve: $\frac{2x + 1}{3} - \frac{3x + 4}{6} = \frac{5x - 9}{9}$

10. Solve: $3x - 2 \geq 6x + 7$

11. What is 0.5% of 8?

12. Solve: $4x - 1 > 5$ or $2 - 3x < 8$

13. Solve: $4 - 3x \geq 7$ and $2x + 3 \geq 7$

14. Solve: $|3 - 5x| = 12$

15. Solve: $2 - |2x - 5| = -7$

16. Solve: $|3x - 5| \leq 4$

17. Solve: $|4x - 3| > 5$

Copyright © Houghton Mifflin Company. All rights reserved.

18. **Consumerism** Gambelli Agency rents cars for \$12 per day plus 10¢ for every mile driven. McDougal Rental rents cars for \$24 per day with unlimited mileage. How many miles a day can you drive a Gambelli Agency car if it is to cost you less than a McDougal Rental car?

19. **Mechanics** A machinist must make a bushing that has a tolerance of 0.002 in. The diameter of the bushing is 2.65 in. Find the lower and upper limits of the diameter of the bushing.

20. **Integers** The sum of two integers is fifteen. Eight times the smaller integer is one less than three times the larger integer. Find the integers.

21. **Mixtures** How many gallons of water must be mixed with 5 gal of a 20% salt solution to make a 16% salt solution?

22. **Mixtures** A butcher combines 100 lb of hamburger that costs \$2.10 per pound with 60 lb of hamburger that costs \$3.70 per pound. Find the cost of the hamburger mixture.

23. **Uniform Motion** A jogger runs a distance at a speed of 8 mph and returns the same distance running at a speed of 6 mph. Find the total distance that the jogger ran if the total time running was 1 hour and 45 minutes.

24. **Uniform Motion** Two trains are 250 mi apart and are traveling toward each other. One train is traveling 5 mph faster than the other train. The trains pass each other in 2 h. Find the speed of each train.

25. **Mixtures** How many ounces of pure water must be added to 60 oz of an 8% salt solution to make a 3% salt solution?

Copyright © Houghton Mifflin Company. All rights reserved.

Cumulative Review Exercises

1. Subtract: $-6 - (-20) - 8$

2. Multiply: $(-2)(-6)(-4)$

3. Subtract: $-\frac{5}{6} - \left(-\frac{7}{16}\right)$

4. Simplify: $-4^2 \cdot \left(-\frac{3}{2}\right)^3$

5. Simplify: $\frac{5}{8} - \left(\frac{1}{2}\right)^2 \div \left(\frac{1}{3} - \frac{3}{4}\right)$

6. Evaluate $3(a - c) - 2ab$ when $a = 2$, $b = 3$, and $c = -4$.

7. Simplify: $3x - 8x + (-12x)$

8. Simplify: $2a - (-b) - 7a - 5b$

9. Simplify: $(16x)\left(\frac{1}{8}\right)$

10. Simplify: $-4(-9y)$

11. Simplify: $-2(-x^2 - 3x + 2)$

12. Simplify: $-2(x - 3) + 2(4 - x)$

13. Simplify: $-3[2x - 4(x - 3)] + 2$

14. Find $A \cap B$ given $A = \{-4, -2, 0, 2\}$ and $B = \{-4, 0, 4, 8\}$.

15. Graph: $\{x \mid x < 3\} \cap \{x \mid x > -2\}$

−5 −4 −3 −2 −1 0 1 2 3 4 5

16. Is -3 a solution of $x^2 + 6x + 9 = x + 3$?

17. Solve: $\frac{3}{5}x = -15$

18. Solve: $7x - 8 = -29$

Copyright © Houghton Mifflin Company. All rights reserved.

19. Solve: $13 - 9x = -14$

20. Solve: $5x - 8 = 12x + 13$

21. Solve: $11 - 4x = 2x + 8$

22. Solve: $8x - 3(4x - 5) = -2x - 11$

23. Solve: $3 - 2(2x - 1) \geq 3(2x - 2) + 1$

24. Solve: $3x + 2 \leq 5$ and $x + 5 \geq 1$

25. Solve: $|3 - 2x| = 5$

26. Solve: $|3x - 1| > 5$

27. Write 55% as a fraction.

28. Write 1.03 as a percent.

29. 25% of what number is 30?

30. **Integers** Translate "the sum of six times a number and thirteen is five less than the product of three and the number" into an equation and solve.

31. **Mixtures** How many pounds of an oat flour that costs \$.80 per pound must be mixed with 40 lb of a wheat flour that costs \$.50 per pound to make a blend that costs \$.60 per pound?

32. **Mixtures** How many grams of pure gold must be added to 100 g of a 20% gold alloy to make an alloy that is 36% gold?

33. **Uniform Motion** A sprinter ran to the end of a track at an average rate of 8 m/s and then jogged back to the starting point at an average rate of 3 m/s. The sprinter took 55 s to run to the end of the track and jog back. Find the length of the track.

Copyright © Houghton Mifflin Company. All rights reserved.

chapter 3

Geometry

This is an aerial view of the house of William Paca, who was a Maryland Patriot and a signer of the Declaration of Independence. The house's large, formal garden has been restored to its original splendor. The best way to appreciate the shapes sculpted in the garden is to view it from above, like in this photo. Each geometric shape combines with the others to form the entire garden. **Exercise 98 on page 193** shows you how to use a geometric formula first to determine the size of an area, and then to calculate how much grass seed is needed for an area of that size.

Need help? For online student resources, such as section quizzes, visit this textbook's website at **math.college.hmco.com/students.**

OBJECTIVES

Section 3.1

A To solve problems involving lines and angles

B To solve problems involving angles formed by intersecting lines

C To solve problems involving the angles of a triangle

Section 3.2

A To solve problems involving the perimeter of a geometric figure

B To solve problems involving the area of a geometric figure

Section 3.3

A To solve problems involving the volume of a solid

B To solve problems involving the surface area of a solid

Copyright © Houghton Mifflin Company. All rights reserved.

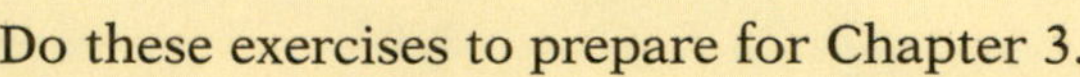

PREP TEST

Do these exercises to prepare for Chapter 3.

1. Solve: $x + 47 = 90$

2. Solve: $32 + (97 + x) = 180$

3. Simplify: $2(18) + 2(10)$

4. Evaluate abc when $a = 2$, $b = 3.14$, and $c = 9$.

5. Evaluate xyz^3 when $x = \frac{4}{3}$, $y = 3.14$, and $z = 3$.

6. Evaluate $\frac{1}{2}a(b + c)$ when $a = 6$, $b = 25$, and $c = 15$.

GO FIGURE

In a school election, one candidate for class president received more than 94%, but less than 100%, of the votes cast. What is the least possible number of votes cast?

Copyright © Houghton Mifflin Company. All rights reserved.

3.1 Introduction to Geometry

Objective A **To solve problems involving lines and angles**

Study Tip

Before you begin a new chapter, you should take some time to review previously learned skills. One way to do this is to complete the Prep Test. See page 160. This test focuses on the particular skills that will be required for the new chapter.

The word *geometry* comes from the Greek words for "earth" and "measure." The original purpose of geometry was to measure land. Today geometry is used in many fields, such as physics, medicine, and geology, and is applied in such areas as mechanical drawing and astronomy. Geometric forms are also used in art and design.

Three basic concepts of geometry are the point, line, and plane. A **point** is symbolized by drawing a dot. A **line** is determined by two distinct points and extends indefinitely in both directions, as the arrows on the line shown at the right indicate. This line contains points A and B and is represented by $\overleftrightarrow{AB}$. A line can also be represented by a single letter, such as ℓ.

A **ray** starts at a point and extends indefinitely in *one* direction. The point at which a ray starts is called the **endpoint** of the ray. The ray shown at the right is denoted by $\overrightarrow{AB}$. Point A is the endpoint of the ray.

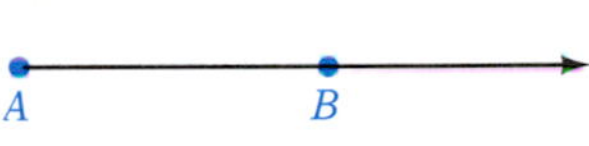

A **line segment** is part of a line and has two endpoints. The line segment shown at the right is denoted by $\overline{AB}$.

The distance between the endpoints of $\overline{AC}$ is denoted by AC. If B is a point on $\overline{AC}$, then AC (the distance from A to C) is the sum of AB (the distance from A to B) and BC (the distance from B to C).

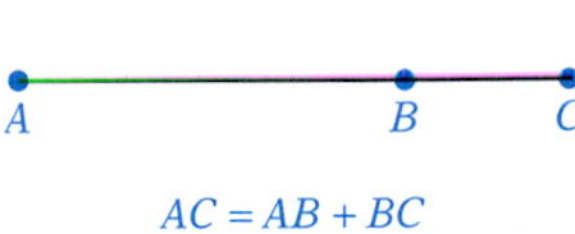

HOW TO Given $AB = 22$ cm and $AC = 31$ cm, find BC.

$AC = AB + BC$ • Write an equation for the distances between points on the line segment.

$31 = 22 + BC$ • Substitute the given distances for AB and AC into the equation.

$9 = BC$ • Solve for BC.

$BC = 9$ cm

In this section we will be discussing figures that lie in a plane. A **plane** is a flat surface and can be pictured as a table top or blackboard that extends in all directions. Figures that lie in a plane are called **plane figures.**

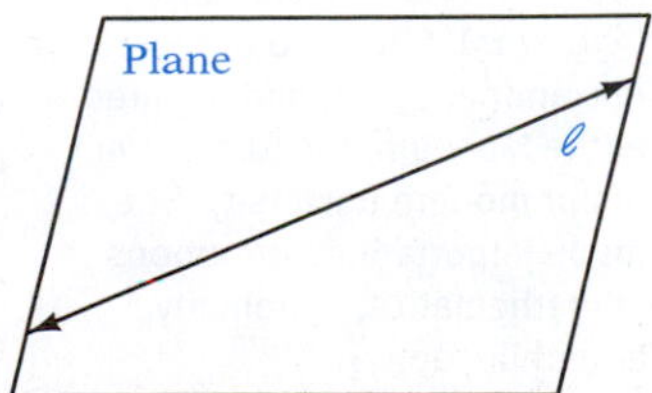

Copyright © Houghton Mifflin Company. All rights reserved.

Point of Interest

Geometry is one of the oldest branches of mathematics. Around 350 B.C., the Greek mathematician Euclid wrote the *Elements*, which contained all of the known concepts of geometry. Euclid's contribution was to unify various concepts into a single deductive system that was based on a set of axioms.

Lines in a plane can be intersecting or parallel. **Intersecting lines** cross at a point in the plane. **Parallel lines** never meet. The distance between them is always the same.

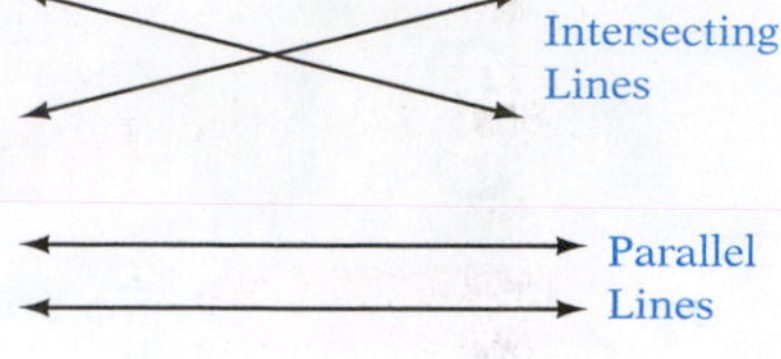

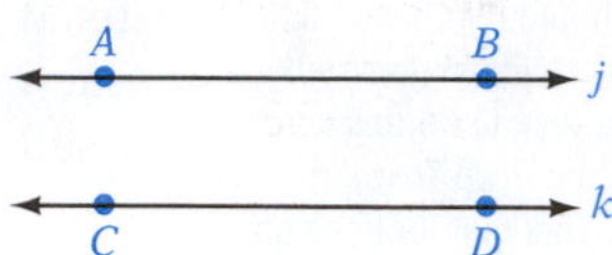

The symbol $\parallel$ means "is parallel to." In the figure at the right, $j \parallel k$ and $\overline{AB} \parallel \overline{CD}$. Note that j contains $\overline{AB}$ and k contains $\overline{CD}$. Parallel lines contain parallel line segments.

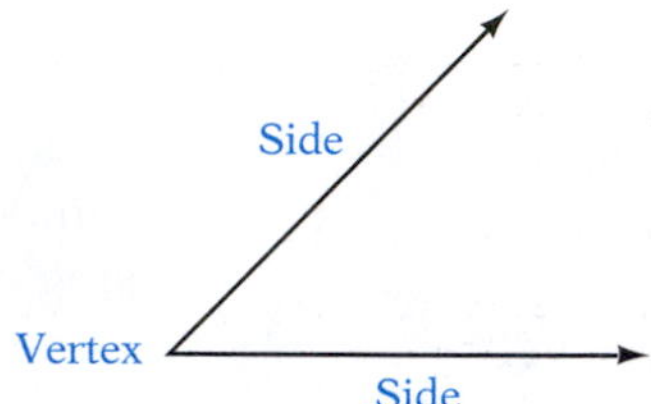

An **angle** is formed by two rays with the same endpoint. The **vertex** of the angle is the point at which the two rays meet. The rays are called the **sides** of the angle.

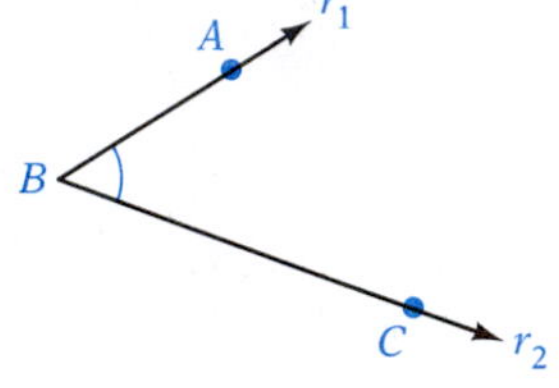

If A and C are points on rays r_1 and r_2, and B is the vertex, then the angle is called $\angle B$ or $\angle ABC$, where $\angle$ is the symbol for angle. Note that either the angle is named by the vertex, or the vertex is the second point listed when the angle is named by giving three points. $\angle ABC$ could also be called $\angle CBA$.

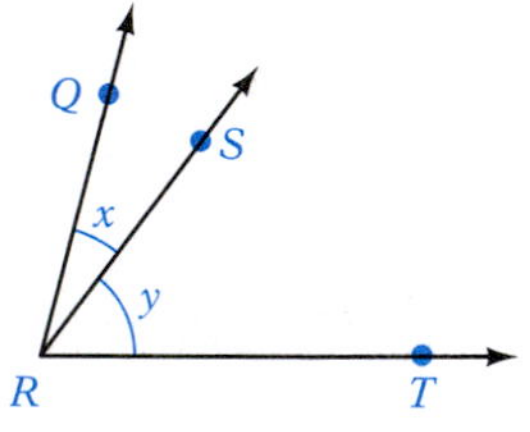

An angle can also be named by a variable written between the rays close to the vertex. In the figure at the right, $\angle x = \angle QRS$ and $\angle y = \angle SRT$. Note that in this figure, more than two rays meet at R. In this case, the vertex cannot be used to name an angle.

Point of Interest

The first woman mathematician for whom documented evidence exists is Hypatia (370–415). She lived in Alexandria, Egypt, and lectured at the Museum, the forerunner of our modern university. She made important contributions in mathematics, astronomy, and philosophy.

An angle is measured in **degrees.** The symbol for degrees is a small raised circle, °. Probably because early Babylonians believed that Earth revolves around the sun in approximately 360 days, the angle formed by a circle has a measure of 360° (360 degrees).

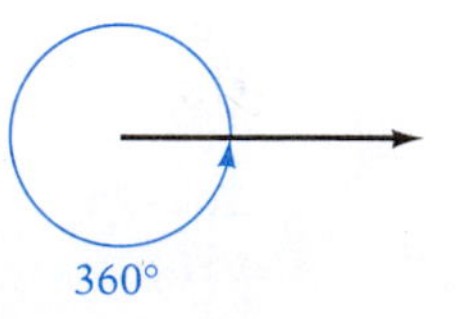

Copyright © Houghton Mifflin Company. All rights reserved.

A **protractor** is used to measure an angle. Place the center of the protractor at the vertex of the angle with the edge of the protractor along a side of the angle. The angle shown in the figure below measures 58°.

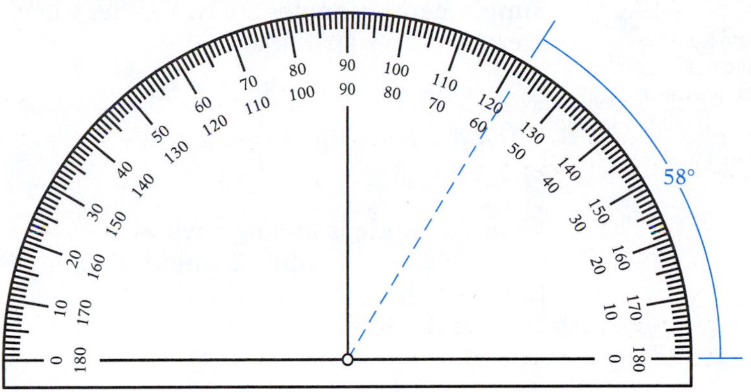

TAKE NOTE
The corner of a page of this book is a good example of a 90° angle.

A 90° angle is called a **right angle.** The symbol ∟ represents a right angle.

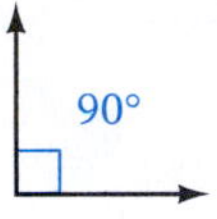

Perpendicular lines are intersecting lines that form right angles.

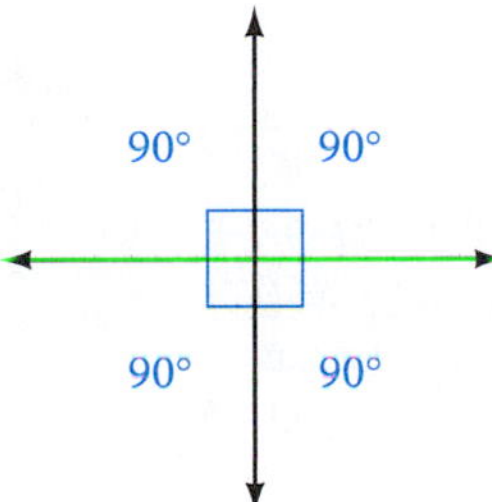

The symbol ⊥ means "is perpendicular to." In the figure at the right, $p \perp q$ and $\overline{AB} \perp \overline{CD}$. Note that line p contains $\overline{AB}$ and line q contains $\overline{CD}$. Perpendicular lines contain perpendicular line segments.

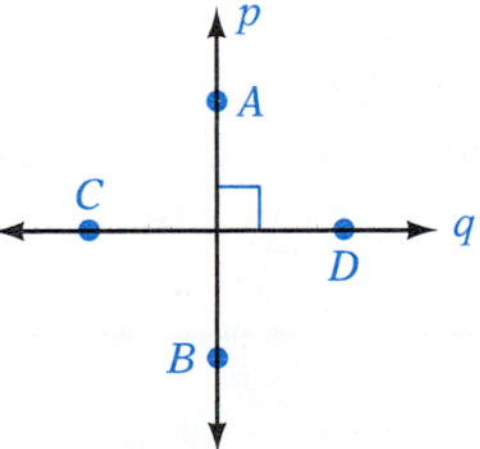

Complementary angles are two angles whose measures have the sum 90°.

$$\angle A + \angle B = 70° + 20° = 90°$$

$\angle A$ and $\angle B$ are complementary angles.

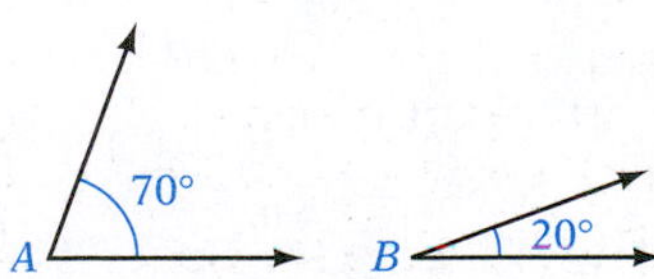

Copyright © Houghton Mifflin Company. All rights reserved.

Study Tip

A great many new vocabulary words are introduced in this chapter. All of these terms are in **bold type.** The bold type indicates that these are concepts you must know to learn the material. Be sure to study each new term as it is presented.

A 180° angle is called a **straight angle.**

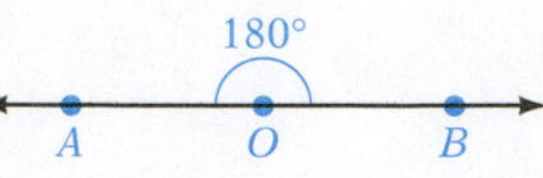

$\angle AOB$ is a straight angle.

Supplementary angles are two angles whose measures have the sum 180°.

130°
A
50°
B

$$\angle A + \angle B = 130° + 50° = 180°$$

$\angle A$ and $\angle B$ are supplementary angles.

An **acute angle** is an angle whose measure is between 0° and 90°. $\angle B$ above is an acute angle. An **obtuse angle** is an angle whose measure is between 90° and 180°. $\angle A$ above is an obtuse angle.

Two angles that share a common side are **adjacent angles.** In the figure at the right, $\angle DAC$ and $\angle CAB$ are adjacent angles. $\angle DAC = 45°$ and $\angle CAB = 55°$.

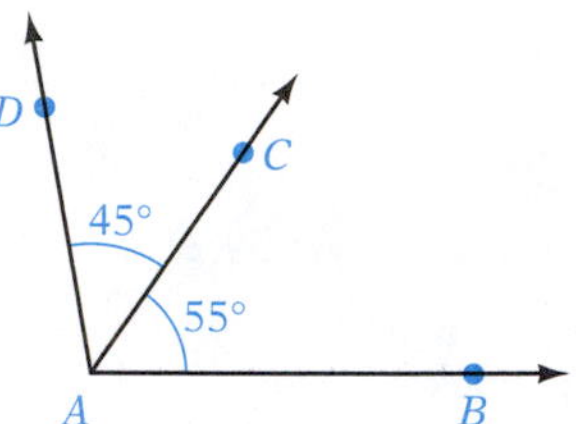

$$\begin{aligned}\angle DAB &= \angle DAC + \angle CAB\\ &= 45° + 55° = 100°\end{aligned}$$

HOW TO In the figure at the right, $\angle EDG = 80°$. $\angle FDG$ is three times the measure of $\angle EDF$. Find the measure of $\angle EDF$.

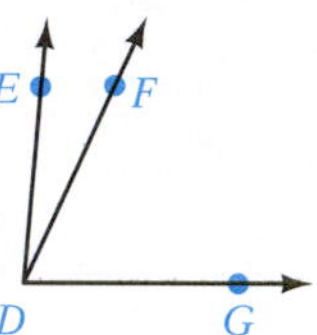

Let x = the measure of $\angle EDF$. Then $3x$ = the measure of $\angle FDG$. Write an equation and solve for x, the measure of $\angle EDF$.

$$\begin{aligned}\angle EDF + \angle FDG &= \angle EDG\\ x + 3x &= 80\\ 4x &= 80\\ x &= 20\end{aligned}$$

$\angle EDF = 20°$

TAKE NOTE

Answers to application problems must have units, such as degrees, feet, dollars, or hours.

Example 1

Given $MN = 15$ mm, $NO = 18$ mm, and $MP = 48$ mm, find OP.

Solution

$$\begin{aligned}MN + NO + OP &= MP\\ 15 + 18 + OP &= 48\\ 33 + OP &= 48\\ OP &= 15\end{aligned}$$

• $MN = 15$, $NO = 18$, $MP = 48$

$OP = 15$ mm

You Try It 1

Given $QR = 24$ cm, $ST = 17$ cm, and $QT = 62$ cm, find RS.

Your solution

Solution on p. S9

Copyright © Houghton Mifflin Company. All rights reserved.

Example 2

Given $XY = 9$ m and YZ is twice XY, find XZ.

Solution

$XZ = XY + YZ$
$XZ = XY + 2(XY)$ • YZ is twice XY.
$XZ = 9 + 2(9)$ • $XY = 9$.
$XZ = 9 + 18$
$XZ = 27$

$XZ = 27$ m

You Try It 2

Given $BC = 16$ ft and $AB = \frac{1}{4}(BC)$, find AC.

Your solution

Example 3

Find the complement of a 38° angle.

Strategy

Complementary angles are two angles whose sum is 90°. To find the complement, let x represent the complement of a 38° angle. Write an equation and solve for x.

Solution

$x + 38° = 90°$
$x = 52°$

The complement of a 38° angle is a 52° angle.

You Try It 3

Find the supplement of a 129° angle.

Your strategy

Your solution

Example 4

Find the measure of $\angle x$.

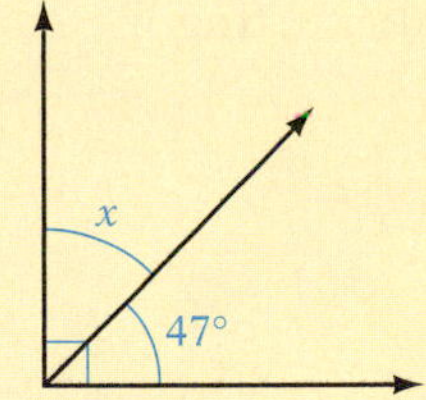

Strategy

To find the measure of $\angle x$, write an equation using the fact that the sum of the measure of $\angle x$ and 47° is 90°. Solve for $\angle x$.

Solution

$\angle x + 47° = 90°$
$\angle x = 43°$

The measure of $\angle x$ is 43°.

You Try It 4

Find the measure of $\angle a$.

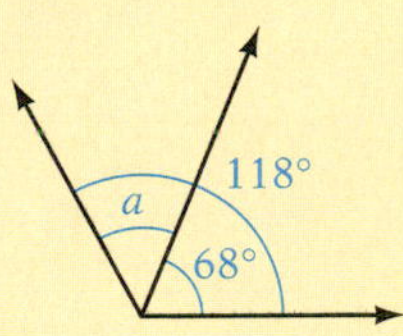

Your strategy

Your solution

Solutions on p. S9

Copyright © Houghton Mifflin Company. All rights reserved.

Objective B To solve problems involving angles formed by intersecting lines

Point of Interest

Many cities in the New World, unlike those in Europe, were designed using rectangular street grids. Washington, D.C. was planned that way except that diagonal avenues were added, primarily for the purpose of enabling quick troop movement in the event the city required defense. As an added precaution, monuments of statuary were constructed at major intersections so that attackers would not have a straight shot down a boulevard.

Four angles are formed by the intersection of two lines. If the two lines are perpendicular, then each of the four angles is a right angle.

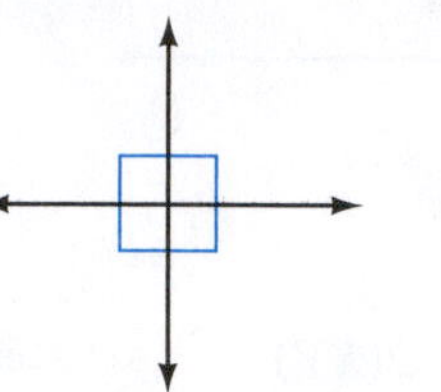

If the two lines are not perpendicular, then two of the angles formed are acute angles and two of the angles are obtuse angles. The two acute angles are always opposite each other, and the two obtuse angles are always opposite each other. In the figure at the right, $\angle w$ and $\angle y$ are acute angles. $\angle x$ and $\angle z$ are obtuse angles.

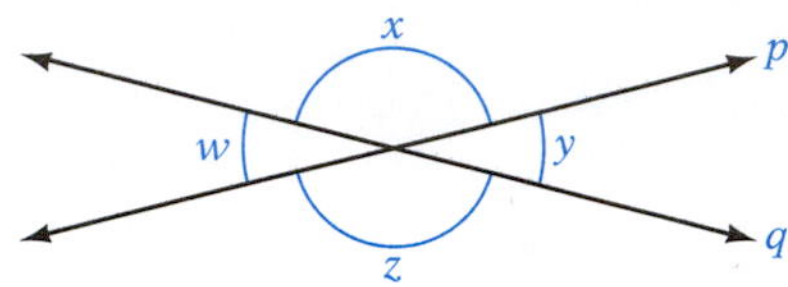

Two angles that are on opposite sides of the intersection of two lines are called **vertical angles.** Vertical angles have the same measure. $\angle w$ and $\angle y$ are vertical angles. $\angle x$ and $\angle z$ are vertical angles.

Vertical angles have the same measure.

$$\angle w = \angle y$$
$$\angle x = \angle z$$

Two angles that share a common side are called **adjacent angles.** For the figure shown above, $\angle x$ and $\angle y$ are adjacent angles, as are $\angle y$ and $\angle z$, $\angle z$ and $\angle w$, and $\angle w$ and $\angle x$. Adjacent angles of intersecting lines are supplementary angles.

Adjacent angles of intersecting lines are supplementary angles.

$$\angle x + \angle y = 180^\circ$$
$$\angle y + \angle z = 180^\circ$$
$$\angle z + \angle w = 180^\circ$$
$$\angle w + \angle x = 180^\circ$$

HOW TO Given that $\angle c = 65^\circ$, find the measures of angles a, b, and d.

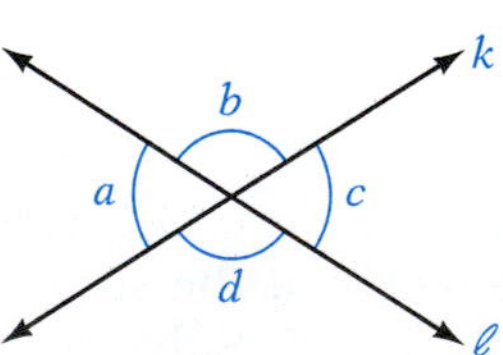

$\angle a = 65^\circ$ • $\angle a = \angle c$ because $\angle a$ and $\angle c$ are vertical angles.

$$\angle b + \angle c = 180^\circ$$
$$\angle b + 65^\circ = 180^\circ$$
$$\angle b = 115^\circ$$

• $\angle b$ is supplementary to $\angle c$ because $\angle b$ and $\angle c$ are adjacent angles of intersecting lines.

$\angle d = 115^\circ$ • $\angle d = \angle b$ because $\angle d$ and $\angle b$ are vertical angles.

Copyright © Houghton Mifflin Company. All rights reserved.

A line that intersects two other lines at different points is called a **transversal.**

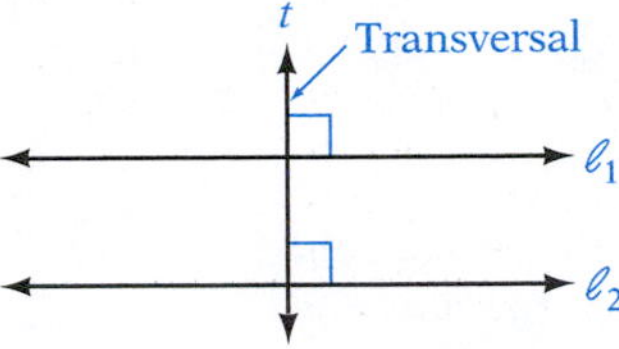

If the lines cut by a transversal *t* are parallel lines and the transversal is perpendicular to the parallel lines, all eight angles formed are right angles.

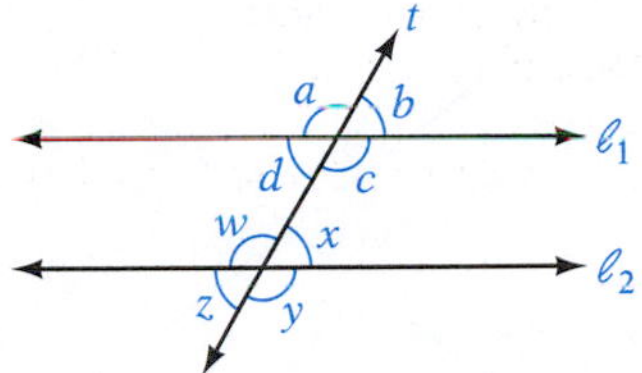

If the lines cut by a transversal *t* are parallel lines and the transversal is not perpendicular to the parallel lines, all four acute angles have the same measure and all four obtuse angles have the same measure. For the figure at the right,

$$\angle b = \angle d = \angle x = \angle z$$
$$\angle a = \angle c = \angle w = \angle y$$

Alternate interior angles are two nonadjacent angles that are on opposite sides of the transversal and lie between the parallel lines. In the figure above, $\angle c$ and $\angle w$ are alternate interior angles; $\angle d$ and $\angle x$ are alternate interior angles. Alternate interior angles have the same measure.

Alternate interior angles have the same measure.

$$\angle c = \angle w$$
$$\angle d = \angle x$$

Alternate exterior angles are two nonadjacent angles that are on opposite sides of the transversal and lie outside the parallel lines. In the figure above, $\angle a$ and $\angle y$ are alternate exterior angles; $\angle b$ and $\angle z$ are alternate exterior angles. Alternate exterior angles have the same measure.

Alternate exterior angles have the same measure.

$$\angle a = \angle y$$
$$\angle b = \angle z$$

Corresponding angles are two angles that are on the same side of the transversal and are both acute angles or are both obtuse angles. For the figure above, the following pairs of angles are corresponding angles: $\angle a$ and $\angle w$, $\angle d$ and $\angle z$, $\angle b$ and $\angle x$, and $\angle c$ and $\angle y$. Corresponding angles have the same measure.

Corresponding angles have the same measure.

$$\angle a = \angle w$$
$$\angle d = \angle z$$
$$\angle b = \angle x$$
$$\angle c = \angle y$$

Copyright © Houghton Mifflin Company. All rights reserved.

HOW TO Given that $\ell_1 \parallel \ell_2$ and $\angle c = 58°$, find the measures of $\angle f$, $\angle h$, and $\angle g$.

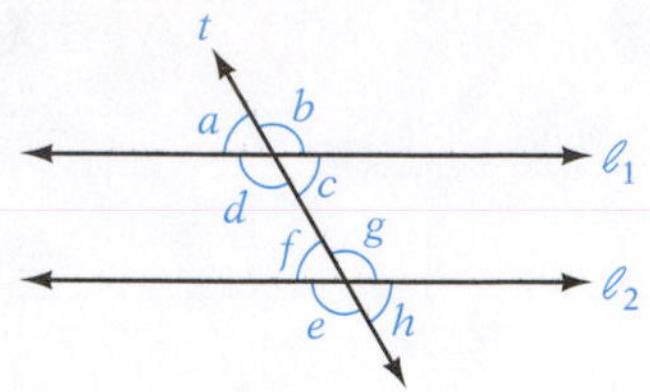

$\angle f = \angle c = 58°$ • $\angle c$ and $\angle f$ are alternate interior angles.

$\angle h = \angle c = 58°$ • $\angle c$ and $\angle h$ are corresponding angles.

$\angle g + \angle h = 180°$ • $\angle g$ is supplementary to $\angle h$.
$\angle g + 58° = 180°$
$\angle g = 122°$

Example 5

Find x.

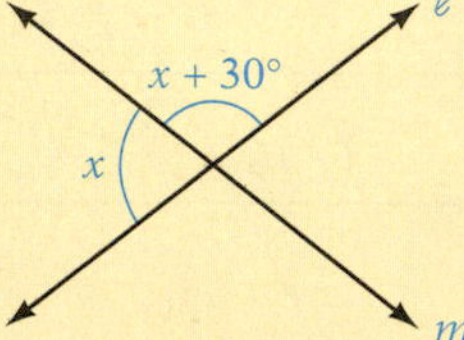

Strategy
The angles labeled are adjacent angles of intersecting lines and are, therefore, supplementary angles. To find x, write an equation and solve for x.

Solution
$x + (x + 30°) = 180°$
$2x + 30° = 180°$
$2x = 150°$
$x = 75°$

You Try It 5

Find x.

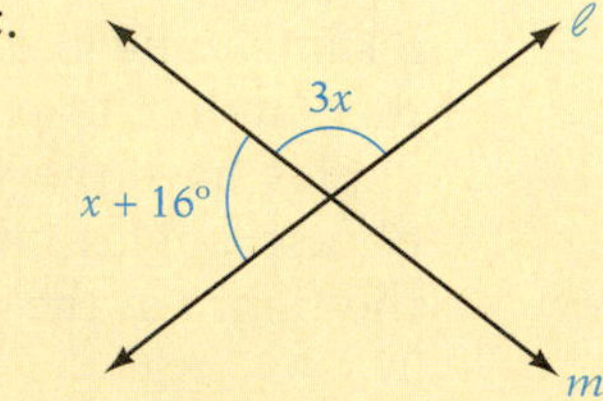

Your strategy

Your solution

Example 6

Given $\ell_1 \parallel \ell_2$, find x.

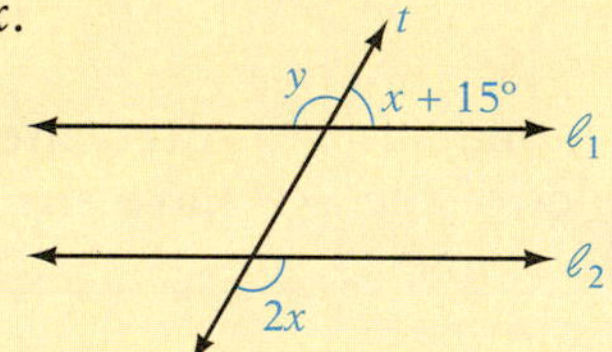

Strategy
$2x = y$ because alternate exterior angles have the same measure. $(x + 15°) + y = 180°$ because adjacent angles of intersecting lines are supplementary angles. Substitute $2x$ for y and solve for x.

Solution
$(x + 15°) + 2x = 180°$
$3x + 15° = 180°$
$3x = 165°$
$x = 55°$

You Try It 6

Given $\ell_1 \parallel \ell_2$, find x.

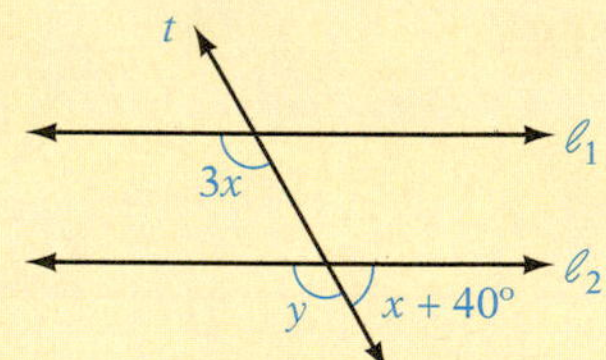

Your strategy

Your solution

Solutions on p. S9

Copyright © Houghton Mifflin Company. All rights reserved.

Objective C To solve problems involving the angles of a triangle

If the lines cut by a transversal are not parallel lines, the three lines will intersect at three points. In the figure at the right, the transversal t intersects lines p and q. The three lines intersect at points A, B, and C. These three points define three line segments: $\overline{AB}$, $\overline{BC}$, and $\overline{AC}$. The plane figure formed by these three line segments is called a **triangle.**

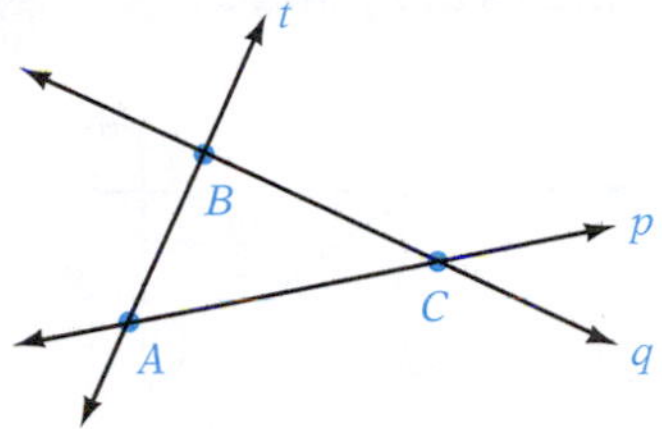

Each of the three points of intersection is the vertex of four angles. The angles within the region enclosed by the triangle are called **interior angles.** In the figure at the right, angles a, b, and c are interior angles. The sum of the measures of the interior angles of a triangle is 180°.

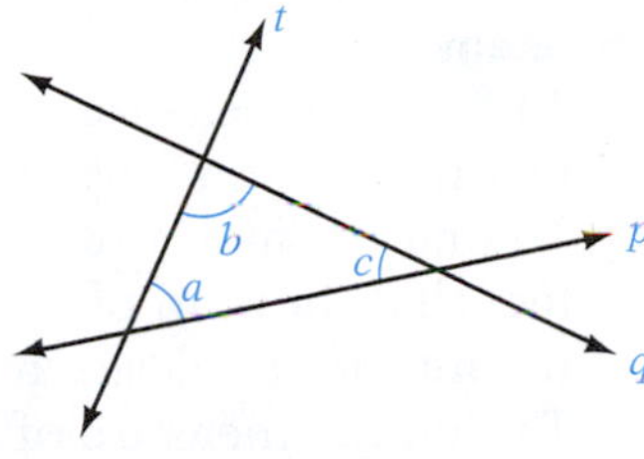

$\angle a + \angle b + \angle c = 180°$

The Sum of the Measures of the Interior Angles of a Triangle

The sum of the measures of the interior angles of a triangle is 180°.

An angle adjacent to an interior angle is an **exterior angle.** In the figure at the right, angles m and n are exterior angles for angle a. The sum of the measures of an interior and an exterior angle is 180°.

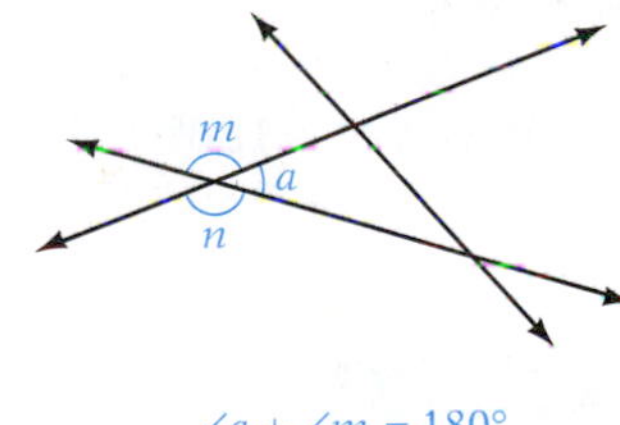

$\angle a + \angle m = 180°$
$\angle a + \angle n = 180°$

HOW TO Given that $\angle c = 40°$ and $\angle d = 100°$, find the measure of $\angle e$.

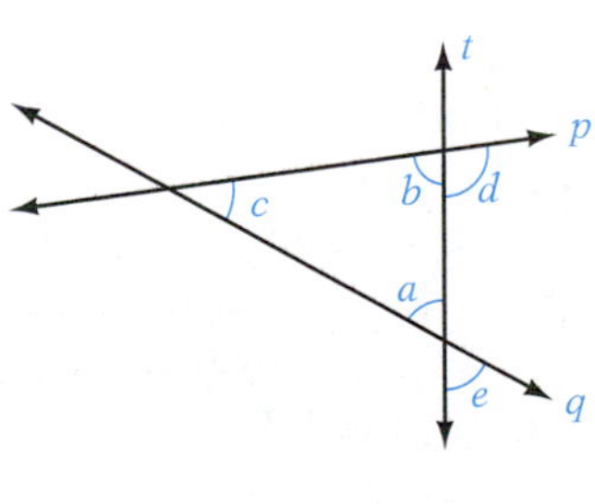

$\angle d$ and $\angle b$ are supplementary angles.

$$\begin{aligned} \angle d + \angle b &= 180° \\ 100° + \angle b &= 180° \\ \angle b &= 80° \end{aligned}$$

The sum of the interior angles is 180°.

$$\begin{aligned} \angle c + \angle b + \angle a &= 180° \\ 40° + 80° + \angle a &= 180° \\ 120° + \angle a &= 180° \\ \angle a &= 60° \end{aligned}$$

$\angle a$ and $\angle e$ are vertical angles.

$$\angle e = \angle a = 60°$$

Copyright © Houghton Mifflin Company. All rights reserved.

Example 7

Given that $\angle y = 55°$, find the measures of angles a, b, and d.

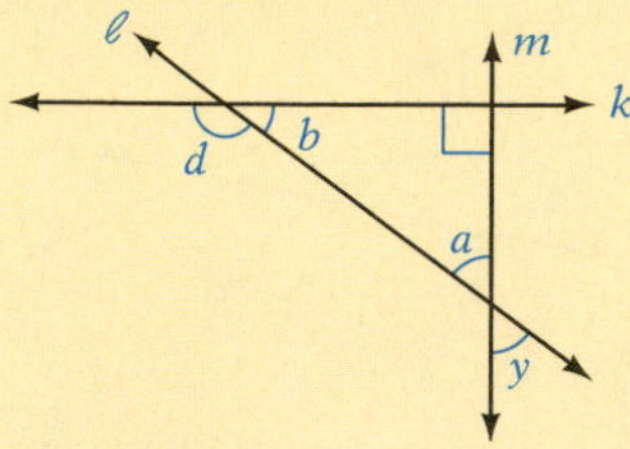

Strategy

- To find the measure of angle a, use the fact that $\angle a$ and $\angle y$ are vertical angles.
- To find the measure of angle b, use the fact that the sum of the measures of the interior angles of a triangle is 180°.
- To find the measure of angle d, use the fact that the sum of an interior and an exterior angle is 180°.

Solution

$\angle a = \angle y = 55°$

$$\angle a + \angle b + 90° = 180°$$
$$55° + \angle b + 90° = 180°$$
$$\angle b + 145° = 180°$$
$$\angle b = 35°$$

$$\angle d + \angle b = 180°$$
$$\angle d + 35° = 180°$$
$$\angle d = 145°$$

You Try It 7

Given that $\angle a = 45°$ and $\angle x = 100°$, find the measures of angles b, c, and y.

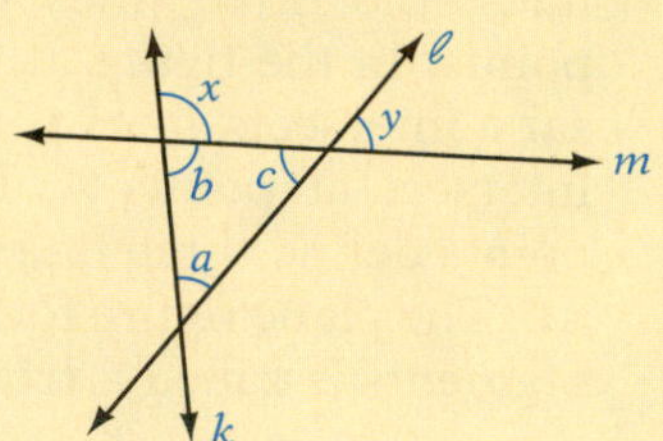

Your strategy

Your solution

Example 8

Two angles of a triangle measure 53° and 78°. Find the measure of the third angle.

Strategy

To find the measure of the third angle, use the fact that the sum of the measures of the interior angles of a triangle is 180°. Write an equation using x to represent the measure of the third angle. Solve the equation for x.

Solution

$$x + 53° + 78° = 180°$$
$$x + 131° = 180°$$
$$x = 49°$$

The measure of the third angle is 49°.

You Try It 8

One angle in a triangle is a right angle, and one angle measures 34°. Find the measure of the third angle.

Your strategy

Your solution

Solutions on pp. S9–S10

Copyright © Houghton Mifflin Company. All rights reserved.

3.1 Exercises

Objective A **To solve problems involving lines and angles**

Use a protractor to measure the angle. State whether the angle is acute, obtuse, or right.

1. 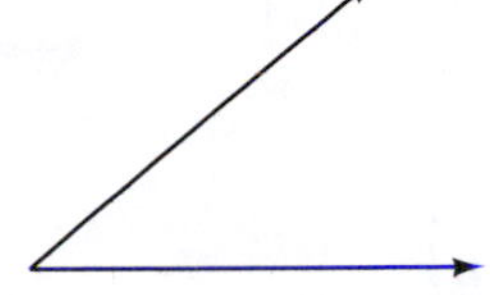

2. 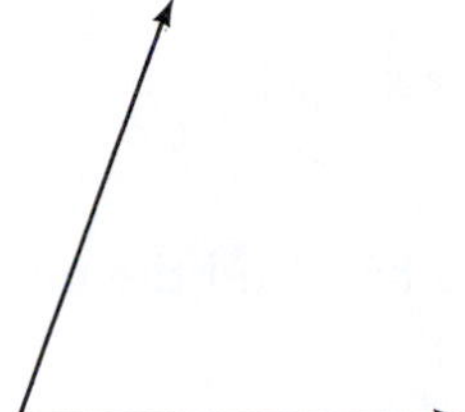

3. 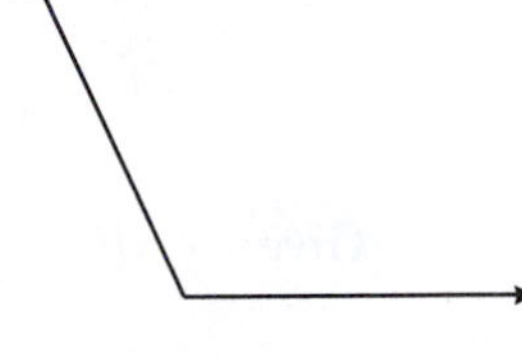

4. 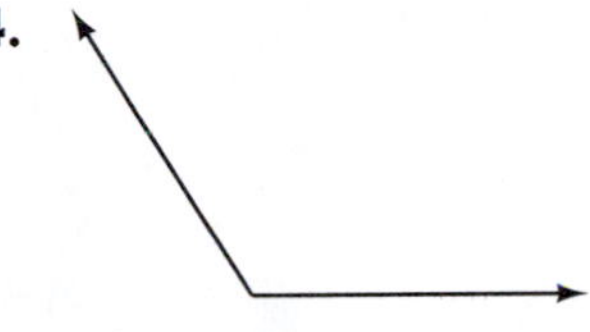

5. 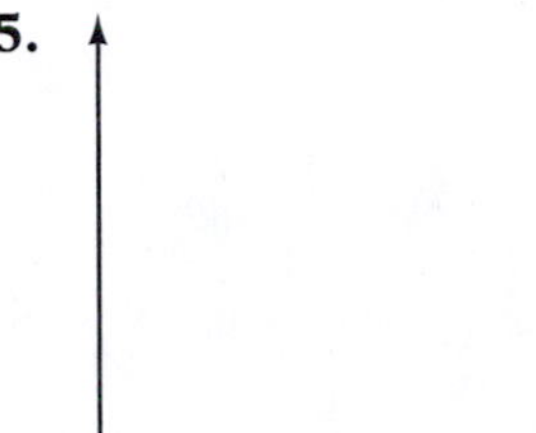

6. 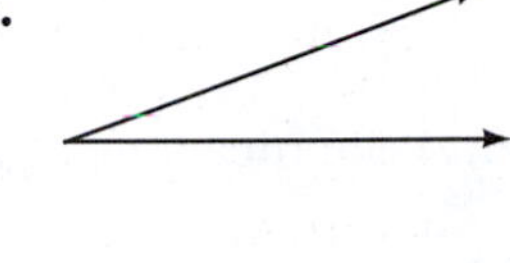

Solve.

7. Find the complement of a 62° angle.

8. Find the complement of a 31° angle.

9. Find the supplement of a 162° angle.

10. Find the supplement of a 72° angle.

11. Given $AB = 12$ cm, $CD = 9$ cm, and $AD = 35$ cm, find the length of BC.

12. Given $AB = 21$ mm, $BC = 14$ mm, and $AD = 54$ mm, find the length of CD.

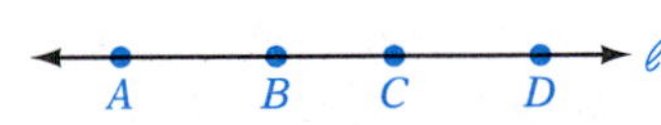

13. Given $QR = 7$ ft and RS is three times the length of QR, find the length of QS.

14. Given $QR = 15$ in. and RS is twice the length of QR, find the length of QS.

15. Given $EF = 20$ m and FG is one-half the length of EF, find the length of EG.

Copyright © Houghton Mifflin Company. All rights reserved.

16. Given $EF = 18$ cm and FG is one-third the length of EF, find the length of EG.

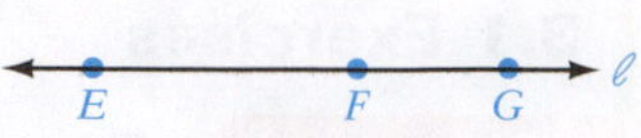

17. Given $\angle LOM = 53°$ and $\angle LON = 139°$, find the measure of $\angle MON$.

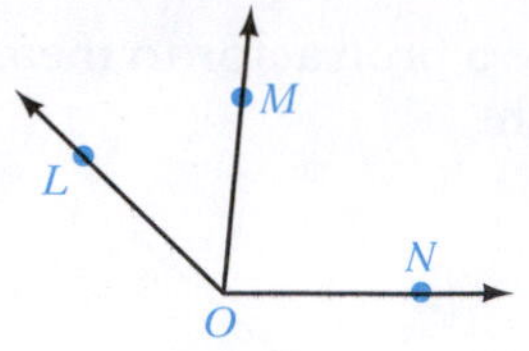

18. Given $\angle MON = 38°$ and $\angle LON = 85°$, find the measure of $\angle LOM$.

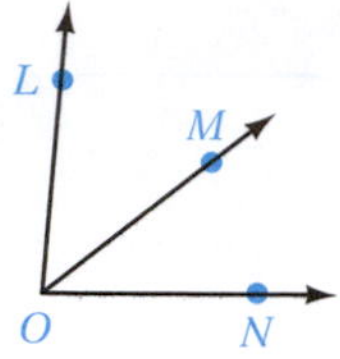

Find the measure of $\angle x$.

19.

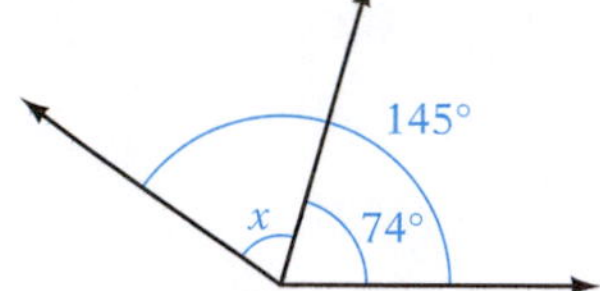

20.

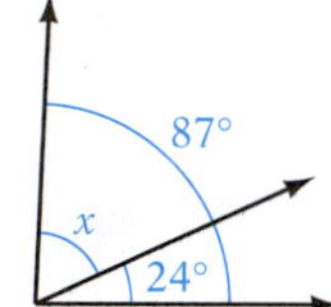

Given that $\angle LON$ is a right angle, find the measure of $\angle x$.

21.

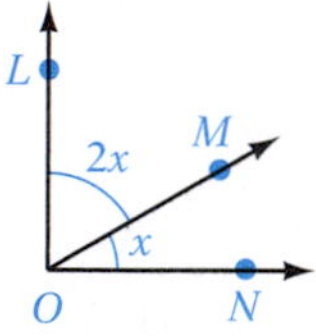

22.

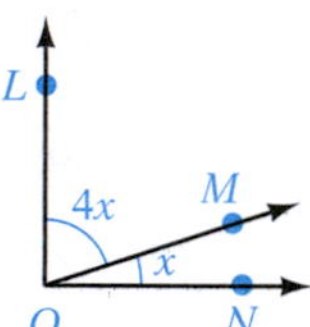

23.

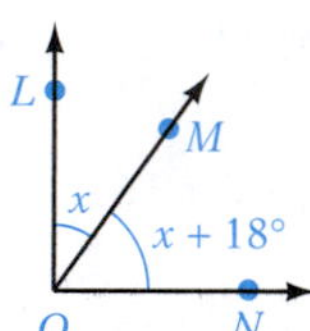

24.

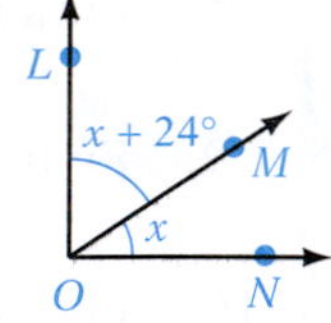

Find the measure of $\angle a$.

25.

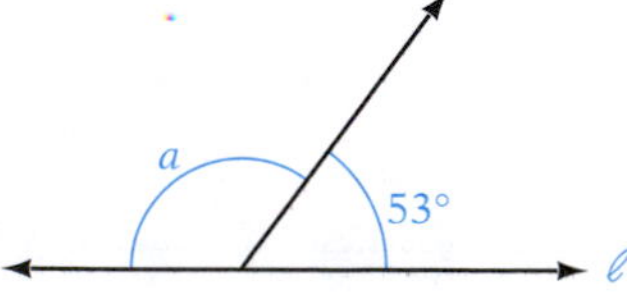

26.

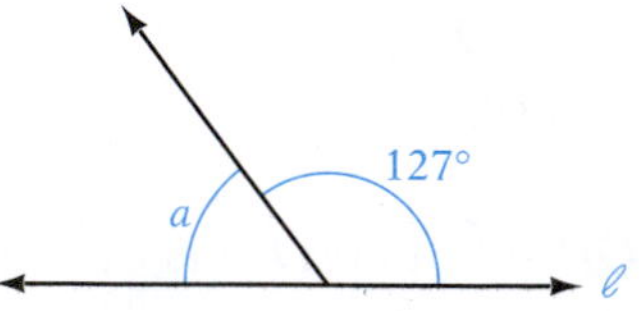

27.

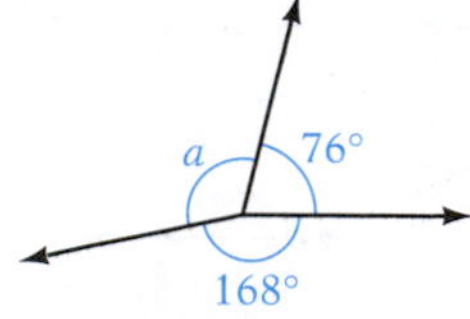

28.

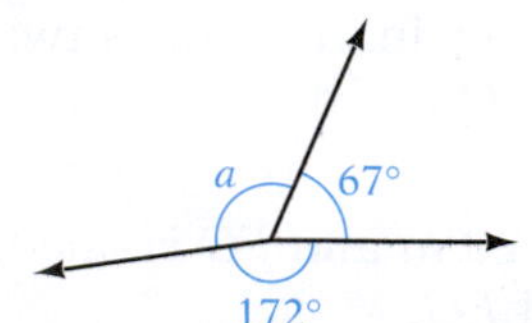

Copyright © Houghton Mifflin Company. All rights reserved.

Find x.

29.

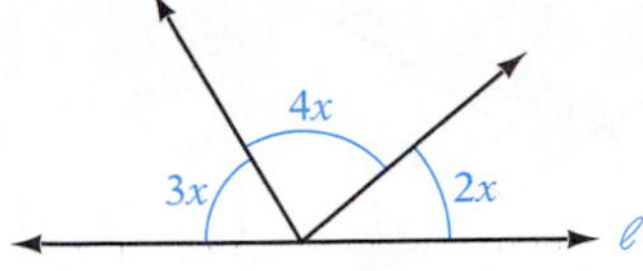

30.

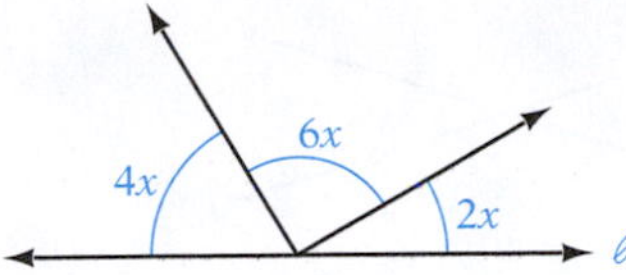

31.

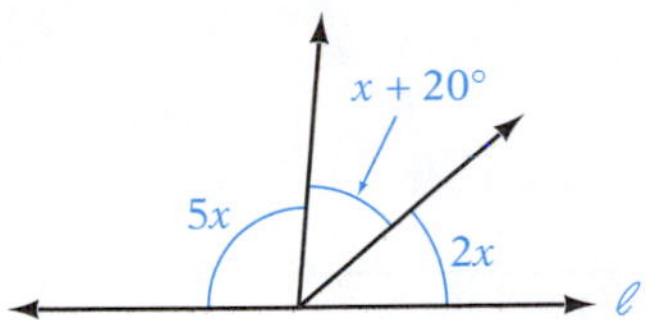

32.

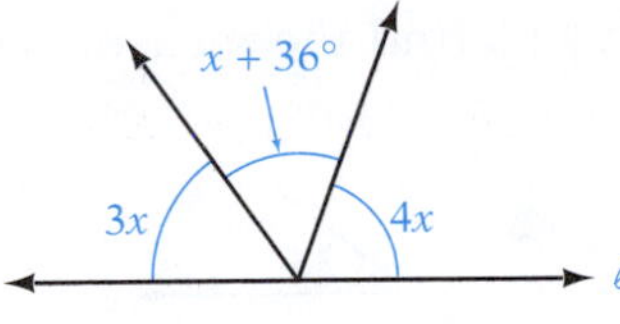

33.

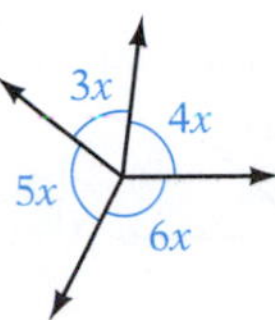

34.

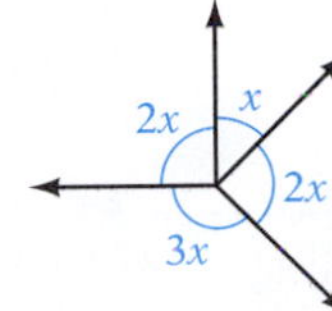

Solve.

35. Given $\angle a = 51°$, find the measure of $\angle b$.

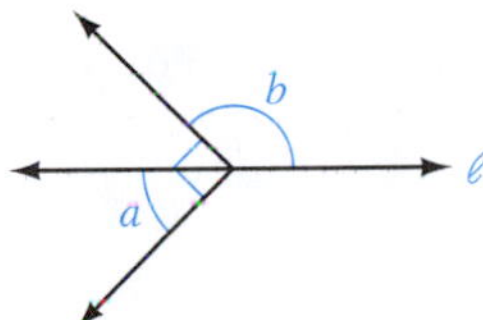

36. Given $\angle a = 38°$, find the measure of $\angle b$.

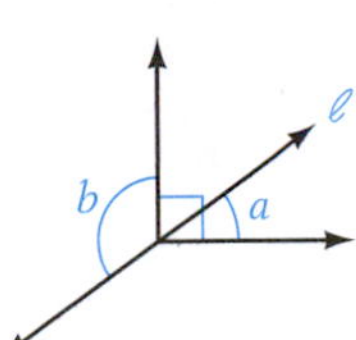

Copyright © Houghton Mifflin Company. All rights reserved.

Objective B To solve problems involving angles formed by intersecting lines

Find the measure of $\angle x$.

37.

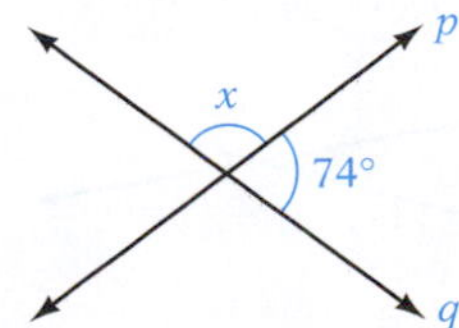

38.

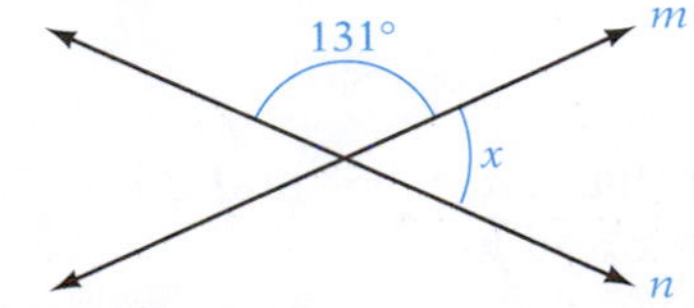

Find x.

39.

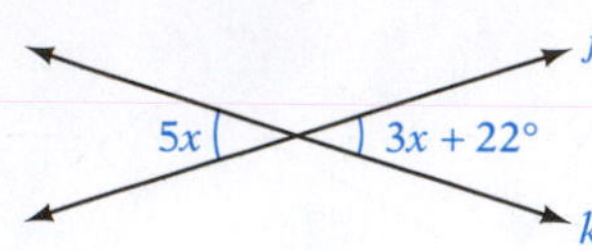

40.

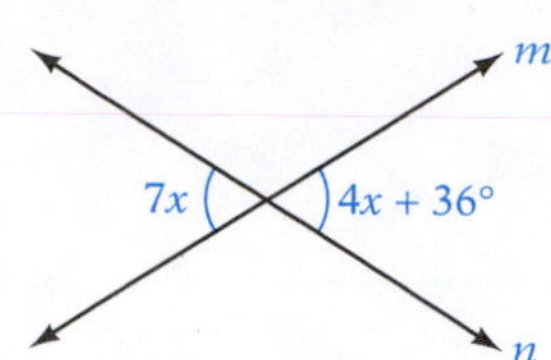

Given that $\ell_1 \parallel \ell_2$, find the measures of angles a and b.

41.

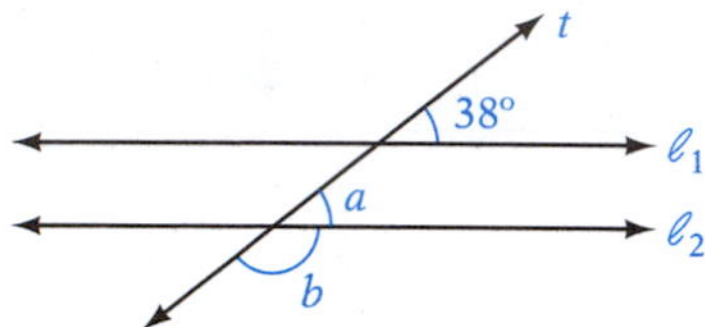

42.

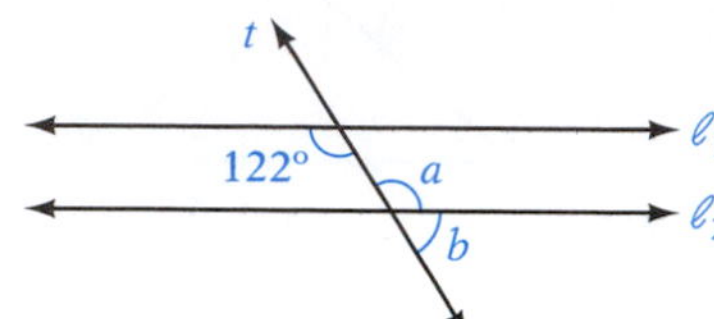

43.

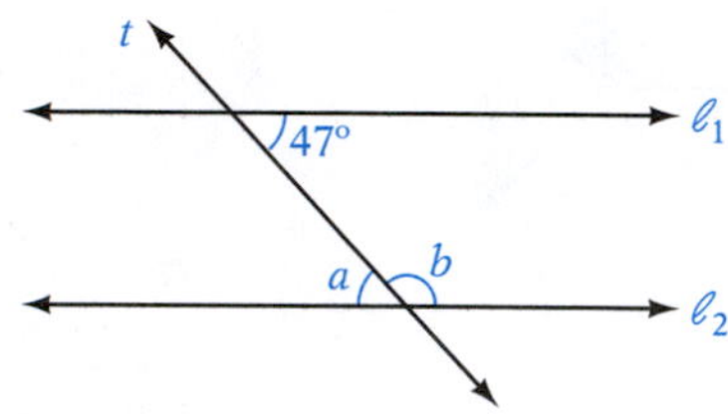

44.

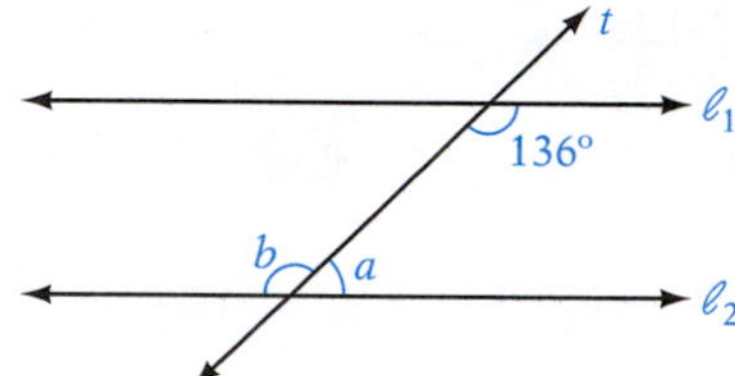

Given that $\ell_1 \parallel \ell_2$, find x.

45.

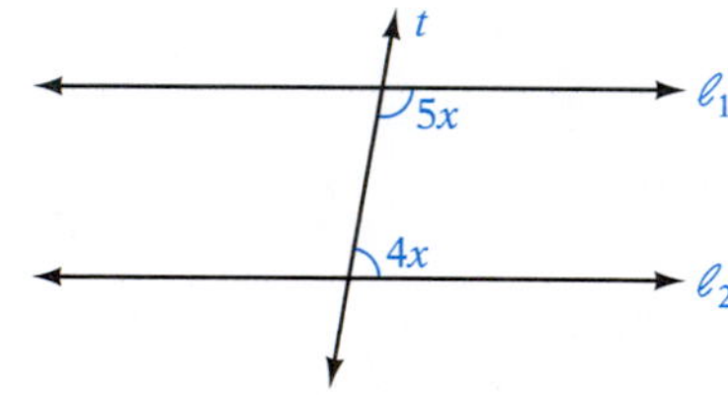

46.

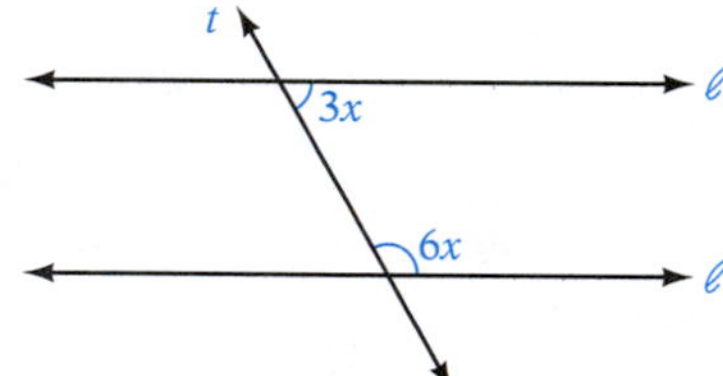

47.

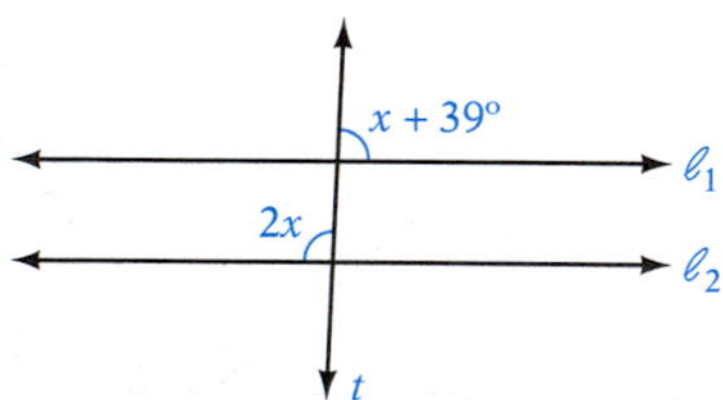

48.

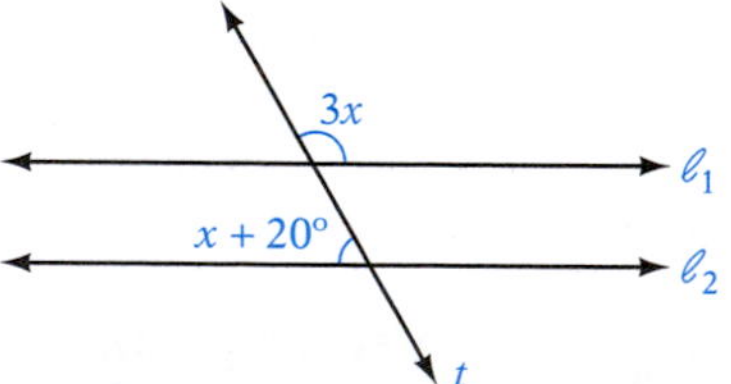

Objective C **To solve problems involving the angles of a triangle**

Solve.

49. Given that $\angle a = 95°$ and $\angle b = 70°$, find the measures of angles x and y.

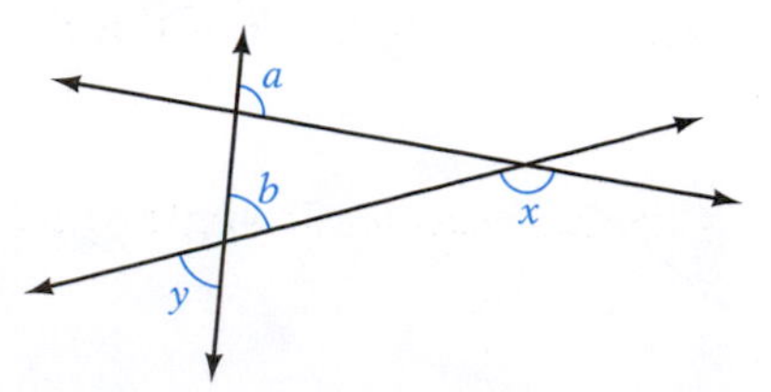

Copyright © Houghton Mifflin Company. All rights reserved.

50. Given that $\angle a = 35°$ and $\angle b = 55°$, find the measures of angles x and y.

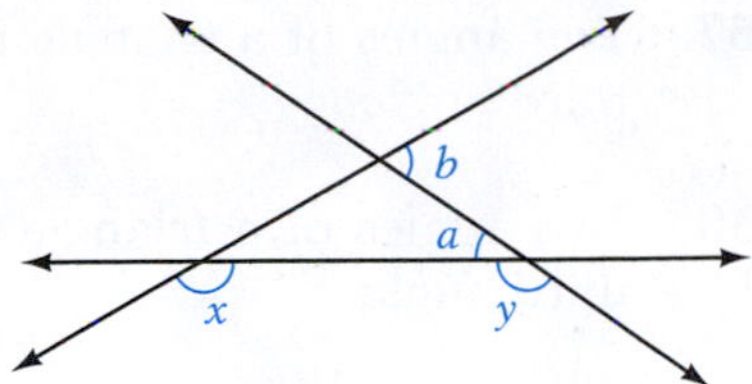

51. Given that $\angle y = 45°$, find the measures of angles a and b.

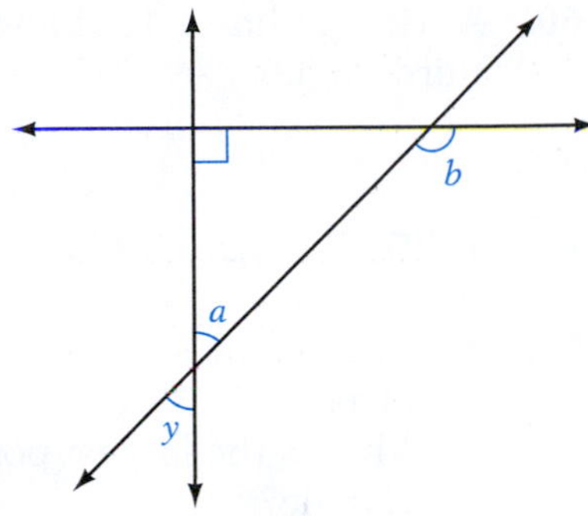

52. Given that $\angle y = 130°$, find the measures of angles a and b.

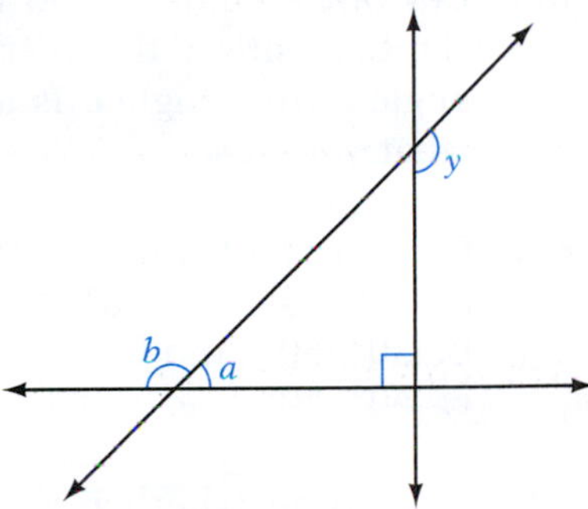

53. Given that $\overline{AO} \perp \overline{OB}$, express in terms of x the number of degrees in $\angle BOC$.

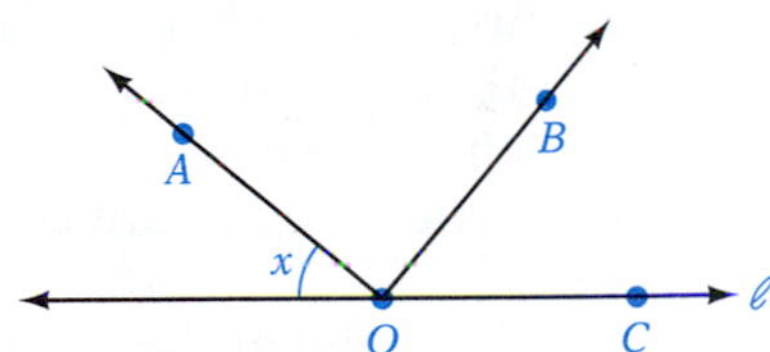

54. Given that $\overline{AO} \perp \overline{OB}$, express in terms of x the number of degrees in $\angle AOC$.

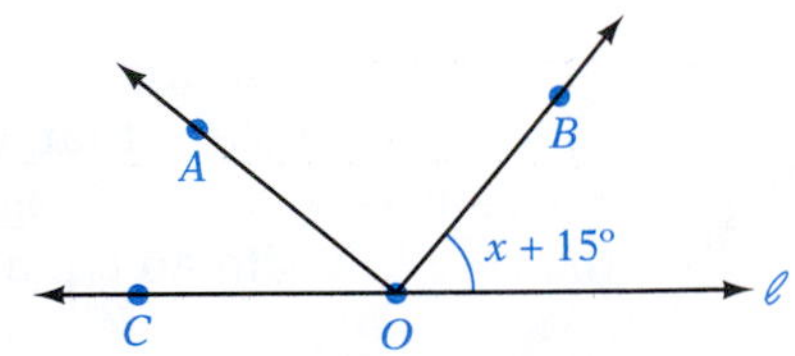

55. One angle in a triangle is a right angle, and one angle is equal to 30°. What is the measure of the third angle?

56. A triangle has a 45° angle and a right angle. Find the measure of the third angle.

Copyright © Houghton Mifflin Company. All rights reserved.

57. Two angles of a triangle measure 42° and 103°. Find the measure of the third angle.

58. Two angles of a triangle measure 62° and 45°. Find the measure of the third angle.

59. A triangle has a 13° angle and a 65° angle. What is the measure of the third angle?

60. A triangle has a 105° angle and a 32° angle. What is the measure of the third angle?

APPLYING THE CONCEPTS

61. **a.** What is the smallest possible whole number of degrees in an angle of a triangle?
b. What is the largest possible whole number of degrees in an angle of a triangle?

62. Cut out a triangle and then tear off two of the angles, as shown at the right. Position the pieces you tore off so that angle a is adjacent to angle b and angle c is adjacent to angle b (on the other side). Describe what you observe. What does this demonstrate?

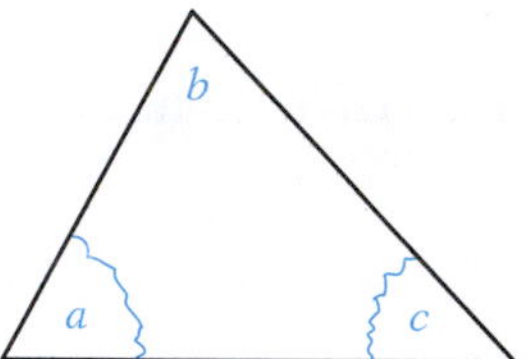

63. Construct a triangle with the given angle measures.
a. 45°, 45°, and 90°
b. 30°, 60°, and 90°
c. 40°, 40°, and 100°

64. Determine whether the statement is always true, sometimes true, or never true.
a. Two lines that are parallel to a third line are parallel to each other.
b. A triangle contains two acute angles.
c. Vertical angles are complementary angles.

65. For the figure at the right, find the sum of the measures of angles x, y, and z.

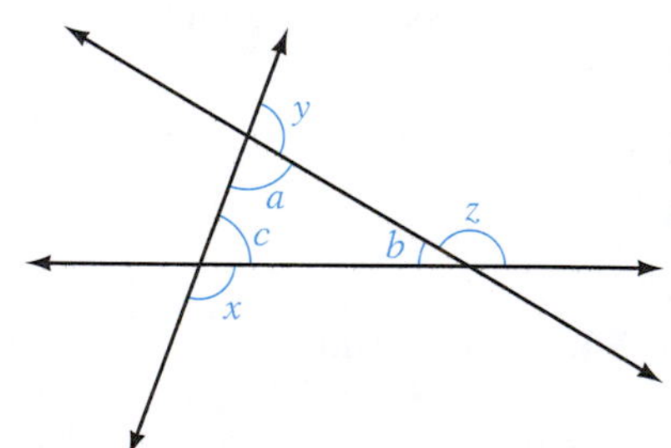

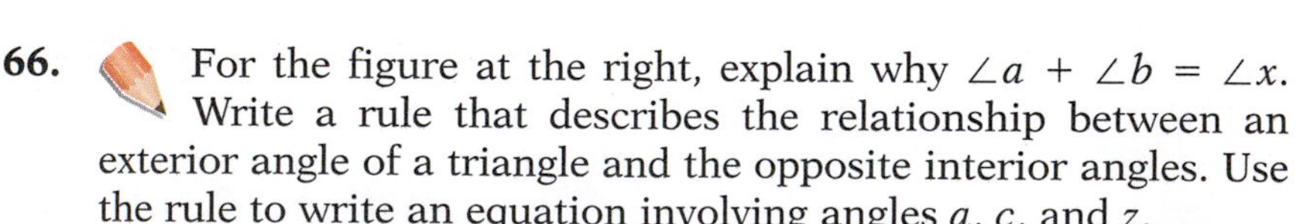

66. For the figure at the right, explain why $\angle a + \angle b = \angle x$. Write a rule that describes the relationship between an exterior angle of a triangle and the opposite interior angles. Use the rule to write an equation involving angles a, c, and z.

67. If $\overline{AB}$ and $\overline{CD}$ intersect at point O, and $\angle AOC = \angle BOC$, explain why $\overline{AB} \perp \overline{CD}$.

68. Do some research on the principle of reflection. Explain how this principle applies to the operation of a periscope and to the game of billiards.

Copyright © Houghton Mifflin Company. All rights reserved.

3.2 Plane Geometric Figures

Objective A

To solve problems involving the perimeter of a geometric figure

A **polygon** is a closed figure determined by three or more line segments that lie in a plane. The line segments that form the polygon are called its **sides.** The figures below are examples of polygons.

A

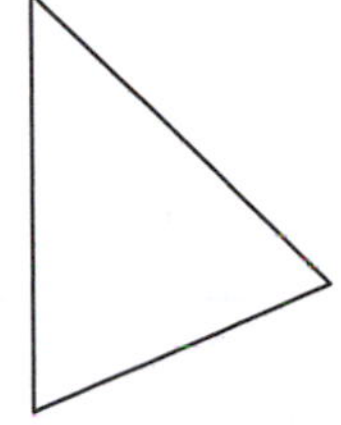
B

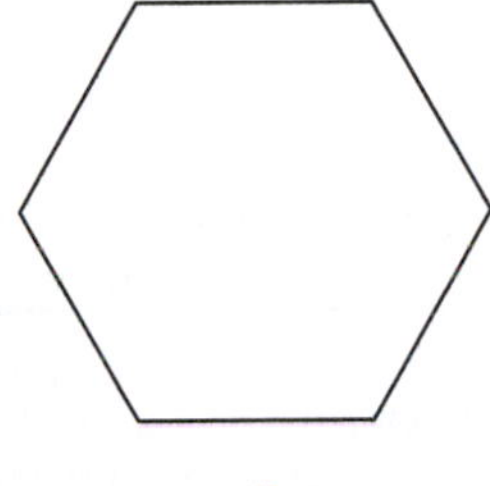
C

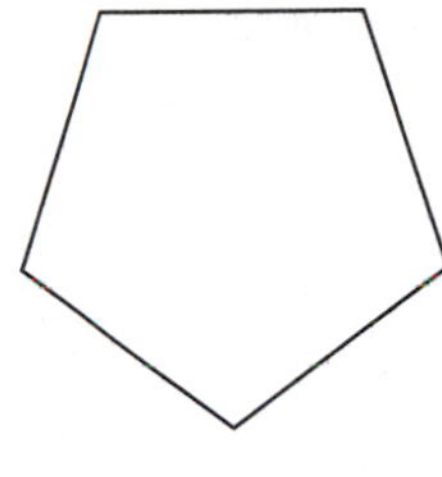
D

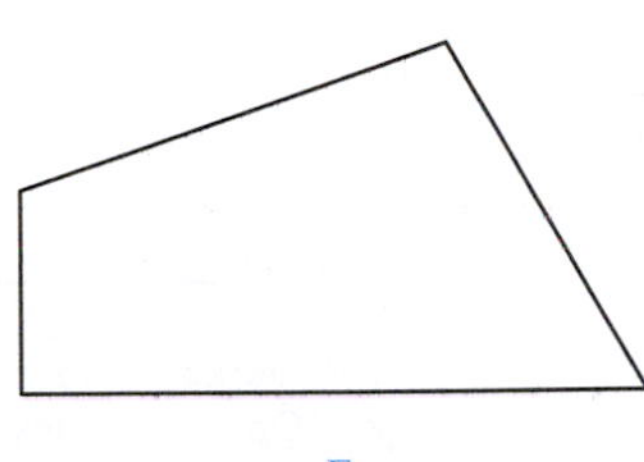
E

Point of Interest

Although a polygon is defined in terms of its *sides* (see the definition above), the word actually comes from the Latin word *polygonum*, which means "having many *angles*." This is certainly the case for a polygon.

The Pentagon in Arlington, Virginia

A **regular polygon** is one in which each side has the same length and each angle has the same measure. The polygons in Figures *A*, *C*, and *D* above are regular polygons.

The name of a polygon is based on the number of its sides. The table below lists the names of polygons that have from 3 to 10 sides.

Number of Sides	*Name of the Polygon*
3	Triangle
4	Quadrilateral
5	Pentagon
6	Hexagon
7	Heptagon
8	Octagon
9	Nonagon
10	Decagon

Triangles and quadrilaterals are two of the most common types of polygons. Triangles are distinguished by the number of equal sides and also by the measures of their angles.

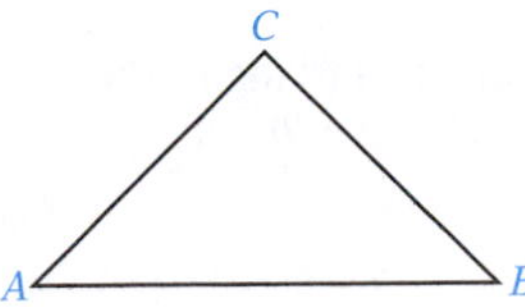

An **isosceles triangle** has two sides of equal length. The angles opposite the equal sides are of equal measure.
$AC = BC$
$\angle A = \angle B$

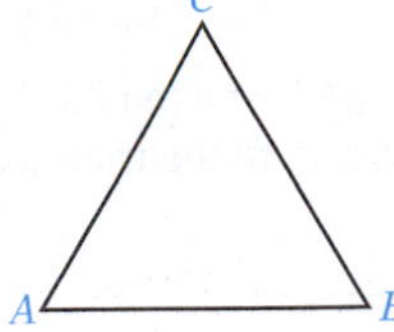

The three sides of an **equilateral triangle** are of equal length. The three angles are of equal measure.
$AB = BC = AC$
$\angle A = \angle B = \angle C$

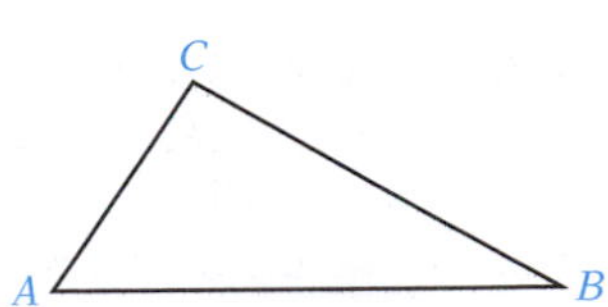

A **scalene triangle** has no two sides of equal length. No two angles are of equal measure.

Copyright © Houghton Mifflin Company. All rights reserved.

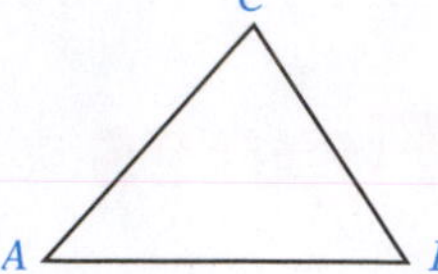

An **acute triangle** has three acute angles.

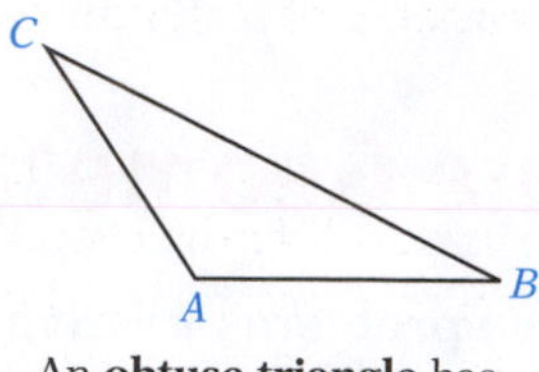

An **obtuse triangle** has one obtuse angle.

A **right triangle** has a right angle.

> **TAKE NOTE**
> The diagram below shows the relationships among all quadrilaterals. The description of each quadrilateral is given within an example of that quadrilateral.

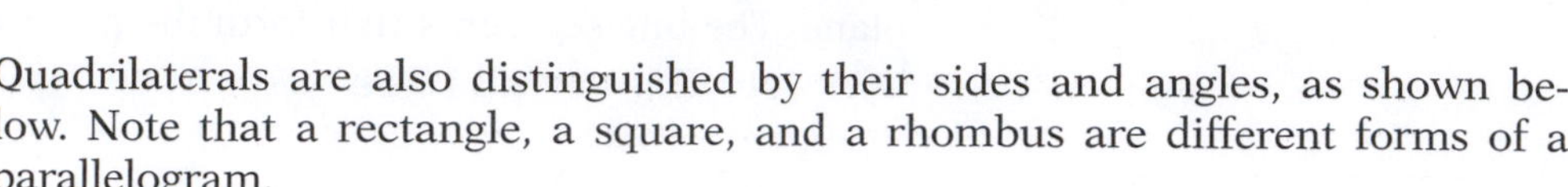

Quadrilaterals are also distinguished by their sides and angles, as shown below. Note that a rectangle, a square, and a rhombus are different forms of a parallelogram.

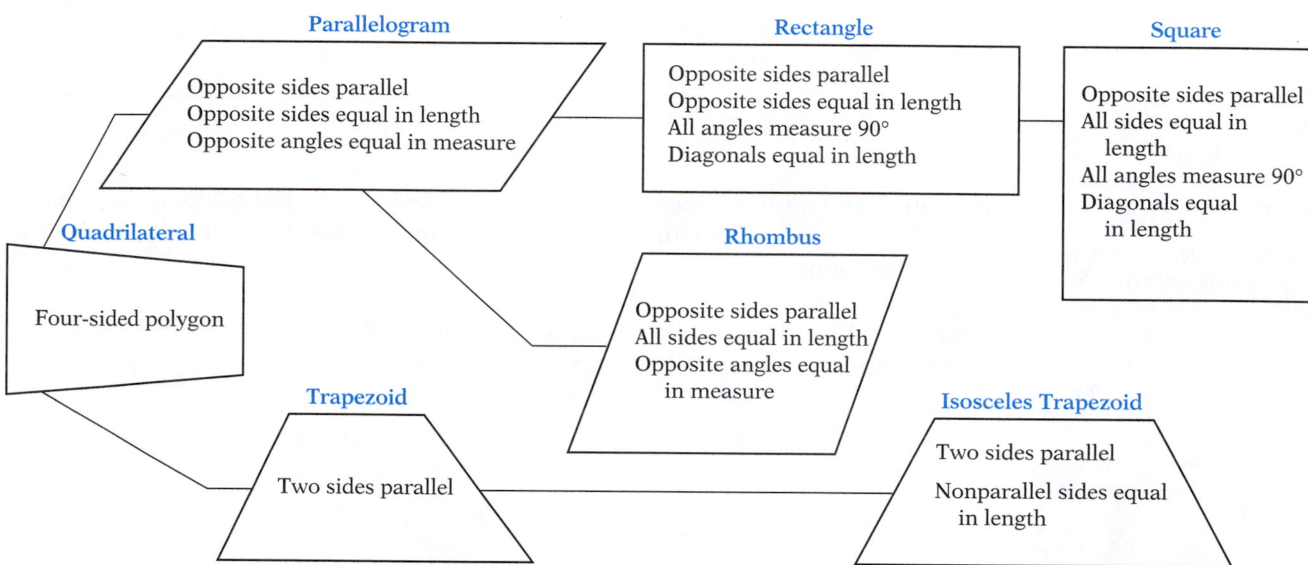

The **perimeter** of a plane geometric figure is a measure of the distance around the figure. Perimeter is used in buying fencing for a lawn or determining how much baseboard is needed for a room.

The perimeter of a triangle is the sum of the lengths of the three sides.

Perimeter of a Triangle

Let a, b, and c be the lengths of the sides of a triangle. The perimeter, P, of the triangle is given by $P = a + b + c$.

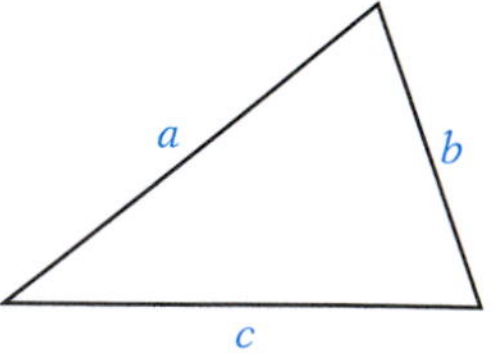

HOW TO Find the perimeter of the triangle shown at the right.

$P = 5 + 7 + 10 = 22$

The perimeter is 22 ft.

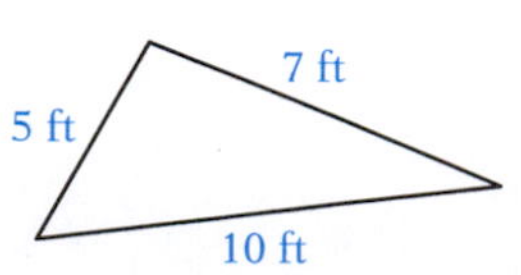

Copyright © Houghton Mifflin Company. All rights reserved.

The perimeter of a quadrilateral is the sum of the lengths of its four sides.

Point of Interest

Leonardo DaVinci painted the *Mona Lisa* on a rectangular canvas whose height was approximately 1.6 times its width. Rectangles with these proportions, called golden rectangles, were used extensively in Renaissance art.

A rectangle is a quadrilateral with opposite sides of equal length. Usually the length, L, of a rectangle refers to the length of one of the longer sides of the rectangle, and the width, W, refers to the length of one of the shorter sides. The perimeter can then be represented by $P = L + W + L + W$.

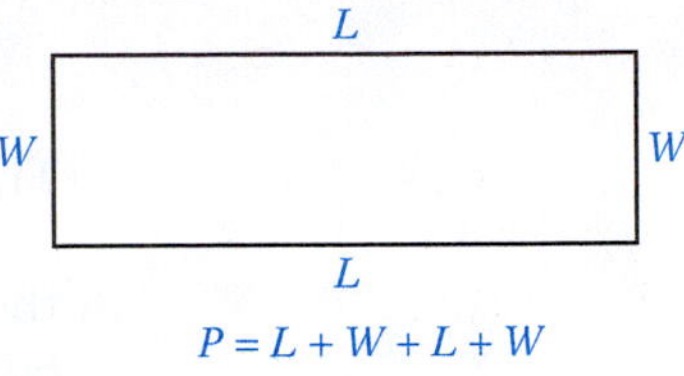

$P = L + W + L + W$

The formula for the perimeter of a rectangle is derived by combining like terms.

$$P = 2L + 2W$$

Perimeter of a Rectangle

Let L represent the length and W the width of a rectangle. The perimeter, P, of the rectangle is given by $P = 2L + 2W$.

HOW TO Find the perimeter of the rectangle shown at the right.

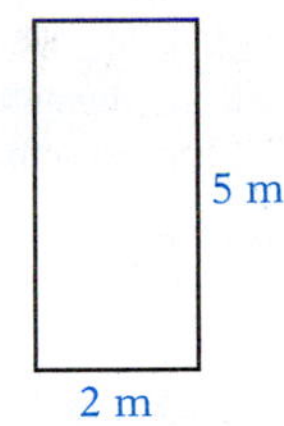

$P = 2L + 2W$

$P = 2(5) + 2(2)$ • The length is 5 m. Substitute 5 for L. The width is 2 m. Substitute 2 for W.

$P = 10 + 4$ • Solve for P.

$P = 14$

The perimeter is 14 m.

A square is a rectangle in which each side has the same length. Let s represent the length of each side of a square. Then the perimeter of the square can be represented by $P = s + s + s + s$.

$P = s + s + s + s$

The formula for the perimeter of a square is derived by combining like terms.

$$P = 4s$$

Perimeter of a Square

Let s represent the length of a side of a square. The perimeter, P, of the square is given by $P = 4s$.

HOW TO Find the perimeter of the square shown at the right.

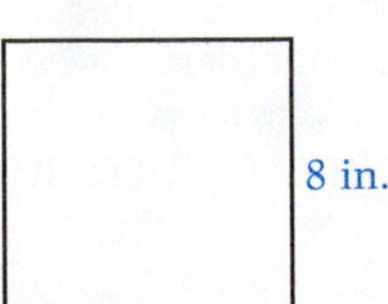

$P = 4s = 4(8) = 32$

The perimeter is 32 in.

Copyright © Houghton Mifflin Company. All rights reserved.

A **circle** is a plane figure in which all points are the same distance from point O, which is called the **center** of the circle.

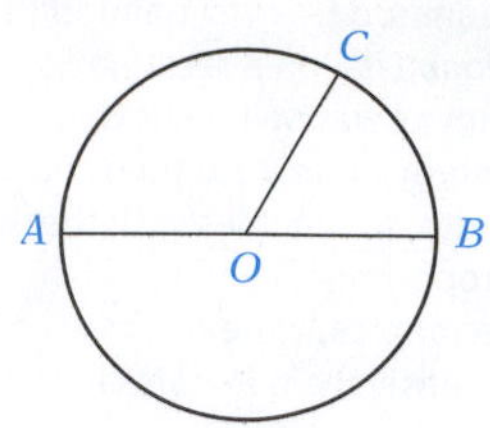

The **diameter** of a circle is a line segment across the circle through point O. AB is a diameter of the circle at the right. The variable d is used to designate the diameter of a circle.

The **radius** of a circle is a line segment from the center of the circle to a point on the circle. OC is a radius of the circle at the right. The variable r is used to designate a radius of a circle.

The length of the diameter of a circle is twice the length of the radius.

$$d = 2r \text{ or } r = \frac{1}{2}d$$

Point of Interest

Archimedes (c. 287–212 B.C.) was the mathematician who gave us the approximate value of π as $\frac{22}{7} = 3\frac{1}{7}$. He actually showed that π was between $3\frac{10}{71}$ and $3\frac{1}{7}$. The approximation $3\frac{10}{71}$ is closer to the exact value of π, but it is more difficult to use.

The distance around a circle is called the **circumference.** The circumference, C, of a circle is equal to the product of π (pi) and the diameter.

$$C = \pi d$$

Because $d = 2r$, the formula for the circumference can be written in terms of r.

$$C = 2\pi r$$

The Circumference of a Circle

The circumference, C, of a circle with diameter d and radius r is given by $C = \pi d$ or $C = 2\pi r$.

The formula for circumference uses the number π, which is an irrational number. The value of π can be approximated by a fraction or by a decimal.

$$\pi \approx \frac{22}{7} \text{ or } \pi \approx 3.14$$

The π key on a scientific calculator gives a closer approximation of π than 3.14. Use a scientific calculator to find approximate values in calculations involving π.

Integrating Technology

The π key on your calculator can be used to find decimal approximations for expressions that contain π. To perform the calculation at the right, enter 6 × π = .

HOW TO Find the circumference of a circle with a diameter of 6 in.

$C = \pi d$

$C = \pi(6)$

- The diameter of the circle is given. Use the circumference formula that involves the diameter. $d = 6$.

$C = 6\pi$

- The exact circumference of the circle is 6π in.

$C \approx 18.85$

- An approximate measure is found by using the π key on a calculator.

The circumference is approximately 18.85 in.

Copyright © Houghton Mifflin Company. All rights reserved.

Example 1

A carpenter is designing a square patio with a perimeter of 44 ft. What is the length of each side?

Strategy
To find the length of each side, use the formula for the perimeter of a square. Substitute 44 for P and solve for s.

Solution
$P = 4s$
$44 = 4s$
$11 = s$

The length of each side of the patio is 11 ft.

You Try It 1

The infield of a softball field is a square with each side of length 60 ft. Find the perimeter of the infield.

Your strategy

Your solution

Example 2

The dimensions of a triangular sail are 18 ft, 11 ft, and 15 ft. What is the perimeter of the sail?

Strategy
To find the perimeter, use the formula for the perimeter of a triangle. Substitute 18 for a, 11 for b, and 15 for c. Solve for P.

Solution
$P = a + b + c$
$P = 18 + 11 + 15$
$P = 44$

The perimeter of the sail is 44 ft.

You Try It 2

What is the perimeter of a standard piece of typing paper that measures $8\frac{1}{2}$ in. by 11 in.?

Your strategy

Your solution

Example 3

Find the circumference of a circle with a radius of 15 cm. Round to the nearest hundredth.

Strategy
To find the circumference, use the circumference formula that involves the radius. An approximation is asked for; use the π key on a calculator. $r = 15$.

Solution
$C = 2\pi r = 2\pi(15) = 30\pi \approx 94.25$

The circumference is approximately 94.25 cm.

You Try It 3

Find the circumference of a circle with a diameter of 9 in. Give the exact measure.

Your strategy

Your solution

Solutions on p. S10

Copyright © Houghton Mifflin Company. All rights reserved.

Objective B To solve problems involving the area of a geometric figure

Area is the amount of surface in a region. Area can be used to describe the size of a rug, a parking lot, a farm, or a national park. Area is measured in square units.

Point of Interest

Polygonal numbers are whole numbers that can be represented as regular geometric figures. For example, a square number is one that can be represented as a square array.

o	o o o o	o o o o o o o o o	o o o o o o o o o o o o o o o o
1	4	9	16

The square numbers are 1, 4, 9, 16, 25, ... They can be represented as $1^2, 2^2, 3^2, 4^2, 5^2, \ldots$.

A square that measures 1 in. on each side has an area of 1 square inch, written 1 in^2.

A square that measures 1 cm on each side has an area of 1 square centimeter, written 1 cm^2.

Larger areas can be measured in square feet (ft^2), square meters (m^2), square miles (mi^2), acres (43,560 ft^2), or any other square unit.

The area of a geometric figure is the number of squares that are necessary to cover the figure. In the figures below, two rectangles have been drawn and covered with squares. In the figure on the left, 12 squares, each of area 1 cm^2, were used to cover the rectangle. The area of the rectangle is 12 cm^2. In the figure on the right, 6 squares, each of area 1 in^2, were used to cover the rectangle. The area of the rectangle is 6 in^2.

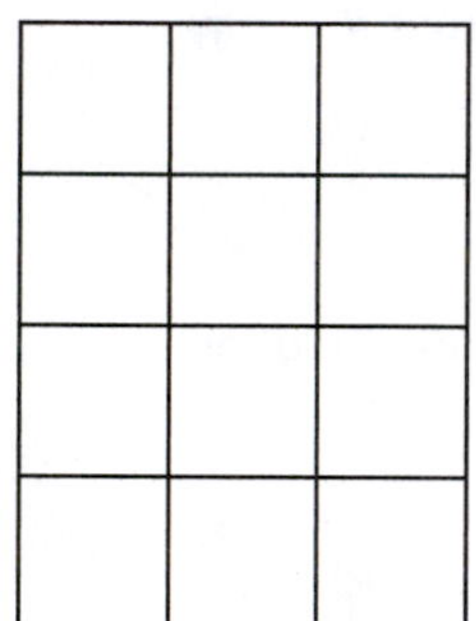

The area of the rectangle is 12 cm^2.

The area of the rectangle is 6 in^2.

Note from the above figures that the area of a rectangle can be found by multiplying the length of the rectangle by its width.

Area of a Rectangle

Let L represent the length and W the width of a rectangle. The area, A, of the rectangle is given by $A = LW$.

HOW TO Find the area of the rectangle shown at the right.

$A = LW = 11(7) = 77$

The area is 77 m^2.

Copyright © Houghton Mifflin Company. All rights reserved.

A square is a rectangle in which all sides are the same length. Therefore, both the length and the width of a square can be represented by s, and $A = LW = s \cdot s = s^2$.

Area of a Square

Let s represent the length of a side of a square. The area, A, of the square is given by $A = s^2$.

$A = s \cdot s = s^2$

HOW TO Find the area of the square shown at the right.

$A = s^2 = 9^2 = 81$

The area is 81 mi^2.

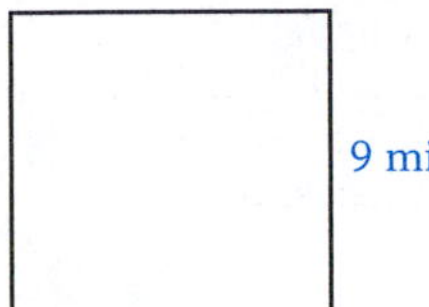

Figure $ABCD$ is a parallelogram. BC is the **base,** b, of the parallelogram. AE, perpendicular to the base, is the **height,** h, of the parallelogram.

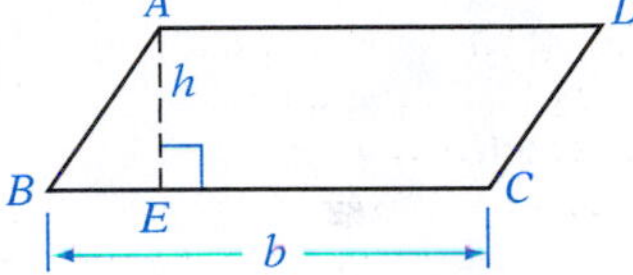

Any side of a parallelogram can be designated as the base. The corresponding height is found by drawing a line segment perpendicular to the base from the opposite side.

A rectangle can be formed from a parallelogram by cutting a right triangle from one end of the parallelogram and attaching it to the other end. The area of the resulting rectangle will equal the area of the original parallelogram.

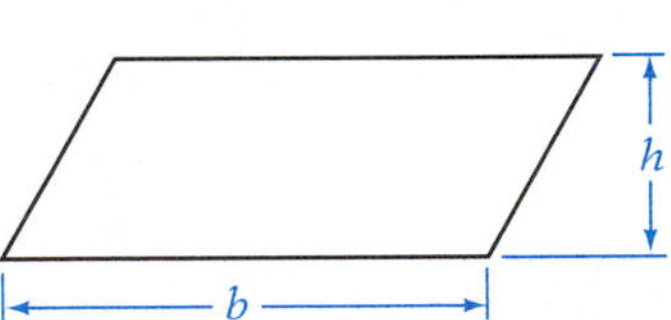

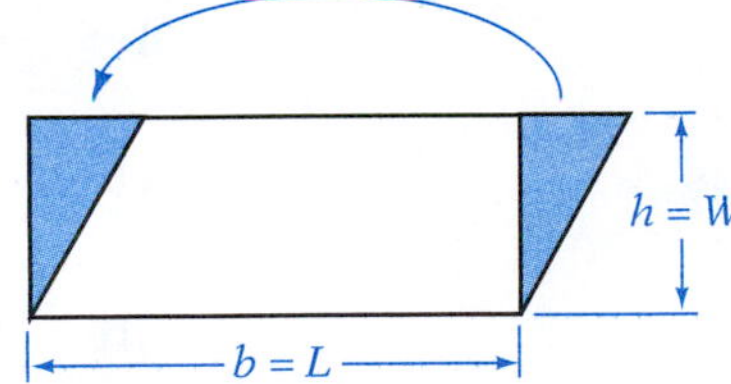

Area of a Parallelogram

Let b represent the length of the base and h the height of a parallelogram. The area, A, of the parallelogram is given by $A = bh$.

HOW TO Find the area of the parallelogram shown at the right.

$A = bh = 12 \cdot 6 = 72$

The area is 72 m^2.

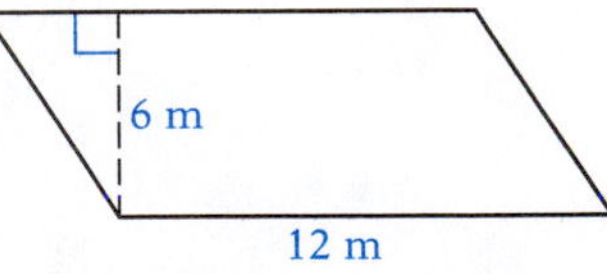

Copyright © Houghton Mifflin Company. All rights reserved.

Figure ABC is a triangle. AB is the **base,** b, of the triangle. CD, perpendicular to the base, is the **height,** h, of the triangle.

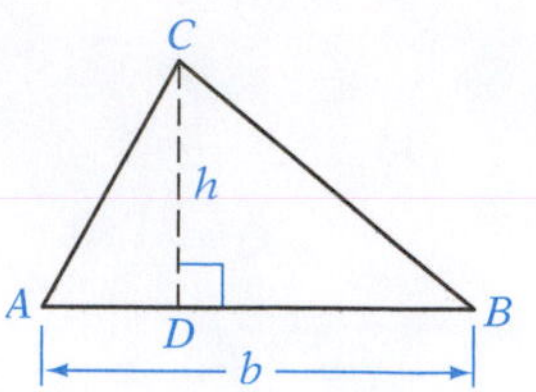

Any side of a triangle can be designated as the base. The corresponding height is found by drawing a line segment perpendicular to the base from the vertex opposite the base.

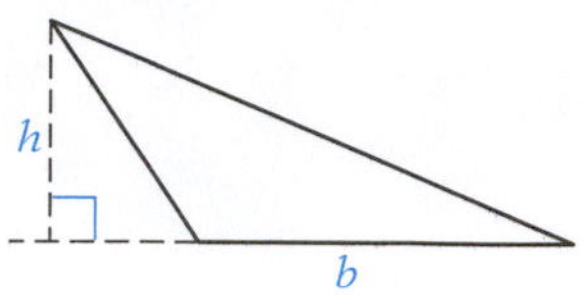

Consider the triangle with base b and height h shown at the right. By extending a line from C parallel to the base AB and equal in length to the base, and extending a line from B parallel to AC and equal in length to AC, a parallelogram is formed. The area of the parallelogram is bh and is twice the area of the triangle. Therefore, the area of the triangle is one-half the area of the parallelogram, or $\frac{1}{2}bh$.

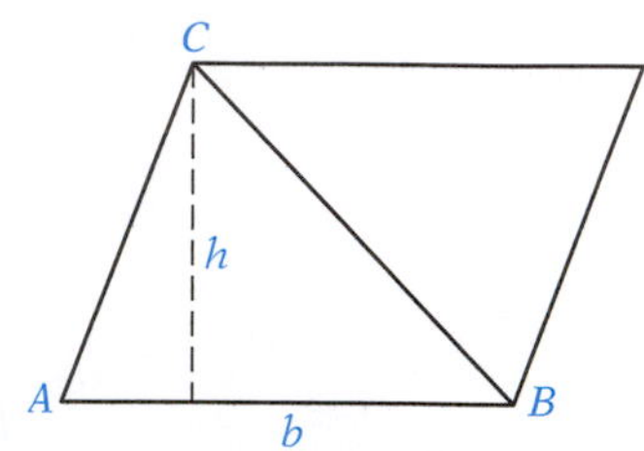

Area of a Triangle

Let b represent the length of the base and h the height of a triangle. The area, A, of the triangle is given by $A = \frac{1}{2}bh$.

Integrating Technology

To calculate the area of the triangle shown at the right, you can enter

18 × 6 ÷ 2 =

or

.5 × 18 × 6 = .

HOW TO Find the area of a triangle with a base of 18 cm and a height of 6 cm.

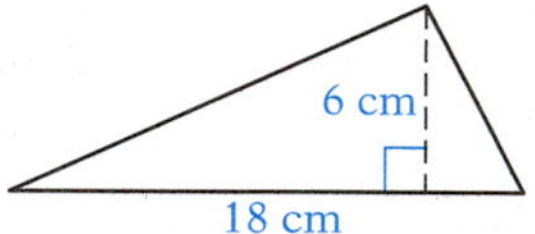

$$A = \frac{1}{2}bh = \frac{1}{2} \cdot 18 \cdot 6 = 54$$

The area is 54 cm².

Figure $ABCD$ is a trapezoid. AB is one **base,** b_1, of the trapezoid, and CD is the other base, b_2. AE, perpendicular to the two bases, is the **height,** h.

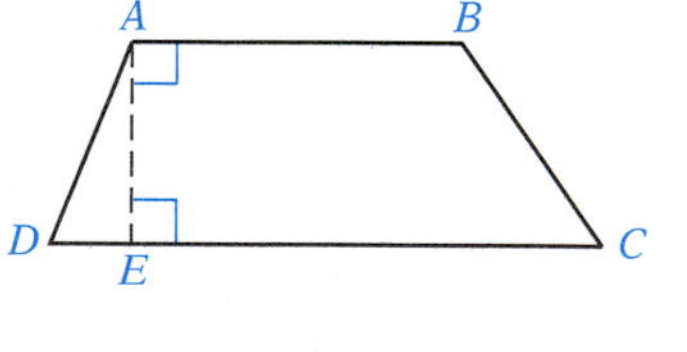

In the trapezoid at the right, the line segment BD divides the trapezoid into two triangles, ABD and BCD. In triangle ABD, b_1 is the base and h is the height. In triangle BCD, b_2 is the base and h is the height. The area of the trapezoid is the sum of the areas of the two triangles.

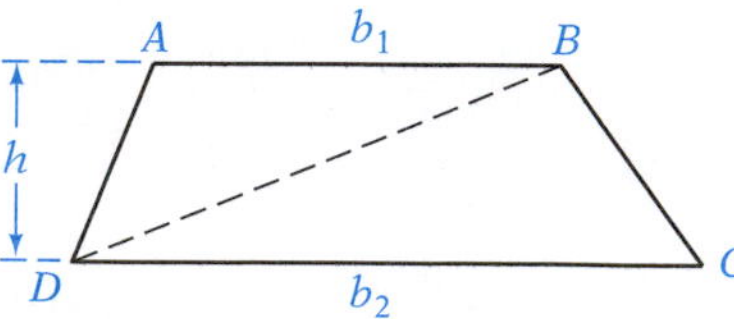

$$\text{Area of trapezoid } ABCD = \text{area of triangle } ABD + \text{area of triangle } BCD$$

$$= \frac{1}{2}b_1h + \frac{1}{2}b_2h = \frac{1}{2}h(b_1 + b_2)$$

Copyright © Houghton Mifflin Company. All rights reserved.

Area of a Trapezoid

Let b_1 and b_2 represent the lengths of the bases and h the height of a trapezoid. The area, A, of the trapezoid is given by $A = \frac{1}{2}h(b_1 + b_2)$.

HOW TO Find the area of a trapezoid that has bases measuring 15 in. and 5 in. and a height of 8 in.

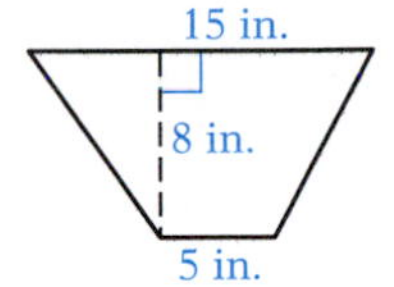

$$A = \frac{1}{2}h(b_1 + b_2)$$

$$= \frac{1}{2} \cdot 8(15 + 5) = 4(20) = 80$$

The area is 80 in^2.

The area of a circle is equal to the product of π and the square of the radius.

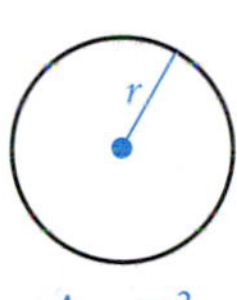

Area of a Circle

The area, A, of a circle with radius r is given by $A = \pi r^2$.

HOW TO Find the area of a circle that has a radius of 6 cm.

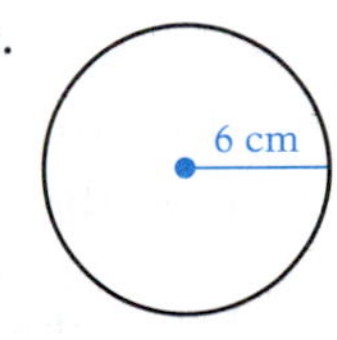

$A = \pi r^2$
$A = \pi(6)^2$
$A = \pi(36)$
- Use the formula for the area of a circle. $r = 6$

$A = 36\pi$
- The exact area of the circle is 36π cm^2.

$A \approx 113.10$
- An approximate measure is found by using the π key on a calculator.

The approximate area of the circle is 113.10 cm^2.

Integrating Technology

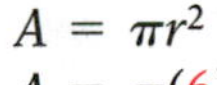

To approximate 36π on your calculator, enter 36 × π =.

For your reference, all of the formulas for the perimeters and areas of the geometric figures presented in this section are listed in the Chapter Summary located at the end of this chapter.

Copyright © Houghton Mifflin Company. All rights reserved.

Example 4

The Parks and Recreation Department of a city plans to plant grass seed in a playground that has the shape of a trapezoid, as shown below. Each bag of grass seed will seed 1500 ft^2. How many bags of grass seed should the department purchase?

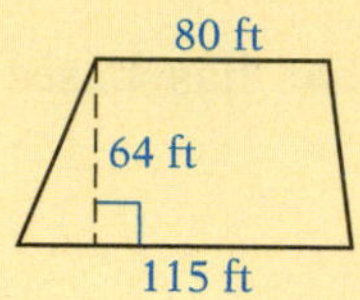

Strategy

To find the number of bags to be purchased:

- Use the formula for the area of a trapezoid to find the area of the playground.
- Divide the area of the playground by the area one bag will seed (1500).

Solution

$$A = \frac{1}{2}h(b_1 + b_2)$$

$$A = \frac{1}{2} \cdot 64(80 + 115)$$

$$A = 6240$$

• **The area of the playground is 6240 ft^2.**

$6240 \div 1500 = 4.16$

Because a portion of a fifth bag is needed, 5 bags of grass seed should be purchased.

You Try It 4

An interior designer decides to wallpaper two walls of a room. Each roll of wallpaper will cover 30 ft^2. Each wall measures 8 ft by 12 ft. How many rolls of wallpaper should be purchased?

Your strategy

Your solution

Example 5

Find the area of a circle with a diameter of 5 ft. Give the exact measure.

Strategy

To find the area:

- Find the radius of the circle.
- Use the formula for the area of a circle. Leave the answer in terms of π.

Solution

$$r = \frac{1}{2}d = \frac{1}{2}(5) = 2.5$$

$$A = \pi r^2 = \pi(2.5)^2 = \pi(6.25) = 6.25\pi$$

The area of the circle is 6.25π ft^2.

You Try It 5

Find the area of a circle with a radius of 11 cm. Round to the nearest hundredth.

Your strategy

Your solution

Solutions on p. S10

Copyright © Houghton Mifflin Company. All rights reserved.

3.2 Exercises

Objective A **To solve problems involving the perimeter of a geometric figure**

Name each polygon.

1. 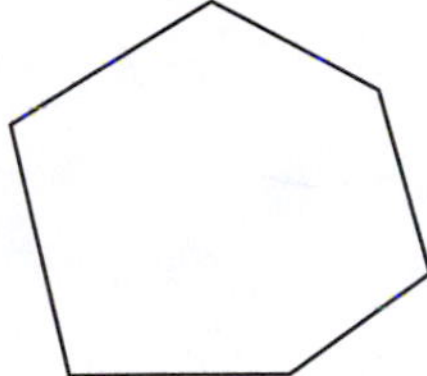

2. 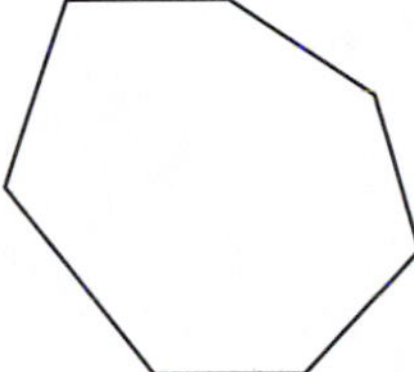

3. 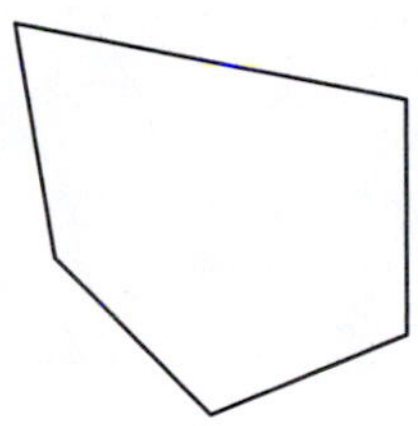

4. 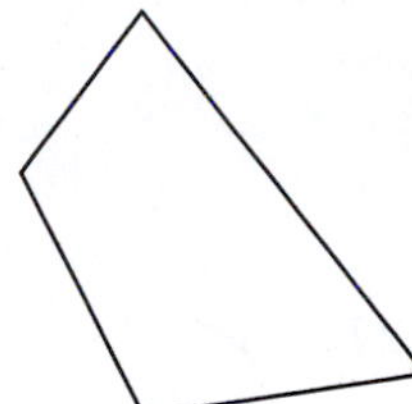

Classify the triangle as isosceles, equilateral, or scalene.

5. 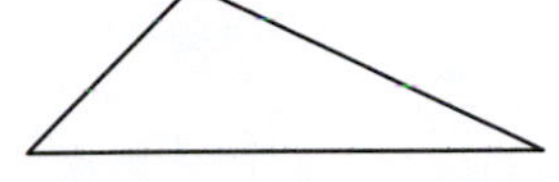

6. 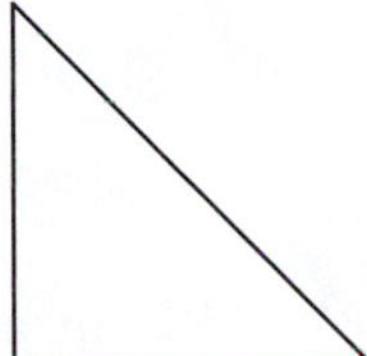

7. 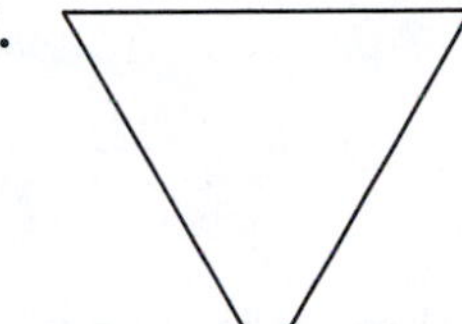

8. 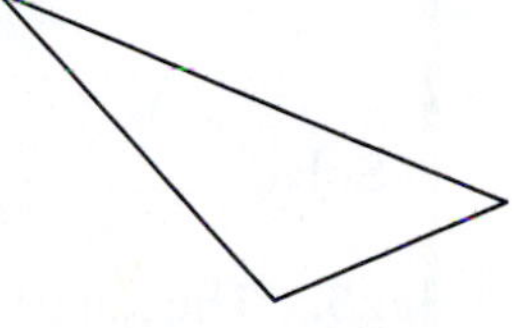

Classify the triangle as acute, obtuse, or right.

9. 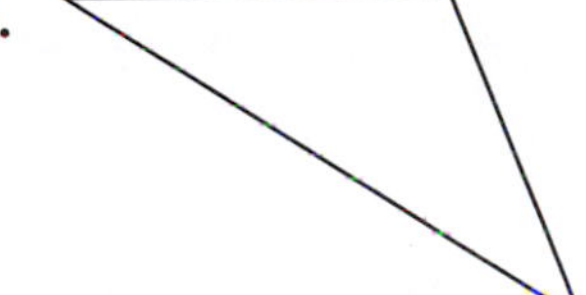

10. 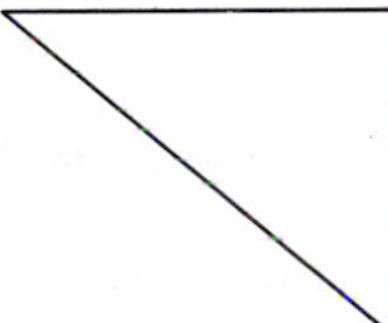

11. 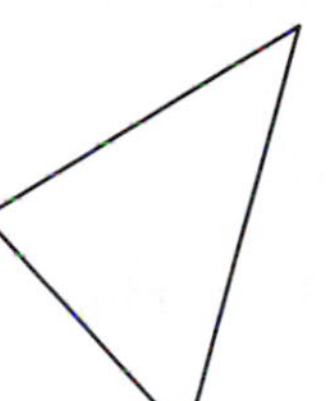

12. 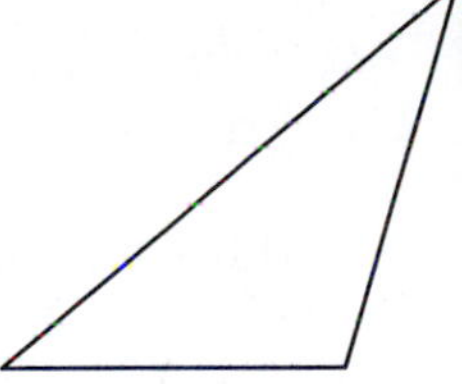

Find the perimeter of the figure.

13. 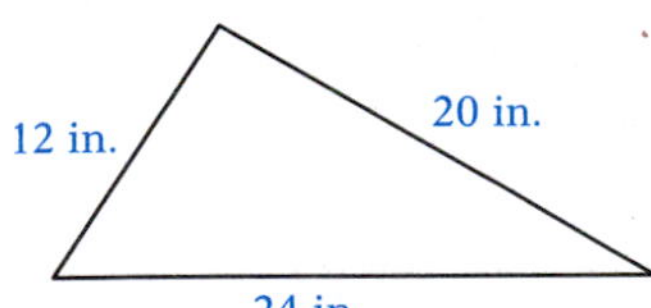

14. 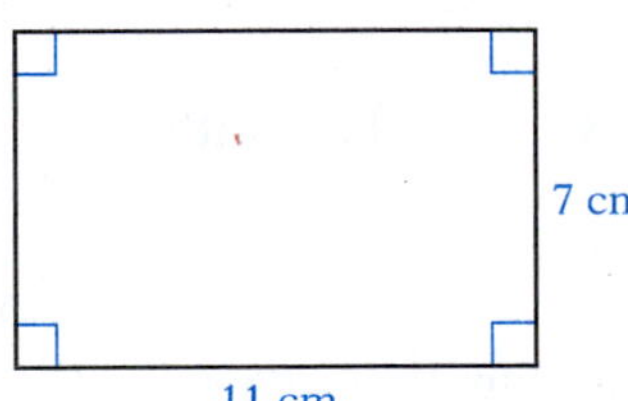

15. 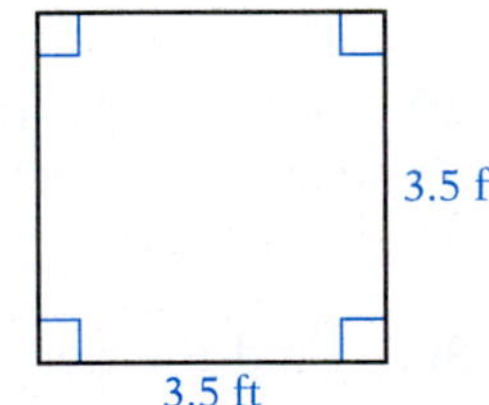

16. 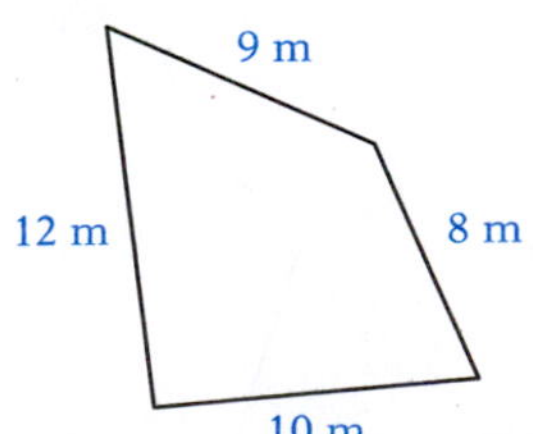

17. 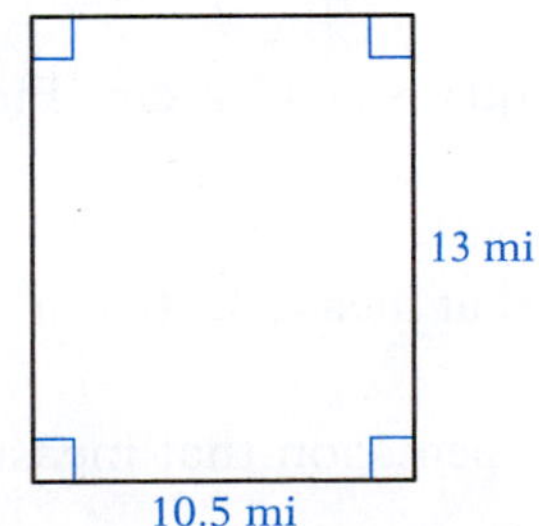

18. 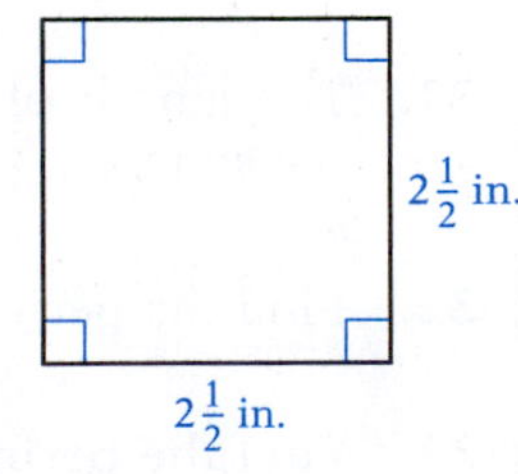

Copyright © Houghton Mifflin Company. All rights reserved.

Find the circumference of the figure. Give both the exact value and an approximation to the nearest hundredth.

19.

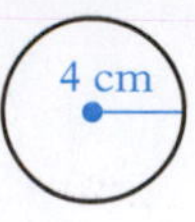

20.

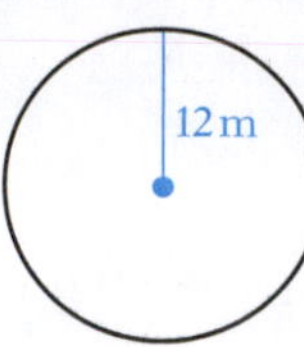

21.

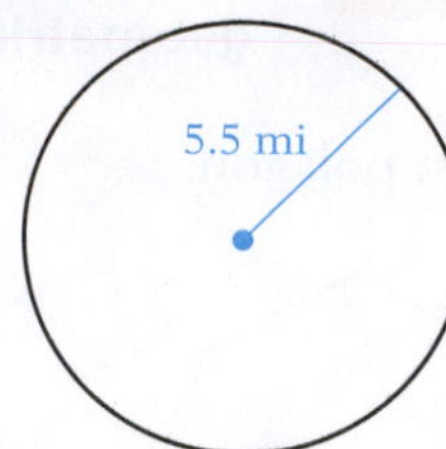

22.

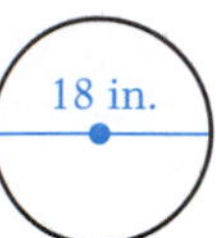

23.

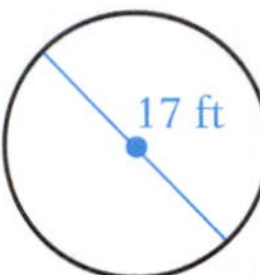

24.

Solve.

25. The lengths of the three sides of a triangle are 3.8 cm, 5.2 cm, and 8.4 cm. Find the perimeter of the triangle.

26. The lengths of the three sides of a triangle are 7.5 m, 6.1 m, and 4.9 m. Find the perimeter of the triangle.

27. The length of each of two sides of an isosceles triangle is $2\frac{1}{2}$ cm. The third side measures 3 cm. Find the perimeter of the triangle.

28. The length of each side of an equilateral triangle is $4\frac{1}{2}$ in. Find the perimeter of the triangle.

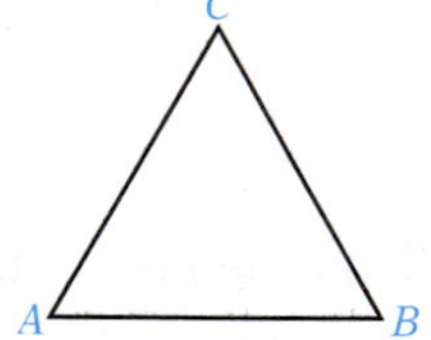

29. A rectangle has a length of 8.5 m and a width of 3.5 m. Find the perimeter of the rectangle.

30. Find the perimeter of a rectangle that has a length of $5\frac{1}{2}$ ft and a width of 4 ft.

31. The length of each side of a square is 12.2 cm. Find the perimeter of the square.

32. Find the perimeter of a square that measures 0.5 m on each side.

33. Find the perimeter of a regular pentagon that measures 3.5 in. on each side.

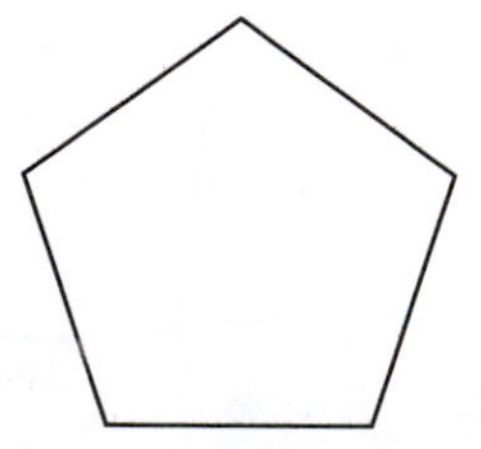

Copyright © Houghton Mifflin Company. All rights reserved.

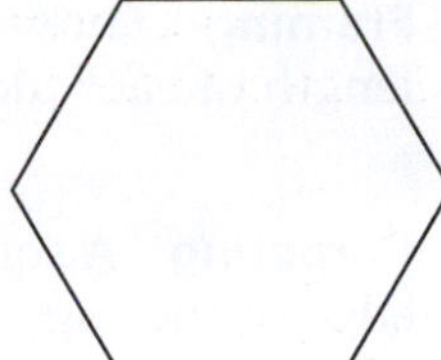

34. What is the perimeter of a regular hexagon that measures 8.5 cm on each side?

35. The radius of a circle is 4.2 cm. Find the length of a diameter of the circle.

36. The diameter of a circle is 0.56 m. Find the length of a radius of the circle.

37. Find the circumference of a circle that has a diameter of 1.5 in. Give the exact value.

38. The diameter of a circle is 4.2 ft. Find the circumference of the circle. Round to the nearest hundredth.

39. The radius of a circle is 36 cm. Find the circumference of the circle. Round to the nearest hundredth.

40. Find the circumference of a circle that has a radius of 2.5 m. Give the exact value.

41. **Fencing** How many feet of fencing should be purchased for a rectangular garden that is 18 ft long and 12 ft wide?

42. **Quilting** How many meters of binding are required to bind the edge of a rectangular quilt that measures 3.5 m by 8.5 m?

43. **Carpeting** Wall-to-wall carpeting is installed in a room that is 12 ft long and 10 ft wide. The edges of the carpet are nailed to the floor. Along how many feet must the carpet be nailed down?

44. **Fencing** The length of a rectangular park is 55 yd. The width is 47 yd. How many yards of fencing are needed to surround the park?

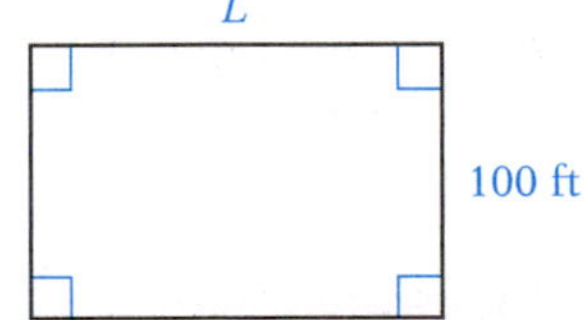

45. **Playgrounds** The perimeter of a rectangular playground is 440 ft. If the width is 100 ft, what is the length of the playground?

46. **Gardens** A rectangular vegetable garden has a perimeter of 64 ft. The length of the garden is 20 ft. What is the width of the garden?

47. **Banners** Each of two sides of a triangular banner measures 18 in. If the perimeter of the banner is 46 in., what is the length of the third side of the banner?

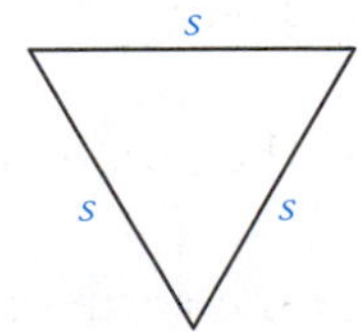

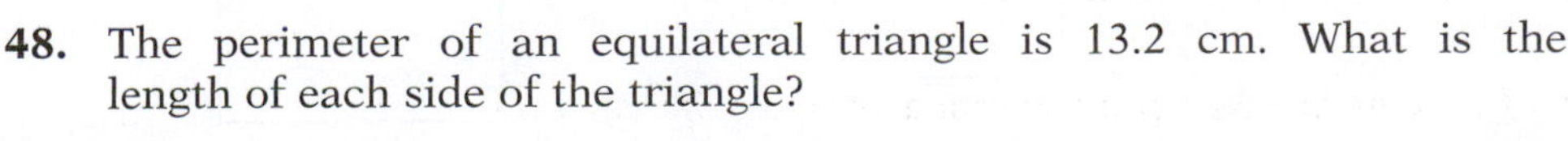

48. The perimeter of an equilateral triangle is 13.2 cm. What is the length of each side of the triangle?

Copyright © Houghton Mifflin Company. All rights reserved.

49. **Framing** The perimeter of a square picture frame is 48 in. Find the length of each side of the frame.

50. **Carpeting** A square rug has a perimeter of 32 ft. Find the length of each edge of the rug.

51. The circumference of a circle is 8 cm. Find the length of a diameter of the circle. Round to the nearest hundredth.

52. The circumference of a circle is 15 in. Find the length of a radius of the circle. Round to the nearest hundredth.

53. **Carpentry** Find the length of molding needed to put around a circular table that is 4.2 ft in diameter. Round to the nearest hundredth.

54. **Carpeting** How much binding is needed to bind the edge of a circular rug that is 3 m in diameter? Round to the nearest hundredth.

55. **Cycling** A bicycle tire has a diameter of 24 in. How many feet does the bicycle travel when the wheel makes eight revolutions? Round to the nearest hundredth.

56. **Cycling** A tricycle tire has a diameter of 12 in. How many feet does the tricycle travel when the wheel makes twelve revolutions? Round to the nearest hundredth.

57. **Earth Science** The distance from the surface of Earth to its center is 6356 km. What is the circumference of Earth? Round to the nearest hundredth.

58. **Sewing** Bias binding is to be sewed around the edge of a rectangular tablecloth measuring 72 in. by 45 in. If the bias binding comes in packages containing 15 ft of binding, how many packages of bias binding are needed for the tablecloth?

Objective B To solve problems involving the area of a geometric figure

Find the area of the figure.

59.

60.

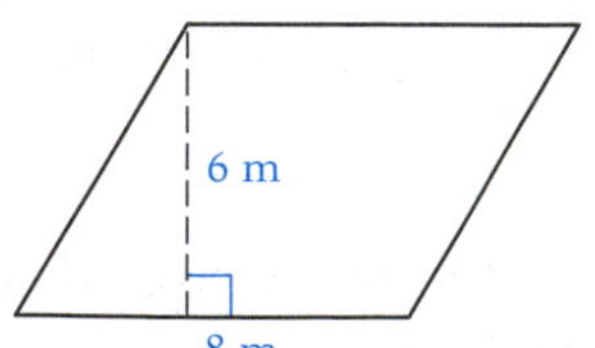

61.

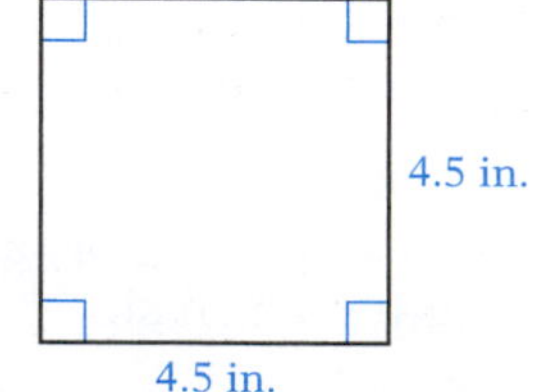

Copyright © Houghton Mifflin Company. All rights reserved.

62.

12 in.

20 in.

63.

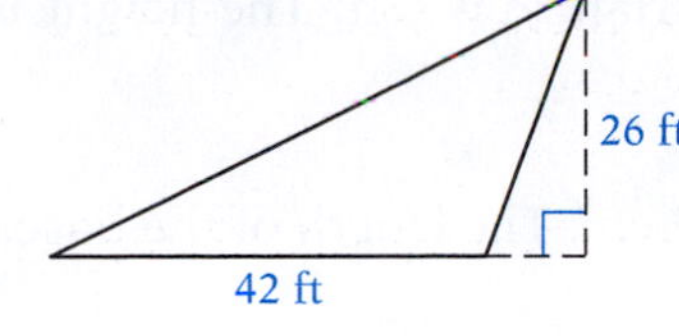

64.

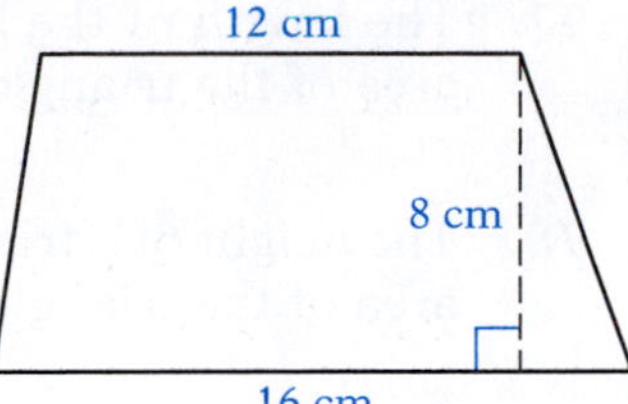

Find the area of the figure. Give both the exact value and an approximation to the nearest hundredth.

65.

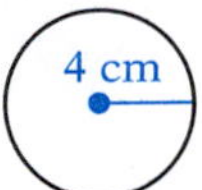

66.

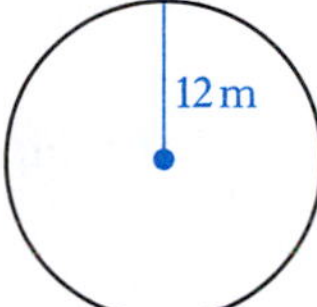

67.

68.

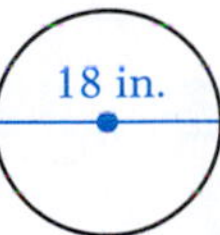

69.

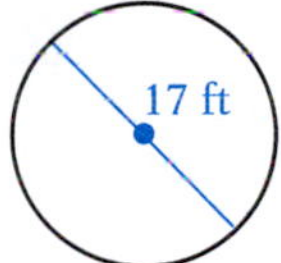

70.

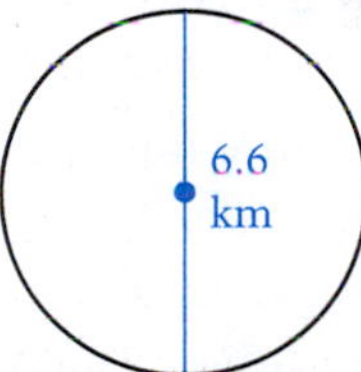

Solve.

71. The length of a side of a square is 12.5 cm. Find the area of the square.

72. Each side of a square measures $3\frac{1}{2}$ in. Find the area of the square.

73. The length of a rectangle is 38 in., and the width is 15 in. Find the area of the rectangle.

74. Find the area of a rectangle that has a length of 6.5 m and a width of 3.8 m.

75. The length of the base of a parallelogram is 16 in., and the height is 12 in. Find the area of the parallelogram.

76. The height of a parallelogram is 3.4 m, and the length of the base is 5.2 m. Find the area of the parallelogram.

Copyright © Houghton Mifflin Company. All rights reserved.

77. The length of the base of a triangle is 6 ft. The height is 4.5 ft. Find the area of the triangle.

78. The height of a triangle is 4.2 cm. The length of the base is 5 cm. Find the area of the triangle.

79. The length of one base of a trapezoid is 35 cm, and the length of the other base is 20 cm. If the height is 12 cm, what is the area of the trapezoid?

80. The height of a trapezoid is 5 in. The bases measure 16 in. and 18 in. Find the area of the trapezoid.

81. The radius of a circle is 5 in. Find the area of the circle. Give the exact value.

82. Find the area of a circle with a radius of 14 m. Round to the nearest hundredth.

83. Find the area of a circle that has a diameter of 3.4 ft. Round to the nearest hundredth.

84. The diameter of a circle is 6.5 m. Find the area of the circle. Give the exact value.

85. **Telescopes** The dome of the Hale telescope at Mount Palomar, California, has a diameter of 200 in. Find the area across the dome. Give the exact value.

86. **Patios** What is the area of a square patio that measures 8.5 m on each side?

87. **Gardens** Find the area of a rectangular flower garden that measures 14 ft by 9 ft.

88. **Irrigation** An irrigation system waters a circular field that has a 50-foot radius. Find the area watered by the irrigation system. Give the exact value.

89. **Athletic Fields** Artificial turf is being used to cover a playing field. If the field is rectangular with a length of 100 yd and a width of 75 yd, how much artificial turf must be purchased to cover the field?

90. **Interior Decorating** A fabric wall hanging is to fill a space that measures 5 m by 3.5 m. Allowing for 0.1 m of the fabric to be folded back along each edge, how much fabric must be purchased for the wall hanging?

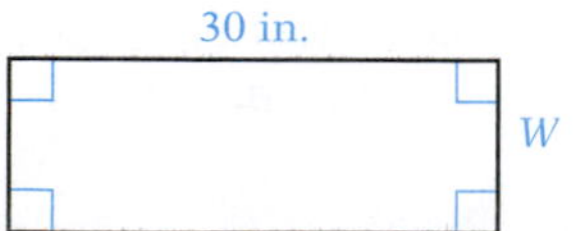

91. The area of a rectangle is 300 in^2. If the length of the rectangle is 30 in., what is the width?

Copyright © Houghton Mifflin Company. All rights reserved.

92. The width of a rectangle is 12 ft. If the area is 312 ft^2, what is the length of the rectangle?

93. The height of a triangle is 5 m. The area of the triangle is 50 m^2. Find the length of the base of the triangle.

94. The area of a parallelogram is 42 m^2. If the height of the parallelogram is 7 m, what is the length of the base?

95. **Home Maintenance** You plan to stain the wooden deck attached to your house. The deck measures 10 ft by 8 ft. If a quart of stain will cover 50 ft^2, how many quarts of stain should you buy?

96. **Flooring** You want to tile your kitchen floor. The floor measures 12 ft by 9 ft. How many tiles, each a square with side $1\frac{1}{2}$ ft, should you purchase for the job?

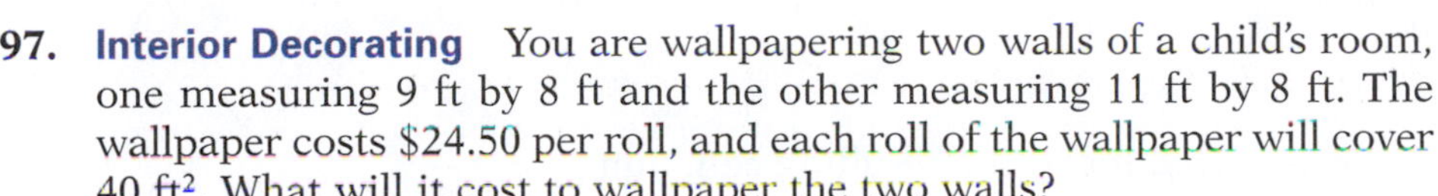

97. **Interior Decorating** You are wallpapering two walls of a child's room, one measuring 9 ft by 8 ft and the other measuring 11 ft by 8 ft. The wallpaper costs $24.50 per roll, and each roll of the wallpaper will cover 40 ft^2. What will it cost to wallpaper the two walls?

98. **Parks** An urban renewal project involves reseeding a park that is in the shape of a square 60 ft on each side. Each bag of grass seed costs $9.75 and will seed 1200 ft^2. How much money should be budgeted for buying grass seed for the park?

99. A circle has a radius of 8 in. Find the increase in area when the radius is increased by 2 in. Round to the nearest hundredth.

100. A circle has a radius of 6 cm. Find the increase in area when the radius is doubled. Round to the nearest hundredth.

101. **Carpeting** You want to install wall-to-wall carpeting in your living room, which measures 15 ft by 24 ft. If the cost of the carpet you would like to purchase is $21.95 per square yard, what will be the cost of the carpeting for your living room? (*Hint:* 9 ft^2 = 1 yd^2)

102. **Interior Decorating** You want to paint the walls of your bedroom. Two walls measure 15 ft by 9 ft, and the other two walls measure 12 ft by 9 ft. The paint you wish to purchase costs $12.98 per gallon, and each gallon will cover 400 ft^2 of wall. Find the total amount you will spend on paint.

103. **Landscaping** A walkway 2 m wide surrounds a rectangular plot of grass. The plot is 30 m long and 20 m wide. What is the area of the walkway?

Copyright © Houghton Mifflin Company. All rights reserved.

104. **Interior Decorating** Pleated draperies for a window must be twice as wide as the width of the window. Draperies are being made for four windows, each 2 ft wide and 4 ft high. Because the drapes will fall slightly below the window sill, and because extra fabric will be needed for hemming the drapes, 1 ft must be added to the height of the window. How much material must be purchased to make the drapes?

105. **Construction** Find the cost of plastering the walls of a room 22 ft long, 25 ft 6 in. wide, and 8 ft high. Subtract 120 ft^2 for windows and doors. The cost is $2.50 per square foot.

APPLYING THE CONCEPTS

106. If both the length and the width of a rectangle are doubled, how many times larger is the area of the resulting rectangle?

107. A **hexagram** is a six-pointed star formed by extending each of the sides of a regular hexagon into an equilateral triangle. A hexagram is shown at the right. Use a pencil, a paper, a protractor, and a ruler to create a hexagram.

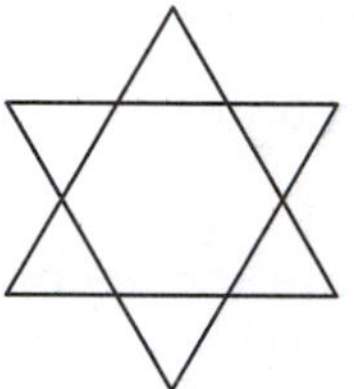

108. If the formula $C = \pi d$ is solved for π, the resulting equation is $\pi = \frac{C}{d}$. Therefore, π is the ratio of the circumference of a circle to the length of its diameter. Use several circular objects, such as coins, plates, tin cans, and wheels, to show that the ratio of the circumference of each object to its diameter is approximately 3.14.

109. Derive a formula for the area of a circle in terms of the diameter of the circle.

110. Determine whether the statement is always true, sometimes true, or never true.
   a. If two triangles have the same perimeter, then they have the same area.
   b. If two rectangles have the same area, then they have the same perimeter.
   c. If two squares have the same area, then the sides of the squares have the same length.
   d. An equilateral triangle is also an isosceles triangle.
   e. All the radii (plural of radius) of a circle are equal.
   f. All the diameters of a circle are equal.

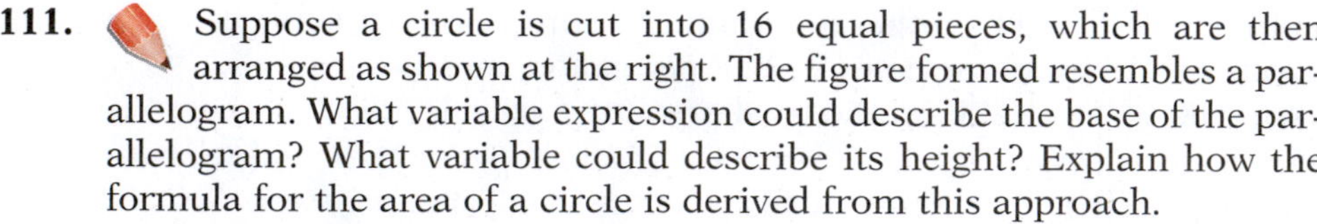

111. Suppose a circle is cut into 16 equal pieces, which are then arranged as shown at the right. The figure formed resembles a parallelogram. What variable expression could describe the base of the parallelogram? What variable could describe its height? Explain how the formula for the area of a circle is derived from this approach.

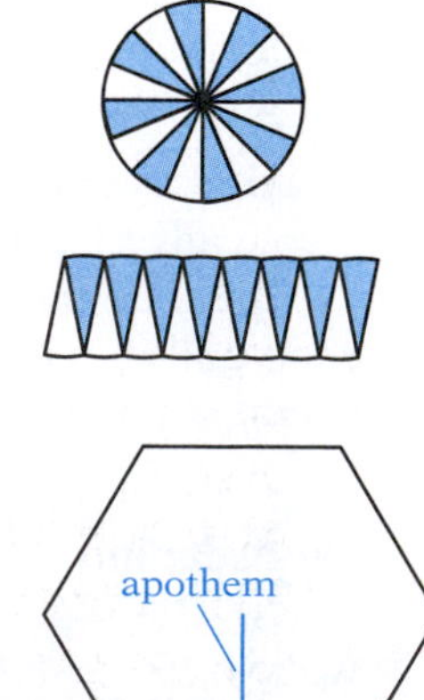

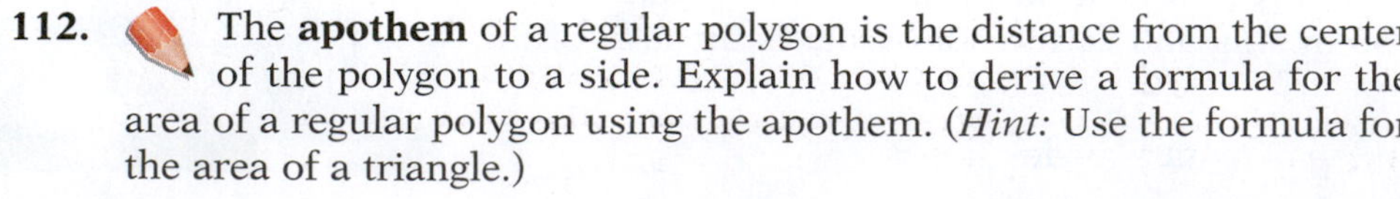

112. The **apothem** of a regular polygon is the distance from the center of the polygon to a side. Explain how to derive a formula for the area of a regular polygon using the apothem. (*Hint:* Use the formula for the area of a triangle.)

Copyright © Houghton Mifflin Company. All rights reserved.

3.3 Solids

Objective A

To solve problems involving the volume of a solid

Geometric solids are figures in space. Five common geometric solids are the rectangular solid, the sphere, the cylinder, the cone, and the pyramid.

A **rectangular solid** is one in which all six sides, called **faces,** are rectangles. The variable L is used to represent the length of a rectangular solid, W its width, and H its height.

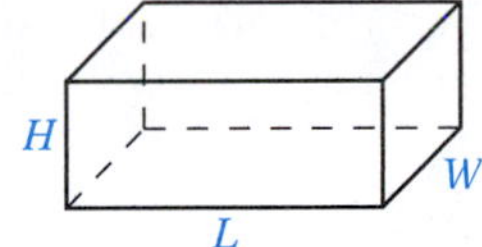

A **sphere** is a solid in which all points are the same distance from point O, which is called the **center** of the sphere. The **diameter,** d, of a sphere is a line across the sphere going through point O. The **radius,** r, is a line from the center to a point on the sphere. AB is a diameter and OC is a radius of the sphere shown at the right.

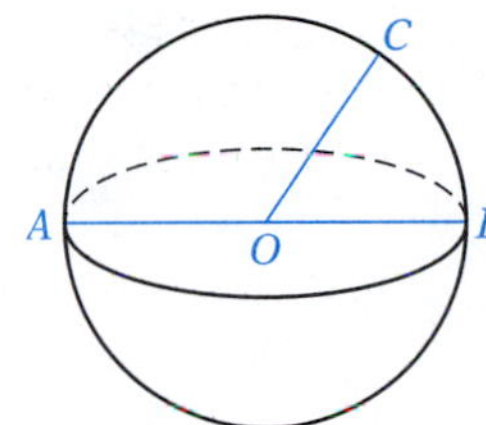

$$d = 2r \text{ or } r = \frac{1}{2}d$$

The most common cylinder, called a **right circular cylinder,** is one in which the bases are circles and are perpendicular to the height of the cylinder. The variable r is used to represent the radius of a base of a cylinder, and h represents the height. In this text, only right circular cylinders are discussed.

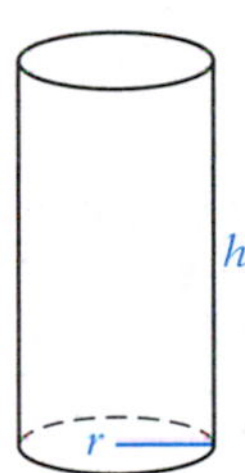

A **right circular cone** is obtained when one base of a right circular cylinder is shrunk to a point, called a **vertex,** V. The variable r is used to represent the radius of the base of the cone, and h represents the height. The variable ℓ is used to represent the **slant height,** which is the distance from a point on the circumference of the base to the vertex. In this text, only right circular cones are discussed.

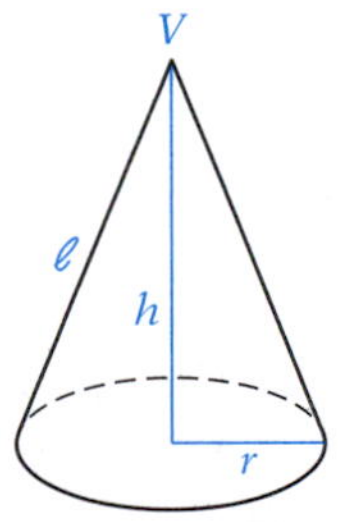

The base of a **regular pyramid** is a regular polygon, and the sides are isosceles triangles. The height, h, is the distance from the vertex, V, to the base and is perpendicular to the base. The variable ℓ is used to represent the **slant height,** which is the height of one of the isosceles triangles on the face of the pyramid. The regular square pyramid at the right has a square base. This is the only type of pyramid discussed in this text.

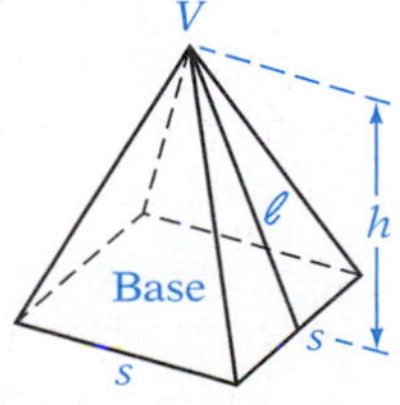

Copyright © Houghton Mifflin Company. All rights reserved.

A **cube** is a special type of rectangular solid. Each of the six faces of a cube is a square. The variable s is used to represent the length of one side of a cube.

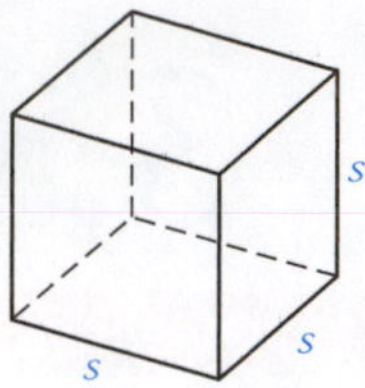

Volume is a measure of the amount of space inside a figure in space. Volume can be used to describe the amount of heating gas used for cooking, the amount of concrete delivered for the foundation of a house, or the amount of water in storage for a city's water supply.

A cube that is 1 ft on each side has a volume of 1 cubic foot, which is written 1 ft^3. A cube that measures 1 cm on each side has a volume of 1 cubic centimeter, which is written 1 cm^3.

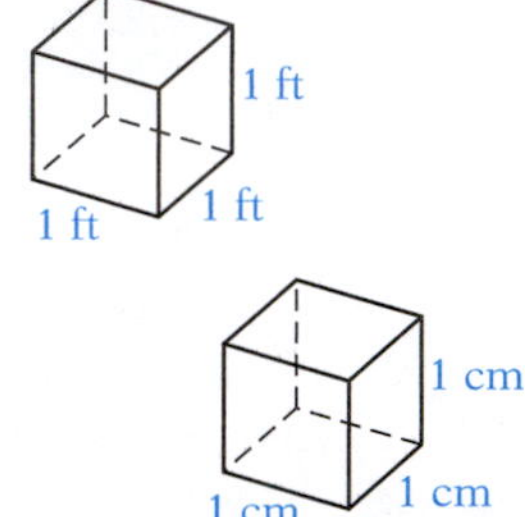

The volume of a solid is the number of cubes that are necessary to exactly fill the solid. The volume of the rectangular solid at the right is 24 cm^3 because it will hold exactly 24 cubes, each 1 cm on a side. Note that the volume can be found by multiplying the length times the width times the height.

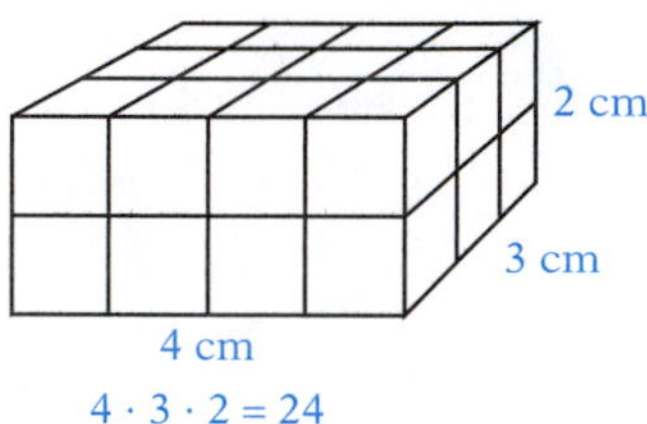

The formulas for the volumes of the geometric solids described above are given below.

Volumes of Geometric Solids

The volume, V, of a **rectangular solid** with length L, width W, and height H is given by $V = LWH$.

The volume, V, of a **cube** with side s is given by $V = s^3$.

The volume, V, of a **sphere** with radius r is given by $V = \frac{4}{3}\pi r^3$.

The volume, V, of a **right circular cylinder** is given by $V = \pi r^2 h$, where r is the radius of the base and h is the height.

The volume, V, of a **right circular cone** is given by $V = \frac{1}{3}\pi r^2 h$, where r is the radius of the circular base and h is the height.

The volume, V, of a **regular square pyramid** is given by $V = \frac{1}{3}s^2 h$, where s is the length of a side of the base and h is the height.

Copyright © Houghton Mifflin Company. All rights reserved.

HOW TO Find the volume of a sphere with a diameter of 6 in.

$r = \frac{1}{2}d = \frac{1}{2}(6) = 3$ • First find the radius of the sphere.

$V = \frac{4}{3}\pi r^3$ • Use the formula for the volume of a sphere.

$V = \frac{4}{3}\pi(3)^3$

$V = \frac{4}{3}\pi(27)$

$V = 36\pi$ • The exact volume of the sphere is 36π in^3.

$V \approx 113.10$ • An approximate measure can be found by using the π key on a calculator.

The approximate volume is 113.10 in^3.

Integrating Technology

To approximate 36π on your calculator, enter 36 × π = .

Example 1

The length of a rectangular solid is 5 m, the width is 3.2 m, and the height is 4 m. Find the volume of the solid.

Strategy
To find the volume, use the formula for the volume of a rectangular solid. $L = 5$, $W = 3.2$, $H = 4$

Solution
$V = LWH = 5(3.2)(4) = 64$

The volume of the rectangular solid is 64 m^3.

You Try It 1

Find the volume of a cube that measures 2.5 m on a side.

Your strategy

Your solution

Example 2

The radius of the base of a cone is 8 cm. The height is 12 cm. Find the volume of the cone. Round to the nearest hundredth.

Strategy
To find the volume, use the formula for the volume of a cone. An approximation is asked for; use the π key on a calculator.
$r = 8, h = 12$

Solution

$V = \frac{1}{3}\pi r^2 h$

$V = \frac{1}{3}\pi(8)^2(12) = \frac{1}{3}\pi(64)(12) = 256\pi$

≈ 804.25

The volume is approximately 804.25 cm^3.

You Try It 2

The diameter of the base of a cylinder is 8 ft. The height of the cylinder is 22 ft. Find the exact volume of the cylinder.

Your strategy

Your solution

Solutions on pp. S10–S11

Copyright © Houghton Mifflin Company. All rights reserved.

Objective B To solve problems involving the surface area of a solid

The **surface area** of a solid is the total area on the surface of the solid.

When a rectangular solid is cut open and flattened out, each face is a rectangle. The surface area, SA, of the rectangular solid is the sum of the areas of the six rectangles:

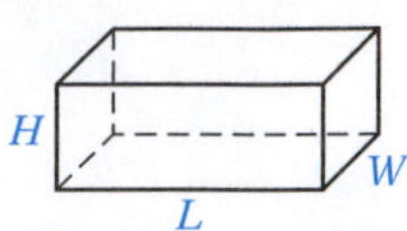

$$SA = LW + LH + WH + LW + WH + LH$$

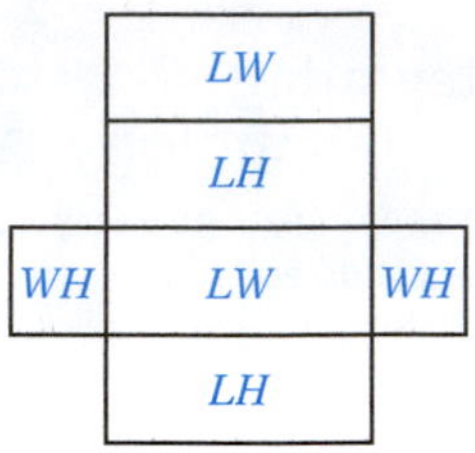

which simplifies to

$$\mathbf{SA = 2LW + 2LH + 2WH}$$

The surface area of a cube is the sum of the areas of the six faces of the cube. The area of each face is s^2. Therefore, the surface area, SA, of a cube is given by the formula $\mathbf{SA = 6s^2}$.

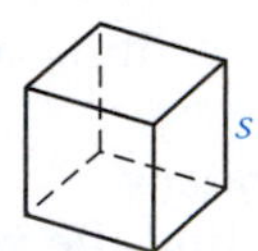

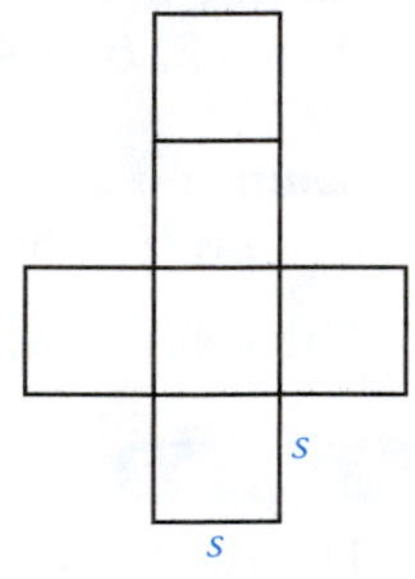

When a cylinder is cut open and flattened out, the top and bottom of the cylinder are circles. The side of the cylinder flattens out to a rectangle. The length of the rectangle is the circumference of the base, which is $2\pi r$; the width is h, the height of the cylinder. Therefore, the area of the rectangle is $2\pi rh$. The surface area, SA, of the cylinder is

$$SA = \pi r^2 + 2\pi rh + \pi r^2$$

which simplifies to

$$\mathbf{SA = 2\pi r^2 + 2\pi rh}$$

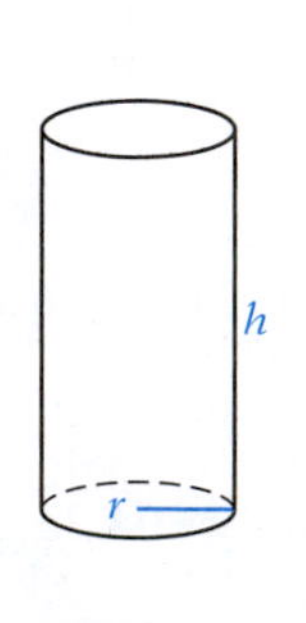

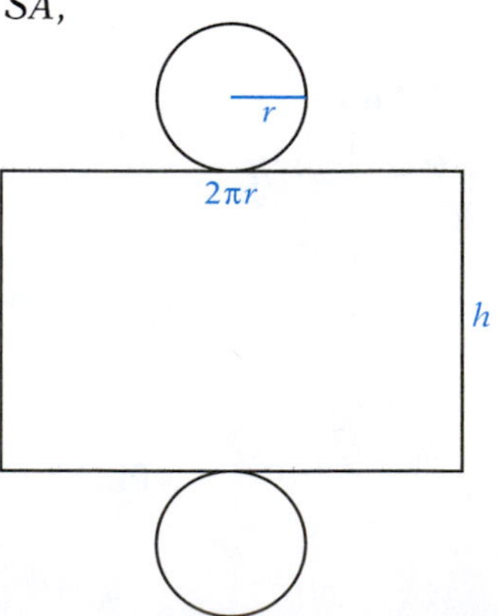

Copyright © Houghton Mifflin Company. All rights reserved.

The surface area of a regular square pyramid is the area of the base plus the area of the four isosceles triangles. A side of the square base is s; therefore, the area of the base is s^2. The slant height, ℓ, is the height of each triangle, and s is the base of each triangle. The surface area, SA, of a pyramid is

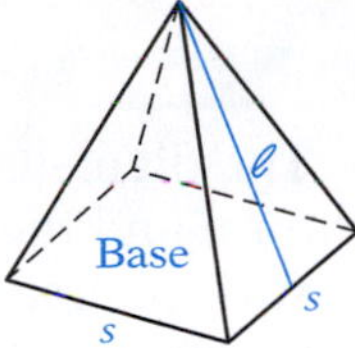

$$SA = s^2 + 4\left(\frac{1}{2}s\ell\right)$$

which simplifies to

$$\mathbf{SA = s^2 + 2s\ell}$$

Formulas for the surface areas of geometric solids are given below.

Surface Areas of Geometric Solids

The surface area, SA, of a **rectangular solid** with length L, width W, and height H is given by $SA = 2LW + 2LH + 2WH$.

The surface area, SA, of a **cube** with side s is given by $SA = 6s^2$.

The surface area, SA, of a **sphere** with radius r is given by $SA = 4\pi r^2$.

The surface area, SA, of a **right circular cylinder** is given by $SA = 2\pi r^2 + 2\pi rh$, where r is the radius of the base and h is the height.

The surface area, SA, of a **right circular cone** is given by $SA = \pi r^2 + \pi r\ell$, where r is the radius of the circular base and ℓ is the slant height.

The surface area, SA, of a **regular square pyramid** is given by $SA = s^2 + 2s\ell$, where s is the length of a side of the base and ℓ is the slant height.

HOW TO Find the surface area of a sphere with a diameter of 18 cm.

$r = \frac{1}{2}d = \frac{1}{2}(18) = 9$ • **First find the radius of the sphere.**

$SA = 4\pi r^2$ • **Use the formula for the surface area of a sphere.**

$SA = 4\pi(9)^2$

$SA = 4\pi(81)$

$SA = 324\pi$ • **The exact surface area of the sphere is 324π cm².**

$SA = 1017.88$ • **An approximate measure can be found by using the π key on a calculator.**

The approximate surface area is 1017.88 cm².

Copyright © Houghton Mifflin Company. All rights reserved.

Example 3

The diameter of the base of a cone is 5 m, and the slant height is 4 m. Find the surface area of the cone. Give the exact measure.

Strategy

To find the surface area of the cone:

- Find the radius of the base of the cone.
- Use the formula for the surface area of a cone. Leave the answer in terms of π.

Solution

$r = \frac{1}{2}d = \frac{1}{2}(5) = 2.5$

$SA = \pi r^2 + \pi r\ell$
$SA = \pi(2.5)^2 + \pi(2.5)(4)$
$SA = \pi(6.25) + \pi(2.5)(4)$
$SA = 6.25\pi + 10\pi$
$SA = 16.25\pi$

The surface area of the cone is 16.25π m^2.

You Try It 3

The diameter of the base of a cylinder is 6 ft, and the height is 8 ft. Find the surface area of the cylinder. Round to the nearest hundredth.

Your strategy

Your solution

Example 4

Find the area of a label used to cover a soup can that has a radius of 4 cm and a height of 12 cm. Round to the nearest hundredth.

Strategy

To find the area of the label, use the fact that the surface area of the sides of a cylinder is given by $2\pi rh$. An approximation is asked for; use the π key on a calculator.
$r = 4, h = 12$

Solution

Area of the label $= 2\pi rh$
Area of the label $= 2\pi(4)(12) = 96\pi$
≈ 301.59

The area is approximately 301.59 cm^2.

You Try It 4

Which has a larger surface area, a cube with a side measuring 10 cm or a sphere with a diameter measuring 8 cm?

Your strategy

Your solution

Solutions on p. S11

Copyright © Houghton Mifflin Company. All rights reserved.

3.3 Exercises

Objective A **To solve problems involving the volume of a solid**

Find the volume of the figure. For calculations involving π, give both the exact value and an approximation to the nearest hundredth.

1. 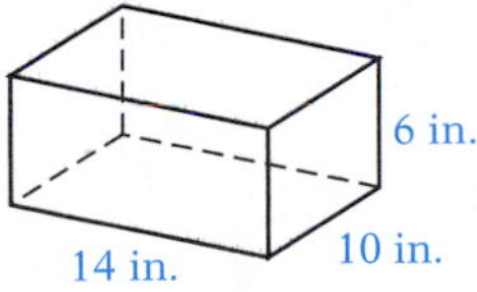

2. 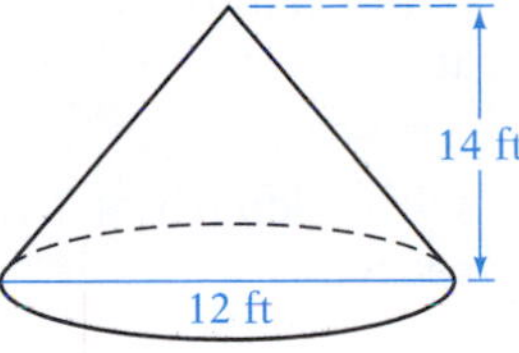

3. 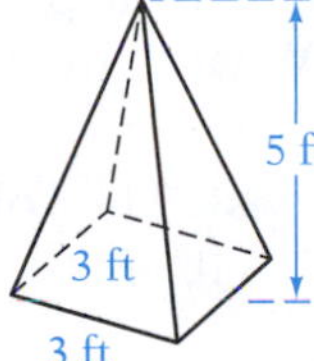

4. 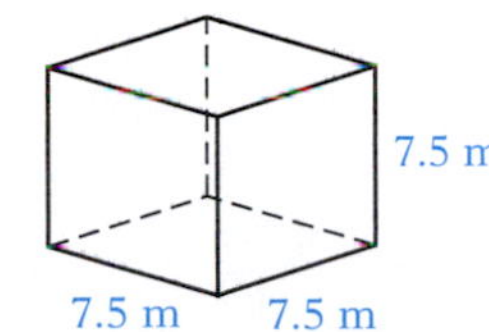

5. 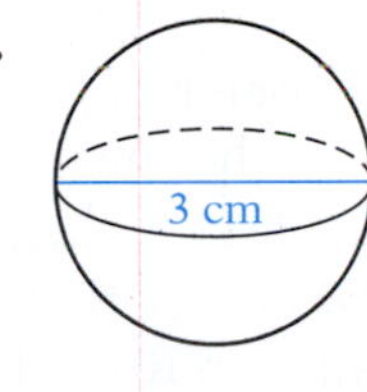

6. 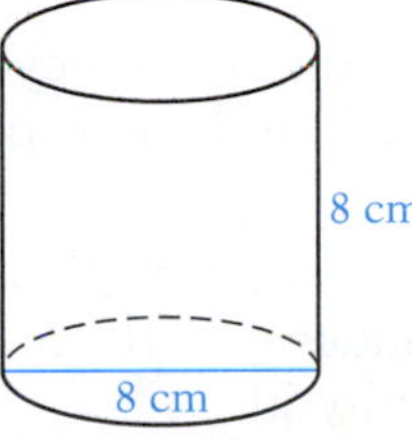

Solve.

7. **Storage Units** A rectangular storage unit has a length of 6.8 m, a width of 2.5 m, and a height of 2 m. Find the volume of the storage unit.

8. **Fish Hatchery** A rectangular tank at a fish hatchery is 9 m long, 3 m wide, and 1.5 m deep. Find the volume of the water in the tank when the tank is full.

9. Find the volume of a cube whose side measures 3.5 in.

10. The length of a side of a cube is 7 cm. Find the volume of the cube.

11. The diameter of a sphere is 6 ft. Find the volume of the sphere. Give the exact measure.

12. Find the volume of a sphere that has a radius of 1.2 m. Round to the nearest tenth.

13. The diameter of the base of a cylinder is 24 cm. The height of the cylinder is 18 cm. Find the volume of the cylinder. Round to the nearest hundredth.

14. The height of a cylinder is 7.2 m. The radius of the base is 4 m. Find the volume of the cylinder. Give the exact measure.

15. The radius of the base of a cone is 5 in. The height of the cone is 9 in. Find the volume of the cone. Give the exact measure.

Copyright © Houghton Mifflin Company. All rights reserved.

16. The height of a cone is 15 cm. The diameter of the cone is 10 cm. Find the volume of the cone. Round to the nearest hundredth.

17. The length of a side of the base of a pyramid is 6 in., and the height is 10 in. Find the volume of the pyramid.

18. The height of a pyramid is 8 m, and the length of a side of the base is 9 m. What is the volume of the pyramid?

19. **Appliances** The volume of a freezer with a length of 7 ft and a height of 3 ft is 52.5 ft^3. Find the width of the freezer.

20. **Aquariums** The length of an aquarium is 18 in., and the width is 12 in. If the volume of the aquarium is 1836 in^3, what is the height of the aquarium?

21. The volume of a cylinder is 502.4 in^3. The diameter of the base is 10 in. Find the height of the cylinder. Round to the nearest hundredth.

22. The diameter of the base of a cylinder is 14 cm. If the volume of the cylinder is 2310 cm^3, find the height of the cylinder. Round to the nearest hundredth.

23. A rectangular solid has a square base and a height of 5 in. If the volume of the solid is 125 in^3, find the length and the width.

24. The volume of a rectangular solid is 864 m^3. The rectangular solid has a square base and a height of 6 m. Find the dimensions of the solid.

25. **Petroleum** An oil storage tank, which is in the shape of a cylinder, is 4 m high and has a diameter of 6 m. The oil tank is two-thirds full. Find the number of cubic meters of oil in the tank. Round to the nearest hundredth.

26. **Agriculture** A silo, which is in the shape of a cylinder, is 16 ft in diameter and has a height of 30 ft. The silo is three-fourths full. Find the volume of the portion of the silo that is not being used for storage. Round to the nearest hundredth.

Objective B To solve problems involving the surface area of a solid

Find the surface area of the figure.

27.

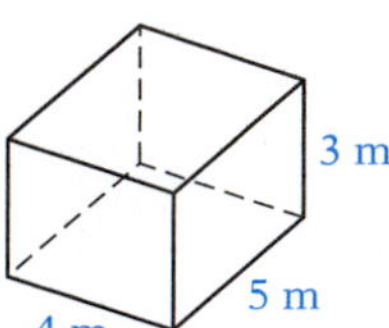

28.

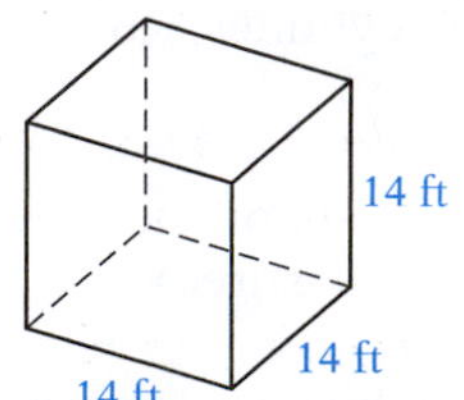

29.

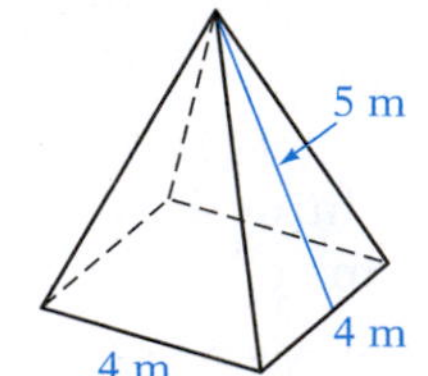

Copyright © Houghton Mifflin Company. All rights reserved.

Find the surface area of the figure. Give both the exact value and an approximation to the nearest hundredth.

30.

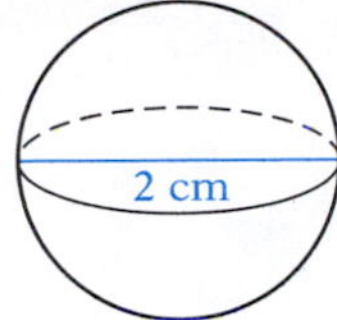

31.

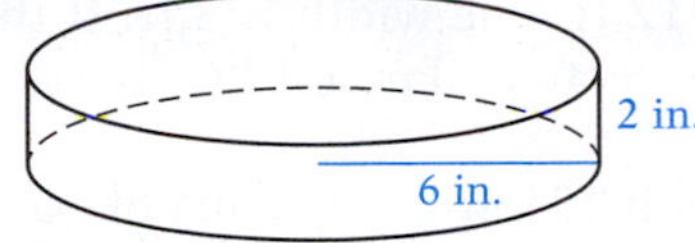

32.

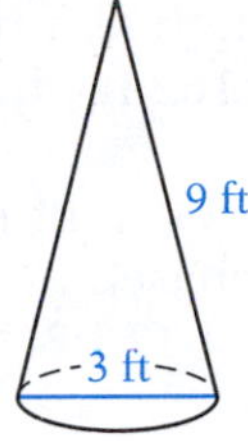

Solve.

33. The height of a rectangular solid is 5 ft. The length is 8 ft, and the width is 4 ft. Find the surface area of the solid.

34. The width of a rectangular solid is 32 cm. The length is 60 cm, and the height is 14 cm. What is the surface area of the solid?

35. The side of a cube measures 3.4 m. Find the surface area of the cube.

36. Find the surface area of a cube that has a side measuring 1.5 in.

37. Find the surface area of a sphere with a diameter of 15 cm. Give the exact value.

38. The radius of a sphere is 2 in. Find the surface area of the sphere. Round to the nearest hundredth.

39. The radius of the base of a cylinder is 4 in. The height of the cylinder is 12 in. Find the surface area of the cylinder. Round to the nearest hundredth.

40. The diameter of the base of a cylinder is 1.8 m. The height of the cylinder is 0.7 m. Find the surface area of the cylinder. Give the exact value.

41. The slant height of a cone is 2.5 ft. The radius of the base is 1.5 ft. Find the surface area of the cone. Give the exact value.

42. The diameter of the base of a cone is 21 in. The slant height is 16 in. What is the surface area of the cone? Round to the nearest hundredth.

43. The length of a side of the base of a pyramid is 9 in., and the slant height is 12 in. Find the surface area of the pyramid.

44. The slant height of a pyramid is 18 m, and the length of a side of the base is 16 m. What is the surface area of the pyramid?

Copyright © Houghton Mifflin Company. All rights reserved.

45. The surface area of a rectangular solid is 108 cm^2. The height of the solid is 4 cm, and the length is 6 cm. Find the width of the rectangular solid.

46. The length of a rectangular solid is 12 ft. The width is 3 ft. If the surface area is 162 ft^2, find the height of the rectangular solid.

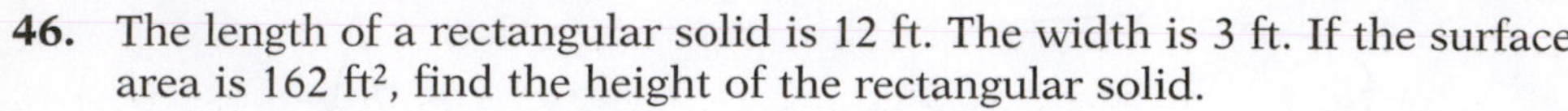

47. **Paint** A can of paint will cover 300 ft^2. How many cans of paint should be purchased in order to paint a cylinder that has a height of 30 ft and a radius of 12 ft?

48. **Ballooning** A hot air balloon is in the shape of a sphere. Approximately how much fabric was used to construct the balloon if its diameter is 32 ft? Round to the nearest whole number.

49. **Aquariums** How much glass is needed to make a fish tank that is 12 in. long, 8 in. wide, and 9 in. high? The fish tank is open at the top.

50. **Packaging** Find the area of a label used to cover a can of juice that has a diameter of 16.5 cm and a height of 17 cm. Round to the nearest hundredth.

51. The length of a side of the base of a pyramid is 5 cm, and the slant height is 8 cm. How much larger is the surface area of this pyramid than the surface area of a cone with a diameter of 5 cm and a slant height of 8 cm? Round to the nearest hundredth.

APPLYING THE CONCEPTS

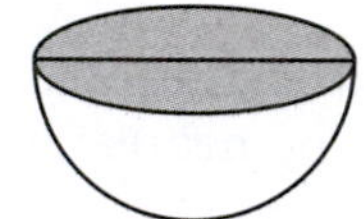

52. Half of a sphere is called a **hemisphere.** Derive formulas for the volume and surface area of a hemisphere.

53. Determine whether the statement is always true, sometimes true, or never true.
    a. The slant height of a regular pyramid is longer than the height.
    b. The slant height of a cone is shorter than the height.
    c. The four triangular faces of a regular pyramid are equilateral triangles.

54. **a.** What is the effect on the surface area of a rectangular solid when the width and height are doubled?
    b. What is the effect on the volume of a rectangular solid when both the length and the width are doubled?
    c. What is the effect on the volume of a cube when the length of each side of the cube is doubled?
    d. What is the effect on the surface area of a cylinder when the radius and height are doubled?

55. Explain how you could cut through a cube so that the face of the resulting solid is
    a. a square
    b. an equilateral triangle
    c. a trapezoid
    d. a hexagon

Copyright © Houghton Mifflin Company. All rights reserved.

Focus on Problem Solving

More on the Trial-and-Error Approach to Problem Solving

Some problems in mathematics are solved by using **trial and error.** The trial-and-error method of arriving at a solution to a problem involves repeated tests or experiments until a satisfactory conclusion is reached.

Many of the Applying the Concepts exercises in this text require a trial-and-error method of solution. For example, an exercise in Section 3 of this chapter reads:

Explain how you could cut through a cube so that the face of the resulting solid is **(a)** a square, **(b)** an equilateral triangle, **(c)** a trapezoid, **(d)** a hexagon.

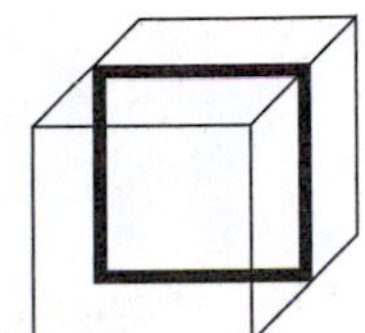

There is no formula to apply to this problem; there is no computation to perform. This problem requires picturing a cube and the results after cutting through it at different places on its surface and at different angles. For part (a), cutting perpendicular to the top and bottom of the cube and parallel to two of its sides will result in a square. The other shapes may prove more difficult.

When solving problems of this type, keep an open mind. Sometimes when using the trial-and-error method, we are hampered by narrowness of vision; we cannot expand our thinking to include other possibilities. Then when we see someone else's solution, it appears so obvious to us! For example, for the Applying the Concepts question above, it is necessary to conceive of cutting through the cube at places other than the top surface; we need to be open to the idea of beginning the cut at one of the corner points of the cube.

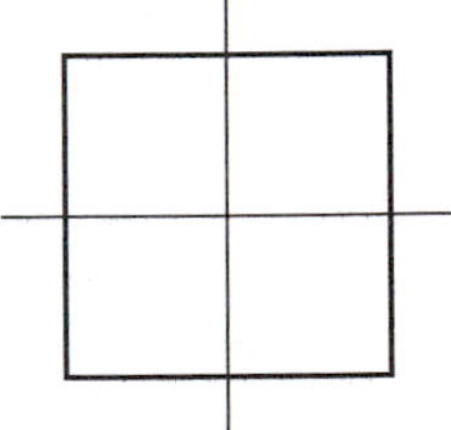

A topic of the Projects and Group Activities in this chapter is symmetry. Here again, trial and error is used to determine the lines of symmetry inherent in an object. For example, in determining lines of symmetry for a square, begin by drawing a square. The horizontal line of symmetry and the vertical line of symmetry may be immediately obvious to you.

But there are two others. Do you see that a line drawn through opposite corners of the square is also a line of symmetry?

Many of the questions in this text that require an answer of "always true, sometimes true, or never true" are best solved by the trial-and-error method. For example, consider the statement presented in Section 2 of this chapter.

If two rectangles have the same area, then they have the same perimeter.

Copyright © Houghton Mifflin Company. All rights reserved.

Try some numbers. Each of two rectangles, one measuring 6 units by 2 units and another measuring 4 units by 3 units, has an area of 12 square units, but the perimeter of the first is 16 units and the perimeter of the second is 14 units. So the answer "always true" has been eliminated. We still need to determine whether there is a case when the statement is true. After experimenting with a lot of numbers, you may come to realize that we are trying to determine if it is possible for two different pairs of factors of a number to have the same sum. Is it?

Don't be afraid to make many experiments, and remember that *errors*, or tests that "don't work," are a part of the trial-and-*error* process.

Projects and Group Activities

Investigating Perimeter

The perimeter of the square at the right is 4 units.

If two squares are joined along one of the sides, the perimeter is 6 units. Note that it does not matter which sides are joined; the perimeter is still 6 units.

If three squares are joined, the perimeter of the resulting figure is 8 units for each possible placement of the squares.

Four squares can be joined in five different ways as shown. There are two possible perimeters, 10 units for A, B, C, and D, and 8 units for E.

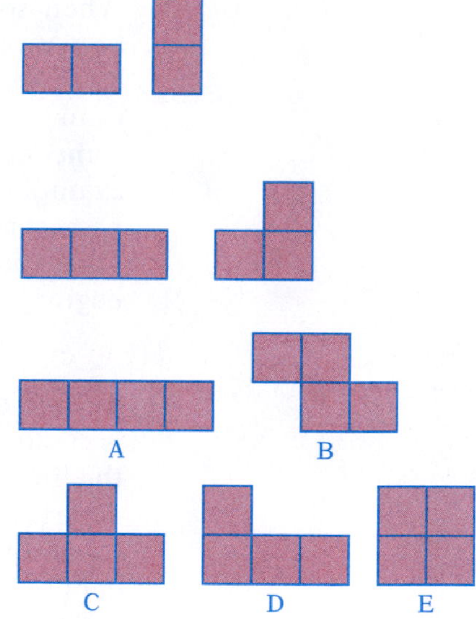

1. If five squares are joined, what is the maximum perimeter possible?

2. If five squares are joined, what is the minimum perimeter possible?

3. If six squares are joined, what is the maximum perimeter possible?

4. If six squares are joined, what is the minimum perimeter possible?

Copyright © Houghton Mifflin Company. All rights reserved.

Symmetry Look at the letter A printed at the left. If the letter were folded along line ℓ, the two sides of the letter would match exactly. This letter has **symmetry** with respect to line ℓ. Line ℓ is called the **axis of symmetry.**

Now consider the letter H printed below at the left. Both lines ℓ_1 and ℓ_2 are axes of symmetry for this letter; the letter could be folded along either line and the two sides would match exactly.

1. Does the letter A have more than one axis of symmetry?
2. Find axes of symmetry for other capital letters of the alphabet.
3. Which lowercase letters have one axis of symmetry?
4. Do any of the lowercase letters have more than one axis of symmetry?
5. Find the number of axes of symmetry for each of the plane geometric figures presented in this chapter.
6. There are other types of symmetry. Look up the meaning of point symmetry and rotational symmetry. Which plane geometric figures provide examples of these types of symmetry?
7. Find examples of symmetry in nature, art, and architecture.

Chapter 3 Summary

Key Words	**Examples**
A *line* extends indefinitely in two directions. A *line segment* is part of a line and has two endpoints. The length of a line segment is the distance between the endpoints of the line segment. [3.1A, p. 161]	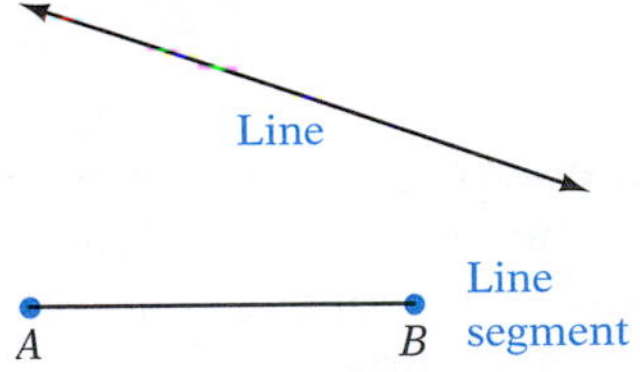
Parallel lines never meet; the distance between them is always the same. The symbol ‖ means "is parallel to." *Intersecting lines* cross at a point in the plane. *Perpendicular lines* are intersecting lines that form right angles. The symbol ⊥ means "is perpendicular to." [3.1A, pp. 162–163]	Parallel lines 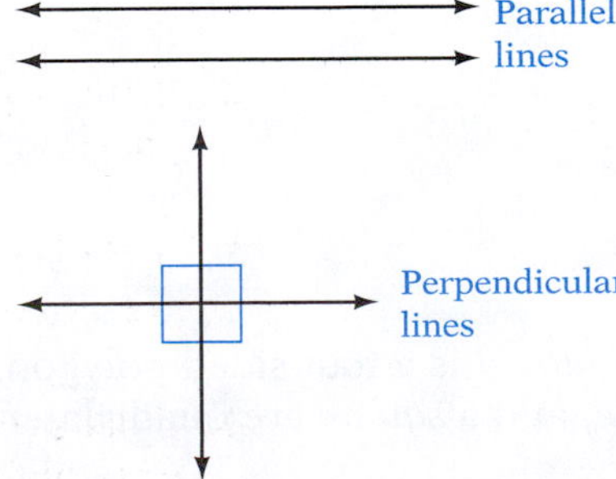

Copyright © Houghton Mifflin Company. All rights reserved.

A *ray* starts at a point and extends indefinitely in one direction. An *angle* is formed when two rays start from the same point. The common point is called the *vertex* of the angle. An angle is measured in *degrees*. A 90° angle is a *right angle*. A 180° angle is a *straight angle*. [3.1A, pp. 161–164]

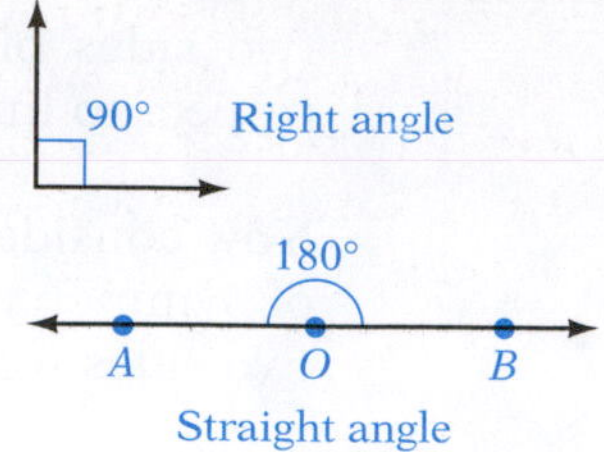

Complementary angles are two angles whose measures have the sum 90°. *Supplementary angles* are two angles whose measures have the sum 180°. [3.1A, pp. 163–164]

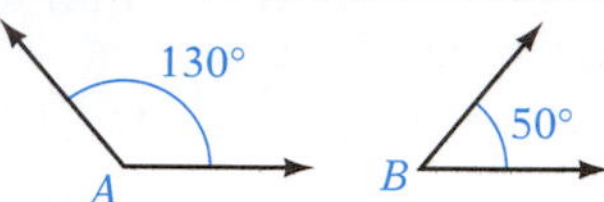

An *acute angle* is an angle whose measure is between 0° and 90°. An *obtuse angle* is an angle whose measure is between 90° and 180°. [3.1A, p. 164]

$\angle A$ above is an obtuse angle.
$\angle B$ above is an acute angle.

Two angles that are on opposite sides of the intersection of two lines are *vertical angles;* vertical angles have the same measure. Two angles that share a common side are *adjacent angles;* adjacent angles of intersecting lines are supplementary angles. [3.1B, p. 166]

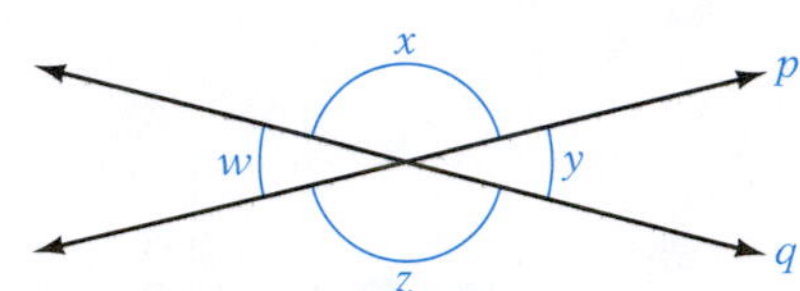

Angles w and y are vertical angles.
Angles x and y are adjacent angles.

A line that intersects two other lines at two different points is a *transversal*. If the lines cut by a transversal are parallel lines, equal angles are formed: *alternate interior angles, alternate exterior angles,* and *corresponding angles*. [3.1B, p. 167]

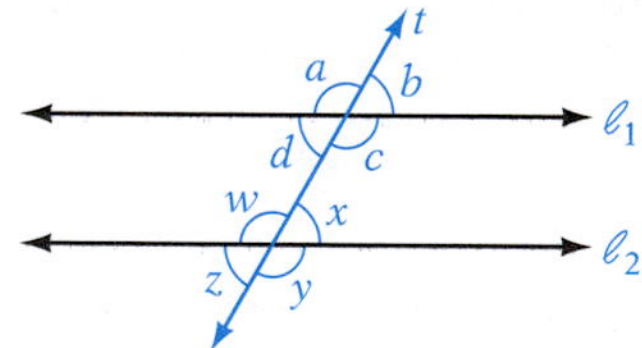

Parallel lines ℓ_1 and ℓ_2 are cut by transversal t. All four acute angles have the same measure. All four obtuse angles have the same measure.

A *quadrilateral* is a four-sided polygon. A *parallelogram*, a *rectangle*, and a *square* are quadrilaterals. [3.2A, pp. 177–178]

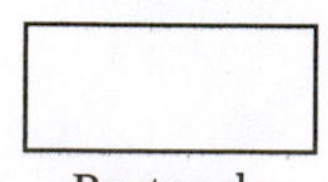

Rectangle Square

Copyright © Houghton Mifflin Company. All rights reserved.

A *polygon* is a closed figure determined by three or more line segments. The line segments that form the polygon are its *sides*. A *regular polygon* is one in which each side has the same length and each angle has the same measure. Polygons are classified by the number of sides. [3.2A, p. 177]

Number of Sides	Name of the Polygon
3	Triangle
4	Quadrilateral
5	Pentagon
6	Hexagon
7	Heptagon
8	Octagon
9	Nonagon
10	Decagon

A *triangle* is a closed, three-sided plane figure. [3.1C, p. 169]

An *isosceles triangle* has two sides of equal length. The three sides of an *equilateral triangle* are of equal length. A *scalene triangle* has no two sides of equal length. An *acute triangle* has three actue angles. An *obtuse triangle* has one obtuse angle. A *right triangle* contains a right angle. [3.2A, pp. 177–178]

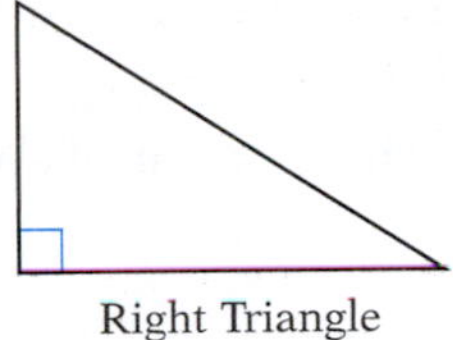

Right Triangle

A *circle* is a plane figure in which all points are the same distance from the center of the circle. A *diameter* of a circle is a line segment across the circle through the center. A *radius* of a circle is a line segment from the center of the circle to a point on the circle. [3.2A, p. 180]

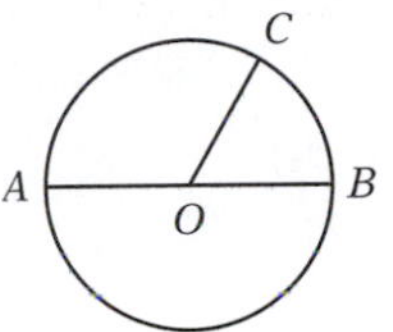

AB is a diameter of the circle.
OC is a radius.

Geometric solids are figures in space. Six common space figures are the rectangular solid, cube, sphere, cylinder, cone, and pyramid. A *rectangular solid* is a solid in which all six faces are rectangles. A *cube* is a rectangular solid in which all six faces are squares. A *sphere* is a solid in which all points on the sphere are the same distance from the center of the sphere. The most common *cylinder* is one in which the bases are circles and are perpendicular to the height. In this text the only types of cones and pyramids discussed are the *right circular cone* and the *regular square pyramid*. [3.3A, pp. 195–196]

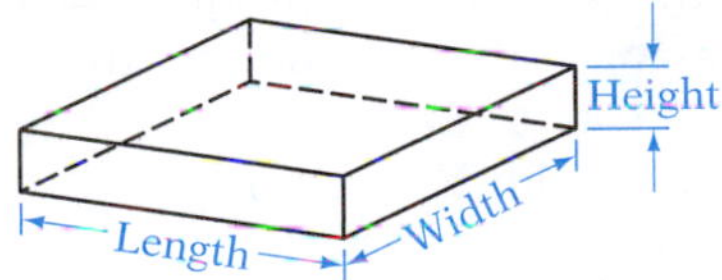

Rectangular Solid

Essential Rules and Procedures

Triangles [3.1C, p. 169]
Sum of the measures of the interior angles = 180°

Examples

Two angles of a triangle measure 32° and 48°. Find the measure of the third angle.

$$\angle A + \angle B + \angle C = 180^\circ$$
$$\angle A + 32^\circ + 48^\circ = 180^\circ$$
$$\angle A + 80^\circ = 180^\circ$$
$$\angle A + 80^\circ - 80^\circ = 180^\circ - 80^\circ$$
$$\angle A = 100^\circ$$

The measure of the third angle is 100°.

Copyright © Houghton Mifflin Company. All rights reserved.

Formulas for Perimeter (the distance around a figure) [3.2A, pp. 178–180]
Triangle: $P = a + b + c$
Rectangle: $P = 2L + 2W$
Square: $P = 4s$
Circumference of a circle: $C = \pi d$ or $C = 2\pi r$

The length of a rectangle is 8 m. The width is 5.5 m. Find the perimeter of the rectangle.
$P = 2L + 2W$
$P = 2(8) + 2(5.5)$
$P = 16 + 11$
$P = 27$
The perimeter is 27 m.

Formulas for Area (the amount of surface in a region) [3.2B, pp. 182–185]
Rectangle: $A = LW$
Square: $A = s^2$
Parallelogram: $A = bh$
Triangle: $A = \frac{1}{2}bh$
Trapezoid: $A = \frac{1}{2}h(b_1 + b_2)$
Circle: $A = \pi r^2$

Find the area of a circle with a radius of 4 cm. Round to the nearest hundredth.
$A = \pi r^2$
$A = \pi(4)^2$
$A = 16\pi$
$A \approx 50.27$
The area is approximately 50.27 cm^2.

Formulas for Volume (the amount of space inside a figure in space) [3.3A, p. 196]
Rectangular solid: $V = LWH$
Cube: $V = s^3$
Sphere: $V = \frac{4}{3}\pi r^3$
Cylinder: $V = \pi r^2 h$
Right circular cone: $V = \frac{1}{3}\pi r^2 h$
Regular square pyramid: $V = \frac{1}{3}s^2 h$

Find the volume of a cube that measures 3 in. on a side.
$V = s^3$
$V = 3^3$
$V = 27$
The volume is 27 in^3.

Formulas for Surface Area (the total area on the surface of a solid) [3.3B, p. 199]
Rectangular solid: $2LW + 2LH + 2WH$
Cube: $SA = 6s^2$
Sphere: $SA = 4\pi r^2$
Cylinder: $SA = 2\pi r^2 + 2\pi rh$
Right circular cone: $SA = \pi r^2 + \pi rl$
Regular square pyramid: $SA = s^2 + 2sl$

Find the surface area of a sphere that has a radius of 3 in. Round to the nearest hundredth.
$SA = 4\pi r^2$
$SA = 4\pi(3)^2$
$SA = 4\pi(9)$
$SA = 36\pi$
$SA \approx 113.10$
The surface area is approximately 113.10 in^3.

Copyright © Houghton Mifflin Company. All rights reserved.

Chapter 3 Review Exercises

1. Given that $\angle a = 74°$ and $\angle b = 52°$, find the measures of angles x and y.

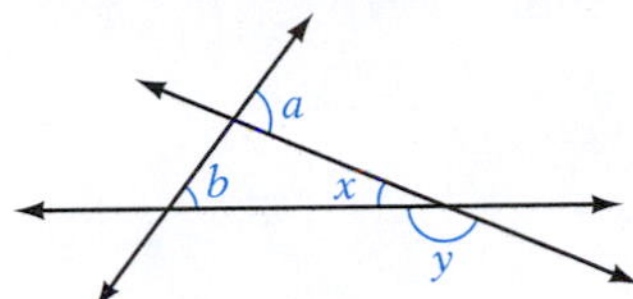

2. Find the measure of $\angle x$.

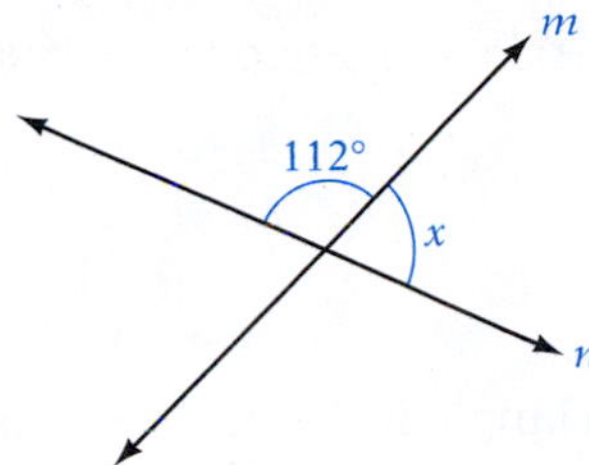

3. Given that $BC = 11$ cm and AB is three times the length of BC, find the length of AC.

4. Find x.

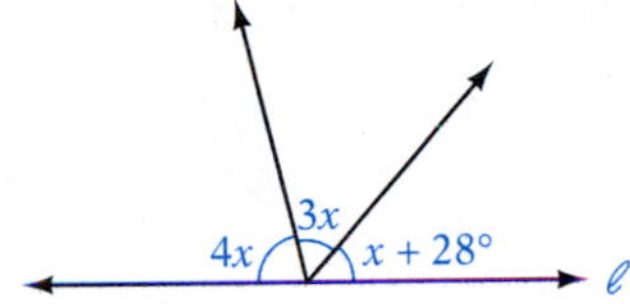

5. Find the volume of the figure.

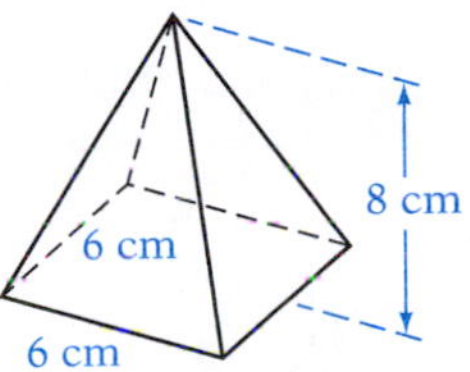

6. Given that $\ell \|_1 \ell_2$, find the measures of angles a and b.

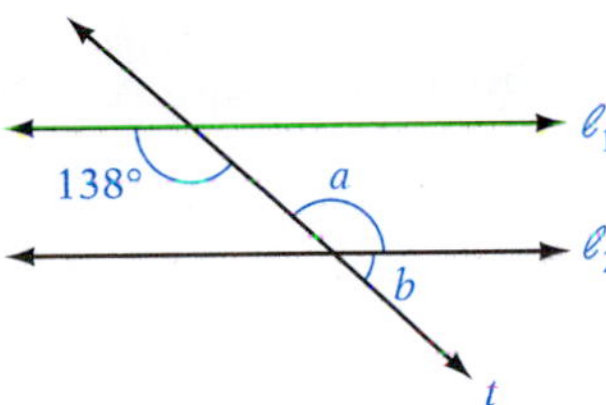

7. Find the surface area of the figure.

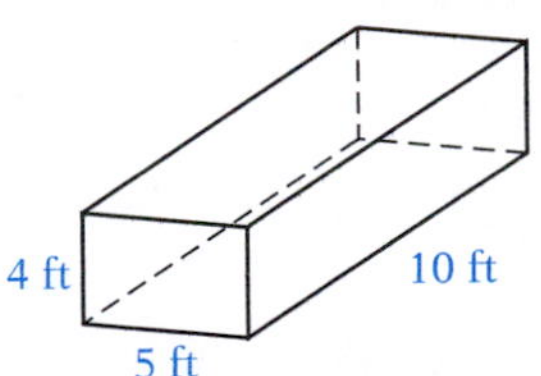

8. Find the supplement of a 32° angle.

9. Determine the area of a rectangle with a length of 12 cm and a width of 6.5 cm.

10. Determine the area of a triangle whose base is 9 m and whose height is 14 m.

Copyright © Houghton Mifflin Company. All rights reserved.

11. Find the volume of a rectangular solid with a length of 6.5 ft, a width of 2 ft, and a height of 3 ft.

12. Two angles of a triangle measure 37° and 48°. Find the measure of the third angle.

13. The height of a triangle is 7 cm. The area of the triangle is 28 cm^2. Find the length of the base of the triangle.

14. Find the volume of a sphere that has a diameter of 12 mm. Find the exact value.

15. Determine the exact volume of a right circular cone whose radius is 7 cm and whose height is 16 cm.

16. **Framing** The perimeter of a square picture frame is 86 cm. Find the length of each side of the frame.

17. **Paint** A can of paint will cover 200 ft^2. How many cans of paint should be purchased in order to paint a cylinder that has a height of 15 ft and a radius of 6 ft?

18. **Parks** The length of a rectangular park is 56 yd. The width is 48 yd. How many yards of fencing are needed to surround the park?

19. **Patios** What is the area of a square patio that measures 9.5 m on each side?

20. **Landscaping** A walkway 2 m wide surrounds a rectangular plot of grass. The plot is 40 m long and 25 m wide. What is the area of the walkway?

Copyright © Houghton Mifflin Company. All rights reserved.

Chapter 3 Test

1. The diameter of a sphere is 1.5 m. Find the radius of the sphere.

2. Find the circumference of a circle with a radius of 5 cm. Round to the nearest hundredth.

3. Find the perimeter of the rectangle in the figure below.

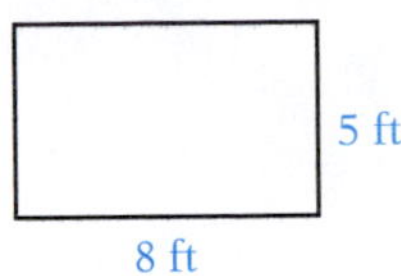

4. Given $AB = 15$, $CD = 6$, and $AD = 24$, find the length of BC.

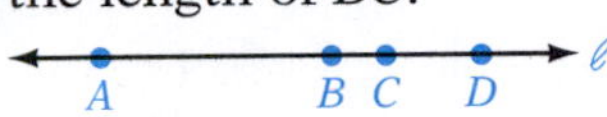

5. Find the volume of a sphere with a diameter of 8 ft. Round to the nearest hundredth.

6. Find the area of the circle shown below. Round to the nearest hundredth.

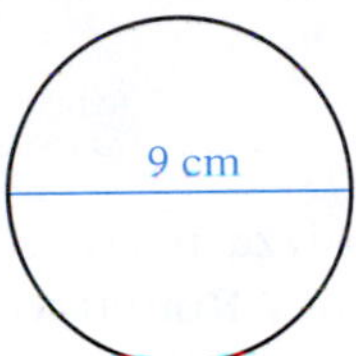

7. Given that $\ell_1 \parallel \ell_2$, find the measures of angles a and b.

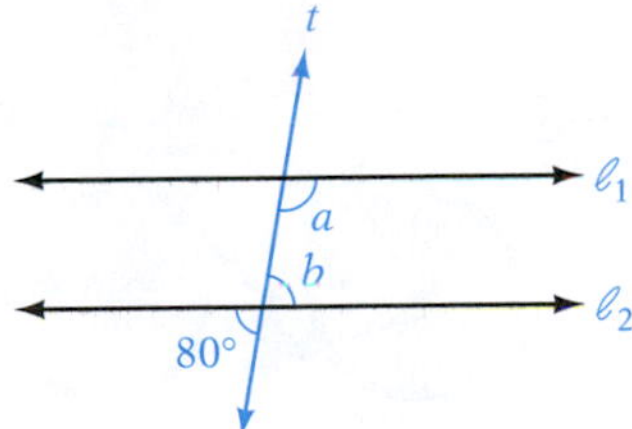

8. Find the supplement of a 105° angle.

9. Given that $\ell_1 \parallel \ell_2$, find the measures of angles a and b.

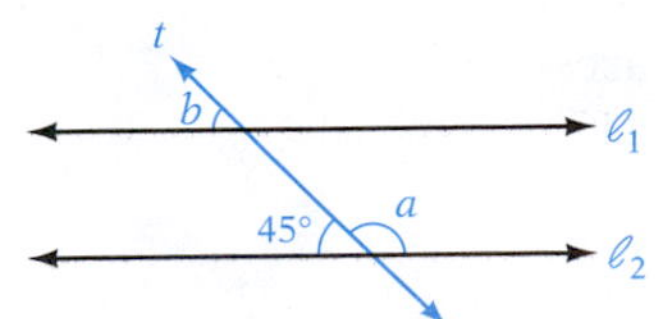

10. Find the area of the rectangle shown below.

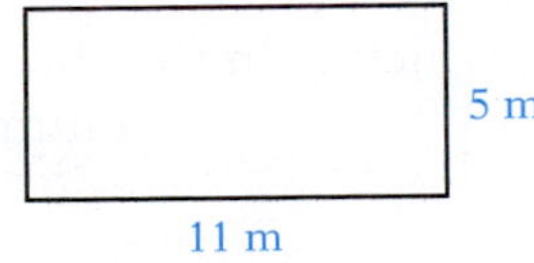

Copyright © Houghton Mifflin Company. All rights reserved.

11. Find the volume of a cylinder with a height of 6 m and a radius of 3 m. Round to the nearest hundredth.

12. Find the perimeter of a rectangle that has a length of 2 m and a width of 1.4 m.

13. Find the complement of a 32° angle.

14. Find the surface area of the figure. Round to the nearest hundredth.

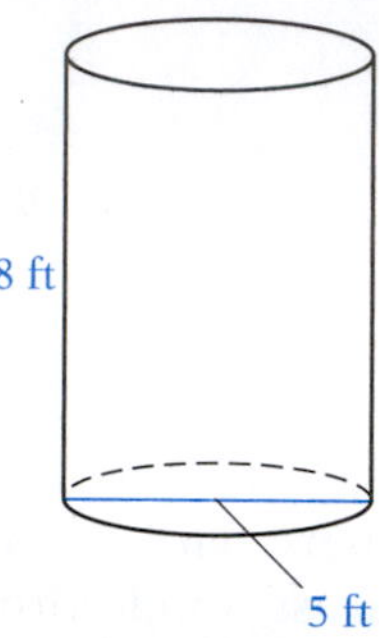

15. **Pizza** How much more pizza is contained in a pizza with radius 10 in. than in one with radius 8 in.? Round to the nearest hundredth.

16. **Triangles** A right triangle has a 32° angle. Find the measures of the other two angles.

17. **Cycling** A bicycle tire has a diameter of 28 in. How many feet does the bicycle travel if the wheel makes 10 revolutions? Round to the nearest tenth.

18. **Carpeting** New carpet is installed in a room measuring 18 ft by 14 ft. Find the area of the room in square yards. ($9 \text{ ft}^2 = 1 \text{ yd}^2$)

19. **Agriculture** A silo, which is in the shape of a cylinder, is 9 ft in diameter and has a height of 18 ft. Find the volume of the silo. Round to the nearest hundredth.

20. **Triangles** Find the area of a right triangle with a base of 8 m and a height of 2.75 m.

Copyright © Houghton Mifflin Company. All rights reserved.

Cumulative Review Exercises

1. Let $x \in \{-3, 0, 1\}$. For what values of x is the inequality $x \leq 1$ a true statement?

2. Write 8.9% as a decimal.

3. Write $\frac{7}{20}$ as a percent.

4. Divide: $-\frac{4}{9} \div \frac{2}{3}$

5. Multiply: $5.7(-4.3)$

6. Simplify: $-\sqrt{125}$

7. Evaluate $5 - 3[10 + (5 - 6)^2]$.

8. Evaluate $a(b - c)^3$ when $a = -1$, $b = -2$, and $c = -4$.

9. Simplify: $5m + 3n - 8m$

10. Simplify: $-7(-3y)$

11. Simplify: $4(3x + 2) - (5x - 1)$

12. Use the roster method to write the set of negative integers greater than or equal to -2.

13. Find $C \cup D$, given $C = \{0, 10, 20, 30\}$ and $D = \{-10, 0, 10\}$.

14. Graph: $x \leq 1$

-5 -4 -3 -2 -1 0 1 2 3 4 5

15. Solve: $4x + 2 = 6x - 8$

16. Solve: $3(2x + 5) = 18$

17. Solve: $4y - 3 \geq 6y + 5$

18. Solve: $8 - 4(3x + 5) \leq 6(x - 8)$

Copyright © Houghton Mifflin Company. All rights reserved.

19. Solve: $2x - 3 > 5$ or $x + 4 < 1$

20. Solve: $-3 \le 2x - 7 \le 5$

21. Solve: $|3x - 1| = 2$

22. Solve: $|x - 8| \le 2$

23. Find the measure of $\angle x$.

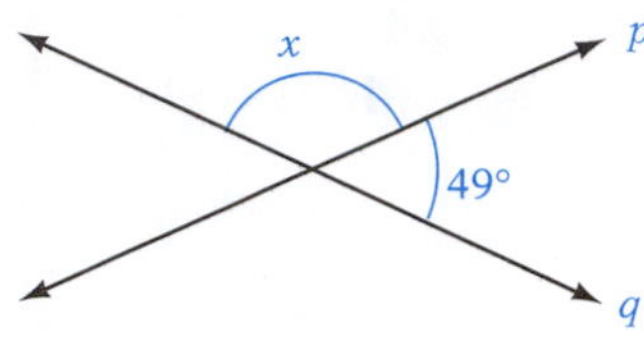

24. Translate "the difference between four times a number and ten is two" into an equation and solve.

25. **Triangles** Two angles of a triangle measure 37° and 21°. Find the measure of the third angle of the triangle.

26. **Investments** Michael deposits \$5000 in an account that earns an annual simple interest rate of 4.5% and \$2500 in an account that earns an annual simple interest rate of 3.5%. How much interest will Michael earn from the two accounts in one year?

27. **Triangles** Two sides of an isosceles triangle measure 7.5 m. The perimeter of the triangle is 19.5 m. Find the measure of the third side of the triangle.

28. **Annual Earnings** According to the Census Bureau, the median annual earnings of a man with a bachelor's degree is \$49,982, and the median earnings of a woman with a bachelor's degree is \$35,408. What percent of the men's median annual earnings is the women's median annual earnings? Round to the nearest tenth of a percent. (Median is a type of average.)

29. Find the exact area of a circle that has a diameter of 9 cm.

30. The volume of a box is 144 ft^3. The length of the box is 12 ft, and the width is 4 ft. Find the height of the box.

Copyright © Houghton Mifflin Company. All rights reserved.

chapter

4 Linear Functions and Inequalities in Two Variables

Do you have a cellular phone? Do you pay a monthly fee plus a charge for each minute you use? In this chapter you will learn how to write a linear function that models the monthly cost of a cell phone in terms of the number of minutes you use it. **Exercise 78 on page 271** asks you to write linear functions for certain cell phone options. Being able to use a linear function to model a relationship between two variables is an important skill in many fields, such as business, economics, and nutrition.

OBJECTIVES

Section 4.1

- **A** To graph points in a rectangular coordinate system
- **B** To determine ordered-pair solutions of an equation in two variables
- **C** To graph a scatter diagram

Section 4.2

- **A** To evaluate a function

Section 4.3

- **A** To graph a linear function
- **B** To graph an equation of the form $Ax + By = C$
- **C** To find the x- and y-intercepts of a straight line
- **D** To solve application problems

Section 4.4

- **A** To find the slope of a line given two points
- **B** To graph a line given a point and the slope

Section 4.5

- **A** To find the equation of a line given a point and the slope
- **B** To find the equation of a line given two points
- **C** To solve application problems

Section 4.6

- **A** To find parallel and perpendicular lines

Section 4.7

- **A** To graph the solution set of an inequality in two variables

Need help? For online student resources, such as section quizzes, visit this textbook's Website at **math.college.hmco.com/students.**

Copyright © Houghton Mifflin Company. All rights reserved.

PREP TEST

Do these exercises to prepare for Chapter 4.

For Exercises 1 to 3, simplify.

1. $-4(x - 3)$

2. $\sqrt{(-6)^2 + (-8)^2}$

3. $\dfrac{3 - (-5)}{2 - 6}$

4. Evaluate $-2x + 5$ for $x = -3$.

5. Evaluate $\dfrac{2r}{r - 1}$ for $r = 5$.

6. Evaluate $2p^3 - 3p + 4$ for $p = -1$.

7. Evaluate $\dfrac{x_1 + x_2}{2}$ for $x_1 = 7$ and $x_2 = -5$.

8. Given $3x - 4y = 12$, find the value of x when $y = 0$.

GO FIGURE

If $\boxed{5} = 4$ and ⑤ $= 6$ and $y = x - 1$, which of the following has the largest value?

$\boxed{x}$ ⓧ $\boxed{y}$ ⓨ

Copyright © Houghton Mifflin Company. All rights reserved.

4.1 The Rectangular Coordinate System

Objective A

To graph points in a rectangular coordinate system

Point of Interest

A rectangular coordinate system is also called a **Cartesian coordinate system,** in honor of Descartes.

Before the 15th century, geometry and algebra were considered separate branches of mathematics. That all changed when René Descartes, a French mathematician who lived from 1596 to 1650, founded analytic geometry. In this geometry, a *coordinate system* is used to study relationships between variables.

A **rectangular coordinate system** is formed by two number lines, one horizontal and one vertical, that intersect at the zero point of each line. The point of intersection is called the **origin.** The two lines are called **coordinate axes,** or simply **axes.**

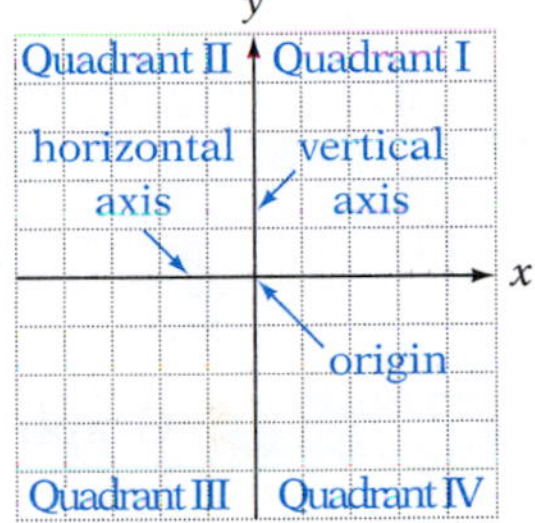

The axes determine a **plane,** which can be thought of as a large, flat sheet of paper. The two axes divide the plane into four regions called **quadrants.** The quadrants are numbered counterclockwise from I to IV.

Point of Interest

Gottfried Leibnitz introduced the words *abscissa* and *ordinate. Abscissa* is from Latin, meaning "to cut off." Originally, Leibnitz used the phrase *abscissa linea,* "cut off a line" (axis). The root of *ordinate* is also a Latin word used to suggest a sense of order.

Each point in the plane can be identified by a pair of numbers called an **ordered pair.** The first number of the pair measures a horizontal distance and is called the **abscissa.** The second number of the pair measures a vertical distance and is called the **ordinate.** The **coordinates of a point** are the numbers in the ordered pair associated with the point. The abscissa is also called the **first coordinate** of the ordered pair, and the ordinate is also called the **second coordinate** of the ordered pair.

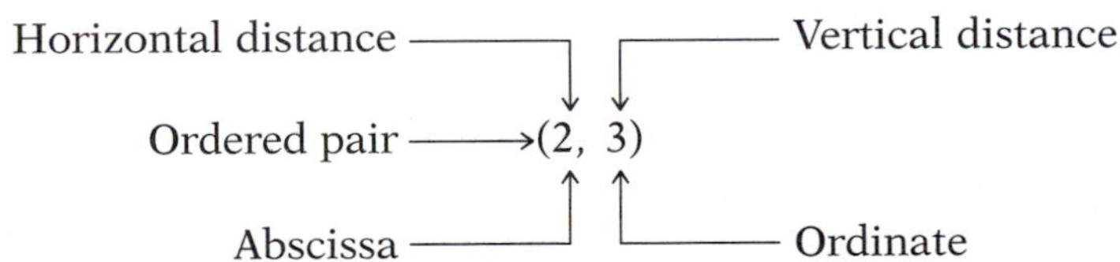

Graphing, or plotting, an ordered pair in the plane means placing a dot at the location given by the ordered pair. The **graph of an ordered pair** is the dot drawn at the coordinates of the point in the plane. The points whose coordinates are (3, 4) and (−2.5, −3) are graphed in the figure at the right.

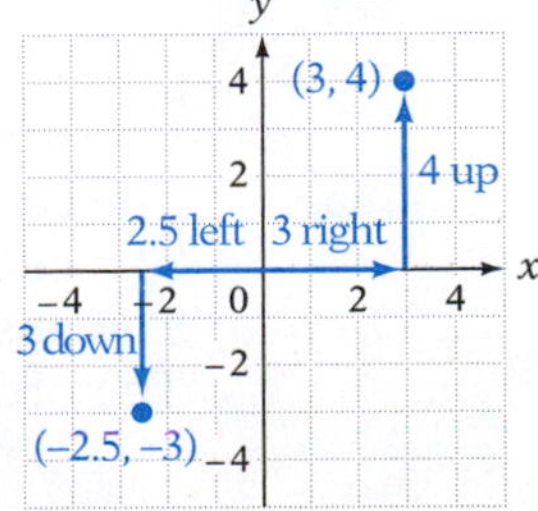

Copyright © Houghton Mifflin Company. All rights reserved.

TAKE NOTE

This is very important. An **ordered pair** is a *pair* of coordinates, and the *order* in which the coordinates appear is crucial.

The points whose coordinates are $(3, -1)$ and $(-1, 3)$ are graphed at the right. Note that the graphed points are in different locations. *The order of the coordinates of an ordered pair is important.*

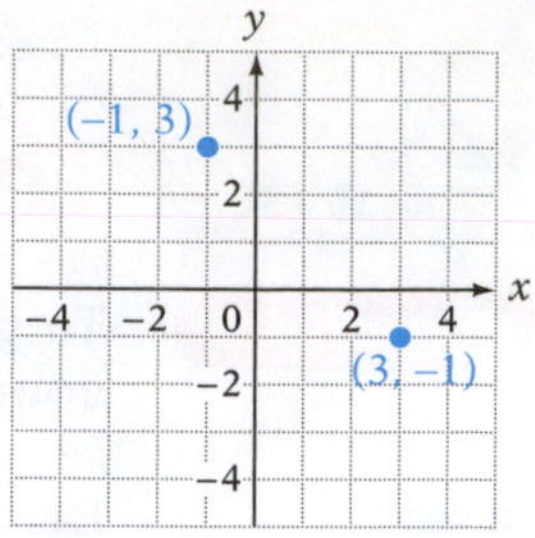

Each point in the plane is associated with an ordered pair, and each ordered pair is associated with a point in the plane. Although only the labels for integers are given on a coordinate grid, the graph of any ordered pair can be approximated. For example, the points whose coordinates are $(-2.3, 4.1)$ and $(\pi, 1)$ are shown on the graph at the right.

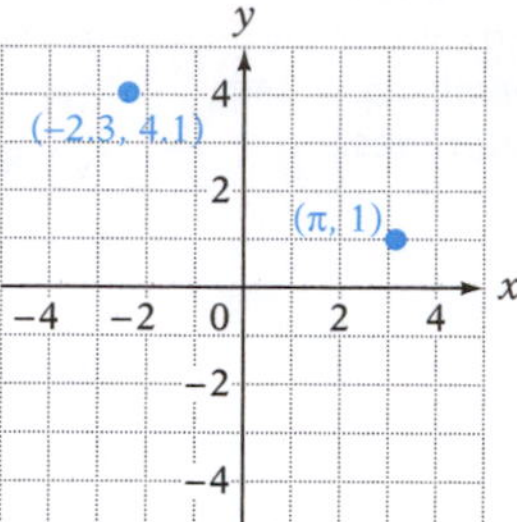

Example 1 Graph the ordered pairs $(-2, -3)$, $(3, -2)$, $(0, -2)$, and $(3, 0)$.

Solution

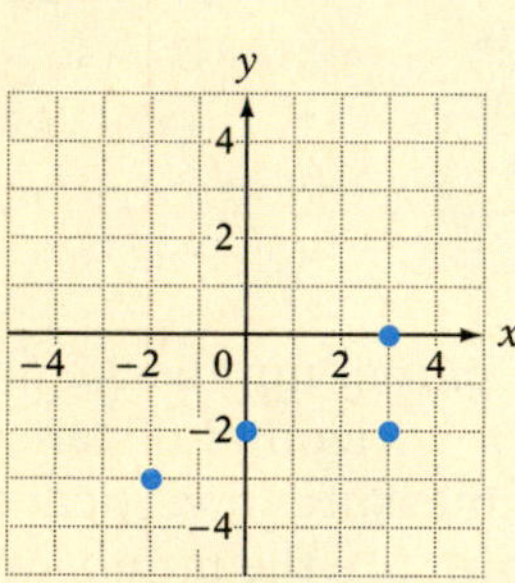

You Try It 1 Graph the ordered pairs $(-4, 1)$, $(3, -3)$, $(0, 4)$, and $(-3, 0)$.

Your solution

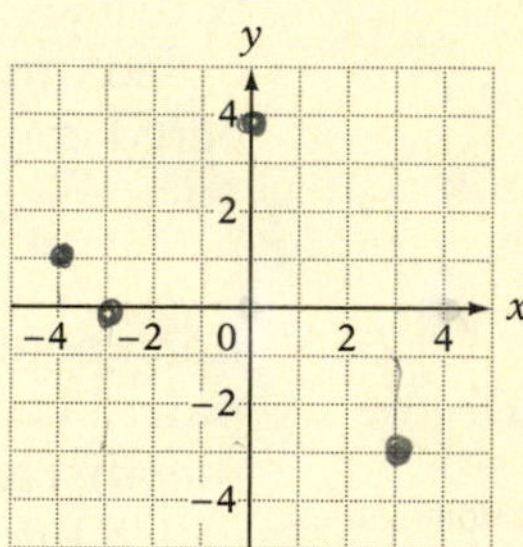

Example 2 Give the coordinates of the points labeled *A* and *B*. Give the abscissa of point *C* and the ordinate of point *D*.

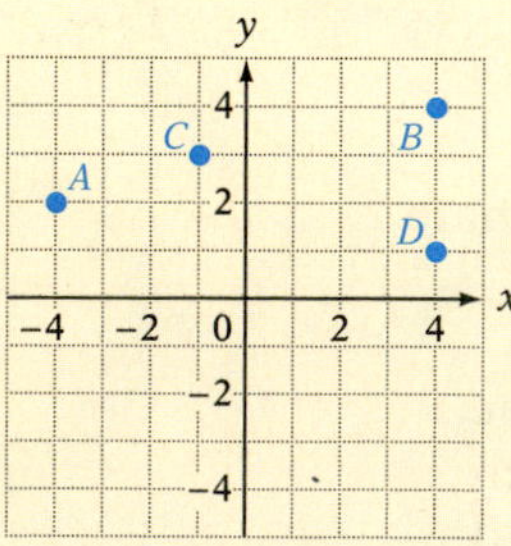

Solution The coordinates of *A* are $(-4, 2)$.
The coordinates of *B* are $(4, 4)$.
The abscissa of *C* is -1.
The ordinate of *D* is 1.

You Try It 2 Give the coordinates of the points labeled *A* and *B*. Give the abscissa of point *D* and the ordinate of point *C*.

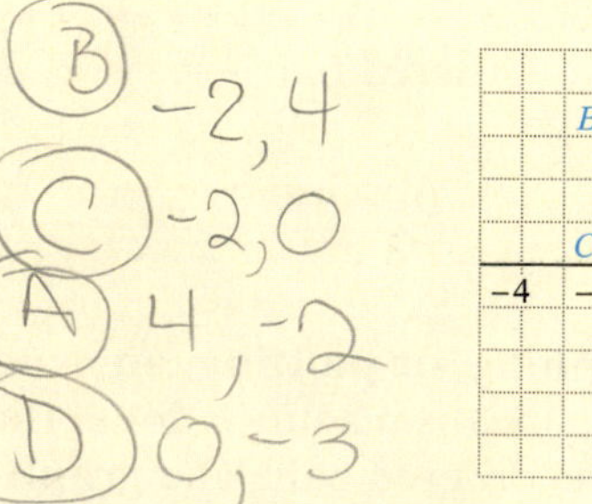

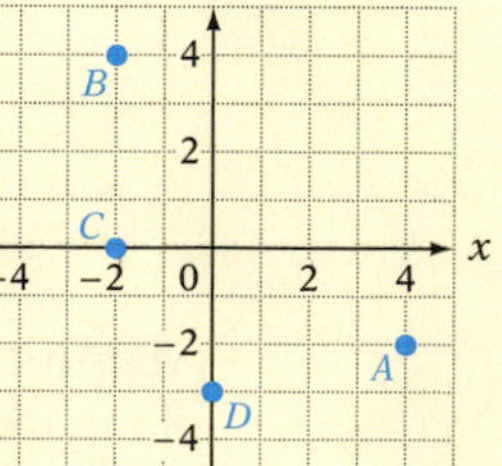

Your solution

Solutions on p. S11

Copyright © Houghton Mifflin Company. All rights reserved.

Objective B To determine ordered-pair solutions of an equation in two variables

When drawing a rectangular coordinate system, we often label the horizontal axis x and the vertical axis y. In this case, the coordinate system is called the ***xy*-coordinate system.** The coordinates of the points are given by ordered pairs (x, y), where the abscissa is called the ***x*-coordinate** and the ordinate is called the ***y*-coordinate**.

A coordinate system is used to study the relationship between two variables. Frequently this relationship is given by an equation. Examples of **equations in two variables** include

$$y = 2x - 3 \qquad 3x + 2y = 6 \qquad x^2 - y = 0$$

A **solution of an equation in two variables** is an ordered pair (x, y) whose coordinates make the equation a true statement.

TAKE NOTE

An ordered pair is of the form (x, y). For the ordered pair $(-3, 7)$, -3 is the x value and 7 is the y value. Substitute -3 for x and 7 for y.

HOW TO Is the ordered pair $(-3, 7)$ a solution of the equation $y = -2x + 1$?

$$\begin{array}{c|l} y & = -2x + 1 \\ \hline 7 & -2(-3) + 1 \\ 7 & 6 + 1 \end{array}$$
$$7 = 7$$

- Replace x by -3 and y by 7.
- Simplify.
- Compare the results. If the resulting equation is a true statement, the ordered pair is a solution of the equation. If it is not a true statement, the ordered pair is not a solution of the equation.

Yes, the ordered pair $(-3, 7)$ is a solution of the equation.

Besides $(-3, 7)$, there are many other ordered-pair solutions of $y = -2x + 1$. For example, $(0, 1)$, $\left(-\frac{3}{2}, 4\right)$, and $(4, -7)$ are also solutions.

In general, an equation in two variables has an infinite number of solutions. By choosing any value of x and substituting that value into the equation, we can calculate a corresponding value of y.

HOW TO Find the ordered-pair solution of $y = \frac{2}{3}x - 3$ that corresponds to $x = 6$.

$$y = \frac{2}{3}x - 3$$
$$= \frac{2}{3}(6) - 3$$
- Replace x by 6.

$$= 4 - 3$$
$$= 1$$
- Solve for y.

The ordered-pair solution is $(6, 1)$.

The solution of an equation in two variables can be graphed in an xy-coordinate system.

Copyright © Houghton Mifflin Company. All rights reserved.

HOW TO Graph the ordered-pair solutions of $y = -2x + 1$ when $x = -2$, -1, 0, 1, and 2.

Use the values of x to determine ordered-pair solutions of the equation. It is convenient to record these in a table.

x	$y = -2x + 1$	y	(x, y)
-2	$-2(-2) + 1$	5	$(-2, 5)$
-1	$-2(-1) + 1$	3	$(-1, 3)$
0	$-2(0) + 1$	1	$(0, 1)$
1	$-2(1) + 1$	-1	$(1, -1)$
2	$-2(2) + 1$	-3	$(2, -3)$

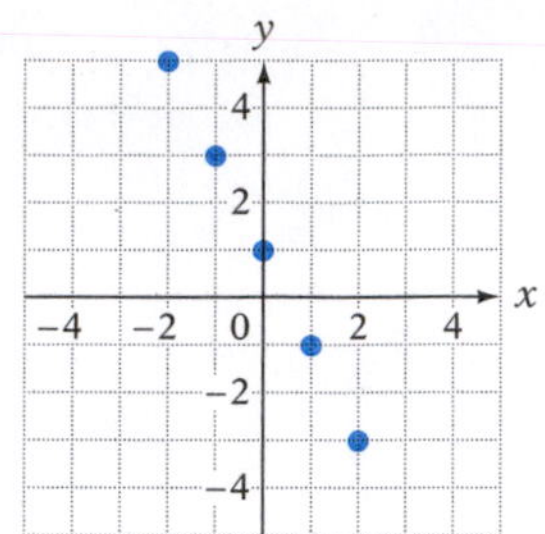

Example 3

Is $(3, -2)$ a solution of $3x - 4y = 15$?

Solution

$3x - 4y$	$= 15$
$3(3) - 4(-2)$	15
$9 + 8$	15
$17 \neq 15$	

• Replace x by 3 and y by -2.

No, $(3, -2)$ is not a solution of $3x - 4y = 15$.

You Try It 3

Is $(-2, 4)$ a solution of $x - 3y = -14$?

Your solution

Example 4

Find the ordered-pair solution of $y = \frac{x}{x - 2}$ corresponding to $x = 4$.

Solution

Replace x by 4 and solve for y.

$$y = \frac{x}{x - 2} = \frac{4}{4 - 2} = \frac{4}{2} = 2$$

The ordered-pair solution is $(4, 2)$.

You Try It 4

Find the ordered-pair solution of $y = \frac{3x}{x + 1}$ corresponding to $x = -2$.

Your solution

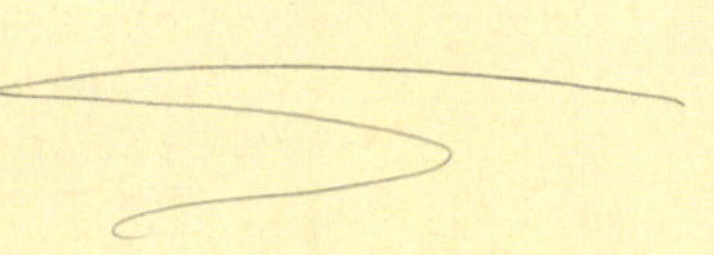

Solutions on p. S11

Copyright © Houghton Mifflin Company. All rights reserved.

Example 5

Graph the ordered-pair solutions of $y = \frac{2}{3}x - 2$ when $x = -3, 0, 3, 6$.

Solution

Replace x in $y = \frac{2}{3}x - 2$ by $-3, 0, 3$, and 6. For each value of x, determine the value of y.

x	$y = \frac{2}{3}x - 2$	y	(x, y)
-3	$\frac{2}{3}(-3) - 2$	-4	$(-3, -4)$
0	$\frac{2}{3}(0) - 2$	-2	$(0, -2)$
3	$\frac{2}{3}(3) - 2$	0	$(3, 0)$
6	$\frac{2}{3}(6) - 2$	2	$(6, 2)$

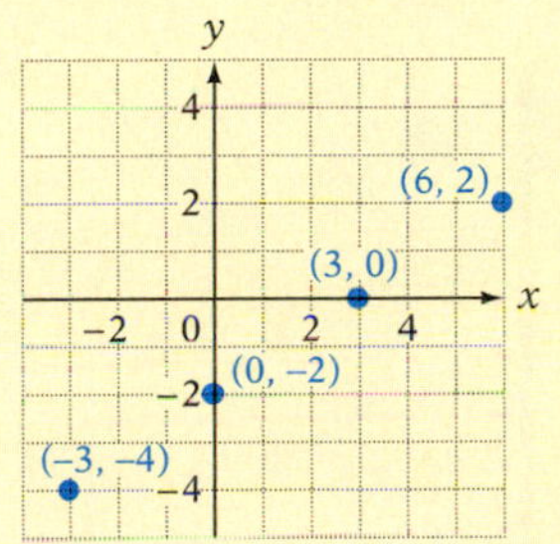

You Try It 5

Graph the ordered-pair solutions of $y = -\frac{1}{2}x + 2$ when $x = -4, -2, 0, 2$.

Your solution

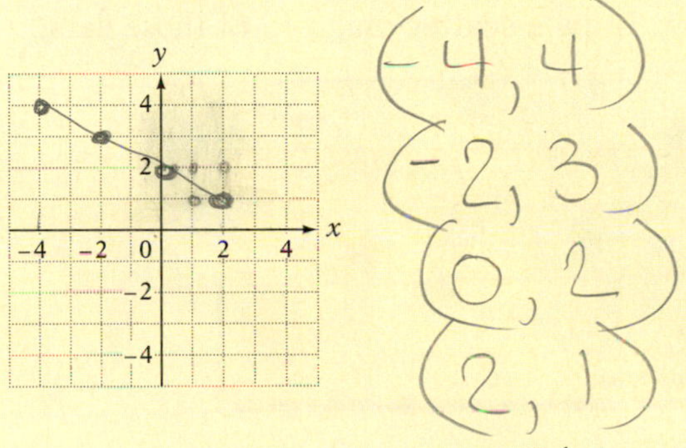

Solution on p. S12

Copyright © Houghton Mifflin Company. All rights reserved.

Objective C To graph a scatter diagram

Discovering a relationship between two variables is an important task in the study of mathematics. These relationships occur in many forms and in a wide variety of applications. Here are some examples.

- A botanist wants to know the relationship between the number of bushels of wheat yielded per acre and the amount of watering per acre.
- An environmental scientist wants to know the relationship between the incidence of skin cancer and the amount of ozone in the atmosphere.
- A business analyst wants to know the relationship between the price of a product and the number of products that are sold at that price.

Integrating Technology

See the appendix Keystroke Guide: *Scatter Diagrams* for instructions on using a graphing calculator to create a scatter diagram.

A researcher may investigate the relationship between two variables by means of *regression analysis,* which is a branch of statistics. The study of the relationship between the two variables may begin with a **scatter diagram,** which is a graph of the ordered pairs of the known data.

The following table shows randomly selected data for a recent Boston Marathon. Ages of participants 40 years old and older and their times (in minutes) are given.

Age (x)	55	46	53	40	40	44	54	44	41	50
Time (y)	254	204	243	194	281	197	238	300	232	216

TAKE NOTE

The jagged portion of the horizontal axis in the figure at the right indicates that the numbers between 0 and 40 are missing.

The scatter diagram for these data is shown at the right. Each ordered pair represents the age and time of a participant. For instance, the ordered pair (53, 243) indicates that a 53-year-old participant ran the marathon in 243 min.

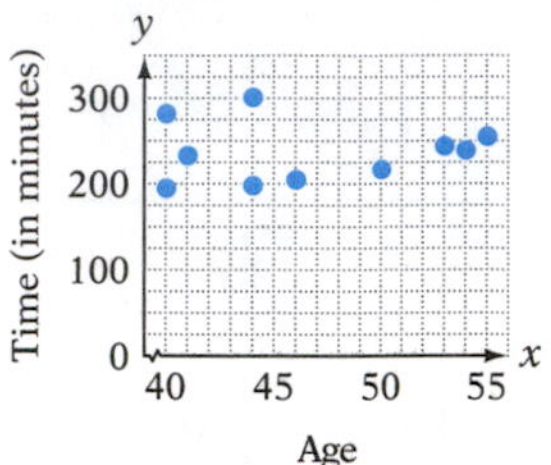

Example 6

The grams of sugar and the grams of fiber in a 1-ounce serving of six breakfast cereals are shown in the table below. Draw a scatter diagram of these data.

	Sugar (x)	*Fiber (y)*
Wheaties	4	3
Rice Krispies	3	0
Total	5	3
Life	6	2
Kix	3	1
Grape-Nuts	7	5

Strategy

To draw a scatter diagram:

- Draw a coordinate grid with the horizontal axis representing the grams of sugar and the vertical axis the grams of fiber.
- Graph the ordered pairs (4, 3), (3, 0), (5, 3), (6, 2), (3, 1), and (7, 5).

Solution

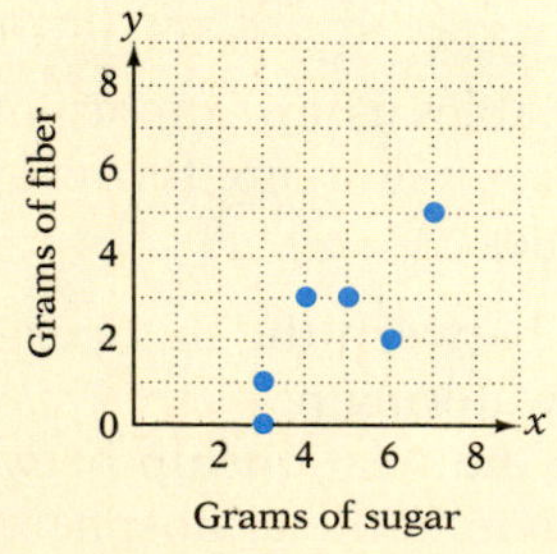

You Try It 6

According to the National Interagency Fire Center, the number of deaths in U.S. wildland fires is as shown in the table below. Draw a scatter diagram of these data.

Year	*Number of Deaths*
1998	14
1999	28
2000	17
2001	18
2002	23
2003	29

Your strategy

Your solution

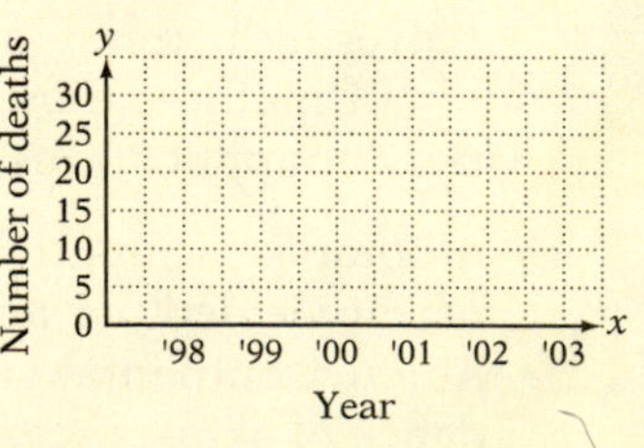

Solution on p. S12

Copyright © Houghton Mifflin Company. All rights reserved.

4.1 Exercises

Objective A To graph points in a rectangular coordinate system

1. Graph $(-2, 1)$, $(3, -5)$, $(-2, 4)$, and $(0, 3)$.

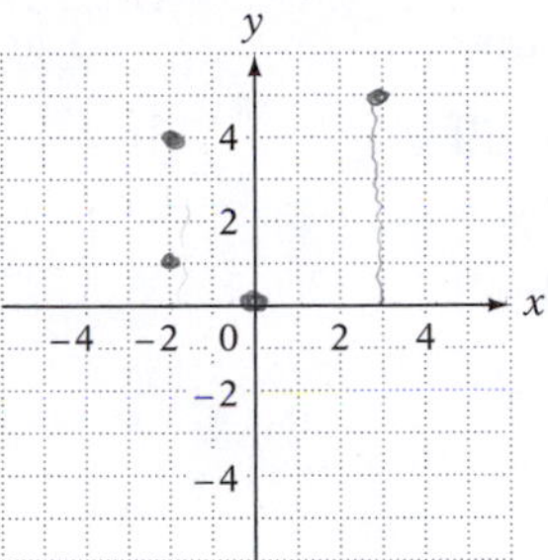

2. Graph $(5, -1)$, $(-3, -3)$, $(-1, 0)$, and $(1, -1)$.

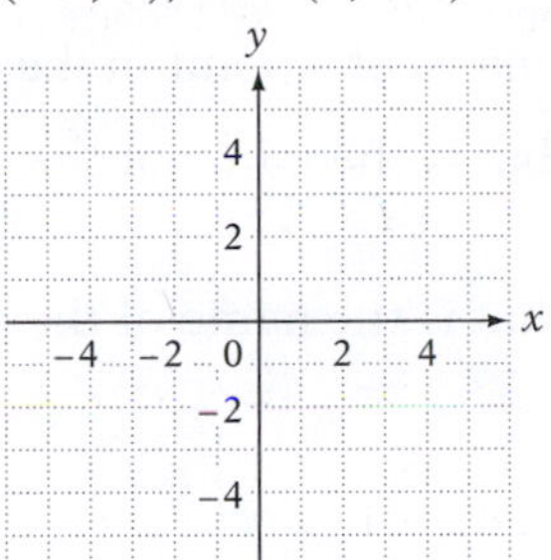

3. Graph $(0, 0)$, $(0, -5)$, $(-3, 0)$, and $(0, 2)$.

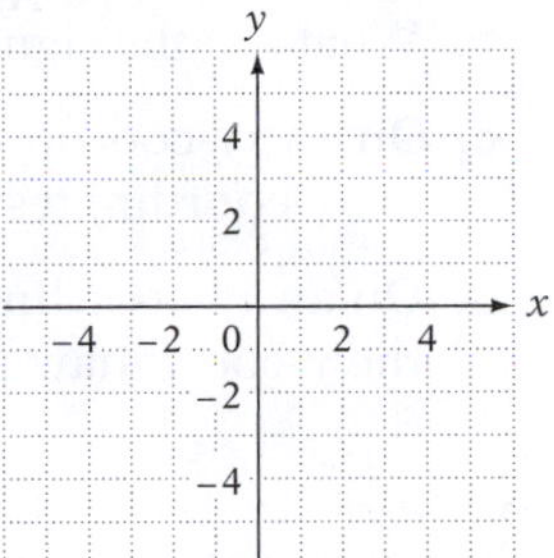

4. Graph $(-4, 5)$, $(-3, 1)$, $(3, -4)$, and $(5, 0)$.

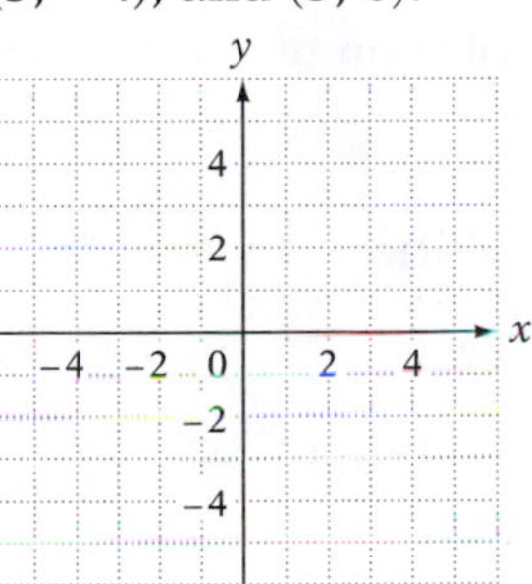

5. Graph $(-1, 4)$, $(-2, -3)$, $(0, 2)$, and $(4, 0)$.

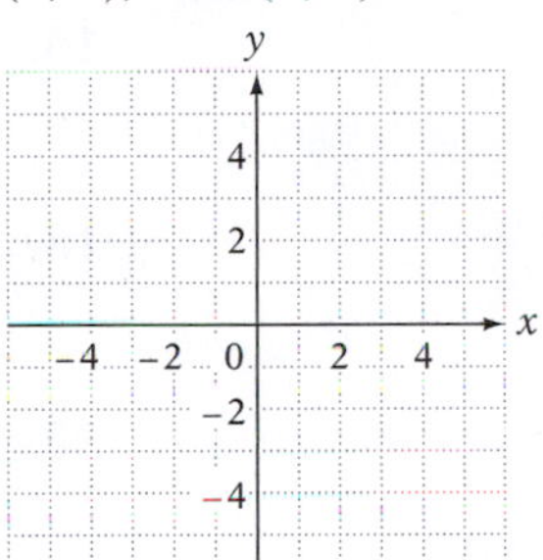

6. Graph $(5, 2)$, $(-4, -1)$, $(0, 0)$, and $(0, 3)$.

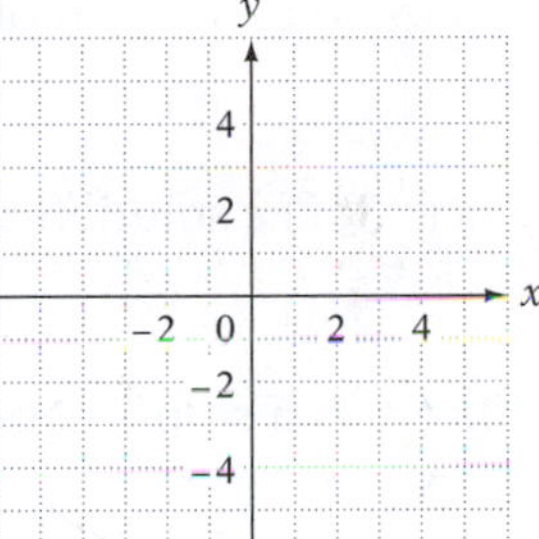

7. Find the coordinates of each of the points.

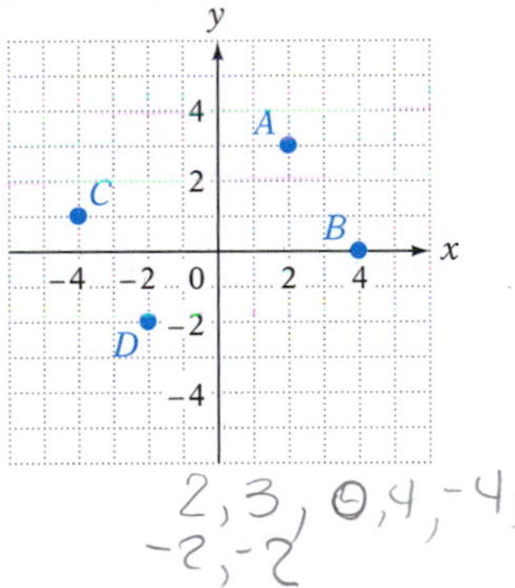

8. Find the coordinates of each of the points.

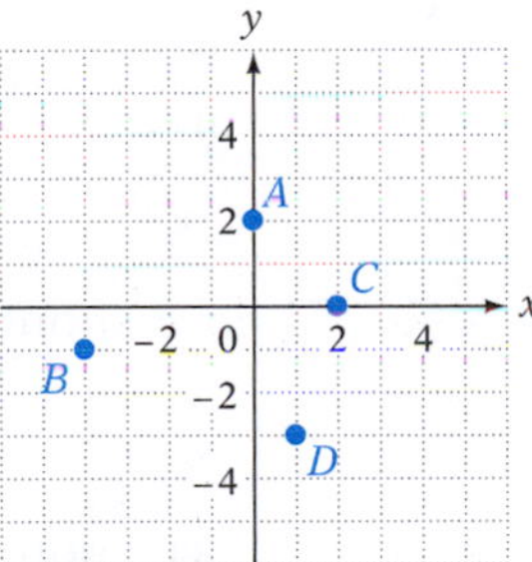

9. Find the coordinates of each of the points.

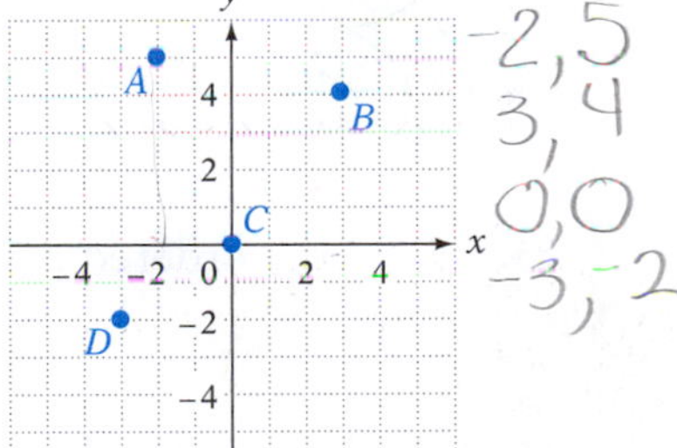

10. Find the coordinates of each of the points.

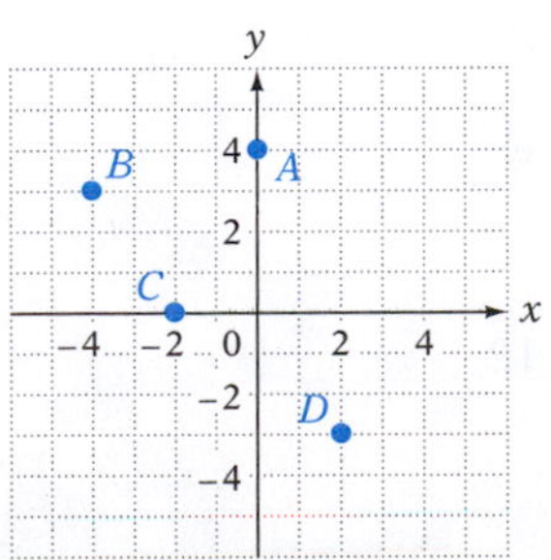

11. **a.** Name the abscissas of points *A* and *C*.
b. Name the ordinates of points *B* and *D*.

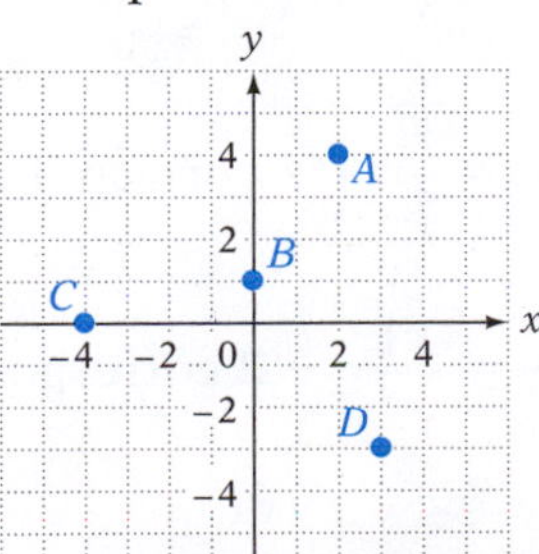

12. **a.** Name the abscissas of points *A* and *C*.
b. Name the ordinates of points *B* and *D*.

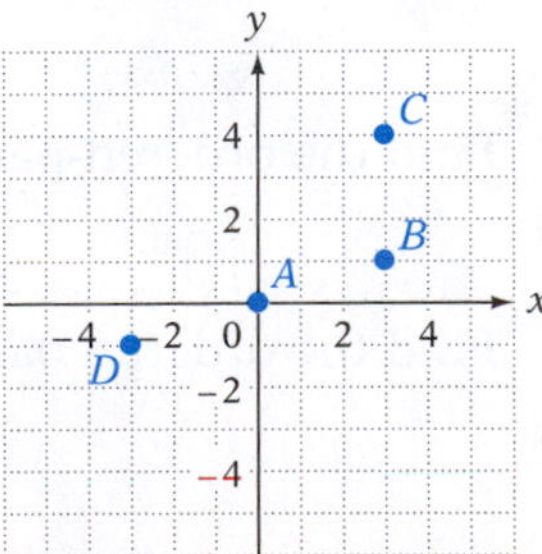

Copyright © Houghton Mifflin Company. All rights reserved.

13. Suppose you are helping a student who is having trouble graphing ordered pairs. The work of the student is at the right. What can you say to this student to correct the error that is being made?

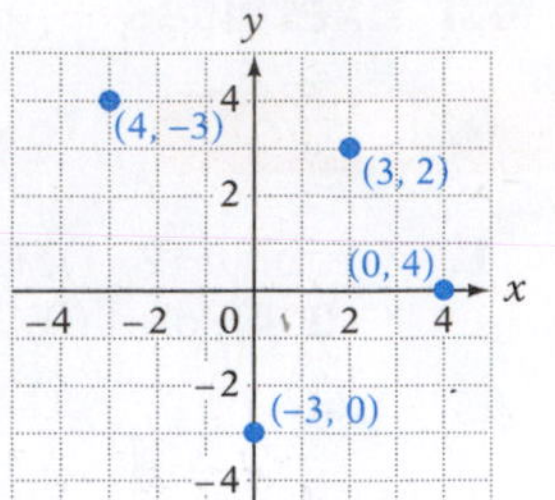

14. **a.** What are the signs of the coordinates of a point in the third quadrant?

    b. What are the signs of the coordinates of a point in the fourth quadrant?

    c. On an xy-coordinate system, what is the name of the axis for which all the x-coordinates are zero?

    d. On an xy-coordinate system, what is the name of the axis for which all the y-coordinates are zero?

Objective B To determine ordered-pair solutions of an equation in two variables

15. Is (3, 4) a solution of $y = -x + 7$?

16. Is (2, −3) a solution of $y = x + 5$?

17. Is (−1, 2) a solution of $y = \frac{1}{2}x - 1$?

18. Is (1, −3) a solution of $y = -2x - 1$?

19. Is (4, 1) a solution of $2x - 5y = 4$?

20. Is (−5, 3) a solution of $3x - 2y = 9$?

21. Is (0, 4) a solution of $3x - 4y = -4$?

22. Is (−2, 0) a solution of $x + 2y = -1$?

23. Find the ordered-pair solution of $y = 3x - 2$ corresponding to $x = 3$.

24. Find the ordered-pair solution of $y = 4x + 1$ corresponding to $x = -1$.

25. Find the ordered-pair solution of $y = \frac{2}{3}x - 1$ corresponding to $x = 6$.

26. Find the ordered-pair solution of $y = \frac{3}{4}x - 2$ corresponding to $x = 4$.

27. Find the ordered-pair solution of $y = -3x + 1$ corresponding to $x = 0$.

28. Find the ordered-pair solution of $y = \frac{2}{5}x - 5$ corresponding to $x = 0$.

29. Find the ordered-pair solution of $y = \frac{2}{5}x + 2$ corresponding to $x = -5$.

30. Find the ordered-pair solution of $y = -\frac{1}{6}x - 2$ corresponding to $x = 12$.

Copyright © Houghton Mifflin Company. All rights reserved.

Graph the ordered-pair solutions for the given values of x.

31. $y = 2x; x = -2, -1, 0, 2$

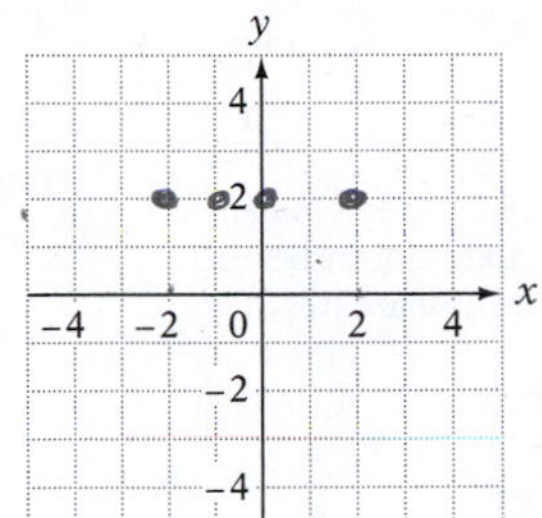

32. $y = -2x; x = -2, -1, 0, 2$

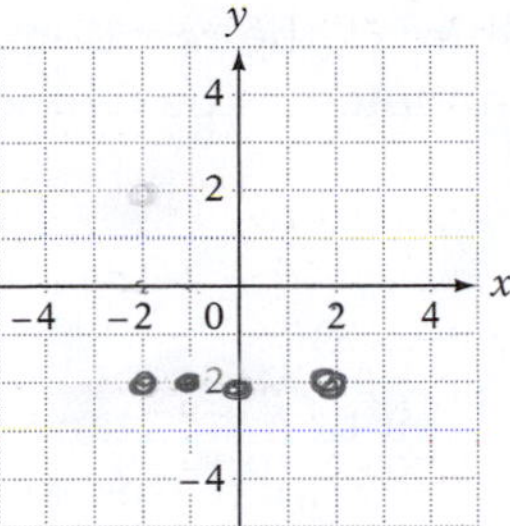

33. $y = x + 2; x = -4, -2, 0, 3$

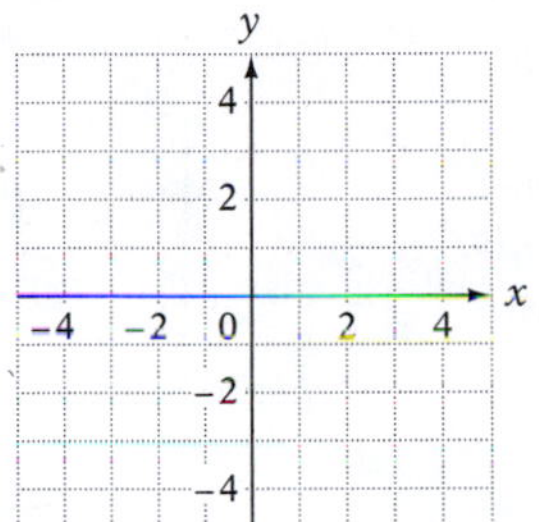

34. $y = \frac{1}{2}x - 1; x = -2, 0, 2, 4$

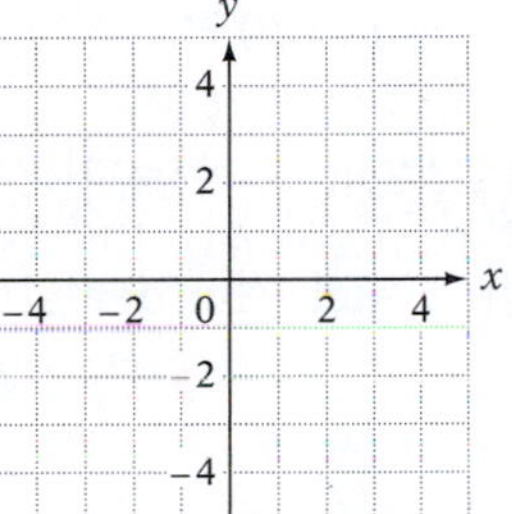

35. $y = \frac{2}{3}x + 1; x = -3, 0, 3$

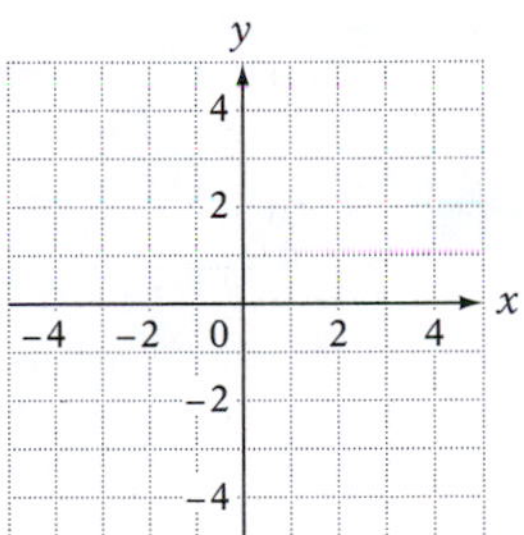

36. $y = -\frac{1}{3}x - 2; x = -3, 0, 3$

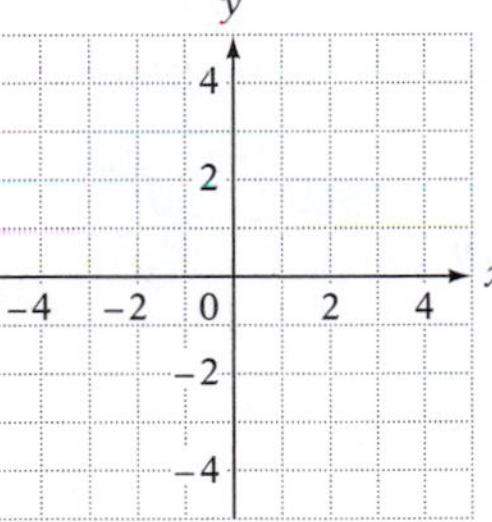

37. $y = x^2; x = -2, -1, 0, 1, 2$

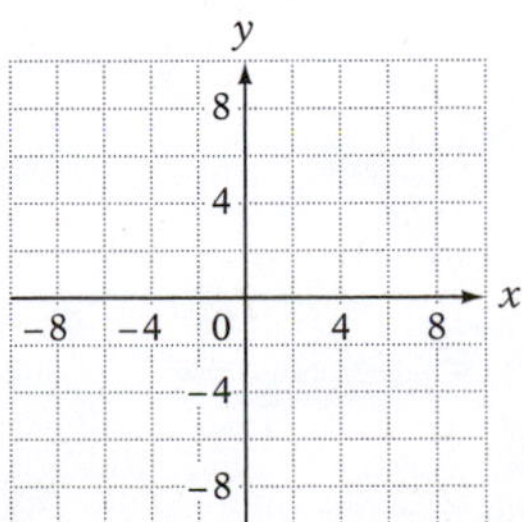

38. $y = |x + 1|; x = -5, -3, 0, 3, 5$

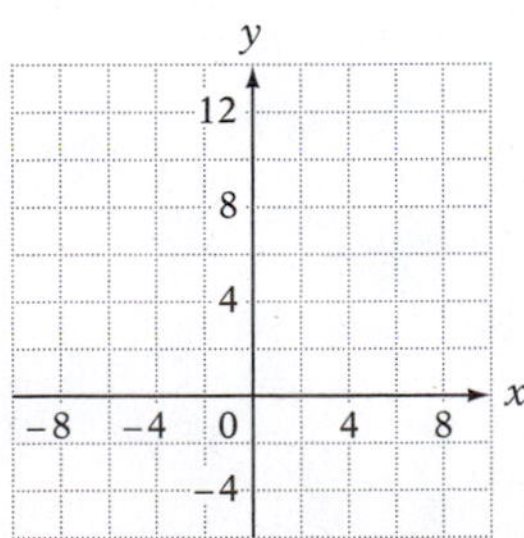

Copyright © Houghton Mifflin Company. All rights reserved.

Objective C To graph a scatter diagram

39. Chemistry The temperature of a chemical reaction is measured at intervals of 10 min and recorded in the scatter diagram at the right.
a. Find the temperature of the reaction after 20 min.
b. After how many minutes is the temperature 160°F?

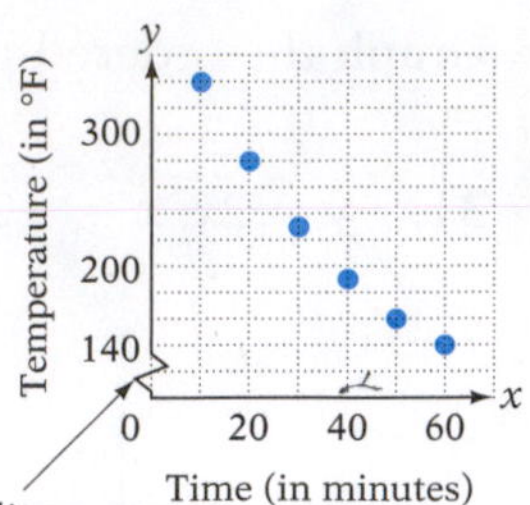

40. Chemistry The amount of a substance that can be dissolved in a fixed amount of water usually increases as the temperature of the water increases. Cerium selenate, however, does not behave in this manner. The graph at the right shows the number of grams of cerium selenate that will dissolve in 100 mg of water for various temperatures, in degrees Celsius.
a. Determine the temperature at which 25 g of cerium selenate will dissolve.
b. Determine the number of grams of cerium selenate that will dissolve when the temperature is 80°C.

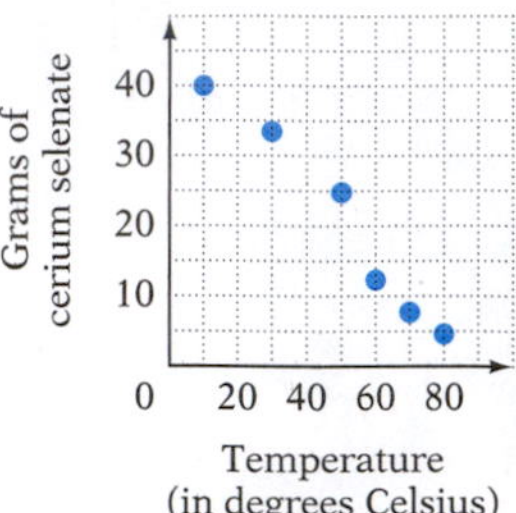

41. Business Past experience of executives of a car company shows that the profit of a dealership will depend on the total income of all the residents of the town in which the dealership is located. The table below shows the profits of several dealerships and the total incomes of the towns. Draw a scatter diagram for these data.

Profit (in thousands of dollars)	65	85	81	77	89	69
Total Income (in billions of dollars)	2.2	2.6	2.5	2.4	2.7	2.3

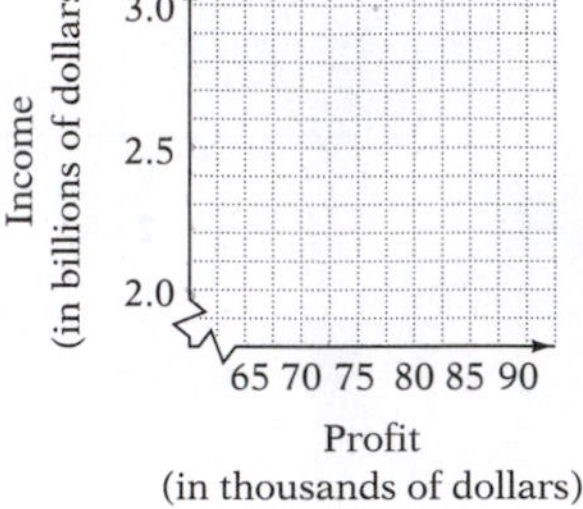

42. Utilities A power company suggests that a larger power plant can produce energy more efficiently and therefore at lower cost to consumers. The table below shows the output and average cost for power plants of various sizes. Draw a scatter diagram for these data.

Output (in millions of watts)	0.7	2.2	2.6	3.2	2.8	3.5
Average Cost (in dollars)	6.9	6.5	6.3	6.4	6.5	6.1

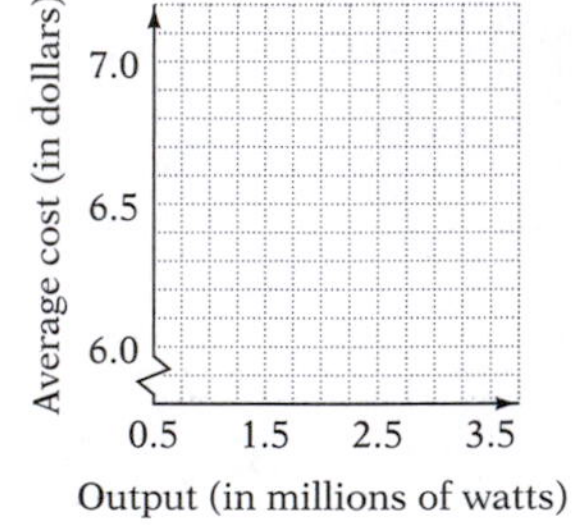

APPLYING THE CONCEPTS

43. Graph the ordered pairs (x, x^2), where $x \in \{-2, -1, 0, 1, 2\}$.

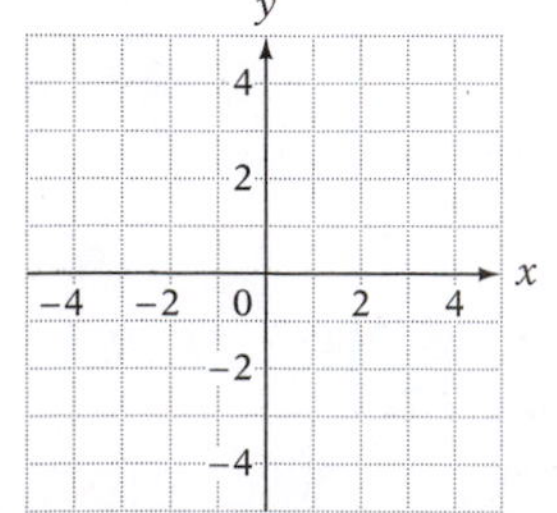

44. Graph the ordered pairs $\left(x, \frac{1}{x}\right)$, where $x \in \left\{-2, -1, -\frac{1}{2}, -\frac{1}{3}, \frac{1}{3}, \frac{1}{2}, 1, 2\right\}$.

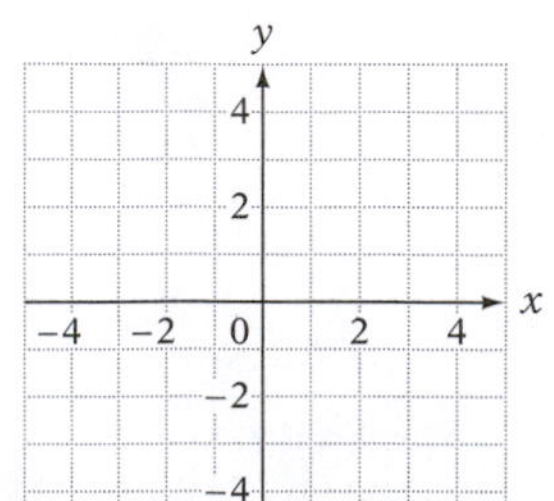

Copyright © Houghton Mifflin Company. All rights reserved.

4.2 Introduction to Functions

Objective A To evaluate a function

In mathematics and its applications, there are many times when it is necessary to investigate a relationship between two quantities. Here is a financial application: Consider a person who is planning to finance the purchase of a car. If the current interest rate for a 5-year loan is 5%, the equation that describes the relationship between the amount that is borrowed B and the monthly payment P is $P = 0.018871B$.

For each amount the purchaser may borrow (B), there is a certain monthly payment (P). The relationship between the amount borrowed and the payment can be recorded as ordered pairs, where the first coordinate is the amount borrowed and the second coordinate is the monthly payment. Some of these ordered pairs are shown at the right.

$$0.018871B = P$$

(6000, 113.23)
(7000, 132.10)
(8000, 150.97)
(9000, 169.84)

A relationship between two quantities is not always given by an equation. The table at the right describes a grading scale that defines a relationship between a score on a test and a letter grade. For each score, the table assigns only one letter grade. The ordered pair (84, B) indicates that a score of 84 receives a letter grade of B.

Score	*Grade*
90–100	A
80–89	B
70–79	C
60–69	D
0–59	F

The graph at the right also shows a relationship between two quantities. It is a graph of the viscosity V of SAE 40 motor oil at various temperatures T. Ordered pairs can be approximated from the graph. The ordered pair (120, 250) indicates that the viscosity of the oil at 120°F is 250 units.

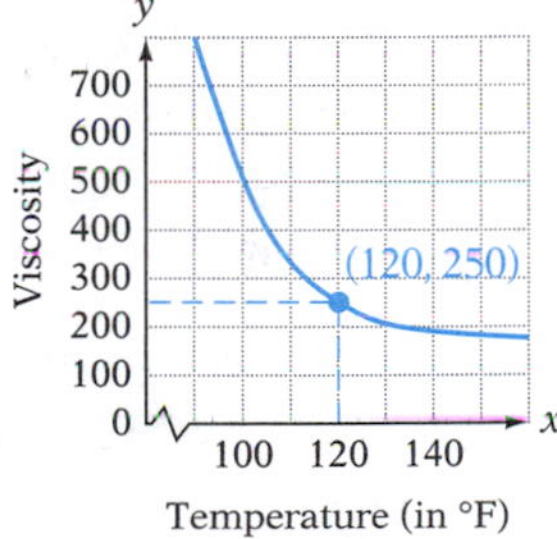

In each of these examples, there is a rule (an equation, a table, or a graph) that determines a certain set of ordered pairs.

Definition of Relation

A **relation** is a set of ordered pairs.

Here are some of the ordered pairs for the relations given above.

Relation	*Some of the Ordered Pairs of the Relation*
Car Payment	(7500, 141.53), (8750, 165.12), (9390, 177.20)
Grading Scale	(78, C), (98, A), (70, C), (81, B), (94, A)
Oil Viscosity	(100, 500), (120, 250), (130, 200), (150, 180)

Copyright © Houghton Mifflin Company. All rights reserved.

Each of these three relations is actually a special type of relation called a *function*. Functions play an important role in mathematics and its applications.

Definition of Function

A **function** is a relation in which no two ordered pairs have the same first coordinate and different second coordinates.

The **domain** of a function is the set of the first coordinates of all the ordered pairs of the function. The **range** is the set of the second coordinates of all the ordered pairs of the function.

For the function defined by the ordered pairs

$$\{(2, 3), (4, 5), (6, 7), (8, 9)\}$$

the domain is {2, 4, 6, 8} and the range is {3, 5, 7, 9}.

HOW TO Find the domain and range of the function {(2, 3), (4, 6), (6, 8), (10, 6)}.

The domain is {2, 4, 6, 10}.
- The domain of the function is the set of the first coordinates of the ordered pairs.

The range is {3, 6, 8}.
- The range of the function is the set of the second coordinates of the ordered pairs.

For each element in the domain of a function there is a corresponding element in the range of the function. A possible diagram for the function above is

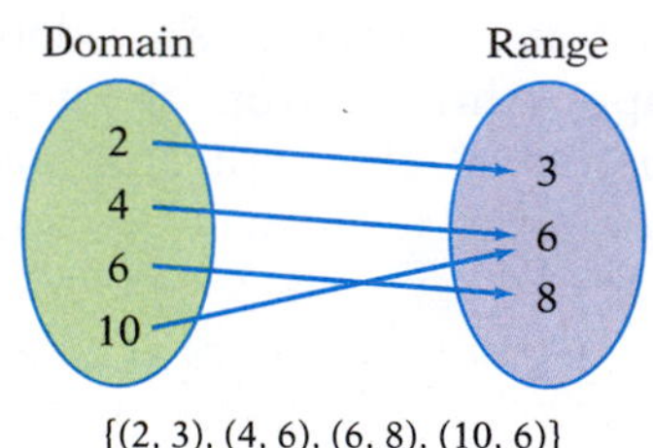

{(2, 3), (4, 6), (6, 8), (10, 6)}

Functions defined by tables or graphs, such as those described at the beginning of this section, have important applications. However, a major focus of this text is functions defined by equations in two variables.

The **square function,** which pairs each real number with its square, can be defined by the equation

$$y = x^2$$

This equation states that for a given value of x in the domain, the corresponding value of y in the range is the square of x. For instance, if $x = 6$, then $y = 36$ and if $x = -7$, then $y = 49$. Because the value of y *depends* on the value of x, y is called the **dependent variable** and x is called the **independent variable.**

Copyright © Houghton Mifflin Company. All rights reserved.

> **TAKE NOTE**
> A pictorial representation of the square function is shown at the right. The function acts as a machine that changes a number from the domain into the square of the number.

A function can be thought of as a rule that pairs one number with another number. For instance, the *square function* pairs a number with its square. The ordered pairs for the values shown at the right are $(-5, 25)$, $\left(\frac{3}{5}, \frac{9}{25}\right)$, $(0, 0)$, and $(3, 9)$. For this function, the second coordinate is the square of the first coordinate. If we let x represent the first coordinate, then the second coordinate is x^2 and we have the ordered pair (x, x^2).

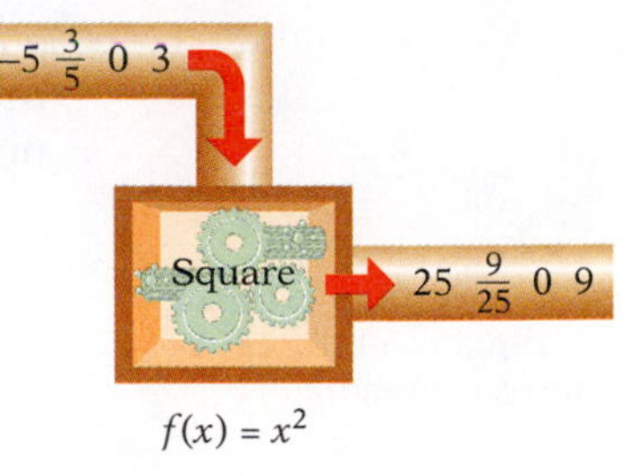

$f(x) = x^2$

A function cannot have two ordered pairs with *different* second coordinates and the same first coordinate. However, a function may contain ordered pairs with the *same* second coordinate. For instance, the square function has the ordered pairs $(-3, 9)$ and $(3, 9)$; the second coordinates are the same but the first coordinates are different.

The **double function** pairs a number with twice that number. The ordered pairs for the values shown at the right are $(-5, -10)$, $\left(\frac{3}{5}, \frac{6}{5}\right)$, $(0, 0)$, and $(3, 6)$. For this function, the second coordinate is twice the first coordinate. If we let x represent the first coordinate, then the second coordinate is $2x$ and we have the ordered pair $(x, 2x)$.

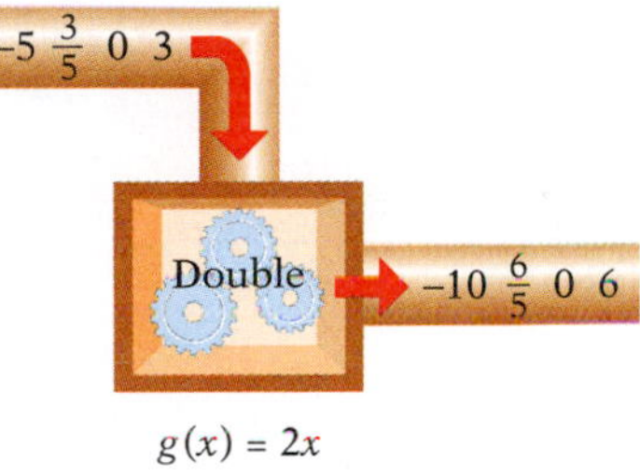

$g(x) = 2x$

Not every equation in two variables defines a function. For instance, consider the equation

$$y^2 = x^2 + 9$$

Because

$$5^2 = 4^2 + 9 \quad \text{and} \quad (-5)^2 = 4^2 + 9$$

the ordered pairs $(4, 5)$ and $(4, -5)$ are both solutions of the equation. Consequently, there are two ordered pairs that have the same first coordinate (4) but *different* second coordinates (5 and -5). Therefore, the equation does not define a function. Other ordered pairs for this equation are $(0, -3)$, $(0, 3)$, $(\sqrt{7}, -4)$, and $(\sqrt{7}, 4)$. A graphical representation of these ordered pairs is shown below.

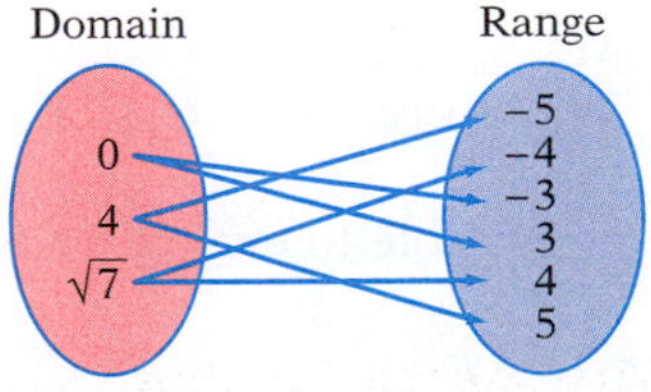

Note from this graphical representation that each element from the domain has two arrows pointing to two different elements in the range. Any time this occurs, the situation does not represent a function. However, this diagram *does* represent a relation. The relation for the values shown is $\{(0, -3), (0, 3), (4, -5), (4, 5), (\sqrt{7}, -4), (\sqrt{7}, 4)\}$.

The phrase "y is a function of x," or the same phrase with different variables, is used to describe an equation in two variables that defines a function. To emphasize that the equation represents a function, **functional notation** is used.

Copyright © Houghton Mifflin Company. All rights reserved.

Just as the variable x is commonly used to represent a number, the letter f is commonly used to name a function. The square function is written in functional notation as follows:

> **TAKE NOTE**
> The dependent variable y and $f(x)$ can be used interchangeably.

This is the value of the function. It is the number that is paired with x.

$$f(x) = x^2$$

The name of the function is f.

This is an algebraic expression that defines the relationship between the dependent and independent variables.

The symbol $f(x)$ is read "the value of f at x" or "f of x."

It is important to note that $f(x)$ does *not* mean f times x. The symbol $f(x)$ is the **value of the function** and represents the value of the dependent variable for a given value of the independent variable. We often write $y = f(x)$ to emphasize the relationship between the independent variable x and the dependent variable y. Remember that y and $f(x)$ are different symbols for the same number.

The letters used to represent a function are somewhat arbitrary. All of the following equations represent the same function.

$$\left.\begin{aligned} f(x) &= x^2 \\ s(t) &= t^2 \\ P(v) &= v^2 \end{aligned}\right\}$$ Each equation represents the square function.

The process of determining $f(x)$ for a given value of x is called **evaluating a function.** For instance, to evaluate $f(x) = x^2$ when $x = 4$, replace x by 4 and simplify.

$$f(x) = x^2$$
$$f(4) = 4^2 = 16$$

The *value* of the function is 16 when $x = 4$. An ordered pair of the function is (4, 16).

> **Integrating Technology**
> See the Projects and Group Activities at the end of this chapter for instructions on using a graphing calculator to evaluate a function. Instructions are also provided in the appendix Keystroke Guide: *Evaluating Functions*.

HOW TO Evaluate $g(t) = 3t^2 - 5t + 1$ when $t = -2$.

$$\begin{aligned} g(t) &= 3t^2 - 5t + 1 \\ g(-2) &= 3(-2)^2 - 5(-2) + 1 \\ &= 3(4) - 5(-2) + 1 \\ &= 12 + 10 + 1 = 23 \end{aligned}$$

- Replace t by -2 and then simplify.

When t is -2, the value of the function is 23.
Therefore, an ordered pair of the function is $(-2, 23)$.

It is possible to evaluate a function for a variable expression.

HOW TO Evaluate $P(z) = 3z - 7$ when $z = 3 + h$.

$$\begin{aligned} P(z) &= 3z - 7 \\ P(3 + h) &= 3(3 + h) - 7 \\ &= 9 + 3h - 7 \\ &= 3h + 2 \end{aligned}$$

- Replace z by $3 + h$ and then simplify.

When z is $3 + h$, the value of the function is $3h + 2$.
Therefore, an ordered pair of the function is $(3 + h, 3h + 2)$.

Copyright © Houghton Mifflin Company. All rights reserved.

Recall that the range of a function is found by applying the function to each element of the domain. If the domain contains an infinite number of elements, it may be difficult to find the range. However, if the domain has a finite number of elements, then the range can be found by evaluating the function for each element in the domain.

HOW TO Find the range of $f(x) = x^3 + x$ if the domain is $\{-2, -1, 0, 1, 2\}$.

$$f(x) = x^3 + x$$
$$f(-2) = (-2)^3 + (-2) = -10$$
$$f(-1) = (-1)^3 + (-1) = -2$$
$$f(0) = 0^3 + 0 = 0$$
$$f(1) = 1^3 + 1 = 2$$
$$f(2) = 2^3 + 2 = 10$$

- **Replace *x* by each member of the domain. The range includes the values of $f(-2)$, $f(-1)$, $f(0)$, $f(1)$, and $f(2)$.**

The range is $\{-10, -2, 0, 2, 10\}$.

When a function is represented by an equation, the domain of the function is all real numbers for which the value of the function is a real number. For instance:

- The domain of $f(x) = x^2$ is all real numbers, because the square of every real number is a real number.
- The domain of $g(x) = \frac{1}{x - 2}$ is all real numbers except 2, because when $x = 2$,

$g(2) = \frac{1}{2 - 2} = \frac{1}{0}$, which is not a real number.

The domain of the grading-scale function is the set of whole numbers from 0 to 100. In set-builder notation, this is written $\{x | 0 \le x \le 100, x \in \text{whole numbers}\}$. The range is {A, B, C, D, F}.

Score	*Grade*
90–100	A
80–89	B
70–79	C
60–69	D
0–59	F

HOW TO What values, if any, are excluded from the domain of $f(x) = 2x^2 - 7x + 1$?

Because the value of $2x^2 - 7x + 1$ is a real number for any value of x, the domain of the function is all real numbers. No values are excluded from the domain of $f(x) = 2x^2 - 7x + 1$.

Example 1

Find the domain and range of the function $\{(5, 3), (9, 7), (13, 7), (17, 3)\}$.

Solution

Domain: $\{5, 9, 13, 17\}$;
Range: $\{3, 7\}$

- **The domain is the set of first coordinates.**

You Try It 1

Find the domain and range of the function $\{(-1, 5), (3, 5), (4, 5), (6, 5)\}$.

Your solution

Solution on p. S12

Copyright © Houghton Mifflin Company. All rights reserved.

Example 2

Given $p(r) = 5r^3 - 6r - 2$, find $p(-3)$.

Solution

$$\begin{aligned} p(r) &= 5r^3 - 6r - 2 \\ p(-3) &= 5(-3)^3 - 6(-3) - 2 \\ &= 5(-27) + 18 - 2 \\ &= -135 + 18 - 2 = -119 \end{aligned}$$

You Try It 2

Evaluate $G(x) = \frac{3x}{x+2}$ when $x = -4$.

Your solution

Example 3

Evaluate $Q(r) = 2r + 5$ when $r = h + 3$.

Solution

$$\begin{aligned} Q(r) &= 2r + 5 \\ Q(h+3) &= 2(h+3) + 5 \\ &= 2h + 6 + 5 \\ &= 2h + 11 \end{aligned}$$

You Try It 3

Evaluate $f(x) = x^2 - 11$ when $x = 3h$.

Your solution

Example 4

Find the range of $f(x) = x^2 - 1$ if the domain is $\{-2, -1, 0, 1, 2\}$.

Solution

To find the range, evaluate the function at each element of the domain.

$$\begin{aligned} f(x) &= x^2 - 1 \\ f(-2) &= (-2)^2 - 1 = 4 - 1 = 3 \\ f(-1) &= (-1)^2 - 1 = 1 - 1 = 0 \\ f(0) &= 0^2 - 1 = 0 - 1 = -1 \\ f(1) &= 1^2 - 1 = 1 - 1 = 0 \\ f(2) &= 2^2 - 1 = 4 - 1 = 3 \end{aligned}$$

The range is $\{-1, 0, 3\}$. Note that 0 and 3 are listed only once.

You Try It 4

Find the range of $h(z) = 3z + 1$ if the domain is $\left\{0, \frac{1}{3}, \frac{2}{3}, 1\right\}$.

Your solution

Example 5

What is the domain of $f(x) = 2x^2 - 7x + 1$?

Solution

Because $2x^2 - 7x + 1$ evaluates to a real number for any value of x, the domain of the function is all real numbers.

You Try It 5

What value is excluded from the domain of $f(x) = \frac{2}{x-5}$?

Your solution

Solutions on p. S12

Copyright © Houghton Mifflin Company. All rights reserved.

4.2 Exercises

Objective A **To evaluate a function**

1. In your own words, explain what a function is.

2. What is the domain of a function? What is the range of a function?

3. Does the diagram below represent a function? Explain your answer.

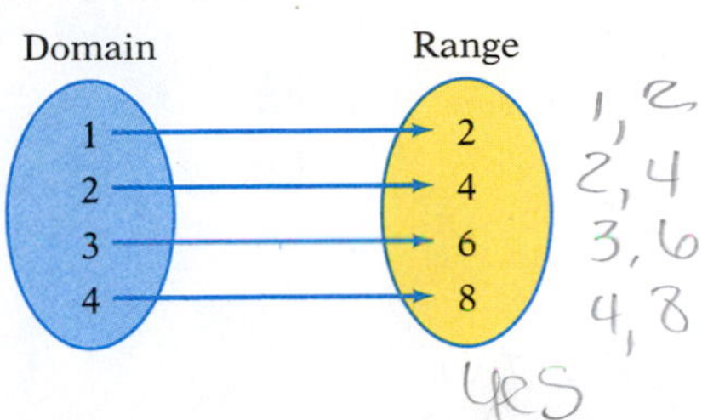

4. Does the diagram below represent a function? Explain your answer.

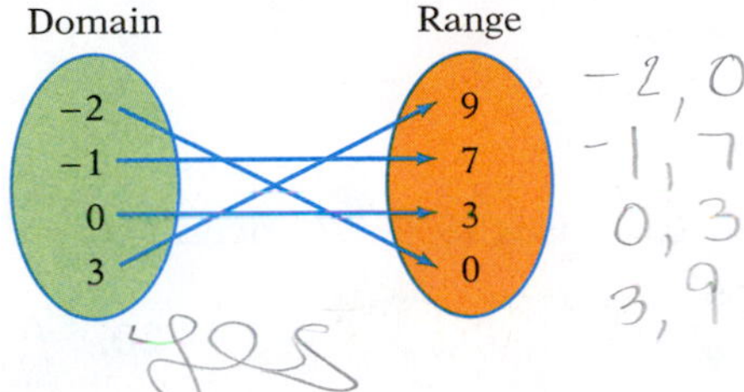

5. Does the diagram below represent a function? Explain your answer.

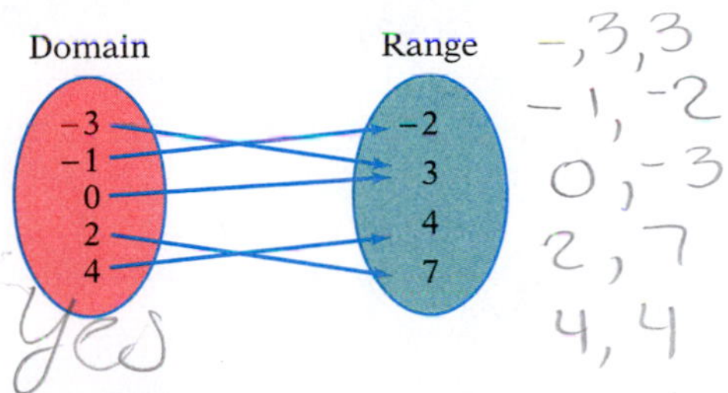

6. Does the diagram below represent a function? Explain your answer.

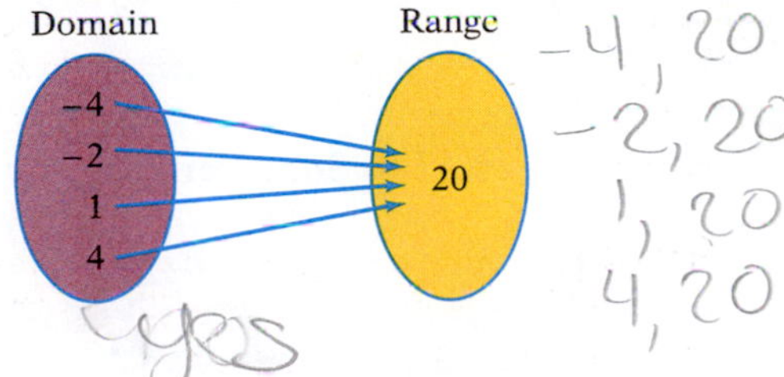

7. Does the diagram below represent a function? Explain your answer.

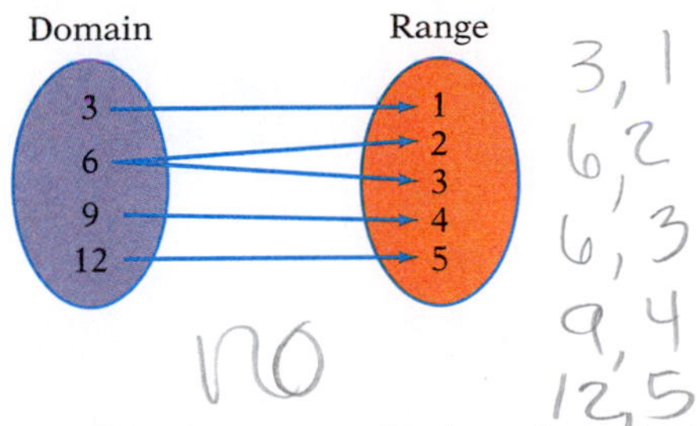

8. Does the diagram below represent a function? Explain your answer.

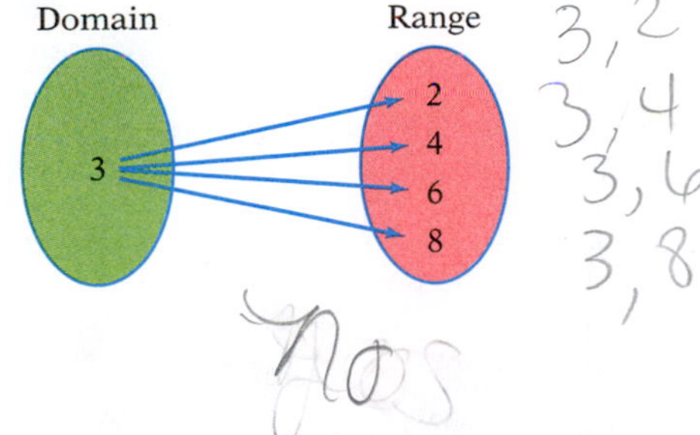

For Exercises 9 to 16, state whether the relation is a function.

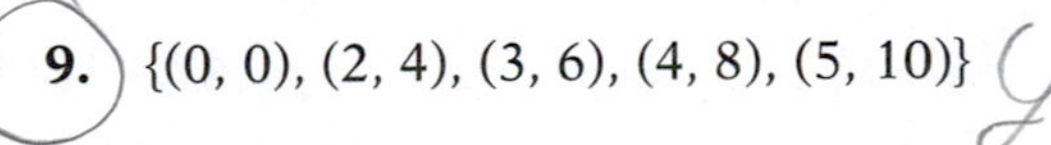

9. {(0, 0), (2, 4), (3, 6), (4, 8), (5, 10)}

10. {(1, 3), (3, 5), (5, 7), (7, 9)}

11. {(−2, −1), (−4, −5), (0, −1), (3, 5)}

12. {(−3, −1), (−1, −1), (0, 1), (2, 6)}

13. {(−2, 3), (−1, 3), (0, −3), (1, 3), (2, 3)}

14. {(0, 0), (1, 0), (2, 0), (3, 0), (4, 0)}

15. {(1, 1), (4, 2), (9, 3), (1, −1), (4, −2)}

16. {(3, 1), (3, 2), (3, 3), (3, 4)}

Copyright © Houghton Mifflin Company. All rights reserved.

17. **Shipping** The table at the right shows the cost to send an overnight package using United Parcel Service.

a. Does this table define a function?

b. Given $x = 2.75$ lb, find y.

Weight in pounds (x)	*Cost (y)*
$0 < x \le 1$	\$28.25
$1 < x \le 2$	\$31.25
$2 < x \le 3$	\$34.75
$3 < x \le 4$	\$37.75
$4 < x \le 5$	\$40.75

18. **Shipping** The table at the right shows the cost to send an "Express Mail" package using the U.S. Postal Service.

a. Does this table define a function?

b. Given $x = 0.5$ lb, find y.

Weight in pounds (x)	*Cost (y)*
$0 < x \le 0.5$	\$13.65
$0.5 < x \le 2$	\$17.85
$2 < x \le 3$	\$21.05
$3 < x \le 4$	\$24.20
$4 < x \le 5$	\$27.30

For Exercises 19 to 22, given $f(x) = 5x - 4$, evaluate:

19. $f(3)$ **20.** $f(-2)$ **21.** $f(0)$ **22.** $f(-1)$

For Exercises 23 to 26, given $G(t) = 4 - 3t$, evaluate:

23. $G(0)$ **24.** $G(-3)$ **25.** $G(-2)$ **26.** $G(4)$

For Exercises 27 to 30, given $q(r) = r^2 - 4$, evaluate:

27. $q(3)$ **28.** $q(4)$ **29.** $q(-2)$ **30.** $q(-5)$

For Exercises 31 to 34, given $F(x) = x^2 + 3x - 4$, evaluate:

31. $F(4)$ **32.** $F(-4)$ **33.** $F(-3)$ **34.** $F(-6)$

For Exercises 35 to 38, given $H(p) = \dfrac{3p}{p + 2}$, evaluate:

35. $H(1)$ **36.** $H(-3)$ **37.** $H(t)$ **38.** $H(v)$

For Exercises 39 to 42, given $s(t) = t^3 - 3t + 4$, evaluate:

39. $s(-1)$ **40.** $s(2)$ **41.** $s(a)$ **42.** $s(w)$

43. Given $P(x) = 4x + 7$, write $P(-2 + h) - P(-2)$ in simplest form.

44. Given $G(t) = 9 - 2t$, write $G(-3 + h) - G(-3)$ in simplest form.

Copyright © Houghton Mifflin Company. All rights reserved.

45. **Business** Game Engineering has just completed the programming and testing for a new computer game. The cost to manufacture and package the game depends on the number of units Game Engineering plans to sell. The table at the right shows the cost per game for packaging various quantities.
a. Evaluate this function when $x = 7000$.
b. Evaluate this function when $x = 20,000$.

Number of Games Manufactured	*Cost to Manufacture One Game*
$0 < x \le 2500$	\$6.00
$2500 < x \le 5000$	\$5.50
$5000 < x \le 10,000$	\$4.75
$10,000 < x \le 20,000$	\$4.00
$20,000 < x \le 40,000$	\$3.00

46. **Airports** Airport administrators have a tendency to price airport parking at a rate that discourages people from using the parking lot for long periods of time. The parking rate structure for an airport is given in the table at the right.
a. Evaluate this function when $t = 2.5$ h.
b. Evaluate this function when $t = 7$ h.

Hours Parked	*Cost*
$0 < t \le 1$	\$1.00
$1 < t \le 2$	\$3.00
$2 < t \le 4$	\$6.50
$4 < t \le 7$	\$10.00
$7 < t \le 12$	\$14.00

47. **Real Estate** A real estate appraiser charges a fee that depends on the estimated value, V, of the property. A table giving the fees charged for various estimated values of the real estate appears at the right.
a. Evaluate this function when $V = \$5,000,000$.
b. Evaluate this function when $V = \$767,000$.

Value of Property	*Appraisal Fee*
$V < 100,000$	\$350
$100,000 \le V < 500,000$	\$525
$500,000 \le V < 1,000,000$	\$950
$1,000,000 \le V < 5,000,000$	\$2500
$5,000,000 \le V < 10,000,000$	\$3000

48. **Shipping** The cost to mail a priority overnight package by Federal Express depends on the weight, w, of the package. A table of the costs for selected weights is given at the right.
a. Evaluate this function when $w = 2$ lb 3 oz.
b. Evaluate this function when $w = 1.9$ lb.

Weight (lb)	*Cost*
$0 < w \le 1$	\$27.00
$1 < w \le 2$	\$29.75
$2 < w \le 3$	\$32.75
$3 < w \le 4$	\$35.75
$4 < w \le 5$	\$39.00

For Exercises 49 to 58, find the domain and range of the function.

49. {(1, 1), (2, 4), (3, 7), (4, 10), (5, 13)}

50. {(2, 6), (4, 18), (6, 38), (8, 66), (10, 102)}

51. {(0, 1), (2, 2), (4, 3), (6, 4)}

52. {(0, 1), (1, 2), (4, 3), (9, 4)}

53. {(1, 0), (3, 0), (5, 0), (7, 0), (9, 0)}

54. {(−2, −4), (2, 4), (−1, 1), (1, 1), (−3, 9), (3, 9)}

Copyright © Houghton Mifflin Company. All rights reserved.

55. $\{(0, 0), (1, 1), (-1, 1), (2, 2), (-2, 2)\}$

56. $\{(0, -5), (5, 0), (10, 5), (15, 10)\}$

57. $\{(-2, -3), (-1, 6), (0, 7), (2, 3), (1, 9)\}$

58. $\{(-8, 0), (-4, 2), (-2, 4), (0, -4), (4, 4)\}$

For Exercises 59 to 76, what values are excluded from the domain of the function?

59. $f(x) = \dfrac{1}{x - 1}$

60. $g(x) = \dfrac{1}{x + 4}$

61. $h(x) = \dfrac{x + 3}{x + 8}$

62. $F(x) = \dfrac{2x - 5}{x - 4}$

63. $f(x) = 3x + 2$

64. $g(x) = 4 - 2x$

65. $G(x) = x^2 + 1$

66. $H(x) = \dfrac{1}{2}x^2$

67. $f(x) = \dfrac{x - 1}{x}$

68. $g(x) = \dfrac{2x + 5}{7}$

69. $H(x) = x^2 - x + 1$

70. $f(x) = 3x^2 + x + 4$

71. $f(x) = \dfrac{2x - 5}{3}$

72. $g(x) = \dfrac{3 - 5x}{5}$

73. $H(x) = \dfrac{x - 2}{x + 2}$

74. $h(x) = \dfrac{3 - x}{6 - x}$

75. $f(x) = \dfrac{x - 2}{2}$

76. $G(x) = \dfrac{2}{x - 2}$

For Exercises 77 to 90, find the range of the function defined by the equation and the given domain.

77. $f(x) = 4x - 3$; domain $= \{0, 1, 2, 3\}$

78. $G(x) = 3 - 5x$; domain $= \{-2, -1, 0, 1, 2\}$

79. $g(x) = 5x - 8$; domain $= \{-3, -1, 0, 1, 3\}$

80. $h(x) = 3x - 7$; domain $= \{-4, -2, 0, 2, 4\}$

81. $h(x) = x^2$; domain $= \{-2, -1, 0, 1, 2\}$

82. $H(x) = 1 - x^2$; domain $= \{-2, -1, 0, 1, 2\}$

Copyright © Houghton Mifflin Company. All rights reserved.

83. $f(x) = 2x^2 - 2x + 2$; domain $= \{-4, -2, 0, 4\}$

84. $G(x) = -2x^2 + 5x - 2$; domain $= \{-3, -1, 0, 1, 3\}$

85. $H(x) = \dfrac{5}{1 - x}$; domain $= \{-2, 0, 2\}$

86. $g(x) = \dfrac{4}{4 - x}$; domain $= \{-5, 0, 3\}$

87. $f(x) = \dfrac{2}{x - 4}$; domain $= \{-2, 0, 2, 6\}$

88. $g(x) = \dfrac{x}{3 - x}$; domain $= \{-2, -1, 0, 1, 2\}$

89. $H(x) = 2 - 3x - x^2$; domain $= \{-5, 0, 5\}$

90. $G(x) = 4 - 3x - x^3$; domain $= \{-3, 0, 3\}$

APPLYING THE CONCEPTS

91. Explain the words *relation* and *function*. Include in your explanation how the meanings of the two words differ.

92. Give a real-world example of a relation that is not a function. Is it possible to give an example of a function that is not a relation? If so, give one. If not, explain why it is not possible.

93. **a.** Find the set of ordered pairs (x, y) determined by the equation $y = x^3$, where $x \in \{-2, -1, 0, 1, 2\}$.
b. Does the set of ordered pairs define a function? Why or why not?

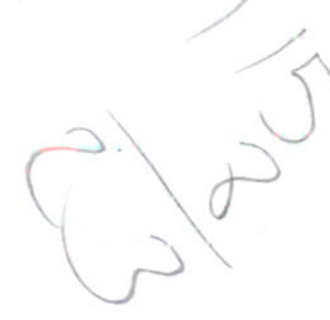

94. **a.** Find the set of ordered pairs (x, y) determined by the equation $|y| = x$, where $x \in \{0, 1, 2, 3\}$.
b. Does the set of ordered pairs define a function? Why or why not?

95. **Energy** The power a windmill can generate is a function of the velocity of the wind. The function can be approximated by $P = f(v) = 0.015v^3$, where P is the power in watts and v is the velocity of the wind in meters per second. How much power will be produced by a windmill when the velocity of the wind is 15 m/s?

Copyright © Houghton Mifflin Company. All rights reserved.

96. Automotive Technology The distance, s (in feet), a car will skid on a certain road surface after the brakes are applied is a function of the car's velocity, v (in miles per hour). The function can be approximated by $s = f(v) = 0.017v^2$. How far will a car skid after its brakes are applied if it is traveling 60 mph?

For Exercises 97 to 100, each graph defines a function. Evaluate the function by estimating the ordinate (which is the value of the function) for the given value of t.

97. Parachuting The graph at the right shows the speed, v, in feet per second that a parachutist is falling during the first 20 seconds after jumping out of a plane.

a. Estimate the speed at which the parachutist is falling when $t = 5$ s.

b. Estimate the speed at which the parachutist is falling when $t = 15$ s.

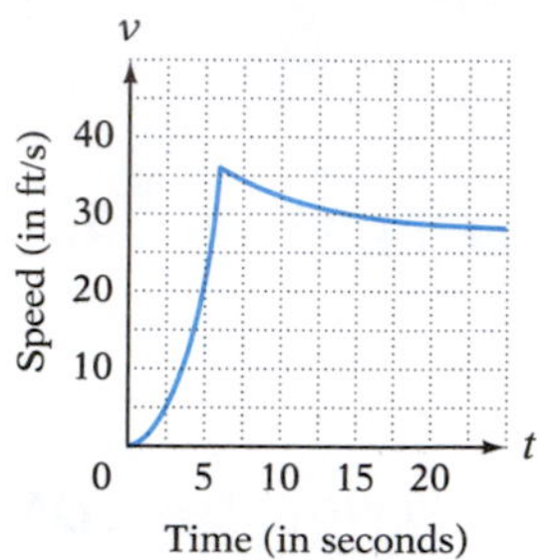

98. Psychology The graph at the right shows what an industrial psychologist has determined to be the average percent score, P, for an employee taking a performance test t weeks after training begins.

a. Estimate the score an employee would receive on this test when $t = 4$ weeks.

b. Estimate the score an employee would receive on this test when $t = 10$ weeks.

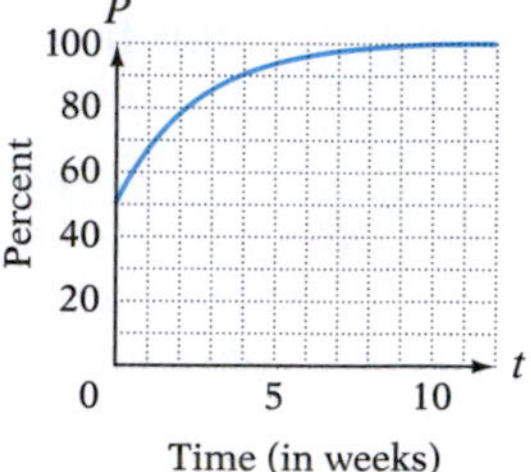

99. Temperature The graph at the right shows the temperature, T, in degrees Fahrenheit, of a can of cola t hours after it is placed in a refrigerator.

a. Use the graph to estimate the temperature of the cola when $t = 5$ h.

b. Use the graph to estimate the temperature of the cola when $t = 15$ h.

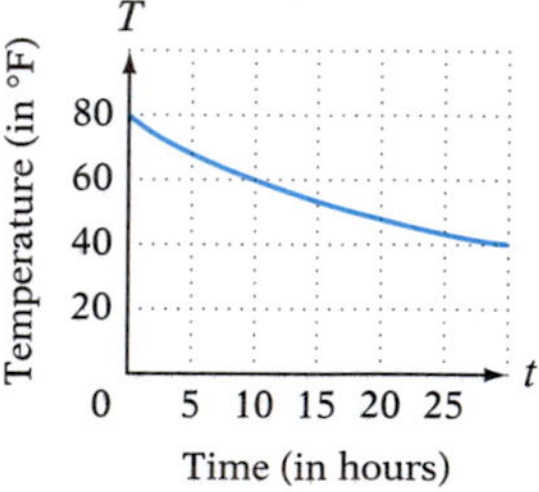

100. Athletics The graph at the right shows the decrease in the heart rate, r, of a runner (in beats per minute) t minutes after the completion of a race.

a. Use the graph to estimate the heart rate of a runner when $t = 5$ min.

b. Use the graph to estimate the heart rate of a runner when $t = 20$ min.

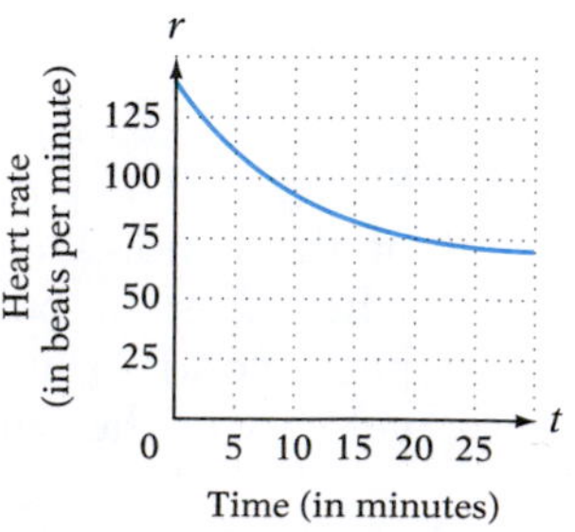

Copyright © Houghton Mifflin Company. All rights reserved.

4.3 Linear Functions

Objective A To graph a linear function

Recall that the ordered pairs of a function can be written as $(x, f(x))$ or (x, y). The **graph of a function** is a graph of the ordered pairs (x, y) that belong to the function. Certain functions have characteristic graphs. A function that can be written in the form $f(x) = mx + b$ (or $y = mx + b$) is called a **linear function** because its graph is a straight line.

Examples of linear functions are shown at the right. Note that the exponent on each variable is 1.

$f(x) = 2x + 5 \quad (m = 2, b = 5)$

$P(t) = 3t - 2 \quad (m = 3, b = -2)$

$y = -2x \quad (m = -2, b = 0)$

$y = -\frac{2}{3}x + 1 \quad \left(m = -\frac{2}{3}, b = 1\right)$

$g(z) = z - 2 \quad (m = 1, b = -2)$

The equation $y = x^2 + 4x + 3$ is not a linear function because it includes a term with a variable squared. The equation $f(x) = \frac{3}{x - 2}$ is not a linear function because a variable occurs in the denominator. Another example of an equation that is not a linear function is $y = \sqrt{x} + 4$; this equation contains a variable within a radical and so is not a linear function.

Consider $f(x) = 2x + 1$. Evaluating the linear function when $x = -2$, -1, 0, 1, and 2 produces some of the ordered pairs of the function. It is convenient to record the results in a table similar to the one at the right. The graph of the ordered pairs is shown in the leftmost figure below.

x	$f(x) = 2x + 1$	y	(x, y)
-2	$2(-2) + 1$	-3	$(-2, -3)$
-1	$2(-1) + 1$	-1	$(-1, -1)$
0	$2(0) + 1$	1	$(0, 1)$
1	$2(1) + 1$	3	$(1, 3)$
2	$2(2) + 1$	5	$(2, 5)$

Evaluating the function when x is not an integer produces more ordered pairs to graph, such as $\left(-\frac{5}{2}, -4\right)$ and $\left(\frac{3}{2}, 4\right)$, as shown in the middle figure below. Evaluating the function for still other values of x would result in more and more ordered pairs being graphed. The result would be so many dots that the graph would look like the straight line shown in the rightmost figure, which is the graph of $f(x) = 2x + 1$.

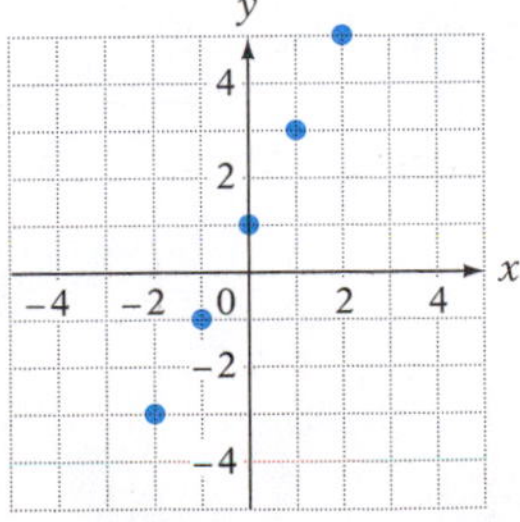

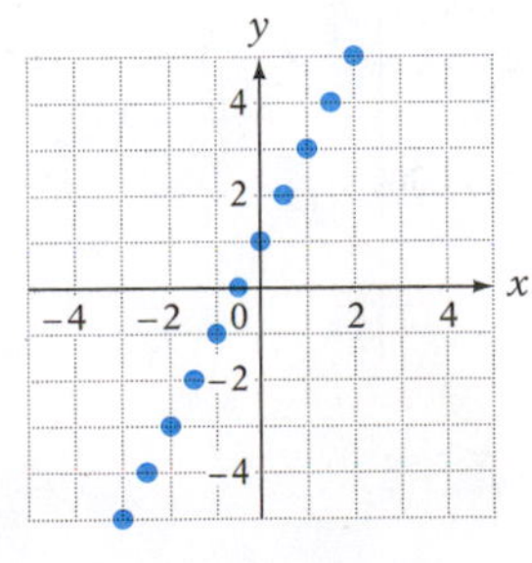

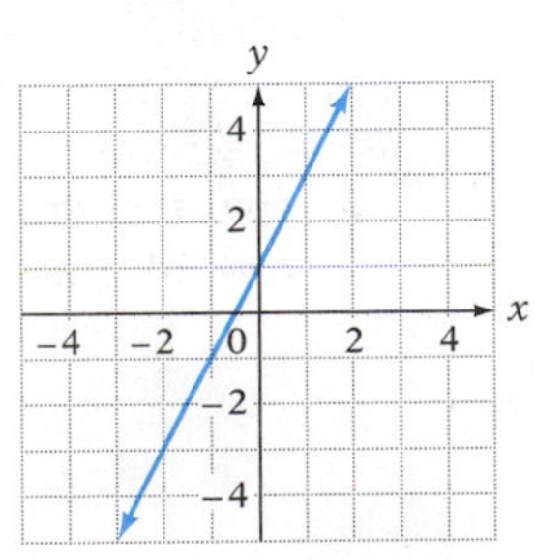

Copyright © Houghton Mifflin Company. All rights reserved.

No matter what value of x is chosen, $2x + 1$ is a real number. This means the domain of $f(x) = 2x + 1$ is all real numbers. Therefore, we can use any real number when evaluating the function. Normally, however, values such as π or $\sqrt{5}$ are not used because it is difficult to graph the resulting ordered pairs.

Note from the graph of $f(x) = 2x + 1$ shown at the right that $(-1.5, -2)$ and $(3, 7)$ are the coordinates of points on the graph and that $f(-1.5) = -2$ and $f(3) = 7$. Note also that the point whose coordinates are $(2, 1)$ is not a point on the graph and that $f(2) \neq 1$. Every point on the graph is an ordered pair that belongs to the function, and every ordered pair that belongs to the function corresponds to a point on the graph.

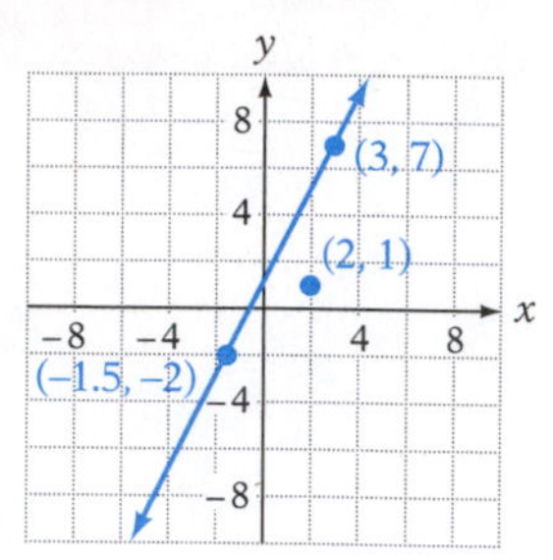

Integrating Technology

See the Projects and Group Activities at the end of this chapter for instructions on using a graphing calculator to graph a linear function. Instructions are also provided in the appendix Keystroke Guide: *Graph*.

Whether an equation is written as $f(x) = mx + b$ or as $y = mx + b$, the equation represents a linear function, and the graph of the equation is a straight line.

Because the graph of a linear function is a straight line, and a straight line is determined by two points, the graph of a linear function can be drawn by finding only two of the ordered pairs of the function. However, it is recommended that you find at least *three* ordered pairs to ensure accuracy.

TAKE NOTE

When the coefficient of x is a fraction, choose values of x that are multiples of the denominator of the fraction. This will result in coordinates that are integers.

HOW TO Graph: $f(x) = -\frac{1}{2}x + 3$

x	$y = f(x)$
−4	5
0	3
2	2

- Find at least three ordered pairs. When the coefficient of x is a fraction, choose values of x that will simplify the calculations. The ordered pairs can be displayed in a table.

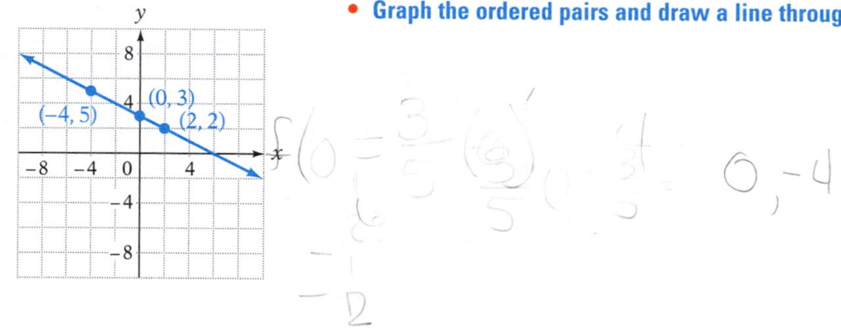

- Graph the ordered pairs and draw a line through the points.

Example 1 Graph: $f(x) = -\frac{3}{2}x - 3$

Solution

x	$y = f(x)$
0	−3
−2	0
−4	3

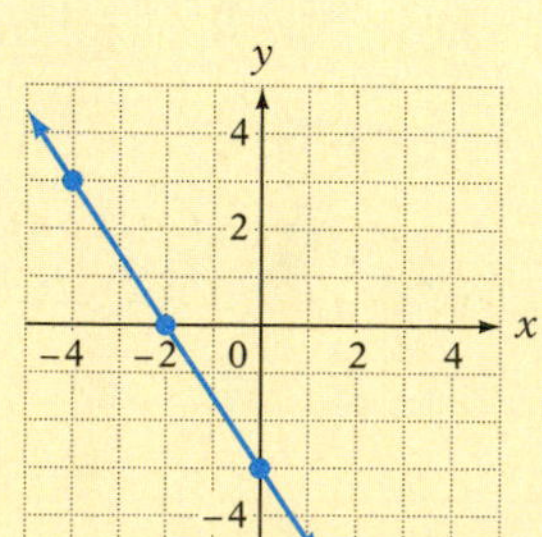

You Try It 1 Graph: $f(x) = \frac{3}{5}x - 4$

Your solution

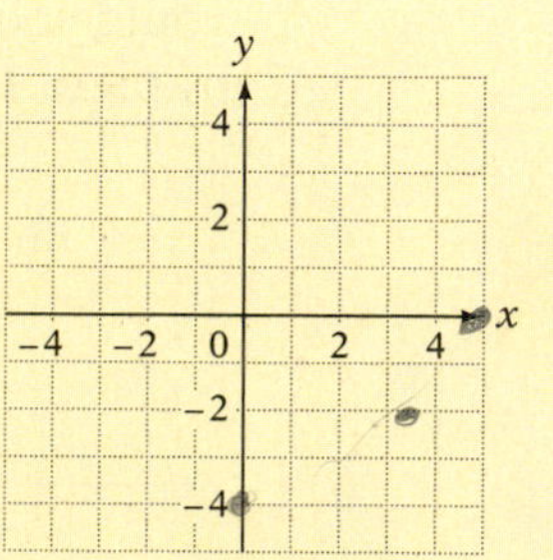

Solution on p. S12

Copyright © Houghton Mifflin Company. All rights reserved.

Example 2 Graph: $y = \frac{2}{3}x$

Solution

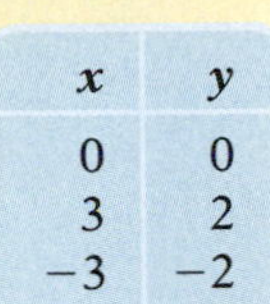

x	y
0	0
3	2
−3	−2

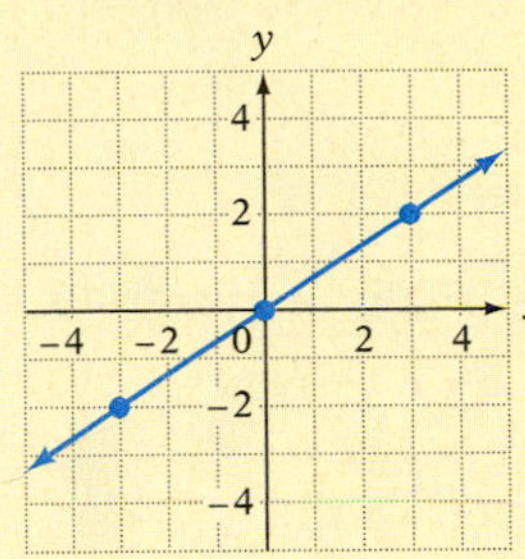

You Try It 2 Graph: $y = -\frac{3}{4}x$

Your solution

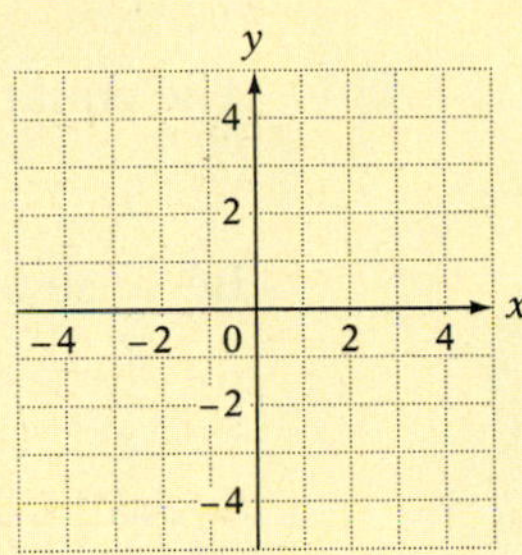

Solution on p. S12

Objective B To graph an equation of the form $Ax + By = C$

An equation of the form $Ax + By = C$, where A and B are coefficients and C is a constant, is also a *linear equation in two variables.* This equation can be written in the form $y = mx + b$.

Study Tip

Be sure to do all you need to do in order to be successful at graphing linear functions: Read through the introductory material, work through the HOW TO examples, study the paired Examples, do the You Try Its and check your solutions against those in the back of the book, and do the exercises in the 4.3 Exercise set. See *AIM for Success,* pages xxvii–xxix.

HOW TO Write $4x - 3y = 6$ in the form $y = mx + b$.

$4x - 3y = 6$

$-3y = -4x + 6$ • Subtract $4x$ from each side of the equation.

$y = \frac{4}{3}x - 2$ • Divide each side of the equation by −3. This is the form $y = mx + b$. $m = \frac{4}{3}$, $b = -2$

To graph an equation of the form $Ax + By = C$, first solve the equation for y. Then follow the same procedure used for graphing an equation of the form $y = mx + b$.

HOW TO Graph: $3x + 2y = 6$

$3x + 2y = 6$

$2y = -3x + 6$ • Solve the equation for y.

$y = -\frac{3}{2}x + 3$

x	y
0	3
2	0
4	−3

• Find at least three solutions.

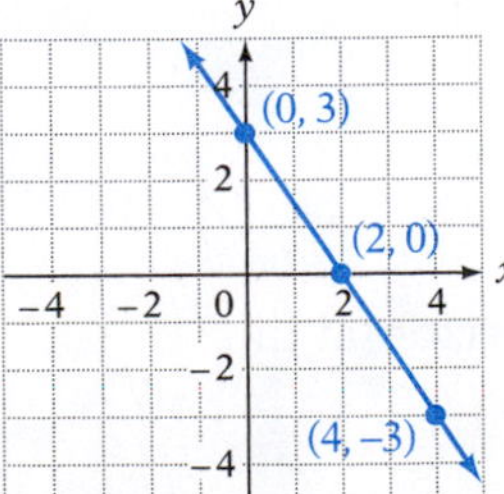

• Graph the ordered pairs in a rectangular coordinate system. Draw a straight line through the points.

Copyright © Houghton Mifflin Company. All rights reserved.

An equation in which one of the variables is missing has a graph that is either a horizontal or a vertical line.

The equation $y = -2$ can be written

$$0 \cdot x + y = -2$$

Because $0 \cdot x = 0$ for all values of x, y is always -2 for every value of x.

Some of the possible ordered pairs are given in the table below. The graph is shown at the right.

x	y
-2	-2
0	-2
3	-2

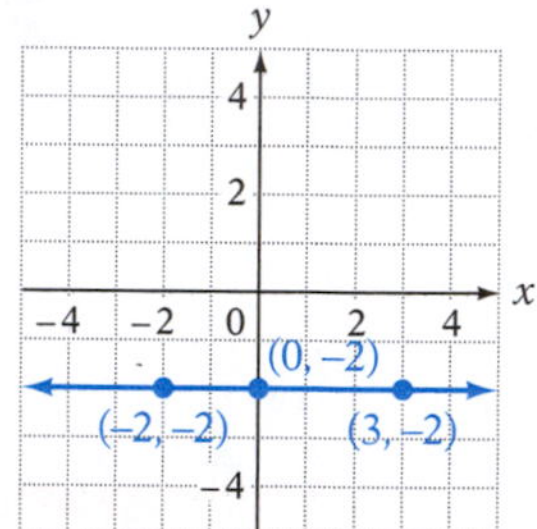

Graph of $y = b$

The **graph of $y = b$** is a horizontal line passing through the point $(0, b)$.

The equation $y = -2$ represents a function. Some of the ordered pairs of this function are $(-2, -2)$, $(0, -2)$, and $(3, -2)$. In functional notation we would write $f(x) = -2$. This function is an example of a constant function. No matter what value of x is selected, $f(x) = -2$.

Graph of a Constant Function

A function given by $f(x) = b$, where b is a constant, is a **constant function.** The graph of the constant function is a horizontal line passing through $(0, b)$.

For each value in the domain of a constant function, the value of the function is the same (constant). For instance, if $f(x) = 4$, then $f(2) = 4$, $f(3) = 4$, $f(\sqrt{3}) = 4$, $f(\pi) = 4$, and so on. The value of $f(x)$ is 4 for all values of x.

HOW TO Graph: $y + 4 = 0$

Solve for y.

$y + 4 = 0$

$y = -4$

The graph of $y = -4$ is a horizontal line passing through $(0, -4)$.

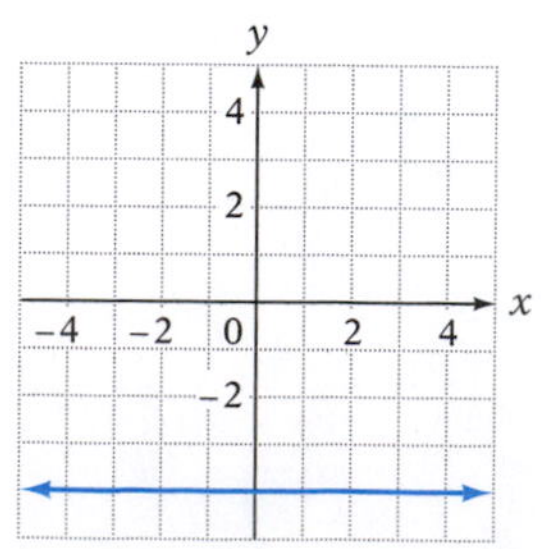

Copyright © Houghton Mifflin Company. All rights reserved.

HOW TO Evaluate $P(t) = -7$ when $t = 6$.

$P(t) = -7$
$P(6) = -7$

• The value of the constant function is the same for all values of the variable.

For the equation $y = -2$, the coefficient of x is zero. For the equation $x = 2$, the coefficient of y is zero. For instance, the equation $x = 2$ can be written

$$x + 0 \cdot y = 2$$

No matter what value of y is chosen, $0 \cdot y = 0$ and therefore x is always 2.

Some of the possible ordered pairs are given in the table below. The graph is shown at the right.

x	y
2	6
2	1
2	−4

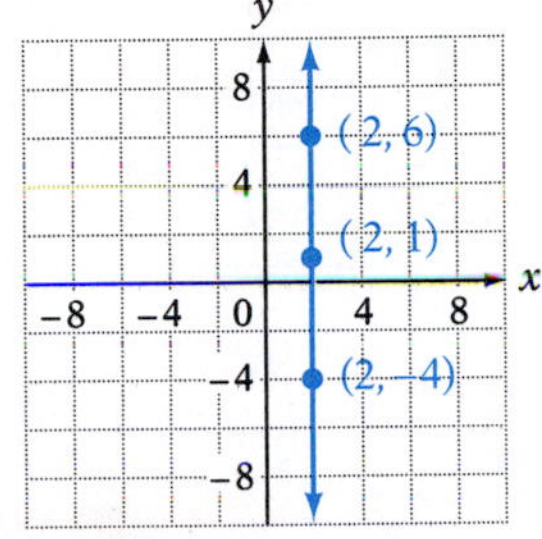

Graph of $x = a$

The **graph of $x = a$** is a vertical line passing through the point $(a, 0)$.

TAKE NOTE

The equation $y = b$ represents a function. The equation $x = a$ does *not* represent a function. Remember, not all equations represent functions.

Recall that a function is a set of ordered pairs in which no two ordered pairs have the same first coordinate and different second coordinates. Because (2, 6), (2, 1), and (2, −4) are ordered pairs belonging to the equation $x = 2$, this equation does not represent a function, and the graph is not the graph of a function.

Example 3 Graph: $2x + 3y = 9$

Solution Solve the equation for y.

$$2x + 3y = 9$$
$$3y = -2x + 9$$
$$y = -\frac{2}{3}x + 3$$

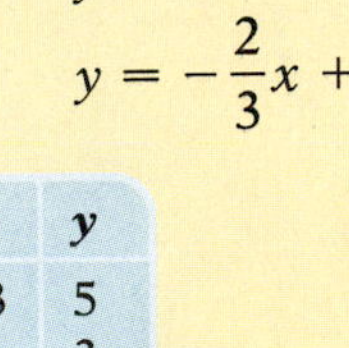

x	y
−3	5
0	3
3	1

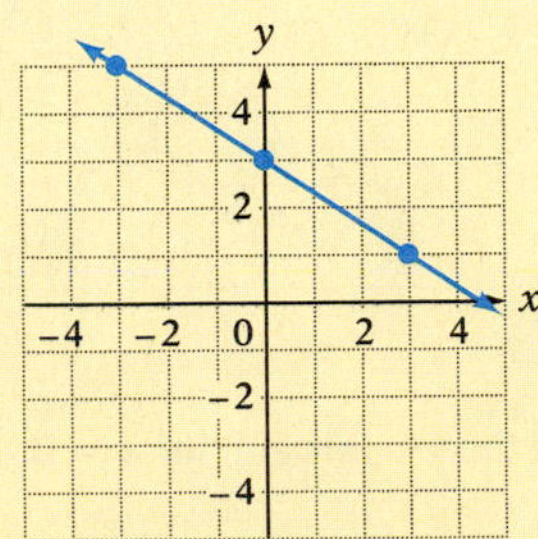

You Try It 3 Graph: $-3x + 2y = 4$

Your solution

y

x

4

2

−4 −2 0 2 4

−2

−4

Solution on p. S12

Copyright © Houghton Mifflin Company. All rights reserved.

Example 4 Graph: $x = -4$

Solution

The graph of an equation of the form $x = a$ is a vertical line passing through the point whose coordinates are $(a, 0)$.

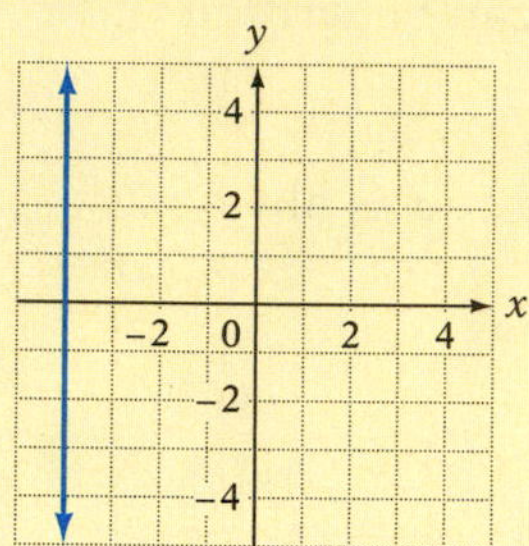

- The graph of $x = -4$ goes through the point $(-4, 0)$.

You Try It 4 Graph: $y - 3 = 0$

Your solution

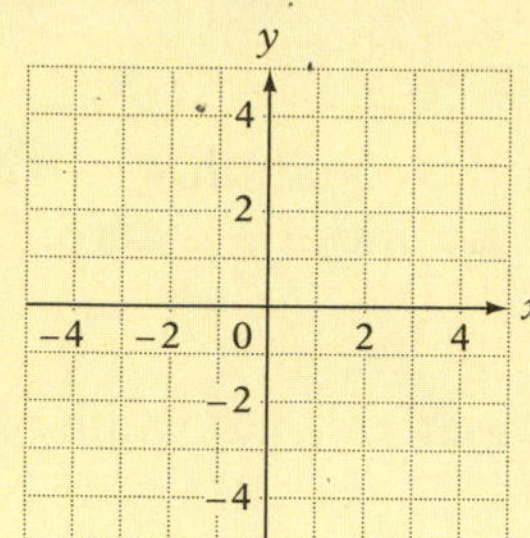

Solution on p. S12

Objective C To find the *x*- and *y*-intercepts of a straight line

VIDEO & DVD CD TUTOR WEB SSM

The graph of the equation $x - 2y = 4$ is shown at the right. The graph crosses the x-axis at the point $(4, 0)$. This point is called the **x-intercept.** The graph also crosses the y-axis at the point $(0, -2)$. This point is called the **y-intercept.**

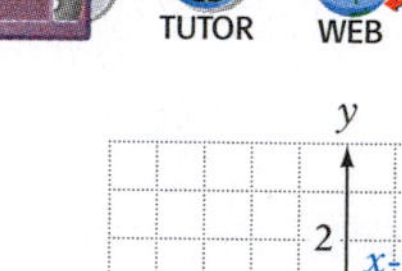

TAKE NOTE

The x-intercept occurs when $y = 0$.

The y-intercept occurs when $x = 0$.

HOW TO Find the x- and y-intercepts of the graph of the equation $3x + 4y = -12$.

To find the x-intercept, let $y = 0$. (Any point on the x-axis has y-coordinate 0.)

$$\begin{aligned} 3x + 4y &= -12 \\ 3x + 4(0) &= -12 \\ 3x &= -12 \\ x &= -4 \end{aligned}$$

The x-intercept is $(-4, 0)$.

To find the y-intercept, let $x = 0$. (Any point on the y-axis has x-coordinate 0.)

$$\begin{aligned} 3x + 4y &= -12 \\ 3(0) + 4y &= -12 \\ 4y &= -12 \\ y &= -3 \end{aligned}$$

The y-intercept is $(0, -3)$.

TAKE NOTE

Note that the y-coordinate of the y-intercept $(0, 7)$ has the same value as the constant term of $y = \frac{2}{3}x + 7$.

As shown below, this is always true:

$$\begin{aligned} y &= mx + b \\ y &= m(0) + b \\ y &= b \end{aligned}$$

Thus, the y-intercept is $(0, b)$.

HOW TO Find the y-intercept of $y = \frac{2}{3}x + 7$.

$$y = \frac{2}{3}x + 7$$

$$y = \frac{2}{3}(0) + 7$$

- To find the y-intercept, let $x = 0$.

$$y = 7$$

The y-intercept is $(0, 7)$.

For any equation of the form $y = mx + b$, the y-intercept is $(0, b)$.

Copyright © Houghton Mifflin Company. All rights reserved.

A linear equation can be graphed by finding the x- and y-intercepts and then drawing a line through the two points.

HOW TO Graph $3x - 2y = 6$ by using the x- and y-intercepts.

$$3x - 2y = 6$$
$$3x - 2(0) = 6$$
$$3x = 6$$
$$x = 2$$

- To find the x-intercept, let $y = 0$. Then solve for x.

$$3x - 2y = 6$$
$$3(0) - 2y = 6$$
$$-2y = 6$$
$$y = -3$$

- To find the y-intercept, let $x = 0$. Then solve for y.

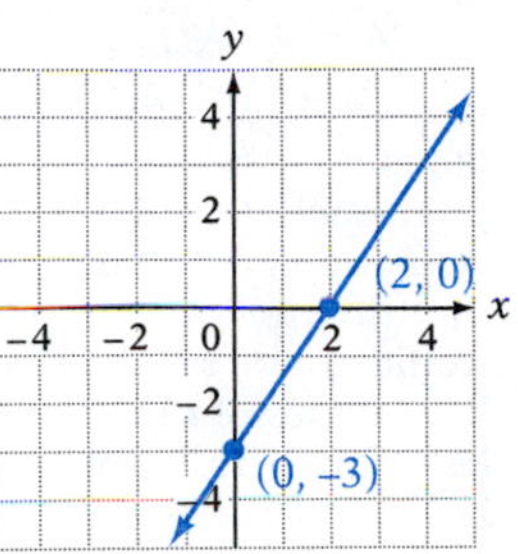

The x-intercept is (2, 0). The y-intercept is (0, −3).

Example 5

Graph $4x - y = 4$ by using the x- and y-intercepts.

Solution

x-intercept:

$$4x - y = 4$$
$$4x - 0 = 4$$
$$4x = 4$$
$$x = 1$$

- Let $y = 0$.

(1, 0)

y-intercept:

$$4x - y = 4$$
$$4(0) - y = 4$$
$$-y = 4$$
$$y = -4$$

- Let $x = 0$.

(0, −4)

You Try It 5

Graph $3x - y = 2$ by using the x- and y-intercepts.

Your solution

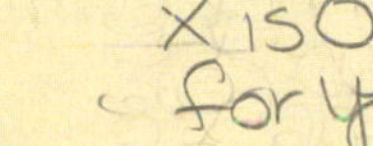

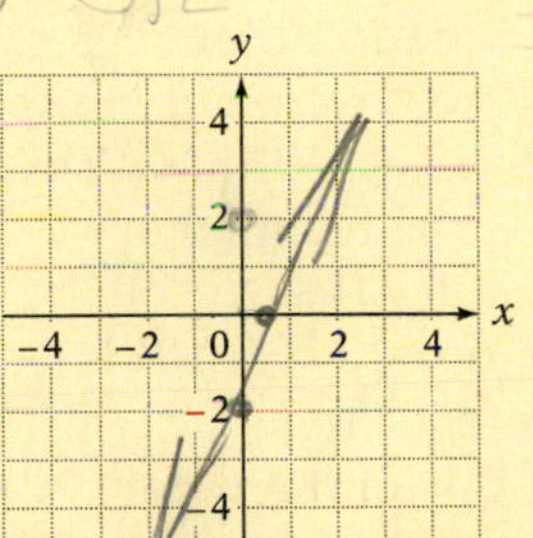

Example 6

Graph $y = \frac{2}{3}x - 2$ by using the x- and y-intercepts.

Solution

x-intercept:

$$y = \frac{2}{3}x - 2$$
$$0 = \frac{2}{3}x - 2$$
$$-\frac{2}{3}x = -2$$
$$x = 3$$

(3, 0)

y-intercept:

(0, b)

$b = -2$

(0, −2)

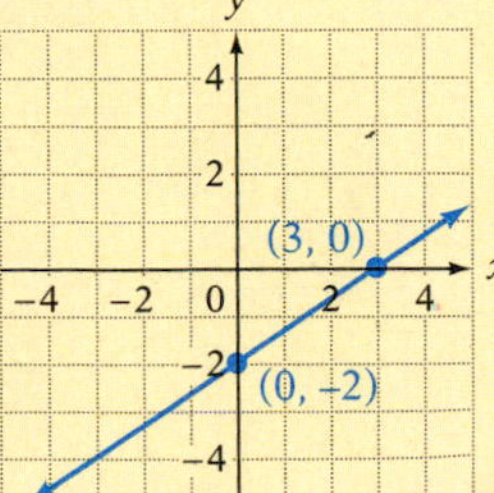

You Try It 6

Graph $y = \frac{1}{4}x + 1$ by using the x- and y-intercepts.

Your solution

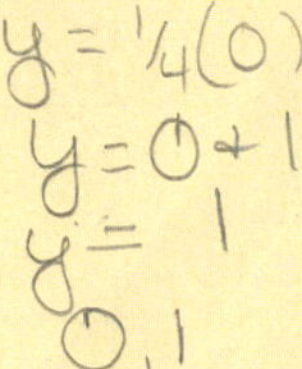

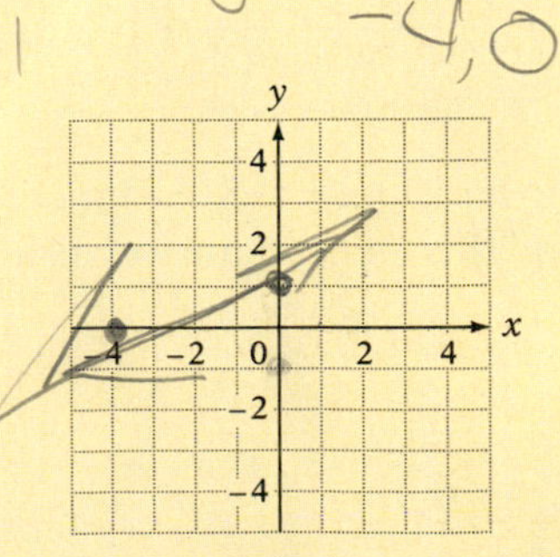

Solutions on pp. S12–S13

Copyright © Houghton Mifflin Company. All rights reserved.

Objective D To solve application problems

There are a variety of applications of linear functions.

TAKE NOTE

In many applications, the domain of the variable is given such that the equation makes sense. For this application, it would not be sensible to have values of t that are less than 0. This would indicate negative time! The number 10 is somewhat arbitrary, but after 10 min most people's heart rates would level off, and a linear function would no longer apply.

HOW TO The heart rate, R, after t minutes for a person taking a brisk walk can be approximated by the equation $R = 2t + 72$.

a. Graph this equation for $0 \le t \le 10$.

b. The point whose coordinates are (5, 82) is on the graph. Write a sentence that describes the meaning of this ordered pair.

a.

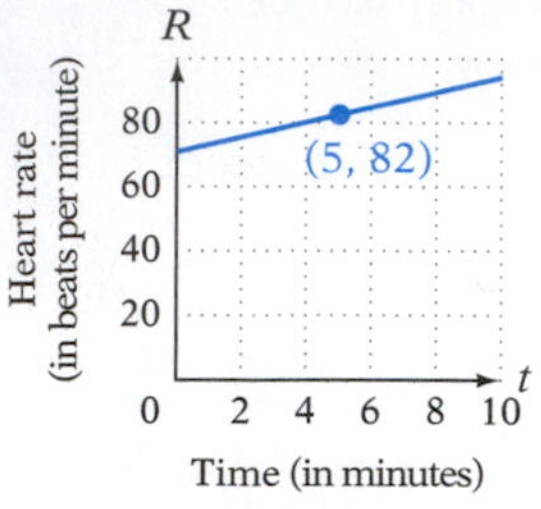

b. The ordered pair (5, 82) means that after 5 min, the person's heart rate is 82 beats per minute.

Example 7

An electronics technician charges \$45 plus \$1 per minute to repair defective wiring in a home or apartment. The equation that describes the total cost, C, to have defective wiring repaired is given by $C = t + 45$, where t is the number of minutes the technician works. Graph this equation for $0 \le t \le 60$. The point whose coordinates are (15, 60) is on this graph. Write a sentence that describes the meaning of this ordered pair.

Solution

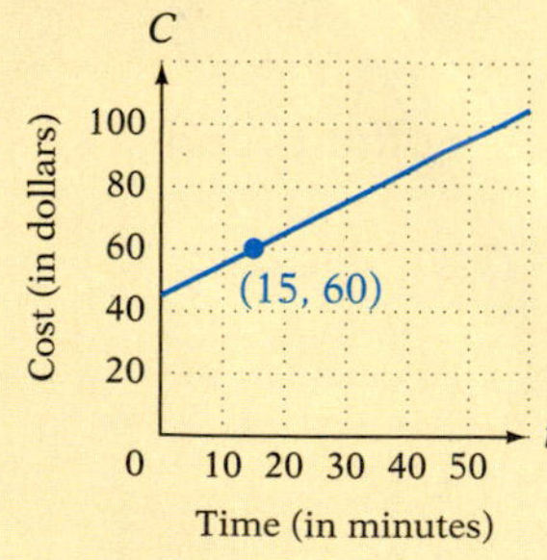

- **Graph $C = t + 45$. When $t = 0$, $C = 45$. When $t = 50$, $C = 95$.**

The ordered pair (15, 60) means that it costs \$60 for the technician to work 15 min.

You Try It 7

The height h (in inches) of a person and the length L (in inches) of that person's stride while walking are related. The equation $h = \frac{3}{4}L + 50$ approximates this relationship. Graph this equation for $15 \le L \le 40$. The point whose coordinates are (32, 74) is on this graph. Write a sentence that describes the meaning of this ordered pair.

Your solution

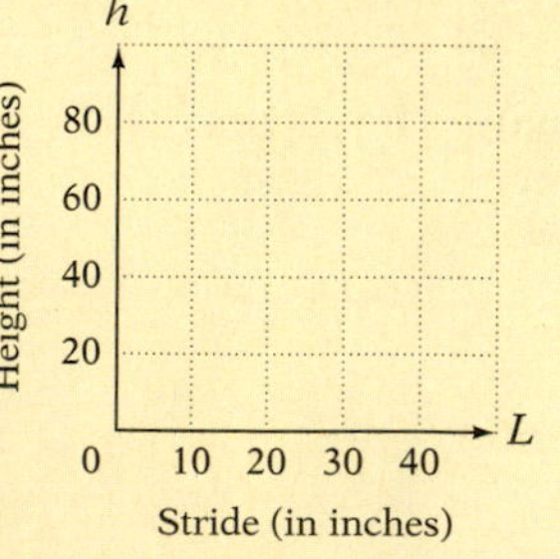

Solution on p. S13

Copyright © Houghton Mifflin Company. All rights reserved.

4.3 Exercises

Objective A To graph a linear function

For Exercises 1 to 9, graph.

1. $y = 3x - 4$

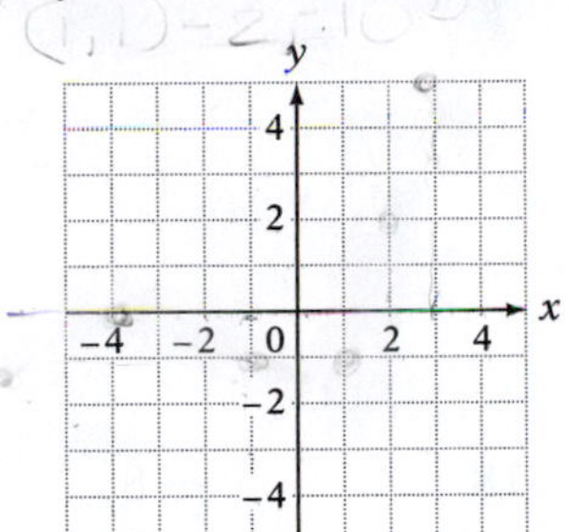

2. $y = -2x + 3$

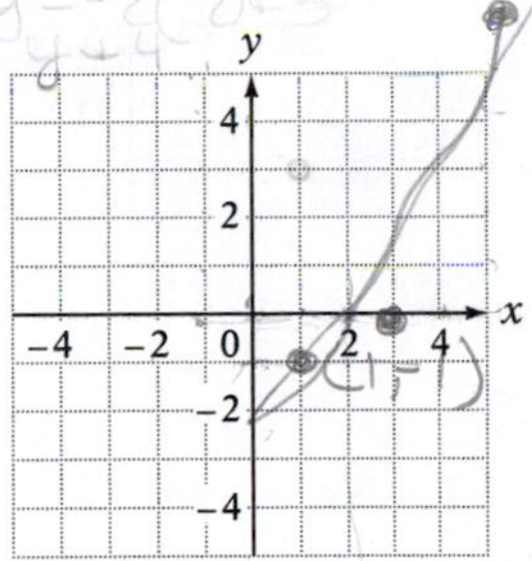

3. $y = -\frac{2}{3}x$

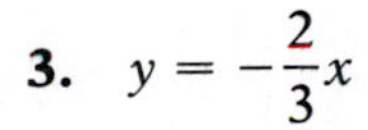

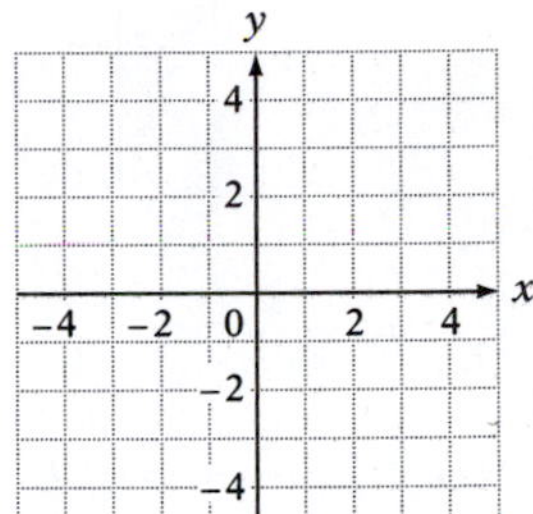

4. $y = \frac{3}{2}x$

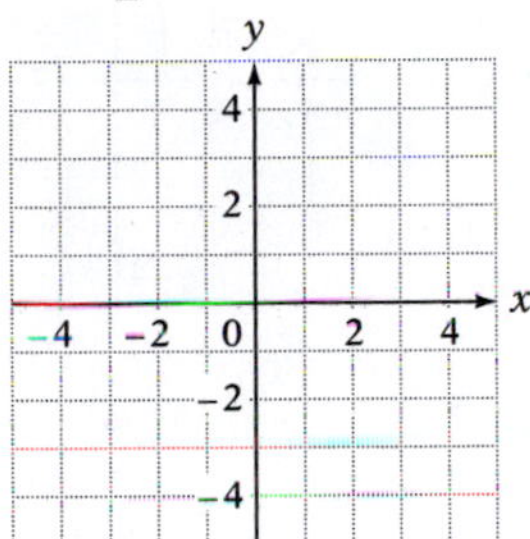

5. $y = \frac{2}{3}x - 4$

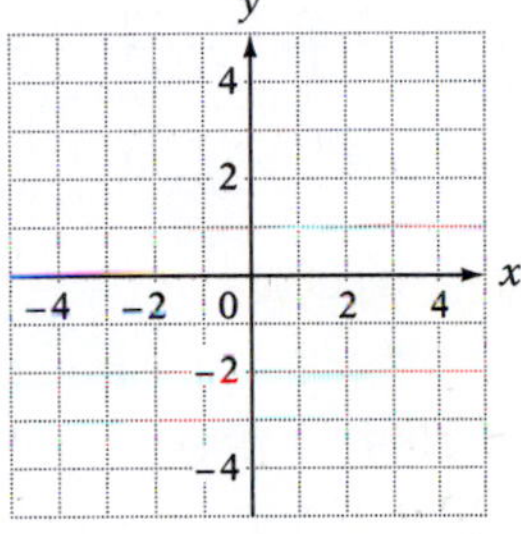

6. $y = \frac{3}{4}x + 2$

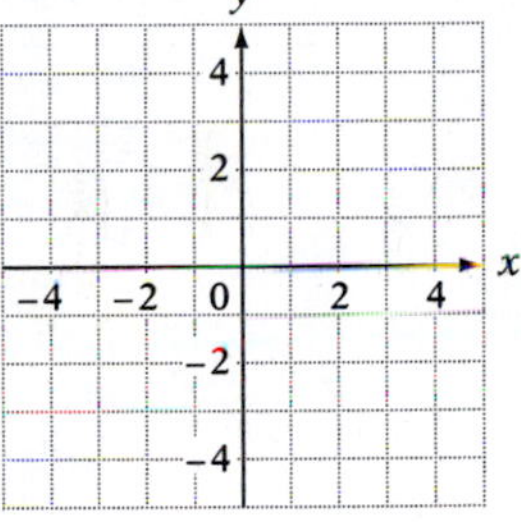

7. $y = -\frac{1}{3}x + 2$

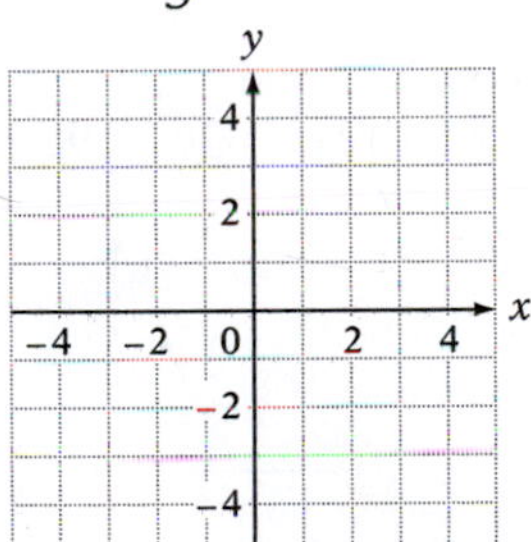

8. $y = -\frac{3}{2}x - 3$

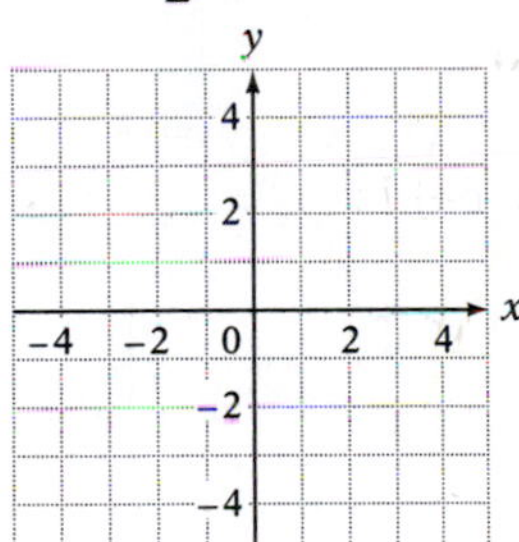

9. $y = \frac{3}{5}x - 1$

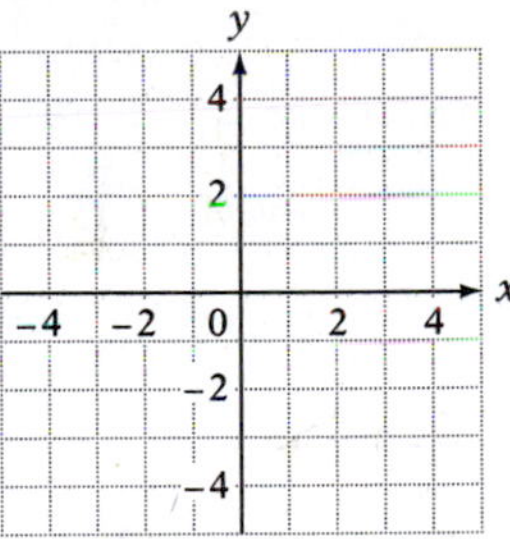

Objective B To graph an equation of the form $Ax + By = C$

For Exercises 10 to 18, graph.

10. $2x - y = 3$

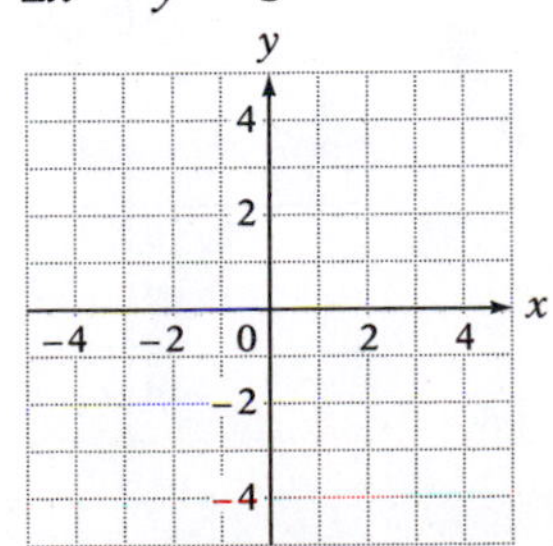

11. $2x + y = -3$

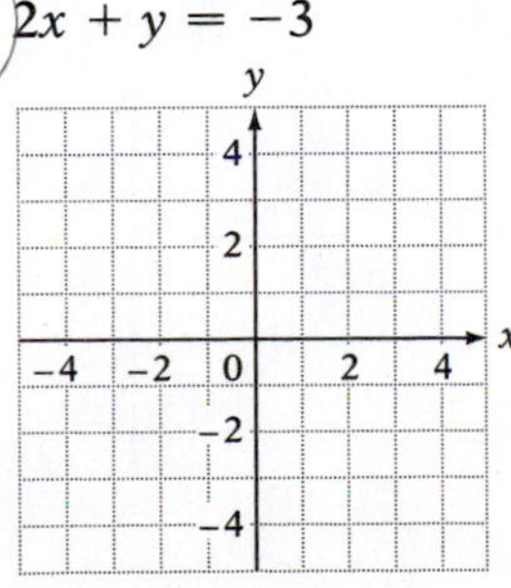

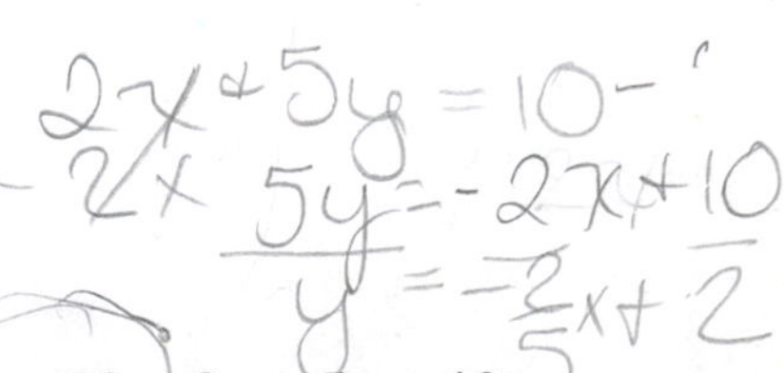

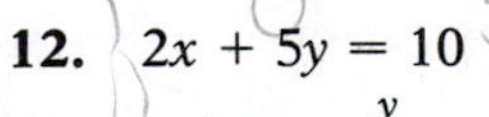

12. $2x + 5y = 10$

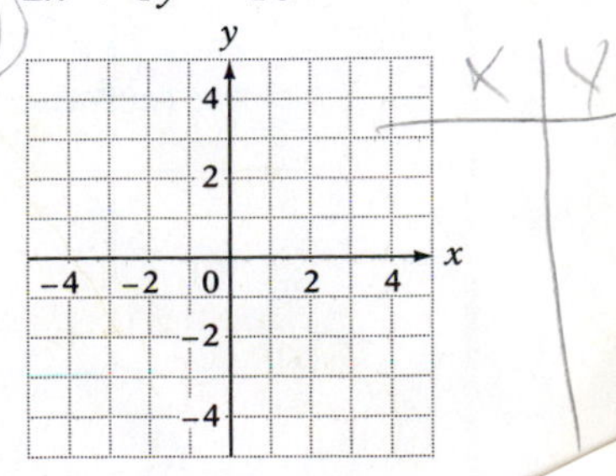

Copyright © Houghton Mifflin Company. All rights reserved.

13. $x - 4y = 8$

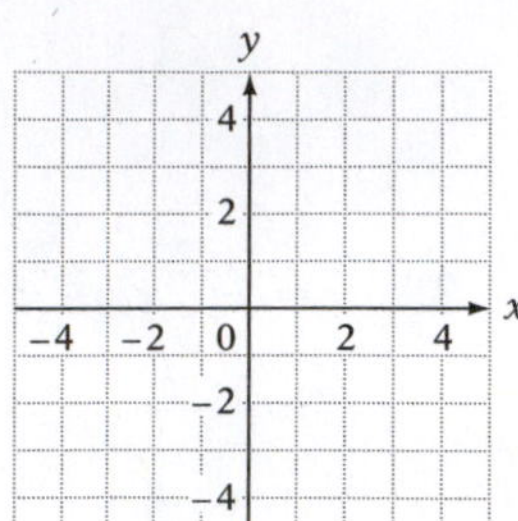

14. $y = -2$

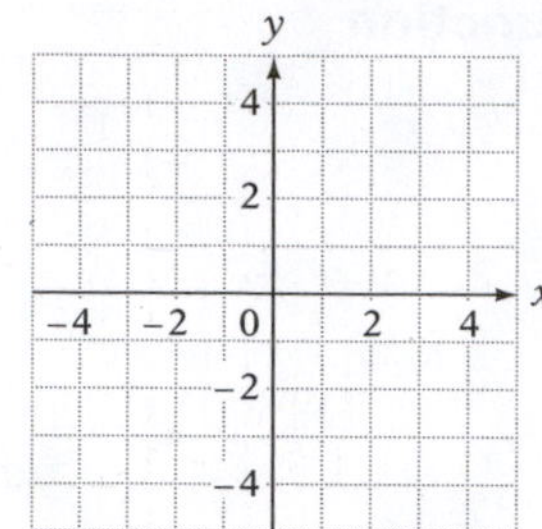

15. $y = \frac{1}{3}x$

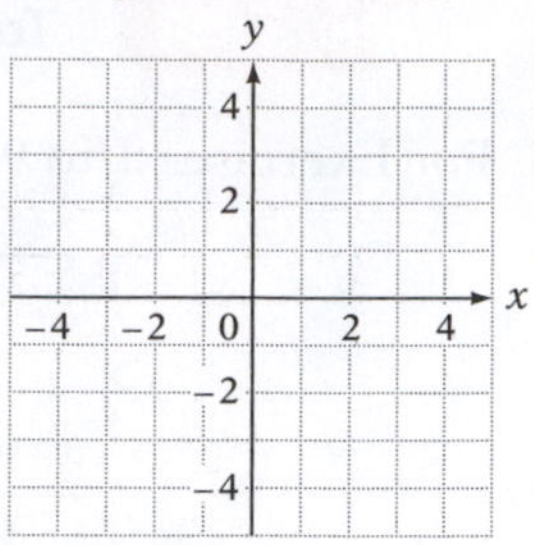

16. $2x - 3y = 12$

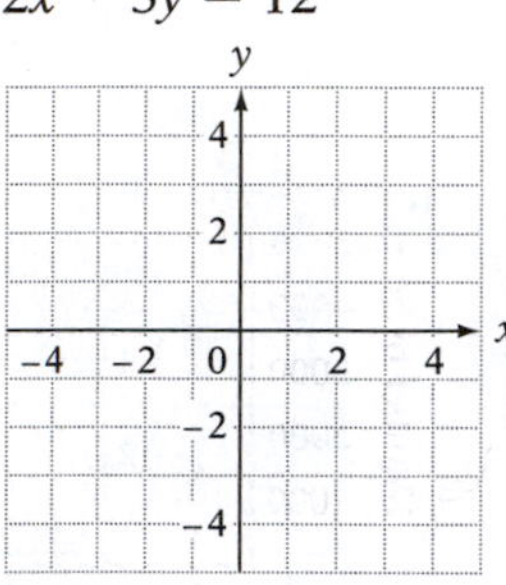

17. $3x - y = -2$

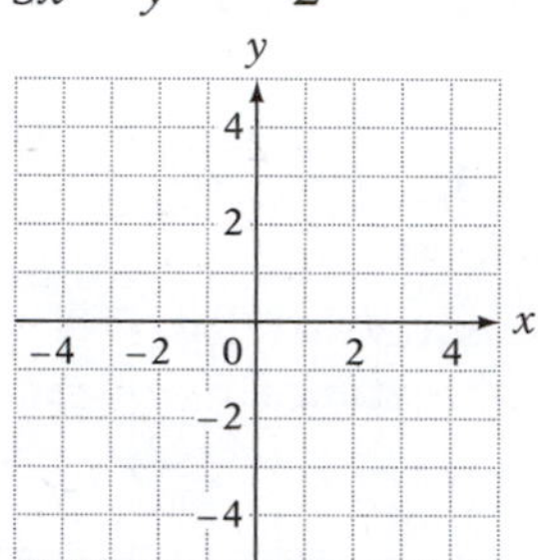

18. $3x - 2y = 8$

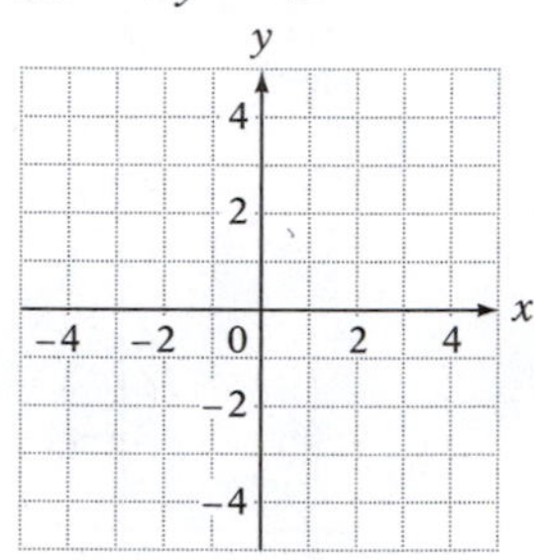

Objective C To find the *x*- and *y*-intercepts of a straight line

For Exercises 19 to 27, find the *x*- and *y*-intercepts and graph.

19. $x - 2y = -4$

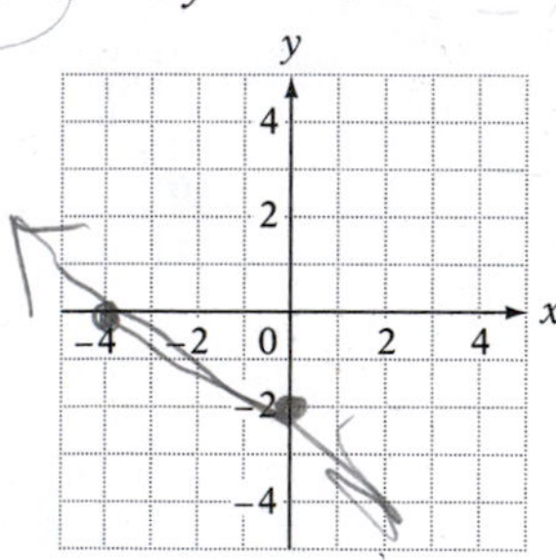

20. $3x + y = 3$

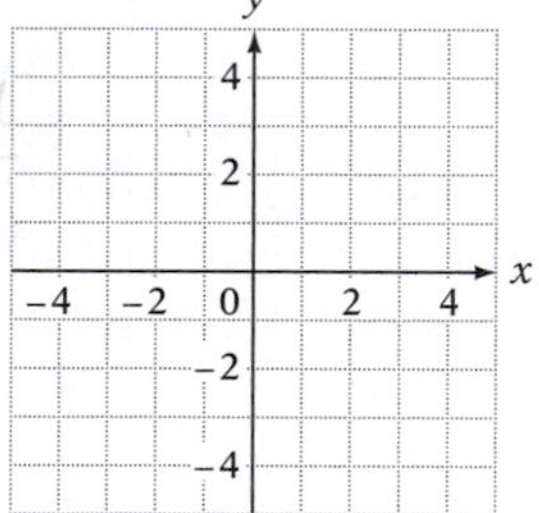

21. $2x - 3y = 9$

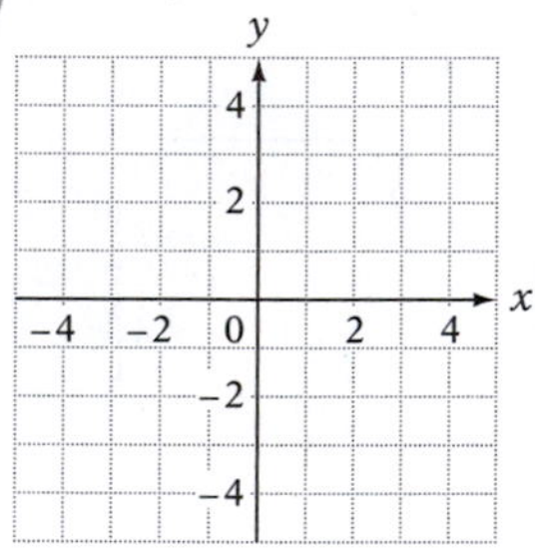

22. $2x - y = 4$

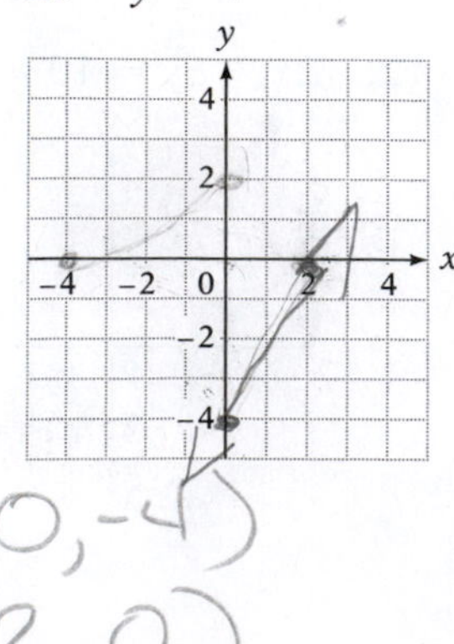

23. $2x + y = 3$

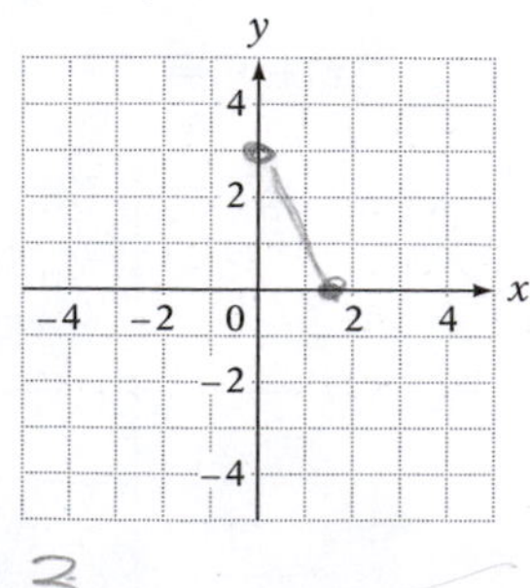

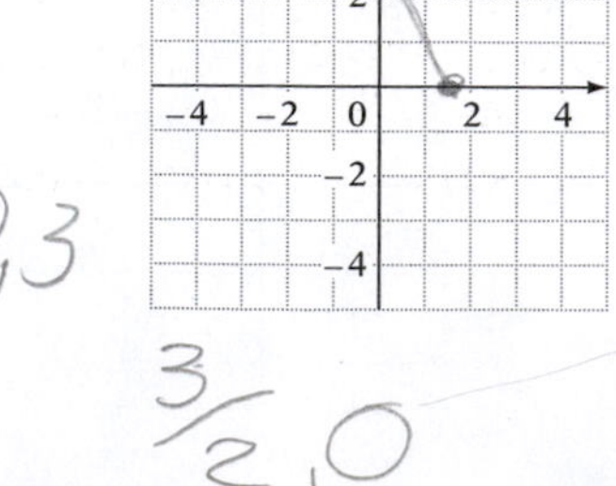

24. $4x - 3y = 8$

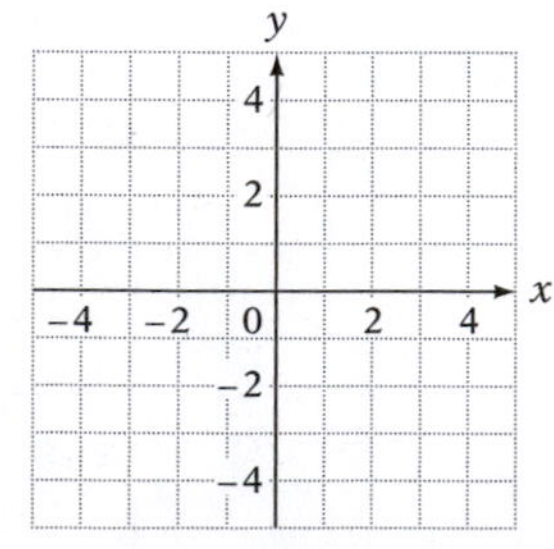

Copyright © Houghton Mifflin Company. All rights reserved.

25. $3x + 2y = 4$

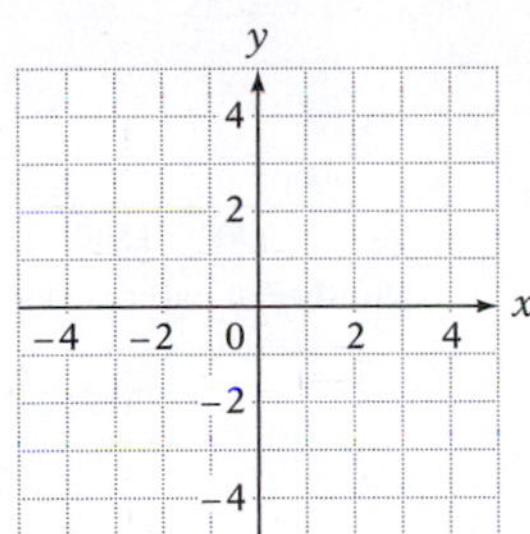

26. $2x - 3y = 4$

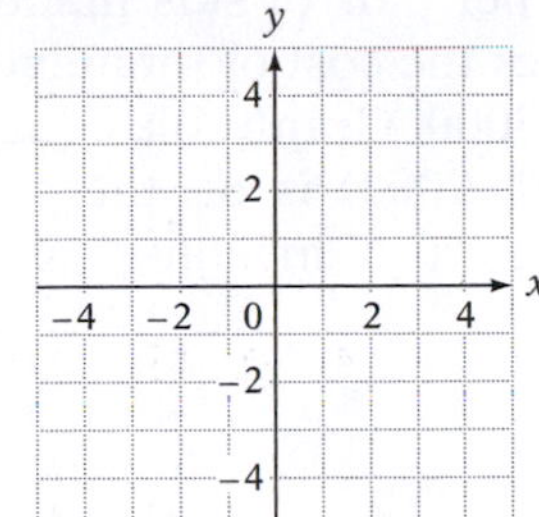

27. $4x - 3y = 6$

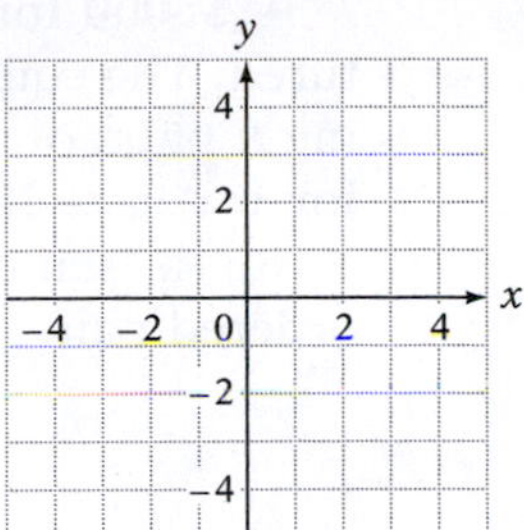

Objective D To solve application problems

28. **Animal Science** A bee beats its wings approximately 100 times per second. The equation that describes the total number of times a bee beats its wings is given by $B = 100t$, where B is the number of beats in t seconds. Graph this equation for $0 \le t \le 60$. The point (35, 3500) is on this graph. Write a sentence that describes the meaning of this ordered pair.

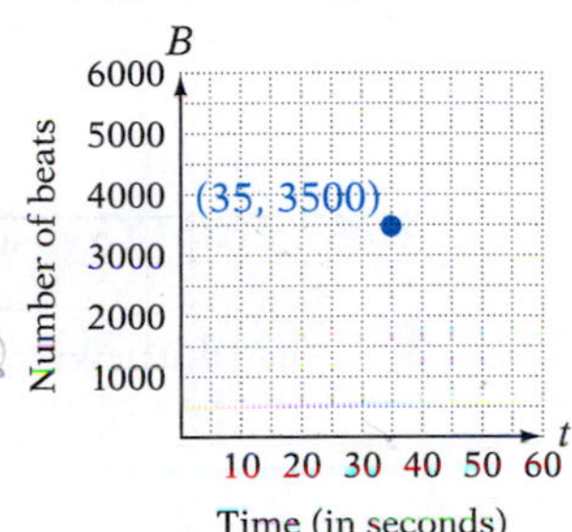

29. **Compensation** Marlys receives $11 per hour as a mathematics department tutor. The equation that describes Marlys's wages is $W = 11t$, where t is the number of hours she spends tutoring. Graph this equation for $0 \le t \le 20$. The ordered pair (15, 165) is on the graph. Write a sentence that describes the meaning of this ordered pair.

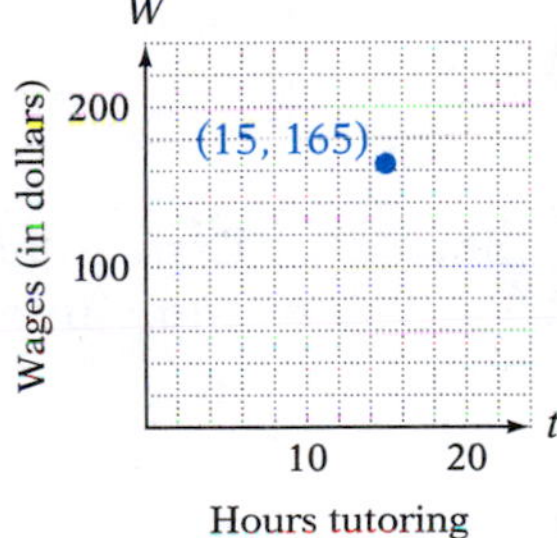

30. **Roller Coasters** The Superman Krypton Coaster at Fiesta Texas® in San Antonio has a maximum speed of approximately 105 ft/s. The equation that describes the total number of feet traveled by the roller coaster in t seconds at this speed is given by $D = 105t$. Graph this equation for $0 \le t \le 10$. The point (5, 525) is on this graph. Write a sentence that describes the meaning of this ordered pair.

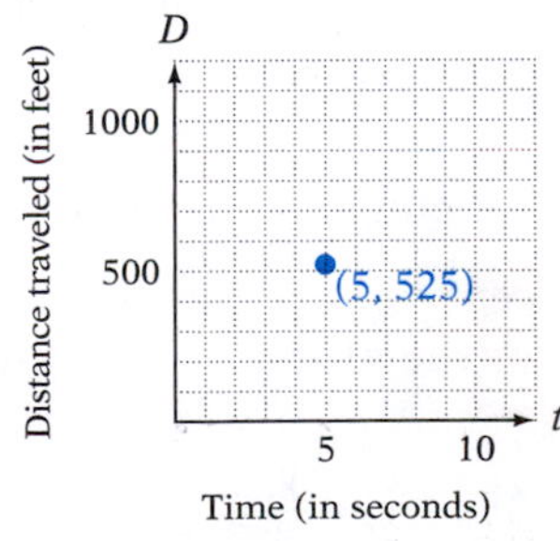

31. **Telecommunications** The monthly cost for receiving messages from a telephone answering service is \$8.00 plus \$.20 a message. The equation that describes the cost is $C = 0.20n + 8.00$, where n is the number of messages received. Graph this equation for $0 \le n \le 40$. The point (32, 14.40) is on the graph. Write a sentence that describes the meaning of this ordered pair.

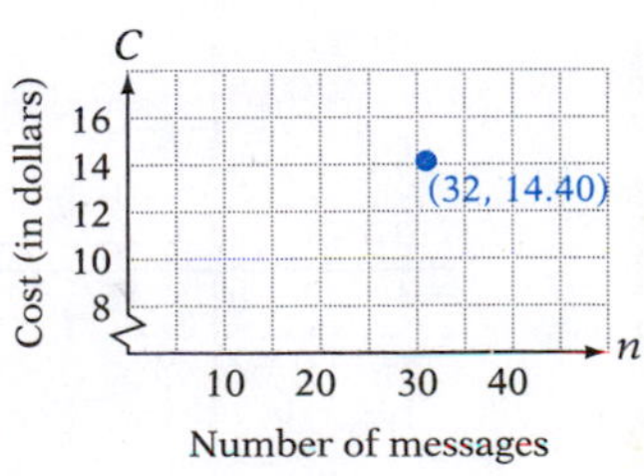

Copyright © Houghton Mifflin Company. All rights reserved.

32. **Manufacturing** The cost of manufacturing skis is \$5000 for startup and \$80 per pair of skis manufactured. The equation that describes the cost of manufacturing n pairs of skis is $C = 80n + 5000$. Graph this equation for $0 \le n \le 2000$. The point (50, 9000) is on the graph. Write a sentence that describes the meaning of this ordered pair.

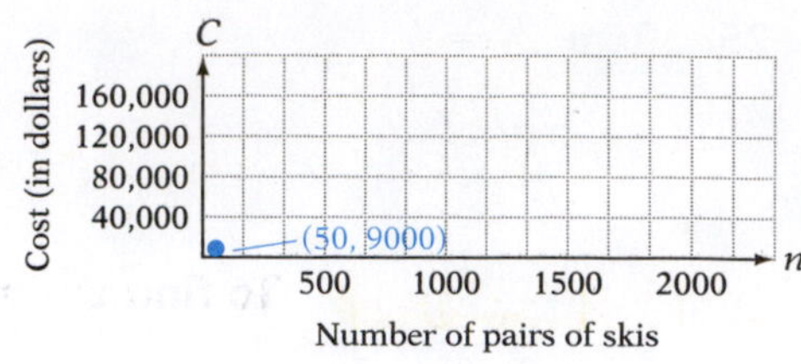

33. **Manufacturing** The cost of manufacturing compact discs is \$3000 for startup and \$5 per disc. The equation that describes the cost of manufacturing n compact discs is $C = 5n + 3000$. Graph this equation for $0 \le n \le 10{,}000$. The point (6000, 33,000) is on the graph. Write a sentence that describes the meaning of this ordered pair.

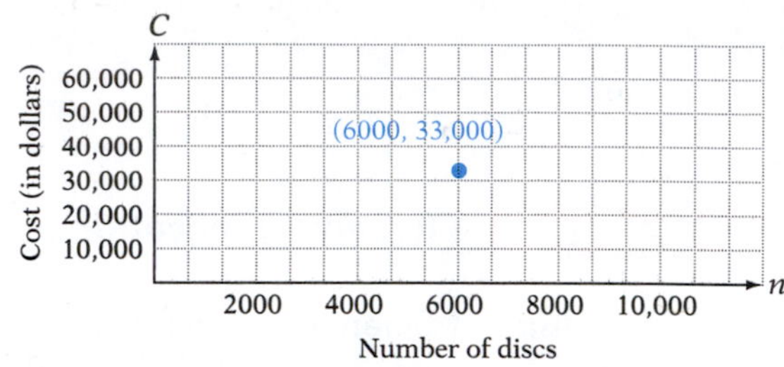

APPLYING THE CONCEPTS

34. Explain what the graph of an equation represents.

35. Explain how to graph the equation of a straight line by plotting points.

36. Explain how to graph the equation of a straight line by using its x- and y-intercepts.

37. Explain why you cannot graph the equation $4x + 3y = 0$ by using just its intercepts.

An equation of the form $\frac{x}{a} + \frac{y}{b} = 1$, where $a \neq 0$ and $b \neq 0$, is called the *intercept form of a straight line* because $(a, 0)$ and $(0, b)$ are the x- and y-intercepts of the graph of the equation. Graph the equations in Exercises 38 to 40.

38. $\frac{x}{3} + \frac{y}{5} = 1$

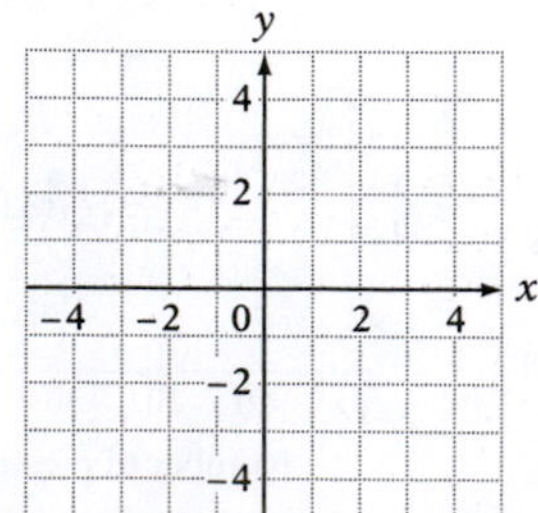

39. $\frac{x}{2} + \frac{y}{3} = 1$

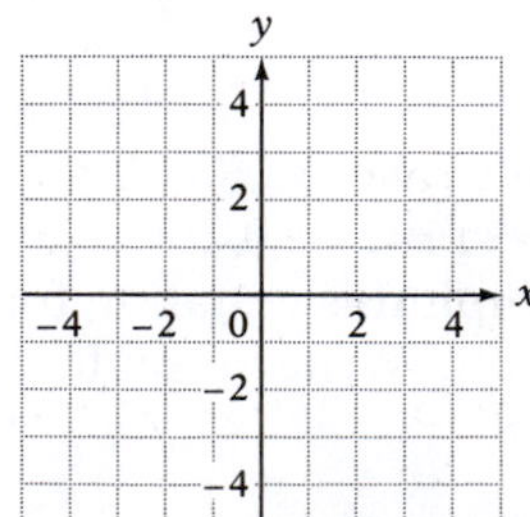

40. $\frac{x}{3} - \frac{y}{2} = 1$

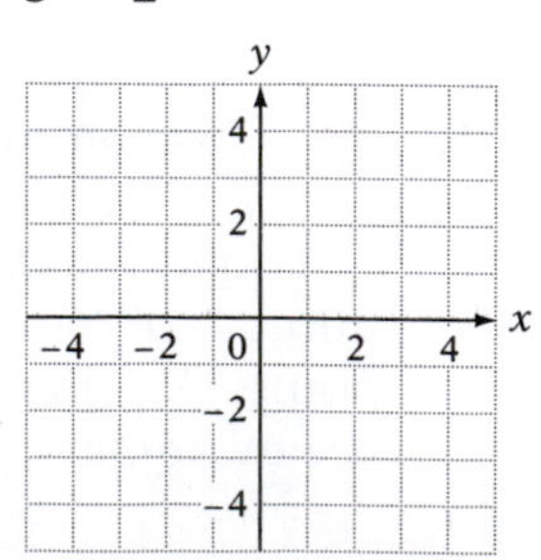

Copyright © Houghton Mifflin Company. All rights reserved.

4.4 Slope of a Straight Line

Objective A To find the slope of a line given two points

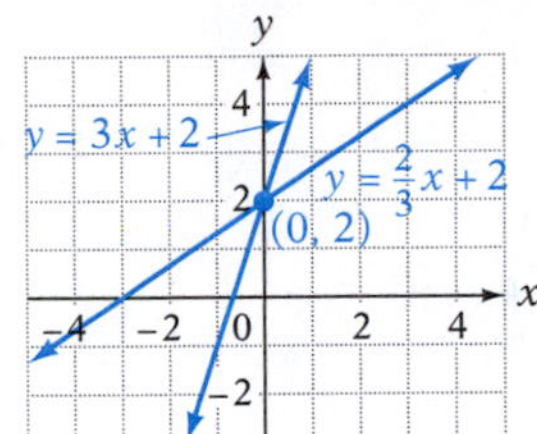

The graphs of $y = 3x + 2$ and $y = \frac{2}{3}x + 2$ are shown at the left. Each graph crosses the y-axis at the point (0, 2), but the graphs have different slants. The **slope** of a line is a measure of the slant of the line. The symbol for slope is m.

The slope of a line containing two points is the ratio of the change in the y values between the two points to the change in the x values. The line containing the points whose coordinates are (−1, −3) and (5, 2) is shown below.

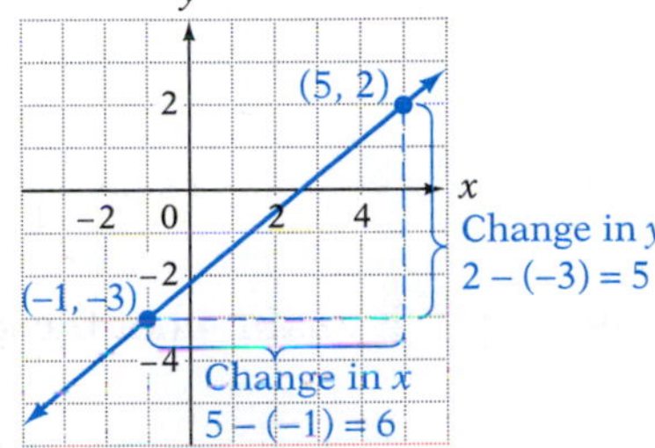

The change in the y values is the difference between the y-coordinates of the two points.

$$\text{Change in } y = 2 - (-3) = 5$$

The change in the x values is the difference between the x-coordinates of the two points.

$$\text{Change in } x = 5 - (-1) = 6$$

The slope of the line between the two points is the ratio of the change in y to the change in x.

$$\text{Slope} = m = \frac{\text{change in } y}{\text{change in } x} = \frac{5}{6} \qquad m = \frac{2 - (-3)}{5 - (-1)} = \frac{5}{6}$$

In general, if $P_1(x_1, y_1)$ and $P_2(x_2, y_2)$ are two points on a line, then

$$\text{Change in } y = y_2 - y_1 \qquad \text{Change in } x = x_2 - x_1$$

Using these ideas, we can state a formula for slope.

Slope Formula

The slope of the line containing the two points $P_1(x_1, y_1)$ and $P_2(x_2, y_2)$ is given by

$$m = \frac{y_2 - y_1}{x_2 - x_1}, x_1 \neq x_2$$

Frequently, the Greek letter Δ (delta) is used to designate the change in a variable. Using this notation, we can write the equations for the change in y and the change in x as follows:

$$\text{Change in } y = \Delta y = y_2 - y_1 \qquad \text{Change in } x = \Delta x = x_2 - x_1$$

Using this notation, the slope formula is written $m = \frac{\Delta y}{\Delta x}$.

Copyright © Houghton Mifflin Company. All rights reserved.

HOW TO Find the slope of the line containing the points whose coordinates are $(-2, 0)$ and $(4, 5)$.

Let $P_1 = (-2, 0)$ and $P_2 = (4, 5)$. (It does not matter which point is named P_1 or P_2; the slope will be the same.)

$$m = \frac{y_2 - y_1}{x_2 - x_1} = \frac{5 - 0}{4 - (-2)} = \frac{5}{6}$$

A line that slants upward to the right always has a **positive slope.**

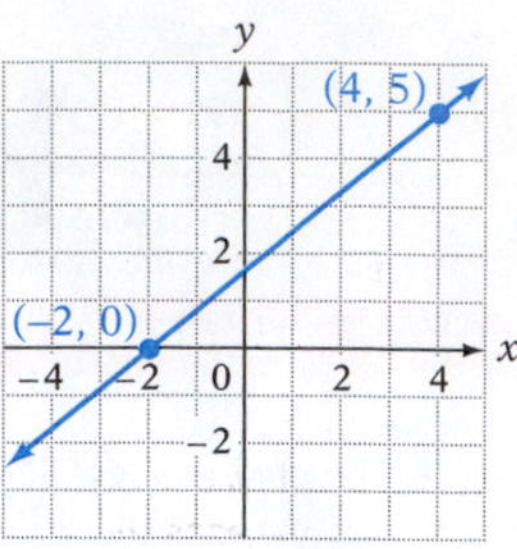

Positive slope

HOW TO Find the slope of the line containing the points whose coordinates are $(-3, 4)$ and $(4, 2)$.

Let $P_1 = (-3, 4)$ and $P_2 = (4, 2)$.

$$m = \frac{y_2 - y_1}{x_2 - x_1} = \frac{2 - 4}{4 - (-3)} = \frac{-2}{7} = -\frac{2}{7}$$

A line that slants downward to the right always has a **negative slope.**

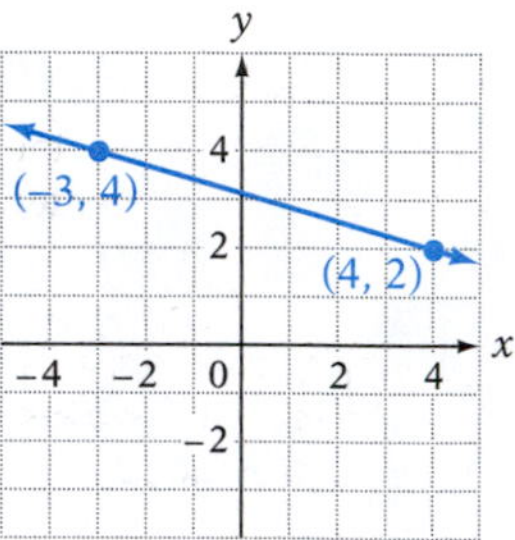

Negative slope

HOW TO Find the slope of the line containing the points whose coordinates are $(-2, 2)$ and $(4, 2)$.

Let $P_1 = (-2, 2)$ and $P_2 = (4, 2)$.

$$m = \frac{y_2 - y_1}{x_2 - x_1} = \frac{2 - 2}{4 - (-2)} = \frac{0}{6} = 0$$

A horizontal line has **zero slope.**

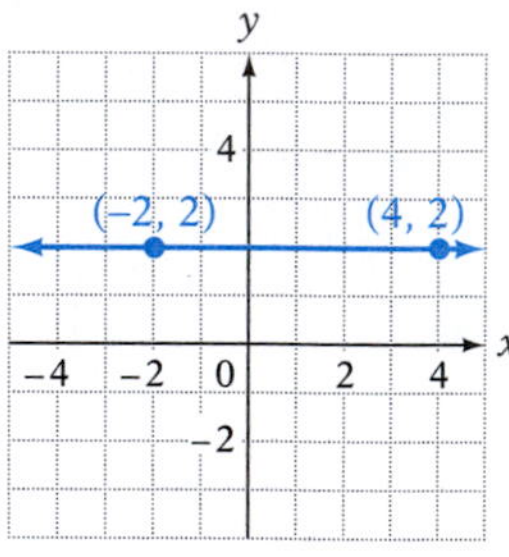

Zero slope

HOW TO Find the slope of the line containing the points whose coordinates are $(1, -2)$ and $(1, 3)$.

Let $P_1 = (1, -2)$ and $P_2 = (1, 3)$.

$$m = \frac{y_2 - y_1}{x_2 - x_1} = \frac{3 - (-2)}{1 - 1} = \frac{5}{0} \quad \text{Not a real number}$$

The slope of a vertical line has **undefined slope.**

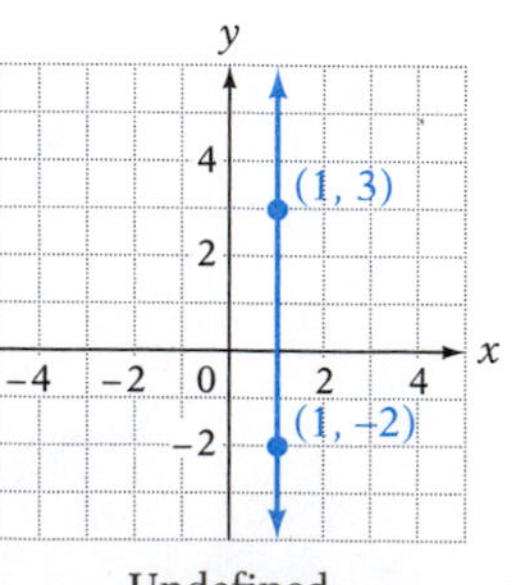

Undefined

Copyright © Houghton Mifflin Company. All rights reserved.

Point of Interest

One of the motivations for the discovery of calculus was the wish to solve a more complicated version of the distance-rate problem at the right.

You may be familiar with twirling a ball on the end of a string. If you release the string, the ball flies off in a path as shown below.

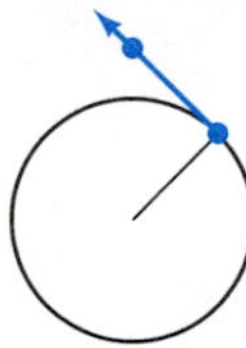

The question that mathematicians tried to answer was essentially, "What is the slope of the line represented by the arrow?"

Answering questions similar to this led to the development of one aspect of calculus.

There are many applications of slope. Here are two examples.

The first record for the one-mile run was recorded in 1865 in England. Richard Webster ran the mile in 4 min 36.5 s. His average speed was approximately 19 feet per second.

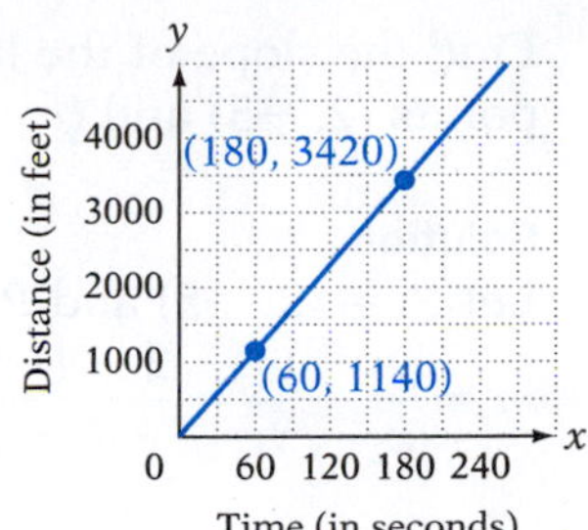

The graph at the right shows the distance Webster ran during that run. From the graph, note that after 60 s (1 min) he had traveled 1140 ft and that after 180 s (3 min) he had traveled 3420 ft.

Let the point (60, 1140) be (x_1, y_1) and the point (180, 3420) be (x_2, y_2). The slope of the line between these two points is

$$m = \frac{y_2 - y_1}{x_2 - x_1}$$

$$= \frac{3420 - 1140}{180 - 60} = \frac{2280}{120} = 19$$

Note that the slope of the line is the same as Webster's average speed, 19 feet per second.

Average speed is related to slope.

The following example is related to agriculture.

The number of farm workers in the United States declined from 10 million in 1950 to 3 million in 2000. (*Source*: U.S. Department of Agriculture)

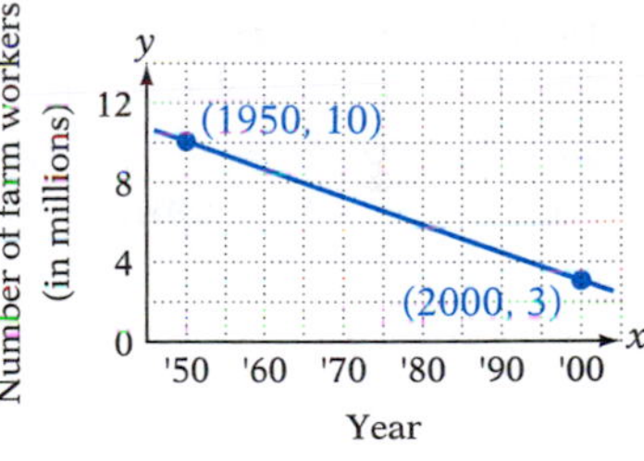

Let the point (1950, 10) be (x_1, y_1) and the point (2000, 3) be (x_2, y_2). The slope of the line between these two points is

$$m = \frac{y_2 - y_1}{x_2 - x_1}$$

$$= \frac{3 - 10}{2000 - 1950} = \frac{-7}{50} = -0.14$$

Note that if we interpret negative slope as decreasing population, then the slope of the line represents the annual rate at which the number of farm workers was decreasing, 0.14 million or 140,000 workers per year.

In general, any quantity that is expressed by using the word *per* is represented mathematically as slope. In the first example, the slope represented the average speed, 19 feet per second. In the second example, the slope represented the rate at which the number of farm workers was decreasing, −0.14 million workers per year.

Copyright © Houghton Mifflin Company. All rights reserved.

Example 1

Find the slope of the line containing the points (2, −5) and (−4, 2).

Solution

Let $P_1 = (2, -5)$ and $P_2 = (-4, 2)$.

$$m = \frac{y_2 - y_1}{x_2 - x_1} = \frac{2 - (-5)}{-4 - 2} = \frac{7}{-6}$$

The slope is $-\frac{7}{6}$.

You Try It 1

Find the slope of the line containing the points (4, −3) and (2, 7).

Your solution

Example 2

Find the slope of the line containing the points (−3, 4) and (5, 4).

Solution

Let $P_1 = (-3, 4)$ and $P_2 = (5, 4)$.

$$m = \frac{y_2 - y_1}{x_2 - x_1} = \frac{4 - 4}{5 - (-3)} = \frac{0}{8} = 0$$

The slope of the line is zero.

You Try It 2

Find the slope of the line containing the points (6, −1) and (6, 7).

Your solution

Example 3

The graph below shows the relationship between the cost of an item and the sales tax. Find the slope of the line between the two points shown on the graph. Write a sentence that states the meaning of the slope.

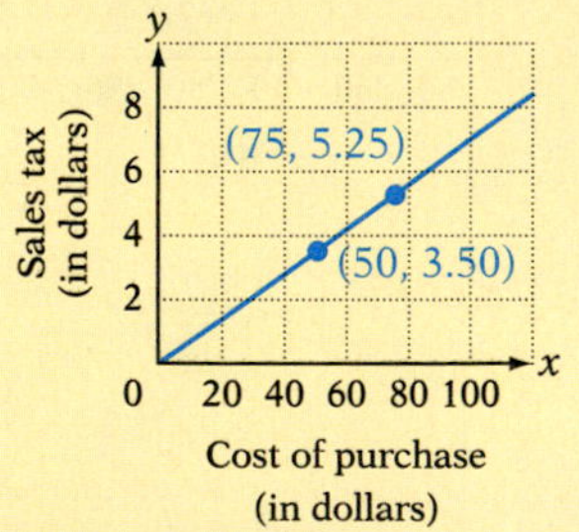

Solution

$$m = \frac{5.25 - 3.50}{75 - 50} = \frac{1.75}{25} = 0.07$$

• $(x_1, y_1) = (50, 3.50)$, $(x_2, y_2) = (75, 5.25)$

A slope of 0.07 means that the sales tax is $.07 per dollar.

You Try It 3

The graph below shows the decrease in the value of a recycling truck for a period of 6 years. Find the slope of the line between the two points shown on the graph. Write a sentence that states the meaning of the slope.

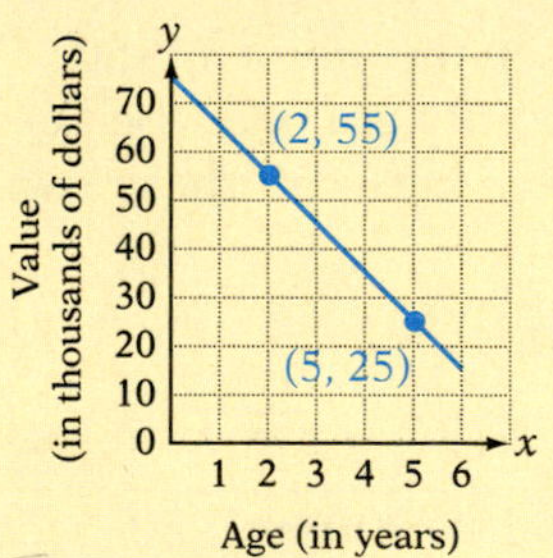

Your solution

Solutions on p. S13

Copyright © Houghton Mifflin Company. All rights reserved.

Objective B To graph a line given a point and the slope

The graph of the equation $y = -\frac{3}{4}x + 4$ is shown at the right. The points $(-4, 7)$ and $(4, 1)$ are on the graph. The slope of the line is

$$m = \frac{7-1}{-4-4} = \frac{6}{-8} = -\frac{3}{4}$$

Note that the slope of the line has the same value as the coefficient of x. The y-intercept is $(0, 4)$.

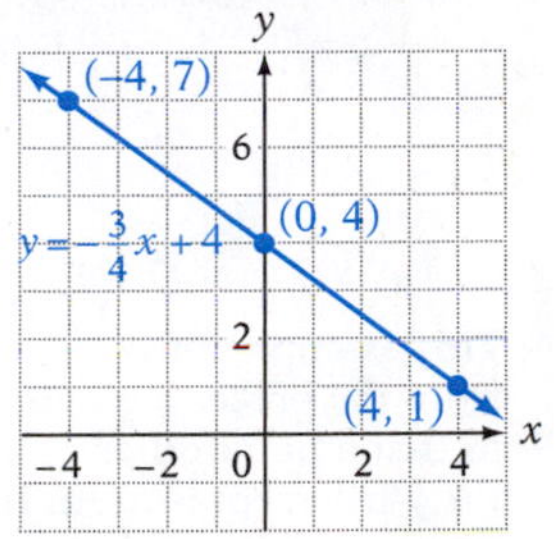

Slope-Intercept Form of a Straight Line

The equation $y = mx + b$ is called the **slope-intercept form of a straight line.** The slope of the line is m, the coefficient of x. The y-intercept is $(0, b)$.

When the equation of a straight line is in the form $y = mx + b$, the graph can be drawn by using the slope and y-intercept. First locate the y-intercept. Use the slope to find a second point on the line. Then draw a line through the two points.

TAKE NOTE

When graphing a line by using its slope and y-intercept, *always* start at the y-intercept.

HOW TO Graph $y = \frac{5}{3}x - 4$ by using the slope and y-intercept.

The slope is the coefficient of x:

$$m = \frac{5}{3} = \frac{\text{change in } y}{\text{change in } x}$$

The y-intercept is $(0, -4)$.

Beginning at the y-intercept $(0, -4)$, move right 3 units (change in x) and then up 5 units (change in y).

The point whose coordinates are $(3, 1)$ is a second point on the graph. Draw a line through the points $(0, -4)$ and $(3, 1)$.

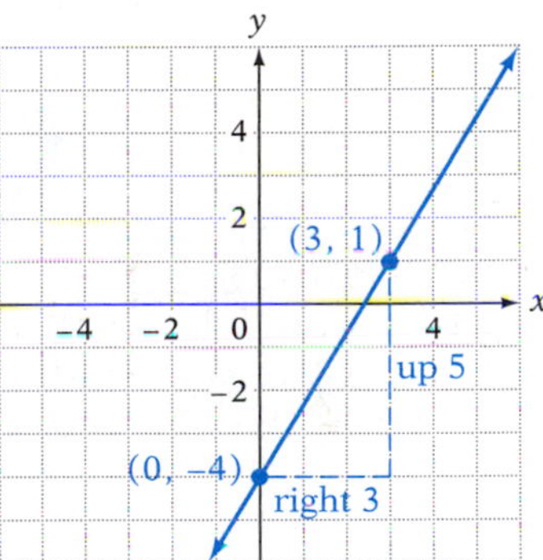

HOW TO Graph $x + 2y = 4$ by using the slope and y-intercept.

Solve the equation for y.

$$x + 2y = 4$$

$$2y = -x + 4$$

$$y = -\frac{1}{2}x + 2$$

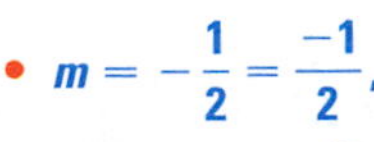

Beginning at the y-intercept $(0, 2)$, move right 2 units (change in x) and then down 1 unit (change in y).

The point whose coordinates are $(2, 1)$ is a second point on the graph. Draw a line through the points $(0, 2)$ and $(2, 1)$.

Copyright © Houghton Mifflin Company. All rights reserved.

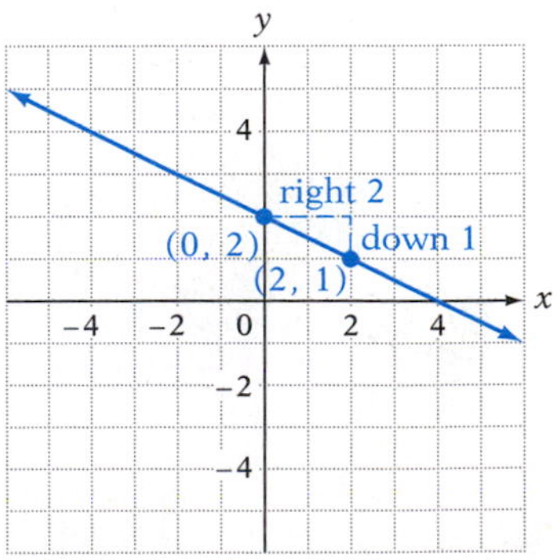

The graph of a line can be drawn when any point on the line and the slope of the line are given.

TAKE NOTE
This example differs from the preceding two in that a point other than the y-intercept is used. In this case, start at the given point.

HOW TO Graph the line that passes through the point $(-4, -4)$ and has slope 2.

When the slope is an integer, write it as a fraction with denominator 1.

$$m = 2 = \frac{2}{1} = \frac{\text{change in } y}{\text{change in } x}$$

Locate $(-4, -4)$ in the coordinate plane. Beginning at that point, move right 1 unit (change in x) and then up 2 units (change in y).

The point whose coordinates are $(-3, -2)$ is a second point on the graph. Draw a line through the points $(-4, -4)$ and $(-3, -2)$.

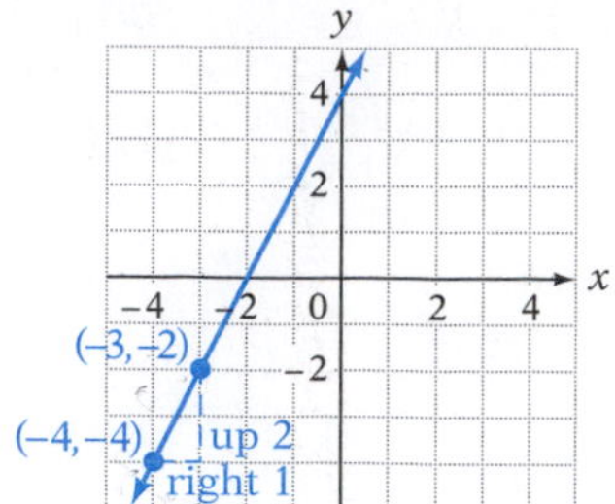

Example 4 Graph $y = -\frac{3}{2}x + 4$ by using the slope and y-intercept.

Solution $m = -\frac{3}{2} = \frac{-3}{2}$

y-intercept $= (0, 4)$

(0, 4) right 2 down 3 (2, 1)

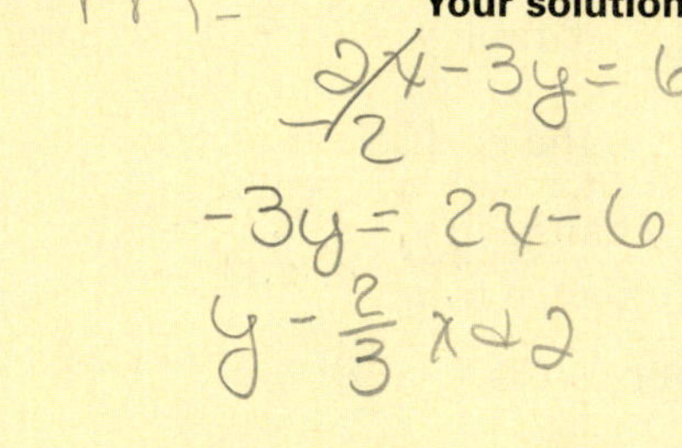

You Try It 4 Graph $2x + 3y = 6$ by using the slope and y-intercept.

Your solution

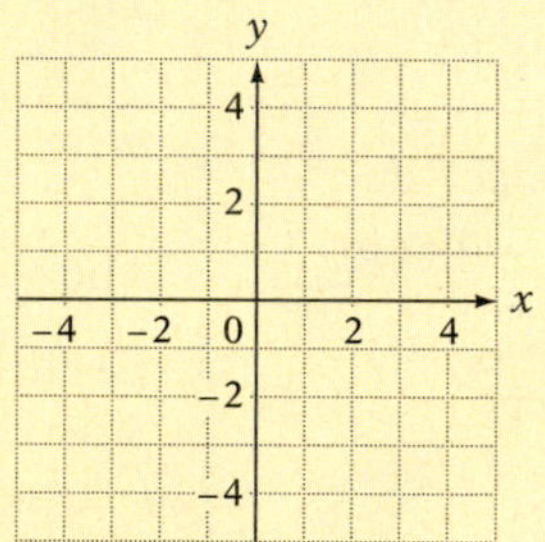

Example 5 Graph the line that passes through the point $(-2, 3)$ and has slope $-\frac{4}{3}$.

Solution $(x_1, y_1) = (-2, 3)$

$m = -\frac{4}{3} = \frac{-4}{3}$

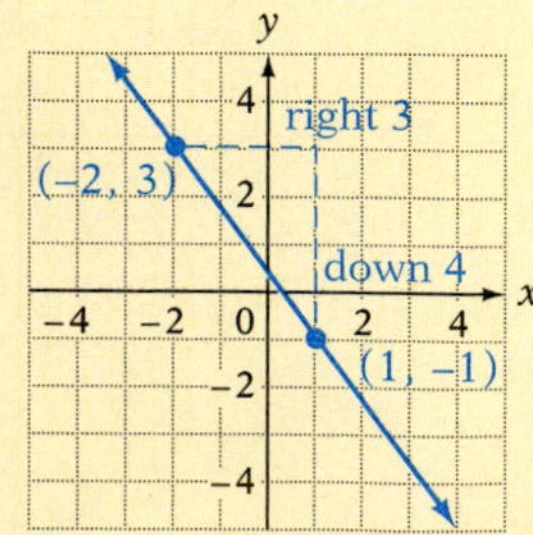

You Try It 5 Graph the line that passes through the point $(-3, -2)$ and has slope 3.

Your solution

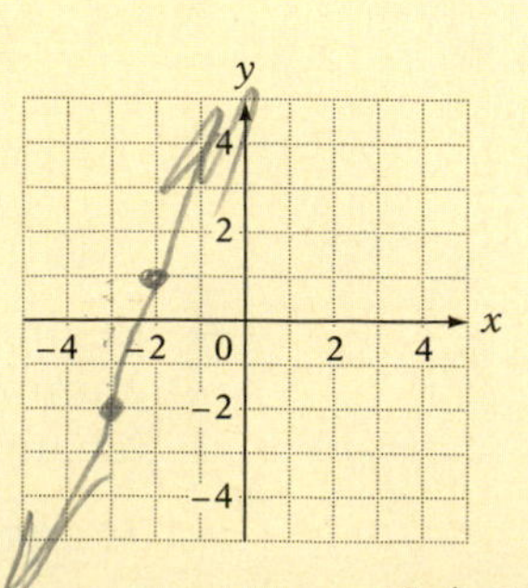

Solutions on p. S13

Copyright © Houghton Mifflin Company. All rights reserved.

4.4 Exercises

Objective A **To find the slope of a line given two points**

For Exercises 1 to 18, find the slope of the line containing the points.

1. $P_1(1, 3), P_2(3, 1)$
2. $P_1(2, 3), P_2(5, 1)$
3. $P_1(-1, 4), P_2(2, 5)$
4. $P_1(3, -2), P_2(1, 4)$
5. $P_1(-1, 3), P_2(-4, 5)$
6. $P_1(-1, -2), P_2(-3, 2)$
7. $P_1(0, 3), P_2(4, 0)$
8. $P_1(-2, 0), P_2(0, 3)$
9. $P_1(2, 4), P_2(2, -2)$
10. $P_1(4, 1), P_2(4, -3)$
11. $P_1(2, 5), P_2(-3, -2)$
12. $P_1(4, 1), P_2(-1, -2)$
13. $P_1(2, 3), P_2(-1, 3)$
14. $P_1(3, 4), P_2(0, 4)$
15. $P_1(0, 4), P_2(-2, 5)$
16. $P_1(-2, 3), P_2(-2, 5)$
17. $P_1(-3, -1), P_2(-3, 4)$
18. $P_1(-2, -5), P_2(-4, -1)$

19. **Travel** The graph below shows the relationship between the distance traveled by a motorist and the time of travel. Find the slope of the line between the two points shown on the graph. Write a sentence that states the meaning of the slope.

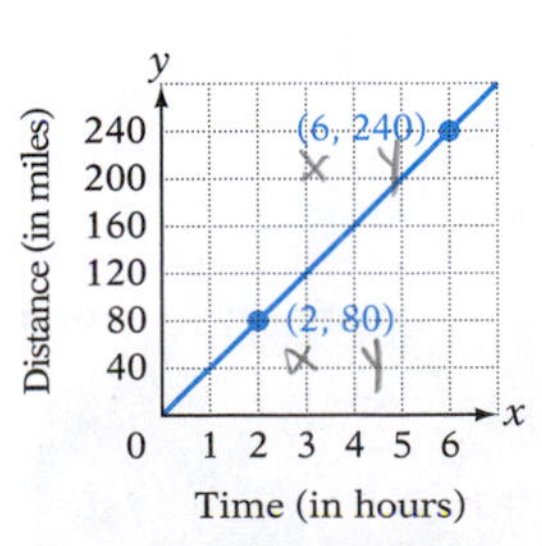

20. **Media** The graph below shows the number of people subscribing to a sports magazine of increasing popularity. Find the slope of the line between the two points shown on the graph. Write a sentence that states the meaning of the slope.

y
900
700
500
('07, 850)
('02, 580)
'02 '04 '06
x
Number of subscriptions (in thousands)
Year

Copyright © Houghton Mifflin Company. All rights reserved.

21. **Temperature** The graph below shows the relationship between the temperature inside an oven and the time since the oven was turned off. Find the slope of the line. Write a sentence that states the meaning of the slope.

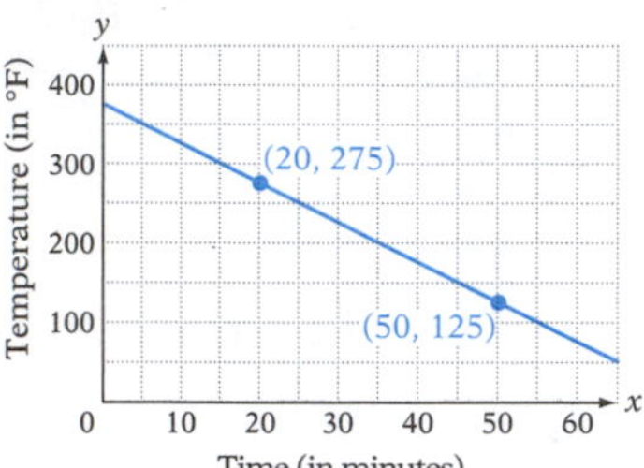

22. **Home Maintenance** The graph below shows the number of gallons of water remaining in a pool x minutes after a valve is opened to drain the pool. Find the slope of the line. Write a sentence that states the meaning of the slope.

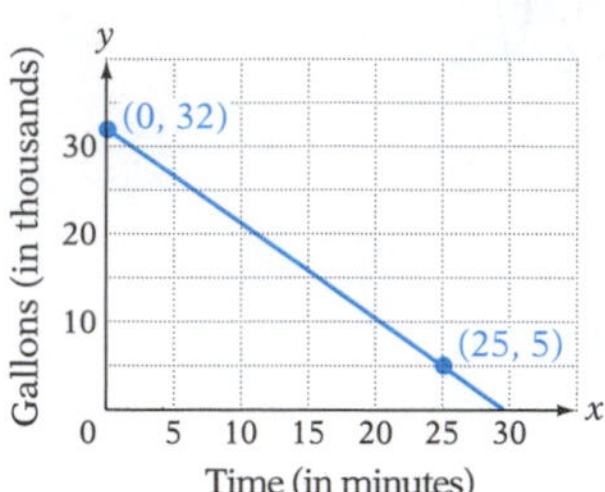

23. **Fuel Consumption** The graph below shows how the amount of gas in the tank of a car decreases as the car is driven. Find the slope of the line. Write a sentence that states the meaning of the slope.

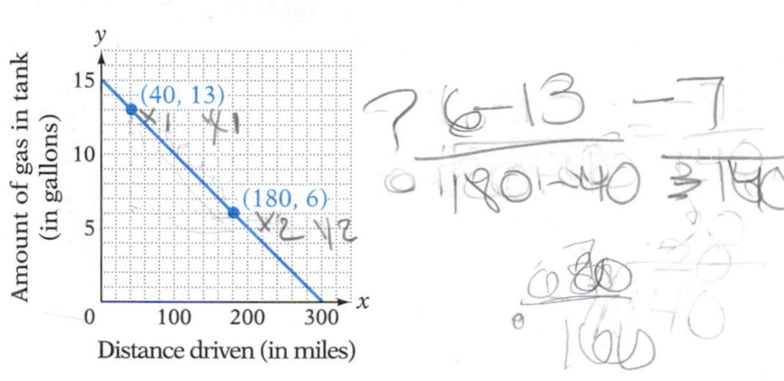

24. **Meteorology** The troposphere extends from Earth's surface to an elevation of about 11 km. The graph below shows the decrease in the temperature of the troposphere as altitude increases. Find the slope of the line. Write a sentence that states the meaning of the slope.

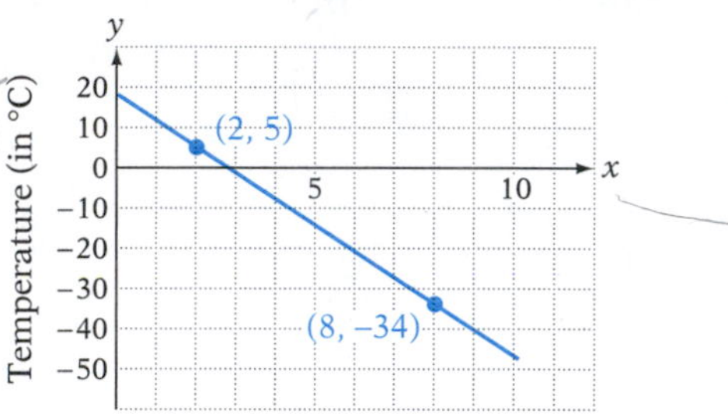

25. **Sports** The graph below shows the relationship between distance and time for the world-record 5000-meter run by Deena Drossin in 2002. Find the slope of the line between the two points shown on the graph. Round to the nearest tenth. Write a sentence that states the meaning of the slope.

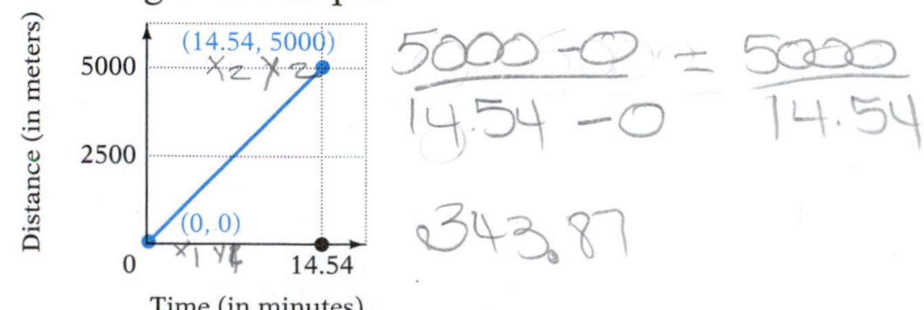

26. **Sports** The graph below shows the relationship between distance and time for the world-record 10,000-meter run by Sammy Kipketer in 2002. Find the slope of the line between the two points shown on the graph. Round to the nearest tenth. Write a sentence that states the meaning of the slope.

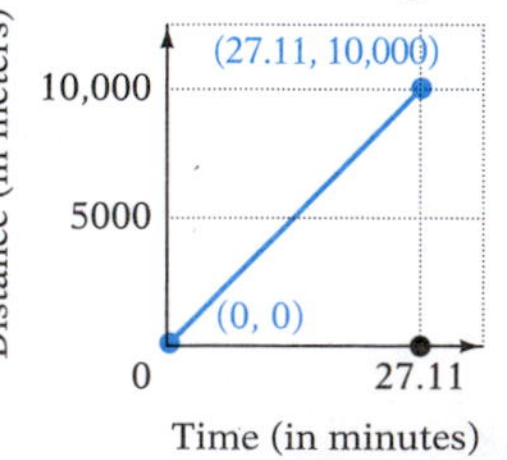

Construction For Exercises 27 and 28, use the fact that the American National Standards Institute (ANSI) states that the slope for a wheelchair ramp must not exceed $\frac{1}{12}$.

27. Does a ramp that is 6 in. high and 5 ft long meet the requirements of ANSI?

28. Does a ramp that is 12 in. high and 170 in. long meet the requirements of ANSI?

Copyright © Houghton Mifflin Company. All rights reserved.

Objective B To graph a line given a point and the slope

For Exercises 29 to 32, complete the table.

	Equation	Value of m	Value of b	Slope	y-intercept
29.	$y = -3x + 5$				
30.	$y = \frac{2}{5}x - 8$				
31.	$y = 4x$				
32.	$y = -7$				

For Exercises 33 to 44, graph by using the slope and the y-intercept.

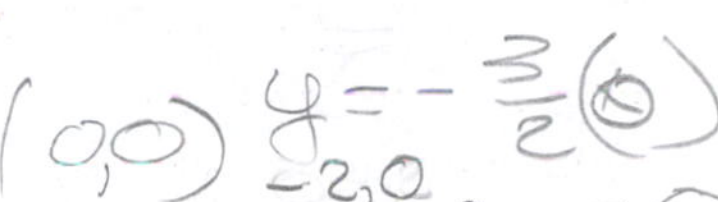

33. $y = \frac{1}{2}x + 2$

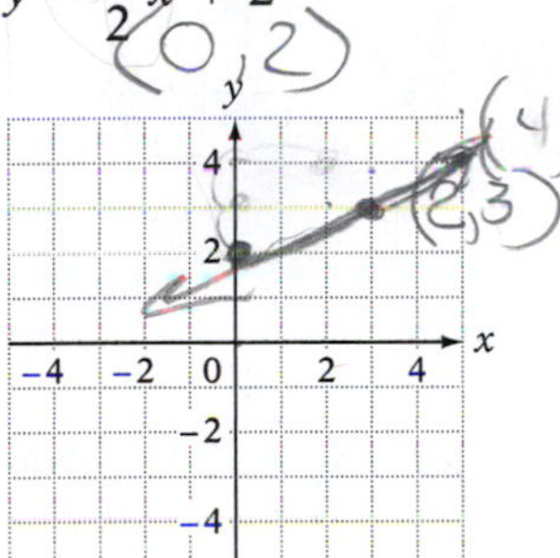

34. $y = \frac{2}{3}x - 3$

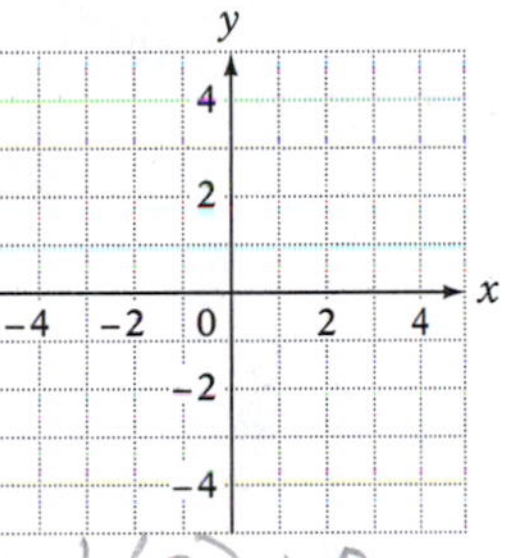

35. $y = -\frac{3}{2}x$

36. $y = \frac{3}{4}x$

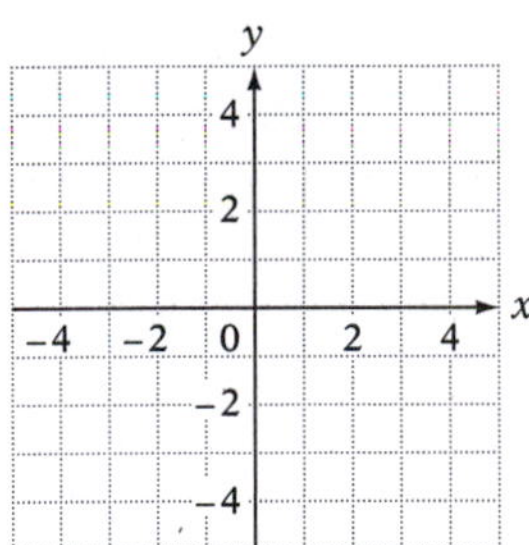

37. $y = -\frac{1}{2}x + 2$

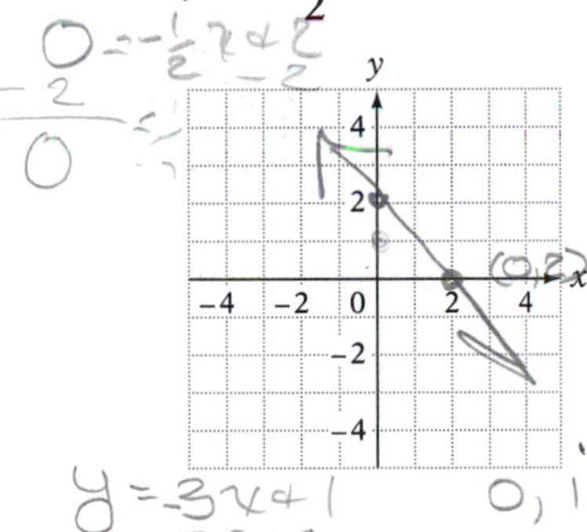

38. $y = \frac{2}{3}x - 1$

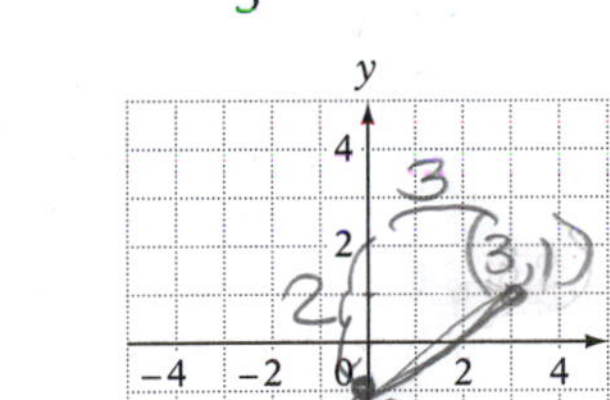

39. $y = 2x - 4$

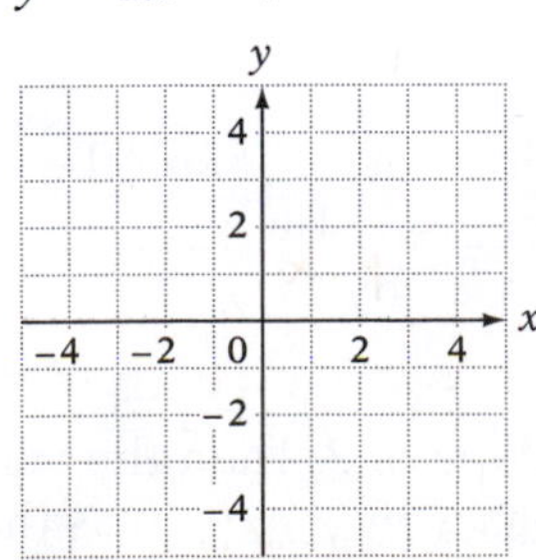

40. $y = -3x + 1$

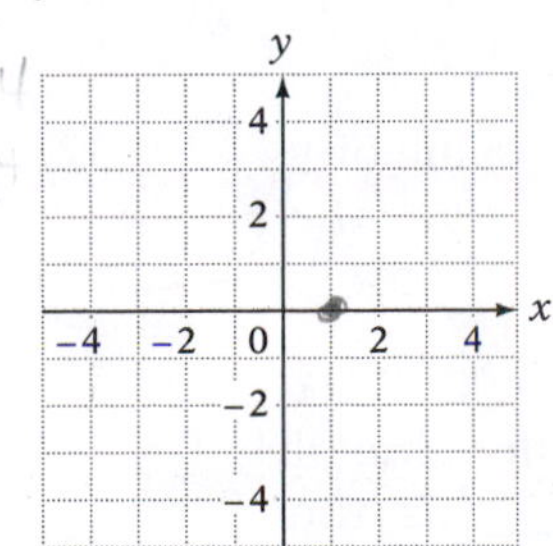

41. $4x - y = 1$

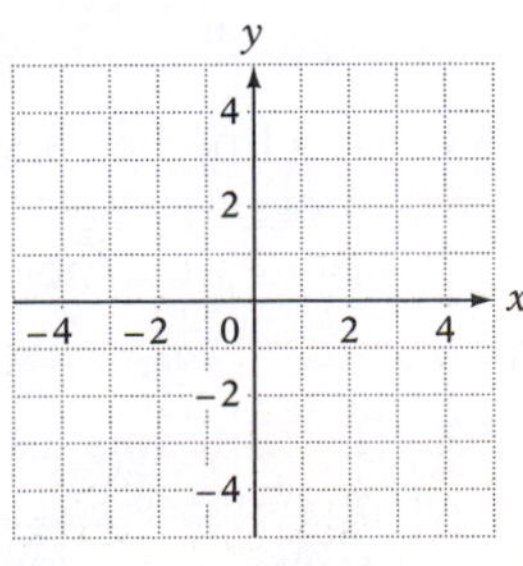

Copyright © Houghton Mifflin Company. All rights reserved.

42. $4x + y = 2$

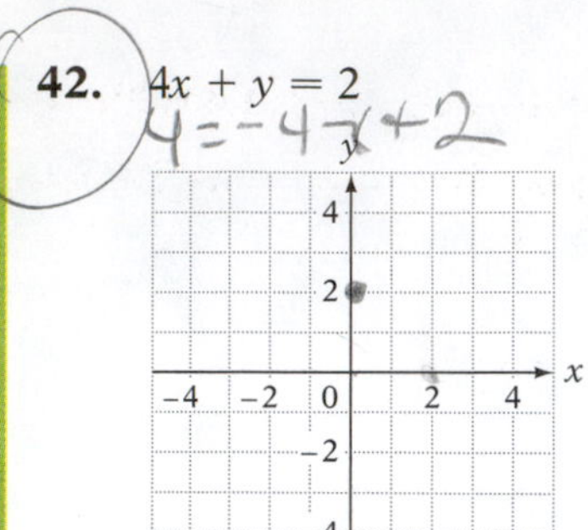

43. $x - 3y = 3$

44. $3x + 2y = 8$

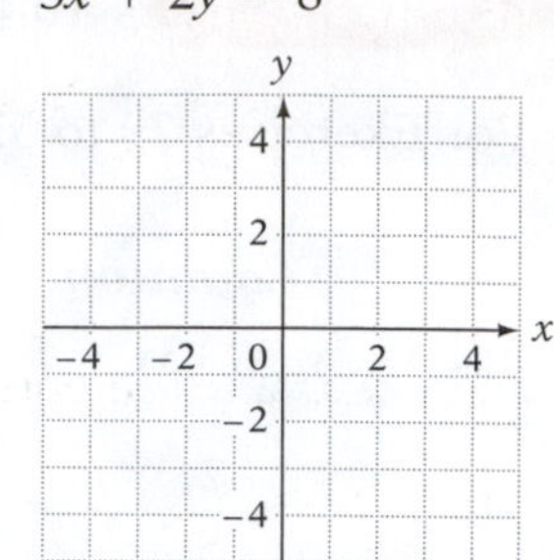

45. Graph the line that passes through the point $(-1, -3)$ and has slope $\frac{4}{3}$.

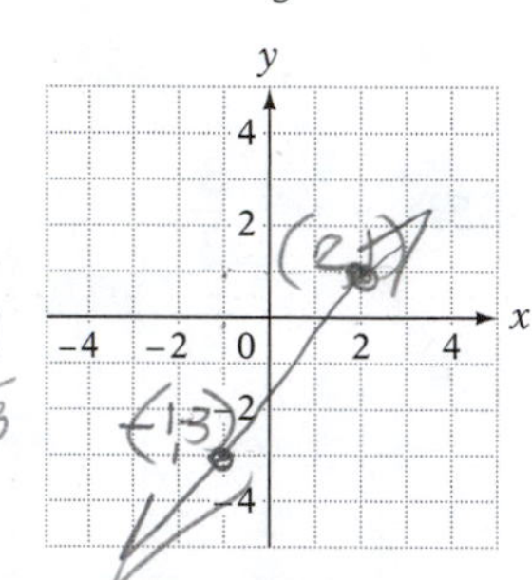

46. Graph the line that passes through the point $(-2, -3)$ and has slope $\frac{5}{4}$.

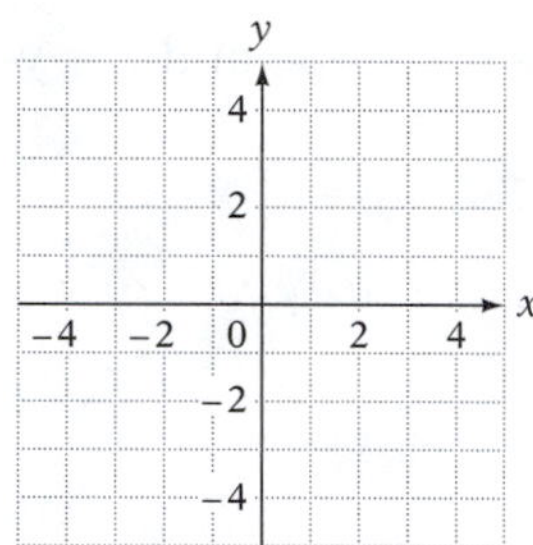

47. Graph the line that passes through the point $(-3, 0)$ and has slope -3.

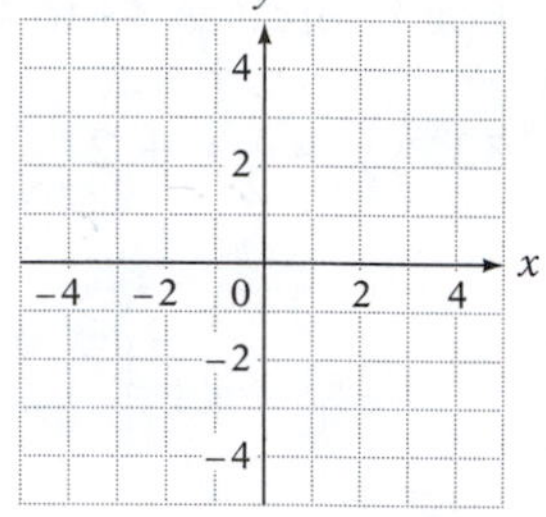

48. Graph the line that passes through the point $(2, -4)$ and has slope $-\frac{1}{2}$.

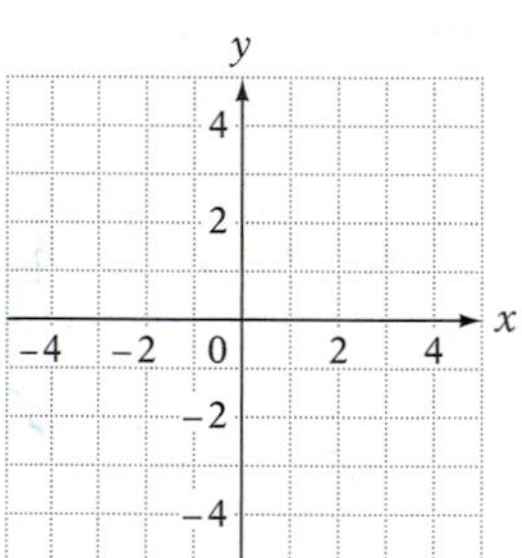

49. Graph the line that passes through the point $(-4, 1)$ and has slope $\frac{2}{3}$.

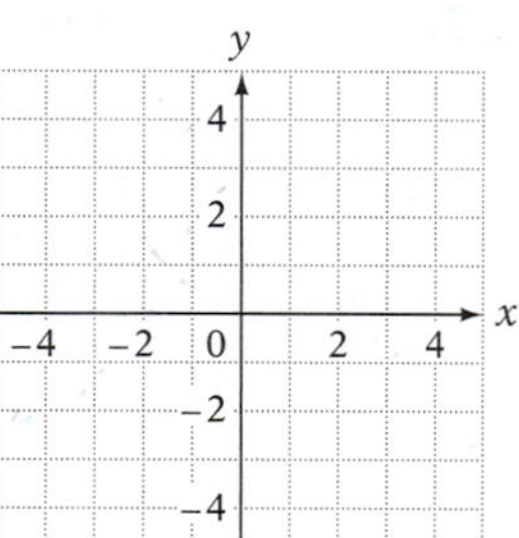

50. Graph the line that passes through the point $(1, 5)$ and has slope -4.

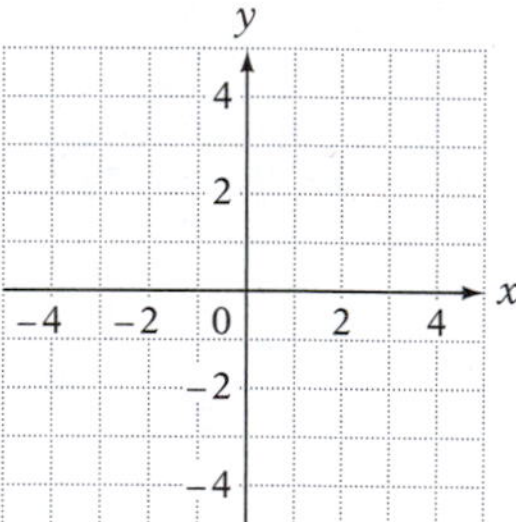

APPLYING THE CONCEPTS

Complete the following sentences.

51. If a line has a slope of 2, then the value of y <u>increases/decreases</u> by _______ as the value of x increases by 1.

52. If a line has a slope of -3, then the value of y <u>increases/decreases</u> by _______ as the value of x increases by 1.

53. If a line has a slope of -2, then the value of y <u>increases/decreases</u> by _______ as the value of x decreases by 1.

54. If a line has a slope of 3, then the value of y <u>increases/decreases</u> by _______ as the value of x decreases by 1.

Copyright © Houghton Mifflin Company. All rights reserved.

55. If a line has a slope of $-\frac{2}{3}$, then the value of y increases/decreases by _______ as the value of x increases by 1.

56. If a line has a slope of $\frac{1}{2}$, then the value of y increases/decreases by _______ as the value of x increases by 1.

57. Match each equation with its graph.

i. $y = -2x + 4$

ii. $y = 2x - 4$

iii. $y = 2$

iv. $2x + 4y = 0$

v. $y = \frac{1}{2}x + 4$

vi. $y = -\frac{1}{4}x - 2$

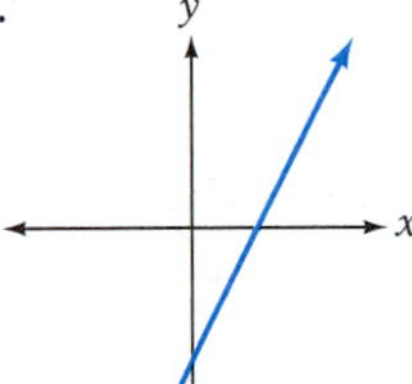

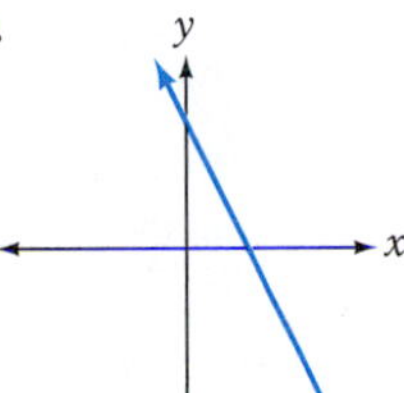

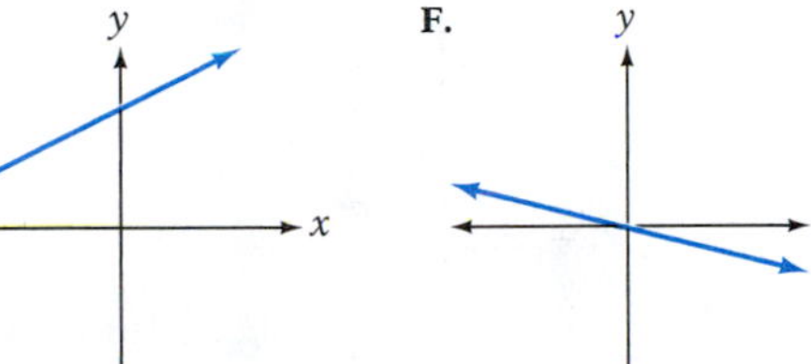

58. Match each equation with its graph.

i. $y = 3x - 2$

ii. $y = -3x + 2$

iii. $y = -3x$

iv. $y = -3$

v. $3x - 2y = 0$

vi. $y = -\frac{1}{3}x + 2$

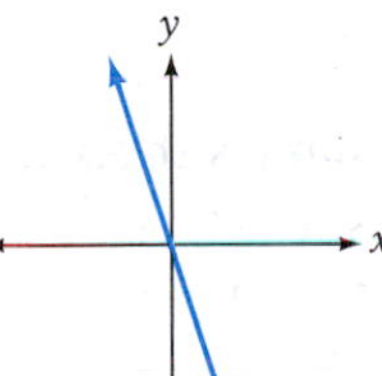

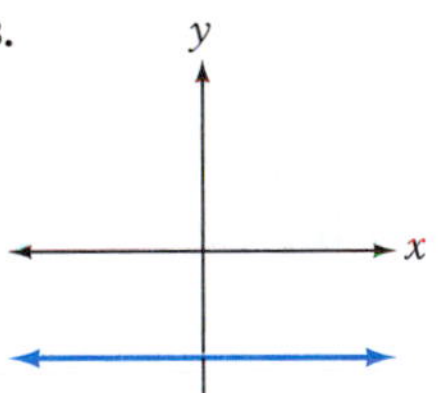

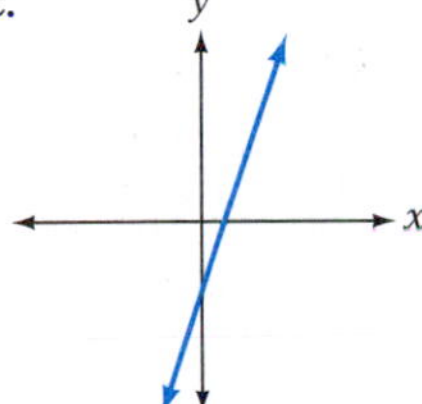

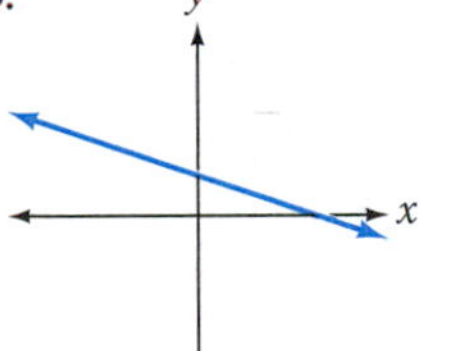

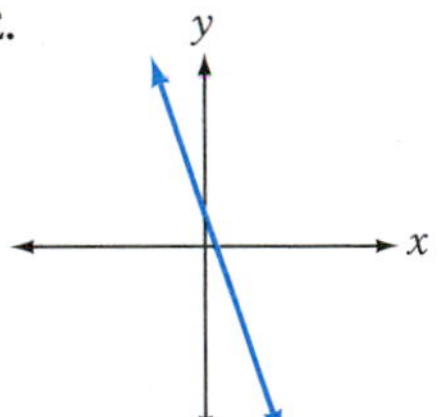

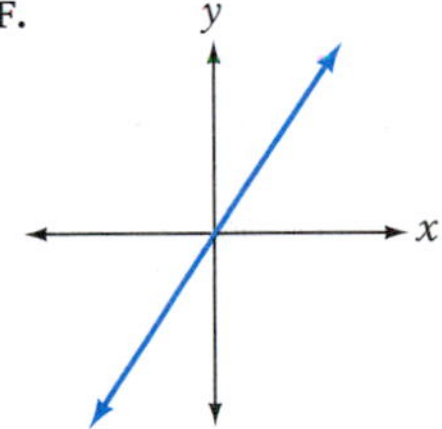

59. Explain how you can use the slope of a line to determine whether three given points lie on the same line. Then use your procedure to determine whether each of the following sets of points lie on the same line.
a. $(2, 5), (-1, -1), (3, 7)$ **b.** $(-1, 5), (0, 3), (-3, 4)$

For Exercises 60 to 63, determine the value of k such that the points whose coordinates are given lie on the same line.

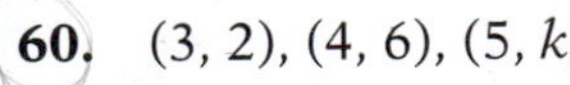

60. $(3, 2), (4, 6), (5, k)$

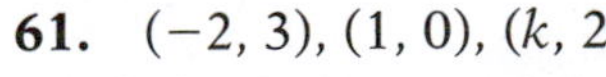

61. $(-2, 3), (1, 0), (k, 2)$

62. $(k, 1), (0, -1), (2, -2)$

63. $(4, -1), (3, -4), (k, k)$

Copyright © Houghton Mifflin Company. All rights reserved.

4.5 Finding Equations of Lines

Objective A To find the equation of a line given a point and the slope

When the slope of a line and a point on the line are known, the equation of the line can be determined. If the particular point is the y-intercept, use the slope-intercept form, $y = mx + b$, to find the equation.

HOW TO Find the equation of the line that contains the point (0, 3) and has slope $\frac{1}{2}$.

The known point is the y-intercept, (0, 3).

$y = mx + b$ • Use the slope-intercept form.

$y = \frac{1}{2}x + 3$ • Replace m with $\frac{1}{2}$, the given slope. Replace b with 3, the y-coordinate of the y-intercept.

The equation of the line is $y = \frac{1}{2}x + 3$.

One method of finding the equation of a line when the slope and *any* point on the line are known involves using the *point-slope* formula. This formula is derived from the formula for the slope of a line as follows.

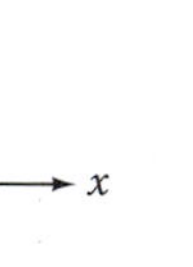

Let (x_1, y_1) be the given point on the line, and let (x, y) be any other point on the line. See the graph at the left.

$\frac{y - y_1}{x - x_1} = m$ • Use the formula for the slope of a line.

$\frac{y - y_1}{x - x_1}(x - x_1) = m(x - x_1)$ • Multiply each side by $(x - x_1)$.

$y - y_1 = m(x - x_1)$ • Simplify.

Point-Slope Formula

Let m be the slope of a line, and let (x_1, y_1) be the coordinates of a point on the line. The equation of the line can be found from the **point-slope formula:**

$$y - y_1 = m(x - x_1)$$

HOW TO Find the equation of the line that contains the point whose coordinates are (4, −1) and has slope $-\frac{3}{4}$.

$y - y_1 = m(x - x_1)$ • Use the point-slope formula.

$y - (-1) = \left(-\frac{3}{4}\right)(x - 4)$ • $m = -\frac{3}{4}$, $(x_1, y_1) = (4, -1)$

$y + 1 = -\frac{3}{4}x + 3$ • Simplify.

$y = -\frac{3}{4}x + 2$ • Write the equation in the form $y = mx + b$.

The equation of the line is $y = -\frac{3}{4}x + 2$.

$y = -\frac{3}{4}x + 2$

Copyright © Houghton Mifflin Company. All rights reserved.

HOW TO Find the equation of the line that passes through the point whose coordinates are (4, 3) and whose slope is undefined.

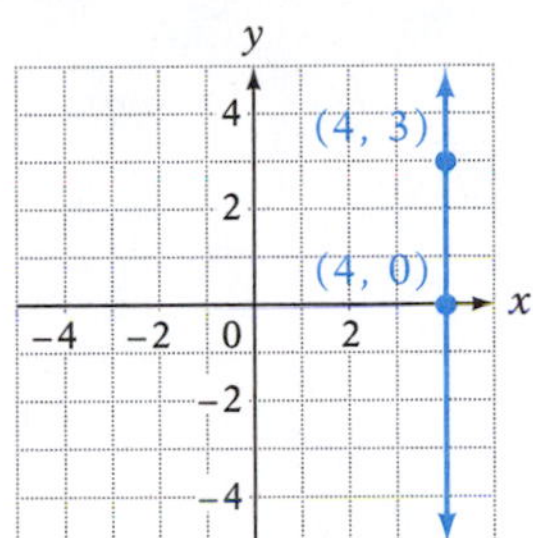

Because the slope is undefined, the point-slope formula cannot be used to find the equation. Instead, recall that when the slope is undefined, the line is vertical, and that the equation of a vertical line is $x = a$, where a is the x-coordinate of the x-intercept. Because the line is vertical and passes through (4, 3), the x-intercept is (4, 0). The equation of the line is $x = 4$.

Example 1

Find the equation of the line that contains the point (3, 0) and has slope −4.

Solution

$m = -4 \qquad (x_1, y_1) = (3, 0)$

$$y - y_1 = m(x - x_1)$$
$$y - 0 = -4(x - 3)$$
$$y = -4x + 12$$

The equation of the line is $y = -4x + 12$.

You Try It 1

Find the equation of the line that contains the point (−3, −2) and has slope $-\frac{1}{3}$.

Your solution

Example 2

Find the equation of the line that contains the point (−2, 4) and has slope 2.

Solution

$m = 2 \qquad (x_1, y_1) = (-2, 4)$

$$y - y_1 = m(x - x_1)$$
$$y - 4 = 2[x - (-2)]$$
$$y - 4 = 2(x + 2)$$
$$y - 4 = 2x + 4$$
$$y = 2x + 8$$

The equation of the line is $y = 2x + 8$.

You Try It 2

Find the equation of the line that contains the point (4, −3) and has slope −3.

Your solution

Solutions on pp. S13–S14

Copyright © Houghton Mifflin Company. All rights reserved.

Objective B To find the equation of a line given two points

The point-slope formula and the formula for slope are used to find the equation of a line when two points are known.

HOW TO Find the equation of the line containing the points (3, 2) and (−5, 6).

To use the point-slope formula, we must know the slope. Use the formula for slope to determine the slope of the line between the two given points. Let $(x_1, y_1) = (3, 2)$ and $(x_2, y_2) = (-5, 6)$.

$$m = \frac{y_2 - y_1}{x_2 - x_1} = \frac{6 - 2}{-5 - 3} = \frac{4}{-8} = -\frac{1}{2}$$

Now use the point-slope formula with $m = -\frac{1}{2}$ and $(x_1, y_1) = (3, 2)$.

$y - y_1 = m(x - x_1)$ • **Use the point-slope formula.**

$y - 2 = \left(-\frac{1}{2}\right)(x - 3)$ • **$m = -\frac{1}{2}$, $(x_1, y_1) = (3, 2)$**

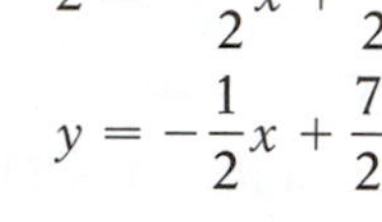

$y - 2 = -\frac{1}{2}x + \frac{3}{2}$ • **Simplify.**

$y = -\frac{1}{2}x + \frac{7}{2}$

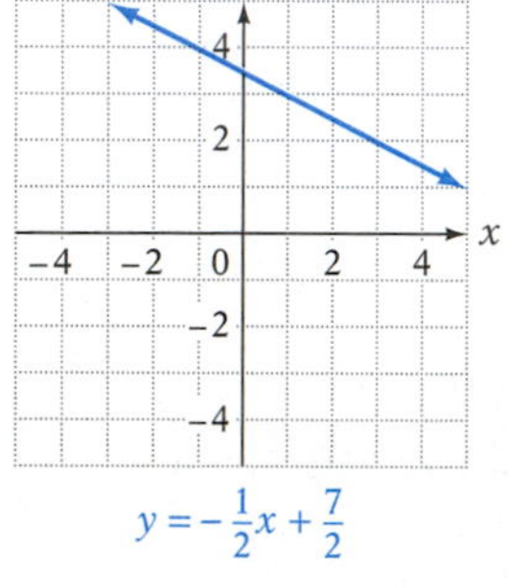

$y = -\frac{1}{2}x + \frac{7}{2}$

The equation of the line is $y = -\frac{1}{2}x + \frac{7}{2}$.

Example 3

Find the equation of the line containing the points (2, 3) and (4, 1).

Solution

Let $(x_1, y_1) = (2, 3)$ and $(x_2, y_2) = (4, 1)$.

$$m = \frac{y_2 - y_1}{x_2 - x_1} = \frac{1 - 3}{4 - 2} = \frac{-2}{2} = -1$$

$$y - y_1 = m(x - x_1)$$
$$y - 3 = -1(x - 2)$$
$$y - 3 = -x + 2$$
$$y = -x + 5$$

The equation of the line is $y = -x + 5$.

You Try It 3

Find the equation of the line containing the points (2, 0) and (5, 3).

Your solution

Example 4

Find the equation of the line containing the points (2, −3) and (2, 5).

Solution

Let $(x_1, y_1) = (2, -3)$ and $(x_2, y_2) = (2, 5)$.

$$m = \frac{y_2 - y_1}{x_2 - x_1} = \frac{5 - (-3)}{2 - 2} = \frac{8}{0}$$

The slope is undefined, so the graph of the line is vertical.

The equation of the line is $x = 2$.

You Try It 4

Find the equation of the line containing the points (2, 3) and (−5, 3).

Your solution

Solutions on p. S14

Copyright © Houghton Mifflin Company. All rights reserved.

Objective C To solve application problems

Linear functions can be used to model a variety of applications in science and business. For each application, data are collected and the independent and dependent variables are selected. Then a linear function is determined that models the data.

Example 5

Suppose a manufacturer has determined that at a price of \$115, consumers will purchase 1 million portable CD players and that at a price of \$90, consumers will purchase 1.25 million portable CD players. Describe this situation with a linear function. Use this function to predict how many portable CD players consumers will purchase if the price is \$80.

Strategy

- Select the independent and dependent variables. Because you are trying to determine the number of CD players, that quantity is the *dependent* variable, y. The price of the CD players is the *independent* variable, x.
- From the given data, two ordered pairs are (115, 1) and (90, 1.25). (The ordinates are in millions of units.) Use these ordered pairs to determine the linear function.
- Evaluate this function when $x = 80$ to predict how many CD players consumers will purchase if the price is \$80.

Solution

Let $(x_1, y_1) = (115, 1)$ and $(x_2, y_2) = (90, 1.25)$.

$$m = \frac{y_2 - y_1}{x_2 - x_1} = \frac{1.25 - 1}{90 - 115} = -\frac{0.25}{25} = -0.01$$

$$y - y_1 = m(x - x_1)$$
$$y - 1 = -0.01(x - 115)$$
$$y = -0.01x + 2.15$$

The linear function is $f(x) = -0.01x + 2.15$.

$$f(80) = -0.01(80) + 2.15 = 1.35$$

Consumers will purchase 1.35 million CD players at a price of \$80.

You Try It 5

Gabriel Daniel Fahrenheit invented the mercury thermometer in 1717. In terms of readings on this thermometer, water freezes at 32°F and boils at 212°F. In 1742, Anders Celsius invented the Celsius temperature scale. On this scale, water freezes at 0°C and boils at 100°C. Determine a linear function that can be used to predict the Celsius temperature when the Fahrenheit temperature is known.

Your strategy

Your solution

Solution on p. S14

Copyright © Houghton Mifflin Company. All rights reserved.

4.5 Exercises

Objective A To find the equation of a line given a point and the slope

1. Explain how to find the equation of a line given its slope and its y-intercept.

2. What is the point-slope formula and how is it used?

For Exercises 3 to 38, find the equation of the line that contains the given point and has the given slope.

3. Point $(0, 5)$, $m = 2$

4. Point $(0, 3)$, $m = 1$

5. Point $(2, 3)$, $m = \frac{1}{2}$

6. Point $(5, 1)$, $m = \frac{2}{3}$

7. Point $(-1, 4)$, $m = \frac{5}{4}$

8. Point $(-2, 1)$, $m = \frac{3}{2}$

9. Point $(3, 0)$, $m = -\frac{5}{3}$

10. Point $(-2, 0)$, $m = \frac{3}{2}$

11. Point $(2, 3)$, $m = -3$

12. Point $(1, 5)$, $m = -\frac{4}{5}$

13. Point $(-1, 7)$, $m = -3$

14. Point $(-2, 4)$, $m = -4$

15. Point $(-1, -3)$, $m = \frac{2}{3}$

16. Point $(-2, -4)$, $m = \frac{1}{4}$

17. Point $(0, 0)$, $m = \frac{1}{2}$

18. Point $(0, 0)$, $m = \frac{3}{4}$

19. Point $(2, -3)$, $m = 3$

20. Point $(4, -5)$, $m = 2$

21. Point $(3, 5)$, $m = -\frac{2}{3}$

22. Point $(5, 1)$, $m = -\frac{4}{5}$

23. Point $(0, -3)$, $m = -1$

24. Point $(2, 0)$, $m = \frac{5}{6}$

25. Point $(1, -4)$, $m = \frac{7}{5}$

26. Point $(3, 5)$, $m = -\frac{3}{7}$

Copyright © Houghton Mifflin Company. All rights reserved.

27. Point $(4, -1)$, $m = -\frac{2}{5}$

28. Point $(-3, 5)$, $m = -\frac{1}{4}$

29. Point $(3, -4)$, slope is undefined

30. Point $(-2, 5)$, slope is undefined

31. Point $(-2, -5)$, $m = -\frac{5}{4}$

32. Point $(-3, -2)$, $m = -\frac{2}{3}$

33. Point $(-2, -3)$, $m = 0$

34. Point $(-3, -2)$, $m = 0$

35. Point $(4, -5)$, $m = -2$

36. Point $(-3, 5)$, $m = 3$

37. Point $(-5, -1)$, slope is undefined

38. Point $(0, 4)$, slope is undefined

Objective B To find the equation of a line given two points

For Exercises 39 to 74, find the equation of the line that contains the given points.

39. $P_1(0, 2)$, $P_2(3, 5)$

40. $P_1(0, 4)$, $P_2(1, 5)$

41. $P_1(0, -3)$, $P_2(-4, 5)$

42. $P_1(0, -2)$, $P_2(-3, 4)$

43. $P_1(2, 3)$, $P_2(5, 5)$

44. $P_1(4, 1)$, $P_2(6, 3)$

45. $P_1(-1, 3)$, $P_2(2, 4)$

46. $P_1(-1, 1)$, $P_2(4, 4)$

47. $P_1(-1, -2)$, $P_2(3, 4)$

48. $P_1(-3, -1)$, $P_2(2, 4)$

49. $P_1(0, 3)$, $P_2(2, 0)$

50. $P_1(0, 4)$, $P_2(2, 0)$

51. $P_1(-3, -1)$, $P_2(2, -1)$

52. $P_1(-3, -5)$, $P_2(4, -5)$

53. $P_1(-2, -3)$, $P_2(-1, -2)$

Copyright © Houghton Mifflin Company. All rights reserved.

54. $P_1(4, 1), P_2(3, -2)$

55. $P_1(-2, 3), P_2(2, -1)$

56. $P_1(3, 1), P_2(-3, -2)$

57. $P_1(2, 3), P_2(5, -5)$

58. $P_1(7, 2), P_2(4, 4)$

59. $P_1(2, 0), P_2(0, -1)$

60. $P_1(0, 4), P_2(-2, 0)$

61. $P_1(3, -4), P_2(-2, -4)$

62. $P_1(-3, 3), P_2(-2, 3)$

63. $P_1(0, 0), P_2(4, 3)$

64. $P_1(2, -5), P_2(0, 0)$

65. $P_1(2, -1), P_2(-1, 3)$

66. $P_1(3, -5), P_2(-2, 1)$

67. $P_1(-2, 5), P_2(-2, -5)$

68. $P_1(3, 2), P_2(3, -4)$

69. $P_1(2, 1), P_2(-2, -3)$

70. $P_1(-3, -2), P_2(1, -4)$

71. $P_1(-4, -3), P_2(2, 5)$

72. $P_1(4, 5), P_2(-4, 3)$

73. $P_1(0, 3), P_2(3, 0)$

74. $P_1(1, -3), P_2(-2, 4)$

Objective C To solve application problems

75. **Aviation** The pilot of a Boeing 747 jet takes off from Boston's Logan Airport, which is at sea level, and climbs to a cruising altitude of 32,000 ft at a constant rate of 1200 ft/min.
   a. Write a linear equation for the height of the plane in terms of the time after takeoff.
   b. Use your equation to find the height of the plane 11 min after takeoff.

76. **Calories** A jogger running at 9 mph burns approximately 14 Calories per minute.
   a. Write a linear equation for the number of Calories burned by the jogger in terms of the number of minutes run.
   b. Use your equation to find the number of Calories that the jogger has burned after jogging for 32 min.

Copyright © Houghton Mifflin Company. All rights reserved.

77. **Construction** A building contractor estimates that the cost to build a home is \$30,000 plus \$85 for each square foot of floor space in the house.
  a. Determine a linear function that will give the cost of building a house that contains a given number of square feet.
  b. Use this model to determine the cost to build a house that contains 1800 square feet.

78. **Telecommunications** A cellular phone company offers several different options for using a cellular telephone. One option, for people who plan on using the phone only in emergencies, costs the user \$4.95 per month plus \$.59 per minute for each minute the phone is used.
  a. Write a linear function for the monthly cost of the phone in terms of the number of minutes the phone is used.
  b. Use your equation to find the cost of using the cellular phone for 13 minutes in one month.

79. **Fuel Consumption** The gas tank of a certain car contains 16 gal when the driver of the car begins a trip. Each mile driven by the driver decreases the amount of gas in the tank by 0.032 gal.
  a. Write a linear function for the number of gallons of gas in the tank in terms of the number of miles driven.
  b. Use your equation to find the number of gallons in the tank after 150 mi are driven.

80. **Boiling Points** At sea level, the boiling point of water is 100°C. At an altitude of 2 km, the boiling point of water is 93°C.
  a. Write a linear function for the boiling point of water in terms of the altitude above sea level.
  b. Use your equation to predict the boiling point of water on top of Mount Everest, which is approximately 8.85 km above sea level. Round to the nearest degree.

Mt. Everest

81. **Business** A manufacturer of economy cars has determined that 50,000 cars per month can be sold at a price of \$9000. At a price of \$8750, the number of cars sold per month would increase to 55,000.
  a. Determine a linear function that will predict the number of cars that would be sold each month at a given price.
  b. Use this model to predict the number of cars that would be sold at a price of \$8500.

82. **Business** A manufacturer of graphing calculators has determined that 10,000 calculators per week will be sold at a price of \$95. At a price of \$90, it is estimated that 12,000 calculators would be sold.
  a. Determine a linear function that will predict the number of calculators that would be sold at a given price.
  b. Use this model to predict the number of calculators that would be sold each week at a price of \$75.

83. **Calories** There are approximately 126 Calories in a 2-ounce serving of lean hamburger and approximately 189 Calories in a 3-ounce serving.
  a. Determine a linear function for the number of Calories in lean hamburger in terms of the size of the serving.
  b. Use your equation to estimate the number of Calories in a 5-ounce serving of lean hamburger.

84. **Compensation** An account executive receives a base salary plus a commission. On \$20,000 in monthly sales, the account executive receives \$1800. On \$50,000 in monthly sales, the account executive receives \$3000.
  a. Determine a linear function that will yield the compensation of the sales executive for a given amount of monthly sales.
  b. Use this model to determine the account executive's compensation for \$85,000 in monthly sales.

Copyright © Houghton Mifflin Company. All rights reserved.

85. Let f be a linear function. If $f(2) = 5$ and $f(0) = 3$, find $f(x)$.

86. Let f be a linear function. If $f(-3) = 4$ and $f(1) = -8$, find $f(x)$.

87. Given that f is a linear function for which $f(1) = 3$ and $f(-1) = 5$, determine $f(4)$.

88. Given that f is a linear function for which $f(-3) = 2$ and $f(2) = 7$, determine $f(0)$.

89. A line with slope $\frac{4}{3}$ passes through the point (3, 2).

a. What is y when $x = -6$?
b. What is x when $y = 6$?

90. A line with slope $-\frac{3}{4}$ passes through the point (8, −2).

a. What is y when $x = -4$?
b. What is x when $y = 1$?

APPLYING THE CONCEPTS

91. Explain the similarities and differences between the point-slope formula and the slope-intercept form of a straight line.

92. Explain why the point-slope formula cannot be used to find the equation of a line that is parallel to the y-axis.

93. Refer to Example 5 in this section for each of the following.

a. Explain the meaning of the slope of the graph of the linear function given in the example.
b. Explain the meaning of the y-intercept.
c. Explain the meaning of the x-intercept.

94. For an equation of the form $y = mx + b$, how does the graph of this equation change if the value of b changes and the value of m remains constant?

95. A line contains the points (−3, 6) and (6, 0). Find the coordinates of three other points that are on this line.

96. A line contains the points (4, −1) and (2, 1). Find the coordinates of three other points that are on this line.

Copyright © Houghton Mifflin Company. All rights reserved.

4.6 Parallel and Perpendicular Lines

Objective A — To find parallel and perpendicular lines

Two lines that have the same slope do not intersect and are called **parallel lines.**

The slope of each of the lines at the right is $\frac{2}{3}$.

The lines are parallel.

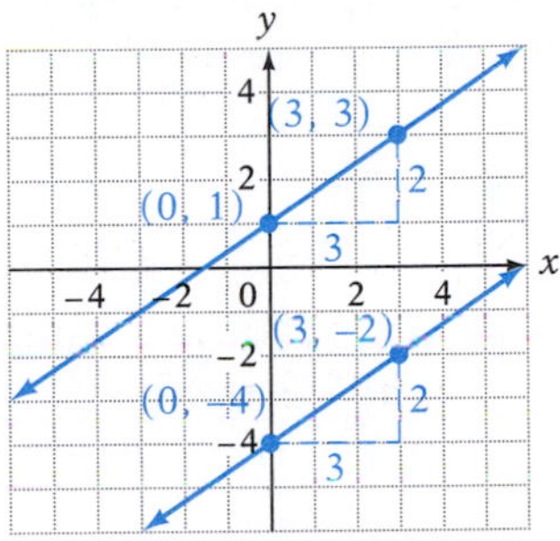

Slopes of Parallel Lines

Two nonvertical lines with slopes of m_1 and m_2 are parallel if and only if $m_1 = m_2$. Any two vertical lines are parallel.

HOW TO Is the line containing the points $(-2, 1)$ and $(-5, -1)$ parallel to the line that contains the points $(1, 0)$ and $(4, 2)$?

$$m_1 = \frac{-1 - 1}{-5 - (-2)} = \frac{-2}{-3} = \frac{2}{3}$$

• Find the slope of the line through $(-2, 1)$ and $(-5, -1)$.

$$m_2 = \frac{2 - 0}{4 - 1} = \frac{2}{3}$$

• Find the slope of the line through $(1, 0)$ and $(4, 2)$.

Because $m_1 = m_2$, the lines are parallel.

HOW TO Find the equation of the line that contains the point $(2, 3)$ and is parallel to the line $y = \frac{1}{2}x - 4$.

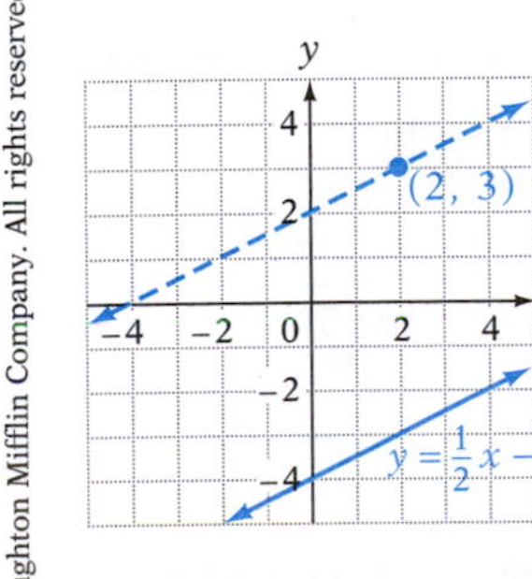

The slope of the given line is $\frac{1}{2}$. Because parallel lines have the same slope, the slope of the unknown line is also $\frac{1}{2}$.

$$y - y_1 = m(x - x_1)$$

• Use the point-slope formula.

$$y - 3 = \frac{1}{2}(x - 2)$$

• $m = \frac{1}{2}$, $(x_1, y_1) = (2, 3)$

$$y - 3 = \frac{1}{2}x - 1$$

• Simplify.

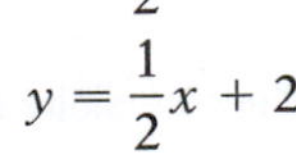

$$y = \frac{1}{2}x + 2$$

• Write the equation in the form $y = mx + b$.

The equation of the line is $y = \frac{1}{2}x + 2$.

Copyright © Houghton Mifflin Company. All rights reserved.

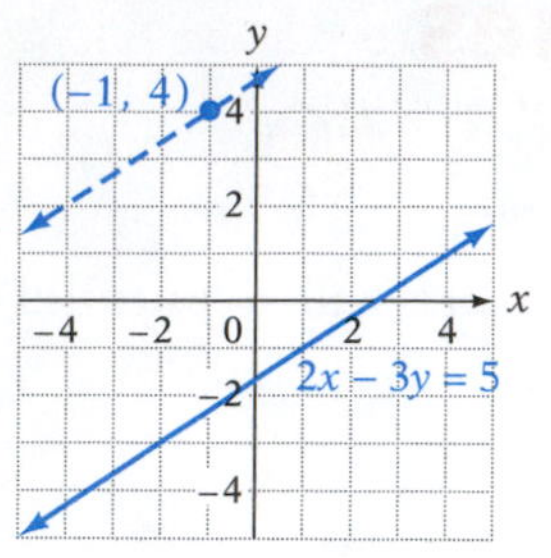

HOW TO Find the equation of the line that contains the point (−1, 4) and is parallel to the line $2x - 3y = 5$.

Because the lines are parallel, the slope of the unknown line is the same as the slope of the given line. Solve $2x - 3y = 5$ for y and determine its slope.

$$2x - 3y = 5$$
$$-3y = -2x + 5$$
$$y = \frac{2}{3}x - \frac{5}{3}$$

The slope of the given line is $\frac{2}{3}$. Because the lines are parallel, this is the slope of the unknown line. Use the point-slope formula to determine the equation.

$y - y_1 = m(x - x_1)$ • **Use the point-slope formula.**

$y - 4 = \frac{2}{3}[x - (-1)]$ • $m = \frac{2}{3}, (x_1, y_1) = (-1, 4)$

$y - 4 = \frac{2}{3}x + \frac{2}{3}$ • **Simplify.**

$y = \frac{2}{3}x + \frac{14}{3}$ • **Write the equation in the form $y = mx + b$.**

The equation of the line is $y = \frac{2}{3}x + \frac{14}{3}$.

Two lines that intersect at right angles are **perpendicular lines.**

Any horizontal line is perpendicular to any vertical line. For example, $x = 3$ is perpendicular to $y = -2$.

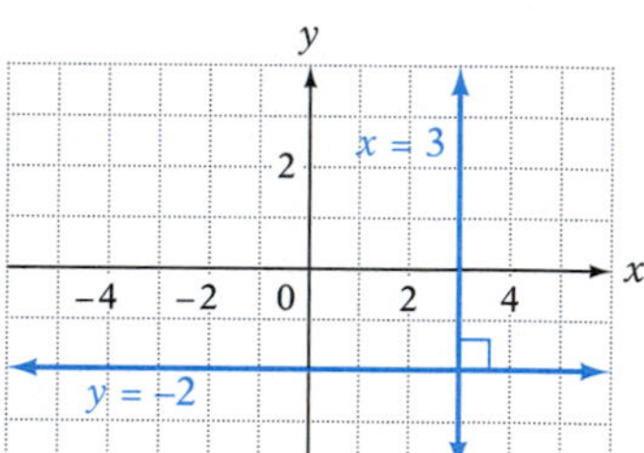

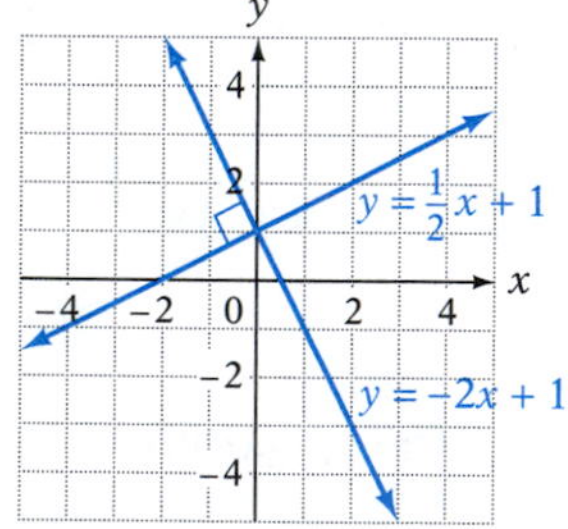

Slopes of Perpendicular Lines

If m_1 and m_2 are the slopes of two lines, neither of which is vertical, then the lines are perpendicular if and only if $m_1 \cdot m_2 = -1$.

A vertical line is perpendicular to a horizontal line.

Solving $m_1 \cdot m_2 = -1$ for m_1 gives $m_1 = -\frac{1}{m_2}$. This last equation states that the slopes of perpendicular lines are **negative reciprocals** of each other.

HOW TO Is the line that contains the points (4, 2) and (−2, 5) perpendicular to the line that contains the points (−4, 3) and (−3, 5)?

$m_1 = \frac{5-2}{-2-4} = \frac{3}{-6} = -\frac{1}{2}$ • **Find the slope of the line through (4, 2) and (−2, 5).**

$m_2 = \frac{5-3}{-3-(-4)} = \frac{2}{1} = 2$ • **Find the slope of the line through (−4, 3) and (−3, 5).**

$m_1 \cdot m_2 = -\frac{1}{2}(2) = -1$ • **Find the product of the two slopes.**

Because $m_1 \cdot m_2 = -1$, the lines are perpendicular.

Copyright © Houghton Mifflin Company. All rights reserved.

HOW TO Are the graphs of the lines whose equations are $3x + 4y = 8$ and $8x + 6y = 5$ perpendicular?

To determine whether the lines are perpendicular, solve each equation for y and find the slope of each line. Then use the equation $m_1 \cdot m_2 = -1$.

$$3x + 4y = 8 \qquad\qquad 8x + 6y = 5$$
$$4y = -3x + 8 \qquad\qquad 6y = -8x + 5$$
$$y = -\frac{3}{4}x + 2 \qquad\qquad y = -\frac{4}{3}x + \frac{5}{6}$$
$$m_1 = -\frac{3}{4} \qquad\qquad m_2 = -\frac{4}{3}$$
$$m_1 \cdot m_2 = \left(-\frac{3}{4}\right)\left(-\frac{4}{3}\right) = 1$$

Because $m_1 \cdot m_2 = 1 \neq -1$, the lines are not perpendicular.

HOW TO Find the equation of the line that contains the point $(-2, 1)$ and is perpendicular to the line $y = -\frac{2}{3}x + 2$.

The slope of the given line is $-\frac{2}{3}$. The slope of the line perpendicular to the given line is the negative reciprocal of $-\frac{2}{3}$, which is $\frac{3}{2}$. Substitute this slope and the coordinates of the given point, $(-2, 1)$, into the point-slope formula.

$$y - y_1 = m(x - x_1)$$ • The point-slope formula

$$y - 1 = \frac{3}{2}[x - (-2)]$$ • $m = \frac{3}{2}, (x_1, y_1) = (-2, 1)$

$$y - 1 = \frac{3}{2}x + 3$$ • Simplify.

$$y = \frac{3}{2}x + 4$$ • Write the equation in the form $y = mx + b$.

The equation of the perpendicular line is $y = \frac{3}{2}x + 4$.

HOW TO Find the equation of the line that contains the point $(3, -4)$ and is perpendicular to the line $2x - y = -3$.

$$2x - y = -3$$
$$-y = -2x - 3$$ • Determine the slope of the given line by solving the equation for y.
$$y = 2x + 3$$ • The slope is 2.

The slope of the line perpendicular to the given line is $-\frac{1}{2}$, the negative reciprocal of 2. Now use the point-slope formula to find the equation of the line.

$$y - y_1 = m(x - x_1)$$ • The point-slope formula

$$y - (-4) = -\frac{1}{2}(x - 3)$$ • $m = -\frac{1}{2}, (x_1, y_1) = (3, -4)$

$$y + 4 = -\frac{1}{2}x + \frac{3}{2}$$ • Simplify.

$$y = -\frac{1}{2}x - \frac{5}{2}$$ • Write the equation in the form $y = mx + b$.

The equation of the perpendicular line is $y = -\frac{1}{2}x - \frac{5}{2}$.

Copyright © Houghton Mifflin Company. All rights reserved.

Example 1

Is the line that contains the points (−4, 2) and (1, 6) parallel to the line that contains the points (2, −4) and (7, 0)?

Solution

$$m_1 = \frac{6-2}{1-(-4)} = \frac{4}{5}$$ • $(x_1, y_1) = (-4, 2)$, $(x_2, y_2) = (1, 6)$

$$m_2 = \frac{0-(-4)}{7-2} = \frac{4}{5}$$ • $(x_1, y_1) = (2, -4)$, $(x_2, y_2) = (7, 0)$

$$m_1 = m_2 = \frac{4}{5}$$

The lines are parallel.

You Try It 1

Is the line that contains the points (−2, −3) and (7, 1) perpendicular to the line that contains the points (4, 1) and (6, −5)?

Your solution

Example 2

Are the lines $4x - y = -2$ and $x + 4y = -12$ perpendicular?

Solution

$$\begin{aligned} 4x - y &= -2 \\ -y &= -4x - 2 \\ y &= 4x + 2 \\ m_1 &= 4 \end{aligned}$$

$$\begin{aligned} x + 4y &= -12 \\ 4y &= -x - 12 \\ y &= -\frac{1}{4}x - 3 \\ m_2 &= -\frac{1}{4} \end{aligned}$$

$$m_1 \cdot m_2 = 4\left(-\frac{1}{4}\right) = -1$$

The lines are perpendicular.

You Try It 2

Are the lines $5x + 2y = 2$ and $5x + 2y = -6$ parallel?

Your solution

Example 3

Find the equation of the line that contains the point (3, −1) and is parallel to the line $3x - 2y = 4$.

Solution

$$\begin{aligned} 3x - 2y &= 4 \\ -2y &= -3x + 4 \\ y &= \frac{3}{2}x - 2 \end{aligned}$$ • $m = \frac{3}{2}$

$$y - y_1 = m(x - x_1)$$

$$y - (-1) = \frac{3}{2}(x - 3)$$ • $(x_1, y_1) = (3, -1)$

$$\begin{aligned} y + 1 &= \frac{3}{2}x - \frac{9}{2} \\ y &= \frac{3}{2}x - \frac{11}{2} \end{aligned}$$

The equation of the line is $y = \frac{3}{2}x - \frac{11}{2}$.

You Try It 3

Find the equation of the line that contains the point (−2, 2) and is perpendicular to the line $x - 4y = 3$.

Your solution

Solutions on p. S14

Copyright © Houghton Mifflin Company. All rights reserved.

4.6 Exercises

Objective A To find parallel and perpendicular lines

1. Explain how to determine whether the graphs of two lines are parallel.

2. Explain how to determine whether the graphs of two lines are perpendicular.

3. The slope of a line is -5. What is the slope of any line parallel to this line?

4. The slope of a line is $\frac{3}{2}$. What is the slope of any line parallel to this line?

5. The slope of a line is 4. What is the slope of any line perpendicular to this line?

6. The slope of a line is $-\frac{4}{5}$. What is the slope of any line perpendicular to this line?

7. Is the line $x = -2$ perpendicular to the line $y = 3$?

8. Is the line $y = \frac{1}{2}$ perpendicular to the line $y = -4$?

9. Is the line $x = -3$ parallel to the line $y = \frac{1}{3}$?

10. Is the line $x = 4$ parallel to the line $x = -4$?

11. Is the line $y = \frac{2}{3}x - 4$ parallel to the line $y = -\frac{3}{2}x - 4$?

12. Is the line $y = -2x + \frac{2}{3}$ parallel to the line $y = -2x + 3$?

13. Is the line $y = \frac{4}{3}x - 2$ perpendicular to the line $y = -\frac{3}{4}x + 2$?

14. Is the line $y = \frac{1}{2}x + \frac{3}{2}$ perpendicular to the line $y = -\frac{1}{2}x + \frac{3}{2}$?

15. Are the lines $2x + 3y = 2$ and $2x + 3y = -4$ parallel?

16. Are the lines $2x - 4y = 3$ and $2x + 4y = -3$ parallel?

17. Are the lines $x - 4y = 2$ and $4x + y = 8$ perpendicular?

18. Are the lines $4x - 3y = 2$ and $4x + 3y = -7$ perpendicular?

19. Is the line that contains the points (3, 2) and (1, 6) parallel to the line that contains the points (−1, 3) and (−1, −1)?

20. Is the line that contains the points (4, −3) and (2, 5) parallel to the line that contains the points (−2, −3) and (−4, 1)?

21. Is the line that contains the points (−3, 2) and (4, −1) perpendicular to the line that contains the points (1, 3) and (−2, −4)?

22. Is the line that contains the points (−1, 2) and (3, 4) perpendicular to the line that contains the points (−1, 3) and (−4, 1)?

Copyright © Houghton Mifflin Company. All rights reserved.

23. Find the equation of the line containing the point $(-2, -4)$ and parallel to the line $2x - 3y = 2$.

24. Find the equation of the line containing the point $(3, 2)$ and parallel to the line $3x + y = -3$.

25. Find the equation of the line containing the point $(4, 1)$ and perpendicular to the line $y = -3x + 4$.

26. Find the equation of the line containing the point $(2, -5)$ and perpendicular to the line $y = \frac{5}{2}x - 4$.

27. Find the equation of the line containing the point $(-1, -3)$ and perpendicular to the line $3x - 5y = 2$.

28. Find the equation of the line containing the point $(-1, 3)$ and perpendicular to the line $2x + 4y = -1$.

APPLYING THE CONCEPTS

Physics For Exercises 29 and 30, suppose that a ball is being twirled at the end of a string and that the center of rotation is the origin of a coordinate system. If the string breaks, the initial path of the ball is on a line that is perpendicular to the radius of the circle.

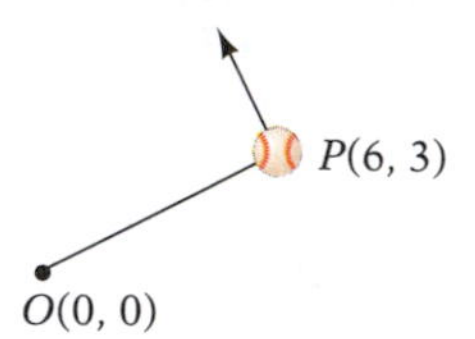

29. Suppose the string breaks when the ball is at the point whose coordinates are $P(6, 3)$. Find the equation of the line on which the initial path lies.

30. Suppose the string breaks when the ball is at the point whose coordinates are $P(2, 8)$. Find the equation of the line on which the initial path lies.

31. If the graphs of $A_1x + B_1y = C_1$ and $A_2x + B_2y = C_2$ are parallel, express $\frac{A_1}{B_1}$ in terms of A_2 and B_2.

32. If the graphs of $A_1x + B_1y = C_1$ and $A_2x + B_2y = C_2$ are perpendicular, express $\frac{A_1}{B_1}$ in terms of A_2 and B_2.

33. The graphs of $y = -\frac{1}{2}x + 2$ and $y = \frac{2}{3}x - 5$ intersect at the point whose coordinates are $(6, -1)$. Find the equation of a line whose graph intersects the graphs of the given lines to form a right triangle. (*Hint:* There is more than one answer to this question.)

34. **Geometry** A theorem from geometry states that a line passing through the center of a circle and through a point P on the circle is perpendicular to the tangent line at P. (See the figure at the right.)
   a. If the coordinates of P are $(5, 4)$ and the coordinates of C are $(3, 2)$, what is the equation of the tangent line?
   b. What is the x-intercept of the tangent line?
   c. What is the y-intercept of the tangent line?

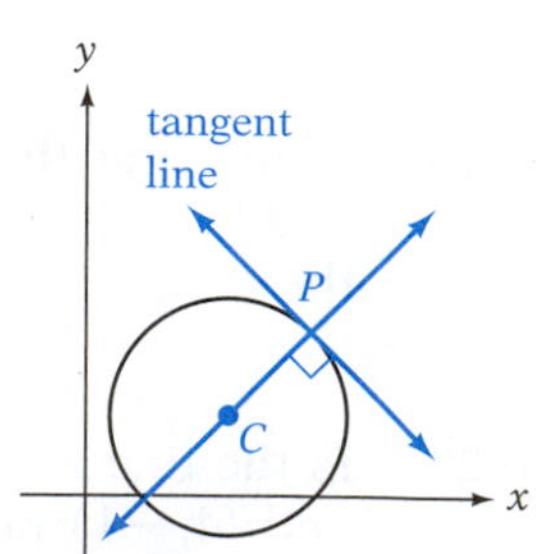

Copyright © Houghton Mifflin Company. All rights reserved.

4.7 Inequalities in Two Variables

Objective A

To graph the solution set of an inequality in two variables

The graph of the linear equation $y = x - 1$ separates the plane into three sets: the set of points on the line, the set of points above the line, and the set of points below the line.

The point whose coordinates are (2, 1) is a solution of $y = x - 1$ and is a point on the line.

The point whose coordinates are (2, 4) is a solution of $y > x - 1$ and is a point above the line.

The point whose coordinates are (2, −2) is a solution of $y < x - 1$ and is a point below the line.

The set of points on the line are the solutions of the equation $y = x - 1$. The set of points above the line are the solutions of the inequality $y > x - 1$. These points form a **half-plane.** The set of points below the line are solutions of the inequality $y < x - 1$. These points also form a half-plane.

An inequality of the form $y > mx + b$ or $Ax + By > C$ is a **linear inequality in two variables.** (The inequality symbol $>$ could be replaced by $\geq$, $<$, or $\leq$.) The solution set of a linear inequality in two variables is a half-plane.

The following illustrates the procedure for graphing the solution set of a linear inequality in two variables.

TAKE NOTE

When solving the inequality at the right for y, both sides of the inequality are divided by −4, so the inequality symbol must be reversed.

HOW TO Graph the solution set of $3x - 4y < 12$.

$$3x - 4y < 12$$
$$-4y < -3x + 12$$
$$y > \frac{3}{4}x - 3$$

• **Solve the inequality for *y*.**

Change the inequality $y > \frac{3}{4}x - 3$ to the equality $y = \frac{3}{4}x - 3$, and graph the line.

If the inequality contains $\leq$ or $\geq$, the line belongs to the solution set and is shown by a *solid line.* If the inequality contains $<$ or $>$, the line is not part of the solution set and is shown by a *dotted line.*

If the inequality contains $>$ or $\geq$, shade the upper half-plane. If the inequality contains $<$ or $\leq$, shade the lower half-plane.

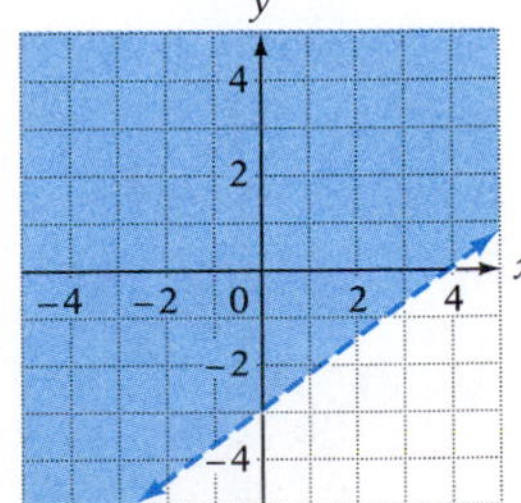

As a check, use the ordered pair (0, 0) to determine whether the correct region of the plane has been shaded. If (0, 0) is a solution of the inequality, then (0, 0) should be in the shaded region. If (0, 0) is not a solution of the inequality, then (0, 0) should not be in the shaded region.

TAKE NOTE

As shown below, (0, 0) is a solution of the inequality in the example at the right.

$$y > \frac{3}{4}x - 3$$
$$0 > \frac{3}{4}(0) - 3$$
$$0 > 0 - 3$$
$$0 > -3$$

Because (0, 0) is a solution of the inequality, (0, 0) should be in the shaded region. The solution set as graphed is correct.

Copyright © Houghton Mifflin Company. All rights reserved.

Integrating Technology

See the appendix Keystroke Guide: *Graphing Inequalities* for instructions on using a graphing calculator to graph the solution set of an inequality in two variables.

If the line passes through the point (0, 0), another point, such as (0, 1), must be used as a check.

From the graph of $y > \frac{3}{4}x - 3$, note that for a given value of x, more than one value of y can be paired with that value of x. For instance, (4, 1), (4, 3), (5, 1), and $\left(5, \frac{9}{4}\right)$ are all ordered pairs that belong to the graph. Because there are ordered pairs with the same first coordinate and different second coordinates, the inequality does not represent a function. The inequality is a relation but not a function.

Example 1 Graph the solution set of $x + 2y \le 4$.

Solution

$$x + 2y \le 4$$
$$2y \le -x + 4$$
$$y \le -\frac{1}{2}x + 2$$

Graph $y = -\frac{1}{2}x + 2$ as a solid line.

Shade the lower half-plane.

Check:

$y \le -\frac{1}{2}x + 2$

$0 \le -\frac{1}{2}(0) + 2$

$0 \le 0 + 2$

$0 \le 2$

The point (0, 0) should be in the shaded region.

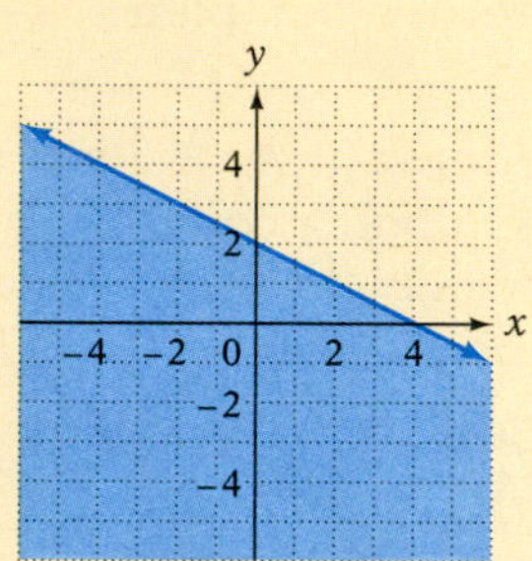

You Try It 1 Graph the solution set of $x + 3y > 6$.

Your solution

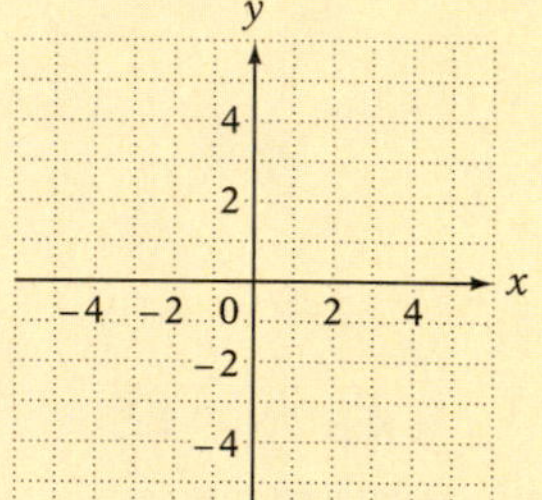

Example 2 Graph the solution set of $x \ge -1$.

Solution Graph $x = -1$ as a solid line.

Shade the half-plane to the right of the line.

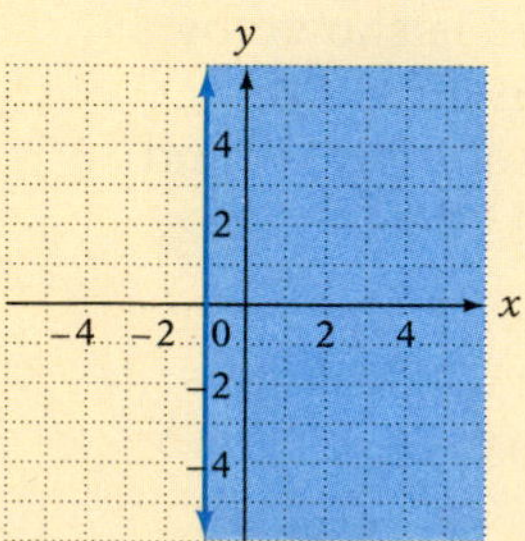

You Try It 2 Graph the solution set of $y < 2$.

Your solution

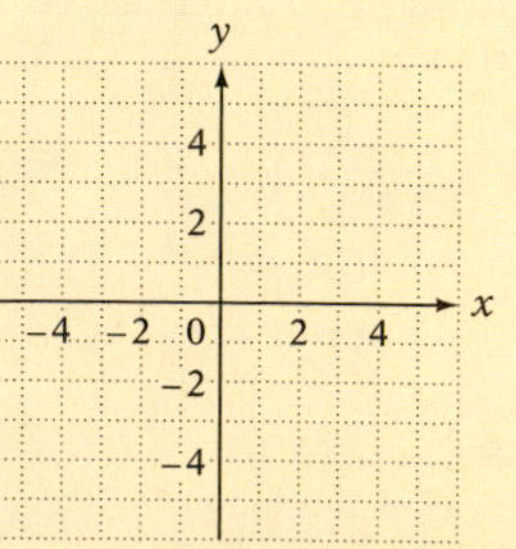

Solutions on pp. S14–S15

Copyright © Houghton Mifflin Company. All rights reserved.

4.7 Exercises

Objective A **To graph the solution set of an inequality in two variables**

1. What is a half-plane?

2. Explain a method you can use to check that the graph of a linear inequality in two variables has been shaded correctly.

3. Is (0, 0) a solution of $y > 2x - 7$?

4. Is (0, 0) a solution of $y < 5x + 3$?

5. Is (0, 0) a solution of $y \le -\frac{2}{3}x - 8$?

6. Is (0, 0) a solution of $y \ge -\frac{3}{4}x + 9$?

For Exercises 7 to 24, graph the solution set.

7. $3x - 2y \ge 6$

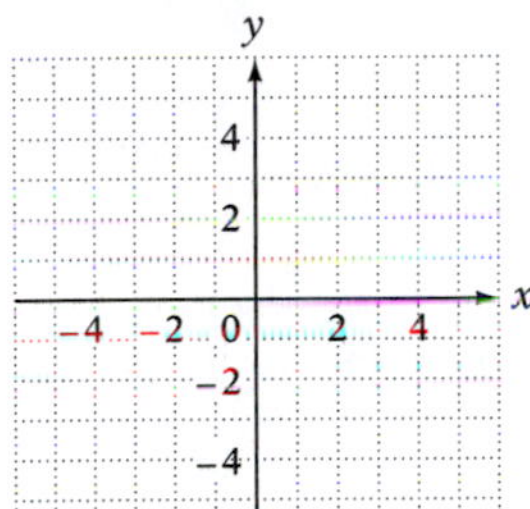

8. $4x - 3y \le 12$

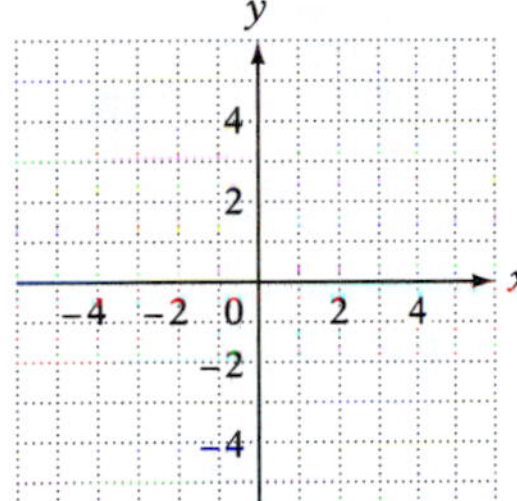

9. $x + 3y < 4$

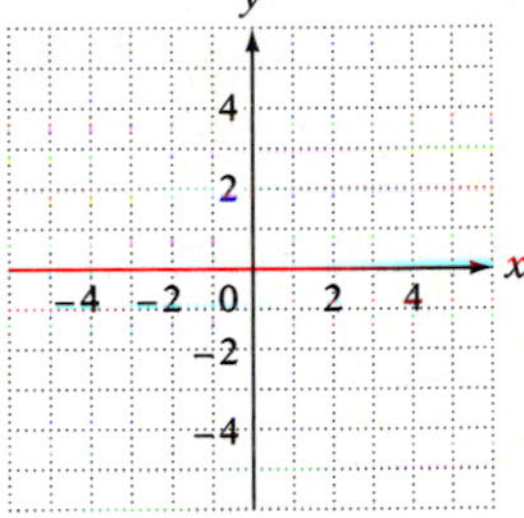

10. $3x - 5y > 15$

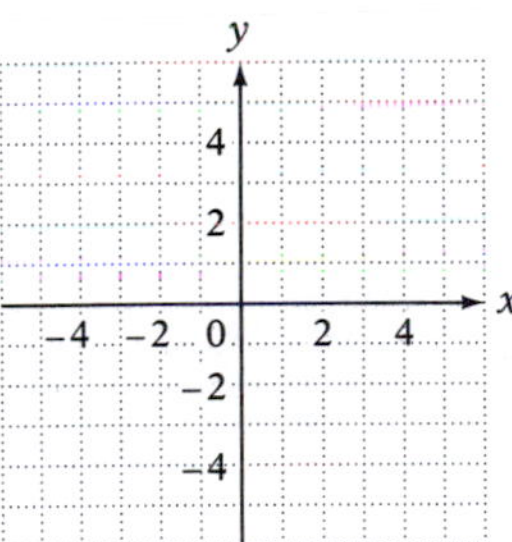

11. $4x - 5y > 10$

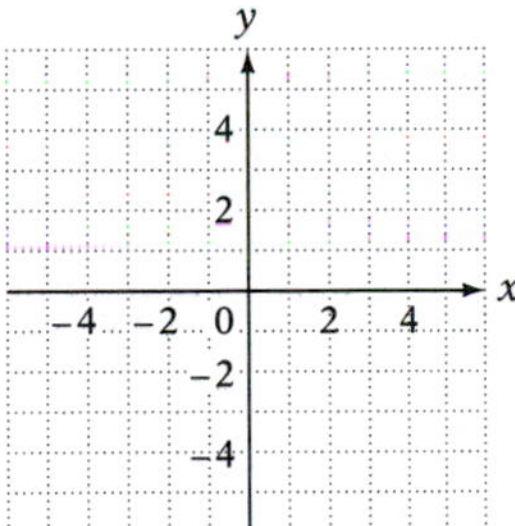

12. $4x + 3y < 9$

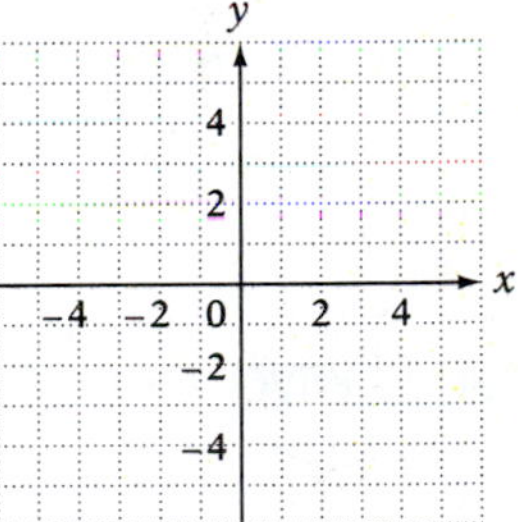

13. $x + 3y < 6$

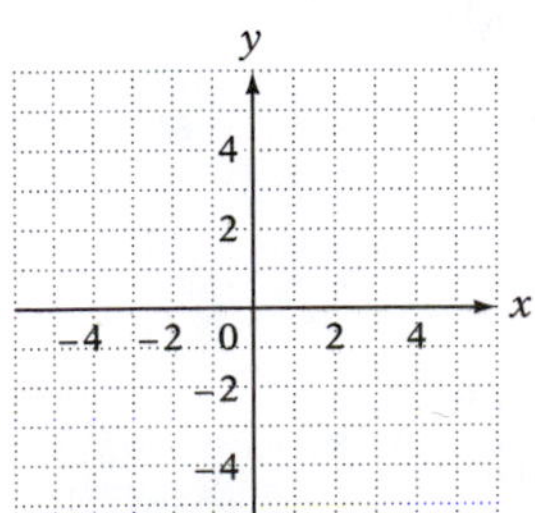

14. $2x - 5y \le 10$

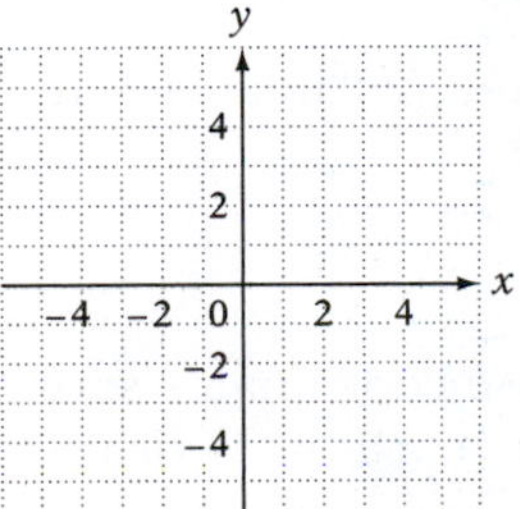

15. $2x + 3y \ge 6$

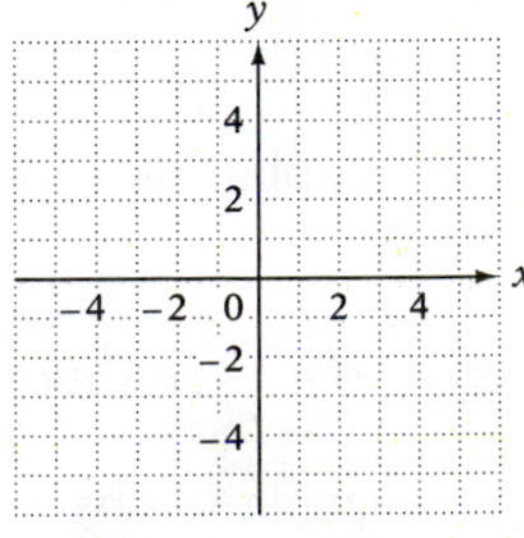

Copyright © Houghton Mifflin Company. All rights reserved.

16. $3x + 2y < 4$

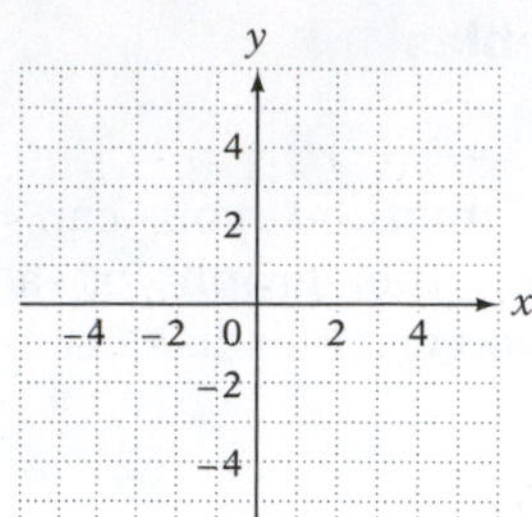
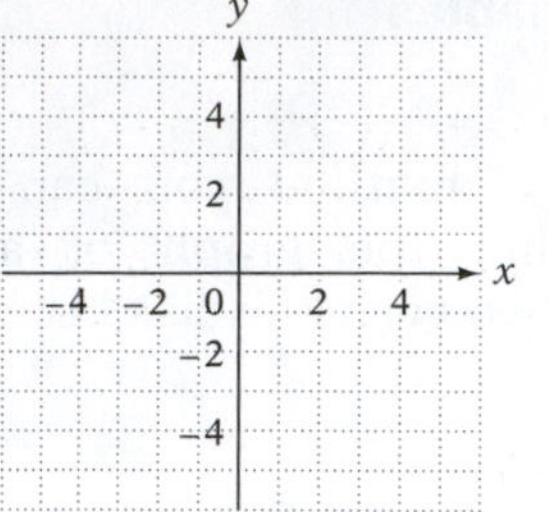

17. $-x + 2y > -8$

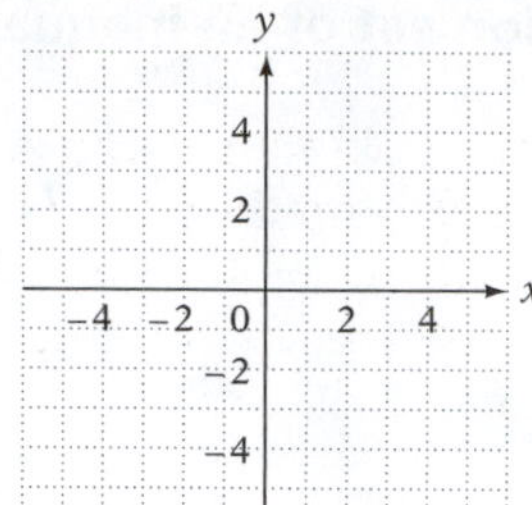

18. $-3x + 2y > 2$

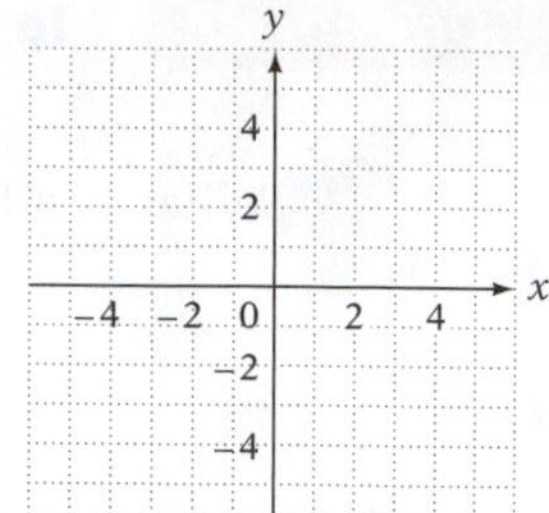

19. $y - 4 < 0$

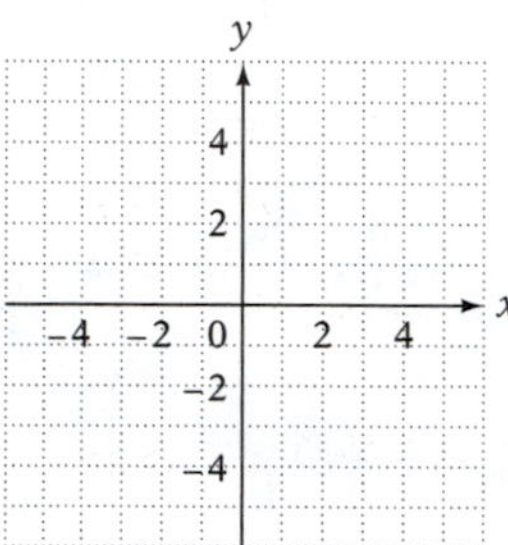

20. $x + 2 \geq 0$

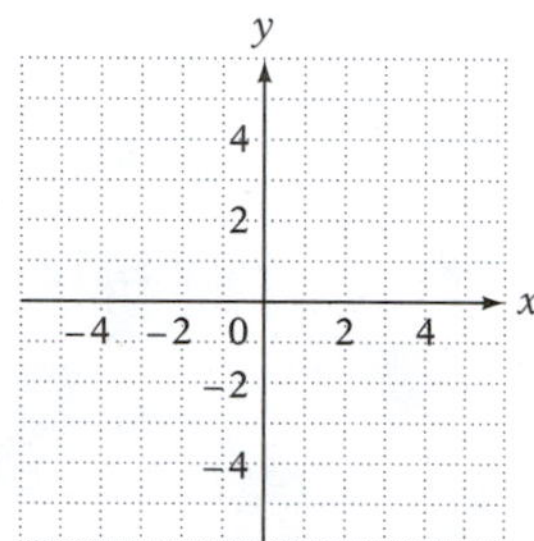

21. $6x + 5y < 15$

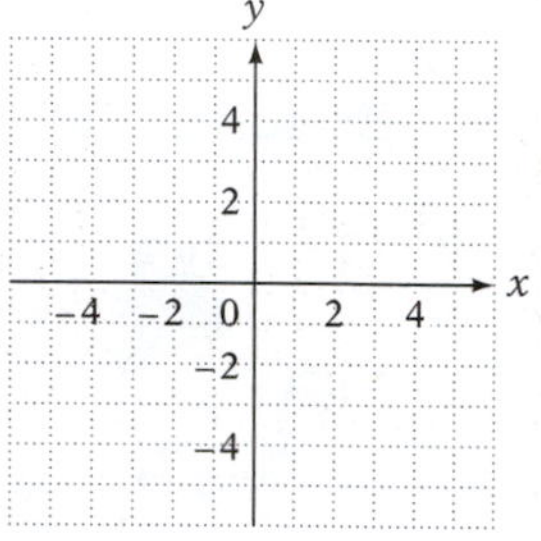

22. $3x - 5y < 10$

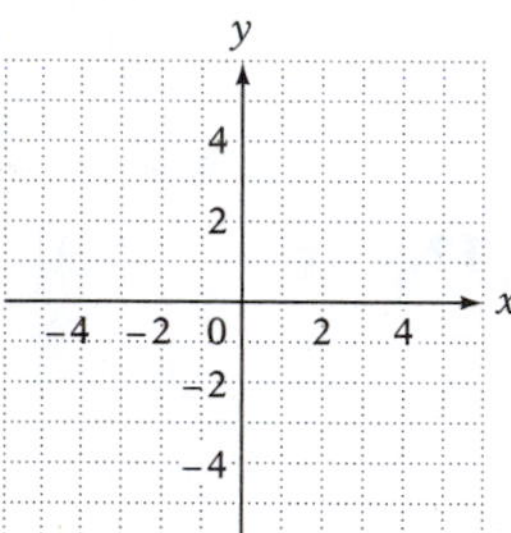

23. $-5x + 3y \geq -12$

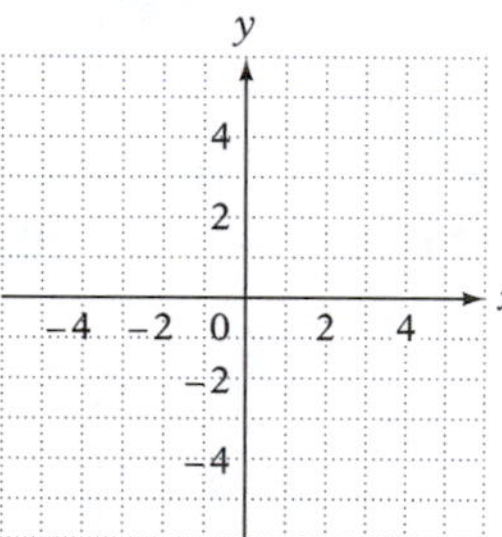

24. $3x + 4y \geq 12$

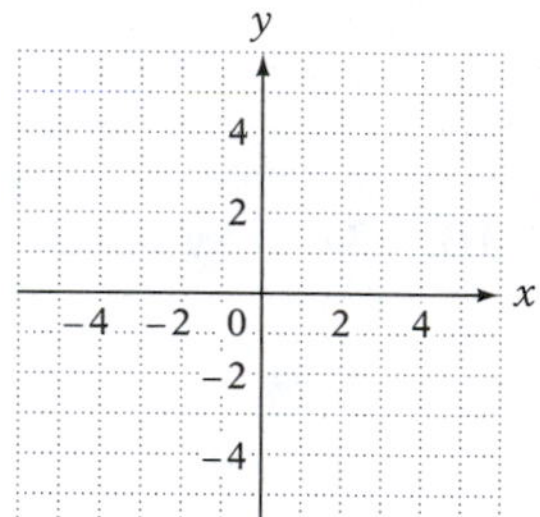

APPLYING THE CONCEPTS

25. Does the inequality $y < 3x - 1$ represent a function? Explain your answer.

26. Are there any points whose coordinates satisfy both $y \leq x + 3$ and $y \geq -\frac{1}{2}x + 1$? If so, give the coordinates of three such points. If not, explain why not.

27. Are there any points whose coordinates satisfy both $y \leq x - 1$ and $y \geq x + 2$? If so, give the coordinates of three such points. If not, explain why not.

Copyright © Houghton Mifflin Company. All rights reserved.

Focus on Problem Solving

Find a Pattern

One approach to problem solving is to try to find a pattern. Karl Friedrich Gauss supposedly used this method to solve a problem that was given to his math class when he was in elementary school. As the story goes, his teacher wanted to grade some papers while the class worked on a math problem. The problem given to the class was to find the sum

$$1 + 2 + 3 + 4 + \cdots + 100$$

Gauss quickly solved the problem by seeing a pattern. Here is what he saw.

101
101
101
101

$$1 + 2 + 3 + 4 + \cdots + 97 + 98 + 99 + 100$$

Note that

$$1 + 100 = 101$$
$$2 + 99 = 101$$
$$3 + 98 = 101$$
$$4 + 97 = 101$$

Gauss noted that there were 50 sums of 101. Therefore, the sum of the first 100 natural numbers is

$$1 + 2 + 3 + 4 + \cdots + 97 + 98 + 99 + 100 = 50(101) = 5050$$

Point of Interest

Karl Friedrich Gauss

Karl Friedrich Gauss (1777–1855) has been called the "Prince of Mathematicians" by some historians. He applied his genius to many areas of mathematics and science. A unit of magnetism, the *gauss*, is named in his honor. Some types of electronic equipment (televisions, for instance) contain a *degausser* that controls magnetic fields.

Try to solve Exercises 1 to 6 by finding a pattern.

1. Find the sum $2 + 4 + 6 + \cdots + 96 + 98 + 100$.

2. Find the sum $1 + 3 + 5 + \cdots + 97 + 99 + 101$.

3. Find another method of finding the sum $1 + 3 + 5 + \cdots + 97 + 99 + 101$ given in the preceding exercise.

4. Find the sum $\frac{1}{1 \cdot 2} + \frac{1}{2 \cdot 3} + \frac{1}{3 \cdot 4} + \cdots + \frac{1}{49 \cdot 50}$.

   Hint: $\frac{1}{1 \cdot 2} = \frac{1}{2}$, $\frac{1}{1 \cdot 2} + \frac{1}{2 \cdot 3} = \frac{2}{3}$, $\frac{1}{1 \cdot 2} + \frac{1}{2 \cdot 3} + \frac{1}{3 \cdot 4} = \frac{3}{4}$

5. A *polynomial number* is a number that can be represented by arranging that number of dots in rows to form a geometric figure such as a triangle, square, pentagon, or hexagon. For instance, the first four *triangular* numbers, 3, 6, 10, and 15, are shown below. What are the next two triangular numbers?

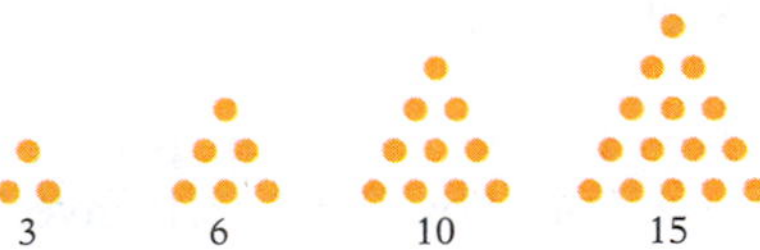

6. The following problem shows that checking a few cases does not always result in a conjecture that is true for *all* cases. Select any two points on a circle (see the drawing in the left margin) and draw a *chord*, a line connecting the points. The chord divides the circle into two regions. Now select three different points and draw chords connecting each of the three points with every other point. The chords divide the circle into four regions. Now select four points and connect each of the points with every other point. Make a conjecture as to the relationship between the number of regions and the number of points on the circle. Does your conjecture work for five points? six points?

2 points, 2 regions

3 points, 4 regions

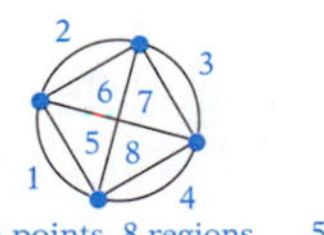
4 points, 8 regions

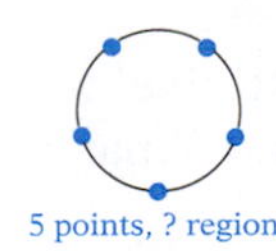
5 points, ? regions

Copyright © Houghton Mifflin Company. All rights reserved.

Projects and Group Activities

Introduction to Graphing Calculators

There are a variety of computer programs and calculators that can graph an equation. A computer or graphing calculator screen is divided into pixels. Depending on the computer or calculator, there are approximately 6000 to 790,000 pixels available on the screen. The greater the number of pixels, the smoother the graph will appear. A portion of a screen is shown at the left. Each little rectangle represents one pixel.

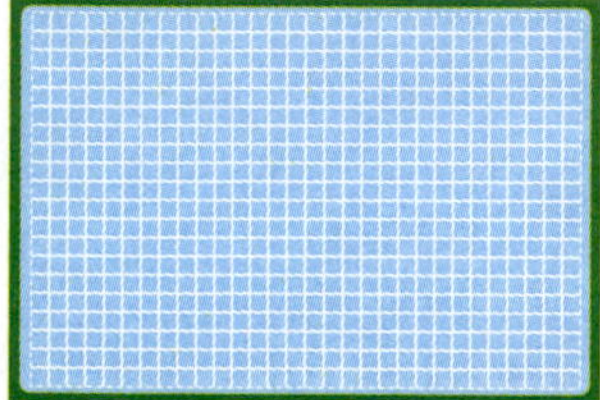

Calculator Screen

A graphing calculator draws a graph in a manner similar to the method we have used in this chapter. Values of x are chosen and ordered pairs calculated. Then a graph is drawn through those points by illuminating pixels (an abbreviation for "picture element") on the screen.

Ymax

Yscl

Xmin Xmax

Xscl

Ymin

Graphing utilities can display only a portion of the xy-plane, called a window. The window [Xmin, Xmax] by [Ymin, Ymax] consists of those points (x, y) that satisfy both of the following inequalities:

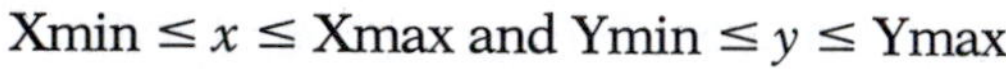

$$\text{Xmin} \leq x \leq \text{Xmax and Ymin} \leq y \leq \text{Ymax}$$

The user sets these values before a graph is drawn.

The numbers Xscl and Yscl are the distances between the tick marks that are drawn on the x- and y-axes. If you do not want tick marks on the axes, set Xscl = 0 and Yscl = 0.

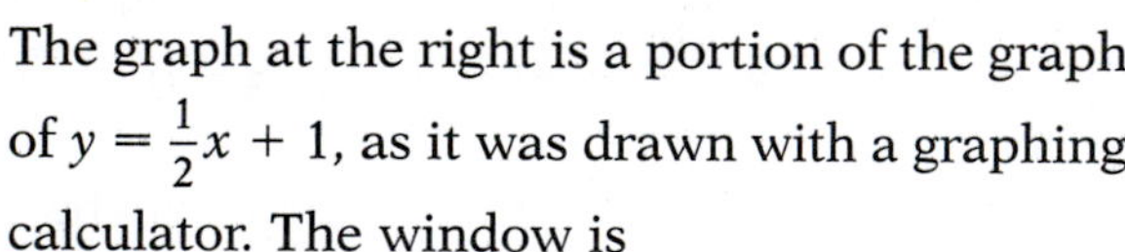

The graph at the right is a portion of the graph of $y = \frac{1}{2}x + 1$, as it was drawn with a graphing calculator. The window is

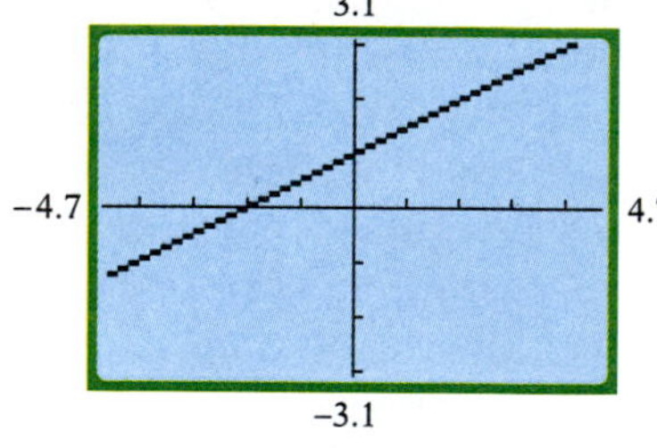

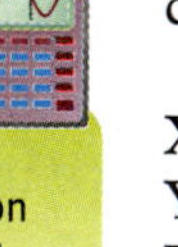

Integrating Technology

See the Keystroke Guide: *Windows* for instructions on changing the viewing window.

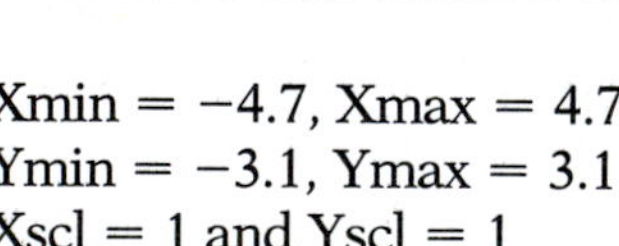

Xmin = −4.7, Xmax = 4.7
Ymin = −3.1, Ymax = 3.1
Xscl = 1 and Yscl = 1

This is often written using the notation [−4.7, 4.7] by [−3.1, 3.1]. The window [−4.7, 4.7] by [−3.1, 3.1] gives "nice" coordinates in the sense that each time the ◀ or the ▶ is pressed, the change in x is 0.1. The reason for this is that the horizontal distance from the middle of the first pixel to the middle of the last pixel is 94 units. By using Xmin = −4.7 and Xmax = 4.7, we have

$$\text{Change in } x = \frac{\text{Xmax} - \text{Xmin}}{94} = \frac{4.7 - (-4.7)}{94} = \frac{9.4}{94} = 0.1$$

Similarly, the vertical distance from the middle of the first pixel to the middle of the last pixel is 62 units. Therefore, using Ymin = −3.1 and Ymax = 3.1 will give nice coordinates in the vertical direction.

Copyright © Houghton Mifflin Company. All rights reserved.

Study Tip

Three important features of this text that can be used to prepare for a test are the:
- Chapter Summary
- Chapter Review Exercises
- Chapter Test

See *AIM for Success*, page xxx.

Graph the equations in Exercises 1 to 6 by using a graphing calculator.

1. $y = 2x + 1$ For $2x$, you may enter $2 \times x$ or just $2x$. The times sign $\times$ is not necessary on many graphing calculators.

2. $y = -x + 2$ Many calculators use the $(-)$ key to enter a negative sign.

3. $3x + 2y = 6$ Solve for y. Then enter the equation.

4. $y = 50x$ You must adjust the viewing window. Try the window $[-4.7, 4.7]$ by $[-250, 250]$ with Yscl = 50.

5. $y = \frac{2}{3}x - 3$ You may enter $\frac{2}{3}x$ as $2x/3$ or $(2/3)x$. Although entering $2/3x$ works on some calculators, it is not recommended.

6. $4x + 3y = 75$ You must adjust the viewing window.

Chapter 4 Summary

Key Words	Examples

A *rectangular coordinate system* is formed by two number lines, one horizontal and one vertical, that intersect at the zero point of each line. The point of intersection is called the *origin*. The number lines that make up a rectangular coordinate system are called *coordinate axes*. A rectangular coordinate system divides the plane into four regions called *quadrants*. [4.1A, p. 219]

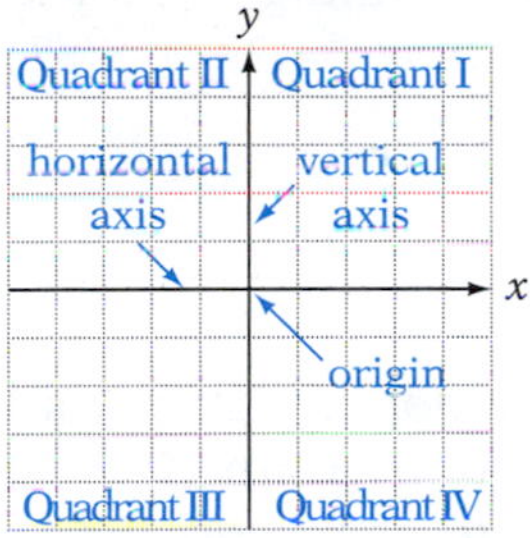

An *ordered pair* (x, y) is used to locate a point in a rectangular coordinate system. The first number of the pair measures a horizontal distance and is called the *abscissa* or *x-coordinate*. The second number of the pair measures a vertical distance and is called the *ordinate* or *y-coordinate*. The *coordinates* of the point are the numbers in the ordered pair associated with the point. To *graph*, or *plot*, a point in the plane, place a dot at the location given by the ordered pair. The *graph of an ordered pair* is the dot drawn at the coordinates of the point in the plane. [4.1A, 4.1B, pp. 219–221]

(3, 4) is an ordered pair.
3 is the abscissa.
4 is the ordinate.
The graph of (3, 4) is shown below.

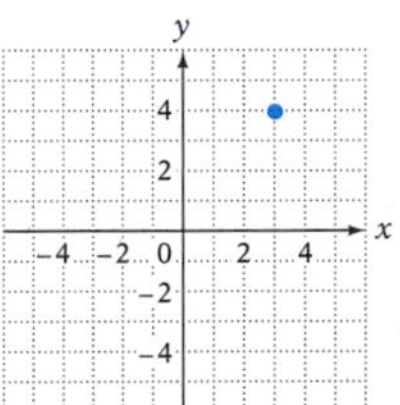

An equation of the form $y = mx + b$, where m and b are constants, is an equation in two variables. *A solution of an equation in two variables* is an ordered pair (x, y) whose coordinates make the equation a true statement. [4.1B, p. 221]

$y = 3x + 2$ is an equation in two variables. Ordered-pair solutions of $y = 3x + 2$ are shown below, along with the graph of the equation.

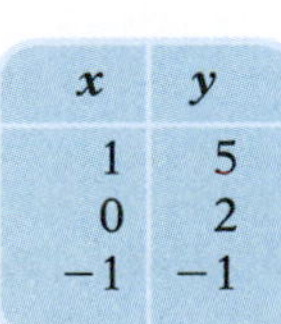

x	y
1	5
0	2
−1	−1

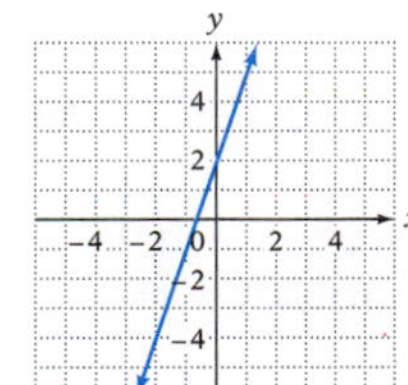

Copyright © Houghton Mifflin Company. All rights reserved.

A *scatter diagram* is a graph of ordered-pair data. [4.1C, p. 224]

A *relation* is any set of ordered pairs. [4.2A, p. 229]

{(2, 3), (2, 4), (3, 4), (5, 7)}

A *function* is a relation in which no two ordered pairs have the same first coordinate and different second coordinates. The *domain* of a function is the set of the first coordinates of all the ordered pairs of the function. The *range* is the set of the second coordinates of all the ordered pairs of the function. [4.2A, p. 230]

{(2, 3), (3, 5), (5, 7), (6, 9)}
The domain is {2, 3, 5, 6}.
The range is {3, 5, 7, 9}.

Functional notation is used for those equations that represent functions. For the equation at the right, x is the *independent variable* and y is the *dependent variable*. The symbol $f(x)$ is the *value of the function* and represents the value of the dependent variable for a given value of the independent variable. [4.2A, pp. 230–232]

In functional notation, $y = 3x + 7$ is written as $f(x) = 3x + 7$.

The process of determining $f(x)$ for a given value of x is called *evaluating the function*. [4.2A, p. 232]

Evaluate $f(x) = 2x - 3$ when $x = 4$.

$f(x) = 2x - 3$
$f(4) = 2(4) - 3$
$f(4) = 5$

The *graph of a function* is a graph of the ordered pairs (x, y) that belong to the function. A function that can be written in the form $f(x) = mx + b$ (or $y = mx + b$) is a *linear function* because its graph is a straight line. [4.3A, p. 241]

$f(x) = -\frac{2}{3}x + 3$ is an example of a linear function. In this case, $m = -\frac{2}{3}$ and $b = 3$.

The point at which a graph crosses the x-axis is called the *x-intercept*, and the point at which a graph crosses the y-axis is called the *y-intercept*. [4.3C, p. 246]

The x-intercept of $x + y = 4$ is (4, 0).
The y-intercept of $x + y = 4$ is (0, 4).

The *slope* of a line is a measure of the slant, or tilt, of the line. The symbol for slope is m. A line that slants upward to the right has a *positive slope*, and a line that slants downward to the right has a *negative slope*. A horizontal line has *zero slope*. The slope of a vertical line is *undefined*. [4.4A, pp. 253–254]

The line $y = 2x - 3$ has a slope of 2 and slants upward to the right.
The line $y = -5x + 2$ has a slope of -5 and slants downward to the right.
The line $y = 4$ has a slope of 0.

An inequality of the form $y > mx + b$ or of the form $Ax + By > C$ is a *linear inequality in two variables*. (The symbol $>$ can be replaced by $\geq$, $<$, or $\leq$.) The solution set of an inequality in two variables is a *half-plane*. [4.7A, p. 279]

$4x - 3y < 12$ and $y \geq 2x + 6$ are linear inequalities in two variables.

Essential Rules and Procedures

Examples

Graph of $y = b$ [4.3B, p. 244]
The graph of $y = b$ is a horizontal line passing through the point $(0, b)$.

The graph of $y = -5$ is a horizontal line passing through the point $(0, -5)$.

Copyright © Houghton Mifflin Company. All rights reserved.

Graph of a Constant Function [4.3B, p. 244]
A function given by $f(x) = b$, where b is a constant, is a *constant function*. The graph of the constant function is a horizontal line passing through $(0, b)$.

The graph of $f(x) = -5$ is a horizontal line passing through the point $(0, -5)$. Note that this is the same as the graph of $y = -5$.

Graph of $x = a$ [4.3B, p. 245]
The graph of $x = a$ is a vertical line passing through the point $(a, 0)$.

The graph of $x = 4$ is a vertical line passing through the point $(4, 0)$.

Finding Intercepts of Graphs of Linear Equations [4.3C, p. 246]
To find the x-intercept, let $y = 0$.
To find the y-intercept, let $x = 0$.
For any equation of the form $y = mx + b$, the y-intercept is $(0, b)$.

$3x + 4y = 12$

Let $y = 0$:
$$3x + 4(0) = 12$$
$$3x = 12$$
$$x = 4$$
The x-intercept is $(4, 0)$.

Let $x = 0$:
$$3(0) + 4y = 12$$
$$4y = 12$$
$$y = 3$$
The y-intercept is $(0, 3)$.

Slope Formula [4.4A, p. 253]
The slope of the line containing the two points $P_1(x_1, y_1)$ and $P_2(x_2, y_2)$ is given by $m = \frac{y_2 - y_1}{x_2 - x_1}$, $x_1 \neq x_2$.

$(x_1, y_1) = (-3, 2)$, $(x_2, y_2) = (1, 4)$
$$m = \frac{y_2 - y_1}{x_2 - x_1} = \frac{4 - 2}{1 - (-3)} = \frac{1}{2}$$
The slope of the line through the points $(-3, 2)$ and $(1, 4)$ is $\frac{1}{2}$.

Slope-Intercept Form of a Straight Line [4.4B, p. 257]
The equation $y = mx + b$ is called the *slope-intercept form* of a straight line. The slope of the line is m, the coefficient of x. The y-intercept is $(0, b)$.

For the equation $y = -3x + 2$, the slope is -3 and the y-intercept is $(0, 2)$.

Point-Slope Formula [4.5A, p. 264]
Let m be the slope of a line, and let (x_1, y_1) be the coordinates of a point on the line. The equation of the line can be found from the point-slope formula: $y - y_1 = m(x - x_1)$.

The equation of the line that passes through the point $(4, 2)$ and has slope -3 is
$$y - y_1 = m(x - x_1)$$
$$y - 2 = -3(x - 4)$$
$$y - 2 = -3x + 12$$
$$y = -3x + 14$$

Slopes of Parallel Lines [4.6A, p. 273]
Two nonvertical lines with slopes of m_1 and m_2 are parallel if and only if $m_1 = m_2$. Any two vertical lines are parallel.

$y = 3x - 4, \quad m_1 = 3$
$y = 3x + 2, \quad m_2 = 3$
Because $m_1 = m_2$, the lines are parallel.

Slopes of Perpendicular Lines [4.6A, p. 274]
If m_1 and m_2 are the slopes of two lines, neither of which is vertical, then the lines are perpendicular if and only if $m_1 \cdot m_2 = -1$. This states that the slopes of perpendicular lines are *negative reciprocals* of each other. A vertical line is perpendicular to a horizontal line.

$y = \frac{1}{2}x - 1, \quad m_1 = \frac{1}{2}$
$y = -2x + 2, \quad m_2 = -2$
Because $m_1 \cdot m_2 = -1$, the lines are perpendicular.

Copyright © Houghton Mifflin Company. All rights reserved.

Chapter 4 Review Exercises

1. Determine the ordered-pair solution of $y = \frac{x}{x - 2}$ that corresponds to $x = 4$.

2. Given $P(x) = 3x + 4$, evaluate $P(-2)$ and $P(a)$.

3. Graph the ordered-pair solutions of $y = 2x^2 - 5$ when $x = -2, -1, 0, 1$, and 2.

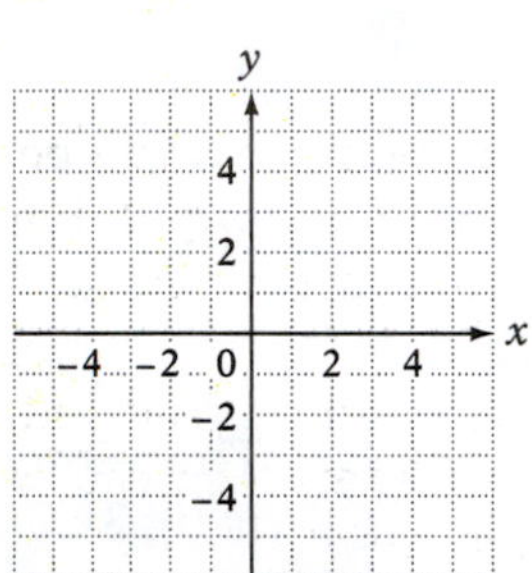

4. Draw a line through all the points with an abscissa of -3.

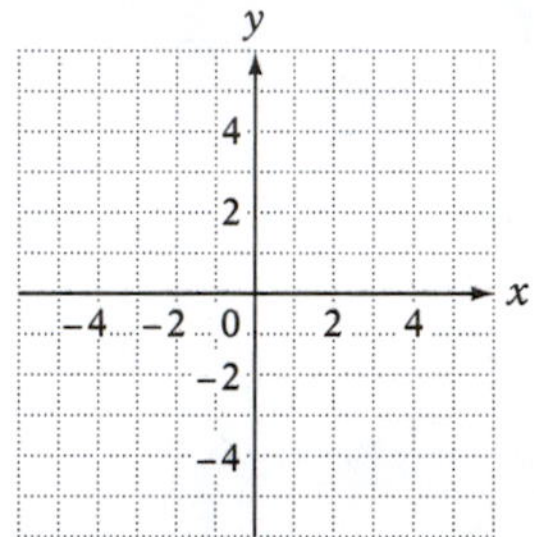

5. Find the range of $f(x) = x^2 + x - 1$ if the domain is $\{-2, -1, 0, 1, 2\}$.

6. Find the domain and range of the function $\{(-1, 0), (0, 2), (1, 2), (5, 4)\}$.

7. Is the line that passes through the points $(-2, 3)$ and $(3, 7)$ parallel to the line that passes through the points $(1, -4)$ and $(6, 0)$?

8. What value of x is excluded from the domain of $f(x) = \frac{x}{x + 4}$?

9. Find the x- and y-intercepts and graph $y = -\frac{2}{3}x - 2$.

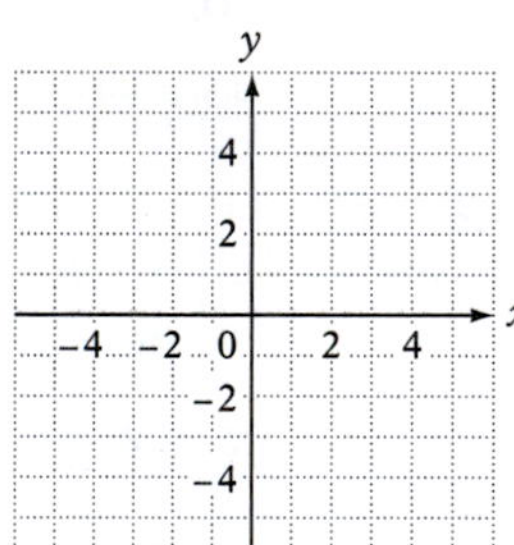

10. Graph $3x + 2y = -6$ by using the x- and y-intercepts.

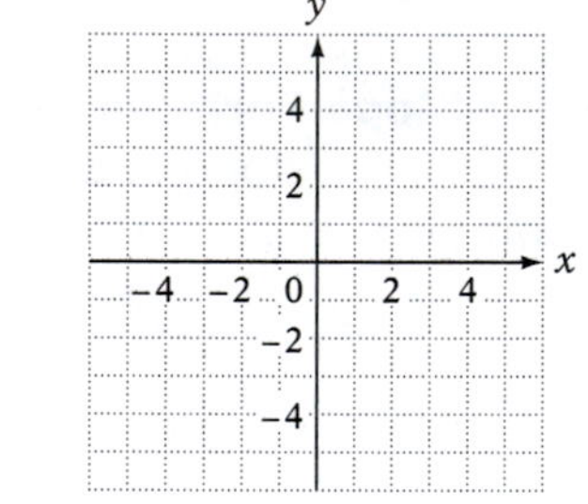

11. Graph: $y = -2x + 2$

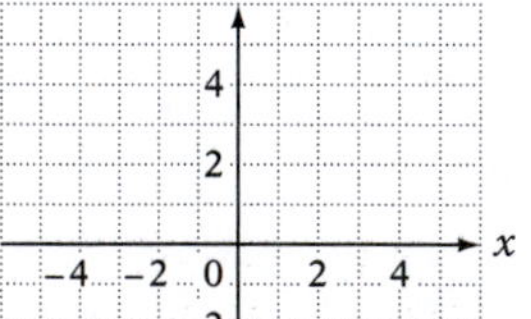

12. Graph: $4x - 3y = 12$

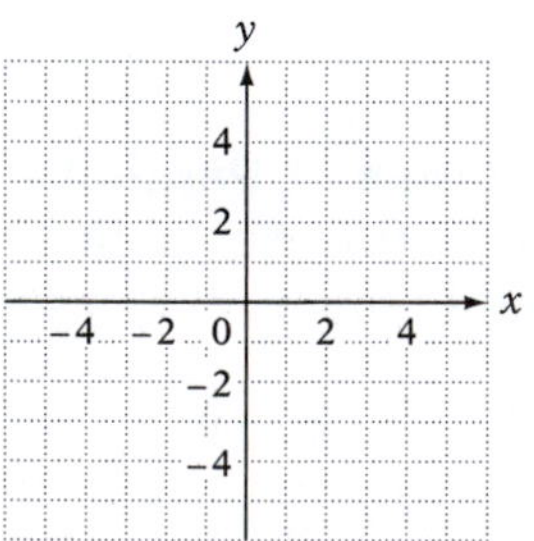

13. Find the slope of the line that contains the ordered pairs $(3, -2)$ and $(-1, 2)$.

14. Find the equation of the line that contains the ordered pair $(-3, 4)$ and has slope $\frac{5}{2}$.

Copyright © Houghton Mifflin Company. All rights reserved.

15. Draw a line through all points with an ordinate of −2.

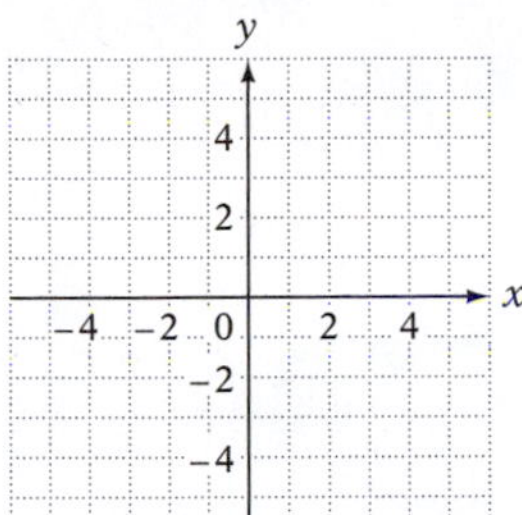

16. Graph the line that passes through the point (−2, 3) and has slope $-\frac{1}{4}$.

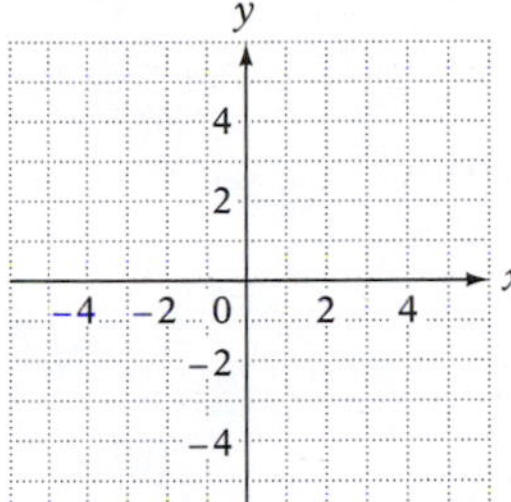

17. Find the range of $f(x) = x^2 - 2$ if the domain is {−2, −1, 0, 1, 2}.

18. **The Hospitality Industry** The manager of a hotel determines that 200 rooms will be occupied if the rate is \$95 per night. For each \$10 increase in the rate, 10 fewer rooms will be occupied.
   a. Determine a linear function that will predict the number of rooms that will be occupied at a given rate.
   b. Use the model to predict occupancy when the rate is \$120.

19. Find the equation of the line that contains the point (−2, 3) and is parallel to the line $y = -4x + 3$.

20. Find the equation of the line that contains the point (−2, 3) and is perpendicular to the line $y = -\frac{2}{5}x - 3$.

21. Graph $y = 1$.

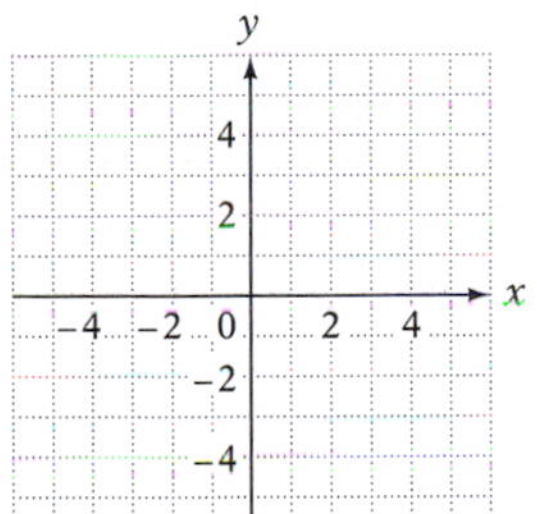

22. Graph $x = -1$.

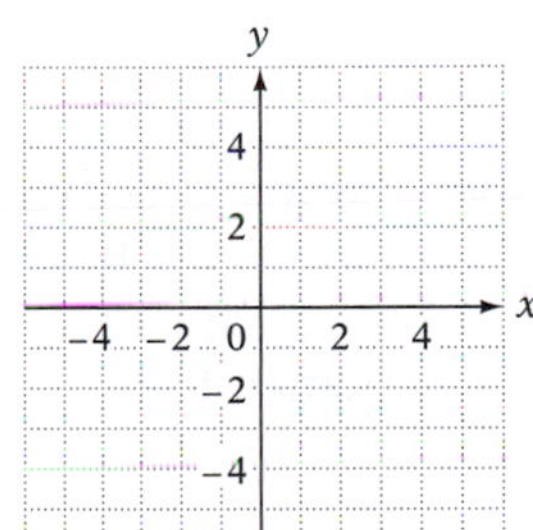

23. Find the equation of the line that contains the ordered pair (−3, 3) and has slope $-\frac{2}{3}$.

24. Find the equation of the line that contains the ordered pairs (−8, 2) and (4, 5).

25. Find the x- and y-intercepts of $3x - 2y = -12$.

26. Find the ordered-pair solution of $y = 3x - 4$ corresponding to $x = -1$.

27. **Life Expectancy** The table below shows the average number of years that women and men were estimated to live beyond age 65 in 1950, 1960, 1970, 1980, 1990, and 2000. (*Source:* 2000 Social Security Trustees Report) Figures are rounded to the nearest whole number. Graph the scatter diagram for these data.

Women, x	15	16	17	18	19	19
Men, y	13	13	13	14	15	16

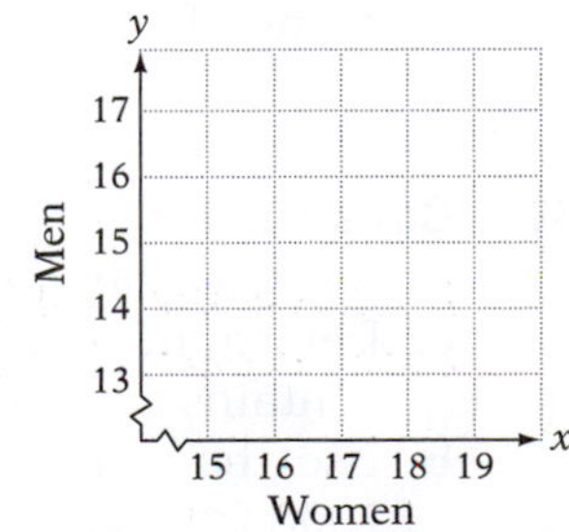

Copyright © Houghton Mifflin Company. All rights reserved.

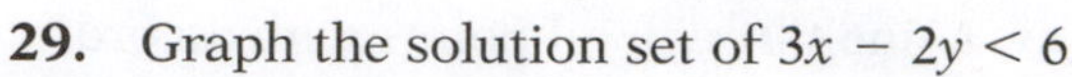

28. Graph the solution set of $y \geq 2x - 3$.

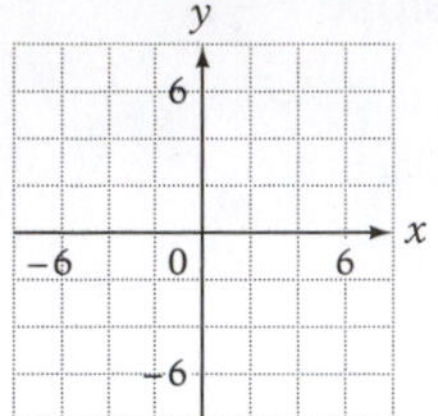

29. Graph the solution set of $3x - 2y < 6$.

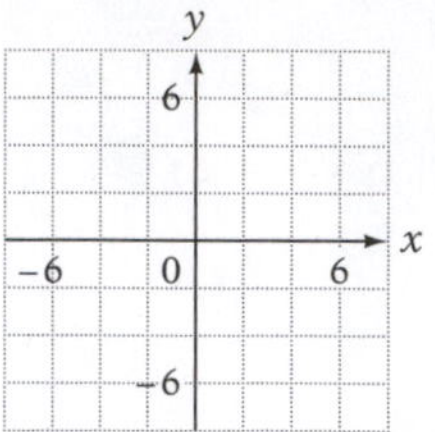

30. Find the equation of the line that contains the ordered pairs $(-2, 4)$ and $(4, -3)$.

31. Find the equation of the line that contains the ordered pair $(-2, -4)$ and is parallel to the graph of $4x - 2y = 7$.

32. Find the equation of the line that contains the ordered pair $(3, -2)$ and is parallel to the graph of $y = -3x + 4$.

33. Find the equation of the line that contains the ordered pair $(2, 5)$ and is perpendicular to the graph of the line $y = -\frac{2}{3}x + 6$.

34. Graph the line that passes through the point whose coordinates are $(-1, 4)$ and has slope $-\frac{1}{3}$.

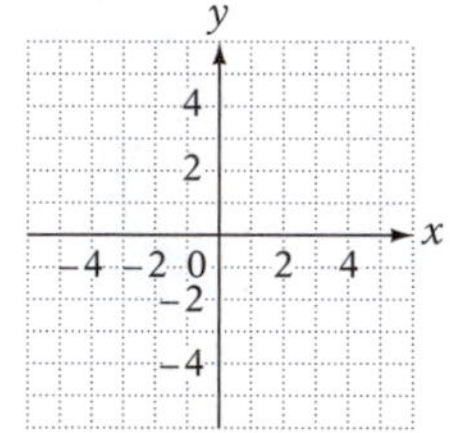

35. **Travel** A car is traveling at 55 mph. The equation that describes the distance traveled is $d = 55t$. Graph this equation for $0 \leq t \leq 6$. The point whose coordinates are $(4, 220)$ is on the graph. Write a sentence that explains the meaning of this ordered pair.

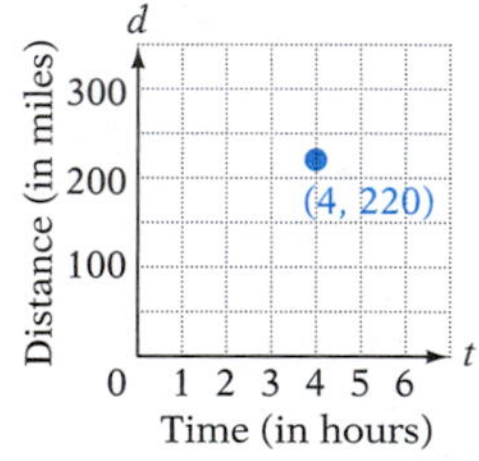

36. **Manufacturing** The graph at the right shows the relationship between the cost of manufacturing calculators and the number of calculators manufactured. Find the slope of the line between the two points shown on the graph. Write a sentence that states the meaning of the slope.

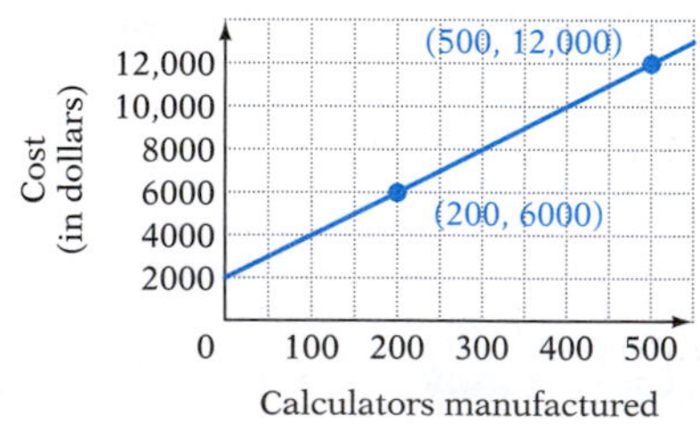

37. **Construction** A building contractor estimates that the cost to build a new home is \$25,000 plus \$80 for each square foot of floor space.

a. Determine a linear function that will give the cost to build a house that contains a given number of square feet.

b. Use the model to determine the cost to build a house that contains 2000 ft^2.

Copyright © Houghton Mifflin Company. All rights reserved.

Chapter 4 Test

1. Graph the ordered-pair solutions of $P(x) = 2 - x^2$ when $x = -2, -1, 0, 1,$ and 2.

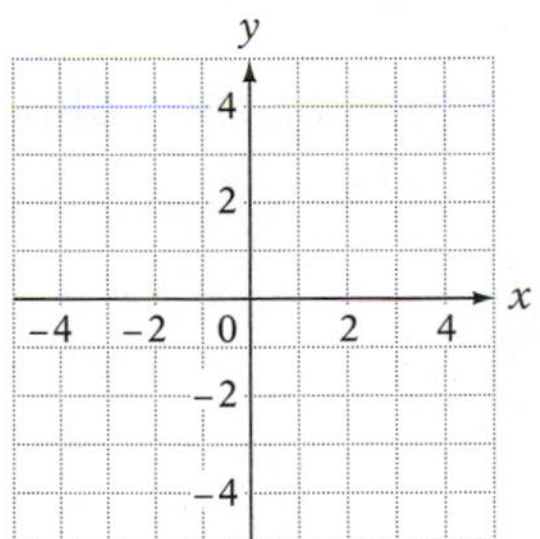

2. Find the ordered-pair solution of $y = 2x + 6$ that corresponds to $x = -3$.

3. Graph: $y = \frac{2}{3}x - 4$

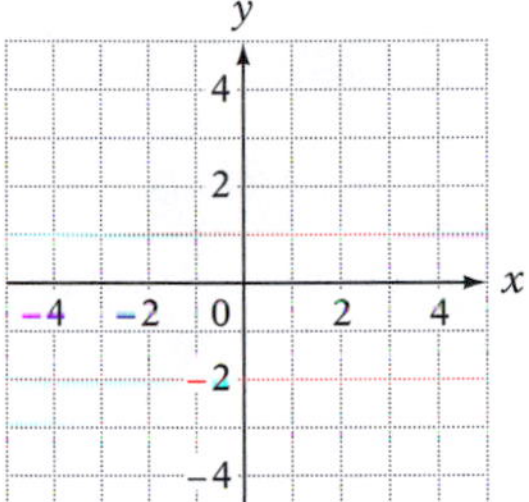

4. Graph: $2x + 3y = -3$

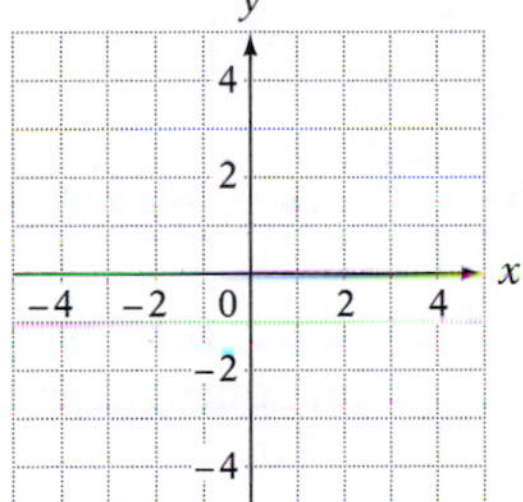

5. Find the equation of the vertical line that contains the point $(-2, 3)$.

6. Is the line that passes through the points $(-3, 1)$ and $(2, 5)$ perpendicular to the line that passes through the points $(-4, 1)$ and $(0, 6)$?

7. Find the slope of the line that contains the points $(-2, 3)$ and $(4, 2)$.

8. Given $P(x) = 3x^2 - 2x + 1$, evaluate $P(2)$.

9. Graph $2x - 3y = 6$ by using the x- and y-intercepts.

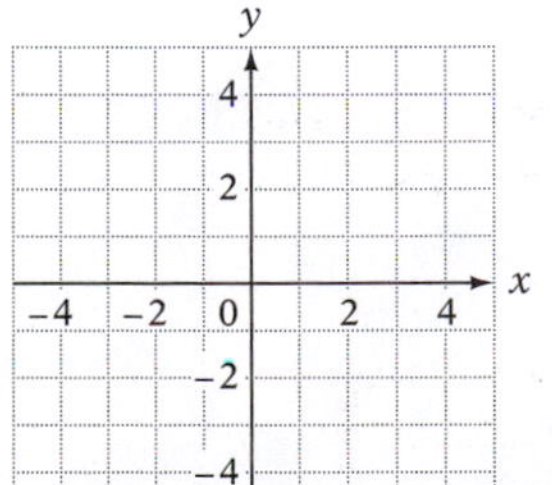

10. Graph the line that passes through the point $(-2, 3)$ and has slope $-\frac{3}{2}$.

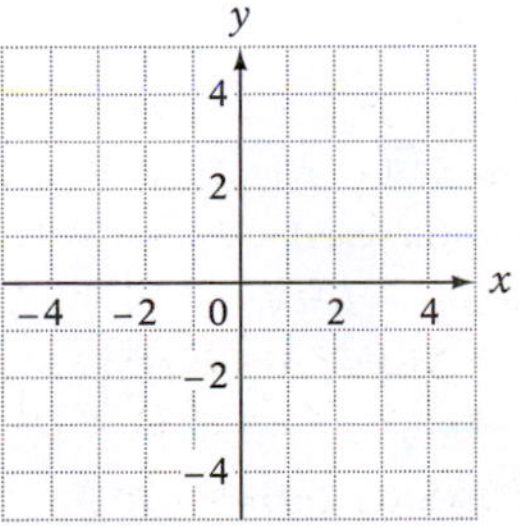

Copyright © Houghton Mifflin Company. All rights reserved.

11. Find the equation of the line that contains the point $(-5, 2)$ and has slope $\frac{2}{5}$.

12. What value of x is excluded from the domain of $f(x) = \frac{2x + 1}{x}$?

13. Find the equation of the line that contains the points $(3, -4)$ and $(-2, 3)$.

14. Find the equation of the horizontal line that contains the point $(4, -3)$.

15. Find the domain and range of the function $\{(-4, 2), (-2, 2), (0, 0), (3, 5)\}$.

16. Find the equation of the line that contains the point $(1, 2)$ and is parallel to the line $y = -\frac{3}{2}x - 6$.

17. Find the equation of the line that contains the point $(-2, -3)$ and is perpendicular to the line $y = -\frac{1}{2}x - 3$.

18. Graph the solution set of $3x - 4y > 8$.

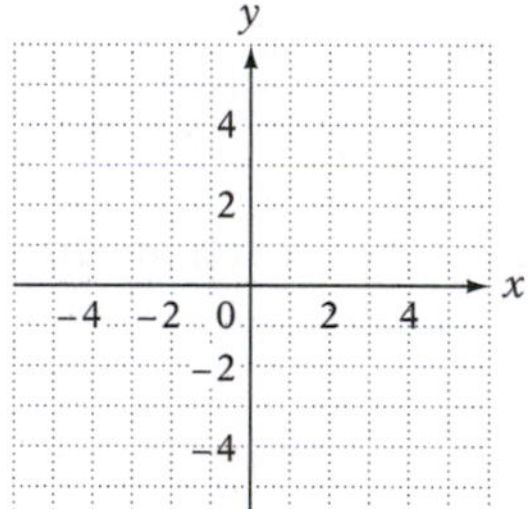

19. **Depreciation** The graph below shows the relationship between the cost of a rental house and the depreciation allowed for income tax purposes. Find the slope between the two points shown on the graph. Write a sentence that states the meaning of the slope.

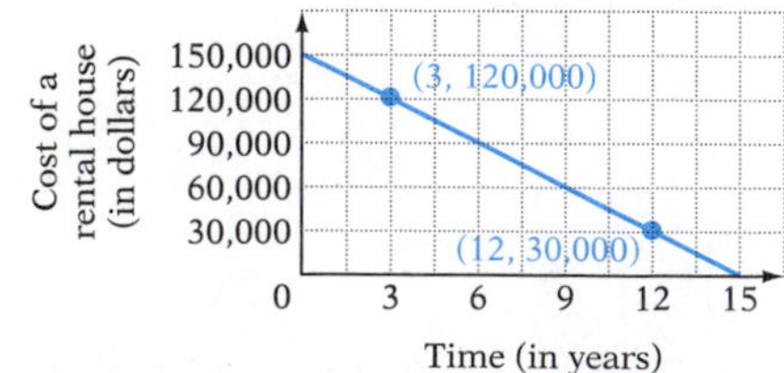

20. **Summer Camp** The director of a baseball camp estimates that 100 students will enroll if the tuition is \$250. For each \$20 increase in tuition, six fewer students will enroll.

a. Determine a linear function that will predict the number of students who will enroll at a given tuition.

b. Use this model to predict enrollment when the tuition is \$300.

Copyright © Houghton Mifflin Company. All rights reserved.

Cumulative Review Exercises

1. Let $x \in \{-5, -3, -1\}$. For what values of x is the inequality $x \le -3$ a true statement?

2. Write $\frac{17}{20}$ as a decimal.

3. Simplify: $3\sqrt{45}$

4. Simplify: $12 - 18 \div 3(-2)^2$

5. Evaluate $\frac{a-b}{a^2-c}$ when $a = -2$, $b = 3$, and $c = -4$.

6. Simplify: $3d - 9 - 7d$

7. Simplify: $4(-8z)$

8. Simplify: $2(x + y) - 5(3x - y)$

9. Graph: $\{x | x < -2\} \cup \{x | x > 0\}$

−5 −4 −3 −2 −1 0 1 2 3 4 5

10. Solve: $2x - \frac{2}{3} = \frac{7}{3}$

11. Solve: $3x - 2(10x - 6) = x - 6$

12. Solve: $4x - 3 < 9x + 2$

13. Solve: $3x - 1 < 4$ and $x - 2 > 2$

14. Solve: $|3x - 5| < 5$

15. Given $f(t) = t^2 + t$, find $f(2)$.

16. Find the slope of the line containing the points $(2, -3)$ and $(4, 1)$.

Copyright © Houghton Mifflin Company. All rights reserved.

17. Graph $y = 3x + 1$.

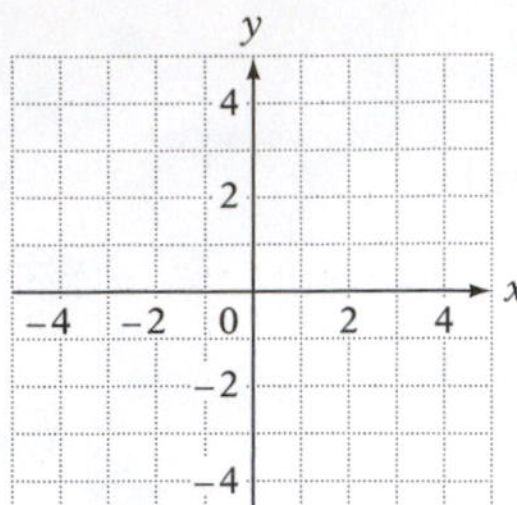

18. Graph $x = -4$.

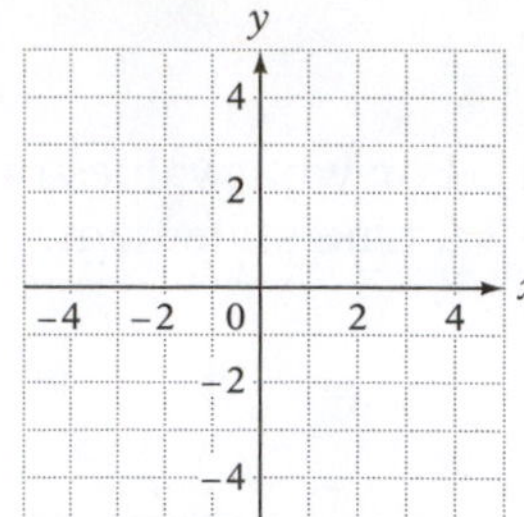

19. Graph the line that has slope $\frac{1}{2}$ and y-intercept $(0, -1)$.

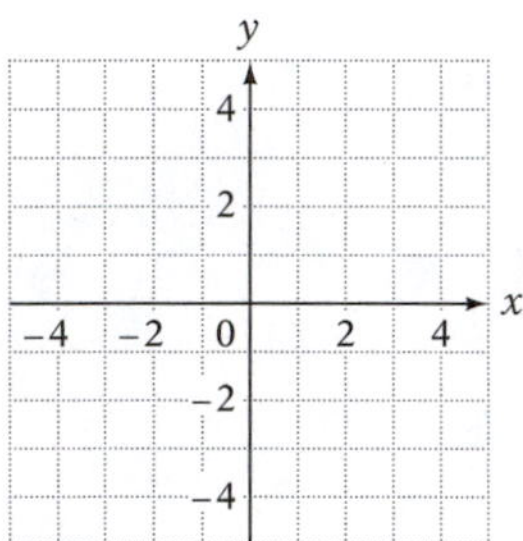

20. Graph the solution set of $3x - 2y \geq 6$.

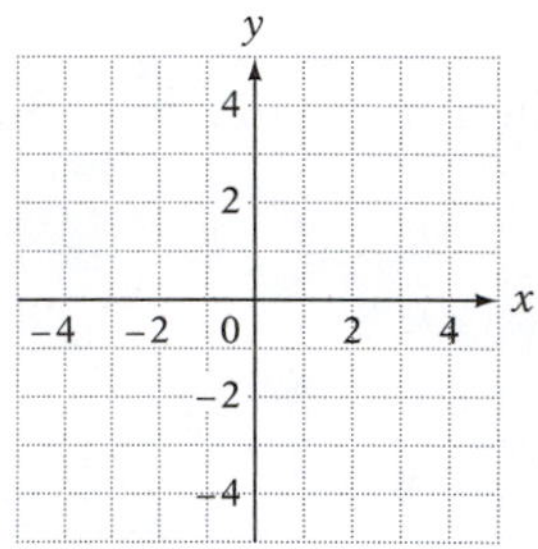

21. Find the equation of the line that passes through the points $(6, -4)$ and $(-3, -1)$.

22. Find the equation of the line that contains the point $(2, 4)$ and is parallel to the line $y = -\frac{3}{2}x + 2$.

23. Travel Two planes are 1800 mi apart and traveling toward each other. The first plane is traveling at twice the speed of the second plane. The planes meet in 3 h. Find the speed of each plane.

24. Mixtures A grocer combines coffee that costs \$9 per pound with coffee that costs \$6 per pound. How many pounds of each should be used to make 60 lb of a blend that costs \$8 per pound?

25. Depreciation The graph at the right shows the relationship between the cost of a backhoe loader and the depreciation allowed for income tax purposes. Find the slope of the line between the two points on the graph. Write a sentence that states the meaning of the slope.

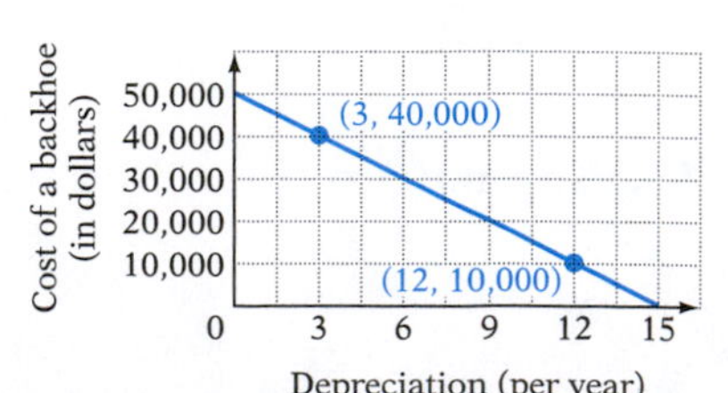

Copyright © Houghton Mifflin Company. All rights reserved.

chapter

5 Systems of Linear Equations and Inequalities

Soccer is one of the most popular team sports in the country, and is played by people of all ages and at every skill level. Children all over the country play in community leagues like AYSO, students play for their high schools and colleges, and the U.S. has an Olympic team. Suppose you volunteer to be the manager of your community's soccer league, which consists of 12 teams. You will need to create a schedule in which each team plays every other team once. How many games must be scheduled? The **Focus on Problem Solving on page 333** discusses methods of solving difficult problems, such as trying an easier version of the problem first, drawing diagrams, and identifying patterns.

OBJECTIVES

Section 5.1

A To solve a system of linear equations by graphing

B To solve a system of linear equations by the substitution method

C To solve investment problems

Section 5.2

A To solve a system of two linear equations in two variables by the addition method

B To solve a system of three linear equations in three variables by the addition method

Section 5.3

A To solve rate-of-wind or rate-of-current problems

B To solve application problems

Section 5.4

A To graph the solution set of a system of linear inequalities

Need help? For online student resources, such as section quizzes, visit this textbook's website at **math.college.hmco.com/students.**

Copyright © Houghton Mifflin Company. All rights reserved.

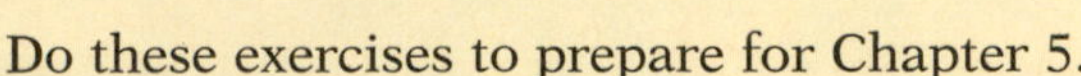

PREP TEST

Do these exercises to prepare for Chapter 5.

1. Simplify: $10\left(\frac{3}{5}x + \frac{1}{2}y\right)$

2. Evaluate $3x + 2y - z$ for $x = -1$, $y = 4$, and $z = -2$.

3. Given $3x - 2z = 4$, find the value of x when $z = -2$.

4. Solve: $3x + 4(-2x - 5) = -5$

5. Solve: $0.45x + 0.06(-x + 4000) = 630$

6. Graph: $y = \frac{1}{2}x - 4$

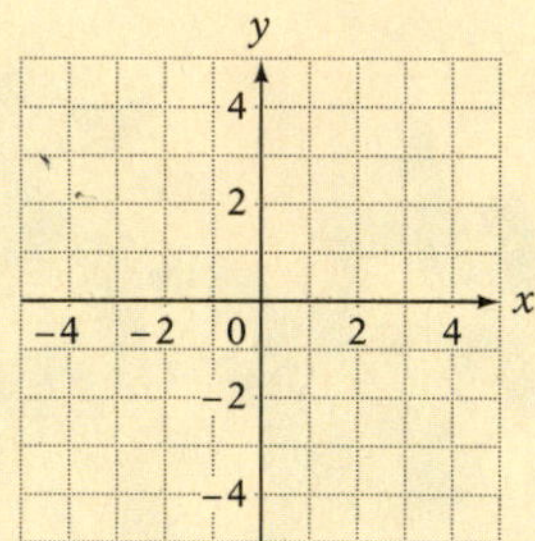

7. Graph: $3x - 2y = 6$

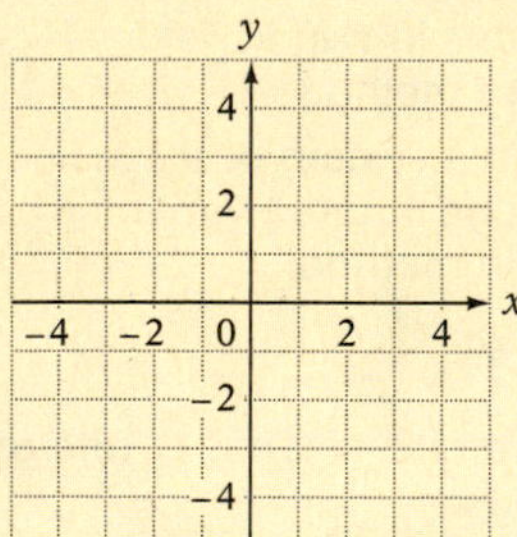

8. Graph: $y > -\frac{3}{5}x + 1$

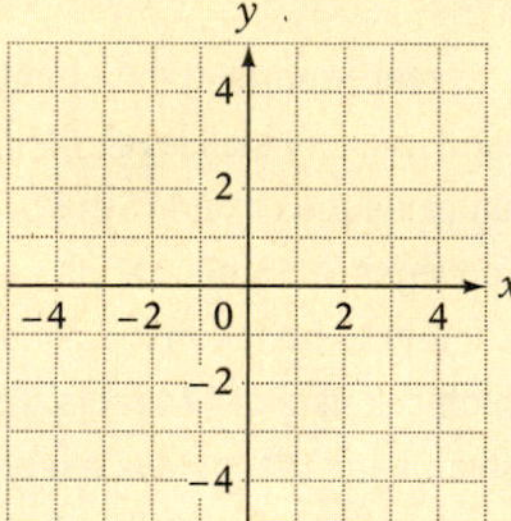

GO FIGURE

Chris can beat Pat by 100 m in a 1000-meter race. Pat can beat Leslie by 10 m in a 100-meter race. If both run at this given rate, by how many meters will Chris beat Leslie in a 1000-meter race?

Copyright © Houghton Mifflin Company. All rights reserved.

5.1 Solving Systems of Linear Equations by Graphing and by the Substitution Method

Objective A To solve a system of linear equations by graphing

A **system of equations** is two or more equations considered together. The system at the right is a system of two linear equations in two variables. The graphs of the equations are straight lines.

$$3x + 4y = 7$$
$$2x - 3y = 6$$

A **solution of a system of equations in two variables** is an ordered pair that is a solution of each equation of the system.

HOW TO Is $(3, -2)$ a solution of the system

$$2x - 3y = 12$$
$$5x + 2y = 11?$$

$2x - 3y = 12$		$5x + 2y = 11$	
$2(3) - 3(-2)$	12	$5(3) + 2(-2)$	11
$6 - (-6)$	12	$15 + (-4)$	11
$12 = 12$		$11 = 11$	

• Replace x by 3 and y by -2.

Yes, because $(3, -2)$ is a solution of each equation, it is a solution of the system of equations.

A solution of a system of linear equations can be found by graphing the lines of the system on the same set of coordinate axes. Three examples of linear equations in two variables are shown below, along with the graphs of the equations of the systems.

System I
$x + 2y = 4$
$2x + y = -1$

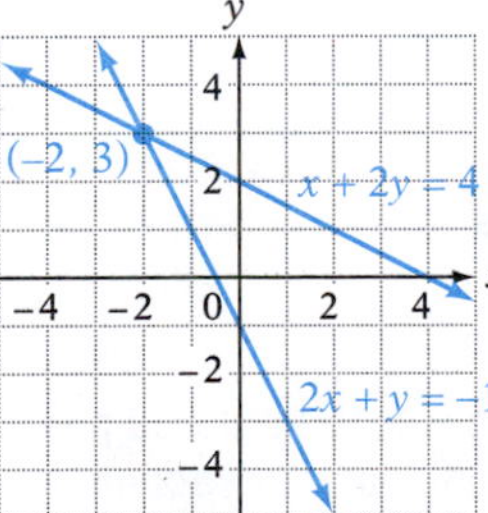

System II
$2x + 3y = 6$
$4x + 6y = -12$

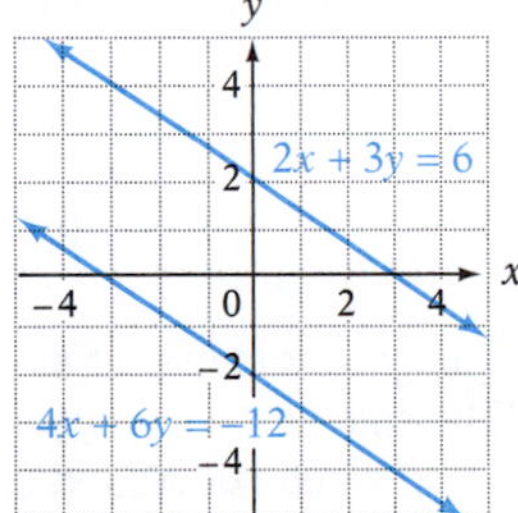

System III
$x - 2y = 4$
$2x - 4y = 8$

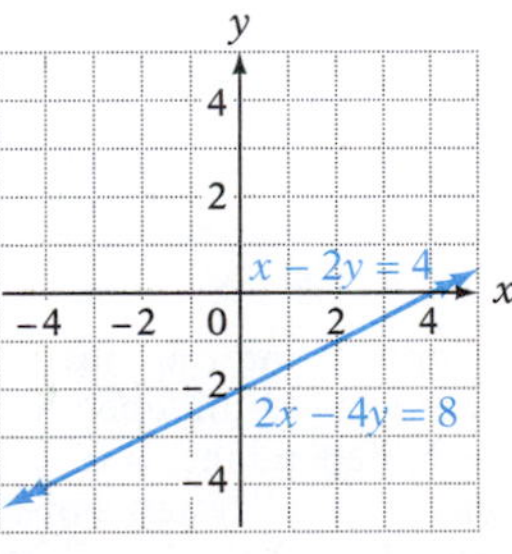

Check:

$x + 2y = 4$	
$-2 + 2(3)$	4
$-2 + 6$	4
$4 = 4$	

$2x + y = -1$	
$2(-2) + 3$	-1
$-4 + 3$	-1
$-1 = -1$	

In System I, the two lines intersect at a single point whose coordinates are $(-2, 3)$. Because this point lies on both lines, it is a solution of each equation of the system of equations. We can check this by replacing x by -2 and y by 3. The check is shown at the left. The ordered pair $(-2, 3)$ is a solution of System I.

When the graphs of a system of equations intersect at only one point, the system is called an **independent system of equations.** System I is an independent system of equations.

Copyright © Houghton Mifflin Company. All rights reserved.

System II from the preceding page and the graph of the equations of that system are shown again at the right. Note in this case that the graphs of the lines are parallel and do not intersect. Since the graphs do not intersect, there is no point that is on both lines. Therefore, the system of equations has no solution.

$$2x + 3y = 6$$
$$4x + 6y = -12$$

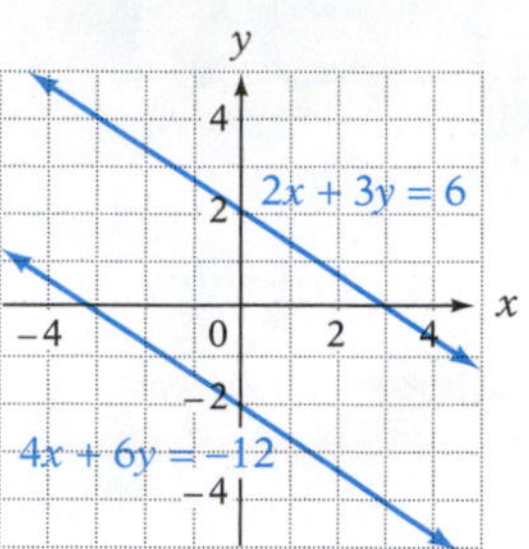

When a system of equations has no solution, it is called an **inconsistent system of equations.** System II is an inconsistent system of equations.

System III from the preceding page and the graph of the equations of that system are shown again at the right. Note that the graph of $x - 2y = 4$ lies directly on top of the graph of $2x - 4y = 8$. Thus the two lines intersect at an infinite number of points. Because the graphs intersect at an infinite number of points, there are an infinite number of solutions of this system of equations. Since each equation represents the same set of points, the solutions of the system of equations can be stated by using the ordered pairs of either one of the equations. Therefore, we can say, "The solutions are the ordered pairs that satisfy $x - 2y = 4$," or we can solve the equation for y and say, "The solutions are the ordered pairs that satisfy $y = \frac{1}{2}x - 2$." We normally state this solution using ordered pairs—"The solutions are the ordered pairs $\left(x, \frac{1}{2}x - 2\right)$."

$$x - 2y = 4$$
$$2x - 4y = 8$$

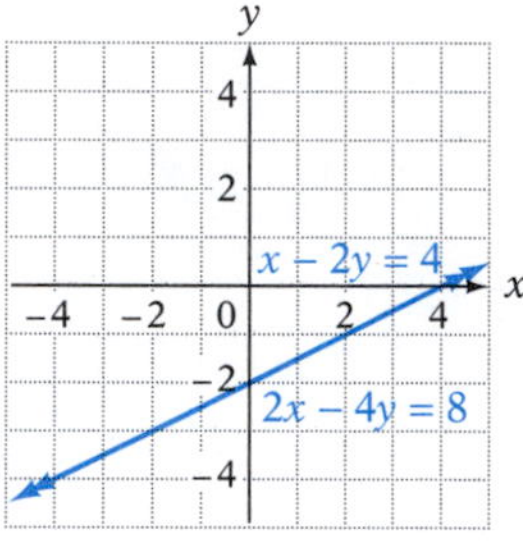

When the two equations in a system of equations represent the same line, the system is called a **dependent system of equations.** System III is a dependent system of equations.

> **TAKE NOTE**
> Keep in mind the differences among independent, dependent, and inconsistent systems of equations. You should be able to express your understanding of these terms by using graphs.

Summary of the Three Possibilities for a System of Linear Equations in Two Variables

1. The graphs intersect at one point. The solution of the system of equations is the ordered pair (x, y) whose coordinates name the point of intersection. The system of equations is independent.
2. The lines are parallel and never intersect. There is no solution of the system of equations. The system of equations is inconsistent.
3. The graphs are the same line, and they intersect at infinitely many points. There are infinitely many solutions of the system of equations. The system of equations is dependent.

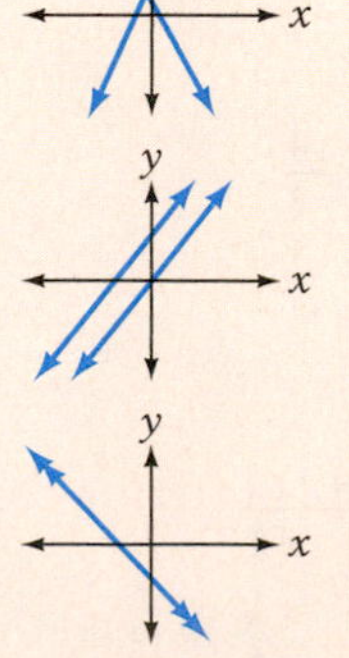

Copyright © Houghton Mifflin Company. All rights reserved.

Integrating Technology

See the Projects and Group Activities at the end of this chapter for instructions on using a graphing calculator to solve a system of equations. Instructions are also provided in the appendix Keystroke Guide: *Intersect.*

HOW TO Solve by graphing: $2x - y = 3$
$4x - 2y = 6$

Graph each line.
The system of equations is dependent.
Solve one of the equations for y.

$$2x - y = 3$$
$$-y = -2x + 3$$
$$y = 2x - 3$$

The solutions are the ordered pairs $(x, 2x - 3)$.

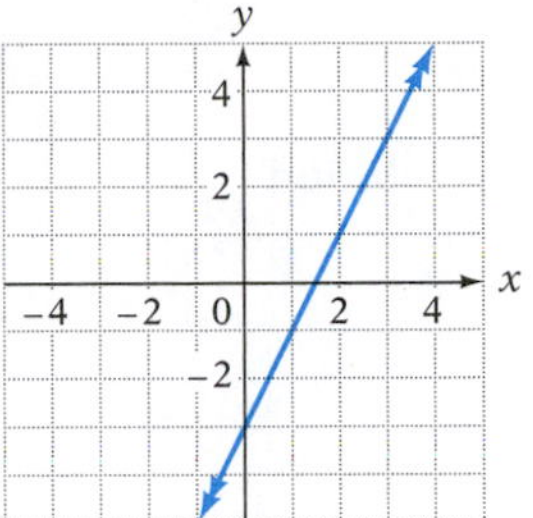

Example 1 Solve by graphing:
$2x - y = 3$
$3x + y = 2$

Solution

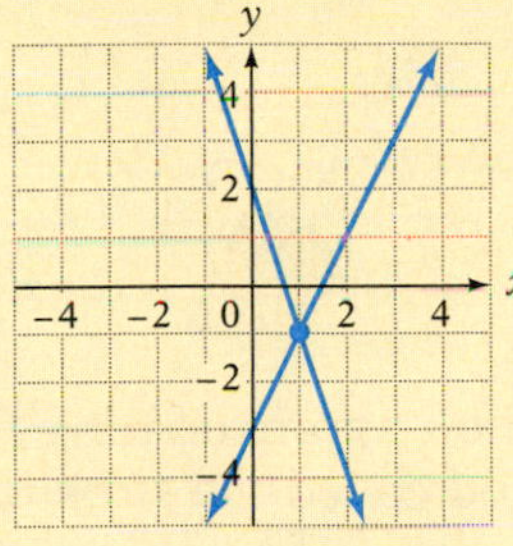

• Find the point of intersection of the graphs of the equations.

The solution is $(1, -1)$.

You Try It 1 Solve by graphing:
$x + y = 1$
$2x + y = 0$

Your solution

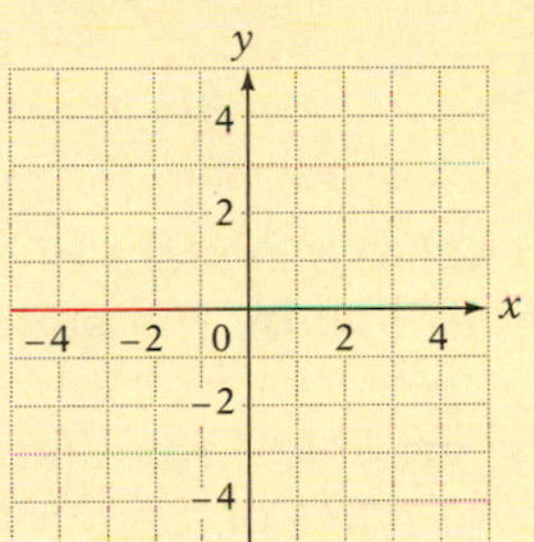

Example 2 Solve by graphing:
$2x + 3y = 6$
$y = -\frac{2}{3}x + 1$

Solution

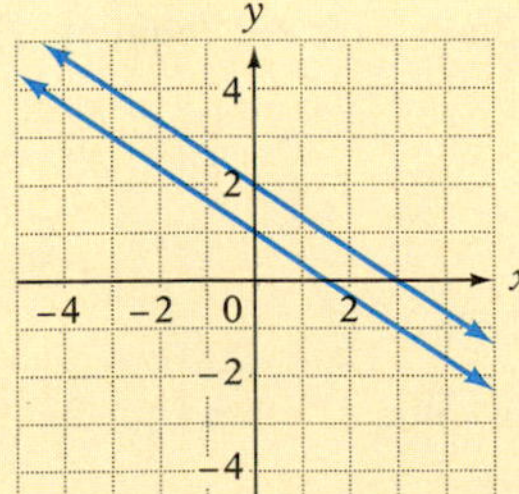

• Graph the two equations.

The lines are parallel and therefore do not intersect. The system of equations has no solution. The system of equations is inconsistent.

You Try It 2 Solve by graphing:
$2x + 5y = 10$
$y = -\frac{2}{5}x - 2$

Your solution

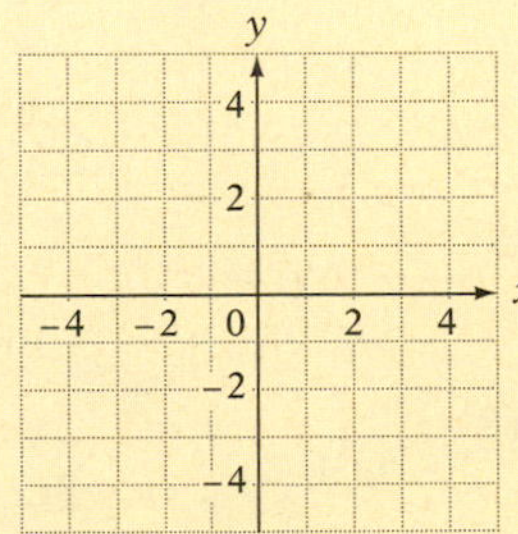

Solutions on p. S15

Copyright © Houghton Mifflin Company. All rights reserved.

Example 3 Solve by graphing:
$x - 2y = 6$
$y = \frac{1}{2}x - 3$

Solution

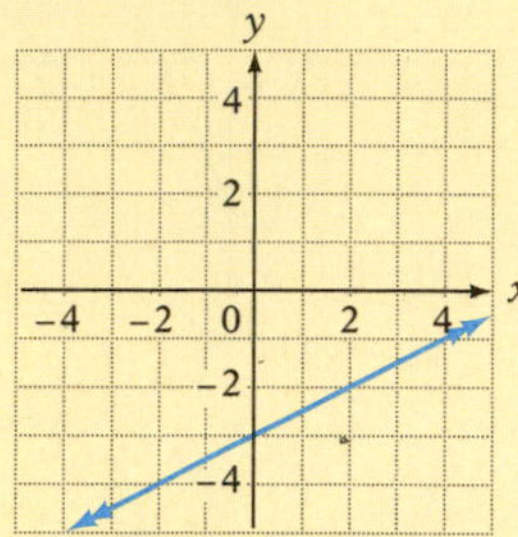

• Graph the two equations.

The system of equations is dependent. The solutions are the ordered pairs $\left(x, \frac{1}{2}x - 3\right)$.

You Try It 3 Solve by graphing:
$3x - 4y = 12$
$y = \frac{3}{4}x - 3$

Your solution

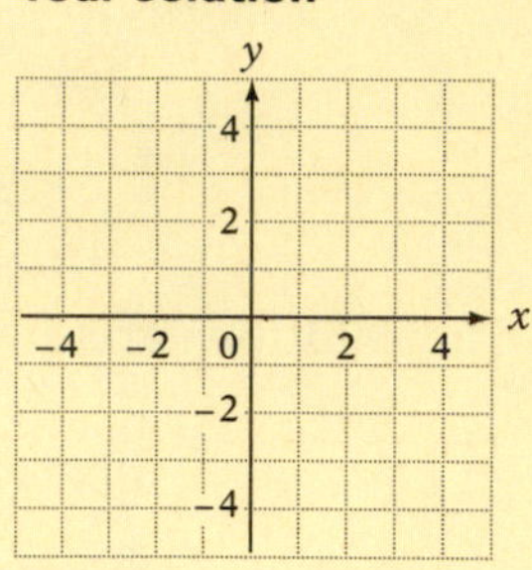

Solution on p. S15

Objective B To solve a system of linear equations by the substitution method

The graphical solution of a system of equations is based on approximating the coordinates of a point of intersection. An algebraic method called the **substitution method** can be used to find an exact solution of a system of equations. To use the substitution method, we must write one of the equations of the system in terms of x or in terms of y.

HOW TO Solve by the substitution method: (1) $3x + y = 5$
(2) $4x + 5y = 3$

$3x + y = 5$
(3) $y = -3x + 5$

• Solve Equation (1) for y. This is Equation (3).

(2) $4x + 5y = 3$
$4x + 5(-3x + 5) = 3$

• This is Equation (2).
• Equation (3) states that $y = -3x + 5$. Substitute $-3x + 5$ for y in Equation (2).

$4x - 15x + 25 = 3$
$-11x + 25 = 3$
$-11x = -22$
$x = 2$

• Solve for x.

(3) $y = -3x + 5$
$y = -3(2) + 5$
$y = -6 + 5$
$y = -1$

• Substitute the value of x into Equation (3) and solve for y.

The solution is the ordered pair $(2, -1)$.

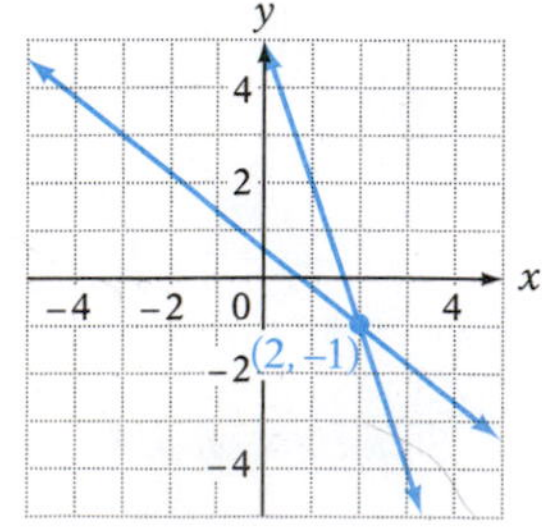

The graph of the system of equations is shown at the left. Note that the graphs intersect at the point whose coordinates are $(2, -1)$, the solution of the system of equations.

Copyright © Houghton Mifflin Company. All rights reserved.

HOW TO Solve by the substitution method: (1) $6x + 2y = 8$
(2) $3x + y = 2$

$3x + y = 2$
(3) $y = -3x + 2$

- We will solve Equation (2) for y.
- This is Equation (3).

(1) $6x + 2y = 8$
$6x + 2(-3x + 2) = 8$

- This is Equation (1).
- Equation (3) states that $y = -3x + 2$. Substitute $-3x + 2$ for y in Equation (1).
- Solve for x.

$6x - 6x + 4 = 8$
$0x + 4 = 8$
$4 = 8$

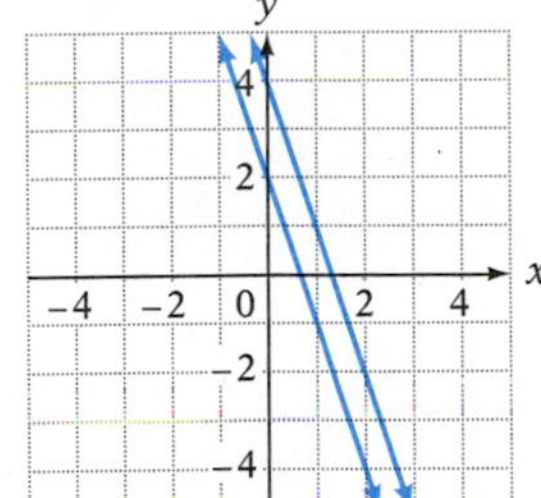

This is not a true equation.
The system of equations has no solution.
The system of equations is inconsistent.

The graph of the system of equations is shown at the left. Note that the lines are parallel.

Example 4 Solve by substitution:
(1) $3x - 2y = 4$
(2) $-x + 4y = -3$

Solution Solve Equation (2) for x.

$-x + 4y = -3$
$-x = -4y - 3$
$x = 4y + 3$ • Equation (3)

Substitute $4y + 3$ for x in Equation (1).

$3x - 2y = 4$ • Equation (1)
$3(4y + 3) - 2y = 4$ • $x = 4y + 3$
$12y + 9 - 2y = 4$
$10y + 9 = 4$
$10y = -5$
$y = -\frac{5}{10} = -\frac{1}{2}$

Substitute the value of y into Equation (3).

$x = 4y + 3$ • Equation (3)
$= 4\left(-\frac{1}{2}\right) + 3$ • $y = -\frac{1}{2}$
$= -2 + 3 = 1$

The solution is $\left(1, -\frac{1}{2}\right)$.

You Try It 4 Solve by substitution:
$3x - y = 3$
$6x + 3y = -4$

Your solution

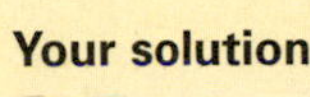

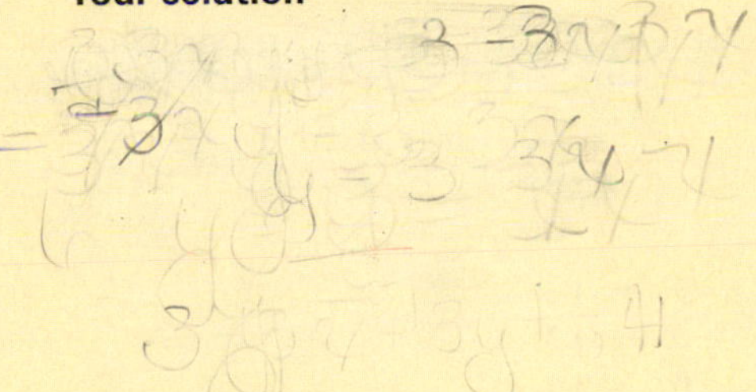

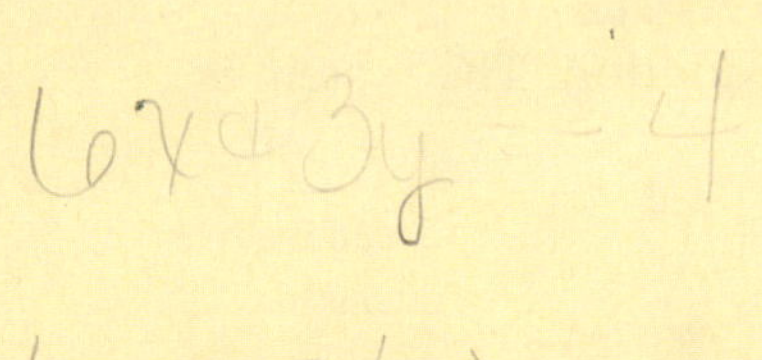

Solution on p. S15

Copyright © Houghton Mifflin Company. All rights reserved.

Example 5 Solve by substitution and graph:

$$3x - 3y = 2$$
$$y = x + 2$$

Solution

$$3x - 3y = 2$$
$$3x - 3(x + 2) = 2$$
$$3x - 3x - 6 = 2$$
$$-6 = 2$$

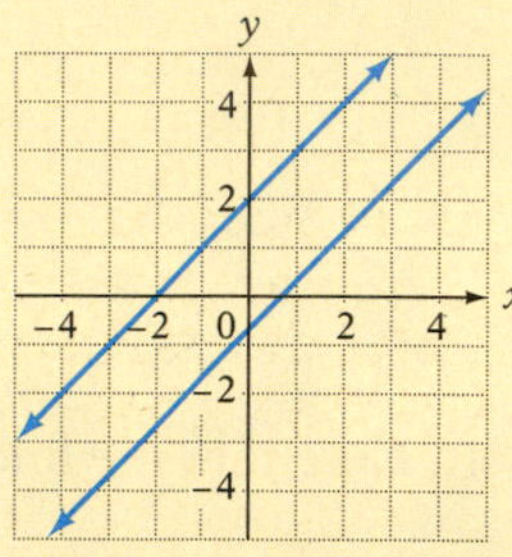

• **Graph the two equations.**

This is not a true equation. The lines are parallel, so the system is inconsistent. The system does not have a solution.

You Try It 5 Solve by substitution and graph:

$$y = 2x - 3$$
$$3x - 2y = 6$$

Your solution

Example 6 Solve by substitution and graph:

$$9x + 3y = 12$$
$$y = -3x + 4$$

Solution

$$9x + 3y = 12$$
$$9x + 3(-3x + 4) = 12$$
$$9x - 9x + 12 = 12$$
$$12 = 12$$

This is a true equation. The system is dependent.

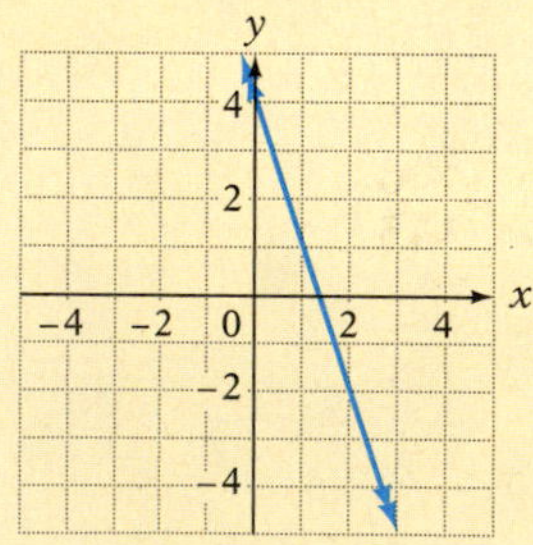

• **Graph the two equations.**

The solutions are the ordered pairs $(x, -3x + 4)$.

You Try It 6 Solve by substitution and graph:

$$6x - 3y = 6$$
$$2x - y = 2$$

Your solution

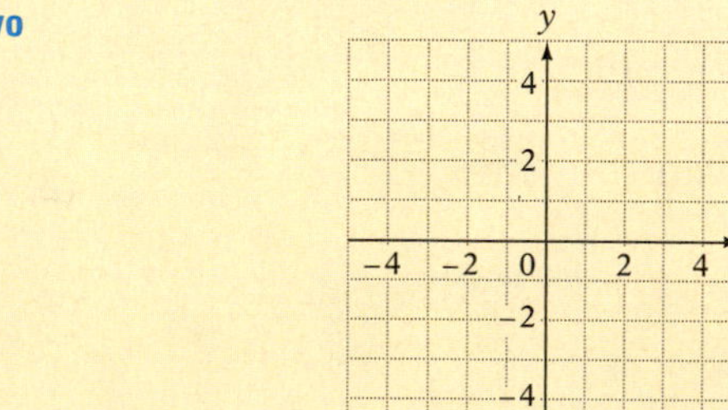

Solutions on pp. S15–S16

Copyright © Houghton Mifflin Company. All rights reserved.

Objective C To solve investment problems

The annual simple interest that an investment earns is given by the equation $Pr = I$, where P is the principal, or the amount invested, r is the simple interest rate, and I is the simple interest.

For instance, if you invest \$500 at a simple interest rate of 5%, then the interest earned after one year is calculated as follows:

$$Pr = I$$
$$500(0.05) = I \quad \text{• Replace } P \text{ by 500 and } r \text{ by 0.05 (5\%).}$$
$$25 = I \quad \text{• Simplify.}$$

The amount of interest earned is \$25.

Study Tip

Note that solving a word problem includes stating a strategy and using the strategy to find a solution. If you have difficulty with a word problem, write down the known information. Be very specific. Write out a phrase or sentence that states what you are trying to find. See *AIM for Success,* page xxix.

HOW TO You have a total of \$5000 to invest in two simple interest accounts. On one account, a money market fund, the annual simple interest rate is 3.5%. On the second account, a bond fund, the annual simple interest rate is 7.5%. If you earn \$245 per year from these two investments, how much do you have invested in each account?

Strategy for Solving Simple-Interest Investment Problems

1. For each amount invested, use the equation $Pr = I$. Write a numerical or variable expression for the principal, the interest rate, and the interest earned.

Amount invested at 3.5%: x
Amount invested at 7.5%: y

	Principal, P	·	*Interest rate, r*	=	*Interest earned, I*
Amount at 3.5%	x	·	0.035	=	$0.035x$
Amount at 7.5%	y	·	0.075	=	$0.075y$

2. Write a system of equations. One equation will express the relationship between the amounts invested. The second equation will express the relationship between the amounts of interest earned by the investments.

The total amount invested is \$5000: $x + y = 5000$
The total annual interest earned is \$245: $0.035x + 0.075y = 245$

Solve the system of equations.
$$(1) \quad x + y = 5000$$
$$(2) \quad 0.035x + 0.075y = 245$$

Solve Equation (1) for y: $(3) \quad y = -x + 5000$

Substitute into Equation (2):
$$(2) \quad 0.035x + 0.075(-x + 5000) = 245$$
$$0.035x - 0.075x + 375 = 245$$
$$-0.04x = -130$$
$$x = 3250$$

Substitute the value of x into Equation (3) and solve for y.

$$y = -x + 5000$$
$$y = -3250 + 5000 = 1750$$

The amount invested at 3.5% is \$3250.
The amount invested at 7.5% is \$1750.

Copyright © Houghton Mifflin Company. All rights reserved.

Example 7

An investment of \$4000 is made at an annual simple interest rate of 4.9%. How much additional money must be invested at an annual simple interest rate of 7.4% so that the total interest earned is 6.4% of the total investment?

Strategy

- Amount invested at 4.9%: 4000
  Amount invested at 7.4%: x
  Amount invested at 6.4%: y

	Principal	*Rate*	*Interest*
Amount at 4.9%	4000	0.049	0.049(4000)
Amount at 7.4%	x	0.074	$0.074x$
Amount at 6.4%	y	0.064	$0.064y$

- The amount invested at 6.4% (y) is \$4000 more than the amount invested at 7.4% (x): $y = x + 4000$
- The sum of the interest earned at 4.9% and the interest earned at 7.4% equals the interest earned at 6.4%: $0.049(4000) + 0.074x = 0.064y$

Solution

$$y = x + 4000 \qquad (1)$$
$$0.049(4000) + 0.074x = 0.064y \qquad (2)$$

Replace y in Equation (2) by $x + 4000$ from Equation (1). Then solve for x.

$$0.049(4000) + 0.074x = 0.064(x + 4000)$$
$$196 + 0.074x = 0.064x + 256$$
$$0.01x = 60$$
$$x = 6000$$

\$6000 must be invested at an annual simple interest rate of 7.4%.

You Try It 7

An investment club invested \$13,600 in two simple interest accounts. On one account, the annual simple interest rate is 4.2%. On the other, the annual simple interest rate is 6%. How much should be invested in each account so that both accounts earn the same annual interest?

Your strategy

Your solution

Solution on p. S16

Copyright © Houghton Mifflin Company. All rights reserved.

5.1 Exercises

Objective A **To solve a system of linear equations by graphing**

For Exercises 1 to 4, determine whether the ordered pair is a solution of the system of equations.

1. $(0, -1)$
$3x - 2y = 2$
$x + 2y = 6$

2. $(2, 1)$
$x + y = 3$
$2x - 3y = 1$

3. $(-3, -5)$
$x + y = -8$
$2x + 5y = -31$

4. $(1, -1)$
$3x - y = 4$
$7x + 2y = -5$

For Exercises 5 to 8, state whether the system of equations is independent, inconsistent, or dependent.

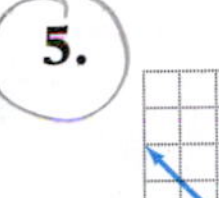

5.

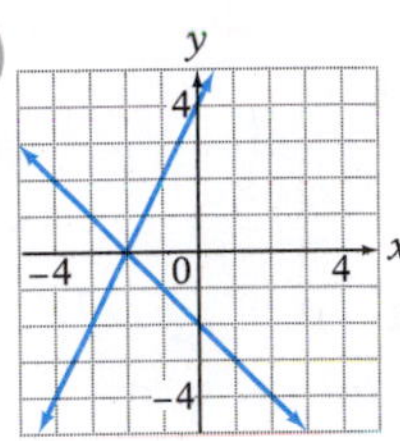

6.

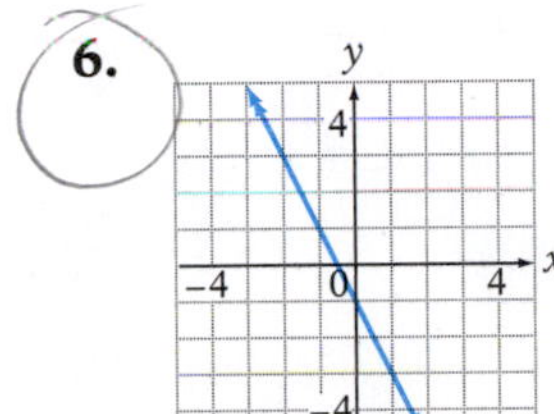

7.

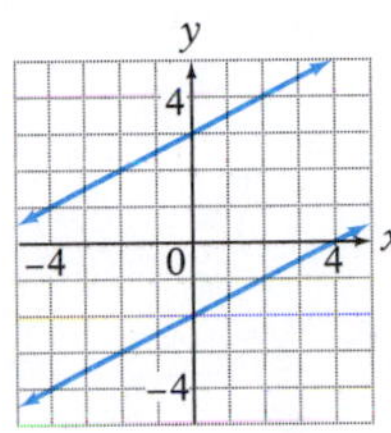

8.

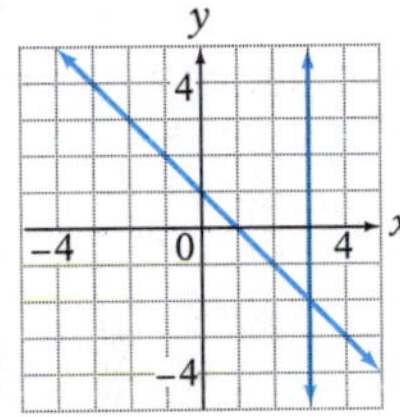

For Exercises 9 to 26, solve by graphing.

9. $x + y = 2$
$x - y = 4$

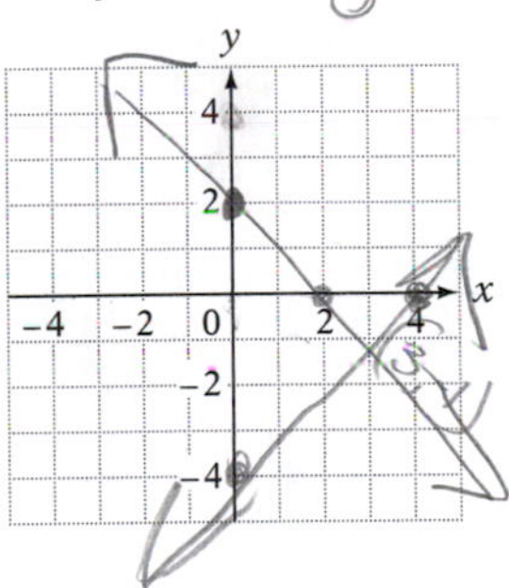

10. $x + y = 1$
$3x - y = -5$

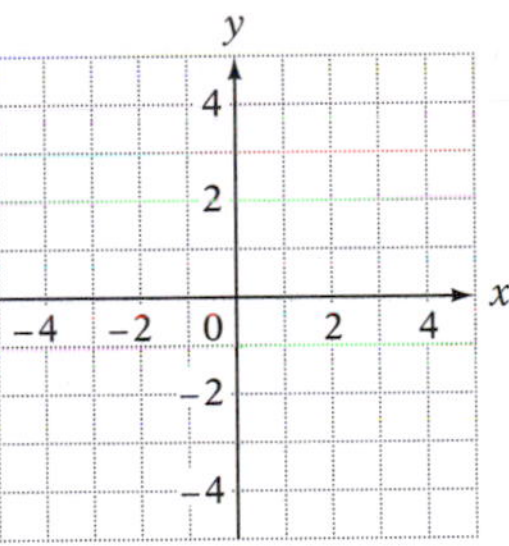

11. $x - y = -2$
$x + 2y = 10$

12. $2x - y = 5$
$3x + y = 5$

13. $3x - 2y = 6$
$y = 3$

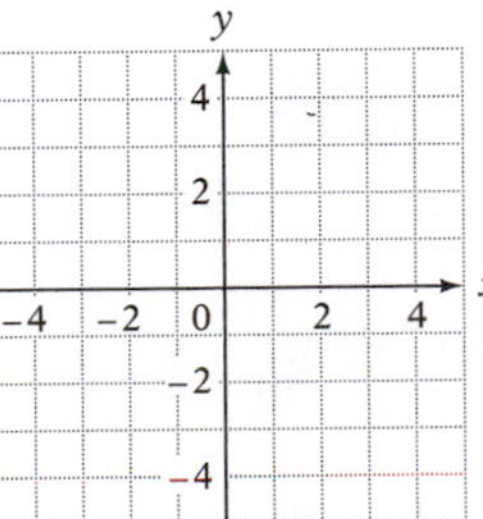

14. $x = 4$
$3x - 2y = 4$

Copyright © Houghton Mifflin Company. All rights reserved.

15. $x = 4$
$y = -1$

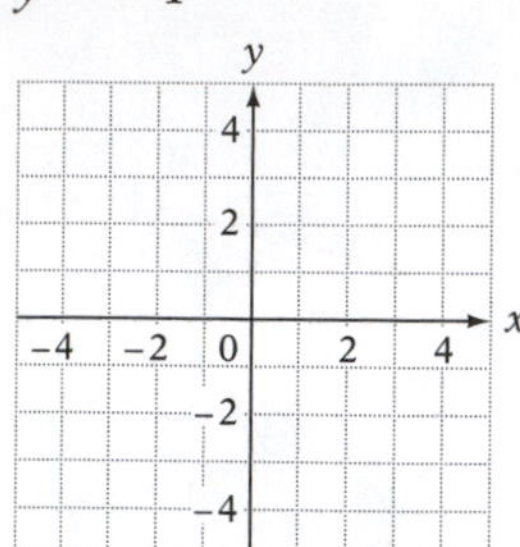

16. $x + 2 = 0$
$y - 1 = 0$

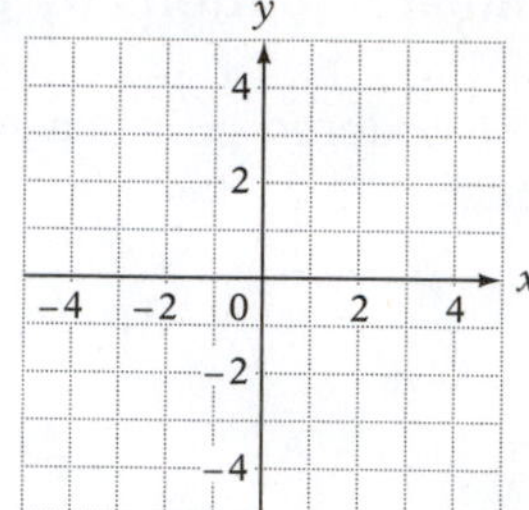

17. $2x + y = 3$
$x - 2 = 0$

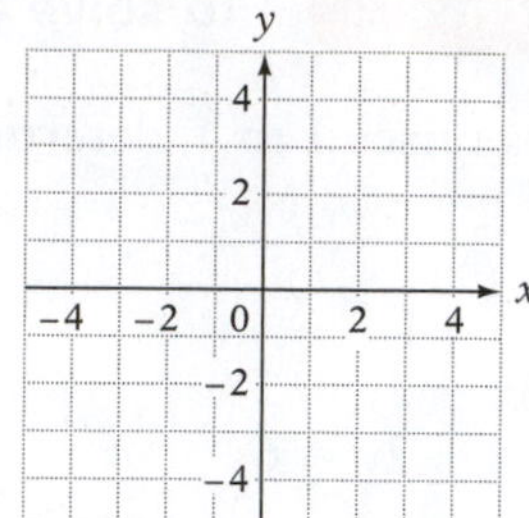

18. $x - 3y = 6$
$y + 3 = 0$

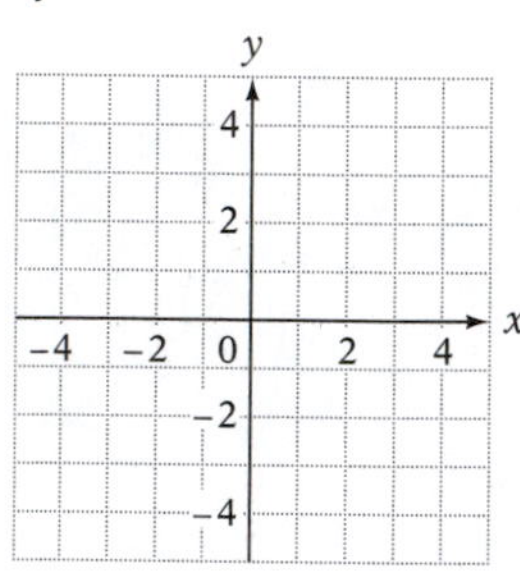

19. $x - y = 6$
$x + y = 2$

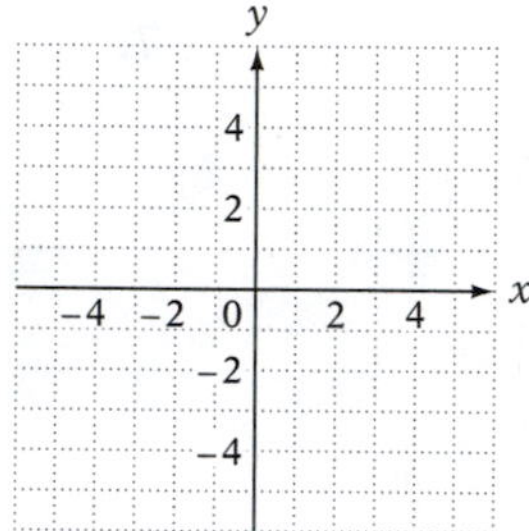

20. $2x + y = 2$
$-x + y = 5$

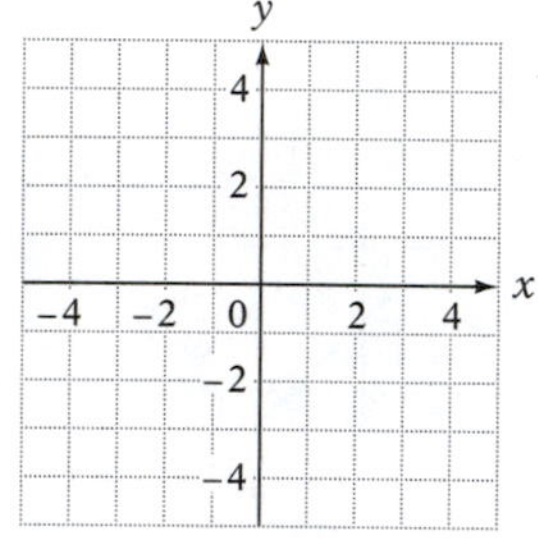

21. $y = x - 5$
$2x + y = 4$

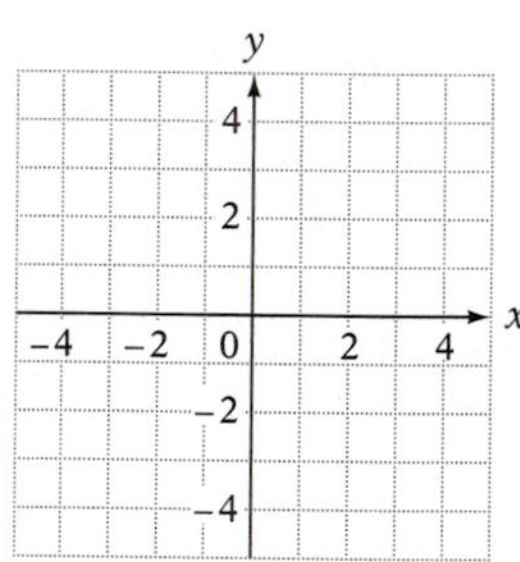

22. $2x - 5y = 4$
$y = x + 1$

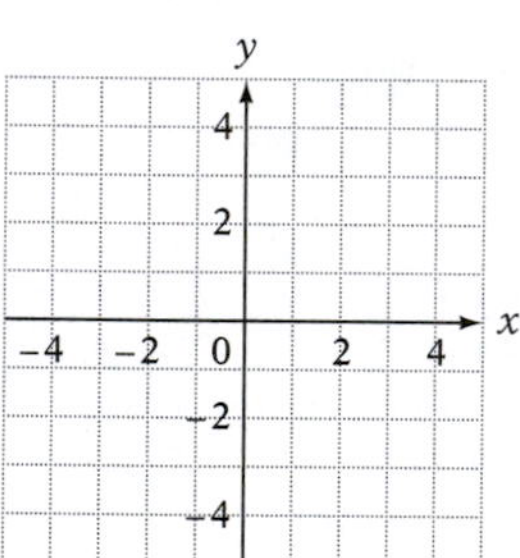

23. $y = \frac{1}{2}x - 2$
$x - 2y = 8$

24. $2x + 3y = 6$
$y = -\frac{2}{3}x + 1$

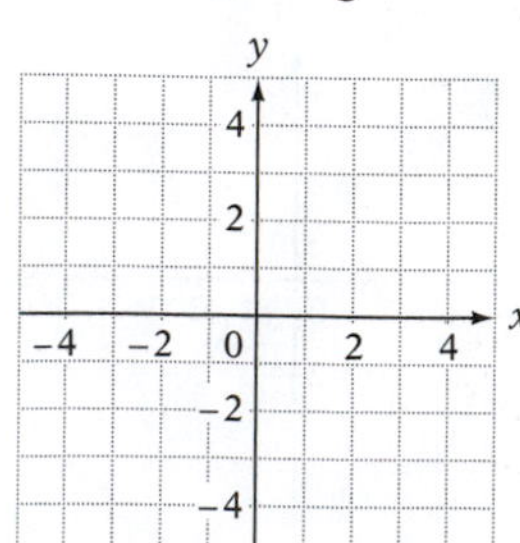

25. $2x - 5y = 10$
$y = \frac{2}{5}x - 2$

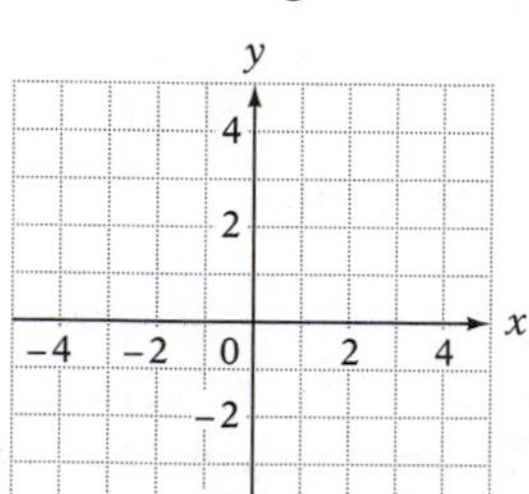

26. $3x - 2y = 6$
$y = \frac{3}{2}x - 3$

Copyright © Houghton Mifflin Company. All rights reserved.

Objective B To solve a system of linear equations by the substitution method

For Exercises 27 to 56, solve by the substitution method.

27. $\begin{aligned} y &= -x + 1 \\ 2x - y &= 5 \end{aligned}$

28. $\begin{aligned} x &= 3y + 1 \\ x - 2y &= 6 \end{aligned}$

29. $\begin{aligned} x &= 2y - 3 \\ 3x + y &= 5 \end{aligned}$

30. $\begin{aligned} 4x - 3y &= 5 \\ y &= 2x - 3 \end{aligned}$

31. $\begin{aligned} 3x + 5y &= -1 \\ y &= 2x - 8 \end{aligned}$

32. $\begin{aligned} 5x - 2y &= 9 \\ y &= 3x - 4 \end{aligned}$

33. $\begin{aligned} 4x - 3y &= 2 \\ y &= 2x + 1 \end{aligned}$

34. $\begin{aligned} x &= 2y + 4 \\ 4x + 3y &= -17 \end{aligned}$

35. $\begin{aligned} 3x - 2y &= -11 \\ x &= 2y - 9 \end{aligned}$

36. $\begin{aligned} 5x + 4y &= -1 \\ y &= 2 - 2x \end{aligned}$

37. $\begin{aligned} 3x + 2y &= 4 \\ y &= 1 - 2x \end{aligned}$

38. $\begin{aligned} 2x - 5y &= -9 \\ y &= 9 - 2x \end{aligned}$

39. $\begin{aligned} 5x + 2y &= 15 \\ x &= 6 - y \end{aligned}$

40. $\begin{aligned} 7x - 3y &= 3 \\ x &= 2y + 2 \end{aligned}$

41. $\begin{aligned} 3x - 4y &= 6 \\ x &= 3y + 2 \end{aligned}$

42. $\begin{aligned} 2x + 2y &= 7 \\ y &= 4x + 1 \end{aligned}$

43. $\begin{aligned} 3x + 7y &= -5 \\ y &= 6x - 5 \end{aligned}$

44. $\begin{aligned} 3x + y &= 5 \\ 2x + 3y &= 8 \end{aligned}$

45. $\begin{aligned} 3x - y &= 10 \\ 6x - 2y &= 5 \end{aligned}$

46. $\begin{aligned} 6x - 4y &= 3 \\ 3x - 2y &= 9 \end{aligned}$

47. $\begin{aligned} 3x + 4y &= 14 \\ 2x + y &= 1 \end{aligned}$

48. $\begin{aligned} 5x + 3y &= 8 \\ 3x + y &= 8 \end{aligned}$

49. $\begin{aligned} 3x + 5y &= 0 \\ x - 4y &= 0 \end{aligned}$

50. $\begin{aligned} 2x - 7y &= 0 \\ 3x + y &= 0 \end{aligned}$

51. $\begin{aligned} 2x - 4y &= 16 \\ -x + 2y &= -8 \end{aligned}$

52. $\begin{aligned} 3x - 12y &= -24 \\ -x + 4y &= 8 \end{aligned}$

53. $\begin{aligned} y &= 3x + 2 \\ y &= 2x + 3 \end{aligned}$

54. $\begin{aligned} y &= 3x - 7 \\ y &= 2x - 5 \end{aligned}$

55. $\begin{aligned} y &= 3x + 1 \\ y &= 6x - 1 \end{aligned}$

56. $\begin{aligned} y &= 2x - 3 \\ y &= 4x - 4 \end{aligned}$

Objective C To solve investment problems

57. The Community Relief Charity Group is earning 3.5% simple interest on the $2800 it invested in a savings account. It also earns 4.2% simple interest on an insured bond fund. The annual interest earned from both accounts is $329. How much is invested in the insured bond fund?

Copyright © Houghton Mifflin Company. All rights reserved.

58. Two investments earn an annual income of \$575. One investment earns an annual simple interest rate of 8.5%, and the other investment earns an annual simple interest rate of 6.4%. The total amount invested is \$8000. How much is invested in each account?

59. An investment club invested \$6000 at an annual simple interest rate of 4.0%. How much additional money must be invested at an annual simple interest rate of 6.5% so that the total annual interest earned will be 5% of the total investment?

60. A small company invested \$30,000 by putting part of it into a savings account that earned 3.2% annual simple interest and the remainder in a risky stock fund that earned 12.6% annual simple interest. If the company earned \$1665 annually from the investments, how much was in each account?

61. An account executive divided \$42,000 between two simple interest accounts. On the tax-free account the annual simple interest rate is 3.5%, and on the money market fund the annual simple interest rate is 4.5%. How much should be invested in each account so that both accounts earn the same annual interest?

62. The Ridge Investment Club placed \$33,000 into two simple interest accounts. On one account, the annual simple interest rate is 6.5%. On the other, the annual simple interest rate is 4.5%. How much should be invested in each account so that both accounts earn the same annual interest?

63. The Cross Creek Investment Club has \$20,000 to invest. The members of the club decided to invest \$16,000 of their money in two bond funds. The first, a mutual bond fund, earns annual simple interest of 4.5%. The second account, a corporate bond fund, earns 8% annual simple interest. If the members earned \$1070 from these two accounts, how much was invested in the mutual bond fund?

64. Cabin Financial Service Group recommends that a client purchase for \$10,000 a corporate bond that earns 5% annual simple interest. How much additional money must be placed in U.S. government securities that earn a simple interest rate of 3.5% so that the total annual interest earned from the two investments is 4% of the total investment?

APPLYING THE CONCEPTS

For Exercises 65 to 67, use a graphing calculator to estimate the solution to the system of equations. Round coordinates to the nearest hundredth. See the Projects and Group Activities at the end of this chapter for assistance.

65. $y = -\frac{1}{2}x + 2$
$y = 2x - 1$

66. $y = \sqrt{2}x - 1$
$y = -\sqrt{3}x + 1$

67. $y = \pi x - \frac{2}{3}$
$y = -x + \frac{\pi}{2}$

Copyright © Houghton Mifflin Company. All rights reserved.

5.2 Solving Systems of Linear Equations by the Addition Method

Objective A

To solve a system of two linear equations in two variables by the addition method

The **addition method** is an alternative method for solving a system of equations. This method is based on the Addition Property of Equations. Use the addition method when it is not convenient to solve one equation for one variable in terms of another variable.

Note for the system of equations at the right the effect of adding Equation (2) to Equation (1). Because $-3y$ and $3y$ are additive inverses, adding the equations results in an equation with only one variable.

$$\begin{aligned} (1)\ 5x - 3y &= 14 \\ (2)\ 2x + 3y &= -7 \\ 7x + 0y &= 7 \\ 7x &= 7 \end{aligned}$$

The solution of the resulting equation is the first coordinate of the ordered-pair solution of the system.

$$\begin{aligned} 7x &= 7 \\ x &= 1 \end{aligned}$$

The second coordinate is found by substituting the value of x into Equation (1) or (2) and then solving for y. Equation (1) is used here.

$$\begin{aligned} (1)\quad 5x - 3y &= 14 \\ 5(1) - 3y &= 14 \\ 5 - 3y &= 14 \\ -3y &= 9 \\ y &= -3 \end{aligned}$$

The solution is $(1, -3)$.

Point of Interest

There are records of Babylonian mathematicians solving systems of equations 3600 years ago. Here is a system of equations from that time (in our modern notation):

$$\begin{aligned} \frac{2}{3}x &= \frac{1}{2}y + 500 \\ x + y &= 1800 \end{aligned}$$

We say *modern notation* for many reasons. Foremost is the fact that using variables did not become widespread until the 17th century. There are many other reasons, however. The equals sign had not been invented, 2 and 3 did not look like they do today, and zero had not even been considered as a possible number.

Sometimes each equation of the system of equations must be multiplied by a constant so that the coefficients of one of the variable terms are opposites.

HOW TO Solve by the addition method: (1) $3x + 4y = 2$
(2) $2x + 5y = -1$

To eliminate x, multiply Equation (1) by 2 and Equation (2) by -3. Note at the right how the constants are chosen.

$$\begin{aligned} 2(3x + 4y) &= 2 \cdot 2 \\ -3(2x + 5y) &= -3(-1) \end{aligned}$$

• The negative is used so that the coefficients will be opposites.

$$\begin{aligned} 6x + 8y &= 4 && \text{• 2 times Equation (1).} \\ -6x - 15y &= 3 && \text{• } -3 \text{ times Equation (2).} \\ -7y &= 7 && \text{• Add the equations.} \\ y &= -1 && \text{• Solve for } y. \end{aligned}$$

Substitute the value of y into Equation (1) or Equation (2) and solve for x. Equation (1) will be used here.

$$\begin{aligned} (1)\quad 3x + 4y &= 2 \\ 3x + 4(-1) &= 2 && \text{• Substitute } -1 \text{ for } y. \\ 3x - 4 &= 2 && \text{• Solve for } x. \\ 3x &= 6 \\ x &= 2 \end{aligned}$$

The solution is $(2, -1)$.

Study Tip

Always check the proposed solution of a system of equations. For the equation at the right:

$$\begin{array}{r|l} 3x + 4y = 2 & \\ \hline 3(2) + 4(-1) & 2 \\ 6 - 4 & 2 \\ 2 = 2 & \end{array}$$

$$\begin{array}{r|l} 2x + 5y = -1 & \\ \hline 2(2) + 5(-1) & -1 \\ 4 - 5 & -1 \\ -1 = -1 & \end{array}$$

The solution checks.

Copyright © Houghton Mifflin Company. All rights reserved.

HOW TO Solve by the addition method: (1) $\frac{2}{3}x + \frac{1}{2}y = 4$
(2) $\frac{1}{4}x - \frac{3}{8}y = -\frac{3}{4}$

Clear fractions. Multiply each equation by the LCM of the denominators.

$$6\left(\frac{2}{3}x + \frac{1}{2}y\right) = 6(4)$$
$$8\left(\frac{1}{4}x - \frac{3}{8}y\right) = 8\left(-\frac{3}{4}\right)$$

$$4x + 3y = 24$$
$$2x - 3y = -6$$
$$6x = 18$$
$$x = 3$$

• Eliminate y. Add the equations. Then solve for x.

$$\frac{2}{3}x + \frac{1}{2}y = 4$$

• This is Equation (1).

$$\frac{2}{3}(3) + \frac{1}{2}y = 4$$

• Substitute $x = 3$ into Equation (1) and solve for y.

$$2 + \frac{1}{2}y = 4$$
$$\frac{1}{2}y = 2$$
$$y = 4$$

The solution is (3, 4).

HOW TO Solve by the addition method: (1) $2x - y = 3$
(2) $4x - 2y = 6$

Eliminate y. Multiply Equation (1) by -2.

(1) $-2(2x - y) = -2(3)$ • -2 times Equation (1).
(3) $-4x + 2y = -6$ • This is Equation (3).

Add Equation (3) to Equation (2).

(2) $4x - 2y = 6$
(3) $-4x + 2y = -6$

$$0 = 0$$

TAKE NOTE

The result of adding Equations (3) and (2) is $0 = 0$. It is not $x = 0$ and it is not $y = 0$. There is no variable in the equation $0 = 0$. This result does not indicate that the solution is (0, 0); rather, it indicates a dependent system of equations.

The equation $0 = 0$ indicates that the system of equations is dependent. This means that the graphs of the two lines are the same. Therefore, the solutions of the system of equations are the ordered-pair solutions of the equation of the line. Solve Equation (1) for y.

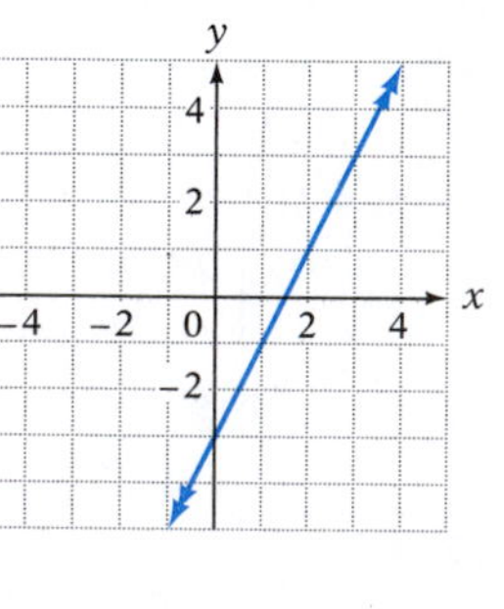

$$2x - y = 3$$
$$-y = -2x + 3$$
$$y = 2x - 3$$

The ordered-pair solutions are (x, y), where $y = 2x - 3$. These ordered pairs are usually written $(x, 2x - 3)$, where $2x - 3$ is substituted for y.

Copyright © Houghton Mifflin Company. All rights reserved.

Example 1 Solve by the addition method:
(1) $3x - 2y = 2x + 5$
(2) $2x + 3y = -4$

Solution
Write Equation (1) in the form $Ax + By = C$.

$$3x - 2y = 2x + 5$$
$$x - 2y = 5$$

Solve the system:

$$x - 2y = 5$$
$$2x + 3y = -4$$

Eliminate x.

$$-2(x - 2y) = -2(5)$$
$$2x + 3y = -4$$

$$-2x + 4y = -10$$
$$2x + 3y = -4$$

$$7y = -14$$ • **Add the equations.**
$$y = -2$$ • **Solve for *y*.**

Replace y in Equation (2).

$$2x + 3y = -4$$
$$2x + 3(-2) = -4$$
$$2x - 6 = -4$$
$$2x = 2$$
$$x = 1$$

The solution is $(1, -2)$.

You Try It 1 Solve by the addition method:
$2x + 5y = 6$
$3x - 2y = 6x + 2$

Your solution

Example 2 Solve by the addition method:
(1) $4x - 8y = 36$
(2) $3x - 6y = 27$

Solution
Eliminate x.

$$3(4x - 8y) = 3(36)$$
$$-4(3x - 6y) = -4(27)$$

$$12x - 24y = 108$$
$$-12x + 24y = -108$$

$$0 = 0$$ • **Add the equations.**

The system of equations is dependent. The solutions are the ordered pairs $\left(x, \frac{1}{2}x - \frac{9}{2}\right)$.

You Try It 2 Solve by the addition method:
$2x + y = 5$
$4x + 2y = 6$

Your solution

Solutions on p. S16

Copyright © Houghton Mifflin Company. All rights reserved.

Objective B To solve a system of three linear equations in three variables by the addition method

An equation of the form $Ax + By + Cz = D$, where A, B, and C are the coefficients of the variables and D is a constant, is a **linear equation in three variables.** Examples of this type of equation are shown at the right.

$$2x + 4y - 3z = 7$$
$$x - 6y + z = -3$$

Graphing an equation in three variables requires a third coordinate axis perpendicular to the xy-plane. The third axis is commonly called the z-axis. The result is a three-dimensional coordinate system called the ***xyz*-coordinate system.** To help visualize a three-dimensional coordinate system, think of a corner of a room: The floor is the xy-plane, one wall is the yz-plane, and the other wall is the xz-plane. A three-dimensional coordinate system is shown at the right.

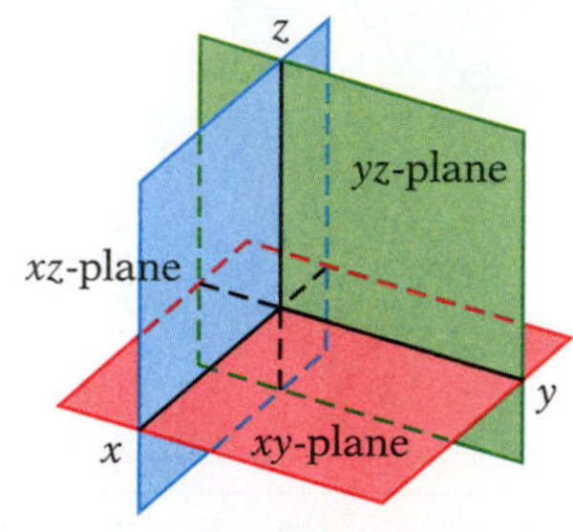

The graph of a point in an xyz-coordinate system is an **ordered triple** (x, y, z). Graphing an ordered triple requires three moves, the first in the direction of the x-axis, the second in the direction of the y-axis, and the third in the direction of the z-axis. The graphs of the points $(-4, 2, 3)$ and $(3, 4, -2)$ are shown at the right.

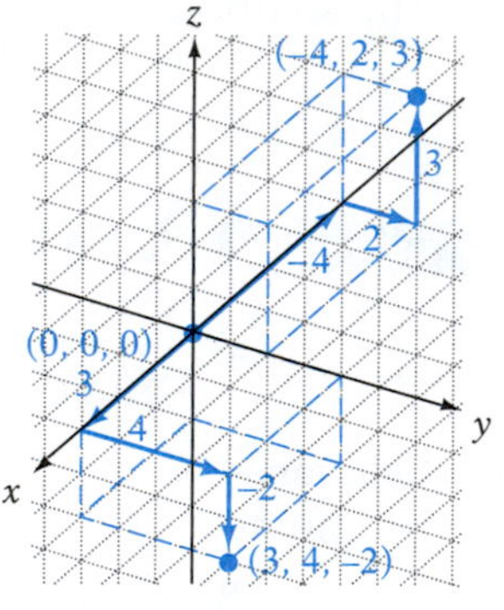

The graph of a linear equation in three variables is a plane. That is, if all the solutions of a linear equation in three variables were plotted in an xyz-coordinate system, the graph would look like a large piece of paper extending infinitely. The graph of $x + y + z = 3$ is shown at the right.

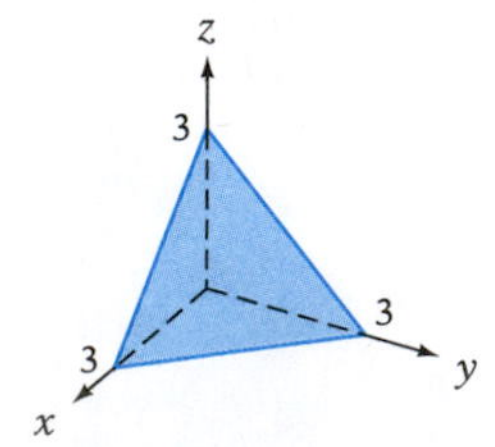

Copyright © Houghton Mifflin Company. All rights reserved.

There are different ways in which three planes can be oriented in an *xyz*-coordinate system. The systems of equations represented by the planes below are inconsistent.

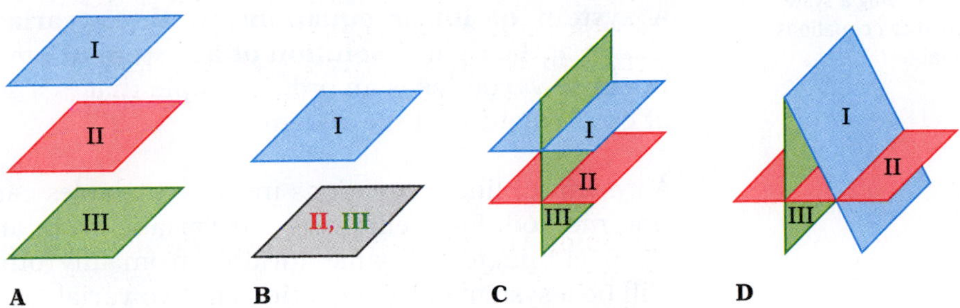

Graphs of Inconsistent Systems of Equations

For a system of three equations in three variables to have a solution, the graphs of the planes must intersect at a single point, they must intersect along a common line, or all equations must have a graph that is the same plane. These situations are shown in the figures below.

The three planes shown in Figure E intersect at a point. A system of equations represented by planes that intersect at a point is independent.

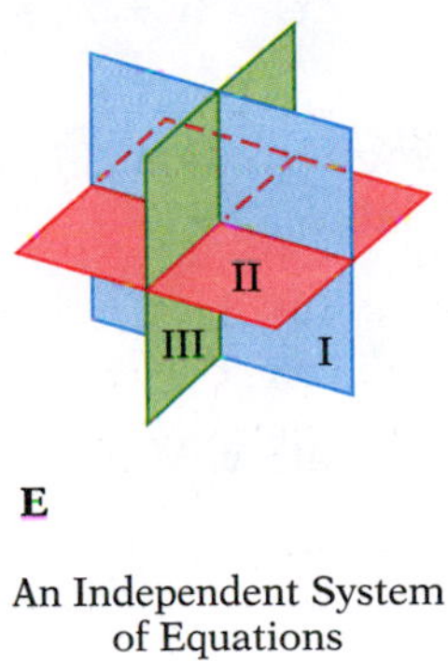

An Independent System of Equations

The planes shown in Figures F and G intersect along a common line. The system of equations represented by the planes in Figure H has a graph that is the same plane. The systems of equations represented by the graphs below are dependent.

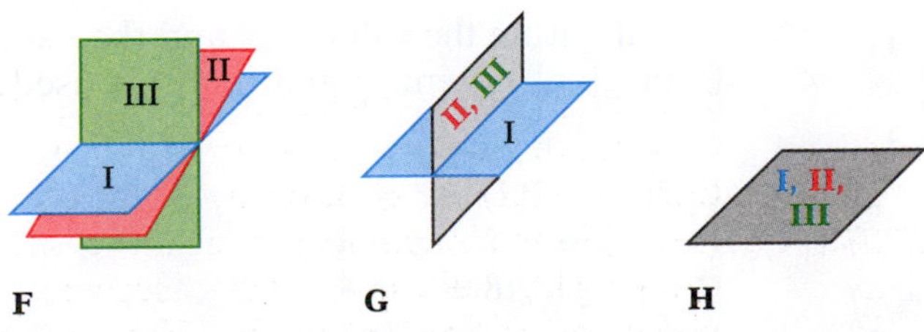

Dependent Systems of Equations

Copyright © Houghton Mifflin Company. All rights reserved.

Point of Interest

In the early 1980s, Stephen Hoppe became interested in winning Monopoly strategies. Finding these strategies required solving a system that contained 123 equations in 123 variables!

Just as a solution of an equation in two variables is an ordered pair (x, y), a **solution of an equation in three variables** is an ordered triple (x, y, z). For example, $(2, 1, -3)$ is a solution of the equation $2x - y - 2z = 9$. The ordered triple $(1, 3, 2)$ is not a solution.

A **system of linear equations in three variables** is shown at the right. A **solution of a system of equations in three variables** is an ordered triple that is a solution of each equation of the system.

$$\begin{aligned} x - 2y + z &= 6 \\ 3x + y - 2z &= 2 \\ 2x - 3y + 5z &= 1 \end{aligned}$$

A system of linear equations in three variables can be solved by using the addition method. First, eliminate one variable from any two of the given equations. Then eliminate the same variable from any other two equations. The result will be a system of two equations in two variables. Solve this system by the addition method.

HOW TO Solve:

$$\begin{aligned} &(1) & x + 4y - z &= 10 \\ &(2) & 3x + 2y + z &= 4 \\ &(3) & 2x - 3y + 2z &= -7 \end{aligned}$$

Eliminate z from Equations (1) and (2) by adding the two equations.

$$\begin{aligned} x + 4y - z &= 10 \\ 3x + 2y + z &= 4 \\ (4) \quad 4x + 6y &= 14 \end{aligned}$$

- Add the equations.

Eliminate z from Equations (1) and (3). Multiply Equation (1) by 2 and add to Equation (3).

$$\begin{aligned} 2x + 8y - 2z &= 20 \\ 2x - 3y + 2z &= -7 \\ (5) \quad 4x + 5y &= 13 \end{aligned}$$

- 2 times Equation (1).
- This is Equation (3).
- Add the equations.

Using Equations (4) and (5), solve the system of two equations in two variables.

$$\begin{aligned} &(4) & 4x + 6y &= 14 \\ &(5) & 4x + 5y &= 13 \end{aligned}$$

Eliminate x. Multiply Equation (5) by -1 and add to Equation (4).

$$\begin{aligned} 4x + 6y &= 14 \\ -4x - 5y &= -13 \\ y &= 1 \end{aligned}$$

- This is Equation (4).
- -1 times Equation (5).
- Add the equations.

Substitute the value of y into Equation (4) or Equation (5) and solve for x. Equation (4) is used here.

$$\begin{aligned} 4x + 6y &= 14 \\ 4x + 6(1) &= 14 \\ 4x &= 8 \\ x &= 2 \end{aligned}$$

- This is Equation (4).
- $y = 1$
- Solve for x.

Substitute the value of y and the value of x into one of the equations in the original system. Equation (2) is used here.

$$\begin{aligned} 3x + 2y + z &= 4 \\ 3(2) + 2(1) + z &= 4 \\ 6 + 2 + z &= 4 \\ 8 + z &= 4 \\ z &= -4 \end{aligned}$$

- $x = 2, y = 1$

The solution is $(2, 1, -4)$.

Study Tip

Always check the proposed solution of a system of equations. For the system at the right:

$$\begin{array}{r|l} x + 4y - z = 10 & \\ \hline 2 + 4(1) - (-4) & 10 \\ 10 = 10 & \end{array}$$

$$\begin{array}{r|l} 3x + 2y + z = 4 & \\ \hline 3(2) + 2(1) + (-4) & 4 \\ 4 = 4 & \end{array}$$

$$\begin{array}{r|l} 2x - 3y + 2z = -7 & \\ \hline 2(2) - 3(1) + 2(-4) & -7 \\ -7 = -7 & \end{array}$$

The solution checks.

Copyright © Houghton Mifflin Company. All rights reserved.

HOW TO Solve:

$$\begin{aligned} (1)\quad 2x - 3y - z &= 1 \\ (2)\quad x + 4y + 3z &= 2 \\ (3)\quad 4x - 6y - 2z &= 5 \end{aligned}$$

Eliminate x from Equations (1) and (2).

$2x - 3y - z = 1$ • This is Equation (1).
$-2x - 8y - 6z = -4$ • −2 times Equation (2).
$-11y - 7z = -3$ • Add the equations.

Eliminate x from Equations (1) and (3).

$-4x + 6y + 2z = -2$ • −2 times Equation (1).
$4x - 6y - 2z = 5$ • This is Equation (3).
$0 = 3$ • Add the equations.

The equation $0 = 3$ is not a true equation. The system of equations is inconsistent and therefore has no solution.

HOW TO Solve:

$$\begin{aligned} (1)\quad 3x - z &= -1 \\ (2)\quad 2y - 3z &= 10 \\ (3)\quad x + 3y - z &= 7 \end{aligned}$$

Eliminate x from Equations (1) and (3). Multiply Equation (3) by −3 and add to Equation (1).

$3x - z = -1$ • This is Equation (1).
$-3x - 9y + 3z = -21$ • −3 times Equation (3).
(4) $-9y + 2z = -22$ • Add the equations.

Use Equations (2) and (4) to form a system of equations in two variables.

(2) $2y - 3z = 10$
(4) $-9y + 2z = -22$

Eliminate z. Multiply Equation (2) by 2 and Equation (4) by 3.

$4y - 6z = 20$ • 2 times Equation (2).
$-27y + 6z = -66$ • 3 times Equation (4).
$-23y = -46$ • Add the equations.
$y = 2$ • Solve for y.

Substitute the value of y into Equation (2) or Equation (4) and solve for z. Equation (2) is used here.

(2) $2y - 3z = 10$ • This is Equation (2).
$2(2) - 3z = 10$ • $y = 2$
$4 - 3z = 10$ • Solve for z.
$-3z = 6$
$z = -2$

Substitute the value of z into Equation (1) and solve for x.

(1) $3x - z = -1$ • This is Equation (1).
$3x - (-2) = -1$ • $z = -2$
$3x + 2 = -1$ • Solve for x.
$3x = -3$
$x = -1$

The solution is $(-1, 2, -2)$.

Copyright © Houghton Mifflin Company. All rights reserved.

Example 3 Solve:
(1) $3x - y + 2z = 1$
(2) $2x + 3y + 3z = 4$
(3) $x + y - 4z = -9$

Solution Eliminate y. Add Equations (1) and (3).

$$3x - y + 2z = 1$$
$$x + y - 4z = -9$$
$$4x - 2z = -8$$

Multiply each side of the equation by $\frac{1}{2}$.

$$2x - z = -4$$ • Equation (4)

Multiply Equation (1) by 3 and add to Equation (2).

$$9x - 3y + 6z = 3$$
$$2x + 3y + 3z = 4$$
$$11x + 9z = 7$$ • Equation (5)

Solve the system of two equations.

(4) $2x - z = -4$
(5) $11x + 9z = 7$

Multiply Equation (4) by 9 and add to Equation (5).

$$18x - 9z = -36$$
$$11x + 9z = 7$$
$$29x = -29$$
$$x = -1$$

Replace x by -1 in Equation (4).

$$2x - z = -4$$
$$2(-1) - z = -4$$
$$-2 - z = -4$$
$$-z = -2$$
$$z = 2$$

Replace x by -1 and z by 2 in Equation (3).

$$x + y - 4z = -9$$
$$-1 + y - 4(2) = -9$$
$$-9 + y = -9$$
$$y = 0$$

The solution is $(-1, 0, 2)$.

You Try It 3 Solve:
$x - y + z = 6$
$2x + 3y - z = 1$
$x + 2y + 2z = 5$

Your solution

Solution on pp. S16–S17

Copyright © Houghton Mifflin Company. All rights reserved.

5.2 Exercises

Objective A **To solve a system of two linear equations in two variables by the addition method**

For Exercises 1 to 42, solve by the addition method.

1. $x - y = 5$, $x + y = 7$

2. $x + y = 1$, $2x - y = 5$

3. $3x + y = 4$, $x + y = 2$

4. $x - 3y = 4$, $x + 5y = -4$

5. $3x + y = 7$, $x + 2y = 4$

6. $x - 2y = 7$, $3x - 2y = 9$

7. $2x + 3y = -1$, $x + 5y = 3$

8. $x + 5y = 7$, $2x + 7y = 8$

9. $3x - y = 4$, $6x - 2y = 8$

10. $x - 2y = -3$, $-2x + 4y = 6$

11. $2x + 5y = 9$, $4x - 7y = -16$

12. $8x - 3y = 21$, $4x + 5y = -9$

13. $4x - 6y = 5$, $2x - 3y = 7$

14. $3x + 6y = 7$, $2x + 4y = 5$

15. $3x - 5y = 7$, $x - 2y = 3$

16. $3x + 4y = 25$, $2x + y = 10$

17. $x + 3y = 7$, $-2x + 3y = 22$

18. $2x - 3y = 14$, $5x - 6y = 32$

19. $3x + 2y = 16$, $2x - 3y = -11$

20. $2x - 5y = 13$, $5x + 3y = 17$

21. $4x + 4y = 5$, $2x - 8y = -5$

Copyright © Houghton Mifflin Company. All rights reserved.

22. $3x + 7y = 16$
$4x - 3y = 9$

23. $5x + 4y = 0$
$3x + 7y = 0$

24. $3x - 4y = 0$
$4x - 7y = 0$

25. $5x + 2y = 1$
$2x + 3y = 7$

26. $3x + 5y = 16$
$5x - 7y = -4$

27. $3x - 6y = 6$
$9x - 3y = 8$

28. $\frac{2}{3}x - \frac{1}{2}y = 3$
$\frac{1}{3}x - \frac{1}{4}y = \frac{3}{2}$

29. $\frac{3}{4}x + \frac{1}{3}y = -\frac{1}{2}$
$\frac{1}{2}x - \frac{5}{6}y = -\frac{7}{2}$

30. $\frac{2}{5}x - \frac{1}{3}y = 1$
$\frac{3}{5}x + \frac{2}{3}y = 5$

31. $\frac{5x}{6} + \frac{y}{3} = \frac{4}{3}$
$\frac{2x}{3} - \frac{y}{2} = \frac{11}{6}$

32. $\frac{3x}{4} + \frac{2y}{5} = -\frac{3}{20}$
$\frac{3x}{2} - \frac{y}{4} = \frac{3}{4}$

33. $\frac{2x}{5} - \frac{y}{2} = \frac{13}{2}$
$\frac{3x}{4} - \frac{y}{5} = \frac{17}{2}$

34. $\frac{x}{2} + \frac{y}{3} = \frac{5}{12}$
$\frac{x}{2} - \frac{y}{3} = \frac{1}{12}$

35. $\frac{3x}{2} - \frac{y}{4} = -\frac{11}{12}$
$\frac{x}{3} - y = -\frac{5}{6}$

36. $\frac{3x}{4} - \frac{2y}{3} = 0$
$\frac{5x}{4} - \frac{y}{3} = \frac{7}{12}$

37. $4x - 5y = 3y + 4$
$2x + 3y = 2x + 1$

38. $5x - 2y = 8x - 1$
$2x + 7y = 4y + 9$

39. $2x + 5y = 5x + 1$
$3x - 2y = 3y + 3$

40. $4x - 8y = 5$
$8x + 2y = 1$

41. $5x + 2y = 2x + 1$
$2x - 3y = 3x + 2$

42. $3x + 3y = y + 1$
$x + 3y = 9 - x$

Copyright © Houghton Mifflin Company. All rights reserved.

Objective B To solve a system of three linear equations in three variables by the addition method

For Exercises 43 to 66, solve by the addition method.

43. $x + 2y - z = 1$
$2x - y + z = 6$
$x + 3y - z = 2$

44. $x + 3y + z = 6$
$3x + y - z = -2$
$2x + 2y - z = 1$

45. $2x - y + 2z = 7$
$x + y + z = 2$
$3x - y + z = 6$

46. $x - 2y + z = 6$
$x + 3y + z = 16$
$3x - y - z = 12$

47. $3x + y = 5$
$3y - z = 2$
$x + z = 5$

48. $2y + z = 7$
$2x - z = 3$
$x - y = 3$

49. $x - y + z = 1$
$2x + 3y - z = 3$
$-x + 2y - 4z = 4$

50. $2x + y - 3z = 7$
$x - 2y + 3z = 1$
$3x + 4y - 3z = 13$

51. $2x + 3z = 5$
$3y + 2z = 3$
$3x + 4y = -10$

52. $3x + 4z = 5$
$2y + 3z = 2$
$2x - 5y = 8$

53. $2x + 4y - 2z = 3$
$x + 3y + 4z = 1$
$x + 2y - z = 4$

54. $x - 3y + 2z = 1$
$x - 2y + 3z = 5$
$2x - 6y + 4z = 3$

55. $2x + y - z = 5$
$x + 3y + z = 14$
$3x - y + 2z = 1$

56. $3x - y - 2z = 11$
$2x + y - 2z = 11$
$x + 3y - z = 8$

57. $3x + y - 2z = 2$
$x + 2y + 3z = 13$
$2x - 2y + 5z = 6$

58. $4x + 5y + z = 6$
$2x - y + 2z = 11$
$x + 2y + 2z = 6$

59. $2x - y + z = 6$
$3x + 2y + z = 4$
$x - 2y + 3z = 12$

60. $3x + 2y - 3z = 8$
$2x + 3y + 2z = 10$
$x + y - z = 2$

Copyright © Houghton Mifflin Company. All rights reserved.

61. $3x - 2y + 3z = -4$
$2x + y - 3z = 2$
$3x + 4y + 5z = 8$

62. $3x - 3y + 4z = 6$
$4x - 5y + 2z = 10$
$x - 2y + 3z = 4$

63. $3x - y + 2z = 2$
$4x + 2y - 7z = 0$
$2x + 3y - 5z = 7$

64. $2x + 2y + 3z = 13$
$-3x + 4y - z = 5$
$5x - 3y + z = 2$

65. $2x - 3y + 7z = 0$
$x + 4y - 4z = -2$
$3x + 2y + 5z = 1$

66. $5x + 3y - z = 5$
$3x - 2y + 4z = 13$
$4x + 3y + 5z = 22$

APPLYING THE CONCEPTS

67. Explain, graphically, the following situations when they are related to a system of three linear equations in three variables.
a. The system of equations has no solution.
b. The system of equations has exactly one solution.
c. The system of equations has infinitely many solutions.

68. Describe the graph of each of the following equations in an xyz-coordinate system.
a. $x = 3$ **b.** $y = 4$ **c.** $z = 2$ **d.** $y = x$

In Exercises 69 to 72, the systems are not linear systems of equations. However, they can be solved by using a modification of the addition method. Solve each system of equations.

69. $\frac{1}{x} - \frac{2}{y} = 3$
$\frac{2}{x} + \frac{3}{y} = -1$

70. $\frac{1}{x} + \frac{2}{y} = 3$
$\frac{1}{x} - \frac{3}{y} = -2$

71. $\frac{3}{x} + \frac{2}{y} = 1$
$\frac{2}{x} + \frac{4}{y} = -2$

72. $\frac{3}{x} - \frac{5}{y} = -\frac{3}{2}$
$\frac{1}{x} - \frac{2}{y} = -\frac{2}{3}$

73. For Exercise 69, solve each equation of the system for y. Then use a graphing calculator to verify the solution you determined algebraically. Note that the graphs are not straight lines.

74. For Exercise 70, solve each equation of the system for y. Then use a graphing calculator to verify the solution you determined algebraically. Note that the graphs are not straight lines.

Copyright © Houghton Mifflin Company. All rights reserved.

5.3 Application Problems

Objective A To solve rate-of-wind or rate-of-current problems

Motion problems that involve an object moving with or against a wind or current normally require two variables to solve.

A motorboat traveling with the current can go 24 mi in 2 h. Against the current, it takes 3 h to go the same distance. Find the rate of the motorboat in calm water and the rate of the current.

Strategy for Solving Rate-of-Wind or Rate-of-Current Problems

1. Choose one variable to represent the rate of the object in calm conditions and a second variable to represent the rate of the wind or current. Using these variables, express the rate of the object with and against the wind or current. Use the equation $rt = d$ to write expressions for the distance traveled by the object. The results can be recorded in a table.

Rate of the boat in calm water: x
Rate of the current: y

	Rate	·	Time	=	Distance
With the current	$x + y$	·	2	=	$2(x + y)$
Against the current	$x - y$	·	3	=	$3(x - y)$

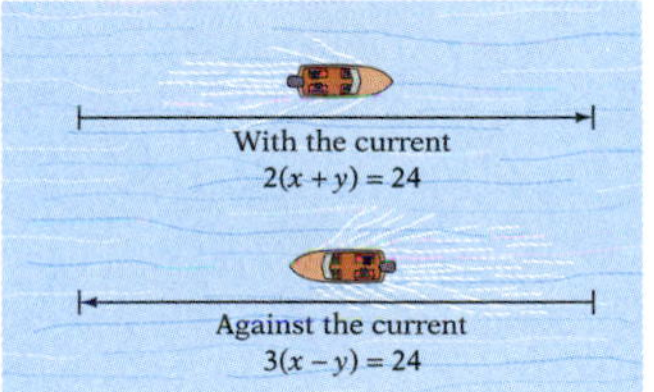

2. Determine how the expressions for distance are related.

The distance traveled with the current is 24 mi: $2(x + y) = 24$
The distance traveled against the current is 24 mi: $3(x - y) = 24$

Solve the system of equations.

$$2(x + y) = 24 \xrightarrow{\text{Multiply by } \frac{1}{2}} \frac{1}{2} \cdot 2(x + y) = \frac{1}{2} \cdot 24 \longrightarrow x + y = 12$$

$$3(x - y) = 24 \xrightarrow{\text{Multiply by } \frac{1}{3}} \frac{1}{3} \cdot 3(x - y) = \frac{1}{3} \cdot 24 \longrightarrow x - y = 8$$

$$2x = 20 \quad \text{• Add the equations.}$$
$$x = 10$$

Replace x by 10 in the equation $x + y = 12$. Solve for y.

$$x + y = 12$$
$$10 + y = 12$$
$$y = 2$$

The rate of the boat in calm water is 10 mph.
The rate of the current is 2 mph.

Copyright © Houghton Mifflin Company. All rights reserved.

Example 1

Flying with the wind, a plane flew 1000 mi in 5 h. Flying against the wind, the plane could fly only 500 mi in the same amount of time. Find the rate of the plane in calm air and the rate of the wind.

Strategy

- Rate of the plane in still air: p
  Rate of the wind: w

	Rate	*Time*	*Distance*
With wind	$p + w$	5	$5(p + w)$
Against wind	$p - w$	5	$5(p - w)$

- The distance traveled with the wind is 1000 mi.
  The distance traveled against the wind is 500 mi.

$$5(p + w) = 1000$$
$$5(p - w) = 500$$

Solution

$$5(p + w) = 1000 \qquad \frac{1}{5} \cdot 5(p + w) = \frac{1}{5} \cdot 1000$$
$$5(p - w) = 500 \qquad \frac{1}{5} \cdot 5(p - w) = \frac{1}{5} \cdot 500$$
$$p + w = 200$$
$$p - w = 100$$
$$2p = 300$$
$$p = 150$$

$$p + w = 200$$
$$150 + w = 200 \qquad \text{• Substitute 150 for } p.$$
$$w = 50$$

The rate of the plane in calm air is 150 mph. The rate of the wind is 50 mph.

You Try It 1

A rowing team rowing with the current traveled 18 mi in 2 h. Against the current, the team rowed 10 mi in 2 h. Find the rate of the rowing team in calm water and the rate of the current.

Your strategy

Your solution

Solution on p. S17

Copyright © Houghton Mifflin Company. All rights reserved.

Objective B To solve application problems

The application problems in this section are varieties of problems solved earlier in the text. Each of the strategies for the problems in this section will result in a system of equations.

A store owner purchased twenty 60-watt light bulbs and 30 fluorescent bulbs for a total cost of $80. A second purchase, at the same prices, included thirty 60-watt light bulbs and 10 fluorescent bulbs for a total cost of $50. Find the cost of a 60-watt bulb and that of a fluorescent bulb.

Strategy for Solving Application Problems

1. Choose a variable to represent each of the unknown quantities. Write numerical or variable expressions for all the remaining quantities. These results may be recorded in tables, one for each condition.

Cost of 60-watt bulb: b
Cost of fluorescent bulb: f

First Purchase

	Amount	·	*Unit Cost*	=	*Value*
60-watt	20	·	b	=	$20b$
Fluorescent	30	·	f	=	$30f$

Second Purchase

	Amount	·	*Unit Cost*	=	*Value*
60-watt	30	·	b	=	$30b$
Fluorescent	10	·	f	=	$10f$

2. Determine a system of equations. The strategies presented in the chapter on First-Degree Equations and Inequalities can be used to determine the relationships among the expressions in the tables. Each table will give one equation of the system of equations.

The total of the first purchase was \$80: $20b + 30f = 80$
The total of the second purchase was \$50: $30b + 10f = 50$

Solve the system of equations: (1) $20b + 30f = 80$
(2) $30b + 10f = 50$

$$\begin{aligned} 60b + 90f &= 240 \\ -60b - 20f &= -100 \\ \hline 70f &= 140 \\ f &= 2 \end{aligned}$$

- **3 times Equation (1).**
- **−2 times Equation (2).**

Replace f by 2 in Equation (1) and solve for b.

$$\begin{aligned} 20b + 30f &= 80 \\ 20b + 30(2) &= 80 \\ 20b + 60 &= 80 \\ 20b &= 20 \\ b &= 1 \end{aligned}$$

The cost of a 60-watt bulb is \$1.00.
The cost of a fluorescent bulb is \$2.00.

Some application problems may require more than two variables, as shown in Example 2 and You Try It 2 on the next page.

Copyright © Houghton Mifflin Company. All rights reserved.

Example 2

An investor has a total of \$20,000 deposited in three different accounts, which earn annual interest rates of 9%, 7%, and 5%. The amount deposited in the 9% account is twice the amount in the 7% account. If the total annual interest earned for the three accounts is \$1300, how much is invested in each account?

Strategy

- Amount invested at 9%: x
  Amount invested at 7%: y
  Amount invested at 5%: z

	Principal	*Rate*	*Interest*
Amount at 9%	x	0.09	$0.09x$
Amount at 7%	y	0.07	$0.07y$
Amount at 5%	z	0.05	$0.05z$

- The amount invested at 9% (x) is twice the amount invested at 7% (y): $x = 2y$
  The sum of the interest earned for all three accounts is \$1300:
  $0.09x + 0.07y + 0.05z = 1300$
  The total amount invested is \$20,000:
  $x + y + z = 20{,}000$

Solution

(1) $x = 2y$
(2) $0.09x + 0.07y + 0.05z = 1300$
(3) $x + y + z = 20{,}000$

Solve the system of equations. Substitute $2y$ for x in Equation (2) and Equation (3).

$0.09(2y) + 0.07y + 0.05z = 1300$
$2y + y + z = 20{,}000$

(4) $0.25y + 0.05z = 1300$ • $0.09(2y) + 0.07y = 0.25y$
(5) $3y + z = 20{,}000$ • $2y + y = 3y$

Solve the system of equations in two variables by multiplying Equation (5) by -0.05 and adding to Equation (4).

$$\begin{aligned} 0.25y + 0.05z &= 1300 \\ -0.15y - 0.05z &= -1000 \\ \hline 0.10y &= 300 \\ y &= 3000 \end{aligned}$$

Substituting the value of y into Equation (1), $x = 6000$.

Substituting the values of x and y into Equation (3), $z = 11{,}000$.

The investor placed \$6000 in the 9% account, \$3000 in the 7% account, and \$11,000 in the 5% account.

You Try It 2

A citrus grower purchased 25 orange trees and 20 grapefruit trees for \$290. The next week, at the same prices, the grower bought 20 orange trees and 30 grapefruit trees for \$330. Find the cost of an orange tree and the cost of a grapefruit tree.

Your strategy

Your solution

Solution on p. S17

Copyright © Houghton Mifflin Company. All rights reserved.

5.3 Exercises

Objective A To solve rate-of-wind or rate-of-current problems

1. A motorboat traveling with the current went 36 mi in 2 h. Against the current, it took 3 h to travel the same distance. Find the rate of the boat in calm water and the rate of the current.

2. A cabin cruiser traveling with the current went 45 mi in 3 h. Against the current, it took 5 h to travel the same distance. Find the rate of the cabin cruiser in calm water and the rate of the current.

3. A jet plane flying with the wind went 2200 mi in 4 h. Against the wind, the plane could fly only 1820 mi in the same amount of time. Find the rate of the plane in calm air and the rate of the wind.

4. Flying with the wind, a small plane flew 300 mi in 2 h. Against the wind, the plane could fly only 270 mi in the same amount of time. Find the rate of the plane in calm air and the rate of the wind.

5. A rowing team rowing with the current traveled 20 km in 2 h. Rowing against the current, the team rowed 12 km in the same amount of time. Find the rate of the team in calm water and the rate of the current.

6. A motorboat traveling with the current went 72 km in 3 h. Against the current, the boat could go only 48 km in the same amount of time. Find the rate of the boat in calm water and the rate of the current.

7. A turboprop plane flying with the wind flew 800 mi in 4 h. Flying against the wind, the plane required 5 h to travel the same distance. Find the rate of the plane in calm air and the rate of the wind.

8. Flying with the wind, a pilot flew 600 mi between two cities in 4 h. The return trip against the wind took 5 h. Find the rate of the plane in calm air and the rate of the wind.

9. A plane flying with a tailwind flew 600 mi in 5 h. Against the wind, the plane required 6 h to fly the same distance. Find the rate of the plane in calm air and the rate of the wind.

Copyright © Houghton Mifflin Company. All rights reserved.

10. Flying with the wind, a plane flew 720 mi in 3 h. Against the wind, the plane required 4 h to fly the same distance. Find the rate of the plane in calm air and the rate of the wind.

Objective B To solve application problems

11. **Flour Mixtures** A baker purchased 12 lb of wheat flour and 15 lb of rye flour for a total cost of $18.30. A second purchase, at the same prices, included 15 lb of wheat flour and 10 lb of rye flour. The cost of the second purchase was $16.75. Find the cost per pound of the wheat flour and of the rye flour.

12. **Business** A merchant mixed 10 lb of cinnamon tea with 5 lb of spice tea. The 15-pound mixture cost $40. A second mixture included 12 lb of cinnamon tea and 8 lb of spice tea. The 20-pound mixture cost $54. Find the cost per pound of the cinnamon tea and the spice tea.

13. **Purchasing** A carpenter purchased 60 ft of redwood and 80 ft of pine for a total cost of $286. A second purchase, at the same prices, included 100 ft of redwood and 60 ft of pine for a total cost of $396. Find the cost per foot of redwood and pine.

14. **Internet Services** A computer online service charges one hourly price for regular use but a higher hourly rate for designated "premium" areas. One customer was charged $28 after spending 2 h in premium areas and 9 regular hours; another spent 3 h in premium areas and 6 regular hours and was charged $27. What does the online service charge per hour for regular and premium services?

15. **Purchasing** A contractor buys 16 yd of nylon carpet and 20 yd of wool carpet for $1840. A second purchase, at the same prices, includes 18 yd of nylon carpet and 25 yd of wool carpet for $2200. Find the cost per yard of the wool carpet.

16. **Finances** During one month, a homeowner used 500 units of electricity and 100 units of gas for a total cost of $352. The next month, 400 units of electricity and 150 units of gas were used for a total cost of $304. Find the cost per unit of gas.

17. **Manufacturing** A company manufactures both mountain bikes and trail bikes. The cost of materials for a mountain bike is $70, and the cost of materials for a trail bike is $50. The cost of labor to manufacture a mountain bike is $80, and the cost of labor to manufacture a trail bike is $40. During a week in which the company has budgeted $2500 for materials and $2600 for labor, how many mountain bikes does the company plan to manufacture?

Copyright © Houghton Mifflin Company. All rights reserved.

18. **Manufacturing** A company manufactures both liquid crystal display (LCD) and cathode ray tube (CRT) color monitors. The cost of materials for a CRT monitor is \$50, whereas the cost of materials for an LCD monitor is \$150. The cost of labor to manufacture a CRT monitor is \$80, whereas the cost of labor to manufacture an LCD monitor is \$130. During a week when the company has budgeted \$9600 for materials and \$8760 for labor, how many LCD monitors does the company plan to manufacture?

19. **Chemistry** A chemist has two alloys, one of which is 10% gold and 15% lead, and the other of which is 30% gold and 40% lead. How many grams of each of the two alloys should be used to make an alloy that contains 60 g of gold and 88 g of lead?

20. **Health Science** A pharmacist has two vitamin-supplement powders. The first powder is 20% vitamin B1 and 10% vitamin B2. The second is 15% vitamin B1 and 20% vitamin B2. How many milligrams of each of the two powders should the pharmacist use to make a mixture that contains 130 mg of vitamin B1 and 80 mg of vitamin B2?

21. **Business** On Monday, a computer manufacturing company sent out three shipments. The first order, which contained a bill for \$114,000, was for four Model I, six Model V, and 10 Model X computers. The second shipment, which contained a bill for \$72,000, was for eight Model I, three Model V, and five Model X computers. The third shipment, which contained a bill for \$81,000, was for two Model I, nine Model V, and five Model X computers. What does the manufacturer charge for each Model V computer?

22. **Purchasing** A relief organization supplies blankets, cots, and lanterns to victims of fires, floods, and other natural disasters. One week the organization purchased 15 blankets, 5 cots, and 10 lanterns for a total cost of \$1250. The next week, at the same prices, the organization purchased 20 blankets, 10 cots, and 15 lanterns for a total cost of \$2000. The next week, at the same prices, the organization purchased 10 blankets, 15 cots, and 5 lanterns for a total cost of \$1625. Find the cost of one blanket, the cost of one cot, and the cost of one lantern.

23. **Ticket Sales** A science museum charges \$10 for a regular admission ticket, but members receive a discount of \$3 and students are admitted for \$5. Last Saturday, 750 tickets were sold for a total of \$5400. If 20 more student tickets than regular tickets were sold, how many of each type of ticket were sold?

24. **Investments** An investor owned 300 shares of an oil company and 200 shares of a movie company. The quarterly dividend from the two stocks was \$165. After the investor sold 100 shares of the oil company and bought an additional 100 shares of the movie company, the quarterly dividend became \$185. Find the dividend per share for each stock.

Copyright © Houghton Mifflin Company. All rights reserved.

25. **Investments** An investor has a total of $25,000 deposited in three different accounts, which earn annual interest rates of 8%, 6%, and 4%. The amount deposited in the 8% account is twice the amount in the 6% account. If the three accounts earn total annual interest of $1520, how much money is deposited in each account?

APPLYING THE CONCEPTS

26. **Geometry** Two angles are complementary. The measure of the larger angle is 9° more than eight times the measure of the smaller angle. Find the measures of the two angles. (Complementary angles are two angles whose sum is 90°.)

27. **Geometry** Two angles are supplementary. The measure of the larger angle is 40° more than three times the measure of the smaller angle. Find the measures of the two angles. (Supplementary angles are two angles whose sum is 180°.)

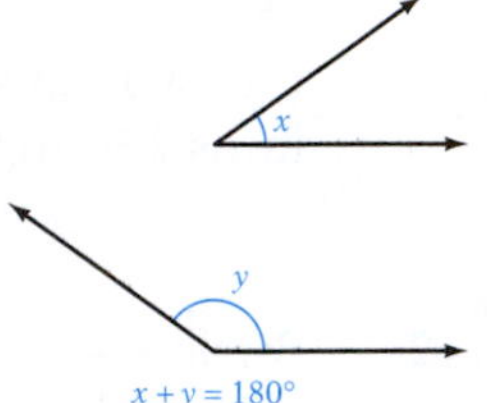

28. **Coins** The sum of the ages of a gold coin and a silver coin is 75 years. The age of the gold coin 10 years from now will be 5 years less than the age of the silver coin 10 years ago. Find the present ages of the two coins.

29. **Art** The difference between the ages of an oil painting and a watercolor is 35 years. The age of the oil painting 5 years from now will be twice the age that the watercolor was 5 years ago. Find the present age of each.

30. **Coins** A coin bank contains only dimes and quarters. The total value of all the coins is $1. How many of each type of coin are in the bank? (*Hint:* There is more than one answer to this problem.)

31. **Investments** An investor has a total of $25,000 deposited in three different accounts, which earn annual interest rates of 8%, 6%, and 4%. The amount deposited in the 8% account is $1000 more than the amount in the 4% account. If the investor wishes to earn $1520 from the three accounts, show that there is more than one way the investor can allocate money to each of the accounts to reach the goal. Explain why this happens here, whereas in Exercise 25 there is only one way to allocate the $25,000. *Suggestion:* Suppose the amount invested at 4% is $8000 and then find the other amounts. Now suppose $10,000 is invested at 4% and then find the other amounts.

Copyright © Houghton Mifflin Company. All rights reserved.

5.4 Solving Systems of Linear Inequalities

Objective A

To graph the solution set of a system of linear inequalities

Point of Interest

Large systems of linear inequalities containing over 100 inequalities have been used to solve application problems in such diverse areas as providing health care and hardening a nuclear missile silo.

Two or more inequalities considered together are called a **system of inequalities.** The **solution set of a system of inequalities** is the intersection of the solution sets of the individual inequalities. To graph the solution set of a system of inequalities, first graph the solution set of each inequality. The solution set of the system of inequalities is the region of the plane represented by the intersection of the shaded areas.

HOW TO Graph the solution set: $2x - y \le 3$
$3x + 2y > 8$

Solve each inequality for y.

$$\begin{aligned} 2x - y &\le 3 \\ -y &\le -2x + 3 \\ y &\ge 2x - 3 \end{aligned} \qquad \begin{aligned} 3x + 2y &> 8 \\ 2y &> -3x + 8 \\ y &> -\frac{3}{2}x + 4 \end{aligned}$$

Graph $y = 2x - 3$ as a solid line. Because the inequality is $\ge$, shade above the line.

Graph $y = -\frac{3}{2}x + 4$ as a dotted line. Because the inequality is $>$, shade above the line.

The solution set of the system is the region of the plane represented by the intersection of the solution sets of the individual inequalities.

TAKE NOTE

You can use a test point to check that the correct region has been denoted as the solution set. We can see from the graph that the point (2, 4) is in the solution set and, as shown below, it is a solution of each inequality in the system. This indicates that the solution set as graphed is correct.

$$\begin{aligned} 2x - y &\le 3 \\ 2(2) - (4) &\le 3 \\ 0 &\le 3 \quad \text{True} \end{aligned}$$

$$\begin{aligned} 3x + 2y &> 8 \\ 3(2) + 2(4) &> 8 \\ 14 &> 8 \quad \text{True} \end{aligned}$$

HOW TO Graph the solution set: $-x + 2y \ge 4$
$x - 2y \ge 6$

Solve each inequality for y.

$$\begin{aligned} -x + 2y &\ge 4 \\ 2y &\ge x + 4 \\ y &\ge \frac{1}{2}x + 2 \end{aligned} \qquad \begin{aligned} x - 2y &\ge 6 \\ -2y &\ge -x + 6 \\ y &\le \frac{1}{2}x - 3 \end{aligned}$$

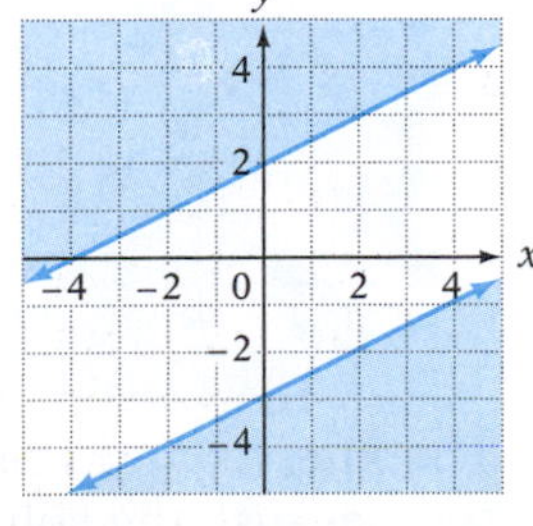

Shade above the solid line $y = \frac{1}{2}x + 2$.

Shade below the solid line $y = \frac{1}{2}x - 3$.

Because the solution sets of the two inequalities do not intersect, the solution of the system is the empty set.

Integrating Technology

See the appendix Keystroke Guide: *Graphing Inequalities* for instructions on using a graphing calculator to graph the solution set of a system of inequalities.

Copyright © Houghton Mifflin Company. All rights reserved.

Example 1

Graph the solution set: $y \geq x - 1$
$y < -2x$

Solution

Shade above the solid line $y = x - 1$.
Shade below the dotted line $y = -2x$.

The solution of the system is the intersection of the solution sets of the individual inequalities.

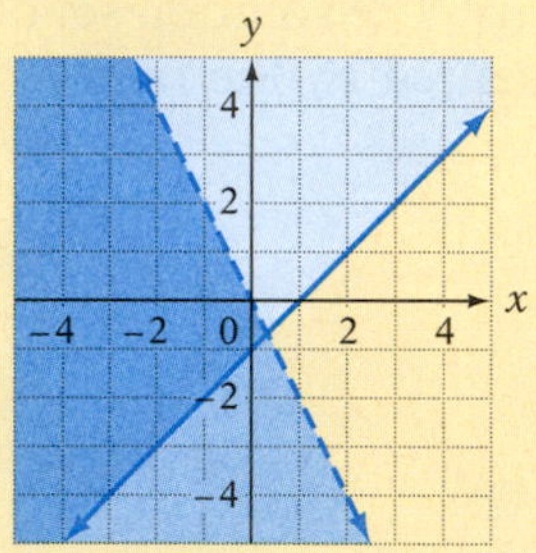

You Try It 1

Graph the solution set: $y \geq 2x - 3$
$y > -3x$

Your solution

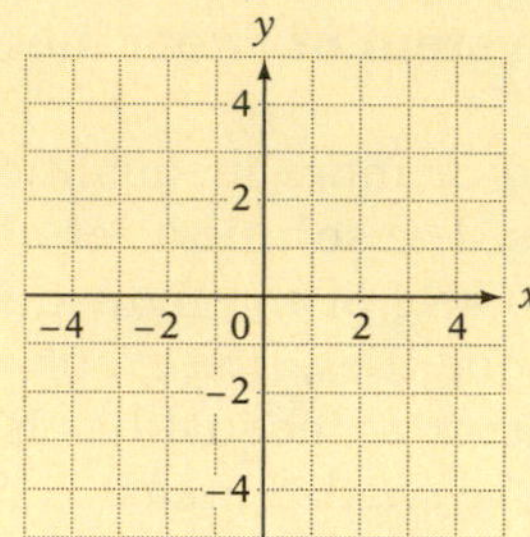

Example 2

Graph the solution set: $2x + 3y > 9$
$y < -\frac{2}{3}x + 1$

Solution

$2x + 3y > 9$
$3y > -2x + 9$
$y > -\frac{2}{3}x + 3$

Graph above the dotted line $y = -\frac{2}{3}x + 3$.

Graph below the dotted line $y = -\frac{2}{3}x + 1$.

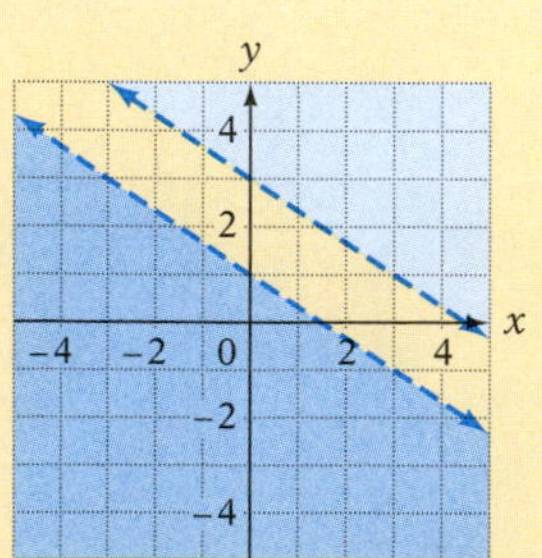

The intersection of the system is the empty set, because the solution sets of the two inequalities do not intersect.

You Try It 2

Graph the solution set: $3x + 4y > 12$
$y < \frac{3}{4}x - 1$

Your solution

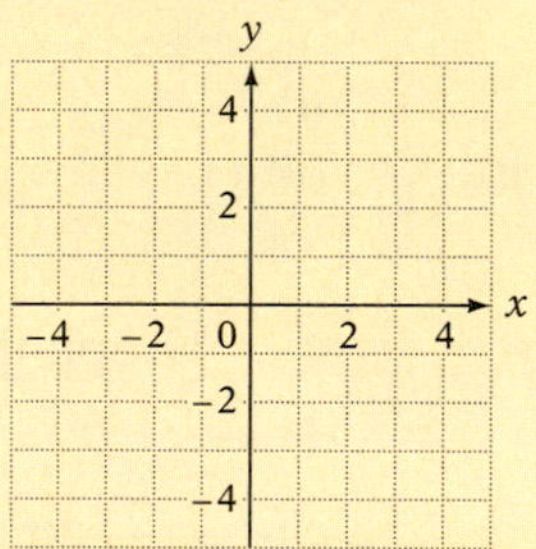

Solutions on pp. S17–S18

Copyright © Houghton Mifflin Company. All rights reserved.

5.4 Exercises

Objective A **To graph the solution set of a system of linear inequalities**

For Exercises 1 to 18, graph the solution set.

1. $x - y \geq 3$
$x + y \leq 5$

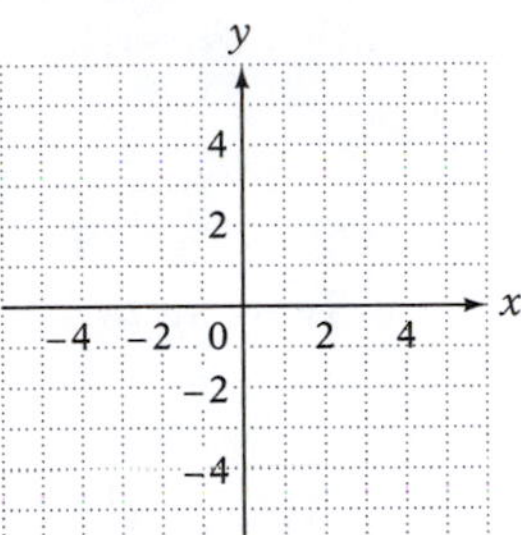

2. $2x - y < 4$
$x + y < 5$

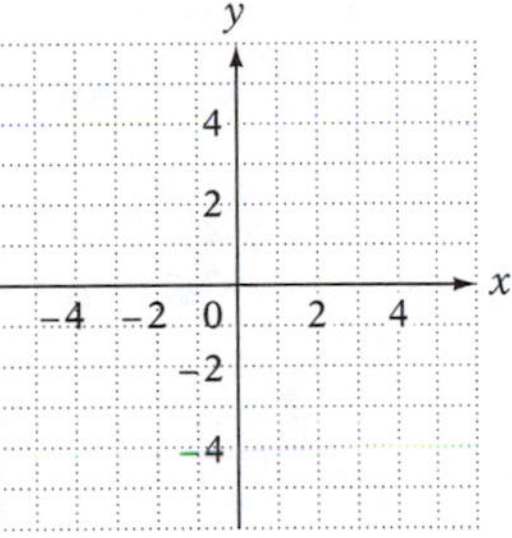

3. $3x - y < 3$
$2x + y \geq 2$

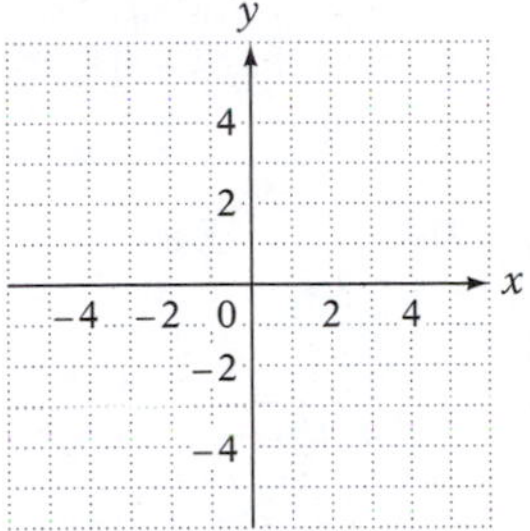

4. $x + 2y \leq 6$
$x - y \leq 3$

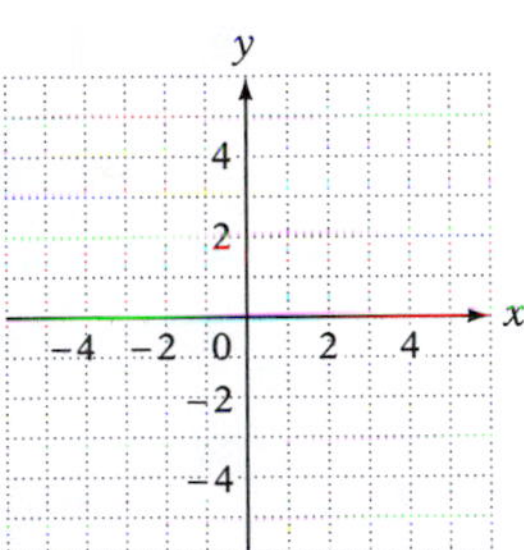

5. $2x + y \geq -2$
$6x + 3y \leq 6$

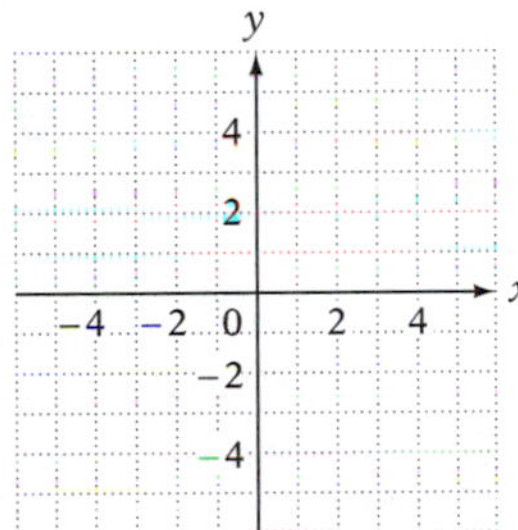

6. $x + y \geq 5$
$3x + 3y \leq 6$

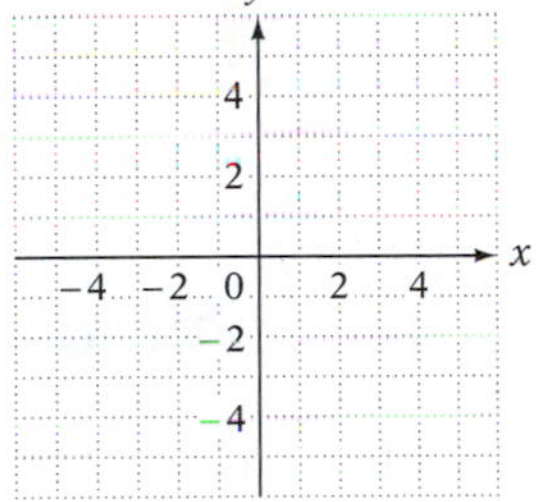

7. $3x - 2y < 6$
$y \leq 3$

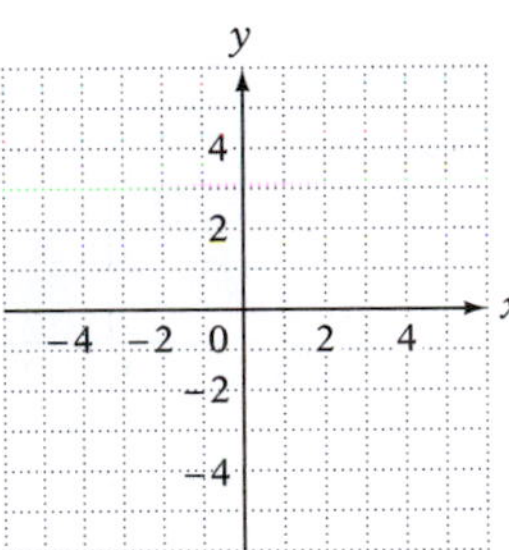

8. $x \leq 2$
$3x + 2y > 4$

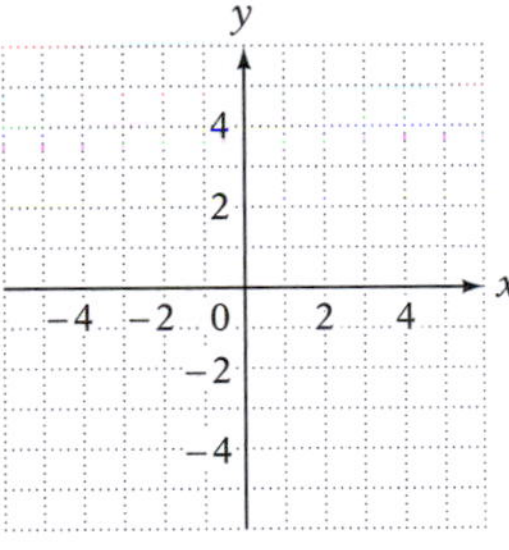

9. $y > 2x - 6$
$x + y < 0$

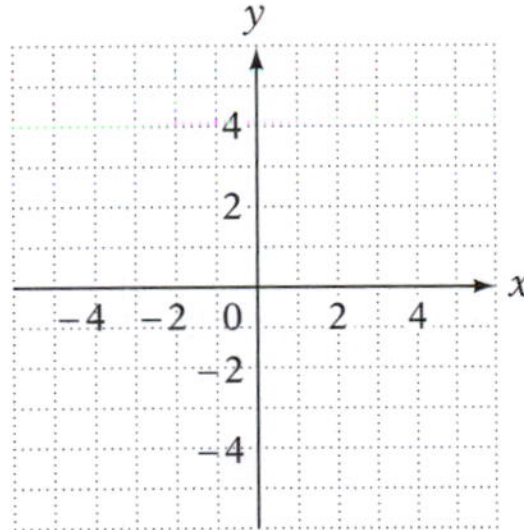

10. $x < 3$
$y < -2$

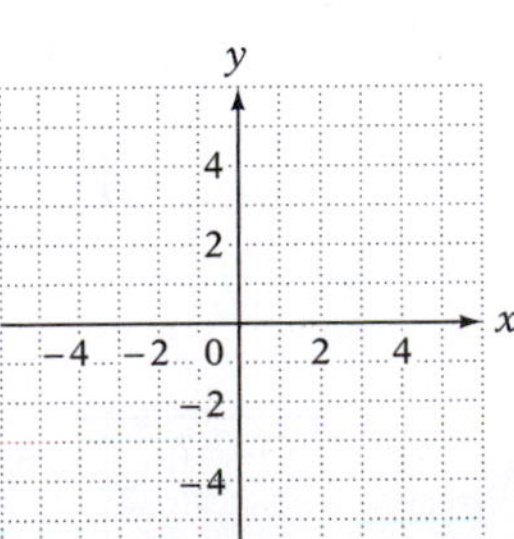

11. $x + 1 \geq 0$
$y - 3 \leq 0$

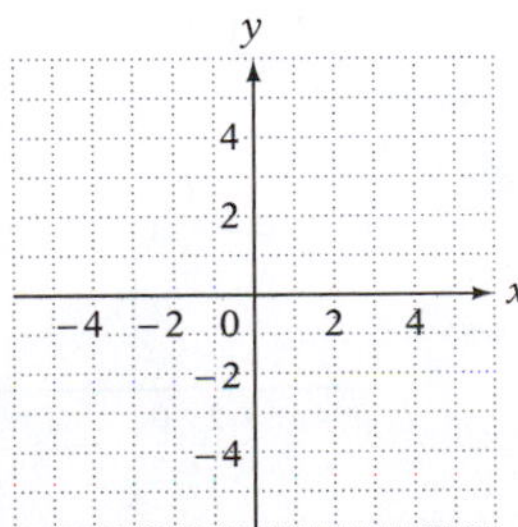

12. $5x - 2y \geq 10$
$3x + 2y \geq 6$

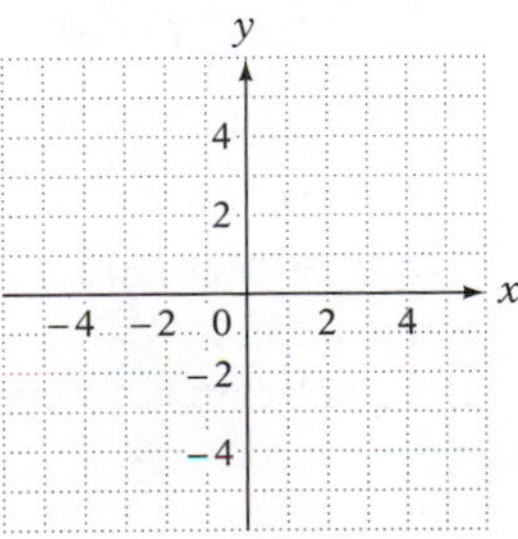

Copyright © Houghton Mifflin Company. All rights reserved.

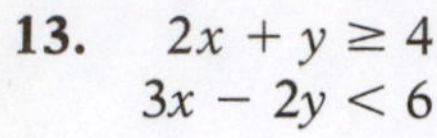

13. $2x + y \geq 4$
$3x - 2y < 6$

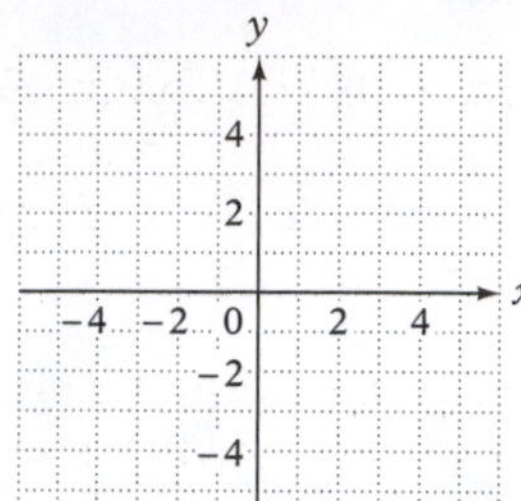

14. $3x - 4y < 12$
$x + 2y < 6$

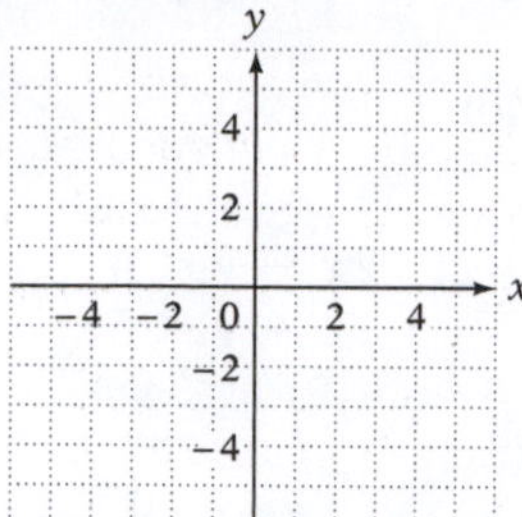

15. $x - 2y \leq 6$
$2x + 3y \leq 6$

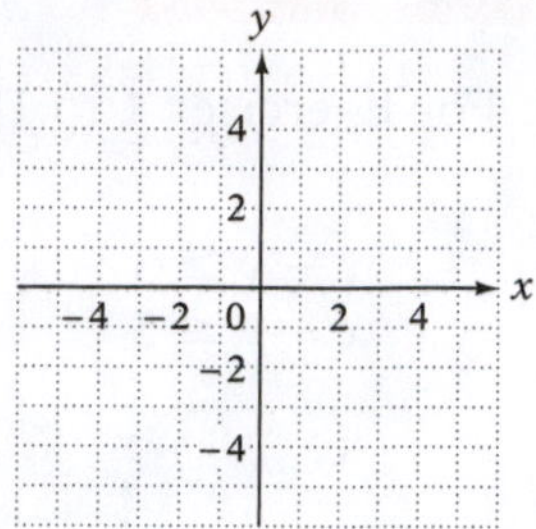

16. $x - 3y > 6$
$2x + y > 5$

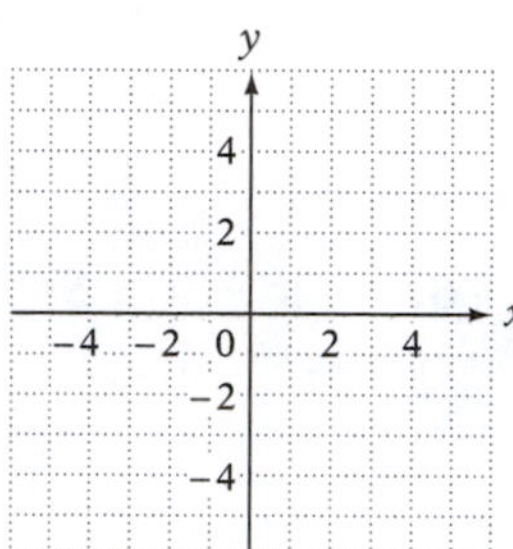

17. $x - 2y \leq 4$
$3x + 2y \leq 8$

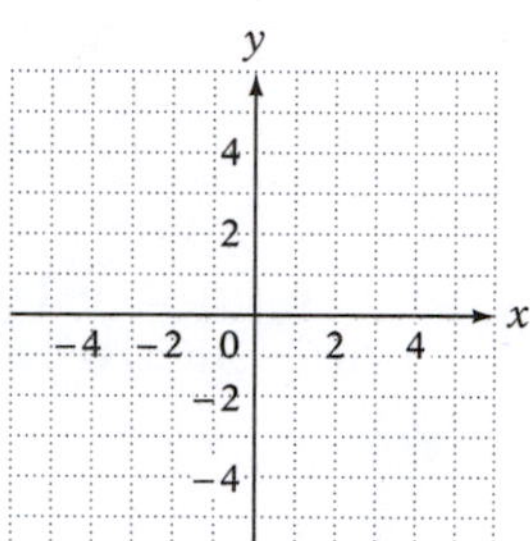

18. $3x - 2y < 0$
$5x + 3y > 9$

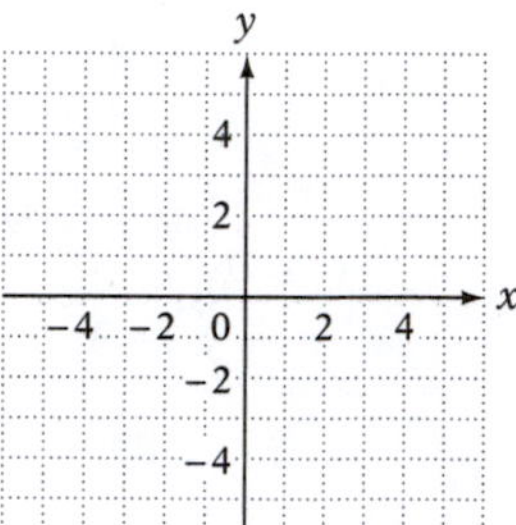

APPLYING THE CONCEPTS

For Exercises 19 to 24, graph the solution set.

19. $2x + 3y \leq 15$
$3x - y \leq 6$
$y \geq 0$

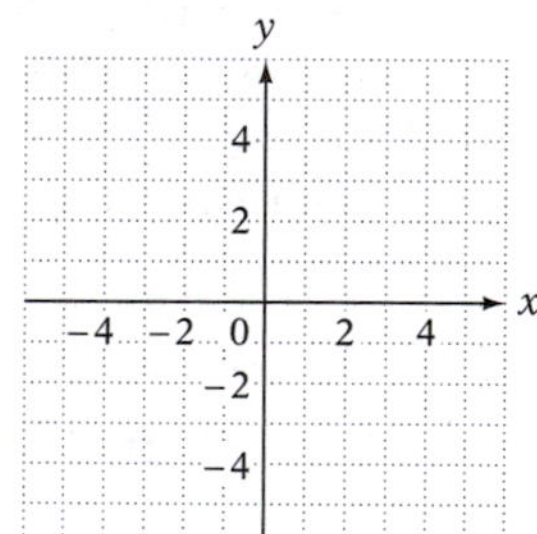

20. $x + y \leq 6$
$x - y \leq 2$
$x \geq 0$

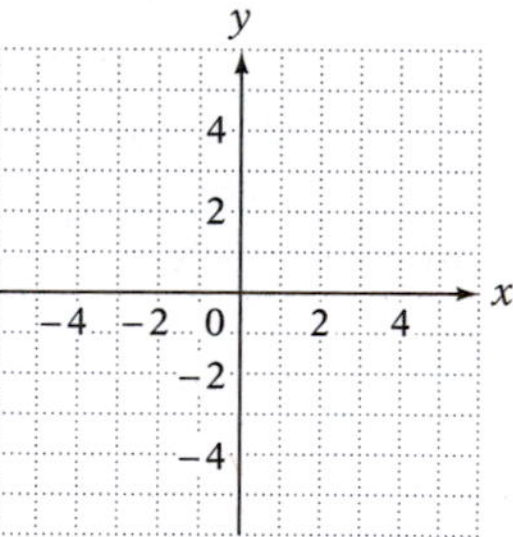

21. $x - y \leq 5$
$2x - y \geq 6$
$y \geq 0$

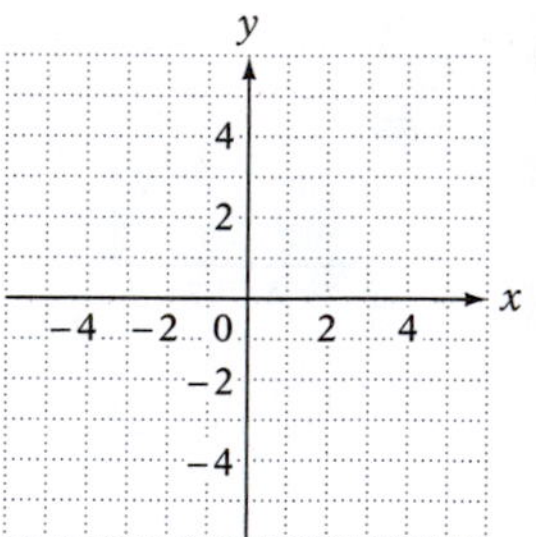

22. $x - 3y \leq 6$
$5x - 2y \geq 4$
$y \geq 0$

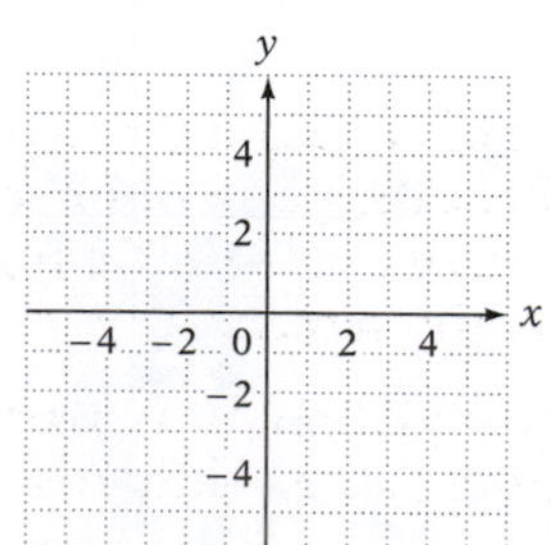

23. $2x - y \leq 4$
$3x + y < 1$
$y \leq 0$

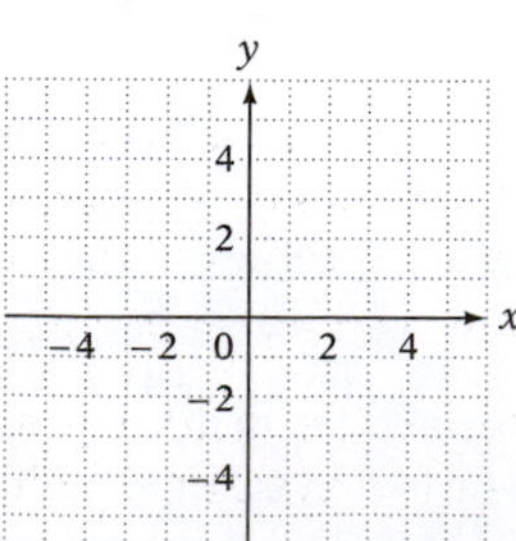

24. $x - y \leq 4$
$2x + 3y > 6$
$x \geq 0$

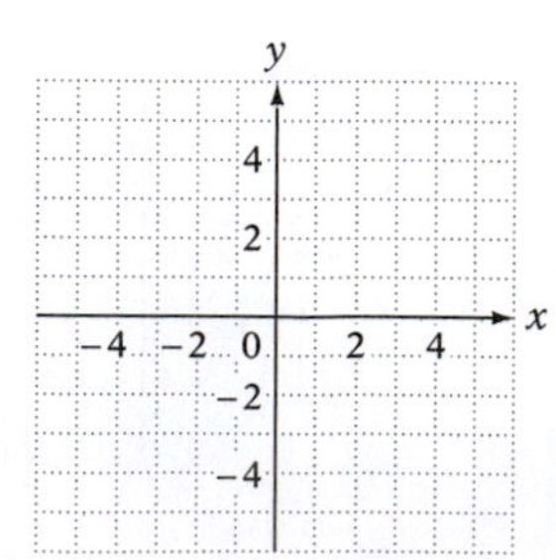

Copyright © Houghton Mifflin Company. All rights reserved.

Focus on Problem Solving

Solve an Easier Problem

One approach to problem solving is to try to solve an easier problem. Suppose you are in charge of your softball league, which consists of 15 teams. You must devise a schedule in which each team plays every other team once. How many games must be scheduled?

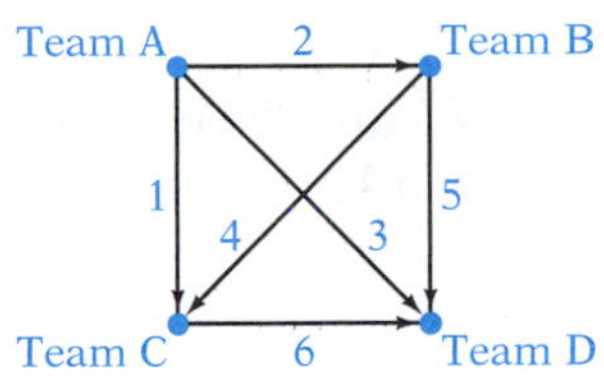

To solve this problem, we will attempt an easier problem first. Suppose that your league contains only a small number of teams. For instance, if there were only 1 team, you would schedule 0 games. If there were 2 teams, you would schedule 1 game. If there were 3 teams, you would schedule 3 games. The diagram at the left shows that 6 games must be scheduled when there are 4 teams in the league.

Here is a table of our results so far. (Remember that making a table is another strategy to be used in problem solving.)

Number of Teams	*Number of Games*	*Possible Pattern*
1	0	0
2	1	1
3	3	1 + 2
4	6	1 + 2 + 3

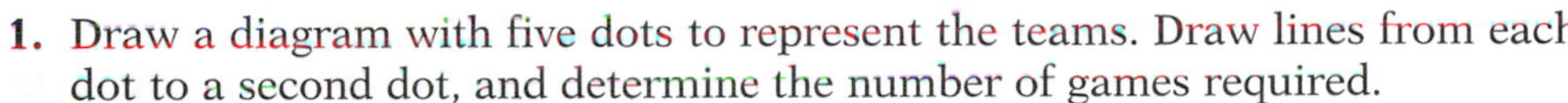

1. Draw a diagram with five dots to represent the teams. Draw lines from each dot to a second dot, and determine the number of games required.
2. What is the apparent pattern for the number of games required?
3. Assuming that the pattern continues, how many games must be scheduled for the 15 teams of the original problem?

After solving a problem, good problem solvers ask whether it is possible to solve the problem in a different manner. Here is a possible alternative method of solving the scheduling problem.

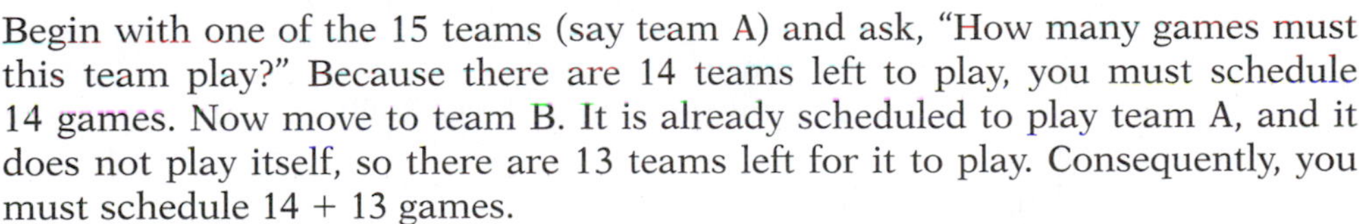

Begin with one of the 15 teams (say team A) and ask, "How many games must this team play?" Because there are 14 teams left to play, you must schedule 14 games. Now move to team B. It is already scheduled to play team A, and it does not play itself, so there are 13 teams left for it to play. Consequently, you must schedule 14 + 13 games.

4. Continue this reasoning for the remaining teams and determine the number of games that must be scheduled. Does this answer correspond to the answer you obtained using the first method?

Copyright © Houghton Mifflin Company. All rights reserved.

Projects and Group Activities

Using a Graphing Calculator to Solve a System of Equations

A graphing calculator can be used to solve a system of equations. For this procedure to work on most calculators, it is necessary that the point of intersection be on the screen. This means that you may have to experiment with Xmin, Xmax, Ymin, and Ymax values until the graphs intersect on the screen.

To solve a system of equations graphically, solve each equation for y. Then graph the equations of the system. Their point of intersection is the solution.

For instance, to solve the system of equations $\begin{aligned} 4x - 3y &= 7 \\ 5x + 4y &= 2 \end{aligned}$

first solve each equation for y. $4x - 3y = 7 \Rightarrow y = \frac{4}{3}x - \frac{7}{3}$

$$5x + 4y = 2 \Rightarrow y = -\frac{5}{4}x + \frac{1}{2}$$

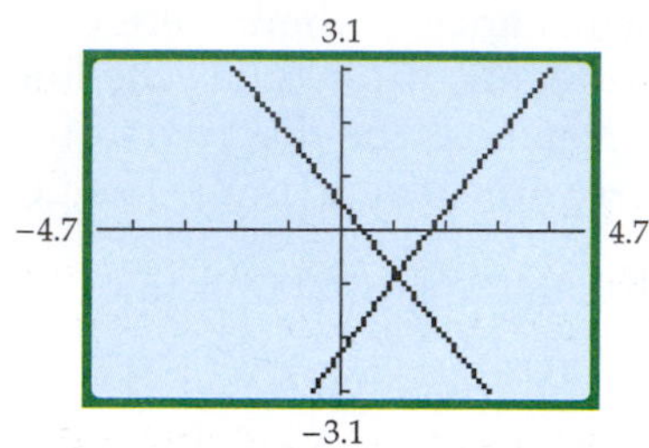

The keystrokes needed to solve this system using a TI-83 are given below. We are using a viewing window of [−4.7, 4.7] by [−3.1, 3.1]. The approximate solution is (1.096774, −0.870968).

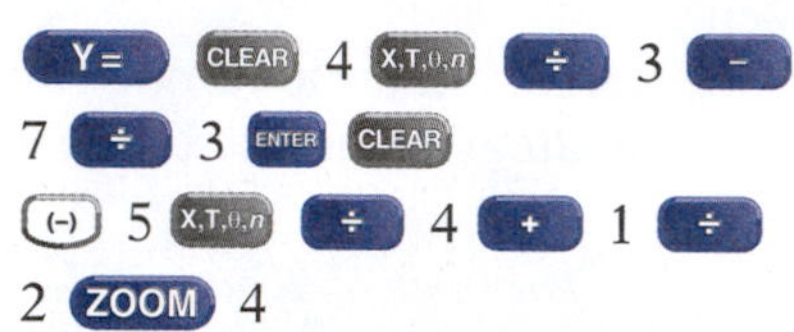

Once the calculator has drawn the graphs, use the following keystrokes to determine the point of intersection.

Here is an example of using a graphing calculator to solve an investment problem.

TAKE NOTE

As shown on the calculator screen in the example at the right, at the point of intersection, $y = 157.5$. This is the interest earned, $157.50, on $4500 invested at 3.5%. It is also the interest earned on $3500 invested at 4.5%. In other words, it is the interest earned on each account when both accounts earn the same interest.

HOW TO A marketing manager deposited $8000 in two simple interest accounts, one with an interest rate of 3.5% and the other with an interest rate of 4.5%. How much is deposited in each account if both accounts earn the same interest?

Amounted invested at 3.5%: x
Amount invested at 4.5%: $8000 - x$

Interest earned on the 3.5% account: $0.035x$
Interest earned on the 4.5% account: $0.045(8000 - x)$

Enter $0.035x$ into Y1. Enter $0.045(8000 - x)$ into Y2. Graph the equations. (We used a window of Xmin = 0, Xmax = 8000, Ymin = 0, Ymax = 400.) Use the intersect feature to find the point of intersection.

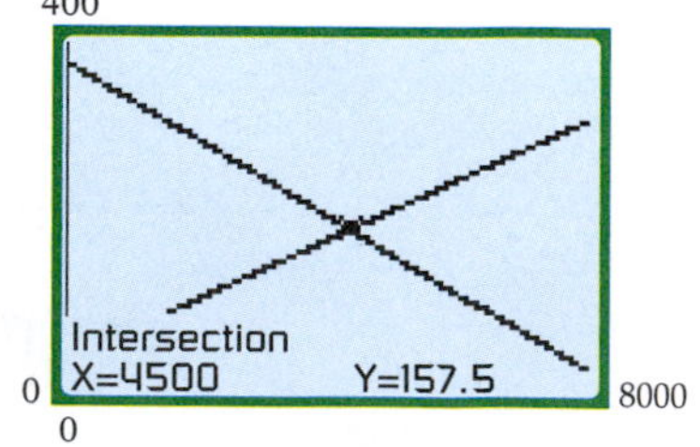

At the point of intersection, $x = 4500$. This is the amount in the 3.5% account.
The amount in the 4.5% account is $8000 - x = 8000 - 4500 = 3500$.

$4500 is invested at 3.5%. $3500 is invested at 4.5%.

Copyright © Houghton Mifflin Company. All rights reserved.

For Exercises 1 to 3, solve using a graphing calculator.

1. **Finances** Suppose that a breadmaker costs \$190, and that the ingredients and electricity needed to make one loaf of bread cost \$.95. If a comparable loaf of bread at a grocery store costs \$1.98, how many loaves of bread must you make before the breadmaker pays for itself?

2. **Finances** Suppose a natural gas clothes dryer costs \$260 and uses \$.40 of gas to dry a load of clothes for 1 h. The laundromat charges \$1.75 to use a dryer for 1 h.
   a. How many loads of clothes must you dry before the gas dryer purchase becomes more economical?
   b. What is the y-coordinate of the point of intersection? What does this represent in the context of the problem?

3. **Investments** When Mitch Deerfield changed jobs, he rolled over the \$7500 in his retirement account into two simple interest accounts. On one account, the annual simple interest rate is 6.25%; on the second account, the annual simple interest rate is 5.75%.
   a. How much is invested in each account if the accounts earn the same amount of annual interest?
   b. What is the y-coordinate of the point of intersection? What does this represent in the context of the problem?

Chapter 5 Summary

Key Words	**Examples**
A *system of equations* is two or more equations considered together. A *solution of a system of equations in two variables* is an ordered pair that is a solution of each equation in the system. [5.1A, p. 297]	The solution of the system $x + y = 2$, $x - y = 4$ is the ordered pair $(3, -1)$. $(3, -1)$ is the only ordered pair that is a solution of both equations.
When the graphs of a system of equations intersect at only one point, the system is called an *independent system of equations*. [5.1A, p. 297]	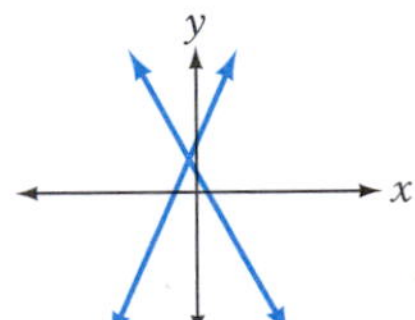
When the graphs of a system of equations do not intersect, the system has no solution and is called an *inconsistent system of equations*. [5.1A, p. 298]	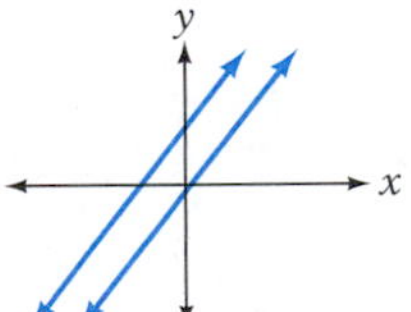
When the graphs of a system of equations coincide, the system is called a *dependent system of equations*. [5.1A, p. 298]	

Copyright © Houghton Mifflin Company. All rights reserved.

An equation of the form $Ax + By + Cz = D$, where A, B, and C are coefficients of the variables and D is a constant, is a *linear equation in three variables*. A *solution of an equation in three variables* is an *ordered triple* (x, y, z). [5.2B, p. 312]

$3x + 2y - 5z = 12$ is a linear equation in three variables. One solution of this equation is the ordered triple $(0, 1, -2)$.

A *solution of a system of equations in three variables* is an ordered triple that is a solution of each equation of the system. [5.2B, p. 314]

The solution of the system

$$\begin{aligned} 3x + y - 3z &= 2 \\ -x + 2y + 3z &= 6 \\ 2x + 2y - 2z &= 4 \end{aligned}$$

is the ordered triple (1, 2, 1). (1, 2, 1) is the only ordered triple that is a solution of all three equations.

Two or more inequalities considered together are called a *system of inequalities*. The *solution set of a system of inequalities* is the intersection of the solution sets of the individual inequalities. [5.4A, p. 329]

$x + y > 3$
$x - y > -2$

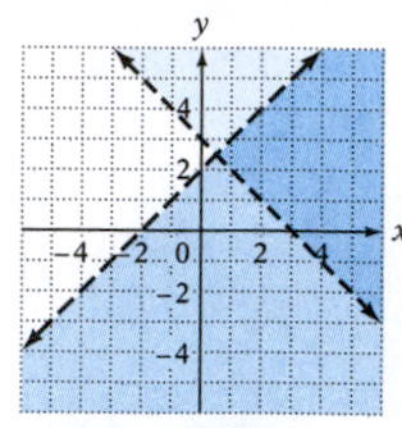

Essential Rules and Procedures

Examples

Solving Systems of Equations

A system of equations can be solved by:

a. *Graphing* [5.1A, p. 297]

$$y = \frac{1}{2}x + 2$$

$$y = \frac{5}{2}x - 2$$

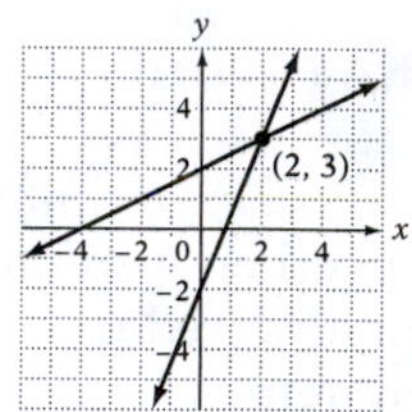

b. *The substitution method* [5.1B, p. 300]

(1) $2x - 3y = 4$
(2) $y = -x + 2$

Substitute the value of y into equation (1).

$2x - 3(-x + 2) = 4$

c. *The addition method* [5.2A, p. 309]

$$\begin{aligned} -2x + 3y &= 7 \\ 2x - 5y &= 2 \\ -2y &= 9 \end{aligned}$$

• **Add the equations.**

Annual Simple Interest Equation [5.1C, p. 303]

Principal · simple interest rate = simple interest

$$Pr = I$$

You have a total of \$10,000 to invest in two simple interest accounts, one earning 4.5% annual simple interest and the other earning 5% annual simple interest. If you earn \$485 per year in interest from these two investments, how much do you have invested in each account?

$$\begin{aligned} x + y &= 10{,}000 \\ 0.045x + 0.05y &= 485 \end{aligned}$$

Copyright © Houghton Mifflin Company. All rights reserved.

Chapter 5 Review Exercises

1. Solve by substitution: $2x - 6y = 15$
$x = 4y + 8$

2. Solve by the addition method: $3x + 2y = 2$
$x + y = 3$

3. Solve by graphing: $x + y = 3$
$3x - 2y = -6$

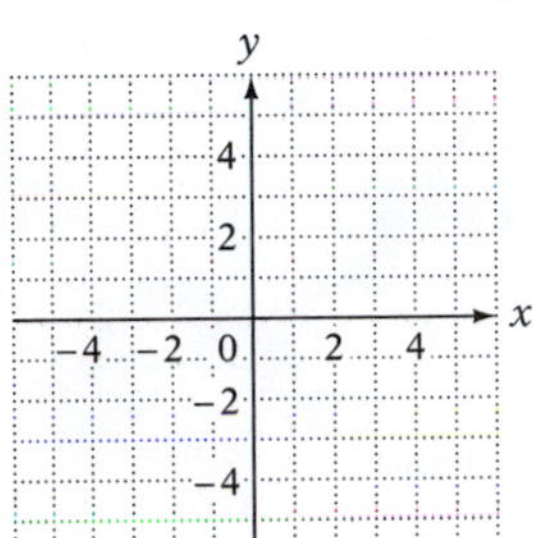

4. Solve by graphing: $2x - y = 4$
$y = 2x - 4$

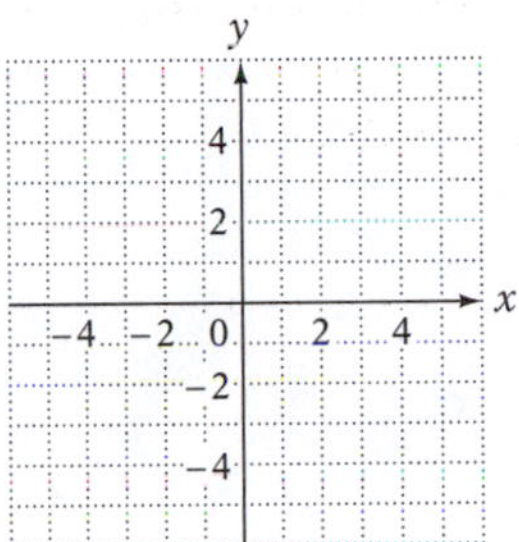

5. Solve by substitution: $3x + 12y = 18$
$x + 4y = 6$

6. Solve by the addition method: $5x - 15y = 30$
$x - 3y = 6$

7. Solve:
$3x - 4y - 2z = 17$
$4x - 3y + 5z = 5$
$5x - 5y + 3z = 14$

8. Solve:
$3x + y = 13$
$2y + 3z = 5$
$x + 2z = 11$

9. Is $(1, -2)$ a solution of the system of equations?
$6x + y = 4$
$2x - 5y = 12$

10. Solve by substitution:
$2x - 4y = 11$
$y = 3x - 4$

Copyright © Houghton Mifflin Company. All rights reserved.

11. Solve by substitution:
$2x - y = 7$
$3x + 2y = 7$

12. Solve by the addition method:
$3x - 4y = 1$
$2x + 5y = 16$

13. Solve:
$x + y + z = 0$
$x + 2y + 3z = 5$
$2x + y + 2z = 3$

14. Solve:
$x + 3y + z = 6$
$2x + y - z = 12$
$x + 2y - z = 13$

15. Graph the solution set:
$x + 3y \leq 6$
$2x - y \geq 4$

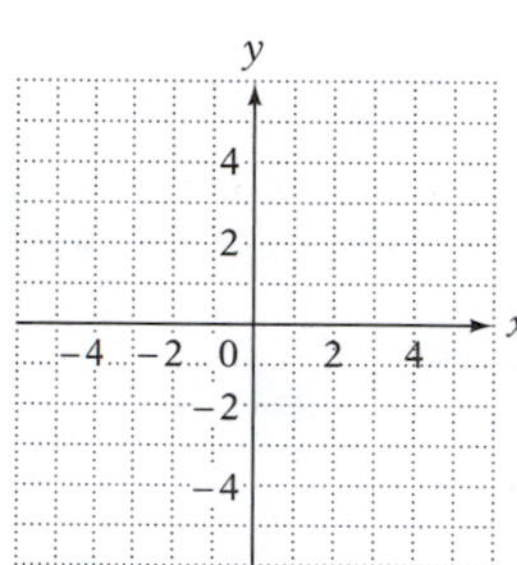

16. Graph the solution set:
$2x + 4y \geq 8$
$x + y \leq 3$

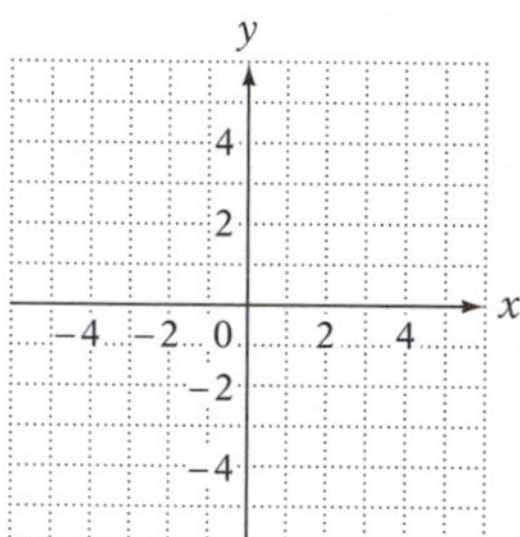

17. Boating A cabin cruiser traveling with the current went 60 mi in 3 h. Against the current, it took 5 h to travel the same distance. Find the rate of the cabin cruiser in calm water and the rate of the current.

18. Aeronautics A pilot flying with the wind flew 600 mi in 3 h. Flying against the wind, the pilot required 4 h to travel the same distance. Find the rate of the plane in calm air and the rate of the wind.

19. Ticket Sales At a movie theater, admission tickets are \$5 for children and \$8 for adults. The receipts for one Friday evening were \$2500. The next day, there were three times as many children as the preceding evening and only half the number of adults as the night before, yet the receipts were still \$2500. Find the number of children who attended on Friday evening.

20. Investments A trust administrator divides \$20,000 between two accounts. One account earns an annual simple interest rate of 3%, and a second account earns an annual simple interest rate of 7%. The total annual income from the two accounts is \$1200. How much is invested in each account?

Copyright © Houghton Mifflin Company. All rights reserved.

Chapter 5 Test

1. Solve by graphing: $2x - 3y = -6$
$2x - y = 2$

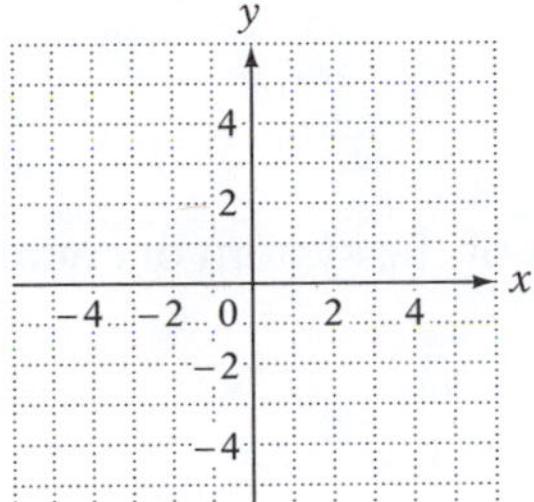

2. Solve by graphing: $x - 2y = -5$
$3x + 4y = -15$

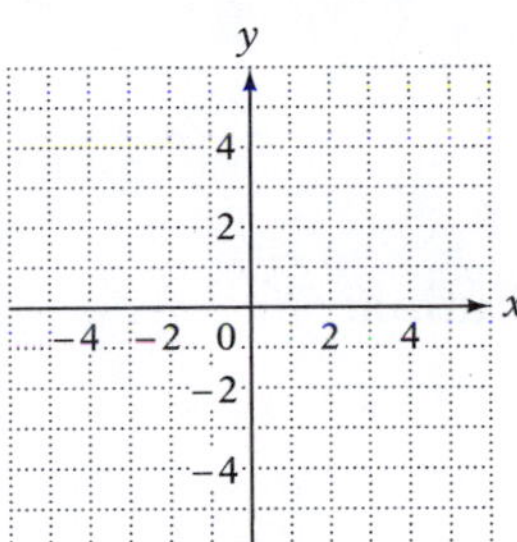

3. Graph the solution set: $2x - y < 3$
$4x + 3y < 11$

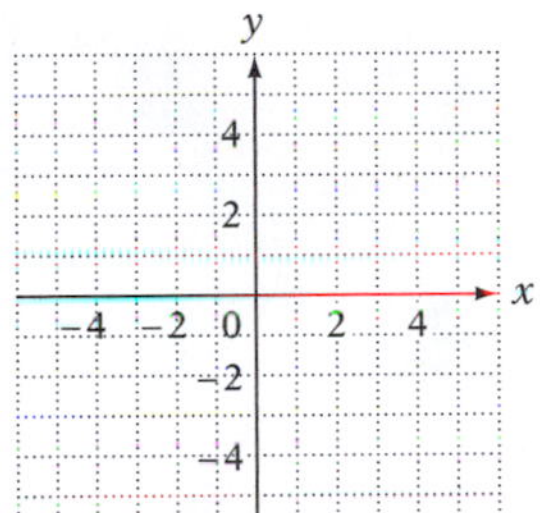

4. Graph the solution set: $x + y > 2$
$2x - y < -1$

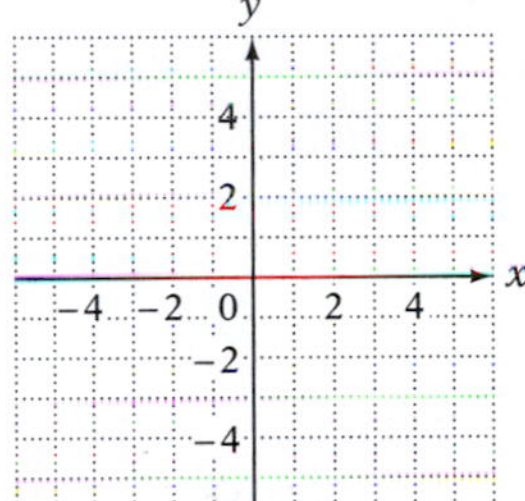

5. Solve by substitution: $3x + 2y = 4$
$x = 2y - 1$

6. Solve by substitution: $5x + 2y = -23$
$2x + y = -10$

7. Solve by substitution: $y = 3x - 7$
$y = -2x + 3$

8. Solve by the addition method:
$3x + 4y = -2$
$2x + 5y = 1$

9. Solve by the addition method:
$4x - 6y = 5$
$6x - 9y = 4$

10. Solve by the addition method:
$3x - y = 2x + y - 1$
$5x + 2y = y + 6$

Copyright © Houghton Mifflin Company. All rights reserved.

11. Solve:
$$2x + 4y - z = 3$$
$$x + 2y + z = 5$$
$$4x + 8y - 2z = 7$$

12. Solve:
$$x - y - z = 5$$
$$2x + z = 2$$
$$3y - 2z = 1$$

13. Solve by substitution:
$$x - y = 3$$
$$2x + y = -4$$

14. Is $(2, -2)$ a solution of the system of equations?
$$5x + 2y = 6$$
$$3x + 5y = -4$$

15. Solve:
$$x - y + z = 2$$
$$2x - y - z = 1$$
$$x + 2y - 3z = -4$$

16. **Aeronautics** A plane flying with the wind went 350 mi in 2 h. The return trip, flying against the wind, took 2.8 h. Find the rate of the plane in calm air and the rate of the wind.

17. **Purchasing** A clothing manufacturer purchased 60 yd of cotton and 90 yd of wool for a total cost of \$1800. Another purchase, at the same prices, included 80 yd of cotton and 20 yd of wool for a total cost of \$1000. Find the cost per yard of the cotton and the wool.

18. **Investments** The annual interest earned on two investments is \$549. One investment is in a 2.7% tax-free annual simple interest account, and the other investment is in a 5.1% annual simple interest CD. The total amount invested is \$15,000. How much is invested in each account?

Copyright © Houghton Mifflin Company. All rights reserved.

Cumulative Review Exercises

1. Simplify: $-2\sqrt{90}$

2. Solve: $3(x - 5) = 2x + 7$

3. Simplify: $3[x - 2(5 - 2x) - 4x] + 6$

4. Evaluate $a + bc \div 2$ when $a = 4$, $b = 8$, and $c = -2$.

5. Solve: $2x - 3 < 9$ or $5x - 1 < 4$

6. Solve: $|x - 2| - 4 < 2$

7. Solve: $|2x - 3| > 5$

8. Given $F(x) = x^2 - 3$, find $F(2)$.

9. Graph the solution set of $\{x|x \leq 2\} \cap \{x|x > -3\}$.

   −5 −4 −3 −2 −1 0 1 2 3 4 5

10. Find the equation of the line that contains the point $(-2, 3)$ and has slope $-\frac{2}{3}$.

11. Find the equation of the line that contains the points $(2, -1)$ and $(3, 4)$.

12. Find the equation of the line that contains the point $(-2, 2)$ and is perpendicular to the line $2x - 3y = 6$.

13. Graph $2x - 5y = 10$.

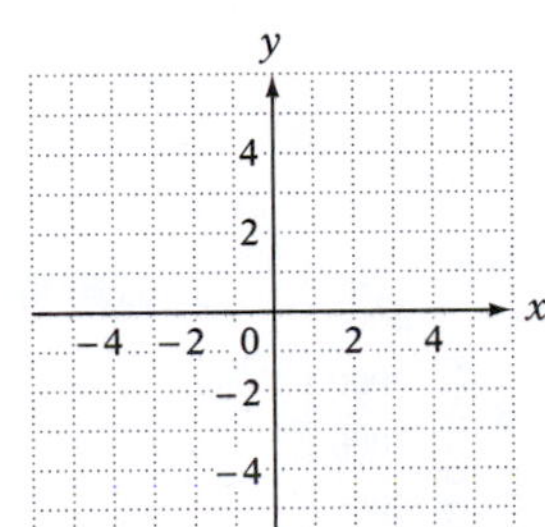

14. Graph the solution set of $3x - 4y \geq 8$.

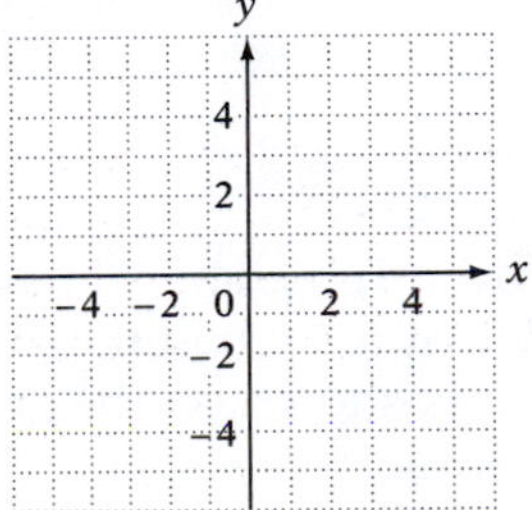

Copyright © Houghton Mifflin Company. All rights reserved.

15. Solve by graphing.
$5x - 2y = 10$
$3x + 2y = 6$

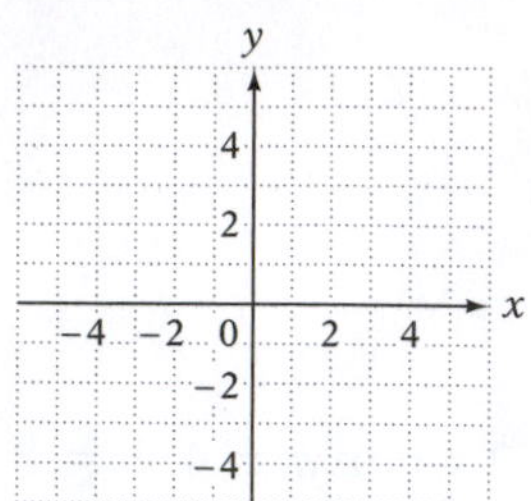

16. Solve by graphing.
$3x - 2y \geq 4$
$x + y < 3$

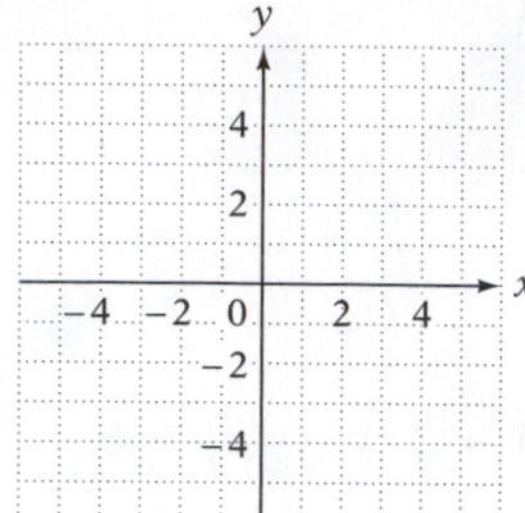

17. Solve:
$3x + 2z = 1$
$2y - z = 1$
$x + 2y = 1$

18. Solve:
$2x - y + z = 2$
$3x + y + 2z = 5$
$3x - y + 4z = 1$

19. Solve by the addition method:
$4x - 3y = 17$
$3x - 2y = 12$

20. Solve by substitution:
$3x - 2y = 7$
$y = 2x - 1$

21. **Mixtures** How many milliliters of pure water must be added to 100 ml of a 4% salt solution to make a 2.5% salt solution?

22. **Travel** Flying with the wind, a small plane required 2 h to fly 150 mi. Against the wind, it took 3 h to fly the same distance. Find the rate of the wind.

23. **Purchasing** A restaurant manager buys 100 lb of hamburger and 50 lb of steak for a total cost of \$980. A second purchase, at the same prices, includes 150 lb of hamburger and 100 lb of steak. The total cost is \$1720. Find the price of 1 lb of steak.

24. **Electronics** Find the lower and upper limits of a 12,000-ohm resistor with a 15% tolerance.

25. **Compensation** The graph shows the relationship between the monthly income and the sales of an account executive. Find the slope of the line between the two points shown on the graph. Write a sentence that states the meaning of the slope.

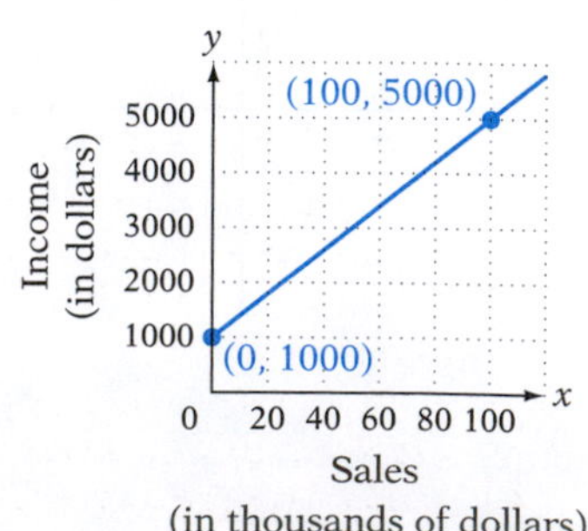

Copyright © Houghton Mifflin Company. All rights reserved.

chapter

6 Polynomials

The twin robots Spirit and Opportunity are part of the Mars Exploration Rover mission. They were launched toward Mars in 2003 to see if the Red Planet once had water. Since the robots' landing in 2004, the NASA Deep Space Network, which is an international network of antennas, has provided the communication links between the scientists on Earth and the rovers on Mars. The communications transmitted travel a distance of over 100 million miles. Distances this great, as well as very small measurements, are generally expressed in scientific notation. **Exercises 115 to 125 on pages 355 and 356** ask you to express distances and other measurements in scientific notation.

Need help? For online student resources, such as section quizzes, visit this textbook's website at **math.college.hmco.com/students.**

OBJECTIVES

Section 6.1

A To multiply monomials
B To divide monomials and simplify expressions with negative exponents
C To write a number using scientific notation
D To solve application problems

Section 6.2

A To evaluate polynomial functions
B To add or subtract polynomials

Section 6.3

A To multiply a polynomial by a monomial
B To multiply two polynomials
C To multiply polynomials that have special products
D To solve application problems

Section 6.4

A To divide a polynomial by a monomial
B To divide polynomials
C To divide polynomials by using synthetic division
D To evaluate a polynomial function using synthetic division

Copyright © Houghton Mifflin Company. All rights reserved.

PREP TEST

Do these exercises to prepare for Chapter 6.

1. Subtract: $-2 - (-3)$

2. Multiply: $-3(6)$

For Exercises 3 to 8, simplify.

3. $-\frac{24}{-36}$

4. $-4(3y)$

5. $(-2)^3$

6. $-4a - 8b + 7a$

7. $3x - 2[y - 4(x + 1) + 5]$

8. $-4y + 4y$

9. Are $2x^2$ and $2x$ like terms?

GO FIGURE

Two perpendicular lines are drawn through the interior of a rectangle, dividing it into four smaller rectangles. The areas of the smaller rectangles are x, 2, 3, and 6. Find the possible values of x.

Copyright © Houghton Mifflin Company. All rights reserved.

6.1 Exponential Expressions

Objective A To multiply monomials

A **monomial** is a number, a variable, or a product of a number and variables.

Point of Interest

Around A.D. 250, the monomial $3x^2$ shown at the right would have been written $\Delta^Y 3$, or at least approximately like that. In A.D. 250, the symbol for 3 was not the one we use today.

The examples at the right are monomials. The **degree of a monomial** is the sum of the exponents of the variables.

x degree 1 $(x = x^1)$
$3x^2$ degree 2
$4x^2y$ degree 3
$6x^3y^4z^2$ degree 9

In this chapter, the variable n is considered a positive integer when used as an exponent.

x^n degree n

The degree of a nonzero constant term is zero.

6 degree 0

The expression $5\sqrt{x}$ is not a monomial because $\sqrt{x}$ cannot be written as a product of variables. The expression $\frac{x}{y}$ is not a monomial because it is a quotient of variables.

The expression x^4 is an exponential expression. The exponent, 4, indicates the number of times the base, x, occurs as a factor.

The product of exponential expressions with the *same* base can be simplified by writing each expression in factored form and writing the result with an exponent.

$$x^3 \cdot x^4 = \underbrace{\overbrace{(x \cdot x \cdot x)}^{3 \text{ factors}} \cdot \overbrace{(x \cdot x \cdot x \cdot x)}^{4 \text{ factors}}}_{7 \text{ factors}}$$
$$= x^7$$

Note that adding the exponents results in the same product.

$$x^3 \cdot x^4 = x^{3+4} = x^7$$

Rule for Multiplying Exponential Expressions

If m and n are positive integers, then $x^m \cdot x^n = x^{m+n}$.

Study Tip

Remember that the HOW TO feature indicates a worked-out example. Using paper and pencil, work through the example. See *AIM for Success*, pages xxvii–xxviii.

HOW TO Simplify: $(-4x^5y^3)(3xy^2)$

$(-4x^5y^3)(3xy^2) = (-4 \cdot 3)(x^5 \cdot x)(y^3 \cdot y^2)$

- Use the Commutative and Associative Properties of Multiplication to rearrange and group factors.

$= -12(x^{5+1})(y^{3+2})$

- Multiply variables with the same base by adding their exponents.

$= -12x^6y^5$

- Simplify.

Copyright © Houghton Mifflin Company. All rights reserved.

As shown below, the power of a monomial can be simplified by writing the power in factored form and then using the Rule for Multiplying Exponential Expressions. It can also be simplified by multiplying each exponent inside the parentheses by the exponent outside the parentheses.

$$(a^2)^3 = a^2 \cdot a^2 \cdot a^2 = a^{2+2+2} = a^6 \qquad (x^3y^4)^2 = (x^3y^4)(x^3y^4) = x^{3+3}y^{4+4} = x^6y^8$$

• Write in factored form. Then use the Rule for Multiplying Exponential Expressions.

$$(a^2)^3 = a^{2\cdot 3} = a^6 \qquad (x^3y^4)^2 = x^{3\cdot 2}y^{4\cdot 2} = x^6y^8$$

• Multiply each exponent inside the parentheses by the exponent outside the parentheses.

Rule for Simplifying the Power of an Exponential Expression

If m and n are positive integers, then $(x^m)^n = x^{mn}$.

Rule for Simplifying Powers of Products

If m, n, and p are positive integers, then $(x^my^n)^p = x^{mp}y^{np}$.

HOW TO Simplify: $(x^4)^5$

$$(x^4)^5 = x^{4\cdot 5} = x^{20}$$

• Use the Rule for Simplifying the Power of an Exponential Expression to multiply the exponents.

HOW TO Simplify: $(2a^3b^4)^3$

$$(2a^3b^4)^3 = 2^{1\cdot 3}a^{3\cdot 3}b^{4\cdot 3} = 2^3a^9b^{12} = 8a^9b^{12}$$

• Use the Rule for Simplifying Powers of Products to multiply each exponent inside the parentheses by the exponent outside the parentheses.

Example 1 Simplify: $(2xy^2)(-3xy^4)^3$

Solution $(2xy^2)(-3xy^4)^3 = (2xy^2)[(-3)^3x^3y^{12}] = (2xy^2)(-27x^3y^{12}) = -54x^4y^{14}$

You Try It 1 Simplify: $(-3a^2b^4)(-2ab^3)^4$

Your solution

Example 2 Simplify: $(x^{n+2})^5$

Solution $(x^{n+2})^5 = x^{5n+10}$ • Multiply the exponents.

You Try It 2 Simplify: $(y^{n-3})^2$

Your solution

Example 3 Simplify: $[(2xy^2)^2]^3$

Solution $[(2xy^2)^2]^3 = [2^2x^2y^4]^3 = [4x^2y^4]^3 = 4^3x^6y^{12} = 64x^6y^{12}$

You Try It 3 Simplify: $[(ab^3)^3]^4$

Your solution

Solutions on p. S18

Copyright © Houghton Mifflin Company. All rights reserved.

Objective B To divide monomials and simplify expressions with negative exponents

The quotient of two exponential expressions with the same base can be simplified by writing each expression in factored form, dividing by the common factors, and then writing the result with an exponent.

$$\frac{x^5}{x^2} = \frac{\overset{1}{\cancel{x}} \cdot \overset{1}{\cancel{x}} \cdot x \cdot x \cdot x}{\underset{1}{\cancel{x}} \cdot \underset{1}{\cancel{x}}} = x^3$$

Note that subtracting the exponents gives the same result.

$$\frac{x^5}{x^2} = x^{5-2} = x^3$$

To divide two monomials with the same base, subtract the exponents of the like bases.

HOW TO Simplify: $\frac{z^8}{z^2}$

$$\frac{z^8}{z^2} = z^{8-2}$$

• **The bases are the same. Subtract the exponents.**

$$= z^6$$

HOW TO Simplify: $\frac{a^5b^9}{a^4b}$

$$\frac{a^5b^9}{a^4b} = a^{5-4}b^{9-1}$$

• **Subtract the exponents of the like bases.**

$$= ab^8$$

Consider the expression $\frac{x^4}{x^4}$, $x \neq 0$. This expression can be simplified, as shown below, by subtracting exponents or dividing by common factors.

$$\frac{x^4}{x^4} = x^{4-4} = x^0 \qquad \frac{x^4}{x^4} = \frac{\overset{1}{\cancel{x}} \cdot \overset{1}{\cancel{x}} \cdot \overset{1}{\cancel{x}} \cdot \overset{1}{\cancel{x}}}{\underset{1}{\cancel{x}} \cdot \underset{1}{\cancel{x}} \cdot \underset{1}{\cancel{x}} \cdot \underset{1}{\cancel{x}}} = 1$$

The equations $\frac{x^4}{x^4} = x^0$ and $\frac{x^4}{x^4} = 1$ suggest the following definition of x^0.

TAKE NOTE

In the example at the right, we indicated that $z \neq 0$. If we try to evaluate $(16z^5)^0$ when $z = 0$, we have $[16(0)^5]^0 = [16(0)]^0 = 0^0$. However, 0^0 is not defined, so we must assume that $z \neq 0$. To avoid stating this for every example or exercise, we will assume that variables cannot take on values that result in the expression 0^0.

Definition of Zero as an Exponent

If $x \neq 0$, then $x^0 = 1$. The expression 0^0 is not defined.

HOW TO Simplify: $(16z^5)^0$, $z \neq 0$

$$(16z^5)^0 = 1$$

• **Any nonzero expression to the zero power is 1.**

HOW TO Simplify: $-(7x^4y^3)^0$

$$-(7x^4y^3)^0 = -(1) = -1$$

• **The negative outside the parentheses is not affected by the exponent.**

Copyright © Houghton Mifflin Company. All rights reserved.

Point of Interest

In the 15th century, the expression $12^{2\overline{m}}$ was used to mean $12x^{-2}$. The use of $\overline{m}$ reflects an Italian influence, where *m* was used for minus and *p* was used for plus. It was understood that $2\overline{m}$ referred to an unnamed variable. Isaac Newton, in the 17th century, advocated the use of a negative exponent, the symbol we use today.

Consider the expression $\frac{x^4}{x^6}$, $x \neq 0$. This expression can be simplified, as shown below, by subtracting exponents or by dividing by common factors.

$$\frac{x^4}{x^6} = x^{4-6} = x^{-2} \qquad \frac{x^4}{x^6} = \frac{\overset{1}{\cancel{x}} \cdot \overset{1}{\cancel{x}} \cdot \overset{1}{\cancel{x}} \cdot \overset{1}{\cancel{x}}}{\underset{1}{\cancel{x}} \cdot \underset{1}{\cancel{x}} \cdot \underset{1}{\cancel{x}} \cdot \underset{1}{\cancel{x}} \cdot x \cdot x} = \frac{1}{x^2}$$

The equations $\frac{x^4}{x^6} = x^{-2}$ and $\frac{x^4}{x^6} = \frac{1}{x^2}$ suggest that $x^{-2} = \frac{1}{x^2}$.

Definition of a Negative Exponent

If $x \neq 0$ and n is a positive integer, then

$$x^{-n} = \frac{1}{x^n} \quad \text{and} \quad \frac{1}{x^{-n}} = x^n$$

TAKE NOTE

Note from the example at the right that 2^{-4} is a *positive* number. A negative exponent does not change the sign of a number.

HOW TO Evaluate: 2^{-4}

$2^{-4} = \frac{1}{2^4}$ • Use the Definition of a Negative Exponent.

$= \frac{1}{16}$ • Evaluate the expression.

The expression $\left(\frac{x^3}{y^4}\right)^2$, $y \neq 0$, can be simplified by squaring $\frac{x^3}{y^4}$ or by multiplying each exponent in the quotient by the exponent outside the parentheses.

$$\left(\frac{x^3}{y^4}\right)^2 = \left(\frac{x^3}{y^4}\right)\left(\frac{x^3}{y^4}\right) = \frac{x^3 \cdot x^3}{y^4 \cdot y^4} = \frac{x^{3+3}}{y^{4+4}} = \frac{x^6}{y^8} \qquad \left(\frac{x^3}{y^4}\right)^2 = \frac{x^{3\cdot 2}}{y^{4\cdot 2}} = \frac{x^6}{y^8}$$

Rule for Simplifying Powers of Quotients

If m, n, and p are integers and $y \neq 0$, then $\left(\frac{x^m}{y^n}\right)^p = \frac{x^{mp}}{y^{np}}$.

HOW TO Simplify: $\left(\frac{a^2}{b^3}\right)^{-2}$

$\left(\frac{a^2}{b^3}\right)^{-2} = \frac{a^{2(-2)}}{b^{3(-2)}}$ • Use the Rule for Simplifying Powers of Quotients.

$= \frac{a^{-4}}{b^{-6}} = \frac{b^6}{a^4}$ • Use the Definition of a Negative Exponent to write the expression with positive exponents.

Copyright © Houghton Mifflin Company. All rights reserved.

The preceding example suggests the following rule.

Rule for Negative Exponents on Fractional Expressions

If $a \neq 0$, $b \neq 0$, and n is a positive integer, then $\left(\frac{a}{b}\right)^{-n} = \left(\frac{b}{a}\right)^{n}$.

TAKE NOTE

The exponent on d is -5 (negative 5). The d^{-5} is written in the denominator as d^5. The exponent on 7 is 1 (positive 1). The 7 remains in the numerator.

Also, note that we indicated $d \neq 0$. This is necessary because division by zero is not defined. In this textbook, we will assume that values of the variables are chosen such that division by zero does not occur.

An exponential expression is in simplest form when it is written with only positive exponents.

HOW TO Simplify: $7d^{-5}$, $d \neq 0$

$$7d^{-5} = 7 \cdot \frac{1}{d^5} = \frac{7}{d^5}$$

- Use the Definition of a Negative Exponent to rewrite the expression with a positive exponent.

HOW TO Simplify: $\frac{2}{5a^{-4}}$

$$\frac{2}{5a^{-4}} = \frac{2}{5} \cdot \frac{1}{a^{-4}} = \frac{2}{5} \cdot a^4 = \frac{2a^4}{5}$$

- Use the Definition of a Negative Exponent to rewrite the expression with a positive exponent.

Now that zero as an exponent and negative exponents have been defined, a rule for dividing exponential expressions can be stated.

Rule for Dividing Exponential Expressions

If m and n are integers and $x \neq 0$, then $\frac{x^m}{x^n} = x^{m-n}$.

HOW TO Simplify: $\frac{x^4}{x^9}$

$$\frac{x^4}{x^9} = x^{4-9}$$

- Use the Rule for Dividing Exponential Expressions.

$$= x^{-5}$$

- Subtract the exponents.

$$= \frac{1}{x^5}$$

- Use the Definition of a Negative Exponent to rewrite the expression with a positive exponent.

The rules for simplifying exponential expressions and powers of exponential expressions are true for all integers. These rules are restated here for convenience.

Rules of Exponents

If m, n, and p are integers, then

$x^m \cdot x^n = x^{m+n}$	$(x^m)^n = x^{mn}$	$(x^m y^n)^p = x^{mp} y^{np}$
$\frac{x^m}{x^n} = x^{m-n}, x \neq 0$	$\left(\frac{x^m}{y^n}\right)^p = \frac{x^{mp}}{y^{np}}, y \neq 0$	$x^{-n} = \frac{1}{x^n}, x \neq 0$
$x^0 = 1, x \neq 0$		

Copyright © Houghton Mifflin Company. All rights reserved.

HOW TO Simplify: $(3ab^{-4})(-2a^{-3}b^{7})$

$$(3ab^{-4})(-2a^{-3}b^{7}) = [3 \cdot (-2)](a^{1+(-3)}b^{-4+7}) = -6a^{-2}b^{3} = -\frac{6b^3}{a^2}$$

- When multiplying expressions, add the exponents on like bases.

HOW TO Simplify: $\frac{4a^{-2}b^5}{6a^5b^2}$

$$\frac{4a^{-2}b^5}{6a^5b^2} = \frac{\overset{1}{\cancel{2}} \cdot 2a^{-2}b^5}{\underset{1}{\cancel{2}} \cdot 3a^5b^2} = \frac{2a^{-2}b^5}{3a^5b^2}$$

- Divide the coefficients by their common factor.

$$= \frac{2a^{-2-5}b^{5-2}}{3}$$

- Use the Rule for Dividing Exponential Expressions.

$$= \frac{2a^{-7}b^3}{3} = \frac{2b^3}{3a^7}$$

- Use the Definition of a Negative Exponent to rewrite the expression with a positive exponent.

HOW TO Simplify: $\left[\frac{6m^2n^3}{8m^7n^2}\right]^{-3}$

$$\left[\frac{6m^2n^3}{8m^7n^2}\right]^{-3} = \left[\frac{3m^{2-7}n^{3-2}}{4}\right]^{-3}$$

- Simplify inside the brackets.

$$= \left[\frac{3m^{-5}n}{4}\right]^{-3}$$

- Subtract the exponents.

$$= \frac{3^{-3}m^{15}n^{-3}}{4^{-3}}$$

- Use the Rule for Simplifying Powers of Quotients.

$$= \frac{4^3m^{15}}{3^3n^3} = \frac{64m^{15}}{27n^3}$$

- Use the Definition of a Negative Exponent to rewrite the expression with positive exponents. Then simplify.

Example 4

Simplify: $\frac{-28x^6z^{-3}}{42x^{-1}z^4}$

Solution

$$\frac{-28x^6z^{-3}}{42x^{-1}z^4} = -\frac{14 \cdot 2x^{6-(-1)}z^{-3-4}}{14 \cdot 3} = -\frac{2x^7z^{-7}}{3} = -\frac{2x^7}{3z^7}$$

You Try It 4

Simplify: $\frac{20r^{-2}t^{-5}}{-16r^{-3}s^{-2}}$

Your solution

Example 5

Simplify: $\frac{(3a^{-1}b^4)^{-3}}{(6^{-1}a^{-3}b^{-4})^3}$

Solution

$$\frac{(3a^{-1}b^4)^{-3}}{(6^{-1}a^{-3}b^{-4})^3} = \frac{3^{-3}a^3b^{-12}}{6^{-3}a^{-9}b^{-12}} = 3^{-3} \cdot 6^3a^{12}b^0 = \frac{6^3a^{12}}{3^3} = \frac{216a^{12}}{27} = 8a^{12}$$

You Try It 5

Simplify: $\frac{(9u^{-6}v^4)^{-1}}{(6u^{-3}v^{-2})^{-2}}$

Your solution

Solutions on p. S18

Copyright © Houghton Mifflin Company. All rights reserved.

Example 6

Simplify: $\dfrac{x^{4n-2}}{x^{2n-5}}$

Solution $\dfrac{x^{4n-2}}{x^{2n-5}} = x^{4n-2-(2n-5)}$ • Subtract the exponents.

$= x^{4n-2-2n+5}$

$= x^{2n+3}$

You Try It 6

Simplify: $\dfrac{a^{2n+1}}{a^{n+3}}$

Your solution

Solution on p. S18

Objective C To write a number using scientific notation

Point of Interest

Astronomers measure the distance of some stars by using the parsec. One parsec is approximately 1.91×10^{13} mi.

Integer exponents are used to represent the very large and very small numbers encountered in the fields of science and engineering. For example, the mass of the electron is 0.0000000000000000000000000000009 g. Numbers such as this are difficult to read and write, so a more convenient system for writing such numbers has been developed. It is called **scientific notation.**

To express a number in scientific notation, write the number as the product of a number between 1 and 10 and a power of 10. The form for scientific notation is $\boldsymbol{a \times 10^n}$, where $1 \le a < 10$.

For numbers greater than 10, move the decimal point to the right of the first digit. The exponent n is positive and equal to the number of places the decimal point has been moved.

$965{,}000 = 9.65 \times 10^5$

$3{,}600{,}000 = 3.6 \times 10^6$

$92{,}000{,}000{,}000 = 9.2 \times 10^{10}$

For numbers less than 1, move the decimal point to the right of the first nonzero digit. The exponent n is negative. The absolute value of the exponent is equal to the number of places the decimal point has been moved.

$0.0002 = 2 \times 10^{-4}$

$0.0000000974 = 9.74 \times 10^{-8}$

$0.000000000086 = 8.6 \times 10^{-11}$

TAKE NOTE

There are two steps involved in writing a number in scientific notation: (1) determine the number between 1 and 10, and (2) determine the exponent on 10.

Converting a number written in scientific notation to decimal notation requires moving the decimal point.

When the exponent is positive, move the decimal point to the right the same number of places as the exponent.

$1.32 \times 10^4 = 13{,}200$

$1.4 \times 10^8 = 140{,}000{,}000$

When the exponent is negative, move the decimal point to the left the same number of places as the absolute value of the exponent.

$1.32 \times 10^{-2} = 0.0132$

$1.4 \times 10^{-4} = 0.00014$

Numerical calculations involving numbers that have more digits than a hand-held calculator is able to handle can be performed using scientific notation.

Integrating Technology

See the appendix Keystroke Guide: *Scientific Notation* for instructions on entering a number that is in scientific notation into a graphing calculator.

HOW TO Simplify: $\dfrac{220{,}000 \times 0.000000092}{0.0000011}$

$\dfrac{220{,}000 \times 0.000000092}{0.0000011} = \dfrac{2.2 \times 10^5 \times 9.2 \times 10^{-8}}{1.1 \times 10^{-6}}$ • Write the numbers in scientific notation.

$= \dfrac{(2.2)(9.2) \times 10^{5+(-8)-(-6)}}{1.1}$ • Simplify.

$= 18.4 \times 10^3 = 18{,}400$

Copyright © Houghton Mifflin Company. All rights reserved.

Example 7 Write 0.000041 in scientific notation.

Solution $0.000041 = 4.1 \times 10^{-5}$

You Try It 7 Write 942,000,000 in scientific notation.

Your solution

Example 8 Write 3.3×10^7 in decimal notation.

Solution $3.3 \times 10^7 = 33{,}000{,}000$

You Try It 8 Write 2.7×10^{-5} in decimal notation.

Your solution

Example 9 Simplify: $\dfrac{2{,}400{,}000{,}000 \times 0.0000063}{0.00009 \times 480}$

Solution

$$\frac{2{,}400{,}000{,}000 \times 0.0000063}{0.00009 \times 480}$$
$$= \frac{2.4 \times 10^9 \times 6.3 \times 10^{-6}}{9 \times 10^{-5} \times 4.8 \times 10^2}$$
$$= \frac{(2.4)(6.3) \times 10^{9+(-6)-(-5)-2}}{(9)(4.8)}$$
$$= 0.35 \times 10^6 = 350{,}000$$

You Try It 9 Simplify: $\dfrac{5{,}600{,}000 \times 0.000000081}{900 \times 0.000000028}$

Your solution

Solutions on p. S18

Objective D To solve application problems

Example 10

How many miles does light travel in 1 day? The speed of light is 186,000 mi/s. Write the answer in scientific notation.

Strategy

To find the distance traveled:

- Write the speed of light in scientific notation.
- Write the number of seconds in 1 day in scientific notation.
- Use the equation $d = rt$, where r is the speed of light and t is the number of seconds in 1 day.

Solution

$r = 186{,}000 = 1.86 \times 10^5$

$t = 24 \cdot 60 \cdot 60 = 86{,}400 = 8.64 \times 10^4$

$d = rt$

$d = (1.86 \times 10^5)(8.64 \times 10^4)$

$= 1.86 \times 8.64 \times 10^9$

$= 16.0704 \times 10^9$

$= 1.60704 \times 10^{10}$

Light travels 1.60704×10^{10} mi in 1 day.

You Try It 10

A computer can do an arithmetic operation in 1×10^{-7} s. In scientific notation, how many arithmetic operations can the computer perform in 1 min?

Your strategy

Your solution

Solution on p. S18

Copyright © Houghton Mifflin Company. All rights reserved.

6.1 Exercises

Objective A To multiply monomials

For Exercises 1 to 37, simplify.

1. $(ab^3)(a^3b)$
2. $(-2ab^4)(-3a^2b^4)$
3. $(9xy^2)(-2x^2y^2)$
4. $(x^2y)^2$
5. $(x^2y^4)^4$
6. $(-2ab^2)^3$
7. $(-3x^2y^3)^4$
8. $(2^2a^2b^3)^3$
9. $(3^3a^5b^3)^2$
10. $(xy)(x^2y)^4$
11. $(x^2y^2)(xy^3)^3$
12. $[(2x)^4]^2$
13. $[(3x)^3]^2$
14. $[(x^2y)^4]^5$
15. $[(ab)^3]^6$
16. $[(2ab)^3]^2$
17. $[(2xy)^3]^4$
18. $[(3x^2y^3)^2]^2$
19. $[(2a^4b^3)^3]^2$
20. $y^n \cdot y^{2n}$
21. $x^n \cdot x^{n+1}$
22. $y^{2n} \cdot y^{4n+1}$
23. $y^{3n} \cdot y^{3n-2}$
24. $(a^n)^{2n}$
25. $(a^{n-3})^{2n}$
26. $(y^{2n-1})^3$
27. $(x^{3n+2})^5$
28. $(b^{2n-1})^n$
29. $(2xy)(-3x^2yz)(x^2y^3z^3)$
30. $(x^2z^4)(2xyz^4)(-3x^3y^2)$
31. $(3b^5)(2ab^2)(-2ab^2c^2)$
32. $(-c^3)(-2a^2bc)(3a^2b)$
33. $(-2x^2y^3z)(3x^2yz^4)$
34. $(2a^2b)^3(-3ab^4)^2$
35. $(-3ab^3)^3(-2^2a^2b)^2$
36. $(4ab)^2(-2ab^2c^3)^3$
37. $(-2ab^2)(-3a^4b^5)^3$

Objective B To divide monomials and simplify expressions with negative exponents

For Exercises 38 to 88, simplify.

38. 2^{-3}
39. $\frac{1}{3^{-5}}$
40. $\frac{1}{x^{-4}}$
41. $\frac{1}{y^{-3}}$

Copyright © Houghton Mifflin Company. All rights reserved.

42. $\dfrac{2x^{-2}}{y^4}$ **43.** $\dfrac{a^3}{4b^{-2}}$ **44.** $x^{-3}y$ **45.** xy^{-4}

46. $-5x^0$ **47.** $\dfrac{1}{2x^0}$ **48.** $\dfrac{(2x)^0}{-2^3}$ **49.** $\dfrac{-3^{-2}}{(2y)^0}$

50. $\dfrac{y^{-7}}{y^{-8}}$ **51.** $\dfrac{y^{-2}}{y^6}$ **52.** $(x^2y^{-4})^2$ **53.** $(x^3y^5)^{-2}$

54. $\dfrac{x^{-2}y^{-11}}{xy^{-2}}$ **55.** $\dfrac{x^4y^3}{x^{-1}y^{-2}}$ **56.** $\dfrac{a^{-1}b^{-3}}{a^4b^{-5}}$ **57.** $\dfrac{a^6b^{-4}}{a^{-2}b^5}$

58. $(2a^{-1})^{-2}(2a^{-1})^4$ **59.** $(3a)^{-3}(9a^{-1})^{-2}$ **60.** $(x^{-2}y)^2(xy)^{-2}$ **61.** $(x^{-1}y^2)^{-3}(x^2y^{-4})^{-3}$

62. $\dfrac{50b^{10}}{70b^5}$ **63.** $\dfrac{x^3y^6}{x^6y^2}$ **64.** $\dfrac{x^{17}y^5}{-x^7y^{10}}$ **65.** $\dfrac{-6x^2y}{12x^4y}$

66. $\dfrac{2x^2y^4}{(3xy^2)^3}$ **67.** $\dfrac{-3ab^2}{(9a^2b^4)^3}$ **68.** $\left(\dfrac{-12a^2b^3}{9a^5b^9}\right)^3$ **69.** $\left(\dfrac{12x^3y^2z}{18xy^3z^4}\right)^4$

70. $\dfrac{(4x^2y)^2}{(2xy^3)^3}$ **71.** $\dfrac{(3a^2b)^3}{(-6ab^3)^2}$ **72.** $\dfrac{(-4x^2y^3)^2}{(2xy^2)^3}$ **73.** $\dfrac{(-3a^2b^3)^2}{(-2ab^4)^3}$

74. $\dfrac{(-4xy^3)^3}{(-2x^7y)^4}$ **75.** $\dfrac{(-8x^2y^2)^4}{(16x^3y^7)^2}$ **76.** $\dfrac{a^{5n}}{a^{3n}}$ **77.** $\dfrac{b^{6n}}{b^{10n}}$

78. $\dfrac{-x^{5n}}{x^{2n}}$ **79.** $\dfrac{y^{2n}}{-y^{8n}}$ **80.** $\dfrac{x^{2n-1}}{x^{n-3}}$ **81.** $\dfrac{y^{3n+2}}{y^{2n+4}}$

82. $\dfrac{a^{3n}b^n}{a^nb^{2n}}$ **83.** $\dfrac{x^ny^{3n}}{x^ny^{5n}}$ **84.** $\dfrac{a^{3n-2}b^{n+1}}{a^{2n+1}b^{2n+2}}$ **85.** $\dfrac{x^{2n-1}y^{n-3}}{x^{n+4}y^{n+3}}$

86. $\left(\dfrac{4^{-2}xy^{-3}}{x^{-3}y}\right)^3\left(\dfrac{8^{-1}x^{-2}y}{x^4y^{-1}}\right)^{-2}$ **87.** $\left(\dfrac{9ab^{-2}}{8a^{-2}b}\right)^{-2}\left(\dfrac{3a^{-2}b}{2a^2b^{-2}}\right)^3$ **88.** $\left(\dfrac{2ab^{-1}}{ab}\right)^{-1}\left(\dfrac{3a^{-2}b}{a^2b^2}\right)^{-2}$

Copyright © Houghton Mifflin Company. All rights reserved.

Objective C To write a number using scientific notation

For Exercises 89 to 94, write in scientific notation.

89. 0.00000467

90. 0.00000005

91. 0.00000000017

92. 4,300,000

93. 200,000,000,000

94. 9,800,000,000

For Exercises 95 to 100, write in decimal notation.

95. 1.23×10^{-7}

96. 6.2×10^{-12}

97. 8.2×10^{15}

98. 6.34×10^{5}

99. 3.9×10^{-2}

100. 4.35×10^{9}

For Exercises 101 to 114, simplify. Write the answer in decimal notation.

101. $(3 \times 10^{-12})(5 \times 10^{16})$

102. $(8.9 \times 10^{-5})(3.2 \times 10^{-6})$

103. $(0.0000065)(3{,}200{,}000{,}000{,}000)$

104. $(480{,}000)(0.0000000096)$

105. $\dfrac{9 \times 10^{-3}}{6 \times 10^{5}}$

106. $\dfrac{2.7 \times 10^{4}}{3 \times 10^{-6}}$

107. $\dfrac{0.0089}{500{,}000{,}000}$

108. $\dfrac{4800}{0.00000024}$

109. $\dfrac{0.00056}{0.000000000004}$

110. $\dfrac{0.000000346}{0.0000005}$

111. $\dfrac{(3.2 \times 10^{-11})(2.9 \times 10^{15})}{8.1 \times 10^{-3}}$

112. $\dfrac{(6.9 \times 10^{27})(8.2 \times 10^{-13})}{4.1 \times 10^{15}}$

113. $\dfrac{(0.00000004)(84{,}000)}{(0.0003)(1{,}400{,}000)}$

114. $\dfrac{(720)(0.0000000039)}{(26{,}000{,}000{,}000)(0.018)}$

Objective D To solve application problems

For Exercises 115 to 125, solve. Write the answer in scientific notation.

115. **Astronomy** Our galaxy is estimated to be 5.6×10^{19} mi across. How long (in hours) would it take a spaceship to cross the galaxy traveling at 25,000 mph?

116. **Astronomy** How long does it take light to travel to Earth from the sun? The sun is 9.3×10^{7} mi from Earth, and light travels 1.86×10^{5} mi/s.

The Milky Way

Copyright © Houghton Mifflin Company. All rights reserved.

117. **Astronomy** The distance from Earth to Saturn is 8.86×10^8 mi. A satellite leaves Earth traveling at a constant rate of 1×10^5 mph. How long does it take for the satellite to reach Saturn?

118. **The Federal Government** In 2004, the gross national debt was approximately 7×10^{12} dollars. How much would each American have to pay in order to pay off the debt? Use 3×10^8 as the number of citizens.

119. **Physics** The mass of an electron is 9.109×10^{-31} kg. The mass of a proton is 1.673×10^{-27} kg. How many times heavier is a proton than an electron?

120. **Geology** The mass of Earth is 5.9×10^{24} kg. The mass of the sun is 2×10^{30} kg. How many times heavier is the sun than Earth?

121. **Physics** How many meters does light travel in 8 h? The speed of light is 3×10^8 m/s.

122. **Physics** How many meters does light travel in 1 day? The speed of light is 3×10^8 m/s.

Sojourner

123. **Astronomy** It took 11 min for the commands from a computer on Earth to travel to the rover Sojourner on Mars, a distance of 119 million miles. How fast did the signals from Earth to Mars travel?

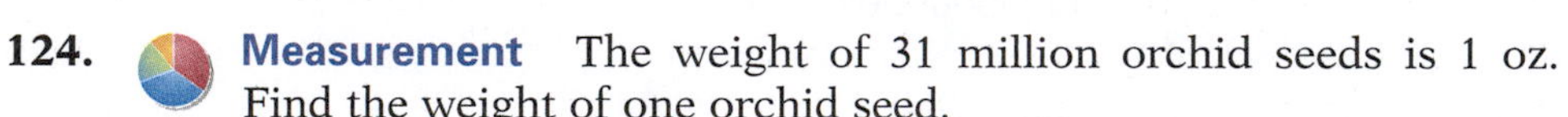

124. **Measurement** The weight of 31 million orchid seeds is 1 oz. Find the weight of one orchid seed.

Centrifuge

125. **Physics** A high-speed centrifuge makes 4×10^8 revolutions each minute. Find the time in seconds for the centrifuge to make one revolution.

APPLYING THE CONCEPTS

126. Correct the error in each of the following expressions. Explain which rule or property was used incorrectly.

a. $x^0 = 0$

b. $(x^4)^5 = x^9$

c. $x^2 \cdot x^3 = x^6$

127. Simplify.

a. $1 + [1 + (1 + 2^{-1})^{-1}]^{-1}$

b. $2 - [2 - (2 - 2^{-1})^{-1}]^{-1}$

Copyright © Houghton Mifflin Company. All rights reserved.

6.2 Introduction to Polynomial Functions

Objective A To evaluate polynomial functions

Study Tip

A great many new vocabulary words are introduced in this chapter. All of these terms are in **boldface type.** The bold type indicates that these are concepts you must know to learn the material. Be sure to study each new term as it is presented.

A **polynomial** is a variable expression in which the terms are monomials.

A polynomial of one term is a **monomial.** $5x$

A polynomial of two terms is a **binomial.** $5x^2y + 6x$

A polynomial of three terms is a **trinomial.** $3x^2 + 9xy - 5y$

Polynomials with more than three terms do not have special names.

The **degree of a polynomial** is the greatest of the degrees of any of its terms.

$3x + 2$	degree 1
$3x^2 + 2x - 4$	degree 2
$4x^3y^2 + 6x^4$	degree 5
$3x^{2n} - 5x^n - 2$	degree $2n$

The terms of a polynomial in one variable are usually arranged so that the exponents of the variable decrease from left to right. This is called **descending order.**

$2x^2 - x + 8$

$3y^3 - 3y^2 + y - 12$

For a polynomial in more than one variable, descending order may refer to any one of the variables.

The polynomial at the right is shown first in descending order of the x variable and then in descending order of the y variable.

$2x^2 + 3xy + 5y^2$

$5y^2 + 3xy + 2x^2$

Polynomial functions have many applications in mathematics. In general, a **polynomial function** is an expression whose terms are monomials. The **linear function** given by $f(x) = mx + b$ is an example of a polynomial function. It is a polynomial function of degree 1. A second-degree polynomial function, called a **quadratic function,** is given by the equation $f(x) = ax^2 + bx + c, a \neq 0$. A third-degree polynomial function is called a **cubic function.**

To **evaluate a polynomial function,** replace the variable by its value and simplify.

Integrating Technology

See the appendix Keystroke Guide: *Evaluating Functions* for instructions on using a graphing calculator to evaluate a function.

HOW TO Given $P(x) = x^3 - 3x^2 + 4$, evaluate $P(-3)$.

$$\begin{aligned} P(x) &= x^3 - 3x^2 + 4 \\ P(-3) &= (-3)^3 - 3(-3)^2 + 4 \\ &= -27 - 27 + 4 \\ &= -50 \end{aligned}$$

• Substitute −3 for x and simplify.

The **leading coefficient** of a polynomial function is the coefficient of the variable with the largest exponent. The **constant term** is the term without a variable.

Copyright © Houghton Mifflin Company. All rights reserved.

HOW TO Find the leading coefficient, the constant term, and the degree of the polynomial function $P(x) = 7x^4 - 3x^2 + 2x - 4$.

The leading coefficient is 7, the constant term is -4, and the degree is 4.

The three equations below do not represent polynomial functions.

$f(x) = 3x^2 + 2x^{-1}$	A polynomial function does not have a variable raised to a negative power.
$g(x) = 2\sqrt{x} - 3$	A polynomial function does not have a variable expression within a radical.
$h(x) = \dfrac{x}{x-1}$	A polynomial function does not have a variable in the denominator of a fraction.

The graph of a linear function is a straight line and can be found by plotting just two points. The graph of a polynomial function of degree greater than 1 is a curve. Consequently, many points may have to be found before an accurate graph can be drawn.

Evaluating the quadratic function given by the equation $f(x) = x^2 - x - 6$ when $x = -3, -2, -1, 0, 1, 2, 3$, and 4 gives the points shown in Figure 1 below. For instance, $f(-3) = 6$, so $(-3, 6)$ is graphed; $f(2) = -4$, so $(2, -4)$ is graphed; and $f(4) = 6$, so $(4, 6)$ is graphed. Evaluating the function when x is not an integer, such as when $x = -\frac{3}{2}$ and $x = \frac{5}{2}$, produces more points to graph, as shown in Figure 2. Connecting the points with a smooth curve results in Figure 3, which is the graph of f.

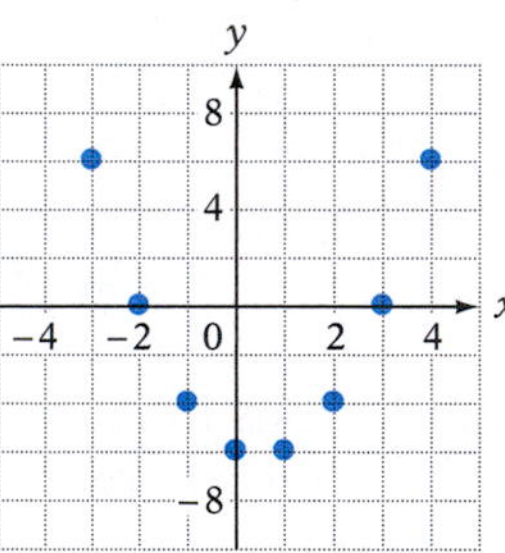

Figure 1

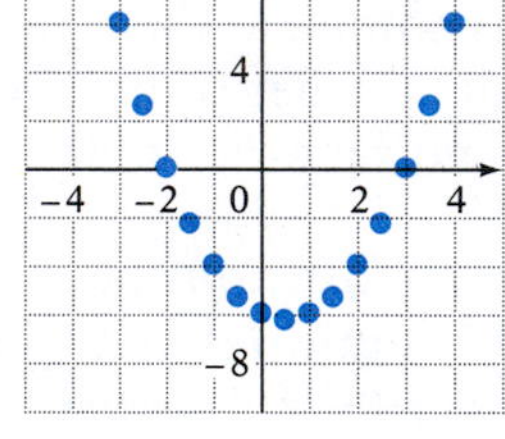

Figure 2

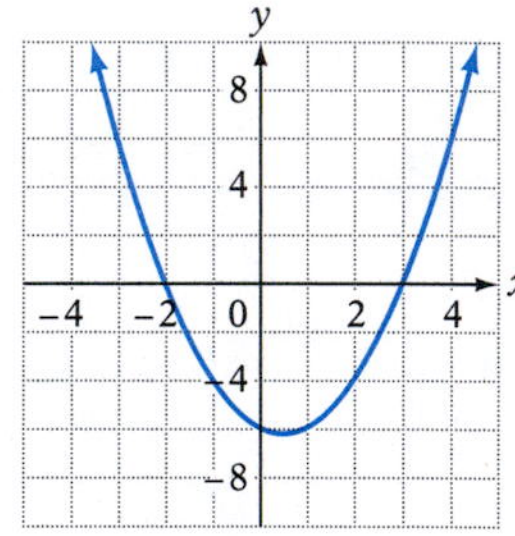

Figure 3

Integrating Technology

You can verify the graphs of these polynomial functions by using a graphing calculator. See the appendix Keystroke Guide: *Graph* for instructions on using a graphing calculator to graph a function.

Here is an example of graphing a cubic function, $P(x) = x^3 - 2x^2 - 5x + 6$. Evaluating the function when $x = -2, -1, 0, 1, 2, 3$, and 4 gives the graph in Figure 4 below. Evaluating for some noninteger values gives the graph in Figure 5. Finally, connecting the dots with a smooth curve gives the graph in Figure 6.

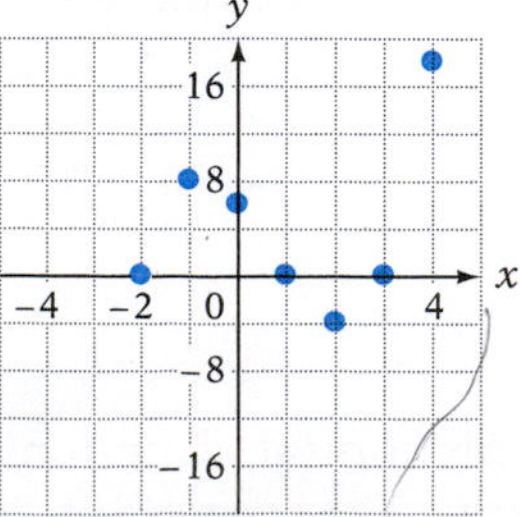

Figure 4

Figure 5

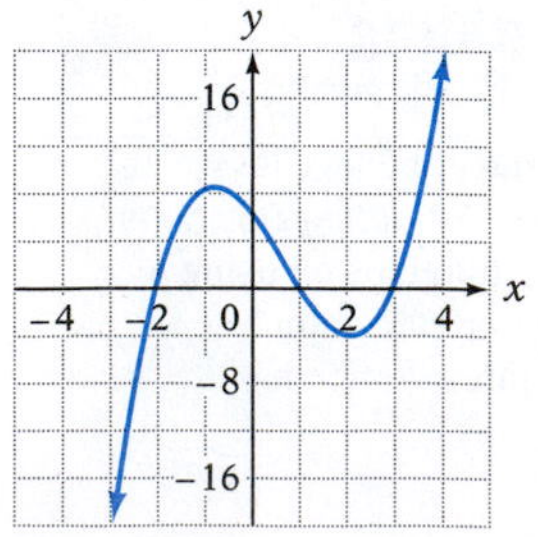

Figure 6

Copyright © Houghton Mifflin Company. All rights reserved.

Example 1

Given $P(x) = x^3 + 3x^2 - 2x + 8$, evaluate $P(-2)$.

Solution

$$P(x) = x^3 + 3x^2 - 2x + 8$$
$$P(-2) = (-2)^3 + 3(-2)^2 - 2(-2) + 8$$
$$= (-8) + 3(4) + 4 + 8$$
$$= -8 + 12 + 4 + 8$$
$$= 16$$

• Replace x by -2. Simplify.

You Try It 1

Given $R(x) = -2x^4 - 5x^3 + 2x - 8$, evaluate $R(2)$.

Your solution

Example 2

Find the leading coefficient, the constant term, and the degree of the polynomial.
$P(x) = 5x^6 - 4x^5 - 3x^2 + 7$

Solution

The leading coefficient is 5, the constant term is 7, and the degree is 6.

You Try It 2

Find the leading coefficient, the constant term, and the degree of the polynomial.
$r(x) = -3x^4 + 3x^3 + 3x^2 - 2x - 12$

Your solution

Example 3

Which of the following is a polynomial function?

a. $P(x) = 3x^{\frac{1}{2}} + 2x^2 - 3$
b. $T(x) = 3\sqrt{x} - 2x^2 - 3x + 2$
c. $R(x) = 14x^3 - \pi x^2 + 3x + 2$

Solution

a. This is not a polynomial function. A polynomial function does not have a variable raised to a fractional power.
b. This is not a polynomial function. A polynomial function does not have a variable expression within a radical.
c. This is a polynomial function.

You Try It 3

Which of the following is a polynomial function?

a. $R(x) = 5x^{14} - 5$
b. $V(x) = -x^{-1} + 2x - 7$
c. $P(x) = 2x^4 - 3\sqrt{x} - 3$

Your solution

Example 4

Graph $f(x) = x^2 - 2$.

Solution

x	$y = f(x)$
−3	7
−2	2
−1	−1
0	−2
1	−1
2	2
3	7

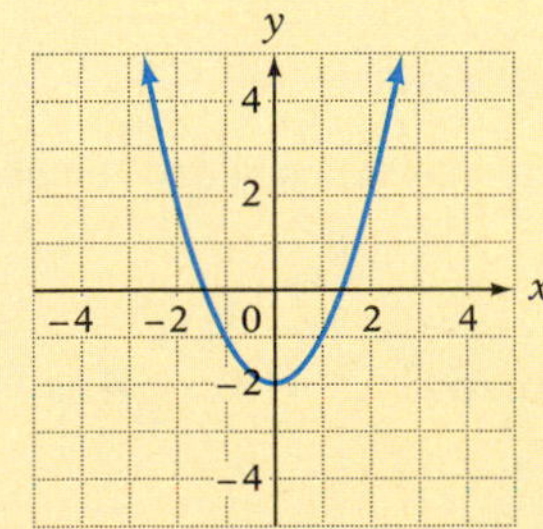

You Try It 4

Graph $f(x) = x^2 + 2x - 3$.

Your solution

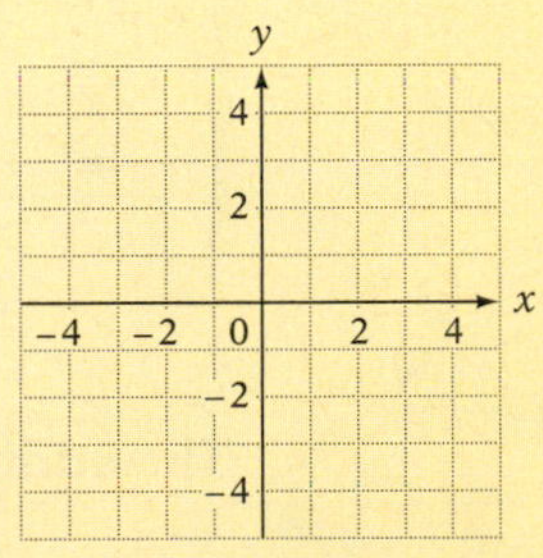

Solutions on pp. S18–S19

Copyright © Houghton Mifflin Company. All rights reserved.

Example 5

Graph $f(x) = x^3 - 1$.

Solution

x	$y = f(x)$
-2	-9
-1	-2
0	-1
1	0
2	7

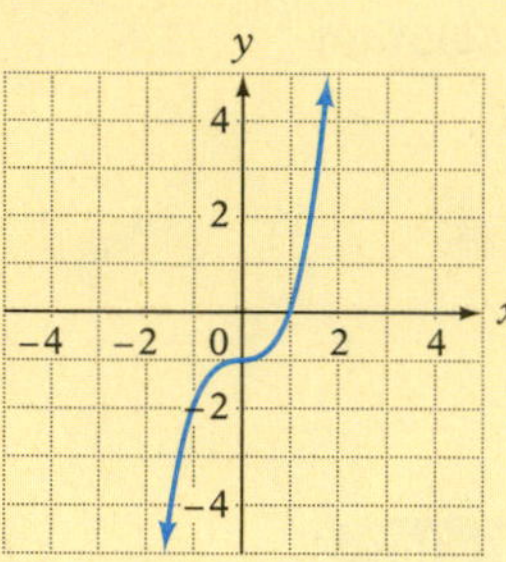

You Try It 5

Graph $f(x) = -x^3 + 1$.

Your solution

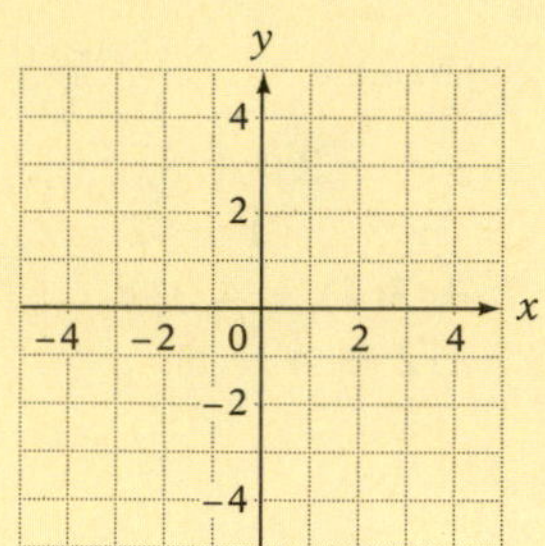

Solution on p. S19

Objective B To add or subtract polynomials

Polynomials can be added by combining like terms. Either a vertical or a horizontal format can be used.

HOW TO Simplify $(3x^2 + 2x - 7) + (7x^3 - 3 + 4x^2)$. Use a horizontal format.

Use the Commutative and Associative Properties of Addition to rearrange and group like terms.

$(3x^2 + 2x - 7) + (7x^3 - 3 + 4x^2)$
$= 7x^3 + (3x^2 + 4x^2) + 2x + (-7 - 3)$
$= 7x^3 + 7x^2 + 2x - 10$ • **Combine like terms.**

HOW TO Simplify $(4x^2 + 5x - 3) + (7x^3 - 7x + 1) + (2x - 3x^2 + 4x^3 + 1)$. Use a vertical format.

Arrange the terms of each polynomial in descending order with like terms in the same column.

$$\begin{array}{rrrr} & 4x^2 & + 5x & - 3 \\ 7x^3 & & - 7x & + 1 \\ 4x^3 & - 3x^2 & + 2x & + 1 \\ \hline 11x^3 & + x^2 & & - 1 \end{array}$$

• **Add the terms in each column.**

TAKE NOTE
The additive inverse of a polynomial is that polynomial with the sign of every term changed.

The **additive inverse of the polynomial** $x^2 + 5x - 4$ is $-(x^2 + 5x - 4)$.

To simplify the additive inverse of a polynomial, change the sign of every term inside the parentheses. $-(x^2 + 5x - 4) = -x^2 - 5x + 4$

Copyright © Houghton Mifflin Company. All rights reserved.

TAKE NOTE
This is the same definition used for subtraction of integers: subtraction is addition of the opposite.

To subtract two polynomials, add the additive inverse of the second polynomial to the first.

HOW TO Simplify $(3x^2 - 7xy + y^2) - (-4x^2 + 7xy - 3y^2)$. Use a horizontal format.

Rewrite the subtraction as addition of the additive inverse.

$$\begin{aligned}(3x^2 - 7xy + y^2) &- (-4x^2 + 7xy - 3y^2)\\ &= (3x^2 - 7xy + y^2) + (4x^2 - 7xy + 3y^2)\\ &= 7x^2 - 14xy + 4y^2\end{aligned}$$

• Combine like terms.

HOW TO Simplify $(6x^3 - 3x + 7) - (3x^2 - 5x + 12)$. Use a vertical format.

Rewrite subtraction as addition of the additive inverse.

$$(6x^3 - 3x + 7) - (3x^2 - 5x + 12) = (6x^3 - 3x + 7) + (-3x^2 + 5x - 12)$$

Arrange the terms of each polynomial in descending order with like terms in the same column.

$$\begin{array}{rrrr} 6x^3 & & -\ 3x & +\ 7 \\ & -\ 3x^2 & +\ 5x & -\ 12 \\ \hline 6x^3 & -\ 3x^2 & +\ 2x & -\ 5 \end{array}$$

• Combine the terms in each column.

Functional notation can be used when adding or subtracting polynomials.

HOW TO Given $P(x) = 3x^2 - 2x + 4$ and $R(x) = -5x^3 + 4x + 7$, find $P(x) + R(x)$.

$$\begin{aligned}P(x) + R(x) &= (3x^2 - 2x + 4) + (-5x^3 + 4x + 7)\\ &= -5x^3 + 3x^2 + 2x + 11\end{aligned}$$

HOW TO Given $P(x) = -5x^2 + 8x - 4$ and $R(x) = -3x^2 - 5x + 9$, find $P(x) - R(x)$.

$$\begin{aligned}P(x) - R(x) &= (-5x^2 + 8x - 4) - (-3x^2 - 5x + 9)\\ &= (-5x^2 + 8x - 4) + (3x^2 + 5x - 9)\\ &= -2x^2 + 13x - 13\end{aligned}$$

HOW TO Given $P(x) = 3x^2 - 5x + 6$ and $R(x) = 2x^2 - 5x - 7$, find $S(x)$, the sum of the two polynomials.

$$\begin{aligned}S(x) = P(x) + R(x) &= (3x^2 - 5x + 6) + (2x^2 - 5x - 7)\\ &= 5x^2 - 10x - 1\end{aligned}$$

Note from the preceding example that evaluating $P(x) = 3x^2 - 5x + 6$ and $R(x) = 2x^2 - 5x - 7$ at, for example, $x = 3$ and then adding the values is the same as evaluating $S(x) = 5x^2 - 10x - 1$ at 3.

$P(3) = 3(3)^2 - 5(3) + 6 = 27 - 15 + 6 = 18$
$R(3) = 2(3)^2 - 5(3) - 7 = 18 - 15 - 7 = -4$

$P(3) + R(3) = 18 + (-4) = 14$

$S(3) = 5(3)^2 - 10(3) - 1 = 45 - 30 - 1 = 14$

Copyright © Houghton Mifflin Company. All rights reserved.

Example 6

Simplify:
$(4x^2 - 3xy + 7y^2) + (-3x^2 + 7xy + y^2)$
Use a vertical format.

Solution

$$\begin{array}{r} 4x^2 - 3xy + 7y^2 \\ -3x^2 + 7xy + y^2 \\ \hline x^2 + 4xy + 8y^2 \end{array}$$

You Try It 6

Simplify:
$(-3x^2 - 4x + 9) + (-5x^2 - 7x + 1)$
Use a vertical format.

Your solution

Example 7

Simplify:
$(3x^2 - 2x + 4) - (7x^2 + 3x - 12)$
Use a vertical format.

Solution

Add the additive inverse of $7x^2 + 3x - 12$ to $3x^2 - 2x + 4$.

$$\begin{array}{r} 3x^2 - 2x + 4 \\ -7x^2 - 3x + 12 \\ \hline -4x^2 - 5x + 16 \end{array}$$

You Try It 7

Simplify:
$(-5x^2 + 2x - 3) - (6x^2 + 3x - 7)$
Use a vertical format.

Your solution

Example 8

Given $P(x) = -3x^2 + 2x - 6$ and $R(x) = 4x^3 - 3x + 4$, find $S(x) = P(x) + R(x)$. Evaluate $S(-2)$.

Solution

$$\begin{aligned} S(x) &= (-3x^2 + 2x - 6) + (4x^3 - 3x + 4) \\ &= 4x^3 - 3x^2 - x - 2 \end{aligned}$$

$$\begin{aligned} S(-2) &= 4(-2)^3 - 3(-2)^2 - (-2) - 2 \\ &= 4(-8) - 3(4) + 2 - 2 \\ &= -44 \end{aligned}$$

You Try It 8

Given $P(x) = 4x^3 - 3x^2 + 2$ and $R(x) = -2x^2 + 2x - 3$, find $S(x) = P(x) + R(x)$. Evaluate $S(-1)$.

Your solution

Example 9

Given $P(x) = (2x^{2n} - 3x^n + 7)$ and $R(x) = (3x^{2n} + 3x^n + 5)$, find $D(x) = P(x) - R(x)$.

Solution

$$\begin{aligned} D(x) &= P(x) - R(x) \\ D(x) &= (2x^{2n} - 3x^n + 7) - (3x^{2n} + 3x^n + 5) \\ &= (2x^{2n} - 3x^n + 7) + (-3x^{2n} - 3x^n - 5) \\ &= -x^{2n} - 6x^n + 2 \end{aligned}$$

You Try It 9

Given $P(x) = (5x^{2n} - 3x^n - 7)$ and $R(x) = (-2x^{2n} - 5x^n + 8)$, find $D(x) = P(x) - R(x)$.

Your solution

Solutions on p. S19

Copyright © Houghton Mifflin Company. All rights reserved.

6.2 Exercises

Objective A **To evaluate polynomial functions**

1. Given $P(x) = 3x^2 - 2x - 8$, evaluate $P(3)$.

2. Given $P(x) = -3x^2 - 5x + 8$, evaluate $P(-5)$.

3. Given $R(x) = 2x^3 - 3x^2 + 4x - 2$, evaluate $R(2)$.

4. Given $R(x) = -x^3 + 2x^2 - 3x + 4$, evaluate $R(-1)$.

5. Given $f(x) = x^4 - 2x^2 - 10$, evaluate $f(-1)$.

6. Given $f(x) = x^5 - 2x^3 + 4x$, evaluate $f(2)$.

In Exercises 7 to 18, indicate which equations define a polynomial function. For those that are polynomial functions: **a.** Identify the leading coefficient. **b.** Identify the constant term. **c.** State the degree.

7. $P(x) = -x^2 + 3x + 8$

8. $P(x) = 3x^4 - 3x - 7$

9. $R(x) = \dfrac{x}{x + 1}$

10. $R(x) = \dfrac{3x^2 - 2x + 1}{x}$

11. $f(x) = \sqrt{x} - x^2 + 2$

12. $f(x) = x^2 - \sqrt{x + 2} - 8$

13. $g(x) = 3x^5 - 2x^2 + \pi$

14. $g(x) = -4x^5 - 3x^2 + x - \sqrt{7}$

15. $P(x) = 3x^2 - 5x^3 + 2$

16. $P(x) = x^2 - 5x^4 - x^6$

17. $R(x) = 14$

18. $R(x) = \dfrac{1}{x} + 2$

For Exercises 19 to 24, graph.

19. $P(x) = x^2 - 1$

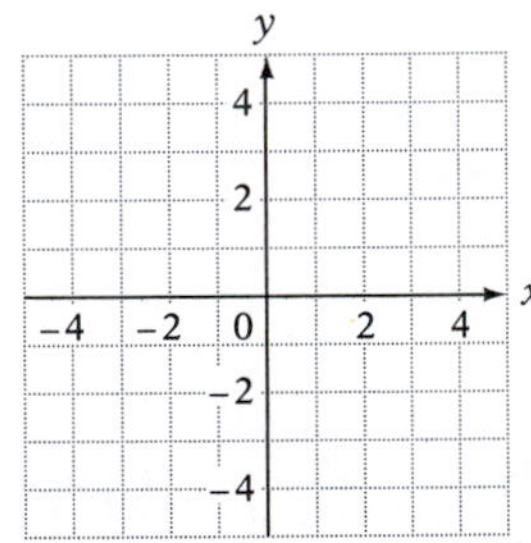

20. $P(x) = 2x^2 + 3$

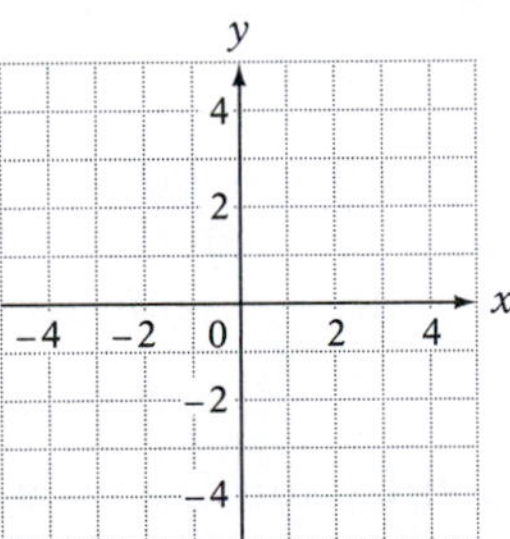

21. $R(x) = x^3 + 2$

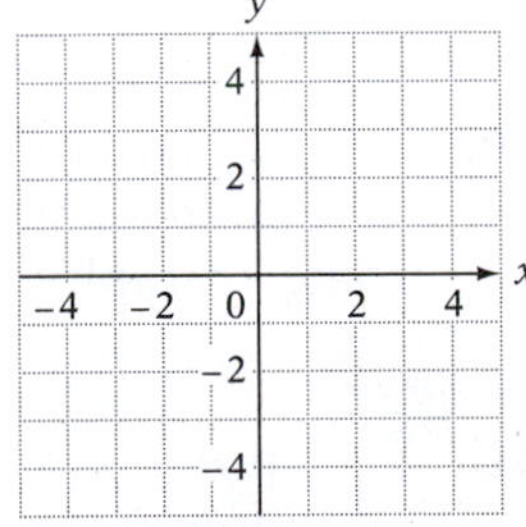

22. $R(x) = x^4 + 1$

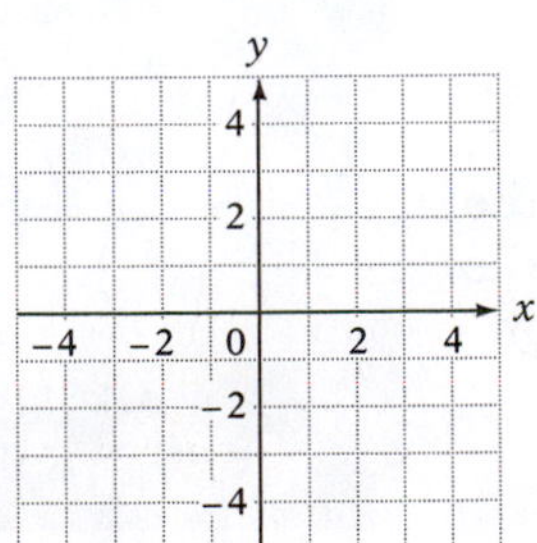

23. $f(x) = x^3 - 2x$

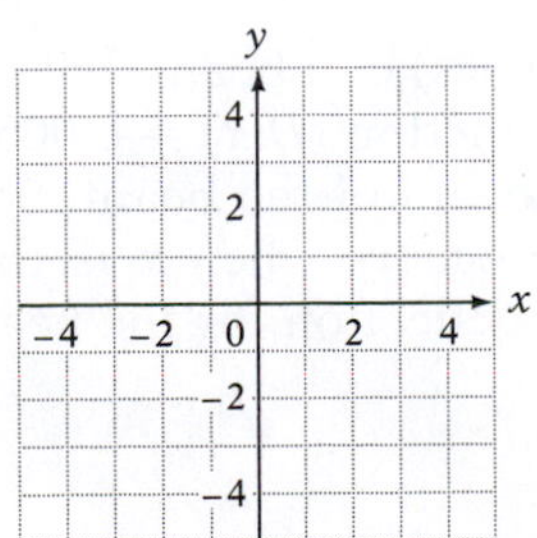

24. $f(x) = x^2 - x - 2$

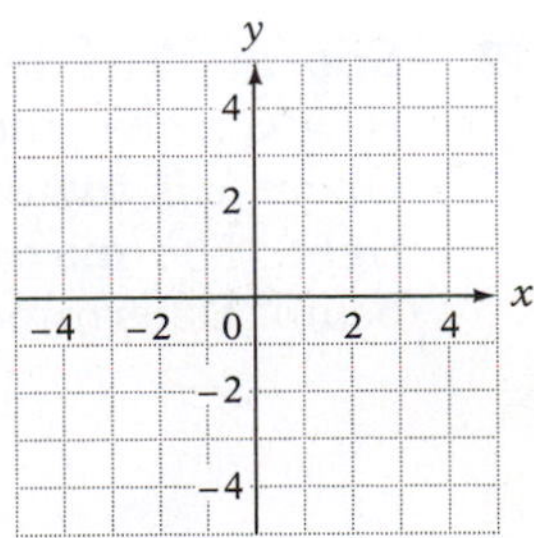

Copyright © Houghton Mifflin Company. All rights reserved.

Objective B To add or subtract polynomials

Simplify. Use a vertical format for Exercises 25 to 28.

25. $(5x^2 + 2x - 7) + (x^2 - 8x + 12)$

26. $(3x^2 - 2x + 7) + (-3x^2 + 2x - 12)$

27. $(x^2 - 3x + 8) - (2x^2 - 3x + 7)$

28. $(2x^2 + 3x - 7) - (5x^2 - 8x - 1)$

Simplify. Use a horizontal format for Exercises 29 to 32.

29. $(3y^2 - 7y) + (2y^2 - 8y + 2)$

30. $(-2y^2 - 4y - 12) + (5y^2 - 5y)$

31. $(2a^2 - 3a - 7) - (-5a^2 - 2a - 9)$

32. $(3a^2 - 9a) - (-5a^2 + 7a - 6)$

33. Given $P(x) = x^2 - 3xy + y^2$ and $R(x) = 2x^2 - 3y^2$, find $P(x) + R(x)$.

34. Given $P(x) = x^{2n} + 7x^n - 3$ and $R(x) = -x^{2n} + 2x^n + 8$, find $P(x) + R(x)$.

35. Given $P(x) = 3x^2 + 2y^2$ and $R(x) = -5x^2 + 2xy - 3y^2$, find $P(x) - R(x)$.

36. Given $P(x) = 2x^{2n} - x^n - 1$ and $R(x) = 5x^{2n} + 7x^n + 1$, find $P(x) - R(x)$.

37. Given $P(x) = 3x^4 - 3x^3 - x^2$ and $R(x) = 3x^3 - 7x^2 + 2x$, find $S(x) = P(x) + R(x)$. Evaluate $S(2)$.

38. Given $P(x) = 3x^4 - 2x + 1$ and $R(x) = 3x^5 - 5x - 8$, find $S(x) = P(x) + R(x)$. Evaluate $S(-1)$.

APPLYING THE CONCEPTS

39. For what value of k is the given equation an identity?
a. $(2x^3 + 3x^2 + kx + 5) - (x^3 + 2x^2 + 3x + 7) = x^3 + x^2 + 5x - 2$
b. $(6x^3 + kx^2 - 2x - 1) - (4x^3 - 3x^2 + 1) = 2x^3 - x^2 - 2x - 2$

40. If $P(x)$ is a third-degree polynomial and $Q(x)$ is a fourth-degree polynomial, what can be said about the degree of $P(x) + Q(x)$? Give some examples of polynomials that support your answer.

41. If $P(x)$ is a fifth-degree polynomial and $Q(x)$ is a fourth-degree polynomial, what can be said about the degree of $P(x) - Q(x)$? Give some examples of polynomials that support your answer.

42. **Sports** The deflection D (in inches) of a beam that is uniformly loaded is given by the polynomial function $D(x) = 0.005x^4 - 0.1x^3 + 0.5x^2$, where x is the distance from one end of the beam. See the figure at the right. The maximum deflection occurs when x is the midpoint of the beam. Determine the maximum deflection for the beam in the diagram.

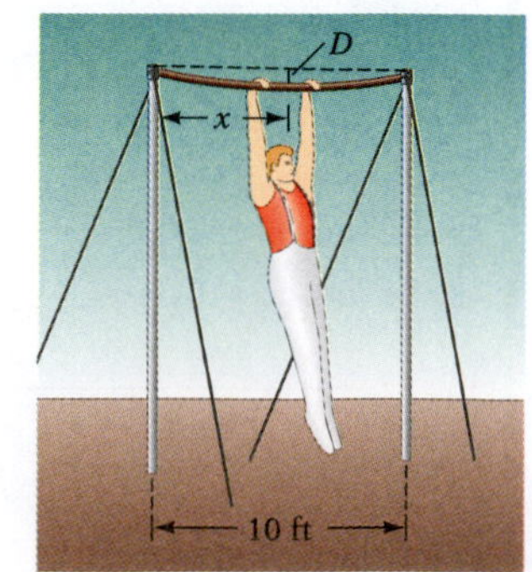

Copyright © Houghton Mifflin Company. All rights reserved.

6.3 Multiplication of Polynomials

Objective A To multiply a polynomial by a monomial

To multiply a polynomial by a monomial, use the Distributive Property and the Rule for Multiplying Exponential Expressions.

HOW TO Multiply: $-3x^2(2x^2 - 5x + 3)$

$-3x^2(2x^2 - 5x + 3)$

$= -3x^2(2x^2) - (-3x^2)(5x) + (-3x^2)(3)$ • Use the Distributive Property.

$= -6x^4 + 15x^3 - 9x^2$ • Use the Rule for Multiplying Exponential Expressions.

HOW TO Simplify: $2x^2 - 3x[2 - x(4x + 1) + 2]$

$2x^2 - 3x[2 - x(4x + 1) + 2]$

$= 2x^2 - 3x[2 - 4x^2 - x + 2]$ • Use the Distributive Property to remove the parentheses.

$= 2x^2 - 3x[-4x^2 - x + 4]$ • Simplify.

$= 2x^2 + 12x^3 + 3x^2 - 12x$ • Use the Distributive Property to remove the brackets.

$= 12x^3 + 5x^2 - 12x$ • Simplify.

HOW TO Multiply: $x^n(x^n - x^2 + 3)$

$x^n(x^n - x^2 + 3)$

$= x^n(x^n) - x^n(x^2) + x^n(3)$ • Use the Distributive Property.

$= x^{2n} - x^{n+2} + 3x^n$

Example 1

Multiply: $(3a^2 - 2a + 4)(-3a)$

Solution

$(3a^2 - 2a + 4)(-3a)$

$= 3a^2(-3a) - 2a(-3a) + 4(-3a)$ • Use the Distributive Property.

$= -9a^3 + 6a^2 - 12a$

You Try It 1

Multiply: $(2b^2 - 7b - 8)(-5b)$

Your solution

Solution on p. S19

Copyright © Houghton Mifflin Company. All rights reserved.

Example 2

Simplify: $y - 3y[y - 2(3y - 6) + 2]$

Solution

$y - 3y[y - 2(3y - 6) + 2]$
$= y - 3y[y - 6y + 12 + 2]$
$= y - 3y[-5y + 14]$
$= y + 15y^2 - 42y$
$= 15y^2 - 41y$

You Try It 2

Simplify: $x^2 - 2x[x - x(4x - 5) + x^2]$

Your solution

Example 3

Multiply: $x^{n+2}(x^{n-1} + 2x - 1)$

Solution

$x^{n+2}(x^{n-1} + 2x - 1)$
$= x^{n+2}(x^{n-1}) + (x^{n+2})(2x) - (x^{n+2})(1)$
$= x^{n+2+(n-1)} + 2x^{n+2+1} - x^{n+2}$
$= x^{2n+1} + 2x^{n+3} - x^{n+2}$

• Use the Distributive Property.

You Try It 3

Multiply: $y^{n+3}(y^{n-2} - 3y^2 + 2)$

Your solution

Solutions on p. S19

Objective B To multiply two polynomials

The product of two polynomials is the polynomial obtained by multiplying each term of one polynomial by each term of the other polynomial and then combining like terms.

HOW TO Multiply: $(2x^2 - 2x + 1)(3x + 2)$

Use the Distributive Property to multiply the trinomial by each term of the binomial.

$$(2x^2 - 2x + 1)(3x + 2) = (2x^2 - 2x + 1)(3x) + (2x^2 - 2x + 1)(2)$$
$$= (6x^3 - 6x^2 + 3x) + (4x^2 - 4x + 2)$$
$$= 6x^3 - 2x^2 - x + 2$$

A convenient method of multiplying two polynomials is to use a vertical format similar to that used for multiplication of whole numbers.

HOW TO Multiply: $(3x^2 - 4x + 8)(2x - 7)$

$$\begin{array}{rrrrl} & 3x^2 & - 4x & + 8 & \\ & & 2x & - 7 & \\ \hline & -21x^2 & + 28x & - 56 & = -7(3x^2 - 4x + 8) \\ 6x^3 & - 8x^2 & + 16x & & = 2x(3x^2 - 4x + 8) \\ \hline 6x^3 & - 29x^2 & + 44x & - 56 & \end{array}$$

• Like terms are in the same column.

• Combine like terms.

Copyright © Houghton Mifflin Company. All rights reserved.

TAKE NOTE

FOIL is not really a different way of multiplying. It is based on the Distributive Property.

$(3x - 2)(2x + 5)$
$= 3x(2x + 5) - 2(2x + 5)$
$= 6x^2 + 15x - 4x - 10$
$= 6x^2 + 11x - 10$

FOIL is an efficient way to remember how to do binomial multiplication.

It is frequently necessary to find the product of two binomials. The product can be found by using a method called **FOIL,** which is based on the Distributive Property. The letters of FOIL stand for **F**irst, **O**uter, **I**nner, and **L**ast.

Multiply: $(3x - 2)(2x + 5)$

Multiply the First terms.	$(3x - 2)(2x + 5)$	$3x \cdot 2x = 6x^2$
Multiply the Outer terms.	$(3x - 2)(2x + 5)$	$3x \cdot 5 = 15x$
Multiply the Inner terms.	$(3x - 2)(2x + 5)$	$-2 \cdot 2x = -4x$
Multiply the Last terms.	$(3x - 2)(2x + 5)$	$-2 \cdot 5 = -10$

F O I L

Add the products. $(3x - 2)(2x + 5) = 6x^2 + 15x - 4x - 10$

Combine like terms. $= 6x^2 + 11x - 10$

HOW TO Multiply: $(6x - 5)(3x - 4)$

$$(6x - 5)(3x - 4) = 6x(3x) + 6x(-4) + (-5)(3x) + (-5)(-4)$$
$$= 18x^2 - 24x - 15x + 20$$
$$= 18x^2 - 39x + 20$$

Example 4

Multiply: $(4a^3 - 3a + 7)(a - 5)$

Solution

$$\begin{array}{rrrrr} 4a^3 & & -\ 3a & +\ 7 \\ & & a & -\ 5 \\ \hline & -\ 20a^3 & & +\ 15a & -\ 35 \\ 4a^4 & & -\ 3a^2 & +\ 7a & \\ \hline 4a^4 & -\ 20a^3 & -\ 3a^2 & +\ 22a & -\ 35 \end{array}$$

• $-5(4a^3 - 3a + 7)$
• $a(4a^3 - 3a + 7)$

You Try It 4

Multiply: $(-2b^2 + 5b - 4)(-3b + 2)$

Your solution

Example 5

Multiply: $(5a - 3b)(2a + 7b)$

Solution

$(5a - 3b)(2a + 7b)$
$= 10a^2 + 35ab - 6ab - 21b^2$ • FOIL
$= 10a^2 + 29ab - 21b^2$

You Try It 5

Multiply: $(3x - 4)(2x - 3)$

Your solution

Example 6

Multiply: $(a^n - 2b^n)(3a^n - b^n)$

Solution

$(a^n - 2b^n)(3a^n - b^n)$
$= 3a^{2n} - a^nb^n - 6a^nb^n + 2b^{2n}$ • FOIL
$= 3a^{2n} - 7a^nb^n + 2b^{2n}$

You Try It 6

Multiply: $(2x^n + y^n)(x^n - 4y^n)$

Your solution

Solutions on p. S19

Copyright © Houghton Mifflin Company. All rights reserved.

Objective C To multiply polynomials that have special products

Using FOIL, a pattern can be found for the **product of the sum and difference of two terms** [that is, a polynomial that can be expressed in the form $(a + b)(a - b)$] and for the **square of a binomial** [that is, a polynomial that can be expressed in the form $(a + b)^2$].

The Product of the Sum and Difference of Two Terms

$$\mathbf{(a + b)(a - b)} = a^2 - ab + ab - b^2$$
$$= \mathbf{a^2 - b^2}$$

Square of the first term → a^2
Square of the second term → b^2

The Square of a Binomial

$$\mathbf{(a + b)^2} = (a + b)(a + b) = a^2 + ab + ab + b^2$$
$$= \mathbf{a^2 + 2ab + b^2}$$

Square of the first term → a^2
Twice the product of the two terms → $2ab$
Square of the second term → b^2

HOW TO Multiply: $(4x + 3)(4x - 3)$

$(4x + 3)(4x - 3)$ is the sum and difference of the same two terms. The product is the difference of the squares of the terms.

$$(4x + 3)(4x - 3) = (4x)^2 - 3^2$$
$$= 16x^2 - 9$$

HOW TO Expand: $(2x - 3y)^2$

$(2x - 3y)^2$ is the square of a binomial.

$$(2x - 3y)^2 = (2x)^2 + 2(2x)(-3y) + (-3y)^2$$
$$= 4x^2 - 12xy + 9y^2$$

Example 7

Multiply: $(2a - 3)(2a + 3)$

Solution

$(2a - 3)(2a + 3) = 4a^2 - 9$ • The sum and difference of two terms

You Try It 7

Multiply: $(3x - 7)(3x + 7)$

Your solution

Example 8

Multiply: $(x^n + 5)(x^n - 5)$

Solution

$(x^n + 5)(x^n - 5) = x^{2n} - 25$ • The sum and difference of two terms

You Try It 8

Multiply: $(2x^n + 3)(2x^n - 3)$

Your solution

Solutions on p. S19

Copyright © Houghton Mifflin Company. All rights reserved.

Example 9

Expand: $(2x + 7y)^2$

Solution

$(2x + 7y)^2 = 4x^2 + 28xy + 49y^2$ • The square of a binomial

You Try It 9

Expand: $(3x - 4y)^2$

Your solution

Example 10

Expand: $(x^{2n} - 2)^2$

Solution

$(x^{2n} - 2)^2 = x^{4n} - 4x^{2n} + 4$ • The square of a binomial

You Try It 10

Expand: $(2x^n - 8)^2$

Your solution

Solutions on p. S19

Objective D To solve application problems    

Example 11

The length of a rectangle is $(2x + 3)$ feet. The width is $(x - 5)$ feet. Find the area of the rectangle in terms of the variable x.

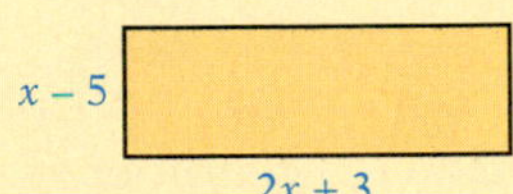

Strategy

To find the area, replace the variables L and W in the equation $A = L \cdot W$ by the given values and solve for A.

Solution

$A = L \cdot W$
$A = (2x + 3)(x - 5)$
$= 2x^2 - 10x + 3x - 15$ • FOIL
$= 2x^2 - 7x - 15$

The area is $(2x^2 - 7x - 15)$ square feet.

You Try It 11

The base of a triangle is $(2x + 6)$ feet. The height is $(x - 4)$ feet. Find the area of the triangle in terms of the variable x.

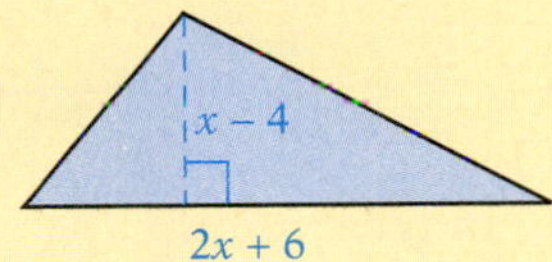

Your strategy

Your solution

Solution on p. S19

Copyright © Houghton Mifflin Company. All rights reserved.

Example 12

The corners are cut from a rectangular piece of cardboard measuring 8 in. by 12 in. The sides are folded up to make a box. Find the volume of the box in terms of the variable x, where x is the length of the side of the square cut from each corner of the rectangle.

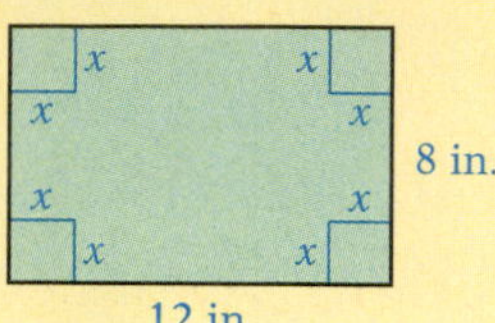

Strategy

Length of the box: $12 - 2x$
Width of the box: $8 - 2x$
Height of the box: x
To find the volume, replace the variables L, W, and H in the equation $V = L \cdot W \cdot H$ and solve for V.

Solution

$V = L \cdot W \cdot H$
$V = (12 - 2x)(8 - 2x)x$
$= (96 - 24x - 16x + 4x^2)x$ • **FOIL**
$= (96 - 40x + 4x^2)x$
$= 96x - 40x^2 + 4x^3$
$= 4x^3 - 40x^2 + 96x$

The volume is $(4x^3 - 40x^2 + 96x)$ cubic inches.

You Try It 12

Find the volume of the rectangular solid shown in the diagram below. All dimensions are in feet.

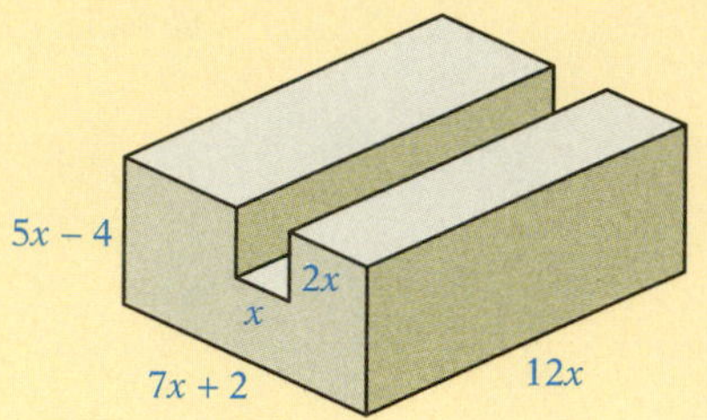

Your strategy

Your solution

Example 13

The radius of a circle is $(3x - 2)$ centimeters. Find the area of the circle in terms of the variable x. Use 3.14 for π.

Strategy

To find the area, replace the variable r in the equation $A = \pi r^2$ by the given value and solve for A.

Solution

$A = \pi r^2$
$A \approx 3.14(3x - 2)^2$
$= 3.14(9x^2 - 12x + 4)$
$= 28.26x^2 - 37.68x + 12.56$

The area is $(28.26x^2 - 37.68x + 12.56)$ square centimeters.

You Try It 13

The radius of a circle is $(2x + 3)$ centimeters. Find the area of the circle in terms of the variable x. Use 3.14 for π.

Your strategy

Your solution

Solutions on pp. S19–S20

Copyright © Houghton Mifflin Company. All rights reserved.

6.3 Exercises

Objective A To multiply a polynomial by a monomial

For Exercises 1 to 28, simplify.

1. $2x(x - 3)$

2. $2a(2a + 4)$

3. $3x^2(2x^2 - x)$

4. $-4y^2(4y - 6y^2)$

5. $3xy(2x - 3y)$

6. $-4ab(5a - 3b)$

7. $x^n(x + 1)$

8. $y^n(y^{2n} - 3)$

9. $x^n(x^n + y^n)$

10. $x - 2x(x - 2)$

11. $2b + 4b(2 - b)$

12. $-2y(3 - y) + 2y^2$

13. $-2a^2(3a^2 - 2a + 3)$

14. $4b(3b^3 - 12b^2 - 6)$

15. $(-3y^2 - 4y + 2)(y^2)$

16. $(6b^4 - 5b^2 - 3)(-2b^3)$

17. $-5x^2(4 - 3x + 3x^2 + 4x^3)$

18. $-2y^2(3 - 2y - 3y^2 + 2y^3)$

19. $-2x^2y(x^2 - 3xy + 2y^2)$

20. $3ab^2(3a^2 - 2ab + 4b^2)$

21. $x^n(x^{2n} + x^n + x)$

22. $x^{2n}(x^{2n-2} + x^{2n} + x)$

23. $a^{n+1}(a^n - 3a + 2)$

24. $a^{n+4}(a^{n-2} + 5a^2 - 3)$

25. $2y^2 - y[3 - 2(y - 4) - y]$

26. $3x^2 - x[x - 2(3x - 4)]$

27. $2y - 3[y - 2y(y - 3) + 4y]$

28. $4a^2 - 2a[3 - a(2 - a + a^2)]$

29. Given $P(b) = 3b$ and $Q(b) = 3b^4 - 3b^2 + 8$, find $P(b) \cdot Q(b)$.

30. Given $P(x) = -2x^2$ and $Q(x) = 2x^2 - 3x - 7$, find $P(x) \cdot Q(x)$.

Copyright © Houghton Mifflin Company. All rights reserved.

Objective B To multiply two polynomials

For Exercises 31 to 62, multiply.

31. $(x - 2)(x + 7)$

32. $(y + 8)(y + 3)$

33. $(2y - 3)(4y + 7)$

34. $(5x - 7)(3x - 8)$

35. $2(2x - 3y)(2x + 5y)$

36. $-3(7x - 3y)(2x - 9y)$

37. $(xy + 4)(xy - 3)$

38. $(xy - 5)(2xy + 7)$

39. $(2x^2 - 5)(x^2 - 5)$

40. $(x^2 - 4)(x^2 - 6)$

41. $(5x^2 - 5y)(2x^2 - y)$

42. $(x^2 - 2y^2)(x^2 + 4y^2)$

43. $(x^n + 2)(x^n - 3)$

44. $(x^n - 4)(x^n - 5)$

45. $(2a^n - 3)(3a^n + 5)$

46. $(5b^n - 1)(2b^n + 4)$

47. $(2a^n - b^n)(3a^n + 2b^n)$

48. $(3x^n + b^n)(x^n + 2b^n)$

49. $(x + 5)(x^2 - 3x + 4)$

50. $(a + 2)(a^2 - 3a + 7)$

51. $(2a - 3b)(5a^2 - 6ab + 4b^2)$

52. $(3a + b)(2a^2 - 5ab - 3b^2)$

53. $(2x - 5)(2x^4 - 3x^3 - 2x + 9)$

54. $(2a - 5)(3a^4 - 3a^2 + 2a - 5)$

55. $(x^2 + 2x - 3)(x^2 - 5x + 7)$

56. $(x^2 - 3x + 1)(x^2 - 2x + 7)$

57. $(a - 2)(2a - 3)(a + 7)$

58. $(b - 3)(3b - 2)(b - 1)$

59. $(x^n + 1)(x^{2n} + x^n + 1)$

60. $(a^{2n} - 3)(a^{5n} - a^{2n} + a^n)$

61. $(x^n + y^n)(x^n - 2x^ny^n + 3y^n)$

62. $(x^n - y^n)(x^{2n} - 3x^ny^n - y^{2n})$

63. Given $P(y) = 2y^2 - 1$ and $Q(y) = y^3 - 5y^2 - 3$, find $P(y) \cdot Q(y)$.

64. Given $P(b) = 2b^2 - 3$ and $Q(b) = 3b^2 - 3b + 6$, find $P(b) \cdot Q(b)$.

Copyright © Houghton Mifflin Company. All rights reserved.

Objective C To multiply polynomials that have special products

For Exercises 65 to 88, simplify or expand.

65. $(3x - 2)(3x + 2)$

66. $(4y + 1)(4y - 1)$

67. $(6 - x)(6 + x)$

68. $(10 + b)(10 - b)$

69. $(2a - 3b)(2a + 3b)$

70. $(5x - 7y)(5x + 7y)$

71. $(x^2 + 1)(x^2 - 1)$

72. $(x^2 + y^2)(x^2 - y^2)$

73. $(x^n + 3)(x^n - 3)$

74. $(x^n + y^n)(x^n - y^n)$

75. $(x - 5)^2$

76. $(y + 2)^2$

77. $(3a + 5b)^2$

78. $(5x - 4y)^2$

79. $(x^2 - 3)^2$

80. $(x^2 + y^2)^2$

81. $(2x^2 - 3y^2)^2$

82. $(x^n - 1)^2$

83. $(a^n - b^n)^2$

84. $(2x^n + 5y^n)^2$

85. $y^2 - (x - y)^2$

86. $a^2 + (a + b)^2$

87. $(x - y)^2 - (x + y)^2$

88. $(a + b)^2 + (a - b)^2$

Objective D To solve application problems

89. **Geometry** The length of a rectangle is $(3x - 2)$ feet. The width is $(x + 4)$ feet. Find the area of the rectangle in terms of the variable x.

90. **Geometry** The base of a triangle is $(x - 4)$ feet. The height is $(3x + 2)$ feet. Find the area of the triangle in terms of the variable x.

91. **Geometry** Find the area of the figure shown below. All dimensions given are in meters.

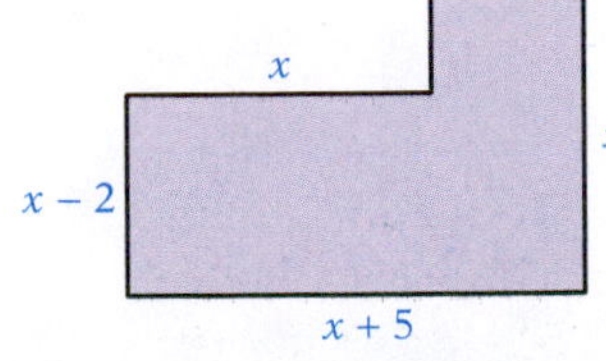

92. **Geometry** Find the area of the figure shown below. All dimensions given are in feet.

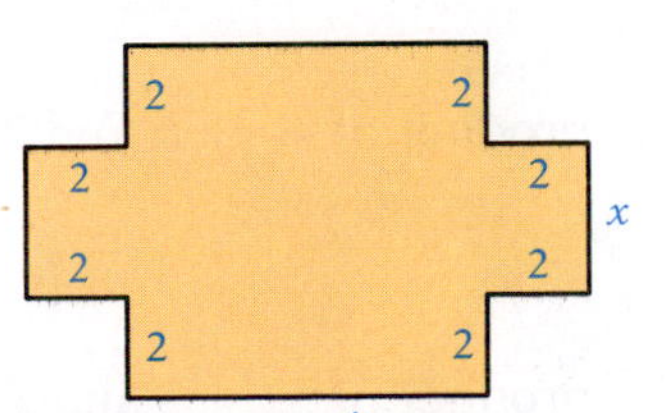

Copyright © Houghton Mifflin Company. All rights reserved.

93. **Geometry** The length of the side of a cube is $(x + 3)$ centimeters. Find the volume of the cube in terms of the variable x.

94. **Geometry** The length of a box is $(3x + 2)$ centimeters, the width is $(x - 4)$ centimeters, and the height is x centimeters. Find the volume of the box in terms of the variable x.

95. **Geometry** Find the volume of the figure shown below. All dimensions given are in inches.

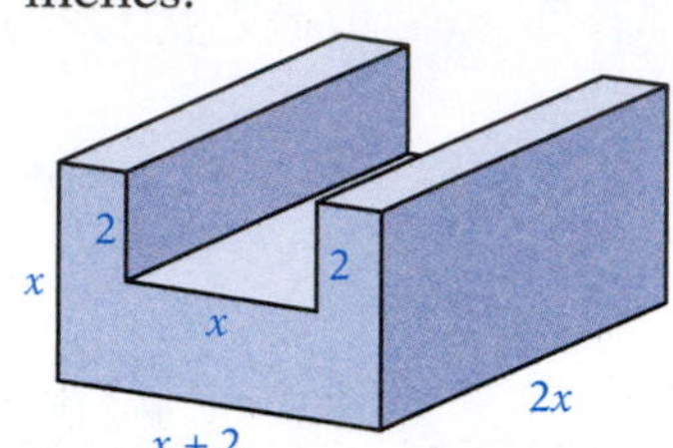

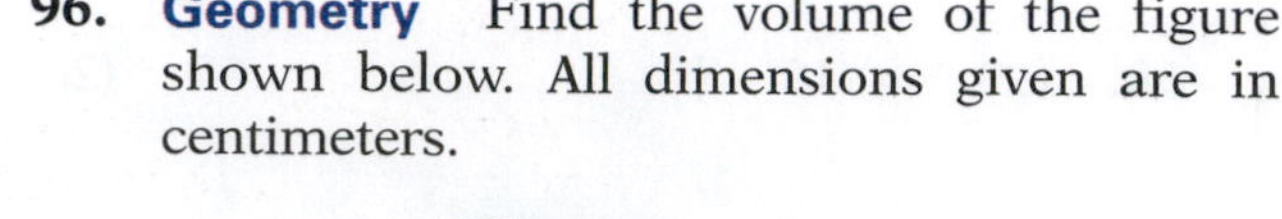

96. **Geometry** Find the volume of the figure shown below. All dimensions given are in centimeters.

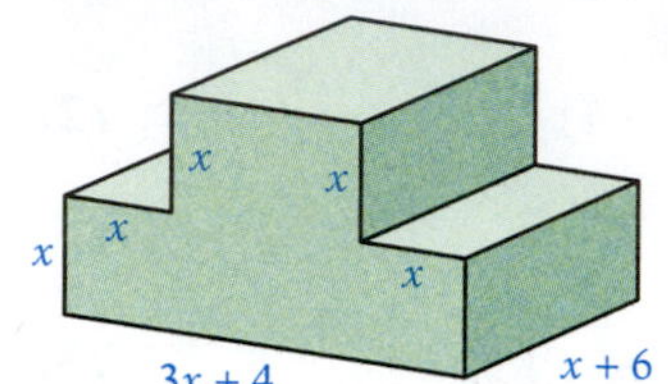

97. **Geometry** The radius of a circle is $(5x + 4)$ inches. Find the area of the circle in terms of the variable x. Use 3.14 for π.

98. **Geometry** The radius of a circle is $(x - 2)$ inches. Find the area of the circle in terms of the variable x. Use 3.14 for π.

APPLYING THE CONCEPTS

99. Find the product.
   a. $(a - b)(a^2 + ab + b^2)$
   b. $(x + y)(x^2 - xy + y^2)$

100. Correct the error in each of the following.
   a. $(x + 3)^2 = x^2 + 9$
   b. $(a - b)^2 = a^2 - b^2$

101. For what value of k is the given equation an identity?
   a. $(3x - k)(2x + k) = 6x^2 + 5x - k^2$
   b. $(4x + k)^2 = 16x^2 + 8x + k^2$

102. Complete.
   a. If $m = n + 1$, then $\frac{a^m}{a^n} =$ _____.
   b. If $m = n + 2$, then $\frac{a^m}{a^n} =$ _____.

103. Subtract the product of $4a + b$ and $2a - b$ from $9a^2 - 2ab$.

104. Subtract the product of $5x - y$ and $x + 3y$ from $6x^2 + 12xy - 2y^2$.

Copyright © Houghton Mifflin Company. All rights reserved.

6.4 Division of Polynomials

Objective A To divide a polynomial by a monomial

As shown below, $\frac{6+4}{2}$ can be simplified by first adding the terms in the numerator and then dividing the result by the denominator. It can also be simplified by first dividing each term in the numerator by the denominator and then adding the results.

$$\frac{6+4}{2} = \frac{10}{2} = 5 \qquad \frac{6+4}{2} = \frac{6}{2} + \frac{4}{2} = 3 + 2 = 5$$

It is this second method that is used to divide a polynomial by a monomial: Divide each term in the numerator by the denominator, and then write the sum of the quotients.

To divide $\frac{6x^2 + 4x}{2x}$, divide each term of the polynomial $6x^2 + 4x$ by the monomial $2x$. Then simplify each quotient.

> **TAKE NOTE**
> Recall that the fraction bar can be read "divided by."

$$\frac{6x^2 + 4x}{2x} = \frac{6x^2}{2x} + \frac{4x}{2x}$$ • Divide each term in the numerator by the denominator.

$$= 3x + 2$$ • Simplify each quotient.

We can check this quotient by multiplying it by the divisor.

$2x(3x + 2) = 6x^2 + 4x$ • The product is the dividend. The quotient checks.

HOW TO Divide and check: $\frac{16x^5 - 8x^3 + 4x}{2x}$

$$\frac{16x^5 - 8x^3 + 4x}{2x} = \frac{16x^5}{2x} - \frac{8x^3}{2x} + \frac{4x}{2x}$$ • Divide each term in the numerator by the denominator.

$$= 8x^4 - 4x^2 + 2$$ • Simplify each quotient.

Check:

$2x(8x^4 - 4x^2 + 2) = 16x^5 - 8x^3 + 4x$ • The quotient checks.

Example 1

Divide and check: $\frac{6x^3 - 3x^2 + 9x}{3x}$

Solution

$$\frac{6x^3 - 3x^2 + 9x}{3x}$$

$$= \frac{6x^3}{3x} - \frac{3x^2}{3x} + \frac{9x}{3x}$$ • Divide each term in the numerator by the denominator.

$$= 2x^2 - x + 3$$ • Simplify each quotient.

Check: $3x(2x^2 - x + 3) = 6x^3 - 3x^2 + 9x$

You Try It 1

Divide and check: $\frac{4x^3y + 8x^2y^2 - 4xy^3}{2xy}$

Your solution

Solution on p. S20

Copyright © Houghton Mifflin Company. All rights reserved.

Objective B To divide polynomials

SSM

The division method illustrated in Objective A is appropriate only when the divisor is a monomial. To divide two polynomials in which the divisor is not a monomial, use a method similar to that used for division of whole numbers.

To check division of polynomials, use

Dividend = (quotient × divisor) + remainder

HOW TO Divide: $(x^2 + 5x - 7) \div (x + 3)$

Step 1

$$\begin{array}{r} x \phantom{{}+5x-7} \\ x+3\,\overline{)\,x^2 + 5x - 7} \\ \underline{x^2 + 3x} \;\downarrow \\ 2x - 7 \end{array}$$

Think: $x\overline{)x^2} = \dfrac{x^2}{x} = x$

Multiply: $x(x + 3) = x^2 + 3x$

Subtract: $(x^2 + 5x) - (x^2 + 3x) = 2x$

Step 2

$$\begin{array}{r} x + 2 \\ x+3\,\overline{)\,x^2 + 5x - 7} \\ \underline{x^2 + 3x} \phantom{{}-7} \\ 2x - 7 \\ \underline{2x + 6} \\ -13 \end{array}$$

Think: $x\overline{)2x} = \dfrac{2x}{x} = 2$

Multiply: $2(x + 3) = 2x + 6$

Subtract: $(2x - 7) - (2x + 6) = -13$

The remainder is -13.

Check: $(x + 2)(x + 3) + (-13) = x^2 + 3x + 2x + 6 - 13 = x^2 + 5x - 7$

$$(x^2 + 5x - 7) \div (x + 3) = x + 2 - \frac{13}{x + 3}$$

HOW TO Divide: $\dfrac{6 - 6x^2 + 4x^3}{2x + 3}$

Arrange the terms in descending order. Note that there is no term of x in $4x^3 - 6x^2 + 6$. Insert a zero for the missing term so that like terms will be in the same columns.

$$\begin{array}{r} 2x^2 - 6x + 9 \\ 2x+3\,\overline{)\,4x^3 - 6x^2 + 0x + 6} \\ \underline{4x^3 + 6x^2} \phantom{{}+0x+6} \\ -12x^2 + 0x \phantom{{}+6} \\ \underline{-12x^2 - 18x} \phantom{{}+6} \\ 18x + 6 \\ \underline{18x + 27} \\ -21 \end{array}$$

$$\frac{4x^3 - 6x^2 + 6}{2x + 3} = 2x^2 - 6x + 9 - \frac{21}{2x + 3}$$

Copyright © Houghton Mifflin Company. All rights reserved.

Example 2

Divide: $\dfrac{12x^2 - 11x + 10}{4x - 5}$

Solution

$$\begin{array}{r}
3x + 1 \\
4x - 5\overline{)12x^2 - 11x + 10} \\
\underline{12x^2 - 15x} \quad\quad \\
4x + 10 \\
\underline{4x - 5} \\
15
\end{array}$$

$$\frac{12x^2 - 11x + 10}{4x - 5} = 3x + 1 + \frac{15}{4x - 5}$$

You Try It 2

Divide: $\dfrac{15x^2 + 17x - 20}{3x + 4}$

Your solution

Example 3

Divide: $\dfrac{x^3 + 1}{x + 1}$

Solution

$$\begin{array}{r}
x^2 - x + 1 \\
x + 1\overline{)x^3 + 0x^2 + 0x + 1} \\
\underline{x^3 + x^2} \quad\quad\quad\quad \\
-x^2 + 0x \quad\quad \\
\underline{-x^2 - x} \quad\quad \\
x + 1 \\
\underline{x + 1} \\
0
\end{array}$$

• Insert zeros for the missing terms.

$$\frac{x^3 + 1}{x + 1} = x^2 - x + 1$$

You Try It 3

Divide: $\dfrac{3x^3 + 8x^2 - 6x + 2}{3x - 1}$

Your solution

Example 4

Divide:
$(2x^4 - 7x^3 + 3x^2 + 4x - 5) \div (x^2 - 2x - 2)$

Solution

$$\begin{array}{r}
2x^2 - 3x + 1 \\
x^2 - 2x - 2\overline{)2x^4 - 7x^3 + 3x^2 + 4x - 5} \\
\underline{2x^4 - 4x^3 - 4x^2} \quad\quad\quad\quad \\
-3x^3 + 7x^2 + 4x \quad\; \\
\underline{-3x^3 + 6x^2 + 6x} \quad\; \\
x^2 - 2x - 5 \\
\underline{x^2 - 2x - 2} \\
-3
\end{array}$$

$(2x^4 - 7x^3 + 3x^2 + 4x - 5) \div (x^2 - 2x - 2)$

$$= 2x^2 - 3x + 1 - \frac{3}{x^2 - 2x - 2}$$

You Try It 4

Divide:
$(3x^4 - 11x^3 + 16x^2 - 16x + 8) \div (x^2 - 3x + 2)$

Your solution

Solutions on p.S20

Copyright © Houghton Mifflin Company. All rights reserved.

Objective C

To divide polynomials by using synthetic division

Study Tip

An important element of success is practice. We cannot do anything well if we do not practice it repeatedly. Practice is crucial to success in mathematics. In this objective you are learning a new procedure, synthetic division. You will need to practice this procedure in order to be successful at it.

Synthetic division is a shorter method of dividing a polynomial by a binomial of the form $x - a$.

Divide $(3x^2 - 4x + 6) \div (x - 2)$ by using long division.

$$\begin{array}{r} 3x + 2 \\ x - 2\overline{)3x^2 - 4x + 6} \\ \underline{3x^2 - 6x} \\ 2x + 6 \\ \underline{2x - 4} \\ 10 \end{array}$$

$(3x^2 - 4x + 6) \div (x - 2) = 3x + 2 + \frac{10}{x - 2}$

The variables can be omitted because the position of a term indicates the power of the term.

$$\begin{array}{r} 3 \quad 2 \\ -2\overline{)3 \quad -4 \quad 6} \\ \underline{3 \quad -6} \\ 2 \quad 6 \\ \underline{2 \quad -4} \\ 10 \end{array}$$

Each number shown in color is exactly the same as the number above it. Removing the colored numbers condenses the vertical spacing.

$$\begin{array}{r} 3 \quad 2 \\ -2\overline{)3 \quad -4 \quad 6} \\ \underline{-6 \quad -4} \\ 2 \quad 10 \end{array}$$

The number in color in the top row is the same as the one in the bottom row. Writing the 3 from the top row in the bottom row allows the spacing to be condensed even further.

$$\begin{array}{r|rrr} -2 & 3 & -4 & 6 \\ & & -6 & -4 \\ \hline & 3 & 2 & 10 \end{array}$$

Terms of the quotient (3, 2) — Remainder (10)

Because the degree of the dividend $(3x^2 - 4x + 6)$ is 2 and the degree of the divisor $(x - 2)$ is 1, the degree of the quotient is $2 - 1 = 1$. This means that, using the terms of the quotient given above, that quotient is $3x + 2$. The remainder is 10.

In general, the degree of the quotient of two polynomials is the difference between the degree of the dividend and the degree of the divisor.

By replacing the constant term in the divisor by its additive inverse, we may add rather than subtract terms. This is illustrated in the following example.

Copyright © Houghton Mifflin Company. All rights reserved.

HOW TO Divide: $(3x^3 + 6x^2 - x - 2) \div (x + 3)$

The additive inverse of the binomial constant

Coefficients of the polynomial

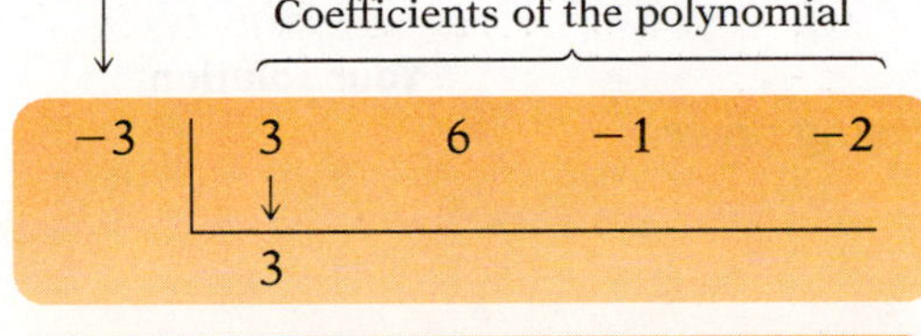

$$\begin{array}{r|rrrr} -3 & 3 & 6 & -1 & -2 \\ & \downarrow & & & \\ \hline & 3 & & & \end{array}$$

- Bring down the 3.

$$\begin{array}{r|rrrr} -3 & 3 & 6 & -1 & -2 \\ & & -9 & & \\ \hline & 3 & -3 & & \end{array}$$

- Multiply $-3(3)$ and add the product to 6.

$$\begin{array}{r|rrrr} -3 & 3 & 6 & -1 & -2 \\ & & -9 & 9 & \\ \hline & 3 & -3 & 8 & \end{array}$$

- Multiply $-3(-3)$ and add the product to -1.

$$\begin{array}{r|rrrr} -3 & 3 & 6 & -1 & -2 \\ & & -9 & 9 & -24 \\ \hline & 3 & -3 & 8 & -26 \end{array}$$

- Multiply $-3(8)$ and add the product to -2.

Terms of the quotient — Remainder

The degree of the dividend is 3 and the degree of the divisor is 1. Therefore, the degree of the quotient is $3 - 1 = 2$.

$$(3x^3 + 6x^2 - x - 2) \div (x + 3) = 3x^2 - 3x + 8 - \frac{26}{x + 3}$$

HOW TO Divide: $(2x^3 - x + 2) \div (x - 2)$

The additive inverse of the binomial constant

Coefficients of the polynomial

$$\begin{array}{r|rrrr} 2 & 2 & 0 & -1 & 2 \\ & \downarrow & & & \\ \hline & 2 & & & \end{array}$$

- Insert a 0 for the missing term and bring down the 2.

$$\begin{array}{r|rrrr} 2 & 2 & 0 & -1 & 2 \\ & & 4 & & \\ \hline & 2 & 4 & & \end{array}$$

- Multiply 2(2) and add the product to 0.

$$\begin{array}{r|rrrr} 2 & 2 & 0 & -1 & 2 \\ & & 4 & 8 & \\ \hline & 2 & 4 & 7 & \end{array}$$

- Multiply 2(4) and add the product to -1.

$$\begin{array}{r|rrrr} 2 & 2 & 0 & -1 & 2 \\ & & 4 & 8 & 14 \\ \hline & 2 & 4 & 7 & 16 \end{array}$$

- Multiply 2(7) and add the product to 2.

Terms of the quotient — Remainder

$$(2x^3 - x + 2) \div (x - 2) = 2x^2 + 4x + 7 + \frac{16}{x - 2}$$

Copyright © Houghton Mifflin Company. All rights reserved.

Example 5

Divide: $(7 - 3x + 5x^2) \div (x - 1)$

Solution

Arrange the coefficients in decreasing powers of x.

$$\begin{array}{r|rrr} 1 & 5 & -3 & 7 \\ & & 5 & 2 \\ \hline & 5 & 2 & 9 \end{array}$$

$$(5x^2 - 3x + 7) \div (x - 1) = 5x + 2 + \frac{9}{x - 1}$$

You Try It 5

Divide: $(6x^2 + 8x - 5) \div (x + 2)$

Your solution

Example 6

Divide: $(2x^3 + 4x^2 - 3x + 12) \div (x + 4)$

Solution

$$\begin{array}{r|rrrr} -4 & 2 & 4 & -3 & 12 \\ & & -8 & 16 & -52 \\ \hline & 2 & -4 & 13 & -40 \end{array}$$

$$(2x^3 + 4x^2 - 3x + 12) \div (x + 4)$$
$$= 2x^2 - 4x + 13 - \frac{40}{x + 4}$$

You Try It 6

Divide: $(5x^3 - 12x^2 - 8x + 16) \div (x - 2)$

Your solution

Example 7

Divide: $(3x^4 - 8x^2 + 2x + 1) \div (x + 2)$

Solution

Insert a zero for the missing term.

$$\begin{array}{r|rrrrr} -2 & 3 & 0 & -8 & 2 & 1 \\ & & -6 & 12 & -8 & 12 \\ \hline & 3 & -6 & 4 & -6 & 13 \end{array}$$

$$(3x^4 - 8x^2 + 2x + 1) \div (x + 2)$$
$$= 3x^3 - 6x^2 + 4x - 6 + \frac{13}{x + 2}$$

You Try It 7

Divide: $(2x^4 - 3x^3 - 8x^2 - 2) \div (x - 3)$

Your solution

Solutions on p. S20

Objective D **To evaluate a polynomial function using synthetic division**

VIDEO & DVD  

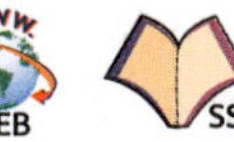

A polynomial can be evaluated by using synthetic division. Consider the polynomial $P(x) = 2x^4 - 3x^3 + 4x^2 - 5x + 1$. One way to evaluate the polynomial when $x = 2$ is to replace x by 2 and then simplify the numerical expression.

$$P(x) = 2x^4 - 3x^3 + 4x^2 - 5x + 1$$

$$\begin{aligned} P(2) &= 2(2)^4 - 3(2)^3 + 4(2)^2 - 5(2) + 1 \\ &= 2(16) - 3(8) + 4(4) - 5(2) + 1 \\ &= 32 - 24 + 16 - 10 + 1 \\ &= 15 \end{aligned}$$

Copyright © Houghton Mifflin Company. All rights reserved.

Now use synthetic division to divide $(2x^4 - 3x^3 + 4x^2 - 5x + 1) \div (x - 2)$.

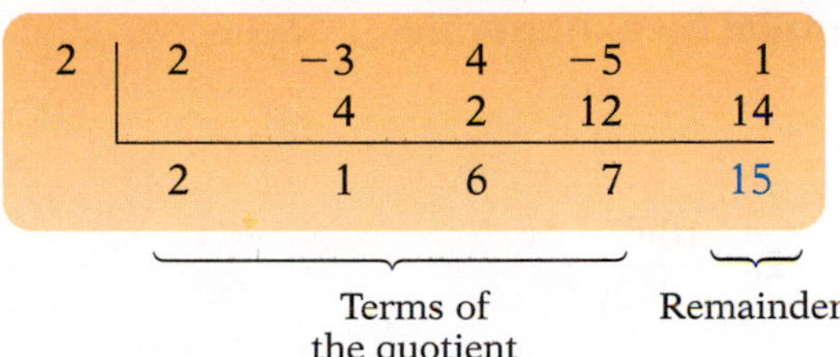

2	2	−3	4	−5	1
		4	2	12	14
	2	1	6	7	15

Note that the remainder is 15, which is the same value as $P(2)$. This is not a coincidence. The following theorem states that this situation is always true.

Remainder Theorem

If the polynomial $P(x)$ is divided by $x - a$, the remainder is $P(a)$.

HOW TO Evaluate $P(x) = x^4 - 3x^2 + 4x - 5$ when $x = -2$ by using the Remainder Theorem.

The value at which the polynomial is evaluated

−2	1	0	−3	4	−5
		−2	4	−2	−4
	1	−2	1	2	−9

- A 0 is inserted for the x^3 term.

← The remainder

$P(-2) = -9$

Example 8

Use synthetic division to evaluate $P(x) = x^2 - 6x + 4$ when $x = 3$.

Solution

3	1	−6	4
		3	−9
	1	−3	−5

$P(3) = -5$

You Try It 8

Use synthetic division to evaluate $P(x) = 2x^2 - 3x - 5$ when $x = 2$.

Your solution

Example 9

Use synthetic division to evaluate $P(x) = -x^4 + 3x^3 + 2x^2 - x - 5$ when $x = -2$.

Solution

−2	−1	3	2	−1	−5
		2	−10	16	−30
	−1	5	−8	15	−35

$P(-2) = -35$

You Try It 9

Use synthetic division to evaluate $P(x) = 2x^3 - 5x^2 + 7$ when $x = -3$.

Your solution

Solutions on p. S20

Copyright © Houghton Mifflin Company. All rights reserved.

6.4 Exercises

Objective A To divide a polynomial by a monomial

For Exercises 1 to 12, divide and check.

1. $\dfrac{3x^2 - 6x}{3x}$

2. $\dfrac{10y^2 - 6y}{2y}$

3. $\dfrac{5x^2 - 10x}{-5x}$

4. $\dfrac{3y^2 - 27y}{-3y}$

5. $\dfrac{5x^2y^2 + 10xy}{5xy}$

6. $\dfrac{8x^2y^2 - 24xy}{8xy}$

7. $\dfrac{x^3 + 3x^2 - 5x}{x}$

8. $\dfrac{a^3 - 5a^2 + 7a}{a}$

9. $\dfrac{9b^5 + 12b^4 + 6b^3}{3b^2}$

10. $\dfrac{a^8 - 5a^5 - 3a^3}{a^2}$

11. $\dfrac{a^5b - 6a^3b + ab}{ab}$

12. $\dfrac{5c^3d + 10c^2d^2 - 15cd^3}{5cd}$

Objective B To divide polynomials

For Exercises 13 to 34, divide by using long division.

13. $(x^2 + 3x - 40) \div (x - 5)$

14. $(x^2 - 14x + 24) \div (x - 2)$

15. $(x^3 - 3x + 2) \div (x - 3)$

16. $(x^3 + 4x^2 - 8) \div (x + 4)$

17. $(6x^2 + 13x + 8) \div (2x + 1)$

18. $(12x^2 + 13x - 14) \div (3x - 2)$

19. $(10x^2 + 9x - 5) \div (2x - 1)$

20. $(18x^2 - 3x + 2) \div (3x + 2)$

Copyright © Houghton Mifflin Company. All rights reserved.

21. $(8x^3 - 9) \div (2x - 3)$

22. $(64x^3 + 4) \div (4x + 2)$

23. $(6x^4 - 13x^2 - 4) \div (2x^2 - 5)$

24. $(12x^4 - 11x^2 + 10) \div (3x^2 + 1)$

25. $\dfrac{-10 - 33x + 3x^3 - 8x^2}{3x + 1}$

26. $\dfrac{10 - 49x + 38x^2 - 8x^3}{1 - 4x}$

27. $\dfrac{x^3 - 5x^2 + 7x - 4}{x - 3}$

28. $\dfrac{2x^3 - 3x^2 + 6x + 4}{2x + 1}$

29. $\dfrac{16x^2 - 13x^3 + 2x^4 + 20 - 9x}{x - 5}$

30. $\dfrac{x - x^2 + 5x^3 + 3x^4 - 2}{x + 2}$

31. $\dfrac{2x^3 + 4x^2 - x + 2}{x^2 + 2x - 1}$

32. $\dfrac{3x^3 - 2x^2 + 5x - 4}{x^2 - x + 3}$

33. $\dfrac{x^4 + 2x^3 - 3x^2 - 6x + 2}{x^2 - 2x - 1}$

34. $\dfrac{x^4 - 3x^3 + 4x^2 - x + 1}{x^2 + x - 3}$

35. Given $Q(x) = 2x + 1$ and $P(x) = 2x^3 + x^2 + 8x + 7$, find $\frac{P(x)}{Q(x)}$.

36. Given $Q(x) = 3x - 2$ and $P(x) = 3x^3 - 2x^2 + 3x - 5$, find $\frac{P(x)}{Q(x)}$.

Copyright © Houghton Mifflin Company. All rights reserved.

Objective C — To divide polynomials by using synthetic division

For Exercises 37 to 52, divide by using synthetic division.

37. $(2x^2 - 6x - 8) \div (x + 1)$

38. $(3x^2 + 19x + 20) \div (x + 5)$

39. $(3x^2 - 14x + 16) \div (x - 2)$

40. $(4x^2 - 23x + 28) \div (x - 4)$

41. $(3x^2 - 4) \div (x - 1)$

42. $(4x^2 - 8) \div (x - 2)$

43. $(2x^3 - x^2 + 6x + 9) \div (x + 1)$

44. $(3x^3 + 10x^2 + 6x - 4) \div (x + 2)$

45. $(18 + x - 4x^3) \div (2 - x)$

46. $(12 - 3x^2 + x^3) \div (x + 3)$

47. $(2x^3 + 5x^2 - 5x + 20) \div (x + 4)$

48. $(5x^3 + 3x^2 - 17x + 6) \div (x + 2)$

49. $\dfrac{5 + 5x - 8x^2 + 4x^3 - 3x^4}{2 - x}$

50. $\dfrac{3 - 13x - 5x^2 + 9x^3 - 2x^4}{3 - x}$

51. $\dfrac{3x^4 + 3x^3 - x^2 + 3x + 2}{x + 1}$

52. $\dfrac{4x^4 + 12x^3 - x^2 - x + 2}{x + 3}$

53. Given $Q(x) = x - 2$ and $P(x) = 3x^2 - 5x + 6$, find $\frac{P(x)}{Q(x)}$.

54. Given $Q(x) = x + 5$ and $P(x) = 2x^2 + 7x - 12$, find $\frac{P(x)}{Q(x)}$.

Objective D **To evaluate a polynomial function using synthetic division**

For Exercises 55 to 72, use the Remainder Theorem to evaluate the polynomial function.

55. $P(x) = 2x^2 - 3x - 1$; $P(3)$

56. $Q(x) = 3x^2 - 5x - 1$; $Q(2)$

Copyright © Houghton Mifflin Company. All rights reserved.

57. $R(x) = x^3 - 2x^2 + 3x - 1; R(4)$

58. $F(x) = x^3 + 4x^2 - 3x + 2; F(3)$

59. $P(z) = 2z^3 - 4z^2 + 3z - 1; P(-2)$

60. $R(t) = 3t^3 + t^2 - 4t + 2; R(-3)$

61. $Z(p) = 2p^3 - p^2 + 3; Z(-3)$

62. $P(y) = 3y^3 + 2y^2 - 5; P(-2)$

63. $Q(x) = x^4 + 3x^3 - 2x^2 + 4x - 9; Q(2)$

64. $Y(z) = z^4 - 2z^3 - 3z^2 - z + 7; Y(3)$

65. $F(x) = 2x^4 - x^3 + 2x - 5; F(-3)$

66. $Q(x) = x^4 - 2x^3 + 4x - 2; Q(-2)$

67. $P(x) = x^3 - 3; P(5)$

68. $S(t) = 4t^3 + 5; S(-4)$

69. $R(t) = 4t^4 - 3t^2 + 5; R(-3)$

70. $P(z) = 2z^4 + z^2 - 3; P(-4)$

71. $Q(x) = x^5 - 4x^3 - 2x^2 + 5x - 2; Q(2)$

72. $R(x) = 2x^5 - x^3 + 4x - 1; R(-2)$

APPLYING THE CONCEPTS

73. Divide by using long division.

a. $\dfrac{a^3 + b^3}{a + b}$ **b.** $\dfrac{x^5 + y^5}{x + y}$ **c.** $\dfrac{x^6 - y^6}{x + y}$

74. For what value of k will the remainder be zero?

a. $(x^3 - x^2 - 3x + k) \div (x + 3)$ **b.** $(2x^3 - x + k) \div (x - 1)$

75. Show how synthetic division can be modified so that the divisor can be of the form $ax + b$.

Copyright © Houghton Mifflin Company. All rights reserved.

Focus on Problem Solving

Dimensional Analysis

In solving application problems, it may be useful to include the units in order to organize the problem so that the answer is in the proper units. Using units to organize and check the correctness of an application is called **dimensional analysis.** We use the operations of multiplying units and dividing units in applying dimensional analysis to application problems.

The Rule for Multiplying Exponential Expressions states that we multiply two expressions with the same base by adding the exponents.

$$x^4 \cdot x^6 = x^{4+6} = x^{10}$$

In calculations that involve quantities, the units are operated on algebraically.

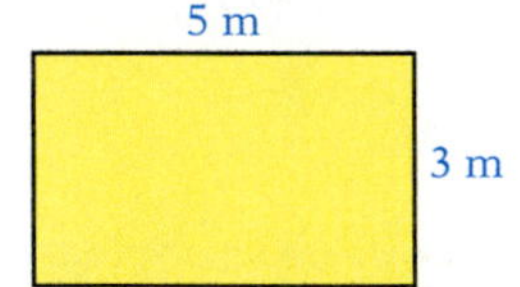

HOW TO A rectangle measures 3 m by 5 m. Find the area of the rectangle.

$A = LW = (3 \text{ m})(5 \text{ m}) = (3 \cdot 5)(\text{m} \cdot \text{m}) = 15 \text{ m}^2$

The area of the rectangle is 15 m^2 (square meters).

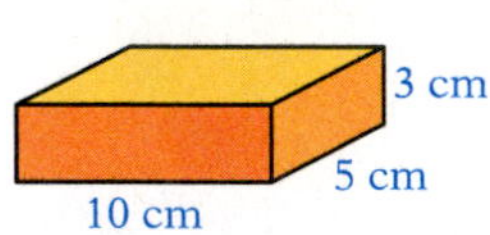

HOW TO A box measures 10 cm by 5 cm by 3 cm. Find the volume of the box.

$V = LWH = (10 \text{ cm})(5 \text{ cm})(3 \text{ cm}) = (10 \cdot 5 \cdot 3)(\text{cm} \cdot \text{cm} \cdot \text{cm}) = 150 \text{ cm}^3$

The volume of the box is 150 cm^3 (cubic centimeters).

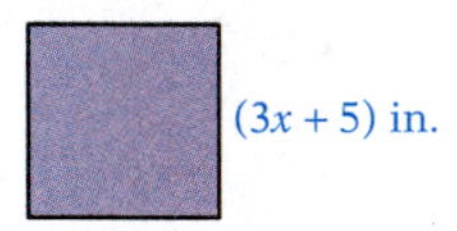

HOW TO Find the area of a square whose side measures $(3x + 5)$ in.

$A = s^2 = [(3x + 5) \text{ in.}]^2 = (3x + 5)^2 \text{ in}^2 = (9x^2 + 30x + 25) \text{ in}^2$

The area of the square is $(9x^2 + 30x + 25)$ in^2 (square inches).

Dimensional analysis is used in the conversion of units.

1 mi
5280 ft

The following example converts the unit miles to feet. The equivalent measures 1 mi = 5280 ft are used to form the following rates, which are called *conversion factors:* $\frac{1 \text{ mi}}{5280 \text{ ft}}$ and $\frac{5280 \text{ ft}}{1 \text{ mi}}$. Because 1 mi = 5280 ft, both of the conversion factors $\frac{1 \text{ mi}}{5280 \text{ ft}}$ and $\frac{5280 \text{ ft}}{1 \text{ mi}}$ are equal to 1.

To convert 3 mi to feet, multiply 3 mi by the conversion factor $\frac{5280 \text{ ft}}{1 \text{ mi}}$.

3 mi
15,840 ft

$$3 \text{ mi} = 3 \text{ mi} \cdot 1 = \frac{3 \text{ mi}}{1} \cdot \frac{5280 \text{ ft}}{1 \text{ mi}} = \frac{3 \cancel{\text{mi}} \cdot 5280 \text{ ft}}{1 \cancel{\text{mi}}} = 3 \cdot 5280 \text{ ft} = 15{,}840 \text{ ft}$$

There are two important points in the above illustration. First, you can think of dividing the numerator and denominator by the common unit "mile" just as you would divide the numerator and denominator of a fraction by a common factor. Second, the conversion factor $\frac{5280 \text{ ft}}{1 \text{ mi}}$ is equal to 1, and multiplying an expression by 1 does not change the value of the expression.

Copyright © Houghton Mifflin Company. All rights reserved.

In the application problem that follows, the units are kept in the problem while the problem is worked.

In 2003, a horse named Funny Cide ran a 1.25-mile race in 2.01 min. Find Funny Cide's average speed for that race in miles per hour. Round to the nearest tenth.

Strategy To find the average speed, use the formula $r = \frac{d}{t}$, where r is the speed, d is the distance, and t is the time. Use the conversion factor $\frac{60 \text{ min}}{1 \text{ h}}$.

Solution

$$r = \frac{d}{t} = \frac{1.25 \text{ mi}}{2.01 \text{ min}} = \frac{1.25 \text{ mi}}{2.01 \text{ min}} \cdot \frac{60 \text{ min}}{1 \text{ h}}$$

$$= \frac{75 \text{ mi}}{2.01 \text{ h}} \approx 37.3 \text{ mph}$$

Funny Cide's average speed was 37.3 mph.

Try each of the following problems. Round to the nearest tenth or nearest cent.

1. Convert 88 ft/s to miles per hour.
2. Convert 8 m/s to kilometers per hour (1 km = 1000 m).
3. A carpet is to be placed in a meeting hall that is 36 ft wide and 80 ft long. At $21.50 per square yard, how much will it cost to carpet the meeting hall?
4. A carpet is to be placed in a room that is 20 ft wide and 30 ft long. At $22.25 per square yard, how much will it cost to carpet the area?
5. Find the number of gallons of water in a fish tank that is 36 in. long and 24 in. wide and is filled to a depth of 16 in. (1 gal = 231 in^3).
6. Find the number of gallons of water in a fish tank that is 24 in. long and 18 in. wide and is filled to a depth of 12 in. (1 gal = 231 in^3).
7. A $\frac{1}{4}$-acre commercial lot is on sale for $2.15 per square foot. Find the sale price of the commercial lot (1 acre = 43,560 ft^2).
8. A 0.75-acre industrial parcel was sold for $98,010. Find the parcel's price per square foot (1 acre = 43,560 ft^2).
9. A new driveway will require 800 ft^3 of concrete. Concrete is ordered by the cubic yard. How much concrete should be ordered?

10. A piston-engined dragster traveled 440 yd in 4.936 s at Ennis, Texas, on October 9, 1988. Find the average speed of the dragster in miles per hour.
11. The Marianas Trench in the Pacific Ocean is the deepest part of the ocean. Its depth is 6.85 mi. The speed of sound under water is 4700 ft/s. Find the time it takes sound to travel from the surface to the bottom of the Marianas Trench and back.

Copyright © Houghton Mifflin Company. All rights reserved.

Projects and Group Activities

Astronomical Distances and Scientific Notation

Astronomers have units of measurement that are useful for measuring vast distances in space. Two of these units are the astronomical unit and the light-year. An **astronomical unit** is the average distance between Earth and the sun. A **light-year** is the distance a ray of light travels in 1 year.

1. Light travels at a speed of 1.86×10^5 mi/s. Find the measure of 1 light-year in miles. Use a 365-day year.

2. The distance between Earth and the star Alpha Centauri is approximately 25 trillion miles. Find the distance between Earth and Alpha Centauri in light-years. Round to the nearest hundredth.

3. The Coma cluster of galaxies is approximately 2.8×10^8 light-years from Earth. Find the distance, in miles, from the Coma cluster to Earth. Write the answer in scientific notation.

4. One astronomical unit (A.U.) is 9.3×10^7 mi. The star Pollux in the constellation Gemini is 1.8228×10^{12} mi from Earth. Find the distance from Pollux to Earth in astronomical units.

5. One light-year is equal to approximately how many astronomical units? Round to the nearest thousand.

Shown below are data on the planets in our solar system. The planets are listed in alphabetical order.

Planet	*Distance from the Sun (in kilometers)*	*Mass (in kilograms)*
Earth	1.50×10^8	5.97×10^{24}
Jupiter	7.79×10^8	1.90×10^{27}
Mars	2.28×10^8	6.42×10^{23}
Mercury	5.79×10^7	3.30×10^{23}
Neptune	4.50×10^9	1.02×10^{26}
Pluto	5.87×10^9	1.25×10^{22}
Saturn	1.43×10^9	5.68×10^{26}
Uranus	2.87×10^9	8.68×10^{25}
Venus	1.08×10^8	4.87×10^{24}

Point of Interest

In November 2001, the Hubble Space Telescope took photos of the atmosphere of a planet orbiting a star 150 light-years from Earth in the constellation Pegasus. The planet is about the size of Jupiter and orbits close to the star HD209458. It was the first discovery of an atmosphere around a planet outside our solar system.

Jupiter

6. Arrange the planets in order from closest to the sun to farthest from the sun.

7. Arrange the planets in order from the one with the greatest mass to the one with the least mass.

8. Write a rule for ordering numbers written in scientific notation.

Copyright © Houghton Mifflin Company. All rights reserved.

Chapter 6 Summary

Key Words	Examples
A *monomial* is a number, a variable, or a product of numbers and variables. [6.1A, p. 345]	5 is a number, y is a variable. $8a^2b^2$ is a product of numbers and variables. 5, y, and $8a^2b^2$ are monomials.
The *degree of a monomial* is the sum of the exponents of the variables. [6.1A, p. 345]	The degree of $8x^4y^5z$ is 10.
A *polynomial* is a variable expression in which the terms are monomials. [6.2A, p. 357]	$x^4 - 2xy - 32x + 8$ is a polynomial. The terms are x^4, $-2xy$, $-32x$, and 8.
A polynomial of one term is a *monomial*, a polynomial of two terms is a *binomial*, and a polynomial of three terms is a *trinomial*. [6.2A, p. 357]	$5x^4$ is a monomial. $6y^3 - 2y$ is a binomial. $2x^2 - 5x + 3$ is a trinomial.
The *degree of a polynomial* is the greatest of the degrees of any of its terms. [6.2A, p. 357]	The degree of the polynomial $x^3 + 3x^2y^2 - 4xy - 3$ is 4.
The terms of a polynomial in one variable are usually arranged so that the exponents of the variable decrease from left to right. This is called *descending order*. [6.2A, p. 357]	The polynomial $4x^3 + 5x^2 - x + 7$ is written in descending order.
A *polynomial function* is an expression whose terms are monomials. Polynomial functions include the *linear function* given by $f(x) = mx + b$; the *quadratic function* given by $f(x) = ax^2 + bx + c$, $a \neq 0$; and the *cubic function*, which is a third-degree polynomial function. The *leading coefficient* of a polynomial function is the coefficient of the variable with the largest exponent. The *constant term* is the term without a variable. [6.2A, p. 357]	$f(x) = 5x - 4$ is a linear function. $f(x) = 3x^2 - 2x + 1$ is a quadratic function. 3 is the leading coefficient, and 1 is the constant term. $f(x) = x^3 - 1$ is a cubic function.

Essential Rules and Procedures	Examples
Rule for Multiplying Exponential Expressions [6.1A, p. 345] $x^m \cdot x^n = x^{m+n}$	$b^5 \cdot b^4 = b^{5+4} = b^9$

Copyright © Houghton Mifflin Company. All rights reserved.

Rule for Simplifying the Power of an Exponential Expression [6.1A, p. 346]

$(x^m)^n = x^{mn}$

$(y^3)^7 = y^{3(7)} = y^{21}$

Rule for Simplifying Powers of Products [6.1A, p. 346]

$(x^m y^n)^p = x^{mp} y^{np}$

$(x^6 y^4 z^5)^2 = x^{6(2)} y^{4(2)} z^{5(2)} = x^{12} y^8 z^{10}$

Definition of Zero as an Exponent [6.1B, p. 347]

For $x \neq 0$, $x^0 = 1$. The expression 0^0 is not defined.

$17^0 = 1$
$(5y)^0 = 1, y \neq 0$

Definition of a Negative Exponent [6.1B, p. 348]

For $x \neq 0$, $x^{-n} = \frac{1}{x^n}$ and $\frac{1}{x^{-n}} = x^n$.

$x^{-6} = \frac{1}{x^6}$ and $\frac{1}{x^{-6}} = x^6$

Rule for Simplifying Powers of Quotients [6.1B, p. 348]

For $y \neq 0$, $\left(\frac{x^m}{y^n}\right)^p = \frac{x^{mp}}{y^{np}}$.

$\left(\frac{x^2}{y^4}\right)^5 = \frac{x^{2 \cdot 5}}{y^{4 \cdot 5}} = \frac{x^{10}}{y^{20}}$

Rule for Negative Exponents on Fractional Expressions [6.1B, p. 349]

For $a \neq 0$, $b \neq 0$, $\left(\frac{a}{b}\right)^{-n} = \left(\frac{b}{a}\right)^n$.

$\left(\frac{3}{8}\right)^{-4} = \left(\frac{8}{3}\right)^4$

Rule for Dividing Exponential Expressions [6.1B, p. 349]

For $x \neq 0$, $\frac{x^m}{x^n} = x^{m-n}$.

$\frac{y^8}{y^3} = y^{8-3} = y^5$

Scientific Notation [6.1C, p. 351]

To express a number in scientific notation, write it in the form $a \times 10^n$, where a is a number between 1 and 10 and n is an integer.

If the number is greater than 10, the exponent on 10 will be positive.

$367{,}000{,}000 = 3.67 \times 10^8$

If the number is less than 1, the exponent on 10 will be negative.

$0.0000059 = 5.9 \times 10^{-6}$

To change a number in scientific notation to decimal notation, move the decimal point to the right if the exponent on 10 is positive and to the left if the exponent on 10 is negative. Move the decimal point the same number of places as the absolute value of the exponent on 10.

$2.418 \times 10^7 = 24{,}180{,}000$
$9.06 \times 10^{-5} = 0.0000906$

Copyright © Houghton Mifflin Company. All rights reserved.

To write the additive inverse of a polynomial, change the sign of every term of the polynomial. [6.2B, p. 360]

The additive inverse of $-y^2 + 4y - 5$ is $y^2 - 4y + 5$.

To add polynomials, combine like terms, which means to add the coefficients of the like terms. [6.2B, p. 360]

$$\begin{aligned}&(8x^2 + 2x - 9) + (-3x^2 + 5x - 7)\\ &\quad= (8x^2 - 3x^2) + (2x + 5x) + (-9 - 7)\\ &\quad= 5x^2 + 7x - 16\end{aligned}$$

To subtract two polynomials, add the additive inverse of the second polynomial to the first polynomial. [6.2B, p. 361]

$$\begin{aligned}&(3y^2 - 8y + 6) - (-y^2 + 4y - 5)\\ &\quad= (3y^2 - 8y + 6) + (y^2 - 4y + 5)\\ &\quad= 4y^2 - 12y + 11\end{aligned}$$

To multiply a polynomial by a monomial, use the Distributive Property and the Rule for Multiplying Exponential Expressions. [6.3A, p. 365]

$$\begin{aligned}&-2x^3(4x^2 + 5x - 1)\\ &\quad= -8x^5 - 10x^4 + 2x^3\end{aligned}$$

The FOIL Method [6.3B, p. 367]
The product of two binomials can be found by adding the products of the **F**irst terms, the **O**uter terms, the **I**nner terms, and the **L**ast terms.

$$\begin{aligned}&(4x + 3)(2x - 5)\\ &\quad= (4x)(2x) + (4x)(-5)\\ &\qquad+ (3)(2x) + (3)(-5)\\ &\quad= 8x^2 - 20x + 6x - 15\\ &\quad= 8x^2 - 14x - 15\end{aligned}$$

To divide a polynomial by a monomial, divide each term of the polynomial by the monomial. [6.4A, p. 375]

$$\begin{aligned}&\frac{12x^5 + 8x^3 - 6x}{4x^2}\\ &\quad= \frac{12x^5}{4x^2} + \frac{8x^3}{4x^2} - \frac{6x}{4x^2} = 3x^3 + 2x - \frac{3}{2x}\end{aligned}$$

Synthetic Division [6.4C, p. 378]
Synthetic division is a shorter method of dividing a polynomial by a binomial of the form $x - a$. This method uses only the coefficients of the variable terms.

$(3x^3 - 9x - 5) \div (x - 2)$

2	3	0	−9	−5
		6	12	6
	3	6	3	1

$$\begin{aligned}&(3x^3 - 9x - 5) \div (x - 2)\\ &\quad= 3x^2 + 6x + 3 + \frac{1}{x - 2}\end{aligned}$$

Remainder Theorem [6.4D, p. 381]
If the polynomial $P(x)$ is divided by $x - a$, the remainder is $P(a)$.

$P(x) = x^3 - x^2 + x - 1$

−2	1	−1	1	−1
		−2	6	−14
	1	−3	7	−15

$P(-2) = -15$

Copyright © Houghton Mifflin Company. All rights reserved.

Chapter 6 Review Exercises

1. Add: $(12y^2 + 17y - 4) + (9y^2 - 13y + 3)$

2. Simplify: $\dfrac{15x^2 + 2x - 2}{3x - 2}$

3. Simplify: $(2x^{-1}y^2z^5)^4(-3x^3yz^{-3})^2$

4. Expand: $(5y - 7)^2$

5. Simplify: $\dfrac{a^{-1}b^3}{a^3b^{-3}}$

6. Use the Remainder Theorem to evaluate $P(x) = x^3 - 2x^2 + 3x - 5$ when $x = 2$.

7. Simplify: $(5x^2 - 8xy + 2y^2) - (x^2 - 3y^2)$

8. Divide: $\dfrac{12b^7 + 36b^5 - 3b^3}{3b^3}$

9. Simplify: $\dfrac{3ab^4}{-6a^2b^4}$

10. Simplify: $(-2a^2b^4)(3ab^2)$

11. Simplify: $\dfrac{8x^{12}}{12x^9}$

12. Simplify: $\dfrac{4x^3 + 27x^2 + 10x + 2}{x + 6}$

13. Given $P(x) = 2x^3 - x + 7$, evaluate $P(-2)$.

14. Subtract: $(13y^3 - 7y - 2) - (12y^2 - 2y - 1)$

15. Divide: $(b^3 - 2b^2 - 33b - 7) \div (b - 7)$

16. Multiply: $4x^2y(3x^3y^2 + 2xy - 7y^3)$

17. Multiply: $(2a - b)(x - 2y)$

18. Multiply: $(2b - 3)(4b + 5)$

19. Simplify: $5x^2 - 4x[x - 3(3x + 2) + x]$

20. Simplify: $(xy^5z^3)(x^3y^3z)$

21. Expand: $(4x - 3y)^2$

22. Simplify: $\dfrac{x^4 - 4}{x - 4}$

Copyright © Houghton Mifflin Company. All rights reserved.

23. Add: $(3x^2 - 2x - 6) + (-x^2 - 3x + 4)$

24. Multiply: $(5x^2yz^4)(2xy^3z^{-1})(7x^{-2}y^{-2}z^3)$

25. Simplify: $\dfrac{3x^4yz^{-1}}{-12xy^3z^2}$

26. Write 948,000,000 in scientific notation.

27. Simplify: $\dfrac{3 \times 10^{-3}}{15 \times 10^2}$

28. Use the Remainder Theorem to evaluate $P(x) = -2x^3 + 2x^2 - 4$ when $x = -3$.

29. Divide: $\dfrac{16x^5 - 8x^3 + 20x}{4x}$

30. Divide: $\dfrac{12x^2 - 16x - 7}{6x + 1}$

31. Multiply: $a^{2n+3}(a^n - 5a + 2)$

32. Multiply: $(x + 6)(x^3 - 3x^2 - 5x + 1)$

33. Multiply: $-2x(4x^2 + 7x - 9)$

34. Multiply: $(3y^2 + 4y - 7)(2y + 3)$

35. Simplify: $(-2u^3v^4)^4$

36. Add: $(2x^3 + 7x^2 + x) + (2x^2 - 4x - 12)$

37. Subtract: $(5x^2 - 2x - 1) - (3x^2 - 5x + 7)$

38. Multiply: $(a + 7)(a - 7)$

39. Simplify: $(5a^7b^6)^2(4ab)$

40. Write 1.46×10^7 in decimal notation.

41. Simplify: $(-2x^3)^2(-3x^4)^3$

42. Divide: $(6y^2 - 35y + 36) \div (3y - 4)$

43. Evaluate: -4^{-2}

44. Multiply: $(5a - 7)(2a + 9)$

45. Divide: $\dfrac{7 - x - x^2}{x + 3}$

46. Write 0.000000127 in scientific notation.

Copyright © Houghton Mifflin Company. All rights reserved.

47. Divide: $\dfrac{16y^2 - 32y}{-4y}$

48. Simplify: $\dfrac{(2a^4b^{-3}c^2)^3}{(2a^3b^2c^{-1})^4}$

49. Multiply: $(x - 4)(3x + 2)(2x - 3)$

50. Simplify: $(-3x^{-2}y^{-3})^{-2}$

51. Simplify: $(2a^{12}b^3)(-9b^2c^6)(3ac)$

52. Multiply: $(5a + 2b)(5a - 2b)$

53. Write 2.54×10^{-3} in decimal notation.

54. Multiply: $2ab^3(4a^2 - 2ab + 3b^2)$

55. Graph $y = x^2 + 1$.

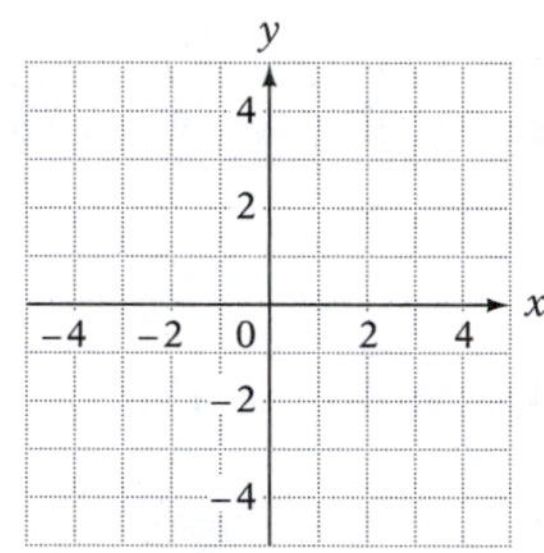

56. For the polynomial $P(x) = 3x^5 - 6x^2 + 7x + 8$:
a. Identify the leading coefficient.
b. Identify the constant term.
c. State the degree.

57. **Physics** The mass of the moon is 3.7×10^{-8} times the mass of the sun. The mass of the sun is 2.19×10^{27} tons. Find the mass of the moon. Write the answer in scientific notation.

The Moon

58. **Geometry** The side of a checkerboard is $(3x - 2)$ inches. Express the area of the checkerboard in terms of the variable x.

59. **Astronomy** The most distant object visible from Earth without the aid of a telescope is the Great Galaxy of Andromeda. It takes light from the Great Galaxy of Andromeda 2.2×10^6 years to travel to Earth. Light travels about 6.7×10^8 mph. How far from Earth is the Great Galaxy of Andromeda? Use a 365-day year.

60. **Geometry** The length of a rectangle is $(5x + 3)$ centimeters. The width is $(2x - 7)$ centimeters. Find the area of the rectangle in terms of the variable x.

Copyright © Houghton Mifflin Company. All rights reserved.

Chapter 6 Test

1. Multiply: $2x(2x^2 - 3x)$

2. Use the Remainder Theorem to evaluate $P(x) = -x^3 + 4x - 8$ when $x = -2$.

3. Simplify: $\dfrac{12x^2}{-3x^8}$

4. Simplify: $(-2xy^2)(3x^2y^4)$

5. Divide: $(x^2 + 1) \div (x + 1)$

6. Multiply: $(x - 3)(x^2 - 4x + 5)$

7. Simplify: $(-2a^2b)^3$

8. Simplify: $\dfrac{(3x^{-2}y^3)^3}{3x^4y^{-1}}$

9. Multiply: $(a - 2b)(a + 5b)$

10. Given $P(x) = 3x^2 - 8x + 1$, evaluate $P(2)$.

11. Divide: $(x^2 + 6x - 7) \div (x - 1)$

12. Multiply: $-3y^2(-2y^2 + 3y - 6)$

13. Multiply: $(-2x^3 + x^2 - 7)(2x - 3)$

14. Simplify: $(4y - 3)(4y + 3)$

Copyright © Houghton Mifflin Company. All rights reserved.

15. Divide: $\dfrac{18x^5 + 9x^4 - 6x^3}{3x^2}$

16. Simplify: $\dfrac{2a^{-1}b}{2^{-2}a^{-2}b^{-3}}$

17. Simplify: $\dfrac{(2a^{-4}b^2)^3}{4a^{-2}b^{-1}}$

18. Subtract: $(3a^2 - 2a - 7) - (5a^3 + 2a - 10)$

19. Simplify: $(2x - 5)^2$

20. Divide: $\dfrac{x^3 - 2x^2 - 5x + 7}{x + 3}$

21. Multiply: $(2x - 7y)(5x - 4y)$

22. Add: $(3x^3 - 2x^2 - 4) + (8x^2 - 8x + 7)$

23. Write 0.00000000302 in scientific notation.

24. Write the number of seconds in 10 weeks in scientific notation.

25. **Geometry** The radius of a circle is $(x - 5)$ meters. Use the equation $A = \pi r^2$, where r is the radius, to find the area of the circle in terms of the variable x. Leave the answer in terms of π.

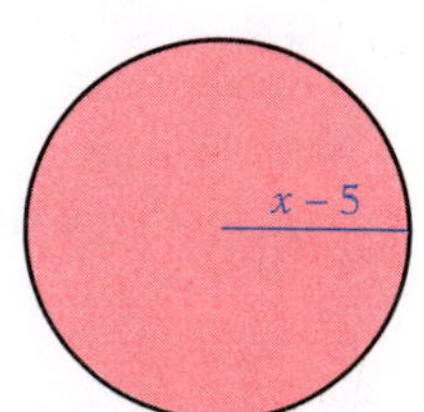

Copyright © Houghton Mifflin Company. All rights reserved.

Cumulative Review Exercises

1. Let $x \in \{-8, -3, 3\}$. For what values of x is the inequality $x \geq -3$ a true statement?

2. Find the additive inverse of 83.

3. Simplify: $8 - 2[-3 - (-1)]^2 \div 4$

4. Evaluate $\frac{2a - b}{b - c}$ when $a = 4$, $b = -2$, and $c = 6$.

5. Simplify: $-5\sqrt{300}$

6. Identify the property that justifies the statement $2x + (-2x) = 0$.

7. Simplify: $2x - 4[x - 2(3 - 23x) + 4]$

8. Solve: $\frac{2}{3} - y = \frac{5}{6}$

9. Solve: $8x - 3 - x = -6 + 3x - 8$

10. Solve: $3 - |2 - 3x| = -2$

11. Given $P(x) = 3x^2 - 2x + 2$, find $P(-2)$.

12. Is the relation $\{(-1, 0), (0, 0), (1, 0)\}$ a function?

13. Find the slope of the line containing the points $(-2, 3)$ and $(4, 2)$.

14. Find the equation of the line that contains the point $(-1, 2)$ and has slope $-\frac{3}{2}$.

15. Find the equation of the line that contains the point $(-2, 4)$ and is perpendicular to the line $3x + 2y = 4$.

16. Solve by substitution:
$$2x - 3y = 4$$
$$x + y = -3$$

17. Solve by the addition method:
$$x - y + z = 0$$
$$2x + y - 3z = -7$$
$$-x + 2y + 2z = 5$$

18. Simplify: $-2x - (-xy) + 7x - 4xy$

19. Multiply: $(2x + 3)(2x^2 - 3x + 1)$

20. Write the number 0.00000501 in scientific notation.

Copyright © Houghton Mifflin Company. All rights reserved.

21. Graph: $3x - 4y = 12$

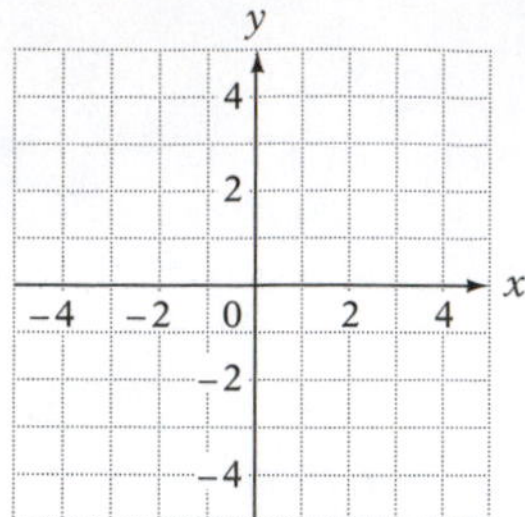

22. Graph: $-3x + 2y < 6$

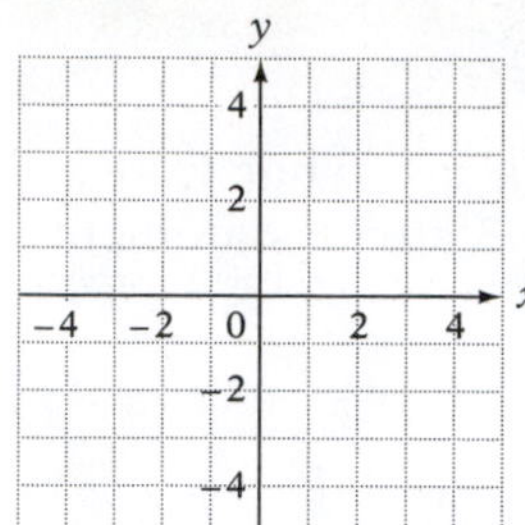

23. Solve by graphing:
$x - 2y = 3$
$-2x + y = -3$

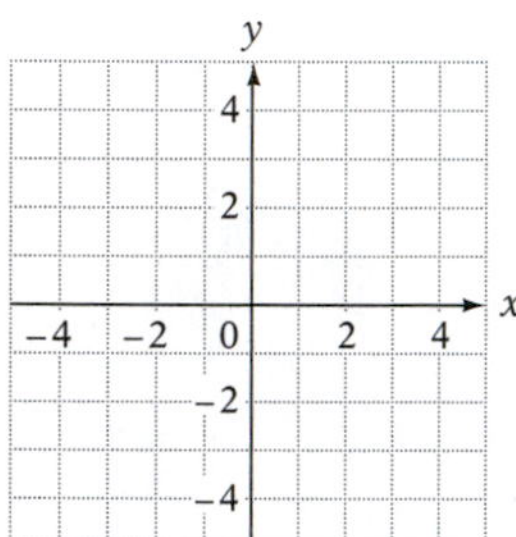

24. Graph the solution set:
$2x + y < 2$
$-6x + 3y \geq 6$

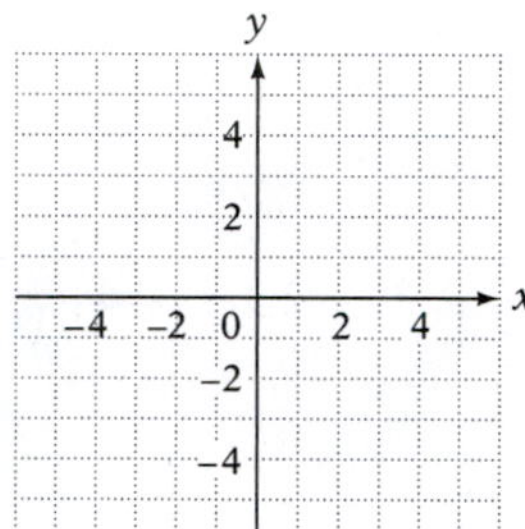

25. Simplify: $(4a^{-2}b^{3})(2a^{3}b^{-1})^{-2}$

26. Simplify: $\dfrac{(5x^{3}y^{-3}z)^{-2}}{y^{4}z^{-2}}$

27. **Integers** The sum of two integers is twenty-four. The difference between four times the smaller integer and nine is three less than twice the larger integer. Find the integers.

28. **Mixtures** How many ounces of pure gold that costs $360 per ounce must be mixed with 80 oz of an alloy that costs $120 per ounce to make a mixture that costs $200 per ounce?

29. **Travel** Two bicycles are 25 mi apart and traveling toward each other. One cyclist is traveling at 1.5 times the rate of the other cyclist. They meet in 2 h. Find the rate of each cyclist.

30. **Investments** An investor has a total of $12,000 invested in two simple interest accounts. On one account, the annual simple interest rate is 4%. On the second account, the annual simple interest rate is 4.5%. How much is invested in the 4% account if the total annual interest earned is $530?

31. **Travel** The graph shows the relationship between the distance traveled and the time of travel. Find the slope of the line between the two points on the graph. Write a sentence that states the meaning of the slope.

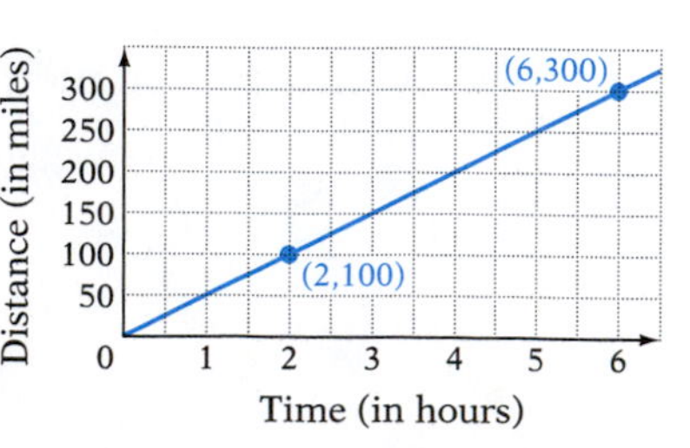

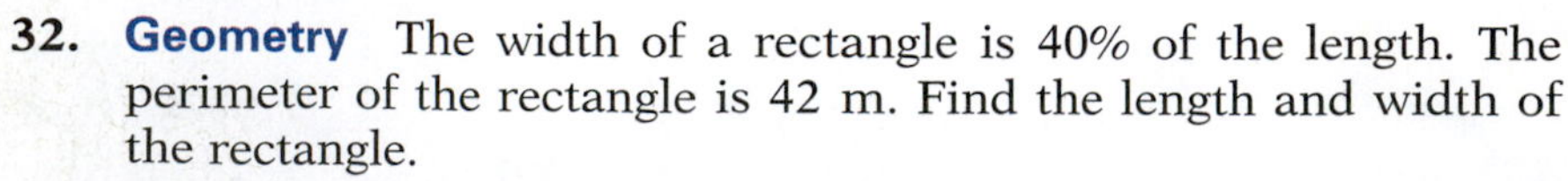

32. **Geometry** The width of a rectangle is 40% of the length. The perimeter of the rectangle is 42 m. Find the length and width of the rectangle.

33. **Geometry** The length of a side of a square is $(2x + 3)$ meters. Find the area of the square in terms of the variable x.

Copyright © Houghton Mifflin Company. All rights reserved.

chapter 7

Factoring

Coordinating the position of each skydiver in a jump involves a lot of careful planning. Position within the "pattern" is determined by when and where each chute is deployed, which is further dependent on each diver's velocity during free fall. Factors such as the initial altitude of the jump, the size of the parachute, and the weight and body position of the diver affect velocity. The velocity of a falling object can be modeled by a quadratic equation, as shown in **Exercises 73 and 74 on page 439.**

OBJECTIVES

Section 7.1

A To factor a monomial from a polynomial

B To factor by grouping

Section 7.2

A To factor a trinomial of the form $x^2 + bx + c$

B To factor completely

Section 7.3

A To factor a trinomial of the form $ax^2 + bx + c$ by using trial factors

B To factor a trinomial of the form $ax^2 + bx + c$ by grouping

Section 7.4

A To factor the difference of two perfect squares or a perfect-square trinomial

B To factor the sum or difference of two perfect cubes

C To factor a trinomial that is quadratic in form

D To factor completely

Section 7.5

A To solve equations by factoring

B To solve application problems

Need help? For online student resources, such as section quizzes, visit this textbook's website at **math.college.hmco.com/students.**

Copyright © Houghton Mifflin Company. All rights reserved.

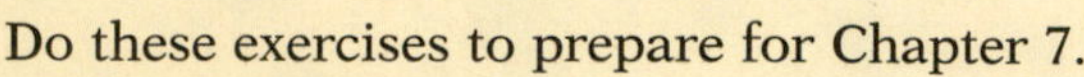

PREP TEST

Do these exercises to prepare for Chapter 7.

1. Write 30 as a product of prime numbers.

2. Simplify: $-3(4y - 5)$

3. Simplify: $-(a - b)$

4. Simplify: $2(a - b) - 5(a - b)$

5. Solve: $4x = 0$

6. Solve: $2x + 1 = 0$

7. Multiply: $(x + 4)(x - 6)$

8. Multiply: $(2x - 5)(3x + 2)$

9. Simplify: $\frac{x^5}{x^2}$

10. Simplify: $\frac{6x^4y^3}{2xy^2}$

GO FIGURE

Find the values of x and y if $x + y$, xy, and $\frac{x}{y}$ all equal the same number.

Copyright © Houghton Mifflin Company. All rights reserved.

7.1 Common Factors

Objective A To factor a monomial from a polynomial

The **greatest common factor (GCF) of two or more numbers** is the largest common factor of the numbers. For example, the GCF of 9 and 12 is 3. The **GCF of two or more monomials** is the product of the GCF of the coefficients and the common variable factors.

$6x^3y = 2 \cdot 3 \cdot x \cdot x \cdot x \cdot y$
$8x^2y^2 = 2 \cdot 2 \cdot 2 \cdot x \cdot x \cdot y \cdot y$
$\text{GCF} = 2 \cdot x \cdot x \cdot y = 2x^2y$

Note that the exponent of each variable in the GCF is the same as the *smallest* exponent of that variable in either of the monomials.

The GCF of $6x^3y$ and $8x^2y^2$ is $2x^2y$.

HOW TO Find the GCF of $12a^4b$ and $18a^2b^2c$.

The common variable factors are a^2 and b; c is not a common variable factor.

$12a^4b = 2 \cdot 2 \cdot 3 \cdot a^4 \cdot b$
$18a^2b^2c = 2 \cdot 3 \cdot 3 \cdot a^2 \cdot b^2 \cdot c$
$\text{GCF} = 2 \cdot 3 \cdot a^2 \cdot b = 6a^2b$

To **factor a polynomial** means to write the polynomial as a product of other polynomials. In the example at the right, $2x$ is the GCF of the terms $2x^2$ and $10x$.

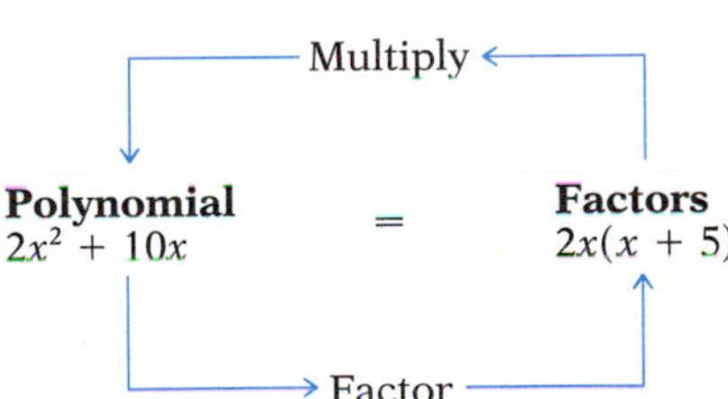

HOW TO Factor: $5x^3 - 35x^2 + 10x$

Find the GCF of the terms of the polynomial.

$5x^3 = 5 \cdot x^3$

$35x^2 = 5 \cdot 7 \cdot x^2$

$10x = 2 \cdot 5 \cdot x$

The GCF is $5x$.

Rewrite the polynomial, expressing each term as a product with the GCF as one of the factors.

$$5x^3 - 35x^2 + 10x = 5x(x^2) + 5x(-7x) + 5x(2)$$
$$= 5x(x^2 - 7x + 2)$$

• Use the Distributive Property to write the polynomial as a product of factors.

> **TAKE NOTE**
> At the right, the factors in parentheses are determined by dividing each term of the trinomial by the GCF, $5x$.
> $\frac{5x^3}{5x} = x^2$, $\frac{-35x^2}{5x} = -7x$, and $\frac{10x}{5x} = 2$

Copyright © Houghton Mifflin Company. All rights reserved.

HOW TO Factor: $21x^2y^3 - 6xy^5 + 15x^4y^2$

Find the GCF of the terms of the polynomial.

$$21x^2y^3 = 3 \cdot 7 \cdot x^2 \cdot y^3$$
$$6xy^5 = 2 \cdot 3 \cdot x \cdot y^5$$
$$15x^4y^2 = 3 \cdot 5 \cdot x^4 \cdot y^2$$

The GCF is $3xy^2$.

Rewrite the polynomial, expressing each term as a product with the GCF as one of the factors.

$$21x^2y^3 - 6xy^5 + 15x^4y^2 = 3xy^2(7xy) + 3xy^2(-2y^3) + 3xy^2(5x^3)$$
$$= 3xy^2(7xy - 2y^3 + 5x^3)$$

• **Use the Distributive Property to write the polynomial as a product of factors.**

Example 1

Factor: $8x^2 + 2xy$

Solution

The GCF is $2x$.

$$8x^2 + 2xy = 2x(4x) + 2x(y)$$
$$= 2x(4x + y)$$

You Try It 1

Factor: $14a^2 - 21a^4b$

Your solution

$7a^2(2-3a^2b)$

Example 2

Factor: $n^3 - 5n^2 + 2n$

Solution

The GCF is n.

$$n^3 - 5n^2 + 2n = n(n^2) + n(-5n) + n(2)$$
$$= n(n^2 - 5n + 2)$$

You Try It 2

Factor: $27b^2 + 18b + 9$

Your solution

$9(3b^2+2b+1)$

Example 3

Factor: $16x^2y + 8x^4y^2 - 12x^4y^5$

Solution

The GCF is $4x^2y$.

$$16x^2y + 8x^4y^2 - 12x^4y^5$$
$$= 4x^2y(4) + 4x^2y(2x^2y) + 4x^2y(-3x^2y^4)$$
$$= 4x^2y(4 + 2x^2y - 3x^2y^4)$$

You Try It 3

Factor: $6x^4y^2 - 9x^3y^2 + 12x^2y^4$

Your solution

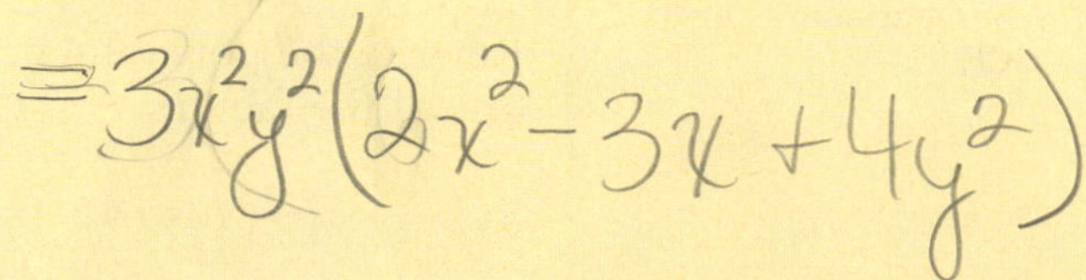

Solutions on p. S21

Copyright © Houghton Mifflin Company. All rights reserved.

Objective B To factor by grouping

VIDEO & DVD   

A factor that has two terms is called a **binomial factor.** In the examples at the right, the binomials in parentheses are binomial factors.

$2a(a + b)^2$
$3xy(x - y)$

The Distributive Property is used to factor a common binomial factor from an expression.

The common binomial factor of the expression $6x(x - 3) + y(x - 3)$ is $(x - 3)$. To factor that expression, use the Distributive Property to write the expression as a product of factors.

$$6x(x - 3) + y(x - 3) = (x - 3)(6x + y)$$

Consider the following simplification of $-(a - b)$.

$$-(a - b) = -1(a - b) = -a + b = b - a$$

Thus $\qquad b - a = -(a - b)$

This equation is sometimes used to factor a common binomial from an expression.

HOW TO Factor: $2x(x - y) + 5(y - x)$

$$\begin{aligned} 2x(x - y) + 5(y - x) &= 2x(x - y) - 5(x - y) \\ &= (x - y)(2x - 5) \end{aligned}$$

- $5(y - x) = 5[(-1)(x - y)] = -5(x - y)$

A polynomial can be **factored by grouping** if its terms can be grouped and factored in such a way that a common binomial factor is found.

HOW TO Factor: $ax + bx - ay - by$

$$\begin{aligned} ax + bx - ay - by &= (ax + bx) - (ay + by) \\ &= x(a + b) - y(a + b) \\ &= (a + b)(x - y) \end{aligned}$$

- Group the first two terms and the last two terms. Note that $-ay - by = -(ay + by)$.
- Factor each group.
- Factor the GCF, $(a + b)$, from each group.

HOW TO Factor: $6x^2 - 9x - 4xy + 6y$

$$\begin{aligned} 6x^2 - 9x - 4xy + 6y &= (6x^2 - 9x) - (4xy - 6y) \\ &= 3x(2x - 3) - 2y(2x - 3) \\ &= (2x - 3)(3x - 2y) \end{aligned}$$

- Group the first two terms and the last two terms. Note that $-4xy + 6y = -(4xy - 6y)$.
- Factor each group.
- Factor the GCF, $(2x - 3)$, from each group.

Copyright © Houghton Mifflin Company. All rights reserved.

Example 4

Factor: $4x(3x - 2) - 7(3x - 2)$

Solution

$4x(3x - 2) - 7(3x - 2)$ • $3x - 2$ is the common binomial factor.

$= (3x - 2)(4x - 7)$

You Try It 4

Factor: $2y(5x - 2) - 3(2 - 5x)$

Your solution

Example 5

Factor: $9x^2 - 15x - 6xy + 10y$

Solution

$9x^2 - 15x - 6xy + 10y$

$= (9x^2 - 15x) - (6xy - 10y)$ • $-6xy + 10y = -(6xy - 10y)$

$= 3x(3x - 5) - 2y(3x - 5)$ • $3x - 5$ is the common factor.

$= (3x - 5)(3x - 2y)$

You Try It 5

Factor: $a^2 - 3a + 2ab - 6b$

Your solution

Example 6

Factor: $3x^2y - 4x - 15xy + 20$

Solution

$3x^2y - 4x - 15xy + 20$

$= (3x^2y - 4x) - (15xy - 20)$ • $-15xy + 20 = -(15xy - 20)$

$= x(3xy - 4) - 5(3xy - 4)$ • $3xy - 4$ is the common factor.

$= (3xy - 4)(x - 5)$

You Try It 6

Factor: $2mn^2 - n + 8mn - 4$

Your solution

Example 7

Factor: $4ab - 6 + 3b - 2ab^2$

Solution

$4ab - 6 + 3b - 2ab^2$

$= (4ab - 6) + (3b - 2ab^2)$

$= 2(2ab - 3) + b(3 - 2ab)$

$= 2(2ab - 3) - b(2ab - 3)$ • $3 - 2ab = -(2ab - 3)$

$= (2ab - 3)(2 - b)$ • $2ab - 3$ is the common factor.

You Try It 7

Factor: $3xy - 9y - 12 + 4x$

Your solution

Solutions on p. S21

Copyright © Houghton Mifflin Company. All rights reserved.

7.1 Exercises

Objective A To factor a monomial from a polynomial

1. Explain the meaning of "a common monomial factor of a polynomial."

2. Explain the meaning of "a factor" and the meaning of "to factor."

For Exercises 3 to 41, factor.

3. $5a + 5$

4. $7b - 7$

5. $16 - 8a^2$

6. $12 + 12y^2$

7. $8x + 12$

8. $16a - 24$

9. $30a - 6$

10. $20b + 5$

11. $7x^2 - 3x$

12. $12y^2 - 5y$

13. $3a^2 + 5a^5$

14. $9x - 5x^2$

15. $14y^2 + 11y$

16. $6b^3 - 5b^2$

17. $2x^4 - 4x$

18. $3y^4 - 9y$

19. $10x^4 - 12x^2$

20. $12a^5 - 32a^2$

21. $8a^8 - 4a^5$

22. $16y^4 - 8y^7$

23. $x^2y^2 - xy$

24. $a^2b^2 + ab$

25. $3x^2y^4 - 6xy$

26. $12a^2b^5 - 9ab$

27. $x^2y - xy^3$

28. $3x^3 + 6x^2 + 9x$

29. $5y^3 - 20y^2 + 5y$

30. $2x^4 - 4x^3 + 6x^2$

31. $3y^4 - 9y^3 - 6y^2$

32. $2x^3 + 6x^2 - 14x$

33. $3y^3 - 9y^2 + 24y$

34. $2y^5 - 3y^4 + 7y^3$

35. $6a^5 - 3a^3 - 2a^2$

36. $x^3y - 3x^2y^2 + 7xy^3$

37. $2a^2b - 5a^2b^2 + 7ab^2$

38. $5y^3 + 10y^2 - 25y$

39. $4b^5 + 6b^3 - 12b$

40. $3a^2b^2 - 9ab^2 + 15b^2$

41. $8x^2y^2 - 4x^2y + x^2$

Objective B To factor by grouping

For Exercises 42 to 68, factor.

42. $x(b + 4) + 3(b + 4)$

43. $y(a + z) + 7(a + z)$

44. $a(y - x) - b(y - x)$

45. $3r(a - b) + s(a - b)$

46. $x(x - 2) + y(2 - x)$

47. $t(m - 7) + 7(7 - m)$

Copyright © Houghton Mifflin Company. All rights reserved.

48. $2x(7 + b) - y(b + 7)$

49. $2y(4a - b) - (b - 4a)$

50. $8c(2m - 3n) + (3n - 2m)$

51. $x^2 + 2x + 2xy + 4y$

52. $x^2 - 3x + 4ax - 12a$

53. $p^2 - 2p - 3rp + 6r$

54. $t^2 + 4t - st - 4s$

55. $ab + 6b - 4a - 24$

56. $xy - 5y - 2x + 10$

57. $2z^2 - z + 2yz - y$

58. $2y^2 - 10y + 7xy - 35x$

59. $8v^2 - 12vy + 14v - 21y$

60. $21x^2 + 6xy - 49x - 14y$

61. $2x^2 - 5x - 6xy + 15y$

62. $4a^2 + 5ab - 10b - 8a$

63. $3y^2 - 6y - ay + 2a$

64. $2ra + a^2 - 2r - a$

65. $3xy - y^2 - y + 3x$

66. $2ab - 3b^2 - 3b + 2a$

67. $3st + t^2 - 2t - 6s$

68. $4x^2 + 3xy - 12y - 16x$

APPLYING THE CONCEPTS

69. **Number Sense** A natural number is a *perfect number* if it is the sum of all its factors less than itself. For example, 6 is a perfect number because all the factors of 6 that are less than 6 are 1, 2, and 3, and $1 + 2 + 3 = 6$.
a. Find the one perfect number between 20 and 30.
b. Find the one perfect number between 490 and 500.

70. **Geometry** In the equation $P = 2L + 2W$, what is the effect on P when the quantity $L + W$ doubles?

71. **Geometry** Write an expression in factored form for the shaded portions in the following diagrams. Use the equation for the area of a rectangle ($A = LW$) and the equation for the area of a circle ($A = \pi r^2$).

a.

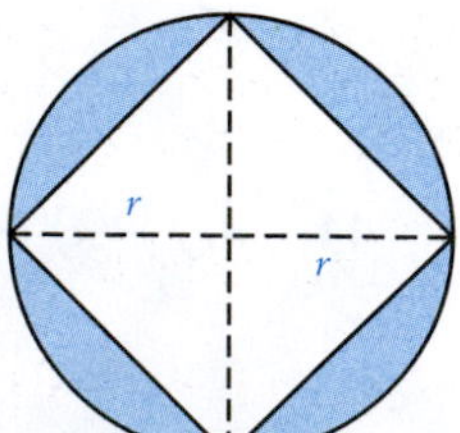

b.

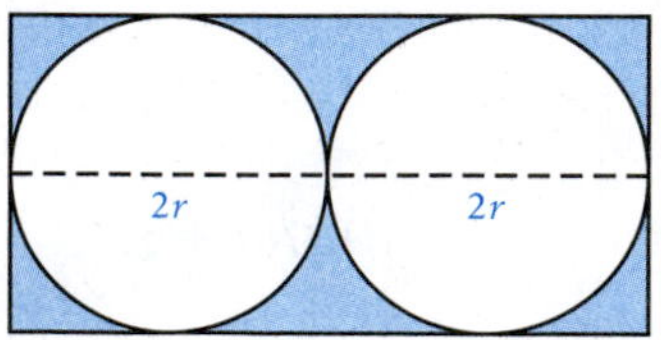

c.

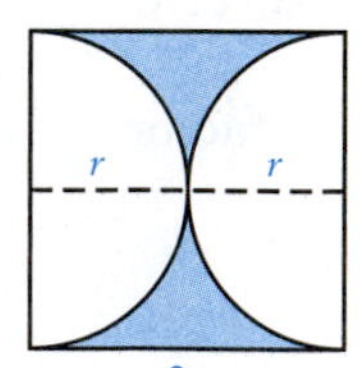

Copyright © Houghton Mifflin Company. All rights reserved.

7.2 Factoring Polynomials of the Form $x^2 + bx + c$

Objective A To factor a trinomial of the form $x^2 + bx + c$

Trinomials of the form $x^2 + bx + c$, where b and c are integers, are shown at the right.

$x^2 + 8x + 12; b = 8, c = 12$
$x^2 - 7x + 12; b = -7, c = 12$
$x^2 - 2x - 15; b = -2, c = -15$

To factor a trinomial of this form means to express the trinomial as the product of two binomials.

Trinomials expressed as the product of binomials are shown at the right.

$x^2 + 8x + 12 = (x + 6)(x + 2)$
$x^2 - 7x + 12 = (x - 3)(x - 4)$
$x^2 - 2x - 15 = (x + 3)(x - 5)$

The method by which factors of a trinomial are found is based on FOIL. Consider the following binomial products, noting the relationship between the constant terms of the binomials and the terms of the trinomials.

The signs in the binomials are the same.

$(x + 6)(x + 2) = x^2 + 2x + 6x + (6)(2) = x^2 + 8x + 12$
Sum of 6 and 2 (→ 8)
Product of 6 and 2 (→ 12)

$(x - 3)(x - 4) = x^2 - 4x - 3x + (-3)(-4) = x^2 - 7x + 12$
Sum of -3 and -4 (→ -7)
Product of -3 and -4 (→ 12)

The signs in the binomials are opposites.

$(x + 3)(x - 5) = x^2 - 5x + 3x + (3)(-5) = x^2 - 2x - 15$
Sum of 3 and -5 (→ -2)
Product of 3 and -5 (→ -15)

$(x - 4)(x + 6) = x^2 + 6x - 4x + (-4)(6) = x^2 + 2x - 24$
Sum of -4 and 6 (→ 2)
Product of -4 and 6 (→ -24)

Factoring $x^2 + bx + c$: IMPORTANT RELATIONSHIPS

1. When the constant term of the trinomial is positive, the constant terms of the binomials have the same sign. They are both positive when the coefficient of the x term in the trinomial is positive. They are both negative when the coefficient of the x term in the trinomial is negative.
2. When the constant term of the trinomial is negative, the constant terms of the binomials have opposite signs.
3. In the trinomial, the coefficient of x is the sum of the constant terms of the binomials.
4. In the trinomial, the constant term is the product of the constant terms of the binomials.

Copyright © Houghton Mifflin Company. All rights reserved.

HOW TO Factor: $x^2 - 7x + 10$

Because the constant term is positive and the coefficient of x is negative, the binomial constants will be negative. Find two negative factors of 10 whose sum is -7. The results can be recorded in a table.

Negative Factors of 10	*Sum*
$-1, -10$	-11
$-2, -5$	-7

• These are the correct factors.

$x^2 - 7x + 10 = (x - 2)(x - 5)$ • Write the trinomial as a product of its factors.

TAKE NOTE
Always check your proposed factorization to ensure accuracy.

You can check the proposed factorization by multiplying the two binomials.

Check: $(x - 2)(x - 5) = x^2 - 5x - 2x + 10$
$= x^2 - 7x + 10$

HOW TO Factor: $x^2 - 9x - 36$

The constant term is negative. The binomial constants will have opposite signs. Find two factors of -36 whose sum is -9.

Factors of -36	*Sum*
$+1, -36$	-35
$-1, +36$	35
$+2, -18$	-16
$-2, +18$	16
$+3, -12$	-9

• Once the correct factors are found, it is not necessary to try the remaining factors.

$x^2 - 9x - 36 = (x + 3)(x - 12)$ • Write the trinomial as a product of its factors.

For some trinomials it is not possible to find integer factors of the constant term whose sum is the coefficient of the middle term. A polynomial that does not factor using only integers is **nonfactorable over the integers.**

TAKE NOTE
Just as 17 is a prime number, $x^2 + 7x + 8$ is a **prime polynomial.** Binomials of the form $x - a$ and $x + a$ are also prime polynomials.

HOW TO Factor: $x^2 + 7x + 8$

The constant term is positive and the coefficient of x is positive. The binomial constants will be positive. Find two positive factors of 8 whose sum is 7.

Positive Factors of 8	*Sum*
1, 8	9
2, 4	6

• There are no positive integer factors of 8 whose sum is 7.

$x^2 + 7x + 8$ is nonfactorable over the integers.

Example 1 Factor: $x^2 - 8x + 15$

Solution
Find two negative factors of 15 whose sum is -8.

Factors	*Sum*
$-1, -15$	-16
$-3, -5$	-8

$x^2 - 8x + 15 = (x - 3)(x - 5)$

You Try It 1 Factor: $x^2 + 9x + 20$

Your solution

Solution on p. S21

Copyright © Houghton Mifflin Company. All rights reserved.

Example 2 Factor: $x^2 + 6x - 27$

Solution
Find two factors of -27 whose sum is 6.

Factors	Sum
+1, −27	−26
−1, +27	26
+3, −9	−6
−3, +9	6

$x^2 + 6x - 27 = (x - 3)(x + 9)$

You Try It 2 Factor: $x^2 + 7x - 18$

Your solution

Solution on p. S21

Objective B To factor completely

SSM

A polynomial is **factored completely** when it is written as a product of factors that are nonfactorable over the integers.

TAKE NOTE
The first step in *any* factoring problem is to determine whether the terms of the polynomial have a *common factor.* If they do, factor it out first.

HOW TO Factor: $4y^3 - 4y^2 - 24y$

$4y^3 - 4y^2 - 24y = 4y(y^2) - 4y(y) - 4y(6)$
- The GCF is $4y$.

$= 4y(y^2 - y - 6)$
- Use the Distributive Property to factor out the GCF.

$= 4y(y + 2)(y - 3)$
- Factor $y^2 - y - 6$. The two factors of -6 whose sum is -1 are 2 and -3.

It is always possible to check the proposed factorization by multiplying the polynomials. Here is the check for the last example.

Check: $4y(y + 2)(y - 3) = 4y(y^2 - 3y + 2y - 6)$
$= 4y(y^2 - y - 6)$
$= 4y^3 - 4y^2 - 24y$
- This is the original polynomial.

TAKE NOTE
$2y$ and $10y$ are placed in the binomials. This is necessary so that the middle term contains xy and the last term contains y^2.

HOW TO Factor: $5x^2 + 60xy + 100y^2$

$5x^2 + 60xy + 100y^2 = 5(x^2) + 5(12xy) + 5(20y^2)$
- The GCF is 5.

$= 5(x^2 + 12xy + 20y^2)$
- Use the Distributive Property to factor out the GCF.

$= 5(x + 2y)(x + 10y)$
- Factor $x^2 + 12xy + 20y^2$. The two factors of 20 whose sum is 12 are 2 and 10.

Note that $2y$ and $10y$ were placed in the binomials. The following check shows that this was necessary.

Check: $5(x + 2y)(x + 10y) = 5(x^2 + 10xy + 2xy + 20y^2)$
$= 5(x^2 + 12xy + 20y^2)$
$= 5x^2 + 60xy + 100y^2$
- This is the original polynomial.

Copyright © Houghton Mifflin Company. All rights reserved.

TAKE NOTE
When the coefficient of the highest power in a polynomial is negative, consider factoring out a negative GCF. Example 3 is another example of this technique.

HOW TO Factor: $15 - 2x - x^2$

Because the coefficient of x^2 is -1, factor -1 from the trinomial and then write the resulting trinomial in descending order.

$15 - 2x - x^2 = -(x^2 + 2x - 15)$ • $15 - 2x - x^2 = -1(-15 + 2x + x^2) = -(x^2 + 2x - 15)$

$= -(x + 5)(x - 3)$ • Factor $x^2 + 2x - 15$. The two factors of -15 whose sum is 2 are 5 and -3.

Check: $-(x + 5)(x - 3) = -(x^2 + 2x - 15)$
$= -x^2 - 2x + 15$
$= 15 - 2x - x^2$ • This is the original polynomial.

Example 3

Factor: $-3x^3 + 9x^2 + 12x$

Solution

The GCF is $-3x$.
$-3x^3 + 9x^2 + 12x = -3x(x^2 - 3x - 4)$
Factor the trinomial $x^2 - 3x - 4$. Find two factors of -4 whose sum is -3.

Factors	*Sum*
+1, −4	−3

$-3x^3 + 9x^2 + 12x = -3x(x + 1)(x - 4)$

You Try It 3

Factor: $-2x^3 + 14x^2 - 12x$

Your solution

Example 4

Factor: $4x^2 - 40xy + 84y^2$

Solution

The GCF is 4.
$4x^2 - 40xy + 84y^2 = 4(x^2 - 10xy + 21y^2)$
Factor the trinomial $x^2 - 10xy + 21y^2$. Find two negative factors of 21 whose sum is -10.

Factors	*Sum*
−1, −21	−22
−3, −7	−10

$4x^2 - 40xy + 84y^2 = 4(x - 3y)(x - 7y)$

You Try It 4

Factor: $3x^2 - 9xy - 12y^2$

Your solution

Solutions on p. S21

Copyright © Houghton Mifflin Company. All rights reserved.

7.2 Exercises

Objective A **To factor a trinomial of the form $x^2 + bx + c$**

1. Fill in the blank. In factoring a trinomial, if the constant term is positive, then the signs in both binomial factors will be _______.

2. Fill in the blanks. To factor $x^2 + 8x - 48$, we must find two numbers whose product is _______ and whose sum is _______.

For Exercises 3 to 75, factor.

3. $x^2 + 3x + 2$ **4.** $x^2 + 5x + 6$ **5.** $x^2 - x - 2$ **6.** $x^2 + x - 6$

7. $a^2 + a - 12$ **8.** $a^2 - 2a - 35$ **9.** $a^2 - 3a + 2$ **10.** $a^2 - 5a + 4$

11. $a^2 + a - 2$ **12.** $a^2 - 2a - 3$ **13.** $b^2 - 6b + 9$ **14.** $b^2 + 8b + 16$

15. $b^2 + 7b - 8$ **16.** $y^2 - y - 6$ **17.** $y^2 + 6y - 55$ **18.** $z^2 - 4z - 45$

19. $y^2 - 5y + 6$ **20.** $y^2 - 8y + 15$ **21.** $z^2 - 14z + 45$ **22.** $z^2 - 14z + 49$

23. $z^2 - 12z - 160$ **24.** $p^2 + 2p - 35$ **25.** $p^2 + 12p + 27$ **26.** $p^2 - 6p + 8$

27. $x^2 + 20x + 100$ **28.** $x^2 + 18x + 81$ **29.** $b^2 + 9b + 20$ **30.** $b^2 + 13b + 40$

31. $x^2 - 11x - 42$ **32.** $x^2 + 9x - 70$ **33.** $b^2 - b - 20$ **34.** $b^2 + 3b - 40$

35. $y^2 - 14y - 51$ **36.** $y^2 - y - 72$ **37.** $p^2 - 4p - 21$ **38.** $p^2 + 16p + 39$

39. $y^2 - 8y + 32$ **40.** $y^2 - 9y + 81$ **41.** $x^2 - 20x + 75$ **42.** $x^2 - 12x + 11$

Copyright © Houghton Mifflin Company. All rights reserved.

43. $p^2 + 24p + 63$

44. $x^2 - 15x + 56$

45. $x^2 + 21x + 38$

46. $x^2 + x - 56$

47. $x^2 + 5x - 36$

48. $a^2 - 21a - 72$

49. $a^2 - 7a - 44$

50. $a^2 - 15a + 36$

51. $a^2 - 21a + 54$

52. $z^2 - 9z - 136$

53. $z^2 + 14z - 147$

54. $c^2 - c - 90$

55. $c^2 - 3c - 180$

56. $z^2 + 15z + 44$

57. $p^2 + 24p + 135$

58. $c^2 + 19c + 34$

59. $c^2 + 11c + 18$

60. $x^2 - 4x - 96$

61. $x^2 + 10x - 75$

62. $x^2 - 22x + 112$

63. $x^2 + 21x - 100$

64. $b^2 + 8b - 105$

65. $b^2 - 22b + 72$

66. $a^2 - 9a - 36$

67. $a^2 + 42a - 135$

68. $b^2 - 23b + 102$

69. $b^2 - 25b + 126$

70. $a^2 + 27a + 72$

71. $z^2 + 24z + 144$

72. $x^2 + 25x + 156$

73. $x^2 - 29x + 100$

74. $x^2 - 10x - 96$

75. $x^2 + 9x - 112$

Copyright © Houghton Mifflin Company. All rights reserved.

Objective B To factor completely

For Exercises 76 to 135, factor.

76. $2x^2 + 6x + 4$

77. $3x^2 + 15x + 18$

78. $18 + 7x - x^2$

79. $12 - 4x - x^2$

80. $ab^2 + 2ab - 15a$

81. $ab^2 + 7ab - 8a$

82. $xy^2 - 5xy + 6x$

83. $xy^2 + 8xy + 15x$

84. $z^3 - 7z^2 + 12z$

85. $-2a^3 - 6a^2 - 4a$

86. $-3y^3 + 15y^2 - 18y$

87. $4y^3 + 12y^2 - 72y$

88. $3x^2 + 3x - 36$

89. $2x^3 - 2x^2 + 4x$

90. $5z^2 - 15z - 140$

91. $6z^2 + 12z - 90$

92. $2a^3 + 8a^2 - 64a$

93. $3a^3 - 9a^2 - 54a$

94. $x^2 - 5xy + 6y^2$

95. $x^2 + 4xy - 21y^2$

96. $a^2 - 9ab + 20b^2$

97. $a^2 - 15ab + 50b^2$

98. $x^2 - 3xy - 28y^2$

99. $s^2 + 2st - 48t^2$

100. $y^2 - 15yz - 41z^2$

101. $x^2 + 85xy + 36y^2$

102. $z^4 - 12z^3 + 35z^2$

103. $z^4 + 2z^3 - 80z^2$

104. $b^4 - 22b^3 + 120b^2$

105. $b^4 - 3b^3 - 10b^2$

106. $2y^4 - 26y^3 - 96y^2$

107. $3y^4 + 54y^3 + 135y^2$

108. $-x^4 - 7x^3 + 8x^2$

109. $-x^4 + 11x^3 + 12x^2$

110. $4x^2y + 20xy - 56y$

111. $3x^2y - 6xy - 45y$

Copyright © Houghton Mifflin Company. All rights reserved.

112. $c^3 + 18c^2 - 40c$

113. $-3x^3 + 36x^2 - 81x$

114. $-4x^3 - 4x^2 + 24x$

115. $x^2 - 8xy + 15y^2$

116. $y^2 - 7xy - 8x^2$

117. $a^2 - 13ab + 42b^2$

118. $y^2 + 4yz - 21z^2$

119. $y^2 + 8yz + 7z^2$

120. $y^2 - 16yz + 15z^2$

121. $3x^2y + 60xy - 63y$

122. $4x^2y - 68xy - 72y$

123. $3x^3 + 3x^2 - 36x$

124. $4x^3 + 12x^2 - 160x$

125. $4z^3 + 32z^2 - 132z$

126. $5z^3 - 50z^2 - 120z$

127. $4x^3 + 8x^2 - 12x$

128. $5x^3 + 30x^2 + 40x$

129. $5p^2 + 25p - 420$

130. $4p^2 - 28p - 480$

131. $p^4 + 9p^3 - 36p^2$

132. $p^4 + p^3 - 56p^2$

133. $t^2 - 12ts + 35s^2$

134. $a^2 - 10ab + 25b^2$

135. $a^2 - 8ab - 33b^2$

APPLYING THE CONCEPTS

For Exercises 136 to 138, factor.

136. $2 + c^2 + 9c$

137. $x^2y - 54y - 3xy$

138. $45a^2 + a^2b^2 - 14a^2b$

For Exercises 139 to 141, find all integers k such that the trinomial can be factored over the integers.

139. $x^2 + kx + 35$

140. $x^2 + kx + 18$

141. $x^2 + kx + 21$

For Exercises 142 to 147, determine the positive integer values of k for which the following polynomials are factorable over the integers.

142. $y^2 + 4y + k$

143. $z^2 + 7z + k$

144. $a^2 - 6a + k$

145. $c^2 - 7c + k$

146. $x^2 - 3x + k$

147. $y^2 + 5y + k$

148. In Exercises 142 to 147, there was the stated requirement that $k > 0$. If k is allowed to be any integer, how many different values of k are possible for each polynomial?

Copyright © Houghton Mifflin Company. All rights reserved.

7.3 Factoring Polynomials of the Form $ax^2 + bx + c$

Objective A To factor a trinomial of the form $ax^2 + bx + c$ by using trial factors

Trinomials of the form $ax^2 + bx + c$, where a, b, and c are integers, are shown at the right.

$3x^2 - x + 4$; $a = 3$, $b = -1$, $c = 4$
$6x^2 + 2x - 3$; $a = 6$, $b = 2$, $c = -3$

These trinomials differ from those in the preceding section in that the coefficient of x^2 is not 1. There are various methods of factoring these trinomials. The method described in this objective is factoring polynomials using trial factors.

To reduce the number of trial factors that must be considered, remember the following:

1. Use the signs of the constant term and the coefficient of x in the trinomial to determine the signs of the binomial factors. If the constant term is positive, the signs of the binomial factors will be the same as the sign of the coefficient of x in the trinomial. If the sign of the constant term is negative, the constant terms in the binomials will have opposite signs.
2. If the terms of the trinomial do not have a common factor, then the terms of neither of the binomial factors will have a common factor.

HOW TO Factor: $2x^2 - 7x + 3$

The terms have no common factor. The constant term is positive. The coefficient of x is negative. The binomial constants will be negative.

Positive Factors of 2 **(coefficient of x^2)**	*Negative Factors of 3* **(constant term)**
1, 2	−1, −3

Write trial factors. Use the **O**uter and **I**nner products of FOIL to determine the middle term, $-7x$, of the trinomial.

Trial Factors	*Middle Term*
$(x - 1)(2x - 3)$	$-3x - 2x = -5x$
$(x - 3)(2x - 1)$	$-x - 6x = -7x$

Write the factors of the trinomial. $2x^2 - 7x + 3 = (x - 3)(2x - 1)$

HOW TO Factor: $3x^2 + 14x + 15$

The terms have no common factor. The constant term is positive. The coefficient of x is positive. The binomial constants will be positive.

Positive Factors of 3 **(coefficient of x^2)**	*Positive Factors of 15* **(constant term)**
1, 3	1, 15 3, 5

Write trial factors. Use the **O**uter and **I**nner products of FOIL to determine the middle term, $14x$, of the trinomial.

Trial Factors	*Middle Term*
$(x + 1)(3x + 15)$	Common factor
$(x + 15)(3x + 1)$	$x + 45x = 46x$
$(x + 3)(3x + 5)$	$5x + 9x = 14x$
$(x + 5)(3x + 3)$	Common factor

Write the factors of the trinomial. $3x^2 + 14x + 15 = (x + 3)(3x + 5)$

Copyright © Houghton Mifflin Company. All rights reserved.

HOW TO Factor: $6x^3 + 14x^2 - 12x$

Factor the GCF, $2x$, from the terms.

$$6x^3 + 14x^2 - 12x = 2x(3x^2 + 7x - 6)$$

Factor the trinomial. The constant term is negative. The binomial constants will have opposite signs.

Positive Factors of 3	*Factors of −6*
1, 3	1, −6
	−1, 6
	2, −3
	−2, 3

Write trial factors. Use the **O**uter and **I**nner products of FOIL to determine the middle term, $7x$, of the trinomial.

It is not necessary to test trial factors that have a common factor.

Trial Factors	*Middle Term*
$(x + 1)(3x - 6)$	Common factor
$(x - 6)(3x + 1)$	$x - 18x = -17x$
$(x - 1)(3x + 6)$	Common factor
$(x + 6)(3x - 1)$	$-x + 18x = 17x$
$(x + 2)(3x - 3)$	Common factor
$(x - 3)(3x + 2)$	$2x - 9x = -7x$
$(x - 2)(3x + 3)$	Common factor
$(x + 3)(3x - 2)$	$-2x + 9x = 7x$

Write the factors of the trinomial.

$$6x^3 + 14x^2 - 12x = 2x(x + 3)(3x - 2)$$

For this example, all the trial factors were listed. Once the correct factors have been found, however, the remaining trial factors can be omitted. For the examples and solutions in this text, all trial factors except those that have a common factor will be listed.

Example 1 Factor: $3x^2 + x - 2$

Solution

Positive factors of 3: 1, 3

Factors of −2: 1, −2
−1, 2

Trial Factors	*Middle Term*
$(x + 1)(3x - 2)$	$-2x + 3x = x$
$(x - 2)(3x + 1)$	$x - 6x = -5x$
$(x - 1)(3x + 2)$	$2x - 3x = -x$
$(x + 2)(3x - 1)$	$-x + 6x = 5x$

$3x^2 + x - 2 = (x + 1)(3x - 2)$

You Try It 1 Factor: $2x^2 - x - 3$

Your solution

Example 2 Factor: $-12x^3 - 32x^2 + 12x$

Solution

The GCF is $-4x$.

$-12x^3 - 32x^2 + 12x = -4x(3x^2 + 8x - 3)$

Factor the trinomial.

Positive factors of 3: 1, 3

Factors of −3: 1, −3
−1, 3

Trial Factors	*Middle Term*
$(x - 3)(3x + 1)$	$x - 9x = -8x$
$(x + 3)(3x - 1)$	$-x + 9x = 8x$

$-12x^3 - 32x^2 + 12x = -4x(x + 3)(3x - 1)$

You Try It 2 Factor: $-45y^3 + 12y^2 + 12y$

Your solution

Solutions on pp. S21–S22

Copyright © Houghton Mifflin Company. All rights reserved.

Objective B

To factor a trinomial of the form $ax^2 + bx + c$ by grouping

In the preceding objective, trinomials of the form $ax^2 + bx + c$ were factored by using trial factors. In this objective, these trinomials will be factored by grouping.

To factor $ax^2 + bx + c$, first find two factors of $a \cdot c$ whose sum is b. Then use factoring by grouping to write the factorization of the trinomial.

HOW TO Factor: $2x^2 + 13x + 15$

Find two positive factors of 30 $(2 \cdot 15)$ whose sum is 13.

Positive Factors of 30	*Sum*
1, 30	31
2, 15	17
3, 10	13

- **Once the required sum has been found, the remaining factors need not be checked.**

$$2x^2 + 13x + 15 = 2x^2 + 3x + 10x + 15$$

- **Use the factors of 30 whose sum is 13 to write $13x$ as $3x + 10x$.**

$$= (2x^2 + 3x) + (10x + 15)$$
$$= x(2x + 3) + 5(2x + 3)$$
$$= (2x + 3)(x + 5)$$

- **Factor by grouping.**

Check: $(2x + 3)(x + 5) = 2x^2 + 10x + 3x + 15$
$= 2x^2 + 13x + 15$

HOW TO Factor: $6x^2 - 11x - 10$

Find two factors of -60 $[6(-10)]$ whose sum is -11.

Factors of −60	*Sum*
1, −60	−59
−1, 60	59
2, −30	−28
−2, 30	28
3, −20	−17
−3, 20	17
4, −15	−11

$$6x^2 - 11x - 10 = 6x^2 + 4x - 15x - 10$$

- **Use the factors of -60 whose sum is -11 to write $-11x$ as $4x - 15x$.**

$$= (6x^2 + 4x) - (15x + 10)$$
$$= 2x(3x + 2) - 5(3x + 2)$$
$$= (3x + 2)(2x - 5)$$

- **Factor by grouping. Recall that $-15x - 10 = -(15x + 10)$.**

Check: $(3x + 2)(2x - 5) = 6x^2 - 15x + 4x - 10$
$= 6x^2 - 11x - 10$

Copyright © Houghton Mifflin Company. All rights reserved.

HOW TO Factor: $3x^2 - 2x - 4$

Find two factors of -12 $[3(-4)]$ whose sum is -2.

Factors of −12	*Sum*
1, −12	−11
−1, 12	11
2, −6	−4
−2, 6	4
3, −4	−1
−3, 4	1

Because no integer factors of -12 have a sum of -2, $3x^2 - 2x - 4$ is nonfactorable over the integers.

TAKE NOTE

$3x^2 - 2x - 4$ is a prime polynomial.

Example 3

Factor: $2x^2 + 19x - 10$

Solution

Factors of −20 [2(−10)]	*Sum*
−1, 20	19

$$\begin{aligned} 2x^2 + 19x - 10 &= 2x^2 - x + 20x - 10 \\ &= (2x^2 - x) + (20x - 10) \\ &= x(2x - 1) + 10(2x - 1) \\ &= (2x - 1)(x + 10) \end{aligned}$$

You Try It 3

Factor: $2a^2 + 13a - 7$

Your solution

Example 4

Factor: $24x^2y - 76xy + 40y$

Solution

The GCF is $4y$.

$24x^2y - 76xy + 40y = 4y(6x^2 - 19x + 10)$

Negative Factors of 60 [6(10)]	*Sum*
−1, −60	−61
−2, −30	−32
−3, −20	−23
−4, −15	−19

$$\begin{aligned} 6x^2 - 19x + 10 &= 6x^2 - 4x - 15x + 10 \\ &= (6x^2 - 4x) - (15x - 10) \\ &= 2x(3x - 2) - 5(3x - 2) \\ &= (3x - 2)(2x - 5) \end{aligned}$$

$$\begin{aligned} 24x^2y - 76xy + 40y &= 4y(6x^2 - 19x + 10) \\ &= 4y(3x - 2)(2x - 5) \end{aligned}$$

You Try It 4

Factor: $15x^3 + 40x^2 - 80x$

Your solution

Solutions on p. S22

Copyright © Houghton Mifflin Company. All rights reserved.

7.3 Exercises

Objective A **To factor a trinomial of the form $ax^2 + bx + c$ by using trial factors**

For Exercises 1 to 70, factor by using trial factors.

1. $2x^2 + 3x + 1$

2. $5x^2 + 6x + 1$

3. $2y^2 + 7y + 3$

4. $3y^2 + 7y + 2$

5. $2a^2 - 3a + 1$

6. $3a^2 - 4a + 1$

7. $2b^2 - 11b + 5$

8. $3b^2 - 13b + 4$

9. $2x^2 + x - 1$

10. $4x^2 - 3x - 1$

11. $2x^2 - 5x - 3$

12. $3x^2 + 5x - 2$

13. $2t^2 - t - 10$

14. $2t^2 + 5t - 12$

15. $3p^2 - 16p + 5$

16. $6p^2 + 5p + 1$

17. $12y^2 - 7y + 1$

18. $6y^2 - 5y + 1$

19. $6z^2 - 7z + 3$

20. $9z^2 + 3z + 2$

21. $6t^2 - 11t + 4$

22. $10t^2 + 11t + 3$

23. $8x^2 + 33x + 4$

24. $7x^2 + 50x + 7$

25. $5x^2 - 62x - 7$

26. $9x^2 - 13x - 4$

27. $12y^2 + 19y + 5$

28. $5y^2 - 22y + 8$

29. $7a^2 + 47a - 14$

30. $11a^2 - 54a - 5$

31. $3b^2 - 16b + 16$

32. $6b^2 - 19b + 15$

33. $2z^2 - 27z - 14$

34. $4z^2 + 5z - 6$

35. $3p^2 + 22p - 16$

36. $7p^2 + 19p + 10$

Copyright © Houghton Mifflin Company. All rights reserved.

37. $4x^2 + 6x + 2$ **38.** $12x^2 + 33x - 9$ **39.** $15y^2 - 50y + 35$ **40.** $30y^2 + 10y - 20$

41. $2x^3 - 11x^2 + 5x$ **42.** $2x^3 - 3x^2 - 5x$ **43.** $3a^2b - 16ab + 16b$ **44.** $2a^2b - ab - 21b$

45. $3z^2 + 95z + 10$ **46.** $8z^2 - 36z + 1$ **47.** $36x - 3x^2 - 3x^3$ **48.** $-2x^3 + 2x^2 + 4x$

49. $80y^2 - 36y + 4$ **50.** $24y^2 - 24y - 18$ **51.** $8z^3 + 14z^2 + 3z$ **52.** $6z^3 - 23z^2 + 20z$

53. $6x^2y - 11xy - 10y$ **54.** $8x^2y - 27xy + 9y$ **55.** $10t^2 - 5t - 50$

56. $16t^2 + 40t - 96$ **57.** $3p^3 - 16p^2 + 5p$ **58.** $6p^3 + 5p^2 + p$

59. $26z^2 + 98z - 24$ **60.** $30z^2 - 87z + 30$ **61.** $10y^3 - 44y^2 + 16y$

62. $14y^3 + 94y^2 - 28y$ **63.** $4yz^3 + 5yz^2 - 6yz$ **64.** $12a^3 + 14a^2 - 48a$

65. $42a^3 + 45a^2 - 27a$ **66.** $36p^2 - 9p^3 - p^4$ **67.** $9x^2y - 30xy^2 + 25y^3$

68. $8x^2y - 38xy^2 + 35y^3$ **69.** $9x^3y - 24x^2y^2 + 16xy^3$ **70.** $9x^3y + 12x^2y + 4xy$

Copyright © Houghton Mifflin Company. All rights reserved.

Objective B **To factor a trinomial of the form $ax^2 + bx + c$ by grouping**

For Exercises 71 to 130, factor by grouping.

71. $6x^2 - 17x + 12$

72. $15x^2 - 19x + 6$

73. $5b^2 + 33b - 14$

74. $8x^2 - 30x + 25$

75. $6a^2 + 7a - 24$

76. $14a^2 + 15a - 9$

77. $4z^2 + 11z + 6$

78. $6z^2 - 25z + 14$

79. $22p^2 + 51p - 10$

80. $14p^2 - 41p + 15$

81. $8y^2 + 17y + 9$

82. $12y^2 - 145y + 12$

83. $18t^2 - 9t - 5$

84. $12t^2 + 28t - 5$

85. $6b^2 + 71b - 12$

86. $8b^2 + 65b + 8$

87. $9x^2 + 12x + 4$

88. $25x^2 - 30x + 9$

89. $6b^2 - 13b + 6$

90. $20b^2 + 37b + 15$

91. $33b^2 + 34b - 35$

92. $15b^2 - 43b + 22$

93. $18y^2 - 39y + 20$

94. $24y^2 + 41y + 12$

95. $15a^2 + 26a - 21$

96. $6a^2 + 23a + 21$

97. $8y^2 - 26y + 15$

98. $18y^2 - 27y + 4$

99. $8z^2 + 2z - 15$

100. $10z^2 + 3z - 4$

101. $15x^2 - 82x + 24$

102. $13z^2 + 49z - 8$

103. $10z^2 - 29z + 10$

104. $15z^2 - 44z + 32$

105. $36z^2 + 72z + 35$

106. $16z^2 + 8z - 35$

107. $3x^2 + xy - 2y^2$

108. $6x^2 + 10xy + 4y^2$

109. $3a^2 + 5ab - 2b^2$

110. $2a^2 - 9ab + 9b^2$

Copyright © Houghton Mifflin Company. All rights reserved.

111. $4y^2 - 11yz + 6z^2$ **112.** $2y^2 + 7yz + 5z^2$ **113.** $28 + 3z - z^2$ **114.** $15 - 2z - z^2$

115. $8 - 7x - x^2$ **116.** $12 + 11x - x^2$ **117.** $9x^2 + 33x - 60$ **118.** $16x^2 - 16x - 12$

119. $24x^2 - 52x + 24$ **120.** $60x^2 + 95x + 20$ **121.** $35a^4 + 9a^3 - 2a^2$

122. $15a^4 + 26a^3 + 7a^2$ **123.** $15b^2 - 115b + 70$ **124.** $25b^2 + 35b - 30$

125. $3x^2 - 26xy + 35y^2$ **126.** $4x^2 + 16xy + 15y^2$ **127.** $216y^2 - 3y - 3$

128. $360y^2 + 4y - 4$ **129.** $21 - 20x - x^2$ **130.** $18 + 17x - x^2$

APPLYING THE CONCEPTS

131. In your own words, explain how the signs of the last terms of the two binomial factors of a trinomial are determined.

For Exercises 132 to 137, factor.

132. $(x + 1)^2 - (x + 1) - 6$ **133.** $(x - 2)^2 + 3(x - 2) + 2$ **134.** $(y + 3)^2 - 5(y + 3) + 6$

135. $2(y + 2)^2 - (y + 2) - 3$ **136.** $3(a + 2)^2 - (a + 2) - 4$ **137.** $4(y - 1)^2 - 7(y - 1) - 2$

For Exercises 138 to 143, find all integers k such that the trinomial can be factored over the integers.

138. $2x^2 + kx + 3$ **139.** $2x^2 + kx - 3$ **140.** $3x^2 + kx + 2$

141. $3x^2 + kx - 2$ **142.** $2x^2 + kx + 5$ **143.** $2x^2 + kx - 5$

Copyright © Houghton Mifflin Company. All rights reserved.

7.4 Special Factoring

Objective A

To factor the difference of two perfect squares or a perfect-square trinomial

The product of a term and itself is called a **perfect square.** The exponents on variables of perfect squares are always even numbers.

Term		*Perfect Square*
5	$5 \cdot 5 =$	25
x	$x \cdot x =$	x^2
$3y^4$	$3y^4 \cdot 3y^4 =$	$9y^8$
x^n	$x^n \cdot x^n =$	x^{2n}

The **square root of a perfect square** is one of the two equal factors of the perfect square. "$\sqrt{\ }$" is the symbol for square root. To find the exponent of the square root of a variable term, multiply the exponent by $\frac{1}{2}$.

$\sqrt{25} = 5$
$\sqrt{x^2} = x$
$\sqrt{9y^8} = 3y^4$
$\sqrt{x^{2n}} = x^n$

The **difference of two perfect squares** is the **product of the sum and difference of two terms.** The factors of the difference of two perfect squares are the sum and difference of the square roots of the perfect squares.

Factors of the Difference of Two Perfect Squares

$a^2 - b^2 = (a + b)(a - b)$

The **sum of two perfect squares,** $a^2 + b^2$, is nonfactorable over the integers.

HOW TO Factor: $4x^2 - 81y^2$

Write the binomial as the difference of two perfect squares.

$4x^2 - 81y^2 = (2x)^2 - (9y)^2$

The factors are the sum and difference of the square roots of the perfect squares.

$= (2x + 9y)(2x - 9y)$

A **perfect-square trinomial** is the square of a binomial.

Factors of a Perfect-Square Trinomial

$a^2 + 2ab + b^2 = (a + b)^2$
$a^2 - 2ab + b^2 = (a - b)^2$

In factoring a perfect-square trinomial, remember that the terms of the binomial are the square roots of the perfect squares of the trinomial. The sign in the binomial is the sign of the middle term of the trinomial.

Copyright © Houghton Mifflin Company. All rights reserved.

HOW TO Factor: $4x^2 + 12x + 9$

Because $4x^2$ is a perfect square $[4x^2 = (2x)^2]$ and 9 is a perfect square $(9 = 3^2)$, try factoring $4x^2 + 12x + 9$ as the square of a binomial.

$4x^2 + 12x + 9 \stackrel{?}{=} (2x + 3)^2$

Check: $(2x + 3)^2 = (2x + 3)(2x + 3) = 4x^2 + 6x + 6x + 9 = 4x^2 + 12x + 9$

The check verifies that $4x^2 + 12x + 9 = (2x + 3)^2$.

It is important to check a proposed factorization as we did above. The next example illustrates the importance of this check.

HOW TO Factor: $x^2 + 13x + 36$

Because x^2 is a perfect square and 36 is a perfect square, try factoring $x^2 + 13x + 36$ as the square of a binomial.

$x^2 + 13x + 36 \stackrel{?}{=} (x + 6)^2$

Check: $(x + 6)^2 = (x + 6)(x + 6) = x^2 + 6x + 6x + 36 = x^2 + 12x + 36$

In this case, the proposed factorization of $x^2 + 13x + 36$ does *not* check. Try another factorization. The numbers 4 and 9 are factors of 36 whose sum is 13.

$x^2 + 13x + 36 = (x + 4)(x + 9)$

Example 1

Factor: $25x^2 - 1$

Solution

$25x^2 - 1 = (5x)^2 - (1)^2$
$= (5x + 1)(5x - 1)$

• **Difference of two squares**

You Try It 1

Factor: $x^2 - 36y^4$

Your solution

Example 2

Factor: $4x^2 - 20x + 25$

Solution

$4x^2 - 20x + 25 = (2x - 5)^2$

• **Perfect-square trinomial**

You Try It 2

Factor: $9x^2 + 12x + 4$

Your solution

Example 3

Factor: $(x + y)^2 - 4$

Solution

$(x + y)^2 - 4$
$= (x + y)^2 - (2)^2$
$= (x + y + 2)(x + y - 2)$

• **Difference of two squares**

You Try It 3

Factor: $(a + b)^2 - (a - b)^2$

Your solution

Solutions on p. S22

Copyright © Houghton Mifflin Company. All rights reserved.

Objective B To factor the sum or difference of two perfect cubes

The product of the same three factors is called a **perfect cube.** The exponents on variables of perfect cubes are always divisible by 3.

Term		*Perfect Cube*
2	$2 \cdot 2 \cdot 2 = 2^3 =$	8
$3y$	$3y \cdot 3y \cdot 3y = (3y)^3 =$	$27y^3$
y^2	$y^2 \cdot y^2 \cdot y^2 = (y^2)^3 =$	y^6

The **cube root of a perfect cube** is one of the three equal factors of the perfect cube. "$\sqrt[3]{}$" is the symbol for cube root. To find the exponent of the cube root of a variable term, multiply the exponent by $\frac{1}{3}$.

$$\sqrt[3]{8} = 2$$
$$\sqrt[3]{27y^3} = 3y$$
$$\sqrt[3]{y^6} = y^2$$

The following rules are used to factor the sum or difference of two perfect cubes.

Factoring the Sum or Difference of Two Perfect Cubes

$a^3 + b^3 = (a + b)(a^2 - ab + b^2)$
$a^3 - b^3 = (a - b)(a^2 + ab + b^2)$

To factor $8x^3 - 27$:

Write the binomial as the difference of two perfect cubes.

$$8x^3 - 27 = (2x)^3 - 3^3$$

The terms of the binomial factor are the cube roots of the perfect cubes. The sign of the binomial factor is the same sign as in the given binomial. The trinomial factor is obtained from the binomial factor.

$$= (2x - 3)(4x^2 + 6x + 9)$$

Square of the first term ($4x^2$)
Opposite of the product of the two terms ($6x$)
Square of the last term (9)

HOW TO Factor: $a^3 + 64y^3$

$a^3 + 64y^3 = a^3 + (4y)^3$ • Write the binomial as the sum of two perfect cubes.

$= (a + 4y)(a^2 - 4ay + 16y^2)$ • Factor.

HOW TO Factor: $64y^4 - 125y$

$64y^4 - 125y = y(64y^3 - 125)$ • Factor out y, the GCF.

$= y[(4y)^3 - 5^3]$ • Write the binomial as the difference of two perfect cubes.

$= y(4y - 5)(16y^2 + 20y + 25)$ • Factor.

Copyright © Houghton Mifflin Company. All rights reserved.

Example 4

Factor: $x^3y^3 - 1$

Solution

$x^3y^3 - 1 = (xy)^3 - 1^3$

$= (xy - 1)(x^2y^2 + xy + 1)$ • Difference of two cubes

You Try It 4

Factor: $a^3b^3 - 27$

Your solution

Example 5

Factor: $64c^3 + 8d^3$

Solution

$64c^3 + 8d^3$

$= 8(8c^3 + d^3)$ • GCF

$= 8[(2c)^3 + d^3]$ • Sum of two cubes

$= 8(2c + d)(4c^2 - 2cd + d^2)$

You Try It 5

Factor: $8x^3 + y^3z^3$

Your solution

Example 6

Factor: $(x + y)^3 - x^3$

Solution

$(x + y)^3 - x^3$ • Difference of two cubes

$= [(x + y) - x][(x + y)^2 + x(x + y) + x^2]$

$= y(x^2 + 2xy + y^2 + x^2 + xy + x^2)$

$= y(3x^2 + 3xy + y^2)$

You Try It 6

Factor: $(x - y)^3 + (x + y)^3$

Your solution

Solutions on p. S22

Objective C To factor a trinomial that is quadratic in form

Certain trinomials that are not of the form $ax^2 + bx + c$ can be expressed as such by making suitable variable substitutions. A trinomial is **quadratic in form** if it can be written as $au^2 + bu + c$.

As shown below, the trinomials $x^4 + 5x^2 + 6$ and $2x^2y^2 + 3xy - 9$ are quadratic in form.

	$x^4 + 5x^2 + 6$		$2x^2y^2 + 3xy - 9$
	$(x^2)^2 + 5(x^2) + 6$		$2(xy)^2 + 3(xy) - 9$
Let $u = x^2$.	$u^2 + 5u + 6$	Let $u = xy$.	$2u^2 + 3u - 9$

When we use this method to factor a trinomial that is quadratic in form, the variable part of the first term in each binomial will be u.

Copyright © Houghton Mifflin Company. All rights reserved.

TAKE NOTE
The trinomial $x^4 + 5x^2 + 6$ was shown to be quadratic in form on the preceding page.

HOW TO Factor: $x^4 + 5x^2 + 6$

$x^4 + 5x^2 + 6 = u^2 + 5u + 6$ • Let $u = x^2$.

$= (u + 3)(u + 2)$ • Factor.

$= (x^2 + 3)(x^2 + 2)$ • Replace u by x^2.

Here is an example in which $u = \sqrt{x}$.

HOW TO Factor: $x - 2\sqrt{x} - 15$

$x - 2\sqrt{x} - 15 = u^2 - 2u - 15$ • Let $u = \sqrt{x}$. Then $u^2 = x$.

$= (u - 5)(u + 3)$ • Factor.

$= (\sqrt{x} - 5)(\sqrt{x} + 3)$ • Replace u by $\sqrt{x}$.

Example 7

Factor: $6x^2y^2 - xy - 12$

Solution
Let $u = xy$.

$$6x^2y^2 - xy - 12 = 6u^2 - u - 12$$
$$= (3u + 4)(2u - 3)$$
$$= (3xy + 4)(2xy - 3)$$

You Try It 7

Factor: $3x^4 + 4x^2 - 4$

Your solution

Solution on p. S22

Objective D To factor completely

Study Tip
You now have completed all the lessons on factoring polynomials. You will need to be able to recognize all of the factoring patterns. To test yourself, you might do the exercises in the Chapter 7 Review Exercises.

TAKE NOTE
Remember that you may have to factor more than once in order to write the polynomial as a product of *prime* factors.

General Factoring Strategy

1. Is there a common factor? If so, factor out the GCF.
2. If the polynomial is a binomial, is it the difference of two perfect squares, the sum of two perfect cubes, or the difference of two perfect cubes? If so, factor.
3. If the polynomial is a trinomial, is it a perfect-square trinomial or the product of two binomials? If so, factor.
4. Can the polynomial be factored by grouping? If so, factor.
5. Is each factor nonfactorable over the integers? If not, factor.

Copyright © Houghton Mifflin Company. All rights reserved.

Example 8

Factor: $6a^3 + 15a^2 - 36a$

Solution

$6a^3 + 15a^2 - 36a = 3a(2a^2 + 5a - 12)$ • GCF

$= 3a(2a - 3)(a + 4)$

You Try It 8

Factor: $18x^3 - 6x^2 - 60x$

Your solution

Example 9

Factor: $x^2y + 2x^2 - y - 2$

Solution

$x^2y + 2x^2 - y - 2$

$= (x^2y + 2x^2) - (y + 2)$ • Factor by grouping.

$= x^2(y + 2) - (y + 2)$

$= (y + 2)(x^2 - 1)$

$= (y + 2)(x + 1)(x - 1)$

You Try It 9

Factor: $4x - 4y - x^3 + x^2y$

Your solution

Example 10

Factor: $x^{4n} - y^{4n}$

Solution

$x^{4n} - y^{4n} = (x^{2n})^2 - (y^{2n})^2$ • Difference of two squares

$= (x^{2n} + y^{2n})(x^{2n} - y^{2n})$

$= (x^{2n} + y^{2n})[(x^n)^2 - (y^n)^2]$

$= (x^{2n} + y^{2n})(x^n + y^n)(x^n - y^n)$

You Try It 10

Factor: $x^{4n} - x^{2n}y^{2n}$

Your solution

Example 11

Factor: $x^{n+3} + x^ny^3$

Solution

$x^{n+3} + x^ny^3 = x^n(x^3 + y^3)$ • GCF

$= x^n(x + y)(x^2 - xy + y^2)$ • Sum of two cubes

You Try It 11

Factor: $ax^5 - ax^2y^6$

Your solution

Solutions on pp. S22–S23

Copyright © Houghton Mifflin Company. All rights reserved.

7.4 Exercises

Objective A **To factor the difference of two perfect squares or a perfect-square trinomial**

For Exercises 1 and 2, determine which expressions are perfect squares.

1. $4; 8; 25x^6; 12y^{10}; 100x^4y^4$
2. $9; 18; 15a^8; 49b^{12}; 64a^{16}b^2$

For Exercises 3 to 6, find the square root of the expression.

3. $16z^8$
4. $36d^{10}$
5. $81a^4b^6$
6. $25m^2n^{12}$

For Exercises 7 to 42, factor.

7. $x^2 - 16$
8. $y^2 - 49$
9. $4x^2 - 1$
10. $81x^2 - 4$
11. $16x^2 - 121$
12. $49y^2 - 36$
13. $1 - 9a^2$
14. $16 - 81y^2$
15. $x^2y^2 - 100$
16. $a^2b^2 - 25$
17. $x^2 + 4$
18. $a^2 + 16$
19. $25 - a^2b^2$
20. $64 - x^2y^2$
21. $a^{2n} - 1$
22. $b^{2n} - 16$
23. $x^2 - 12x + 36$
24. $y^2 - 6y + 9$
25. $b^2 - 2b + 1$
26. $a^2 + 14a + 49$
27. $16x^2 - 40x + 25$
28. $49x^2 + 28x + 4$
29. $4a^2 + 4a - 1$
30. $9x^2 + 12x - 4$
31. $b^2 + 7b + 14$
32. $y^2 - 5y + 25$
33. $x^2 + 6xy + 9y^2$
34. $4x^2y^2 + 12xy + 9$
35. $25a^2 - 40ab + 16b^2$
36. $4a^2 - 36ab + 81b^2$

Copyright © Houghton Mifflin Company. All rights reserved.

37. $x^{2n} + 6x^n + 9$

38. $y^{2n} - 16y^n + 64$

39. $(x - 4)^2 - 9$

40. $16 - (a - 3)^2$

41. $(x - y)^2 - (a + b)^2$

42. $(x - 2y)^2 - (x + y)^2$

Objective B To factor the sum or difference of two perfect cubes

For Exercises 43 and 44, determine which expressions are perfect cubes.

43. 4; 8; x^9; a^8b^8; $27c^{15}d^{18}$

44. 9; 27; y^{12}; m^3n^6; $64mn^9$

For Exercises 45 to 48, find the cube root of the expression.

45. $8x^9$

46. $27y^{15}$

47. $64a^6b^{18}$

48. $125c^{12}d^3$

For Exercises 49 to 72, factor.

49. $x^3 - 27$

50. $y^3 + 125$

51. $8x^3 - 1$

52. $64a^3 + 27$

53. $x^3 - y^3$

54. $x^3 - 8y^3$

55. $m^3 + n^3$

56. $27a^3 + b^3$

57. $64x^3 + 1$

58. $1 - 125b^3$

59. $27x^3 - 8y^3$

60. $64x^3 + 27y^3$

61. $x^3y^3 + 64$

62. $8x^3y^3 + 27$

63. $16x^3 - y^3$

64. $27x^3 - 8y^2$

65. $8x^3 - 9y^3$

66. $27a^3 - 16$

67. $(a - b)^3 - b^3$

68. $a^3 + (a + b)^3$

69. $x^{6n} + y^{3n}$

70. $x^{3n} + y^{3n}$

71. $x^{3n} + 8$

72. $a^{3n} + 64$

Copyright © Houghton Mifflin Company. All rights reserved.

Objective C — To factor a trinomial that is quadratic in form

For Exercises 73 to 93, factor.

73. $x^2y^2 - 8xy + 15$

74. $x^2y^2 - 8xy - 33$

75. $x^2y^2 - 17xy + 60$

76. $a^2b^2 + 10ab + 24$

77. $x^4 - 9x^2 + 18$

78. $y^4 - 6y^2 - 16$

79. $b^4 - 13b^2 - 90$

80. $a^4 + 14a^2 + 45$

81. $x^4y^4 - 8x^2y^2 + 12$

82. $a^4b^4 + 11a^2b^2 - 26$

83. $x^{2n} + 3x^n + 2$

84. $a^{2n} - a^n - 12$

85. $3x^2y^2 - 14xy + 15$

86. $5x^2y^2 - 59xy + 44$

87. $6a^2b^2 - 23ab + 21$

88. $10a^2b^2 + 3ab - 7$

89. $2x^4 - 13x^2 - 15$

90. $3x^4 + 20x^2 + 32$

91. $2x^{2n} - 7x^n + 3$

92. $4x^{2n} + 8x^n - 5$

93. $6a^{2n} + 19a^n + 10$

Objective D — To factor completely

For Exercises 94 to 129, factor.

94. $5x^2 + 10x + 5$

95. $12x^2 - 36x + 27$

96. $3x^4 - 81x$

97. $27a^4 - a$

98. $7x^2 - 28$

99. $20x^2 - 5$

100. $y^4 - 10y^3 + 21y^2$

101. $y^5 + 6y^4 - 55y^3$

102. $x^4 - 16$

103. $16x^4 - 81$

104. $8x^5 - 98x^3$

105. $16a - 2a^4$

Copyright © Houghton Mifflin Company. All rights reserved.

106. $x^3y^3 - x^3$

107. $a^3b^6 - b^3$

108. $x^6y^6 - x^3y^3$

109. $8x^4 - 40x^3 + 50x^2$

110. $6x^5 + 74x^4 + 24x^3$

111. $x^4 - y^4$

112. $16a^4 - b^4$

113. $x^6 + y^6$

114. $x^4 - 5x^2 - 4$

115. $a^4 - 25a^2 - 144$

116. $3b^5 - 24b^2$

117. $16a^4 - 2a$

118. $x^4y^2 - 5x^3y^3 + 6x^2y^4$

119. $a^4b^2 - 8a^3b^3 - 48a^2b^4$

120. $16x^3y + 4x^2y^2 - 42xy^3$

121. $24a^2b^2 - 14ab^3 - 90b^4$

122. $x^3 - 2x^2 - x + 2$

123. $x^3 - 2x^2 - 4x + 8$

124. $4x^2y^2 - 4x^2 - 9y^2 + 9$

125. $4x^4 - x^2 - 4x^2y^2 + y^2$

126. $a^{2n+2} - 6a^{n+2} + 9a^2$

127. $x^{2n+1} + 2x^{n+1} + x$

128. $2x^{n+2} - 7x^{n+1} + 3x^n$

129. $3b^{n+2} + 4b^{n+1} - 4b^n$

APPLYING THE CONCEPTS

130. Factor: $x^2(x - 3) - 3x(x - 3) + 2(x - 3)$

131. Given that $(x - 3)$ and $(x + 4)$ are factors of $x^3 + 6x^2 - 7x - 60$, explain how you can find a third *first-degree* factor of $x^3 + 6x^2 - 7x - 60$. Then find the factor.

Copyright © Houghton Mifflin Company. All rights reserved.

7.5 Solving Equations

Objective A To solve equations by factoring

The Multiplication Property of Zero states that the product of a number and zero is zero. This property is stated below.

If a is a real number, then $a \cdot 0 = 0 \cdot a = 0$.

Now consider $x \cdot y = 0$. For this to be a true equation, then either $x = 0$ or $y = 0$.

Principle of Zero Products

If the product of two factors is zero, then at least one of the factors must be zero.

If $a \cdot b = 0$, then $a = 0$ or $b = 0$.

The Principle of Zero Products is used to solve some equations.

HOW TO Solve: $(x - 2)(x - 3) = 0$

$$(x - 2)(x - 3) = 0$$

$x - 2 = 0 \quad x - 3 = 0$ • Let each factor equal zero (the Principle of Zero Products).

$x = 2 \quad x = 3$ • Solve each equation for x.

Check:

$(x - 2)(x - 3) = 0$

$(2 - 2)(2 - 3) \mid 0$

$0(-1) \mid 0$

$0 = 0$ • A true equation

$(x - 2)(x - 3) = 0$

$(3 - 2)(3 - 3) \mid 0$

$(1)(0) \mid 0$

$0 = 0$ • A true equation

The solutions are 2 and 3.

An equation that can be written in the form $ax^2 + bx + c = 0$, $a \neq 0$, is a **quadratic equation.** A quadratic equation is in **standard form** when the polynomial is in descending order and equal to zero. The quadratic equations at the right are in standard form.

$3x^2 + 2x + 1 = 0$
$a = 3, b = 2, c = 1$

$4x^2 - 3x + 2 = 0$
$a = 4, b = -3, c = 2$

Copyright © Houghton Mifflin Company. All rights reserved.

HOW TO Solve: $2x^2 + x = 6$

$$2x^2 + x = 6$$
$$2x^2 + x - 6 = 0$$ • Write the equation in standard form.
$$(2x - 3)(x + 2) = 0$$ • Factor.
$$2x - 3 = 0 \qquad x + 2 = 0$$ • Use the Principle of Zero Products.
$$2x = 3 \qquad x = -2$$ • Solve each equation for *x*.
$$x = \frac{3}{2}$$

Check: $\frac{3}{2}$ and -2 check as solutions.

The solutions are $\frac{3}{2}$ and -2.

Example 1

Solve: $x(x - 3) = 0$

Solution

$$x(x - 3) = 0$$
$$x = 0 \qquad x - 3 = 0$$
$$x = 3$$
• Use the Principle of Zero Products.

The solutions are 0 and 3.

You Try It 1

Solve: $2x(x + 7) = 0$

Your solution

$2x = 0 \quad x + 7 = 0$

$x = 0 \quad x = -7$

Example 2

Solve: $2x^2 - 50 = 0$

Solution

$$2x^2 - 50 = 0$$
$$2(x^2 - 25) = 0$$ • Factor the GCF, 2.
$$2(x + 5)(x - 5) = 0$$ • Factor the difference of two squares.
$$x + 5 = 0 \qquad x - 5 = 0$$
$$x = -5 \qquad x = 5$$
• Use the Principle of Zero Products.

The solutions are -5 and 5.

You Try It 2

Solve: $4x^2 - 9 = 0$

$(2x - 3) \quad 2x + 3 = 0$

Your solution

$2x = 3 \quad 2x = -3$

$x \frac{3}{2} \quad x = -\frac{3}{2}$

Example 3

Solve: $(x - 3)(x - 10) = -10$

Solution

$$(x - 3)(x - 10) = -10$$
$$x^2 - 13x + 30 = -10$$ • Multiply $(x - 3)(x - 10)$.
$$x^2 - 13x + 40 = 0$$ • Add 10 to each side of the equation. The equation is now in standard form.
$$(x - 8)(x - 5) = 0$$
$$x - 8 = 0 \qquad x - 5 = 0$$
$$x = 8 \qquad x = 5$$

The solutions are 8 and 5.

You Try It 3

Solve: $(x + 2)(x - 7) = 52$

Your solution

Solutions on p. S23

Copyright © Houghton Mifflin Company. All rights reserved.

Objective B To solve application problems

Example 4

The sum of the squares of two consecutive positive even integers is equal to 100. Find the two integers.

Strategy

First positive even integer: n

Second positive even integer: $n + 2$

The sum of the square of the first positive even integer and the square of the second positive even integer is 100.

Solution

$$n^2 + (n + 2)^2 = 100$$
$$n^2 + n^2 + 4n + 4 = 100$$
$$2n^2 + 4n + 4 = 100$$
$$2n^2 + 4n - 96 = 0$$
$$2(n^2 + 2n - 48) = 0$$
$$2(n - 6)(n + 8) = 0$$

$n - 6 = 0$ $\quad$ $n + 8 = 0$ $\quad$ • **Principle of Zero Products**

$n = 6$ $\quad$ $n = -8$

Because -8 is not a positive even integer, it is not a solution.

$n = 6$

$n + 2 = 6 + 2 = 8$

The two integers are 6 and 8.

You Try It 4

The sum of the squares of two consecutive positive integers is 61. Find the two integers.

Your strategy

Your solution

Solution on p. S23

Copyright © Houghton Mifflin Company. All rights reserved.

Example 5

A stone is thrown into a well with an initial speed of 4 ft/s. The well is 420 ft deep. How many seconds later will the stone hit the bottom of the well? Use the equation $d = vt + 16t^2$, where d is the distance in feet that the stone travels in t seconds when its initial speed is v feet per second.

Strategy

To find the time for the stone to drop to the bottom of the well, replace the variables d and v by their given values and solve for t.

Solution

$$d = vt + 16t^2$$
$$420 = 4t + 16t^2$$
$$0 = -420 + 4t + 16t^2$$
$$0 = 16t^2 + 4t - 420$$
$$0 = 4(4t^2 + t - 105)$$
$$0 = 4(4t + 21)(t - 5)$$

$4t + 21 = 0$ $\qquad$ $t - 5 = 0$ $\qquad$ • Principle of Zero Products

$4t = -21$ $\qquad$ $t = 5$

$t = -\frac{21}{4}$

Because the time cannot be a negative number, $-\frac{21}{4}$ is not a solution.

The stone will hit the bottom of the well 5 s later.

You Try It 5

The length of a rectangle is 4 in. longer than twice the width. The area of the rectangle is 96 in^2. Find the length and width of the rectangle.

Your strategy

Your solution

Solution on p. S23

Copyright © Houghton Mifflin Company. All rights reserved.

7.5 Exercises

Objective A To solve equations by factoring

1. In your own words, explain the Principle of Zero Products.

2. Fill in the blanks. If $(x + 5)(2x - 7) = 0$, then ______ $= 0$ or ______ $= 0$.

For Exercises 3 to 60, solve.

3. $(y + 3)(y + 2) = 0$
4. $(y - 3)(y - 5) = 0$
5. $(z - 7)(z - 3) = 0$
6. $(z + 8)(z - 9) = 0$
7. $x(x - 5) = 0$
8. $x(x + 2) = 0$
9. $a(a - 9) = 0$
10. $a(a + 12) = 0$
11. $y(2y + 3) = 0$
12. $t(4t - 7) = 0$
13. $2a(3a - 2) = 0$
14. $4b(2b + 5) = 0$
15. $(b + 2)(b - 5) = 0$
16. $(b - 8)(b + 3) = 0$
17. $x^2 - 81 = 0$
18. $x^2 - 121 = 0$
19. $4x^2 - 49 = 0$
20. $16x^2 - 1 = 0$
21. $9x^2 - 1 = 0$
22. $16x^2 - 49 = 0$
23. $x^2 + 6x + 8 = 0$
24. $x^2 - 8x + 15 = 0$
25. $z^2 + 5z - 14 = 0$
26. $z^2 + z - 72 = 0$
27. $2a^2 - 9a - 5 = 0$
28. $3a^2 + 14a + 8 = 0$
29. $6z^2 + 5z + 1 = 0$
30. $6y^2 - 19y + 15 = 0$
31. $x^2 - 3x = 0$
32. $a^2 - 5a = 0$
33. $x^2 - 7x = 0$
34. $2a^2 - 8a = 0$
35. $a^2 + 5a = -4$
36. $a^2 - 5a = 24$
37. $y^2 - 5y = -6$
38. $y^2 - 7y = 8$
39. $2t^2 + 7t = 4$
40. $3t^2 + t = 10$
41. $3t^2 - 13t = -4$
42. $5t^2 - 16t = -12$
43. $x(x - 12) = -27$
44. $x(x - 11) = 12$
45. $y(y - 7) = 18$
46. $y(y + 8) = -15$

Copyright © Houghton Mifflin Company. All rights reserved.

47. $p(p + 3) = -2$

48. $p(p - 1) = 20$

49. $y(y + 4) = 45$

50. $y(y - 8) = -15$

51. $x(x + 3) = 28$

52. $p(p - 14) = 15$

53. $(x + 8)(x - 3) = -30$

54. $(x + 4)(x - 1) = 14$

55. $(z - 5)(z + 4) = 52$

56. $(z - 8)(z + 4) = -35$

57. $(z - 6)(z + 1) = -10$

58. $(a + 3)(a + 4) = 72$

59. $(a - 4)(a + 7) = -18$

60. $(2x + 5)(x + 1) = -1$

Objective B To solve application problems

61. **Number Sense** The square of a positive number is six more than five times the positive number. Find the number.

62. **Number Sense** The square of a negative number is fifteen more than twice the negative number. Find the number.

63. **Number Sense** The sum of two numbers is six. The sum of the squares of the two numbers is twenty. Find the two numbers.

64. **Number Sense** The sum of two numbers is eight. The sum of the squares of the two numbers is thirty-four. Find the two numbers.

65. **Number Sense** The sum of the squares of two consecutive positive integers is forty-one. Find the two integers.

66. **Number Sense** The sum of the squares of two consecutive positive even integers is one hundred. Find the two integers.

67. **Number Sense** The sum of two numbers is ten. The product of the two numbers is twenty-one. Find the two numbers.

68. **Number Sense** The sum of two numbers is thirteen. The product of the two numbers is forty. Find the two numbers.

Copyright © Houghton Mifflin Company. All rights reserved.

Sum of Natural Numbers The formula $S = \frac{n^2 + n}{2}$ gives the sum, S, of the first n natural numbers. Use this formula for Exercises 69 and 70.

69. How many consecutive natural numbers beginning with 1 will give a sum of 78?

70. How many consecutive natural numbers beginning with 1 will give a sum of 171?

Sports The formula $N = \frac{t^2 - t}{2}$ gives the number, N, of football games that must be scheduled in a league with t teams if each team is to play every other team once. Use this formula for Exercises 71 and 72.

71. How many teams are in a league that schedules 15 games in such a way that each team plays every other team once?

72. How many teams are in a league that schedules 45 games in such a way that each team plays every other team once?

Physics The distance s, in feet, that an object will fall (neglecting air resistance) in t seconds is given by $s = vt + 16t^2$, where v is the initial velocity of the object in feet per second. Use this formula for Exercises 73 and 74.

73. An object is released from the top of a building 192 ft high. The object's initial velocity is 16 ft/s, and air resistance is neglected. How many seconds later will the object hit the ground?

74. In October 2003, the world's tallest building, Taipei 101, was completed. The top of the spire is 1667 ft above ground. If an object is released from this building at a point 640 ft above the ground at an initial velocity of 48 ft/s, assuming no air resistance, how many seconds later will the object reach the ground?

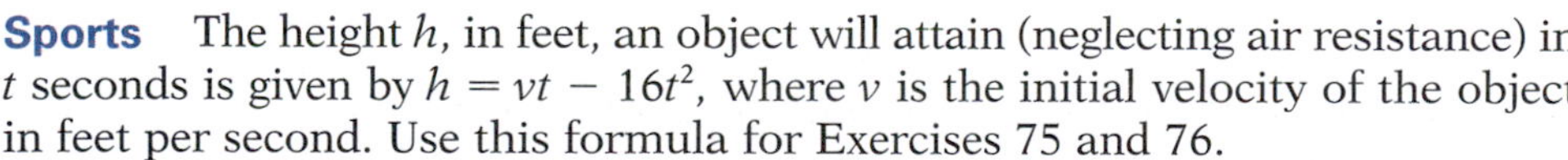

Sports The height h, in feet, an object will attain (neglecting air resistance) in t seconds is given by $h = vt - 16t^2$, where v is the initial velocity of the object in feet per second. Use this formula for Exercises 75 and 76.

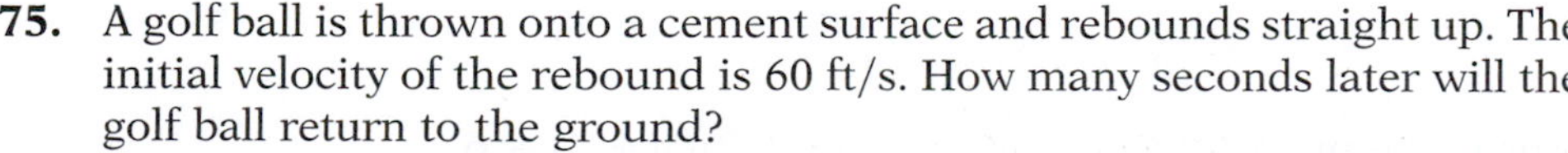

75. A golf ball is thrown onto a cement surface and rebounds straight up. The initial velocity of the rebound is 60 ft/s. How many seconds later will the golf ball return to the ground?

76. A foul ball leaves a bat, hits home plate, and travels straight up with an initial velocity of 64 ft/s. How many seconds later will the ball be 64 ft above the ground?

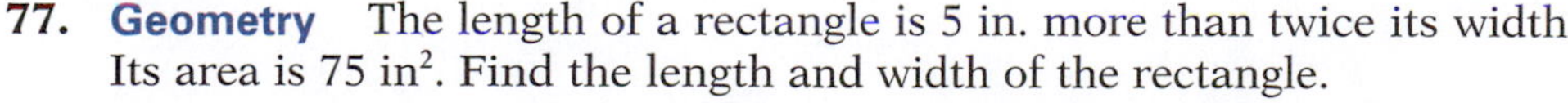

77. **Geometry** The length of a rectangle is 5 in. more than twice its width. Its area is 75 in^2. Find the length and width of the rectangle.

Copyright © Houghton Mifflin Company. All rights reserved.

78. **Geometry** The width of a rectangle is 5 ft less than the length. The area of the rectangle is 176 ft^2. Find the length and width of the rectangle.

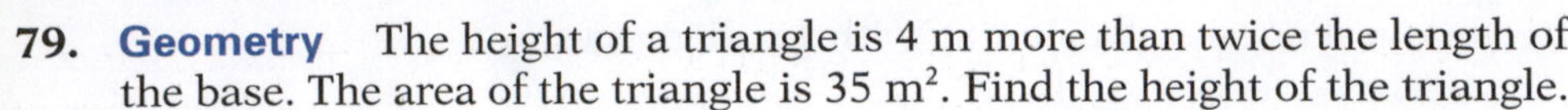

79. **Geometry** The height of a triangle is 4 m more than twice the length of the base. The area of the triangle is 35 m^2. Find the height of the triangle.

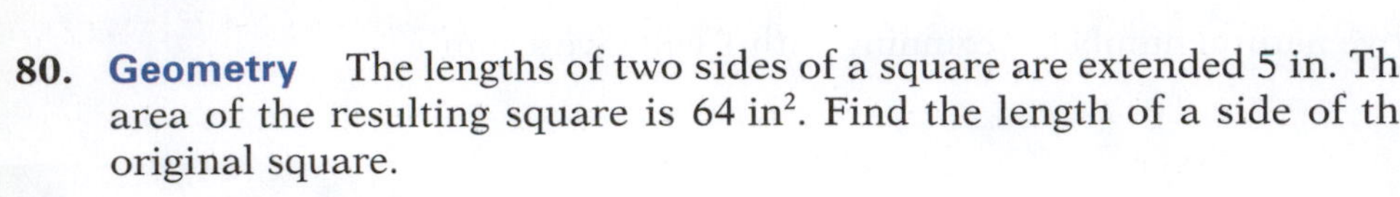

80. **Geometry** The lengths of two sides of a square are extended 5 in. The area of the resulting square is 64 in^2. Find the length of a side of the original square.

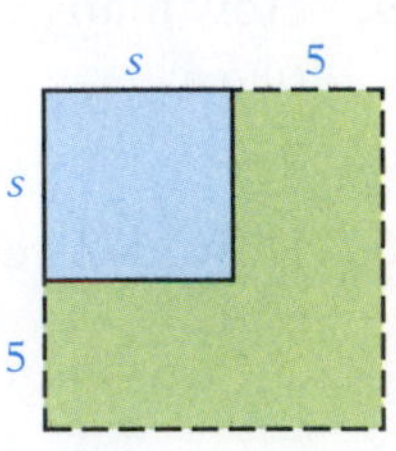

81. **Publishing** The page of a book measures 6 in. by 9 in. A uniform border around the page leaves 28 in^2 for type. What are the dimensions of the type area?

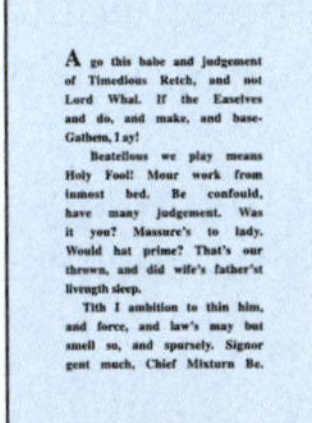

82. **Gardening** A small garden measures 8 ft by 10 ft. A uniform border around the garden increases the total area to 143 ft^2. What is the width of the border?

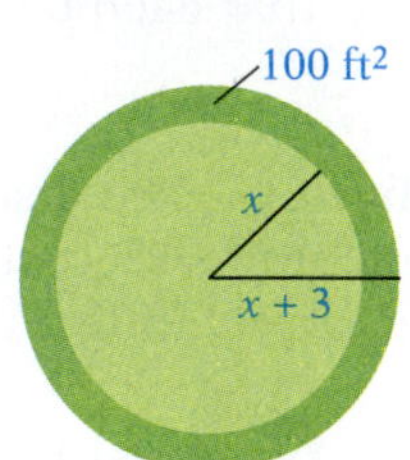

83. **Landscaping** A landscape designer decides to increase the radius of a circular lawn by 3 ft. This increases the area of the lawn by 100 ft^2. Find the radius of the original circular lawn. Round to the nearest hundredth.

84. **Geometry** A circle has a radius of 10 in. Find the increase in area that occurs when the radius is increased by 2 in. Round to the nearest hundredth.

APPLYING THE CONCEPTS

85. Find $3n^2$ if $n(n + 5) = -4$.

86. Find $2n^2$ if $n(n + 3) = 4$.

For Exercises 87 to 90, solve.

87. $2y(y + 4) = -5(y + 3)$

88. $(b + 5)^2 = 16$

89. $p^3 = 9p^2$

90. $(x + 3)(2x - 1) = (3 - x)(5 - 3x)$

91. Explain the error made in solving the equation at the right. Solve the equation correctly.

$$(x + 2)(x - 3) = 6$$
$$x + 2 = 6 \qquad x - 3 = 6$$
$$x = 4 \qquad x = 9$$

92. Explain the error made in solving the equation at the right. Solve the equation correctly.

$$x^2 = x$$
$$\frac{x^2}{x} = \frac{x}{x}$$
$$x = 1$$

Copyright © Houghton Mifflin Company. All rights reserved.

Focus on Problem Solving

Making a Table

There are six students using a gym. The wall on the gym has six lockers that are numbered 1, 2, 3, 4, 5, and 6. After a practice, the first student goes by and opens all the lockers. The second student shuts every second locker, the third student changes every third locker (opens a locker if it is shut, shuts a locker if it is open), the fourth student changes every fourth locker, the fifth student changes every fifth locker, and the sixth student changes every sixth locker. After the sixth student makes changes, which lockers are open?

One method of solving this problem would be to create a table as shown below.

Student / *Locker*	1	2	3	4	5	6
1	O	O	O	O	O	O
2	O	C	C	C	C	C
3	O	O	C	C	C	C
4	O	C	C	O	O	O
5	O	O	O	O	C	C
6	O	C	O	O	O	C

From this table, lockers 1 and 4 are open after the sixth student passes through.

Now extend this to more lockers and students. In each case, the *n*th student changes multiples of the *n*th locker. For instance, the eighth student would change the 8th, 16th, 24th, . . .

1. Suppose there were 10 lockers and 10 students. Which lockers would remain open?
2. Suppose there were 16 lockers and 16 students. Which lockers would remain open?
3. Suppose there were 25 lockers and 25 students. Which lockers would remain open?
4. Suppose there were 40 lockers and 40 students. Which lockers would remain open?
5. Suppose there were 50 lockers and 50 students. Which lockers would remain open?
6. Make a conjecture as to which lockers would be open if there were 100 lockers and 100 students.
7. Give a reason why your conjecture should be true. [*Hint:* Consider how many factors there are for the door numbers that remain open and for those that remain closed. For instance, with 40 lockers and 40 students, locker 36 (which remains open) has factors 1, 2, 3, 4, 6, 9, 12, 18, 36—an odd number of factors. Locker 28, a closed locker, has factors 1, 2, 4, 7, 14, 28—an even number of factors.]

Copyright © Houghton Mifflin Company. All rights reserved.

Projects and Group Activities

Exploring Integers

Number theory is a branch of mathematics that focuses on integers and the relationships that exist among the integers. Some of the results from this field of study have important, practical applications for sending sensitive information such as credit card numbers over the Internet. In this project, you will be asked to discover some of those relationships.

1. If n is an integer, explain why the product $n(n + 1)$ is always an even number.
2. If n is an integer, explain why $2n$ is always an even integer.
3. If n is an integer, explain why $2n + 1$ is always an odd integer.
4. Select any odd integer greater than 1, square it, and then subtract 1. Try this for various odd integers greater than 1. Is the result always evenly divisible by 8?
5. Prove the assertion in Exercise 4. [*Suggestion:* From Exercise 3, an odd integer can be represented as $2n + 1$. Therefore, the assertion in Exercise 4 can be stated "$(2n + 1)^2 - 1$ is evenly divisible by 8." Expand this expression and explain why the result must be divisible by 8. You will need to use the result from Exercise 1.]
6. The integers 2 and 3 are consecutive prime numbers. Are there any other consecutive prime numbers? Why?
7. If n is a positive integer, for what values of n is $n^2 - 1$ a prime number?
8. A *Mersenne* prime number is a prime that can be written in the form $2^n - 1$, where n is also a prime number. For instance, $2^5 - 1 = 32 - 1 = 31$. Because 5 and 31 are prime numbers, 31 is a Mersenne prime number. On the other hand, $2^{11} - 1 = 2048 - 1 = 2047$. In this case, although 11 is a prime number, $2047 = 23 \cdot 89$, and so is not a prime number. Find two Mersenne prime numbers other than 31.

Chapter 7 Summary

Key Words	Examples
The *greatest common factor (GCF) of two or more monomials* is the product of the GCF of the coefficients and the common variable factors. [7.1A, p. 401]	The GCF of $8x^2y$ and $12xyz$ is $4xy$.
To *factor a polynomial* means to write the polynomial as a product of other polynomials. [7.1A, p. 401]	To factor $x^2 + 3x + 2$ means to write it as the product $(x + 1)(x + 2)$.

Copyright © Houghton Mifflin Company. All rights reserved.

A factor that has two terms is called a *binomial factor.* [7.1B, p. 403]	$(x + 1)$ is a binomial factor of $3x(x + 1)$.
A polynomial that does not factor using only integers is *nonfactorable over the integers.* [7.2A, p. 408]	The trinomial $x^2 + x + 4$ is nonfactorable over the integers. There are no integers whose product is 4 and whose sum is 1.
A polynomial is *factored completely* if it is written as a product of factors that are nonfactorable over the integers. [7.2B, p. 409]	The polynomial $3y^3 + 9y^2 - 12y$ is factored completely as $3y(y + 4)(y - 1)$.
The product of a term and itself is called a *perfect square*. The *square root* of a perfect square is one of the two equal factors of the perfect square. [7.4A, p. 423]	$(5x)(5x) = 25x^2$; $25x^2$ is a perfect square. $\sqrt{25x^2} = 5x$
The product of the same three factors is called a *perfect cube*. The *cube root* of a perfect cube is one of the three equal factors of the perfect cube. [7.4B, p. 425]	$(2x)(2x)(2x) = 8x^3$; $8x^3$ is a perfect cube. $\sqrt[3]{8x^3} = 2x$
A trinomial is *quadratic in form* if it can be written as $au^2 + bu + c$. [7.4C, p. 426]	$6x^4 - 5x^2 - 4$ $6(x^2)^2 - 5(x^2) - 4$ $6u^2 - 5u - 4$
An equation that can be written in the form $ax^2 + bx + c = 0$, $a \neq 0$, is a *quadratic equation*. A quadratic equation is in *standard form* when the polynomial is written in descending order and equal to zero. [7.5A, p. 433]	The equation $2x^2 - 3x + 7 = 0$ is a quadratic equation in standard form.

Essential Rules and Procedures	Examples
Factoring $x^2 + bx + c$: IMPORTANT RELATIONSHIPS [7.2A, p. 407]	
1. When the constant term of the trinomial is positive, the constant terms of the binomials have the same sign. They are both positive when the coefficient of the x term in the trinomial is positive. They are both negative when the coefficient of the x term in the trinomial is negative.	$x^2 + 6x + 8 = (x + 4)(x + 2)$ $x^2 - 6x + 5 = (x - 5)(x - 1)$
2. When the constant term of the trinomial is negative, the constant terms of the binomials have opposite signs.	$x^2 - 4x - 21 = (x + 3)(x - 7)$
3. In the trinomial, the coefficient of x is the sum of the constant terms of the binomials.	In the three examples above, note that $6 = 4 + 2$, $-6 = -5 + (-1)$, and $-4 = 3 + (-7)$.
4. In the trinomial, the constant term is the product of the constant terms of the binomials.	In the three examples above, note that $8 = 4 \cdot 2$, $5 = -5(-1)$, and $-21 = 3(-7)$.

Copyright © Houghton Mifflin Company. All rights reserved.

Factoring by Grouping [7.1B, p. 403]

A polynomial can be factored by grouping if its terms can be grouped and factored in such a way that a common binomial factor is found.

$$\begin{aligned} &3a^2 - a - 15ab + 5b \\ &\quad = (3a^2 - a) - (15ab - 5b) \\ &\quad = a(3a - 1) - 5b(3a - 1) \\ &\quad = (3a - 1)(a - 5b) \end{aligned}$$

To factor $ax^2 + bx + c$ by grouping [7.3B, p. 417]

First find two factors of $a \cdot c$ whose sum is b. Then use factoring by grouping to write the factorization of the trinomial.

$3x^2 - 11x - 20$

$a \cdot c = 3(-20) = -60$

The product of 4 and -15 is -60.

The sum of 4 and -15 is -11.

$$\begin{aligned} &3x^2 + 4x - 15x - 20 \\ &\quad = (3x^2 + 4x) - (15x + 20) \\ &\quad = x(3x + 4) - 5(3x + 4) \\ &\quad = (3x + 4)(x - 5) \end{aligned}$$

Factoring the Difference of Two Perfect Squares [7.4A, p. 423]

The difference of two perfect squares factors as the sum and difference of the same terms.

$a^2 - b^2 = (a + b)(a - b)$

$x^2 - 64 = (x + 8)(x - 8)$

$$\begin{aligned} 4x^2 - 81 &= (2x)^2 - 9^2 \\ &= (2x + 9)(2x - 9) \end{aligned}$$

Factoring a Perfect-Square Trinomial [7.4A, p. 423]

A perfect-square trinomial is the square of a binomial.

$a^2 + 2ab + b^2 = (a + b)^2$

$a^2 - 2ab + b^2 = (a - b)^2$

$x^2 + 14x + 49 = (x + 7)^2$

$x^2 - 10x + 25 = (x - 5)^2$

Factoring the Sum or Difference of Two Cubes [7.4B, p. 425]

$a^3 + b^3 = (a + b)(a^2 - ab + b^2)$

$a^3 - b^3 = (a - b)(a^2 + ab + b^2)$

$x^3 + 64 = (x + 4)(x^2 - 4x + 16)$

$8b^3 - 1 = (2b - 1)(4b^2 + 2b + 1)$

To Factor Completely [7.4D, p. 427]

When factoring a polynomial completely, ask the following questions about the polynomial.

1. Is there a common factor? If so, factor out the GCF.
2. If the polynomial is a binomial, is it the difference of two perfect squares, the sum of two perfect cubes, or the difference of two perfect cubes? If so, factor.
3. If the polynomial is a trinomial, is it a perfect-square trinomial or the product of two binomials? If so, factor.
4. Can the polynomial be factored by grouping? If so, factor.
5. Is each factor nonfactorable over the integers? If not, factor.

$$\begin{aligned} 54x^3 - 6x &= 6x(9x^2 - 1) \\ &= 6x(3x + 1)(3x - 1) \end{aligned}$$

Principle of Zero Products [7.5A, p. 433]

If the product of two factors is zero, then at least one of the factors must be zero.

If $ab = 0$, then $a = 0$ or $b = 0$.

$(x - 4)(x + 2) = 0$

$x - 4 = 0 \qquad x + 2 = 0$

Copyright © Houghton Mifflin Company. All rights reserved.

Chapter 7 Review Exercises

1. Factor: $b^2 - 13b + 30$

2. Factor: $4x(x - 3) - 5(3 - x)$

3. Factor $2x^2 - 5x + 6$ by using trial factors.

4. Factor: $21x^4y^4 + 23x^2y^2 + 6$

5. Factor: $14y^9 - 49y^6 + 7y^3$

6. Factor: $y^2 + 5y - 36$

7. Factor $6x^2 - 29x + 28$ by using trial factors.

8. Factor: $12a^2b + 3ab^2$

9. Factor: $a^6 - 100$

10. Factor: $n^4 - 2n^3 - 3n^2$

11. Factor $12y^2 + 16y - 3$ by using trial factors.

12. Factor: $12b^3 - 58b^2 + 56b$

13. Factor: $9y^4 - 25z^2$

14. Factor: $c^2 + 8c + 12$

15. Factor $18a^2 - 3a - 10$ by grouping.

16. Solve: $4x^2 + 27x = 7$

17. Factor: $4x^3 - 20x^2 - 24x$

18. Factor: $64a^3 - 27b^3$

Copyright © Houghton Mifflin Company. All rights reserved.

19. Factor $2a^2 - 19a - 60$ by grouping.

20. Solve: $(x + 1)(x - 5) = 16$

21. Factor: $21ax - 35bx - 10by + 6ay$

22. Factor: $36x^8 - 36x^4 + 5$

23. Factor: $10x^2 + 25x + 4xy + 10y$

24. Factor: $5x^2 - 5x - 30$

25. Factor: $3x^2 + 36x + 108$

26. Factor $3x^2 - 17x + 10$ by grouping.

27. Sports The length of the field in field hockey is 20 yd less than twice the width of the field. The area of the field in field hockey is 6000 yd². Find the length and width of the field.

28. Image Projection The size, S, of an image from a slide projector depends on the distance, d, of the screen from the projector and is given by $S = d^2$. Find the distance between the projector and the screen when the size of the picture is 400 ft².

29. Photography A rectangular photograph has dimensions 15 in. by 12 in. A picture frame around the photograph increases the total area to 270 in². What is the width of the frame?

30. Gardening The length of each side of a square garden plot is extended 4 ft. The area of the resulting square is 576 ft². Find the length of a side of the original garden plot.

Copyright © Houghton Mifflin Company. All rights reserved.

Chapter 7 Test

1. Factor: $ab + 6a - 3b - 18$

2. Factor: $2y^4 - 14y^3 - 16y^2$

3. Factor $8x^2 + 20x - 48$ by grouping.

4. Factor $6x^2 + 19x + 8$ by using trial factors.

5. Factor: $a^2 - 19a + 48$

6. Factor: $6x^3 - 8x^2 + 10x$

7. Factor: $x^2 + 2x - 15$

8. Solve: $4x^2 - 1 = 0$

9. Factor: $5x^2 - 45x - 15$

10. Factor: $p^2 + 12p + 36$

11. Solve: $x(x - 8) = -15$

12. Factor: $3x^2 + 12xy + 12y^2$

13. Factor: $b^2 - 16$

14. Factor $6x^2y^2 + 9xy^2 + 3y^2$ by grouping.

Copyright © Houghton Mifflin Company. All rights reserved.

15. Factor: $27x^3 - 8$

16. Factor: $6a^4 - 13a^2 - 5$

17. Factor: $x(p + 1) - (p + 1)$

18. Factor: $3a^2 - 75$

19. Factor $2x^2 + 4x - 5$ by using trial factors.

20. Factor: $x^2 - 9x - 36$

21. Factor: $4a^2 - 12ab + 9b^2$

22. Factor: $4x^2 - 49y^2$

23. Solve: $(2a - 3)(a + 7) = 0$

24. **Number Sense** The sum of two numbers is ten. The sum of the squares of the two numbers is fifty-eight. Find the two numbers.

25. **Geometry** The length of a rectangle is 3 cm longer than twice its width. The area of the rectangle is 90 cm^2. Find the length and width of the rectangle.

Copyright © Houghton Mifflin Company. All rights reserved.

Cumulative Review Exercises

1. Subtract: $-2 - (-3) - 5 - (-11)$

2. Simplify: $(3 - 7)^2 \div (-2) - 3 \cdot (-4)$

3. Evaluate $-2a^2 \div (2b) - c$ when $a = -4$, $b = 2$, and $c = -1$.

4. Simplify: $-\frac{3}{4}(-20x^2)$

5. Simplify: $-2[4x - 2(3 - 2x) - 8x]$

6. Solve: $-\frac{5}{7}x = -\frac{10}{21}$

7. Solve: $3x - 2 = 12 - 5x$

8. Solve: $-2 + 4[3x - 2(4 - x) - 3] = 4x + 2$

9. 120% of what number is 54?

10. Given $f(x) = -x^2 + 3x - 1$, find $f(2)$.

11. Graph $y = \frac{1}{4}x + 3$.

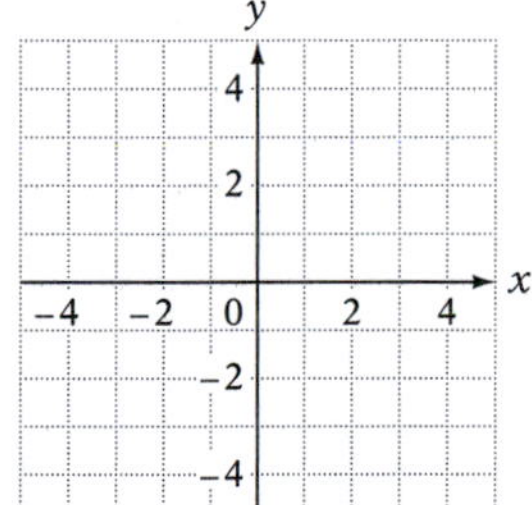

12. Graph $5x + 3y = 15$.

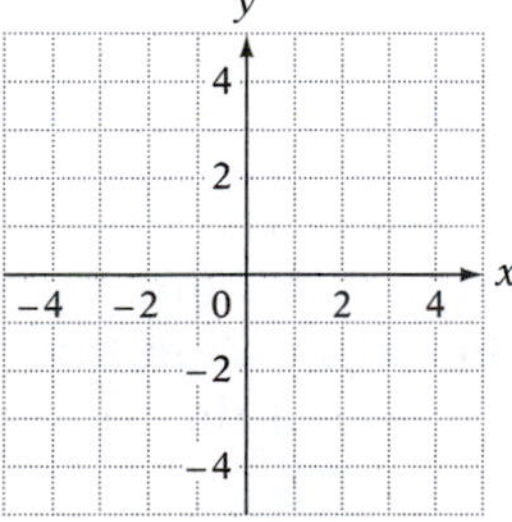

13. Find the equation of the line that contains the point $(-3, 4)$ and has slope $\frac{2}{3}$.

14. Solve by substitution: $8x - y = 2$
$y = 5x + 1$

Copyright © Houghton Mifflin Company. All rights reserved.

15. Solve by the addition method:
$5x + 2y = -9$
$12x - 7y = 2$

16. Simplify: $(-3a^3b^2)^2$

17. Multiply: $(x + 2)(x^2 - 5x + 4)$

18. Divide: $(8x^2 + 4x - 3) \div (2x - 3)$

19. Simplify: $(x^{-4}y^3)^2$

20. Factor: $3a - 3b - ax + bx$

21. Factor: $15xy^2 - 20xy^4$

22. Factor: $x^2 - 5xy - 14y^2$

23. Solve: $3x^2 + 19x - 14 = 0$

24. Solve: $6x^2 + 60 = 39x$

25. **Geometry** A triangle has a 31° angle and a right angle. Find the measure of the third angle.

26. **Gardening** A rectangular flower garden has a perimeter of 86 ft. The length of the garden is 28 ft. What is the width of the garden?

27. **Carpentry** A board 10 ft long is cut into two pieces. Four times the length of the shorter piece is 2 ft less than three times the length of the longer piece. Find the length of each piece.

28. **Investments** An investor has a total of $15,000 invested in two simple interest accounts. On one account, the annual simple interest rate is 4.5%. On the second account, the annual simple interest rate is 4%. How much is invested in the 4% account if the total annual interest earned is $635?

29. **Travel** A family drove to a resort at an average speed of 42 mph and later returned over the same road at an average speed of 56 mph. Find the distance to the resort if the total driving time was 7 h.

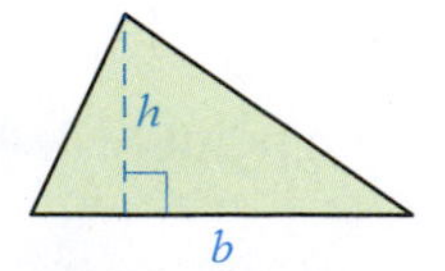

30. **Geometry** The length of the base of a triangle is three times the height. The area of the triangle is 24 in^2. Find the length of the base of the triangle.

Copyright © Houghton Mifflin Company. All rights reserved.

chapter

8 Rational Expressions

In order to monitor species that are or are becoming endangered, scientists need to determine the present population of that species. Scientists catch and tag a certain number of the animals and then release them. Later, a group of the animals from that same habitat is caught and the number tagged is counted. A proportion is used to estimate the total population size in that region, as shown in **Exercise 27 on page 482**. Tracking the tagged animals also assists scientists in learning more about the habits of that species.

OBJECTIVES

Section 8.1

A To simplify a rational expression
B To multiply rational expressions
C To divide rational expressions

Section 8.2

A To rewrite rational expressions in terms of a common denominator
B To add or subtract rational expressions

Section 8.3

A To simplify a complex fraction

Section 8.4

A To solve an equation containing fractions

Section 8.5

A To solve a proportion
B To solve application problems
C To solve problems involving similar triangles

Section 8.6

A To solve a literal equation for one of the variables

Section 8.7

A To solve work problems
B To use rational expressions to solve uniform motion problems

Section 8.8

A To solve variation problems

Need help? For online student resources, such as section quizzes, visit this textbook's website at **math.college.hmco.com/students.**

Copyright © Houghton Mifflin Company. All rights reserved.

PREP TEST

Do these exercises to prepare for Chapter 8.

1. Find the LCM of 10 and 25.

For Exercises 2 to 5, add, subtract, multiply, or divide.

2. $-\frac{3}{8} \cdot \frac{4}{9}$

3. $-\frac{4}{5} \div \frac{8}{15}$

4. $-\frac{5}{6} + \frac{7}{8}$

5. $-\frac{3}{8} - \left(-\frac{7}{12}\right)$

6. Evaluate $\frac{2x - 3}{x^2 - x + 1}$ for $x = 2$.

7. Solve: $4(2x + 1) = 3(x - 2)$

8. Solve: $10\left(\frac{t}{2} + \frac{t}{5}\right) = 10(1)$

9. **Travel** Two planes start from the same point and fly in opposite directions. The first plane is flying 20 mph slower than the second plane. In 2 h, the planes are 480 mi apart. Find the rate of each plane.

GO FIGURE

If 6 machines can fill 12 boxes of cereal in 7 min, how many boxes of cereal can 14 machines fill in 12 min?

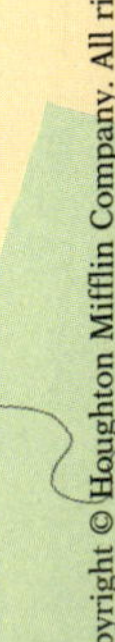
Copyright © Houghton Mifflin Company. All rights reserved.

8.1 Multiplication and Division of Rational Expressions

Objective A

To simplify a rational expression

SSM

A fraction in which the numerator and denominator are polynomials is called a **rational expression.** Examples of rational expressions are shown at the right.

$$\frac{5}{z}, \quad \frac{x^2+1}{2x-1}, \quad \frac{y^2+y-1}{4y^2+1}$$

Care must be exercised with a rational expression to ensure that when the variables are replaced with numbers, the resulting denominator is not zero. Consider the rational expression at the right. The value of x cannot be 3 because the denominator would then be zero.

$$\frac{4x^2-9}{2x-6}$$

$$\frac{4(3)^2-9}{2(3)-6} = \frac{27}{0} \quad \text{Not a real number}$$

TAKE NOTE

Recall that division by zero is not defined. Therefore, the denominator of a fraction cannot be zero.

In the **simplest form of a rational expression,** the numerator and denominator have no common factors. The Multiplication Property of One is used to write a rational expression in simplest form.

HOW TO Simplify: $\frac{x^2-4}{x^2-2x-8}$

$$\frac{x^2-4}{x^2-2x-8} = \frac{(x-2)(x+2)}{(x-4)(x+2)}$$

- Factor the numerator and denominator.

$$= \frac{x-2}{x-4} \cdot \boxed{\frac{x+2}{x+2}} = \frac{x-2}{x-4} \cdot 1$$

$$= \frac{x-2}{x-4}, x \neq -2, 4$$

- The restrictions, $x \neq -2$ or 4, are necessary to prevent division by zero.

This simplification is usually shown with slashes through the common factors:

$$\frac{x^2-4}{x^2-2x-8} = \frac{(x-2)\overset{1}{\cancel{(x+2)}}}{(x-4)\underset{1}{\cancel{(x+2)}}}$$

- Factor the numerator and denominator.

$$= \frac{x-2}{x-4}, x \neq -2, 4$$

- Divide by the common factors. The restrictions, $x \neq -2$ or 4, are necessary to prevent division by zero.

In summary, to simplify a rational expression, factor the numerator and denominator. Then divide the numerator and denominator by the common factors.

HOW TO Simplify: $\frac{10+3x-x^2}{x^2-4x-5}$

$$\frac{10+3x-x^2}{x^2-4x-5} = \frac{-(x^2-3x-10)}{x^2-4x-5}$$

- Because the coefficient of x^2 in the numerator is -1, factor -1 from the numerator.

$$= \frac{-\overset{1}{\cancel{(x-5)}}(x+2)}{\underset{1}{\cancel{(x-5)}}(x+1)}$$

- Factor the numerator and denominator. Divide by the common factors.

$$= -\frac{x+2}{x+1}, x \neq -1, 5$$

Copyright © Houghton Mifflin Company. All rights reserved.

For the remaining examples, we will omit the restrictions on the variables that prevent division by zero and assume the values of the variables are such that division by zero is not possible.

Example 1

Simplify: $\dfrac{4x^3y^4}{6x^4y}$

Solution

$\dfrac{4x^3y^4}{6x^4y} = \dfrac{2y^3}{3x}$ • Use rules of exponents.

You Try It 1

Simplify: $\dfrac{6x^5y}{12x^2y^3}$

Your solution

Example 2

Simplify: $\dfrac{9 - x^2}{x^2 + x - 12}$

Solution

$$\frac{9 - x^2}{x^2 + x - 12} = \frac{\overset{-1}{\cancel{(3 - x)}}(3 + x)}{\underset{1}{\cancel{(x - 3)}}(x + 4)} \quad \bullet\ (3 - x) = -1(x - 3)$$

$$= -\frac{x + 3}{x + 4}$$

You Try It 2

Simplify: $\dfrac{x^2 + 2x - 24}{16 - x^2}$

Your solution

Example 3

Simplify: $\dfrac{x^2 + 2x - 15}{x^2 - 7x + 12}$

Solution

$$\frac{x^2 + 2x - 15}{x^2 - 7x + 12} = \frac{(x + 5)\overset{1}{\cancel{(x - 3)}}}{\underset{1}{\cancel{(x - 3)}}(x - 4)} = \frac{x + 5}{x - 4}$$

You Try It 3

Simplify: $\dfrac{x^2 + 4x - 12}{x^2 - 3x + 2}$

Your solution

Solutions on pp. S23–S24

Objective B To multiply rational expressions

The product of two fractions is a fraction whose numerator is the product of the numerators of the two fractions and whose denominator is the product of the denominators of the two fractions.

Multiplying Rational Expressions

Multiply the numerators.
Multiply the denominators. $\dfrac{a}{b} \cdot \dfrac{c}{d} = \dfrac{ac}{bd}$

$$\frac{2}{3} \cdot \frac{4}{5} = \frac{8}{15} \qquad \frac{3x}{y} \cdot \frac{2}{z} = \frac{6x}{yz} \qquad \frac{x + 2}{x} \cdot \frac{3}{x - 2} = \frac{3x + 6}{x^2 - 2x}$$

Copyright © Houghton Mifflin Company. All rights reserved.

HOW TO Multiply: $\frac{x^2 + 3x}{x^2 - 3x - 4} \cdot \frac{x^2 - 5x + 4}{x^2 + 2x - 3}$

$$\frac{x^2 + 3x}{x^2 - 3x - 4} \cdot \frac{x^2 - 5x + 4}{x^2 + 2x - 3}$$

$$= \frac{x(x + 3)}{(x - 4)(x + 1)} \cdot \frac{(x - 4)(x - 1)}{(x + 3)(x - 1)}$$

- Factor the numerator and denominator of each fraction.

$$= \frac{x\overset{1}{\cancel{(x + 3)}}\overset{1}{\cancel{(x - 4)}}\overset{1}{\cancel{(x - 1)}}}{\underset{1}{\cancel{(x - 4)}}(x + 1)\underset{1}{\cancel{(x + 3)}}\underset{1}{\cancel{(x - 1)}}}$$

- Multiply. Then divide by the common factors.

$$= \frac{x}{x + 1}$$

- Write the answer in simplest form.

Example 4

Multiply: $\frac{10x^2 - 15x}{12x - 8} \cdot \frac{3x - 2}{20x - 25}$

Solution

$$\frac{10x^2 - 15x}{12x - 8} \cdot \frac{3x - 2}{20x - 25}$$

$$= \frac{5x(2x - 3)}{4(3x - 2)} \cdot \frac{(3x - 2)}{5(4x - 5)}$$

- Factor.

$$= \frac{\overset{1}{\cancel{5}}x(2x - 3)\overset{1}{\cancel{(3x - 2)}}}{4\underset{1}{\cancel{(3x - 2)}}\underset{1}{\cancel{5}}(4x - 5)}$$

- Divide by common factors.

$$= \frac{x(2x - 3)}{4(4x - 5)}$$

You Try It 4

Multiply: $\frac{12x^2 + 3x}{10x - 15} \cdot \frac{8x - 12}{9x + 18}$

Your solution

Example 5

Multiply: $\frac{x^2 + x - 6}{x^2 + 7x + 12} \cdot \frac{x^2 + 3x - 4}{4 - x^2}$

Solution

$$\frac{x^2 + x - 6}{x^2 + 7x + 12} \cdot \frac{x^2 + 3x - 4}{4 - x^2}$$

$$= \frac{(x + 3)(x - 2)}{(x + 3)(x + 4)} \cdot \frac{(x + 4)(x - 1)}{(2 - x)(2 + x)}$$

- Factor.

$$= \frac{\overset{1}{\cancel{(x + 3)}}\overset{-1}{\cancel{(x - 2)}}\overset{1}{\cancel{(x + 4)}}(x - 1)}{\underset{1}{\cancel{(x + 3)}}\underset{1}{\cancel{(x + 4)}}\underset{1}{\cancel{(2 - x)}}(2 + x)}$$

- Divide by common factors.

$$= -\frac{x - 1}{x + 2}$$

You Try It 5

Multiply: $\frac{x^2 + 2x - 15}{9 - x^2} \cdot \frac{x^2 - 3x - 18}{x^2 - 7x + 6}$

Your solution

Solutions on p. S24

Copyright © Houghton Mifflin Company. All rights reserved.

Objective C To divide rational expressions

The **reciprocal of a rational expression** is the rational expression with the numerator and denominator interchanged.

Fraction		Reciprocal
$\frac{a}{b}$	$\frac{b}{a}$	
$x^2 = \frac{x^2}{1}$	$\frac{1}{x^2}$	
$\frac{x+2}{x}$	$\frac{x}{x+2}$	

Dividing Rational Expressions

Multiply the dividend by the reciprocal of the divisor. $\frac{a}{b} \div \frac{c}{d} = \frac{a}{b} \cdot \frac{d}{c} = \frac{ad}{bc}$

$$\frac{4}{x} \div \frac{y}{5} = \frac{4}{x} \cdot \frac{5}{y} = \frac{20}{xy} \qquad \frac{x+4}{x} \div \frac{x-2}{4} = \frac{x+4}{x} \cdot \frac{4}{x-2} = \frac{4(x+4)}{x(x-2)}$$

The basis for the division rule is shown at the right.

$$\frac{a}{b} \div \frac{c}{d} = \frac{\frac{a}{b}}{\frac{c}{d}} = \frac{\frac{a}{b} \cdot \frac{d}{c}}{\frac{c}{d} \cdot \frac{d}{c}} = \frac{\frac{a}{b} \cdot \frac{d}{c}}{1} = \frac{a}{b} \cdot \frac{d}{c}$$

Example 6

Divide: $\frac{xy^2 - 3x^2y}{z^2} \div \frac{6x^2 - 2xy}{z^3}$

Solution

$$\frac{xy^2 - 3x^2y}{z^2} \div \frac{6x^2 - 2xy}{z^3}$$

$$= \frac{xy^2 - 3x^2y}{z^2} \cdot \frac{z^3}{6x^2 - 2xy}$$ • Multiply by the reciprocal.

$$= \frac{xy(y - 3x) \cdot z^3}{z^2 \cdot 2x(3x - y)} = -\frac{yz}{2}$$

You Try It 6

Divide: $\frac{a^2}{4bc^2 - 2b^2c} \div \frac{a}{6bc - 3b^2}$

Your solution

Example 7

Divide: $\frac{2x^2 + 5x + 2}{2x^2 + 3x - 2} \div \frac{3x^2 + 13x + 4}{2x^2 + 7x - 4}$

Solution

$$\frac{2x^2 + 5x + 2}{2x^2 + 3x - 2} \div \frac{3x^2 + 13x + 4}{2x^2 + 7x - 4}$$

$$= \frac{2x^2 + 5x + 2}{2x^2 + 3x - 2} \cdot \frac{2x^2 + 7x - 4}{3x^2 + 13x + 4}$$ • Multiply by the reciprocal.

$$= \frac{(2x+1)(x+2) \cdot (2x-1)(x+4)}{(2x-1)(x+2) \cdot (3x+1)(x+4)} = \frac{2x+1}{3x+1}$$

You Try It 7

Divide: $\frac{3x^2 + 26x + 16}{3x^2 - 7x - 6} \div \frac{2x^2 + 9x - 5}{x^2 + 2x - 15}$

Your solution

Solutions on p. S24

Copyright © Houghton Mifflin Company. All rights reserved.

8.1 Exercises

Objective A To simplify a rational expression

1. Explain the procedure for writing a rational expression in simplest form.

2. Explain why the following simplification is incorrect.

$$\frac{x+3}{x} = \frac{\overset{1}{\cancel{x}} + 3}{\underset{1}{\cancel{x}}} = 4$$

For Exercises 3 to 29, simplify.

3. $\frac{9x^3}{12x^4}$

4. $\frac{16x^2y}{24xy^3}$

5. $\frac{(x+3)^2}{(x+3)^3}$

6. $\frac{(2x-1)^5}{(2x-1)^4}$

7. $\frac{3n-4}{4-3n}$

8. $\frac{5-2x}{2x-5}$

9. $\frac{6y(y+2)}{9y^2(y+2)}$

10. $\frac{12x^2(3-x)}{18x(3-x)}$

11. $\frac{6x(x-5)}{8x^2(5-x)}$

12. $\frac{14x^3(7-3x)}{21x(3x-7)}$

13. $\frac{a^2+4a}{ab+4b}$

14. $\frac{x^2-3x}{2x-6}$

15. $\frac{4-6x}{3x^2-2x}$

16. $\frac{5xy-3y}{9-15x}$

17. $\frac{y^2-3y+2}{y^2-4y+3}$

18. $\frac{x^2+5x+6}{x^2+8x+15}$

19. $\frac{x^2+3x-10}{x^2+2x-8}$

20. $\frac{a^2+7a-8}{a^2+6a-7}$

21. $\frac{x^2+x-12}{x^2-6x+9}$

22. $\frac{x^2+8x+16}{x^2-2x-24}$

23. $\frac{x^2-3x-10}{25-x^2}$

24. $\frac{4-y^2}{y^2-3y-10}$

25. $\frac{2x^3+2x^2-4x}{x^3+2x^2-3x}$

26. $\frac{3x^3-12x}{6x^3-24x^2+24x}$

27. $\frac{6x^2-7x+2}{6x^2+5x-6}$

28. $\frac{2n^2-9n+4}{2n^2-5n-12}$

29. $\frac{x^2+3x-28}{24-2x-x^2}$

Copyright © Houghton Mifflin Company. All rights reserved.

Objective B To multiply rational expressions

For Exercises 30 to 55, multiply.

30. $\frac{8x^2}{9y^3} \cdot \frac{3y^2}{4x^3}$

31. $\frac{14a^2b^3}{15x^5y^2} \cdot \frac{25x^3y}{16ab}$

32. $\frac{12x^3y^4}{7a^2b^3} \cdot \frac{14a^3b^4}{9x^2y^2}$

33. $\frac{18a^4b^2}{25x^2y^3} \cdot \frac{50x^5y^6}{27a^6b^2}$

34. $\frac{3x-6}{5x-20} \cdot \frac{10x-40}{27x-54}$

35. $\frac{8x-12}{14x+7} \cdot \frac{42x+21}{32x-48}$

36. $\frac{3x^2+2x}{2xy-3y} \cdot \frac{2xy^3-3y^3}{3x^3+2x^2}$

37. $\frac{4a^2x-3a^2}{2by+5b} \cdot \frac{2b^3y+5b^3}{4ax-3a}$

38. $\frac{x^2+5x+4}{x^3y^2} \cdot \frac{x^2y^3}{x^2+2x+1}$

39. $\frac{x^2+x-2}{xy^2} \cdot \frac{x^3y}{x^2+5x+6}$

40. $\frac{x^4y^2}{x^2+3x-28} \cdot \frac{x^2-49}{xy^4}$

41. $\frac{x^5y^3}{x^2+13x+30} \cdot \frac{x^2+2x-3}{x^7y^2}$

42. $\frac{2x^2-5x}{2xy+y} \cdot \frac{2xy^2+y^2}{5x^2-2x^3}$

43. $\frac{3a^3+4a^2}{5ab-3b} \cdot \frac{3b^3-5ab^3}{3a^2+4a}$

44. $\frac{x^2-2x-24}{x^2-5x-6} \cdot \frac{x^2+5x+6}{x^2+6x+8}$

45. $\frac{x^2-8x+7}{x^2+3x-4} \cdot \frac{x^2+3x-10}{x^2-9x+14}$

46. $\frac{x^2+2x-35}{x^2+4x-21} \cdot \frac{x^2+3x-18}{x^2+9x+18}$

47. $\frac{y^2+y-20}{y^2+2y-15} \cdot \frac{y^2+4y-21}{y^2+3y-28}$

Copyright © Houghton Mifflin Company. All rights reserved.

48. $\dfrac{x^2 - 3x - 4}{x^2 + 6x + 5} \cdot \dfrac{x^2 + 5x + 6}{8 + 2x - x^2}$

49. $\dfrac{25 - n^2}{n^2 - 2n - 35} \cdot \dfrac{n^2 - 8n - 20}{n^2 - 3n - 10}$

50. $\dfrac{12x^2 - 6x}{x^2 + 6x + 5} \cdot \dfrac{2x^4 + 10x^3}{4x^2 - 1}$

51. $\dfrac{8x^3 + 4x^2}{x^2 - 3x + 2} \cdot \dfrac{x^2 - 4}{16x^2 + 8x}$

52. $\dfrac{16 + 6x - x^2}{x^2 - 10x - 24} \cdot \dfrac{x^2 - 6x - 27}{x^2 - 17x + 72}$

53. $\dfrac{x^2 - 11x + 28}{x^2 - 13x + 42} \cdot \dfrac{x^2 + 7x + 10}{20 - x - x^2}$

54. $\dfrac{2x^2 + 5x + 2}{2x^2 + 7x + 3} \cdot \dfrac{x^2 - 7x - 30}{x^2 - 6x - 40}$

55. $\dfrac{x^2 - 4x - 32}{x^2 - 8x - 48} \cdot \dfrac{3x^2 + 17x + 10}{3x^2 - 22x - 16}$

Objective C To divide rational expressions

56. 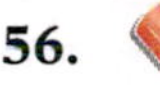 What is the reciprocal of a rational expression?

57. Explain how to divide rational expressions.

For Exercises 58 to 77, divide.

58. $\dfrac{4x^2y^3}{15a^2b^3} \div \dfrac{6xy}{5a^3b^5}$

59. $\dfrac{9x^3y^4}{16a^4b^2} \div \dfrac{45x^4y^2}{14a^7b}$

60. $\dfrac{6x - 12}{8x + 32} \div \dfrac{18x - 36}{10x + 40}$

61. $\dfrac{28x + 14}{45x - 30} \div \dfrac{14x + 7}{30x - 20}$

62. $\dfrac{6x^3 + 7x^2}{12x - 3} \div \dfrac{6x^2 + 7x}{36x - 9}$

63. $\dfrac{5a^2y + 3a^2}{2x^3 + 5x^2} \div \dfrac{10ay + 6a}{6x^3 + 15x^2}$

64. $\dfrac{x^2 + 4x + 3}{x^2y} \div \dfrac{x^2 + 2x + 1}{xy^2}$

65. $\dfrac{x^3y^2}{x^2 - 3x - 10} \div \dfrac{xy^4}{x^2 - x - 20}$

66. $\dfrac{x^2 - 49}{x^4y^3} \div \dfrac{x^2 - 14x + 49}{x^4y^3}$

67. $\dfrac{x^2y^5}{x^2 - 11x + 30} \div \dfrac{xy^6}{x^2 - 7x + 10}$

Copyright © Houghton Mifflin Company. All rights reserved.

68. $\dfrac{4ax - 8a}{c^2} \div \dfrac{2y - xy}{c^3}$

69. $\dfrac{3x^2y - 9xy}{a^2b} \div \dfrac{3x^2 - x^3}{ab^2}$

70. $\dfrac{x^2 - 5x + 6}{x^2 - 9x + 18} \div \dfrac{x^2 - 6x + 8}{x^2 - 9x + 20}$

71. $\dfrac{x^2 + 3x - 40}{x^2 + 2x - 35} \div \dfrac{x^2 + 2x - 48}{x^2 + 3x - 18}$

72. $\dfrac{x^2 + 2x - 15}{x^2 - 4x - 45} \div \dfrac{x^2 + x - 12}{x^2 - 5x - 36}$

73. $\dfrac{y^2 - y - 56}{y^2 + 8y + 7} \div \dfrac{y^2 - 13y + 40}{y^2 - 4y - 5}$

74. $\dfrac{8 + 2x - x^2}{x^2 + 7x + 10} \div \dfrac{x^2 - 11x + 28}{x^2 - x - 42}$

75. $\dfrac{x^2 - x - 2}{x^2 - 7x + 10} \div \dfrac{x^2 - 3x - 4}{40 - 3x - x^2}$

76. $\dfrac{2x^2 - 3x - 20}{2x^2 - 7x - 30} \div \dfrac{2x^2 - 5x - 12}{4x^2 + 12x + 9}$

77. $\dfrac{6n^2 + 13n + 6}{4n^2 - 9} \div \dfrac{6n^2 + n - 2}{4n^2 - 1}$

APPLYING THE CONCEPTS

78. Given the expression $\frac{9}{x^2 + 1}$, choose some values of x and evaluate the expression for those values. Is it possible to choose a value of x for which the value of the expression is greater than 10? If so, what is that value of x? If not, explain why it is not possible.

79. Given the expression $\frac{1}{y - 3}$, choose some values of y and evaluate the expression for those values. Is it possible to choose a value of y for which the value of the expression is greater than 10,000,000? If so, what is that value of y? If not, explain why it is not possible.

Geometry For Exercises 80 and 81, write in simplest form the ratio of the shaded area of the figure to the total area of the figure.

80.

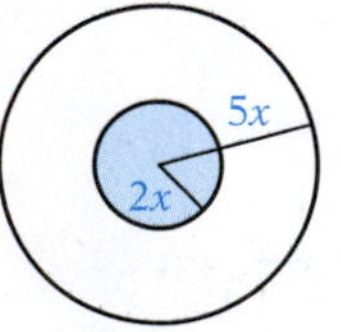

81.

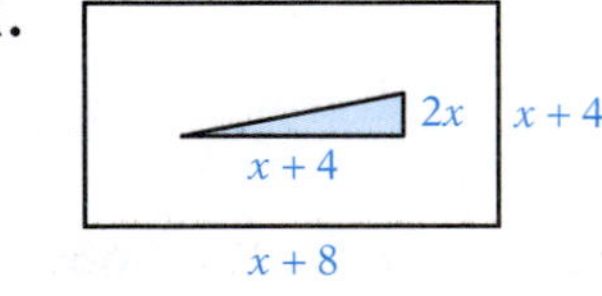

For Exercises 82 to 84, complete the simplification.

82. $\dfrac{8x}{9y} \div \dfrac{?}{} = \dfrac{10y}{3}$

83. $\dfrac{n}{n + 3} \div \dfrac{?}{} = \dfrac{n}{n - 2}$

84. $\dfrac{?}{} \div \dfrac{n - 1}{4n^3} = 2n^2(n + 1)$

Copyright © Houghton Mifflin Company. All rights reserved.

8.2 Addition and Subtraction of Rational Expressions

Objective A

To rewrite rational expressions in terms of a common denominator

Study Tip

As you know, often in mathematics you learn one skill in order to perform another. This is true of this objective. You are learning to rewrite rational expressions in terms of a common denominator in order to add and subtract rational expressions in the next objective. To ensure success, be certain you understand this lesson before studying the next.

In adding or subtracting rational expressions, it is frequently necessary to express the rational expressions in terms of a common denominator. This common denominator is the least common multiple (LCM) of the denominators.

The **least common multiple (LCM) of two or more polynomials** is the simplest polynomial that contains the factors of each polynomial. To find the LCM, first factor each polynomial completely. The LCM is the product of each factor the greatest number of times it occurs in any one factorization.

HOW TO Find the LCM of $3x^2 + 15x$ and $6x^4 + 24x^3 - 30x^2$.

Factor each polynomial.

$$3x^2 + 15x = 3x(x + 5)$$
$$6x^4 + 24x^3 - 30x^2 = 6x^2(x^2 + 4x - 5)$$
$$= 6x^2(x - 1)(x + 5)$$

The LCM is the product of the LCM of the numerical coefficients and each variable factor the greatest number of times it occurs in any one factorization.

LCM $= 6x^2(x - 1)(x + 5)$

TAKE NOTE

This is the only section in this text in which the numerator in the final answer is in unfactored form. In building fractions, the fraction is not in simplest form, and leaving the fractions in unfactored form removes the temptation to cancel.

HOW TO Write the fractions $\frac{5}{6x^2y}$ and $\frac{a}{9xy^3}$ in terms of the LCM of the denominators.

The LCM is $18x^2y^3$.
- Find the LCM of the denominators.

$$\frac{5}{6x^2y} = \frac{5}{6x^2y} \cdot \frac{3y^2}{3y^2} = \frac{15y^2}{18x^2y^3}$$

$$\frac{a}{9xy^3} = \frac{a}{9xy^3} \cdot \frac{2x}{2x} = \frac{2ax}{18x^2y^3}$$

- For each fraction, multiply the numerator and denominator by the factor whose product with the denominator is the LCM.

TAKE NOTE

$x^2 - 2x = x(x - 2)$
$3x - 6 = 3(x - 2)$
The LCM of $x(x - 2)$ and $3(x - 2)$ is $3x(x - 2)$.

HOW TO Write the fractions $\frac{x+2}{x^2 - 2x}$ and $\frac{5x}{3x - 6}$ in terms of the LCM of the denominators.

The LCM is $3x(x - 2)$.
- Find the LCM of the denominators.

$$\frac{x+2}{x^2 - 2x} = \frac{x+2}{x(x-2)} \cdot \frac{3}{3} = \frac{3x+6}{3x(x-2)}$$

$$\frac{5x}{3x - 6} = \frac{5x}{3(x-2)} \cdot \frac{x}{x} = \frac{5x^2}{3x(x-2)}$$

- For each fraction, multiply the numerator and denominator by the factor whose product with the denominator is the LCM.

Copyright © Houghton Mifflin Company. All rights reserved.

Example 1

Write the fractions $\frac{3x}{x-1}$ and $\frac{4}{2x+5}$ in terms of the LCM of the denominators.

Solution

The LCM of $x - 1$ and $2x + 5$ is $(x - 1)(2x + 5)$.

$$\frac{3x}{x-1} = \frac{3x}{x-1} \cdot \frac{2x+5}{2x+5} = \frac{6x^2 + 15x}{(x-1)(2x+5)}$$

$$\frac{4}{2x+5} = \frac{4}{2x+5} \cdot \frac{x-1}{x-1} = \frac{4x-4}{(x-1)(2x+5)}$$

You Try It 1

Write the fractions $\frac{2x}{2x-5}$ and $\frac{3}{x+4}$ in terms of the LCM of the denominators.

Your solution

Example 2

Write the fractions $\frac{2a-3}{a^2-2a}$ and $\frac{a+1}{2a^2-a-6}$ in terms of the LCM of the denominators.

Solution

$a^2 - 2a = a(a - 2)$;
$2a^2 - a - 6 = (2a + 3)(a - 2)$

The LCM is $a(a - 2)(2a + 3)$.

$$\frac{2a-3}{a^2-2a} = \frac{2a-3}{a(a-2)} \cdot \frac{2a+3}{2a+3} = \frac{4a^2-9}{a(a-2)(2a+3)}$$

$$\frac{a+1}{2a^2-a-6} = \frac{a+1}{(a-2)(2a+3)} \cdot \frac{a}{a} = \frac{a^2+a}{a(a-2)(2a+3)}$$

You Try It 2

Write the fractions $\frac{3x}{2x^2-11x+15}$ and $\frac{x-2}{x^2-3x}$ in terms of the LCM of the denominators.

Your solution

Example 3

Write the fractions $\frac{2x-3}{3x-x^2}$ and $\frac{3x}{x^2-4x+3}$ in terms of the LCM of the denominators.

Solution

$3x - x^2 = x(3 - x) = -x(x - 3)$;
$x^2 - 4x + 3 = (x - 3)(x - 1)$

The LCM is $x(x - 3)(x - 1)$.

$$\frac{2x-3}{3x-x^2} = -\frac{2x-3}{x(x-3)} \cdot \frac{x-1}{x-1} = -\frac{2x^2-5x+3}{x(x-3)(x-1)}$$

$$\frac{3x}{x^2-4x+3} = \frac{3x}{(x-3)(x-1)} \cdot \frac{x}{x} = \frac{3x^2}{x(x-3)(x-1)}$$

You Try It 3

Write the fractions $\frac{2x-7}{2x-x^2}$ and $\frac{3x-2}{3x^2-5x-2}$ in terms of the LCM of the denominators.

Your solution

Solutions on p. S24

Copyright © Houghton Mifflin Company. All rights reserved.

Objective B To add or subtract rational expressions

$$\frac{a}{c} + \frac{b}{c} = \frac{a+b}{c}$$

When adding rational expressions in which the denominators are the same, add the numerators. The denominator of the sum is the common denominator.

$$\frac{4x}{15} + \frac{8x}{15} = \frac{4x+8x}{15} = \frac{12x}{15} = \frac{4x}{5}$$

• Note that the sum is written in simplest form.

$$\frac{a}{a^2-b^2} + \frac{b}{a^2-b^2} = \frac{a+b}{a^2-b^2} = \frac{a+b}{(a-b)(a+b)} = \frac{\overset{1}{\cancel{(a+b)}}}{(a-b)\underset{1}{\cancel{(a+b)}}} = \frac{1}{a-b}$$

$$\frac{a}{c} - \frac{b}{c} = \frac{a-b}{c}$$

When subtracting rational expressions in which the denominators are the same, subtract the numerators. The denominator of the difference is the common denominator. Write the answer in simplest form.

$$\frac{7x-12}{2x^2+5x-12} - \frac{3x-6}{2x^2+5x-12} = \frac{(7x-12)-(3x-6)}{2x^2+5x-12} = \frac{4x-6}{2x^2+5x-12}$$

$$= \frac{2\overset{1}{\cancel{(2x-3)}}}{\underset{1}{\cancel{(2x-3)}}(x+4)} = \frac{2}{x+4}$$

Before two rational expressions with *different* denominators can be added or subtracted, each rational expression must be expressed in terms of a common denominator. This common denominator is the LCM of the denominators of the rational expressions.

TAKE NOTE

Note the steps involved in adding or subtracting rational expressions:

1. Find the LCM of the denominators.
2. Rewrite each fraction in terms of the common denominator.
3. Add or subtract the rational expressions.
4. Simplify the resulting sum or difference.

HOW TO Simplify: $\frac{x}{x-3} - \frac{x+1}{x-2}$

The LCM is $(x-3)(x-2)$. • Find the LCM of the denominators.

$$\frac{x}{x-3} - \frac{x+1}{x-2} = \frac{x}{x-3}\cdot\frac{x-2}{x-2} - \frac{x+1}{x-2}\cdot\frac{x-3}{x-3}$$

• Express each fraction in terms of the LCM.

$$= \frac{x(x-2)-(x+1)(x-3)}{(x-3)(x-2)}$$

• Subtract the fractions.

$$= \frac{(x^2-2x)-(x^2-2x-3)}{(x-3)(x-2)}$$

$$= \frac{3}{(x-3)(x-2)}$$

• Simplify.

HOW TO Simplify: $\frac{3x}{2x-3} + \frac{3x+6}{2x^2+x-6}$

The LCM of $2x-3$ and $2x^2+x-6$ is $(2x-3)(x+2)$. • Find the LCM of the denominators.

$$\frac{3x}{2x-3} + \frac{3x+6}{2x^2+x-6} = \frac{3x}{2x-3}\cdot\frac{x+2}{x+2} + \frac{3x+6}{(2x-3)(x+2)}$$

• Express each fraction in terms of the LCM.

$$= \frac{3x(x+2)+(3x+6)}{(2x-3)(x+2)}$$

• Add the fractions.

$$= \frac{(3x^2+6x)+(3x+6)}{(2x-3)(x+2)}$$

$$= \frac{3x^2+9x+6}{(2x-3)(x+2)} = \frac{3\cancel{(x+2)}(x+1)}{(2x-3)\cancel{(x+2)}}$$

• Simplify.

$$= \frac{3(x+1)}{2x-3}$$

Copyright © Houghton Mifflin Company. All rights reserved.

Example 4

Simplify: $\frac{2}{x} - \frac{3}{x^2} + \frac{1}{xy}$

Solution

The LCM is x^2y.

$$\frac{2}{x} - \frac{3}{x^2} + \frac{1}{xy} = \frac{2}{x} \cdot \frac{xy}{xy} - \frac{3}{x^2} \cdot \frac{y}{y} + \frac{1}{xy} \cdot \frac{x}{x}$$
$$= \frac{2xy}{x^2y} - \frac{3y}{x^2y} + \frac{x}{x^2y}$$
$$= \frac{2xy - 3y + x}{x^2y}$$

You Try It 4

Simplify: $\frac{2}{b} - \frac{1}{a} + \frac{4}{ab}$

Your solution

Example 5

Simplify: $\frac{x}{2x - 4} - \frac{4 - x}{x^2 - 2x}$

Solution

$2x - 4 = 2(x - 2)$; $x^2 - 2x = x(x - 2)$

The LCM is $2x(x - 2)$.

$$\frac{x}{2x - 4} - \frac{4 - x}{x^2 - 2x} = \frac{x}{2(x - 2)} \cdot \frac{x}{x} - \frac{4 - x}{x(x - 2)} \cdot \frac{2}{2}$$
$$= \frac{x^2 - (4 - x)2}{2x(x - 2)}$$
$$= \frac{x^2 - (8 - 2x)}{2x(x - 2)} = \frac{x^2 + 2x - 8}{2x(x - 2)}$$
$$= \frac{(x + 4)(x - 2)}{2x(x - 2)} = \frac{(x + 4)\overset{1}{\cancel{(x - 2)}}}{2x\underset{1}{\cancel{(x - 2)}}}$$
$$= \frac{x + 4}{2x}$$

You Try It 5

Simplify: $\frac{a - 3}{a^2 - 5a} + \frac{a - 9}{a^2 - 25}$

Your solution

Example 6

Simplify: $\frac{x}{x + 1} - \frac{2}{x - 2} - \frac{3}{x^2 - x - 2}$

Solution

The LCM is $(x + 1)(x - 2)$.

$$\frac{x}{x + 1} - \frac{2}{x - 2} - \frac{3}{x^2 - x - 2}$$
$$= \frac{x}{x + 1} \cdot \frac{x - 2}{x - 2} - \frac{2}{x - 2} \cdot \frac{x + 1}{x + 1} - \frac{3}{(x + 1)(x - 2)}$$
$$= \frac{x(x - 2) - 2(x + 1) - 3}{(x + 1)(x - 2)}$$
$$= \frac{x^2 - 2x - 2x - 2 - 3}{(x + 1)(x - 2)}$$
$$= \frac{x^2 - 4x - 5}{(x + 1)(x - 2)} = \frac{\overset{1}{\cancel{(x + 1)}}(x - 5)}{\underset{1}{\cancel{(x + 1)}}(x - 2)} = \frac{x - 5}{x - 2}$$

You Try It 6

Simplify: $\frac{2x}{x - 4} - \frac{x - 1}{x + 1} + \frac{2}{x^2 - 3x - 4}$

Your solution

Solutions on p. S24

Copyright © Houghton Mifflin Company. All rights reserved.

8.2 Exercises

Objective A **To rewrite rational expressions in terms of a common denominator**

For Exercises 1 to 25, write each fraction in terms of the LCM of the denominators.

1. $\dfrac{3}{4x^2y}, \dfrac{17}{12xy^4}$

2. $\dfrac{5}{16a^3b^3}, \dfrac{7}{30a^5b}$

3. $\dfrac{x-2}{3x(x-2)}, \dfrac{3}{6x^2}$

4. $\dfrac{5x-1}{4x(2x+1)}, \dfrac{2}{5x^3}$

5. $\dfrac{3x-1}{2x^2-10x}, -3x$

6. $\dfrac{4x-3}{3x(x-2)}, 2x$

7. $\dfrac{3x}{2x-3}, \dfrac{5x}{2x+3}$

8. $\dfrac{2}{7y-3}, \dfrac{-3}{7y+3}$

9. $\dfrac{2x}{x^2-9}, \dfrac{x+1}{x-3}$

10. $\dfrac{3x}{16-x^2}, \dfrac{2x}{16-4x}$

11. $\dfrac{3}{3x^2-12y^2}, \dfrac{5}{6x-12y}$

12. $\dfrac{2x}{x^2-36}, \dfrac{x-1}{6x-36}$

13. $\dfrac{3x}{x^2-1}, \dfrac{5x}{x^2-2x+1}$

14. $\dfrac{x^2+2}{x^3-1}, \dfrac{3}{x^2+x+1}$

15. $\dfrac{x-3}{8-x^3}, \dfrac{2}{4+2x+x^2}$

16. $\dfrac{2x}{x^2+x-6}, \dfrac{-4x}{x^2+5x+6}$

17. $\dfrac{2x}{x^2+2x-3}, \dfrac{-x}{x^2+6x+9}$

18. $\dfrac{3x}{2x^2-x-3}, \dfrac{-2x}{2x^2-11x+12}$

19. $\dfrac{-4x}{4x^2-16x+15}, \dfrac{3x}{6x^2-19x+10}$

20. $\dfrac{3}{2x^2+5x-12}, \dfrac{2x}{3-2x}, \dfrac{3x-1}{x+4}$

21. $\dfrac{5}{6x^2-17x+12}, \dfrac{2x}{4-3x}, \dfrac{x+1}{2x-3}$

Copyright © Houghton Mifflin Company. All rights reserved.

22. $\frac{3x}{x-4}, \frac{4}{x+5}, \frac{x+2}{20-x-x^2}$

23. $\frac{2x}{x-3}, \frac{-2}{x+5}, \frac{x-1}{15-2x-x^2}$

24. $\frac{2}{x^{2n}-1}, \frac{5}{x^{2n}+2x^n+1}$

25. $\frac{x-5}{x^{2n}+3x^n+2}, \frac{2x}{x^n+2}$

Objective B **To add or subtract rational expressions**

For Exercises 26 to 81, simplify.

26. $\frac{3}{2xy} - \frac{7}{2xy} - \frac{9}{2xy}$

27. $-\frac{3}{4x^2} + \frac{8}{4x^2} - \frac{3}{4x^2}$

28. $\frac{x}{x^2-3x+2} - \frac{2}{x^2-3x+2}$

29. $\frac{3x}{3x^2+x-10} - \frac{5}{3x^2+x-10}$

30. $\frac{3}{2x^2y} - \frac{8}{5x} - \frac{9}{10xy}$

31. $\frac{2}{5ab} - \frac{3}{10a^2b} + \frac{4}{15ab^2}$

32. $\frac{2}{3x} - \frac{3}{2xy} + \frac{4}{5xy} - \frac{5}{6x}$

33. $\frac{3}{4ab} - \frac{2}{5a} + \frac{3}{10b} - \frac{5}{8ab}$

34. $\frac{2x-1}{12x} - \frac{3x+4}{9x}$

35. $\frac{3x-4}{6x} - \frac{2x-5}{4x}$

36. $\frac{3x+2}{4x^2y} - \frac{y-5}{6xy^2}$

37. $\frac{2y-4}{5xy^2} + \frac{3-2x}{10x^2y}$

38. $\frac{2x}{x-3} - \frac{3x}{x-5}$

39. $\frac{3a}{a-2} - \frac{5a}{a+1}$

40. $\frac{3}{2a-3} + \frac{2a}{3-2a}$

41. $\frac{x}{2x-5} - \frac{2}{5x-2}$

42. $\frac{1}{x+h} - \frac{1}{h}$

43. $\frac{1}{a-b} + \frac{1}{b}$

44. $\frac{2}{x} - 3 - \frac{10}{x-4}$

45. $\frac{6a}{a-3} - 5 + \frac{3}{a}$

46. $\frac{1}{2x-3} - \frac{5}{2x} + 1$

Copyright © Houghton Mifflin Company. All rights reserved.

47. $\frac{5}{x} - \frac{5x}{5 - 6x} + 2$

48. $\frac{3}{x^2 - 1} + \frac{2x}{x^2 + 2x + 1}$

49. $\frac{1}{x^2 - 6x + 9} - \frac{1}{x^2 - 9}$

50. $\frac{x}{x + 3} - \frac{3 - x}{x^2 - 9}$

51. $\frac{1}{x + 2} - \frac{3x}{x^2 + 4x + 4}$

52. $\frac{2x - 3}{x + 5} - \frac{x^2 - 4x - 19}{x^2 + 8x + 15}$

53. $\frac{-3x^2 + 8x + 2}{x^2 + 2x - 8} - \frac{2x - 5}{x + 4}$

54. $\frac{x^n}{x^{2n} - 1} - \frac{2}{x^n + 1}$

55. $\frac{2}{x^n - 1} + \frac{x^n}{x^{2n} - 1}$

56. $\frac{2}{x^n - 1} - \frac{6}{x^{2n} + x^n - 2}$

57. $\frac{2x^n - 6}{x^{2n} - x^n - 6} + \frac{x^n}{x^n + 2}$

58. $\frac{2x - 2}{4x^2 - 9} - \frac{5}{3 - 2x}$

59. $\frac{x^2 + 4}{4x^2 - 36} - \frac{13}{x + 3}$

60. $\frac{x - 2}{x + 1} - \frac{3 - 12x}{2x^2 - x - 3}$

61. $\frac{3x - 4}{4x + 1} + \frac{3x + 6}{4x^2 + 9x + 2}$

62. $\frac{x + 1}{x^2 + x - 6} - \frac{x + 2}{x^2 + 4x + 3}$

63. $\frac{x + 1}{x^2 + x - 12} - \frac{x - 3}{x^2 + 7x + 12}$

64. $\frac{x^2 + 6x}{x^2 + 3x - 18} - \frac{2x - 1}{x + 6} + \frac{x - 2}{3 - x}$

65. $\frac{2x^2 - 2x}{x^2 - 2x - 15} - \frac{2}{x + 3} + \frac{x}{5 - x}$

66. $\frac{7 - 4x}{2x^2 - 9x + 10} + \frac{x - 3}{x - 2} - \frac{x + 1}{2x - 5}$

67. $\frac{x}{3x + 4} + \frac{3x + 2}{x - 5} - \frac{7x^2 + 24x + 28}{3x^2 - 11x - 20}$

68. $\frac{32x - 9}{2x^2 + 7x - 15} + \frac{x - 2}{3 - 2x} + \frac{3x + 2}{x + 5}$

69. $\frac{x + 1}{1 - 2x} - \frac{x + 3}{4x - 3} + \frac{10x^2 + 7x - 9}{8x^2 - 10x + 3}$

Copyright © Houghton Mifflin Company. All rights reserved.

70. $\dfrac{x^2}{x^3 - 8} - \dfrac{x + 2}{x^2 + 2x + 4}$

71. $\dfrac{2x}{4x^2 + 2x + 1} + \dfrac{4x + 1}{8x^3 - 1}$

72. $\dfrac{2x^2}{x^4 - 1} - \dfrac{1}{x^2 - 1} + \dfrac{1}{x^2 + 1}$

73. $\dfrac{x^2 - 12}{x^4 - 16} + \dfrac{1}{x^2 - 4} - \dfrac{1}{x^2 + 4}$

74. $\left(\dfrac{x + 8}{4} + \dfrac{4}{x}\right) \div \dfrac{x + 4}{16x^2}$

75. $\left(\dfrac{a - 3}{a^2} - \dfrac{a - 3}{9}\right) \div \dfrac{a^2 - 9}{3a}$

76. $\dfrac{3}{x - 2} - \dfrac{x^2 + x}{2x^3 + 3x^2} \cdot \dfrac{2x^2 + x - 3}{x^2 + 3x + 2}$

77. $\dfrac{x^2 - 4x + 4}{2x + 1} \cdot \dfrac{2x^2 + x}{x^3 - 4x} - \dfrac{3x - 2}{x + 1}$

78. $\left(\dfrac{x - y}{x^2} - \dfrac{x - y}{y^2}\right) \div \dfrac{x^2 - y^2}{xy}$

79. $\left(\dfrac{a - 2b}{b} + \dfrac{b}{a}\right) \div \left(\dfrac{b + a}{a} - \dfrac{2a}{b}\right)$

80. $\dfrac{2}{x - 3} - \dfrac{x}{x^2 - x - 6} \cdot \dfrac{x^2 - 2x - 3}{x^2 - x}$

81. $\dfrac{2x}{x^2 - x - 6} - \dfrac{6x - 6}{2x^2 - 9x + 9} \div \dfrac{x^2 + x - 2}{2x - 3}$

APPLYING THE CONCEPTS

82. Correct the following expressions.

a. $\dfrac{3}{4} + \dfrac{x}{5} = \dfrac{3 + x}{4 + 5}$ **b.** $\dfrac{4x + 5}{4} = x + 5$ **c.** $\dfrac{1}{x} + \dfrac{1}{y} = \dfrac{1}{x + y}$

83. Simplify.

a. $\left(\dfrac{b}{6} - \dfrac{6}{b}\right) \div \left(\dfrac{6}{b} - 4 + \dfrac{b}{2}\right)$ **b.** $\left(\dfrac{x + 1}{2x - 1} - \dfrac{x - 1}{2x + 1}\right) \cdot \left(\dfrac{2x - 1}{x} - \dfrac{2x - 1}{x^2}\right)$

84. Rewrite each fraction as the sum of two fractions in simplest form.

a. $\dfrac{3x + 6y}{xy}$ **b.** $\dfrac{4a^2 + 3ab}{a^2b^2}$ **c.** $\dfrac{3m^2n + 2mn^2}{12m^3n^2}$

85. Let $f(x) = \dfrac{x}{x + 2}$, $g(x) = \dfrac{4}{x - 3}$, and $S(x) = \dfrac{x^2 + x + 8}{x^2 - x - 6}$. **a.** Evaluate $f(4)$, $g(4)$, and $S(4)$. Does $f(4) + g(4) = S(4)$? **b.** Let a be a real number ($a \neq -2, a \neq 3$). Express $S(a)$ in terms of $f(a)$ and $g(a)$.

Copyright © Houghton Mifflin Company. All rights reserved.

8.3 Complex Fractions

Objective A To simplify a complex fraction

A **complex fraction** is a fraction whose numerator or denominator contains one or more fractions. Examples of complex fractions are shown below.

$$\frac{5}{2+\frac{1}{2}}, \quad \frac{5+\frac{1}{y}}{5-\frac{1}{y}}, \quad \frac{x+4+\frac{1}{x+2}}{x-2+\frac{1}{x+2}}$$

HOW TO Simplify: $\dfrac{\frac{1}{x}+\frac{1}{y}}{\frac{1}{x}-\frac{1}{y}}$

The LCM of x and y is xy.

$$\frac{\frac{1}{x}+\frac{1}{y}}{\frac{1}{x}-\frac{1}{y}} = \frac{\frac{1}{x}+\frac{1}{y}}{\frac{1}{x}-\frac{1}{y}} \cdot \frac{xy}{xy}$$

$$= \frac{\frac{1}{x}\cdot xy + \frac{1}{y}\cdot xy}{\frac{1}{x}\cdot xy - \frac{1}{y}\cdot xy} = \frac{y+x}{y-x}$$

- Find the LCM of the denominators of the fractions in the numerator and denominator.
- Multiply the numerator and denominator of the complex fraction by the LCM.

The method shown above of simplifying a complex fraction by multiplying the numerator and denominator by the LCM of the denominators is used in Examples 1 and 2 on the next page. However, a different approach is to rewrite the numerator and denominator of the complex fraction as single fractions and then divide the numerator by the denominator. The example shown above is simplified below by using this alternative method.

$$\frac{\frac{1}{x}+\frac{1}{y}}{\frac{1}{x}-\frac{1}{y}} = \frac{\frac{1}{x}\cdot\frac{y}{y}+\frac{1}{y}\cdot\frac{x}{x}}{\frac{1}{x}\cdot\frac{y}{y}-\frac{1}{y}\cdot\frac{x}{x}} = \frac{\frac{y}{xy}+\frac{x}{xy}}{\frac{y}{xy}-\frac{x}{xy}} = \frac{\frac{y+x}{xy}}{\frac{y-x}{xy}}$$

$$= \frac{y+x}{xy} \div \frac{y-x}{xy} = \frac{y+x}{xy}\cdot\frac{xy}{y-x}$$

$$= \frac{(y+x)xy}{xy(y-x)} = \frac{y+x}{y-x}$$

- Rewrite the numerator and denominator of the complex fraction as single fractions.
- Divide the numerator of the complex fraction by the denominator.
- Multiply the fractions. Simplify.

Note that this is the same result as shown above.

Copyright © Houghton Mifflin Company. All rights reserved.

Example 1

Simplify: $\dfrac{2x - 1 + \dfrac{7}{x+4}}{3x - 8 + \dfrac{17}{x+4}}$

Solution

The LCM of $x + 4$ and $x + 4$ is $x + 4$.

$$\frac{2x - 1 + \frac{7}{x+4}}{3x - 8 + \frac{17}{x+4}} = \frac{2x - 1 + \frac{7}{x+4}}{3x - 8 + \frac{17}{x+4}} \cdot \frac{x+4}{x+4}$$

$$= \frac{(2x-1)(x+4) + \frac{7}{x+4}(x+4)}{(3x-8)(x+4) + \frac{17}{x+4}(x+4)}$$

$$= \frac{2x^2 + 7x - 4 + 7}{3x^2 + 4x - 32 + 17}$$

$$= \frac{2x^2 + 7x + 3}{3x^2 + 4x - 15}$$

$$= \frac{(2x+1)(x+3)}{(3x-5)(x+3)}$$

$$= \frac{(2x+1)\overset{1}{\cancel{(x+3)}}}{(3x-5)\underset{1}{\cancel{(x+3)}}} = \frac{2x+1}{3x-5}$$

You Try It 1

Simplify: $\dfrac{2x + 5 + \dfrac{14}{x-3}}{4x + 16 + \dfrac{49}{x-3}}$

Your solution

Example 2

Simplify: $1 + \dfrac{a}{2 + \dfrac{1}{a}}$

Solution

The LCM of the denominators is a.

$$1 + \frac{a}{2 + \frac{1}{a}} = 1 + \frac{a}{2 + \frac{1}{a}} \cdot \frac{a}{a}$$

$$= 1 + \frac{a \cdot a}{2 \cdot a + \frac{1}{a} \cdot a} = 1 + \frac{a^2}{2a+1}$$

The LCM of 1 and $2a + 1$ is $2a + 1$.

$$1 + \frac{a^2}{2a+1} = 1 \cdot \frac{2a+1}{2a+1} + \frac{a^2}{2a+1}$$

$$= \frac{2a+1}{2a+1} + \frac{a^2}{2a+1} = \frac{2a + 1 + a^2}{2a+1}$$

$$= \frac{a^2 + 2a + 1}{2a+1} = \frac{(a+1)^2}{2a+1}$$

You Try It 2

Simplify: $2 - \dfrac{1}{2 - \dfrac{1}{x}}$

Your solution

Solutions on p. S25

Copyright © Houghton Mifflin Company. All rights reserved.

8.3 Exercises

Objective A **To simplify a complex fraction**

1. What is a complex fraction?

2. What is the general goal of simplifying a complex fraction?

For Exercises 3 to 46, simplify.

3. $\dfrac{2-\dfrac{1}{3}}{4+\dfrac{11}{3}}$

4. $\dfrac{3+\dfrac{5}{2}}{8-\dfrac{3}{2}}$

5. $\dfrac{3-\dfrac{2}{3}}{5+\dfrac{5}{6}}$

6. $\dfrac{5-\dfrac{3}{4}}{2+\dfrac{1}{2}}$

7. $\dfrac{1+\dfrac{1}{x}}{1-\dfrac{1}{x^2}}$

8. $\dfrac{\dfrac{1}{y^2}-1}{1+\dfrac{1}{y}}$

9. $\dfrac{a-2}{\dfrac{4}{a}-a}$

10. $\dfrac{\dfrac{25}{a}-a}{5+a}$

11. $\dfrac{\dfrac{1}{a^2}-\dfrac{1}{a}}{\dfrac{1}{a^2}+\dfrac{1}{a}}$

12. $\dfrac{\dfrac{1}{b}+\dfrac{1}{2}}{\dfrac{4}{b^2}-1}$

13. $\dfrac{2-\dfrac{4}{x+2}}{5-\dfrac{10}{x+2}}$

14. $\dfrac{4+\dfrac{12}{2x-3}}{5+\dfrac{15}{2x-3}}$

15. $\dfrac{\dfrac{3}{2a-3}+2}{\dfrac{-6}{2a-3}-4}$

16. $\dfrac{\dfrac{-5}{b-5}-3}{\dfrac{10}{b-5}+6}$

17. $\dfrac{\dfrac{x}{x+1}-\dfrac{1}{x}}{\dfrac{x}{x+1}+\dfrac{1}{x}}$

18. $\dfrac{\dfrac{2a}{a-1}-\dfrac{3}{a}}{\dfrac{1}{a-1}+\dfrac{2}{a}}$

19. $\dfrac{1-\dfrac{1}{x}-\dfrac{6}{x^2}}{1-\dfrac{4}{x}+\dfrac{3}{x^2}}$

20. $\dfrac{1-\dfrac{3}{x}-\dfrac{10}{x^2}}{1+\dfrac{11}{x}+\dfrac{18}{x^2}}$

21. $\dfrac{1+\dfrac{1}{x}-\dfrac{12}{x^2}}{\dfrac{9}{x^2}+\dfrac{3}{x}-2}$

22. $\dfrac{\dfrac{15}{x^2}-\dfrac{2}{x}-1}{\dfrac{4}{x^2}-\dfrac{5}{x}+4}$

Copyright © Houghton Mifflin Company. All rights reserved.

23. $a + \dfrac{a}{a + \dfrac{1}{a}}$

24. $4 - \dfrac{2}{2 - \dfrac{3}{x}}$

25. $\dfrac{1 - \dfrac{1}{x - 4}}{1 - \dfrac{6}{x + 1}}$

26. $\dfrac{1 + \dfrac{3}{x + 2}}{1 + \dfrac{6}{x - 1}}$

27. $\dfrac{x - \dfrac{1}{x}}{x + \dfrac{1}{x}}$

28. $\dfrac{a - \dfrac{1}{a}}{\dfrac{1}{a} + a}$

29. $\dfrac{\dfrac{1}{x + h} - \dfrac{1}{x}}{h}$

30. $\dfrac{\dfrac{1}{(x + h)^2} - \dfrac{1}{x^2}}{h}$

31. $\dfrac{1 - \dfrac{2}{x - 3}}{1 + \dfrac{3}{2 - x}}$

32. $\dfrac{1 + \dfrac{x}{x + 1}}{1 + \dfrac{x - 1}{x + 2}}$

33. $\dfrac{x - 4 + \dfrac{9}{2x + 3}}{x + 3 - \dfrac{5}{2x + 3}}$

34. $\dfrac{2x - 3 - \dfrac{10}{4x - 5}}{3x + 2 + \dfrac{11}{4x - 5}}$

35. $\dfrac{3x - 2 - \dfrac{5}{2x - 1}}{x - 6 + \dfrac{9}{2x - 1}}$

36. $\dfrac{x + 4 - \dfrac{7}{2x - 5}}{2x + 7 - \dfrac{28}{2x - 5}}$

37. $\dfrac{\dfrac{1}{a} - \dfrac{3}{a - 2}}{\dfrac{2}{a} + \dfrac{5}{a - 2}}$

38. $\dfrac{\dfrac{2}{b} - \dfrac{5}{b + 3}}{\dfrac{3}{b} + \dfrac{3}{b + 3}}$

39. $\dfrac{\dfrac{1}{y^2} - \dfrac{1}{xy} - \dfrac{2}{x^2}}{\dfrac{1}{y^2} - \dfrac{3}{xy} + \dfrac{2}{x^2}}$

40. $\dfrac{\dfrac{2}{b^2} - \dfrac{5}{ab} - \dfrac{3}{a^2}}{\dfrac{2}{b^2} + \dfrac{7}{ab} + \dfrac{3}{a^2}}$

41. $\dfrac{\dfrac{x - 1}{x + 1} - \dfrac{x + 1}{x - 1}}{\dfrac{x - 1}{x + 1} + \dfrac{x + 1}{x - 1}}$

42. $\dfrac{\dfrac{y}{y + 2} - \dfrac{y}{y - 2}}{\dfrac{y}{y + 2} + \dfrac{y}{y - 2}}$

43. $a - \dfrac{a}{1 - \dfrac{a}{1 - a}}$

44. $3 - \dfrac{3}{3 - \dfrac{3}{3 - x}}$

45. $3 - \dfrac{2}{1 - \dfrac{2}{3 - \dfrac{2}{x}}}$

46. $a + \dfrac{a}{2 + \dfrac{1}{1 - \dfrac{2}{a}}}$

APPLYING THE CONCEPTS

47. **Integers** Find the sum of the reciprocals of three consecutive even integers.

48. If $a = \dfrac{b^2 + 4b + 4}{b^2 - 4}$ and $b = \dfrac{1}{c}$, express a in terms of c.

49. Simplify.

a. $\dfrac{x^{-1}}{x^{-1} + 2^{-1}}$ **b.** $\dfrac{-x^{-1} + x}{x^{-1} - x}$ **c.** $\dfrac{x^{-1} - x^{-2} - 6x^{-3}}{x^{-1} - 4x^{-3}}$

Copyright © Houghton Mifflin Company. All rights reserved.

8.4 Solving Equations Containing Fractions

Objective A To solve an equation containing fractions

Recall that to solve an equation containing fractions, clear denominators by multiplying each side of the equation by the LCM of the denominators. Then solve for the variable.

HOW TO Solve: $\frac{3x-1}{4}+\frac{2}{3}=\frac{7}{6}$

$$\frac{3x-1}{4}+\frac{2}{3}=\frac{7}{6}$$

$$12\left(\frac{3x-1}{4}+\frac{2}{3}\right)=12\cdot\frac{7}{6}$$

- The LCM is 12. To clear denominators, multiply each side of the equation by the LCM.

$$12\left(\frac{3x-1}{4}\right)+12\cdot\frac{2}{3}=12\cdot\frac{7}{6}$$

- Simplify using the Distributive Property and the Properties of Fractions.

$$\frac{\overset{3}{\cancel{12}}}{1}\left(\frac{3x-1}{\underset{1}{\cancel{4}}}\right)+\frac{\overset{4}{\cancel{12}}}{1}\cdot\frac{2}{\underset{1}{\cancel{3}}}=\frac{\overset{2}{\cancel{12}}}{1}\cdot\frac{7}{\underset{1}{\cancel{6}}}$$

$$9x-3+8=14$$

- Solve for x.

$$9x+5=14$$

$$9x=9$$

$$x=1$$

1 checks as a solution. The solution is 1.

Occasionally, a value of the variable that appears to be a solution of an equation will make one of the denominators zero. In this case, that value is not a solution of the equation.

HOW TO Solve: $\frac{2x}{x-2}=1+\frac{4}{x-2}$

$$\frac{2x}{x-2}=1+\frac{4}{x-2}$$

$$(x-2)\frac{2x}{x-2}=(x-2)\left(1+\frac{4}{x-2}\right)$$

- The LCM is $x-2$. Multiply each side of the equation by the LCM.

$$(x-2)\frac{2x}{x-2}=(x-2)\cdot 1+(x-2)\frac{4}{x-2}$$

- Simplify using the Distributive Property and the Properties of Fractions.

$$\frac{\overset{1}{\cancel{(x-2)}}}{1}\cdot\frac{2x}{\underset{1}{\cancel{x-2}}}=(x-2)\cdot 1+\frac{\overset{1}{\cancel{(x-2)}}}{1}\cdot\frac{4}{\underset{1}{\cancel{x-2}}}$$

$$2x=x-2+4$$

- Solve for x.

$$2x=x+2$$

$$x=2$$

When x is replaced by 2, the denominators of $\frac{2x}{x-2}$ and $\frac{4}{x-2}$ are zero. Therefore, the equation has no solution.

Copyright © Houghton Mifflin Company. All rights reserved.

Example 1

Solve: $\frac{x}{x+4} = \frac{2}{x}$

Solution

The LCM is $x(x+4)$.

$$\frac{x}{x+4} = \frac{2}{x}$$

$$x(x+4)\left(\frac{x}{x+4}\right) = x(x+4)\left(\frac{2}{x}\right)$$ • Multiply by the LCM.

$$\frac{\overset{1}{\cancel{x(x+4)}}}{1} \cdot \frac{x}{\underset{1}{\cancel{x+4}}} = \frac{\overset{1}{\cancel{x}}(x+4)}{1} \cdot \frac{2}{\underset{1}{\cancel{x}}}$$ • Divide by the common factors.

$$x^2 = (x+4)2$$ • Simplify.

$$x^2 = 2x + 8$$

Solve the quadratic equation by factoring.

$$x^2 - 2x - 8 = 0$$ • Write in standard form.

$$(x-4)(x+2) = 0$$ • Factor.

$$x - 4 = 0 \qquad x + 2 = 0$$ • Principle of Zero Products

$$x = 4 \qquad x = -2$$

Both 4 and -2 check as solutions.

The solutions are 4 and -2.

You Try It 1

Solve: $\frac{x}{x+6} = \frac{3}{x}$

Your solution

Example 2

Solve: $\frac{3x}{x-4} = 5 + \frac{12}{x-4}$

Solution

The LCM is $x - 4$.

$$\frac{3x}{x-4} = 5 + \frac{12}{x-4}$$

$$(x-4)\left(\frac{3x}{x-4}\right) = (x-4)\left(5 + \frac{12}{x-4}\right)$$ • Clear denominators.

$$\frac{\overset{1}{\cancel{(x-4)}}}{1} \cdot \frac{3x}{\underset{1}{\cancel{x-4}}} = (x-4)5 + \frac{\overset{1}{\cancel{(x-4)}}}{1} \cdot \frac{12}{\underset{1}{\cancel{x-4}}}$$

$$3x = (x-4)5 + 12$$ • Solve for x.

$$3x = 5x - 20 + 12$$

$$3x = 5x - 8$$

$$-2x = -8$$

$$x = 4$$

4 does not check as a solution.

The equation has no solution.

You Try It 2

Solve: $\frac{5x}{x+2} = 3 - \frac{10}{x+2}$

Your solution

Solutions on p. S25

Copyright © Houghton Mifflin Company. All rights reserved.

8.4 Exercises

Objective A **To solve an equation containing fractions**

1. Can 2 be a solution of the equation $\frac{6x}{x+1} - \frac{x}{x-2} = 4$? Explain your answer.

2. After multiplying each side of an equation by a variable expression, why must we check the solution?

For Exercises 3 to 35, solve.

3. $\frac{2x}{3} - \frac{5}{2} = -\frac{1}{2}$

4. $\frac{x}{3} - \frac{1}{4} = \frac{1}{12}$

5. $\frac{x}{3} - \frac{1}{4} = \frac{x}{4} - \frac{1}{6}$

6. $\frac{2y}{9} - \frac{1}{6} = \frac{y}{9} + \frac{1}{6}$

7. $\frac{2x-5}{8} + \frac{1}{4} = \frac{x}{8} + \frac{3}{4}$

8. $\frac{3x+4}{12} - \frac{1}{3} = \frac{5x+2}{12} - \frac{1}{2}$

9. $\frac{6}{2a+1} = 2$

10. $\frac{12}{3x-2} = 3$

11. $\frac{9}{2x-5} = -2$

12. $\frac{6}{4-3x} = 3$

13. $2 + \frac{5}{x} = 7$

14. $3 + \frac{8}{n} = 5$

15. $1 - \frac{9}{x} = 4$

16. $3 - \frac{12}{x} = 7$

17. $\frac{2}{y} + 5 = 9$

18. $\frac{6}{x} + 3 = 11$

19. $\frac{3}{x-2} = \frac{4}{x}$

20. $\frac{5}{x+3} = \frac{3}{x-1}$

21. $\frac{2}{3x-1} = \frac{3}{4x+1}$

22. $\frac{5}{3x-4} = \frac{-3}{1-2x}$

23. $\frac{-3}{2x+5} = \frac{2}{x-1}$

Copyright © Houghton Mifflin Company. All rights reserved.

24. $\frac{4}{5y-1} = \frac{2}{2y-1}$

25. $\frac{4x}{x-4} + 5 = \frac{5x}{x-4}$

26. $\frac{2x}{x+2} - 5 = \frac{7x}{x+2}$

27. $2 + \frac{3}{a-3} = \frac{a}{a-3}$

28. $\frac{x}{x+4} = 3 - \frac{4}{x+4}$

29. $\frac{x}{x-1} = \frac{8}{x+2}$

30. $\frac{x}{x+12} = \frac{1}{x+5}$

31. $\frac{2x}{x+4} = \frac{3}{x-1}$

32. $\frac{5}{3n-8} = \frac{n}{n+2}$

33. $x + \frac{6}{x-2} = \frac{3x}{x-2}$

34. $x - \frac{6}{x-3} = \frac{2x}{x-3}$

35. $\frac{8}{y} = \frac{2}{y-2} + 1$

APPLYING THE CONCEPTS

36. Explain the procedure for solving an equation containing fractions. Include in your discussion how the LCM of the denominators is used to eliminate fractions in the equation.

For Exercises 37 to 42, solve.

37. $\frac{3}{5}y - \frac{1}{3}(1 - y) = \frac{2y - 5}{15}$

38. $\frac{3}{4}a = \frac{1}{2}(3 - a) + \frac{a - 2}{4}$

39. $\frac{b+2}{5} = \frac{1}{4}b - \frac{3}{10}(b - 1)$

40. $\frac{x}{2x^2 - x - 1} = \frac{3}{x^2 - 1} + \frac{3}{2x + 1}$

41. $\frac{x+1}{x^2 + x - 2} = \frac{x+2}{x^2 - 1} + \frac{3}{x + 2}$

42. $\frac{y+2}{y^2 - y - 2} + \frac{y+1}{y^2 - 4} = \frac{1}{y + 1}$

Copyright © Houghton Mifflin Company. All rights reserved.

8.5 Ratio and Proportion

Objective A To solve a proportion

Quantities such as 4 meters, 15 seconds, and 8 gallons are number quantities written with units. In these examples the units are meters, seconds, and gallons.

A **ratio** is the quotient of two quantities that have the same unit.

The length of a living room is 16 ft and the width is 12 ft. The ratio of the length to the width is written

$\frac{16 \text{ ft}}{12 \text{ ft}} = \frac{16}{12} = \frac{4}{3}$ A ratio is in simplest form when the two numbers do not have a common factor. Note that the units are not written.

A **rate** is the quotient of two quantities that have different units.

There are 2 lb of salt in 8 gal of water. The salt-to-water rate is

$\frac{2 \text{ lb}}{8 \text{ gal}} = \frac{1 \text{ lb}}{4 \text{ gal}}$ A rate is in simplest form when the two numbers do not have a common factor. The units are written as part of the rate.

A **proportion** is an equation that states the equality of two ratios or rates. Examples of proportions are shown at the right.

$\frac{30 \text{ mi}}{4 \text{ h}} = \frac{15 \text{ mi}}{2 \text{ h}}$ $\frac{4}{6} = \frac{8}{12}$ $\frac{3}{4} = \frac{x}{8}$

Study Tip

Always check the solution of an equation. For the equation at the right:

$$\frac{4}{x} = \frac{2}{3}$$
$$\frac{4}{6} \,\Big|\, \frac{2}{3}$$
$$\frac{2}{3} = \frac{2}{3}$$

The solution checks.

HOW TO Solve the proportion $\frac{4}{x} = \frac{2}{3}$.

$$\frac{4}{x} = \frac{2}{3}$$
$$3x\left(\frac{4}{x}\right) = 3x\left(\frac{2}{3}\right)$$
• The LCM of the denominators is $3x$. To clear denominators, multiply each side of the proportion by the LCM.
$$12 = 2x$$
• Solve the equation.
$$6 = x$$

The solution is 6.

Example 1 Solve: $\frac{8}{x+3} = \frac{4}{x}$

Solution

$$\frac{8}{x+3} = \frac{4}{x}$$
$$x(x+3)\frac{8}{x+3} = x(x+3)\frac{4}{x}$$
• Clear denominators.
$$8x = 4(x+3)$$
• Solve for x.
$$8x = 4x + 12$$
$$4x = 12$$
$$x = 3$$

The solution is 3.

You Try It 1 Solve: $\frac{2}{x+3} = \frac{6}{5x+5}$

Your solution

Solution on p. S25

Copyright © Houghton Mifflin Company. All rights reserved.

Objective B To solve application problems

Example 2

The monthly loan payment for a car is \$28.35 for each \$1000 borrowed. At this rate, find the monthly payment for a \$6000 car loan.

Strategy

To find the monthly payment, write and solve a proportion, using P to represent the monthly car payment.

Solution

$$\frac{28.35}{1000} = \frac{P}{6000}$$ • Write a proportion.

$$6000\left(\frac{28.35}{1000}\right) = 6000\left(\frac{P}{6000}\right)$$ • Clear denominators.

$$170.10 = P$$

The monthly payment is \$170.10.

You Try It 2

Sixteen ceramic tiles are needed to tile a 9-square-foot area. At this rate, how many square feet can be tiled using 256 ceramic tiles?

Your strategy

Your solution

Solution on p. S25

Objective C To solve problems involving similar triangles

Similar objects have the same shape but not necessarily the same size. A tennis ball is similar to a basketball. A model ship is similar to an actual ship.

Similar objects have corresponding parts; for example, the rudder on the model ship corresponds to the rudder on the actual ship. The relationship between the sizes of the corresponding parts can be written as a ratio, and each ratio will be the same. If the rudder on the model ship is $\frac{1}{100}$ the size of the rudder on the actual ship, then the model wheelhouse is $\frac{1}{100}$ the size of the actual wheelhouse, the width of the model is $\frac{1}{100}$ the width of the actual ship, and so on.

The two triangles ABC and DEF shown at the right are similar. Side AB corresponds to DE, side BC corresponds to EF, and side AC corresponds to DF. The height CH corresponds to the height FK. The ratios of corresponding parts of similar triangles are equal.

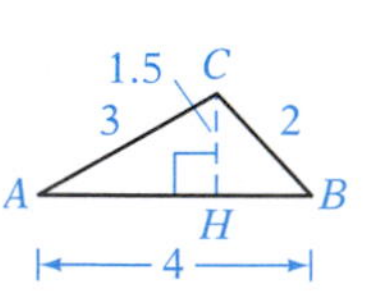

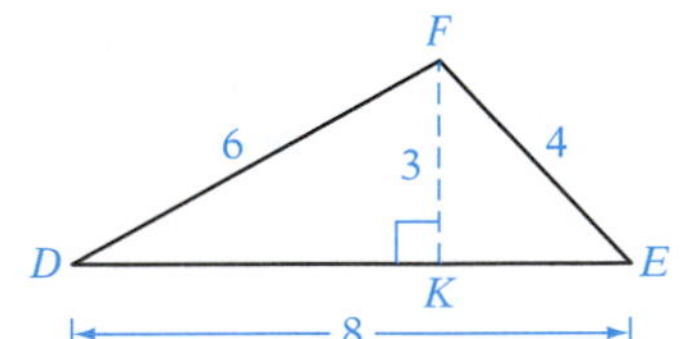

$$\frac{AB}{DE} = \frac{4}{8} = \frac{1}{2}, \quad \frac{AC}{DF} = \frac{3}{6} = \frac{1}{2}, \quad \frac{BC}{EF} = \frac{2}{4} = \frac{1}{2}, \quad \text{and} \quad \frac{CH}{FK} = \frac{1.5}{3} = \frac{1}{2}$$

Because the ratios of corresponding parts are equal, three proportions can be formed using the sides of the triangles.

$$\frac{AB}{DE} = \frac{AC}{DF}, \quad \frac{AB}{DE} = \frac{BC}{EF}, \quad \text{and} \quad \frac{AC}{DF} = \frac{BC}{EF}$$

Copyright © Houghton Mifflin Company. All rights reserved.

Three proportions can also be formed by using the sides and height of the triangles.

$$\frac{AB}{DE} = \frac{CH}{FK}, \quad \frac{AC}{DF} = \frac{CH}{FK}, \quad \text{and} \quad \frac{BC}{EF} = \frac{CH}{FK}$$

The measures of the corresponding angles in similar triangles are equal. Therefore,

$$m\angle A = m\angle D, \quad m\angle B = m\angle E, \quad \text{and} \quad m\angle C = m\angle F$$

It is also true that if the measures of the three angles of one triangle are equal, respectively, to the measures of the three angles of another triangle, then the two triangles are similar.

TAKE NOTE

Vertical angles of intersecting lines, corresponding angles of parallel lines, and angles of a triangle are discussed in Section 3.1.

A line DE is drawn parallel to the base AB in the triangle at the right. $m\angle x = m\angle m$ and $m\angle y = m\angle n$ because corresponding angles are equal. $m\angle C = m\angle C$; thus the measures of the three angles of triangle DEC are equal, respectively, to the measures of the three angles of triangle ABC. Triangle DEC is similar to triangle ABC.

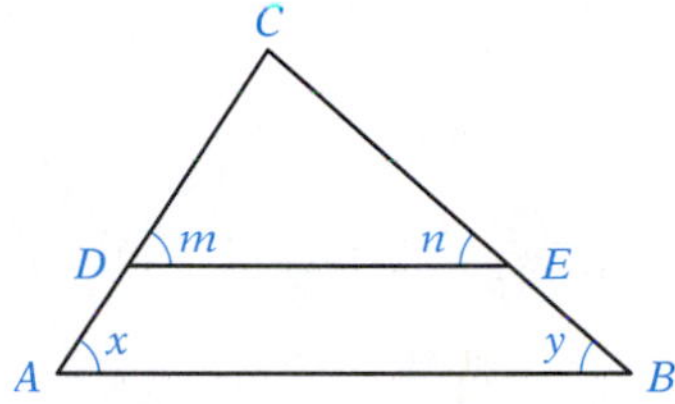

The sum of the measures of the three angles of a triangle is 180°. If two angles of one triangle are equal in measure to two angles of another triangle, then the third angles must be equal in measure. Thus we can say that if two angles of one triangle are equal in measure to two angles of another triangle, then the two triangles are similar.

HOW TO The line segments AB and CD intersect at point O in the figure at the right. Angles C and D are right angles. Find the length of DO.

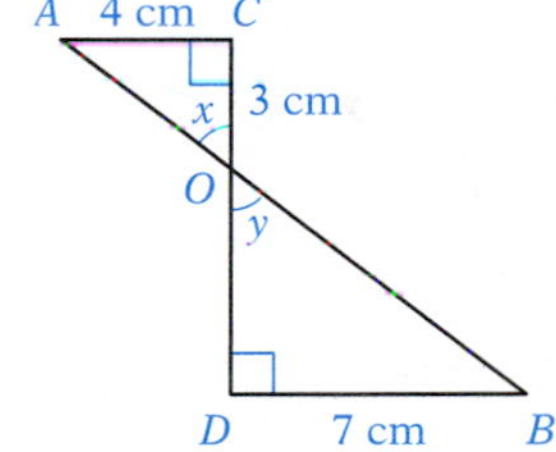

First we must determine whether triangle AOC is similar to triangle BOD.

$m\angle C = m\angle D$ because they are right angles.

$m\angle x = m\angle y$ because they are vertical angles.

Triangle AOC is similar to triangle BOD because two angles of one triangle are equal in measure to two angles of the other triangle.

$$\frac{AC}{DB} = \frac{CO}{DO}$$
• Use a proportion to find the length of the unknown side.

$$\frac{4}{7} = \frac{3}{DO}$$
• $AC = 4$, $CO = 3$, and $DB = 7$.

$$7(DO)\frac{4}{7} = 7(DO)\frac{3}{DO}$$
• To clear denominators, multiply each side of the proportion by $7(DO)$.

$$4(DO) = 7(3)$$
• Solve for DO.

$$4(DO) = 21$$

$$DO = 5.25 \text{ cm}$$

Copyright © Houghton Mifflin Company. All rights reserved.

HOW TO Triangles ABC and DEF at the right are similar. Find the area of triangle ABC.

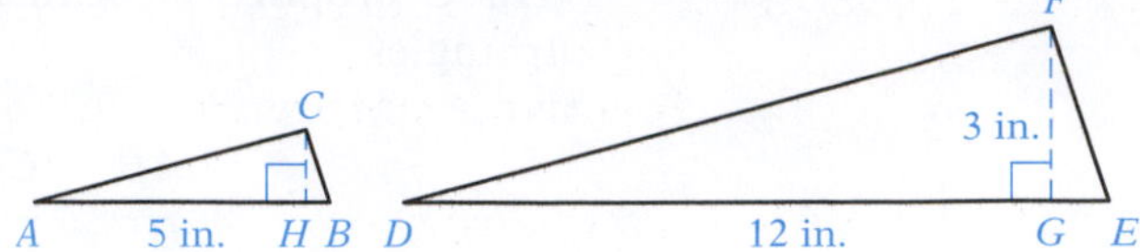

$$\frac{AB}{DE} = \frac{CH}{FG}$$
• Solve a proportion to find the height of triangle ABC.

$$\frac{5}{12} = \frac{CH}{3}$$
• $AB = 5$, $DE = 12$, and $FG = 3$.

$$12 \cdot \frac{5}{12} = 12 \cdot \frac{CH}{3}$$
• To clear denominators, multiply each side of the proportion by 12.

$$5 = 4(CH)$$
• Solve for CH.

$$1.25 = CH$$
• The height is 1.25 in. The base is 5 in.

$$A = \frac{1}{2}bh = \frac{1}{2}(5)(1.25) = 3.125$$
• Use the formula for the area of a triangle.

The area of triangle ABC is 3.125 in^2.

Example 3

In the figure below, AB is parallel to DC, and angles B and D are right angles. $AB = 12$ m, $DC = 4$ m, and $AC = 18$ m. Find the length of CO.

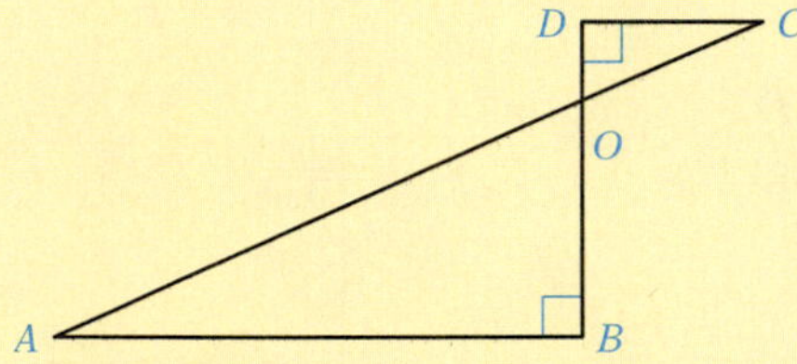

Strategy

Triangle AOB is similar to triangle COD. Solve a proportion to find the length of CO. Let x represent the length of CO and $18 - x$ represent the length of AO.

Solution

$$\frac{DC}{AB} = \frac{CO}{AO}$$
• Write a proportion.

$$\frac{4}{12} = \frac{x}{18 - x}$$
• Substitute.

$$12(18 - x) \cdot \frac{4}{12} = 12(18 - x) \cdot \frac{x}{18 - x}$$
• Clear denominators.

$$4(18 - x) = 12x$$
• Solve for x.

$$72 - 4x = 12x$$
$$72 = 16x$$
$$4.5 = x$$
• x is the length of CO.

The length of CO is 4.5 m.

You Try It 3

In the figure below, AB is parallel to DC, and angles A and D are right angles. $AB = 10$ cm, $CD = 4$ cm, and $DO = 3$ cm. Find the area of triangle AOB.

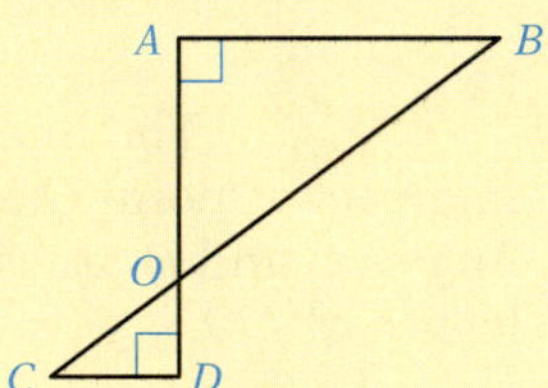

Your strategy

Your solution

Solution on p. S26

Copyright © Houghton Mifflin Company. All rights reserved.

8.5 Exercises

Objective A To solve a proportion

1. What is a proportion?

2. Explain a method for solving a proportion.

For Exercises 3 to 17, solve.

3. $\frac{x}{12} = \frac{3}{4}$

4. $\frac{6}{x} = \frac{2}{3}$

5. $\frac{4}{9} = \frac{x}{27}$

6. $\frac{16}{9} = \frac{64}{x}$

7. $\frac{x+3}{12} = \frac{5}{6}$

8. $\frac{3}{5} = \frac{x-4}{10}$

9. $\frac{18}{x+4} = \frac{9}{5}$

10. $\frac{2}{11} = \frac{20}{x-3}$

11. $\frac{2}{x} = \frac{4}{x+1}$

12. $\frac{16}{x-2} = \frac{8}{x}$

13. $\frac{x+3}{4} = \frac{x}{8}$

14. $\frac{x-6}{3} = \frac{x}{5}$

15. $\frac{2}{x-1} = \frac{6}{2x+1}$

16. $\frac{9}{x+2} = \frac{3}{x-2}$

17. $\frac{2x}{7} = \frac{x-2}{14}$

Objective B To solve application problems

18. **Cooking** Simple syrup used in making some desserts requires 2 c of sugar for every $\frac{2}{3}$ c of boiling water. At this rate, how many cups of sugar are required for 2 c of boiling water?

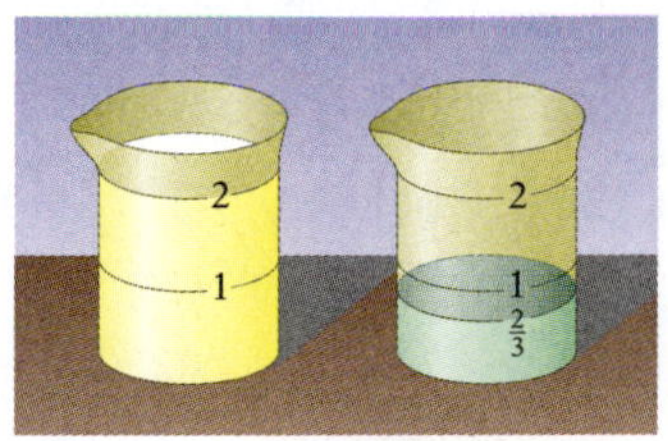

19. **Surveys** An exit poll survey showed that 4 out of every 7 voters cast a ballot in favor of an amendment to a city charter. At this rate, how many voters voted in favor of the amendment if 35,000 people voted?

20. **Surveys** In a city of 25,000 homes, a survey was taken to determine the number with cable television. Of the 300 homes surveyed, 210 had cable television. Estimate the number of homes in the city that have cable television.

21. **Cartography** On a map, two cities are $2\frac{5}{8}$ in. apart. If $\frac{3}{8}$ in. on the map represents 25 mi, find the number of miles between the two cities.

Copyright © Houghton Mifflin Company. All rights reserved.

22. **Business** A company decides to accept a large shipment of 10,000 computer chips if there are 2 or fewer defects in a sample of 100 randomly chosen chips. Assuming that there are 300 defective chips in the shipment and that the rate of defective chips in the sample is the same as the rate in the shipment, will the shipment be accepted?

23. **Taxes** The sales tax on a car that sold for \$12,000 is \$780. At this rate, how much higher is the sales tax on a car that sells for \$13,500?

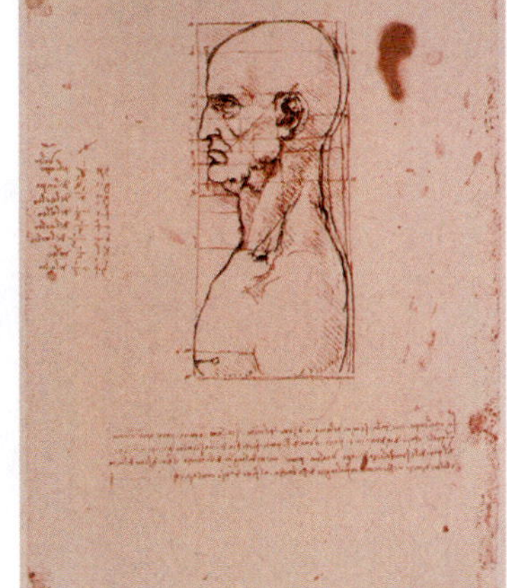

24. **Art** Leonardo da Vinci measured various distances on the human body in order to make accurate drawings. He determined that generally, the ratio of the kneeling height of a person to the standing height of that person was $\frac{3}{4}$. Using this ratio, determine how tall a person is who has a kneeling height of 48 in.

25. **Art** In one of Leonardo da Vinci's notebooks, he wrote that "...from the top to the bottom of the chin is the sixth part of a face, and it is the fifty-fourth part of the man." Suppose the distance from the top to the bottom of the chin of a person is 1.25 in. Using da Vinci's measurements, find the height of this person.

26. **Conservation** As part of a conservation effort for a lake, 40 fish are caught, tagged, and then released. Later 80 fish are caught. Four of the 80 fish are found to have tags. Estimate the number of fish in the lake.

27. **Conservation** In a wildlife preserve, 10 elk are captured, tagged, and then released. Later 15 elk are captured and 2 are found to have tags. Estimate the number of elk in the preserve.

28. **Rocketry** The engine of a small rocket burns 170,000 lb of fuel in 1 min. At this rate, how many pounds of fuel does the rocket burn in 45 s?

Objective C To solve problems involving similar triangles

Triangles *ABC* and *DEF* in Exercises 29 to 36 are similar. Round answers to the nearest tenth.

29. Find side *AC*.

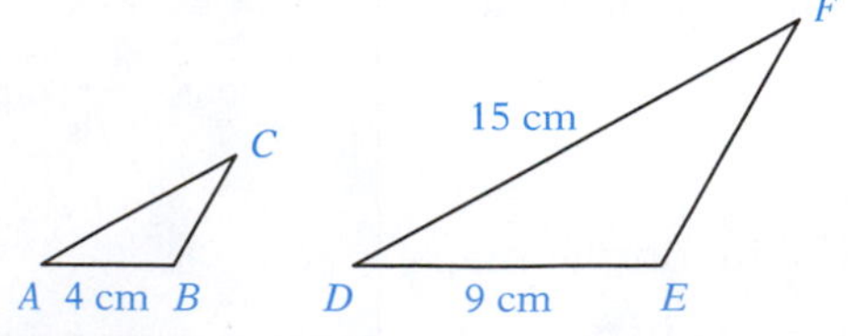

30. Find side *DE*.

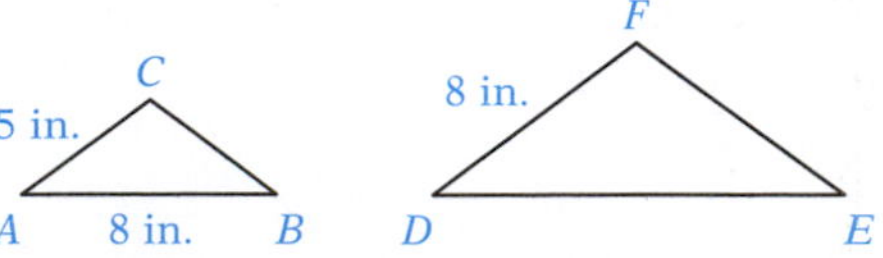

Copyright © Houghton Mifflin Company. All rights reserved.

31. Find the height of triangle *ABC*.

32. Find the height of triangle *DEF*.

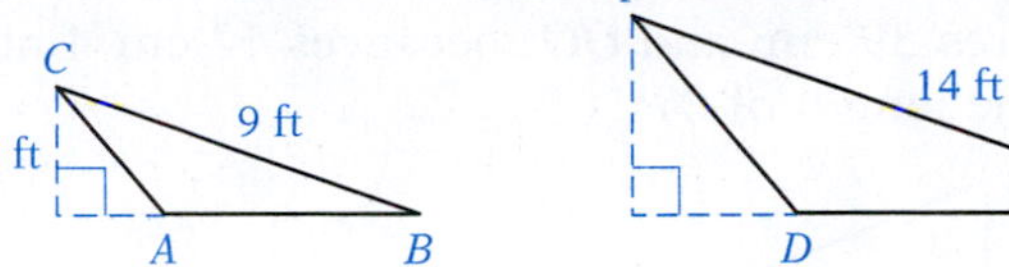

33. Find the perimeter of triangle *DEF*.

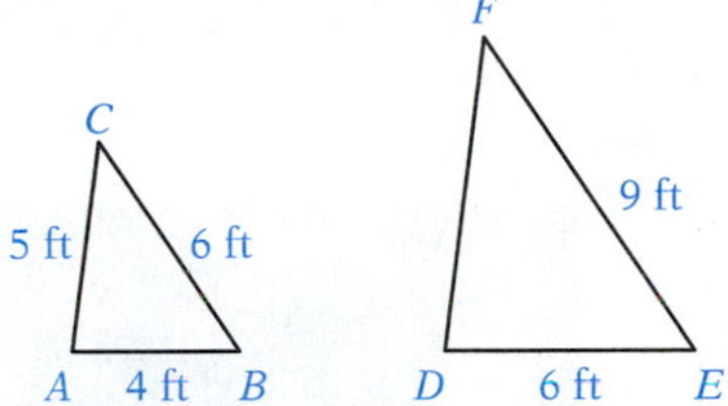

34. Find the perimeter of triangle *ABC*.

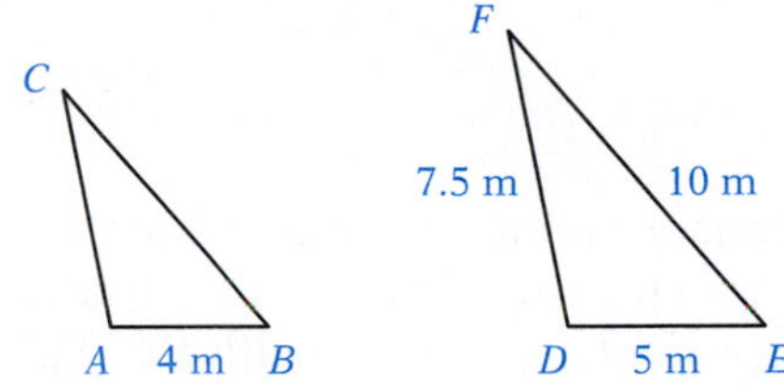

35. Find the area of triangle *ABC*.

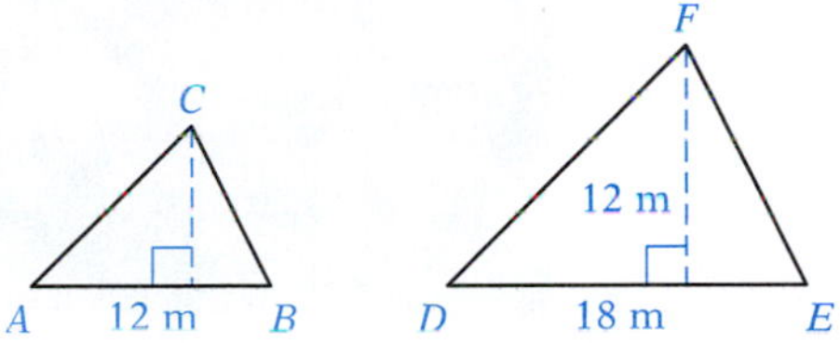

36. Find the area of triangle *ABC*.

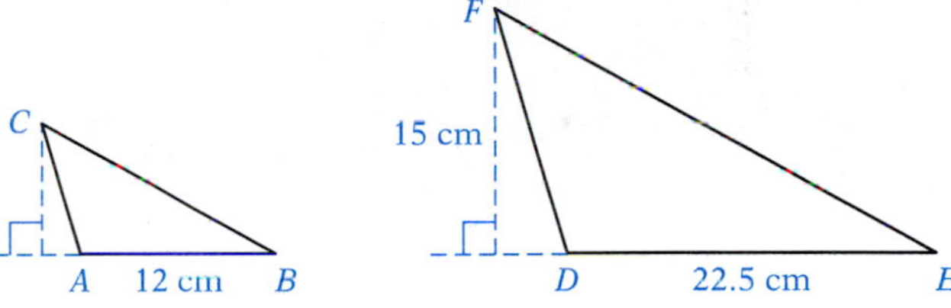

37. Given $BD \parallel AE$, *BD* measures 5 cm, *AE* measures 8 cm, and *AC* measures 10 cm, find the length of *BC*.

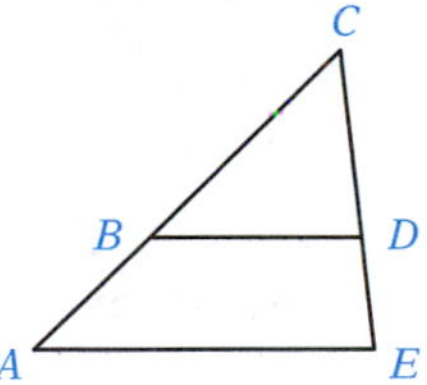

38. Given $AC \parallel DE$, *BD* measures 8 m, *AD* measures 12 m, and *BE* measures 6 m, find the length of *BC*.

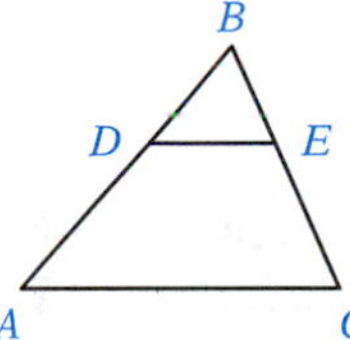

39. Given $DE \parallel AC$, *DE* measures 6 in., *AC* measures 10 in., and *AB* measures 15 in., find the length of *DA*.

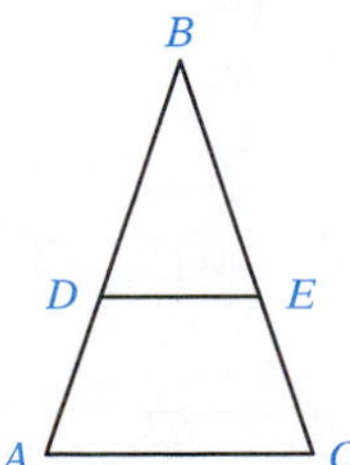

40. Given *MP* and *NQ* intersect at *O*, *NO* measures 25 ft, *MO* measures 20 ft, and *PO* measures 8 ft, find the length of *QO*.

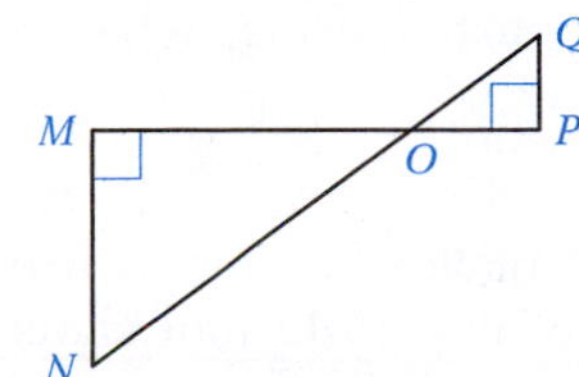

Copyright © Houghton Mifflin Company. All rights reserved.

41. Given MP and NQ intersect at O, NO measures 24 cm, MN measures 10 cm, MP measures 39 cm, and QO measures 12 cm, find the length of OP.

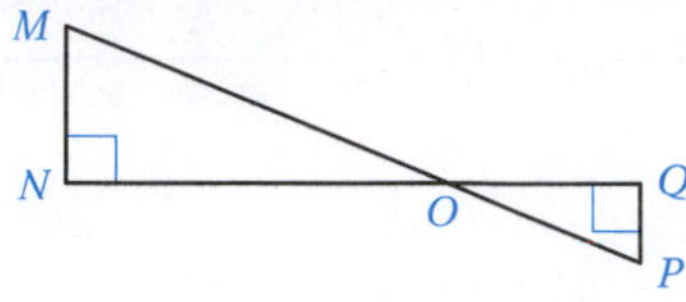

42. Given MQ and NP intersect at O, NO measures 12 m, MN measures 9 m, PQ measures 3 m, and MQ measures 20 m, find the perimeter of triangle OPQ.

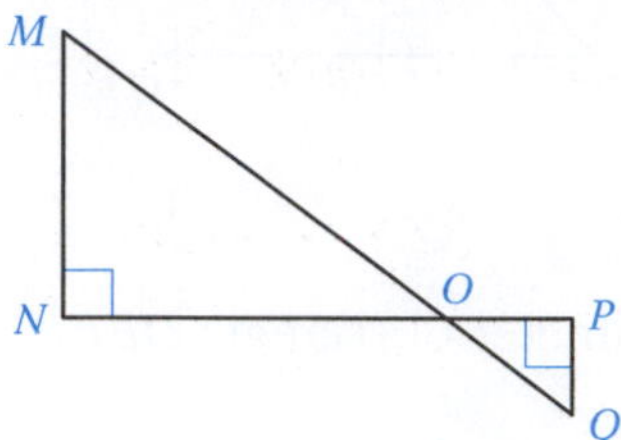

43. **Indirect Measurement** Similar triangles can be used as an indirect way to measure inaccessible distances. The diagram at the right represents a river of width DC. The triangles AOB and DOC are similar. The distances AB, BO, and OC can be measured. Find the width of the river.

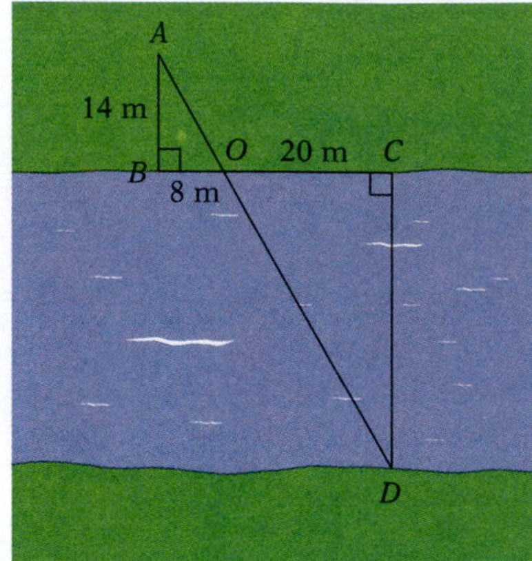

44. **Indirect Measurement** The sun's rays cast a shadow as shown in the diagram at the right. Find the height of the flagpole. Write the answer in terms of feet.

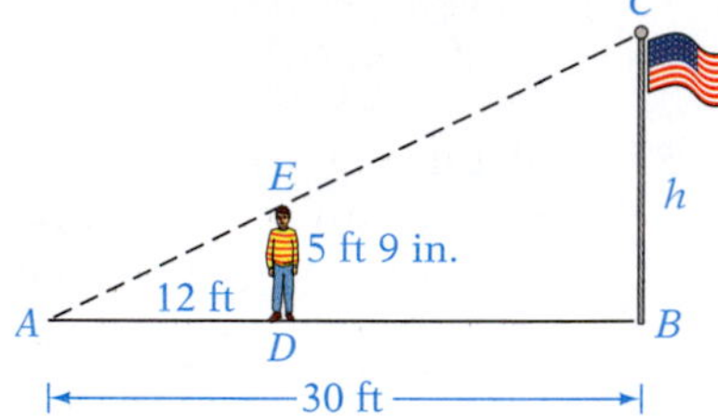

APPLYING THE CONCEPTS

45. **Lottery Tickets** Three people put their money together to buy lottery tickets. The first person put in \$25, the second person put in \$30, and the third person put in \$35. One of their tickets was a winning ticket. If they won \$4.5 million, what was the first person's share of the winnings?

46. **Clubs** No one belongs to both the Math Club and the Photography Club, but the two clubs join to hold a car wash. Ten members of the Math Club and six members of the Photography Club participate. The profits from the car wash are \$120. If each club's profits are proportional to the number of members participating, what share of the profits does the Math Club receive?

47. **Sports** A basketball player has made five out of every six foul shots attempted in 1 year of play. If 42 foul shots were missed that year, how many foul shots did the basketball player make?

Copyright © Houghton Mifflin Company. All rights reserved.

8.6 Literal Equations

Objective A To solve a literal equation for one of the variables

A **literal equation** is an equation that contains more than one variable. Examples of literal equations are shown at the right.

$$2x + 3y = 6$$
$$4w - 2x + z = 0$$

Formulas are used to express relationships among physical quantities. A **formula** is a literal equation that states a rule about measurements. Examples of formulas are shown at the right.

$$\frac{1}{R_1} + \frac{1}{R_2} = \frac{1}{R} \quad \text{(Physics)}$$
$$s = a + (n - 1)d \quad \text{(Mathematics)}$$
$$A = P + Prt \quad \text{(Business)}$$

The Addition and Multiplication Properties can be used to solve a literal equation for one of the variables. The goal is to rewrite the equation so that the variable being solved for is alone on one side of the equation and all the other numbers and variables are on the other side.

HOW TO Solve $A = P(1 + i)$ for i.

The goal is to rewrite the equation so that i is on one side of the equation and all other variables are on the other side.

$$A = P(1 + i)$$
$$A = P + Pi$$ • Use the Distributive Property to remove parentheses.
$$A - P = P - P + Pi$$ • Subtract P from each side of the equation.
$$A - P = Pi$$
$$\frac{A - P}{P} = \frac{Pi}{P}$$ • Divide each side of the equation by P.
$$\frac{A - P}{P} = i$$

Example 1

Solve $3x - 4y = 12$ for y.

Solution

$$3x - 4y = 12$$
$$3x - 3x - 4y = -3x + 12$$ • Subtract $3x$.
$$-4y = -3x + 12$$
$$\frac{-4y}{-4} = \frac{-3x + 12}{-4}$$ • Divide by -4.
$$y = \frac{3}{4}x - 3$$

You Try It 1

Solve $5x - 2y = 10$ for y.

Your solution

Solution on p. S26

Copyright © Houghton Mifflin Company. All rights reserved.

Example 2

Solve $I = \dfrac{E}{R + r}$ for R.

Solution

$$I = \frac{E}{R + r}$$

$$(R + r)I = (R + r)\frac{E}{R + r}$$ • Multiply by $(R + r)$.

$$RI + rI = E$$

$$RI + rI - rI = E - rI$$ • Subtract rI.

$$RI = E - rI$$

$$\frac{RI}{I} = \frac{E - rI}{I}$$ • Divide by I.

$$R = \frac{E - rI}{I}$$

You Try It 2

Solve $s = \dfrac{A + L}{2}$ for L.

Your solution

Example 3

Solve $L = a(1 + ct)$ for c.

Solution

$$L = a(1 + ct)$$

$$L = a + act$$ • Distributive Property

$$L - a = a - a + act$$ • Subtract a.

$$L - a = act$$

$$\frac{L - a}{at} = \frac{act}{at}$$ • Divide by at.

$$\frac{L - a}{at} = c$$

You Try It 3

Solve $S = a + (n - 1)d$ for n.

Your solution

Example 4

Solve $S = C - rC$ for C.

Solution

$$S = C - rC$$

$$S = (1 - r)C$$ • Factor.

$$\frac{S}{1 - r} = \frac{(1 - r)C}{1 - r}$$ • Divide by $(1 - r)$.

$$\frac{S}{1 - r} = C$$

You Try It 4

Solve $S = rS + C$ for S.

Your solution

Solutions on p. S26

Copyright © Houghton Mifflin Company. All rights reserved.

8.6 Exercises

Objective A **To solve a literal equation for one of the variables**

For Exercises 1 to 15, solve for y.

1. $3x + y = 10$

2. $2x + y = 5$

3. $4x - y = 3$

4. $5x - y = 7$

5. $3x + 2y = 6$

6. $2x + 3y = 9$

7. $2x - 5y = 10$

8. $5x - 2y = 4$

9. $2x + 7y = 14$

10. $6x - 5y = 10$

11. $x + 3y = 6$

12. $x + 2y = 8$

13. $y - 2 = 3(x + 2)$

14. $y + 4 = -2(x - 3)$

15. $y - 1 = -\frac{2}{3}(x + 6)$

For Exercises 16 to 24, solve for x.

16. $x + 3y = 6$

17. $x + 6y = 10$

18. $3x - y = 3$

19. $2x - y = 6$

20. $2x + 5y = 10$

21. $4x + 3y = 12$

22. $x - 2y + 1 = 0$

23. $x - 4y - 3 = 0$

24. $5x + 4y + 20 = 0$

For Exercises 25 to 40, solve the formula for the given variable.

25. $d = rt$; t (Physics)

26. $E = IR$; R (Physics)

27. $PV = nRT$; T (Chemistry)

28. $A = bh$; h (Geometry)

Copyright © Houghton Mifflin Company. All rights reserved.

29. $P = 2l + 2w; l$ (Geometry)

30. $F = \frac{9}{5}C + 32; C$ (Temperature conversion)

31. $A = \frac{1}{2}h(b_1 + b_2); b_1$ (Geometry)

32. $C = \frac{5}{9}(F - 32); F$ (Temperature conversion)

33. $V = \frac{1}{3}Ah; h$ (Geometry)

34. $P = R - C; C$ (Business)

35. $R = \frac{C - S}{t}; S$ (Business)

36. $P = \frac{R - C}{n}; R$ (Business)

37. $A = P + Prt; P$ (Business)

38. $T = fm - gm; m$ (Engineering)

39. $A = Sw + w; w$ (Physics)

40. $a = S - Sr; S$ (Mathematics)

APPLYING THE CONCEPTS

Business Break-even analysis is a method used to determine the sales volume required for a company to break even, or experience neither a profit nor a loss on the sale of a product. The break-even point represents the number of units that must be made and sold for income from sales to equal the cost of producing the product. The break-even point can be calculated using the formula $B = \frac{F}{S - V}$, where F is the fixed costs, S is the selling price per unit, and V is the variable costs per unit. Use this information for Exercise 41.

41. **a.** Solve the formula $B = \frac{F}{S - V}$ for S.

b. Use your answer to part **a** to find the selling price per unit required for a company to break even. The fixed costs are \$20,000, the variable costs per unit are \$80, and the company plans to make and sell 200 desks.

c. Use your answer to part **a** to find the selling price per unit required for a company to break even. The fixed costs are \$15,000, the variable costs per unit are \$50, and the company plans to make and sell 600 cameras.

Copyright © Houghton Mifflin Company. All rights reserved.

8.7 Application Problems

Objective A To solve work problems

If a painter can paint a room in 4 h, then in 1 h the painter can paint $\frac{1}{4}$ of the room. The painter's rate of work is $\frac{1}{4}$ of the room each hour. The **rate of work** is the part of a task that is completed in 1 unit of time.

A pipe can fill a tank in 30 min. This pipe can fill $\frac{1}{30}$ of the tank in 1 min. The rate of work is $\frac{1}{30}$ of the tank each minute. If a second pipe can fill the tank in x min, the rate of work for the second pipe is $\frac{1}{x}$ of the tank each minute.

In solving a work problem, the goal is to determine the time it takes to complete a task. The basic equation that is used to solve work problems is

Rate of work × time worked = part of task completed

For example, if a faucet can fill a sink in 6 min, then in 5 min the faucet will fill $\frac{1}{6} \times 5 = \frac{5}{6}$ of the sink. In 5 min the faucet completes $\frac{5}{6}$ of the task.

Study Tip

Note in the examples in this section that solving a word problem includes stating a strategy and using the strategy to find a solution. If you have difficulty with a word problem, write down the known information. Be very specific. Write out a phrase or sentence that states what you are trying to find. See *AIM for Success*, page xxix.

HOW TO A painter can paint a wall in 20 min. The painter's apprentice can paint the same wall in 30 min. How long will it take them to paint the wall when they work together?

Strategy for Solving a Work Problem

1. For each person or machine, write a numerical or variable expression for the rate of work, the time worked, and the part of the task completed. The results can be recorded in a table.

Unknown time to paint the wall working together: t

TAKE NOTE

Use the information given in the problem to fill in the "Rate" and "Time" columns of the table. Fill in the "Part Completed" column by multiplying the two expressions you wrote in each row.

	Rate of Work	·	*Time Worked*	=	*Part of Task Completed*
Painter	$\frac{1}{20}$	·	t	=	$\frac{t}{20}$
Apprentice	$\frac{1}{30}$	·	t	=	$\frac{t}{30}$

2. Determine how the parts of the task completed are related. Use the fact that the sum of the parts of the task completed must equal 1, the complete task.

$$\frac{t}{20} + \frac{t}{30} = 1$$
• The sum of the part of the task completed by the painter and the part of the task completed by the apprentice is 1.

$$60\left(\frac{t}{20} + \frac{t}{30}\right) = 60 \cdot 1$$
• Multiply by the LCM of 20 and 30.

$$3t + 2t = 60$$
• Distributive Property

$$5t = 60$$
$$t = 12$$

Working together, they will paint the wall in 12 min.

Copyright © Houghton Mifflin Company. All rights reserved.

Example 1

A small water pipe takes three times longer to fill a tank than does a large water pipe. With both pipes open it takes 4 h to fill the tank. Find the time it would take the small pipe, working alone, to fill the tank.

Strategy

- Time for large pipe to fill the tank: t
  Time for small pipe to fill the tank: $3t$

Fills tank in $3t$ hours

Fills tank in t hours

Fills $\frac{4}{3t}$ of the tank in 4 hours

Fills $\frac{4}{t}$ of the tank in 4 hours

	Rate	*Time*	*Part*
Small pipe	$\frac{1}{3t}$	4	$\frac{4}{3t}$
Large pipe	$\frac{1}{t}$	4	$\frac{4}{t}$

- The sum of the parts of the task completed by each pipe must equal 1.

Solution

$$\frac{4}{3t} + \frac{4}{t} = 1$$

$$3t\left(\frac{4}{3t} + \frac{4}{t}\right) = 3t \cdot 1$$ • Multiply by the LCM of $3t$ and t.

$$4 + 12 = 3t$$ • Distributive Property

$$16 = 3t$$

$$\frac{16}{3} = t$$

$$3t = 3\left(\frac{16}{3}\right) = 16$$

The small pipe working alone takes 16 h to fill the tank.

You Try It 1

Two computer printers that work at the same rate are working together to print the payroll checks for a large corporation. After they work together for 2 h, one of the printers quits. The second requires 3 h more to complete the payroll checks. Find the time it would take one printer, working alone, to print the payroll.

Your strategy

Your solution

Solution on p. S26

Copyright © Houghton Mifflin Company. All rights reserved.

Objective B To use rational expressions to solve uniform motion problems

A car that travels constantly in a straight line at 30 mph is in uniform motion. **Uniform motion** means that the speed or direction of an object does not change.

The basic equation used to solve uniform motion problems is

Distance = rate × time

An alternative form of this equation can be written by solving the equation for time.

$$\frac{\textbf{Distance}}{\textbf{Rate}} = \textbf{time}$$

This form of the equation is useful when the total time of travel for two objects or the time of travel between two points is known.

HOW TO The speed of a boat in still water is 20 mph. The boat traveled 75 mi down a river in the same amount of time it took to travel 45 mi up the river. Find the rate of the river's current.

Strategy for Solving a Uniform Motion Problem

1. For each object, write a numerical or variable expression for the distance, rate, and time. The results can be recorded in a table.

The unknown rate of the river's current: r

	Distance	÷	*Rate*	=	*Time*
Down river	75	÷	$20 + r$	=	$\frac{75}{20 + r}$
Up river	45	÷	$20 - r$	=	$\frac{45}{20 - r}$

TAKE NOTE

Use the information given in the problem to fill in the "Distance" and "Rate" columns of the table. Fill in the "Time" column by dividing the two expressions you wrote in each row.

2. Determine how the times traveled by each object are related. For example, it may be known that the times are equal, or the total time may be known.

$$\frac{75}{20 + r} = \frac{45}{20 - r}$$

• The time down the river is equal to the time up the river.

$$(20 + r)(20 - r)\frac{75}{20 + r} = (20 + r)(20 - r)\frac{45}{20 - r}$$

• Multiply by the LCM.

$$(20 - r)75 = (20 + r)45$$
$$1500 - 75r = 900 + 45r$$

• Distributive Property

$$-120r = -600$$
$$r = 5$$

The rate of the river's current is 5 mph.

Copyright © Houghton Mifflin Company. All rights reserved.

Example 2

A cyclist rode the first 20 mi of a trip at a constant rate. For the next 16 mi, the cyclist reduced the speed by 2 mph. The total time for the 36 mi was 4 h. Find the rate of the cyclist for each leg of the trip.

Strategy

- Rate for the first 20 mi: r
  Rate for the next 16 mi: $r - 2$

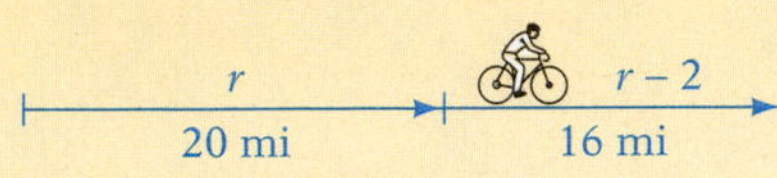

	Distance	*Rate*	*Time*
First 20 mi	20	r	$\frac{20}{r}$
Next 16 mi	16	$r - 2$	$\frac{16}{r-2}$

- The total time for the trip was 4 h.

Solution

$$\frac{20}{r} + \frac{16}{r-2} = 4$$ • The total time was 4 h.

$$r(r-2)\left[\frac{20}{r} + \frac{16}{r-2}\right] = r(r-2) \cdot 4$$ • Multiply by the LCM.

$$(r-2)20 + 16r = 4r^2 - 8r$$ • Distributive Property

$$20r - 40 + 16r = 4r^2 - 8r$$

$$36r - 40 = 4r^2 - 8r$$

Solve the quadratic equation by factoring.

$$0 = 4r^2 - 44r + 40$$ • Standard form

$$0 = 4(r^2 - 11r + 10)$$

$$0 = 4(r - 10)(r - 1)$$ • Factor.

$$r - 10 = 0 \qquad r - 1 = 0$$ • Principle of Zero Products

$$r = 10 \qquad r = 1$$

The solution $r = 1$ mph is not possible, because the rate on the last 16 mi would then be -1 mph.

10 mph was the rate for the first 20 mi.
8 mph was the rate for the next 16 mi.

You Try It 2

The total time it took for a sailboat to sail back and forth across a lake 6 km wide was 2 h. The rate sailing back was three times the rate sailing across. Find the rate sailing out across the lake.

Your strategy

Your solution

Solution on p. S26

Copyright © Houghton Mifflin Company. All rights reserved.

8.7 Exercises

Objective A To solve work problems

1.  Explain the meaning of the phrase "rate of work."

2. If $\frac{2}{5}$ of a room can be painted in 1 h, what is the rate of work? At the same rate, how long will it take to paint the entire room?

3. A park has two sprinklers that are used to fill a fountain. One sprinkler can fill the fountain in 3 h, whereas the second sprinkler can fill the fountain in 6 h. How long will it take to fill the fountain with both sprinklers operating?

4. One grocery clerk can stock a shelf in 20 min, whereas a second clerk requires 30 min to stock the same shelf. How long would it take to stock the shelf if the two clerks worked together?

5. One person with a skiploader requires 12 h to remove a large quantity of earth. A second, larger skiploader can remove the same amount of earth in 4 h. How long would it take to remove the earth with both skiploaders working together?

6. An experienced painter can paint a fence twice as fast as an inexperienced painter. Working together, the painters require 4 h to paint the fence. How long would it take the experienced painter, working alone, to paint the fence?

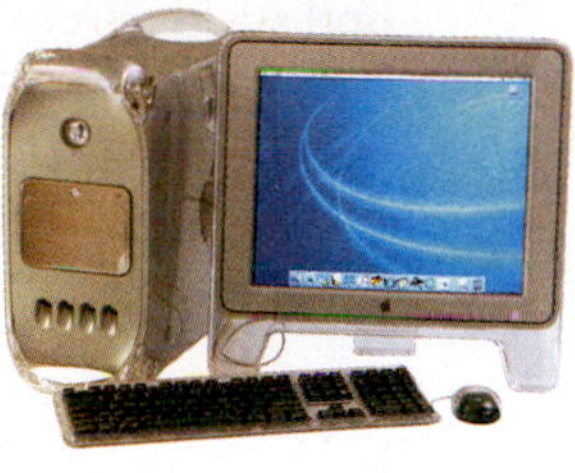

7. One computer can solve a complex prime factorization problem in 75 h. A second computer can solve the same problem in 50 h. How long would it take both computers, working together, to solve the problem?

8. A new machine can make 10,000 aluminum cans three times faster than an older machine. With both machines working, 10,000 cans can be made in 9 h. How long would it take the new machine, working alone, to make the 10,000 cans?

9. A small air conditioner can cool a room 5° in 75 min. A larger air conditioner can cool the room 5° in 50 min. How long would it take to cool the room 5° with both air conditioners working?

10. One printing press can print the first edition of a book in 55 min, whereas a second printing press requires 66 min to print the same number of copies. How long would it take to print the first edition with both presses operating?

11. Two oil pipelines can fill a small tank in 30 min. One of the pipelines would require 45 min to fill the tank. How long would it take the second pipeline, working alone, to fill the tank?

Copyright © Houghton Mifflin Company. All rights reserved.

12. Working together, two dock workers can load a crate in 6 min. One dock worker, working alone, can load the crate in 15 min. How long would it take the second dock worker, working alone, to load the crate?

13. A mason can construct a retaining wall in 10 h. With the mason's apprentice assisting, the task takes 6 h. How long would it take the apprentice, working alone, to construct the wall?

14. A mechanic requires 2 h to repair a transmission, whereas an apprentice requires 6 h to make the same repairs. The mechanic worked alone for 1 h and then stopped. How long will it take the apprentice, working alone, to complete the repairs?

15. One computer technician can wire a modem in 4 h, whereas it takes 6 h for a second technician to do the same job. After working alone for 2 h, the first technician quits. How long will it take the second technician to complete the wiring?

16. A wallpaper hanger requires 2 h to hang the wallpaper on one wall of a room. A second wallpaper hanger requires 4 h to hang the same amount of paper. The first wallpaper hanger worked alone for 1 h and then quit. How long will it take the second wallpaper hanger, working alone, to complete the wall?

17. Two welders who work at the same rate are welding the girders of a building. After they work together for 10 h, one of the welders quits. The second welder requires 20 more hours to complete the welds. Find the time it would have taken one of the welders, working alone, to complete the welds.

18. A large and a small heating unit are being used to heat the water of a pool. The larger unit, working alone, requires 8 h to heat the pool. After both units have been operating for 2 h, the larger unit is turned off. The small unit requires 9 h more to heat the pool. How long would it take the small unit, working alone, to heat the pool?

19. Two machines that fill cereal boxes work at the same rate. After they work together for 7 h, one machine breaks down. The second machine requires 14 h more to finish filling the boxes. How long would it have taken one of the machines, working alone, to fill the boxes?

20. A large and a small drain are opened to drain a pool. The large drain can empty the pool in 6 h. After both drains have been open for 1 h, the large drain becomes clogged and is closed. The smaller drain remains open and requires 9 h more to empty the pool. How long would it have taken the small drain, working alone, to empty the pool?

Copyright © Houghton Mifflin Company. All rights reserved.

Objective B **To use rational expressions to solve uniform motion problems**

21. Running at a constant speed, a jogger ran 24 mi in 3 h. How far did the jogger run in 2 h?

22. For uniform motion, distance = rate · time. How is time related to distance and rate? How is rate related to distance and time?

23. Commuting from work to home, a lab technician traveled 10 mi at a constant rate through congested traffic. On reaching the expressway, the technician increased her speed by 20 mph. An additional 20 mi was traveled at the increased speed. The total time for the trip was 1 h. Find the rate of travel through the congested traffic.

10 mi 20 mi
r $r + 20$

24. The president of a company traveled 1800 mi by jet and 300 mi on a prop plane. The rate of the jet was four times the rate of the prop plane. The entire trip took a total of 5 h. Find the rate of the jet plane.

25. As part of a conditioning program, a jogger ran 8 mi in the same amount of time a cyclist rode 20 mi. The rate of the cyclist was 12 mph faster than the rate of the jogger. Find the rate of the jogger and that of the cyclist.

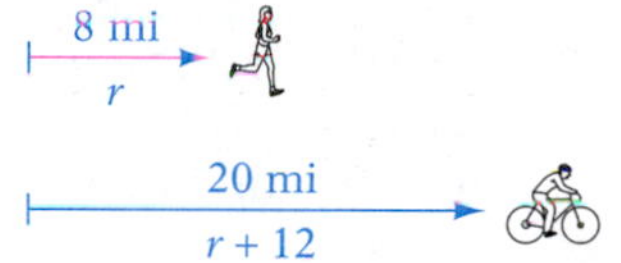

26. An express train travels 600 mi in the same amount of time it takes a freight train to travel 360 mi. The rate of the express train is 20 mph faster than that of the freight train. Find the rate of each train.

27. To assess the damage done by a fire, a forest ranger traveled 1080 mi by jet and then an additional 180 mi by helicopter. The rate of the jet was four times the rate of the helicopter. The entire trip took a total of 5 h. Find the rate of the jet.

28. A twin-engine plane can fly 800 mi in the same time that it takes a single-engine plane to fly 600 mi. The rate of the twin-engine plane is 50 mph faster than that of the single-engine plane. Find the rate of the twin-engine plane.

29. As part of an exercise plan, Camille Ellison walked for 40 min and then ran for 20 min. If Camille runs 3 mph faster than she walks and covered 5 mi during the 1-hour exercise period, what is her walking speed?

30. A car and a bus leave a town at 1 P.M. and head for a town 300 mi away. The rate of the car is twice the rate of the bus. The car arrives 5 h ahead of the bus. Find the rate of the car.

Copyright © Houghton Mifflin Company. All rights reserved.

31. A car is traveling at a rate that is 36 mph faster than the rate of a cyclist. The car travels 384 mi in the same time it takes the cyclist to travel 96 mi. Find the rate of the car.

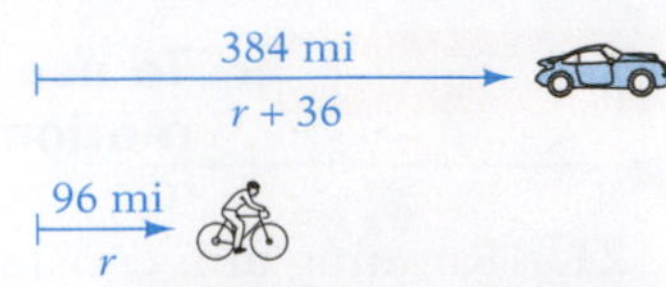

32. A backpacker hiking into a wilderness area walked 9 mi at a constant rate and then reduced this rate by 1 mph. Another 4 mi was hiked at this reduced rate. The time required to hike the 4 mi was 1 h less than the time required to walk the 9 mi. Find the rate at which the hiker walked the first 9 mi.

33. A plane can fly 180 mph in calm air. Flying with the wind, the plane can fly 600 mi in the same amount of time it takes to fly 480 mi against the wind. Find the rate of the wind.

34. A commercial jet can fly 550 mph in calm air. Traveling with the jet stream, the plane flew 2400 mi in the same amount of time it takes to fly 2000 mi against the jet stream. Find the rate of the jet stream.

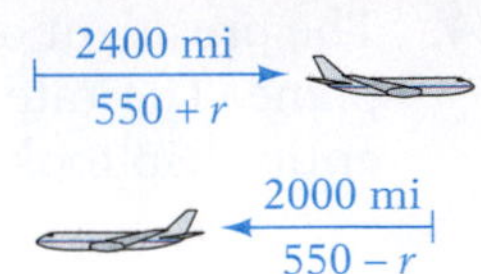

35. A cruise ship can sail at 28 mph in calm water. Sailing with the gulf current, the ship can sail 170 mi in the same amount of time that it can sail 110 mi against the gulf current. Find the rate of the gulf current.

36. Rowing with the current of a river, a rowing team can row 25 mi in the same amount of time it takes to row 15 mi against the current. The rate of the rowing team in calm water is 20 mph. Find the rate of the current.

37. On a recent trip, a trucker traveled 330 mi at a constant rate. Because of road construction, the trucker then had to reduce the speed by 25 mph. An additional 30 mi was traveled at the reduced rate. The total time for the entire trip was 7 h. Find the rate of the trucker for the first 330 mi.

APPLYING THE CONCEPTS

38. **Work** One pipe can fill a tank in 2 h, a second pipe can fill the tank in 4 h, and a third pipe can fill the tank in 5 h. How long will it take to fill the tank with all three pipes working?

39. **Transportation** Because of bad weather, a bus driver reduced the usual speed along a 150-mile bus route by 10 mph. The bus arrived only 30 min later than its usual arrival time. How fast does the bus usually travel?

Copyright © Houghton Mifflin Company. All rights reserved.

8.8 Variation

Objective A To solve variation problems

Direct variation is a special function that can be expressed as the equation $y = kx$, where k is a constant. The equation $y = kx$ is read "y varies directly as x" or "y is directly proportional to x." The constant k is called the **constant of variation** or the **constant of proportionality.**

The circumference (C) of a circle varies directly as the diameter (d). The direct variation equation is written $C = \pi d$. The constant of variation is π.

A nurse earns \$30 per hour. The total wage (w) of the nurse is directly proportional to the number of hours (h) worked. The equation of variation is $w = 30h$. The constant of proportionality is 30.

A direct variation equation can be written in the form $y = kx^n$, where n is a positive number. For example, the equation $y = kx^2$ is read "y varies directly as the square of x."

The area (A) of a circle varies directly as the square of the radius (r) of the circle. The direct variation equation is $A = \pi r^2$. The constant of variation is π.

HOW TO Given that V varies directly as r and that $V = 20$ when $r = 4$, find the constant of variation and the variation equation.

$V = kr$ • Write the basic direct variation equation.

$20 = k \cdot 4$ • Replace V and r by the given values. Then solve for k.

$5 = k$ • This is the constant of variation.

$V = 5r$ • Write the direct variation equation by substituting the value of k into the basic direct variation equation.

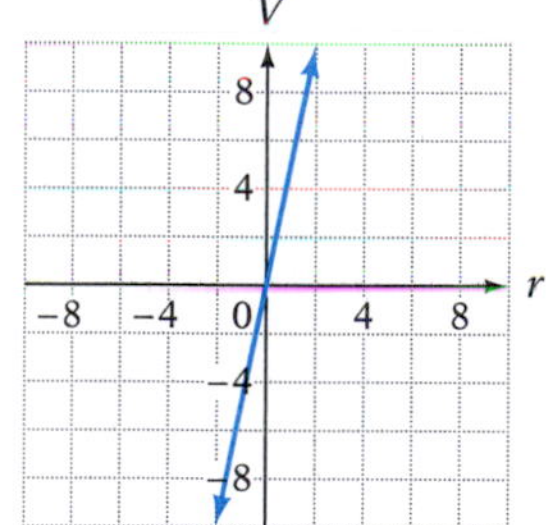

The graph of $V = 5r$

HOW TO The tension (T) in a spring varies directly as the distance (x) it is stretched. If $T = 8$ lb when $x = 2$ in., find T when $x = 4$ in.

$T = kx$ • Write the basic direct variation equation.

$8 = k \cdot 2$ • Replace T and x by the given values.

$4 = k$ • Solve for the constant of variation.

$T = 4x$ • Write the direct variation equation.

To find T when $x = 4$ in., substitute 4 for x in the equation and solve for T.

$T = 4x$

$T = 4 \cdot 4 = 16$

The tension is 16 lb.

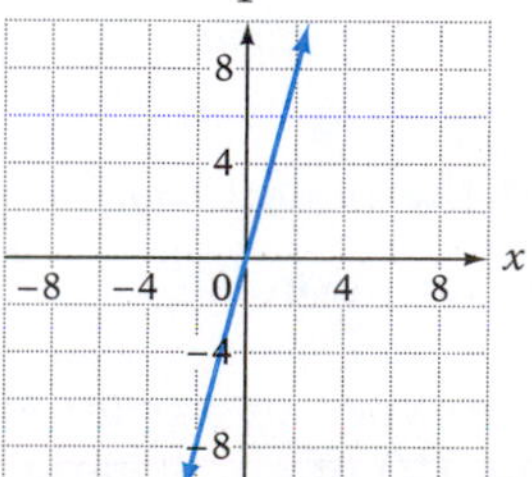

The graph of $T = 4x$

Copyright © Houghton Mifflin Company. All rights reserved.

Inverse variation is a function that can be expressed as the equation $y = \frac{k}{x}$, where k is a constant. The equation $y = \frac{k}{x}$ is read "y varies inversely as x" or "y is inversely proportional to x." In general, an inverse variation equation can be written $y = \frac{k}{x^n}$, where n is a positive number. For example, the equation $y = \frac{k}{x^2}$ is read "y varies inversely as the square of x."

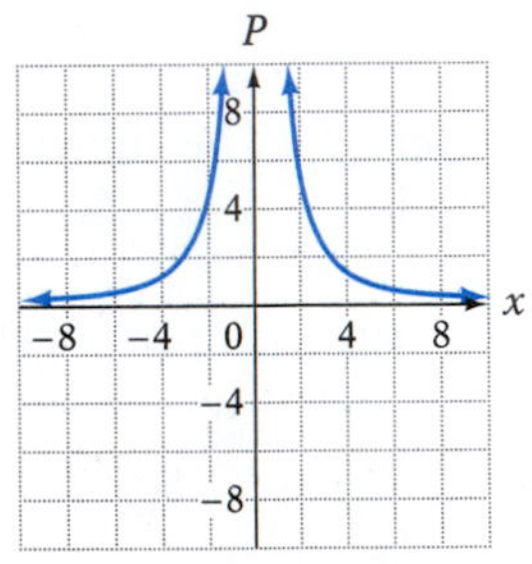

The graph of $P = \frac{20}{x^2}$

HOW TO Given that P varies inversely as the square of x and that $P = 5$ when $x = 2$, find the constant of variation and the variation equation.

$P = \frac{k}{x^2}$ • Write the basic inverse variation equation.

$5 = \frac{k}{2^2}$ • Replace P and x by the given values. Then solve for k.

$20 = k$ • This is the constant of variation.

$P = \frac{20}{x^2}$ • Write the inverse variation equation by substituting the value of k into the basic inverse variation equation.

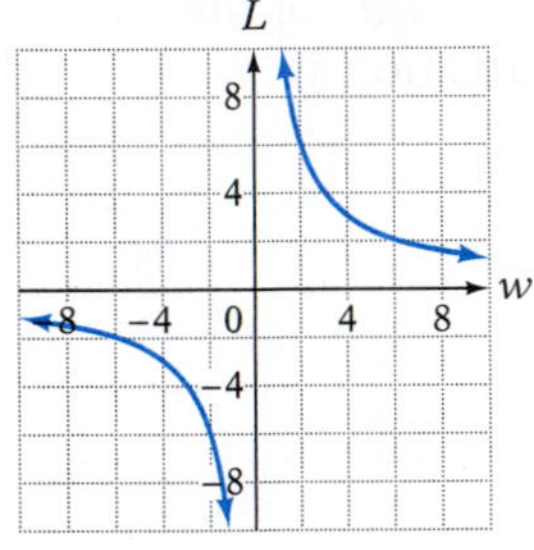

The graph of $L = \frac{12}{w}$

HOW TO The length (L) of a rectangle with fixed area is inversely proportional to the width (w). If $L = 6$ ft when $w = 2$ ft, find L when $w = 3$ ft.

$L = \frac{k}{w}$ • Write the basic inverse variation equation.

$6 = \frac{k}{2}$ • Replace L and w by the given values.

$12 = k$ • Solve for the constant of variation.

$L = \frac{12}{w}$ • Write the inverse variation equation.

To find L when $w = 3$ ft, substitute 3 for w in the equation and solve for L.

$$L = \frac{12}{w}$$

$$L = \frac{12}{3} = 4$$

The length is 4 ft.

Joint variation is a variation in which a variable varies directly as the product of two or more other variables. A joint variation can be expressed as the equation $z = kxy$, where k is a constant. The equation $z = kxy$ is read "z varies jointly as x and y."

The area (A) of a triangle varies jointly as the base (b) and the height (h). The joint variation equation is written $A = \frac{1}{2}bh$. The constant of variation is $\frac{1}{2}$.

A **combined variation** is a variation in which two or more types of variation occur at the same time. For example, in physics, the volume (V) of a gas varies directly as the temperature (T) and inversely as the pressure (P). This combined variation is written $V = \frac{kT}{P}$.

Copyright © Houghton Mifflin Company. All rights reserved.

HOW TO A ball is being twirled on the end of a string. The tension (T) in the string is directly proportional to the square of the speed (v) of the ball and inversely proportional to the length (r) of the string. If the tension is 96 lb when the length of the string is 0.5 ft and the speed is 4 ft/s, find the tension when the length of the string is 1 ft and the speed is 5 ft/s.

$T = \dfrac{kv^2}{r}$ • Write the basic combined variation equation.

$96 = \dfrac{k \cdot 4^2}{0.5}$ • Replace T, v, and r by the given values.

$96 = \dfrac{k \cdot 16}{0.5}$

$96 = k \cdot 32$ • Solve for the constant of variation.

$3 = k$

$T = \dfrac{3v^2}{r}$ • Write the combined variation equation.

To find T when $r = 1$ ft and $v = 5$ ft/s, substitute 1 for r and 5 for v and solve for T.

$$T = \frac{3v^2}{r} = \frac{3 \cdot 5^2}{1} = 3 \cdot 25 = 75$$

The tension is 75 lb.

Example 1

The amount (A) of medication prescribed for a person is directly related to the person's weight (W). For a 50-kilogram person, 2 ml of medication is prescribed. How many milliliters of medication are required for a person who weighs 75 kg?

Strategy

To find the required amount of medication:

- Write the basic direct variation equation, replace the variables by the given values, and solve for k.
- Write the direct variation equation, replacing k by its value. Substitute 75 for W and solve for A.

Solution

$A = kW$ • Direct variation equation

$2 = k \cdot 50$ • Replace A by 2 and W by 50.

$\dfrac{1}{25} = k$

$A = \dfrac{1}{25}W = \dfrac{1}{25} \cdot 75 = 3$ • $k = \dfrac{1}{25}$, $W = 75$

The required amount of medication is 3 ml.

You Try It 1

The distance (s) a body falls from rest varies directly as the square of the time (t) of the fall. An object falls 64 ft in 2 s. How far will it fall in 5 s?

Your strategy

Your solution

Solution on p. S27

Copyright © Houghton Mifflin Company. All rights reserved.

Example 2

A company that produces personal computers has determined that the number of computers it can sell (s) is inversely proportional to the price (P) of the computer. Two thousand computers can be sold when the price is \$2500. How many computers can be sold when the price of a computer is \$2000?

Strategy
To find the number of computers:
- Write the basic inverse variation equation, replace the variables by the given values, and solve for k.
- Write the inverse variation equation, replacing k by its value. Substitute 2000 for P and solve for s.

Solution

$$s = \frac{k}{P}$$ • Inverse variation equation

$$2000 = \frac{k}{2500}$$ • Replace s by 2000 and P by 2500.

$$5{,}000{,}000 = k$$

$$s = \frac{5{,}000{,}000}{P} = \frac{5{,}000{,}000}{2000} = 2500$$ • $k = 5{,}000{,}000$, $P = 2000$

At \$2000 each, 2500 computers can be sold.

You Try It 2

The resistance (R) to the flow of electric current in a wire of fixed length is inversely proportional to the square of the diameter (d) of the wire. If a wire of diameter 0.01 cm has a resistance of 0.5 ohm, what is the resistance in a wire that is 0.02 cm in diameter?

Your strategy

Your solution

Example 3

The pressure (P) of a gas varies directly as the temperature (T) and inversely as the volume (V). When $T = 50°$ and $V = 275$ in^3, $P = 20$ lb/in^2. Find the pressure of a gas when $T = 60°$ and $V = 250$ in^3.

Strategy
To find the pressure:
- Write the basic combined variation equation, replace the variables by the given values, and solve for k.
- Write the combined variation equation, replacing k by its value. Substitute 60 for T and 250 for V, and solve for P.

Solution

$$P = \frac{kT}{V}$$ • Combined variation equation

$$20 = \frac{k \cdot 50}{275}$$ • Replace P by 20, T by 50, and V by 275.

$$110 = k$$

$$P = \frac{110T}{V} = \frac{110 \cdot 60}{250} = 26.4$$ • $k = 110$, $T = 60$, $V = 250$

The pressure is 26.4 lb/in^2.

You Try It 3

The strength (s) of a rectangular beam varies jointly as its width (W) and the square of its depth (d) and inversely as its length (L). If the strength of a beam 2 in. wide, 12 in. deep, and 12 ft long is 1200 lb, find the strength of a beam 4 in. wide, 8 in. deep, and 16 ft long.

Your strategy

Your solution

Solutions on p. S27

Copyright © Houghton Mifflin Company. All rights reserved.

8.8 Exercises

Objective A To solve variation problems

1. **Business** The profit (P) realized by a company varies directly as the number of products it sells (s). If a company makes a profit of \$4000 on the sale of 250 products, what is the profit when the company sells 5000 products?

2. **Compensation** The income (I) of a computer analyst varies directly as the number of hours (h) worked. If the analyst earns \$336 for working 8 h, how much will the analyst earn for working 36 h?

3. **Recreation** The pressure (p) on a diver in the water varies directly as the depth (d) below the surface. If the pressure is 4.5 lb/in^2 when the depth is 10 ft, what is the pressure when the depth is 15 ft?

4. **Physics** The distance (d) that a spring will stretch varies directly as the force (f) applied to the spring. If a force of 6 lb is required to stretch a spring 3 in., what force is required to stretch the spring 4 in.?

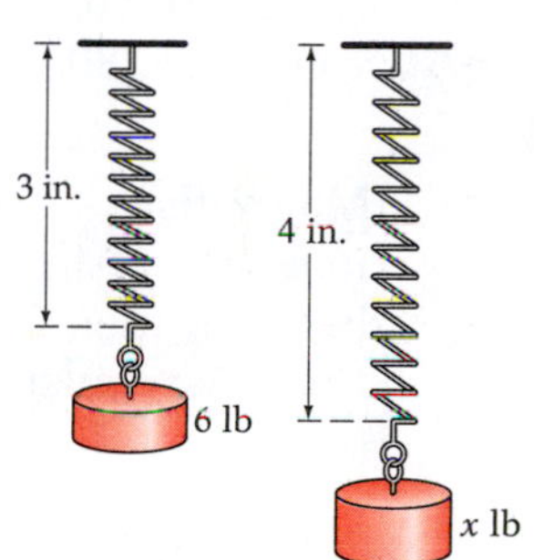

5. **Physics** The distance (d) an object will fall is directly proportional to the square of the time (t) of the fall. If an object falls 144 ft in 3 s, how far will the object fall in 10 s?

6. **Physics** The period (p) of a pendulum, or the time it takes the pendulum to make one complete swing, varies directly as the square root of the length (L) of the pendulum. If the period of a pendulum is 1.5 s when the length is 2 ft, find the period when the length is 5 ft. Round to the nearest hundredth.

7. **Physics** The distance (s) a ball will roll down an inclined plane is directly proportional to the square of the time (t). If the ball rolls 6 ft in 1 s, how far will it roll in 3 s?

8. **Safety** The stopping distance (s) of a car varies directly as the square of its speed (v). If a car traveling 30 mph requires 63 ft to stop, find the stopping distance for a car traveling 55 mph.

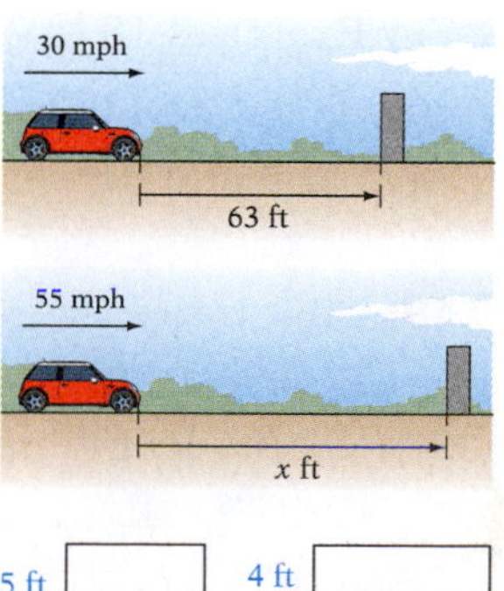

9. **Geometry** The length (L) of a rectangle of fixed area varies inversely as the width (w). If the length of a rectangle is 8 ft when the width is 5 ft, find the length of the rectangle when the width is 4 ft.

5 ft 8 ft 4 ft x ft

Copyright © Houghton Mifflin Company. All rights reserved.

10. **Travel** The time (t) for a car to travel between two cities is inversely proportional to the rate (r) of travel. If it takes the car 5 h to travel between the cities at a rate of 55 mph, find the time to travel between the two cities at a rate of 65 mph. Round to the nearest tenth.

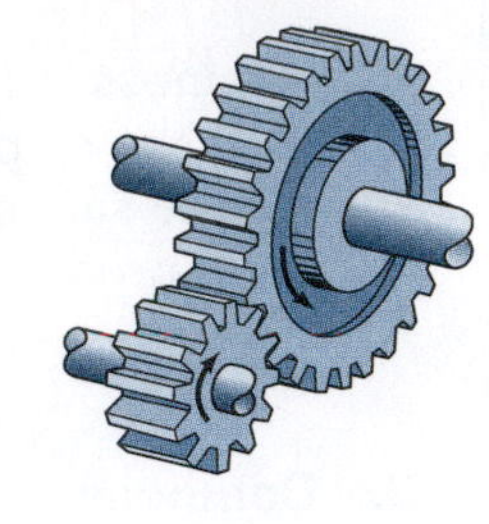

11. **Mechanics** The speed (v) of a gear varies inversely as the number of teeth (t). If a gear that has 45 teeth makes 24 revolutions per minute, how many revolutions per minute will a gear that has 36 teeth make?

12. **Physics** The pressure (p) of a liquid varies directly as the product of the depth (d) and the density (D) of the liquid. If the pressure is 150 lb/in^2 when the depth is 100 in. and the density is 1.2, find the pressure when the density remains the same and the depth is 75 in.

13. **Electronics** The current (I) in a wire varies directly as the voltage (v) and inversely as the resistance (r). If the current is 10 amps when the voltage is 110 volts and the resistance is 11 ohms, find the current when the voltage is 180 volts and the resistance is 24 ohms.

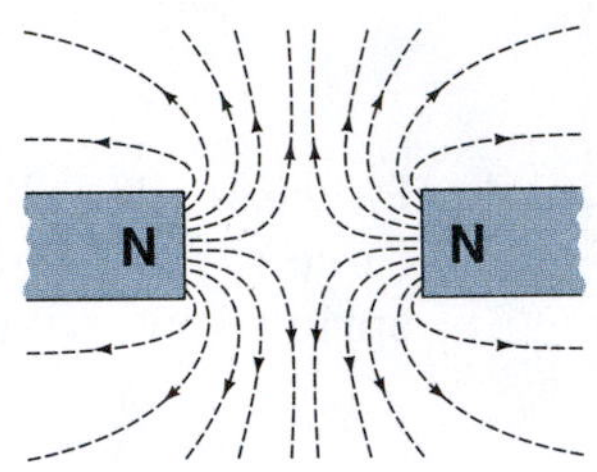

14. **Magnetism** The repulsive force (f) between the north poles of two magnets is inversely proportional to the square of the distance (d) between them. If the repulsive force is 20 lb when the distance is 4 in., find the repulsive force when the distance is 2 in.

15. **Light** The intensity (I) of a light source is inversely proportional to the square of the distance (d) from the source. If the intensity is 12 foot-candles at a distance of 10 ft, what is the intensity when the distance is 5 ft?

APPLYING THE CONCEPTS

16. In the inverse variation equation $y = \frac{k}{x}$, what is the effect on x when y is doubled?

17. In the direct variation equation $y = kx$, what is the effect on y when x is doubled?

For Exercises 18 to 21, complete using the word *directly* or *inversely*.

18. If a varies directly as b and inversely as c, then c varies __________ as b and __________ as a.

19. If a varies __________ as b and c, then abc is constant.

20. If the length of a rectangle is held constant, the area of the rectangle varies __________ as the width.

21. If the area of a rectangle is held constant, the length of the rectangle varies __________ as the width.

Copyright © Houghton Mifflin Company. All rights reserved.

Focus on Problem Solving

Implication

Sentences that are constructed using "If ..., then ..." occur frequently in problem-solving situations. These sentences are called **implications.** The sentence "If it rains, then I will stay home" is an implication. The phrase *it rains* is called the **antecedent** of the implication. The phrase *I will stay home* is the **consequent** of the implication.

The sentence "If $x = 4$, then $x^2 = 16$" is a true sentence. The **contrapositive** of an implication is formed by switching the antecedent and the consequent and then negating each one. The contrapositive of "If $x = 4$, then $x^2 = 16$" is "If $x^2 \neq 16$, then $x \neq 4$." This sentence is also true. It is a principle of logic that an implication and its contrapositive are either both true or both false.

The **converse** of an implication is formed by switching the antecedent and the consequent. The converse of "If $x = 4$, then $x^2 = 16$" is "If $x^2 = 16$, then $x = 4$." Note that the converse is not a true statement, because if $x^2 = 16$, then x could equal -4. The converse of an implication may or may not be true.

Those statements for which the implication and the converse are both true are very important. They can be stated in the form "x if and only if y." For instance, a number is divisible by 5 if and only if the last digit of the number is 0 or 5. The implication "If a number is divisible by 5, then it ends in 0 or 5" and the converse "If a number ends in 0 or 5, then it is divisible by 5" are both true.

For Exercises 1 to 14, state the contrapositive and the converse of the implication. If the converse and the implication are both true statements, write a sentence using the phrase *if and only if.*

1. If I live in Chicago, then I live in Illinois.
2. If today is June 1, then yesterday was May 31.
3. If today is not Thursday, then tomorrow is not Friday.
4. If a number is divisible by 8, then it is divisible by 4.
5. If a number is an even number, then it is divisible by 2.
6. If a number is a multiple of 6, then it is a multiple of 3.
7. If $4z = 20$, then $z = 5$.
8. If an angle measures 90°, then it is a right angle.
9. If p is a prime number greater than 2, then p is an odd number.
10. If the equation of a graph can be written in the form $y = mx + b$, then the graph of the equation is a straight line.
11. If $a = 0$ or $b = 0$, then $ab = 0$.
12. If the coordinates of a point are $(5, 0)$, then the point is on the x-axis.
13. If a quadrilateral is a square, then the quadrilateral has four sides of equal length.
14. If $x = y$, then $x^2 = y^2$.

Copyright © Houghton Mifflin Company. All rights reserved.

Projects and Group Activities

Intensity of Illumination

You are already aware that the standard unit of length in the metric system is the meter (m) and that the standard unit of mass in the metric system is the gram (g). You may not know that the standard unit of light intensity is the **candela (cd).**

The rate at which light falls on a 1-square-unit area of surface is called the **intensity of illumination.** Intensity of illumination is measured in **lumens (lm).** A lumen is defined in the following illustration.

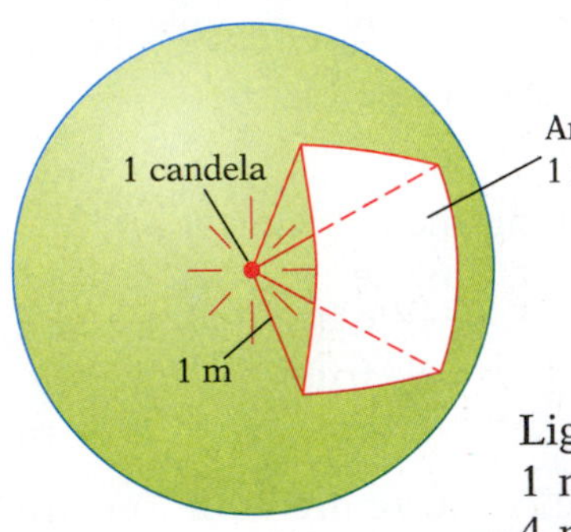

Picture a source of light equal to 1 cd positioned at the center of a hollow sphere that has a radius of 1 m. The rate at which light falls on 1 m^2 of the inner surface of the sphere is equal to 1 lm. If a light source equal to 4 cd is positioned at the center of the sphere, each square meter of the inner surface receives four times as much illumination, or 4 lm.

Light rays diverge as they leave a light source. The light that falls on an area of 1 m^2 at a distance of 1 m from the source of light spreads out over an area of 4 m^2 when it is 2 m from the source. The same light spreads out over an area of 9 m^2 when it is 3 m from the light source and over an area of 16 m^2 when it is 4 m from the light source. Therefore, as a surface moves farther away from the source of light, the intensity of illumination on the surface decreases from its value at 1 m to $\left(\frac{1}{2}\right)^2$, or $\frac{1}{4}$, that value at 2 m; to $\left(\frac{1}{3}\right)^2$, or $\frac{1}{9}$, that value at 3 m; and to $\left(\frac{1}{4}\right)^2$, or $\frac{1}{16}$, that value at 4 m.

The formula for the intensity of illumination is

$$I = \frac{s}{r^2}$$

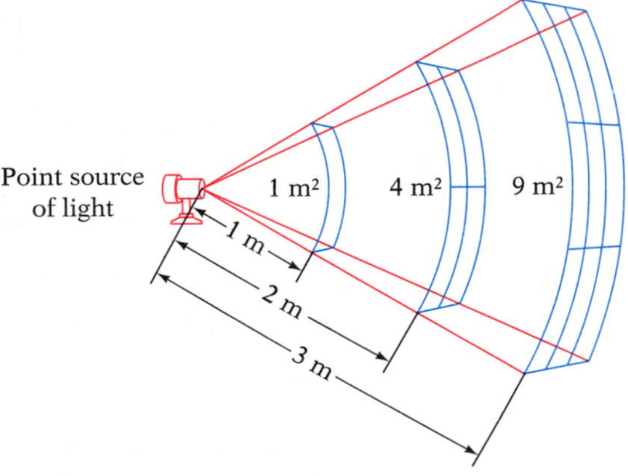

where I is the intensity of illumination in lumens, s is the strength of the light source in candelas, and r is the distance in meters between the light source and the illuminated surface.

A 30-candela lamp is 0.5 m above a desk. Find the illumination on the desk.

$$I = \frac{s}{r^2}$$

$$I = \frac{30}{(0.5)^2} = 120$$

The illumination on the desk is 120 lm.

Copyright © Houghton Mifflin Company. All rights reserved.

Solve.

1. A 100-candela light is hanging 5 m above a floor. What is the intensity of the illumination on the floor beneath it?
2. A 25-candela source of light is 2 m above a desk. Find the intensity of illumination on the desk.
3. How strong a light source is needed to cast 20 lm of light on a surface 4 m from the source?
4. How strong a light source is needed to cast 80 lm of light on a surface 5 m from the source?
5. How far from the desk surface must a 40-candela light source be positioned if the desired intensity of illumination is 10 lm?
6. Find the distance between a 36-candela light source and a surface if the intensity of illumination on the surface is 0.01 lm.
7. Two lights cast the same intensity of illumination on a wall. One light is 6 m from the wall and has a rating of 36 cd. The second light is 8 m from the wall. Find the candela rating of the second light.
8. A 40-candela light source and a 10-candela light source both throw the same intensity of illumination on a wall. The 10-candela light is 6 m from the wall. Find the distance from the 40-candela light to the wall.

Chapter 8 Summary

Key Words	**Examples**
A *rational expression* is a fraction in which the numerator and denominator are polynomials. A rational expression is in *simplest form* when the numerator and denominator have no common factors. [8.1A, p. 453]	$\frac{2x+1}{x^2+4}$ is a rational expression in simplest form.
The *reciprocal of a rational expression* is the rational expression with the numerator and denominator interchanged. [8.1C, p. 456]	The reciprocal of $\frac{3x-y}{x+4}$ is $\frac{x+4}{3x-y}$.
The *least common multiple (LCM) of two or more polynomials* is the polynomial of least degree that contains all the factors of each polynomial. [8.2A, p. 461]	The LCM of $3x^2 - 6x$ and $x^2 - 4$ is $3x(x-2)(x+2)$, because it contains the factors of $3x^2 - 6x = 3x(x-2)$ and the factors of $x^2 - 4 = (x-2)(x+2)$.
A *complex fraction* is a fraction whose numerator or denominator contains one or more fractions. [8.3A, p. 469]	$\dfrac{x - \frac{2}{x+1}}{1 - \frac{4}{x}}$ is a complex fraction.
A *ratio* is the quotient of two quantities that have the same unit. A *rate* is the quotient of two quantities that have different units. [8.5A, p. 477]	$\frac{9}{4}$ is a ratio. $\frac{60 \text{ m}}{12 \text{ s}}$ is a rate.

Copyright © Houghton Mifflin Company. All rights reserved.

A *proportion* is an equation that states the equality of two ratios or rates. [8.5A, p. 477]

$\frac{3}{8} = \frac{12}{32}$ and $\frac{x \text{ ft}}{12 \text{ s}} = \frac{15 \text{ ft}}{160 \text{ s}}$ are proportions.

A *literal equation* is an equation that contains more than one variable. A *formula* is a literal equation that states a rule about measurements. [8.6A, p. 485]

$3x - 4y = 12$ is a literal equation. $A = LW$ is a literal equation that is also the formula for the area of a rectangle.

Direct variation is a special function that can be expressed as the equation $y = kx^n$, where k is a constant called the *constant of variation* or the *constant of proportionality*. [8.8A, p. 497]

$E = mc^2$ is Einstein's formula relating energy and mass. c^2 is the constant of proportionality.

Inverse variation is a function that can be expressed as the equation $y = \frac{k}{x^n}$, where k is a constant. [8.8A, p. 498]

$I = \frac{k}{d^2}$ gives the intensity of a light source at a distance d from the source.

A *joint variation* is a variation in which a variable varies directly as the product of two or more variables. A joint variation can be expressed as the equation $z = kxy$, where k is a constant. [8.8A, p. 498]

$C = kAT$ is a formula for the cost of insulation, where A is the area to be insulated and T is the thickness of the insulation.

A *combined variation* is a variation in which two or more types of variation occur at the same time. [8.8A, p. 498]

$V = \frac{kT}{P}$ is a formula that states that the volume of a gas is directly proportional to the temperature and inversely proportional to the pressure.

Essential Rules and Procedures

Examples

Simplifying Rational Expressions [8.1A, p. 453]
Factor the numerator and denominator. Divide the numerator and denominator by the common factors.

$$\frac{x^2 - 3x - 10}{x^2 - 5} = \frac{(x+2)(x-5)}{(x+5)(x-5)} = \frac{x+2}{x+5}$$

Multiplying Rational Expressions [8.1B, p. 454]
Multiply the numerators. Multiply the denominators. Write the answer in simplest form.

$$\frac{a}{b} \cdot \frac{c}{d} = \frac{ac}{bd}$$

$$\frac{x^2 - 3x}{x^2 + x} \cdot \frac{x^2 + 5x + 4}{x^2 - 4x + 3}$$
$$= \frac{x(x-3)}{x(x+1)} \cdot \frac{(x+1)(x+4)}{(x-3)(x-1)}$$
$$= \frac{x(x-3)(x+1)(x+4)}{x(x+1)(x-3)(x-1)}$$
$$= \frac{x+4}{x-1}$$

Copyright © Houghton Mifflin Company. All rights reserved.

Dividing Rational Expressions [8.1C, p. 456]

Multiply the dividend by the reciprocal of the divisor. Write the answer in simplest form.

$$\frac{a}{b} \div \frac{c}{d} = \frac{a}{b} \cdot \frac{d}{c} = \frac{ad}{bc}$$

$$\frac{4x + 16}{3x - 6} \div \frac{x^2 + 6x + 8}{x^2 - 4}$$
$$= \frac{4x + 16}{3x - 6} \cdot \frac{x^2 - 4}{x^2 + 6x + 8}$$
$$= \frac{4(x + 4)}{3(x - 2)} \cdot \frac{(x - 2)(x + 2)}{(x + 4)(x + 2)}$$
$$= \frac{4}{3}$$

Adding and Subtracting Rational Expressions [8.2B, p. 463]

1. Find the LCM of the denominators.
2. Write each fraction as an equivalent fraction using the LCM as the denominator.
3. Add or subtract the numerators and place the result over the common denominator.
4. Write the answer in simplest form.

$$\frac{a}{b} + \frac{c}{b} = \frac{a + c}{b} \qquad \frac{a}{b} - \frac{c}{b} = \frac{a - c}{b}$$

$$\frac{x}{x + 1} - \frac{x + 3}{x - 2}$$
$$= \frac{x}{x + 1} \cdot \frac{x - 2}{x - 2} - \frac{x + 3}{x - 2} \cdot \frac{x + 1}{x + 1}$$
$$= \frac{x(x - 2)}{(x + 1)(x - 2)} - \frac{(x + 3)(x + 1)}{(x + 1)(x - 2)}$$
$$= \frac{x(x - 2) - (x + 3)(x + 1)}{(x + 1)(x - 2)}$$
$$= \frac{(x^2 - 2x) - (x^2 + 4x + 3)}{(x + 1)(x - 2)}$$
$$= \frac{-6x - 3}{(x + 1)(x - 2)}$$

Simplifying Complex Fractions [8.3A, p. 469]

Method 1: Multiply by 1 in the form $\frac{\text{LCM}}{\text{LCM}}$.

1. Determine the LCM of the denominators of the fractions in the numerator and denominator of the complex fraction.
2. Multiply the numerator and denominator of the complex fraction by the LCM.
3. Simplify.

Method 1: $$\frac{\frac{1}{x} + \frac{1}{y}}{\frac{1}{x} - \frac{1}{y}} = \frac{\frac{1}{x} + \frac{1}{y}}{\frac{1}{x} - \frac{1}{y}} \cdot \frac{xy}{xy}$$
$$= \frac{\frac{1}{x} \cdot xy + \frac{1}{y} \cdot xy}{\frac{1}{x} \cdot xy - \frac{1}{y} \cdot xy}$$
$$= \frac{y + x}{y - x}$$

Method 2: Multiply the numerator by the reciprocal of the denominator.

1. Simplify the numerator to a single fraction and simplify the denominator to a single fraction.
2. Using the definition for dividing fractions, multiply the numerator by the reciprocal of the denominator.
3. Simplify.

Method 2: $$\frac{\frac{1}{x} + \frac{1}{y}}{\frac{1}{x} - \frac{1}{y}} = \frac{\frac{y + x}{xy}}{\frac{y - x}{xy}}$$
$$= \frac{y + x}{xy} \cdot \frac{xy}{y - x}$$
$$= \frac{y + x}{y - x}$$

Copyright © Houghton Mifflin Company. All rights reserved.

Solving Equations Containing Fractions [8.4A, p. 473]

Clear denominators by multiplying each side of the equation by the LCM of the denominators. Then solve for the variable.

$$\frac{1}{2a} = \frac{2}{a} - \frac{3}{8}$$

$$8a\left(\frac{1}{2a}\right) = 8a\left(\frac{2}{a}\right) - 8a\left(\frac{3}{8}\right)$$

$$4 = 16 - 3a$$

$$-12 = -3a$$

$$4 = a$$

Similar Triangles [8.5C, pp. 478–479]

Similar triangles have the same shape but not necessarily the same size. The ratios of corresponding parts of similar triangles are equal. The measures of the corresponding angles of similar triangles are equal.

Triangles ABC and DFE are similar triangles. The ratios of corresponding parts are equal to $\frac{2}{3}$.

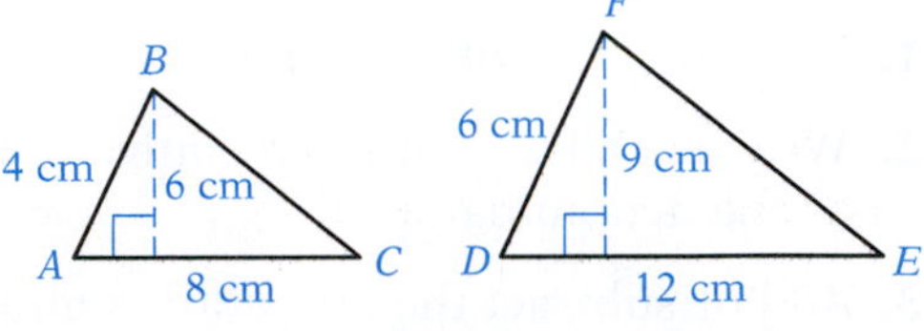

If two angles of one triangle are equal in measure to two angles of another triangle, then the two triangles are similar.

Triangles AOB and COD are similar because $m\angle AOB = m\angle COD$ and $m\angle B = m\angle D$.

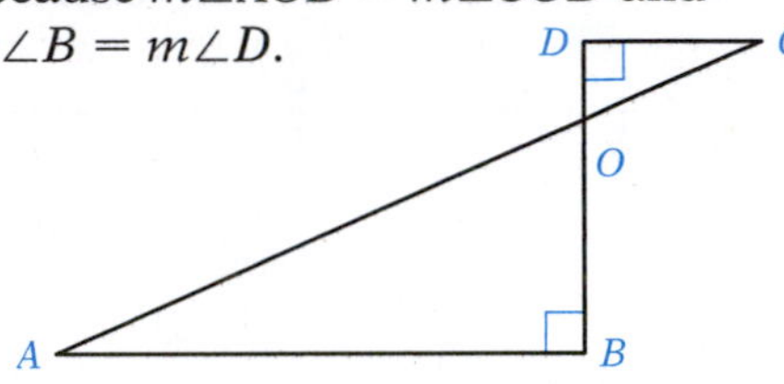

Solving Literal Equations [8.6A, p. 485]

Rewrite the equation so that the letter being solved for is alone on one side of the equation and all numbers and other variables are on the other side.

Solve $2x + ax = 5$ for x.

$$2x + ax = 5$$

$$x(2 + a) = 5$$

$$\frac{x(2 + a)}{2 + a} = \frac{5}{2 + a}$$

$$x = \frac{5}{2 + a}$$

Work Problems [8.7A, p. 489]

Rate of work × time worked = part of task completed

Pat can do a certain job in 3 h. Chris can do the same job in 5 h. How long would it take them, working together, to get the job done?

$$\frac{t}{3} + \frac{t}{5} = 1$$

Uniform Motion Problems with Rational Expressions [8.7B, p. 491]

$$\frac{\text{Distance}}{\text{Rate}} = \text{time}$$

Train A's speed is 15 mph faster than train B's speed. Train A travels 150 mi in the same amount of time it takes train B to travel 120 mi. Find the rate of train B.

$$\frac{120}{r} = \frac{150}{r + 15}$$

Copyright © Houghton Mifflin Company. All rights reserved.

Chapter 8 Review Exercises

1. Divide: $\dfrac{6a^2b^7}{25x^3y} \div \dfrac{12a^3b^4}{5x^2y^2}$

2. Add: $\dfrac{x+7}{15x} + \dfrac{x-2}{20x}$

3. Simplify: $\dfrac{x - \dfrac{16}{5x-2}}{3x - 4 - \dfrac{88}{5x-2}}$

4. Simplify: $\dfrac{x^2 + x - 30}{15 + 2x - x^2}$

5. Simplify: $\dfrac{16x^5y^3}{24xy^{10}}$

6. Solve: $\dfrac{20}{x+2} = \dfrac{5}{16}$

7. Divide: $\dfrac{10 - 23y + 12y^2}{6y^2 - y - 5} \div \dfrac{4y^2 - 13y + 10}{18y^2 + 3y - 10}$

8. Multiply: $\dfrac{8ab^2}{15x^3y} \cdot \dfrac{5xy^4}{16a^2b}$

9. Simplify: $\dfrac{1 - \dfrac{1}{x}}{1 - \dfrac{8x-7}{x^2}}$

10. Write each fraction in terms of the LCM of the denominators.

$\dfrac{x}{12x^2 + 16x - 3}, \dfrac{4x^2}{6x^2 + 7x - 3}$

11. Solve $T = 2(ab + bc + ca)$ for a.

12. Solve: $\dfrac{5}{7} + \dfrac{x}{2} = 2 - \dfrac{x}{7}$

13. Solve $i = \dfrac{100m}{c}$ for c.

14. Solve: $\dfrac{x+8}{x+4} = 1 + \dfrac{5}{x+4}$

15. Divide: $\dfrac{20x^2 - 45x}{6x^3 + 4x^2} \div \dfrac{40x^3 - 90x^2}{12x^2 + 8x}$

16. Add: $\dfrac{2y}{5y-7} + \dfrac{3}{7-5y}$

Copyright © Houghton Mifflin Company. All rights reserved.

17. Subtract: $\dfrac{5x+3}{2x^2+5x-3} - \dfrac{3x+4}{2x^2+5x-3}$

18. Find the LCM of $10x^2 - 11x + 3$ and $20x^2 - 17x + 3$.

19. Solve $4x + 9y = 18$ for y.

20. Multiply: $\dfrac{24x^2-94x+15}{12x^2-49x+15} \cdot \dfrac{24x^2+7x-5}{4-27x+18x^2}$

21. Solve: $\dfrac{20}{2x+3} = \dfrac{17x}{2x+3} - 5$

22. Add: $\dfrac{x-1}{x+2} + \dfrac{3x-2}{5-x} + \dfrac{5x^2+15x-11}{x^2-3x-10}$

23. Solve: $\dfrac{6}{x-7} = \dfrac{8}{x-6}$

24. Solve: $\dfrac{3}{20} = \dfrac{x}{80}$

25. Triangles ABC and DEF are similar. Find the perimeter of triangle ABC.

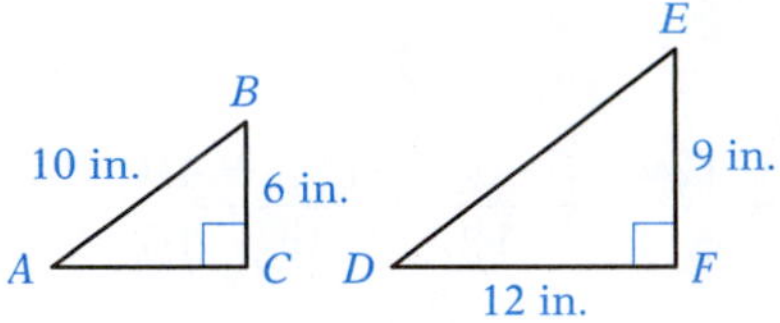

26. **Work** One hose can fill a pool in 15 h. The second hose can fill the pool in 10 h. How long would it take to fill the pool using both hoses?

27. **Travel** A car travels 315 mi in the same amount of time that a bus travels 245 mi. The rate of the car is 10 mph greater than that of the bus. Find the rate of the car.

28. **Travel** The rate of a jet is 400 mph in calm air. Traveling with the wind, the jet can fly 2100 mi in the same amount of time it takes to fly 1900 mi against the wind. Find the rate of the wind.

29. **Baseball** A pitcher's earned run average (ERA) is the average number of runs allowed in nine innings of pitching. If a pitcher allows 15 runs in 100 innings, find the pitcher's ERA.

30. **Electronics** The current (I) in an electric circuit varies inversely as the resistance (R). If the current in the circuit is 4 amps when the resistance is 50 ohms, find the current in the circuit when the resistance is 100 ohms.

Copyright © Houghton Mifflin Company. All rights reserved.

Chapter 8 Test

1. Simplify: $\dfrac{16x^5y}{24x^2y^4}$

2. Simplify: $\dfrac{x^2 + 4x - 5}{1 - x^2}$

3. Multiply: $\dfrac{x^3y^4}{x^2 - 4x + 4} \cdot \dfrac{x^2 - x - 2}{x^6y^4}$

4. Multiply: $\dfrac{x^2 + 2x - 3}{x^2 + 6x + 9} \cdot \dfrac{2x^2 - 11x + 5}{2x^2 + 3x - 5}$

5. Divide: $\dfrac{x^2 + 3x + 2}{x^2 + 5x + 4} \div \dfrac{x^2 - x - 6}{x^2 + 2x - 15}$

6. Find the LCM of $6x - 3$ and $2x^2 + x - 1$.

7. Write each fraction in terms of the LCM of the denominators.
$\dfrac{3}{x^2 - 2x}, \dfrac{x}{x^2 - 4}$

8. Subtract: $\dfrac{2x}{x^2 + 3x - 10} - \dfrac{4}{x^2 + 3x - 10}$

9. Subtract: $\dfrac{2}{2x - 1} - \dfrac{3}{3x + 1}$

10. Subtract: $\dfrac{x}{x + 3} - \dfrac{2x - 5}{x^2 + x - 6}$

11. Simplify: $\dfrac{1 + \dfrac{1}{x} - \dfrac{12}{x^2}}{1 + \dfrac{2}{x} - \dfrac{8}{x^2}}$

12. Solve: $\dfrac{6}{x} - 2 = 1$

Copyright © Houghton Mifflin Company. All rights reserved.

13. Solve: $\frac{2x}{x + 1} - 3 = \frac{-2}{x + 1}$

14. Solve: $\frac{3}{x + 4} = \frac{5}{x + 6}$

15. Triangles *ABC* and *DEF* are similar. Find the area of triangle *DEF*.

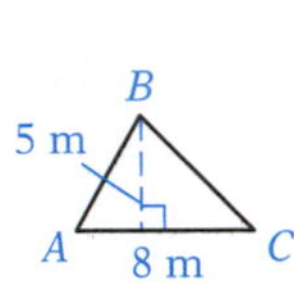

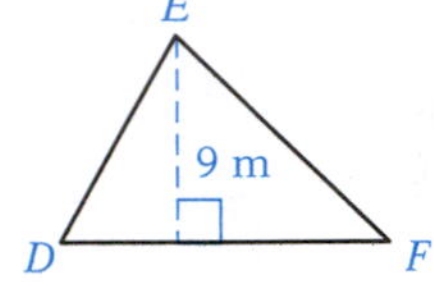

16. Solve $d = s + rt$ for t.

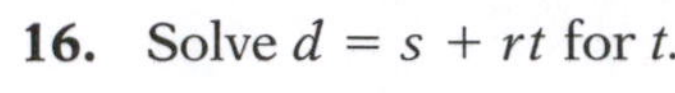

17. **Interior Design** An interior designer uses two rolls of wallpaper for every 45 ft^2 of wall space in an office. At this rate, how many rolls of wallpaper are needed for an office that has 315 ft^2 of wall space?

18. **Landscaping** One landscaper can till the soil for a lawn in 30 min, whereas it takes a second landscaper 15 min to do the same job. How long would it take to till the soil for the lawn with both landscapers working together?

19. **Travel** A cyclist travels 20 mi in the same amount of time that it takes a hiker to walk 6 mi. The rate of the cyclist is 7 mph faster than the rate of the hiker. Find the rate of the cyclist.

20. **Electronics** The electrical resistance (r) of a cable varies directly as its length (l) and inversely as the square of its diameter (d). If a cable 16,000 ft long and $\frac{1}{4}$ in. in diameter has a resistance of 3.2 ohms, what is the resistance of a cable that is 8000 ft long and $\frac{1}{2}$ in. in diameter?

Copyright © Houghton Mifflin Company. All rights reserved.

Cumulative Review Exercises

1. Simplify: $\left(\frac{2}{3}\right)^2 \div \left(\frac{3}{2} - \frac{2}{3}\right) + \frac{1}{2}$

2. Evaluate $-a^2 + (a - b)^2$ when $a = -2$ and $b = 3$.

3. Simplify: $-2x - (-3y) + 7x - 5y$

4. Simplify: $2[3x - 7(x - 3) - 8]$

5. Solve: $4 - \frac{2}{3}x = 7$

6. Solve: $3[x - 2(x - 3)] = 2(3 - 2x)$

7. Find $16\frac{2}{3}\%$ of 60.

8. Solve: $x - 3(1 - 2x) \geq 1 - 4(3 - 2x)$

9. Find the volume of the rectangular solid shown in the figure.

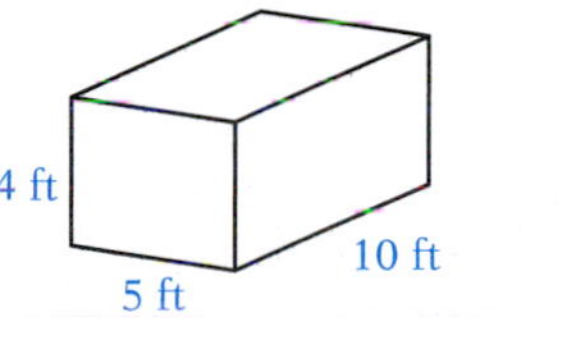

10. 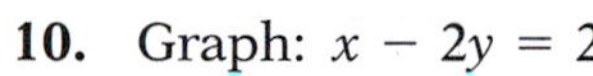
Graph: $x - 2y = 2$

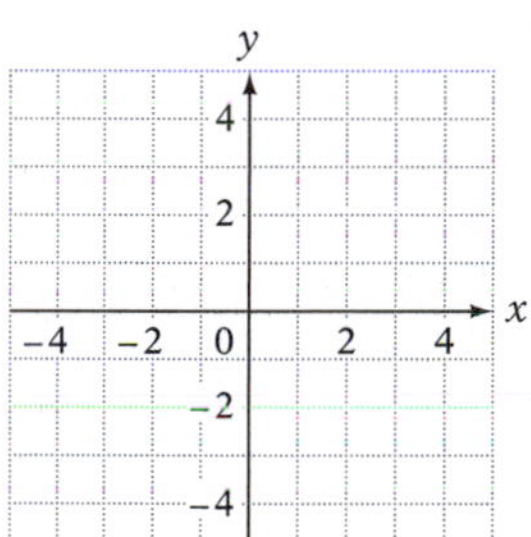

11. Given $P(x) = \frac{x - 1}{2x - 3}$, find $P(-2)$.

12. Find the equation of the line that contains the point $(-2, -1)$ and is parallel to the line $3x - 2y = 6$.

13. Solve:
$$\begin{aligned} 2x - y + z &= 2 \\ 3x + y - 2z &= 9 \\ x - y + z &= 0 \end{aligned}$$

14. Multiply: $(a^2b^5)(ab^2)$

15. Simplify: $\frac{(2a^{-2}b^3)^{-2}}{(4a)^{-1}}$

16. Write 0.000000035 in scientific notation.

17. Multiply: $(2a^2 - 3a + 1)(-2a^2)$

18. Multiply: $(a - 3b)(a + 4b)$

Copyright © Houghton Mifflin Company. All rights reserved.

19. Divide: $(x^3 - 8) \div (x - 2)$

20. Factor: $y^2 - 7y + 6$

21. Factor: $12x^2 - x - 1$

22. Factor: $2a^3 + 7a^2 - 15a$

23. Factor: $4b^2 - 100$

24. Solve: $(x + 3)(2x - 5) = 0$

25. Simplify: $\dfrac{12x^4y^2}{18xy^7}$

26. Simplify: $\dfrac{x^2 - 7x + 10}{25 - x^2}$

27. Divide: $\dfrac{x^2 - x - 56}{x^2 + 8x + 7} \div \dfrac{x^2 - 13x + 40}{x^2 - 4x - 5}$

28. Subtract: $\dfrac{2}{2x - 1} - \dfrac{1}{x + 1}$

29. Simplify: $\dfrac{1 - \dfrac{2}{x} - \dfrac{15}{x^2}}{1 - \dfrac{25}{x^2}}$

30. Solve: $\dfrac{3x}{x - 3} - 2 = \dfrac{10}{x - 3}$

31. **Mixtures** A silversmith mixes 60 g of an alloy that is 40% silver with 120 g of another silver alloy. The resulting alloy is 60% silver. Find the percent of silver in the 120 g of alloy.

32. **Insurance** A life insurance policy costs \$32 for every \$1000 of coverage. At this rate, how much money would a policy of \$5000 cost?

33. **Work** One water pipe can fill a tank in 9 min, whereas a second pipe requires 18 min to fill the tank. How long would it take both pipes, working together, to fill the tank?

Copyright © Houghton Mifflin Company. All rights reserved.

chapter

9 Exponents and Radicals

Thousands of man-made satellites have been launched into space and are now orbiting Earth. These satellites do hundreds of different tasks, from collecting data on our atmosphere to facilitating telecommunications. The speed at which a satellite orbits Earth is a function of its height above Earth's surface. The height at which it orbits Earth is dependent on the satellite's purpose. The relationship between a satellite's speed and its height above Earth can be modeled by a radical equation, such as the one used in **Exercise 26 on page 542.**

OBJECTIVES

Section 9.1

A To simplify expressions with rational exponents

B To write exponential expressions as radical expressions and to write radical expressions as exponential expressions

C To simplify radical expressions that are roots of perfect powers

Section 9.2

A To simplify radical expressions

B To add or subtract radical expressions

C To multiply radical expressions

D To divide radical expressions

Section 9.3

A To solve a radical equation

B To solve application problems

Section 9.4

A To simplify a complex number

B To add or subtract complex numbers

C To multiply complex numbers

D To divide complex numbers

Copyright © Houghton Mifflin Company. All rights reserved.

Need help? For online student resources, such as section quizzes, visit this textbook's website at **math.college.hmco.com/students.**

PREP TEST

Do these exercises to prepare for Chapter 9.

1. Complete: $48 = ? \cdot 3$

For Exercises 2 to 6, simplify.

2. 2^5

3. $6\left(\frac{3}{2}\right)$

4. $\frac{1}{2} - \frac{2}{3} + \frac{1}{4}$

5. $(3 - 7x) - (4 - 2x)$

6. $\frac{3x^5y^6}{12x^4y}$

7. Expand: $(3x - 2)^2$

For Exercises 8 and 9, multiply.

8. $(2 + 4x)(5 - 3x)$

9. $(6x - 1)(6x + 1)$

10. Solve: $x^2 - 14x - 5 = 10$

GO FIGURE

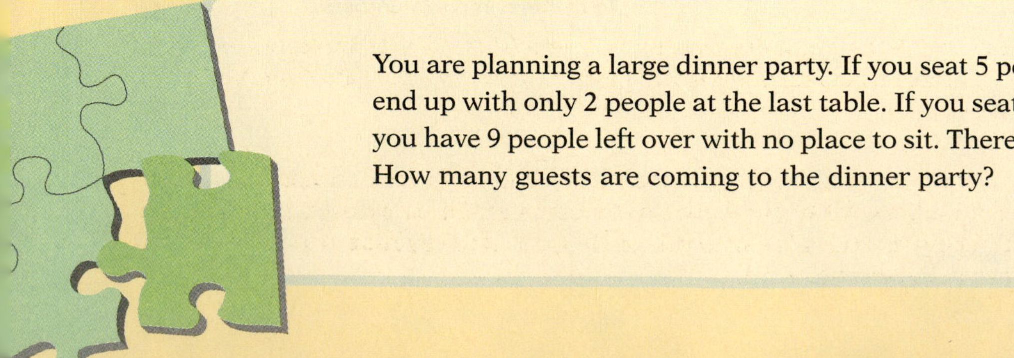

You are planning a large dinner party. If you seat 5 people at each table, you end up with only 2 people at the last table. If you seat 3 people at each table, you have 9 people left over with no place to sit. There are less than 10 tables. How many guests are coming to the dinner party?

Copyright © Houghton Mifflin Company. All rights reserved.

9.1 Rational Exponents and Radical Expressions

Objective A To simplify expressions with rational exponents

Point of Interest

Nicolas Chuquet (c. 1475), a French physician, wrote an algebra text in which he used a notation for expressions with fractional exponents. He wrote $R^2 6$ to mean $6^{1/2}$ and $R^3 15$ to mean $15^{1/3}$. This was an improvement over earlier notations that used words for these expressions.

In this section, the definition of an exponent is extended beyond integers so that any rational number can be used as an exponent. The definition is expressed in such a way that the Rules of Exponents hold true for rational exponents.

Consider the expression $(a^{1/n})^n$ for $a > 0$ and n a positive integer. Now simplify, assuming that the Rule for Simplifying the Power of an Exponential Expression is true.

$$(a^{1/n})^n = a^{\frac{1}{n} \cdot n} = a^1 = a$$

Because $(a^{1/n})^n = a$, the number $a^{1/n}$ is the number whose nth power is a.

If $a > 0$ and n is a positive number, then $a^{1/n}$ is called the nth root of a.

$25^{1/2} = 5$ because $(5)^2 = 25$.

$8^{1/3} = 2$ because $(2)^3 = 8$.

In the expression $a^{1/n}$, if a is a negative number and n is a positive even integer, then $a^{1/n}$ is not a real number.

$(-4)^{1/2}$ is not a real number, because there is no real number whose second power is -4.

When n is a positive odd integer, a can be a positive or a negative number.

$(-27)^{1/3} = -3$ because $(-3)^3 = -27$.

Integrating Technology

A calculator can be used to evaluate expressions with rational exponents. For example, to evaluate the expression at the right, press

3) ENTER. The display reads -3.

Using the definition of $a^{1/n}$ and the Rules of Exponents, it is possible to define any exponential expression that contains a rational exponent.

> **Rule for Rational Exponents**
>
> If m and n are positive integers and $a^{1/n}$ is a real number, then
>
> $$a^{m/n} = (a^{1/n})^m$$

The expression $a^{m/n}$ can also be written $a^{m/n} = a^{m \cdot \frac{1}{n}} = (a^m)^{1/n}$. However, rewriting $a^{m/n}$ as $(a^m)^{1/n}$ is not as useful as rewriting it as $(a^{1/n})^m$. See the Take Note at the top of the next page.

As shown above, expressions that contain rational exponents do not always represent real numbers when the base of the exponential expression is a negative number. For this reason, **all variables in this chapter represent positive numbers unless otherwise stated.**

Copyright © Houghton Mifflin Company. All rights reserved.

TAKE NOTE

Although we can simplify an expression by rewriting $a^{m/n}$ in the form $(a^m)^{1/n}$, it is usually easier to simplify the form $(a^{1/n})^m$. For instance, simplifying $(27^{1/3})^2$ is easier than simplifying $(27^2)^{1/3}$.

HOW TO Simplify: $27^{2/3}$

$$\begin{aligned} 27^{2/3} &= (3^3)^{2/3} && \text{• Rewrite 27 as } 3^3. \\ &= 3^{3(2/3)} && \text{• Multiply the exponents.} \\ &= 3^2 && \text{• Simplify.} \\ &= 9 \end{aligned}$$

TAKE NOTE

Note that $32^{-2/5} = \frac{1}{4}$, a positive number. A negative exponent does not affect the sign of a number.

HOW TO Simplify: $32^{-2/5}$

$$\begin{aligned} 32^{-2/5} &= (2^5)^{-2/5} && \text{• Rewrite 32 as } 2^5. \\ &= 2^{-2} && \text{• Multiply the exponents.} \\ &= \frac{1}{2^2} && \text{• Use the Rule of Negative Exponents.} \\ &= \frac{1}{4} && \text{• Simplify.} \end{aligned}$$

HOW TO Simplify: $a^{1/2} \cdot a^{2/3} \cdot a^{-1/4}$

$$\begin{aligned} a^{1/2} \cdot a^{2/3} \cdot a^{-1/4} &= a^{1/2 + 2/3 - 1/4} && \text{• Use the Rule for Multiplying Exponential Expressions.} \\ &= a^{6/12 + 8/12 - 3/12} \\ &= a^{11/12} && \text{• Simplify.} \end{aligned}$$

HOW TO Simplify: $(x^6y^4)^{3/2}$

$$\begin{aligned} (x^6y^4)^{3/2} &= x^{6(3/2)}y^{4(3/2)} && \text{• Use the Rule for Simplifying Powers of Products.} \\ &= x^9y^6 && \text{• Simplify.} \end{aligned}$$

Study Tip

Remember that a HOW TO indicates a worked-out example. Using paper and pencil, work through the example. See *AIM for Success*, pages xxvii and xxviii.

HOW TO Simplify: $\left(\frac{8a^3b^{-4}}{64a^{-9}b^2}\right)^{2/3}$

$$\begin{aligned} \left(\frac{8a^3b^{-4}}{64a^{-9}b^2}\right)^{2/3} &= \left(\frac{2^3a^3b^{-4}}{2^6a^{-9}b^2}\right)^{2/3} && \text{• Rewrite 8 as } 2^3 \text{ and 64 as } 2^6. \\ &= (2^{-3}a^{12}b^{-6})^{2/3} && \text{• Use the Rule for Dividing Exponential Expressions.} \\ &= 2^{-2}a^8b^{-4} && \text{• Use the Rule for Simplifying Powers of Products.} \\ &= \frac{a^8}{2^2b^4} = \frac{a^8}{4b^4} && \text{• Use the Rule of Negative Exponents and simplify.} \end{aligned}$$

Copyright © Houghton Mifflin Company. All rights reserved.

Example 1 Simplify: $64^{-2/3}$

Solution $64^{-2/3} = (2^6)^{-2/3} = 2^{-4}$ • $64 = 2^6$

$= \frac{1}{2^4} = \frac{1}{16}$

You Try It 1 Simplify: $16^{-3/4}$

Your solution

Example 2 Simplify: $(-49)^{3/2}$

Solution The base of the exponential expression is a negative number, while the denominator of the exponent is a positive even number.

Therefore, $(-49)^{3/2}$ is not a real number.

You Try It 2 Simplify: $(-81)^{3/4}$

Your solution

Example 3 Simplify: $(x^{1/2}y^{-3/2}z^{1/4})^{-3/2}$

Solution $(x^{1/2}y^{-3/2}z^{1/4})^{-3/2}$

$= x^{-3/4}y^{9/4}z^{-3/8}$

$= \frac{y^{9/4}}{x^{3/4}z^{3/8}}$

• Use the Rule for Simplifying Powers of Products.

You Try It 3 Simplify: $(x^{3/4}y^{1/2}z^{-2/3})^{-4/3}$

Your solution

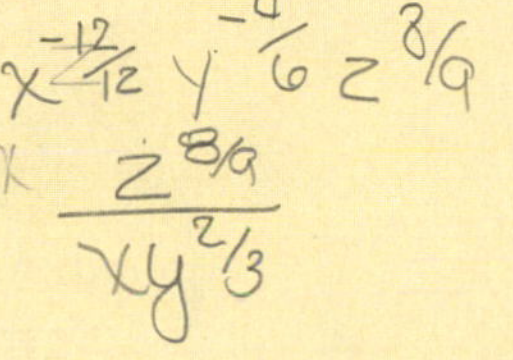

Example 4 Simplify: $\frac{x^{1/2}y^{-5/4}}{x^{-4/3}y^{1/3}}$

Solution $\frac{x^{1/2}y^{-5/4}}{x^{-4/3}y^{1/3}}$

$= x^{3/6-(-8/6)}y^{-15/12-4/12}$

$= x^{11/6}y^{-19/12} = \frac{x^{11/6}}{y^{19/12}}$

• Use the Rule for Dividing Exponential Expressions.

You Try It 4 Simplify: $\left(\frac{16a^{-2}b^{4/3}}{9a^4b^{-2/3}}\right)^{-1/2}$

Your solution

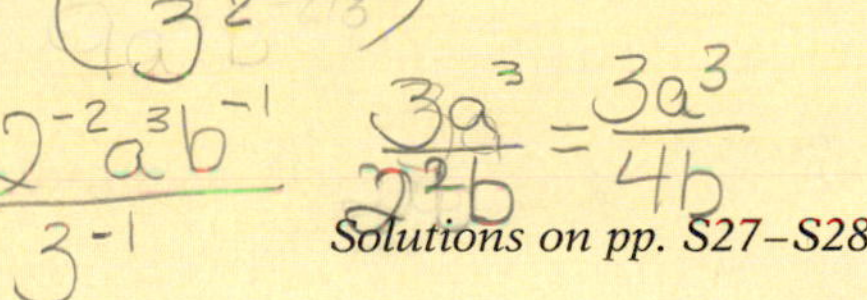

Solutions on pp. S27–S28

Objective B To write exponential expressions as radical expressions and to write radical expressions as exponential expressions

Point of Interest

The radical sign was introduced in 1525 in a book by Christoff Rudolff called *Coss*. He modified the symbol to indicate square roots, cube roots, and fourth roots. The idea of using an index, as we do in our modern notation, did not occur until some years later.

Recall that $a^{1/n}$ is the nth root of a. The expression $\sqrt[n]{a}$ is another symbol for the nth root of a.

If a is a real number, then $a^{1/n} = \sqrt[n]{a}$.

In the expression $\sqrt[n]{a}$, the symbol $\sqrt{}$ is called a **radical,** n is the **index** of the radical, and a is the **radicand.** When $n = 2$, the radical expression represents a square root and the index 2 is usually not written.

An exponential expression with a rational exponent can be written as a radical expression.

Writing Exponential Expressions as Radical Expressions

If $a^{1/n}$ is a real number, then $a^{1/n} = \sqrt[n]{a}$ and $a^{m/n} = a^{m \cdot 1/n} = (a^m)^{1/n} = \sqrt[n]{a^m}$.

The expression $a^{m/n}$ can also be written $a^{m/n} = (a^{1/n})^m = (\sqrt[n]{a})^m$.

Copyright © Houghton Mifflin Company. All rights reserved.

The exponential expression at the right has been written as a radical expression.

$y^{2/3} = (y^2)^{1/3}$
$= \sqrt[3]{y^2}$

The radical expressions at the right have been written as exponential expressions.

$\sqrt[5]{x^6} = (x^6)^{1/5} = x^{6/5}$
$\sqrt{17} = (17)^{1/2} = 17^{1/2}$

HOW TO Write $(5x)^{2/5}$ as a radical expression.

$(5x)^{2/5} = \sqrt[5]{(5x)^2}$ • The denominator of the rational exponent is the index of the radical. The numerator is the power of the radicand.

$= \sqrt[5]{25x^2}$ • Simplify.

HOW TO Write $\sqrt[3]{x^4}$ as an exponential expression with a rational exponent.

$\sqrt[3]{x^4} = (x^4)^{1/3}$ • The index of the radical is the denominator of the rational exponent. The power of the radicand is the numerator of the rational exponent.

$= x^{4/3}$

HOW TO Write $\sqrt[3]{a^3 + b^3}$ as an exponential expression with a rational exponent.

$\sqrt[3]{a^3 + b^3} = (a^3 + b^3)^{1/3}$

Note that $(a^3 + b^3)^{1/3} \neq a + b$.

Example 5 Write $(3x)^{5/4}$ as a radical expression.

Solution $(3x)^{5/4} = \sqrt[4]{(3x)^5} = \sqrt[4]{243x^5}$

You Try It 5 Write $(2x^3)^{3/4}$ as a radical expression.

Your solution

Example 6 Write $-2x^{2/3}$ as a radical expression.

Solution $-2x^{2/3} = -2(x^2)^{1/3} = -2\sqrt[3]{x^2}$

You Try It 6 Write $-5a^{5/6}$ as a radical expression.

Your solution

Example 7 Write $\sqrt[4]{3a}$ as an exponential expression.

Solution $\sqrt[4]{3a} = (3a)^{1/4}$

You Try It 7 Write $\sqrt[3]{3ab}$ as an exponential expression.

Your solution

Example 8 Write $\sqrt{a^2 - b^2}$ as an exponential expression.

Solution $\sqrt{a^2 - b^2} = (a^2 - b^2)^{1/2}$

You Try It 8 Write $\sqrt[4]{x^4 + y^4}$ as an exponential expression.

Your solution

Solutions on p. S28

Copyright © Houghton Mifflin Company. All rights reserved.

Objective C

To simplify radical expressions that are roots of perfect powers

Every positive number has two square roots, one a positive number and one a negative number. For example, because $(5)^2 = 25$ and $(-5)^2 = 25$, there are two square roots of 25: 5 and −5.

The symbol $\sqrt{\ }$ is used to indicate the positive square root, or **principal square root.** To indicate the negative square root of a number, a negative sign is placed in front of the radical.

$\sqrt{25} = 5$

$-\sqrt{25} = -5$

The square root of zero is zero. $\sqrt{0} = 0$

The square root of a negative number is not a real number, because the square of a real number must be positive. $\sqrt{-25}$ is not a real number.

Note that

$$\sqrt{(-5)^2} = \sqrt{25} = 5 \quad \text{and} \quad \sqrt{5^2} = \sqrt{25} = 5$$

This is true for all real numbers and is stated as the following result.

For any real number a, $\sqrt{a^2} = |a|$ and $-\sqrt{a^2} = -|a|$. If a is a positive real number, then $\sqrt{a^2} = a$ and $(\sqrt{a})^2 = a$.

Integrating Technology

See the appendix Keystroke Guide: *Radical Expressions* for instructions on using a graphing calculator to evaluate a numerical radical expression.

Besides square roots, we can also determine cube roots, fourth roots, and so on.

$\sqrt[3]{8} = 2$, because $2^3 = 8$. • The cube root of a positive number is positive.

$\sqrt[3]{-8} = -2$, because $(-2)^3 = -8$. • The cube root of a negative number is negative.

$\sqrt[4]{625} = 5$, because $5^4 = 625$.

$\sqrt[5]{243} = 3$, because $3^5 = 243$.

The following properties hold true for finding the nth root of a real number.

If n is an even integer, then $\sqrt[n]{a^n} = |a|$ and $-\sqrt[n]{a^n} = -|a|$. If n is an odd integer, then $\sqrt[n]{a^n} = a$.

For example,

$$\sqrt[6]{y^6} = |y| \qquad -\sqrt[12]{x^{12}} = -|x| \qquad \sqrt[5]{b^5} = b$$

TAKE NOTE

Note that when the index is an even natural number, the nth root requires absolute value symbols.

$\sqrt[6]{y^6} = |y|$ but $\sqrt[5]{y^5} = y$

Because we stated that variables within radicals represent *positive* numbers, we will omit the absolute value symbols when writing an answer.

For the remainder of this chapter, we will assume that variable expressions inside a radical represent positive numbers. Therefore, it is not necessary to use the absolute value signs.

HOW TO Simplify: $\sqrt[4]{x^4y^8}$

$\sqrt[4]{x^4y^8} = (x^4y^8)^{1/4}$ • The radicand is a perfect fourth power because the exponents on the variables are divisible by 4. Write the radical expression as an exponential expression.

$= xy^2$ • Use the Rule for Simplifying Powers of Products.

Copyright © Houghton Mifflin Company. All rights reserved.

HOW TO Simplify: $\sqrt[3]{125c^9d^6}$

$\sqrt[3]{125c^9d^6} = (5^3c^9d^6)^{1/3}$

- The radicand is a perfect cube because 125 is a perfect cube ($125 = 5^3$) and all the exponents on the variables are divisible by 3.

$= 5c^3d^2$

- Use the Rule for Simplifying Powers of Products.

Note that **a variable expression is a perfect power if the exponents on the factors are evenly divisible by the index of the radical.**

The chart below shows roots of perfect powers. Knowledge of these roots is very helpful when simplifying radical expressions.

Square Roots		*Cube Roots*	*Fourth Roots*	*Fifth Roots*
$\sqrt{1} = 1$	$\sqrt{36} = 6$	$\sqrt[3]{1} = 1$	$\sqrt[4]{1} = 1$	$\sqrt[5]{1} = 1$
$\sqrt{4} = 2$	$\sqrt{49} = 7$	$\sqrt[3]{8} = 2$	$\sqrt[4]{16} = 2$	$\sqrt[5]{32} = 2$
$\sqrt{9} = 3$	$\sqrt{64} = 8$	$\sqrt[3]{27} = 3$	$\sqrt[4]{81} = 3$	$\sqrt[5]{243} = 3$
$\sqrt{16} = 4$	$\sqrt{81} = 9$	$\sqrt[3]{64} = 4$	$\sqrt[4]{256} = 4$	
$\sqrt{25} = 5$	$\sqrt{100} = 10$	$\sqrt[3]{125} = 5$	$\sqrt[4]{625} = 5$	

TAKE NOTE

From the chart, $\sqrt[5]{243} = 3$, which means that $3^5 = 243$. From this we know that $(-3)^5 = -243$, which means $\sqrt[5]{-243} = -3$.

HOW TO Simplify: $\sqrt[5]{-243x^5y^{15}}$

$\sqrt[5]{-243x^5y^{15}} = -3xy^3$

- From the chart, 243 is a perfect fifth power, and each exponent is divisible by 5. Therefore, the radicand is a perfect fifth power.

Example 9 Simplify: $\sqrt[3]{-125a^6b^9}$

Solution The radicand is a perfect cube.

$\sqrt[3]{-125a^6b^9} = -5a^2b^3$

- Divide each exponent by 3.

You Try It 9 Simplify: $\sqrt[3]{-8x^{12}y^3}$

Your solution

Example 10 Simplify: $-\sqrt[4]{16a^4b^8}$

Solution The radicand is a perfect fourth power.

$-\sqrt[4]{16a^4b^8} = -2ab^2$

- Divide each exponent by 4.

You Try It 10 Simplify: $-\sqrt[4]{81x^{12}y^8}$

Your solution

Solutions on p. S28

Copyright © Houghton Mifflin Company. All rights reserved.

9.1 Exercises

Objective A **To simplify expressions with rational exponents**

For Exercises 1 to 80, simplify.

1. $8^{1/3}$ **2.** $16^{1/2}$ **3.** $9^{3/2}$ **4.** $25^{3/2}$ **5.** $27^{-2/3}$

6. $64^{-1/3}$ **7.** $32^{2/5}$ **8.** $16^{3/4}$ **9.** $(-25)^{5/2}$ **10.** $(-36)^{1/4}$

11. $\left(\frac{25}{49}\right)^{-3/2}$ **12.** $\left(\frac{8}{27}\right)^{-2/3}$ **13.** $x^{1/2}x^{1/2}$ **14.** $a^{1/3}a^{5/3}$ **15.** $y^{-1/4}y^{3/4}$

16. $x^{2/5} \cdot x^{-4/5}$ **17.** $x^{-2/3} \cdot x^{3/4}$ **18.** $x \cdot x^{-1/2}$ **19.** $a^{1/3} \cdot a^{3/4} \cdot a^{-1/2}$ **20.** $y^{-1/6} \cdot y^{2/3} \cdot y^{1/2}$

21. $\frac{a^{1/2}}{a^{3/2}}$ **22.** $\frac{b^{1/3}}{b^{4/3}}$ **23.** $\frac{y^{-3/4}}{y^{1/4}}$ **24.** $\frac{x^{-3/5}}{x^{1/5}}$ **25.** $\frac{y^{2/3}}{y^{-5/6}}$

26. $\frac{b^{3/4}}{b^{-3/2}}$ **27.** $(x^2)^{-1/2}$ **28.** $(a^8)^{-3/4}$ **29.** $(x^{-2/3})^6$ **30.** $(y^{-5/6})^{12}$

31. $(a^{-1/2})^{-2}$ **32.** $(b^{-2/3})^{-6}$ **33.** $(x^{-3/8})^{-4/5}$ **34.** $(y^{-3/2})^{-2/9}$

35. $(a^{1/2} \cdot a)^2$ **36.** $(b^{2/3} \cdot b^{1/6})^6$ **37.** $(x^{-1/2} \cdot x^{3/4})^{-2}$ **38.** $(a^{1/2} \cdot a^{-2})^3$

39. $(y^{-1/2} \cdot y^{2/3})^{2/3}$ **40.** $(b^{-2/3} \cdot b^{1/4})^{-4/3}$ **41.** $(x^8y^2)^{1/2}$ **42.** $(a^3b^9)^{2/3}$

Copyright © Houghton Mifflin Company. All rights reserved.

43. $(x^4y^2z^6)^{3/2}$

44. $(a^8b^4c^4)^{3/4}$

45. $(x^{-3}y^6)^{-1/3}$

46. $(a^2b^{-6})^{-1/2}$

47. $(x^{-2}y^{1/3})^{-3/4}$

48. $(a^{-2/3}b^{2/3})^{3/2}$

49. $\left(\dfrac{x^{1/2}}{y^2}\right)^4$

50. $\left(\dfrac{b^{-3/4}}{a^{-1/2}}\right)^8$

51. $\dfrac{x^{1/4} \cdot x^{-1/2}}{x^{2/3}}$

52. $\dfrac{b^{1/2} \cdot b^{-3/4}}{b^{1/4}}$

53. $\left(\dfrac{y^{2/3} \cdot y^{-5/6}}{y^{1/9}}\right)^9$

54. $\left(\dfrac{a^{1/3} \cdot a^{-2/3}}{a^{1/2}}\right)^4$

55. $\left(\dfrac{b^2 \cdot b^{-3/4}}{b^{-1/2}}\right)^{-1/2}$

56. $\dfrac{(x^{-5/6} \cdot x^3)^{-2/3}}{x^{4/3}}$

57. $(a^{2/3}b^2)^6(a^3b^3)^{1/3}$

58. $(x^3y^{-1/2})^{-2}(x^{-3}y^2)^{1/6}$

59. $(16m^{-2}n^4)^{-1/2}(mn^{1/2})$

60. $(27m^3n^{-6})^{1/3}(m^{-1/3}n^{5/6})^6$

61. $\left(\dfrac{x^{1/2}y^{-3/4}}{y^{2/3}}\right)^{-6}$

62. $\left(\dfrac{x^{1/2}y^{-5/4}}{y^{-3/4}}\right)^{-4}$

63. $\left(\dfrac{2^{-6}b^{-3}}{a^{-1/2}}\right)^{-2/3}$

64. $\left(\dfrac{49c^{5/3}}{a^{-1/4}b^{5/6}}\right)^{-3/2}$

65. $y^{3/2}(y^{1/2} - y^{-1/2})$

66. $y^{3/5}(y^{2/5} + y^{-3/5})$

67. $a^{-1/4}(a^{5/4} - a^{9/4})$

68. $x^{4/3}(x^{2/3} + x^{-1/3})$

69. $x^n \cdot x^{3n}$

70. $a^{2n} \cdot a^{-5n}$

71. $x^n \cdot x^{n/2}$

72. $a^{n/2} \cdot a^{-n/3}$

73. $\dfrac{y^{n/2}}{y^{-n}}$

74. $\dfrac{b^{m/3}}{b^m}$

75. $(x^{2n})^n$

76. $(x^{5n})^{2n}$

77. $(x^{n/4}y^{n/8})^8$

78. $(x^{n/2}y^{n/3})^6$

79. $(x^{n/5}y^{n/10})^{20}$

80. $(x^{n/2}y^{n/5})^{10}$

Copyright © Houghton Mifflin Company. All rights reserved.

Objective B **To write exponential expressions as radical expressions and to write radical expressions as exponential expressions**

For Exercises 81 to 96, rewrite the exponential expression as a radical expression.

81. $3^{1/4}$ **82.** $5^{1/2}$ **83.** $a^{3/2}$ **84.** $b^{4/3}$

85. $(2t)^{5/2}$ **86.** $(3x)^{2/3}$ **87.** $-2x^{2/3}$ **88.** $-3a^{2/5}$

89. $(a^2b)^{2/3}$ **90.** $(x^2y^3)^{3/4}$ **91.** $(a^2b^4)^{3/5}$ **92.** $(a^3b^7)^{3/2}$

93. $(4x - 3)^{3/4}$ **94.** $(3x - 2)^{1/3}$ **95.** $x^{-2/3}$ **96.** $b^{-3/4}$

For Exercises 97 to 112, rewrite the radical expression as an exponential expression.

97. $\sqrt{14}$ **98.** $\sqrt{7}$ **99.** $\sqrt[3]{x}$ **100.** $\sqrt[4]{y}$

101. $\sqrt[3]{x^4}$ **102.** $\sqrt[4]{a^3}$ **103.** $\sqrt[5]{b^3}$ **104.** $\sqrt[4]{b^5}$

105. $\sqrt[3]{2x^2}$ **106.** $\sqrt[5]{4y^7}$ **107.** $-\sqrt{3x^5}$ **108.** $-\sqrt[4]{4x^5}$

109. $3x\sqrt[3]{y^2}$ **110.** $2y\sqrt{x^3}$ **111.** $\sqrt{a^2 - 2}$ **112.** $\sqrt{3 - y^2}$

Copyright © Houghton Mifflin Company. All rights reserved.

Objective C To simplify radical expressions that are roots of perfect powers

For Exercises 113 to 136, simplify.

113. $\sqrt{x^{16}}$

114. $\sqrt{y^{14}}$

115. $-\sqrt{x^8}$

116. $-\sqrt{a^6}$

117. $\sqrt[3]{x^3y^9}$

118. $\sqrt[3]{a^6b^{12}}$

119. $-\sqrt[3]{x^{15}y^3}$

120. $-\sqrt[3]{a^9b^9}$

121. $\sqrt{16a^4b^{12}}$

122. $\sqrt{25x^8y^2}$

123. $\sqrt{-16x^4y^2}$

124. $\sqrt{-9a^4b^8}$

125. $\sqrt[3]{27x^9}$

126. $\sqrt[3]{8a^{21}b^6}$

127. $\sqrt[3]{-64x^9y^{12}}$

128. $\sqrt[3]{-27a^3b^{15}}$

129. $-\sqrt[4]{x^8y^{12}}$

130. $-\sqrt[4]{a^{16}b^4}$

131. $\sqrt[5]{x^{20}y^{10}}$

132. $\sqrt[5]{a^5b^{25}}$

133. $\sqrt[4]{81x^4y^{20}}$

134. $\sqrt[4]{16a^8b^{20}}$

135. $\sqrt[5]{32a^5b^{10}}$

136. $\sqrt[5]{-32x^{15}y^{20}}$

APPLYING THE CONCEPTS

137. Determine whether the following statements are true or false. If the statement is false, correct the right-hand side of the equation.

a. $\sqrt{(-2)^2} = -2$ **b.** $\sqrt[3]{(-3)^3} = -3$ **c.** $\sqrt[n]{a} = a^{1/n}$

d. $\sqrt[n]{a^n + b^n} = a + b$ **e.** $(a^{1/2} + b^{1/2})^2 = a + b$ **f.** $\sqrt[m]{a^n} = a^{mn}$

138. Simplify.

a. $\sqrt[3]{\sqrt{x^6}}$ **b.** $\sqrt[4]{\sqrt{a^8}}$ **c.** $\sqrt{\sqrt{81y^8}}$

d. $\sqrt{\sqrt[n]{a^{4n}}}$ **e.** $\sqrt[n]{\sqrt{b^{6n}}}$ **f.** $\sqrt{\sqrt[3]{x^{12}y^{24}}}$

139. If x is any real number, is $\sqrt{x^2} = x$ always true? Show why or why not.

Copyright © Houghton Mifflin Company. All rights reserved.

9.2 Operations on Radical Expressions

Objective A To simplify radical expressions

Point of Interest

The Latin expression for irrational numbers was *numerus surdus*, which literally means "inaudible number." A prominent 16th-century mathematician wrote of irrational numbers, ". . . just as an infinite number is not a number, so an irrational number is not a true number, but lies hidden in some sort of cloud of infinity." In 1872, Richard Dedekind wrote a paper that established the first logical treatment of irrational numbers.

If a number is not a perfect power, its root can only be approximated; examples include $\sqrt{5}$ and $\sqrt[3]{3}$. These numbers are irrational numbers. Their decimal representations never terminate or repeat.

$$\sqrt{5} = 2.2360679\ldots \qquad \sqrt[3]{3} = 1.4422495\ldots$$

A radical expression is in simplest form when the radicand contains no factor that is a perfect power. The Product Property of Radicals is used to simplify radical expressions whose radicands are not perfect powers.

The Product Property of Radicals

If $\sqrt[n]{a}$ and $\sqrt[n]{b}$ are positive real numbers, then $\sqrt[n]{ab} = \sqrt[n]{a} \cdot \sqrt[n]{b}$ and $\sqrt[n]{a} \cdot \sqrt[n]{b} = \sqrt[n]{ab}$.

HOW TO Simplify: $\sqrt{48}$

$\sqrt{48} = \sqrt{16 \cdot 3}$
- Write the radicand as the product of a perfect square and a factor that does not contain a perfect square.

$= \sqrt{16}\sqrt{3}$
- Use the Product Property of Radicals to write the expression as a product.

$= 4\sqrt{3}$
- Simplify $\sqrt{16}$.

Note that 48 must be written as the product of a perfect square and *a factor that does not contain a perfect square*. Therefore, it would not be correct to rewrite $\sqrt{48}$ as $\sqrt{4 \cdot 12}$ and simplify the expression as shown at the right. Although 4 is a perfect square factor of 48, 12 contains a perfect square $(12 = 4 \cdot 3)$ and thus $\sqrt{12}$ can be simplified further. Remember to find the *largest* perfect power that is a factor of the radicand.

$$\begin{aligned}\sqrt{48} &= \sqrt{4 \cdot 12} \\ &= \sqrt{4}\sqrt{12} \\ &= 2\sqrt{12}\end{aligned}$$

Not in simplest form

HOW TO Simplify: $\sqrt{18x^2y^3}$

$\sqrt{18x^2y^3} = \sqrt{9x^2y^2 \cdot 2y}$
- Write the radicand as the product of a perfect square and factors that do not contain a perfect square.

$= \sqrt{9x^2y^2}\sqrt{2y}$
- Use the Product Property of Radicals to write the expression as a product.

$= 3xy\sqrt{2y}$
- Simplify.

Copyright © Houghton Mifflin Company. All rights reserved.

HOW TO Simplify: $\sqrt[3]{x^7}$

$\sqrt[3]{x^7} = \sqrt[3]{x^6 \cdot x}$
- Write the radicand as the product of a perfect cube and a factor that does not contain a perfect cube.

$= \sqrt[3]{x^6}\,\sqrt[3]{x}$
- Use the Product Property of Radicals to write the expression as a product.

$= x^2\sqrt[3]{x}$
- Simplify.

HOW TO Simplify: $\sqrt[4]{32x^7}$

$\sqrt[4]{32x^7} = \sqrt[4]{16x^4(2x^3)}$
- Write the radicand as the product of a perfect fourth power and factors that do not contain a perfect fourth power.

$= \sqrt[4]{16x^4}\,\sqrt[4]{2x^3}$
- Use the Product Property of Radicals to write the expression as a product.

$= 2x\sqrt[4]{2x^3}$
- Simplify.

Example 1 Simplify: $\sqrt[4]{x^9}$

Solution

$\sqrt[4]{x^9} = \sqrt[4]{x^8 \cdot x}$
$= \sqrt[4]{x^8}\,\sqrt[4]{x}$
$= x^2\sqrt[4]{x}$

- x^8 is a perfect fourth power.

You Try It 1 Simplify: $\sqrt[5]{x^7}$

Your solution

Example 2 Simplify: $\sqrt[3]{-27a^5b^{12}}$

Solution

$\sqrt[3]{-27a^5b^{12}}$
$= \sqrt[3]{-27a^3b^{12}(a^2)}$
$= \sqrt[3]{-27a^3b^{12}}\,\sqrt[3]{a^2}$
$= -3ab^4\sqrt[3]{a^2}$

- $-27a^3b^{12}$ is a perfect third power.

You Try It 2 Simplify: $\sqrt[3]{-64x^8y^{18}}$

Your solution

Solutions on p. S28

Objective B To add or subtract radical expressions

The Distributive Property is used to simplify the sum or difference of radical expressions that have the same radicand and the same index. For example,

$$3\sqrt{5} + 8\sqrt{5} = (3+8)\sqrt{5} = 11\sqrt{5}$$
$$2\sqrt[3]{3x} - 9\sqrt[3]{3x} = (2-9)\sqrt[3]{3x} = -7\sqrt[3]{3x}$$

Radical expressions that are in simplest form and have unlike radicands or different indices cannot be simplified by the Distributive Property. The expressions below cannot be simplified by the Distributive Property.

$$3\sqrt[4]{2} - 6\sqrt[4]{3} \qquad 2\sqrt[4]{4x} + 3\sqrt[3]{4x}$$

Copyright © Houghton Mifflin Company. All rights reserved.

HOW TO Simplify: $3\sqrt{32x^2} - 2x\sqrt{2} + \sqrt{128x^2}$

$3\sqrt{32x^2} - 2x\sqrt{2} + \sqrt{128x^2}$
$= 3\sqrt{16x^2}\sqrt{2} - 2x\sqrt{2} + \sqrt{64x^2}\sqrt{2}$
$= 3 \cdot 4x\sqrt{2} - 2x\sqrt{2} + 8x\sqrt{2}$
$= 12x\sqrt{2} - 2x\sqrt{2} + 8x\sqrt{2}$
$= 18x\sqrt{2}$

- First simplify each term. Then combine like terms by using the Distributive Property.

Example 3

Simplify: $5b\sqrt[4]{32a^7b^5} - 2a\sqrt[4]{162a^3b^9}$

Solution

$5b\sqrt[4]{32a^7b^5} - 2a\sqrt[4]{162a^3b^9}$
$= 5b\sqrt[4]{16a^4b^4 \cdot 2a^3b} - 2a\sqrt[4]{81b^8 \cdot 2a^3b}$
$= 5b \cdot 2ab\sqrt[4]{2a^3b} - 2a \cdot 3b^2\sqrt[4]{2a^3b}$
$= 10ab^2\sqrt[4]{2a^3b} - 6ab^2\sqrt[4]{2a^3b}$
$= 4ab^2\sqrt[4]{2a^3b}$

You Try It 3

Simplify: $3xy\sqrt[3]{81x^5y} - \sqrt[3]{192x^8y^4}$

Your solution

Solution on p. S28

Objective C To multiply radical expressions

The Product Property of Radicals is used to multiply radical expressions with the same index.

$\sqrt{3x} \cdot \sqrt{5y} = \sqrt{3x \cdot 5y} = \sqrt{15xy}$

HOW TO Simplify: $\sqrt[3]{2a^5b}\sqrt[3]{16a^2b^2}$

$\sqrt[3]{2a^5b}\sqrt[3]{16a^2b^2} = \sqrt[3]{32a^7b^3}$

- Use the Product Property of Radicals to multiply the radicands.

$= \sqrt[3]{8a^6b^3 \cdot 4a}$

- Simplify.

$= 2a^2b\sqrt[3]{4a}$

HOW TO Simplify: $\sqrt{2x}(\sqrt{8x} - \sqrt{3})$

$\sqrt{2x}(\sqrt{8x} - \sqrt{3}) = \sqrt{2x}(\sqrt{8x}) - \sqrt{2x}(\sqrt{3})$

- Use the Distributive Property.

$= \sqrt{16x^2} - \sqrt{6x}$

- Simplify.

$= 4x - \sqrt{6x}$

HOW TO Simplify: $(2\sqrt{5} - 3)(3\sqrt{5} + 4)$

$(2\sqrt{5} - 3)(3\sqrt{5} + 4) = 6(\sqrt{5})^2 + 8\sqrt{5} - 9\sqrt{5} - 12$

- Use the FOIL method to multiply the numbers.

$= 30 + 8\sqrt{5} - 9\sqrt{5} - 12$
$= 18 - \sqrt{5}$

- Combine like terms.

Copyright © Houghton Mifflin Company. All rights reserved.

HOW TO Simplify: $(4\sqrt{a} - \sqrt{b})(2\sqrt{a} + 5\sqrt{b})$

$(4\sqrt{a} - \sqrt{b})(2\sqrt{a} + 5\sqrt{b})$
$= 8(\sqrt{a})^2 + 20\sqrt{ab} - 2\sqrt{ab} - 5(\sqrt{b})^2$ • Use the FOIL method.
$= 8a + 18\sqrt{ab} - 5b$

TAKE NOTE

The concept of conjugate is used in a number of different instances. Make sure you understand this idea.

The conjugate of $\sqrt{3} - 4$ is $\sqrt{3} + 4$.

The conjugate of $\sqrt{3} + 4$ is $\sqrt{3} - 4$.

The conjugate of $\sqrt{5a} + \sqrt{b}$ is $\sqrt{5a} - \sqrt{b}$.

The expressions $a + b$ and $a - b$ are **conjugates** of each other: binomial expressions that differ only in the sign of a term. Recall that $(a + b)(a - b) = a^2 - b^2$. This identity is used to simplify conjugate radical expressions.

HOW TO Simplify: $(\sqrt{11} - 3)(\sqrt{11} + 3)$

$(\sqrt{11} - 3)(\sqrt{11} + 3) = (\sqrt{11})^2 - 3^2 = 11 - 9$
$= 2$ • The radical expressions are conjugates.

Example 4

Simplify: $\sqrt{3x}(\sqrt{27x^2} - \sqrt{3x})$

Solution

$\sqrt{3x}(\sqrt{27x^2} - \sqrt{3x})$
$= \sqrt{81x^3} - \sqrt{9x^2}$ • The Distributive Property
$= \sqrt{81x^2 \cdot x} - \sqrt{9x^2}$ • Simplify each radical expression.
$= 9x\sqrt{x} - 3x$

You Try It 4

Simplify: $\sqrt{5b}(\sqrt{3b} - \sqrt{10})$

Your solution $\sqrt{15b^2} - \sqrt{50b}$
$b\sqrt{15} - 5\sqrt{2b}$

Example 5

Simplify: $(2\sqrt[3]{x} - 3)(3\sqrt[3]{x} - 4)$

Solution

$(2\sqrt[3]{x} - 3)(3\sqrt[3]{x} - 4)$
$= 6\sqrt[3]{x^2} - 8\sqrt[3]{x} - 9\sqrt[3]{x} + 12$ • The FOIL method
$= 6\sqrt[3]{x^2} - 17\sqrt[3]{x} + 12$

You Try It 5

Simplify: $(2\sqrt[3]{2x} - 3)(\sqrt[3]{2x} - 5)$

Your solution

Example 6

Simplify: $(2\sqrt{x} - \sqrt{2y})(2\sqrt{x} + \sqrt{2y})$

Solution

$(2\sqrt{x} - \sqrt{2y})(2\sqrt{x} + \sqrt{2y})$ • The expressions are conjugates.
$= (2\sqrt{x})^2 - (\sqrt{2y})^2$
$= 4x - 2y$

You Try It 6

Simplify: $(\sqrt{a} - 3\sqrt{y})(\sqrt{a} + 3\sqrt{y})$

Your solution

Solutions on p. S28

Copyright © Houghton Mifflin Company. All rights reserved.

Objective D To divide radical expressions

The Quotient Property of Radicals is used to divide radical expressions with the same index.

The Quotient Property of Radicals

If $\sqrt[n]{a}$ and $\sqrt[n]{b}$ are real numbers, and $b \neq 0$, then

$$\sqrt[n]{\frac{a}{b}} = \frac{\sqrt[n]{a}}{\sqrt[n]{b}} \quad \text{and} \quad \frac{\sqrt[n]{a}}{\sqrt[n]{b}} = \sqrt[n]{\frac{a}{b}}$$

HOW TO Simplify: $\sqrt[3]{\dfrac{81x^5}{y^6}}$

$$\sqrt[3]{\frac{81x^5}{y^6}} = \frac{\sqrt[3]{81x^5}}{\sqrt[3]{y^6}}$$
• Use the Quotient Property of Radicals.

$$= \frac{\sqrt[3]{27x^3 \cdot 3x^2}}{\sqrt[3]{y^6}} = \frac{3x\sqrt[3]{3x^2}}{y^2}$$
• Simplify each radical expression.

HOW TO Simplify: $\dfrac{\sqrt{5a^4b^7c^2}}{\sqrt{ab^3c}}$

$$\frac{\sqrt{5a^4b^7c^2}}{\sqrt{ab^3c}} = \sqrt{\frac{5a^4b^7c^2}{ab^3c}}$$
• Use the Quotient Property of Radicals.

$$= \sqrt{5a^3b^4c}$$
• Simplify the radicand.

$$= \sqrt{a^2b^4 \cdot 5ac} = ab^2\sqrt{5ac}$$

A radical expression is in simplest form when no radical remains in the denominator of the radical expression. The procedure used to remove a radical from the denominator is called **rationalizing the denominator.**

HOW TO Simplify: $\dfrac{5}{\sqrt{2}}$

$$\frac{5}{\sqrt{2}} = \frac{5}{\sqrt{2}} \cdot 1 = \frac{5}{\sqrt{2}} \cdot \frac{\sqrt{2}}{\sqrt{2}}$$
• Multiply by $\frac{\sqrt{2}}{\sqrt{2}}$, which equals 1.

$$= \frac{5\sqrt{2}}{2}$$
• $\sqrt{2} \cdot \sqrt{2} = (\sqrt{2})^2 = 2$

HOW TO Simplify: $\dfrac{3}{\sqrt[3]{4x}}$

$$\frac{3}{\sqrt[3]{4x}} = \frac{3}{\sqrt[3]{4x}} \cdot \frac{\sqrt[3]{2x^2}}{\sqrt[3]{2x^2}}$$
• Because $4x \cdot 2x^2 = 8x^3$, a perfect cube, multiply the expression by $\frac{\sqrt[3]{2x^2}}{\sqrt[3]{2x^2}}$, which equals 1.

$$= \frac{3\sqrt[3]{2x^2}}{\sqrt[3]{8x^3}} = \frac{3\sqrt[3]{2x^2}}{2x}$$
• Simplify.

TAKE NOTE

Multiplying by $\frac{\sqrt[3]{4x}}{\sqrt[3]{4x}}$ will not rationalize the denominator of $\frac{3}{\sqrt[3]{4x}}$.

$$\frac{3}{\sqrt[3]{4x}} \cdot \frac{\sqrt[3]{4x}}{\sqrt[3]{4x}} = \frac{3\sqrt[3]{4x}}{\sqrt[3]{16x^2}}$$

Because $16x^2$ is not a perfect cube, the denominator still contains a radical expression.

Copyright © Houghton Mifflin Company. All rights reserved.

To simplify a fraction that has a square-root radical expression with two terms in the denominator, multiply the numerator and denominator by the conjugate of the denominator. Then simplify.

TAKE NOTE
Here is an example of using a conjugate to simplify a radical expression.

HOW TO Simplify: $\dfrac{\sqrt{x}-\sqrt{y}}{\sqrt{x}+\sqrt{y}}$

$$\frac{\sqrt{x}-\sqrt{y}}{\sqrt{x}+\sqrt{y}} = \frac{\sqrt{x}-\sqrt{y}}{\sqrt{x}+\sqrt{y}} \cdot \frac{\sqrt{x}-\sqrt{y}}{\sqrt{x}-\sqrt{y}}$$

• Multiply by $\dfrac{\sqrt{x}-\sqrt{y}}{\sqrt{x}-\sqrt{y}}$.

$$= \frac{(\sqrt{x})^2 - \sqrt{xy} - \sqrt{xy} + (\sqrt{y})^2}{(\sqrt{x})^2 - (\sqrt{y})^2} = \frac{x - 2\sqrt{xy} + y}{x - y}$$

• Simplify.

Example 7

Simplify: $\dfrac{5}{\sqrt{5x}}$

Solution

$$\frac{5}{\sqrt{5x}} = \frac{5}{\sqrt{5x}} \cdot \frac{\sqrt{5x}}{\sqrt{5x}} = \frac{5\sqrt{5x}}{(\sqrt{5x})^2}$$

• Rationalize the denominator.

$$= \frac{5\sqrt{5x}}{5x} = \frac{\sqrt{5x}}{x}$$

You Try It 7

Simplify: $\dfrac{y}{\sqrt{3y}}$

Your solution

Example 8

Simplify: $\dfrac{3x}{\sqrt[4]{2x}}$

Solution

$$\frac{3x}{\sqrt[4]{2x}} = \frac{3x}{\sqrt[4]{2x}} \cdot \frac{\sqrt[4]{8x^3}}{\sqrt[4]{8x^3}}$$

• Rationalize the denominator.

$$= \frac{3x\sqrt[4]{8x^3}}{\sqrt[4]{16x^4}} = \frac{3x\sqrt[4]{8x^3}}{2x}$$

$$= \frac{3\sqrt[4]{8x^3}}{2}$$

You Try It 8

Simplify: $\dfrac{3x}{\sqrt[3]{3x^2}}$

Your solution

Example 9

Simplify: $\dfrac{3}{5 - 2\sqrt{3}}$

Solution

$$\frac{3}{5-2\sqrt{3}} = \frac{3}{5-2\sqrt{3}} \cdot \frac{5+2\sqrt{3}}{5+2\sqrt{3}}$$

• Rationalize the denominator.

$$= \frac{15 + 6\sqrt{3}}{5^2 - (2\sqrt{3})^2}$$

$$= \frac{15 + 6\sqrt{3}}{25 - 12} = \frac{15 + 6\sqrt{3}}{13}$$

You Try It 9

Simplify: $\dfrac{3 + \sqrt{6}}{2 - \sqrt{6}}$

Your solution

Solutions on p. S28

Copyright © Houghton Mifflin Company. All rights reserved.

9.2 Exercises

Objective A To simplify radical expressions

For Exercises 1 to 16, simplify.

1. $\sqrt{x^4y^3z^5}$
2. $\sqrt{x^3y^6z^9}$
3. $\sqrt{8a^3b^8}$
4. $\sqrt{24a^9b^6}$
5. $\sqrt{45x^2y^3z^5}$
6. $\sqrt{60xy^7z^{12}}$
7. $\sqrt{-9x^3}$
8. $\sqrt{-x^2y^5}$
9. $\sqrt[3]{a^{16}b^8}$
10. $\sqrt[3]{a^5b^8}$
11. $\sqrt[3]{-125x^2y^4}$
12. $\sqrt[3]{-216x^5y^9}$
13. $\sqrt[3]{a^4b^5c^6}$
14. $\sqrt[3]{a^8b^{11}c^{15}}$
15. $\sqrt[4]{16x^9y^5}$
16. $\sqrt[4]{64x^8y^{10}}$

Objective B To add or subtract radical expressions

For Exercises 17 to 42, simplify.

17. $2\sqrt{x} - 8\sqrt{x}$
18. $3\sqrt{y} + 12\sqrt{y}$
19. $\sqrt{8} - \sqrt{32}$
20. $\sqrt{27a} - \sqrt{8a}$
21. $\sqrt{18b} + \sqrt{75b}$
22. $2\sqrt{2x^3} + 4x\sqrt{8x}$
23. $3\sqrt{8x^2y^3} - 2x\sqrt{32y^3}$
24. $2\sqrt{32x^2y^3} - xy\sqrt{98y}$
25. $2a\sqrt{27ab^5} + 3b\sqrt{3a^3b}$
26. $\sqrt[3]{128} + \sqrt[3]{250}$
27. $\sqrt[3]{16} - \sqrt[3]{54}$
28. $2\sqrt[3]{3a^4} - 3a\sqrt[3]{81a}$
29. $2b\sqrt[3]{16b^2} + \sqrt[3]{128b^5}$
30. $3\sqrt[3]{x^5y^7} - 8xy\sqrt[3]{x^2y^4}$

Copyright © Houghton Mifflin Company. All rights reserved.

31. $3\sqrt[4]{32a^5} - a\sqrt[4]{162a}$

32. $2a\sqrt[4]{16ab^5} + 3b\sqrt[4]{256a^5b}$

33. $2\sqrt{50} - 3\sqrt{125} + \sqrt{98}$

34. $3\sqrt{108} - 2\sqrt{18} - 3\sqrt{48}$

35. $\sqrt{9b^3} - \sqrt{25b^3} + \sqrt{49b^3}$

36. $\sqrt{4x^7y^5} + 9x^2\sqrt{x^3y^5} - 5xy\sqrt{x^5y^3}$

37. $2x\sqrt{8xy^2} - 3y\sqrt{32x^3} + \sqrt{4x^3y^3}$

38. $5a\sqrt{3a^3b} + 2a^2\sqrt{27ab} - 4\sqrt{75a^5b}$

39. $\sqrt[3]{54xy^3} - 5\sqrt[3]{2xy^3} + y\sqrt[3]{128x}$

40. $2\sqrt[3]{24x^3y^4} + 4x\sqrt[3]{81y^4} - 3y\sqrt[3]{24x^3y}$

41. $2a\sqrt[4]{32b^5} - 3b\sqrt[4]{162a^4b} + \sqrt[4]{2a^4b^5}$

42. $6y\sqrt[4]{48x^5} - 2x\sqrt[4]{243xy^4} - 4\sqrt[4]{3x^5y^4}$

Objective C **To multiply radical expressions**

For Exercises 43 to 73, simplify.

43. $\sqrt{8}\,\sqrt{32}$

44. $\sqrt{14}\,\sqrt{35}$

45. $\sqrt[3]{4}\,\sqrt[3]{8}$

46. $\sqrt[3]{6}\,\sqrt[3]{36}$

47. $\sqrt{x^2y^5}\,\sqrt{xy}$

48. $\sqrt{a^3b}\,\sqrt{ab^4}$

49. $\sqrt{2x^2y}\,\sqrt{32xy}$

50. $\sqrt{5x^3y}\,\sqrt{10x^3y^4}$

51. $\sqrt[3]{x^2y}\,\sqrt[3]{16x^4y^2}$

52. $\sqrt[3]{4a^2b^3}\,\sqrt[3]{8ab^5}$

53. $\sqrt[4]{12ab^3}\,\sqrt[4]{4a^5b^2}$

54. $\sqrt[4]{36a^2b^4}\,\sqrt[4]{12a^5b^3}$

55. $\sqrt{3}\,(\sqrt{27} - \sqrt{3})$

56. $\sqrt{10}\,(\sqrt{10} - \sqrt{5})$

57. $\sqrt{x}\,(\sqrt{x} - \sqrt{2})$

58. $\sqrt{y}\,(\sqrt{y} - \sqrt{5})$

59. $\sqrt{2x}\,(\sqrt{8x} - \sqrt{32})$

60. $\sqrt{3a}\,(\sqrt{27a^2} - \sqrt{a})$

Copyright © Houghton Mifflin Company. All rights reserved.

61. $(\sqrt{x} - 3)^2$

62. $(\sqrt{2x} + 4)^2$

63. $(4\sqrt{5} + 2)^2$

64. $2\sqrt{3x^2} \cdot 3\sqrt{12xy^3} \cdot \sqrt{6x^3y}$

65. $2\sqrt{14xy} \cdot 4\sqrt{7x^2y} \cdot 3\sqrt{8xy^2}$

66. $\sqrt[3]{8ab}\,\sqrt[3]{4a^2b^3}\,\sqrt[3]{9ab^4}$

67. $\sqrt[3]{2a^2b}\,\sqrt[3]{4a^3b^2}\,\sqrt[3]{8a^5b^6}$

68. $(\sqrt{2} - 3)(\sqrt{2} + 4)$

69. $(\sqrt{5} - 5)(2\sqrt{5} + 2)$

70. $(\sqrt{y} - 2)(\sqrt{y} + 2)$

71. $(\sqrt{x} - y)(\sqrt{x} + y)$

72. $(\sqrt{2x} - 3\sqrt{y})(\sqrt{2x} + 3\sqrt{y})$

73. $(2\sqrt{3x} - \sqrt{y})(2\sqrt{3x} + \sqrt{y})$

Objective D **To divide radical expressions**

74. When is a radical expression in simplest form?

75. Explain what it means to rationalize the denominator of a radical expression and how to do so.

For Exercises 76 to 112, simplify.

76. $\frac{\sqrt{32x^2}}{\sqrt{2x}}$

77. $\frac{\sqrt{60y^4}}{\sqrt{12y}}$

78. $\frac{\sqrt{42a^3b^5}}{\sqrt{14a^2b}}$

79. $\frac{\sqrt{65ab^4}}{\sqrt{5ab}}$

80. $\frac{1}{\sqrt{5}}$

81. $\frac{1}{\sqrt{2}}$

82. $\frac{1}{\sqrt{2x}}$

83. $\frac{2}{\sqrt{3y}}$

84. $\frac{5}{\sqrt{5x}}$

85. $\frac{9}{\sqrt{3a}}$

86. $\sqrt{\frac{x}{5}}$

87. $\sqrt{\frac{y}{2}}$

88. $\frac{3}{\sqrt[3]{2}}$

89. $\frac{5}{\sqrt[3]{9}}$

90. $\frac{3}{\sqrt[3]{4x^2}}$

91. $\frac{5}{\sqrt[3]{3y}}$

Copyright © Houghton Mifflin Company. All rights reserved.

92. $\dfrac{\sqrt{40x^3y^2}}{\sqrt{80x^2y^3}}$ **93.** $\dfrac{\sqrt{15a^2b^5}}{\sqrt{30a^5b^3}}$ **94.** $\dfrac{\sqrt{24a^2b}}{\sqrt{18ab^4}}$ **95.** $\dfrac{\sqrt{12x^3y}}{\sqrt{20x^4y}}$

96. $\dfrac{5}{\sqrt{3}-2}$ **97.** $\dfrac{-2}{1-\sqrt{2}}$ **98.** $\dfrac{-3}{2-\sqrt{3}}$ **99.** $\dfrac{-4}{3-\sqrt{2}}$

100. $\dfrac{2}{\sqrt{5}+2}$ **101.** $\dfrac{5}{2-\sqrt{7}}$ **102.** $\dfrac{3}{\sqrt{y}-2}$ **103.** $\dfrac{-7}{\sqrt{x}-3}$

104. $\dfrac{\sqrt{2}-\sqrt{3}}{\sqrt{2}+\sqrt{3}}$ **105.** $\dfrac{\sqrt{3}+\sqrt{4}}{\sqrt{2}+\sqrt{3}}$ **106.** $\dfrac{2+3\sqrt{7}}{5-2\sqrt{7}}$

107. $\dfrac{2+3\sqrt{5}}{1-\sqrt{5}}$ **108.** $\dfrac{2\sqrt{3}-1}{3\sqrt{3}+2}$ **109.** $\dfrac{2\sqrt{a}-\sqrt{b}}{4\sqrt{a}+3\sqrt{b}}$

110. $\dfrac{2\sqrt{x}-4}{\sqrt{x}+2}$ **111.** $\dfrac{3\sqrt{y}-y}{\sqrt{y}+2y}$ **112.** $\dfrac{3\sqrt{x}-4\sqrt{y}}{3\sqrt{x}-2\sqrt{y}}$

APPLYING THE CONCEPTS

113. Determine whether the following statements are true or false. If the statement is false, correct the right-hand side of the equation.

a. $\sqrt[2]{3} \cdot \sqrt[3]{4} = \sqrt[5]{12}$ **b.** $\sqrt{3} \cdot \sqrt{3} = 3$ **c.** $\sqrt[3]{x} \cdot \sqrt[3]{x} = x$

d. $\sqrt{x} + \sqrt{y} = \sqrt{x+y}$ **e.** $\sqrt[2]{2} + \sqrt[3]{3} = \sqrt[5]{2+3}$ **f.** $8\sqrt[5]{a} - 2\sqrt[5]{a} = 6\sqrt[5]{a}$

114. Multiply: $(\sqrt[3]{a} + \sqrt[3]{b})(\sqrt[3]{a^2} - \sqrt[3]{ab} + \sqrt[3]{b^2})$

115. Rewrite $\dfrac{\sqrt[4]{(a+b)^3}}{\sqrt{a+b}}$ as an expression with a single radical.

Copyright © Houghton Mifflin Company. All rights reserved.

9.3 Solving Equations Containing Radical Expressions

Objective A To solve a radical equation

An equation that contains a variable expression in a radicand is a **radical equation.**

$$\left.\begin{array}{r}\sqrt[3]{2x-5}+x=7\\ \sqrt{x+1}-\sqrt{x}=4\end{array}\right\}\text{ Radical equations}$$

The following property is used to solve a radical equation.

> **The Property of Raising Each Side of an Equation to a Power**
>
> If two numbers are equal, then the same powers of the numbers are equal.
>
> If $a = b$, then $a^n = b^n$.

HOW TO Solve: $\sqrt{x-2}-6=0$

$\sqrt{x-2}-6=0$

$\sqrt{x-2}=6$ • Isolate the radical by adding 6 to each side of the equation.

$(\sqrt{x-2})^2=6^2$ • Square each side of the equation.

$x-2=36$ • Simplify and solve for x.

$x=38$

Check:

$$\begin{array}{r|l}\multicolumn{2}{c}{\sqrt{x-2}-6=0}\\ \hline \sqrt{38-2}-6 & 0\\ \sqrt{36}-6 & 0\\ 6-6 & 0\\ \multicolumn{2}{r}{0=0}\end{array}$$

38 checks as a solution. The solution is 38.

HOW TO Solve: $\sqrt[3]{x+2}=-3$

$\sqrt[3]{x+2}=-3$

$(\sqrt[3]{x+2})^3=(-3)^3$ • Cube each side of the equation.

$x+2=-27$ • Solve the resulting equation.

$x=-29$

Check:

$$\begin{array}{r|l}\multicolumn{2}{c}{\sqrt[3]{x+2}=-3}\\ \hline \sqrt[3]{-29+2} & -3\\ \sqrt[3]{-27} & -3\\ \multicolumn{2}{r}{-3=-3}\end{array}$$

-29 checks as a solution. The solution is -29.

Copyright © Houghton Mifflin Company. All rights reserved.

Raising each side of an equation to an even power may result in an equation that has a solution that is not a solution of the original equation. This is called an **extraneous solution.** Here is an example:

TAKE NOTE

Note that

$(2 - \sqrt{x})^2$
$= (2 - \sqrt{x})(2 - \sqrt{x})$
$= 4 - 4\sqrt{x} + x$

HOW TO Solve: $\sqrt{2x-1} + \sqrt{x} = 2$

$$\sqrt{2x-1} + \sqrt{x} = 2$$

$$\sqrt{2x-1} = 2 - \sqrt{x}$$ • Solve for one of the radical expressions.

$$(\sqrt{2x-1})^2 = (2 - \sqrt{x})^2$$ • Square each side. Recall that $(a-b)^2 = a^2 - 2ab + b^2$.

$$2x - 1 = 4 - 4\sqrt{x} + x$$

$$x - 5 = -4\sqrt{x}$$

$$(x-5)^2 = (-4\sqrt{x})^2$$

$$x^2 - 10x + 25 = 16x$$

$$x^2 - 26x + 25 = 0$$ • Solve the quadratic equation by factoring.

$$(x-25)(x-1) = 0$$

$$x = 25 \quad \text{or} \quad x = 1$$

Check:

$\sqrt{2x-1} + \sqrt{x}$	$= 2$
$\sqrt{2(25)-1} + \sqrt{25}$	2
$7 + 5$	2
$12 \neq 2$	

$\sqrt{2x-1} + \sqrt{x}$	$= 2$
$\sqrt{2(1)-1} + \sqrt{1}$	2
$1 + 1$	2
$2 = 2$	

25 does not check as a solution. 1 checks as a solution. The solution is 1. Here 25 is an extraneous solution.

TAKE NOTE

You must always check the proposed solutions to radical equations.

The proposed solutions of the equation on the right were 1 and 25. However, 25 did not check as a solution.

Example 1

Solve: $\sqrt{x-1} + \sqrt{x+4} = 5$

Solution

$$\sqrt{x-1} + \sqrt{x+4} = 5$$

$$\sqrt{x+4} = 5 - \sqrt{x-1}$$ • Subtract $\sqrt{x-1}$.

$$(\sqrt{x+4})^2 = (5 - \sqrt{x-1})^2$$ • Square each side.

$$x + 4 = 25 - 10\sqrt{x-1} + x - 1$$

$$-20 = -10\sqrt{x-1}$$

$$2 = \sqrt{x-1}$$

$$2^2 = (\sqrt{x-1})^2$$ • Square each side.

$$4 = x - 1$$

$$5 = x$$

The solution checks. The solution is 5.

You Try It 1

Solve: $\sqrt{x} - \sqrt{x+5} = 1$

Your solution

Example 2

Solve: $\sqrt[3]{3x-1} = -4$

Solution

$$\sqrt[3]{3x-1} = -4$$

$$(\sqrt[3]{3x-1})^3 = (-4)^3$$ • Cube each side.

$$3x - 1 = -64$$

$$3x = -63$$

$$x = -21$$

The solution checks. The solution is −21.

You Try It 2

Solve: $\sqrt[4]{x-8} = 3$

Your solution

Solutions on pp. S28–S29

Copyright © Houghton Mifflin Company. All rights reserved.

Objective B To solve application problems

SSM

Pythagoras
(c. 580 B.C.–529 B.C.)

A right triangle contains one 90° angle. The side opposite the 90° angle is called the **hypotenuse.** The other two sides are called **legs.**

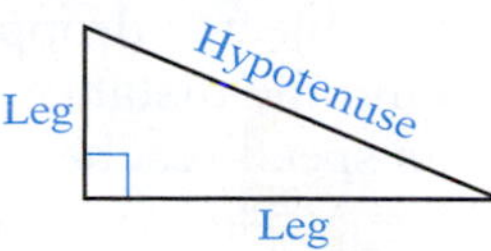

Pythagoras, a Greek mathematician, discovered that the square of the hypotenuse of a right triangle is equal to the sum of the squares of the two legs. This is called the **Pythagorean Theorem.**

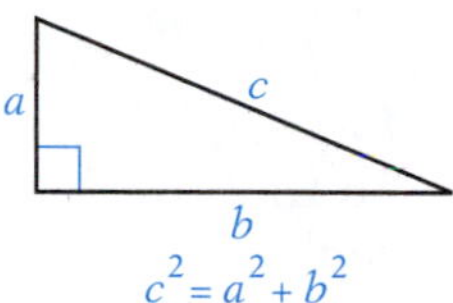

Example 3

A ladder 20 ft long is leaning against a building. How high on the building will the ladder reach when the bottom of the ladder is 8 ft from the building? Round to the nearest tenth.

Strategy

To find the distance, use the Pythagorean Theorem. The hypotenuse is the length of the ladder. One leg is the distance from the bottom of the ladder to the base of the building. The distance along the building from the ground to the top of the ladder is the unknown leg.

Solution

$c^2 = a^2 + b^2$ • Pythagorean Theorem

$20^2 = 8^2 + b^2$ • Replace c by 20 and a by 8.

$400 = 64 + b^2$ • Solve for b.

$336 = b^2$

$(336)^{1/2} = (b^2)^{1/2}$ • Raise each side to the $\frac{1}{2}$ power.

$\sqrt{336} = b$ • $a^{1/2} = \sqrt{a}$

$18.3 \approx b$

The distance is approximately 18.3 ft.

You Try It 3

Find the diagonal of a rectangle that is 6 cm long and 3 cm wide. Round to the nearest tenth.

Your strategy

Your solution

Solution on p. S29

Copyright © Houghton Mifflin Company. All rights reserved.

Example 4

An object is dropped from a high building. Find the distance the object has fallen when its speed reaches 96 ft/s. Use the equation $v = \sqrt{64d}$, where v is the speed of the object in feet per second and d is the distance in feet.

Strategy

To find the distance the object has fallen, replace v in the equation with the given value and solve for d.

Solution

$$v = \sqrt{64d}$$

$$96 = \sqrt{64d}$$ • Replace v by 96.

$$(96)^2 = (\sqrt{64d})^2$$ • Square each side.

$$9216 = 64d$$

$$144 = d$$

The object has fallen 144 ft.

You Try It 4

How far would a submarine periscope have to be above the water to locate a ship 5.5 mi away? The equation for the distance in miles that the lookout can see is $d = \sqrt{1.5h}$, where h is the height in feet above the surface of the water. Round to the nearest hundredth.

Your strategy

Your solution

Example 5

Find the length of a pendulum that makes one swing in 1.5 s. The equation for the time of one swing is given by $T = 2\pi\sqrt{\frac{L}{32}}$, where T is the time in seconds and L is the length in feet. Use 3.14 for π. Round to the nearest hundredth.

Strategy

To find the length of the pendulum, replace T in the equation with the given value and solve for L.

Solution

$$T = 2\pi\sqrt{\frac{L}{32}}$$

$$1.5 = 2(3.14)\sqrt{\frac{L}{32}}$$ • Replace T by 1.5 and π by 3.14.

$$\frac{1.5}{2(3.14)} = \sqrt{\frac{L}{32}}$$ • Divide each side by 2(3.14).

$$\left[\frac{1.5}{2(3.14)}\right]^2 = \left(\sqrt{\frac{L}{32}}\right)^2$$ • Square each side.

$$\left(\frac{1.5}{6.28}\right)^2 = \frac{L}{32}$$ • Solve for L. Multiply each side by 32.

$$1.83 \approx L$$

The length of the pendulum is approximately 1.83 ft.

You Try It 5

Find the distance required for a car to reach a velocity of 88 ft/s when the acceleration is 22 ft/s^2. Use the equation $v = \sqrt{2as}$, where v is the velocity in feet per second, a is the acceleration, and s is the distance in feet.

Your strategy

Your solution

Solutions on p. S29

Copyright © Houghton Mifflin Company. All rights reserved.

9.3 Exercises

Objective A To solve a radical equation

For Exercises 1 to 21, solve.

1. $\sqrt[3]{4x} = -2$
2. $\sqrt[3]{6x} = -3$
3. $\sqrt{3x - 2} = 5$
4. $\sqrt{3x + 9} - 12 = 0$
5. $\sqrt{4x - 3} - 5 = 0$
6. $\sqrt{4x - 2} = \sqrt{3x + 9}$
7. $\sqrt{2x + 4} = \sqrt{5x - 9}$
8. $\sqrt[3]{x - 2} = 3$
9. $\sqrt[3]{2x - 6} = 4$
10. $\sqrt[3]{3x - 9} = \sqrt[3]{2x + 12}$
11. $\sqrt[3]{x - 12} = \sqrt[3]{5x + 16}$
12. $\sqrt[4]{4x + 1} = 2$
13. $\sqrt[3]{2x - 3} + 5 = 2$
14. $\sqrt[3]{x - 4} + 7 = 5$
15. $\sqrt{x} + \sqrt{x - 5} = 5$
16. $\sqrt{x + 3} + \sqrt{x - 1} = 2$
17. $\sqrt{2x + 5} - \sqrt{2x} = 1$
18. $\sqrt{3x} - \sqrt{3x - 5} = 1$
19. $\sqrt{2x} - \sqrt{x - 1} = 1$
20. $\sqrt{2x - 5} + \sqrt{x + 1} = 3$
21. $\sqrt{2x + 2} + \sqrt{x} = 3$

Objective B To solve application problems

22. **Physics** An object is dropped from a bridge. Find the distance the object has fallen when its speed reaches 100 ft/s. Use the equation $v = \sqrt{64d}$, where v is the speed in feet per second and d is the distance in feet.

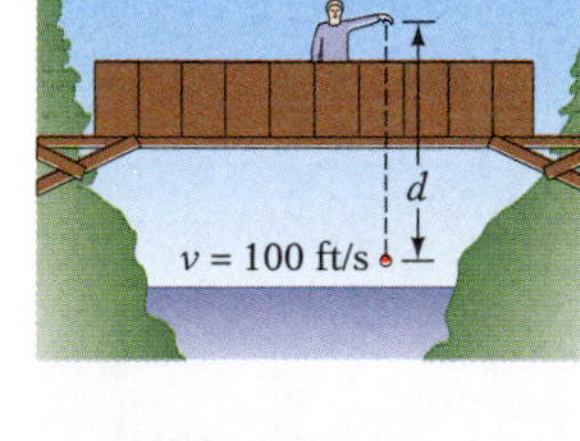

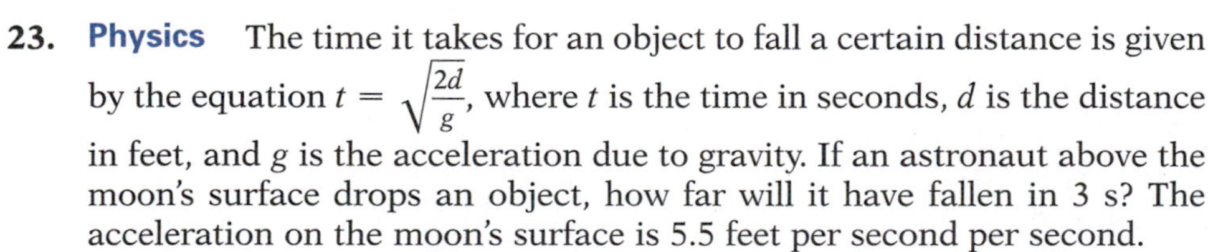

23. **Physics** The time it takes for an object to fall a certain distance is given by the equation $t = \sqrt{\frac{2d}{g}}$, where t is the time in seconds, d is the distance in feet, and g is the acceleration due to gravity. If an astronaut above the moon's surface drops an object, how far will it have fallen in 3 s? The acceleration on the moon's surface is 5.5 feet per second per second.

24. **Physics** The time it takes for an object to fall a certain distance is given by the equation $t = \sqrt{\frac{2d}{g}}$, where t is the time in seconds, d is the distance in feet, and g is the acceleration due to gravity. The acceleration due to gravity on Earth is 32 feet per second per second. If an object is dropped from an airplane, how far will it fall in 6 s?

Copyright © Houghton Mifflin Company. All rights reserved.

25. Television High definition television (HDTV) gives consumers a wider viewing area, more like a film screen in a theater. A regular television with a 27-inch diagonal measurement has a screen 16.2 in. tall. An HDTV screen with the same 16.2-inch height would have a diagonal measuring 33 in. How many inches wider is the HDTV screen? Round to the nearest hundredth.

26. Satellites At what height above Earth's surface is a satellite in orbit if it is traveling at a speed of 7500 m/s? Use the equation $v = \sqrt{\dfrac{4 \times 10^{14}}{h + 6.4 \times 10^6}}$, where v is the speed of the satellite in meters per second and h is the height above Earth's surface in meters. Round to the nearest thousand.

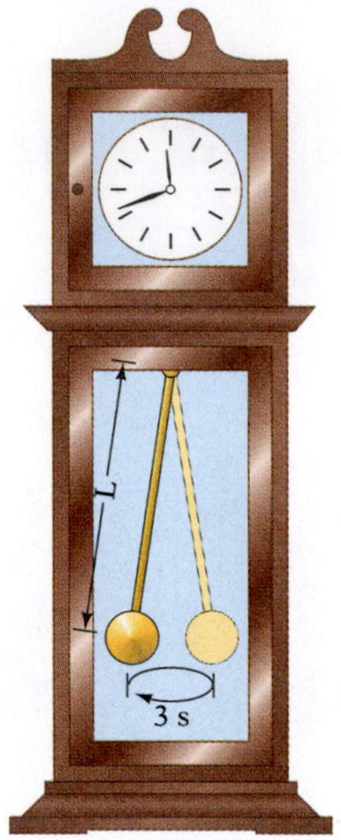

27. Pendulums Find the length of a pendulum that makes one swing in 3 s. The equation for the time of one swing of a pendulum is $T = 2\pi\sqrt{\dfrac{L}{32}}$, where T is the time in seconds and L is the length in feet. Round to the nearest hundredth.

APPLYING THE CONCEPTS

28. Solve the following equations. Describe the solution by using the following terms: integer, rational number, irrational number, and real number. Note that more than one term may be used to describe the answer.

a. $x^2 + 3 = 7$

b. $x^{3/4} = 8$

29. Solve: $\sqrt{3x - 2} = \sqrt{2x - 3} + \sqrt{x - 1}$

30. Solve $a^2 + b^2 = c^2$ for a.

31. Geometry Solve $V = \frac{4}{3}\pi r^3$ for r.

32. Geometry Find the length of the side labeled x.

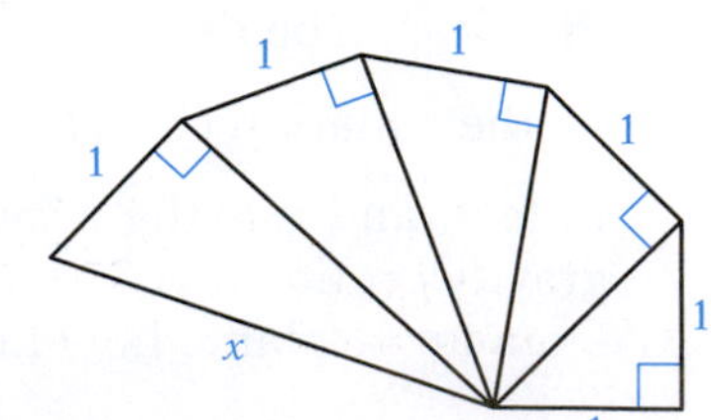

Copyright © Houghton Mifflin Company. All rights reserved.

9.4 Complex Numbers

Objective A To simplify a complex number

The radical expression $\sqrt{-4}$ is not a real number because there is no real number whose square is -4. However, the solution of an algebraic equation is sometimes the square root of a negative number.

For example, the equation $x^2 + 1 = 0$ does not have a real number solution, because there is no real number whose square is a negative number.

$$x^2 + 1 = 0$$
$$x^2 = -1$$

Around the 17th century, a new number, called an **imaginary number,** was defined so that a negative number would have a square root. The letter i was chosen to represent the number whose square is -1.

$$\mathbf{i^2 = -1}$$

An imaginary number is defined in terms of i.

Point of Interest

The first written occurrence of an imaginary number was in a book published in 1545 by Hieronimo Cardan, where he wrote (in our modern notation) $5 + \sqrt{-15}$. He went on to say that the number "is as refined as it is useless." It was not until the 20th century that applications of complex numbers were found.

Definition of $\sqrt{-a}$

If a is a positive real number, then the principal square root of negative a is the imaginary number $i\sqrt{a}$.

$$\sqrt{-a} = i\sqrt{a}$$

Here are some examples.

$$\sqrt{-16} = i\sqrt{16} = 4i$$
$$\sqrt{-12} = i\sqrt{12} = 2i\sqrt{3}$$
$$\sqrt{-21} = i\sqrt{21}$$
$$\sqrt{-1} = i\sqrt{1} = i$$

It is customary to write i in front of a radical to avoid confusing $\sqrt{a}\,i$ with $\sqrt{ai}$.

The real numbers and imaginary numbers make up the complex numbers.

Complex Number

A **complex number** is a number of the form $\boldsymbol{a + bi}$, where a and b are real numbers and $i = \sqrt{-1}$. The number a is the **real part** of $a + bi$, and the number b is the **imaginary part.**

TAKE NOTE

The *imaginary part* of a complex number is a real number. As another example, the imaginary part of $6 - 8i$ is -8.

Examples of complex numbers are shown at the right.

Real Part	*Imaginary Part*
a	$+ bi$
3	$+ 2i$
8	$- 10i$

Copyright © Houghton Mifflin Company. All rights reserved.

Complex numbers $a + bi$
- Real Numbers $a + 0i$ — A *real number* is a complex number in which $b = 0$.
- Imaginary Numbers $0 + bi$ — An *imaginary number* is a complex number in which $a = 0$.

Study Tip

Be sure you understand how to simplify expressions such as those in Example 1 and You Try It 1, as it is a prerequisite for solving quadratic equations in Chapter 10.

HOW TO Simplify: $\sqrt{20} - \sqrt{-50}$

$$\sqrt{20} - \sqrt{-50} = \sqrt{20} - i\sqrt{50}$$

- Write the complex number in the form $a + bi$.

$$= \sqrt{4 \cdot 5} - i\sqrt{25 \cdot 2}$$

- Use the Product Property of Radicals to simplify each radical.

$$= 2\sqrt{5} - 5i\sqrt{2}$$

Example 1

Simplify: $\sqrt{-80}$

Solution

$\sqrt{-80} = i\sqrt{80} = i\sqrt{16 \cdot 5} = 4i\sqrt{5}$

You Try It 1

Simplify: $\sqrt{-45}$

Your solution

Example 2

Simplify: $\sqrt{25} + \sqrt{-40}$

Solution

$$\sqrt{25} + \sqrt{-40} = \sqrt{25} + i\sqrt{40}$$
$$= \sqrt{25} + i\sqrt{4 \cdot 10}$$
$$= 5 + 2i\sqrt{10}$$

You Try It 2

Simplify: $\sqrt{98} - \sqrt{-60}$

Your solution

Solutions on p. S29

Objective B To add or subtract complex numbers

Integrating Technology

See the appendix Keystroke Guide: *Complex Numbers* for instructions on using a graphing calculator to perform operations on complex numbers.

Addition and Subtraction of Complex Numbers

To add two complex numbers, add the real parts and add the imaginary parts. To subtract two complex numbers, subtract the real parts and subtract the imaginary parts.

$$(a + bi) + (c + di) = (a + c) + (b + d)i$$
$$(a + bi) - (c + di) = (a - c) + (b - d)i$$

HOW TO Simplify: $(3 - 7i) - (4 - 2i)$

$$(3 - 7i) - (4 - 2i) = (3 - 4) + [-7 - (-2)]i$$

- Subtract the real parts and subtract the imaginary parts of the complex numbers.

$$= -1 - 5i$$

Copyright © Houghton Mifflin Company. All rights reserved.

HOW TO Simplify: $(3 + \sqrt{-12}) + (7 - \sqrt{-27})$

$(3 + \sqrt{-12}) + (7 - \sqrt{-27})$

$= (3 + i\sqrt{12}) + (7 - i\sqrt{27})$ • Write each complex number in the form $a + bi$.

$= (3 + i\sqrt{4 \cdot 3}) + (7 - i\sqrt{9 \cdot 3})$ • Use the Product Property of Radicals to simplify each radical.

$= (3 + 2i\sqrt{3}) + (7 - 3i\sqrt{3})$

$= 10 - i\sqrt{3}$ • Add the complex numbers.

Example 3

Simplify: $(3 + 2i) + (6 - 5i)$

Solution

$(3 + 2i) + (6 - 5i) = 9 - 3i$

You Try It 3

Simplify: $(-4 + 2i) - (6 - 8i)$

Your solution

Example 4

Simplify: $(9 - \sqrt{-8}) - (5 + \sqrt{-32})$

Solution

$(9 - \sqrt{-8}) - (5 + \sqrt{-32})$

$= (9 - i\sqrt{8}) - (5 + i\sqrt{32})$

$= (9 - i\sqrt{4 \cdot 2}) - (5 + i\sqrt{16 \cdot 2})$

$= (9 - 2i\sqrt{2}) - (5 + 4i\sqrt{2})$

$= 4 - 6i\sqrt{2}$

You Try It 4

Simplify: $(16 - \sqrt{-45}) - (3 + \sqrt{-20})$

Your solution

Example 5

Simplify: $(6 + 4i) + (-6 - 4i)$

Solution

$(6 + 4i) + (-6 - 4i) = 0 + 0i = 0$

This illustrates that the additive inverse of $a + bi$ is $-a - bi$.

You Try It 5

Simplify: $(3 - 2i) + (-3 + 2i)$

Your solution

Solutions on p. S29

Copyright © Houghton Mifflin Company. All rights reserved.

Objective C To multiply complex numbers

When multiplying complex numbers, we often find that the term i^2 is a part of the product. Recall that $i^2 = -1$.

HOW TO Simplify: $2i \cdot 3i$

$2i \cdot 3i = 6i^2$ • Multiply the imaginary numbers.

$= 6(-1)$ • Replace i^2 by -1.

$= -6$ • Simplify.

TAKE NOTE
This example illustrates an important point. When working with an expression that has a square root of a negative number, always rewrite the number as the product of a real number and i before continuing.

HOW TO Simplify: $\sqrt{-6} \cdot \sqrt{-24}$

$\sqrt{-6} \cdot \sqrt{-24} = i\sqrt{6} \cdot i\sqrt{24}$ • Write each radical as the product of a real number and i.

$= i^2\sqrt{144}$ • Multiply the imaginary numbers.

$= -\sqrt{144}$ • Replace i^2 by -1.

$= -12$ • Simplify the radical expression.

Note from the last example that it would have been incorrect to multiply the radicands of the two radical expressions. To illustrate,

$$\sqrt{-6} \cdot \sqrt{-24} = \sqrt{(-6)(-24)} = \sqrt{144} = 12, \textit{ not } -12$$

The Product Property of Radicals does not hold true when both radicands are negative and the index is an even number.

HOW TO Simplify: $4i(3 - 2i)$

$4i(3 - 2i) = 12i - 8i^2$ • Use the Distributive Property to remove parentheses.

$= 12i - 8(-1)$ • Replace i^2 by -1.

$= 8 + 12i$ • Write the answer in the form $a + bi$.

The product of two complex numbers is defined as follows.

Multiplication of Complex Numbers

$(a + bi)(c + di) = (ac - bd) + (ad + bc)i$

One way to remember this rule is to think of the FOIL method.

HOW TO Simplify: $(2 + 4i)(3 - 5i)$

$(2 + 4i)(3 - 5i) = 6 - 10i + 12i - 20i^2$ • Use the FOIL method to find the product.

$= 6 + 2i - 20i^2$

$= 6 + 2i - 20(-1)$ • Replace i^2 by -1.

$= 26 + 2i$ • Write the answer in the form $a + bi$.

The conjugate of $a + bi$ is $a - bi$.

The product of conjugates, $(a + bi)(a - bi)$, is the real number $a^2 + b^2$.

$$\begin{aligned}(a + bi)(a - bi) &= a^2 - b^2i^2\\ &= a^2 - b^2(-1)\\ &= a^2 + b^2\end{aligned}$$

HOW TO Simplify: $(2 + 3i)(2 - 3i)$

$(2 + 3i)(2 - 3i) = 2^2 + 3^2$ • $a = 2, b = 3$. The product of the conjugates is $2^2 + 3^2$.

$= 4 + 9$

$= 13$

Note that the product of a complex number and its conjugate is a real number.

Copyright © Houghton Mifflin Company. All rights reserved.

Example 6

Simplify: $(2i)(-5i)$

Solution

$(2i)(-5i) = -10i^2 = (-10)(-1) = 10$

You Try It 6

Simplify: $(-3i)(-10i)$

Your solution

Example 7

Simplify: $\sqrt{-10} \cdot \sqrt{-5}$

Solution

$$\begin{aligned}\sqrt{-10} \cdot \sqrt{-5} &= i\sqrt{10} \cdot i\sqrt{5} \\ &= i^2\sqrt{50} = -\sqrt{25 \cdot 2} = -5\sqrt{2}\end{aligned}$$

You Try It 7

Simplify: $-\sqrt{-8} \cdot \sqrt{-5}$

Your solution

Example 8

Simplify: $3i(2 - 4i)$

Solution

$$\begin{aligned}3i(2 - 4i) &= 6i - 12i^2 \\ &= 6i - 12(-1) \\ &= 12 + 6i\end{aligned}$$

• **The Distributive Property**

You Try It 8

Simplify: $-6i(3 + 4i)$

Your solution

Example 9

Simplify: $\sqrt{-8}(\sqrt{6} - \sqrt{-2})$

Solution

$$\begin{aligned}\sqrt{-8}(\sqrt{6} - \sqrt{-2}) &= i\sqrt{8}(\sqrt{6} - i\sqrt{2}) \\ &= i\sqrt{48} - i^2\sqrt{16} \\ &= i\sqrt{16 \cdot 3} - (-1)\sqrt{16} \\ &= 4i\sqrt{3} + 4 = 4 + 4i\sqrt{3}\end{aligned}$$

You Try It 9

Simplify: $\sqrt{-3}(\sqrt{27} - \sqrt{-6})$

Your solution

Example 10

Simplify: $(3 - 4i)(2 + 5i)$

Solution

$$\begin{aligned}(3 - 4i)(2 + 5i) &= 6 + 15i - 8i - 20i^2 \\ &= 6 + 7i - 20i^2 \\ &= 6 + 7i - 20(-1) = 26 + 7i\end{aligned}$$

• **FOIL**

You Try It 10

Simplify: $(4 - 3i)(2 - i)$

Your solution

Example 11

Simplify: $(4 + 5i)(4 - 5i)$

Solution

$$\begin{aligned}(4 + 5i)(4 - 5i) &= 4^2 + 5^2 \\ &= 16 + 25 = 41\end{aligned}$$

• **Conjugates**

You Try It 11

Simplify: $(3 + 6i)(3 - 6i)$

Your solution

Solutions on p. S29

Copyright © Houghton Mifflin Company. All rights reserved.

Objective D To divide complex numbers

A rational expression containing one or more complex numbers is in simplest form when no imaginary number remains in the denominator.

HOW TO Simplify: $\dfrac{2-3i}{2i}$

$$\frac{2-3i}{2i} = \frac{2-3i}{2i}\cdot\frac{i}{i}$$
- Multiply the numerator and denominator by $\frac{i}{i}$.

$$= \frac{2i-3i^2}{2i^2}$$

$$= \frac{2i-3(-1)}{2(-1)}$$
- Replace i^2 by -1.

$$= \frac{3+2i}{-2} = -\frac{3}{2} - i$$
- Simplify. Write the answer in the form $a + bi$.

HOW TO Simplify: $\dfrac{3+2i}{1+i}$

$$\frac{3+2i}{1+i} = \frac{3+2i}{1+i}\cdot\frac{1-i}{1-i}$$
- Multiply the numerator and denominator by the conjugate of $1 + i$.

$$= \frac{3-3i+2i-2i^2}{1^2+1^2}$$

$$= \frac{3-i-2(-1)}{1+1}$$
- Replace i^2 by -1 and simplify.

$$= \frac{5-i}{2} = \frac{5}{2} - \frac{1}{2}i$$
- Write the answer in the form $a + bi$.

Example 12

Simplify: $\dfrac{5+4i}{3i}$

Solution

$$\frac{5+4i}{3i} = \frac{5+4i}{3i}\cdot\frac{i}{i} = \frac{5i+4i^2}{3i^2}$$

$$= \frac{5i+4(-1)}{3(-1)} = \frac{-4+5i}{-3} = \frac{4}{3} - \frac{5}{3}i$$

You Try It 12

Simplify: $\dfrac{2-3i}{4i}$

Your solution

Example 13

Simplify: $\dfrac{5-3i}{4+2i}$

Solution

$$\frac{5-3i}{4+2i} = \frac{5-3i}{4+2i}\cdot\frac{4-2i}{4-2i}$$

$$= \frac{20-10i-12i+6i^2}{4^2+2^2}$$

$$= \frac{20-22i+6(-1)}{16+4}$$

$$= \frac{14-22i}{20} = \frac{7-11i}{10} = \frac{7}{10} - \frac{11}{10}i$$

You Try It 13

Simplify: $\dfrac{2+5i}{3-2i}$

Your solution

Solutions on p. S30

Copyright © Houghton Mifflin Company. All rights reserved.

9.4 Exercises

Objective A To simplify a complex number

1. What is an imaginary number? What is a complex number?

2. Are all real numbers also complex numbers? Are all complex numbers also real numbers?

For Exercises 3 to 14, simplify.

3. $\sqrt{-4}$

4. $\sqrt{-64}$

5. $\sqrt{-98}$

6. $\sqrt{-72}$

7. $\sqrt{-27}$

8. $\sqrt{-75}$

9. $\sqrt{16} + \sqrt{-4}$

10. $\sqrt{25} + \sqrt{-9}$

11. $\sqrt{12} - \sqrt{-18}$

12. $\sqrt{60} - \sqrt{-48}$

13. $\sqrt{160} - \sqrt{-147}$

14. $\sqrt{96} - \sqrt{-125}$

Objective B To add or subtract complex numbers

For Exercises 15 to 24, simplify.

15. $(2 + 4i) + (6 - 5i)$

16. $(6 - 9i) + (4 + 2i)$

17. $(-2 - 4i) - (6 - 8i)$

18. $(3 - 5i) + (8 - 2i)$

19. $(8 - \sqrt{-4}) - (2 + \sqrt{-16})$

20. $(5 - \sqrt{-25}) - (11 - \sqrt{-36})$

21. $(12 - \sqrt{-50}) + (7 - \sqrt{-8})$

22. $(5 - \sqrt{-12}) - (9 + \sqrt{-108})$

23. $(\sqrt{8} + \sqrt{-18}) + (\sqrt{32} - \sqrt{-72})$

24. $(\sqrt{40} - \sqrt{-98}) - (\sqrt{90} + \sqrt{-32})$

Objective C To multiply complex numbers

For Exercises 25 to 42, simplify.

25. $(7i)(-9i)$

26. $(-6i)(-4i)$

27. $\sqrt{-2}\sqrt{-8}$

Copyright © Houghton Mifflin Company. All rights reserved.

28. $\sqrt{-5}\sqrt{-45}$

29. $\sqrt{-3}\sqrt{-6}$

30. $\sqrt{-5}\sqrt{-10}$

31. $2i(6 + 2i)$

32. $-3i(4 - 5i)$

33. $\sqrt{-2}(\sqrt{8} + \sqrt{-2})$

34. $\sqrt{-3}(\sqrt{12} - \sqrt{-6})$

35. $(5 - 2i)(3 + i)$

36. $(2 - 4i)(2 - i)$

37. $(6 + 5i)(3 + 2i)$

38. $(4 - 7i)(2 + 3i)$

39. $(1 - i)\left(\frac{1}{2} + \frac{1}{2}i\right)$

40. $\left(\frac{4}{5} - \frac{2}{5}i\right)\left(1 + \frac{1}{2}i\right)$

41. $\left(\frac{6}{5} + \frac{3}{5}i\right)\left(\frac{2}{3} - \frac{1}{3}i\right)$

42. $(2 - i)\left(\frac{2}{5} + \frac{1}{5}i\right)$

Objective D To divide complex numbers

For Exercises 43 to 58, simplify.

43. $\frac{3}{i}$

44. $\frac{4}{5i}$

45. $\frac{2 - 3i}{-4i}$

46. $\frac{16 + 5i}{-3i}$

47. $\frac{4}{5 + i}$

48. $\frac{6}{5 + 2i}$

49. $\frac{2}{2 - i}$

50. $\frac{5}{4 - i}$

51. $\frac{1 - 3i}{3 + i}$

52. $\frac{2 + 12i}{5 + i}$

53. $\frac{\sqrt{-10}}{\sqrt{8} - \sqrt{-2}}$

54. $\frac{\sqrt{-2}}{\sqrt{12} - \sqrt{-8}}$

55. $\frac{2 - 3i}{3 + i}$

56. $\frac{3 + 5i}{1 - i}$

57. $\frac{5 + 3i}{3 - i}$

58. $\frac{3 - 2i}{2i + 3}$

APPLYING THE CONCEPTS

59. **a.** Is $3i$ a solution of $2x^2 + 18 = 0$?
b. Is $3 + i$ a solution of $x^2 - 6x + 10 = 0$?

60. Evaluate i^n for $n = 0, 1, 2, 3, 4, 5, 6,$ and 7. Make a conjecture about the value of i^n for any natural number. Using your conjecture, evaluate i^{76}.

Copyright © Houghton Mifflin Company. All rights reserved.

Focus on Problem Solving

Polya's Four-Step Process

One of the foremost mathematicians to study problem solving was George Polya (1887–1985). The basic structure that Polya advocated for problem solving has four steps: understand the problem, devise a plan, carry out the plan, and review the solution. We illustrate these steps with the following problem.

A = 1
B = 2
C = 3
D = 4
E = 5
F = 6
G = 7
H = 8
I = 9
J = 10
K = 11
L = 12
M = 13
N = 14
O = 15
P = 16
Q = 17
R = 18
S = 19
T = 20
U = 21
V = 22
W = 23
X = 24
Y = 25
Z = 26

Number the letters of the alphabet in sequence from 1 to 26. (See the list at the left.) Find a word for which the product of the numerical values of the letters of the word equals 1,000,000. We will agree that a "word" is any sequence of letters that contains at least one vowel but is not necessarily in the dictionary.

Understand the Problem

Consider REZB. The product of the values of the letters is $18 \cdot 5 \cdot 26 \cdot 2 = 4680$. This "word" is a sequence of letters with at least one vowel. However, the product of the numerical values of the letters is not 1,000,000. Thus this word does not solve our problem.

Devise a Plan

Actually, we should have known that the product of the values of the letters in REZB could not equal 1,000,000. The letter R has a factor of 9, and the letter Z has a factor of 13. Neither of these two numbers is a factor of 1,000,000. Consequently, R and Z cannot be letters in the word we are trying to find. This observation leads to an important consideration: Each of the letters that make up our word must be a factor of 1,000,000. To find these letters, consider the prime factorization of 1,000,000.

$$1{,}000{,}000 = 2^6 \cdot 5^6$$

Looking at the prime factorization, we note that letters that contain only 2 or 5 as factors are possible candidates. These letters are B, D, E, H, J, P, T, and Y. One additional point: Because 1 times any number is the number, the letter A can be part of any word we construct.

Our task is now to construct a word from these letters such that the product is 1,000,000. From the prime factorization above, we must have 2 as a factor six times and 5 as a factor six times.

Carry Out the Plan

We must construct a word with the characteristics described in our plan. Here is a possibility:

THEBEYE

Review the Solution

You should multiply the values of all the letters and verify that the product is 1,000,000. To ensure that you have an understanding of the problem, find other "words" that satisfy the conditions of the problem.

Point of Interest

George Polya was born in Hungary and moved to the United States in 1940. He lived in Providence, Rhode Island, where he taught at Brown University until 1942, when he moved to California. There he taught at Stanford University until his retirement. While at Stanford, he published 10 books and a number of articles for mathematics journals. Of the books Polya published, *How To Solve It* (1945) is one of his best known. In this book, Polya outlines a strategy for solving problems. This strategy, although frequently applied to mathematics, can be used to solve problems from virtually any discipline.

Copyright © Houghton Mifflin Company. All rights reserved.

Projects and Group Activities

Solving Radical Equations with a Graphing Calculator

The radical equation $\sqrt{x-2} - 6 = 0$ was solved algebraically at the beginning of Section 9.3. To solve $\sqrt{x-2} - 6 = 0$ with a graphing calculator, use the left side of the equation and write an equation in two variables.

$$y = \sqrt{x-2} - 6$$

The graph of $y = \sqrt{x-2} - 6$ is shown below. The solution set of the equation $y = \sqrt{x-2} - 6$ is the set of ordered pairs (x, y) whose coordinates make the equation a true statement. The x-coordinate at which $y = 0$ is the solution of the equation $\sqrt{x-2} - 6 = 0$. The solution is the x-intercept of the curve given by the equation $y = \sqrt{x-2} - 6$. The solution is 38.

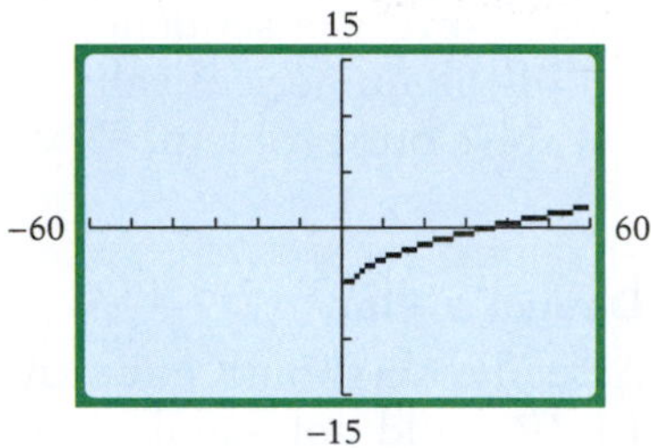

The solution of the radical equation $\sqrt[3]{x-2} = -0.8$ is 1.488. A graphing calculator can be used to find any rational or irrational solution to a predetermined degree of accuracy. The graph of $y = \sqrt[3]{x-2} + 0.8$ is shown below. The graph intersects the x-axis at $x = 1.488$.

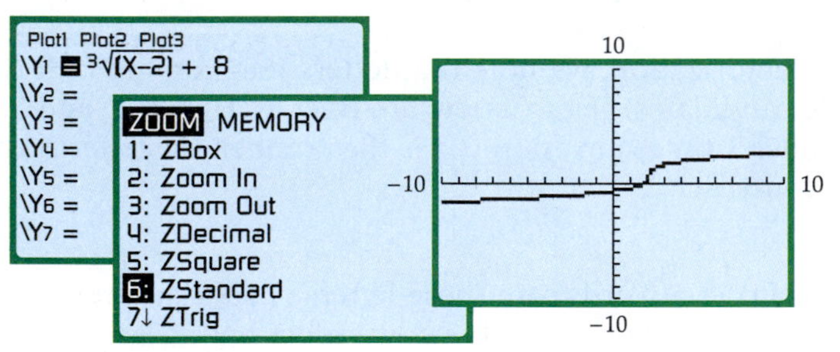

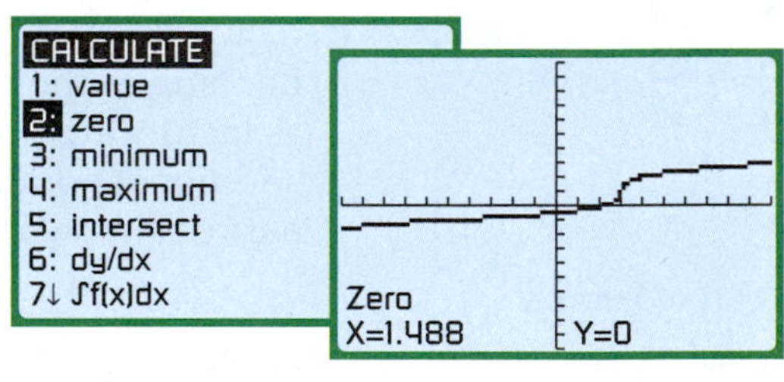

To graph the equation $y = \sqrt[3]{x-2} + 0.8$ on a TI-83, enter

Enter 2nd CALC 2 to find the x-intercept of the equation.

Integrating Technology

See the appendix Keystroke Guide: *Graph* for instructions on using a graphing calculator to graph a function, and *Zero* for instructions on finding the x-intercepts of a function.

You were to solve Exercises 1 to 21 in Section 9.3 by algebraic means. Try some of those exercises again, this time using a graphing calculator.

Use a graphing calculator to find the solutions of the equations in Exercises 1 to 3. Round to the nearest thousandth.

1. $\sqrt{x + 0.3} = 1.3$ **2.** $\sqrt[3]{x + 1.2} = -1.1$ **3.** $\sqrt[4]{3x - 1.5} = 1.4$

Copyright © Houghton Mifflin Company. All rights reserved.

Chapter 9 Summary

Key Words	Examples
$a^{1/n}$ is the *nth root of a*. [9.1A, p. 517]	$16^{1/2} = 4$ because $4^2 = 16$.
The expression $\sqrt[n]{a}$ is another symbol for the *n*th root of *a*. In the expression $\sqrt[n]{a}$, the symbol $\sqrt{\ }$ is called a *radical*, *n* is the *index* of the radical, and *a* is the *radicand*. [9.1B, p. 519]	$125^{1/3} = \sqrt[3]{125} = 5$ The index is 3, and the radicand is 125.
The symbol $\sqrt{\ }$ is used to indicate the positive square root, or *principal square root*, of a number. [9.1C, p. 521]	$\sqrt{16} = 4$ $-\sqrt{16} = -4$
The expressions $a + b$ and $a - b$ are called *conjugates* of each other. The product of conjugates of the form $(a + b)(a - b)$ is $a^2 - b^2$. [9.2C, p. 530]	$(x + 3)(x - 3) = x^2 - 3^2 = x^2 - 9$
The procedure used to remove a radical from the denominator of a radical expression is called *rationalizing the denominator*. [9.2D, p. 531]	$\frac{2}{1-\sqrt{3}} = \frac{2}{1-\sqrt{3}} \cdot \frac{1+\sqrt{3}}{1+\sqrt{3}}$ $= \frac{2(1+\sqrt{3})}{(1-\sqrt{3})(1+\sqrt{3})}$ $= \frac{2+2\sqrt{3}}{1-3} = \frac{2+2\sqrt{3}}{-2}$ $= -1 - \sqrt{3}$
A *radical equation* is an equation that contains a variable expression in a radicand. [9.3A, p. 537]	$\sqrt{x-2} - 3 = 6$ is a radical equation.
A *complex number* is a number of the form $a + bi$, where a and b are real numbers and $i = \sqrt{-1}$. For the complex number $a + bi$, a is the *real part* of the complex number and b is the *imaginary part* of the complex number. [9.4A, p. 543]	$3 + 2i$ is a complex number. 3 is the real part and 2 is the imaginary part of the complex number.

Essential Rules and Procedures	Examples
Rule for Rational Exponents [9.1A, p. 517] If m and n are positive integers and $a^{1/n}$ is a real number, then $a^{m/n} = (a^{1/n})^m$.	$8^{2/3} = (8^{1/3})^2 = 2^2 = 4$

Copyright © Houghton Mifflin Company. All rights reserved.

Definition of *n*th Root of *a* [9.1B, p. 519]

If a is a real number, then $a^{1/n} = \sqrt[n]{a}$.

$x^{1/3} = \sqrt[3]{x}$

Writing Exponential Expressions as Radical Expressions [9.1B, p. 519]

If $a^{1/n}$ is a real number, then $a^{m/n} = a^{m \cdot 1/n} = (a^m)^{1/n} = \sqrt[n]{a^m}$.

The expression $a^{m/n}$ can also be written $(\sqrt[n]{a})^m$.

$b^{3/4} = \sqrt[4]{b^3}$

$8^{2/3} = (\sqrt[3]{8})^2 = 2^2 = 4$

Product Property of Radicals [9.2A, p. 527]

If $\sqrt[n]{a}$ and $\sqrt[n]{b}$ are positive real numbers, then $\sqrt[n]{ab} = \sqrt[n]{a} \cdot \sqrt[n]{b}$ and $\sqrt[n]{a} \cdot \sqrt[n]{b} = \sqrt[n]{ab}$.

$\sqrt{9 \cdot 7} = \sqrt{9} \cdot \sqrt{7} = 3\sqrt{7}$

Quotient Property of Radicals [9.2D, p. 531]

If $\sqrt[n]{a}$ and $\sqrt[n]{b}$ are positive real numbers, and $b \neq 0$, then

$$\sqrt[n]{\frac{a}{b}} = \frac{\sqrt[n]{a}}{\sqrt[n]{b}} \text{ and } \frac{\sqrt[n]{a}}{\sqrt[n]{b}} = \sqrt[n]{\frac{a}{b}}.$$

$$\sqrt[3]{\frac{5}{27}} = \frac{\sqrt[3]{5}}{\sqrt[3]{27}} = \frac{\sqrt[3]{5}}{3}$$

Property of Raising Each Side of an Equation to a Power [9.3A, p. 537]

If $a = b$, then $a^n = b^n$.

If $x = 4$, then $x^2 = 16$.

Pythagorean Theorem [9.3B, p. 539]

The square of the hypotenuse of a right triangle is equal to the sum of the squares of the two legs.

$c^2 = a^2 + b^2$

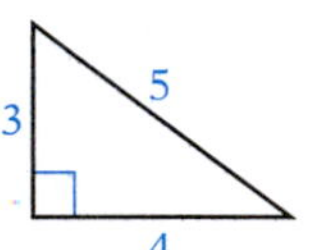

$5^2 = 3^2 + 4^2$

Definition of $\sqrt{-a}$ [9.4A, p. 543]

If a is a positive real number, then the principal square root of negative a is the imaginary number $i\sqrt{a}$: $\sqrt{-a} = i\sqrt{a}$.

$\sqrt{-8} = i\sqrt{8} = 2i\sqrt{2}$

Addition and Subtraction of Complex Numbers [9.4B, p. 544]

$(a + bi) + (c + di) = (a + c) + (b + d)i$

$(a + bi) - (c + di) = (a - c) + (b - d)i$

$(2 + 4i) + (3 + 6i)$
$= (2 + 3) + (4 + 6)i$
$= 5 + 10i$

$(4 + 3i) - (7 + 4i)$
$= (4 - 7) + (3 - 4)i$
$= -3 - i$

Multiplication of Complex Numbers [9.4C, p. 546]

$(a + bi)(c + di) = (ac - bd) + (ad + bc)i$

One way to remember this rule is to think of the FOIL method.

$(2 - 3i)(5 + 4i)$
$= 10 + 8i - 15i - 12i^2$
$= 10 - 7i - 12(-1)$
$= 22 - 7i$

Copyright © Houghton Mifflin Company. All rights reserved.

Chapter 9 Review Exercises

1. Simplify: $(16x^{-4}y^{12})^{1/4}(100x^{6}y^{-2})^{1/2}$

2. Solve: $\sqrt[4]{3x - 5} = 2$

3. Simplify: $(6 - 5i)(4 + 3i)$

4. Rewrite $7y\sqrt[3]{x^2}$ as an exponential expression.

5. Simplify: $(\sqrt{3} + 8)(\sqrt{3} - 2)$

6. Solve: $\sqrt{4x + 9} + 10 = 11$

7. Simplify: $\dfrac{x^{-3/2}}{x^{7/2}}$

8. Simplify: $\dfrac{8}{\sqrt{3y}}$

9. Simplify: $\sqrt[3]{-8a^6b^{12}}$

10. Simplify: $\sqrt{50a^4b^3} - ab\sqrt{18a^2b}$

11. Simplify: $\dfrac{x + 2}{\sqrt{x} + \sqrt{2}}$

12. Simplify: $\dfrac{5 + 2i}{3i}$

13. Simplify: $\sqrt{18a^3b^6}$

14. Simplify: $(\sqrt{50} + \sqrt{-72}) - (\sqrt{162} - \sqrt{-8})$

15. Simplify: $3x\sqrt[3]{54x^8y^{10}} - 2x^2y\sqrt[3]{16x^5y^7}$

16. Simplify: $\sqrt[3]{16x^4y}\sqrt[3]{4xy^5}$

17. Simplify: $i(3 - 7i)$

18. Rewrite $3x^{3/4}$ as a radical expression.

Copyright © Houghton Mifflin Company. All rights reserved.

19. Simplify: $\sqrt[5]{-64a^8b^{12}}$

20. Simplify: $\dfrac{5 + 9i}{1 - i}$

21. Simplify: $\sqrt{-12}\sqrt{-6}$

22. Solve: $\sqrt{x - 5} + \sqrt{x + 6} = 11$

23. Simplify: $\sqrt[4]{81a^8b^{12}}$

24. Simplify: $\sqrt{-50}$

25. Simplify: $(-8 + 3i) - (4 - 7i)$

26. Simplify: $(5 - \sqrt{6})^2$

27. Simplify: $4x\sqrt{12x^2y} + \sqrt{3x^4y} - x^2\sqrt{27y}$

28. Simplify: $81^{-1/4}$

29. Simplify: $(a^{16})^{-5/8}$

30. Simplify: $-\sqrt{49x^6y^{16}}$

31. **Energy** The velocity of the wind determines the amount of power generated by a windmill. A typical equation for this relationship is $v = 4.05\sqrt[3]{P}$, where v is the velocity in miles per hour and P is the power in watts. Find the amount of power generated by a 20-mph wind. Round to the nearest whole number.

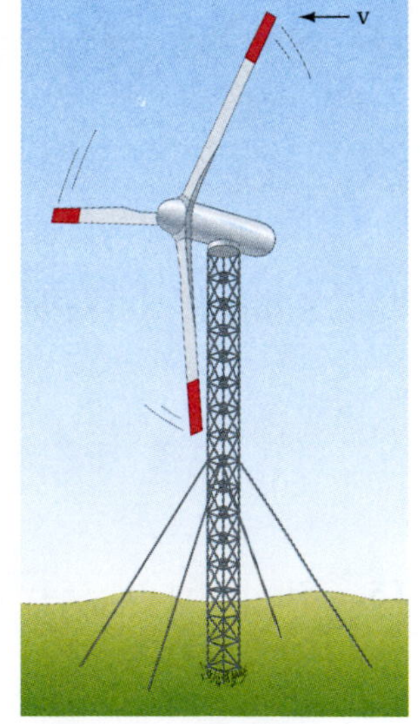

32. **Automotive Technology** Find the distance required for a car to reach a velocity of 88 ft/s when the acceleration is 16 ft/s². Use the equation $v = \sqrt{2as}$, where v is the velocity in feet per second, a is the acceleration, and s is the distance in feet.

33. **Home Maintenance** A 12-foot ladder is leaning against a house in preparation for washing the windows. How far from the house is the bottom of the ladder when the top of the ladder touches the house 10 ft above the ground? Round to the nearest hundredth.

Copyright © Houghton Mifflin Company. All rights reserved.

Chapter 9 Test

1. Write $\frac{1}{2}\sqrt[4]{x^3}$ as an exponential expression.

2. Simplify:
$\sqrt[3]{54x^7y^3} - x\sqrt[3]{128x^4y^3} - x^2\sqrt[3]{2xy^3}$

3. Write $3y^{2/5}$ as a radical expression.

4. Simplify: $(2 + 5i)(4 - 2i)$

5. Simplify: $(2\sqrt{x} + \sqrt{y})^2$

6. Simplify: $\dfrac{r^{2/3}r^{-1}}{r^{-1/2}}$

7. Solve: $\sqrt{x + 12} - \sqrt{x} = 2$

8. Simplify: $\sqrt[3]{8x^3y^6}$

9. Simplify: $\sqrt{3x}(\sqrt{x} - \sqrt{25x})$

10. Simplify: $(5 - 2i) - (8 - 4i)$

11. Simplify: $\sqrt{32x^4y^7}$

12. Simplify: $(2\sqrt{3} + 4)(3\sqrt{3} - 1)$

13. Simplify: $(2 + i) + (2 - i)(3 + 2i)$

14. Simplify: $\dfrac{4 - 2\sqrt{5}}{2 - \sqrt{5}}$

Copyright © Houghton Mifflin Company. All rights reserved.

15. Simplify: $\sqrt{18a^3} + a\sqrt{50a}$

16. Simplify: $(\sqrt{a} - 3\sqrt{b})(2\sqrt{a} + 5\sqrt{b})$

17. Simplify: $\dfrac{(2x^{1/3}y^{-2/3})^6}{(x^{-4}y^8)^{1/4}}$

18. Simplify: $\dfrac{\sqrt{x}}{\sqrt{x} - \sqrt{y}}$

19. Simplify: $\dfrac{2 + 3i}{1 - 2i}$

20. Solve: $\sqrt[3]{2x - 2} + 4 = 2$

21. Simplify: $\left(\dfrac{4a^4}{b^2}\right)^{-3/2}$

22. Simplify: $\sqrt[3]{27a^4b^3c^7}$

23. Simplify: $\dfrac{\sqrt{32x^5y}}{\sqrt{2xy^3}}$

24. Simplify: $(\sqrt{-8})(\sqrt{-2})$

25. **Physics** An object is dropped from a high building. Find the distance the object has fallen when its speed reaches 192 ft/s. Use the equation $v = \sqrt{64d}$, where v is the speed of the object in feet per second and d is the distance in feet.

Copyright © Houghton Mifflin Company. All rights reserved.

Cumulative Review Exercises

1. Simplify: $2^3 \cdot 3 - 4(3 - 4 \cdot 5)$

2. Evaluate $4a^2b - a^3$ when $a = -2$ and $b = 3$.

3. Simplify: $-3(4x - 1) - 2(1 - x)$

4. Solve: $5 - \frac{2}{3}x = 4$

5. Solve: $2[4 - 2(3 - 2x)] = 4(1 - x)$

6. Solve: $6x - 3(2x + 2) > 3 - 3(x + 2)$

7. Solve: $2 + |4 - 3x| = 5$

8. Solve: $|2x + 3| \leq 9$

9. Find the area of the triangle shown in the figure below.

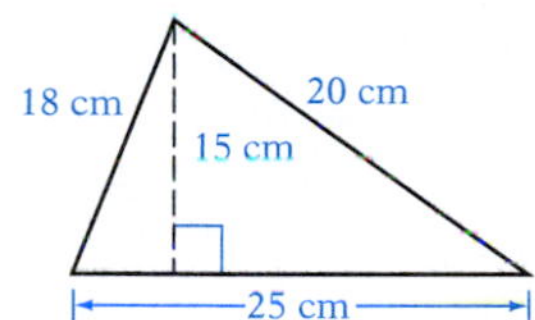

10. Find the volume of a rectangular solid with a length of 3.5 ft, a width of 2 ft, and a height of 2 ft.

11. Graph $3x - 2y = -6$. State the slope and y-intercept.

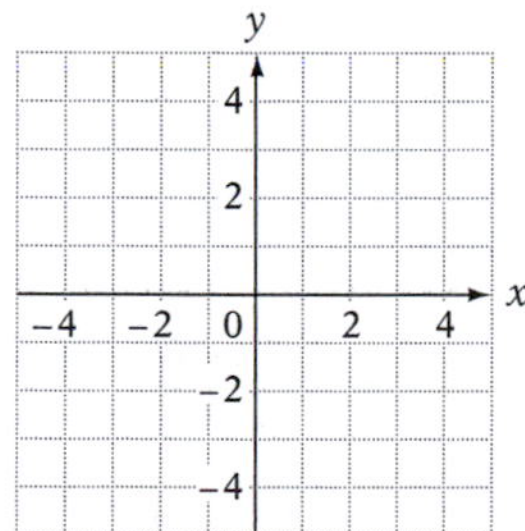

12. Graph the solution set of $3x + 2y \leq 4$.

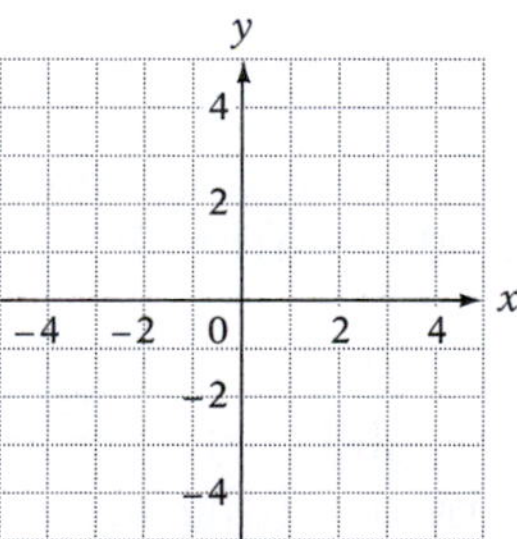

13. Find the equation of the line that passes through the points (2, 3) and (−1, 2).

14. Solve by the addition method:

$$2x - y = 4$$
$$-2x + 3y = 0$$

15. Simplify: $(2^{-1}x^2y^{-6})(2^{-1}y^{-4})^{-2}$

16. Factor: $81x^2 - y^2$

Copyright © Houghton Mifflin Company. All rights reserved.

17. Factor: $x^5 + 2x^3 - 3x$

18. Solve $P = \dfrac{R - C}{n}$ for C.

19. Simplify: $\left(\dfrac{x^{-2/3}y^{1/2}}{y^{-1/3}}\right)^6$

20. Subtract: $\sqrt{40x^3} - x\sqrt{90x}$

21. Multiply: $(\sqrt{3} - 2)(\sqrt{3} - 5)$

22. Simplify: $\dfrac{4}{\sqrt{6} - \sqrt{2}}$

23. Simplify: $\dfrac{2i}{3 - i}$

24. Solve: $\sqrt[3]{3x - 4} + 5 = 1$

25. The two triangles are similar triangles. Find the length of side DE.

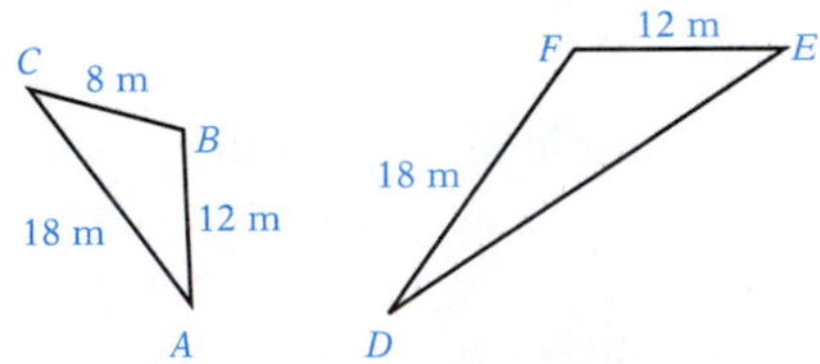

26. **Investments** An investor has a total of \$10,000 deposited in two simple interest accounts. On one account, the annual simple interest rate is 3.5%. On the second account, the annual simple interest rate is 4.5%. How much is invested in the 3.5% account if the total annual interest earned is \$425?

27. **Travel** A sales executive traveled 25 mi by car and then an additional 625 mi by plane. The rate of the plane was five times the rate of the car. The total time of the trip was 3 h. Find the rate of the plane.

28. **Astronomy** How long does it take light to travel to Earth from the moon when the moon is 232,500 mi from Earth? Light travels 1.86×10^5 mi/s.

29. **Investments** The graph shows the amount invested and the annual interest earned on the investment. Find the slope of the line between the two points shown on the graph. Then write a sentence that states the meaning of the slope.

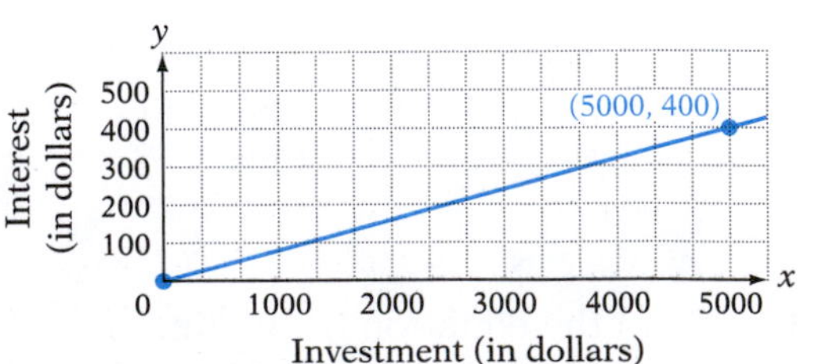

30. **Oceanography** How far would a submarine periscope have to be above the water for the lookout to locate a ship 7 mi away? The equation for the distance in miles that the lookout can see is $d = \sqrt{1.5h}$, where h is the height in feet above the surface of the water. Round to the nearest tenth of a foot.

Copyright © Houghton Mifflin Company. All rights reserved.

chapter 10

Quadratic Equations

This model rocket, just like an airplane, is subject to the forces of weight and thrust as it moves through the air. The study of forces and their effect on the motion of objects traveling through the air is called aerodynamics. The motion of a model rocket through the air can be described and explained by the principles of physics, which were discovered by Sir Isaac Newton over 300 years ago. **Exercises 6 to 8 on page 595** use quadratic equations to mathematically model the rocket's movement, as well as the movements of other projectiles.

Need help? For online student resources, such as section quizzes, visit this textbook's website at **math.college.hmco.com/students.**

OBJECTIVES

Section 10.1

A To solve a quadratic equation by factoring

B To write a quadratic equation given its solutions

C To solve a quadratic equation by taking square roots

Section 10.2

A To solve a quadratic equation by completing the square

Section 10.3

A To solve a quadratic equation by using the quadratic formula

Section 10.4

A To solve an equation that is quadratic in form

B To solve a radical equation that is reducible to a quadratic equation

C To solve a fractional equation that is reducible to a quadratic equation

Section 10.5

A To solve a nonlinear inequality

Section 10.6

A To solve application problems

Copyright © Houghton Mifflin Company. All rights reserved.

PREP TEST

Do these exercises to prepare for Chapter 10.

1. Simplify: $\sqrt{18}$

2. Simplify: $\sqrt{-9}$

3. Simplify: $\dfrac{3x - 2}{x - 1} - 1$

4. Evaluate $b^2 - 4ac$ when $a = 2$, $b = -4$, and $c = 1$.

5. Is $4x^2 + 28x + 49$ a perfect square trinomial?

6. Factor: $4x^2 - 4x + 1$

7. Factor: $9x^2 - 4$

8. Graph $\{x \mid x < -1\} \cap \{x \mid x < 4\}$

−5 −4 −3 −2 −1 0 1 2 3 4 5

9. Solve: $x(x - 1) = x + 15$

10. Solve: $\dfrac{4}{x - 3} = \dfrac{16}{x}$

GO FIGURE

The numeral $0.AAA\ldots$ is a repeating decimal. Given that $A \neq B$ and $\sqrt{0.AAA\ldots} = \sqrt{B}$, find the value of A.

Copyright © Houghton Mifflin Company. All rights reserved.

10.1 Solving Quadratic Equations by Factoring or by Taking Square Roots

Objective A To solve a quadratic equation by factoring

Recall that a *quadratic equation* is an equation of the form $ax^2 + bx + c = 0$, where a and b are coefficients, c is a constant, and $a \neq 0$.

Quadratic Equations
$$\begin{cases} 3x^2 - x + 2 = 0, & a = 3, & b = -1, & c = 2 \\ -x^2 + 4 = 0, & a = -1, & b = 0, & c = 4 \\ 6x^2 - 5x = 0, & a = 6, & b = -5, & c = 0 \end{cases}$$

A quadratic equation is in *standard form* when the polynomial is in descending order and equal to zero. Because the degree of the polynomial $ax^2 + bx + c$ is 2, a quadratic equation is also called a **second-degree equation.**

As we discussed earlier, quadratic equations sometimes can be solved by using the Principle of Zero Products. This method is reviewed here.

The Principle of Zero Products

If a and b are real numbers and $ab = 0$, then $a = 0$ or $b = 0$.

The Principle of Zero Products states that if the product of two factors is zero, then at least one of the factors must be zero.

TAKE NOTE

Recall that the steps involved in solving a quadratic equation by factoring are

1. Write the equation in standard form.
2. Factor.
3. Use the Principle of Zero Products to set each factor equal to 0.
4. Solve each equation.
5. Check the solutions.

HOW TO Solve by factoring: $3x^2 = 2 - 5x$

$$3x^2 = 2 - 5x$$
$$3x^2 + 5x - 2 = 0$$ • Write the equation in standard form.
$$(3x - 1)(x + 2) = 0$$ • Factor.
$$3x - 1 = 0 \qquad x + 2 = 0$$ • Use the Principle of Zero Products to write two equations.
$$3x = 1 \qquad x = -2$$ • Solve each equation.
$$x = \frac{1}{3}$$

$\frac{1}{3}$ and -2 check as solutions. The solutions are $\frac{1}{3}$ and -2.

TAKE NOTE

When a quadratic equation has two solutions that are the same number, the solution is called a **double root** of the equation. 3 is a double root of $x^2 - 6x = -9$.

HOW TO Solve by factoring: $x^2 - 6x = -9$

$$x^2 - 6x = -9$$
$$x^2 - 6x + 9 = 0$$ • Write the equation in standard form.
$$(x - 3)(x - 3) = 0$$ • Factor.
$$x - 3 = 0 \qquad x - 3 = 0$$ • Use the Principle of Zero Products.
$$x = 3 \qquad x = 3$$ • Solve each equation.

3 checks as a solution. The solution is 3.

Copyright © Houghton Mifflin Company. All rights reserved.

Example 1

Solve by factoring: $2x(x - 3) = x + 4$

Solution

$$2x(x - 3) = x + 4$$
$$2x^2 - 6x = x + 4$$
$$2x^2 - 7x - 4 = 0$$ • Write in standard form.
$$(2x + 1)(x - 4) = 0$$ • Solve by factoring.
$$2x + 1 = 0 \qquad x - 4 = 0$$
$$2x = -1 \qquad x = 4$$
$$x = -\frac{1}{2}$$

The solutions are $-\frac{1}{2}$ and 4.

You Try It 1

Solve by factoring: $2x^2 = 7x - 3$

Your solution

Example 2

Solve for x by factoring: $x^2 - 4ax - 5a^2 = 0$

Solution

This is a literal equation. Solve for x in terms of a.

$$x^2 - 4ax - 5a^2 = 0$$
$$(x + a)(x - 5a) = 0$$
$$x + a = 0 \qquad x - 5a = 0$$
$$x = -a \qquad x = 5a$$

The solutions are $-a$ and $5a$.

You Try It 2

Solve for x by factoring: $x^2 - 3ax - 4a^2 = 0$

Your solution

Solutions on p. S30

Objective B To write a quadratic equation given its solutions

As shown below, the solutions of the equation $(x - r_1)(x - r_2) = 0$ are r_1 and r_2.

$$(x - r_1)(x - r_2) = 0$$
$$x - r_1 = 0 \qquad x - r_2 = 0$$
$$x = r_1 \qquad x = r_2$$

Check:

$(x - r_1)(x - r_2) = 0$	
$(r_1 - r_1)(r_1 - r_2)$	0
$0 \cdot (r_1 - r_2)$	0
$0 = 0$	

$(x - r_1)(x - r_2) = 0$	
$(r_2 - r_1)(r_2 - r_2)$	0
$(r_2 - r_1) \cdot 0$	0
$0 = 0$	

Using the equation $(x - r_1)(x - r_2) = 0$ and the fact that r_1 and r_2 are solutions of this equation, it is possible to write a quadratic equation given its solutions.

HOW TO Write a quadratic equation that has solutions 4 and −5.

$$(x - r_1)(x - r_2) = 0$$
$$(x - 4)[x - (-5)] = 0$$ • Replace r_1 by 4 and r_2 by −5.
$$(x - 4)(x + 5) = 0$$ • Simplify.
$$x^2 + x - 20 = 0$$ • Multiply.

Copyright © Houghton Mifflin Company. All rights reserved.

HOW TO Write a quadratic equation with integer coefficients and solutions $\frac{2}{3}$ and $\frac{1}{2}$.

$$(x - r_1)(x - r_2) = 0$$

$$\left(x - \frac{2}{3}\right)\left(x - \frac{1}{2}\right) = 0$$
• Replace r_1 by $\frac{2}{3}$ and r_2 by $\frac{1}{2}$.

$$x^2 - \frac{7}{6}x + \frac{1}{3} = 0$$
• Multiply.

$$6\left(x^2 - \frac{7}{6}x + \frac{1}{3}\right) = 6 \cdot 0$$
• Multiply each side of the equation by the LCM of the denominators.

$$6x^2 - 7x + 2 = 0$$

Example 3

Write a quadratic equation with integer coefficients and solutions $\frac{1}{2}$ and -4.

Solution

$$(x - r_1)(x - r_2) = 0$$

$$\left(x - \frac{1}{2}\right)[x - (-4)] = 0$$
• $r_1 = \frac{1}{2}$, $r_2 = -4$

$$\left(x - \frac{1}{2}\right)(x + 4) = 0$$

$$x^2 + \frac{7}{2}x - 2 = 0$$

$$2\left(x^2 + \frac{7}{2}x - 2\right) = 2 \cdot 0$$

$$2x^2 + 7x - 4 = 0$$

You Try It 3

Write a quadratic equation with integer coefficients and solutions 3 and $-\frac{1}{2}$.

Your solution

Solution on p. S30

Objective C **To solve a quadratic equation by taking square roots**

The solution of the quadratic equation $x^2 = 16$ is shown at the right.

$$x^2 = 16$$
$$x^2 - 16 = 0$$
$$(x - 4)(x + 4) = 0$$
$$x - 4 = 0 \qquad x + 4 = 0$$
$$x = 4 \qquad x = -4$$

The solution can also be found by taking the square root of each side of the equation and writing the positive and negative square roots of the radicand. The notation ± 4 means $x = 4$ or $x = -4$.

$$x^2 = 16$$
$$\sqrt{x^2} = \sqrt{16}$$
$$x = \pm 4$$

The solutions are 4 and −4.

HOW TO Solve by taking square roots: $3x^2 = 54$

$$3x^2 = 54$$

$$x^2 = 18$$
• Solve for x^2.

$$\sqrt{x^2} = \sqrt{18}$$
• Take the square root of each side of the equation.

$$x = \pm\sqrt{18}$$
• Simplify.

$$x = \pm 3\sqrt{2}$$

The solutions are $3\sqrt{2}$ and $-3\sqrt{2}$.
• $3\sqrt{2}$ and $-3\sqrt{2}$ check as solutions.

Copyright © Houghton Mifflin Company. All rights reserved.

Solving a quadratic equation by taking the square root of each side of the equation can lead to solutions that are complex numbers.

HOW TO Solve by taking square roots: $2x^2 + 18 = 0$

$2x^2 + 18 = 0$

$2x^2 = -18$ • Solve for x^2.

$x^2 = -9$

$\sqrt{x^2} = \sqrt{-9}$ • Take the square root of each side of the equation.

$x = \pm\sqrt{-9}$ • Simplify.

$x = \pm 3i$

Check:

$2x^2 + 18 = 0$		$2x^2 + 18 = 0$	
$2(3i)^2 + 18$	0	$2(-3i)^2 + 18$	0
$2(-9) + 18$	0	$2(-9) + 18$	0
$-18 + 18$	0	$-18 + 18$	0
$0 = 0$		$0 = 0$	

The solutions are $3i$ and $-3i$.

Study Tip

Always check the solution of an equation.

An equation containing the square of a binomial can be solved by taking square roots.

HOW TO Solve by taking square roots: $(x + 2)^2 - 24 = 0$

$(x + 2)^2 - 24 = 0$

$(x + 2)^2 = 24$ • Solve for $(x + 2)^2$.

$\sqrt{(x + 2)^2} = \sqrt{24}$ • Take the square root of each side of the equation. Then simplify.

$x + 2 = \pm\sqrt{24}$

$x + 2 = \pm 2\sqrt{6}$

$x + 2 = 2\sqrt{6}$ $\quad$ $x + 2 = -2\sqrt{6}$ • Solve for x.

$x = -2 + 2\sqrt{6}$ $\quad$ $x = -2 - 2\sqrt{6}$

The solutions are $-2 + 2\sqrt{6}$ and $-2 - 2\sqrt{6}$.

Example 4

Solve by taking square roots.
$3(x - 2)^2 + 12 = 0$

Solution

$3(x - 2)^2 + 12 = 0$

$3(x - 2)^2 = -12$ • Solve for $(x - 2)^2$.

$(x - 2)^2 = -4$

$\sqrt{(x - 2)^2} = \sqrt{-4}$ • Take the square root of each side of the equation.

$x - 2 = \pm\sqrt{-4}$

$x - 2 = \pm 2i$

$x - 2 = 2i$ $\quad$ $x - 2 = -2i$ • Solve for x.

$x = 2 + 2i$ $\quad$ $x = 2 - 2i$

The solutions are $2 + 2i$ and $2 - 2i$.

You Try It 4

Solve by taking square roots.
$2(x + 1)^2 - 24 = 0$

Your solution

Solution on p. S30

Copyright © Houghton Mifflin Company. All rights reserved.

10.1 Exercises

Objective A To solve a quadratic equation by factoring

1. Explain why the restriction $a \neq 0$ is necessary in the definition of a quadratic equation.

2. What does the Principle of Zero Products state? How is it used to solve a quadratic equation?

For Exercises 3 to 6, write the quadratic equation in standard form with the coefficient of x^2 positive. Name the values of a, b, and c.

3. $2x^2 - 4x = 5$
4. $x^2 = 3x + 1$
5. $5x = 4x^2 + 6$
6. $3x^2 = 7$

For Exercises 7 to 37, solve by factoring.

7. $x^2 - 4x = 0$
8. $y^2 + 6y = 0$
9. $t^2 - 25 = 0$
10. $p^2 - 81 = 0$
11. $s^2 - s - 6 = 0$
12. $v^2 + 4v - 5 = 0$
13. $y^2 - 6y + 9 = 0$
14. $x^2 + 10x + 25 = 0$
15. $9z^2 - 18z = 0$
16. $4y^2 + 20y = 0$
17. $r^2 - 3r = 10$
18. $p^2 + 5p = 6$
19. $v^2 + 10 = 7v$
20. $t^2 - 16 = 15t$
21. $2x^2 - 9x - 18 = 0$
22. $3y^2 - 4y - 4 = 0$
23. $4z^2 - 9z + 2 = 0$
24. $2s^2 - 9s + 9 = 0$
25. $3w^2 + 11w = 4$
26. $2r^2 + r = 6$
27. $6x^2 = 23x + 18$
28. $6x^2 = 7x - 2$
29. $4 - 15u - 4u^2 = 0$
30. $3 - 2y - 8y^2 = 0$
31. $x + 18 = x(x - 6)$
32. $t + 24 = t(t + 6)$
33. $4s(s + 3) = s - 6$

Copyright © Houghton Mifflin Company. All rights reserved.

34. $3v(v - 2) = 11v + 6$

35. $u^2 - 2u + 4 = (2u - 3)(u + 2)$

36. $(3v - 2)(2v + 1) = 3v^2 - 11v - 10$

37. $(3x - 4)(x + 4) = x^2 - 3x - 28$

For Exercises 38 to 52, solve for x by factoring.

38. $x^2 + 14ax + 48a^2 = 0$

39. $x^2 - 9bx + 14b^2 = 0$

40. $x^2 + 9xy - 36y^2 = 0$

41. $x^2 - 6cx - 7c^2 = 0$

42. $x^2 - ax - 20a^2 = 0$

43. $2x^2 + 3bx + b^2 = 0$

44. $3x^2 - 4cx + c^2 = 0$

45. $3x^2 - 14ax + 8a^2 = 0$

46. $3x^2 - 11xy + 6y^2 = 0$

47. $3x^2 - 8ax - 3a^2 = 0$

48. $3x^2 - 4bx - 4b^2 = 0$

49. $4x^2 + 8xy + 3y^2 = 0$

50. $6x^2 - 11cx + 3c^2 = 0$

51. $6x^2 + 11ax + 4a^2 = 0$

52. $12x^2 - 5xy - 2y^2 = 0$

Objective B To write a quadratic equation given its solutions

For Exercises 53 to 82, write a quadratic equation that has integer coefficients and has as solutions the given pair of numbers.

53. 2 and 5

54. 3 and 1

55. −2 and −4

56. −1 and −3

57. 6 and −1

58. −2 and 5

59. 3 and −3

60. 5 and −5

61. 4 and 4

62. 2 and 2

63. 0 and 5

64. 0 and −2

Copyright © Houghton Mifflin Company. All rights reserved.

65. 0 and 3

66. 0 and -1

67. 3 and $\frac{1}{2}$

68. 2 and $\frac{2}{3}$

69. $-\frac{3}{4}$ and 2

70. $-\frac{1}{2}$ and 5

71. $-\frac{5}{3}$ and -2

72. $-\frac{3}{2}$ and -1

73. $-\frac{2}{3}$ and $\frac{2}{3}$

74. $-\frac{1}{2}$ and $\frac{1}{2}$

75. $\frac{1}{2}$ and $\frac{1}{3}$

76. $\frac{3}{4}$ and $\frac{2}{3}$

77. $\frac{6}{5}$ and $-\frac{1}{2}$

78. $\frac{3}{4}$ and $-\frac{3}{2}$

79. $-\frac{1}{4}$ and $-\frac{1}{2}$

80. $-\frac{5}{6}$ and $-\frac{2}{3}$

81. $\frac{3}{5}$ and $-\frac{1}{10}$

82. $\frac{7}{2}$ and $-\frac{1}{4}$

Objective C To solve a quadratic equation by taking square roots

For Exercises 83 to 112, solve by taking square roots.

83. $y^2 = 49$

84. $x^2 = 64$

85. $z^2 = -4$

86. $v^2 = -16$

87. $s^2 - 4 = 0$

88. $r^2 - 36 = 0$

89. $4x^2 - 81 = 0$

90. $9x^2 - 16 = 0$

91. $y^2 + 49 = 0$

92. $z^2 + 16 = 0$

93. $v^2 - 48 = 0$

94. $s^2 - 32 = 0$

95. $r^2 - 75 = 0$

96. $u^2 - 54 = 0$

97. $z^2 + 18 = 0$

Copyright © Houghton Mifflin Company. All rights reserved.

98. $t^2 + 27 = 0$

99. $(x - 1)^2 = 36$

100. $(x + 2)^2 = 25$

101. $3(y + 3)^2 = 27$

102. $4(s - 2)^2 = 36$

103. $5(z + 2)^2 = 125$

104. $2(y - 3)^2 = 18$

105. $\left(v - \frac{1}{2}\right)^2 = \frac{1}{4}$

106. $\left(r + \frac{2}{3}\right)^2 = \frac{1}{9}$

107. $(x + 5)^2 - 6 = 0$

108. $(t - 1)^2 - 15 = 0$

109. $(v - 3)^2 + 45 = 0$

110. $(x + 5)^2 + 32 = 0$

111. $\left(u + \frac{2}{3}\right)^2 - 18 = 0$

112. $\left(z - \frac{1}{2}\right)^2 - 20 = 0$

APPLYING THE CONCEPTS

For Exercises 113 to 120, write a quadratic equation that has as solutions the given pair of numbers.

113. $\sqrt{2}$ and $-\sqrt{2}$

114. $\sqrt{5}$ and $-\sqrt{5}$

115. i and $-i$

116. $2i$ and $-2i$

117. $2\sqrt{2}$ and $-2\sqrt{2}$

118. $3\sqrt{2}$ and $-3\sqrt{2}$

119. $i\sqrt{2}$ and $-i\sqrt{2}$

120. $2i\sqrt{3}$ and $-2i\sqrt{3}$

For Exercises 121 to 126, solve for x.

121. $4a^2x^2 = 36b^2, a > 0, b > 0$

122. $3y^2x^2 = 27z^2, y > 0, z > 0$

123. $(x + a)^2 - 4 = 0$

124. $(x - b)^2 - 1 = 0$

125. $(2x - 1)^2 = (2x + 3)^2$

126. $(x - 4)^2 = (x + 2)^2$

127. **a.** Show that the solutions of the equation $ax^2 + bx = 0$, $a > 0$, $b > 0$, are 0 and $-\frac{b}{a}$.

b. Show that the solutions of the equation $ax^2 + c = 0$, $a > 0$, $c > 0$, are $\frac{\sqrt{ca}}{a}i$ and $-\frac{\sqrt{ca}}{a}i$.

Copyright © Houghton Mifflin Company. All rights reserved.

10.2 Solving Quadratic Equations by Completing the Square

Objective A To solve a quadratic equation by completing the square

Recall that a perfect-square trinomial is the square of a binomial.

Perfect-Square Trinomial		*Square of a Binomial*
$x^2 + 8x + 16$	=	$(x + 4)^2$
$x^2 - 10x + 25$	=	$(x - 5)^2$
$x^2 + 2ax + a^2$	=	$(x + a)^2$

For each perfect-square trinomial, the square of $\frac{1}{2}$ of the coefficient of x equals the constant term.

$$x^2 + 8x + 16, \quad \left(\frac{1}{2} \cdot 8\right)^2 = 16$$

$$x^2 - 10x + 25, \quad \left[\frac{1}{2}(-10)\right]^2 = 25$$

$$x^2 + 2ax + a^2, \quad \left(\frac{1}{2} \cdot 2a\right)^2 = a^2$$

$$\left(\frac{1}{2}\textbf{ coefficient of } x\right)^2 = \textbf{constant term}$$

Point of Interest

Early attempts to solve quadratic equations were primarily geometric. The Persian mathematician al-Khowarizmi (c. A.D. 800) essentially completed a square of $x^2 + 12x$ as follows.

This relationship can be used to write the constant term for a perfect-square trinomial. Adding to a binomial the constant term that makes it a perfect-square trinomial is called **completing the square.**

HOW TO Complete the square on $x^2 + 12x$. Write the resulting perfect-square trinomial as the square of a binomial.

$\left[\frac{1}{2}(12)\right]^2 = (6)^2 = 36$ • **Find the constant term.**

$x^2 + 12x + 36$ • **Complete the square on $x^2 + 12x$ by adding the constant term.**

$x^2 + 12x + 36 = (x + 6)^2$ • **Write the resulting perfect-square trinomial as the square of a binomial.**

HOW TO Complete the square on $z^2 - 3z$. Write the resulting perfect-square trinomial as the square of a binomial.

$\left[\frac{1}{2} \cdot (-3)\right]^2 = \left(-\frac{3}{2}\right)^2 = \frac{9}{4}$ • **Find the constant term.**

$z^2 - 3z + \frac{9}{4}$ • **Complete the square on $z^2 - 3z$ by adding the constant term.**

$z^2 - 3z + \frac{9}{4} = \left(z - \frac{3}{2}\right)^2$ • **Write the resulting perfect-square trinomial as the square of a binomial.**

Any quadratic equation can be solved by completing the square. Add to each side of the equation the term that completes the square. Rewrite the equation in the form $(x + a)^2 = b$. Then take the square root of each side of the equation.

Copyright © Houghton Mifflin Company. All rights reserved.

Point of Interest

Mathematicians have studied quadratic equations for centuries. Many of the initial equations were a result of trying to solve a geometry problem. One of the most famous, which dates from around 500 B.C., is "squaring the circle." The question was "Is it possible to construct a square whose area is that of a given circle?" For these early mathematicians, to *construct* meant to draw with only a straightedge and a compass. It was approximately 2300 years later that mathematicians were able to prove that such a construction is impossible.

HOW TO Solve by completing the square: $x^2 - 6x - 15 = 0$

$x^2 - 6x - 15 = 0$

$x^2 - 6x = 15$ • Add 15 to each side of the equation.

$x^2 - 6x + 9 = 15 + 9$ • Complete the square. Add $\left[\frac{1}{2}(-6)\right]^2 = (-3)^2 = 9$ to each side of the equation.

$(x - 3)^2 = 24$ • Factor the perfect-square trinomial.

$\sqrt{(x - 3)^2} = \sqrt{24}$ • Take the square root of each side of the equation.

$x - 3 = \pm\sqrt{24}$ • Solve for *x*.

$x - 3 = \pm 2\sqrt{6}$

$x - 3 = 2\sqrt{6}$ $\quad$ $x - 3 = -2\sqrt{6}$

$x = 3 + 2\sqrt{6}$ $\quad$ $x = 3 - 2\sqrt{6}$

Check:

$x^2 - 6x - 15 = 0$	
$(3 + 2\sqrt{6})^2 - 6(3 + 2\sqrt{6}) - 15$	0
$9 + 12\sqrt{6} + 24 - 18 - 12\sqrt{6} - 15$	0
$0 = 0$	

$x^2 - 6x - 15 = 0$	
$(3 - 2\sqrt{6})^2 - 6(3 - 2\sqrt{6}) - 15$	0
$9 - 12\sqrt{6} + 24 - 18 + 12\sqrt{6} - 15$	0
$0 = 0$	

The solutions are $3 + 2\sqrt{6}$ and $3 - 2\sqrt{6}$.

TAKE NOTE

The exact solutions of the equation $x^2 - 6x - 15 = 0$ are $3 + 2\sqrt{6}$ and $3 - 2\sqrt{6}$. 7.899 and −1.899 are approximate solutions of the equation.

In the example above, the solutions of the equation $x^2 - 6x - 15 = 0$ are $3 + 2\sqrt{6}$ and $3 - 2\sqrt{6}$. These are the exact solutions. However, in some situations it may be preferable to have decimal approximations of the solutions of a quadratic equation. Approximate solutions can be found by using a calculator and then rounding to the desired degree of accuracy.

$$3 + 2\sqrt{6} \approx 7.899 \text{ and } 3 - 2\sqrt{6} \approx -1.899$$

To the nearest thousandth, the approximate solutions of the equation $x^2 - 6x - 15 = 0$ are 7.899 and −1.899.

HOW TO Solve $2x^2 - x - 2 = 0$ by completing the square. Find the exact solutions and approximate the solutions to the nearest thousandth.

In order for us to complete the square on an expression, the coefficient of the squared term must be 1. After adding the constant term to each side of the equation, multiply each side of the equation by $\frac{1}{2}$.

$2x^2 - x - 2 = 0$

$2x^2 - x = 2$ • Add 2 to each side of the equation.

$\frac{1}{2}(2x^2 - x) = \frac{1}{2} \cdot 2$ • Multiply each side of the equation by $\frac{1}{2}$.

Copyright © Houghton Mifflin Company. All rights reserved.

Integrating Technology

A graphing calculator can be used to check the proposed solutions of a quadratic equation. For the equation at the right, enter the expression $2x^2 - x - 2$ in Y1. Then evaluate the function for $\frac{1+\sqrt{17}}{4}$ and $\frac{1-\sqrt{17}}{4}$. The value of each should be zero. See the appendix Keystroke Guide: *Evaluating Functions*.

$$x^2 - \frac{1}{2}x = 1$$
• The coefficient of x^2 is now 1.

$$x^2 - \frac{1}{2}x + \frac{1}{16} = 1 + \frac{1}{16}$$
• Complete the square. Add $\left[\frac{1}{2}\left(-\frac{1}{2}\right)\right]^2 = \left(-\frac{1}{4}\right)^2 = \frac{1}{16}$ to each side of the equation.

$$\left(x - \frac{1}{4}\right)^2 = \frac{17}{16}$$
• Factor the perfect-square trinomial.

$$\sqrt{\left(x - \frac{1}{4}\right)^2} = \sqrt{\frac{17}{16}}$$
• Take the square root of each side of the equation.

$$x - \frac{1}{4} = \pm\frac{\sqrt{17}}{4}$$
• Solve for x.

$$x - \frac{1}{4} = \frac{\sqrt{17}}{4} \qquad x - \frac{1}{4} = -\frac{\sqrt{17}}{4}$$

$$x = \frac{1}{4} + \frac{\sqrt{17}}{4} \qquad x = \frac{1}{4} - \frac{\sqrt{17}}{4}$$

The exact solutions are $\frac{1+\sqrt{17}}{4}$ and $\frac{1-\sqrt{17}}{4}$.

$$\frac{1+\sqrt{17}}{4} \approx 1.281 \qquad \frac{1-\sqrt{17}}{4} \approx -0.781$$

To the nearest thousandth, the solutions are 1.281 and −0.781.

Check of exact solutions:

$$\begin{array}{c|c} 2x^2 - x - 2 & 0 \\ \hline 2\left(\frac{1+\sqrt{17}}{4}\right)^2 - \frac{1+\sqrt{17}}{4} - 2 & 0 \\ 2\left(\frac{18+2\sqrt{17}}{16}\right) - \frac{1+\sqrt{17}}{4} - 2 & 0 \\ \frac{18+2\sqrt{17}}{8} - \frac{2+2\sqrt{17}}{8} - 2 & 0 \\ \frac{16}{8} - 2 & 0 \\ 0 = 0 & \end{array} \qquad \begin{array}{c|c} 2x^2 - x - 2 & 0 \\ \hline 2\left(\frac{1-\sqrt{17}}{4}\right)^2 - \frac{1-\sqrt{17}}{4} - 2 & 0 \\ 2\left(\frac{18-2\sqrt{17}}{16}\right) - \frac{1-\sqrt{17}}{4} - 2 & 0 \\ \frac{18-2\sqrt{17}}{8} - \frac{2-2\sqrt{17}}{8} - 2 & 0 \\ \frac{16}{8} - 2 & 0 \\ 0 = 0 & \end{array}$$

The above example illustrates all the steps required in solving a quadratic equation by completing the square.

Study Tip

This is a new skill and one that is difficult for many students. Be sure to do all you need to do in order to be successful at solving equations by completing the square: Read through the introductory material, work through the HOW TO examples, study the paired Examples, do the You Try Its, and check your solutions against the ones in the back of the book. See *AIM for Success*, pages xxvii–xxx.

Procedure for Solving a Quadratic Equation by Completing the Square

1. Write the equation in the form $ax^2 + bx = -c$.
2. Multiply both sides of the equation by $\frac{1}{a}$.
3. Complete the square on $x^2 + \frac{b}{a}x$. Add the number that completes the square to both sides of the equation.
4. Factor the perfect-square trinomial.
5. Take the square root of each side of the equation.
6. Solve the resulting equation for x.
7. Check the solutions.

Copyright © Houghton Mifflin Company. All rights reserved.

Example 1

Solve by completing the square: $4x^2 - 8x + 1 = 0$

Solution

$$4x^2 - 8x + 1 = 0$$

$$4x^2 - 8x = -1$$ • Write in the form $ax^2 + bx = -c$.

$$\frac{1}{4}(4x^2 - 8x) = \frac{1}{4}(-1)$$ • Multiply both sides by $\frac{1}{a}$.

$$x^2 - 2x = -\frac{1}{4}$$

$$x^2 - 2x + 1 = -\frac{1}{4} + 1$$ • Complete the square.

$$(x - 1)^2 = \frac{3}{4}$$ • Factor.

$$\sqrt{(x - 1)^2} = \sqrt{\frac{3}{4}}$$ • Take square roots.

$$x - 1 = \pm\frac{\sqrt{3}}{2}$$

$$x - 1 = \frac{\sqrt{3}}{2} \qquad x - 1 = -\frac{\sqrt{3}}{2}$$ • Solve for x.

$$x = 1 + \frac{\sqrt{3}}{2} \qquad x = 1 - \frac{\sqrt{3}}{2}$$

$$= \frac{2 + \sqrt{3}}{2} \qquad = \frac{2 - \sqrt{3}}{2}$$

The solutions are $\frac{2 + \sqrt{3}}{2}$ and $\frac{2 - \sqrt{3}}{2}$.

You Try It 1

Solve by completing the square: $4x^2 - 4x - 1 = 0$

Your solution

Example 2

Solve by completing the square: $x^2 + 4x + 5 = 0$

Solution

$$x^2 + 4x + 5 = 0$$

$$x^2 + 4x = -5$$

$$x^2 + 4x + 4 = -5 + 4$$ • Complete the square.

$$(x + 2)^2 = -1$$ • Factor.

$$\sqrt{(x + 2)^2} = \sqrt{-1}$$ • Take square roots.

$$x + 2 = \pm i$$

$$x + 2 = i \qquad x + 2 = -i$$ • Solve for x.

$$x = -2 + i \qquad x = -2 - i$$

The solutions are $-2 + i$ and $-2 - i$.

You Try It 2

Solve by completing the square: $x^2 + 4x + 8 = 0$

Your solution

Solutions on pp. S30–S31

Copyright © Houghton Mifflin Company. All rights reserved.

10.2 Exercises

Objective A **To solve a quadratic equation by completing the square**

For Exercises 1 to 48, solve by completing the square.

1. $x^2 - 4x - 5 = 0$

2. $y^2 + 6y + 5 = 0$

3. $v^2 + 8v - 9 = 0$

4. $w^2 - 2w - 24 = 0$

5. $z^2 - 6z + 9 = 0$

6. $u^2 + 10u + 25 = 0$

7. $r^2 + 4r - 7 = 0$

8. $s^2 + 6s - 1 = 0$

9. $x^2 - 6x + 7 = 0$

10. $y^2 + 8y + 13 = 0$

11. $z^2 - 2z + 2 = 0$

12. $t^2 - 4t + 8 = 0$

13. $s^2 - 5s - 24 = 0$

14. $v^2 + 7v - 44 = 0$

15. $x^2 + 5x - 36 = 0$

16. $y^2 - 9y + 20 = 0$

17. $p^2 - 3p + 1 = 0$

18. $r^2 - 5r - 2 = 0$

19. $t^2 - t - 1 = 0$

20. $u^2 - u - 7 = 0$

21. $y^2 - 6y = 4$

22. $w^2 + 4w = 2$

23. $x^2 = 8x - 15$

24. $z^2 = 4z - 3$

25. $v^2 = 4v - 13$

26. $x^2 = 2x - 17$

27. $p^2 + 6p = -13$

28. $x^2 + 4x = -20$

29. $y^2 - 2y = 17$

30. $x^2 + 10x = 7$

31. $z^2 = z + 4$

32. $r^2 = 3r - 1$

33. $x^2 + 13 = 2x$

Copyright © Houghton Mifflin Company. All rights reserved.

34. $6v^2 - 7v = 3$

35. $4x^2 - 4x + 5 = 0$

36. $4t^2 - 4t + 17 = 0$

37. $9x^2 - 6x + 2 = 0$

38. $9y^2 - 12y + 13 = 0$

39. $2s^2 = 4s + 5$

40. $3u^2 = 6u + 1$

41. $2r^2 = 3 - r$

42. $2x^2 = 12 - 5x$

43. $y - 2 = (y - 3)(y + 2)$

44. $8s - 11 = (s - 4)(s - 2)$

45. $6t - 2 = (2t - 3)(t - 1)$

46. $2z + 9 = (2z + 3)(z + 2)$

47. $(x - 4)(x + 1) = x - 3$

48. $(y - 3)^2 = 2y + 10$

For Exerises 49 to 54, solve by completing the square. Approximate the solutions to the nearest thousandth.

49. $z^2 + 2z = 4$

50. $t^2 - 4t = 7$

51. $2x^2 = 4x - 1$

52. $3y^2 = 5y - 1$

53. $4z^2 + 2z = 1$

54. $4w^2 - 8w = 3$

APPLYING THE CONCEPTS

For Exercises 55 to 57, solve for x by completing the square.

55. $x^2 - ax - 2a^2 = 0$

56. $x^2 + 3ax - 4a^2 = 0$

57. $x^2 + 3ax - 10a^2 = 0$

58. **Sports** After a baseball is hit, the height h (in feet) of the ball above the ground t seconds after it is hit can be approximated by the equation $h = -16t^2 + 70t + 4$. Using this equation, determine when the ball will hit the ground. Round to the nearest hundredth. (*Hint:* The ball hits the ground when $h = 0$.)

59. **Sports** After a baseball is hit, there are two equations that can be considered. One gives the height h (in feet) of the ball above the ground t seconds after it is hit. The second is the distance s (in feet) of the ball from home plate t seconds after it is hit. A model of this situation is given by $h = -16t^2 + 70t + 4$ and $s = 44.5t$. Using this model, determine whether the ball will clear a fence 325 ft from home plate.

60. Explain how to complete the square on $x^2 + bx$.

Copyright © Houghton Mifflin Company. All rights reserved.

10.3 Solving Quadratic Equations by Using the Quadratic Formula

Objective A

To solve a quadratic equation by using the quadratic formula

A general formula known as the **quadratic formula** can be derived by applying the method of completing the square to the standard form of a quadratic equation. This formula can be used to solve any quadratic equation. The equation $ax^2 + bx + c = 0$ is solved by completing the square as follows.

$$ax^2 + bx + c = 0$$

Add the opposite of the constant term to each side of the equation.

$$ax^2 + bx + c + (-c) = 0 + (-c)$$

$$ax^2 + bx = -c$$

Multiply each side of the equation by the reciprocal of a, the coefficient of x^2.

$$\frac{1}{a}(ax^2 + bx) = \frac{1}{a}(-c)$$

$$x^2 + \frac{b}{a}x = -\frac{c}{a}$$

Complete the square by adding $\left(\frac{1}{2} \cdot \frac{b}{a}\right)^2 = \frac{b^2}{4a^2}$ to each side of the equation.

$$x^2 + \frac{b}{a}x + \frac{b^2}{4a^2} = \frac{b^2}{4a^2} - \frac{c}{a}$$

Simplify the right side of the equation.

$$x^2 + \frac{b}{a}x + \frac{b^2}{4a^2} = \frac{b^2}{4a^2} - \left(\frac{c}{a} \cdot \frac{4a}{4a}\right)$$

$$x^2 + \frac{b}{a}x + \frac{b^2}{4a^2} = \frac{b^2}{4a^2} - \frac{4ac}{4a^2}$$

$$x^2 + \frac{b}{a}x + \frac{b^2}{4a^2} = \frac{b^2 - 4ac}{4a^2}$$

Factor the perfect-square trinomial on the left side of the equation.

$$\left(x + \frac{b}{2a}\right)^2 = \frac{b^2 - 4ac}{4a^2}$$

Take the square root of each side of the equation.

$$\sqrt{\left(x + \frac{b}{2a}\right)^2} = \sqrt{\frac{b^2 - 4ac}{4a^2}}$$

$$x + \frac{b}{2a} = \pm\frac{\sqrt{b^2 - 4ac}}{2a}$$

Solve for x.

$$x + \frac{b}{2a} = \frac{\sqrt{b^2 - 4ac}}{2a} \qquad x + \frac{b}{2a} = -\frac{\sqrt{b^2 - 4ac}}{2a}$$

$$x = -\frac{b}{2a} + \frac{\sqrt{b^2 - 4ac}}{2a} \qquad x = -\frac{b}{2a} - \frac{\sqrt{b^2 - 4ac}}{2a}$$

$$= \frac{-b + \sqrt{b^2 - 4ac}}{2a} \qquad = \frac{-b - \sqrt{b^2 - 4ac}}{2a}$$

The Quadratic Formula

The solutions of $ax^2 + bx + c = 0$, $a \neq 0$, are

$$\frac{-b + \sqrt{b^2 - 4ac}}{2a} \quad \text{and} \quad \frac{-b - \sqrt{b^2 - 4ac}}{2a}$$

Point of Interest

Although mathematicians have studied quadratic equations since around 500 B.C., it was not until the 18th century that the formula was written as it is today. Of further note, the word *quadratic* has the same Latin root as the word *square*.

Copyright © Houghton Mifflin Company. All rights reserved.

The quadratic formula is frequently written as $x = \dfrac{-b \pm \sqrt{b^2 - 4ac}}{2a}$.

HOW TO Solve by using the quadratic formula: $2x^2 + 5x + 3 = 0$

TAKE NOTE
The solutions of this quadratic equation are rational numbers. When this happens, the equation could have been solved by factoring and using the Principle of Zero Products. This may be easier than applying the quadratic formula.

$$x = \frac{-b \pm \sqrt{b^2 - 4ac}}{2a}$$

- The equation $2x^2 + 5x + 3 = 0$ is in standard form. $a = 2$, $b = 5$, $c = 3$
- Replace a, b, and c in the quadratic formula with these values.

$$= \frac{-(5) \pm \sqrt{(5)^2 - 4(2)(3)}}{2(2)}$$

$$= \frac{-5 \pm \sqrt{25 - 24}}{4}$$

$$= \frac{-5 \pm \sqrt{1}}{4} = \frac{-5 \pm 1}{4}$$

$$x = \frac{-5 + 1}{4} = \frac{-4}{4} = -1 \qquad x = \frac{-5 - 1}{4} = \frac{-6}{4} = -\frac{3}{2}$$

The solutions are -1 and $-\frac{3}{2}$.

HOW TO Solve $3x^2 = 4x + 6$ by using the quadratic formula. Find the exact solutions and approximate the solutions to the nearest thousandth.

$$3x^2 = 4x + 6$$
$$3x^2 - 4x - 6 = 0$$

- Write the equation in standard form. Subtract $4x$ and 6 from each side of the equation. $a = 3$, $b = -4$, $c = -6$

$$x = \frac{-b \pm \sqrt{b^2 - 4ac}}{2a}$$

- Replace a, b, and c in the quadratic formula with these values.

$$= \frac{-(-4) \pm \sqrt{(-4)^2 - 4(3)(-6)}}{2(3)}$$

$$= \frac{4 \pm \sqrt{16 - (-72)}}{6}$$

$$= \frac{4 \pm \sqrt{88}}{6} = \frac{4 \pm 2\sqrt{22}}{6}$$

$$= \frac{2(2 \pm \sqrt{22})}{2 \cdot 3} = \frac{2 \pm \sqrt{22}}{3}$$

Check:

$3x^2$	$4x + 6$
$3\left(\frac{2 + \sqrt{22}}{3}\right)^2$	$4\left(\frac{2 + \sqrt{22}}{3}\right) + 6$
$3\left(\frac{4 + 4\sqrt{22} + 22}{9}\right)$	$\frac{8}{3} + \frac{4\sqrt{22}}{3} + \frac{18}{3}$
$3\left(\frac{26 + 4\sqrt{22}}{9}\right)$	$\frac{26}{3} + \frac{4\sqrt{22}}{3}$
$\frac{26 + 4\sqrt{22}}{3}$	$= \frac{26 + 4\sqrt{22}}{3}$

$3x^2$	$4x + 6$
$3\left(\frac{2 - \sqrt{22}}{3}\right)^2$	$4\left(\frac{2 - \sqrt{22}}{3}\right) + 6$
$3\left(\frac{4 - 4\sqrt{22} + 22}{9}\right)$	$\frac{8}{3} - \frac{4\sqrt{22}}{3} + \frac{18}{3}$
$3\left(\frac{26 - 4\sqrt{22}}{9}\right)$	$\frac{26}{3} - \frac{4\sqrt{22}}{3}$
$\frac{26 - 4\sqrt{22}}{3}$	$= \frac{26 - 4\sqrt{22}}{3}$

The exact solutions are $\frac{2 + \sqrt{22}}{3}$ and $\frac{2 - \sqrt{22}}{3}$.

$$\frac{2 + \sqrt{22}}{3} \approx 2.230 \qquad \frac{2 - \sqrt{22}}{3} \approx -0.897$$

To the nearest thousandth, the solutions are 2.230 and -0.897.

Copyright © Houghton Mifflin Company. All rights reserved.

Integrating Technology

See the Projects and Group Activities at the end of this chapter for instructions on using a graphing calculator to solve a quadratic equation. Instructions are also provided in the appendix Keystroke Guide: *Zero*.

HOW TO Solve by using the quadratic formula: $4x^2 = 8x - 13$

$$4x^2 = 8x - 13$$
$$4x^2 - 8x + 13 = 0$$ • Write the equation in standard form.
$$x = \frac{-b \pm \sqrt{b^2 - 4ac}}{2a}$$ • Use the quadratic formula.
$$= \frac{-(-8) \pm \sqrt{(-8)^2 - 4 \cdot 4 \cdot 13}}{2 \cdot 4}$$ • $a = 4, b = -8, c = 13$
$$= \frac{8 \pm \sqrt{64 - 208}}{8} = \frac{8 \pm \sqrt{-144}}{8}$$
$$= \frac{8 \pm 12i}{8} = \frac{4(2 + 3i)}{4 \cdot 2} = \frac{2 \pm 3i}{2}$$

The solutions are $1 + \frac{3}{2}i$ and $1 - \frac{3}{2}i$.

Of the three preceding examples, the first two had real number solutions; the last one had complex number solutions.

In the quadratic formula, the quantity $b^2 - 4ac$ is called the **discriminant.** When a, b, and c are real numbers, the discriminant determines whether a quadratic equation will have a double root, two real number solutions that are not equal, or two complex number solutions.

The Effect of the Discriminant on the Solutions of a Quadratic Equation

1. If $b^2 - 4ac = 0$, the equation has one real number solution, a double root.
2. If $b^2 - 4ac > 0$, the equation has two unequal real number solutions.
3. If $b^2 - 4ac < 0$, the equation has two complex number solutions.

HOW TO Use the discriminant to determine whether $x^2 - 4x - 5 = 0$ has one real number solution, two real number solutions, or two complex number solutions.

$b^2 - 4ac$ • Evaluate the discriminant.
$(-4)^2 - 4(1)(-5) = 16 + 20 = 36$ $a = 1, b = -4, c = -5$
$36 > 0$

Because $b^2 - 4ac > 0$, the equation has two real number solutions.

Example 1 Solve by using the quadratic formula: $2x^2 - x + 5 = 0$

Solution

$$2x^2 - x + 5 = 0$$
$$a = 2, b = -1, c = 5$$
$$x = \frac{-b \pm \sqrt{b^2 - 4ac}}{2a}$$
$$= \frac{-(-1) \pm \sqrt{(-1)^2 - 4(2)(5)}}{2 \cdot 2}$$
$$= \frac{1 \pm \sqrt{1 - 40}}{4} = \frac{1 \pm \sqrt{-39}}{4}$$
$$= \frac{1 \pm i\sqrt{39}}{4}$$

The solutions are $\frac{1}{4} + \frac{\sqrt{39}}{4}i$ and $\frac{1}{4} - \frac{\sqrt{39}}{4}i$.

You Try It 1 Solve by using the quadratic formula: $x^2 - 2x + 10 = 0$

Your solution

Solution on p. S31

Copyright © Houghton Mifflin Company. All rights reserved.

Example 2

Solve by using the quadratic formula:
$2x^2 = (x - 2)(x - 3)$

Solution

$2x^2 = (x - 2)(x - 3)$

$2x^2 = x^2 - 5x + 6$ • Write in standard form.

$x^2 + 5x - 6 = 0$

$a = 1, b = 5, c = -6$

$$x = \frac{-b \pm \sqrt{b^2 - 4ac}}{2a}$$

$$= \frac{-5 \pm \sqrt{5^2 - 4(1)(-6)}}{2 \cdot 1}$$

$$= \frac{-5 \pm \sqrt{25 + 24}}{2} = \frac{-5 \pm \sqrt{49}}{2}$$

$$= \frac{-5 \pm 7}{2}$$

$$x = \frac{-5 + 7}{2} \qquad x = \frac{-5 - 7}{2}$$

$$= \frac{2}{2} = 1 \qquad = \frac{-12}{2} = -6$$

The solutions are 1 and -6.

You Try It 2

Solve by using the quadratic formula:
$4x^2 = 4x - 1$

Your solution

Example 3

Use the discriminant to determine whether $4x^2 - 2x + 5 = 0$ has one real number solution, two real number solutions, or two complex number solutions.

Solution

$a = 4, b = -2, c = 5$

$$b^2 - 4ac = (-2)^2 - 4(4)(5)$$
$$= 4 - 80$$
$$= -76$$

$-76 < 0$

Because the discriminant is less than zero, the equation has two complex number solutions.

You Try It 3

Use the discriminant to determine whether $3x^2 - x - 1 = 0$ has one real number solution, two real number solutions, or two complex number solutions.

Your solution

Solutions on p. S31

Copyright © Houghton Mifflin Company. All rights reserved.

10.3 Exercises

Objective A **To solve a quadratic equation by using the quadratic formula**

1. Write the quadratic formula. What does each variable in the formula represent?

2. Write the expression that appears under the radical symbol in the quadratic formula. What is this quantity called? What can it be used to determine?

For Exercises 3 to 29, solve by using the quadratic formula.

3. $x^2 - 3x - 10 = 0$

4. $z^2 - 4z - 8 = 0$

5. $y^2 + 5y - 36 = 0$

6. $z^2 - 3z - 40 = 0$

7. $w^2 = 8w + 72$

8. $t^2 = 2t + 35$

9. $v^2 = 24 - 5v$

10. $x^2 = 18 - 7x$

11. $2y^2 + 5y - 1 = 0$

12. $4p^2 - 7p + 1 = 0$

13. $8s^2 = 10s + 3$

14. $12t^2 = 5t + 2$

15. $x^2 = 14x - 4$

16. $v^2 = 12v - 24$

17. $2z^2 - 2z - 1 = 0$

18. $6w^2 = 9w - 1$

19. $z^2 + 2z + 2 = 0$

20. $p^2 - 4p + 5 = 0$

21. $y^2 - 2y + 5 = 0$

22. $x^2 + 6x + 13 = 0$

23. $s^2 - 4s + 13 = 0$

24. $t^2 - 6t + 10 = 0$

25. $2w^2 - 2w - 5 = 0$

26. $2v^2 + 8v + 3 = 0$

27. $2x^2 + 6x + 5 = 0$

28. $2y^2 + 2y + 13 = 0$

29. $4t^2 - 6t + 9 = 0$

Copyright © Houghton Mifflin Company. All rights reserved.

For Exercises 30 to 35, solve by using the quadratic formula. Approximate the solutions to the nearest thousandth.

30. $x^2 - 6x - 6 = 0$ **31.** $p^2 - 8p + 3 = 0$ **32.** $r^2 - 2r = 4$

33. $w^2 + 4w = 1$ **34.** $3t^2 = 7t + 1$ **35.** $2y^2 = y + 5$

For Exercises 36 to 41, use the discriminant to determine whether the quadratic equation has one real number solution, two real number solutions, or two complex number solutions.

36. $2z^2 - z + 5 = 0$ **37.** $3y^2 + y + 1 = 0$ **38.** $9x^2 - 12x + 4 = 0$

39. $4x^2 + 20x + 25 = 0$ **40.** $2v^2 - 3v - 1 = 0$ **41.** $3w^2 + 3w - 2 = 0$

APPLYING THE CONCEPTS

42. **Sports** The height h, in feet, of a baseball above the ground t seconds after it is hit by a Little Leaguer can be approximated by the equation $h = -0.01t^2 + 2t + 3.5$. Does the ball reach a height of 100 ft?

3.5 ft

43. **Sports** The height h, in feet, of an arrow shot upward can be approximated by the equation $h = 128t - 16t^2$, where t is the time in seconds. Does the arrow reach a height of 275 ft?

44. Name the three methods of solving a quadratic equation that have been discussed. What are the advantages and disadvantages of each?

For what values of p do the quadratic equations in Exercises 45 and 46 have two real number solutions that are not equal? Write the answer in set-builder notation.

45. $x^2 - 6x + p = 0$ **46.** $x^2 + 10x + p = 0$

For what values of p do the quadratic equations in Exercises 47 and 48 have two complex number solutions? Write the answer in set-builder notation.

47. $x^2 - 2x + p = 0$ **48.** $x^2 + 4x + p = 0$

49. Find all values of x that satisfy the equation $x^2 + ix + 2 = 0$.

50. Show that the equation $x^2 + bx - 1 = 0$, $b \in$ real numbers, always has real number solutions regardless of the value of b.

51. Show that the equation $2x^2 + bx - 2 = 0$, $b \in$ real numbers, always has real number solutions regardless of the value of b.

Copyright © Houghton Mifflin Company. All rights reserved.

10.4 Solving Equations That Are Reducible to Quadratic Equations

Objective A To solve an equation that is quadratic in form

Certain equations that are not quadratic can be expressed in quadratic form by making suitable substitutions. An equation is **quadratic in form** if it can be written as $au^2 + bu + c = 0$.

The equation $x^4 - 4x^2 - 5 = 0$ is quadratic in form.

$$x^4 - 4x^2 - 5 = 0$$
$$(x^2)^2 - 4(x^2) - 5 = 0$$
$$u^2 - 4u - 5 = 0 \qquad \text{• Let } x^2 = u.$$

The equation $y - y^{1/2} - 6 = 0$ is quadratic in form.

$$y - y^{1/2} - 6 = 0$$
$$(y^{1/2})^2 - (y^{1/2}) - 6 = 0$$
$$u^2 - u - 6 = 0 \qquad \text{• Let } y^{1/2} = u.$$

The key to recognizing equations that are quadratic in form is as follows. When the equation is written in standard form, the exponent on one variable term is $\frac{1}{2}$ the exponent on the other variable term.

HOW TO Solve: $z + 7z^{1/2} - 18 = 0$

$$z + 7z^{1/2} - 18 = 0 \qquad \text{• The equation is quadratic in form.}$$
$$(z^{1/2})^2 + 7(z^{1/2}) - 18 = 0$$
$$u^2 + 7u - 18 = 0 \qquad \text{• Let } z^{1/2} = u.$$
$$(u - 2)(u + 9) = 0 \qquad \text{• Solve by factoring.}$$

$$u - 2 = 0 \qquad u + 9 = 0$$
$$u = 2 \qquad u = -9$$

$$z^{1/2} = 2 \qquad z^{1/2} = -9 \qquad \text{• Replace } u \text{ by } z^{1/2}.$$
$$\sqrt{z} = 2 \qquad \sqrt{z} = -9$$

$$(\sqrt{z})^2 = 2^2 \qquad (\sqrt{z})^2 = (-9)^2 \qquad \text{• Solve for } z.$$
$$z = 4 \qquad z = 81$$

Check each solution.

Check:

$z + 7z^{1/2} - 18 = 0$	
$4 + 7(4)^{1/2} - 18$	0
$4 + 7 \cdot 2 - 18$	0
$4 + 14 - 18$	0
$0 = 0$	

$z + 7z^{1/2} - 18 = 0$	
$81 + 7(81)^{1/2} - 18$	0
$81 + 7 \cdot 9 - 18$	0
$81 + 63 - 18$	0
$126 \neq 0$	

4 checks as a solution, but 81 does not check as a solution.

The solution is 4.

TAKE NOTE

When each side of an equation is squared, the resulting equation may have a solution that is not a solution of the original equation.

Copyright © Houghton Mifflin Company. All rights reserved.

Example 1 Solve: $x^4 + x^2 - 12 = 0$

Solution

$$x^4 + x^2 - 12 = 0$$
$$(x^2)^2 + (x^2) - 12 = 0$$
$$u^2 + u - 12 = 0$$
$$(u - 3)(u + 4) = 0$$

$u - 3 = 0$ $\qquad$ $u + 4 = 0$
$u = 3$ $\qquad$ $u = -4$

Replace u by x^2.

$x^2 = 3$ $\qquad$ $x^2 = -4$
$\sqrt{x^2} = \sqrt{3}$ $\qquad$ $\sqrt{x^2} = \sqrt{-4}$
$x = \pm\sqrt{3}$ $\qquad$ $x = \pm 2i$

The solutions are $\sqrt{3}$, $-\sqrt{3}$, $2i$, and $-2i$.

You Try It 1 Solve: $x - 5x^{1/2} + 6 = 0$

Your solution

Solution on p. S31

Objective B To solve a radical equation that is reducible to a quadratic equation

Certain equations containing radicals can be expressed as quadratic equations.

HOW TO Solve: $\sqrt{x + 2} + 4 = x$

$$\sqrt{x + 2} + 4 = x$$
$$\sqrt{x + 2} = x - 4$$ • Solve for the radical expression.
$$(\sqrt{x + 2})^2 = (x - 4)^2$$ • Square each side of the equation.
$$x + 2 = x^2 - 8x + 16$$ • Simplify.
$$0 = x^2 - 9x + 14$$ • Write the equation in standard form.
$$0 = (x - 7)(x - 2)$$ • Solve for *x*.

$x - 7 = 0$ $\qquad$ $x - 2 = 0$
$x = 7$ $\qquad$ $x = 2$

Check each solution.

Check:

$\sqrt{x + 2} + 4 = x$	
$\sqrt{7 + 2} + 4$	7
$\sqrt{9} + 4$	7
$3 + 4$	7
$7 = 7$	

$\sqrt{x + 2} + 4 = x$	
$\sqrt{2 + 2} + 4$	2
$\sqrt{4} + 4$	2
$2 + 4$	2
$6 \neq 2$	

7 checks as a solution, but 2 does not check as a solution.

The solution is 7.

Copyright © Houghton Mifflin Company. All rights reserved.

Example 2

Solve : $\sqrt{7y-3}+3=2y$

Solution

$\sqrt{7y-3}+3=2y$ • Solve for the radical.

$\sqrt{7y-3}=2y-3$

$(\sqrt{7y-3})^2=(2y-3)^2$ • Square each side.

$7y-3=4y^2-12y+9$ • Write in standard form.

$0=4y^2-19y+12$

$0=(4y-3)(y-4)$ • Solve by factoring.

$4y-3=0 \qquad y-4=0$

$4y=3 \qquad y=4$

$y=\frac{3}{4}$

4 checks as a solution.

$\frac{3}{4}$ does not check as a solution.

The solution is 4.

You Try It 2

Solve: $\sqrt{2x+1}+x=7$

Your solution

Example 3

Solve: $\sqrt{2y+1}-\sqrt{y}=1$

Solution

$\sqrt{2y+1}-\sqrt{y}=1$

Solve for one of the radical expressions.

$\sqrt{2y+1}=\sqrt{y}+1$

$(\sqrt{2y+1})^2=(\sqrt{y}+1)^2$ • Square each side.

$2y+1=y+2\sqrt{y}+1$

$y=2\sqrt{y}$

$y^2=(2\sqrt{y})^2$ • Square each side.

$y^2=4y$

$y^2-4y=0$

$y(y-4)=0$

$y=0 \qquad y-4=0$

$y=4$

0 and 4 check as solutions.

The solutions are 0 and 4.

You Try It 3

Solve: $\sqrt{2x-1}+\sqrt{x}=2$

Your solution

Solutions on p. S31

Copyright © Houghton Mifflin Company. All rights reserved.

Objective C **To solve a fractional equation that is reducible to a quadratic equation**

After each side of a fractional equation has been multiplied by the LCM of the denominators, the resulting equation may be a quadratic equation.

HOW TO Solve: $\frac{1}{r} + \frac{1}{r+1} = \frac{3}{2}$

$$\frac{1}{r} + \frac{1}{r+1} = \frac{3}{2}$$

$$2r(r+1)\left(\frac{1}{r} + \frac{1}{r+1}\right) = 2r(r+1) \cdot \frac{3}{2}$$

• Multiply each side of the equation by the LCM of the denominators.

$$2(r+1) + 2r = r(r+1) \cdot 3$$
$$2r + 2 + 2r = 3r(r+1)$$
$$4r + 2 = 3r^2 + 3r$$
$$0 = 3r^2 - r - 2$$

• Write the equation in standard form.

$$0 = (3r+2)(r-1)$$

• Solve for *r* by factoring.

$$3r + 2 = 0 \qquad r - 1 = 0$$
$$3r = -2 \qquad r = 1$$
$$r = -\frac{2}{3}$$

$-\frac{2}{3}$ and 1 check as solutions.

The solutions are $-\frac{2}{3}$ and 1.

Example 4

Solve: $\frac{9}{x-3} = 2x + 1$

Solution

$$\frac{9}{x-3} = 2x + 1$$

$$(x-3)\frac{9}{x-3} = (x-3)(2x+1)$$

• Clear denominators.

$$9 = 2x^2 - 5x - 3$$
$$0 = 2x^2 - 5x - 12$$

• Write in standard form.

$$0 = (2x+3)(x-4)$$

• Solve by factoring.

$$2x + 3 = 0 \qquad x - 4 = 0$$
$$2x = -3 \qquad x = 4$$
$$x = -\frac{3}{2}$$

• The solutions check.

The solutions are $-\frac{3}{2}$ and 4.

You Try It 4

Solve: $3y + \frac{25}{3y-2} = -8$

Your solution

Solution on p. S32

Copyright © Houghton Mifflin Company. All rights reserved.

10.4 Exercises

Objective A To solve an equation that is quadratic in form

For Exercises 1 to 18, solve.

1. $x^4 - 13x^2 + 36 = 0$
2. $y^4 - 5y^2 + 4 = 0$
3. $z^4 - 6z^2 + 8 = 0$
4. $t^4 - 12t^2 + 27 = 0$
5. $p - 3p^{1/2} + 2 = 0$
6. $v - 7v^{1/2} + 12 = 0$
7. $x - x^{1/2} - 12 = 0$
8. $w - 2w^{1/2} - 15 = 0$
9. $z^4 + 3z^2 - 4 = 0$
10. $y^4 + 5y^2 - 36 = 0$
11. $x^4 + 12x^2 - 64 = 0$
12. $x^4 - 81 = 0$
13. $p + 2p^{1/2} - 24 = 0$
14. $v + 3v^{1/2} - 4 = 0$
15. $y^{2/3} - 9y^{1/3} + 8 = 0$
16. $z^{2/3} - z^{1/3} - 6 = 0$
17. $9w^4 - 13w^2 + 4 = 0$
18. $4y^4 - 7y^2 - 36 = 0$

Objective B To solve a radical equation that is reducible to a quadratic equation

For Exercises 19 to 36, solve.

19. $\sqrt{x + 1} + x = 5$
20. $\sqrt{x - 4} + x = 6$
21. $x = \sqrt{x} + 6$
22. $\sqrt{2y - 1} = y - 2$
23. $\sqrt{3w + 3} = w + 1$
24. $\sqrt{2s + 1} = s - 1$
25. $\sqrt{4y + 1} - y = 1$
26. $\sqrt{3s + 4} + 2s = 12$
27. $\sqrt{10x + 5} - 2x = 1$

Copyright © Houghton Mifflin Company. All rights reserved.

28. $\sqrt{t+8} = 2t + 1$

29. $\sqrt{p+11} = 1 - p$

30. $x - 7 = \sqrt{x-5}$

31. $\sqrt{x-1} - \sqrt{x} = -1$

32. $\sqrt{y} + 1 = \sqrt{y+5}$

33. $\sqrt{2x-1} = 1 - \sqrt{x-1}$

34. $\sqrt{x+6} + \sqrt{x+2} = 2$

35. $\sqrt{t+3} + \sqrt{2t+7} = 1$

36. $\sqrt{5-2x} = \sqrt{2-x} + 1$

Objective C To solve a fractional equation that is reducible to a quadratic equation

For Exercises 37 to 51, solve.

37. $x = \frac{10}{x-9}$

38. $z = \frac{5}{z-4}$

39. $\frac{t}{t+1} = \frac{-2}{t-1}$

40. $\frac{2v}{v-1} = \frac{5}{v+2}$

41. $\frac{y-1}{y+2} + y = 1$

42. $\frac{2p-1}{p-2} + p = 8$

43. $\frac{3r+2}{r+2} - 2r = 1$

44. $\frac{2v+3}{v+4} + 3v = 4$

45. $\frac{2}{2x+1} + \frac{1}{x} = 3$

46. $\frac{3}{s} - \frac{2}{2s-1} = 1$

47. $\frac{16}{z-2} + \frac{16}{z+2} = 6$

48. $\frac{2}{y+1} + \frac{1}{y-1} = 1$

49. $\frac{t}{t-2} + \frac{2}{t-1} = 4$

50. $\frac{4t+1}{t+4} + \frac{3t-1}{t+1} = 2$

51. $\frac{5}{2p-1} + \frac{4}{p+1} = 2$

APPLYING THE CONCEPTS

52. Solve: $(x^2 - 7)^{1/2} = (x - 1)^{1/2}$

53. Solve: $(\sqrt{x} - 2)^2 - 5\sqrt{x} + 14 = 0$ (*Hint:* Let $u = \sqrt{x} - 2$.)

54. Solve: $(\sqrt{x} + 3)^2 - 4\sqrt{x} - 17 = 0$ (*Hint:* Let $u = \sqrt{x} + 3$.)

55. **Mathematics** The fourth power of a number is twenty-five less than ten times the square of the number. Find the number.

Copyright © Houghton Mifflin Company. All rights reserved.

10.5 Quadratic Inequalities and Rational Inequalities

Objective A To solve a nonlinear inequality

A **quadratic inequality** is one that can be written in the form $ax^2 + bx + c < 0$ or $ax^2 + bx + c > 0$, where $a \neq 0$. The symbols $\leq$ and $\geq$ can also be used. The solution set of a quadratic inequality can be found by solving a compound inequality.

To solve $x^2 - 3x - 10 > 0$, first factor the trinomial.

$$x^2 - 3x - 10 > 0$$
$$(x + 2)(x - 5) > 0$$

There are two cases for which the product of the factors will be positive: (1) both factors are positive, or (2) both factors are negative.

$$(1)\ x + 2 > 0 \quad \text{and} \quad x - 5 > 0$$
$$(2)\ x + 2 < 0 \quad \text{and} \quad x - 5 < 0$$

Solve each pair of compound inequalities.

$$(1)\ x + 2 > 0 \quad \text{and} \quad x - 5 > 0$$
$$x > -2 \qquad\qquad x > 5$$
$$\{x \mid x > -2\} \cap \{x \mid x > 5\} = \{x \mid x > 5\}$$

$$(2)\ x + 2 < 0 \quad \text{and} \quad x - 5 < 0$$
$$x < -2 \qquad\qquad x < 5$$
$$\{x \mid x < -2\} \cap \{x \mid x < 5\} = \{x \mid x < -2\}$$

Because the two cases for which the product will be positive are connected by *or*, the solution set is the union of the solution sets of the individual inequalities.

$$\{x \mid x > 5\} \cup \{x \mid x < -2\} = \{x \mid x > 5 \text{ or } x < -2\}$$

Although the solution set of any quadratic inequality can be found by using the method outlined above, a graphical method is often easier to use.

HOW TO Solve and graph the solution set of $x^2 - x - 6 < 0$.

Factor the trinomial.

$$x^2 - x - 6 < 0$$
$$(x - 3)(x + 2) < 0$$

On a number line, draw lines indicating the numbers that make each factor equal to zero.

−5 −4 −3 −2 −1 0 1 2 3 4 5

$$x - 3 = 0 \qquad x + 2 = 0$$
$$x = 3 \qquad\quad x = -2$$

For each factor, place plus signs above the number line for those regions where the factor is positive and minus signs where the factor is negative.

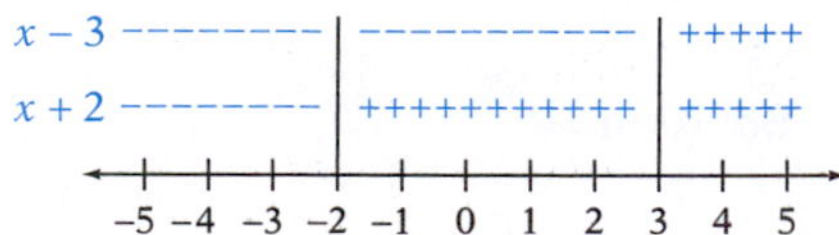

Because $x^2 - x - 6 < 0$, the solution set will be the regions where one factor is positive and the other factor is negative.

Write the solution set.

$$\{x \mid -2 < x < 3\}$$

The graph of the solution set of $x^2 - x - 6 < 0$ is shown at the right.

TAKE NOTE

For each factor, choose a number in each region. For example: when $x = -4$, $x - 3$ is negative; when $x = 1$, $x - 3$ is negative; and when $x = 4$, $x - 3$ is positive. When $x = -4$, $x + 2$ is negative; when $x = 1$, $x + 2$ is positive; and when $x = 4$, $x + 2$ is positive.

Copyright © Houghton Mifflin Company. All rights reserved.

HOW TO Solve and graph the solution set of $(x-2)(x+1)(x-4) > 0$.

On a number line, identify for each factor the regions where the factor is positive and those where the factor is negative.

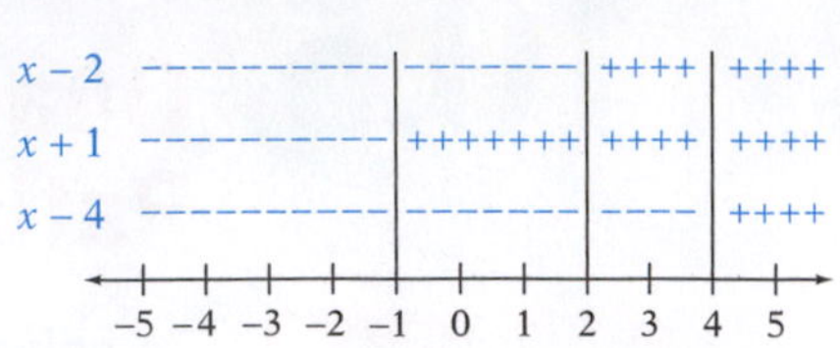

There are two regions where the product of the three factors is positive.

Write the solution set. $\{x \mid -1 < x < 2 \text{ or } x > 4\}$

The graph of the solution set of $(x-2)(x+1)(x-4) > 0$ is shown at the right.

HOW TO Solve: $\dfrac{2x-5}{x-4} \leq 1$

$$\frac{2x-5}{x-4} \leq 1$$

Rewrite the inequality so that 0 appears on the right side of the inequality.

$$\frac{2x-5}{x-4} - 1 \leq 0$$

Simplify.

$$\frac{2x-5}{x-4} - \frac{x-4}{x-4} \leq 0$$

$$\frac{x-1}{x-4} \leq 0$$

On a number line, identify for each factor of the numerator and each factor of the denominator the regions where the factor is positive and those where the factor is negative.

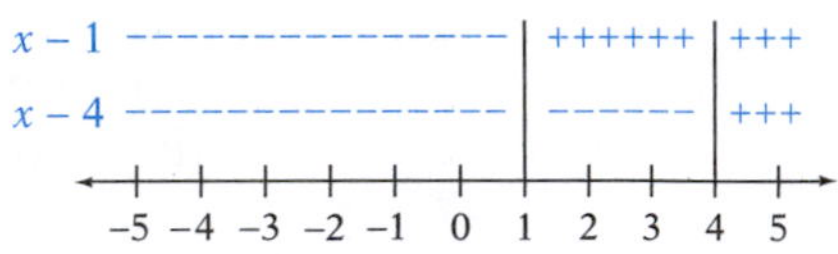

The region where the quotient of the two factors is negative is between 1 and 4.

Write the solution set. $\{x \mid 1 \leq x < 4\}$

Note that 1 is part of the solution set but 4 is not because the denominator of the rational expression is zero when $x = 4$.

Example 1

Solve and graph the solution set of $2x^2 - x - 3 \geq 0$.

Solution

$2x^2 - x - 3 \geq 0$

$(2x-3)(x+1) \geq 0$

$2x-3$ ---|-----|+++

$x+1$ ---|+++++|+++

–2 –1 0 1 2

$\left\{x \mid x \leq -1 \text{ or } x \geq \dfrac{3}{2}\right\}$

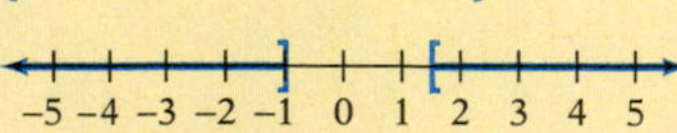

You Try It 1

Solve and graph the solution set of $2x^2 - x - 10 \leq 0$.

Your solution

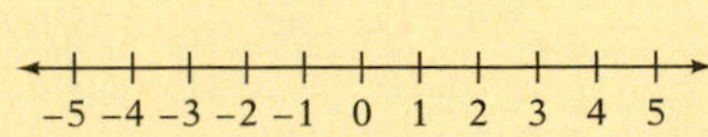

Solution on p. S32

Copyright © Houghton Mifflin Company. All rights reserved.

10.5 Exercises

Objective A **To solve a nonlinear inequality**

1. If $(x - 3)(x - 5) > 0$, what must be true of the values of $x - 3$ and $x - 5$?

2. For the inequality $\frac{x-2}{x-3} \le 1$, which of the values 1, 2, and 3 is not a possible element of the solution set? Why?

For Exercises 3 to 18, solve and graph the solution set.

3. $(x - 4)(x + 2) > 0$

−5 −4 −3 −2 −1 0 1 2 3 4 5

4. $(x + 1)(x - 3) > 0$

−5 −4 −3 −2 −1 0 1 2 3 4 5

5. $x^2 - 3x + 2 \ge 0$

−5 −4 −3 −2 −1 0 1 2 3 4 5

6. $x^2 + 5x + 6 > 0$

−5 −4 −3 −2 −1 0 1 2 3 4 5

7. $x^2 - x - 12 < 0$

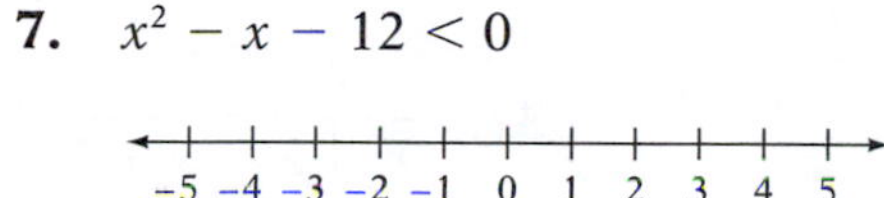

8. $x^2 + x - 20 < 0$

−5 −4 −3 −2 −1 0 1 2 3 4 5

9. $(x - 1)(x + 2)(x - 3) < 0$

−5 −4 −3 −2 −1 0 1 2 3 4 5

10. $(x + 4)(x - 2)(x + 1) > 0$

−5 −4 −3 −2 −1 0 1 2 3 4 5

11. $(x + 4)(x - 2)(x - 1) \ge 0$

−5 −4 −3 −2 −1 0 1 2 3 4 5

12. $(x - 1)(x + 5)(x - 2) \le 0$

−5 −4 −3 −2 −1 0 1 2 3 4 5

13. $\frac{x - 4}{x + 2} > 0$

−5 −4 −3 −2 −1 0 1 2 3 4 5

14. $\frac{x + 2}{x - 3} > 0$

−5 −4 −3 −2 −1 0 1 2 3 4 5

15. $\frac{x - 3}{x + 1} \le 0$

−5 −4 −3 −2 −1 0 1 2 3 4 5

16. $\frac{x - 1}{x} > 0$

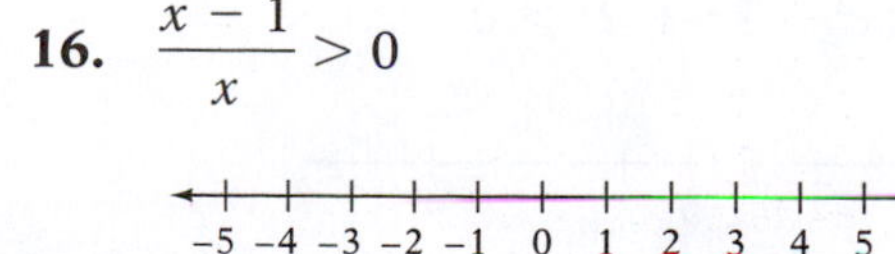

Copyright © Houghton Mifflin Company. All rights reserved.

17. $\dfrac{(x-1)(x+2)}{x-3} \le 0$

−5 −4 −3 −2 −1 0 1 2 3 4 5

18. $\dfrac{(x+3)(x-1)}{x-2} \ge 0$

−5 −4 −3 −2 −1 0 1 2 3 4 5

For Exercises 19 to 33, solve.

19. $x^2 - 16 > 0$

20. $x^2 - 4 \ge 0$

21. $x^2 - 9x \le 36$

22. $x^2 + 4x > 21$

23. $4x^2 - 8x + 3 < 0$

24. $2x^2 + 11x + 12 \ge 0$

25. $\dfrac{3}{x-1} < 2$

26. $\dfrac{x}{(x-1)(x+2)} \ge 0$

27. $\dfrac{x-2}{(x+1)(x-1)} \le 0$

28. $\dfrac{1}{x} < 2$

29. $\dfrac{x}{2x-1} \ge 1$

30. $\dfrac{x}{2x-3} \le 1$

31. $\dfrac{x}{2-x} \le -3$

32. $\dfrac{3}{x-2} > \dfrac{2}{x+2}$

33. $\dfrac{3}{x-5} > \dfrac{1}{x+1}$

APPLYING THE CONCEPTS

For Exercises 34 to 39, graph the solution set.

34. $(x+2)(x-3)(x+1)(x+4) > 0$

−4 −2 0 2 4

35. $(x-1)(x+3)(x-2)(x-4) \ge 0$

−4 −2 0 2 4

36. $(x^2 + 2x - 8)(x^2 - 2x - 3) < 0$

−4 −2 0 2 4

37. $(x^2 + 2x - 3)(x^2 + 3x + 2) \ge 0$

−4 −2 0 2 4

38. $(x^2 + 1)(x^2 - 3x + 2) > 0$

−4 −2 0 2 4

39. $\dfrac{x^2(3-x)(2x+1)}{(x+4)(x+2)} \ge 0$

−4 −2 0 2 4

Copyright © Houghton Mifflin Company. All rights reserved.

10.6 Applications of Quadratic Equations

Objective A To solve application problems

The application problems in this section are similar to problems solved earlier in the text. Each of the strategies for the problems in this section will result in a quadratic equation.

HOW TO A small pipe takes 16 min longer to empty a tank than does a larger pipe. Working together, the pipes can empty the tank in 6 min. How long would it take each pipe working alone to empty the tank?

Empties tank in t minutes

Empties $\frac{1}{t}$ of the tank each minute

Empties tank in $(t + 16)$ minutes

Empties $\frac{1}{t+16}$ of the tank each minute

Strategy for Solving an Application Problem

1. Determine the type of problem. Is it a uniform motion problem, a geometry problem, an integer problem, or a work problem?

The problem is a work problem.

2. Choose a variable to represent the unknown quantity. Write numerical or variable expressions for all the remaining quantities. These results can be recorded in a table.

The unknown time of the larger pipe: t
The unknown time of the smaller pipe: $t + 16$

	Rate of Work	·	*Time Worked*	=	*Part of Task Completed*
Larger pipe	$\frac{1}{t}$	·	6	=	$\frac{6}{t}$
Smaller pipe	$\frac{1}{t+16}$	·	6	=	$\frac{6}{t+16}$

3. Determine how the quantities are related.

$$\frac{6}{t} + \frac{6}{t+16} = 1$$

• The sum of the parts of the task completed must equal 1.

$$t(t+16)\left(\frac{6}{t} + \frac{6}{t+16}\right) = t(t+16) \cdot 1$$
$$(t+16)6 + 6t = t^2 + 16t$$
$$6t + 96 + 6t = t^2 + 16t$$
$$0 = t^2 + 4t - 96$$
$$0 = (t+12)(t-8)$$

$t + 12 = 0 \qquad t - 8 = 0$
$t = -12 \qquad t = 8$

Because time cannot be negative, the solution $t = -12$ is not possible.

The time for the smaller pipe is $t + 16$.

$t + 16 = 8 + 16 = 24$ • Replace t by 8 and evaluate.

The larger pipe requires 8 min to empty the tank.
The smaller pipe requires 24 min to empty the tank.

Copyright © Houghton Mifflin Company. All rights reserved.

Example 1

In 8 h, two campers rowed 15 mi down a river and then rowed back to their campsite. The rate of the river's current was 1 mph. Find the rate at which the campers rowed.

Strategy

- This is a uniform motion problem.
- Unknown rowing rate of the campers: r

	Distance	*Rate*	*Time*
Down river	15	$r + 1$	$\frac{15}{r+1}$
Up river	15	$r - 1$	$\frac{15}{r-1}$

- The total time of the trip was 8 h.

Solution

$$\frac{15}{r+1} + \frac{15}{r-1} = 8$$

$$(r+1)(r-1)\left(\frac{15}{r+1} + \frac{15}{r-1}\right) = (r+1)(r-1)8$$

$$(r-1)15 + (r+1)15 = (r^2 - 1)8$$

$$15r - 15 + 15r + 15 = 8r^2 - 8$$

$$30r = 8r^2 - 8$$

$$0 = 8r^2 - 30r - 8$$

$$0 = 2(4r^2 - 15r - 4)$$

$$0 = 2(4r + 1)(r - 4)$$

$$4r + 1 = 0 \qquad r - 4 = 0$$

$$4r = -1 \qquad r = 4$$

$$r = -\frac{1}{4}$$

The solution $r = -\frac{1}{4}$ is not possible, because the rate cannot be a negative number.

The rowing rate was 4 mph.

You Try It 1

The length of a rectangle is 3 m more than the width. The area is 54 m^2. Find the length of the rectangle.

Your strategy

Your solution

Solution on p. S32

Copyright © Houghton Mifflin Company. All rights reserved.

10.6 Exercises

Objective A **To solve application problems**

1. **Geometry** The length of the base of a triangle is 1 cm less than five times the height of the triangle. The area of the triangle is 21 cm^2. Find the height of the triangle and the length of the base of the triangle.

2. **Geometry** The length of a rectangle is 2 ft less than three times the width of the rectangle. The area of the rectangle is 65 ft^2. Find the length and width of the rectangle.

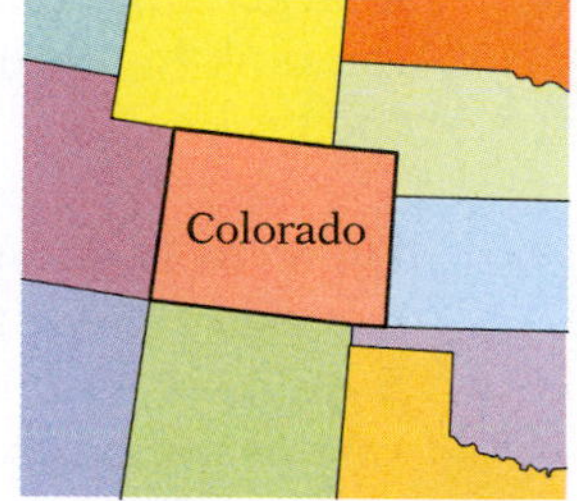

3. **Geography** The state of Colorado is almost perfectly rectangular, with its north border 111 mi longer than its west border. If the state encompasses 104,000 mi^2, estimate the dimensions of Colorado. Round to the nearest mile.

4. **Geometry** A square piece of cardboard is formed into a box by cutting 10-centimeter squares from each of the four corners and then folding up the sides, as shown in the figure. If the volume, V, of the box is to be 49,000 cm^3, what size square piece of cardboard is needed? Recall that $V = LWH$.

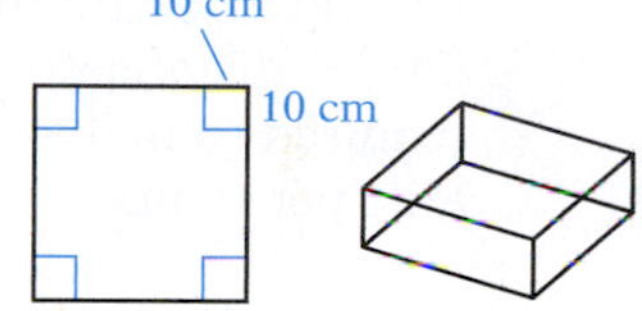

5. **Safety** A car with good tire tread can stop in less distance than a car with poor tread. The formula for the stopping distance d, in feet, of a car with good tread on dry cement is approximated by $d = 0.04v^2 + 0.5v$, where v is the speed of the car. If the driver must be able to stop within 60 ft, what is the maximum safe speed, to the nearest mile per hour, of the car?

6. **Rockets** A model rocket is launched with an initial velocity of 200 ft/s. The height h, in feet, of the rocket t seconds after the launch is given by $h = -16t^2 + 200t$. How many seconds after the launch will the rocket be 300 ft above the ground? Round to the nearest hundredth of a second.

7. **Physics** The height of a projectile fired upward is given by the formula $s = v_0t - 16t^2$, where s is the height in feet, v_0 is the initial velocity, and t is the time in seconds. Find the time for a projectile to return to Earth if it has an initial velocity of 200 ft/s.

8. **Physics** The height of a projectile fired upward is given by the formula $s = v_0t - 16t^2$, where s is the height in feet, v_0 is the initial velocity, and t is the time in seconds. Find the time for a projectile to reach a height of 64 ft if it has an initial velocity of 128 ft/s. Round to the nearest hundredth of a second.

9. **Safety** In Germany, there is no speed limit on some portions of the autobahn (highway). Other portions have a speed limit of 180 km/h (approximately 112 mph). The distance d, in meters, required to stop a car traveling at v kilometers per hour is $d = 0.019v^2 + 0.69v$. Approximate, to the nearest tenth, the maximum speed a driver can be traveling and still be able to stop within 150 m.

German Autobahn System

Copyright © Houghton Mifflin Company. All rights reserved.

10. **Ice Cream** A perfectly spherical scoop of mint chocolate chip ice cream is placed in a cone, as shown in the figure. How far is the bottom of the scoop of ice cream from the bottom of the cone? Round to the nearest tenth. (*Hint:* A line segment from the center of the ice cream to the point at which the ice cream touches the cone is perpendicular to the edge of the cone.)

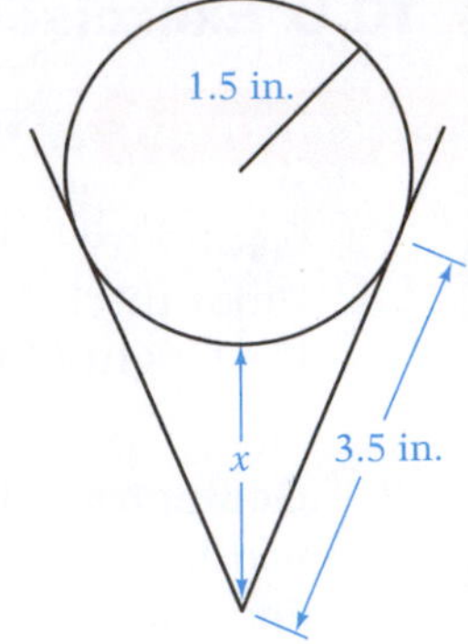

11. **Tanks** A small pipe can fill a tank in 6 min more time than it takes a larger pipe to fill the same tank. Working together, both pipes can fill the tank in 4 min. How long would it take each pipe working alone to fill the tank?

12. **Uniform Motion** A cruise ship made a trip of 100 mi in 8 h. The ship traveled the first 40 mi at a constant rate before increasing its speed by 5 mph. Then it traveled another 60 mi at the increased speed. Find the rate of the cruise ship for the first 40 mi.

13. **Uniform Motion** The Concorde's speed in calm air was 1320 mph. Flying with the wind, the Concorde could fly from New York to London, a distance of approximately 4000 mi, in 0.5 h less than the time required to make the return trip. Find the rate of the wind to the nearest mile per hour.

14. **Uniform Motion** A car travels 120 mi. A second car, traveling 10 mph faster than the first car, makes the same trip in 1 h less time. Find the speed of each car.

15. **Uniform Motion** For a portion of the Green River in Utah, the rate of the river's current is 4 mph. A tour guide can row 5 mi down this river and back in 3 h. Find the rowing rate of the guide in calm water.

16. **Construction** The height h, in feet, of an arch is given by the equation $h(x) = -\frac{3}{64}x^2 + 27$, where $|x|$ is the distance in feet from the center of the arch.

   a. What is the maximum height of the arch?
   b. What is the height of the arch 8 ft to the right of the center?
   c. How far from the center is the arch 8 ft tall? Round to the nearest hundredth.

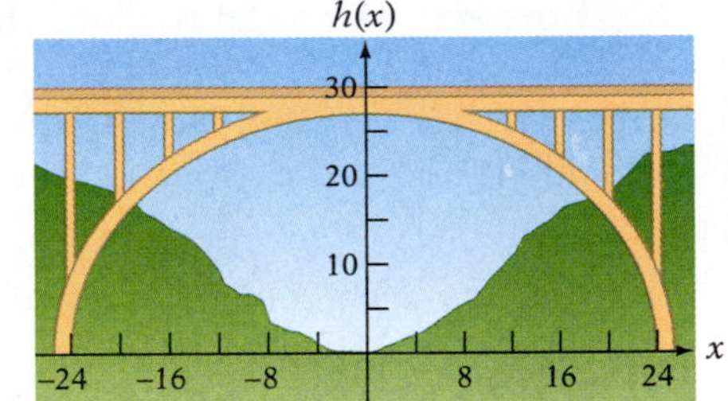

APPLYING THE CONCEPTS

17. **Geometry** The surface area of the ice cream cone shown at the right is given by $A = \pi r^2 + \pi rs$, where r is the radius of the circular top of the cone and s is the slant height of the cone. If the surface area of the cone is 11.25π in^2 and the slant height is 6 in., find the radius of the cone.

18. **Physics** Using Torricelli's Principle, it can be shown that the depth d of a liquid in a bottle with a hole of area 0.5 cm^2 in its side can be approximated by $d = 0.0034t^2 - 0.52518t + 20$, where t is the time in seconds since a stopper was removed from the hole. When will the depth be 10 cm? Round to the nearest tenth of a second.

Copyright © Houghton Mifflin Company. All rights reserved.

Focus on Problem Solving

Using a Variety of Problem-Solving Techniques

We have examined several problem-solving strategies throughout the text. See if you can apply those techniques to the following problems.

1. Eight coins look exactly alike, but one is lighter than the others. Explain how the different coin can be found in two weighings on a balance scale.

2. For the sequence of numbers 1, 1, 2, 3, 5, 8, 13, . . . , identify a possible pattern and then use that pattern to determine the next number in the sequence.

3. Arrange the numbers 1, 2, 3, 4, 5, 6, 7, 8, and 9 in the squares at the right so that the sum of any row, column, or diagonal is 15. (*Suggestion:* Note that 1, 5, 9; 2, 5, 8; 3, 5, 7; and 4, 5, 6 all add to 15. Because 5 is part of each sum, this suggests that 5 be placed in the center of the squares.)

4. A restaurant charges $10.00 for a pizza that has a diameter of 9 in. Determine the selling price of a pizza with a diameter of 18 in. so that the selling price per square inch is the same as that of the 9-inch pizza.

5. You have a balance scale and weights of 1 g, 4 g, 8 g, and 16 g. Using only these weights, can you weigh something that weighs 7 g? 9 g? 12 g? 19 g?

6. Can the checkerboard at the right be covered with dominos (which look like ◻◼) so that every square on the board is covered by a domino? Why or why not? (*Note:* The dominos cannot overlap.)

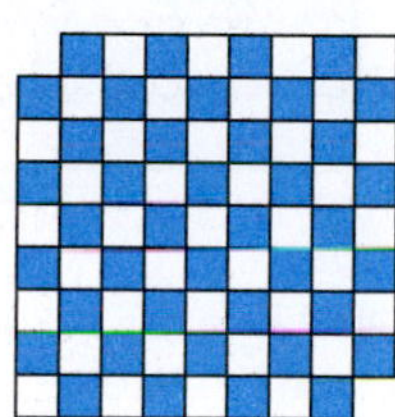

Projects and Group Activities

Using a Graphing Calculator to Solve a Quadratic Equation

Recall that an x-intercept of the graph of an equation is a point at which the graph crosses the x-axis. For the graph in Figure 1, the x-intercepts are $(-2, 0)$ and $(3, 0)$.

Recall also that to find the x-intercept of a graph, set $y = 0$ and then solve for x. For the equation in Figure 1, if we set $y = 0$, the resulting equation is $0 = x^2 - x - 6$, which is a quadratic equation. Solving this equation by factoring, we have

$$0 = x^2 - x - 6$$
$$0 = (x + 2)(x - 3)$$

$$x + 2 = 0 \qquad x - 3 = 0$$
$$x = -2 \qquad x = 3$$

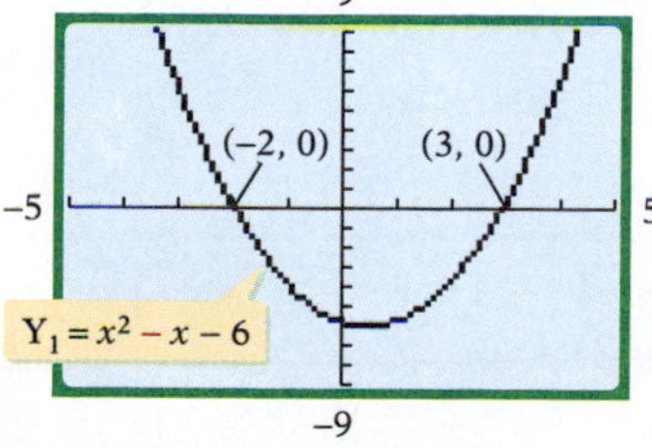

Figure 1

Copyright © Houghton Mifflin Company. All rights reserved.

Thus the solutions of the equation are the x-coordinates of the x-intercepts of the graph.

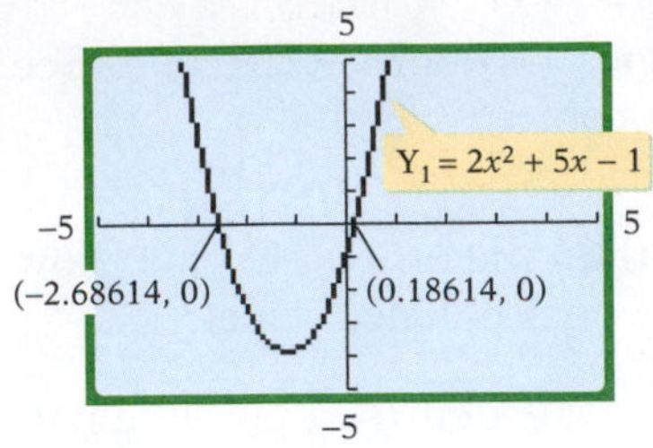

Figure 2

This connection between the solutions of an equation and the x-intercepts of its graph allows us to find approximations of the real number solutions of an equation graphically. For example, to use a TI-83 or TI-84 to approximate the solutions of $2x^2 + 5x - 1 = 0$ graphically, use the following keystrokes to graph $y = 2x^2 + 5x - 1$ and find decimal approximations for the x-coordinates of the x-intercepts. Use the window shown in the graph in Figure 2.

Use the arrow keys to move to the left of the leftmost x-intercept. Press ENTER. Use the arrow keys to move to a point just to the right of the leftmost x-intercept. Press ENTER twice. The x-coordinate at the bottom of the screen is the approximation of one solution of the equation. To find the other solution, use the same procedure but move the cursor first to the left and then to the right of the rightmost x-intercept.

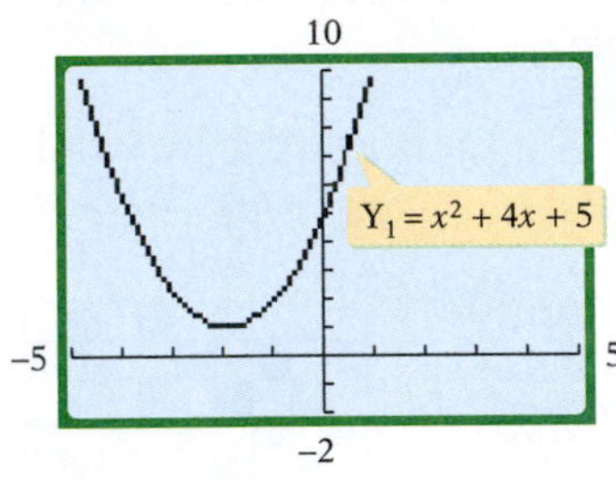

Figure 3

Attempting to find the solutions of an equation graphically will not necessarily yield all the solutions. Because the x-coordinates of the x-intercepts of a graph are *real* numbers, only real number solutions can be found. For instance, consider the equation $x^2 + 4x + 5 = 0$. The graph of $y = x^2 + 4x + 5$ is shown in Figure 3. Note that the graph has no x-intercepts and that consequently it has no real number solutions. However, $x^2 + 4x + 5 = 0$ does have complex number solutions that can be obtained by using the quadratic formula. They are $-2 + i$ and $-2 - i$.

Use a graphing calculator to approximate the solutions of the equations in Exercises 1 to 4.

1. $x^2 + 6x + 1 = 0$

2. $-x^2 + x + 3 = 0$

3. $2x^2 + 4x - 1 = 0$

4. $x^2 + 4x + 5 = 0$

Chapter 10 Summary

Key Words	**Examples**
A *quadratic equation* is an equation of the form $ax^2 + bx + c = 0$, where $a \neq 0$. A quadratic equation is also called a *second-degree equation.* A quadratic equation is in *standard form* when the polynomial is in descending order and equal to zero. [10.1A, p. 563]	$3x^2 + 4x - 7 = 0$, $x^2 - 1 = 0$, and $4x^2 + 8x = 0$ are quadratic equations. They are all written in standard form.
When a quadratic equation has two solutions that are the same number, the solution is called a *double root* of the equation. [10.1A, p. 563]	$x^2 - 4x + 4 = 0$ $(x - 2)(x - 2) = 0$ $x - 2 = 0 \quad x - 2 = 0$ $x = 2 \quad x = 2$ 2 is a double root.

Copyright © Houghton Mifflin Company. All rights reserved.

For an equation of the form $ax^2 + bx + c = 0$, the quantity $b^2 - 4ac$ is called the *discriminant*. [10.3A, p. 579]

$$2x^2 - 3x + 5 = 0$$
$$a = 2, b = -3, c = 5$$
$$b^2 - 4ac = (-3)^2 - 4(2)(5) = 9 - 40 = -31$$

An equation is *quadratic in form* if it can be written as $au^2 + bu + c = 0$. [10.4A, p. 583]

$$6x^4 - 5x^2 - 4 = 0$$
$$6(x^2)^2 - 5(x^2) - 4 = 0$$
$$6u^2 - 5u - 4 = 0$$
The equation is quadratic in form.

A *quadratic inequality* is one that can be written in the form $ax^2 + bx + c < 0$ or $ax^2 + bx + c > 0$, where $a \neq 0$. The symbols $\leq$ and $\geq$ can also be used. [10.5A, p. 589]

$3x^2 + 5x - 8 \leq 0$ is a quadratic inequality.

Essential Rules and Procedures

Examples

Principle of Zero Products [10.1A, p. 563]
If a and b are real numbers and $ab = 0$, then $a = 0$ or $b = 0$.

$$(x - 3)(x + 4) = 0$$
$$x - 3 = 0 \qquad x + 4 = 0$$
$$x = 3 \qquad x = -4$$

To Solve a Quadratic Equation by Factoring [10.1A, p. 563]

1. Write the equation in standard form.
2. Factor the polynomial.
3. Use the Principle of Zero Products to set each factor equal to 0.
4. Solve each equation.
5. Check the solutions.

$$x^2 + 2x = 35$$
$$x^2 + 2x - 35 = 0$$
$$(x - 5)(x + 7) = 0$$
$$x - 5 = 0 \qquad x + 7 = 0$$
$$x = 5 \qquad x = -7$$

To Write a Quadratic Equation Given Its Solutions [10.1B, p. 564]
Use the equation $(x - r_1)(x - r_2) = 0$. Replace r_1 with one solution and r_2 with the other solution. Then multiply the two factors.

Write a quadratic equation that has solutions -3 and 6.

$$(x - r_1)(x - r_2) = 0$$
$$[x - (-3)](x - 6) = 0$$
$$(x + 3)(x - 6) = 0$$
$$x^2 - 3x - 18 = 0$$

Copyright © Houghton Mifflin Company. All rights reserved.

To Solve a Quadratic Equation by Taking Square Roots [10.1C, pp. 565–566]

1. Solve for x^2 or for $(x + a)^2$.
2. Take the square root of each side of the equation.
3. Simplify.
4. Check the solutions.

$$(x+2)^2 - 9 = 0$$
$$(x+2)^2 = 9$$
$$\sqrt{(x+2)^2} = \sqrt{9}$$
$$x + 2 = \pm\sqrt{9}$$
$$x + 2 = \pm 3$$
$$x + 2 = 3 \qquad x + 2 = -3$$
$$x = 1 \qquad x = -5$$

To Complete the Square [10.2A, p. 571]

Add to a binomial of the form $x^2 + bx$ the square of $\frac{1}{2}$ of the coefficient of x, making it a perfect-square trinomial.

To complete the square on $x^2 - 8x$, add $\left[\frac{1}{2}(-8)\right]^2 = 16$: $x^2 - 8x + 16$.

To Solve a Quadratic Equation by Completing the Square [10.2A, p. 573]

1. Write the equation in the form $ax^2 + bx = -c$.
2. Multiply both sides of the equation by $\frac{1}{a}$.
3. Complete the square on $x^2 + \frac{b}{a}x$. Add the number that completes the square to both sides of the equation.
4. Factor the perfect-square trinomial.
5. Take the square root of each side of the equation.
6. Solve the resulting equation for x.
7. Check the solutions.

$$x^2 + 4x - 1 = 0$$
$$x^2 + 4x = 1$$
$$x^2 + 4x + 4 = 1 + 4$$
$$(x+2)^2 = 5$$
$$\sqrt{(x+2)^2} = \sqrt{5}$$
$$x + 2 = \pm\sqrt{5}$$
$$x + 2 = \sqrt{5} \qquad x + 2 = -\sqrt{5}$$
$$x = -2 + \sqrt{5} \qquad x = -2 - \sqrt{5}$$

Quadratic Formula [10.3A, p. 577]

The solutions of $ax^2 + bx + c = 0$, $a \neq 0$, are $x = \frac{-b \pm \sqrt{b^2 - 4ac}}{2a}$.

$$2x^2 - 3x + 4 = 0$$
$$a = 2, b = -3, c = 4$$
$$x = \frac{-(-3) \pm \sqrt{(-3)^2 - 4(2)(4)}}{2(2)}$$
$$= \frac{3 \pm \sqrt{-23}}{4} = \frac{3 \pm i\sqrt{23}}{4}$$
$$= \frac{3}{4} \pm \frac{\sqrt{23}}{4}i$$

The Effect of the Discriminant on the Solutions of a Quadratic Equation [10.3A, p. 579]

1. If $b^2 - 4ac = 0$, the equation has one real number solution, a double root.
2. If $b^2 - 4ac > 0$, the equation has two unequal real number solutions.
3. If $b^2 - 4ac < 0$, the equation has two complex number solutions.

$x^2 + 8x + 16 = 0$ has a double root because $b^2 - 4ac = 8^2 - 4(1)(16) = 0$.
$2x^2 + 3x - 5 = 0$ has two unequal real number solutions because $b^2 - 4ac = 3^2 - 4(2)(-5) = 49$.
$3x^2 + 2x + 4 = 0$ has two complex number solutions because $b^2 - 4ac = 2^2 - 4(3)(4) = -44$.

Copyright © Houghton Mifflin Company. All rights reserved.

Chapter 10 Review Exercises

1. Solve by factoring: $2x^2 - 3x = 0$

2. Solve for x by factoring: $6x^2 + 9cx = 6c^2$

3. Solve by taking square roots:
$x^2 = 48$

4. Solve by taking square roots:
$\left(x + \frac{1}{2}\right)^2 + 4 = 0$

5. Solve by completing the square:
$x^2 + 4x + 3 = 0$

6. Solve by completing the square:
$7x^2 - 14x + 3 = 0$

7. Solve by using the quadratic formula:
$12x^2 - 25x + 12 = 0$

8. Solve by using the quadratic formula:
$x^2 - x + 8 = 0$

9. Write a quadratic equation that has integer coefficients and has solutions 0 and -3.

10. Write a quadratic equation that has integer coefficients and has solutions $\frac{3}{4}$ and $-\frac{2}{3}$.

11. Solve by completing the square:
$x^2 - 2x + 8 = 0$

12. Solve by completing the square:
$(x - 2)(x + 3) = x - 10$

13. Solve by using the quadratic formula:
$3x(x - 3) = 2x - 4$

14. Use the discriminant to determine whether $3x^2 - 5x + 1 = 0$ has one real number solution, two real number solutions, or two complex number solutions.

15. Solve: $(x + 3)(2x - 5) < 0$

16. Solve: $(x - 2)(x + 4)(2x + 3) \le 0$

Copyright © Houghton Mifflin Company. All rights reserved.

17. Solve: $x^{2/3} + x^{1/3} - 12 = 0$

18. Solve: $2(x - 1) + 3\sqrt{x - 1} - 2 = 0$

19. Solve: $3x = \dfrac{9}{x - 2}$

20. Solve: $\dfrac{3x + 7}{x + 2} + x = 3$

21. Solve and graph the solution set:
$\dfrac{x - 2}{2x - 3} \geq 0$

−5 −4 −3 −2 −1 0 1 2 3 4 5

22. Solve and graph the solution set:
$\dfrac{(2x - 1)(x + 3)}{x - 4} \leq 0$

−5 −4 −3 −2 −1 0 1 2 3 4 5

23. Solve: $x = \sqrt{x} + 2$

24. Solve: $2x = \sqrt{5x + 24} + 3$

25. Solve: $\dfrac{x - 2}{2x + 3} - \dfrac{x - 4}{x} = 2$

26. Solve: $1 - \dfrac{x + 4}{2 - x} = \dfrac{x - 3}{x + 2}$

27. **Geometry** The length of a rectangle is 2 cm more than twice the width. The area of the rectangle is 60 cm². Find the length and width of the rectangle.

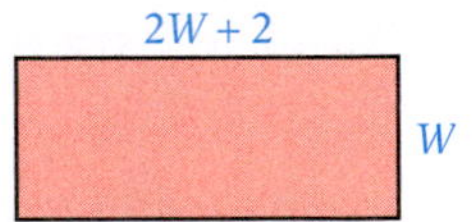

28. **Integers** The sum of the squares of three consecutive even integers is fifty-six. Find the three integers.

29. **Computers** An older computer requires 12 min longer to print the payroll than does a newer computer. Together the computers can print the payroll in 8 min. Find the time for the new computer working alone to print the payroll.

30. **Sports** To prepare for an upcoming race, a sculling crew rowed 16 mi down a river and back in 6 h. If the rate of the river's current is 2 mph, find the sculling crew's rate of rowing in calm water.

Copyright © Houghton Mifflin Company. All rights reserved.

Chapter 10 Test

1. Solve by factoring: $3x^2 + 10x = 8$

2. Solve by factoring: $6x^2 - 5x - 6 = 0$

3. Write a quadratic equation that has integer coefficients and has solutions 3 and -3.

4. Write a quadratic equation that has integer coefficients and has solutions $\frac{1}{2}$ and -4.

5. Solve by taking square roots: $3(x - 2)^2 - 24 = 0$

6. Solve by completing the square: $x^2 - 6x - 2 = 0$

7. Solve by completing the square: $3x^2 - 6x = 2$

8. Solve by using the quadratic formula: $2x^2 - 2x = 1$

9. Solve by using the quadratic formula: $x^2 + 4x + 12 = 0$

10. Use the discriminant to determine whether $3x^2 - 4x = 1$ has one real number solution, two real number solutions, or two complex number solutions.

11. Use the discriminant to determine whether $x^2 - 6x = -15$ has one real number solution, two real number solutions, or two complex number solutions.

12. Solve: $2x + 7x^{1/2} - 4 = 0$

Copyright © Houghton Mifflin Company. All rights reserved.

13. Solve: $x^4 - 4x^2 + 3 = 0$

14. Solve: $\sqrt{2x + 1} + 5 = 2x$

15. Solve: $\sqrt{x - 2} = \sqrt{x} - 2$

16. Solve: $\dfrac{2x}{x - 3} + \dfrac{5}{x - 1} = 1$

17. Solve and graph the solution set of $(x - 2)(x + 4)(x - 4) < 0$.

−5 −4 −3 −2 −1 0 1 2 3 4 5

18. Solve and graph the solution set of $\dfrac{2x - 3}{x + 4} \leq 0$.

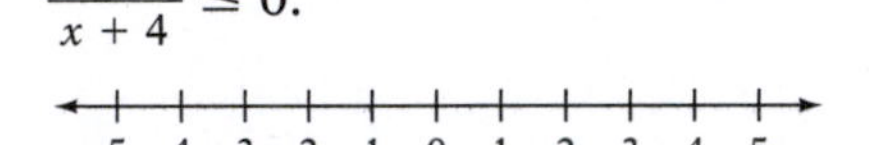

19. **Sports** A basketball player shoots at a basket that is 25 ft away. The height h, in feet, of the ball above the ground after t seconds is given by $h = -16t^2 + 32t + 6.5$. How many seconds after the ball is released does it hit the basket? *Note:* The basket is 10 ft off the ground. Round to the nearest hundredth.

20. **Uniform Motion** The rate of a river's current is 2 mph. A canoe was rowed 6 mi down the river and back in 4 h. Find the rowing rate in calm water.

Copyright © Houghton Mifflin Company. All rights reserved.

Cumulative Review Exercises

1. Evaluate $2a^2 - b^2 \div c^2$ when $a = 3$, $b = -4$, and $c = -2$.

2. Solve: $|3x - 2| < 8$

3. Find the volume of a cylinder with a height of 6 m and a radius of 3 m. Give the exact measure.

4. Given $f(x) = \dfrac{2x - 3}{x^2 - 1}$, find $f(-2)$.

5. Find the slope of the line containing the points $(3, -4)$ and $(-1, 2)$.

6. Find the x- and y-intercepts of the graph of $6x - 5y = 15$.

7. Find the equation of the line that contains the point $(1, 2)$ and is parallel to the line $x - y = 1$.

8. Solve the system of equations.

$$\begin{aligned} x + y + z &= 2 \\ -x + 2y - 3z &= -9 \\ x - 2y - 2z &= -1 \end{aligned}$$

9. Graph the solution set:
$x + y \leq 3$
$2x - y < 4$

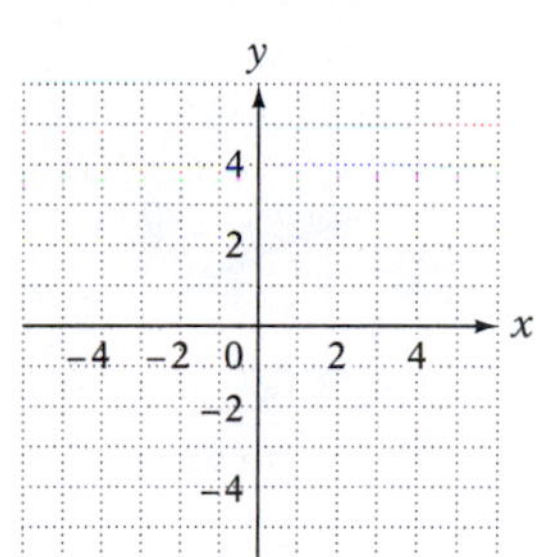

10. Triangles ABC and DEF are similar. Find the height of triangle DEF.

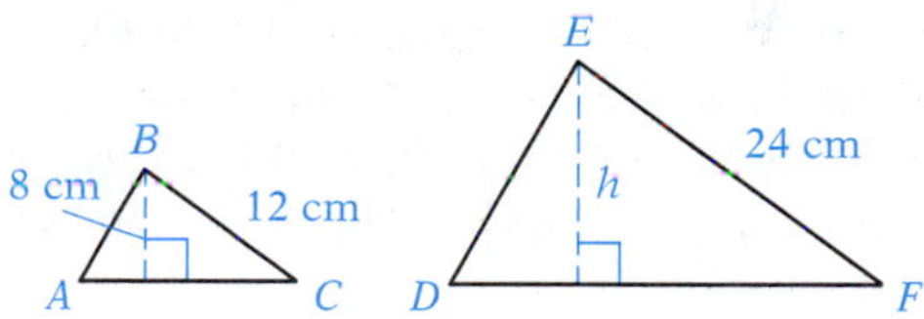

11. Divide: $(3x^3 - 13x^2 + 10) \div (3x - 4)$

12. Factor: $-3x^3y + 6x^2y^2 - 9xy^3$

13. Factor: $6x^2 - 7x - 20$

14. Multiply: $\dfrac{x^2 + 2x + 1}{8x^2 + 8x} \cdot \dfrac{4x^3 - 4x^2}{x^2 - 1}$

Copyright © Houghton Mifflin Company. All rights reserved.

15. Solve: $\frac{x}{x+2} - \frac{4x}{x+3} = 1$

16. Solve $S = \frac{n}{2}(a + b)$ for b.

17. Multiply: $a^{-1/2}(a^{1/2} - a^{3/2})$

18. Multiply: $-2i(7 - 4i)$

19. Solve: $\sqrt{3x + 1} - 1 = x$

20. Solve: $x^4 - 6x^2 + 8 = 0$

21. Mechanics A piston rod for an automobile is $9\frac{3}{8}$ in. with a tolerance of $\frac{1}{64}$ in. Find the lower and upper limits of the length of the piston rod.

22. Geometry The base of a triangle is $(x + 8)$ feet. The height is $(2x - 4)$ feet. Find the area of the triangle in terms of the variable x.

23. Depreciation The graph shows the relationship between the value of a building and the time, in years, since depreciation began. Find the slope of the line between the two points shown on the graph. Write a sentence that states the meaning of the slope.

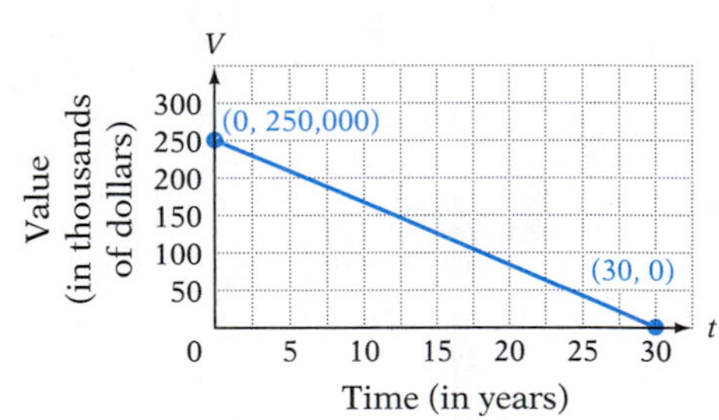

24. Home Maintenance How high on a house will a 17-foot ladder reach when the bottom of the ladder is 8 feet from the house?

25. Use the discriminant to determine whether $2x^2 + 4x + 3 = 0$ has one real number solution, two real number solutions, or two complex number solutions.

Copyright © Houghton Mifflin Company. All rights reserved.

chapter

11 Functions and Relations

Due to constantly changing airfares and schedules, thousands of available vacation packages, and the vast amount of travel information available on the Internet, travel planning can be frustrating and time-consuming. For assistance, tourists and businesspeople often turn to travel agents, who assess their needs and make the appropriate travel arrangements for them. Like any business, the travel agency's main goal is to earn a profit. Profit is the difference between a company's revenue (the total amount of money the company earns by selling its products or services) and its costs (the total amount of money the company spends in doing business). The **Projects and Group Activities on pages 648 and 649** use quadratic equations to determine the maximum profit companies can earn.

OBJECTIVES

Section 11.1

- **A** To graph a quadratic function
- **B** To find the x-intercepts of a parabola
- **C** To find the minimum or maximum of a quadratic function
- **D** To solve application problems

Section 11.2

- **A** To graph functions

Section 11.3

- **A** To perform operations on functions
- **B** To find the composition of two functions

Section 11.4

- **A** To determine whether a function is one-to-one
- **B** To find the inverse of a function

Need help? For online student resources, such as section quizzes, visit this textbook's website at **math.college.hmco.com/students.**

Copyright © Houghton Mifflin Company. All rights reserved.

PREP TEST • • •

Do these exercises to prepare for Chapter 11.

1. Evaluate $-\frac{b}{2a}$ for $b = -4$ and $a = 2$.

2. Given $y = -x^2 + 2x + 1$, find the value of y when $x = -2$.

3. Given $f(x) = x^2 - 3x + 2$, find $f(-4)$.

4. Evaluate $p(r) = r^2 - 5$ when $r = 2 + h$.

5. Solve: $0 = 3x^2 - 7x - 6$

6. Solve by using the quadratic formula: $0 = x^2 - 4x + 1$

7. Solve $x = 2y + 4$ for y.

8. Find the domain and range of the relation $\{(-2, 4), (3, 5), (4, 6), (6, 5)\}$. Is the relation a function?

9. Graph: $x = -2$

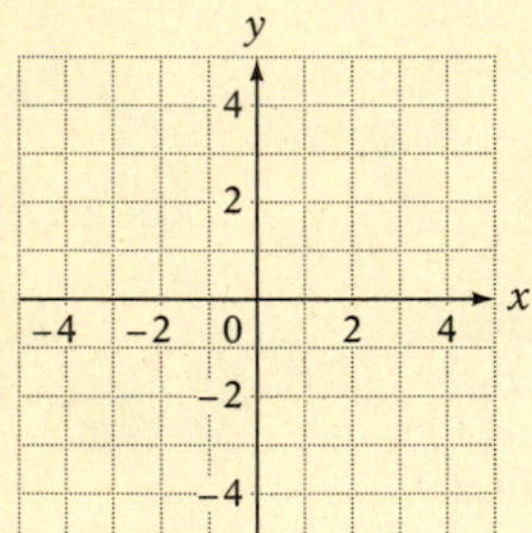

GO FIGURE • • •

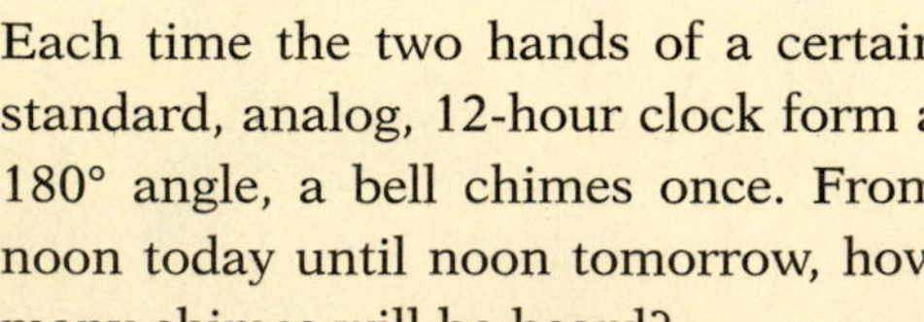
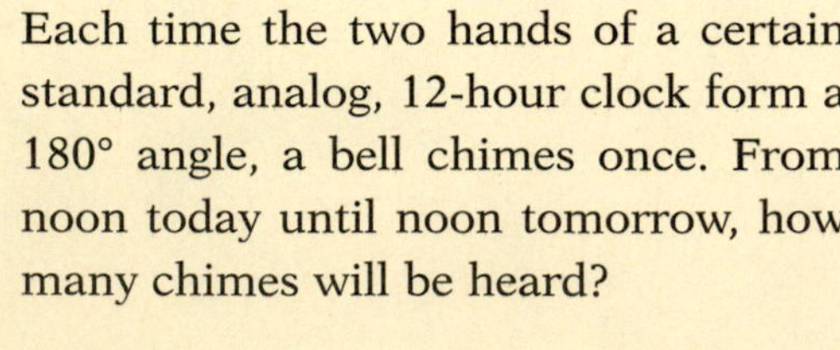

Each time the two hands of a certain standard, analog, 12-hour clock form a 180° angle, a bell chimes once. From noon today until noon tomorrow, how many chimes will be heard?

Copyright © Houghton Mifflin Company. All rights reserved.

11.1 Properties of Quadratic Functions

Objective A To graph a quadratic function

Recall that a linear function is one that can be expressed by the equation $f(x) = mx + b$. The graph of a linear function has certain characteristics. It is a straight line with slope m and y-intercept $(0, b)$. A **quadratic function** is one that can be expressed by the equation $f(x) = ax^2 + bx + c$, $a \neq 0$. The graph of this function, called a **parabola,** also has certain characteristics. The graph of a quadratic function can be drawn by finding ordered pairs that belong to the function.

TAKE NOTE

Sometimes the value of the independent variable is called the **input** because it is *put in* place of the independent variable. The result of evaluating the function is called the **output.**

An **input/output table** shows the results of evaluating a function for various values of the independent variable. An input/output table for $f(x) = x^2 - 2x - 3$ is shown at the right.

HOW TO Graph $f(x) = x^2 - 2x - 3$.

By evaluating the function for various values of x, find enough ordered pairs to determine the shape of the graph.

x	$f(x) = x^2 - 2x - 3$	$f(x)$	(x, y)
-2	$f(-2) = (-2)^2 - 2(-2) - 3$	5	$(-2, 5)$
-1	$f(-1) = (-1)^2 - 2(-1) - 3$	0	$(-1, 0)$
0	$f(0) = (0)^2 - 2(0) - 3$	-3	$(0, -3)$
1	$f(1) = (1)^2 - 2(1) - 3$	-4	$(1, -4)$
2	$f(2) = (2)^2 - 2(2) - 3$	-3	$(2, -3)$
3	$f(3) = (3)^2 - 2(3) - 3$	0	$(3, 0)$
4	$f(4) = (4)^2 - 2(4) - 3$	5	$(4, 5)$

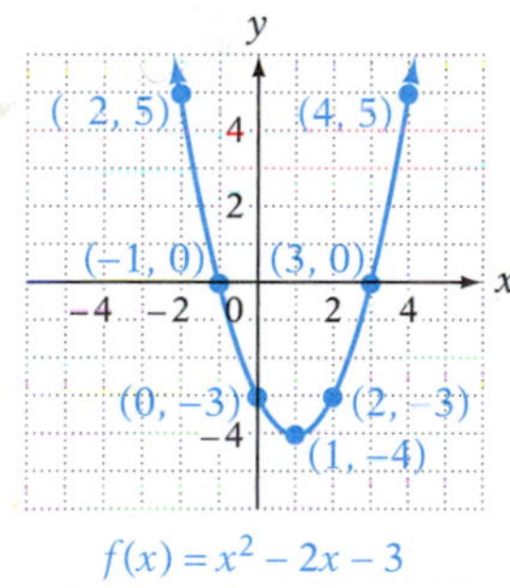

$f(x) = x^2 - 2x - 3$

Because the value of $f(x) = x^2 - 2x - 3$ is a real number for all values of x, the domain of f is all real numbers. From the graph, it appears that no value of y is less than -4. Thus the range is $\{y \mid y \geq -4\}$. The range can also be determined algebraically, as shown below, by completing the square.

TAKE NOTE

In completing the square, 1 is both added and subtracted. Because $1 - 1 = 0$, the expression $x^2 - 2x - 3$ is not changed. Note that

$$\begin{aligned}(x-1)^2 - 4 &= (x^2 - 2x + 1) - 4\\ &= x^2 - 2x - 3\end{aligned}$$

which is the original expression.

$f(x) = x^2 - 2x - 3$

$= (x^2 - 2x) - 3$ • Group the variable terms.

$= (x^2 - 2x + 1) - 1 - 3$ • Complete the square on $x^2 - 2x$. Add and subtract $\left[\frac{1}{2}(-2)\right]^2 = 1$ to and from $x^2 - 2x$.

$= (x - 1)^2 - 4$ • Factor and combine like terms.

Because the square of a real number is always nonnegative, we have

$(x - 1)^2 \geq 0$

$(x - 1)^2 - 4 \geq -4$ • Subtract 4 from each side of the inequality.

$f(x) \geq -4$ • $f(x) = (x - 1)^2 - 4$

$y \geq -4$

From the last inequality, the range is $\{y \mid y \geq -4\}$.

Copyright © Houghton Mifflin Company. All rights reserved.

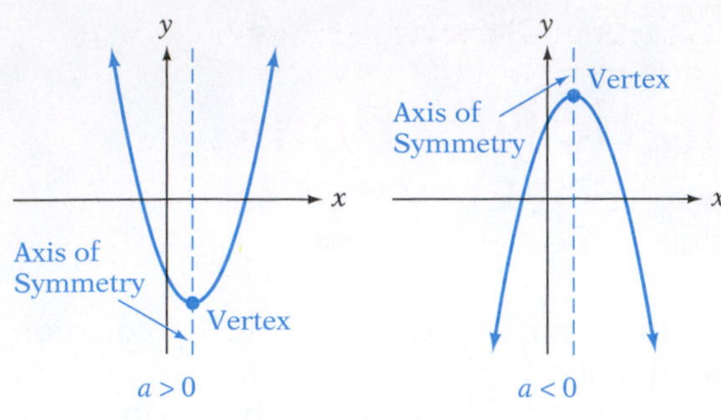

In general, the graph of $f(x) = ax^2 + bx + c$, $a \neq 0$, resembles a "cup" shape, as shown at the left. When $a > 0$, the parabola opens up and the **vertex of the parabola** is the point with the smallest y-coordinate. When $a < 0$, the parabola opens down and the vertex is the point with the largest y-coordinate. The **axis of symmetry of a parabola** is the vertical line that passes through the vertex of the parabola and is parallel to the y-axis. To understand the axis of symmetry, think of folding the graph along that vertical line. The two portions of the graph will match up.

The vertex and axis of symmetry of a parabola can be found by completing the square.

HOW TO Find the vertex and axis of symmetry of the graph of $F(x) = x^2 + 4x + 3$.

To find the coordinates of the vertex, complete the square.

$F(x) = x^2 + 4x + 3$

$= (x^2 + 4x) + 3$ • **Group the variable terms.**

$= (x^2 + 4x + 4) - 4 + 3$ • **Complete the square on $x^2 + 4x$. Add and**

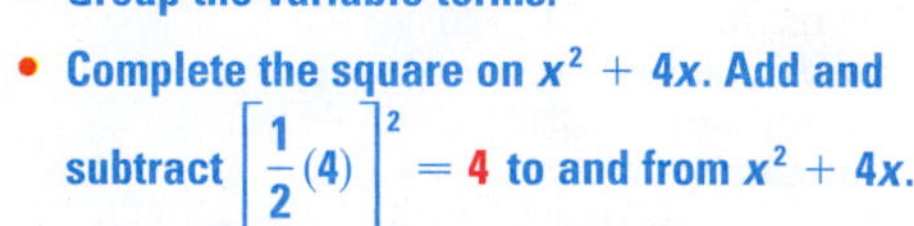
subtract $\left[\frac{1}{2}(4)\right]^2 = 4$ to and from $x^2 + 4x$.

$= (x + 2)^2 - 1$ • **Factor and combine like terms.**

Because a, the coefficient of x^2, is positive ($a = 1$), the parabola opens up and the vertex is the point with the smallest y-coordinate. Because $(x + 2)^2 \geq 0$ for all values of x, the smallest y-coordinate occurs when $(x + 2)^2 = 0$. The quantity $(x + 2)^2$ is equal to zero when $x = -2$. Therefore, the x-coordinate of the vertex is -2.

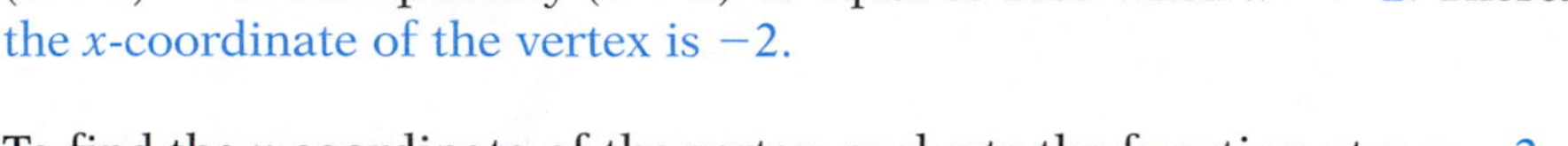

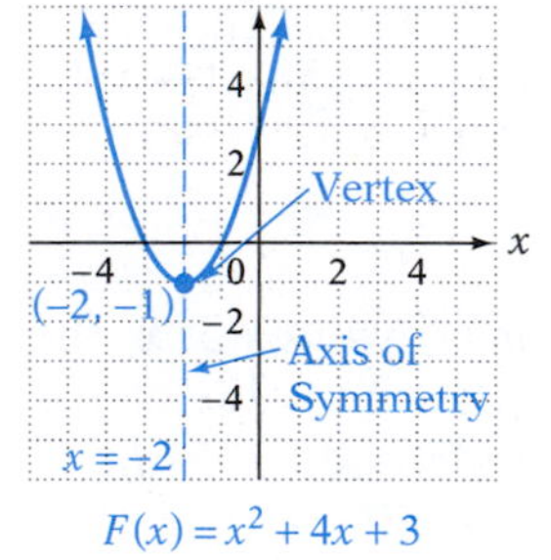

$F(x) = x^2 + 4x + 3$

To find the y-coordinate of the vertex, evaluate the function at $x = -2$.

$F(x) = (x + 2)^2 - 1$

$F(-2) = (-2 + 2)^2 - 1 = -1$

The y-coordinate of the vertex is -1.

From the results above, the coordinates of the vertex are $(-2, -1)$.

The axis of symmetry is the vertical line that passes through the vertex. The equation of the vertical line that passes through the point $(-2, -1)$ is $x = -2$. The axis of symmetry is the line $x = -2$.

By following the process illustrated in the preceding example and completing the square on $f(x) = ax^2 + bx + c$, we can find a formula for the coordinates of the vertex and the equation of the axis of symmetry of a parabola.

Vertex and Axis of Symmetry of a Parabola

Let $f(x) = ax^2 + bx + c$ be the equation of a parabola. The coordinates of the vertex are $\left(-\frac{b}{2a}, f\left(-\frac{b}{2a}\right)\right)$. The equation of the axis of symmetry is $x = -\frac{b}{2a}$.

Copyright © Houghton Mifflin Company. All rights reserved.

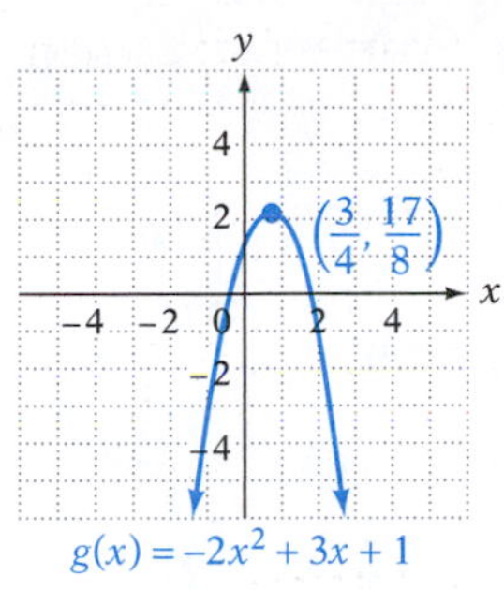

HOW TO Find the vertex of the parabola whose equation is $g(x) = -2x^2 + 3x + 1$. Then graph the equation.

x-coordinate of the vertex: $-\frac{b}{2a} = -\frac{3}{2(-2)} = \frac{3}{4}$

• From the equation $g(x) = -2x^2 + 3x + 1$, $a = -2$, $b = 3$.

y-coordinate of the vertex:

$$g(x) = -2x^2 + 3x + 1$$
$$g\left(\frac{3}{4}\right) = -2\left(\frac{3}{4}\right)^2 + 3\left(\frac{3}{4}\right) + 1$$
$$g\left(\frac{3}{4}\right) = \frac{17}{8}$$

• Evaluate the function at the value of the x-coordinate of the vertex.

The vertex is $\left(\frac{3}{4}, \frac{17}{8}\right)$.

Because a is negative, the graph opens down. Find a few ordered pairs that belong to the function and then sketch the graph, as shown at the left.

TAKE NOTE

Once the coordinates of the vertex are found, the range of the quadratic function can be determined.

Once the y-coordinate of the vertex is known, the range of the function can be determined. Here, the graph of g opens down, so the y-coordinate of the vertex is the largest value of y. Therefore, the range of g is $\left\{y \mid y \le \frac{17}{8}\right\}$. The domain is $\{x \mid x \in \text{ real numbers}\}$.

Example 1 Find the vertex and axis of symmetry of the parabola whose equation is $y = -x^2 + 4x + 1$. Then graph the equation.

Solution x-coordinate of the vertex:

$$-\frac{b}{2a} = -\frac{4}{2(-1)} = 2$$

y-coordinate of vertex:

$$y = -x^2 + 4x + 1$$
$$= -(2)^2 + 4(2) + 1$$
$$= 5$$

Vertex: (2, 5)

Axis of symmetry: $x = 2$

You Try It 1 Find the vertex and axis of symmetry of the parabola whose equation is $y = 4x^2 + 4x + 1$. Then graph the equation.

Your solution

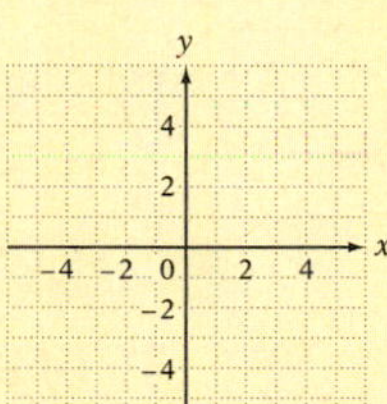

Example 2 Find the domain and range of $f(x) = 0.5x^2 - 3$. Then graph the equation.

Solution x-coordinate of the vertex:

$$-\frac{b}{2a} = -\frac{0}{2(0.5)} = 0$$

y-coordinate of vertex:

$$f(x) = 0.5x^2 - 3$$
$$f(0) = 0.5(0)^2 - 3$$
$$= -3$$

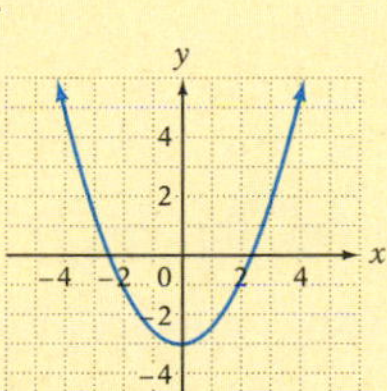

Vertex: (0, −3)

The domain is $\{x \mid x \in \text{ real numbers}\}$.

The range is $\{y \mid y \ge -3\}$.

You Try It 2 Find the domain and range of $f(x) = -x^2 - 2x - 1$. Then graph the equation.

Your solution

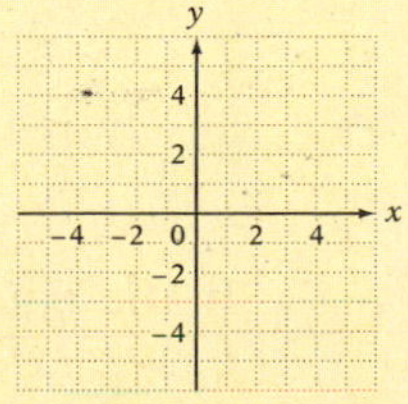

Solutions on p. S32

Copyright © Houghton Mifflin Company. All rights reserved.

Objective B To find the x-intercepts of a parabola

Recall that a point at which a graph crosses the x- or y-axis is called an *intercept* of the graph. The x-intercepts of the graph of an equation occur when $y = 0$; the y-intercepts occur when $x = 0$.

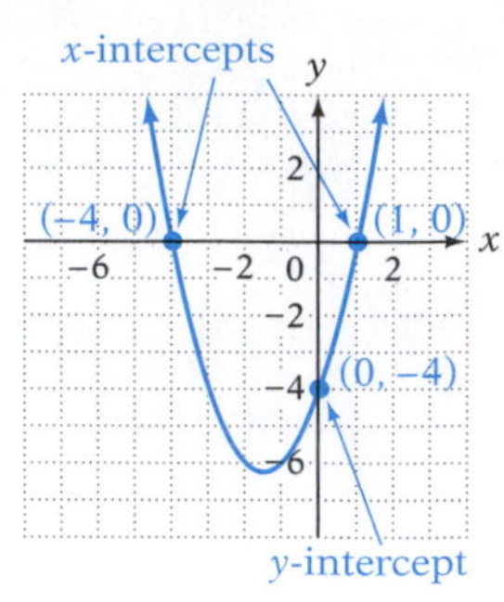

The graph of $y = x^2 + 3x - 4$ is shown at the right. The points whose coordinates are $(-4, 0)$ and $(1, 0)$ are the x-intercepts of the graph. The point whose coordinates are $(0, -4)$ is the y-intercept of the graph.

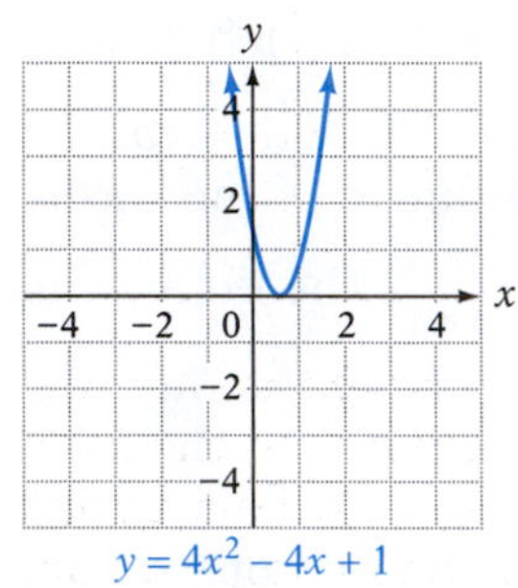

$y = 4x^2 - 4x + 1$

HOW TO Find the x-intercepts of the parabola whose equation is $y = 4x^2 - 4x + 1$.

To find the x-intercepts, let $y = 0$ and then solve for x.

$y = 4x^2 - 4x + 1$

$0 = 4x^2 - 4x + 1$ • Let $y = 0$.

$0 = (2x - 1)(2x - 1)$ • Solve for x by factoring and using the Principle of Zero Products.

$2x - 1 = 0 \qquad 2x - 1 = 0$

$2x = 1 \qquad 2x = 1$

$x = \frac{1}{2} \qquad x = \frac{1}{2}$

The x-intercept is $\left(\frac{1}{2}, 0\right)$.

Integrating Technology

See the appendix Keystroke Guide: *Zero* for instructions on using a graphing calculator to find the x-intercepts of the graph of a parabola.

In the preceding example, the parabola has only one x-intercept. In this case, the parabola is said to be *tangent* to the x-axis at $x = \frac{1}{2}$.

TAKE NOTE

A zero of a function is the x-coordinate of the x-intercept of the graph of the function. Because the x-intercepts of the graph of $f(x) = 2x^2 - x - 6$ are $\left(-\frac{3}{2}, 0\right)$ and $(2, 0)$, the zeros are $-\frac{3}{2}$ and 2.

HOW TO Find the x-intercepts of $y = 2x^2 - x - 6$.

To find the x-intercepts, let $y = 0$ and solve for x.

$y = 2x^2 - x - 6$

$0 = 2x^2 - x - 6$ • Let $y = 0$.

$0 = (2x + 3)(x - 2)$ • Solve for x by factoring and using the Principle of Zero Products.

$2x + 3 = 0 \qquad x - 2 = 0$

$x = -\frac{3}{2} \qquad x = 2$

The x-intercepts are $\left(-\frac{3}{2}, 0\right)$ and $(2, 0)$.

If the equation above, $y = 2x^2 - x - 6$, were written in functional notation as $f(x) = 2x^2 - x - 6$, then to find the x-intercepts you would let $f(x) = 0$ and solve for x. A value of x for which $f(x) = 0$ is a **zero of the function.** Thus $-\frac{3}{2}$ and 2 are *zeros* of $f(x) = 2x^2 - x - 6$.

Copyright © Houghton Mifflin Company. All rights reserved.

Integrating Technology

See the appendix Keystroke Guide: *Zero* for instructions on using a graphing calculator to find the zeros of a function.

HOW TO Find the zeros of $f(x) = x^2 - 2x - 1$.

To find the zeros, let $f(x) = 0$ and solve for x.

$$f(x) = x^2 - 2x - 1$$
$$0 = x^2 - 2x - 1$$
$$x = \frac{-b \pm \sqrt{b^2 - 4ac}}{2a}$$

• **Because $x^2 - 2x - 1$ does not easily factor, use the quadratic formula to solve for x.**

$$= \frac{-(-2) \pm \sqrt{(-2)^2 - 4(1)(-1)}}{2(1)}$$

• $a = 1, b = -2, c = -1$

$$= \frac{2 \pm \sqrt{4 + 4}}{2} = \frac{2 \pm \sqrt{8}}{2}$$
$$= \frac{2 \pm 2\sqrt{2}}{2}$$
$$= 1 \pm \sqrt{2}$$

The zeros of the function are $1 - \sqrt{2}$ and $1 + \sqrt{2}$.

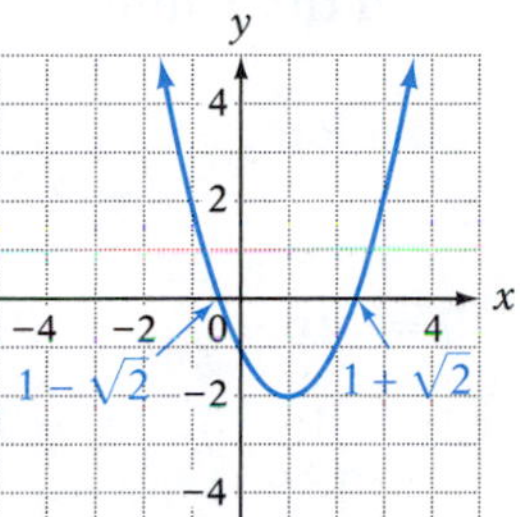

TAKE NOTE

The graph of $f(x) = x^2 - 2x - 1$ is shown at the right. Note that the zeros are the x-intercepts of the graph of f.

The preceding examples suggest that there is a relationship among the x-intercepts of the graph of a function, the zeros of the function, and the solutions of an equation. In fact, these three concepts are different ways of discussing the same number. The choice depends on the focus of the discussion. If we are discussing graphing, then the intercepts are our focus; if we are discussing functions, then the zeros of the function are our focus; and if we are discussing equations, the solution of the equation is our focus.

The graph of a parabola may not have x-intercepts. The graph of $y = -x^2 + 2x - 2$ is shown at the right. Note that the graph does not pass through the x-axis and thus there are no x-intercepts. This means there are no real number zeros of $f(x) = -x^2 + 2x - 2$ and there are no real number solutions of $-x^2 + 2x - 2 = 0$.

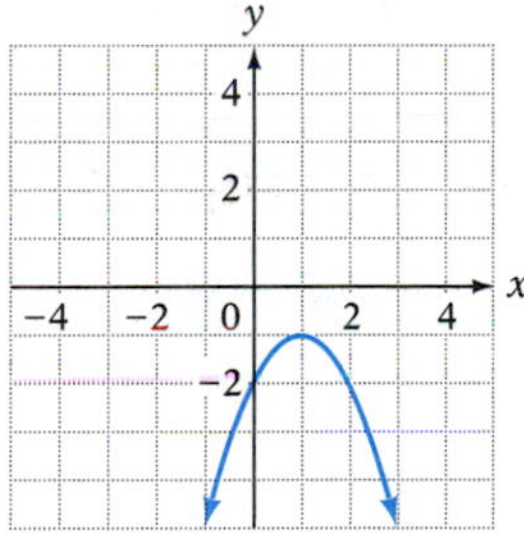

When we apply the quadratic formula, we find that the solutions of the equation $-x^2 + 2x - 2 = 0$ are $1 - i$ and $1 + i$. Thus the zeros of $f(x) = -x^2 + 2x - 2$ are the complex numbers $1 - i$ and $1 + i$.

Study Tip

The paragraph at the right begins with the word "Recall." This signals that the content refers to material presented earlier in the text. The ideas presented here will be more meaningful if you return to the word *discriminant* on page 579 and review the concepts presented there.

Recall that the *discriminant* of $ax^2 + bx + c$ is the expression $b^2 - 4ac$ and that this expression can be used to determine whether $ax^2 + bx + c = 0$ has zero, one, or two real number solutions. Because there is a connection between the solutions of $ax^2 + bx + c = 0$ and the x-intercepts of the graph of $y = ax^2 + bx + c$, the discriminant can be used to determine the number of x-intercepts of a parabola.

The Effect of the Discriminant on the Number of x-Intercepts of a Parabola

1. If $b^2 - 4ac = 0$, the parabola has one x-intercept.
2. If $b^2 - 4ac > 0$, the parabola has two x-intercepts.
3. If $b^2 - 4ac < 0$, the parabola has no x-intercepts.

Copyright © Houghton Mifflin Company. All rights reserved.

HOW TO Use the discriminant to determine the number of x-intercepts of the parabola whose equation is $y = 2x^2 - x + 2$.

$b^2 - 4ac$ • Evaluate the discriminant.

$(-1)^2 - 4(2)(2) = 1 - 16 = -15$ • $a = 2, b = -1, c = 2$

$-15 < 0$

The discriminant is less than zero. Therefore, the parabola has no x-intercepts.

Example 3

Find the x-intercepts of $y = 2x^2 - 5x + 2$.

Solution

$y = 2x^2 - 5x + 2$

$0 = 2x^2 - 5x + 2$ • Let $y = 0$.

$0 = (2x - 1)(x - 2)$ • Solve for x by factoring.

$2x - 1 = 0 \qquad x - 2 = 0$

$2x = 1 \qquad x = 2$

$x = \frac{1}{2}$

The x-intercepts are $\left(\frac{1}{2}, 0\right)$ and $(2, 0)$.

You Try It 3

Find the x-intercepts of $y = x^2 + 3x + 4$.

Your solution

Example 4

Find the zeros of $f(x) = x^2 + 4x + 5$.

Solution

$f(x) = x^2 + 4x + 5$

$0 = x^2 + 4x + 5$ • Let $f(x) = 0$.

$x = \frac{-b \pm \sqrt{b^2 - 4ac}}{2a}$ • Use the quadratic formula.

$= \frac{-4 \pm \sqrt{4^2 - 4(1)(5)}}{2(1)}$ • $a = 1, b = 4, c = 5$

$= \frac{-4 \pm \sqrt{16 - 20}}{2} = \frac{-4 \pm \sqrt{-4}}{2}$

$= \frac{-4 \pm 2i}{2} = -2 \pm i$

The zeros of the function are $-2 + i$ and $-2 - i$.

You Try It 4

Find the zeros of $g(x) = x^2 - x + 6$.

Your solution

Example 5

Use the discriminant to determine the number of x-intercepts of $y = x^2 - 6x + 9$.

Solution

$b^2 - 4ac$ • Evaluate the discriminant.

$(-6)^2 - 4(1)(9)$ • $a = 1, b = -6, c = 9$

$= 36 - 36 = 0$

The discriminant is equal to zero.
The parabola has one x-intercept.

You Try It 5

Use the discriminant to determine the number of x-intercepts of $y = x^2 - x - 6$.

Your solution

Solutions on pp. S32–S33

Copyright © Houghton Mifflin Company. All rights reserved.

Objective C To find the minimum or maximum of a quadratic function

Study Tip

After studying this objective, you should be able to describe in words what the minimum or maximum value of a quadratic function is, draw a graph of a quadratic function with either a maximum or a minimum value, determine from a quadratic equation whether it has a minimum or a maximum value, and calculate algebraically the minimum or maximum value of a quadratic function.

The graph of $f(x) = x^2 - 2x + 3$ is shown at the right. Because a is positive, the parabola opens up. The vertex of the parabola is the lowest point on the parabola. It is the point that has the minimum y-coordinate. This point represents the **minimum value of the function.**

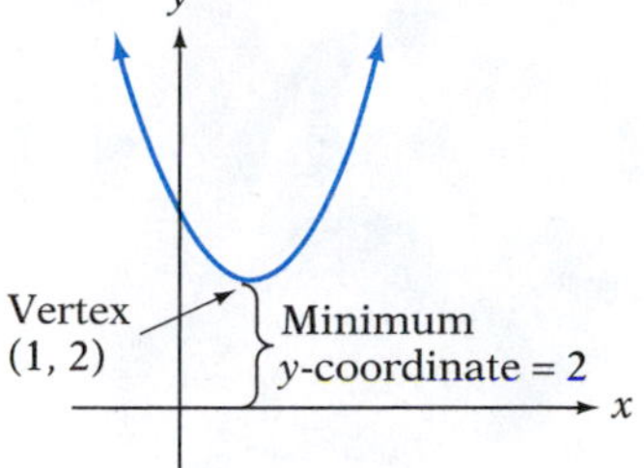

The graph of $f(x) = -x^2 + 2x + 1$ is shown at the right. Because a is negative, the parabola opens down. The vertex of the parabola is the highest point on the parabola. It is the point that has the maximum y-coordinate. This point represents the **maximum value of the function.**

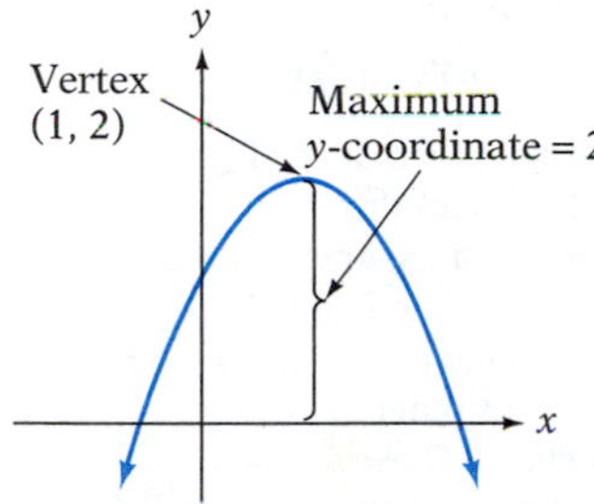

To find the minimum or maximum value of a quadratic function, first find the x-coordinate of the vertex. Then evaluate the function at that value.

Integrating Technology

See the Projects and Group Activities at the end of this chapter for instructions on using a graphing calculator to find the minimum or maximum value of a function.

HOW TO Find the maximum or minimum value of the function $f(x) = -2x^2 + 4x + 3$.

$$x = -\frac{b}{2a} = -\frac{4}{2(-2)} = 1$$

- Find the x-coordinate of the vertex. $a = -2$, $b = 4$

$$\begin{aligned} f(x) &= -2x^2 + 4x + 3 \\ f(1) &= -2(1)^2 + 4(1) + 3 \\ &= -2 + 4 + 3 \\ &= 5 \end{aligned}$$

- Evaluate the function at $x = 1$.

The maximum value of the function is 5.

- Because $a < 0$, the graph of f opens down. Therefore, the function has a maximum value.

Example 6

Find the maximum or minimum value of $f(x) = 2x^2 - 3x + 1$.

Solution

$$x = -\frac{b}{2a} = -\frac{-3}{2(2)} = \frac{3}{4}$$

- The x-coordinate of the vertex

$$\begin{aligned} f(x) &= 2x^2 - 3x + 1 \\ f\left(\frac{3}{4}\right) &= 2\left(\frac{3}{4}\right)^2 - 3\left(\frac{3}{4}\right) + 1 \\ &= \frac{9}{8} - \frac{9}{4} + 1 = -\frac{1}{8} \end{aligned}$$

- $x = \frac{3}{4}$

Because a is positive, the graph opens up. The function has a minimum value.

The minimum value of the function is $-\frac{1}{8}$.

You Try It 6

Find the maximum or minimum value of $f(x) = -3x^2 + 4x - 1$.

Your solution

Solution on p. S33

Copyright © Houghton Mifflin Company. All rights reserved.

Objective D To solve application problems

Point of Interest

Calculus is a branch of mathematics that demonstrates, among other things, how to find the maximum or minimum of functions other than quadratic functions. These are very important problems in applied mathematics. For instance, an automotive engineer wants to design a car whose shape will *minimize* the effects of air flow. The same engineer tries to *maximize* the efficiency of the car's engine. Similarly, an economist may try to determine what business practices will *minimize* cost and *maximize* profit.

HOW TO A carpenter is forming a rectangular floor for a storage shed. The perimeter of the rectangle is 44 ft. What dimensions of the rectangle would give the floor a maximum area? What is the maximum area?

We are given the perimeter of the rectangle, and we want to find the dimensions of the rectangle that will yield the maximum area for the floor. Use the equation for the perimeter of a rectangle.

$P = 2L + 2W$

$44 = 2L + 2W$ • $P = 44$

$22 = L + W$ • Divide both sides of the equation by 2.

$22 - L = W$ • Solve the equation for W.

Now use the equation for the area of a rectangle. Use substitution to express the area in terms of L.

$A = LW$

$A = L(22 - L)$ • From the equation above, $W = 22 - L$. Substitute $22 - L$ for W.

$A = 22L - L^2$ • The area of the rectangle is $22L - L^2$.

To find the length of the rectangle, find the L-coordinate of the vertex of the function $f(L) = -L^2 + 22L$.

$L = -\frac{b}{2a} = -\frac{22}{2(-1)} = 11$ • For the equation $f(L) = -L^2 + 22L$, $a = -1$ and $b = 22$.

The length of the rectangle is 11 ft.

To find the width, replace L in $22 - L$ by the L-coordinate of the vertex and evaluate.

$W = 22 - L$

$W = 22 - 11 = 11$ • Replace L by 11 and evaluate.

The width of the rectangle is 11 ft.

The dimensions of the rectangle that would give the floor a maximum area are 11 ft by 11 ft.

To find the maximum area of the floor, evaluate $f(L) = -L^2 + 22L$ at the L-coordinate of the vertex.

$f(L) = -L^2 + 22L$

$f(11) = -(11)^2 + 22(11)$ • Evaluate the function at 11.

$= -121 + 242 = 121$

The maximum area of the floor is 121 ft^2.

The graph of the function $f(L) = -L^2 + 22L$ is shown at the right. Note that the vertex of the parabola is (11, 121). For any value of L less than 11, the area of the floor will be less than 121 ft^2. For any value of L greater than 11, the area of the floor will be less than 121 ft^2. The maximum value of the function is 121, and the maximum value occurs when $L = 11$.

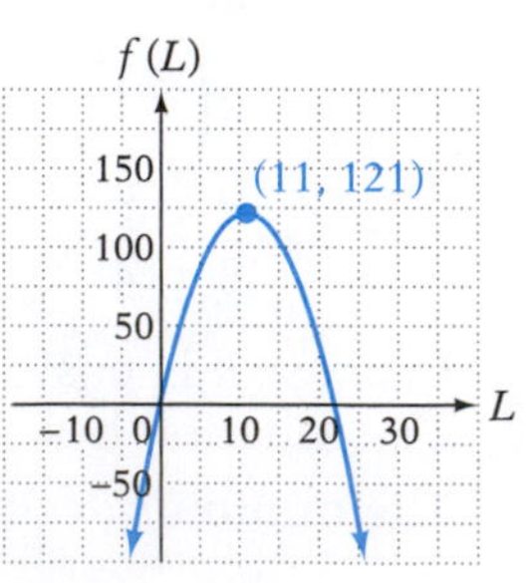

Copyright © Houghton Mifflin Company. All rights reserved.

Example 7

A mining company has determined that the cost c, in dollars per ton, of mining a mineral is given by $c(x) = 0.2x^2 - 2x + 12$, where x is the number of tons of the mineral mined. Find the number of tons of the mineral that should be mined to minimize the cost. What is the minimum cost per ton?

Strategy

- To find the number of tons of the mineral that should be mined to minimize the cost, find the x-coordinate of the vertex.
- To find the minimum cost, evaluate $c(x)$ at the x-coordinate of the vertex.

Solution

$c(x) = 0.2x^2 - 2x + 12$ • $a = 0.2$, $b = -2$, $c = 12$

$x = -\frac{b}{2a} = -\frac{(-2)}{2(0.2)} = 5$ • The x-coordinate of the vertex

To minimize the cost, 5 tons of the mineral should be mined.

$$\begin{aligned} c(x) &= 0.2x^2 - 2x + 12 \\ c(5) &= 0.2(5)^2 - 2(5) + 12 \quad \bullet\ x = 5 \\ &= 5 - 10 + 12 \\ &= 7 \end{aligned}$$

The minimum cost is \$7 per ton.

Note: The graph of the function $c(x) = 0.2x^2 - 2x + 12$ is shown below. The vertex of the parabola is (5, 7). For any value of x less than 5, the cost per ton is greater than \$7. For any value of x greater than 5, the cost per ton is greater than \$7. 7 is the minimum value of the function, and the minimum value occurs when $x = 5$.

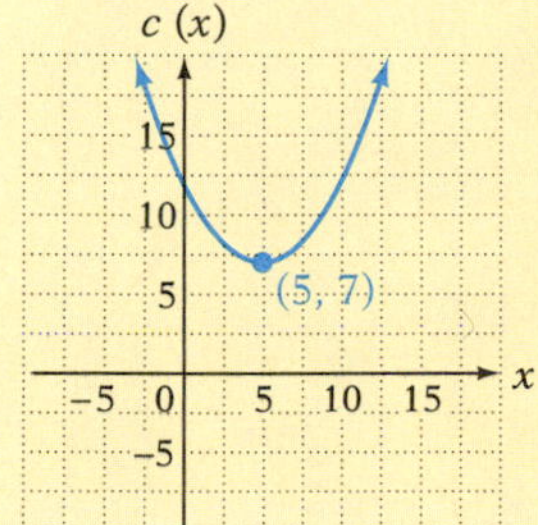

You Try It 7

The height s, in feet, of a ball thrown straight up is given by $s(t) = -16t^2 + 64t$, where t is the time in seconds. Find the time it takes the ball to reach its maximum height. What is the maximum height?

Your strategy

Your solution

Solution on p. S33

Copyright © Houghton Mifflin Company. All rights reserved.

Example 8

Find two numbers whose difference is 10 and whose product is a minimum. What is the minimum product of the two numbers?

Strategy

- Let x represent one number. Because the difference between the two numbers is 10,

$$x + 10$$

represents the other number.
[*Note:* $(x + 10) - (x) = 10$]
Then their product is represented by

$$x(x + 10) = x^2 + 10x$$

- To find one of the two numbers, find the x-coordinate of the vertex of $f(x) = x^2 + 10x$.
- To find the other number, replace x in $x + 10$ by the x-coordinate of the vertex and evaluate.
- To find the minimum product, evaluate the function at the x-coordinate of the vertex.

Solution

$f(x) = x^2 + 10x$ • $a = 1, b = 10, c = 0$

$x = -\frac{b}{2a} = -\frac{10}{2(1)} = -5$ • One number is −5.

$x + 10$
$-5 + 10 = 5$ • The other number is 5.

The numbers are −5 and 5.

$$\begin{aligned} f(x) &= x^2 + 10x \\ f(-5) &= (-5)^2 + 10(-5) \\ &= 25 - 50 \\ &= -25 \end{aligned}$$

The minimum product of the two numbers is −25.

You Try It 8

A rectangular fence is being constructed along a stream to enclose a picnic area. If there are 100 ft of fence available, what dimensions of the rectangle will produce the maximum area for picnicking?

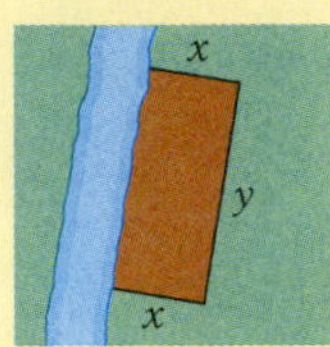

Your strategy

Your solution

Solution on p. S33

Copyright © Houghton Mifflin Company. All rights reserved.

11.1 Exercises

Objective A To graph a quadratic function

1. What is a quadratic function?

2. Describe the graph of a parabola.

3. What is the vertex of a parabola?

4. What is the axis of symmetry of the graph of a parabola?

5. The axis of symmetry of a parabola is the line $x = -5$. What is the x-coordinate of the vertex of the parabola?

6. The axis of symmetry of a parabola is the line $x = 8$. What is the x-coordinate of the vertex of the parabola?

7. The vertex of a parabola is $(7, -9)$. What is the equation of the axis of symmetry of the parabola?

8. The vertex of a parabola is $(-4, 10)$. What is the equation of the axis of symmetry of the parabola?

For Exercises 9 to 23, find the vertex and axis of symmetry of the parabola given by each equation. Then graph the equation.

9. $y = x^2 - 2x - 4$

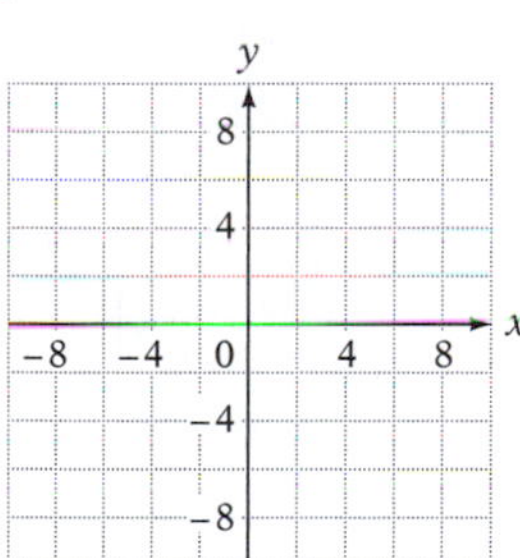

10. $y = x^2 + 4x - 4$

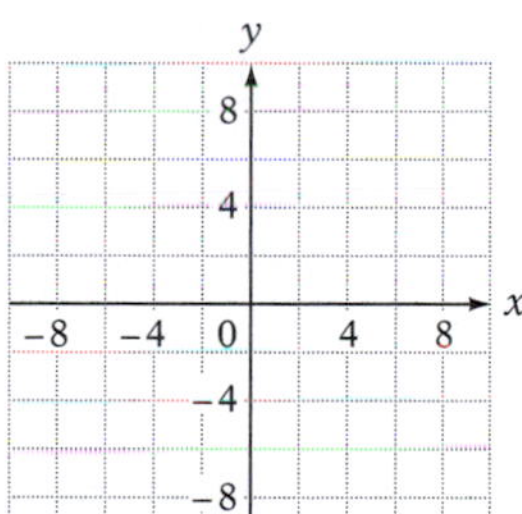

11. $y = -x^2 + 2x - 3$

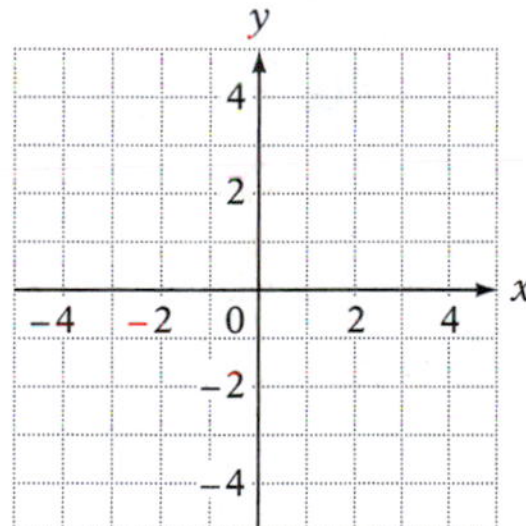

12. $y = -x^2 + 4x - 5$

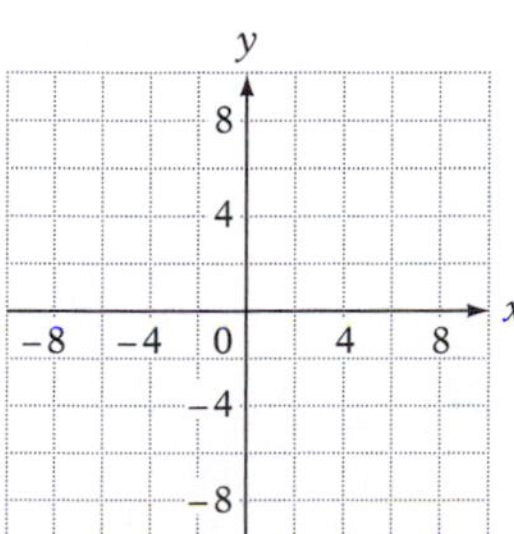

13. $f(x) = x^2 - x - 6$

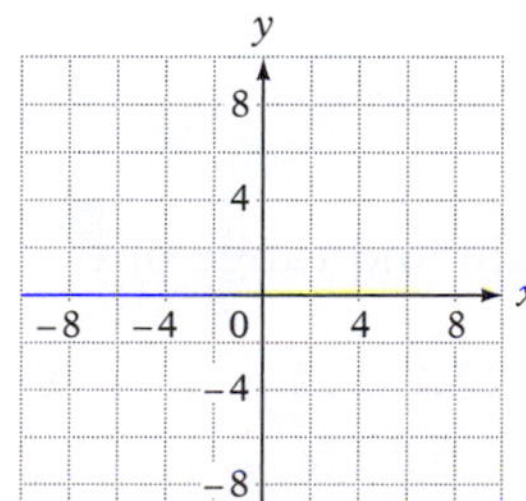

14. $G(x) = x^2 - x - 2$

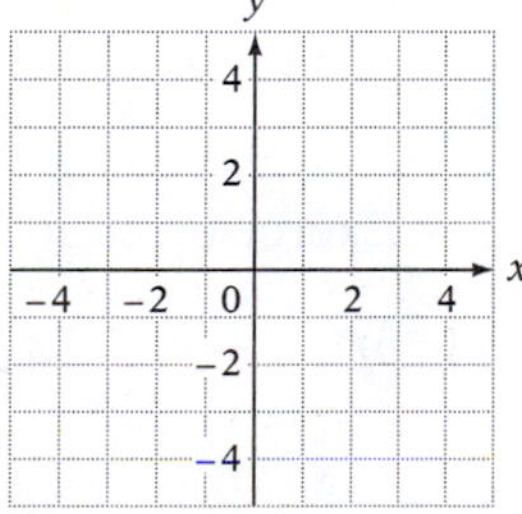

Copyright © Houghton Mifflin Company. All rights reserved.

15. $F(x) = x^2 - 3x + 2$

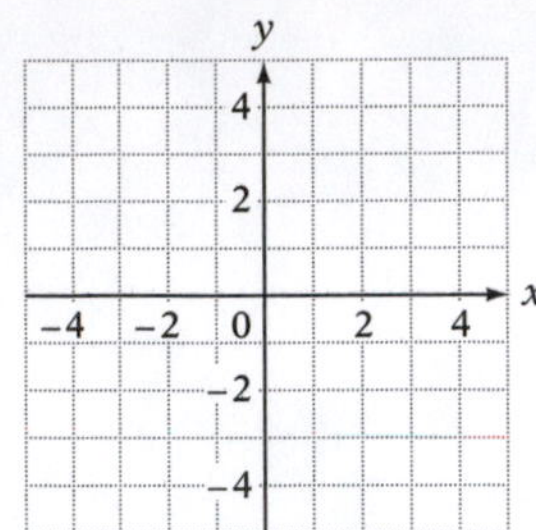

16. $y = 2x^2 - 4x + 1$

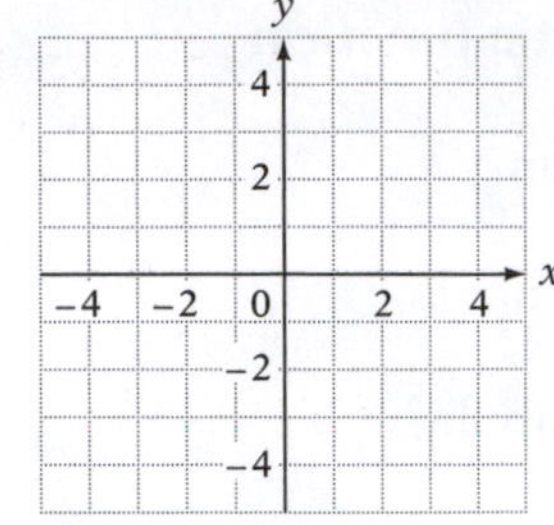

17. $y = -2x^2 + 6x$

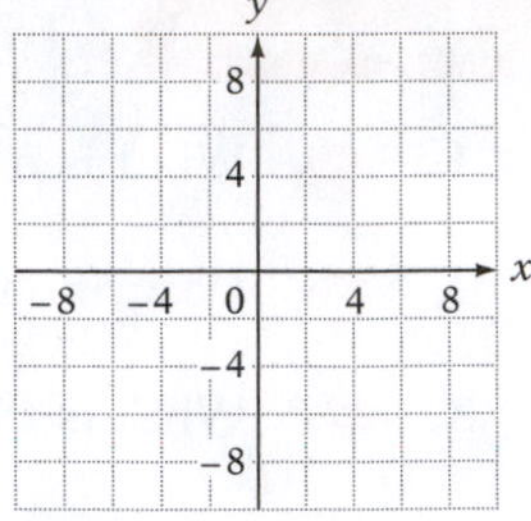

18. $y = \frac{1}{2}x^2 + 4$

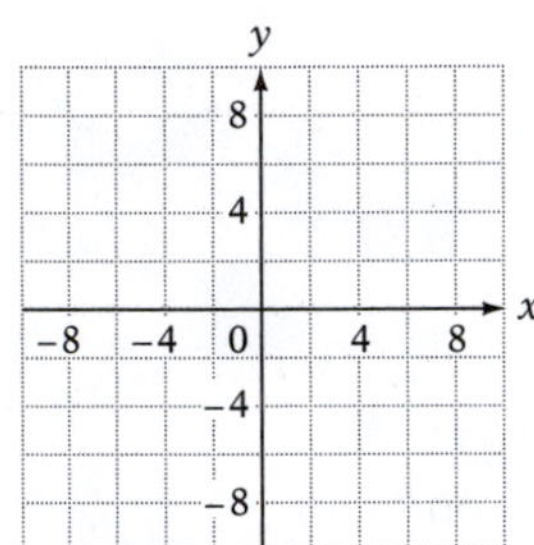

19. $y = -\frac{1}{4}x^2 - 1$

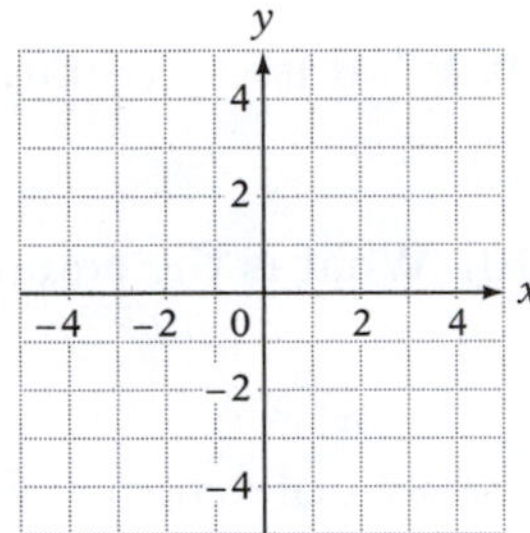

20. $h(x) = \frac{1}{2}x^2 - x + 1$

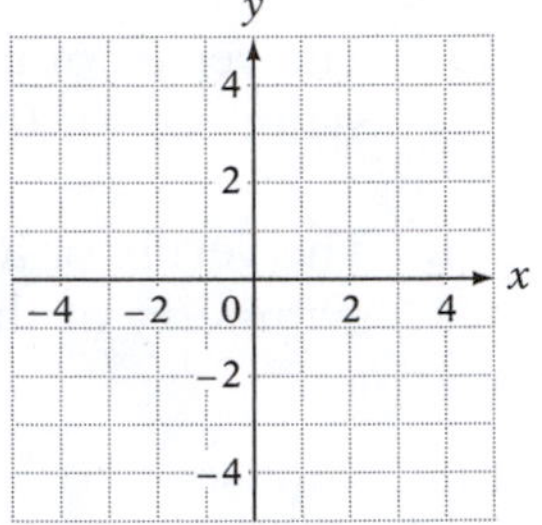

21. $P(x) = -\frac{1}{2}x^2 + 2x - 3$

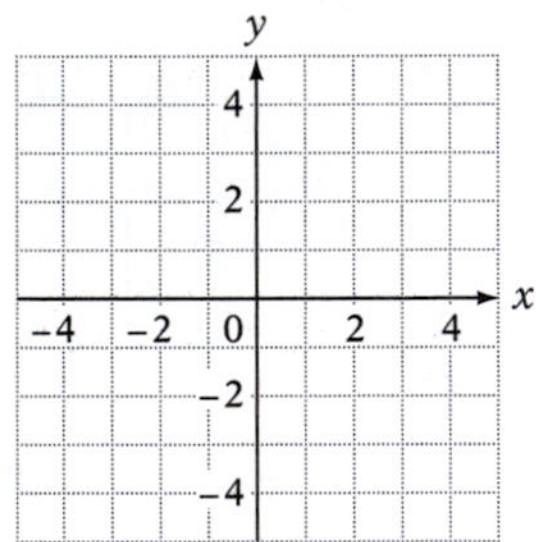

22. $y = \frac{1}{2}x^2 + 2x - 6$

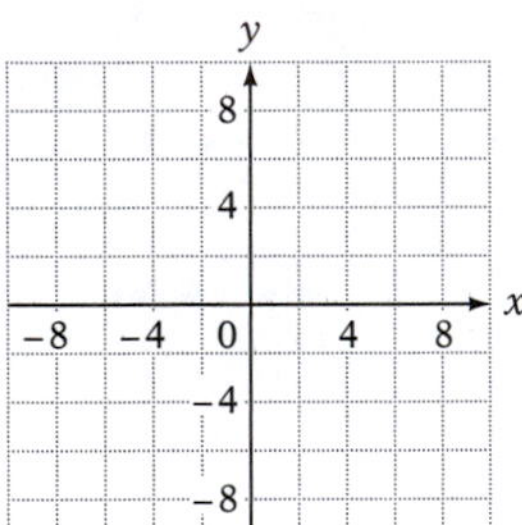

23. $y = -\frac{1}{2}x^2 + x - 3$

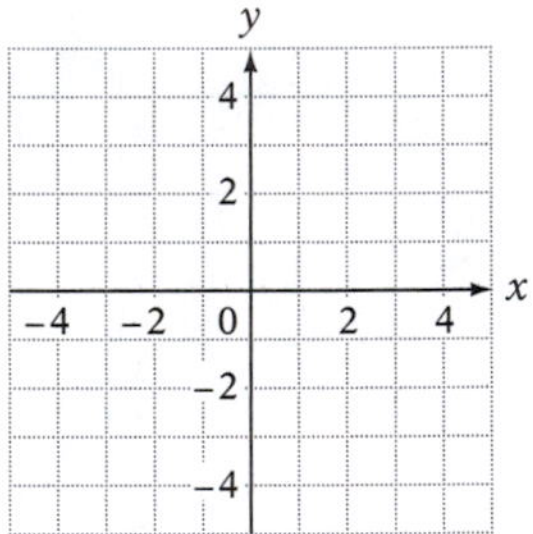

For Exercises 24 to 29, state the domain and range of the function.

24. $f(x) = 2x^2 - 4x - 5$

25. $f(x) = 2x^2 + 8x + 3$

26. $f(x) = -2x^2 - 3x + 2$

27. $f(x) = -x^2 + 6x - 9$

28. $f(x) = -x^2 - 4x - 5$

29. $f(x) = x^2 + 4x - 3$

Copyright © Houghton Mifflin Company. All rights reserved.

Objective B To find the x-intercepts of a parabola

30. **a.** What is an x-intercept of the graph of a parabola?
b. How many x-intercepts can the graph of a parabola have?

31. **a.** What is the y-intercept of the graph of a parabola?
b. How many y-intercepts can the graph of a parabola have?

For Exercises 32 to 43, find the x-intercepts of the parabola given by each equation.

32. $y = x^2 - 4$

33. $y = x^2 - 9$

34. $y = 2x^2 - 4x$

35. $y = 3x^2 + 6x$

36. $y = x^2 - x - 2$

37. $y = x^2 - 2x - 8$

38. $y = 2x^2 - x - 1$

39. $y = 2x^2 - 5x - 3$

40. $y = x^2 + 2x - 1$

41. $y = x^2 + 4x - 3$

42. $y = x^2 + 6x + 10$

43. $y = -x^2 - 4x - 5$

For Exercises 44 to 55, find the zeros of the function.

44. $f(x) = x^2 + 3x + 2$

45. $f(x) = x^2 - 6x + 9$

46. $f(x) = -x^2 + 4x - 5$

47. $f(x) = -x^2 + 3x + 8$

48. $f(x) = 2x^2 - 3x$

49. $f(x) = -3x^2 + 4x$

50. $f(x) = 2x^2 - 4$

51. $f(x) = 3x^2 + 6$

52. $f(x) = 2x^2 + 3x + 2$

53. $f(x) = 3x^2 - x + 4$

54. $f(x) = -3x^2 - 4x + 1$

55. $f(x) = -2x^2 + x + 5$

For Exercises 56 to 67, use the discriminant to determine the number of x-intercepts of the graph of the equation.

56. $y = 2x^2 + 2x - 1$

57. $y = -x^2 - x + 3$

58. $y = x^2 - 8x + 16$

Copyright © Houghton Mifflin Company. All rights reserved.

59. $y = x^2 - 10x + 25$

60. $y = -3x^2 - x - 2$

61. $y = -2x^2 + x - 1$

62. $y = -2x^2 + x + 1$

63. $y = 4x^2 - x - 2$

64. $y = 2x^2 + x + 1$

65. $y = 2x^2 + x + 4$

66. $y = -3x^2 + 2x - 8$

67. $y = 4x^2 + 2x - 5$

68. The zeros of the function $f(x) = x^2 - 2x - 3$ are -1 and 3. What are the x-intercepts of the graph of the equation $y = x^2 - 2x - 3$?

69. The zeros of the function $f(x) = x^2 - x - 20$ are -4 and 5. What are the x-intercepts of the graph of the equation $y = x^2 - x - 20$?

Objective C To find the minimum or maximum of a quadratic function

70. What is the minimum or maximum value of a quadratic function?

71. Describe how to find the minimum or maximum value of a quadratic function.

72. Does the function have a minimum or a maximum value?
a. $f(x) = -x^2 + 6x - 1$ **b.** $f(x) = 2x^2 - 4$ **c.** $f(x) = -5x^2 + x$

73. Does the function have a minimum or a maximum value?
a. $f(x) = 3x^2 - 2x + 4$ **b.** $f(x) = -x^2 + 9$ **c.** $f(x) = 6x^2 - 3x$

For Exercises 74 to 85, find the minimum or maximum value of the quadratic function.

74. $f(x) = x^2 - 2x + 3$

75. $f(x) = 2x^2 + 4x$

76. $f(x) = -2x^2 + 4x - 3$

77. $f(x) = -2x^2 + 4x - 5$

78. $f(x) = -2x^2 - 3x + 4$

79. $f(x) = -2x^2 - 3x$

80. $f(x) = 2x^2 + 3x - 8$

81. $f(x) = 3x^2 + 3x - 2$

82. $f(x) = -3x^2 + x - 6$

83. $f(x) = -x^2 - x + 2$

84. $f(x) = x^2 - 5x + 3$

85. $f(x) = 3x^2 + 5x + 2$

Copyright © Houghton Mifflin Company. All rights reserved.

86. Which of the following parabolas has the greatest minimum value?
a. $y = x^2 - 2x - 3$ **b.** $y = x^2 - 10x + 20$ **c.** $y = 3x^2 - 6$

87. Which of the following parabolas has the greatest maximum value?
a. $y = -2x^2 + 2x - 1$ **b.** $y = -x^2 + 8x - 2$ **c.** $y = -4x^2 + 3$

88. The vertex of a parabola that opens up is $(-4, 7)$. Does the function have a maximum or a minimum value? What is the maximum or minimum value of the function?

89. The vertex of a parabola that opens down is $(3, -5)$. Does the function have a maximum or a minimum value? What is the maximum or minimum value of the function?

Objective D To solve application problems

90. **Physics** The height s, in feet, of a rock thrown upward at an initial speed of 64 ft/s from a cliff 50 ft above an ocean beach is given by the function $s(t) = -16t^2 + 64t + 50$, where t is the time in seconds. Find the maximum height above the beach that the rock will attain.

91. **Business** A manufacturer of microwave ovens believes that the revenue R, in dollars, the company receives is related to the price P, in dollars, of an oven by the function $R(P) = 125P - 0.25P^2$. What price will give the maximum revenue?

92. **Business** A tour operator believes that the profit P, in dollars, from selling x tickets is given by $P(x) = 40x - 0.25x^2$. Using this model, what is the maximum profit the tour operator can expect?

93. **The Olympics** An event in the Summer Olympics is 10-meter springboard diving. In this event, the height s, in meters, of a diver above the water t seconds after jumping is given by $s(t) = -4.9t^2 + 7.8t + 10$. What is the maximum height that the diver will be above the water? Round to the nearest tenth.

94. **Chemistry** A pool is treated with a chemical to reduce the amount of algae in the pool. The amount of algae in the pool t days after the treatment can be approximated by the function $A(t) = 40t^2 - 400t + 500$. How many days after the treatment will the pool have the least amount of algae?

95. **Construction** The suspension cable that supports a small footbridge hangs in the shape of a parabola. The height h, in feet, of the cable above the bridge is given by the function $h(x) = 0.25x^2 - 0.8x + 25$, where x is the distance in feet from one end of the bridge. What is the minimum height of the cable above the bridge?

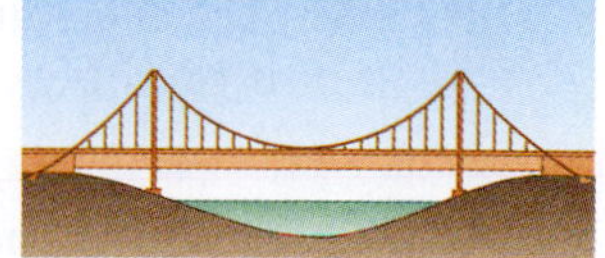

Copyright © Houghton Mifflin Company. All rights reserved.

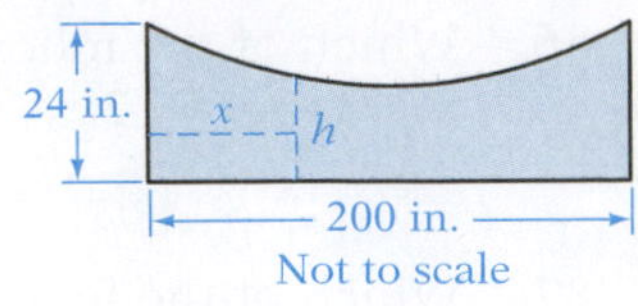

96. **Telescopes** An equation that models the thickness h, in inches, of the mirror in the telescope at the Palomar Mountain Observatory is given by $h(x) = 0.000379x^2 - 0.0758x + 24$, where x is measured in inches from the edge of the mirror. Find the minimum thickness of the mirror.

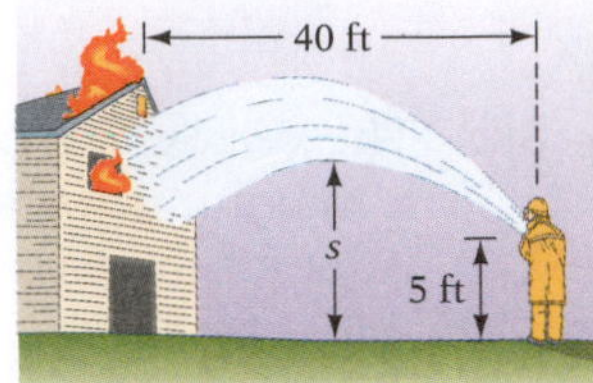

97. **Fire Science** The height s, in feet, of water squirting from a certain fire hose nozzle is given by the equation $s(x) = -\frac{1}{30}x^2 + 2x + 5$, where x is the horizontal distance, in feet, from the nozzle. How high on a building 40 ft from the fire hose will the water land?

98. **Fountains** The Buckingham Fountain in Chicago shoots water from a nozzle at the base of the fountain. The height h, in feet, of the water above the ground t seconds after it leaves the nozzle is given by the function $h(t) = -16t^2 + 90t + 15$. What is the maximum height of the water? Round to the nearest tenth.

99. **Safety** On wet concrete, the stopping distance s, in feet, of a car traveling v miles per hour is given by the function $s(v) = 0.055v^2 + 1.1v$. What is the maximum speed at which a car could be traveling and still stop at a stop sign 44 ft away?

100. **Sports** Some football fields are built in a parabolic mound shape so that water will drain off the field. A model for the shape of such a field is given by $h(x) = -0.00023475x^2 + 0.0375x$, where h is the height of the field, in feet, at a distance of x feet from the sideline. What is the maximum height? Round to the nearest tenth.

101. **Mathematics** Find two numbers whose sum is 20 and whose product is a maximum.

102. **Mathematics** Find two numbers whose difference is 14 and whose product is a minimum.

103. **Ranching** A rancher has 200 ft of fencing to build a rectangular corral alongside an existing fence. Determine the dimensions of the corral that will maximize the enclosed area.

APPLYING THE CONCEPTS

104. One root of the quadratic equation $2x^2 - 5x + k = 0$ is 4. What is the other root?

105. What is the value of k if the vertex of the parabola $y = x^2 - 8x + k$ is a point on the x-axis?

106. The roots of the function $f(x) = mx^2 + nx + 1$ are -2 and 3. What are the roots of the function $g(x) = nx^2 + mx - 1$?

Copyright © Houghton Mifflin Company. All rights reserved.

11.2 Graphs of Functions

Objective A To graph functions

The graphs of the polynomial functions $L(x) = mx + b$ (a straight line) and $Q(x) = ax^2 + bx + c$, $a \neq 0$ (a parabola) were discussed in previous objectives. The graphs of other functions can also be drawn by finding ordered pairs that belong to the function, plotting the points that correspond to the ordered pairs, and then drawing a curve through the points.

> **TAKE NOTE**
> The units along the x-axis are different from those along the y-axis. If the units along the x-axis were the same as those along the y-axis, the graph would appear narrower than the one shown.

HOW TO Graph: $F(x) = x^3$

Select several values of x and evaluate the function.

x	$F(x) = x^3$	$F(x)$	(x, y)
-2	$(-2)^3$	-8	$(-2, -8)$
-1	$(-1)^3$	-1	$(-1, -1)$
0	0^3	0	$(0, 0)$
1	1^3	1	$(1, 1)$
2	2^3	8	$(2, 8)$

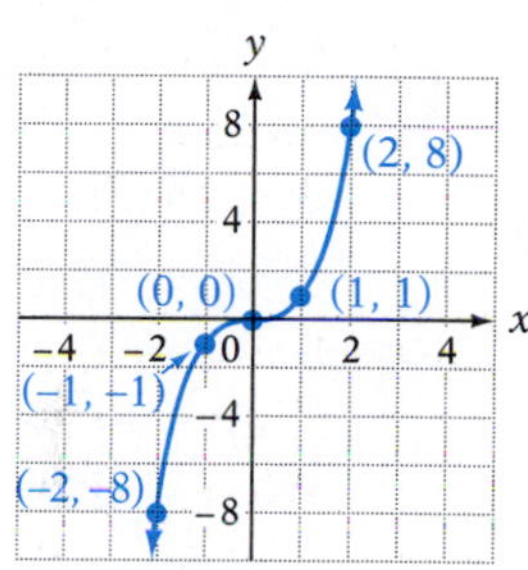

Plot the ordered pairs and draw a graph through the points.

HOW TO Graph: $g(x) = x^3 - 4x + 5$

Select several values of x and evaluate the function.

x	$g(x) = x^3 - 4x + 5$	$g(x)$	(x, y)
-3	$(-3)^3 - 4(-3) + 5$	-10	$(-3, -10)$
-2	$(-2)^3 - 4(-2) + 5$	5	$(-2, 5)$
-1	$(-1)^3 - 4(-1) + 5$	8	$(-1, 8)$
0	$(0)^3 - 4(0) + 5$	5	$(0, 5)$
1	$(1)^3 - 4(1) + 5$	2	$(1, 2)$
2	$(2)^3 - 4(2) + 5$	5	$(2, 5)$

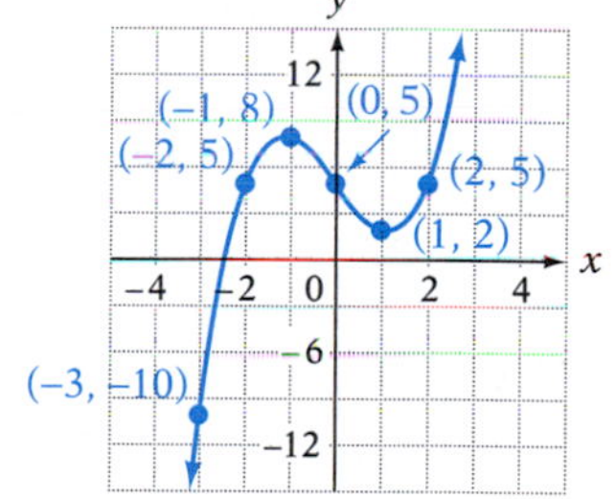

Plot the ordered pairs and draw a graph through the points.

Note from the graphs of the two cubic functions above that the shapes of the graphs can be quite different. The following graphs of typical cubic polynomial functions show the general shape of a cubic polynomial.

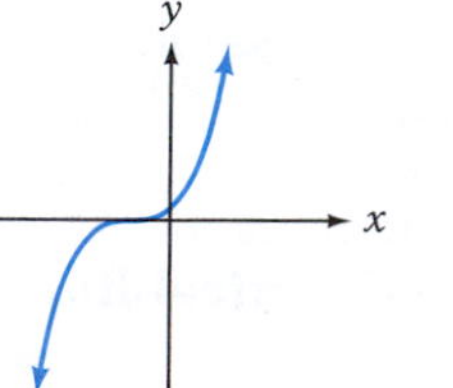

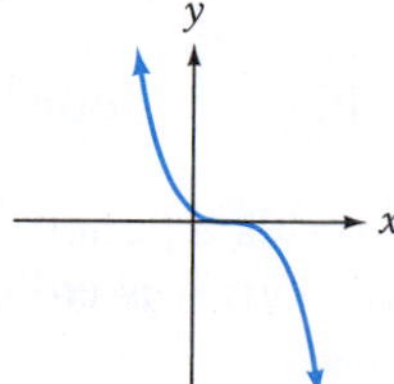

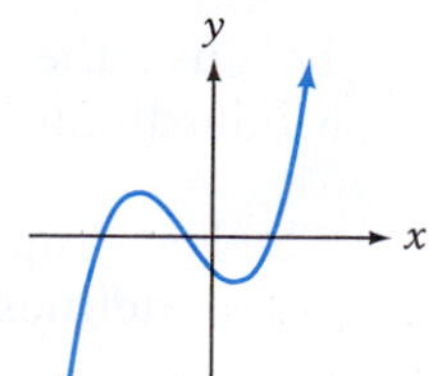

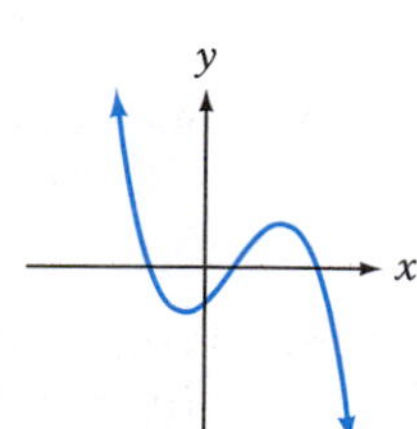

As the degree of a polynomial increases, the graph of the polynomial function can change significantly. In such cases, it may be necessary to plot many points before an accurate graph can be drawn. Only polynomials of degree 3 are considered here.

Copyright © Houghton Mifflin Company. All rights reserved.

Integrating Technology

See the appendix Keystroke Guide: *Graph* for instructions on using a graphing calculator to graph a function.

HOW TO Graph: $f(x) = |x + 2|$

This is an **absolute-value function.**

x	$f(x) = \|x + 2\|$	$f(x)$	(x, y)
-3	$\|-3 + 2\|$	1	$(-3, 1)$
-2	$\|-2 + 2\|$	0	$(-2, 0)$
-1	$\|-1 + 2\|$	1	$(-1, 1)$
0	$\|0 + 2\|$	2	$(0, 2)$
1	$\|1 + 2\|$	3	$(1, 3)$
2	$\|2 + 2\|$	4	$(2, 4)$

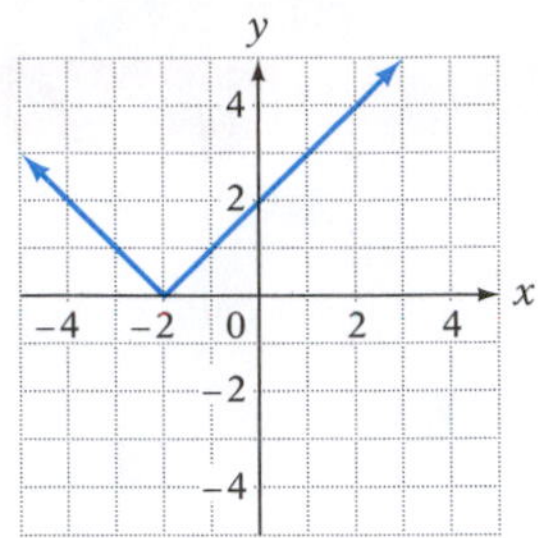

In general, the graph of the absolute value of a linear polynomial is V-shaped.

HOW TO Graph: $R(x) = \sqrt{2x - 4}$

This is a **radical function.** Because the square root of a negative number is not a real number, the domain of this function requires that $2x - 4 \geq 0$. Solve this inequality for x.

$$2x - 4 \geq 0$$
$$2x \geq 4$$
$$x \geq 2$$

The domain is $\{x \mid x \geq 2\}$. This means that only values of x that are greater than or equal to 2 can be chosen as values at which to evaluate the function. In this case, some of the y-coordinates must be approximated.

x	$R(x) = \sqrt{2x - 4}$	$R(x)$	(x, y)
2	$\sqrt{2(2) - 4}$	0	$(2, 0)$
3	$\sqrt{2(3) - 4}$	1.41	$(3, 1.41)$
4	$\sqrt{2(4) - 4}$	2	$(4, 2)$
5	$\sqrt{2(5) - 4}$	2.45	$(5, 2.45)$
6	$\sqrt{2(6) - 4}$	2.83	$(6, 2.83)$

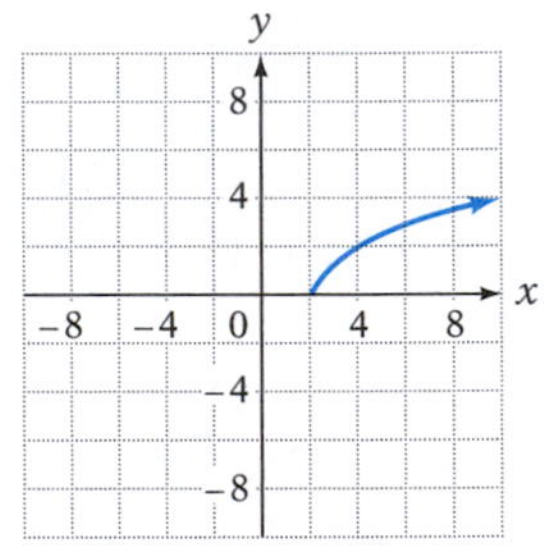

Recall that a function is a special type of relation, one for which no two ordered pairs have the same first coordinate. Graphically, this means that the graph of a function cannot pass through two points that have the same x-coordinate and different y-coordinates. For instance, the graph at the right is not the graph of a function because there are ordered pairs with the same x-coordinate and different y-coordinates.

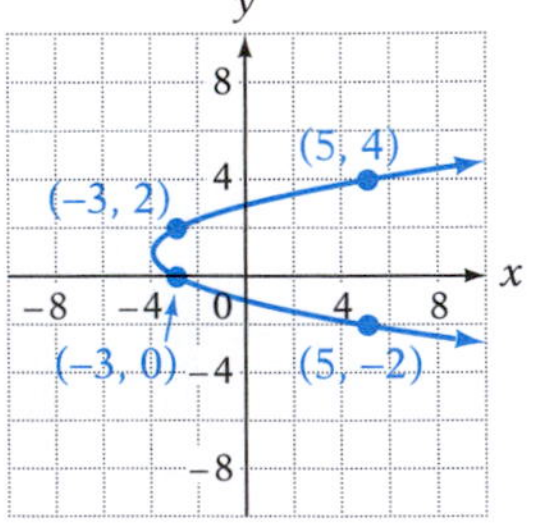

The last graph illustrates a general statement that can be made about whether a graph defines a function. It is called the **vertical-line test.**

Vertical-Line Test

A graph defines a function if any vertical line intersects the graph at no more than one point.

Copyright © Houghton Mifflin Company. All rights reserved.

For example, the graph of a nonvertical straight line is the graph of a function. Any vertical line intersects the graph no more than once. The graph of a circle, however, is not the graph of a function. There are vertical lines that intersect the graph at more than one point.

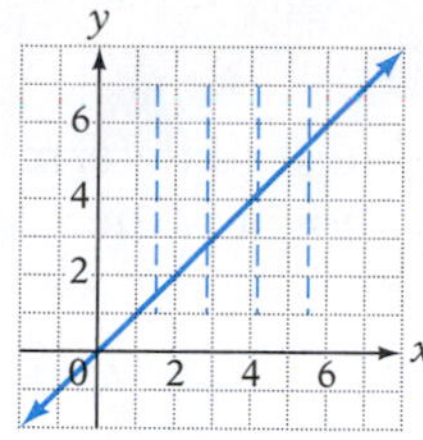

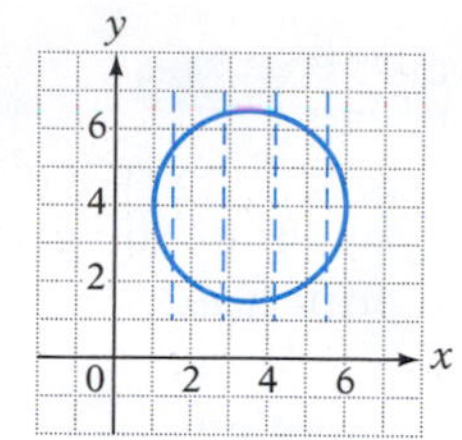

Some practical applications can be modeled by a graph that is not the graph of a function. Below is an example of such an application.

One of the causes of smog is an inversion layer of air where temperatures at higher altitudes are warmer than those at lower altitudes. The graph at the right shows the altitudes at which various temperatures were recorded. As shown by the dashed lines in the graph, there are two altitudes at which the temperature was 25°C. This means that there are two ordered pairs (shown in the graph) with the same first coordinate but different second coordinates. The graph does not define a function.

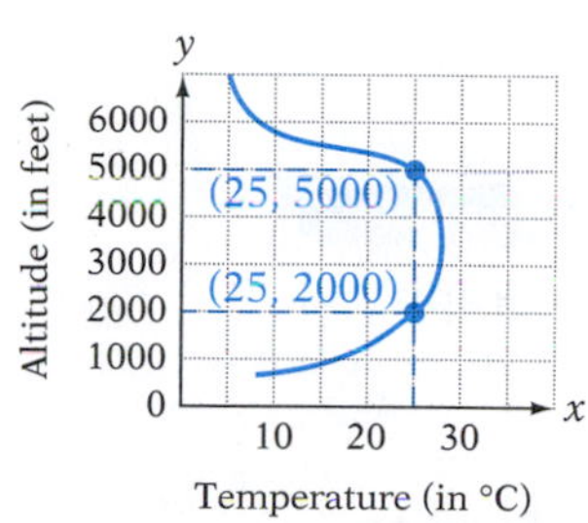

When a graph does define a function, the domain and range can be estimated from the graph.

> **TAKE NOTE**
> To determine the domain, think of collapsing the graph onto the x-axis and determining the interval on the x-axis that the graph covers. To determine the range, think of collapsing the graph onto the y-axis and determining the interval on the y-axis that the graph covers.

HOW TO Determine the domain and range of the function given by the graph at the right. Write the answer in set-builder notation.

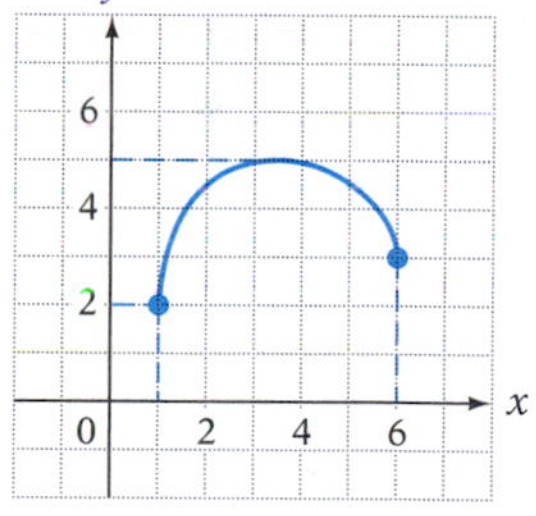

The solid dots on the graph indicate its beginning and ending points.

The domain is the set of x-coordinates.
Domain: $\{x \mid 1 \le x \le 6\}$

The range is the set of y-coordinates.
Range: $\{y \mid 2 \le y \le 5\}$

HOW TO Determine the domain and range of the function given by the graph at the right. Write the answer in set-builder notation.

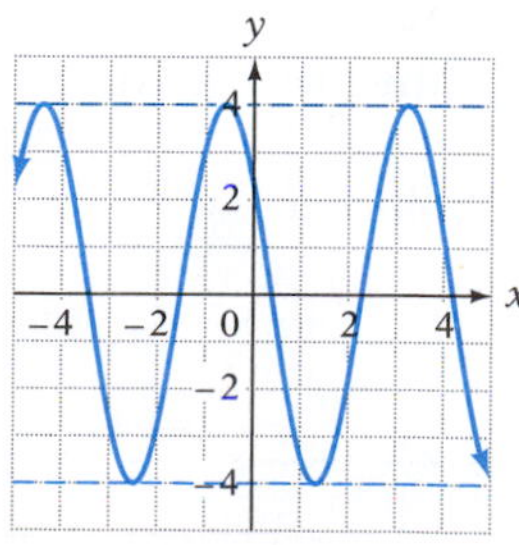

The arrows on the graph indicate that the graph continues in the same manner.

The domain is the set of x-coordinates.
Domain: $\{x \mid x \in \text{real numbers}\}$

The range is the set of y-coordinates.
Range: $\{y \mid -4 \le y \le 4\}$

Copyright © Houghton Mifflin Company. All rights reserved.

Example 1

Use the vertical-line test to determine whether the graph shown is the graph of a function.

Solution

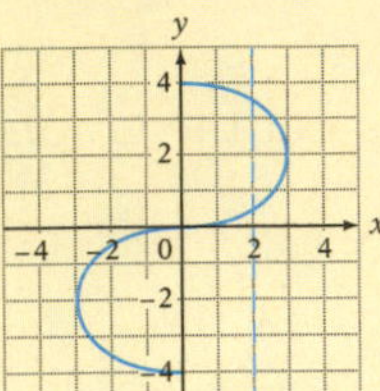

A vertical line intersects the graph more than once. The graph is not the graph of a function.

You Try It 1

Use the vertical-line test to determine whether the graph shown is the graph of a function.

Your solution

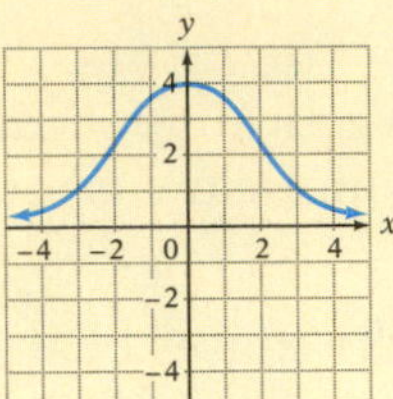

Example 2

Graph $f(x) = x^3 - 3x$. State the domain and range of the function.

Solution

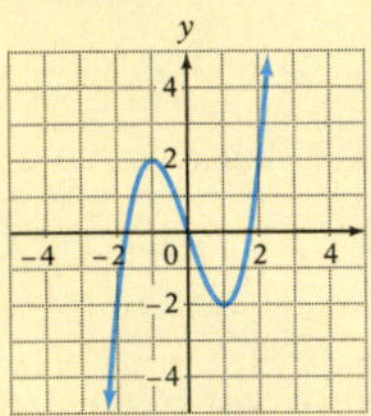

Domain: $\{x|x \in \text{real numbers}\}$
Range: $\{y|y \in \text{real numbers}\}$

You Try It 2

Graph $f(x) = -\frac{1}{2}x^3 + 2x$. State the domain and range of the function.

Your solution

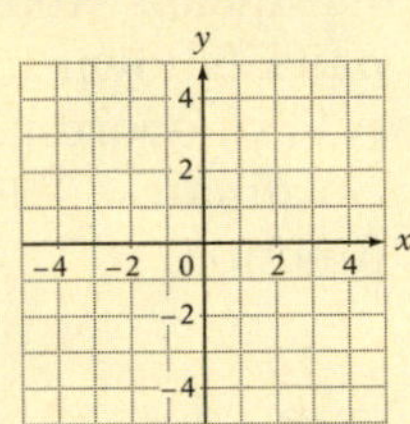

Example 3

Graph $f(x) = |x| + 2$. State the domain and range of the function.

Solution

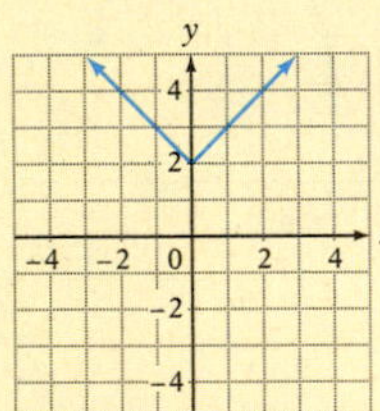

Domain: $\{x|x \in \text{real numbers}\}$
Range: $\{y|y \geq 2\}$

You Try It 3

Graph $f(x) = |x - 3|$. State the domain and range of the function.

Your solution

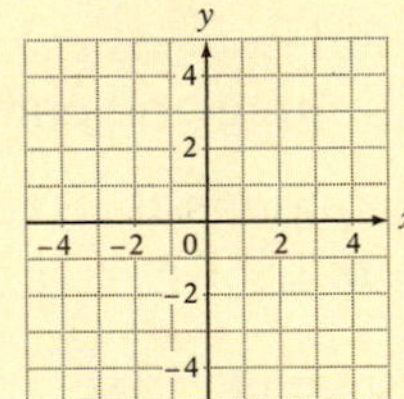

Example 4

Graph $f(x) = \sqrt{2 - x}$. State the domain and range of the function.

Solution

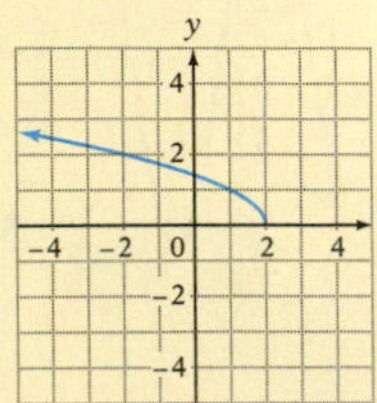

Domain: $\{x|x \leq 2\}$
Range: $\{y|y \geq 0\}$

You Try It 4

Graph $f(x) = -\sqrt{x - 1}$. State the domain and range of the function.

Your solution

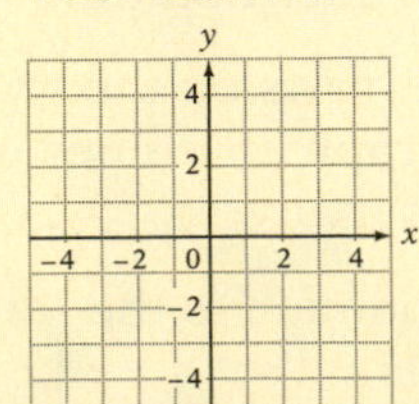

Solutions on pp. S33–S34

Copyright © Houghton Mifflin Company. All rights reserved.

11.2 Exercises

Objective A To graph functions

For Exercises 1 to 6, use the vertical-line test to determine whether the graph is the graph of a function.

1.

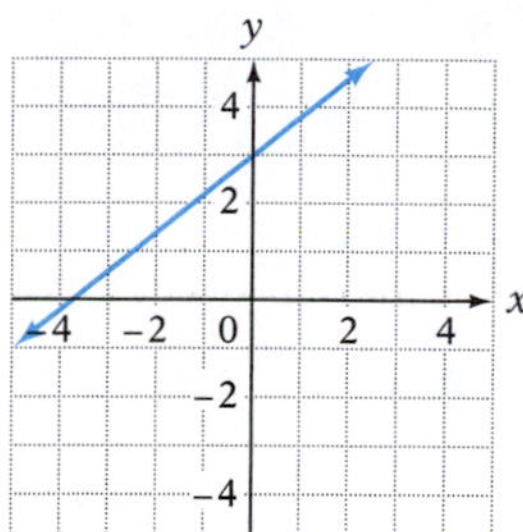

2.

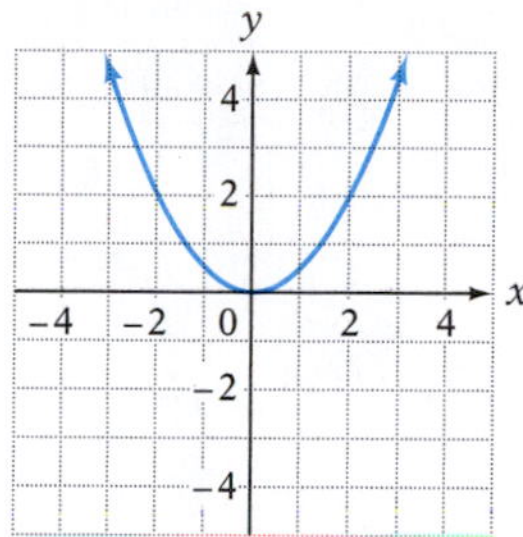

3.

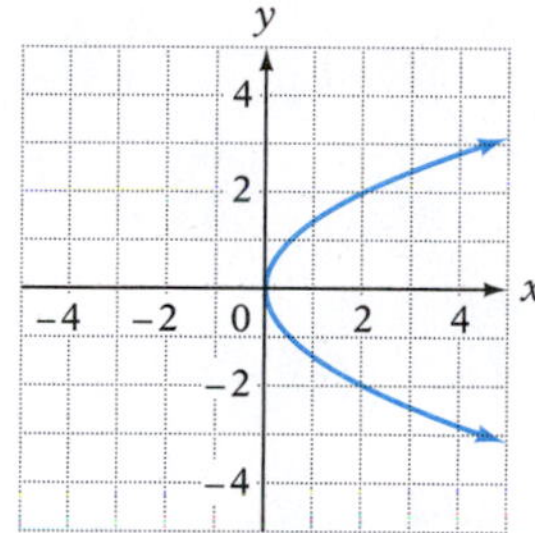

4.

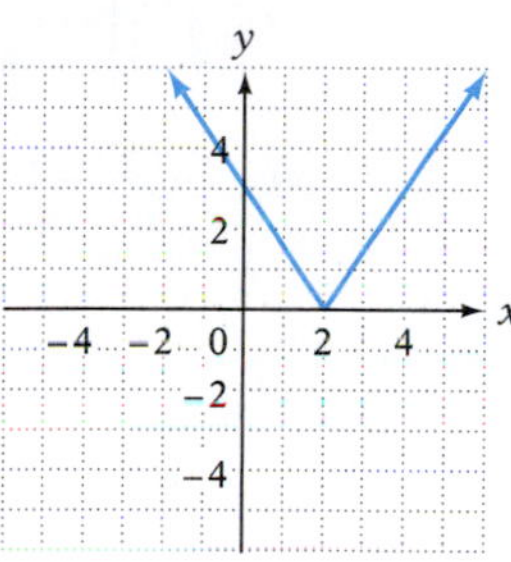

5.

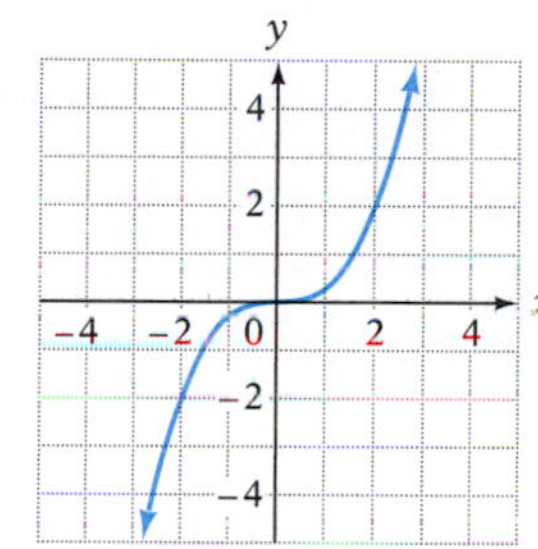

6.

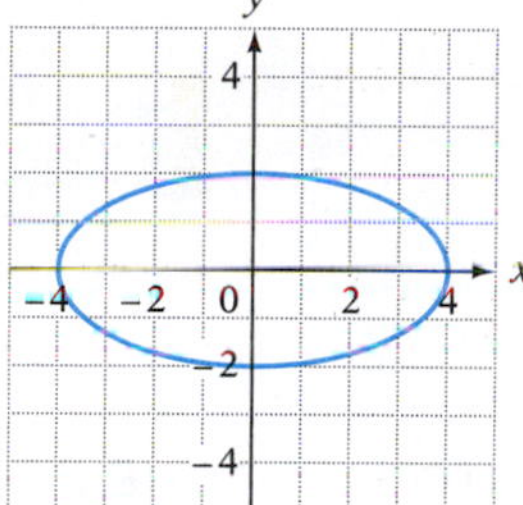

For Exercises 7 to 18, graph the function and state its domain and range.

7. $f(x) = 3|2 - x|$

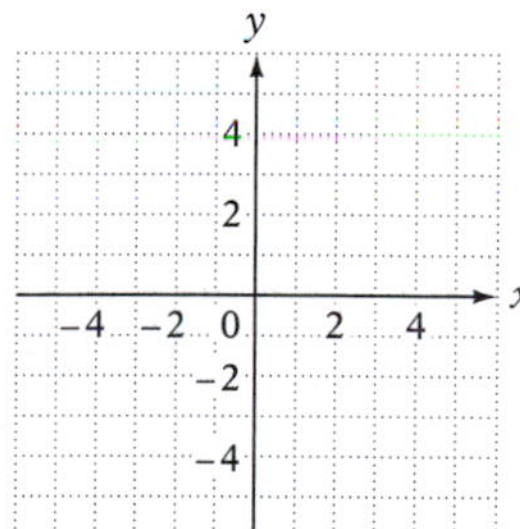

8. $f(x) = x^3 - 1$

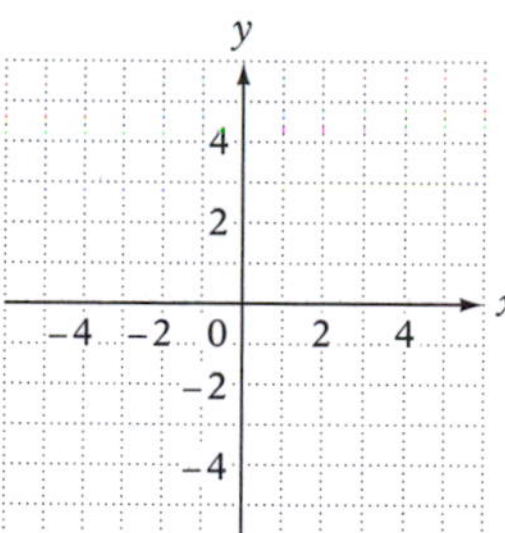

9. $f(x) = 1 - x^3$

10. $f(x) = \sqrt{1 + x}$

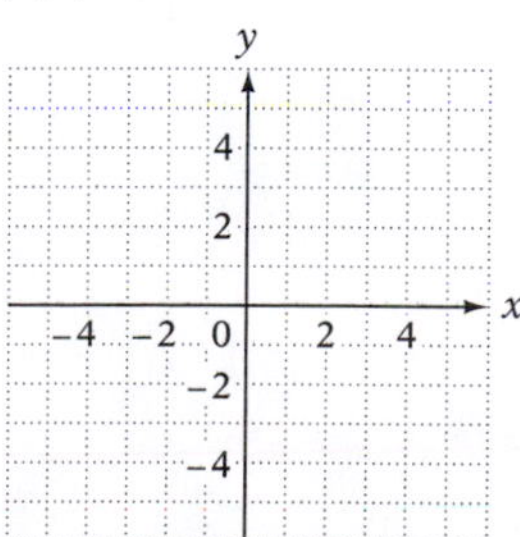

11. $f(x) = \sqrt{4 - x}$

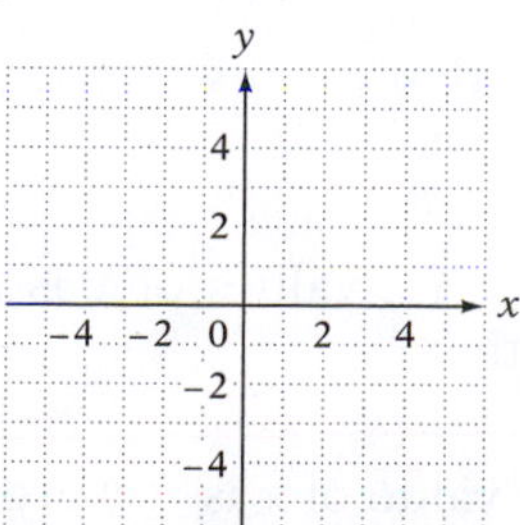

12. $f(x) = |2x - 1|$

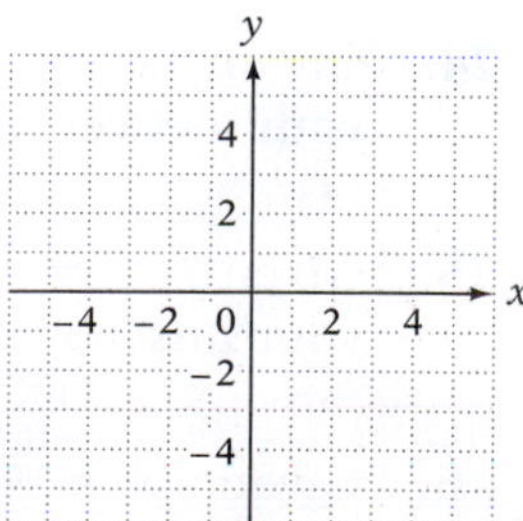

Copyright © Houghton Mifflin Company. All rights reserved.

13. $f(x) = x^3 + 4x^2 + 4x$

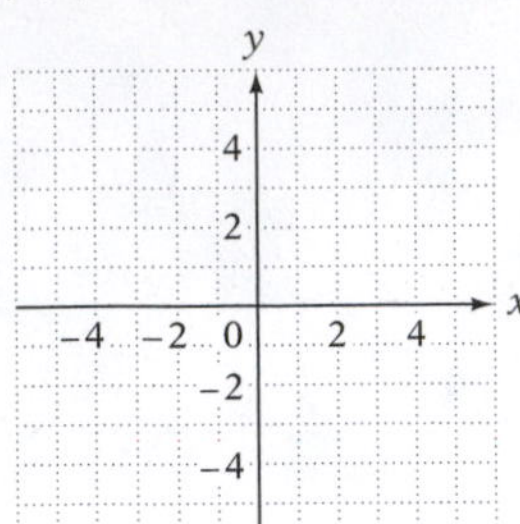

14. $f(x) = x^3 - x^2 - x + 1$

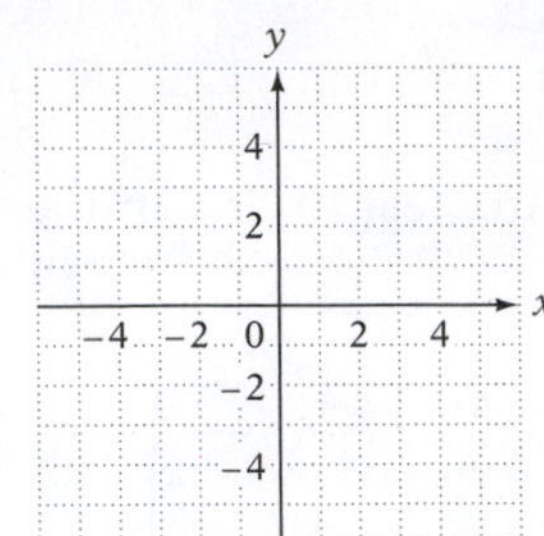

15. $f(x) = -\sqrt{x + 2}$

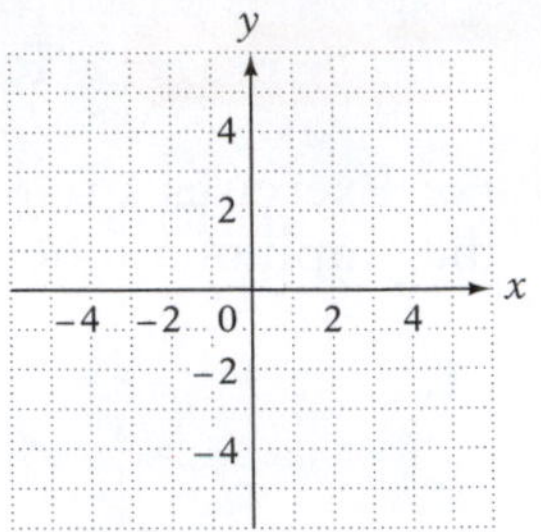

16. $f(x) = -\sqrt{x - 3}$

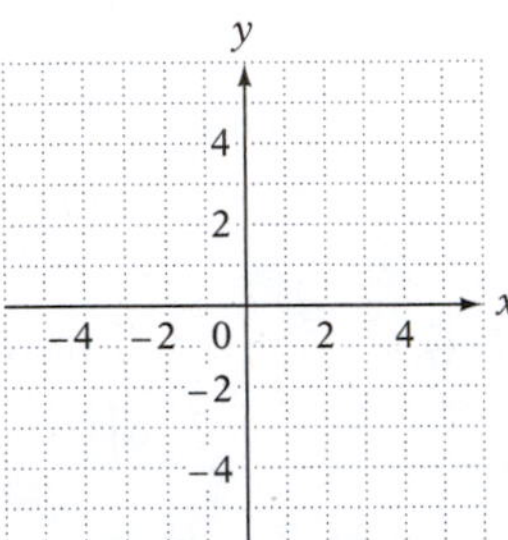

17. $f(x) = |2x + 2|$

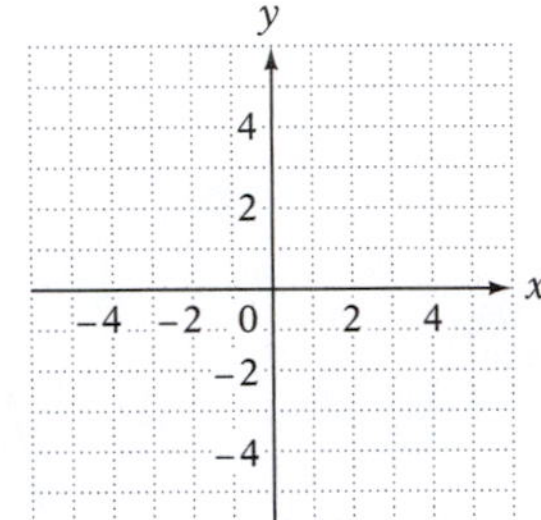

18. $f(x) = 2|x + 1|$

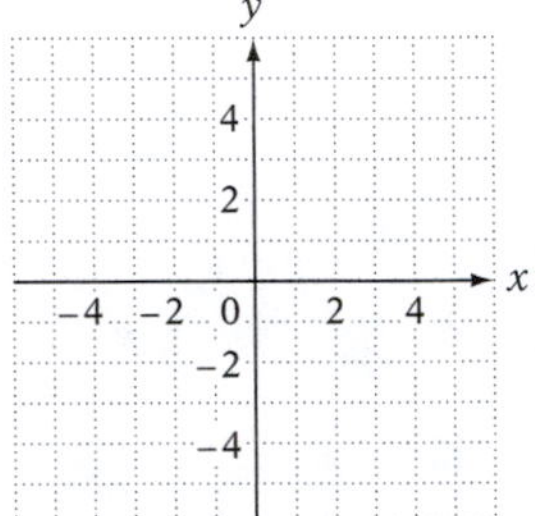

APPLYING THE CONCEPTS

19. If $f(x) = \sqrt{x - 2}$ and $f(a) = 4$, find a.

20. If $f(x) = \sqrt{x + 5}$ and $f(a) = 3$, find a.

21. $f(a, b) =$ the sum of a and b
$g(a, b) =$ the product of a and b
Find $f(2, 5) + g(2, 5)$.

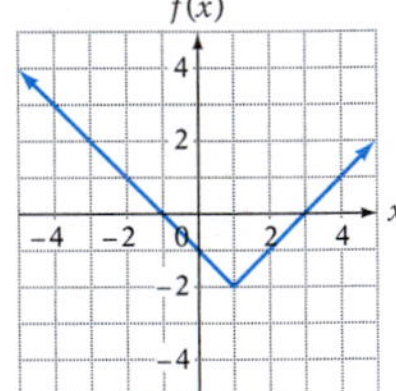

22. The graph of the function f is shown at the right. For this function, which of the following are true?

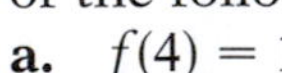

a. $f(4) = 1$ **b.** $f(0) = 3$ **c.** $f(-3) = 2$

23. Let $f(x)$ be the digit in the xth decimal place of the repeating decimal $0.\overline{387}$. For example, $f(3) = 7$ because 7 is the digit in the third decimal place. Find $f(14)$.

24. Given $f(x) = (x + 1)(x - 1)$, for what values of x is $f(x)$ negative? Write your answer in set-builder notation.

25. Given $f(x) = (x + 2)(x - 2)$, for what values of x is $f(x)$ negative? Write your answer in set-builder notation.

26. Given $f(x) = -|x + 3|$, for what value of x is $f(x)$ greatest?

27. Given $f(x) = |2x - 2|$, for what value of x is $f(x)$ smallest?

Copyright © Houghton Mifflin Company. All rights reserved.

11.3 Algebra of Functions

Objective A To perform operations on functions

The operations of addition, subtraction, multiplication, and division of functions are defined as follows.

Operations on Functions

If f and g are functions and x is an element of the domain of each function, then

$(f + g)(x) = f(x) + g(x)$ $\qquad$ $(f \cdot g)(x) = f(x) \cdot g(x)$

$(f - g)(x) = f(x) - g(x)$ $\qquad$ $\left(\dfrac{f}{g}\right)(x) = \dfrac{f(x)}{g(x)}, g(x) \neq 0$

HOW TO Given $f(x) = x^2 + 1$ and $g(x) = 3x - 2$, find $(f + g)(3)$ and $(f \cdot g)(-1)$.

$$\begin{aligned}(f + g)(3) &= f(3) + g(3)\\ &= [(3)^2 + 1] + [3(3) - 2]\\ &= 10 + 7 = 17\end{aligned}$$

$$\begin{aligned}(f \cdot g)(-1) &= f(-1) \cdot g(-1)\\ &= [(-1)^2 + 1] \cdot [3(-1) - 2]\\ &= 2 \cdot (-5) = -10\end{aligned}$$

Consider the functions f and g from the last example. Let $S(x)$ be the sum of the two functions. Then

$S(x) = (f + g)(x) = f(x) + g(x)$ • The definition of addition of functions

$\quad = [x^2 + 1] + [3x - 2]$ • $f(x) = x^2 + 1, g(x) = 3x - 2$

$S(x) = x^2 + 3x - 1$

Now evaluate $S(3)$.

$$\begin{aligned}S(x) &= x^2 + 3x - 1\\ S(3) &= (3)^2 + 3(3) - 1\\ &= 9 + 9 - 1\\ &= 17 = (f + g)(3)\end{aligned}$$

Note that $S(3) = 17$ and $(f + g)(3) = 17$. This shows that adding $f(x) + g(x)$ and then evaluating is the same as evaluating $f(x)$ and $g(x)$ and then adding. The same is true for the other operations on functions. For instance, let $P(x)$ be the product of the functions f and g. Then

$$\begin{aligned}P(x) &= (f \cdot g)(x) = f(x) \cdot g(x)\\ &= (x^2 + 1)(3x - 2)\\ &= 3x^3 - 2x^2 + 3x - 2\\ P(-1) &= 3(-1)^3 - 2(-1)^2 + 3(-1) - 2\\ &= -3 - 2 - 3 - 2\\ &= -10\end{aligned}$$

• Note that $P(-1) = -10$ and $(f \cdot g)(-1) = -10$.

Copyright © Houghton Mifflin Company. All rights reserved.

HOW TO Given $f(x) = 2x^2 - 5x + 3$ and $g(x) = x^2 - 1$, find $\left(\frac{f}{g}\right)(1)$.

$$\left(\frac{f}{g}\right)(1) = \frac{f(1)}{g(1)}$$
$$= \frac{2(1)^2 - 5(1) + 3}{(1)^2 - 1} = \frac{0}{0}$$

• Not a real number

Because $\frac{0}{0}$ is not defined, the expression $\left(\frac{f}{g}\right)(1)$ cannot be evaluated.

Example 1

Given $f(x) = x^2 - x + 1$ and $g(x) = x^3 - 4$, find $(f - g)(3)$.

Solution

$$\begin{aligned}(f - g)(3) &= f(3) - g(3)\\ &= (3^2 - 3 + 1) - (3^3 - 4)\\ &= 7 - 23\\ &= -16\end{aligned}$$

$(f - g)(3) = -16$

You Try It 1

Given $f(x) = x^2 + 2x$ and $g(x) = 5x - 2$, find $(f + g)(-2)$.

Your solution

Example 2

Given $f(x) = x^2 + 2$ and $g(x) = 2x + 3$, find $(f \cdot g)(-2)$.

Solution

$$\begin{aligned}(f \cdot g)(-2) &= f(-2) \cdot g(-2)\\ &= [(-2)^2 + 2] \cdot [2(-2) + 3]\\ &= 6(-1)\\ &= -6\end{aligned}$$

$(f \cdot g)(-2) = -6$

You Try It 2

Given $f(x) = 4 - x^2$ and $g(x) = 3x - 4$, find $(f \cdot g)(3)$.

Your solution

Example 3

Given $f(x) = x^2 + 4x + 4$ and $g(x) = x^3 - 2$, find $\left(\frac{f}{g}\right)(3)$.

Solution

$$\begin{aligned}\left(\frac{f}{g}\right)(3) &= \frac{f(3)}{g(3)}\\ &= \frac{3^2 + 4(3) + 4}{3^3 - 2}\\ &= \frac{25}{25}\\ &= 1\end{aligned}$$

$\left(\frac{f}{g}\right)(3) = 1$

You Try It 3

Given $f(x) = x^2 - 4$ and $g(x) = x^2 + 2x + 1$, find $\left(\frac{f}{g}\right)(4)$.

Your solution

Solutions on p. S34

Copyright © Houghton Mifflin Company. All rights reserved.

Objective B To find the composition of two functions

Composition of functions is another way in which functions can be combined. This method of combining functions uses the output of one function as the input for a second function.

Suppose a forest fire is started by lightning striking a tree, and the spread of the fire can be approximated by a circle whose radius r, in feet, is given by $r(t) = 24\sqrt{t}$, where t is the number of hours after the tree is struck by the lightning. The area of the fire is the area of a circle and is given by the formula $A(r) = \pi r^2$. Because the area of the fire depends on the radius of the circle and the radius depends on the time since the tree was struck, there is a relationship between the area of the fire and time. This relationship can be found by evaluating the formula for the area of a circle using $r(t) = 24\sqrt{t}$.

$$A(r) = \pi r^2$$
$$A[r(t)] = \pi[r(t)]^2 \quad \text{• Replace } r \text{ by } r(t).$$
$$= \pi[24\sqrt{t}]^2 \quad \text{• } r(t) = 24\sqrt{t}$$
$$= 576\pi t \quad \text{• Simplify.}$$

The result is the function $A(t) = 576\pi t$, which gives the area of the fire in terms of the time since the lightning struck. For instance, when $t = 3$, we have

$$A(t) = 576\pi t$$
$$A(3) = 576\pi(3)$$
$$\approx 5429$$

Three hours after the lightning strikes, the area of the fire is approximately 5429 ft^2.

The function produced above, in which one function was evaluated using another function, is referred to as the *composition* of A with r. The notation $A \circ r$ is used to denote this composition of functions. That is,

$$(A \circ r)(t) = 576\pi t$$

Definition of the Composition of Two Functions

Let f and g be two functions such that $g(x)$ is in the domain of f for all x in the domain of g. Then the **composition of the two functions,** denoted by $f \circ g$, is the function whose value at x is given by $(f \circ g)(x) = f[g(x)]$.

The function defined by $(f \circ g)(x)$ is also called the **composite** of f and g and represents a **composite function.** We read $(f \circ g)(x)$ or $f[g(x)]$ as "f of g of x."

Copyright © Houghton Mifflin Company. All rights reserved.

The function machine at the right illustrates the composition of the two functions $g(x) = x^2$ and $f(x) = 2x$. Note that a composite function combines two functions. First one function pairs an input with an output. Then that output is used as the input for a second function, which in turn produces a final output.

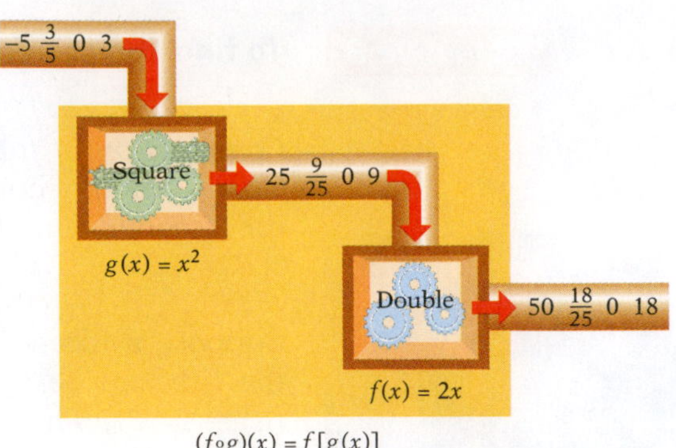

$(f \circ g)(x) = f[g(x)]$

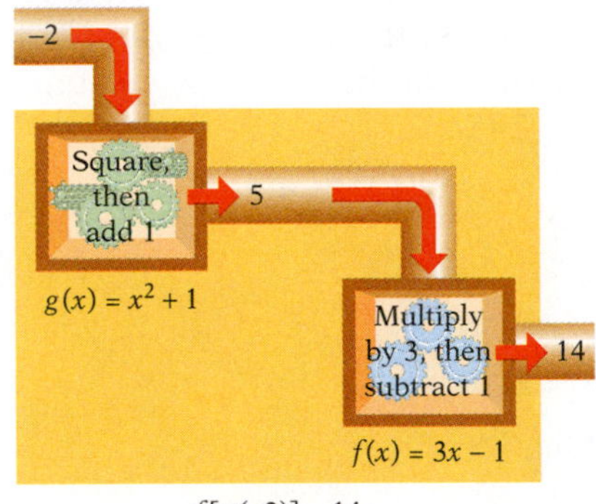

$f[g(-2)] = 14$

Consider $f(x) = 3x - 1$ and $g(x) = x^2 + 1$. The expression $(f \circ g)(-2)$ or, equivalently, $f[g(-2)]$ means to evaluate the function f at $g(-2)$.

$$g(x) = x^2 + 1$$
$$g(-2) = (-2)^2 + 1 \qquad \text{• Evaluate } g \text{ at } -2.$$
$$g(-2) = 5$$
$$f(x) = 3x - 1$$
$$f(5) = 3(5) - 1 = 14 \qquad \text{• Evaluate } f \text{ at } g(-2) = 5.$$

If we apply our function machine analogy, the composition of functions looks something like the figure at the left.

We can find a general expression for $f[g(x)]$ by evaluating f at $g(x)$. For instance, using $f(x) = 3x - 1$ and $g(x) = x^2 + 1$, we have

$$f(x) = 3x - 1$$
$$f[g(x)] = 3[g(x)] - 1 \qquad \text{• Replace } x \text{ by } g(x).$$
$$= 3[x^2 + 1] - 1 \qquad \text{• Replace } g(x) \text{ by } x^2 + 1.$$
$$= 3x^2 + 2 \qquad \text{• Simplify.}$$

The requirement in the definition of the composition of two functions that $g(x)$ be in the domain of f for all x in the domain of g is important. For instance, let

$$f(x) = \frac{1}{x - 1} \qquad \text{and} \qquad g(x) = 3x - 5$$

When $x = 2$,

$$g(2) = 3(2) - 5 = 1$$
$$f[g(2)] = f(1) = \frac{1}{1 - 1} = \frac{1}{0} \qquad \text{• This is not a real number.}$$

In this case, $g(2)$ is not in the domain of f. Thus the composition is not defined at 2.

HOW TO Given $f(x) = x^3 - x + 1$ and $g(x) = 2x^2 - 10$, evaluate $(g \circ f)(2)$.

$$f(2) = (2)^3 - (2) + 1 = 7$$
$$(g \circ f)(2) = g[f(2)]$$
$$= g(7)$$
$$= 2(7)^2 - 10 = 88$$

Copyright © Houghton Mifflin Company. All rights reserved.

HOW TO Given $f(x) = 3x - 2$ and $g(x) = x^2 - 2x$, find $(f \circ g)(x)$.

$$\begin{aligned}(f \circ g)(x) &= f[g(x)] \\ &= 3(x^2 - 2x) - 2 \\ &= 3x^2 - 6x - 2\end{aligned}$$

When we evaluate compositions of functions, the order in which the functions are applied is important. In the two diagrams below, the order of the *square function*, $g(x) = x^2$, and the *double function*, $f(x) = 2x$, is interchanged. Note that the final outputs are different. Therefore, $(f \circ g)(x) \neq (g \circ f)(x)$.

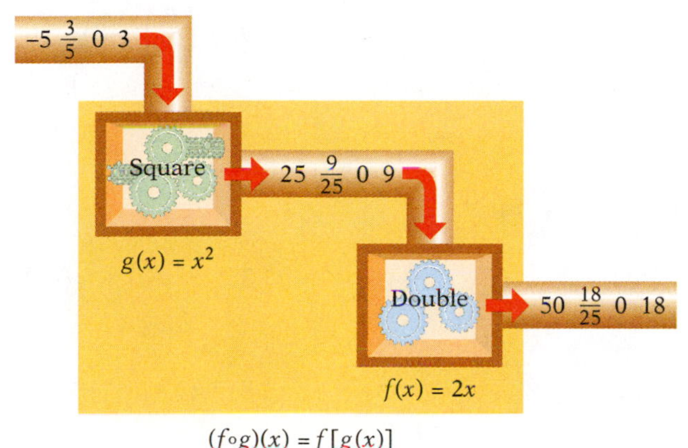

$(f \circ g)(x) = f[g(x)]$

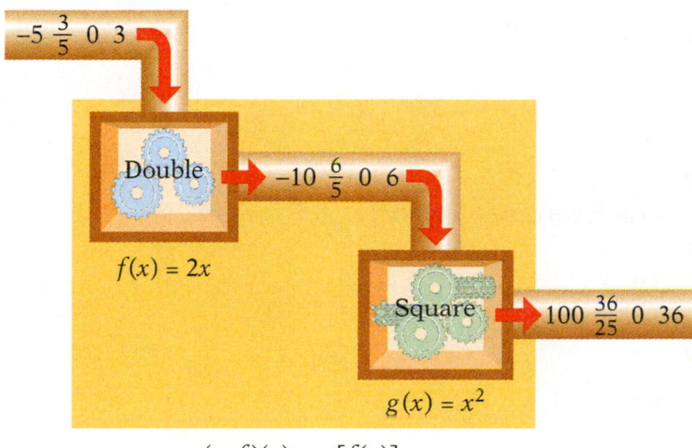

$(g \circ f)(x) = g[f(x)]$

Example 4

Given $f(x) = x^2 - x$ and $g(x) = 3x - 2$, find $f[g(3)]$.

Solution

$g(x) = 3x - 2$

$g(3) = 3(3) - 2 = 9 - 2 = 7$ • **Evaluate g at 3.**

$f(x) = x^2 - x$ • **Evaluate f at $g(3) = 7$.**

$f[g(3)] = f(7) = 7^2 - 7 = 42$

You Try It 4

Given $f(x) = 1 - 2x$ and $g(x) = x^2$, find $f[g(-1)]$.

Your solution

Example 5

Given $s(t) = t^2 + 3t - 1$ and $v(t) = 2t + 1$, determine $s[v(t)]$.

Solution

$s(t) = t^2 + 3t - 1$ • **Evaluate s at $v(t)$.**

$$\begin{aligned}s[v(t)] &= (2t + 1)^2 + 3(2t + 1) - 1 \\ &= (4t^2 + 4t + 1) + 6t + 3 - 1 \\ &= 4t^2 + 10t + 3\end{aligned}$$

You Try It 5

Given $L(s) = s + 1$ and $M(s) = s^3 + 1$, determine $M[L(s)]$.

Your solution

Solutions on p. S34

Copyright © Houghton Mifflin Company. All rights reserved.

11.3 Exercises

Objective A To perform operations on functions

For Exercises 1 to 9, let $f(x) = 2x^2 - 3$ and $g(x) = -2x + 4$. Find:

1. $f(2) - g(2)$

2. $f(3) - g(3)$

3. $f(0) + g(0)$

4. $f(1) + g(1)$

5. $(f \cdot g)(2)$

6. $(f \cdot g)(-1)$

7. $\left(\frac{f}{g}\right)(4)$

8. $\left(\frac{f}{g}\right)(-1)$

9. $\left(\frac{g}{f}\right)(-3)$

For Exercises 10 to 18, let $f(x) = 2x^2 + 3x - 1$ and $g(x) = 2x - 4$. Find:

10. $f(-3) + g(-3)$

11. $f(1) + g(1)$

12. $f(-2) - g(-2)$

13. $f(4) - g(4)$

14. $(f \cdot g)(-2)$

15. $(f \cdot g)(1)$

16. $\left(\frac{f}{g}\right)(2)$

17. $\left(\frac{f}{g}\right)(-3)$

18. $(f \cdot g)\left(\frac{1}{2}\right)$

For Exercises 19 to 21, let $f(x) = x^2 + 3x - 5$ and $g(x) = x^3 - 2x + 3$. Find:

19. $f(2) - g(2)$

20. $(f \cdot g)(-3)$

21. $\left(\frac{f}{g}\right)(-2)$

Objective B To find the composition of two functions

22. Explain the meaning of the notation $f[g(3)]$.

23. Explain the meaning of the expression $(f \circ g)(-2)$.

Given $f(x) = 2x - 3$ and $g(x) = 4x - 1$, evaluate the composite functions in Exercises 24 to 29.

24. $f[g(0)]$

25. $g[f(0)]$

26. $f[g(2)]$

27. $g[f(-2)]$

28. $f[g(x)]$

29. $g[f(x)]$

Copyright © Houghton Mifflin Company. All rights reserved.

Given $h(x) = 2x + 4$ and $f(x) = \frac{1}{2}x + 2$, evaluate the composite functions in Exercises 30 to 35.

30. $h[f(0)]$ **31.** $f[h(0)]$ **32.** $h[f(2)]$

33. $f[h(-1)]$ **34.** $h[f(x)]$ **35.** $f[h(x)]$

Given $g(x) = x^2 + 3$ and $h(x) = x - 2$, evaluate the composite functions in Exercises 36 to 41.

36. $g[h(0)]$ **37.** $h[g(0)]$ **38.** $g[h(4)]$

39. $h[g(-2)]$ **40.** $g[h(x)]$ **41.** $h[g(x)]$

Given $f(x) = x^2 + x + 1$ and $h(x) = 3x + 2$, evaluate the composite functions in Exercises 42 to 47.

42. $f[h(0)]$ **43.** $h[f(0)]$ **44.** $f[h(-1)]$

45. $h[f(-2)]$ **46.** $f[h(x)]$ **47.** $h[f(x)]$

Given $f(x) = x - 2$ and $g(x) = x^3$, evaluate the composite functions in Exercises 48 to 53.

48. $f[g(2)]$ **49.** $f[g(-1)]$ **50.** $g[f(2)]$

51. $g[f(-1)]$ **52.** $f[g(x)]$ **53.** $g[f(x)]$

54. Oil Spills Suppose the spread of an oil leak from a tanker can be approximated by a circle with the tanker at its center and radius r, in feet. The radius of the spill t hours after the beginning of the leak is given by $r(t) = 45t$.

a. Find the area of the spill as a function of time.

b. What is the area of the spill after 3 h? Round to the nearest whole number.

55. Manufacturing Suppose the manufacturing cost, in dollars, per digital camera is given by the function $M(x) = \frac{50x + 10{,}000}{x}$, where x is the number of cameras manufactured. A camera store will sell the cameras by marking up the manufacturing cost per camera, $M(x)$, by 60%.

a. Express the selling price of a camera as a function of the number of cameras to be manufactured. That is, find $S \circ M$.

b. Find $(S \circ M)(5000)$.

c. Explain the meaning of the answer to part **b.**

Copyright © Houghton Mifflin Company. All rights reserved.

56. **Manufacturing** The number of electric scooters e that a factory can produce per day is a function of the number of hours h it operates and is given by $e(h) = 250h$ for $0 \le h \le 10$. The daily cost c to manufacture e electric scooters is given by the function $c(e) = 0.05e^2 + 60e + 1000$.
  a. Find $(c \circ e)(h)$.
  b. Evaluate $(c \circ e)(10)$.
  c.  Write a sentence that explains the meaning of the answer to part **b.**

57. **Automobile Rebates** A car dealership offers a \$1500 rebate and a 10% discount off the price of a new car. Let p be the sticker price of a new car on the dealer's lot, r the price after the rebate, and d the discounted price. Then $r(p) = p - 1500$ and $d(p) = 0.90p$.
  a. Write a composite function for the dealer taking the rebate first and then the discount.
  b. Write a composite function for the dealer taking the discount first and then the rebate.
  c. Which composite function would you prefer the dealer to use when you buy a new car?

APPLYING THE CONCEPTS

The graphs of f and g are shown at the right. Use the graphs to determine the values of the composite functions in Exercises 58 to 63.

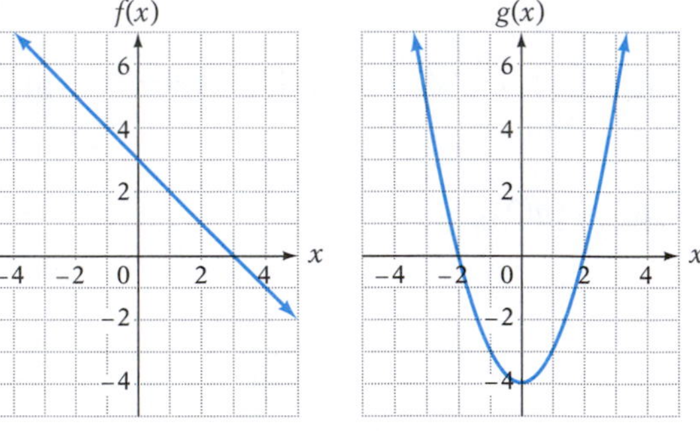

58. $f[g(-1)]$
59. $g[f(1)]$
60. $(g \circ f)(2)$
61. $(f \circ g)(3)$
62. $g[f(3)]$
63. $f[g(0)]$

For Exercises 64 to 69, let $g(x) = x^2 - 1$. Find:

64. $g(2 + h)$
65. $g(3 + h) - g(3)$
66. $g(-1 + h) - g(-1)$
67. $\dfrac{g(1 + h) - g(1)}{h}$
68. $\dfrac{g(-2 + h) - g(-2)}{h}$
69. $\dfrac{g(a + h) - g(a)}{h}$

For Exercises 70 to 75, let $f(x) = 2x$, $g(x) = 3x - 1$, and $h(x) = x - 2$. Find:

70. $f(g[h(2)])$
71. $g(h[f(1)])$
72. $h(g[f(-1)])$
73. $f(h[g(0)])$
74. $f(g[h(x)])$
75. $g(f[h(x)])$

Copyright © Houghton Mifflin Company. All rights reserved.

11.4 One-to-One and Inverse Functions

Objective A To determine whether a function is one-to-one

Recall that a function is a set of ordered pairs in which no two ordered pairs that have the same first coordinate have different second coordinates. This means that given any x, there is only one y that can be paired with that x. A **one-to-one function** satisfies the additional condition that given any y, there is only one x that can be paired with the given y. One-to-one functions are commonly written as 1–1.

> **One-to-One Function**
>
> A function f is a 1–1 function if for any a and b in the domain of f, $f(a) = f(b)$ implies that $a = b$.

This definition states that for a 1–1 function, if the y-coordinates of two ordered pairs are equal, $f(a) = f(b)$, then the x-coordinates must be equal, $a = b$.

The function defined by $f(x) = 2x + 1$ is a 1–1 function. To show this, determine $f(a)$ and $f(b)$. Then form the equation $f(a) = f(b)$.

$$f(a) = 2a + 1 \qquad f(b) = 2b + 1$$

$$f(a) = f(b)$$
$$2a + 1 = 2b + 1$$
$$2a = 2b \qquad \text{• Subtract 1 from each side of the equation.}$$
$$a = b \qquad \text{• Divide each side of the equation by 2.}$$

Because $f(a) = f(b)$ implies that $a = b$, the function is a 1–1 function.

Consider the function defined by $g(x) = x^2 - x$. Evaluate the function at -2 and 3.

$$g(-2) = (-2)^2 - (-2) = 6 \qquad g(3) = 3^2 - 3 = 6$$

From this evaluation, $g(-2) = 6$ and $g(3) = 6$, but $-2 \neq 3$. Thus g is not a 1–1 function.

The graphs of $f(x) = 2x + 1$ and $g(x) = x^2 - x$ are shown below. Note that a horizontal line intersects the graph of f at no more than one point. However, a horizontal line intersects the graph of g at more than one point.

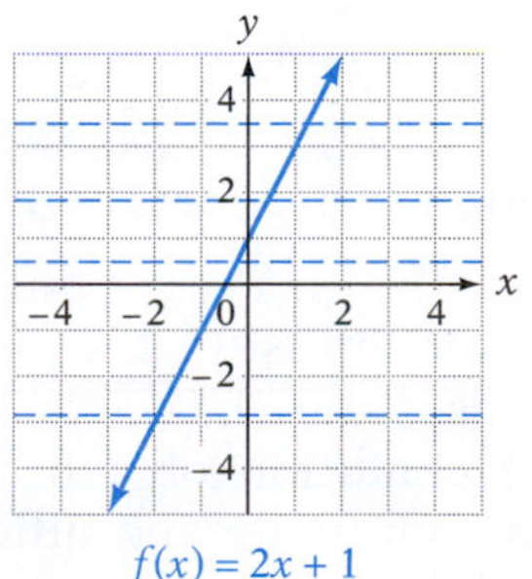

$f(x) = 2x + 1$

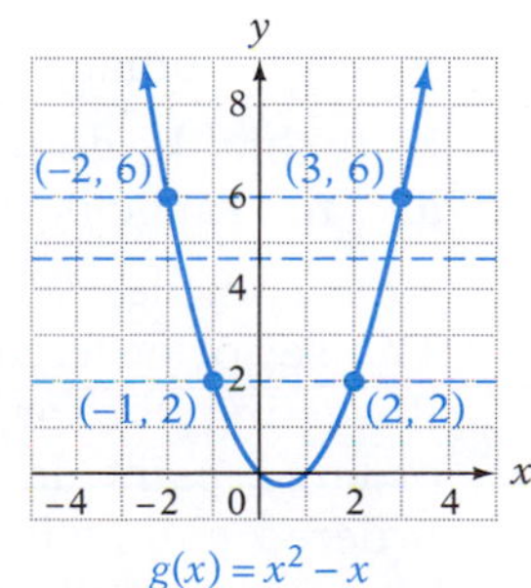

$g(x) = x^2 - x$

Copyright © Houghton Mifflin Company. All rights reserved.

Looking at the graph of f on the previous page, note that for each y-coordinate there is only one x-coordinate. Thus f is a 1–1 function. From the graph of g, however, there are *two* x-coordinates for a given y-coordinate. For instance, $(-2, 6)$ and $(3, 6)$ are the coordinates of two points on the graph for which the y-coordinates are the same and the x-coordinates are different. Therefore, g is not a 1–1 function.

Horizontal-Line Test

The graph of a function represents the graph of a 1–1 function if any horizontal line intersects the graph at no more than one point.

Example 1

Determine whether the graph is the graph of a 1–1 function.

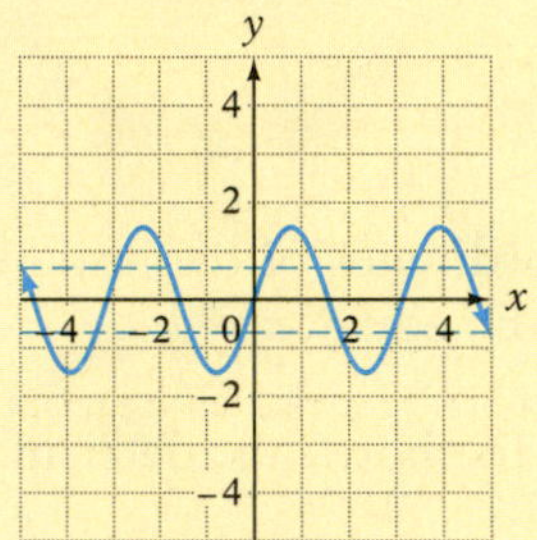

Solution

Because a horizontal line intersects the graph more than once, the graph is not the graph of a 1–1 function.

You Try It 1

Determine whether the graph is the graph of a 1–1 function.

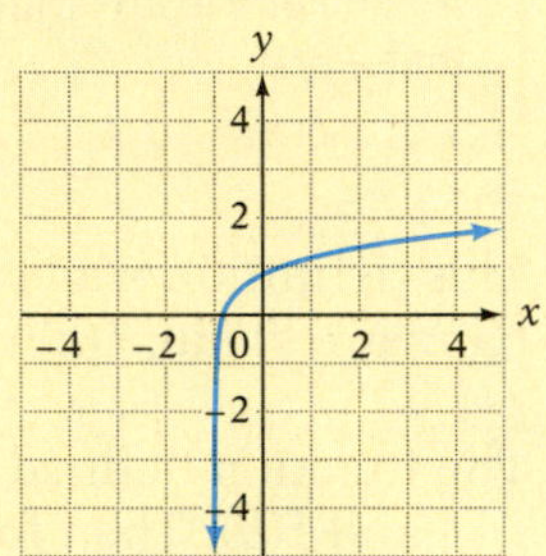

Your solution

Solution on p. S34

Objective B To find the inverse of a function

The **inverse of a function** is the set of ordered pairs formed by reversing the coordinates of each ordered pair of the function.

For example, the set of ordered pairs of the function defined by $f(x) = 2x$ with domain $\{-2, -1, 0, 1, 2\}$ is $\{(-2, -4), (-1, -2), (0, 0), (1, 2), (2, 4)\}$. The set of ordered pairs of the inverse function is $\{(-4, -2), (-2, -1), (0, 0), (2, 1), (4, 2)\}$.

From the ordered pairs of f, we have

$$\text{Domain} = \{-2, -1, 0, 1, 2\} \quad \text{and} \quad \text{range} = \{-4, -2, 0, 2, 4\}$$

From the ordered pairs of the inverse function, we have

$$\text{Domain} = \{-4, -2, 0, 2, 4\} \quad \text{and} \quad \text{range} = \{-2, -1, 0, 1, 2\}$$

Note that the domain of the inverse function is the range of the original function, and the range of the inverse function is the domain of the original function.

Now consider the function defined by $g(x) = x^2$ with domain $\{-2, -1, 0, 1, 2\}$. The set of ordered pairs of this function is $\{(-2, 4), (-1, 1), (0, 0), (1, 1), (2, 4)\}$. Reversing the ordered pairs gives $\{(4, -2), (1, -1), (0, 0), (1, 1), (4, 2)\}$. These ordered pairs do not satisfy the condition of a function, because there are ordered pairs with the same first coordinate and different second coordinates. This example illustrates that not all functions have an inverse function.

Copyright © Houghton Mifflin Company. All rights reserved.

The graphs of $f(x) = 2x$ and $g(x) = x^2$, with the set of real numbers as the domain, are shown below.

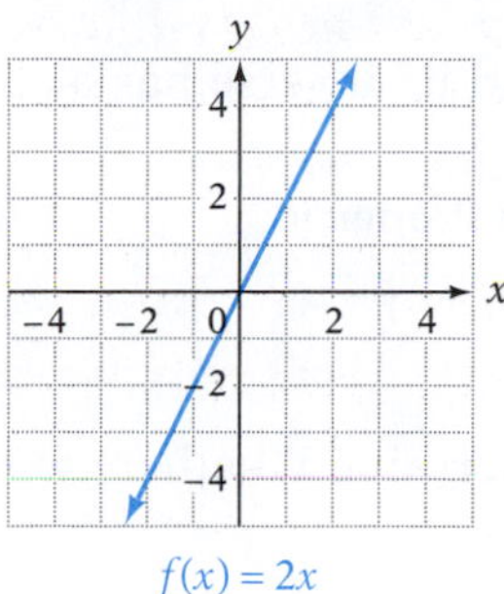

$f(x) = 2x$

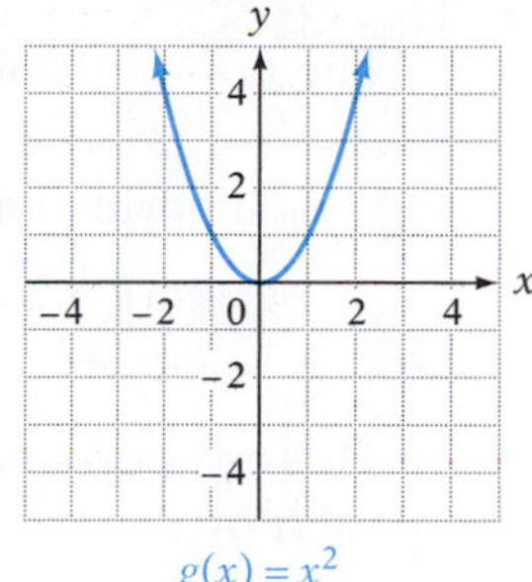

$g(x) = x^2$

By the horizontal-line test, f is a 1–1 function but g is not.

Condition for an Inverse Function

A function f has an inverse function if and only if f is a 1–1 function.

The symbol f^{-1} is used to denote the inverse of the function f. The symbol $f^{-1}(x)$ is read "f inverse of x."

$f^{-1}(x)$ is *not* the reciprocal of $f(x)$ but is, rather, the notation for the inverse of a 1–1 function.

To find the inverse of a function, interchange x and y. Then solve for y.

TAKE NOTE

If the ordered pairs of f are given by (x, y), then the ordered pairs of f^{-1} are given by (y, x). That is, x and y are interchanged. This is the reason for Step 2 at the right.

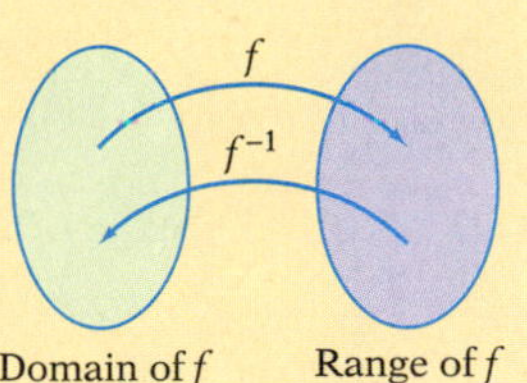

Domain of f / Range of f^{-1} Range of f / Domain of f^{-1}

HOW TO Find the inverse of the function defined by $f(x) = 3x + 6$.

$f(x) = 3x + 6$

$y = 3x + 6$ • **Replace $f(x)$ by y.**

$x = 3y + 6$ • **Interchange x and y.**

$x - 6 = 3y$ • **Solve for y.**

$\frac{1}{3}x - 2 = y$

$f^{-1}(x) = \frac{1}{3}x - 2$ • **Replace y by $f^{-1}(x)$.**

The inverse of the function $f(x) = 3x + 6$ is $f^{-1}(x) = \frac{1}{3}x - 2$.

The fact that the ordered pairs of the inverse of a function are the reverse of those of the original function has a graphical interpretation. In the graph in the middle, the points from the figure on the left are plotted with the coordinates reversed. The inverse function is graphed by drawing a smooth curve through those points, as shown in the figure on the right.

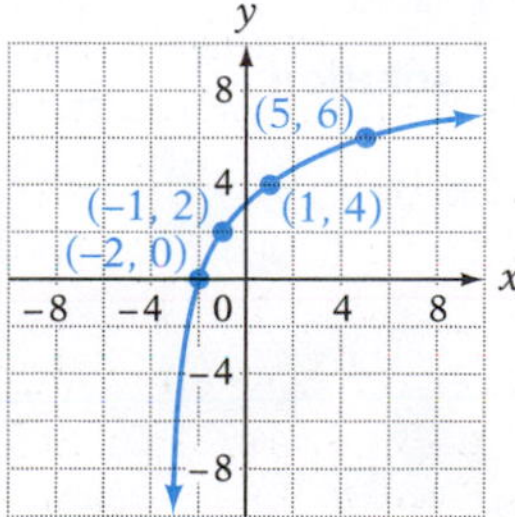

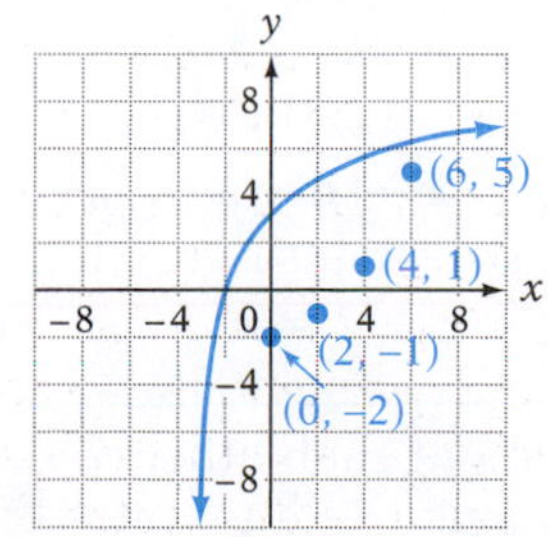

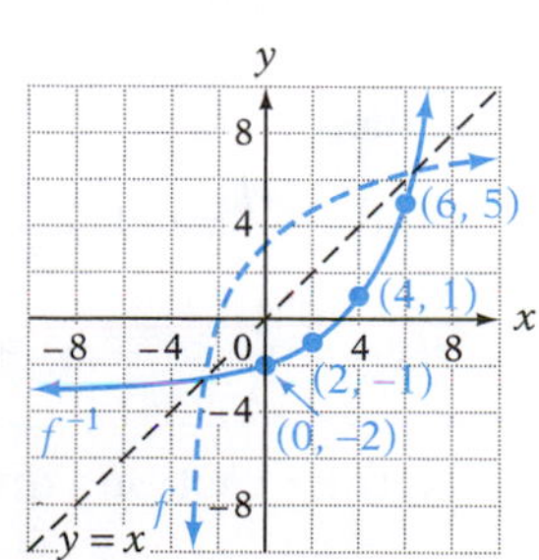

Copyright © Houghton Mifflin Company. All rights reserved.

Note that the dashed graph of $y = x$ is shown in black in the figure on the bottom right of the preceding page. If two functions are inverses of each other, their graphs are mirror images with respect to the graph of the line $y = x$.

The composition of a function and its inverse has a special property.

Composition of Inverse Functions Property

$f^{-1}[f(x)] = x$ and $f[f^{-1}(x)] = x$

This property can be used to determine whether two functions are inverses of each other.

HOW TO Are $f(x) = 2x - 4$ and $g(x) = \frac{1}{2}x + 2$ inverses of each other?

To determine whether the functions are inverses, use the Composition of Inverse Functions Property.

$$\begin{aligned} f[g(x)] &= 2\left(\frac{1}{2}x + 2\right) - 4 \\ &= x + 4 - 4 \\ &= x \end{aligned} \qquad \begin{aligned} g[f(x)] &= \frac{1}{2}(2x - 4) + 2 \\ &= x - 2 + 2 \\ &= x \end{aligned}$$

Because $f[g(x)] = x$ and $g[f(x)] = x$, the functions are inverses of each other.

Example 2

Find the inverse of the function defined by $f(x) = 2x - 3$.

Solution

$$\begin{aligned} f(x) &= 2x - 3 \\ y &= 2x - 3 && \text{• Replace } f(x) \text{ by } y. \\ x &= 2y - 3 && \text{• Interchange } x \text{ and } y. \\ x + 3 &= 2y && \text{• Solve for } y. \\ \frac{x}{2} + \frac{3}{2} &= y \\ f^{-1}(x) &= \frac{x}{2} + \frac{3}{2} && \text{• Replace } y \text{ by } f^{-1}(x). \end{aligned}$$

The inverse of the function is given by $f^{-1}(x) = \frac{x}{2} + \frac{3}{2}$.

You Try It 2

Find the inverse of the function defined by $f(x) = \frac{1}{2}x + 4$.

Your solution

Example 3

Are $f(x) = 3x - 6$ and $g(x) = \frac{1}{3}x + 2$ inverses of each other?

Solution

$$f[g(x)] = 3\left(\frac{1}{3}x + 2\right) - 6 = x + 6 - 6 = x$$

$$g[f(x)] = \frac{1}{3}(3x - 6) + 2 = x - 2 + 2 = x$$

Yes, the functions are inverses of each other.

You Try It 3

Are $f(x) = 2x - 6$ and $g(x) = \frac{1}{2}x - 3$ inverses of each other?

Your solution

Solutions on p. S34

Copyright © Houghton Mifflin Company. All rights reserved.

11.4 Exercises

Objective A **To determine whether a function is one-to-one**

1. What is a 1–1 function?

2. What is the horizontal-line test?

For Exercises 3 to 14, determine whether the graph represents the graph of a 1–1 function.

3. 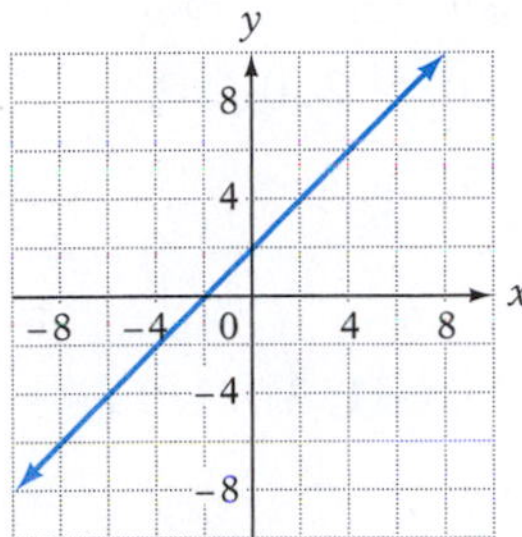

4. 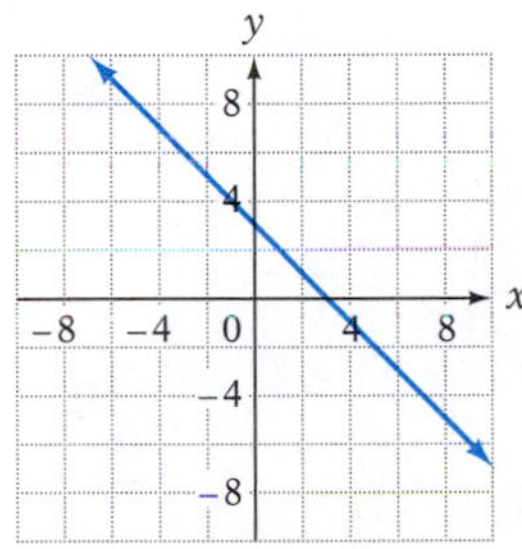

5. 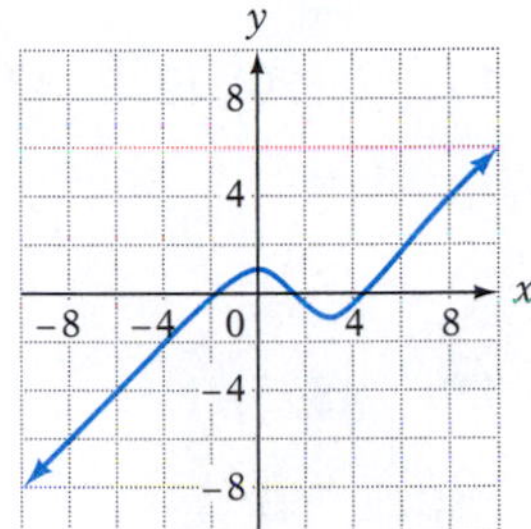

6. 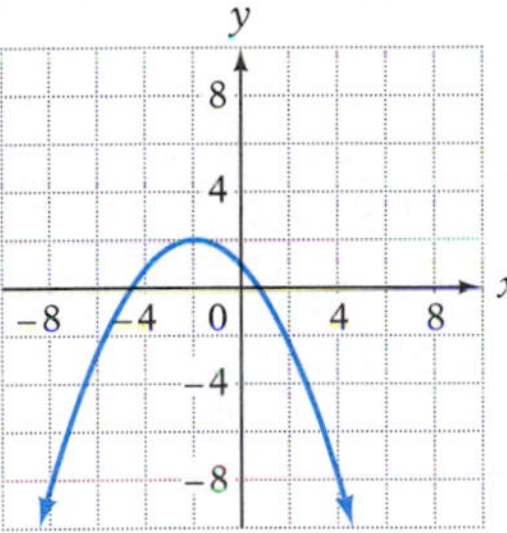

7. 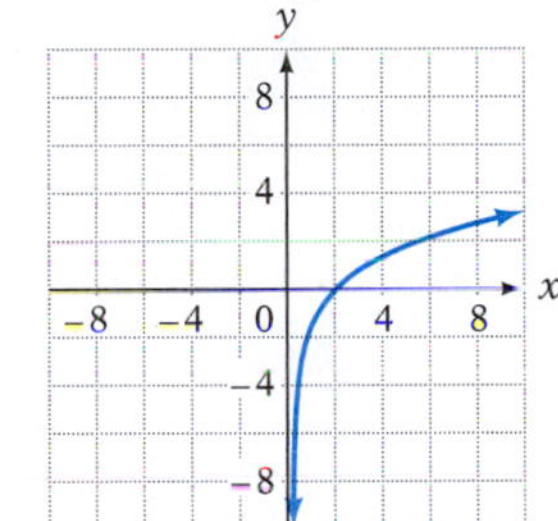

8. 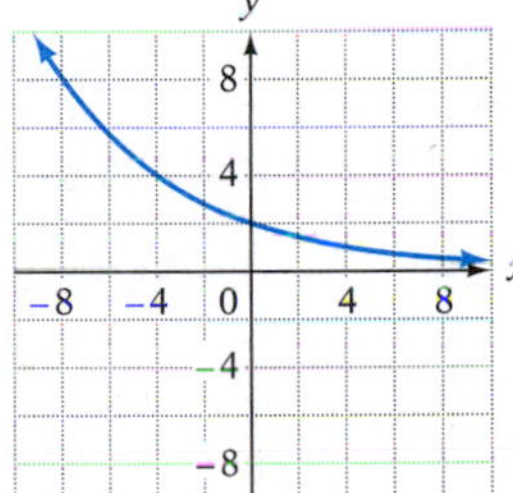

9. 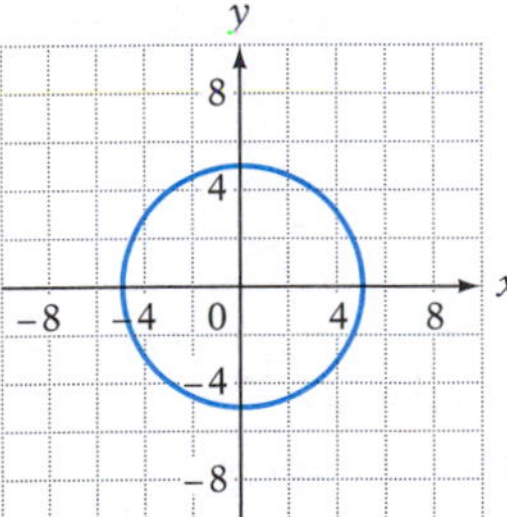

10. 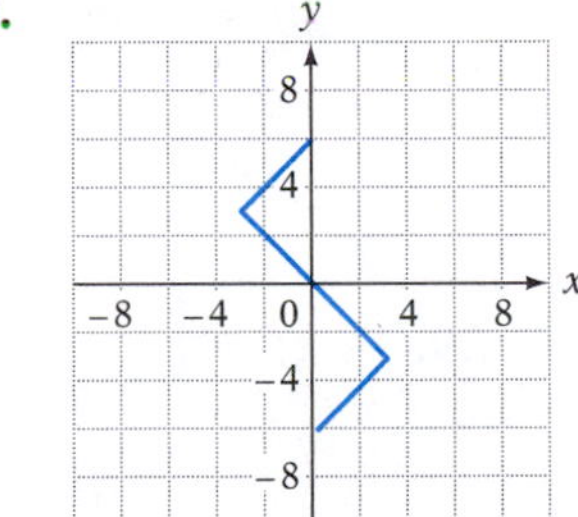

11. 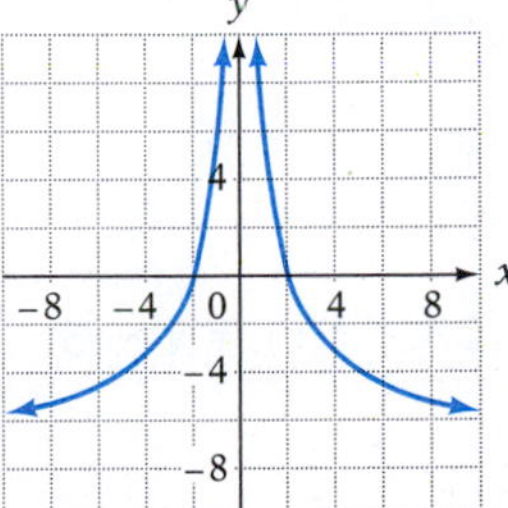

12. 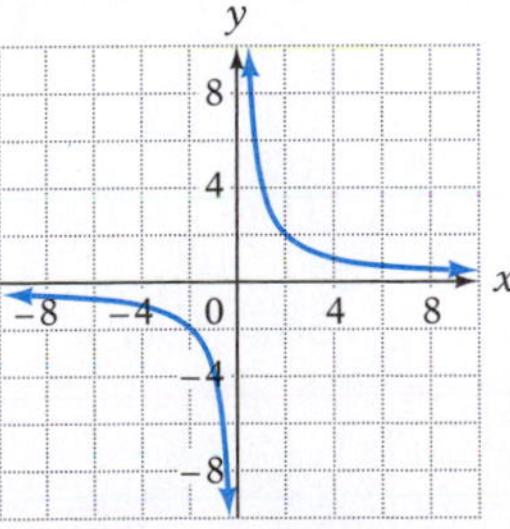

13. 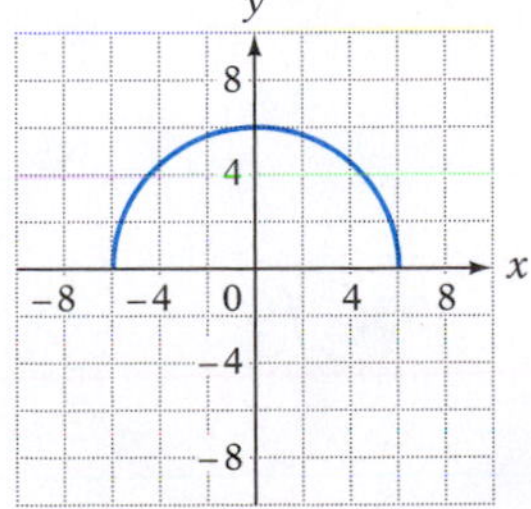

14. 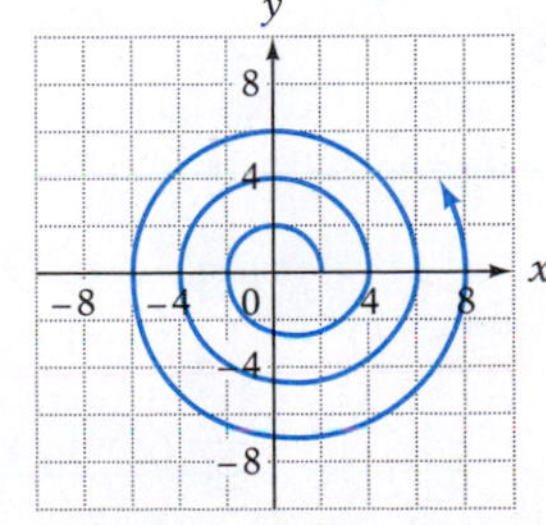

Copyright © Houghton Mifflin Company. All rights reserved.

Objective B To find the inverse of a function

15. How are the ordered pairs of the inverse of a function related to the function?

16. Why is it that not all functions have an inverse function?

For Exercises 17 to 24, find the inverse of the function. If the function does not have an inverse function, write "no inverse."

17. $\{(1, 0), (2, 3), (3, 8), (4, 15)\}$

18. $\{(1, 0), (2, 1), (-1, 0), (-2, 0)\}$

19. $\{(3, 5), (-3, -5), (2, 5), (-2, -5)\}$

20. $\{(-5, -5), (-3, -1), (-1, 3), (1, 7)\}$

21. $\{(0, -2), (-1, 5), (3, 3), (-4, 6)\}$

22. $\{(-2, -2), (0, 0), (2, 2), (4, 4)\}$

23. $\{(-2, -3), (-1, 3), (0, 3), (1, 3)\}$

24. $\{(2, 0), (1, 0), (3, 0), (4, 0)\}$

For Exercises 25 to 42, find $f^{-1}(x)$.

25. $f(x) = 4x - 8$

26. $f(x) = 3x + 6$

27. $f(x) = 2x + 4$

28. $f(x) = x - 5$

29. $f(x) = \frac{1}{2}x - 1$

30. $f(x) = \frac{1}{3}x + 2$

31. $f(x) = -2x + 2$

32. $f(x) = -3x - 9$

33. $f(x) = \frac{2}{3}x + 4$

34. $f(x) = \frac{3}{4}x - 4$

35. $f(x) = -\frac{1}{3}x + 1$

36. $f(x) = -\frac{1}{2}x + 2$

Copyright © Houghton Mifflin Company. All rights reserved.

37. $f(x) = 2x - 5$

38. $f(x) = 3x + 4$

39. $f(x) = 5x - 2$

40. $f(x) = 4x - 2$

41. $f(x) = 6x - 3$

42. $f(x) = -8x + 4$

For Exercises 43 to 45, let $f(x) = 3x - 5$. Find:

43. $f^{-1}(0)$

44. $f^{-1}(2)$

45. $f^{-1}(4)$

For Exercises 46 to 48, state whether the graph is the graph of a function. If it is the graph of a function, does the function have an inverse?

46.

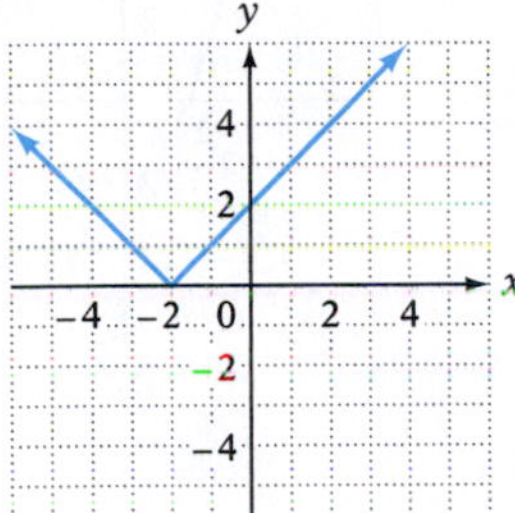

47.

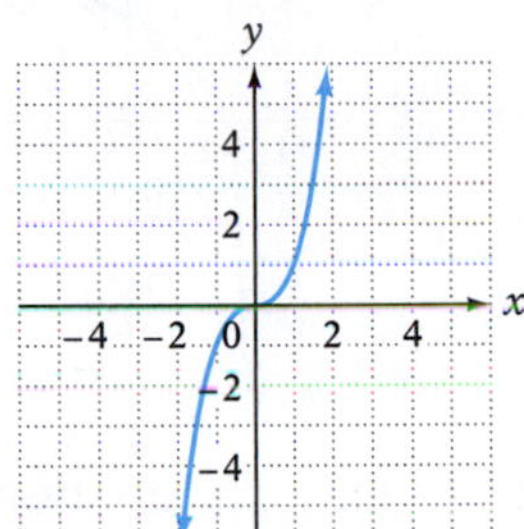

48.

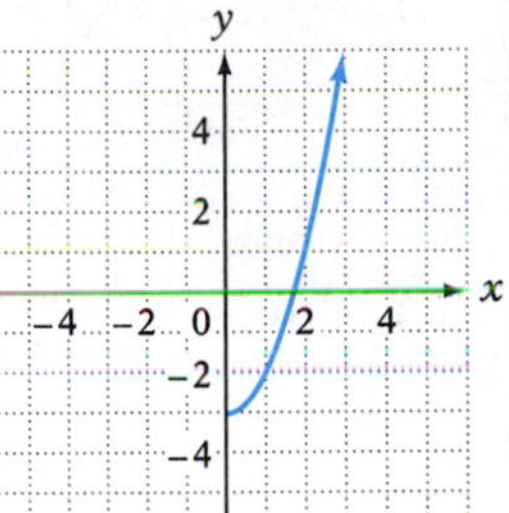

For Exercises 49 to 56, use the Composition of Inverse Functions Property to determine whether the functions are inverses.

49. $f(x) = 4x; g(x) = \frac{x}{4}$

50. $g(x) = x + 5; h(x) = x - 5$

51. $f(x) = 3x; h(x) = \frac{1}{3x}$

52. $h(x) = x + 2; g(x) = 2 - x$

53. $g(x) = 3x + 2; f(x) = \frac{1}{3}x - \frac{2}{3}$

54. $h(x) = 4x - 1; f(x) = \frac{1}{4}x + \frac{1}{4}$

55. $f(x) = \frac{1}{2}x - \frac{3}{2}; g(x) = 2x + 3$

56. $g(x) = -\frac{1}{2}x - \frac{1}{2}; h(x) = -2x + 1$

Copyright © Houghton Mifflin Company. All rights reserved.

APPLYING THE CONCEPTS

Given the graph of the 1–1 function in each of Exercises 57 to 62, draw the graph of the inverse of the function by using the technique shown in Objective B of this section.

57.

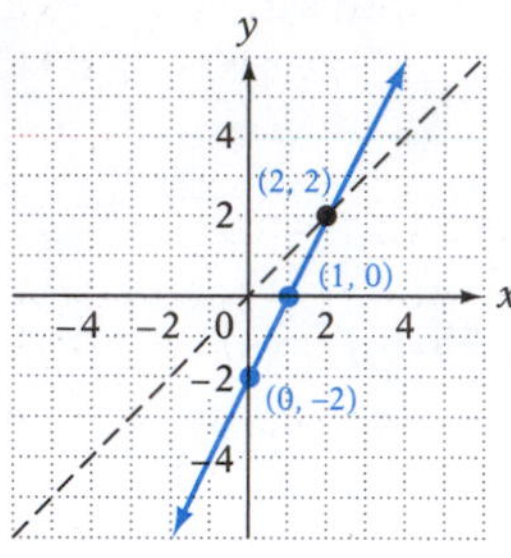

58.

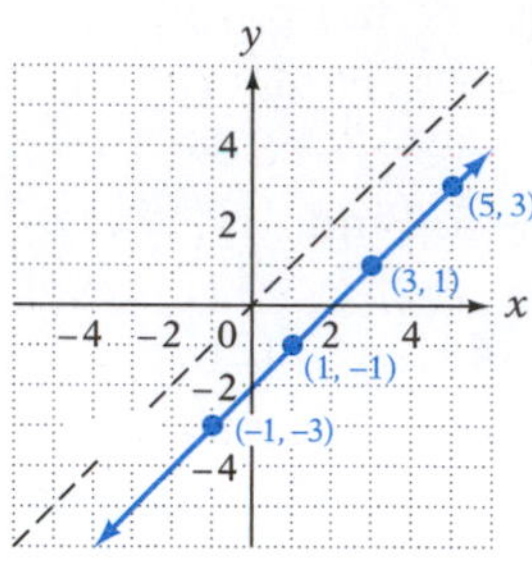

59.

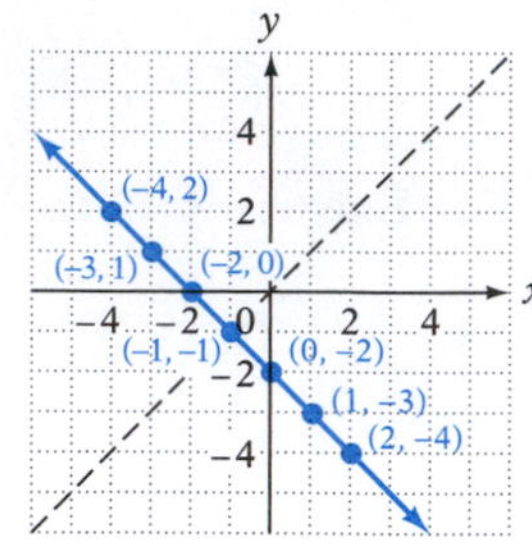

60.

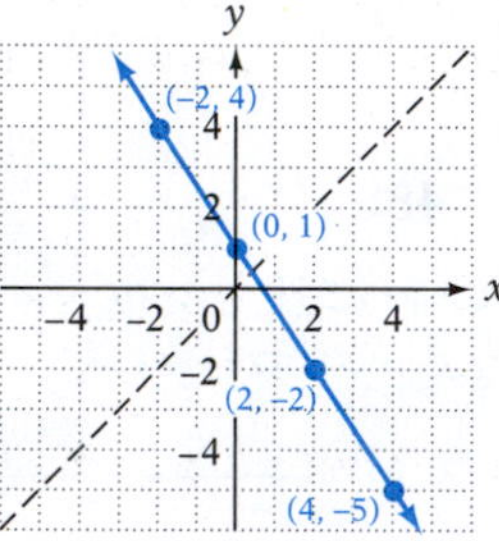

61.

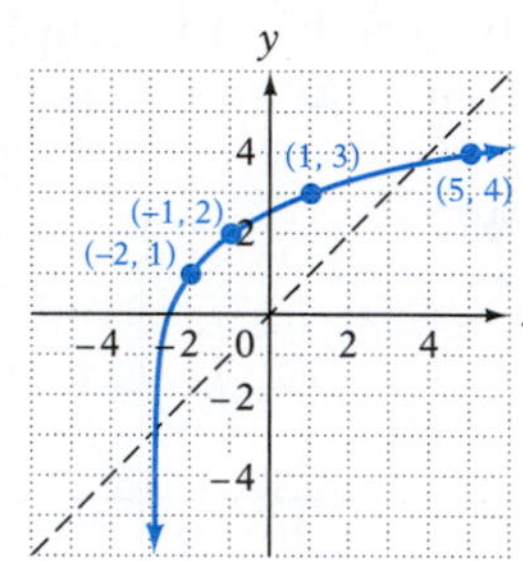

62.

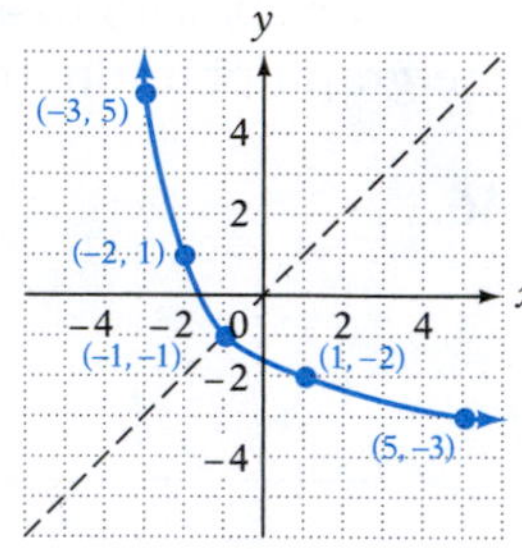

Each of the tables in Exercises 63 and 64 defines a function. Is the inverse of the function a function? Explain your answer.

63.  **Grading Scale**

Score	Grade
90–100	A
80–89	B
70–79	C
60–69	D
0–59	F

64.  **First-Class Postage Rates**

Weight	Cost
$0 < w \le 1$	\$ .37
$1 < w \le 2$	\$ .60
$2 < w \le 3$	\$ .83
$3 < w \le 4$	\$1.06

Given that f is a 1–1 function and that $f(0) = 1$, $f(3) = -1$, and $f(5) = -3$, evaluate the inverse functions in Exercises 65 to 67.

65. $f^{-1}(-3)$ **66.** $f^{-1}(-1)$ **67.** $f^{-1}(1)$

Given that f is a 1–1 function and that $f(-3) = 3$, $f(-4) = 7$, and $f(0) = 8$, evaluate the inverse functions in Exercises 68 to 70.

68. $f^{-1}(3)$ **69.** $f^{-1}(7)$ **70.** $f^{-1}(8)$

71. Is the inverse of a constant function a function? Explain your answer.

72. The graphs of all functions given by $f(x) = mx + b$, $m \neq 0$, are straight lines. Are all of these functions 1–1 functions? If so, explain why. If not, give an example of a linear function of this form that is not 1–1.

Copyright © Houghton Mifflin Company. All rights reserved.

Focus on Problem Solving

Algebraic Manipulation and Graphing Techniques

Problem solving is often easier when we have both algebraic manipulation and graphing techniques at our disposal. Solving quadratic equations and graphing quadratic equations in two variables are the techniques used here to solve problems involving profit.

A company's revenue, R, is the total amount of money the company earned by selling its products. The cost, C, is the total amount of money the company spent to manufacture and sell its products. A company's profit, P, is the difference between the revenue and the cost: $P = R - C$. A company's revenue and cost may be represented by equations.

A company manufactures and sells woodstoves. The total monthly cost, in dollars, to produce n woodstoves is $C = 30n + 2000$. Write a variable expression for the company's monthly profit if the revenue, in dollars, obtained from selling all n woodstoves is $R = 150n - 0.4n^2$.

$$P = R - C$$
$$P = 150n - 0.4n^2 - (30n + 2000)$$
$$P = -0.4n^2 + 120n - 2000$$

• Replace R by $150n - 0.4n^2$ and C by $30n + 2000$. Then simplify.

How many woodstoves must the company manufacture and sell in order to make a profit of \$6000 a month?

$$P = -0.4n^2 + 120n - 2000$$
$$6000 = -0.4n^2 + 120n - 2000$$ • Substitute 6000 for P.
$$0 = -0.4n^2 + 120n - 8000$$ • Write the equation in standard form.
$$0 = n^2 - 300n + 20{,}000$$ • Divide each side of the equation by −0.4.
$$0 = (n - 100)(n - 200)$$ • Factor.
$$n - 100 = 0 \qquad n - 200 = 0$$ • Solve for n.
$$n = 100 \qquad n = 200$$

The company will make a monthly profit of \$6000 if either 100 or 200 woodstoves are manufactured and sold.

The graph of $P = -0.4n^2 + 120n - 2000$ is shown at the right. Note that when $P = 6000$, the values of n are 100 and 200.

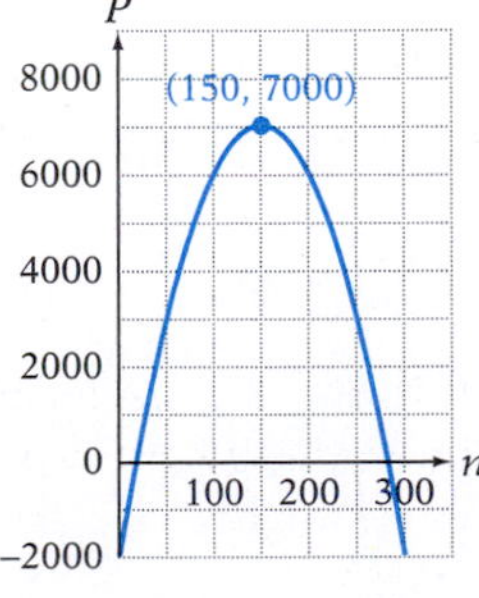

Also note that the coordinates of the highest point on the graph are (150, 7000). This means that the company makes a *maximum* profit of \$7000 per month when 150 woodstoves are manufactured and sold.

1. The total cost, in dollars, for a company to produce and sell n guitars per month is $C = 240n + 1200$. The company's revenue, in dollars, from selling all n guitars is $R = 400n - 2n^2$.
   a. How many guitars must the company produce and sell each month in order to make a monthly profit of \$1200?
   b. Graph the profit equation. What is the maximum monthly profit that the company can make?

Copyright © Houghton Mifflin Company. All rights reserved.

Projects and Group Activities

Finding the Maximum or Minimum of a Function Using a Graphing Calculator

Recall that for a quadratic function, the maximum or minimum value of the function is the y-coordinate of the vertex of the graph of the function. We determined algebraically that the coordinates of the vertex are $\left(-\frac{b}{2a}, f\left(-\frac{b}{2a}\right)\right)$. It is also possible to use a graphing calculator to approximate a maximum or minimum value. The method used to find the value depends on the calculator model. However, it is generally true for most calculators that the point at which the function has a maximum or minimum must be shown on the screen. If it is not, you must select a different viewing window.

For a TI-83/84 graphing calculator, press 2nd CALC 3 to determine a minimum value or 2nd CALC 4 to determine a maximum value. Move the cursor to a point on the curve that is to the left of the minimum (or maximum). Press ENTER. Move the cursor to a point on the curve that is to the right of the minimum (or maximum). Press ENTER twice. The minimum (or maximum) value is shown as the y-coordinate at the bottom of the screen.

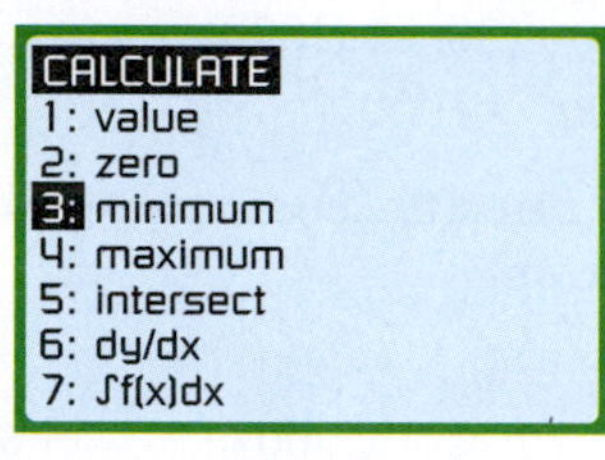

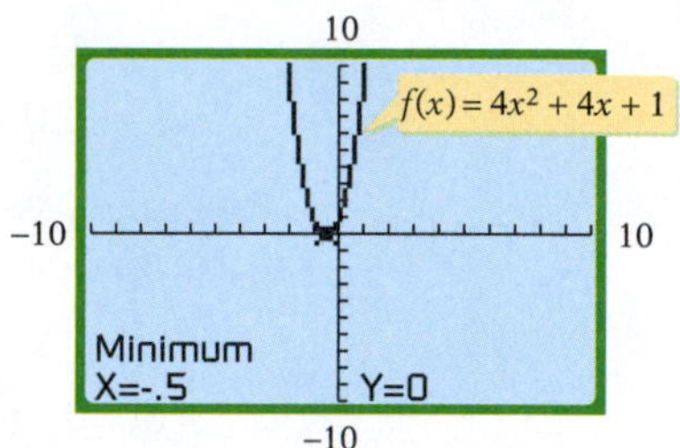

1. Find the maximum or minimum value of $f(x) = -x^2 - 3x + 2$ by using a graphing calculator.
2. Find the maximum or minimum value of $f(x) = x^2 - 4x - 2$ by using a graphing calculator.
3. Find the maximum or minimum value of $f(x) = -\sqrt{2}x^2 - \pi x + 2$ by using a graphing calculator.
4. Find the maximum or minimum value of $f(x) = \frac{x^2}{\sqrt{5}} - \sqrt{7}x - \sqrt{11}$ by using a graphing calculator.
5. **Business** A manufacturer of camera lenses estimates that the average monthly cost C of production is given by the function $C(x) = 0.1x^2 - 20x + 2000$, where x is the number of lenses produced each month. Find the number of lenses the company should produce in order to minimize the average cost.
6. **Compensation** The net annual income I, in dollars, of a family physician can be modeled by the equation $I(x) = -290(x - 48)^2 + 148{,}000$, where x is the age of the physician and $27 \le x \le 70$.
   a. Find the age at which the physician's income will be a maximum.
   b. Find the maximum income.

Business Applications of Maximum and Minimum Values of Quadratic Functions

A company's revenue, R, is the total amount of money the company earned by selling its products. The cost, C, is the total amount of money the company spent to manufacture and sell its products. A company's profit, P, is the difference between the revenue and the cost.

$$P = R - C$$

Copyright © Houghton Mifflin Company. All rights reserved.

A company's revenue and cost may be represented by equations.

HOW TO The owners of a travel agency are selling tickets for a tour. The travel agency must pay a cost per person of \$10. A maximum of 180 people can be accommodated on the tour. For every \$.25 increase in the price of a ticket over the \$10 cost, the owners estimate that they will sell one ticket less. Therefore, the price per ticket can be represented by

$$\begin{aligned} p &= 10 + 0.25(180 - x) \\ &= 55 - 0.25x \end{aligned}$$

where x represents the number of tickets sold. Determine the maximum profit and the cost per ticket that yields the maximum profit.

Solution

The cost is represented by $10x$ and the revenue is represented by $x(55 - 0.25x)$, where x represents the number of tickets sold.

Use the equation $P = R - C$ to represent the profit.

$$P = R - C$$
$$P = x(55 - 0.25x) - 10x$$
$$P = -0.25x^2 + 45x$$

The graph of $P = -0.25x^2 + 45x$ is a parabola that opens down. Find the x-coordinates of the vertex.

$$x = -\frac{b}{2a} = -\frac{45}{2(-0.25)} = 90$$

To find the maximum profit, evaluate $P = -0.25x^2 + 45x$ at 90.

$$P = -0.25x^2 + 45x$$
$$P = -0.25(90^2) + 45(90)$$
$$P = 2025$$

To find the price per ticket, evaluate $p = 55 - 0.25x$ at 90.

$$p = 55 - 0.25x$$
$$p = 55 - 0.25(90)$$
$$p = 32.50$$

The travel agency can expect a maximum profit of \$2025 when 90 people take the tour at a ticket price of \$32.50 per person.

The graph of the profit function is shown at the left.

P
(90, 2025)
2000
1500
1000
500
0 50 100 150 200 x

$P = -0.25x^2 + 45x$

1. **Business** A company's total monthly cost, in dollars, for producing and selling n DVDs per month is $C = 25n + 4000$. The company's revenue, in dollars, from selling all n DVDs is $R = 275n - 0.2n^2$.
   a. How many DVDs must the company produce and sell each month in order to make the maximum monthly profit?
   b. What is the maximum monthly profit?

2. **Business** A shipping company has determined that its cost to deliver x packages per week is $1000 + 6x$. The price per package that the company charges when it sends x packages per week is $24 - 0.02x$. Determine the company's maximum weekly profit and the price per package that yields the maximum profit.

Copyright © Houghton Mifflin Company. All rights reserved.

Chapter 11 Summary

Key Words	Examples
A *quadratic function* is one that can be expressed by the equation $f(x) = ax^2 + bx + c$, $a \neq 0$. [11.1A, p. 609]	$f(x) = 4x^2 + 3x - 5$ is a quadratic function.
The value of the independent variable is sometimes called the *input*. The result of evaluating the function is called the *output*. An *input/output table* shows the result of evaluating a function for various values of the independent variable. [11.1A, p. 609]	$f(x) = x^2 + 4x - 6$ (see table below)

x	y
−2	−10
−1	−9
0	−6
1	−1
2	6

The graph of a quadratic function is a *parabola*. When $a > 0$, the parabola opens up and the *vertex* of the parabola is the point with the smallest y-coordinate. When $a < 0$, the parabola opens down and the *vertex* of the parabola is the point with the largest y-coordinate. The *axis of symmetry* is the vertical line that passes through the vertex of the parabola and is parallel to the y-axis. [11.1A, pp. 609, 610]

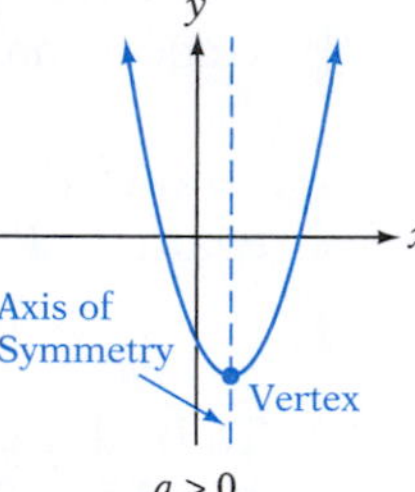

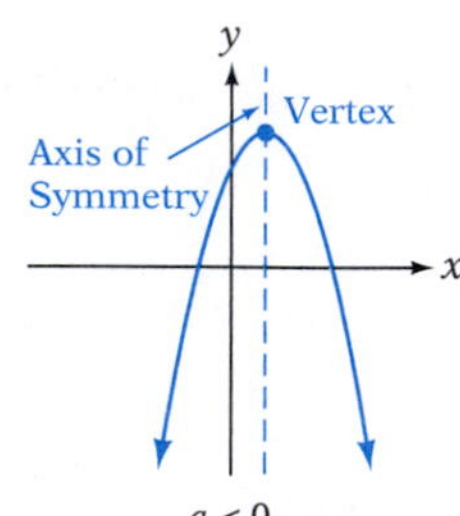

A point at which a graph crosses the x-axis is called an *x-intercept* of the graph. The x-intercepts occur when $y = 0$. [11.1B, p. 612]

$$x^2 - 5x + 6 = 0$$
$$(x - 2)(x - 3) = 0$$
$$x - 2 = 0 \qquad x - 3 = 0$$
$$x = 2 \qquad x = 3$$

The x-intercepts are (2, 0) and (3, 0).

A value of x for which $f(x) = 0$ is a *zero* of the function. [11.1B, p. 612]

$$f(x) = x^2 - 7x + 10$$
$$0 = x^2 - 7x + 10$$
$$0 = (x - 2)(x - 5)$$
$$x - 2 = 0 \qquad x - 5 = 0$$
$$x = 2 \qquad x = 5$$

2 and 5 are zeros of the function.

Copyright © Houghton Mifflin Company. All rights reserved.

The graph of a quadratic function has a *minimum value* if $a > 0$ and a *maximum value* if $a < 0$. [11.1C, p. 615]

The graph of $f(x) = 3x^2 - x + 4$ has a minimum value.
The graph of $f(x) = -x^2 - 6x + 5$ has a maximum value.

A function is a *one-to-one function* if, for any a and b in the domain of f, $f(a) = f(b)$ implies $a = b$. This means that given any y, there is only one x that can be paired with the given y. One-to-one functions are commonly written as 1–1. [11.4A, p. 639]

A linear function of the form $y = mx + b$, $m \neq 0$, is a 1–1 function.

The *inverse of a function* is the set of ordered pairs formed by reversing the coordinates of each ordered pair of the function. [11.4B, p. 640]

The inverse of the function $\{(1, 2), (2, 4), (3, 6), (4, 8), (5, 10)\}$ is $\{(2, 1), (4, 2), (6, 3), (8, 4), (10, 5)\}$.

Essential Rules and Procedures

Examples

Vertex and Axis of Symmetry of a Parabola [11.1A, p. 610]
Let $f(x) = ax^2 + bx + c$ be the equation of a parabola. The coordinates of the vertex are $\left(-\frac{b}{2a}, f\left(-\frac{b}{2a}\right)\right)$. The equation of the axis of symmetry is $x = -\frac{b}{2a}$.

$f(x) = x^2 - 2x - 4$
$a = 1, b = -2$
$-\dfrac{b}{2a} = -\dfrac{-2}{2(1)} = 1$
$f(1) = 1^2 - 2(1) - 4 = -5$

The coordinates of the vertex are $(1, -5)$.
The axis of symmetry is $x = 1$.

Effect of the Discriminant on the Number of x-Intercepts of a Parabola [11.1B, p. 613]
1. If $b^2 - 4ac = 0$, the parabola has one x-intercept.
2. If $b^2 - 4ac > 0$, the parabola has two x-intercepts.
3. If $b^2 - 4ac < 0$, the parabola has no x-intercepts.

$x^2 + 8x + 16 = 0$ has one x-intercept because $b^2 - 4ac = 8^2 - 4(1)(16) = 0$.

$2x^2 + 3x - 5 = 0$ has two x-intercepts because $b^2 - 4ac = 3^2 - 4(2)(-5) = 49$.

$3x^2 + 2x + 4 = 0$ has no x-intercepts because $b^2 - 4ac = 2^2 - 4(3)(4) = -44$.

To Find the Minimum or Maximum Value of a Quadratic Function [11.1C, p. 615]
First find the x-coordinate of the vertex. Then evaluate the function at that value.

$f(x) = x^2 - 2x - 4$
$a = 1, b = -2$
$-\dfrac{b}{2a} = -\dfrac{-2}{2(1)} = 1$
$f(1) = 1^2 - 2(1) - 4 = -5$
$a > 0$. The minimum value of the function is -5.

Copyright © Houghton Mifflin Company. All rights reserved.

Vertical-Line Test [11.2A, p. 626]
A graph defines a function if any vertical line intersects the graph at no more than one point.

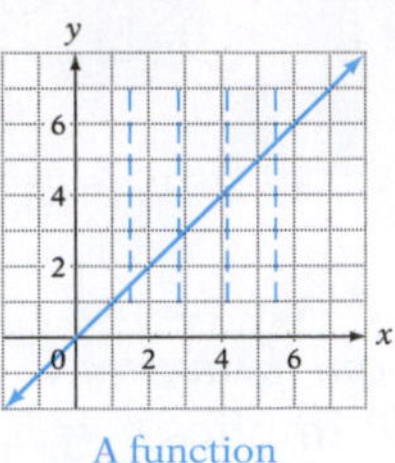
A function

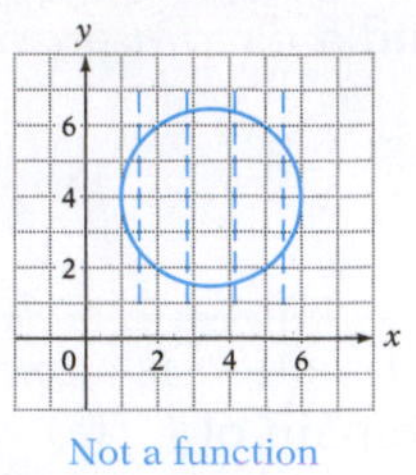
Not a function

Operations on Functions [11.3A, p. 631]
If f and g are functions and x is an element of the domain of each function, then

$(f+g)(x) = f(x) + g(x)$
$(f-g)(x) = f(x) - g(x)$
$(f \cdot g)(x) = f(x) \cdot g(x)$
$\left(\frac{f}{g}\right)(x) = \frac{f(x)}{g(x)}, g(x) \neq 0$

Given $f(x) = x + 2$ and $g(x) = 2x$, then

$$(f+g)(4) = f(4) + g(4) = (4+2) + 2(4) = 6 + 8 = 14$$
$$(f-g)(4) = f(4) - g(4) = (4+2) - 2(4) = 6 - 8 = -2$$
$$(f \cdot g)(4) = f(4) \cdot g(4) = (4+2) \cdot 2(4) = 6 \cdot 8 = 48$$
$$\left(\frac{f}{g}\right)(4) = \frac{f(4)}{g(4)} = \frac{4+2}{2(4)} = \frac{6}{8} = \frac{3}{4}$$

Definition of the Composition of Two Functions [11.3B, p. 633]
The composition of two functions f and g, symbolized by $f \circ g$, is the function whose value at x is given by $(f \circ g)(x) = f[g(x)]$.

Given $f(x) = x - 4$ and $g(x) = 4x$, then

$$(f \circ g)(2) = f[g(2)] = f(8) \quad \text{because } g(2) = 8$$
$$= 8 - 4 = 4$$

Horizontal-Line Test [11.4A, p. 640]
The graph of a function represents the graph of a 1–1 function if any horizontal line intersects the graph at no more than one point.

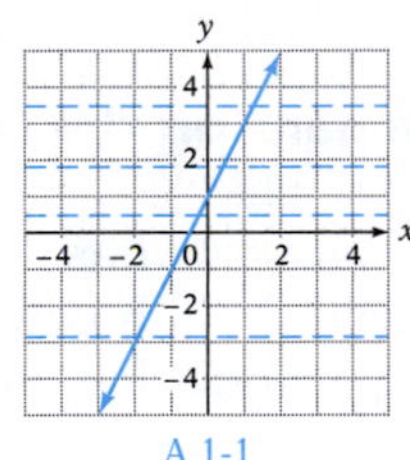
A 1-1 function

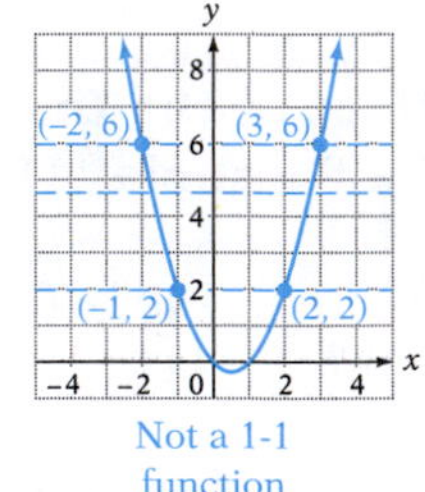

Not a 1-1 function

Condition for an Inverse Function [11.4B, p. 641]
A function f has an inverse function if and only if f is a 1–1 function.

The function $f(x) = x^2$ does not have an inverse function. When $y = 4$, $x = 2$ or -2; therefore, the function $f(x)$ is not a 1–1 function.

Composition of Inverse Functions Property [11.4B, p. 642]
$f^{-1}[f(x)] = x$ and $f[f^{-1}(x)] = x$.

$$f(x) = 2x - 3 \qquad f^{-1}(x) = \frac{1}{2}x + \frac{3}{2}$$
$$f^{-1}[f(x)] = \frac{1}{2}(2x - 3) + \frac{3}{2} = x$$
$$f[f^{-1}(x)] = 2\left(\frac{1}{2}x + \frac{3}{2}\right) - 3 = x$$

Copyright © Houghton Mifflin Company. All rights reserved.

Chapter 11 Review Exercises

1. Is the graph below the graph of a function?

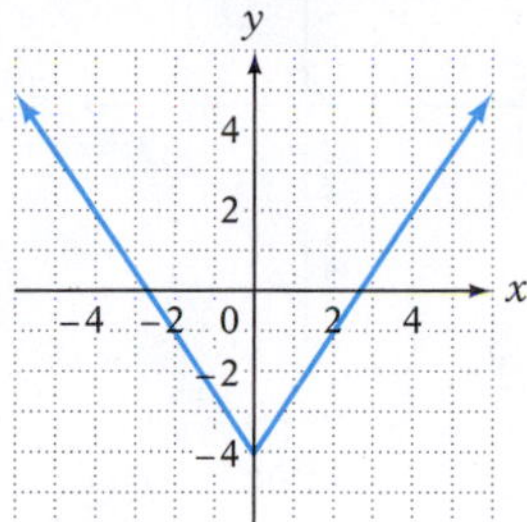

2. Is the graph below the graph of a 1–1 function?

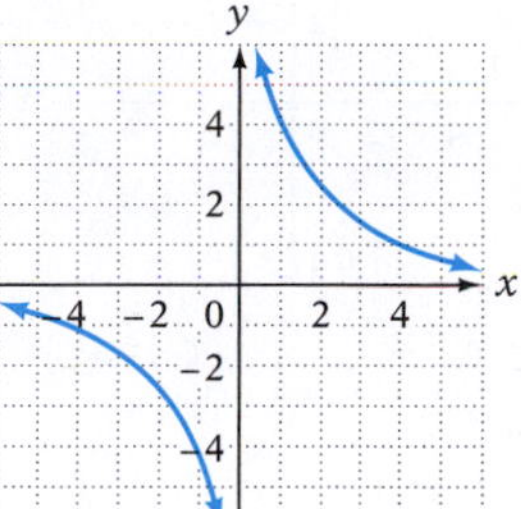

3. Graph $f(x) = 3x^3 - 2$. State the domain and range.

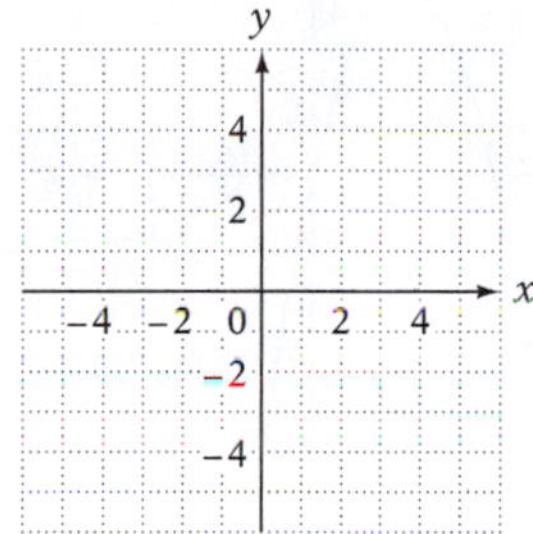

4. Graph $f(x) = \sqrt{x + 4}$. State the domain and range.

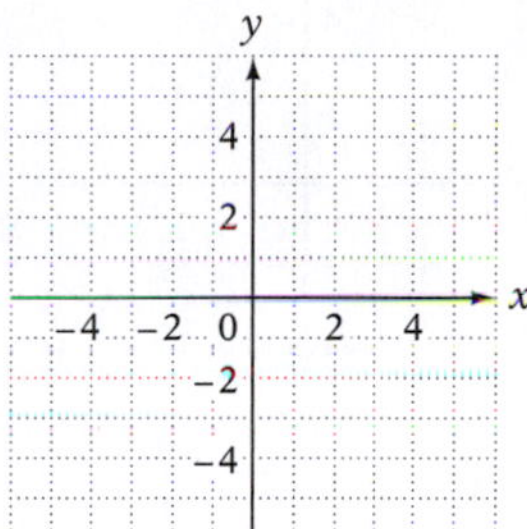

5. Use the discriminant to determine the number of x-intercepts of $y = -3x^2 + 4x + 6$.

6. Find the x-intercepts of $y = 3x^2 + 9x$.

7. Find the zeros of $f(x) = 3x^2 + 2x + 2$.

8. Find the maximum value of $f(x) = -2x^2 + 4x + 1$.

9. Find the minimum value of $f(x) = x^2 - 7x + 8$.

10. Given $f(x) = x^2 + 4$ and $g(x) = 4x - 1$, find $f[g(0)]$.

11. Given $f(x) = 6x + 8$ and $g(x) = 4x + 2$, find $g[f(-1)]$.

12. Given $f(x) = 3x^2 - 4$ and $g(x) = 2x + 1$, find $f[g(x)]$.

13. Are the functions given by $f(x) = -\frac{1}{4}x + \frac{5}{4}$ and $g(x) = -4x + 5$ inverses of each other?

Copyright © Houghton Mifflin Company. All rights reserved.

14. Graph $f(x) = x^2 + 2x - 4$. State the domain and range.

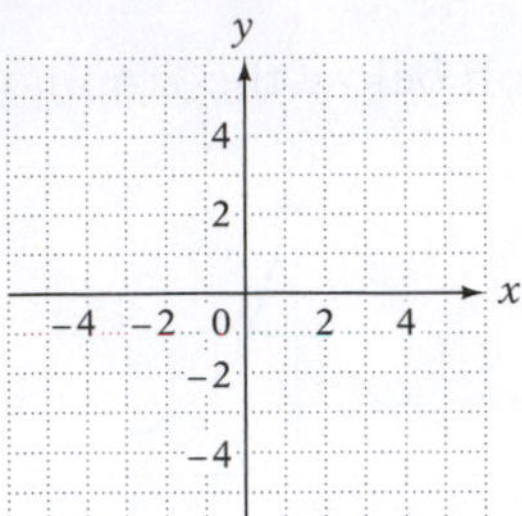

15. Find the vertex and axis of symmetry of the parabola whose equation is $y = x^2 - 2x + 3$. Then graph the equation.

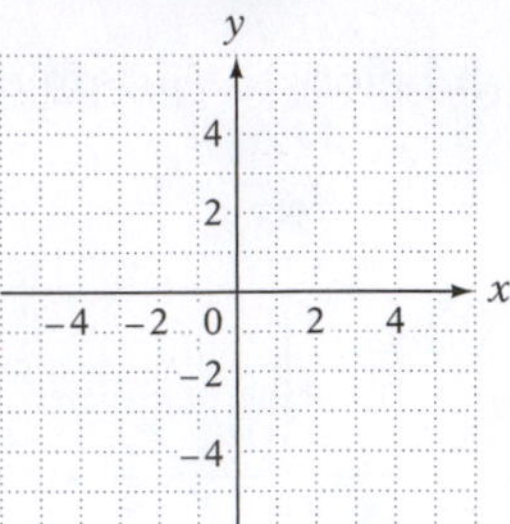

16. Graph $f(x) = |x| - 3$. State the domain and range.

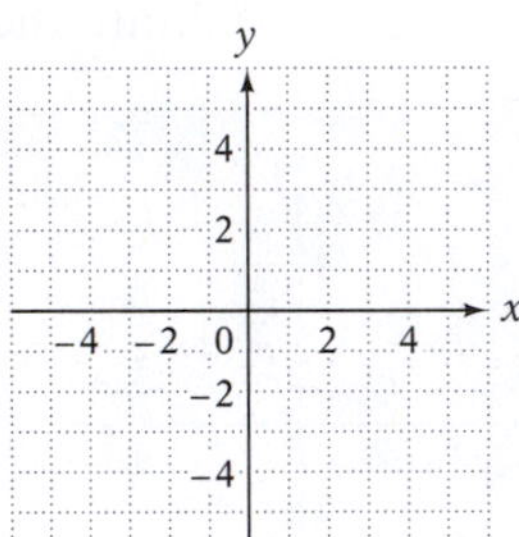

17. Is the graph below the graph of a 1–1 function?

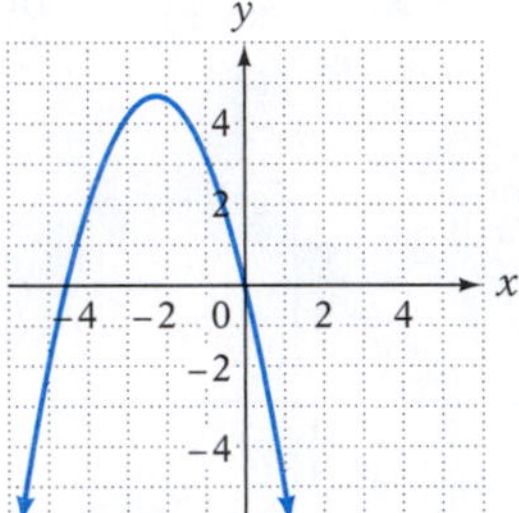

For Exercises 18 to 21, use $f(x) = x^2 + 2x - 3$ and $g(x) = x^2 - 2$.

18. Evaluate: $(f + g)(2)$

19. Evaluate: $(f - g)(-4)$

20. Evaluate: $(f \cdot g)(-4)$

21. Evaluate: $\left(\frac{f}{g}\right)(3)$

22. Given $f(x) = 2x^2 + x - 5$ and $g(x) = 3x - 1$, find $g[f(x)]$.

23. Find the inverse of $f(x) = -6x + 4$.

24. Find the inverse of $f(x) = \frac{2}{3}x - 12$.

25. Find the inverse of $f(x) = \frac{1}{2}x + 8$.

26. **Geometry** The perimeter of a rectangle is 28 ft. What dimensions would give the rectangle a maximum area?

Copyright © Houghton Mifflin Company. All rights reserved.

Chapter 11 Test

1. Find the zeros of the function $f(x) = 2x^2 - 3x + 4$.

2. Find the x-intercepts of $y = x^2 + 3x - 8$.

3. Use the discriminant to determine the number of zeros of the parabola $y = 3x^2 + 2x - 4$.

4. Given $f(x) = x^2 + 2x - 3$ and $g(x) = x^3 - 1$, find $(f - g)(2)$.

5. Given $f(x) = x^3 + 1$ and $g(x) = 2x - 3$, find $(f \cdot g)(-3)$.

6. Given $f(x) = 4x - 5$ and $g(x) = x^2 + 3x + 4$, find $\left(\frac{f}{g}\right)(-2)$.

7. Given $f(x) = x^2 + 4$ and $g(x) = 2x^2 + 2x + 1$, find $(f - g)(-4)$.

8. Given $f(x) = 4x + 2$ and $g(x) = \frac{x}{x + 1}$, find $f[g(3)]$.

9. Given $f(x) = 2x^2 - 7$ and $g(x) = x - 1$, find $f[g(x)]$.

10. Find the maximum value of the function $f(x) = -x^2 + 8x - 7$.

11. Find the inverse of the function $f(x) = 4x - 2$.

Copyright © Houghton Mifflin Company. All rights reserved.

12. Find the inverse of the function $f(x) = \frac{1}{4}x - 4$.

13. Find the inverse of the function $\{(2, 6), (3, 5), (4, 4), (5, 3)\}$.

14. Are the functions $f(x) = \frac{1}{2}x + 2$ and $g(x) = 2x - 4$ inverses of each other?

15. Find the maximum product of two numbers whose sum is 20.

16. Graph $f(x) = -\sqrt{3 - x}$. State the domain and range.

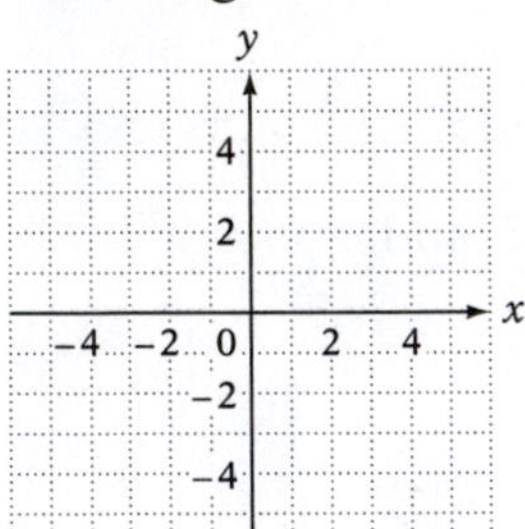

17. Graph $f(x) = \left|\frac{1}{2}x\right| - 2$. State the domain and range.

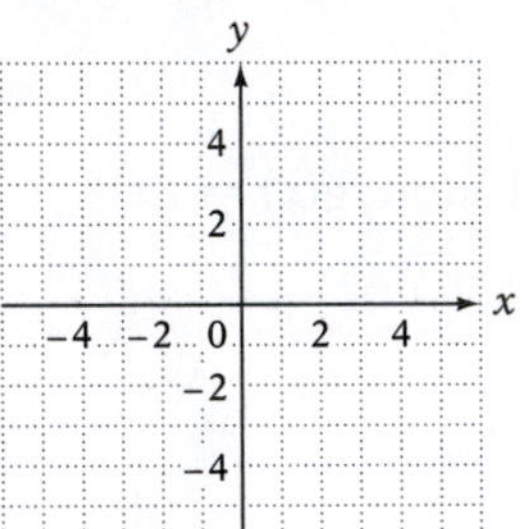

18. Graph $f(x) = x^3 - 3x + 2$. State the domain and range.

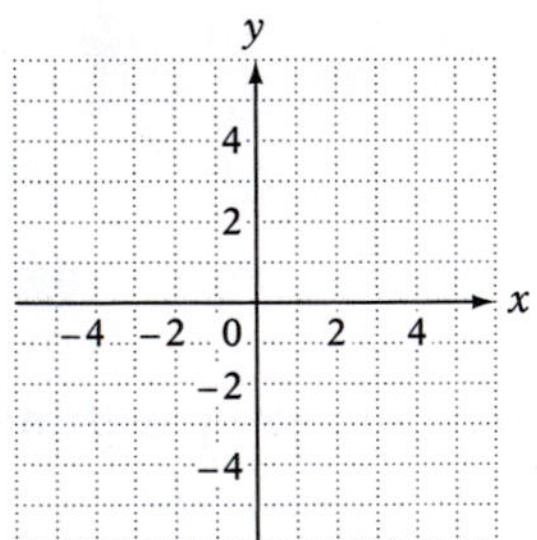

19. Determine whether the graph is the graph of a 1–1 function.

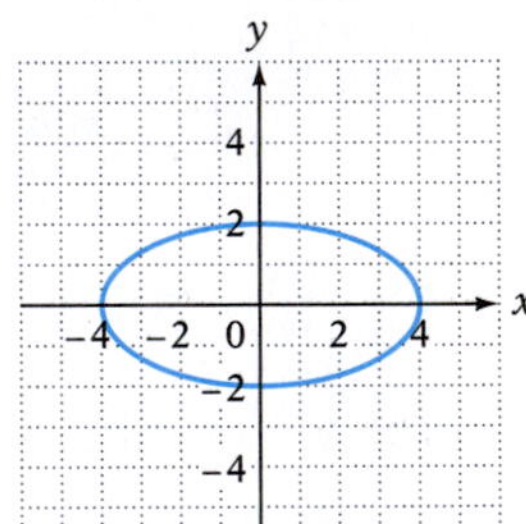

20. **Geometry** The perimeter of a rectangle is 200 cm. What dimensions would give the rectangle a maximum area? What is the maximum area?

Copyright © Houghton Mifflin Company. All rights reserved.

Cumulative Review Exercises

1. Evaluate $-3a + \left|\frac{3b - ab}{3b - c}\right|$ when $a = 2$, $b = 2$, and $c = -2$.

2. Graph $\{x \mid x < -3\} \cap \{x \mid x > -4\}$.

−5 −4 −3 −2 −1 0 1 2 3 4 5

3. Solve: $\frac{3x - 1}{6} - \frac{5 - x}{4} = \frac{5}{6}$

4. Solve: $4x - 2 < -10$ or $3x - 1 > 8$
Write the solution set in set-builder notation.

5. Solve: $|8 - 2x| \geq 0$

6. Simplify: $\left(\frac{3a^3b}{2a}\right)^2\left(\frac{a^2}{-3b^2}\right)^3$

7. Simplify: $(x - 4)(2x^2 + 4x - 1)$

8. Solve by using the addition method:
$6x - 2y = -3$
$4x + y = 5$

9. Factor: $x^3y + x^2y^2 - 6xy^3$

10. Solve: $(b + 2)(b - 5) = 2b + 14$

11. Solve: $x^2 - 2x > 15$

12. Simplify: $\frac{x^2 + 4x - 5}{2x^2 - 3x + 1} - \frac{x}{2x - 1}$

13. Solve: $\frac{5}{x^2 + 7x + 12} = \frac{9}{x + 4} - \frac{2}{x + 3}$

14. Simplify: $\frac{4 - 6i}{2i}$

15. Graph $f(x) = \frac{1}{4}x^2$. Find the vertex and axis of symmetry.

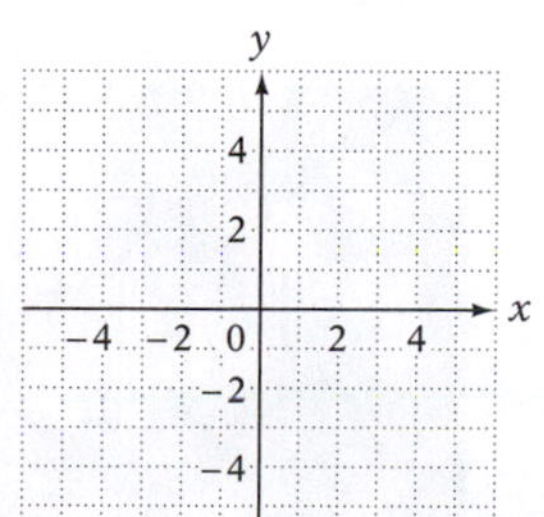

16. Graph the solution set of $3x - 4y \geq 8$.

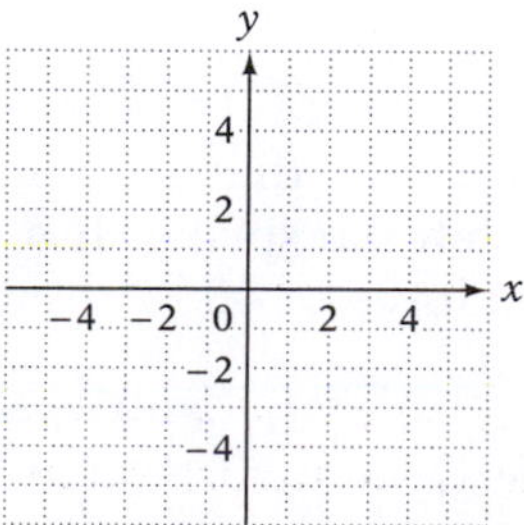

Copyright © Houghton Mifflin Company. All rights reserved.

17. Find the equation of the line containing the points $(-3, 4)$ and $(2, -6)$.

18. Find the equation of the line that contains the point $(-3, 1)$ and is perpendicular to the line $2x - 3y = 6$.

19. Solve: $3x^2 = 3x - 1$

20. Solve: $\sqrt{8x + 1} = 2x - 1$

21. Find the minimum value of the function $f(x) = 2x^2 - 3$.

22. Find the range of $f(x) = |3x - 4|$ if the domain is $\{0, 1, 2, 3\}$.

23. Is this set of ordered pairs a function? $\{(-3, 0), (-2, 0), (-1, 1), (0, 1)\}$

24. Solve: $\sqrt[3]{5x - 2} = 2$

25. Given $g(x) = 3x - 5$ and $h(x) = \frac{1}{2}x + 4$, find $g[h(2)]$.

26. Find the inverse of the function given by $f(x) = -3x + 9$.

27. **Mixtures** Find the cost per pound of a tea mixture made from 30 lb of tea costing $4.50 per pound and 45 lb of tea costing $3.60 per pound.

28. **Mixtures** How many pounds of an 80% copper alloy must be mixed with 50 lb of a 20% copper alloy to make an alloy that is 40% copper?

29. **Mixtures** Six ounces of an insecticide are mixed with 16 gal of water to make a spray for spraying an orange grove. How much additional insecticide is required if it is to be mixed with 28 gal of water?

30. **Tanks** A large pipe can fill a tank in 8 min less time than it takes a smaller pipe to fill the same tank. Working together, both pipes can fill the tank in 3 min. How long would it take the larger pipe, working alone, to fill the tank?

31. **Physics** The distance, d, that a spring stretches varies directly as the force, f, used to stretch the spring. If a force of 50 lb can stretch a spring 30 in., how far can a force of 40 lb stretch the spring?

32. **Music** The frequency of vibration, f, in a pipe in an open pipe organ varies inversely as the length, L, of the pipe. If the air in a pipe 2 m long vibrates 60 times per minute, find the frequency in a pipe that is 1.5 m long.

Copyright © Houghton Mifflin Company. All rights reserved.

chapter

12 Exponential and Logarithmic Functions

This archaeopteryx skeleton is an exciting find for archaeologists. The archaeopteryx is thought to be from the Upper Jurassic period, and is considered by some scientists to be the most primitive of all birds. To determine exactly how old the bones are, archaeologists will use the carbon-dating method, which involves an exponential function. Instruments are used to detect the amount of carbon-14 left in an object. Since carbon-14 occurs naturally in living things and gradually decays after death, the amount of carbon-14 remaining can reveal the object's age. **The HOW TO on page 690** gives an example of the carbon dating process.

OBJECTIVES

Section 12.1

A To evaluate an exponential function
B To graph an exponential function

Section 12.2

A To find the logarithm of a number
B To use the Properties of Logarithms to simplify expressions containing logarithms
C To use the Change-of-Base Formula

Section 12.3

A To graph a logarithmic function

Section 12.4

A To solve an exponential equation
B To solve a logarithmic equation

Section 12.5

A To solve application problems

Need help? For online student resources, such as section quizzes, visit this textbook's website at **math.college.hmco.com/students.**

Copyright © Houghton Mifflin Company. All rights reserved.

PREP TEST • • •

Do these exercises to prepare for Chapter 12.

1. Simplify: 3^{-2}

2. Simplify: $\left(\frac{1}{2}\right)^{-4}$

3. Complete: $\frac{1}{8} = 2^{?}$

4. Evaluate $f(x) = x^4 + x^3$ for $x = -1$ and $x = 3$.

5. Solve: $3x + 7 = x - 5$

6. Solve: $16 = x^2 - 6x$

7. Evaluate $A(1 + i)^n$ for $A = 5000$, $i = 0.04$, and $n = 6$. Round to the nearest hundredth.

8. Graph: $f(x) = x^2 - 1$

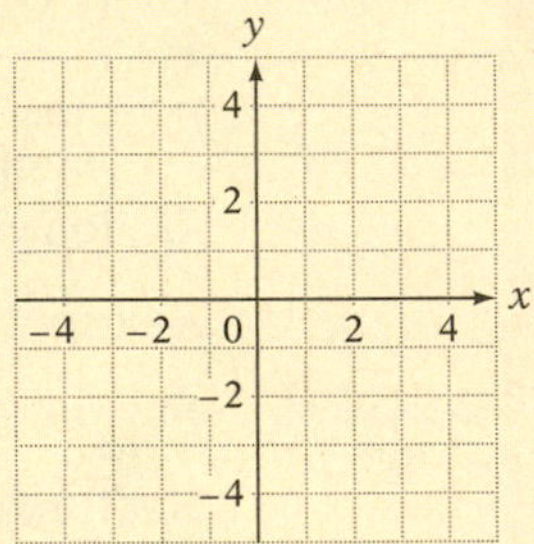

GO FIGURE • • •

What is the ones digit of $1 + 9 + 9^2 + 9^3 + 9^4 + \cdots + 9^{2000} + 9^{2001}$?

Copyright © Houghton Mifflin Company. All rights reserved.

12.1 Exponential Functions

Objective A To evaluate an exponential function

The growth of a \$500 savings account that earns 5% annual interest compounded daily is shown in the graph at the right. In approximately 14 years, the savings account contains approximately \$1000, twice the initial amount. The growth of this savings account is an example of an exponential function.

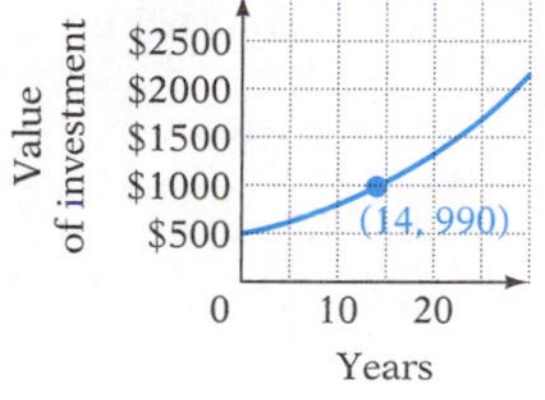

The pressure of the atmosphere at a certain height is shown in the graph at the right. This is another example of an exponential function. From the graph, we read that the air pressure is approximately 6.5 lb/in² at an altitude of 20,000 ft.

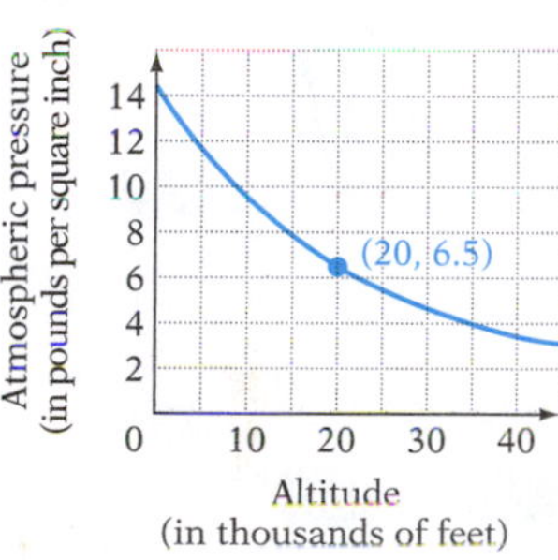

Definition of an Exponential Function

The **exponential function** with base b is defined by

$$f(x) = b^x$$

where $b > 0$, $b \neq 1$, and x is any real number.

In the definition of an exponential function, b, the base, is required to be positive. If the base were a negative number, the value of the function would be a complex number for some values of x. For instance, the value of $f(x) = (-4)^x$ when $x = \frac{1}{2}$ is $f\left(\frac{1}{2}\right) = (-4)^{1/2} = \sqrt{-4} = 2i$. To avoid complex number values of a function, the base of the exponential function is always a positive number.

Integrating Technology

A graphing calculator can be used to evaluate an exponential expression in which the exponent is an irrational number.

```
4^√(2)
            7.102993301
```

HOW TO Evaluate $f(x) = 2^x$ at $x = 3$ and $x = -2$.

$f(3) = 2^3 = 8$ • Substitute 3 for x and simplify.

$f(-2) = 2^{-2} = \frac{1}{2^2} = \frac{1}{4}$ • Substitute -2 for x and simplify.

To evaluate an exponential expression in which the exponent is an irrational number such as $\sqrt{2}$, we obtain an approximation to the value of the function by approximating the irrational number. For instance, the value of $f(x) = 4^x$ when $x = \sqrt{2}$ can be approximated by using an approximation of $\sqrt{2}$.

$$f(\sqrt{2}) = 4^{\sqrt{2}} \approx 4^{1.4142} \approx 7.1029$$

Copyright © Houghton Mifflin Company. All rights reserved.

TAKE NOTE
The natural exponential function is an extremely important function. It is used extensively in applied problems in virtually all disciplines, from archaeology to zoology. Leonhard Euler (1707–1783) was the first to use the letter e as the base of the natural exponential function.

Because $f(x) = b^x$ ($b > 0, b \neq 1$) can be evaluated at both rational and irrational numbers, the domain of f is all real numbers. And because $b^x > 0$ for all values of x, the range of f is the positive real numbers.

A frequently used base in applications of exponential functions is an irrational number designated by e. **The number e is approximately 2.71828183.** It is an irrational number, so it has a nonterminating, nonrepeating decimal representation.

Natural Exponential Function

The function defined by $f(x) = e^x$ is called the **natural exponential function.**

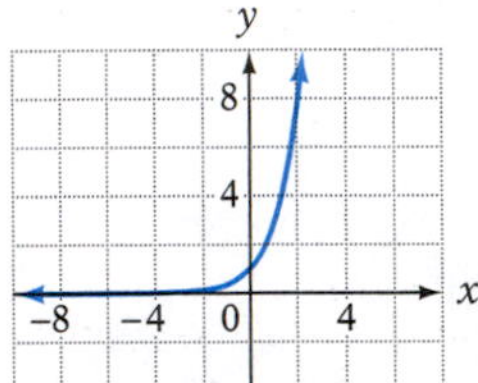

The e^x key on a calculator can be used to evaluate the natural exponential function. The graph of $y = e^x$ is shown at the left.

Example 1

Evaluate $f(x) = \left(\frac{1}{2}\right)^x$ at $x = 2$ and $x = -3$.

Solution

$f(x) = \left(\frac{1}{2}\right)^x$

$f(2) = \left(\frac{1}{2}\right)^2 = \frac{1}{4}$ • $x = 2$

$f(-3) = \left(\frac{1}{2}\right)^{-3} = 2^3 = 8$ • $x = -3$

You Try It 1

Evaluate $f(x) = \left(\frac{2}{3}\right)^x$ at $x = 3$ and $x = -2$.

Your solution

Example 2

Evaluate $f(x) = 2^{3x-1}$ at $x = 1$ and $x = -1$.

Solution

$f(x) = 2^{3x-1}$

$f(1) = 2^{3(1)-1} = 2^2 = 4$ • $x = 1$

$f(-1) = 2^{3(-1)-1} = 2^{-4} = \frac{1}{2^4} = \frac{1}{16}$ • $x = -1$

You Try It 2

Evaluate $f(x) = 2^{2x+1}$ at $x = 0$ and $x = -2$.

Your solution

Example 3

Evaluate $f(x) = e^{2x}$ at $x = 1$ and $x = -1$. Round to the nearest ten-thousandth.

Solution

$f(x) = e^{2x}$

$f(1) = e^{2\cdot 1} = e^2 \approx 7.3891$ • $x = 1$

$f(-1) = e^{2(-1)} = e^{-2} \approx 0.1353$ • $x = -1$

You Try It 3

Evaluate $f(x) = e^{2x-1}$ at $x = 2$ and $x = -2$. Round to the nearest ten-thousandth.

Your solution

Solutions on p. S35

Copyright © Houghton Mifflin Company. All rights reserved.

Objective B To graph an exponential function

SSM

Integrating Technology

See the appendix Keystroke Guide: *Graph* for instructions on using a graphing calculator to graph functions.

Some properties of an exponential function can be seen from its graph.

HOW TO Graph $f(x) = 2^x$.

Think of this as the equation $y = 2^x$.

Choose values of x and find the corresponding values of y. The results can be recorded in a table.

x	$f(x) = y$
-2	$2^{-2} = \frac{1}{4}$
-1	$2^{-1} = \frac{1}{2}$
0	$2^0 = 1$
1	$2^1 = 2$
2	$2^2 = 4$
3	$2^3 = 8$

Graph the ordered pairs on a rectangular coordinate system.

Connect the points with a smooth curve.

Note that any vertical line would intersect the graph at only one point. Therefore, by the vertical-line test, the graph of $f(x) = 2^x$ is the graph of a function. Also note that any horizontal line would intersect the graph at only one point. Therefore, the graph of $f(x) = 2^x$ is the graph of a one-to-one function.

HOW TO Graph $f(x) = \left(\frac{1}{2}\right)^x$.

Think of this as the equation $y = \left(\frac{1}{2}\right)^x$.

Choose values of x and find the corresponding values of y.

Graph the ordered pairs on a rectangular coordinate system.

Connect the points with a smooth curve.

x	$f(x) = y$
-3	$\left(\frac{1}{2}\right)^{-3} = 8$
-2	$\left(\frac{1}{2}\right)^{-2} = 4$
-1	$\left(\frac{1}{2}\right)^{-1} = 2$
0	$\left(\frac{1}{2}\right)^{0} = 1$
1	$\left(\frac{1}{2}\right)^{1} = \frac{1}{2}$
2	$\left(\frac{1}{2}\right)^{2} = \frac{1}{4}$

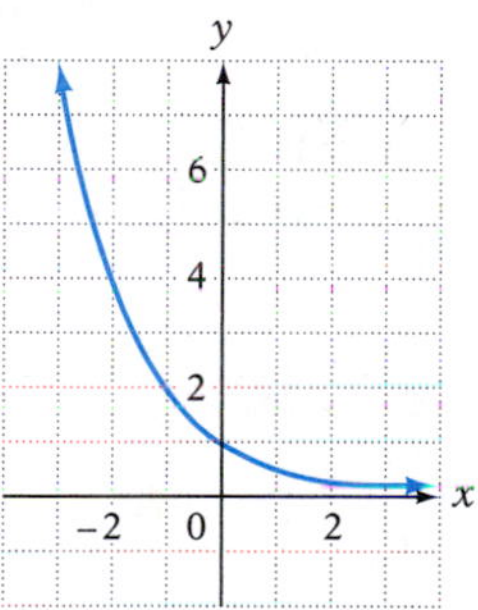

Applying the vertical-line and horizontal-line tests reveals that the graph of $f(x) = \left(\frac{1}{2}\right)^x$ is also the graph of a one-to-one function.

TAKE NOTE

Note that because $2^{-x} = (2^{-1})^x = \left(\frac{1}{2}\right)^x$, the graphs of $f(x) = 2^{-x}$ and $f(x) = \left(\frac{1}{2}\right)^x$ are the same.

HOW TO Graph $f(x) = 2^{-x}$.

Think of this as the equation $y = 2^{-x}$.

Choose values of x and find the corresponding values of y.

Graph the ordered pairs on a rectangular coordinate system.

Connect the points with a smooth curve.

x	y
-3	8
-2	4
-1	2
0	1
1	$\frac{1}{2}$
2	$\frac{1}{4}$

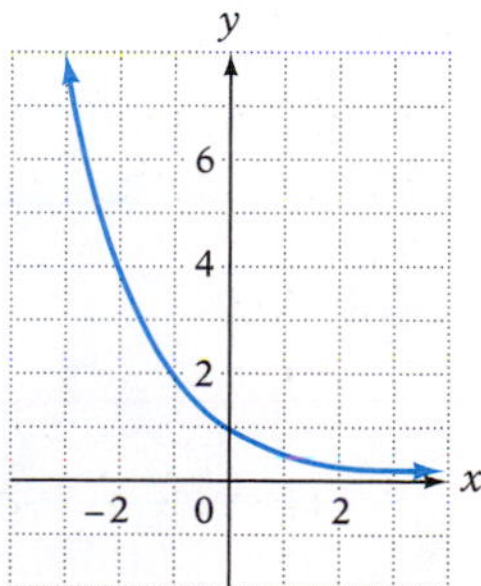

Copyright © Houghton Mifflin Company. All rights reserved.

Example 4 Graph: $f(x) = 3^{\frac{1}{2}x-1}$

Solution

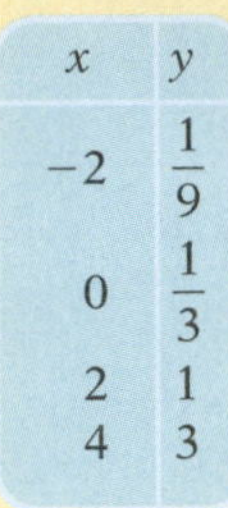

x	y
-2	$\frac{1}{9}$
0	$\frac{1}{3}$
2	1
4	3

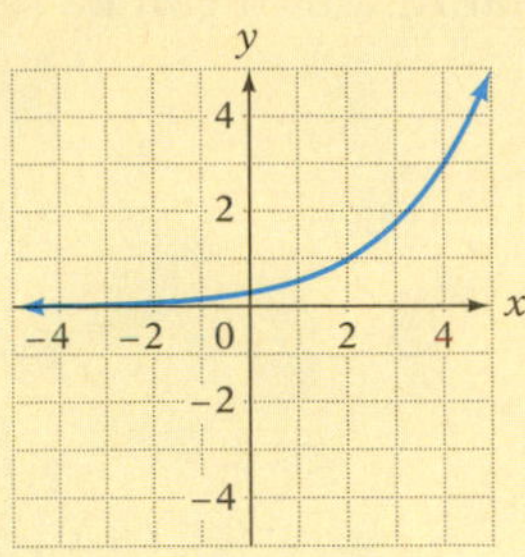

You Try It 4 Graph: $f(x) = 2^{-\frac{1}{2}x}$

Your solution

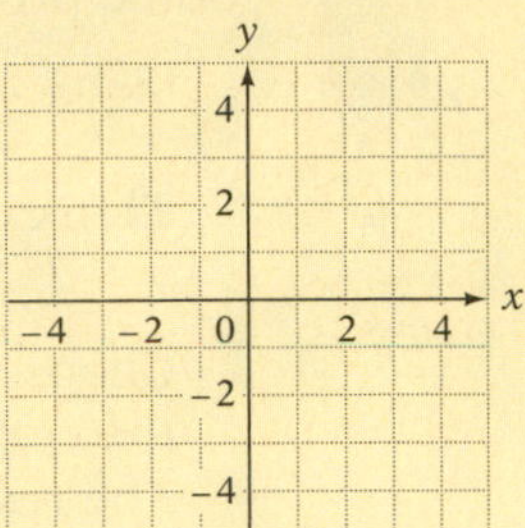

Example 5 Graph: $f(x) = 2^x - 1$

Solution

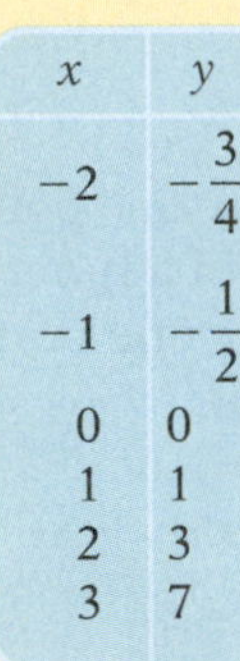

x	y
-2	$-\frac{3}{4}$
-1	$-\frac{1}{2}$
0	0
1	1
2	3
3	7

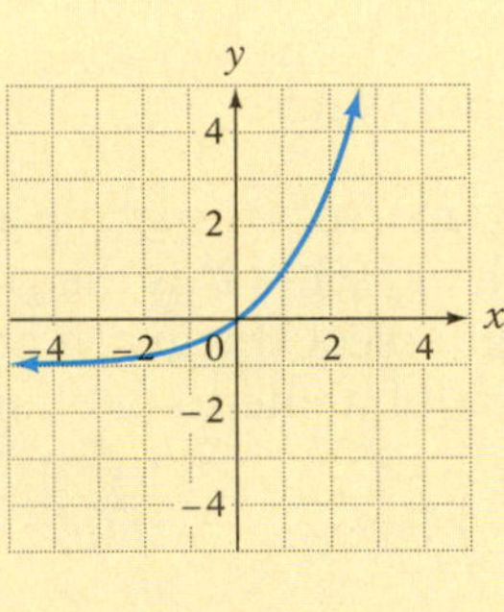

You Try It 5 Graph: $f(x) = 2^x + 1$

Your solution

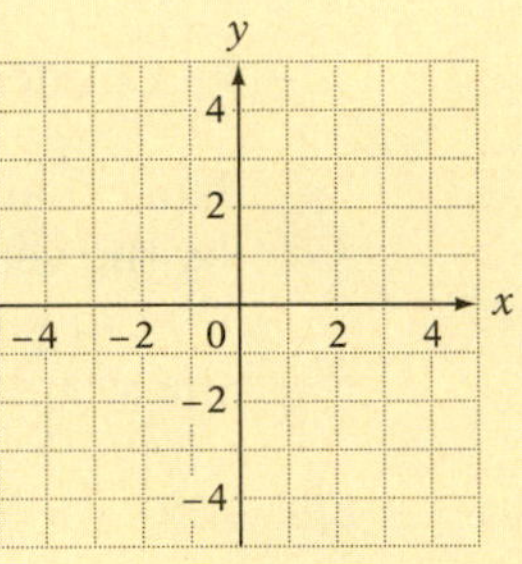

Example 6 Graph: $f(x) = \left(\frac{1}{3}\right)^x - 2$

Solution

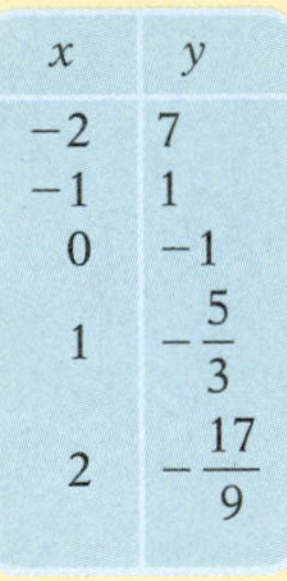

x	y
-2	7
-1	1
0	-1
1	$-\frac{5}{3}$
2	$-\frac{17}{9}$

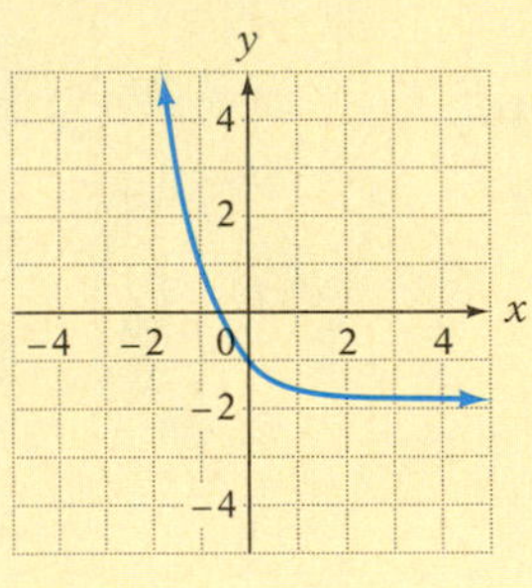

You Try It 6 Graph: $f(x) = 2^{-x} + 2$

Your solution

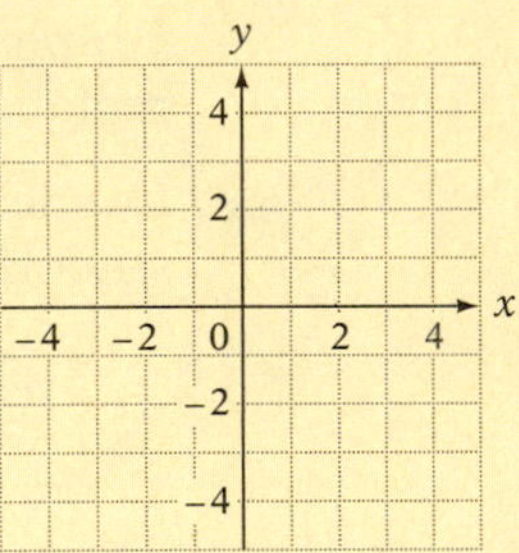

Example 7 Graph: $f(x) = 2^{-\frac{1}{2}x} - 1$

Solution

x	y
-6	7
-4	3
-2	1
0	0
2	$-\frac{1}{2}$
4	$-\frac{3}{4}$

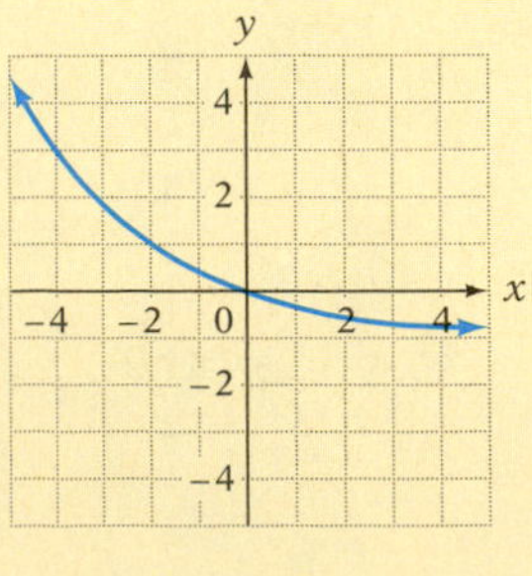

You Try It 7 Graph: $f(x) = \left(\frac{1}{2}\right)^{-\frac{1}{2}x} + 2$

Your solution

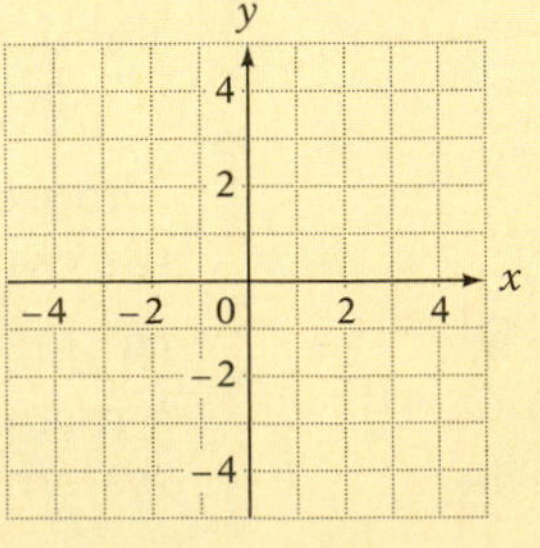

Solutions on p. S35

Copyright © Houghton Mifflin Company. All rights reserved.

12.1 Exercises

Objective A To evaluate an exponential function

1. What is an exponential function?

2. What is the natural exponential function?

3. Which of the following cannot be the base of an exponential function?
 a. 7 **b.** $\frac{1}{4}$ **c.** -5 **d.** 0.01

4. Which of the following cannot be the base of an exponential function?
 a. 0.9 **b.** 476 **c.** 8 **d.** $-\frac{1}{2}$

5. Given $f(x) = 3^x$, evaluate the following.
 a. $f(2)$ **b.** $f(0)$ **c.** $f(-2)$

6. Given $H(x) = 2^x$, evaluate the following.
 a. $H(-3)$ **b.** $H(0)$ **c.** $H(2)$

7. Given $g(x) = 2^{x+1}$, evaluate the following.
 a. $g(3)$ **b.** $g(1)$ **c.** $g(-3)$

8. Given $F(x) = 3^{x-2}$, evaluate the following.
 a. $F(-4)$ **b.** $F(-1)$ **c.** $F(0)$

9. Given $P(x) = \left(\frac{1}{2}\right)^{2x}$, evaluate the following.
 a. $P(0)$ **b.** $P\left(\frac{3}{2}\right)$ **c.** $P(-2)$

10. Given $R(t) = \left(\frac{1}{3}\right)^{3t}$, evaluate the following.
 a. $R\left(-\frac{1}{3}\right)$ **b.** $R(1)$ **c.** $R(-2)$

11. Given $G(x) = e^{x/2}$, evaluate the following. Round to the nearest ten-thousandth.
 a. $G(4)$ **b.** $G(-2)$ **c.** $G\left(\frac{1}{2}\right)$

12. Given $f(x) = e^{2x}$, evaluate the following. Round to the nearest ten-thousandth.
 a. $f(-2)$ **b.** $f\left(-\frac{2}{3}\right)$ **c.** $f(2)$

13. Given $H(r) = e^{-r+3}$, evaluate the following. Round to the nearest ten-thousandth.
 a. $H(-1)$ **b.** $H(3)$ **c.** $H(5)$

14. Given $P(t) = e^{-\frac{1}{2}t}$, evaluate the following. Round to the nearest ten-thousandth.
 a. $P(-3)$ **b.** $P(4)$ **c.** $P\left(\frac{1}{2}\right)$

Copyright © Houghton Mifflin Company. All rights reserved.

15. Given $F(x) = 2^{x^2}$, evaluate the following.
a. $F(2)$ **b.** $F(-2)$ **c.** $F(0)$

16. Given $Q(x) = 2^{-x^2}$, evaluate the following.
a. $Q(3)$ **b.** $Q(-1)$ **c.** $Q(-2)$

17. Given $f(x) = e^{-x^2/2}$, evaluate the following. Round to the nearest ten-thousandth.
a. $f(-2)$ **b.** $f(2)$ **c.** $f(-3)$

18. Given $f(x) = e^{-2x} + 1$, evaluate the following. Round to the nearest ten-thousandth.
a. $f(-1)$ **b.** $f(3)$ **c.** $f(-2)$

Objective B To graph an exponential function

Graph the functions in Exercises 19 to 30.

19. $f(x) = 3^x$

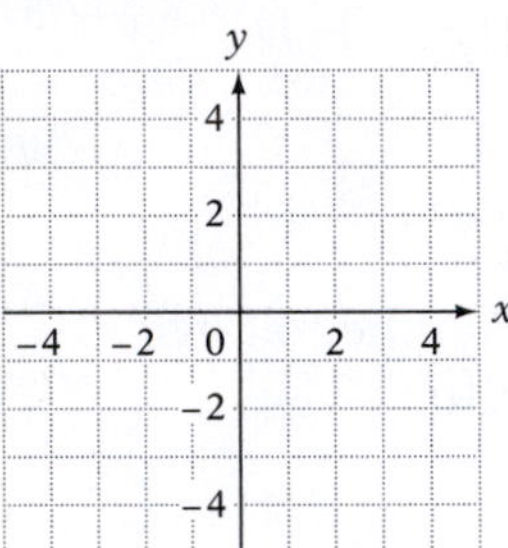

20. $f(x) = 3^{-x}$

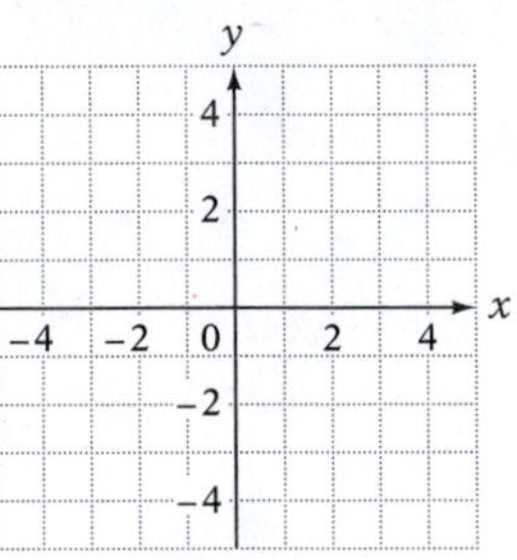

21. $f(x) = 2^{x+1}$

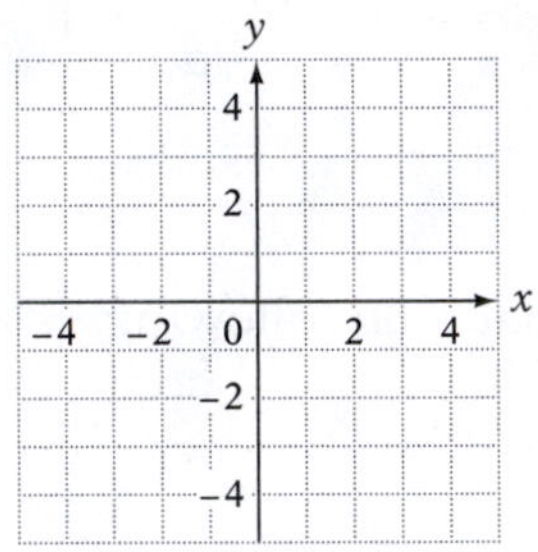

22. $f(x) = 2^{x-1}$

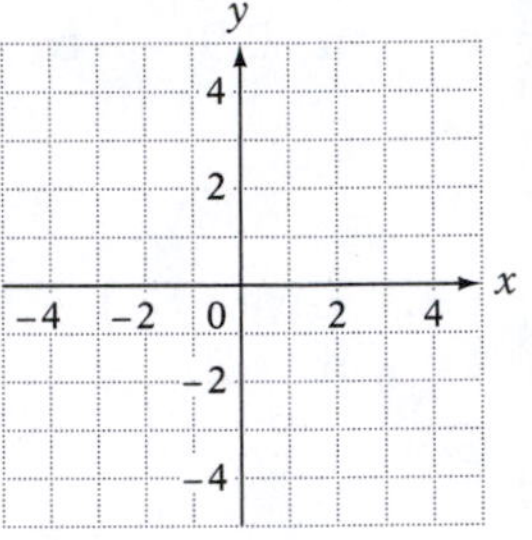

23. $f(x) = \left(\frac{1}{3}\right)^x$

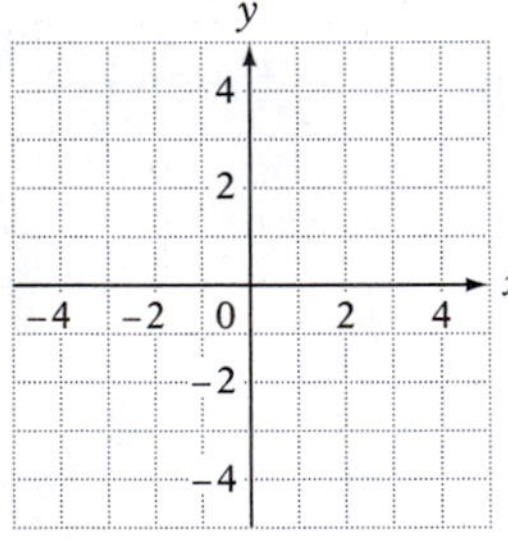

24. $f(x) = \left(\frac{2}{3}\right)^x$

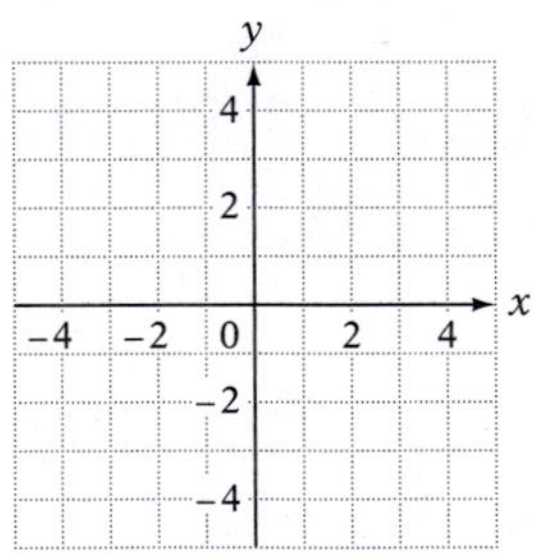

25. $f(x) = 2^{-x} + 1$

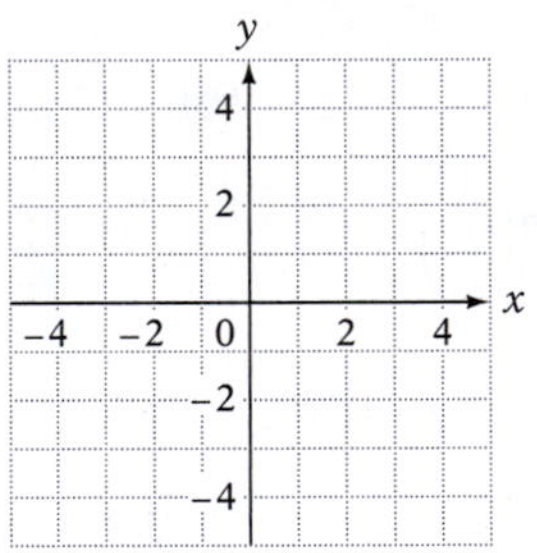

26. $f(x) = 2^x - 3$

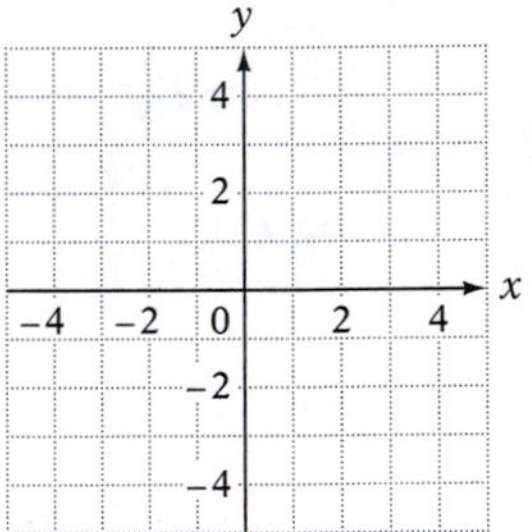

27. $f(x) = \left(\frac{1}{3}\right)^{-x}$

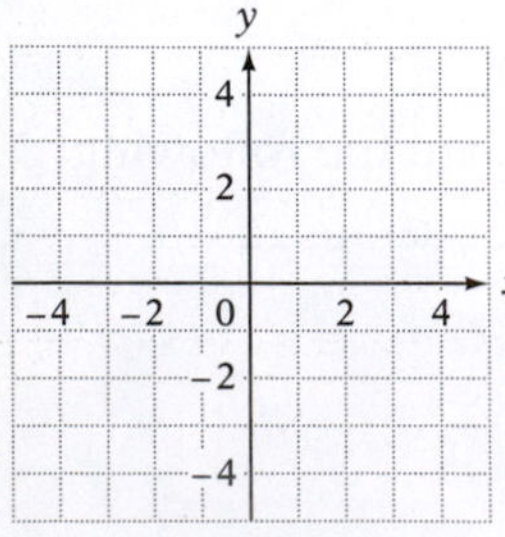

28. $f(x) = \left(\frac{3}{2}\right)^{-x}$

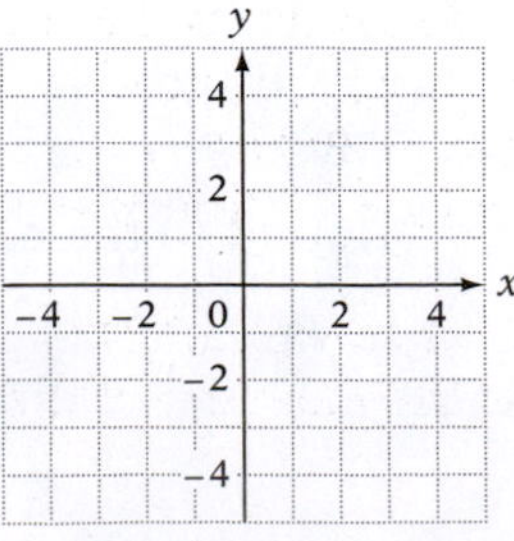

29. $f(x) = \left(\frac{1}{2}\right)^{-x} + 2$

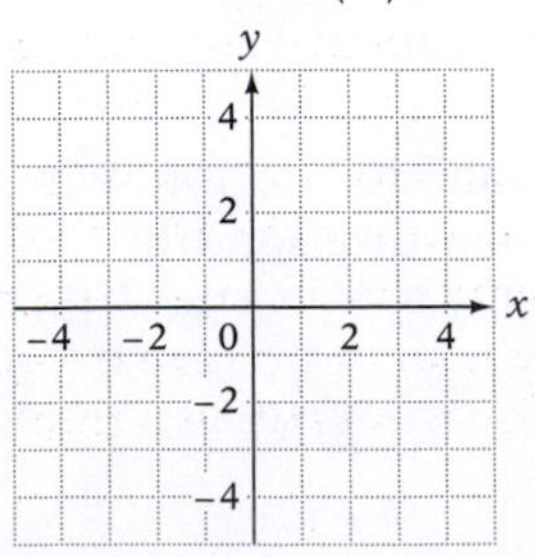

30. $f(x) = \left(\frac{1}{2}\right)^x - 1$

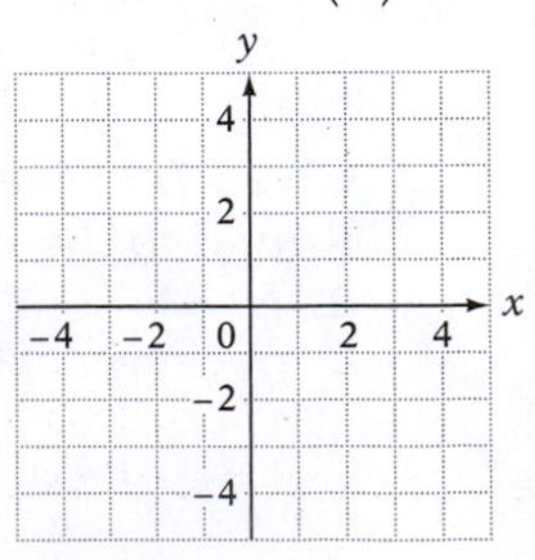

Copyright © Houghton Mifflin Company. All rights reserved.

31. Which of the following functions have the same graph?

a. $f(x) = 3^x$ **b.** $f(x) = \left(\frac{1}{3}\right)^x$ **c.** $f(x) = x^3$ **d.** $f(x) = 3^{-x}$

32. Which of the following functions have the same graph?

a. $f(x) = x^4$ **b.** $f(x) = 4^{-x}$ **c.** $f(x) = 4^x$ **d.** $f(x) = \left(\frac{1}{4}\right)^x$

33. Graph $f(x) = 3^x$ and $f(x) = 3^{-x}$ and find the point of intersection of the two graphs.

34. Graph $f(x) = 2^{x+1}$ and $f(x) = 2^{-x+1}$ and find the point of intersection of the two graphs.

35. Graph $f(x) = \left(\frac{1}{3}\right)^x$. What are the x- and y-intercepts of the graph of the function?

36. Graph $f(x) = \left(\frac{1}{3}\right)^{-x}$. What are the x- and y-intercepts of the graph of the function?

APPLYING THE CONCEPTS

Use a graphing calculator to graph the functions in Exercises 37 to 39.

37. $P(x) = (\sqrt{3})^x$

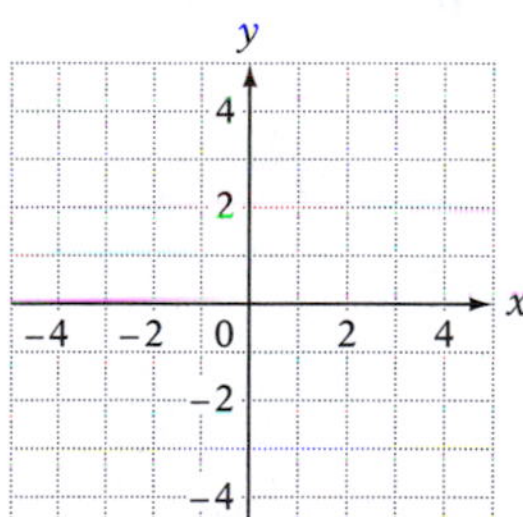

38. $Q(x) = (\sqrt{3})^{-x}$

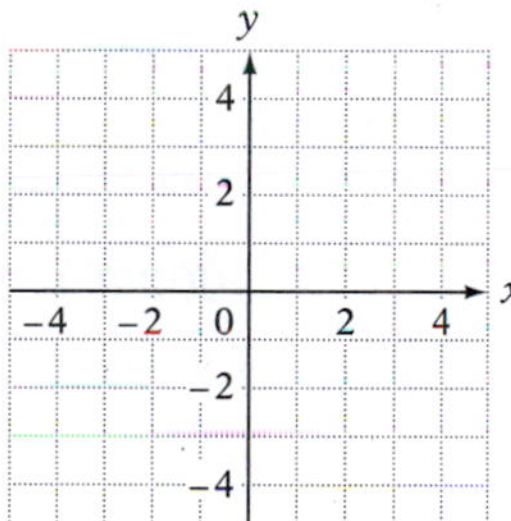

39. $f(x) = \pi^x$

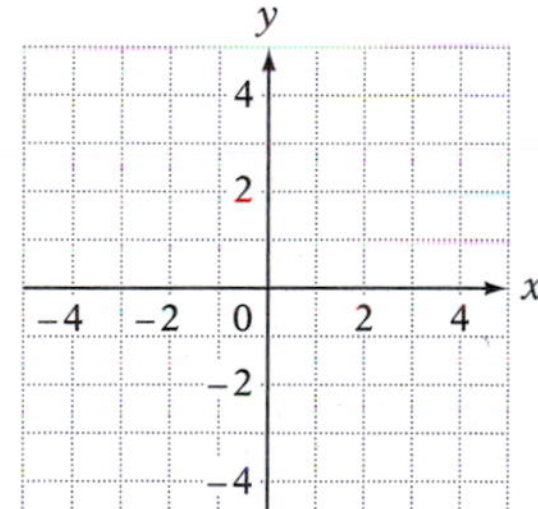

40. Evaluate $\left(1 + \frac{1}{n}\right)^n$ for $n = 100$, 1000, 10,000, and 100,000 and compare the results with the value of e, the base of the natural exponential function. On the basis of your evaluation, complete the following sentence: As n increases, $\left(1 + \frac{1}{n}\right)^n$ becomes closer to ________.

41. Physics If air resistance is ignored, the speed v, in feet per second, of an object t seconds after it has been dropped is given by $v = 32t$. However, if air resistance is considered, then the speed depends on the mass of the object (and on other things). For a certain mass, the speed t seconds after the object has been dropped is given by $v = 32(1 - e^{-t})$.

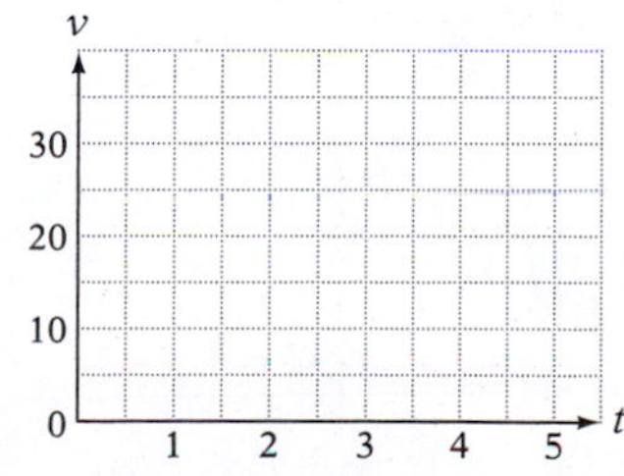

a. Graph this equation. *Suggestion:* Use Xmin = 0, Xmax = 5.5, Ymin = 0, Ymax = 40, and Yscl = 5.

b. The point whose approximate coordinates are (2, 27.7) is on this graph. Write a sentence that explains the meaning of these coordinates.

Copyright © Houghton Mifflin Company. All rights reserved.

12.2 Introduction to Logarithms

Objective A To find the logarithm of a number

Suppose a bacteria colony that originally contained 1000 bacteria doubled in size every hour. Then the table at the right would show the number of bacteria in that colony after 1, 2, and 3 h.

Time (in hours)	*Number of Bacteria*
0	1000
1	2000
2	4000
3	8000

The exponential function $A = 1000(2^t)$, where A is the number of bacteria in the colony at time t, is a model of the growth of the colony. For instance, when $t = 3$ hours, we have

$$A = 1000(2^t)$$

$$A = 1000(2^3) \quad \text{• Replace } t \text{ by } 3.$$

$$A = 1000(8) = 8000$$

After 3 h there are 8000 bacteria in the colony.

Now we ask, "How long will it take for there to be 32,000 bacteria in the colony?" To answer the question, we must solve the *exponential equation* $32{,}000 = 1000(2^t)$. By trial and error, we find that when $t = 5$,

$$A = 1000(2^t)$$

$$A = 1000(2^5) \quad \text{• Replace } t \text{ by } 5.$$

$$A = 1000(32) = 32{,}000$$

After 5 h there will be 32,000 bacteria in the colony.

Now suppose we want to know how long it takes before the colony reaches 50,000 bacteria. To answer that question, we must find t so that $50{,}000 = 1000(2^t)$. Using trial and error again, we find that

$$1000(2^5) = 32{,}000 \quad \text{and} \quad 1000(2^6) = 64{,}000$$

Because 50,000 is between 32,000 and 64,000, we conclude that t is between 5 and 6 h. If we try $t = 5.5$ (halfway between 5 and 6), then

$$A = 1000(2^t)$$

$$A = 1000(2^{5.5}) \quad \text{• Replace } t \text{ by } 5.5.$$

$$A \approx 1000(45.25) = 45{,}250$$

In 5.5 h, there are approximately 45,250 bacteria in the colony. Because this is less than 50,000, the value of t must be a little greater than 5.5.

Copyright © Houghton Mifflin Company. All rights reserved.

We could continue to use trial and error to find the correct value of t, but it would be more efficient if we could just solve the exponential equation $50{,}000 = 1000(2^t)$ for t. If we follow the procedures for solving equations that were discussed earlier in the text, we have

$$50{,}000 = 1000(2^t)$$

$$50 = 2^t$$ • Divide each side of the equation by 1000.

Integrating Technology

Using a calculator, we can verify that $2^{5.644} \approx 50$. On a graphing calculator, press 2 ^ 5.644.

To proceed to the next step, it would be helpful to have a function that would find the power of 2 that produces 50.

Around the mid-sixteenth century, mathematicians created such a function, which we now call a *logarithmic function*. We write the solution of $50 = 2^t$ as $t = \log_2 50$. This is read "t equals the logarithm base 2 of 50" and it means "t equals the power of 2 that produces 50." When logarithms were first introduced, tables were used to find a numerical value of t. Today, a calculator is used. Using a calculator, we can approximate the value of t as 5.644. This means that $2^{5.644} \approx 50$.

TAKE NOTE

Note that earlier, when we tried $t = 5.5$, we stated that the actual value of t must be greater than 5.5; 5.644 is a little greater than 5.5.

The equivalence of the expressions $50 = 2^t$ and $t = \log_2 50$ are described in the following definition of **logarithm.**

TAKE NOTE

Read $\log_b x$ as "the logarithm of x, base b" or "log base b of x."

Definition of Logarithm

For $x > 0$, $b > 0$, $b \neq 1$, $y = \log_b x$ is equivalent to $x = b^y$.

The table at the right shows equivalent statements written in both exponential and logarithmic form.

Exponential Form	*Logarithmic Form*
$2^4 = 16$	$\log_2 16 = 4$
$\left(\frac{2}{3}\right)^2 = \frac{4}{9}$	$\log_{2/3}\left(\frac{4}{9}\right) = 2$
$10^{-1} = 0.1$	$\log_{10}(0.1) = -1$

Study Tip

Be sure you can rewrite an exponential equation as a logarithmic equation and a logarithmic equation as an exponential equation. This relationship is very important.

HOW TO Write $\log_3 81 = 4$ in exponential form.

$\log_3 81 = 4$ is equivalent to $3^4 = 81$.

HOW TO Write $10^{-2} = 0.01$ in logarithmic form.

$10^{-2} = 0.01$ is equivalent to $\log_{10}(0.01) = -2$.

It is important to note that the exponential function is a 1–1 function and thus has an inverse function. **The inverse function of the exponential function is called a logarithm.**

Copyright © Houghton Mifflin Company. All rights reserved.

The 1–1 property of exponential functions can be used to evaluate some logarithms.

1–1 Property of Exponential Functions

For $b > 0$, $b \neq 1$, if $b^u = b^v$, then $u = v$.

HOW TO Evaluate $\log_2 8$.

$\log_2 8 = x$ • Write an equation.

$8 = 2^x$ • Write the equation in its equivalent exponential form.

$2^3 = 2^x$ • Write 8 as 2^3.

$3 = x$ • Use the 1–1 Property of Exponential Functions.

$\log_2 8 = 3$

HOW TO Solve $\log_4 x = -2$ for x.

$\log_4 x = -2$

$4^{-2} = x$ • Write the equation in its equivalent exponential form.

$\frac{1}{16} = x$ • Simplify the expression with a negative exponent.

The solution is $\frac{1}{16}$.

Integrating Technology

The logarithms of most numbers are irrational numbers. Therefore, the value displayed by a calculator is an approximation.

Logarithms to the base 10 are called **common logarithms.** Usually the base, 10, is omitted when writing the common logarithm of a number. Therefore, $\log_{10} x$ is written $\log x$. To find the common logarithm of most numbers, a calculator is necessary. A calculator was used to find the value of log 384, shown below.

$$\log 384 \approx 2.5843312$$

When e (the base of the natural exponential function) is used as the base of a logarithm, the logarithm is referred to as the **natural logarithm** and is abbreviated $\ln x$. This is read "el en x." Use a calculator to approximate natural logarithms.

$$\ln 23 \approx 3.135494216$$

Example 1

Evaluate: $\log_3 \frac{1}{9}$

Solution

$\log_3 \frac{1}{9} = x$ • Write an equation.

$\frac{1}{9} = 3^x$ • Write the equivalent exponential form.

$3^{-2} = 3^x$ • $\frac{1}{9} = 3^{-2}$

$-2 = x$ • The bases are the same. The exponents are equal.

$\log_3 \frac{1}{9} = -2$

You Try It 1

Evaluate: $\log_4 64$

Your solution

Solution on p. S35

Copyright © Houghton Mifflin Company. All rights reserved.

Example 2 Solve for x: $\log_5 x = 2$

Solution

$$\log_5 x = 2$$
$$5^2 = x$$ • Write the equivalent exponential form.
$$25 = x$$

The solution is 25.

You Try It 2 Solve for x: $\log_2 x = -4$

Your solution

Example 3 Solve $\log x = -1.5$ for x. Round to the nearest ten-thousandth.

Solution

$$\log x = -1.5$$
$$10^{-1.5} = x$$ • Write the equivalent exponential form.
$$0.0316 \approx x$$ • Use a calculator.

You Try It 3 Solve $\ln x = 3$ for x. Round to the nearest ten-thousandth.

Your solution

Solutions on p. S35

Objective B To use the Properties of Logarithms to simplify expressions containing logarithms

Because a logarithm is a special kind of exponent, the Properties of Logarithms are similar to the Properties of Exponents.

The property of logarithms that states that the logarithm of the product of two numbers equals the sum of the logarithms of the two numbers is similar to the property of exponents that states that to multiply two exponential expressions with the same base, we add the exponents.

> **TAKE NOTE**
> Pay close attention to this property. Note, for instance, that this property states that $\log_3(4 \cdot p) = \log_3 4 + \log_3 p$. It also states that $\log_5 9 + \log_5 z = \log_5(9z)$. It does *not* state any relationship regarding the expression $\log_b(x + y)$. **This expression cannot be simplified.**

The Logarithm Property of the Product of Two Numbers

For any positive real numbers x, y, and b, $b \neq 1$, $\log_b(xy) = \log_b x + \log_b y$.

A proof of this property can be found in the Appendix.

HOW TO Write $\log_b(6z)$ in expanded form.

$$\log_b(6z) = \log_b 6 + \log_b z$$ • Use the Logarithm Property of Products.

HOW TO Write $\log_b 12 + \log_b r$ as a single logarithm.

$$\log_b 12 + \log_b r = \log_b(12r)$$ • Use the Logarithm Property of Products.

The Logarithm Property of Products can be extended to include the logarithm of the product of more than two factors. For instance,

$$\log_b(xyz) = \log_b x + \log_b y + \log_b z$$
$$\log_b(7rt) = \log_b 7 + \log_b r + \log_b t$$

Copyright © Houghton Mifflin Company. All rights reserved.

A second property of logarithms involves the logarithm of the quotient of two numbers. This property of logarithms is also based on the facts that a logarithm is an exponent and that to divide two exponential expressions with the same base, we subtract the exponents.

TAKE NOTE

This property is used to rewrite expressions such as $\log_5\left(\frac{m}{8}\right) = \log_5 m - \log_5 8$. It does *not* state any relationship regarding the expression $\frac{\log_b x}{\log_b y}$. **This expression cannot be simplified.**

The Logarithm Property of the Quotient of Two Numbers

For any positive real numbers x, y, and b, $b \neq 1$,

$\log_b \frac{x}{y} = \log_b x - \log_b y$.

A proof of this property can be found in the Appendix.

HOW TO Write $\log_b \frac{p}{8}$ in expanded form.

$\log_b \frac{p}{8} = \log_b p - \log_b 8$ • Use the Logarithm Property of Quotients.

HOW TO Write $\log_b y - \log_b v$ as a single logarithm.

$\log_b y - \log_b v = \log_b \frac{y}{v}$ • Use the Logarithm Property of Quotients.

A third property of logarithms is used to simplify powers of a number.

The Logarithm Property of the Power of a Number

For any positive real numbers x and b, $b \neq 1$, and for any real number r, $\log_b x^r = r\log_b x$.

A proof of this property can be found in the Appendix.

HOW TO Rewrite $\log_b x^3$ in terms of $\log_b x$.

$\log_b x^3 = 3 \log_b x$ • Use the Logarithm Property of Powers.

HOW TO Rewrite $\frac{2}{3} \log_b x$ with a coefficient of 1.

$\frac{2}{3} \log_b x = \log_b x^{2/3}$ • Use the Logarithm Property of Powers.

The following table summarizes the properties of logarithms that we have discussed, along with three other properties.

Summary of the Properties of Logarithms

Let x, y, and b be positive real numbers with $b \neq 1$. Then

Product Property	$\log_b(x \cdot y) = \log_b x + \log_b y$
Quotient Property	$\log_b \frac{x}{y} = \log_b x - \log_b y$
Power Property	$\log_b x^r = r \log_b x$, r a real number
Logarithm of One	$\log_b 1 = 0$
Inverse Property	$b^{\log_b x} = x$ and $\log_b b^x = x$
1–1 Property	If $\log_b x = \log_b y$, then $x = y$.

Point of Interest

Logarithms were developed independently by Jobst Burgi (1552–1632) and John Napier (1550–1617) as a means of simplifying the calculations of astronomers. The idea was to devise a method by which two numbers could be multiplied by performing additions. Napier is usually given credit for logarithms because he published his results first.

John Napier

In Napier's original work, the logarithm of 10,000,000 was 0. After this work was published, Napier, in discussions with Henry Briggs (1561–1631), decided that tables of logarithms would be easier to use if the logarithm of 1 were 0. Napier died before new tables could be determined, and Briggs took on the task. His table consisted of logarithms accurate to 30 decimal places, all accomplished without a calculator!

The logarithms Briggs calculated are the common logarithms mentioned earlier.

Copyright © Houghton Mifflin Company. All rights reserved.

HOW TO Write $\log_b \frac{xy}{z}$ in expanded form.

$$\log_b \frac{xy}{z} = \log_b(xy) - \log_b z$$ • Use the Logarithm Property of Quotients.

$$= \log_b x + \log_b y - \log_b z$$ • Use the Logarithm Property of Products.

HOW TO Write $\log_b \frac{x^2}{y^3}$ in expanded form.

$$\log_b \frac{x^2}{y^3} = \log_b x^2 - \log_b y^3$$ • Use the Logarithm Property of Quotients.

$$= 2\log_b x - 3\log_b y$$ • Use the Logarithm Property of Powers.

HOW TO Write $2\log_b x + 4\log_b y$ as a single logarithm with a coefficient of 1.

$$2\log_b x + 4\log_b y = \log_b x^2 + \log_b y^4$$ • Use the Logarithm Property of Powers.

$$= \log_b(x^2y^4)$$ • Use the Logarithm Property of Products.

Example 4

Write $\log \sqrt{x^3y}$ in expanded form.

Solution

$$\log \sqrt{x^3y} = \log(x^3y)^{1/2} = \frac{1}{2}\log(x^3y)$$ • Power Property

$$= \frac{1}{2}(\log x^3 + \log y)$$ • Product Property

$$= \frac{1}{2}(3\log x + \log y)$$ • Power Property

$$= \frac{3}{2}\log x + \frac{1}{2}\log y$$ • Distributive Property

You Try It 4

Write $\log_8 \sqrt[3]{xy^2}$ in expanded form.

Your solution

Example 5

Write $\frac{1}{2}(\log_3 x - 3\log_3 y + \log_3 z)$ as a single logarithm with a coefficient of 1.

Solution

$$\frac{1}{2}(\log_3 x - 3\log_3 y + \log_3 z)$$

$$= \frac{1}{2}(\log_3 x - \log_3 y^3 + \log_3 z)$$

$$= \frac{1}{2}\left(\log_3 \frac{x}{y^3} + \log_3 z\right)$$

$$= \frac{1}{2}\left(\log_3 \frac{xz}{y^3}\right) = \log_3\left(\frac{xz}{y^3}\right)^{1/2} = \log_3 \sqrt{\frac{xz}{y^3}}$$

You Try It 5

Write $\frac{1}{3}(\log_4 x - 2\log_4 y + \log_4 z)$ as a single logarithm with a coefficient of 1.

Your solution

Example 6

Find $\log_4 4^7$.

Solution

$\log_4 4^7 = 7$ • Inverse Property

You Try It 6

Find $\log_9 1$.

Your solution

Solutions on pp. S35–S36

Copyright © Houghton Mifflin Company. All rights reserved.

Objective C To use the Change-of-Base Formula

Although only common logarithms and natural logarithms are programmed into calculators, the logarithms for other positive bases can be found.

HOW TO Evaluate $\log_5 22$.

$\log_5 22 = x$ • **Write an equation.**

$5^x = 22$ • **Write the equation in its equivalent exponential form.**

$\log 5^x = \log 22$ • **Apply the common logarithm to each side of the equation.**

$x \log 5 = \log 22$ • **Use the Power Property of Logarithms.**

$x = \dfrac{\log 22}{\log 5}$ • **Exact answer**

$x \approx 1.92057$ • **Approximate answer**

$\log_5 22 \approx 1.92057$

In the third step above, the natural logarithm, instead of the common logarithm, could have been applied to each side of the equation. The same result would have been obtained.

Using a procedure similar to the one used to evaluate $\log_5 22$, we can derive a formula for changing bases.

Change-of-Base Formula

$$\log_a N = \frac{\log_b N}{\log_b a}$$

HOW TO Evaluate $\log_2 14$.

$$\log_2 14 = \frac{\log 14}{\log 2} \approx 3.80735$$

• **Use the Change-of-Base Formula with $N = 14$, $a = 2$, $b = 10$.**

In the last example, common logarithms were used. Here is the same example using natural logarithms. Note that the answers are the same.

$$\log_2 14 = \frac{\ln 14}{\ln 2} \approx 3.80735$$

Example 7

Evaluate $\log_8 0.137$ by using natural logarithms.

Solution

$$\log_8 0.137 = \frac{\ln 0.137}{\ln 8} \approx -0.95592$$

You Try It 7

Evaluate $\log_3 0.834$ by using natural logarithms.

Your solution

Example 8

Evaluate $\log_2 90.813$ by using common logarithms.

Solution

$$\log_2 90.813 = \frac{\log 90.813}{\log 2} \approx 6.50483$$

You Try It 8

Evaluate $\log_7 6.45$ by using common logarithms.

Your solution

Solutions on p. S36

Copyright © Houghton Mifflin Company. All rights reserved.

12.2 Exercises

Objective A To find the logarithm of a number

1. **a.** What is a common logarithm?
   b. How is the common logarithm of $4z$ written?

2. **a.** What is a natural logarithm?
   b. How is the natural logarithm of $3x$ written?

For Exercises 3 to 10, write the exponential expression in logarithmic form.

3. $5^2 = 25$ **4.** $10^3 = 1000$ **5.** $4^{-2} = \frac{1}{16}$ **6.** $3^{-3} = \frac{1}{27}$

7. $10^y = x$ **8.** $e^y = x$ **9.** $a^x = w$ **10.** $b^y = c$

For Exercises 11 to 18, write the logarithmic expression in exponential form.

11. $\log_3 9 = 2$ **12.** $\log_2 32 = 5$ **13.** $\log 0.01 = -2$ **14.** $\log_5 \frac{1}{5} = -1$

15. $\ln x = y$ **16.** $\log x = y$ **17.** $\log_b u = v$ **18.** $\log_c x = y$

Evaluate the expressions in Exercises 19 to 30.

19. $\log_3 81$ **20.** $\log_7 49$ **21.** $\log_2 128$ **22.** $\log_5 125$

23. $\log 100$ **24.** $\log 0.001$ **25.** $\ln e^3$ **26.** $\ln e^2$

27. $\log_8 1$ **28.** $\log_3 243$ **29.** $\log_5 625$ **30.** $\log_2 64$

For Exercises 31 to 38, solve for x.

31. $\log_3 x = 2$ **32.** $\log_5 x = 1$ **33.** $\log_4 x = 3$ **34.** $\log_2 x = 6$

35. $\log_7 x = -1$ **36.** $\log_8 x = -2$ **37.** $\log_6 x = 0$ **38.** $\log_4 x = 0$

Copyright © Houghton Mifflin Company. All rights reserved.

For Exercises 39 to 46, solve for x. Round to the nearest hundredth.

39. $\log x = 2.5$ **40.** $\log x = 3.2$ **41.** $\log x = -1.75$ **42.** $\log x = -2.1$

43. $\ln x = 2$ **44.** $\ln x = 1.4$ **45.** $\ln x = -\frac{1}{2}$ **46.** $\ln x = -1.7$

Objective B To use the Properties of Logarithms to simplify expressions containing logarithms

47. What is the Product Property of Logarithms?

48. What is the Quotient Property of Logarithms?

For Exercises 49 to 76, express as a single logarithm with a coefficient of 1.

49. $\log_3 x^3 + \log_3 y^2$ **50.** $\log_7 x + \log_7 z^2$ **51.** $\ln x^4 - \ln y^2$

52. $\ln x^2 - \ln y$ **53.** $3 \log_7 x$ **54.** $4 \log_8 y$

55. $3 \ln x + 4 \ln y$ **56.** $2 \ln x - 5 \ln y$ **57.** $2(\log_4 x + \log_4 y)$

58. $3(\log_5 r + \log_5 t)$ **59.** $2 \log_3 x - \log_3 y + 2 \log_3 z$ **60.** $4 \log_5 r - 3 \log_5 s + \log_5 t$

61. $\ln x - (2 \ln y + \ln z)$ **62.** $2 \log_b x - 3(\log_b y + \log_b z)$

63. $\frac{1}{2}(\log_6 x - \log_6 y)$ **64.** $\frac{1}{3}(\log_8 x - \log_8 y)$

65. $2(\log_4 s - 2 \log_4 t + \log_4 r)$ **66.** $3(\log_9 x + 2 \log_9 y - 2 \log_9 z)$

67. $\ln x - 2(\ln y + \ln z)$ **68.** $\ln t - 3(\ln u + \ln v)$

69. $3 \log_2 t - 2(\log_2 r - \log_2 v)$ **70.** $2 \log_{10} x - 3(\log_{10} y - \log_{10} z)$

Copyright © Houghton Mifflin Company. All rights reserved.

71. $\frac{1}{2}(3 \log_4 x - 2 \log_4 y + \log_4 z)$

72. $\frac{1}{3}(4 \log_5 t - 5 \log_5 u - 7 \log_5 v)$

73. $\frac{1}{2}(\ln x - 3 \ln y)$

74. $\frac{1}{3}\ln a + \frac{2}{3}\ln b$

75. $\frac{1}{2}\log_2 x - \frac{2}{3}\log_2 y + \frac{1}{2}\log_2 z$

76. $\frac{2}{3}\log_3 x + \frac{1}{3}\log_3 y - \frac{1}{2}\log_3 z$

For Exercises 77 to 100, write the logarithm in expanded form.

77. $\log_8(xz)$

78. $\log_7(rt)$

79. $\log_3 x^5$

80. $\log_2 y^7$

81. $\log_b \frac{r}{s}$

82. $\log_c \frac{z}{4}$

83. $\log_3(x^2 y^6)$

84. $\log_4(t^4 u^2)$

85. $\log_7 \frac{u^3}{v^4}$

86. $\log_{10} \frac{s^5}{t^2}$

87. $\log_2(rs)^2$

88. $\log_3(x^2 y)^3$

89. $\ln(x^2 yz)$

90. $\ln(xy^2 z^3)$

91. $\log_5 \frac{xy^2}{z^4}$

92. $\log_b \frac{r^2 s}{t^3}$

93. $\log_8 \frac{x^2}{yz^2}$

94. $\log_9 \frac{x}{y^2 z^3}$

95. $\log_4 \sqrt{x^3 y}$

96. $\log_3 \sqrt{x^5 y^3}$

97. $\log_7 \sqrt{\frac{x^3}{y}}$

Copyright © Houghton Mifflin Company. All rights reserved.

98. $\log_b \sqrt[3]{\frac{r^2}{t}}$ **99.** $\log_3 \frac{t}{\sqrt{x}}$ **100.** $\log_4 \frac{x}{\sqrt{y^2 z}}$

Objective C To use the Change-of-Base Formula

Evaluate the expressions in Exercises 101 to 124. Round to the nearest ten-thousandth.

101. $\log_{10}7$ **102.** $\log_{10}9$ **103.** $\log_{10}\frac{3}{5}$ **104.** $\log_{10}\frac{13}{3}$

105. $\ln 4$ **106.** $\ln 6$ **107.** $\ln \frac{17}{6}$ **108.** $\ln \frac{13}{17}$

109. $\log_8 6$ **110.** $\log_4 8$ **111.** $\log_5 30$ **112.** $\log_6 28$

113. $\log_3 0.5$ **114.** $\log_5 0.6$ **115.** $\log_7 1.7$ **116.** $\log_6 3.2$

117. $\log_5 15$ **118.** $\log_3 25$ **119.** $\log_{12} 120$ **120.** $\log_9 90$

121. $\log_4 2.55$ **122.** $\log_8 6.42$ **123.** $\log_5 67$ **124.** $\log_8 35$

APPLYING THE CONCEPTS

125. For each of the following, answer True or False. Assume all variables represent positive numbers.

a. $\log_3(-9) = -2$

b. $x^y = z$ and $\log_x z = y$ are equivalent equations.

c. $\log(x^{-1}) = \frac{1}{\log x}$

d. $\log \frac{x}{y} = \log x - \log y$

e. $\log(x \cdot y) = \log x \cdot \log y$

f. If $\log x = \log y$, then $x = y$.

126. Complete each statement using the equation $\log_a b = c$.

a. $a^c =$ ________

b. $\log_a(a^c) =$ ________

Copyright © Houghton Mifflin Company. All rights reserved.

12.3 Graphs of Logarithmic Functions

Objective A **To graph a logarithmic function**

The graph of a logarithmic function can be drawn by using the relationship between the exponential and logarithmic functions.

Point of Interest

Although logarithms were originally developed to assist with computations, logarithmic functions have a much broader use today. These functions occur in geology, acoustics, chemistry, and economics, for example.

HOW TO Graph: $f(x) = \log_2 x$

Think of $f(x) = \log_2 x$ as the equation $y = \log_2 x$.

$$f(x) = \log_2 x$$
$$y = \log_2 x$$

Write the equivalent exponential equation.

$$x = 2^y$$

Because the equation is solved for x in terms of y, it is easier to choose values of y and find the corresponding values of x. The results can be recorded in a table.

Graph the ordered pairs on a rectangular coordinate system.

Connect the points with a smooth curve.

x	y
$\frac{1}{4}$	-2
$\frac{1}{2}$	-1
1	0
2	1
4	2

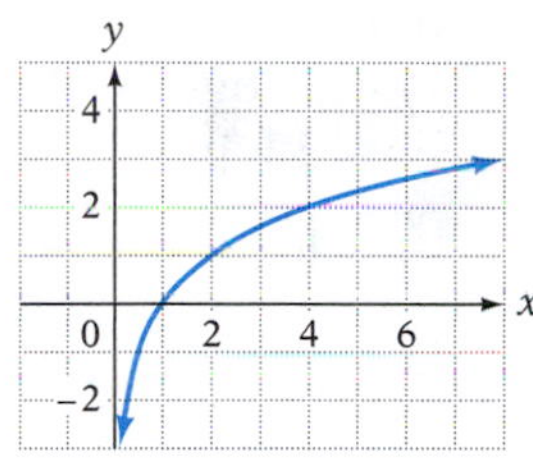

Applying the vertical-line and horizontal-line tests reveals that $f(x) = \log_2 x$ is a one-to-one function.

Integrating Technology

See the Projects and Group Activities at the end of this chapter for suggestions on graphing logarithmic functions using a graphing calculator.

HOW TO Graph: $f(x) = \log_2 x + 1$

Think of $f(x) = \log_2 x + 1$ as the equation $y = \log_2 x + 1$.

$$f(x) = \log_2 x + 1$$
$$y = \log_2 x + 1$$

Solve for $\log_2 x$.

$$y - 1 = \log_2 x$$

Write the equivalent exponential equation.

$$2^{y-1} = x$$

Choose values of y and find the corresponding values of x.

Graph the ordered pairs on a rectangular coordinate system.

Connect the points with a smooth curve.

x	y
$\frac{1}{4}$	-1
$\frac{1}{2}$	0
1	1
2	2
4	3

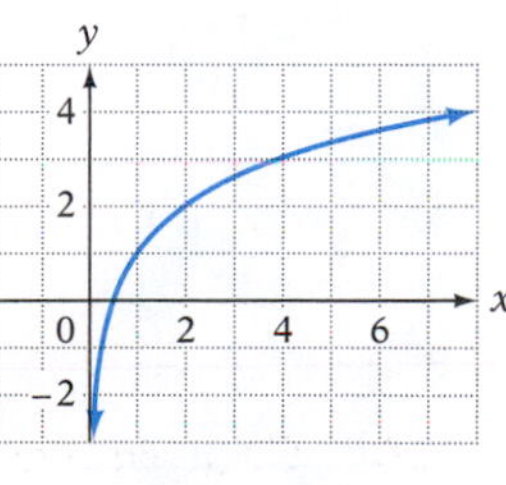

Copyright © Houghton Mifflin Company. All rights reserved.

Example 1 Graph: $f(x) = \log_3 x$

Solution

$f(x) = \log_3 x$

$y = \log_3 x$ • $f(x) = y$

$3^y = x$ • Write the equivalent exponential equation.

x	y
$\frac{1}{9}$	-2
$\frac{1}{3}$	-1
1	0
3	1

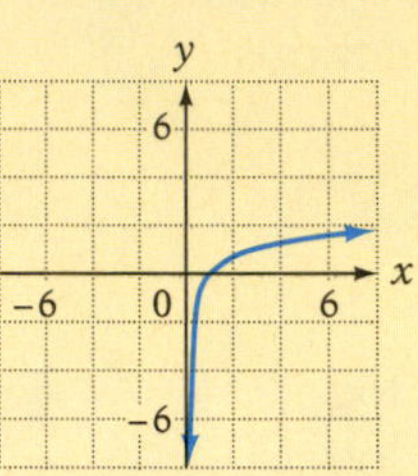

You Try It 1 Graph: $f(x) = \log_2 (x - 1)$

Your solution

Example 2 Graph: $f(x) = 2 \log_3 x$

Solution

$f(x) = 2 \log_3 x$

$y = 2 \log_3 x$ • $f(x) = y$

$\frac{y}{2} = \log_3 x$ • Divide both sides by 2.

$3^{y/2} = x$ • Write the equivalent exponential equation.

x	y
$\frac{1}{9}$	-4
$\frac{1}{3}$	-2
1	0
3	2

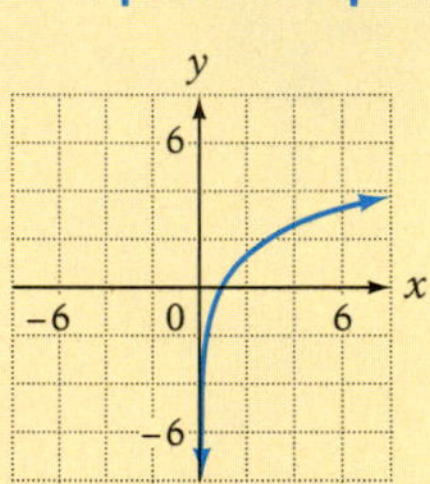

You Try It 2 Graph: $f(x) = \log_3 (2x)$

Your solution

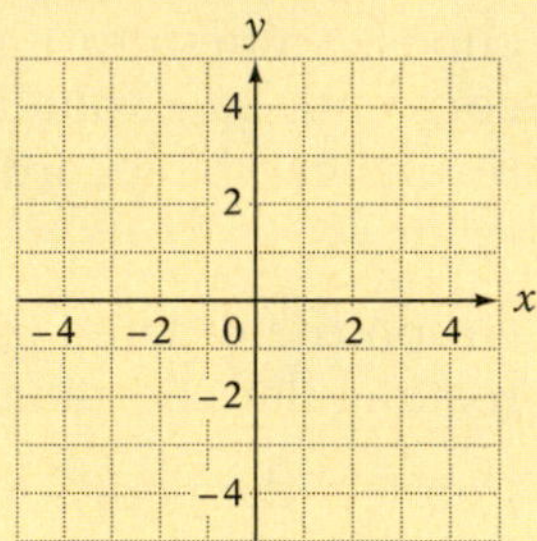

Example 3 Graph: $f(x) = -\log_2 (x - 2)$

Solution

$f(x) = -\log_2 (x - 2)$

$y = -\log_2 (x - 2)$ • $f(x) = y$

$-y = \log_2 (x - 2)$ • Multiply both sides by −1.

$2^{-y} = x - 2$ • Write the equivalent exponential equation.

$2^{-y} + 2 = x$

x	y
6	-2
4	-1
3	0
$\frac{5}{2}$	1
$\frac{9}{4}$	2
$\frac{17}{8}$	3

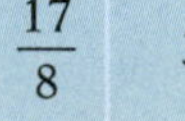

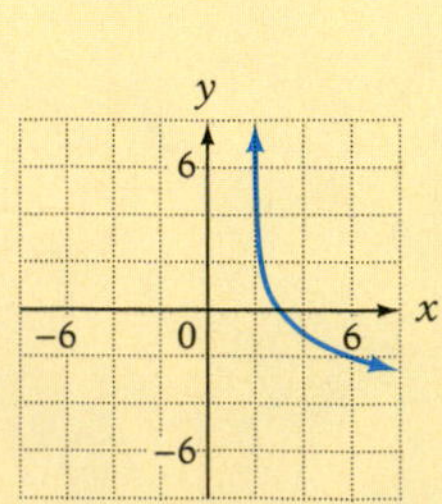

You Try It 3 Graph: $f(x) = -\log_3 (x + 1)$

Your solution

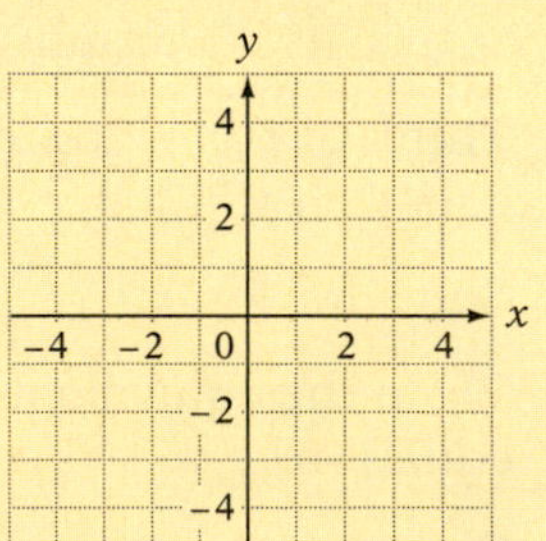

Solutions on p. S36

Copyright © Houghton Mifflin Company. All rights reserved.

12.3 Exercises

Objective A To graph a logarithmic function

1. Is the function $f(x) = \log x$ a 1–1 function? Why or why not?

2. Name two characteristics of the graph of $y = \log_b x$, $b > 1$.

3. What is the relationship between the graphs of $x = 3^y$ and $y = \log_3 x$?

4. What is the relationship between the graphs of $y = 3^x$ and $y = \log_3 x$?

Graph the functions in Exercises 5 to 16.

5. $f(x) = \log_4 x$

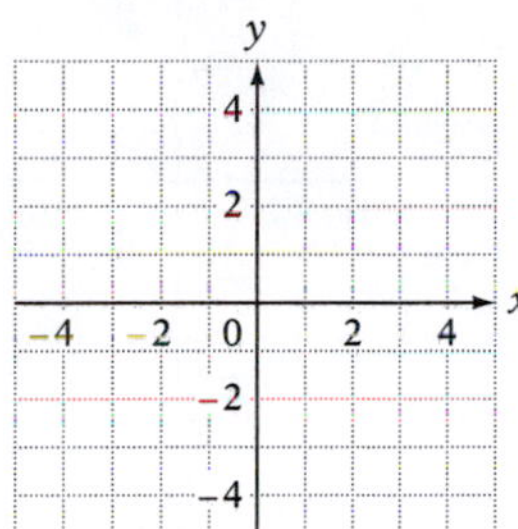

6. $f(x) = \log_2(x + 1)$

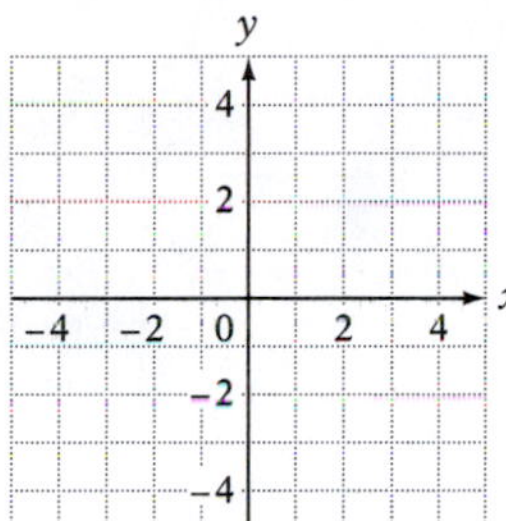

7. $f(x) = \log_3(2x - 1)$

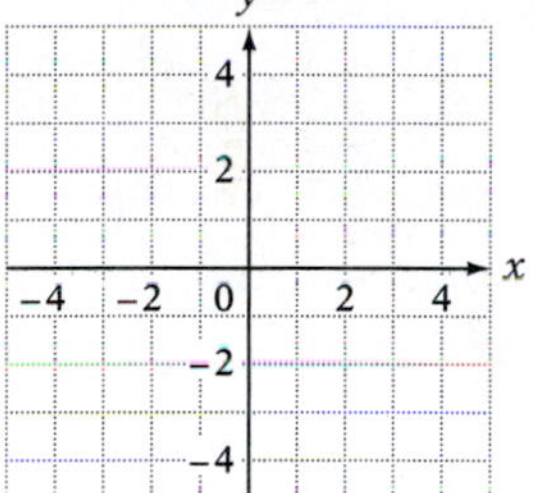

8. $f(x) = \log_2\left(\frac{1}{2}x\right)$

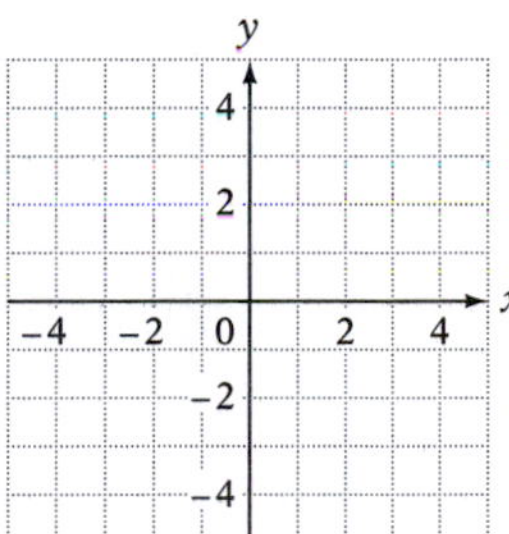

9. $f(x) = 3\log_2 x$

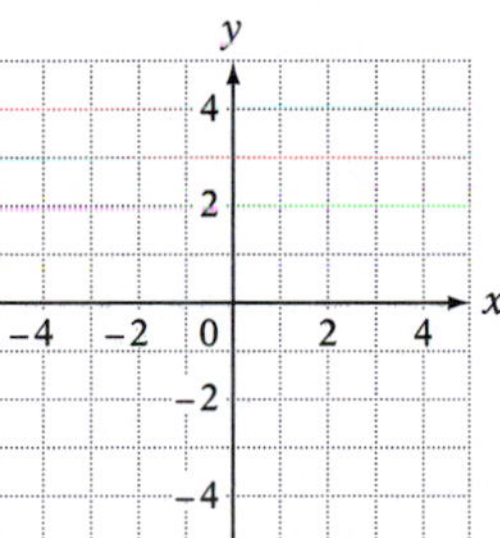

10. $f(x) = \frac{1}{2}\log_2 x$

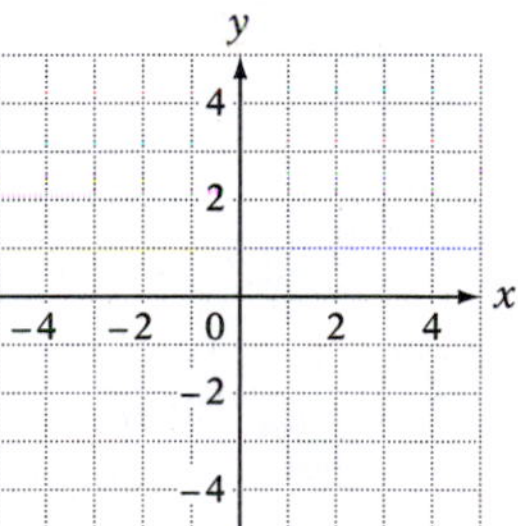

11. $f(x) = -\log_2 x$

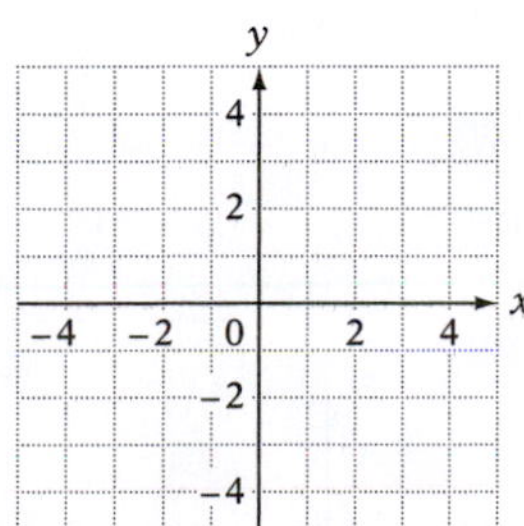

12. $f(x) = -\log_3 x$

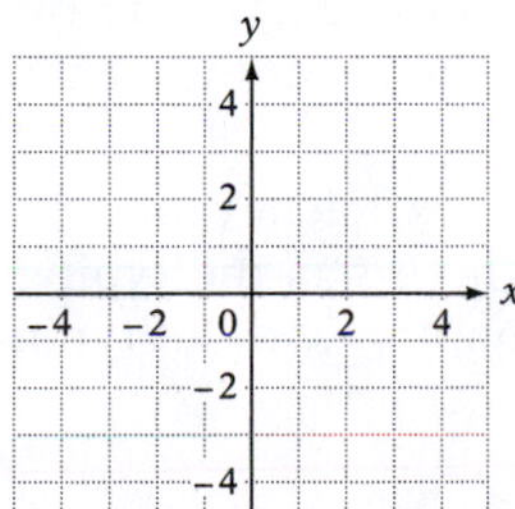

13. $f(x) = \log_2(x - 1)$

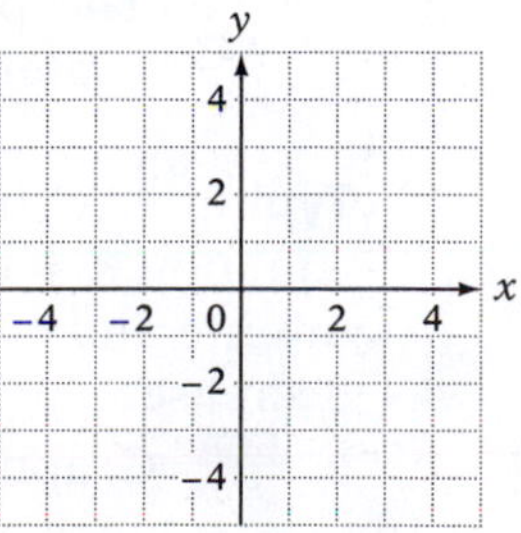

Copyright © Houghton Mifflin Company. All rights reserved.

14. $f(x) = \log_3(2 - x)$

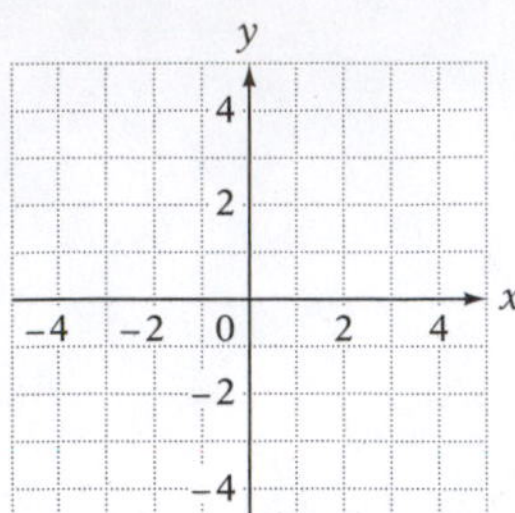

15. $f(x) = -\log_2(x - 1)$

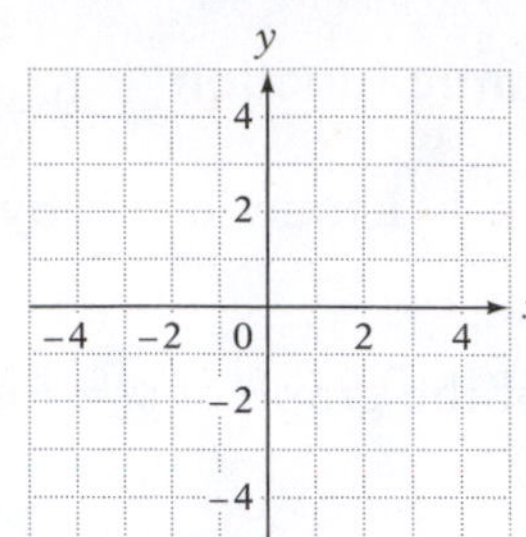

16. $f(x) = -\log_2(1 - x)$

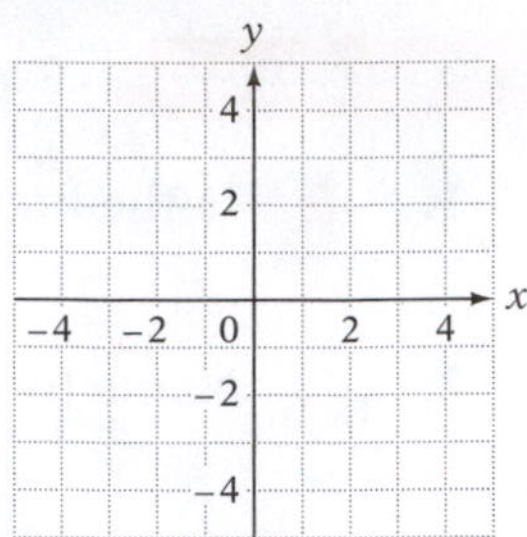

APPLYING THE CONCEPTS

 Use a graphing calculator to graph the functions in Exercises 17 to 22.

17. $f(x) = x - \log_2(1 - x)$

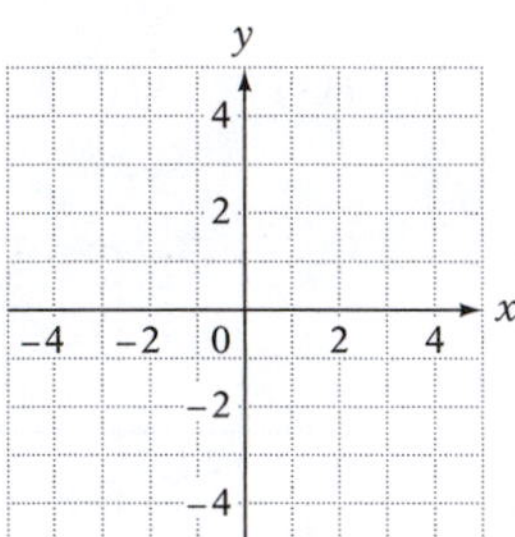

18. $f(x) = -\frac{1}{2}\log_2 x - 1$

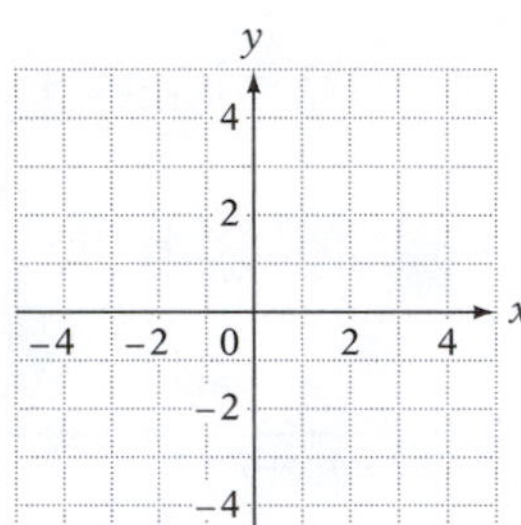

19. $f(x) = \frac{x}{2} - 2\log_2(x + 1)$

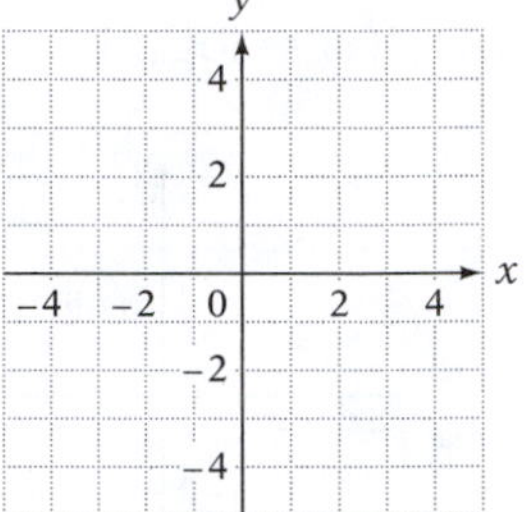

20. $f(x) = x + \log_3(2 - x)$

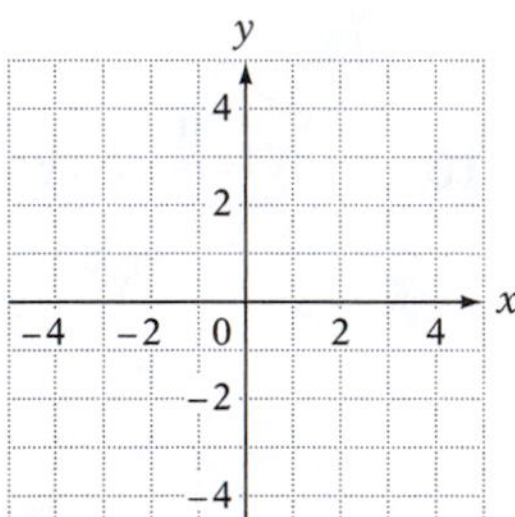

21. $f(x) = x^2 - 10\ln(x - 1)$

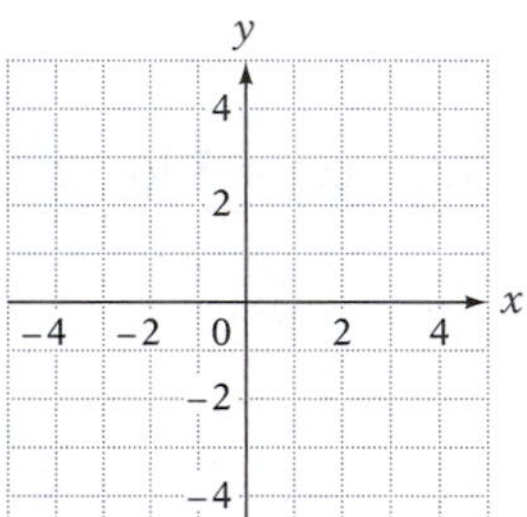

22. $f(x) = \frac{x}{3} - 3\log_2(x + 3)$

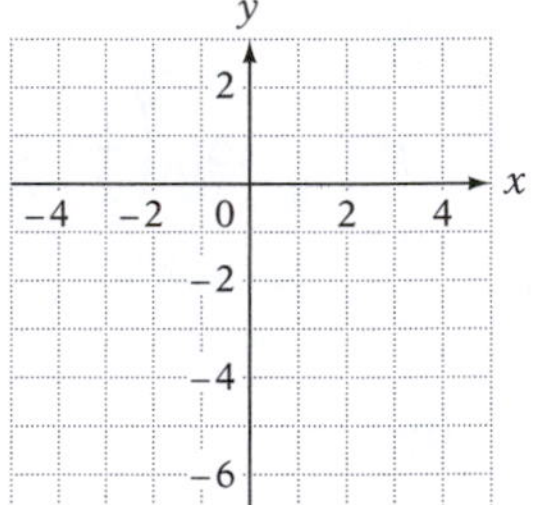

23. **Astronomy** Astronomers use the *distance modulus* of a star as a method of determining the star's distance from Earth. The formula is $M = 5\log(s - 5)$, where M is the distance modulus and s is the star's distance from Earth in parsecs. (One parsec $\approx 1.9 \times 10^{13}$ mi)

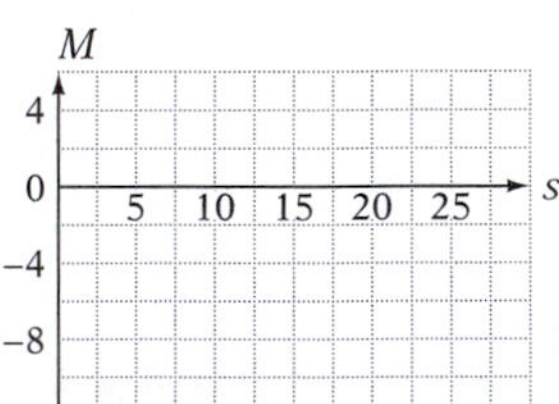

a. 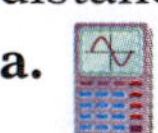 Graph the equation.

b.  The point with coordinates (25.1, 2) is on the graph. Write a sentence that describes the meaning of this ordered pair.

24. **Typing** Without practice, the proficiency of a typist decreases. The equation $s = 60 - 7\ln(t + 1)$, where s is the typing speed in words per minute and t is the number of months without typing, approximates this decrease.

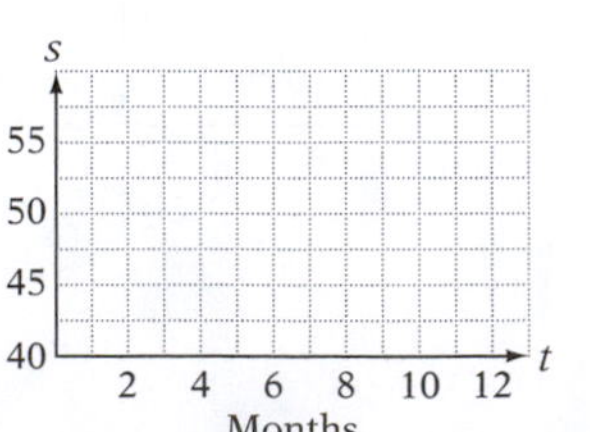

a. 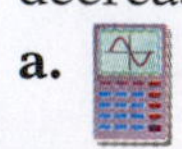 Graph the equation.

b.  The point with coordinates (4, 49) is on the graph. Write a sentence that describes the meaning of this ordered pair.

Copyright © Houghton Mifflin Company. All rights reserved.

12.4 Solving Exponential and Logarithmic Equations

Objective A **To solve an exponential equation**

An **exponential equation** is one in which a variable occurs in the exponent. The equations at the right are exponential equations.

$$6^{2x+1} = 6^{3x-2}$$
$$4^x = 3$$
$$2^{x+1} = 7$$

An exponential equation in which each side of the equation can be expressed in terms of the same base can be solved by using the 1–1 Property of Exponential Functions. Recall that the 1–1 Property of Exponential Functions states that for $b > 0, b \neq 1$,

$$\text{if } b^x = b^y, \text{ then } x = y$$

In the two examples below, this property is used in solving exponential equations.

HOW TO Solve: $10^{3x+5} = 10^{x-3}$

$10^{3x+5} = 10^{x-3}$

$3x + 5 = x - 3$ • Use the 1–1 Property of Exponential Functions to equate the exponents.

$2x + 5 = -3$ • Solve the resulting equation.

$2x = -8$

$x = -4$

Check: $10^{3x+5} = 10^{x-3}$

$10^{3(-4)+5}$	10^{-4-3}
10^{-12+5}	10^{-7}

$10^{-7} = 10^{-7}$

The solution is -4.

HOW TO Solve: $9^{x+1} = 27^{x-1}$

$9^{x+1} = 27^{x-1}$

$(3^2)^{x+1} = (3^3)^{x-1}$ • $3^2 = 9; 3^3 = 27$

$3^{2x+2} = 3^{3x-3}$ • Rule for Simplifying Powers of Exponential Expressions

$2x + 2 = 3x - 3$ • Use the 1–1 Property of Exponential Functions to equate the exponents.

$2 = x - 3$ • Solve for x.

$5 = x$

Check: $9^{x+1} = 27^{x-1}$

9^{5+1}	27^{5-1}
9^6	27^4

$531{,}441 = 531{,}441$

The solution is 5.

Copyright © Houghton Mifflin Company. All rights reserved.

Integrating Technology

To evaluate $\frac{\log 7}{\log 4}$ on many scientific calculators, use the keystrokes

7   4  ENTER

The display should read 1.4036775.

When both sides of an exponential equation cannot easily be expressed in terms of the same base, logarithms are used to solve the exponential equation.

HOW TO Solve: $4^x = 7$

$$4^x = 7$$
$$\log 4^x = \log 7$$
- Take the common logarithm of each side of the equation.

$$x \log 4 = \log 7$$
- Rewrite the equation using the Properties of Logarithms.

$$x = \frac{\log 7}{\log 4} \approx 1.4037$$
- Solve for x.

The solution is approximately 1.4037.
- Note that $\frac{\log 7}{\log 4} \neq \log 7 - \log 4$.

Study Tip

Always check the solution of an equation, even when the solution is an approximation. For the equation at the right:

$3^{x+1} = 5$	
$3^{0.4650+1}$	5
$3^{1.4650}$	5

$5.000\overline{145} \approx 5$

HOW TO Solve: $3^{x+1} = 5$

$$3^{x+1} = 5$$
$$\log 3^{x+1} = \log 5$$
- Take the common logarithm of each side of the equation.

$$(x + 1)\log 3 = \log 5$$
- Rewrite the equation using the Properties of Logarithms.

$$x + 1 = \frac{\log 5}{\log 3}$$
$$x + 1 \approx 1.4650$$
- Solve for x.

$$x \approx 0.4650$$

The solution is approximately 0.4650.

Example 1

Solve for n: $(1.1)^n = 2$

Solution

$$(1.1)^n = 2$$
$$\log (1.1)^n = \log 2$$
- Take the log of each side.

$$n \log 1.1 = \log 2$$
- Power Property

$$n = \frac{\log 2}{\log 1.1}$$
- Divide both sides by log 1.1.

$$n \approx 7.2725$$

The solution is approximately 7.2725.

You Try It 1

Solve for n: $(1.06)^n = 1.5$

Your solution

Example 2

Solve for x: $3^{2x} = 4$

Solution

$$3^{2x} = 4$$
$$\log 3^{2x} = \log 4$$
- Take the log of each side.

$$2x \log 3 = \log 4$$
- Power Property

$$2x = \frac{\log 4}{\log 3}$$
- Divide both sides by log 3.

$$2x \approx 1.26185$$
$$x \approx 0.6309$$
- Divide both sides by 2.

The solution is approximately 0.6309.

You Try It 2

Solve for x: $4^{3x} = 25$

Your solution

Solutions on p. S36

Copyright © Houghton Mifflin Company. All rights reserved.

Objective B To solve a logarithmic equation

A logarithmic equation can be solved by using the Properties of Logarithms. Here are two examples.

HOW TO Solve: $\log_9 x + \log_9(x - 8) = 1$

$$\log_9 x + \log_9(x - 8) = 1$$

$$\log_9[x(x - 8)] = 1$$

- Use the Logarithm Property of Products to rewrite the left side of the equation.

$$9^1 = x(x - 8)$$

- Write the equation in exponential form.

$$9 = x^2 - 8x$$

$$0 = x^2 - 8x - 9$$

- Write in standard form.

$$0 = (x - 9)(x + 1)$$

- Factor and use the Principle of Zero Products.

$$x - 9 = 0 \qquad x + 1 = 0$$

$$x = 9 \qquad x = -1$$

Replacing x by 9 in the original equation reveals that 9 checks as a solution. Replacing x by -1 in the original equation results in the expression $\log_9(-1)$. Because the logarithm of a negative number is not a real number, -1 does not check as a solution.

The solution of the equation is 9.

HOW TO Solve: $\log_3 6 - \log_3(2x + 3) = \log_3(x + 1)$

$$\log_3 6 - \log_3(2x + 3) = \log_3(x + 1)$$

$$\log_3 \frac{6}{2x + 3} = \log_3(x + 1)$$

- Use the Quotient Property of Logarithms to rewrite the left side of the equation.

$$\frac{6}{2x + 3} = x + 1$$

- Use the 1–1 Property of Logarithms.

$$6 = (2x + 3)(x + 1)$$

$$6 = 2x^2 + 5x + 3$$

$$0 = 2x^2 + 5x - 3$$

- Write in standard form.

$$0 = (2x - 1)(x + 3)$$

- Solve for x.

$$2x - 1 = 0 \qquad x + 3 = 0$$

$$x = \frac{1}{2} \qquad x = -3$$

Replacing x by $\frac{1}{2}$ in the original equation reveals that $\frac{1}{2}$ checks as a solution. Replacing x by -3 in the original equation results in the expression $\log_3(-2)$. Because the logarithm of a negative number is not a real number, -3 does not check as a solution.

The solution of the equation is $\frac{1}{2}$.

Copyright © Houghton Mifflin Company. All rights reserved.

Example 3

Solve for x: $\log_3(2x - 1) = 2$

Solution

$\log_3(2x - 1) = 2$

$3^2 = 2x - 1$ • Write in exponential form.

$9 = 2x - 1$

$10 = 2x$

$5 = x$

5 checks as a solution. The solution is 5.

You Try It 3

Solve for x: $\log_4(x^2 - 3x) = 1$

Your solution

Example 4

Solve for x: $\log_2 x - \log_2(x - 1) = \log_2 2$

Solution

$\log_2 x - \log_2(x - 1) = \log_2 2$

$\log_2\left(\frac{x}{x-1}\right) = \log_2 2$ • Quotient Property

$\frac{x}{x-1} = 2$ • 1–1 Property of Logarithms

$(x-1)\left(\frac{x}{x-1}\right) = (x-1)2$

$x = 2x - 2$

$-x = -2$

$x = 2$

2 checks as a solution. The solution is 2.

You Try It 4

Solve for x: $\log_3 x + \log_3(x + 3) = \log_3 4$

Your solution

Example 5

Solve for x:
$\log_2(3x + 8) = \log_2(2x + 2) + \log_2(x - 2)$

Solution

$\log_2(3x + 8) = \log_2(2x + 2) + \log_2(x - 2)$

$\log_2(3x + 8) = \log_2[(2x + 2)(x - 2)]$

$\log_2(3x + 8) = \log_2(2x^2 - 2x - 4)$

$3x + 8 = 2x^2 - 2x - 4$ • 1–1 Property of Logarithms

$0 = 2x^2 - 5x - 12$

$0 = (2x + 3)(x - 4)$ • Solve by factoring.

$2x + 3 = 0$ $\qquad x - 4 = 0$

$x = -\frac{3}{2}$ $\qquad x = 4$

$-\frac{3}{2}$ does not check as a solution;

4 checks as a solution. The solution is 4.

You Try It 5

Solve for x: $\log_3 x + \log_3(x + 6) = 3$

Your solution

Solutions on p. S36

Copyright © Houghton Mifflin Company. All rights reserved.

12.4 Exercises

Objective A To solve an exponential equation

1. What is an exponential equation?

2. **a.** What does the 1–1 Property of Exponential Functions state?
b. Provide an example of when you would use this property.

For Exercises 3 to 26, solve for x. For Exercises 15 to 26, round to the nearest ten-thousandth.

3. $5^{4x-1} = 5^{x-2}$

4. $7^{4x-3} = 7^{2x+1}$

5. $8^{x-4} = 8^{5x+8}$

6. $10^{4x-5} = 10^{x+4}$

7. $9^x = 3^{x+1}$

8. $2^{x-1} = 4^x$

9. $8^{x+2} = 16^x$

10. $9^{3x} = 81^{x-4}$

11. $16^{2-x} = 32^{2x}$

12. $27^{2x-3} = 81^{4-x}$

13. $25^{3-x} = 125^{2x-1}$

14. $8^{4x-7} = 64^{x-3}$

15. $5^x = 6$

16. $7^x = 10$

17. $e^x = 3$

18. $e^x = 2$

19. $10^x = 21$

20. $10^x = 37$

21. $2^{-x} = 7$

22. $3^{-x} = 14$

23. $2^{x-1} = 6$

24. $4^{x+1} = 9$

25. $3^{2x-1} = 4$

26. $4^{-x+2} = 12$

Objective B To solve a logarithmic equation

27. What is a logarithmic equation?

28. What does the 1–1 Property of Logarithms state?

For Exercises 29 to 46, solve for x.

29. $\log_2(2x - 3) = 3$

30. $\log_4(3x + 1) = 2$

31. $\log_2(x^2 + 2x) = 3$

32. $\log_3(x^2 + 6x) = 3$

33. $\log_5 \frac{2x}{x - 1} = 1$

34. $\log_6 \frac{3x}{x + 1} = 1$

Copyright © Houghton Mifflin Company. All rights reserved.

35. $\log x = \log (1 - x)$

36. $\ln(3x - 2) = \ln(x + 1)$

37. $\ln 5 = \ln(4x - 13)$

38. $\log_3(x - 2) = \log_3(2x)$

39. $\ln(3x + 2) = 4$

40. $\ln(2x + 3) = -1$

41. $\log_2(8x) - \log_2(x^2 - 1) = \log_2 3$

42. $\log_5(3x) - \log_5(x^2 - 1) = \log_5 2$

43. $\log_9 x + \log_9(2x - 3) = \log_9 2$

44. $\log_6 x + \log_6(3x - 5) = \log_6 2$

45. $\log_8(6x) = \log_8 2 + \log_8(x - 4)$

46. $\log_7(5x) = \log_7 3 + \log_7(2x + 1)$

APPLYING THE CONCEPTS

For Exercises 47 to 52, solve for x. Round to the nearest ten-thousandth.

47. $8^{\frac{x}{2}} = 6$

48. $4^{\frac{x}{3}} = 2$

49. $5^{\frac{3x}{2}} = 7$

50. $9^{\frac{2x}{3}} = 8$

51. $1.2^{\frac{x}{2} - 1} = 1.4$

52. $5.6^{\frac{x}{3} + 1} = 7.8$

53. **Physics** A model for the distance s, in feet, that an object experiencing air resistance will fall in t seconds is given by $s = 312.5 \ln \dfrac{e^{0.32t} + e^{-0.32t}}{2}$.

a. Graph this equation. *Suggestion:* Use Xmin = 0, Xmax = 4.5, Ymin = 0, Ymax = 140, and Yscl = 20.

b. Determine, to the nearest hundredth of a second, the time it takes the object to travel 100 ft.

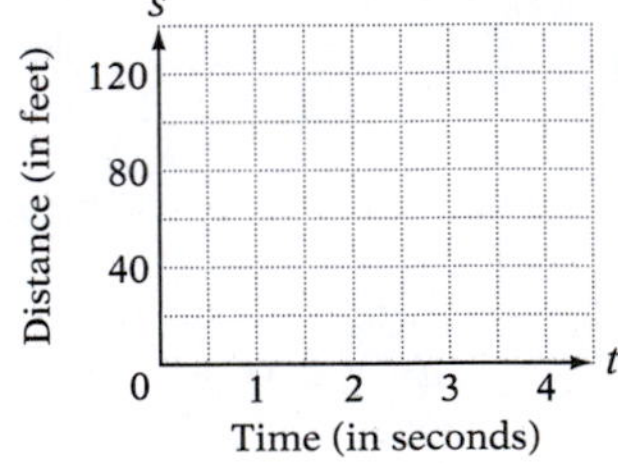

54. **Physics** A model for the distance s, in feet, that an object experiencing air resistance will fall in t seconds is given by $s = 78 \ln \dfrac{e^{0.8t} + e^{-0.8t}}{2}$.

a. Graph this equation. *Suggestion:* Use Xmin = 0, Xmax = 4.5, Ymin = 0, Ymax = 140, and Yscl = 20.

b. Determine, to the nearest hundredth of a second, the time it takes the object to travel 125 ft.

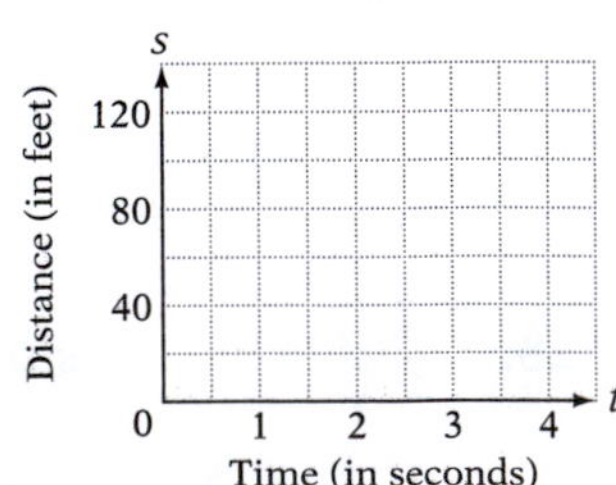

55. The following "proof" shows that $0.5 < 0.25$. Explain the error.

$$
\begin{aligned}
1 &< 2 \\
1 \cdot \log 0.5 &< 2 \cdot \log 0.5 \\
\log 0.5 &< \log (0.5)^2 \\
0.5 &< (0.5)^2 \\
0.5 &< 0.25
\end{aligned}
$$

Copyright © Houghton Mifflin Company. All rights reserved.

12.5 Applications of Exponential and Logarithmic Functions

Objective A

To solve application problems

A biologist places one single-celled bacterium in a culture, and each hour that particular species of bacteria divides into two bacteria. After 1 h there will be two bacteria. After 2 h, each of these two bacteria will divide and there will be four bacteria. After 3 h, each of the four bacteria will divide and there will be eight bacteria.

Point of Interest

C. Northcote Parkinson

Parkinson's Law, named after C. Northcote Parkinson, is sometimes stated as "A job will expand to fill the time alloted for the job." However, Parkinson actually said that in any new government administration, administrative employees will be added at the rate of about 5% to 6% per year. This is an example of exponential growth and means that a staff of 500 will grow to approximately 630 by the end of a 4-year term.

The table at the right shows the number of bacteria in the culture after various intervals of time t, in hours. Values in this table could also be found by using the exponential equation $N = 2^t$.

Time, t	*Number of Bacteria, N*
0	1
1	2
2	4
3	8
4	16

The equation $N = 2^t$ is an example of an **exponential growth equation.** In general, any equation that can be written in the form $A = A_0 b^{kt}$, where A is the size at time t, A_0 is the initial size, $b > 1$, and k is a positive real number, is an exponential growth equation. These equations are important not only in population growth studies but also in physics, chemistry, psychology, and economics.

Recall that interest is the amount of money that one pays (or receives) when borrowing (or investing) money. **Compound interest** is interest that is computed not only on the original principal but also on the interest already earned. The compound interest formula is an exponential equation.

The **compound interest formula** is $A = P(1 + i)^n$, where P is the original value of an investment, i is the interest rate per compounding period, n is the total number of compounding periods, and A is the value of the investment after n periods.

HOW TO An investment broker deposits \$1000 into an account that earns 8% annual interest compounded quarterly. What is the value of the investment after 3 years?

$$i = \frac{8\%}{4} = \frac{0.08}{4} = 0.02$$

- **Find i, the interest rate per quarter. The quarterly rate is the annual rate divided by 4, the number of quarters in 1 year.**

$$n = 4 \cdot 3 = 12$$

- **Find n, the number of compounding periods. The investment is compounded quarterly, 4 times a year, for 3 years.**

$$A = P(1 + i)^n$$

- **Use the compound interest formula.**

$$A = 1000(1 + 0.02)^{12}$$

- **Replace P, i, and n by their values.**

$$A \approx 1268$$

- **Solve for A.**

The value of the investment after 3 years is approximately \$1268.

Copyright © Houghton Mifflin Company. All rights reserved.

Exponential decay offers another example of an exponential equation. One of the most common illustrations of exponential decay is the decay of a radioactive substance. For instance, tritium, a radioactive nucleus of hydrogen that has been used in luminous watch dials, has a half-life of approximately 12 years. This means that one-half of any given amount of tritium will disintegrate in 12 years.

The table at the right indicates the amount of an initial 10-microgram sample of tritium that remains after various intervals of time, t, in years. Values in this table could also be found by using the exponential equation $A = 10(0.5)^{t/12}$.

Time, t	*Amount, A*
0	10
12	5
24	2.5
36	1.25
48	0.625

The equation $A = 10(0.5)^{t/12}$ is an example of an **exponential decay equation.** Comparing this equation to the exponential growth equation, note that for exponential growth, the base of the exponential equation is greater than 1, whereas for exponential decay, the base is between 0 and 1.

A method by which an archaeologist can measure the age of a bone is called carbon dating. Carbon dating is based on a radioactive isotope of carbon called carbon-14, which has a half-life of approximately 5570 years. The exponential decay equation is given by $A = A_0(0.5)^{t/5570}$, where A_0 is the original amount of carbon-14 present in the bone, t is the age of the bone in years, and A is the amount of carbon-14 present after t years.

HOW TO A bone that originally contained 100 mg of carbon-14 now has 70 mg of carbon-14. What is the approximate age of the bone?

$$A = A_0(0.5)^{t/5570}$$
• Use the exponential decay equation.

$$70 = 100(0.5)^{t/5570}$$
• Replace A by 70 and A_0 by 100 and solve for t.

$$0.7 = (0.5)^{t/5570}$$
• Divide each side by 100.

$$\log 0.7 = \log(0.5)^{t/5570}$$
• Take the common logarithm of each side of the equation.

$$\log 0.7 = \frac{t}{5570}\log 0.5$$
• Power Property

$$\frac{5570 \log 0.7}{\log 0.5} = t$$
• Multiply by 5570 and divide by log 0.5.

$$2866 \approx t$$

The bone is approximately 2866 years old.

Point of Interest

Søren Sørensen

The pH scale was created by the Danish biochemist Søren Sørensen in 1909 to measure the acidity of water used in the brewing of beer. pH is an abbreviation for *pondus hydrogenii*, which translates as "potential hydrogen."

A chemist measures the acidity or alkalinity of a solution by measuring the concentration of hydrogen ions, H^+, in the solution using the formula $pH = -\log(H^+)$. A neutral solution such as distilled water has a pH of 7, acids have a pH less than 7, and alkaline solutions (also called basic solutions) have a pH greater than 7.

HOW TO Find the pH of orange juice that has a hydrogen ion concentration, H^+, of 2.9×10^{-4}. Round to the nearest tenth.

$$pH = -\log(H^+)$$
$$= -\log(2.9 \times 10^{-4})$$
• $H^+ = 2.9 \times 10^{-4}$
$$\approx 3.5376$$

The pH of the orange juice is approximately 3.5.

Copyright © Houghton Mifflin Company. All rights reserved.

Logarithmic functions are used to scale very large or very small numbers into numbers that are easier to comprehend. For instance, the *Richter scale magnitude* of an earthquake uses a logarithmic function to convert the intensity of shock waves I into a number M, which for most earthquakes is in the range of 0 to 10. The intensity I of an earthquake is often given in terms of the constant I_0, where I_0 is the intensity of the smallest earthquake, called a *zero-level earthquake,* that can be measured on a seismograph near the earthquake's epicenter. An earthquake with an intensity I has a Richter scale magnitude of $M = \log \frac{I}{I_0}$, where I_0 is the measure of the intensity of a zero-level earthquake.

TAKE NOTE

Note that we do not need to know the value of I_0 to determine the Richter scale magnitude of the earthquake.

HOW TO Find the Richter scale magnitude of the 2003 Amazonas, Brazil, earthquake that had an intensity I of $12{,}589{,}254I_0$. Round to the nearest tenth.

$$M = \log \frac{I}{I_0}$$

$$M = \log \frac{12{,}589{,}254I_0}{I_0}$$ • $I = 12{,}589{,}254I_0$

$$M = \log 12{,}589{,}254$$ • Divide the numerator and denominator by I_0.

$$M \approx 7.1$$ • Evaluate log 12,589,254.

The 2003 Amazonas, Brazil, earthquake had a Richter scale magnitude of about 7.1.

If you know the Richter scale magnitude of an earthquake, you can determine the intensity of the earthquake.

HOW TO  Find the intensity of the 2003 Colima, Mexico, earthquake that measured 7.6 on the Richter scale. Write the answer in terms of I_0. Round to the nearest thousand.

$$M = \log \frac{I}{I_0}$$

$$7.6 = \log \frac{I}{I_0}$$ • Replace M by 7.6.

$$10^{7.6} = \frac{I}{I_0}$$ • Write in exponential form.

$$10^{7.6}I_0 = I$$ • Multiply both sides by I_0.

$$39{,}810{,}717I_0 \approx I$$ • Evaluate $10^{7.6}$.

The 2003 Colima, Mexico, earthquake had an intensity that was approximately 39,811,000 times the intensity of a zero-level earthquake.

Point of Interest

Charles F. Richter

The Richter scale was created by seismologist Charles F. Richter in 1935. Note that a tenfold increase in the intensity level of an earthquake increases the Richter scale magnitude of the earthquake by only 1.

The percent of light that will pass through a substance is given by $\log P = -kd$, where P is the percent of light passing through the substance, k is a constant depending on the substance, and d is the thickness of the substance in meters.

HOW TO For certain parts of the ocean, $k = 0.03$. Using this value, at what depth will the percent of light be 50% of the light at the surface of the ocean? Round to the nearest meter.

$$\log P = -kd$$

$$\log(0.5) = -0.03d$$ • Replace P by 0.5 (50%) and k by 0.03.

$$\frac{\log(0.5)}{-0.03} = d$$ • Solve for d.

$$10.0343 \approx d$$

At a depth of about 10 m, the light will be 50% of the light at the surface.

Copyright © Houghton Mifflin Company. All rights reserved.

Example 1

An investment of \$3000 is placed into an account that earns 12% annual interest compounded monthly. In approximately how many years will the investment be worth twice the original amount?

Strategy

To find the time, solve the compound interest formula for n. Use $A = 6000$, $P = 3000$, and $i = \frac{12\%}{12} = \frac{0.12}{12} = 0.01$.

Solution

$$A = P(1 + i)^n$$
$$6000 = 3000(1 + 0.01)^n$$
$$6000 = 3000(1.01)^n$$
$$2 = (1.01)^n$$
$$\log 2 = \log(1.01)^n$$ • Take the log of each side.
$$\log 2 = n \log 1.01$$ • Power Property
$$\frac{\log 2}{\log 1.01} = n$$ • Divide both sides by log 1.01.
$$70 \approx n$$

70 months ÷ 12 ≈ 5.8 years

In approximately 6 years, the investment will be worth \$6000.

You Try It 1

Find the hydrogen ion concentration, H^+, of vinegar that has a pH of 2.9.

Your strategy

Your solution

Example 2

The number of words per minute that a student can type will increase with practice and can be approximated by the equation $N = 100[1 - (0.9)^t]$, where N is the number of words typed per minute after t days of instruction. Find the number of words a student will type per minute after 8 days of instruction.

Strategy

To find the number of words per minute, replace t in the equation by its given value and solve for N.

Solution

$$N = 100[1 - (0.9)^t]$$
$$= 100[1 - (0.9)^8]$$ • $t = 8$
$$\approx 56.95$$

After 8 days of instruction, a student will type approximately 57 words per minute.

You Try It 2

On April 29, 2003, an earthquake measuring 4.6 on the Richter scale struck Fort Payne, Alabama. Find the intensity of the quake in terms of I_0.

Your strategy

Your solution

Solutions on p. S37

Copyright © Houghton Mifflin Company. All rights reserved.

12.5 Exercises

Objective A To solve application problems

Compound Interest For Exercises 1 to 4, use the compound interest formula $A = P(1 + i)^n$, where P is the original value of an investment, i is the interest rate per compounding period, n is the total number of compounding periods, and A is the value of the investment after n periods.

1. An investment broker deposits $1000 into an account that earns 8% annual interest compounded quarterly. What is the value of the investment after 2 years? Round to the nearest dollar.

2. A financial advisor recommends that a client deposit $2500 into a fund that earns 7.5% annual interest compounded monthly. What will be the value of the investment after 3 years? Round to the nearest cent.

3. To save for college tuition, the parents of a preschooler invest $5000 in a bond fund that earns 6% annual interest compounded monthly. In approximately how many years will the investment be worth $15,000?

4. A hospital administrator deposits $10,000 into an account that earns 6% annual interest compounded monthly. In approximately how many years will the investment be worth $15,000?

Radioactivity For Exercises 5 to 8, use the exponential decay equation $A = A_0\left(\frac{1}{2}\right)^{t/k}$, where A is the amount of a radioactive material present after time t, k is the half-life of the radioactive substance, and A_0 is the original amount of the radioactive substance. Round to the nearest tenth.

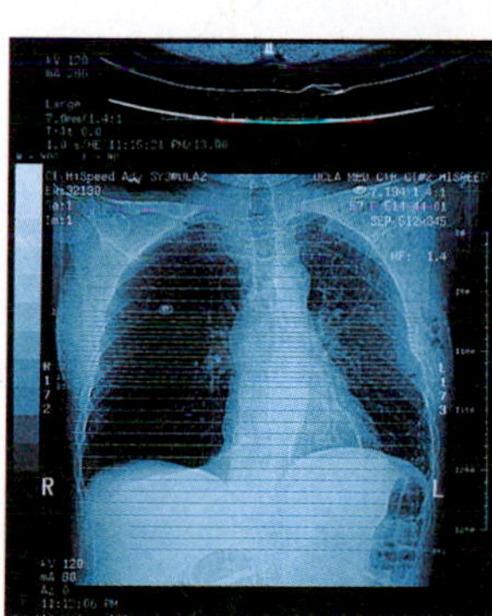

5. An isotope of technetium is used to prepare images of internal body organs. This isotope has a half-life of approximately 6 h. A patient is injected with 30 mg of this isotope.

a. What is the technetium level in the patient after 3 h?
b. How long (in hours) will it take for the technetium level to reach 20 mg?

6. Iodine-131 is an isotope that is used to study the functioning of the thyroid gland. This isotope has a half-life of approximately 8 days. A patient is given an injection that contains 8 micrograms of iodine-131.

a. What is the iodine level in the patient after 5 days?
b. How long (in days) will it take for the iodine level to reach 5 micrograms?

7. A sample of promethium-147 (used in some luminous paints) weighs 25 mg. One year later, the sample weighs 18.95 mg. What is the half-life of promethium-147, in years?

Copyright © Houghton Mifflin Company. All rights reserved.

8. Francium-223 is a very rare radioactive isotope discovered in 1939 by Marguerite Percy. A 3-microgram sample of francium-223 decays to 2.54 micrograms in 5 min. What is the half-life of francium-223, in minutes?

9. **Earth Science** Earth's atmospheric pressure changes as you rise above its surface. At an altitude of h kilometers, where $0 < h < 80$, the pressure P in newtons per square centimeter is approximately modeled by the equation $P(h) = 10.13e^{-0.116h}$.
   a. What is the approximate pressure at 40 km above Earth's surface?
   b. What is the approximate pressure on Earth's surface?
   c. Does atmospheric pressure increase or decrease as you rise above Earth's surface?

10. **Demography** The U.S. Census Bureau provides information about various segments of the population in the United States. The following table gives the number of people, in millions, age 80 and older at the beginning of each decade from 1910 to 2000. An equation that approximately models the data is $y = 0.18808(1.0365)^x$, where x is the number of years since 1900 and y is the population, in millions, of people age 80 and over.

Year	1910	1920	1930	1940	1950	1960	1970	1980	1990	2000
Number of people age 80 and over (in millions)	0.3	0.4	0.5	0.8	1.1	1.6	2.3	2.9	3.9	9.3

   a. According to the model, what is the predicted population of this age group in the year 2020? Round to the nearest tenth of a million. (*Hint:* You will need to determine the x-value for the year 2020.)
   b. In what year does this model predict that the population of this age group will be 15 million? Round to the nearest year.

Chemistry For Exercises 11 and 12, use the equation $\text{pH} = -\log(\text{H}^+)$, where H^+ is the hydrogen ion concentration of a solution. Round to the nearest tenth.

11. Find the pH of milk, which has a hydrogen ion concentration of 3.97×10^{-7}.

12. Find the pH of a baking soda solution for which the hydrogen ion concentration is 3.98×10^{-9}.

Light For Exercises 13 and 14, use the equation $\log P = -kd$, which gives the relationship between the percent P, as a decimal, of light passing through a substance and the thickness d, in meters, of the substance.

13. The value of k for a swimming pool is approximately 0.05. At what depth, in meters, will the percent of light be 75% of the light at the surface of the pool? Round to the nearest tenth.

14. The constant k for a piece of blue stained glass is 20. What percent of light will pass through a piece of this glass that is 0.005 m thick?

Copyright © Houghton Mifflin Company. All rights reserved.

Sound For Exercises 15 and 16, use the equation $D = 10(\log I + 16)$, where D is the number of decibels of a sound and I is the power of the sound measured in watts. Round to the nearest whole number.

15. Find the number of decibels of normal conversation. The power of the sound of normal conversation is approximately 3.2×10^{-10} watts.

16. The loudest sound made by any animal is made by the blue whale and can be heard more than 500 mi away. The power of the sound is 630 watts. Find the number of decibels of sound emitted by the blue whale.

17. **Sports** During the 1980s and 1990s, the average time T to play a Major League baseball game increased each year. If the year 1981 is represented by $x = 1$, then the function $T(x) = 149.57 + 7.63 \ln x$ approximates the time T, in minutes, to play a Major League baseball game for the years $x = 1$ to $x = 19$. By how many minutes did the average time of a Major League baseball game increase from 1981 to 1999? Round to the nearest tenth.

18. **Postage** In 1962, the cost of a first-class postage stamp was \$.04. In 2004, the cost was \$.37. The increase in cost can be modeled by the equation $C = 0.04e^{0.057t}$, where C is the cost and t is the number of years after 1962. According to this model, in what year did a first-class postage stamp cost \$.22?

19. **Chemistry** The intensity I of an x-ray after it has passed through a material that is x centimeters thick is given by $I = I_0 e^{-kx}$, where I_0 is the initial intensity of the x-ray and k is a number that depends on the material. The constant k for copper is 3.2. Find the thickness of copper that is needed so that the intensity of an x-ray after passing through the copper is 25% of the original intensity. Round to the nearest tenth.

20. **Natural Resources** One model for the time it will take for the world's oil supply to be depleted is given by the equation $T = 14.29 \ln (0.00411r + 1)$, where r is the estimated world oil reserves in billions of barrels and T is the time, in years, before that amount of oil is depleted. Use this equation to determine how many barrels of oil are necessary to meet the world demand for 20 years. Round to the nearest tenth of a billion barrels.

Seismology For Exercises 21 to 24, use the Richter scale equation $M = \log \frac{I}{I_0}$, where M is the magnitude of an earthquake, I is the intensity of the shock waves, and I_0 is the measure of the intensity of a zero-level earthquake.

21. On May 21, 2003, an earthquake struck Northern Algeria. The earthquake had an intensity of $I = 6{,}309{,}573I_0$. Find the Richter scale magnitude of the earthquake. Round to the nearest tenth.

Copyright © Houghton Mifflin Company. All rights reserved.

22. The earthquake on November 17, 2003, in the Aleutian Islands of Moska had an intensity of $I = 63{,}095{,}734I_0$. Find the Richter scale magnitude of the earthquake. Round to the nearest tenth.

23. An earthquake in Japan on March 2, 1933, measured 8.9 on the Richter scale. Find the intensity of the earthquake in terms of I_0. Round to the nearest whole number.

24. An earthquake that occurred in China in 1978 measured 8.2 on the Richter scale. Find the intensity of the earthquake in terms of I_0. Round to the nearest whole number.

Seismology Shown at the right is a *seismogram,* which is used to measure the magnitude of an earthquake. The magnitude is determined by the amplitude A of a shock wave and the difference in time t, in seconds, between the occurrences of two types of waves called *primary waves* and *secondary waves*. As you can see on the graph, a primary wave is abbreviated *p-wave* and a secondary wave is abbreviated *s-wave*. The amplitude A of a wave is one-half the difference between its highest and lowest points. For this graph, A is 23 mm. The equation is $M = \log A + 3 \log 8t - 2.92$. Use this information for Exercises 25 to 27. Round to the nearest tenth.

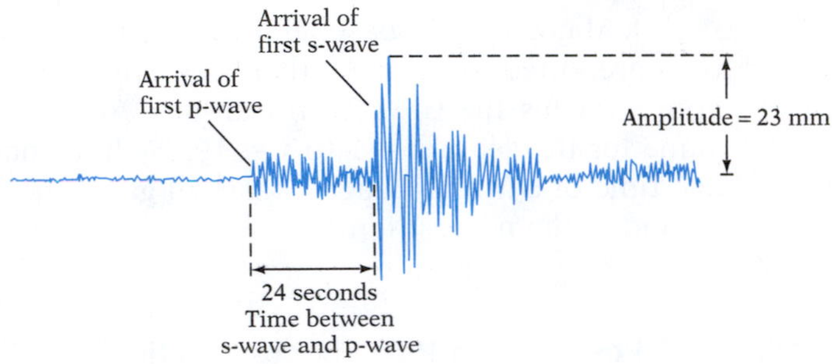

25. Determine the magnitude of the earthquake for the seismogram given in the figure.

26. Find the magnitude of an earthquake that has a seismogram with an amplitude of 30 mm and for which t is 21 s.

27. Find the magnitude of an earthquake that has a seismogram with an amplitude of 28 mm and for which t is 28 s.

APPLYING THE CONCEPTS

28. **Investments** The value of an investment in an account that earns an annual interest rate of 7% compounded daily grows according to the equation $A = A_0\left(1 + \frac{0.07}{365}\right)^{365t}$, where A_0 is the original value of an investment and t is the time in years. Find the time for the investment to double in value. Round to the nearest year.

29. **Investments** Some banks now use continuous compounding of an amount invested. In this case, the equation that relates the value of an initial investment of P dollars after t years earning an annual interest rate r is given by $A = Pe^{rt}$. Using this equation, find the value after 5 years of an investment of \$2500 in an account that earns 5% annual interest.

Copyright © Houghton Mifflin Company. All rights reserved.

Focus on Problem Solving

Proof by Contradiction

The four-step plan for solving problems that we have used before is restated here.

1. Understand the problem.
2. Devise a plan.
3. Carry out the plan.
4. Review the solution.

One of the techniques that can be used in the second step is a method called *proof by contradiction.* In this method you assume that the conditions of the problem you are trying to solve can be met and then show that your assumption leads to a condition you already know is not true.

To illustrate this method, suppose we try to prove that $\sqrt{2}$ is a rational number. We begin by recalling that a rational number is one that can be written as the quotient of integers. Therefore, let a and b be two integers with no common factors. If $\sqrt{2}$ is a rational number, then

$$\sqrt{2} = \frac{a}{b}$$

$$(\sqrt{2})^2 = \left(\frac{a}{b}\right)^2 \qquad \text{• Square each side.}$$

$$2 = \frac{a^2}{b^2}$$

$$2b^2 = a^2 \qquad \text{• Multiply each side by } b^2.$$

From the last equation, a^2 is an even number. Because a^2 is an even number, a is an even number. Now divide each side of the last equation by 2.

$$2b^2 = a^2$$

$$2b^2 = a \cdot a$$

$$b^2 = a \cdot x \qquad \text{• } x = \frac{a}{2}$$

Because a is an even number, $a \cdot x$ is an even number. Because $a \cdot x$ is an even number, b^2 is an even number, and this in turn means that b is an even number. This result, however, contradicts the assumption that a and b are two integers with no common factors. Because this assumption is now known not to be true, we conclude that our original assumption, that $\sqrt{2}$ is a rational number, is false. This proves that $\sqrt{2}$ is an irrational number.

Try a proof by contradiction for the following problem: "Is it possible to write numbers using each of the digits 0, 1, 2, 3, 4, 5, 6, 7, 8, and 9 exactly once such that the sum of the numbers is exactly 100?"* Here are some suggestions. First note that the sum of the 10 digits is 45. This means that some of the digits used must be tens digits. Let x be the sum of those digits.

a. What is the sum of the remaining units digits?

b. Express "the sum of the units digits and the tens digits equals 100" as an equation.

c. Solve the equation for x.

d. Explain why this result means that it is impossible to satisfy both conditions of the problem.

*Problem adapted from G. Polya, *How to Solve It: A New Aspect of Mathematical Method.* Copyright © 1945 Princeton University Press, 1973 renewed PUP. Reprinted by permission of Princeton University Press.

Copyright © Houghton Mifflin Company. All rights reserved.

Projects and Group Activities

Solving Exponential and Logarithmic Equations Using a Graphing Calculator

A graphing calculator can be used to draw the graphs of logarithmic functions. Note that there are two logarithmic keys, LOG and LN, on a graphing calculator. The first key gives the values of common logarithms (base 10), and the second gives the values of natural logarithms (base e).

To graph $y = \ln x$, press the Y= key. Clear any equations already entered. Press LN X,T,θ,n) GRAPH. The graph is shown at the left with a viewing window of Xmin = −1, Xmax = 8, Ymin = −5, and Ymax = 5.

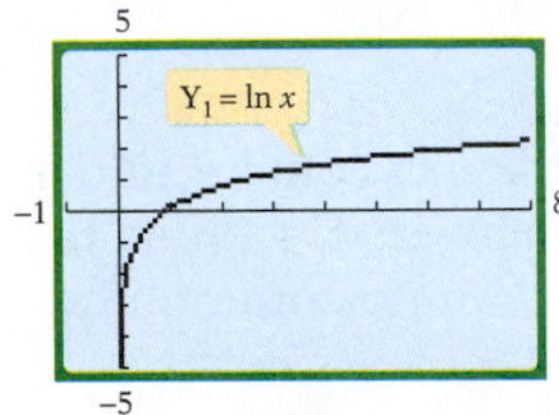

Some exponential and logarithmic equations cannot be solved algebraically. In these cases, a graphical approach may be appropriate. Here is an example.

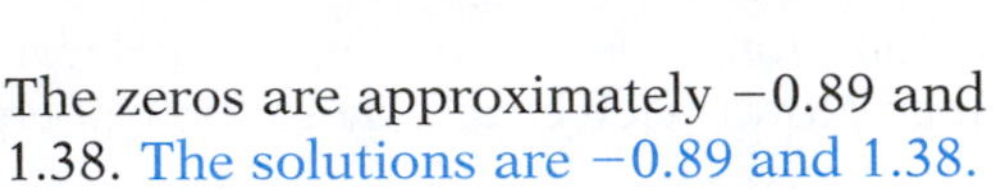

HOW TO Solve $\ln (2x + 4) = x^2$ for x. Round to the nearest hundredth.

Rewrite the equation by subtracting x^2 from each side.

$$\ln (2x + 4) = x^2$$
$$\ln (2x + 4) - x^2 = 0$$

The zeros of $f(x) = \ln (2x + 4) - x^2$ are the solutions of $\ln (2x + 4) - x^2 = 0$.

Graph f and use the zero feature of the graphing calculator to estimate the solutions to the nearest hundredth.

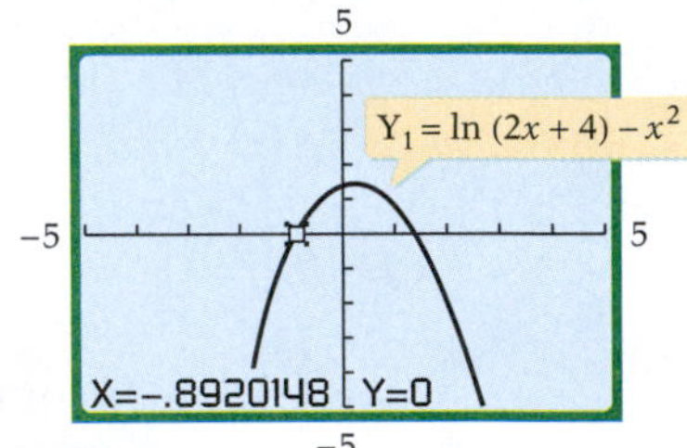

The zeros are approximately −0.89 and 1.38. The solutions are −0.89 and 1.38.

For Exercises 1 to 8, solve for x by graphing. Round to the nearest hundredth.

1. $2^x = 2x + 4$
2. $3^x = -x - 1$
3. $e^x = -2x - 2$
4. $e^x = 3x + 4$
5. $\log (2x - 1) = -x + 3$
6. $\log (x + 4) = -2x + 1$
7. $\ln (x + 2) = x^2 - 3$
8. $\ln x = -x^2 + 1$

To graph a logarithmic function with a base other than base 10 or base e, the Change-of-Base Formula is used. Here is the Change-of-Base Formula, discussed earlier:

$$\log_a N = \frac{\log_b N}{\log_b a}$$

To graph $y = \log_2 x$, first change from base 2 logarithms to the equivalent base 10 logarithms or natural logarithms. Base 10 is shown here.

$$y = \log_2 x = \frac{\log_{10} x}{\log_{10} 2} \qquad \bullet\ a = 2, b = 10, N = x$$

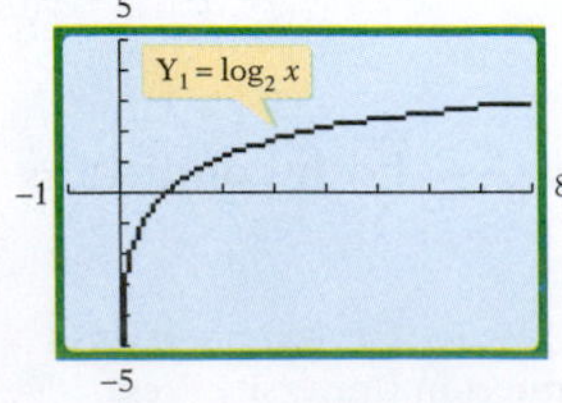

Then graph the equivalent equation $y = \frac{\log x}{\log 2}$. The graph is shown at the left with a viewing window of Xmin = −1, Xmax = 8, Ymin = −5, and Ymax = 5.

Copyright © Houghton Mifflin Company. All rights reserved.

Credit Reports and FICO® Scores*

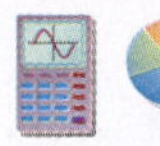
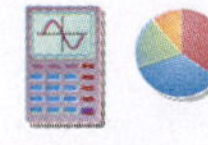

When a consumer applies for a loan, the lender generally wants to know the consumer's credit history. For this, the lender turns to a credit reporting agency. These agencies maintain files on millions of borrowers. They provide the lender with a credit report, which lists information such as the types of credit the consumer uses, the lengths of time the consumer's credit accounts have been open, the amounts owed by the consumer, and whether the consumer has paid his or her bills on time. Along with the credit report, the lender can buy a credit score based on the information in the report. The credit score gives the lender a quick measure of the consumer's credit risk, or how likely the consumer is to repay a debt. It answers the lender's question: "If I lend this person money (or give this person a credit card), what is the probability that I will be paid back in a timely manner?"

The most widely used credit bureau score is the FICO® score, so named because scores are produced from software developed by Fair Isaac Corporation (FICO). FICO scores range from 300 to 850. The higher the score, the lower the predicted credit risk for lenders.

The following graph shows the delinquency rate, or credit risk, associated with ranges of FICO scores. For example, the delinquency rate of consumers in the 550–599 range of scores is 51%. This means that within the next 2 years, for every 100 borrowers in this range, approximately 51 will default on a loan, file for bankruptcy, or fail to pay a credit card bill within 90 days of the due date.

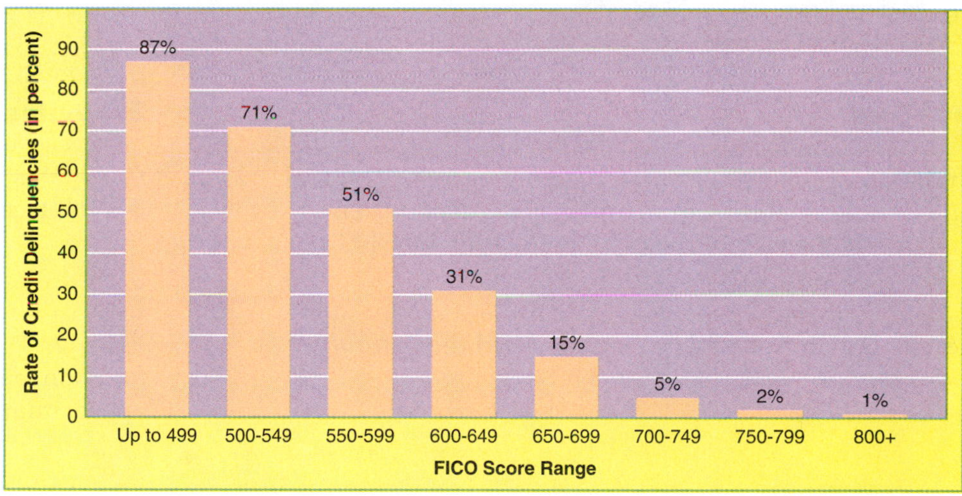

Because the deliquency rate depends on the FICO score, we can let the independent variable x represent the FICO score and the dependent variable y represent the delinquency rate. We will use the middle of each range for the values of x. The resulting ordered pairs are recorded in the table at the right.

x	y
400	87
525	71
575	51
625	31
675	15
725	5
775	2
825	1

*Source: **www.myfico.com**. Reprinted by permission of Fair Isaac Corporation.

Copyright © Houghton Mifflin Company. All rights reserved.

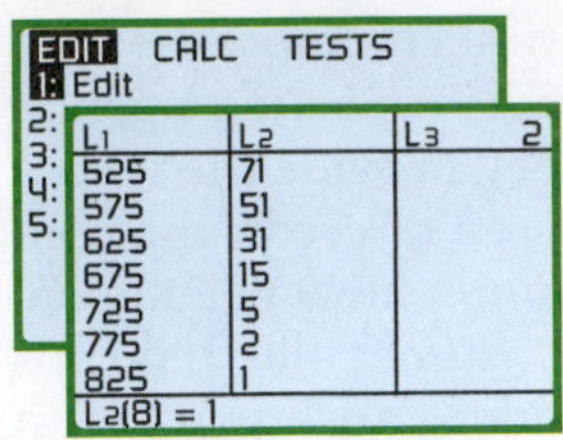

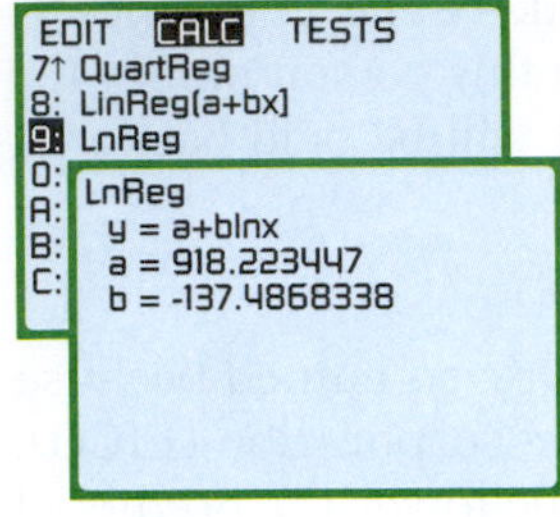

1. Use your graphing calculator to find a logarithmic equation that approximates these data. Here are instructions for the TI-83/84:

   Press STAT ENTER. This will bring up a table into which you can enter data. Enter the values of x in L1. Enter the values of y in L2.

   Press STAT again, press the right arrow key to highlight CALC, and arrow down to LnReg (ln for the natural logarithm). Press ENTER twice. The values for a and b in the equation $y = a + b \ln x$ will appear on the screen.

2. Use your equation to predict the delinquency rate of a consumer with a score of 500. Round to the nearest whole number.

3. The equation pairs a delinquency rate of 36% with what score?

4. Use the graph on page 699 to determine the highest FICO score that will result in a delinquency rate of 71% or more.

5. Use the graph on page 699 to determine the lowest FICO score that will result in a delinquency rate of 5% or lower.

6. Use the Internet to find the name of at least one of the major credit reporting agencies.

Chapter 12 Summary

Key Words	Examples
A function of the form $f(x) = b^x$,where b is a positive real number not equal to 1, is an *exponential function*. The number b is the *base* of the exponential function. [12.1A, p. 661]	$f(x) = 3^x$ is an exponential function. 3 is the base of the function.
The function defined by $f(x) = e^x$ is called the *natural exponential function*. [12.1A, p. 662]	$f(x) = 2e^{x-1}$ is a natural exponential function. e is an irrational number approximately equal to 2.71828183.
Because the exponential function is a 1–1 function, it has an inverse function that is called a *logarithm*. The definition of logarithm is: For $x > 0$, $b > 0$, $b \neq 1$, $y = \log_b x$ is equivalent to $x = b^y$. [12.2A, p. 669]	$\log_2 8 = 3$ is equivalent to $8 = 2^3$.
Logarithms with base 10 are called *common logarithms*. We usually omit the base, 10, when writing the common logarithm of a number. [12.2A, p. 670]	$\log_{10} 100 = 2$ is usually written $\log 100 = 2$.
When e (the base of the natural exponential function) is used as the base of a logarithm, the logarithm is referred to as the *natural logarithm* and is abbreviated $\ln x$. [12.2A, p. 670]	$\log_e 100 \approx 4.61$ is usually written $\ln 100 \approx 4.61$.

Copyright © Houghton Mifflin Company. All rights reserved.

An *exponential equation* is one in which a variable occurs in the exponent. [12.4A, p. 683]

$2^x = 12$ is an exponential equation.

An *exponential growth equation* is an equation that can be written in the form $A = A_0 b^{kt}$, where A is the size at time t, A_0 is the initial size, $b > 1$, and k is a positive real number. In an *exponential decay equation*, the base is between 0 and 1. [12.5A, pp. 689–690]

$P = 1000(1.03)^n$ is an exponential growth equation.
$A = 10(0.5)^x$ is an exponential decay equation.

Essential Rules and Procedures

Examples

1–1 Property of Exponential Functions [12.2A, p. 670]
For $b > 0$, $b \neq 1$, if $b^u = b^v$, then $u = v$.

If $b^x = b^5$, then $x = 5$.

Logarithm Property of the Product of Two Numbers [12.2B, p. 671]
For any positive real numbers x, y, and b, $b \neq 1$, $\log_b(xy) = \log_b x + \log_b y$.

$\log_b(3x) = \log_b 3 + \log_b x$

Logarithm Property of the Quotient of Two Numbers [12.2B, p. 672]
For any positive real numbers x, y, and b, $b \neq 1$, $\log_b \frac{x}{y} = \log_b x - \log_b y$.

$\log_b \frac{x}{20} = \log_b x - \log_b 20$

Logarithm Property of the Power of a Number [12.2B, p. 672]
For any positive real numbers x and b, $b \neq 1$, and for any real number r, $\log_b x^r = r \log_b x$.

$\log_b x^5 = 5 \log_b x$

Logarithm Property of One [12.2B, p. 672]
For any positive real numbers x, y, and b, $b \neq 1$, $\log_b 1 = 0$.

$\log_6 1 = 0$

Inverse Property of Logarithms [12.2B, p. 672]
For any positive real numbers x, y, and b, $b \neq 1$, $b^{\log_b x} = x$ and $\log_b b^x = x$.

$\log_3 3^4 = 4$

1–1 Property of Logarithms [12.2B, p. 672]
For any positive real numbers x, y, and b, $b \neq 1$, if $\log_b x = \log_b y$, then $x = y$.

If $\log_5(x - 2) = \log_5 3$, then $x - 2 = 3$.

Change-of-Base Formula [12.2C, p. 674]
$$\log_a N = \frac{\log_b N}{\log_b a}$$

$\log_3 12 = \frac{\log 12}{\log 3}$ $\quad \log_6 16 = \frac{\ln 16}{\ln 6}$

Copyright © Houghton Mifflin Company. All rights reserved.

Chapter 12 Review Exercises

1. Evaluate $f(x) = e^{x-2}$ at $x = 2$.

2. Write $\log_5 25 = 2$ in exponential form.

3. Graph: $f(x) = 3^{-x} + 2$

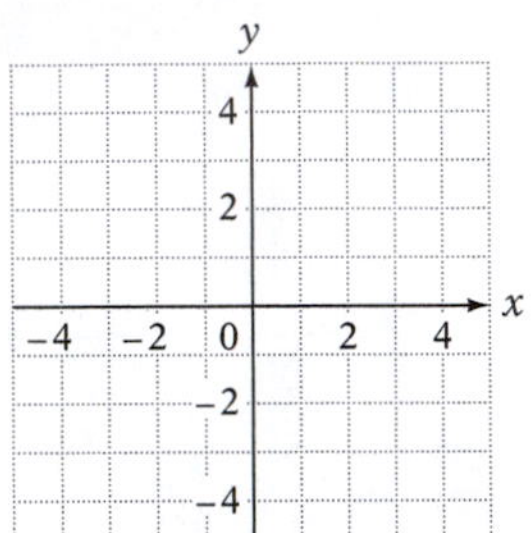

4. Graph: $f(x) = \log_3 (x - 1)$

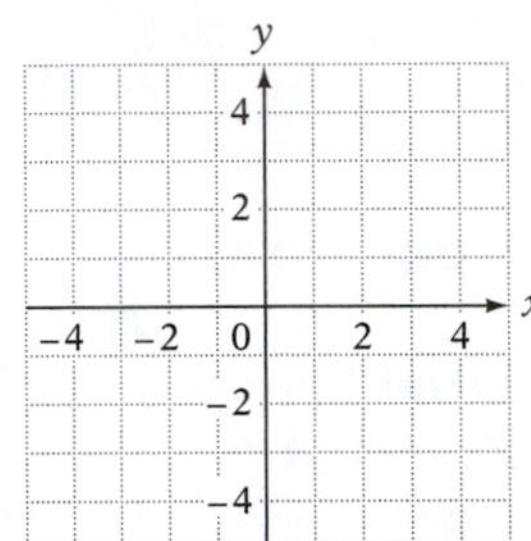

5. Write $\log_3 \sqrt[5]{x^2y^4}$ in expanded form.

6. Write $2 \log_3 x - 5 \log_3 y$ as a single logarithm with a coefficient of 1.

7. Solve: $27^{2x+4} = 81^{x-3}$

8. Solve: $\log_5 \dfrac{7x + 2}{3x} = 1$

9. Find $\log_6 22$. Round to the nearest ten-thousandth.

10. Solve: $\log_2 x = 5$

11. Solve: $\log_3(x + 2) = 4$

12. Solve: $\log_{10} x = 3$

13. Write $\frac{1}{3}(\log_7 x + 4 \log_7 y)$ as a single logarithm with a coefficient of 1.

14. Write $\log_8 \sqrt{\dfrac{x^5}{y^3}}$ in expanded form.

15. Write $2^5 = 32$ in logarithmic form.

16. Find $\log_3 1.6$. Round to the nearest ten-thousandth.

Copyright © Houghton Mifflin Company. All rights reserved.

17. Solve $3^{x+2} = 5$. Round to the nearest thousandth.

18. Evaluate $f(x) = \left(\frac{2}{3}\right)^{x+2}$ at $x = -3$.

19. Solve: $\log_8(x + 2) - \log_8 x = \log_8 4$

20. Solve: $\log_6(2x) = \log_6 2 + \log_6(3x - 4)$

21. Graph: $f(x) = \left(\frac{2}{3}\right)^{x+1}$

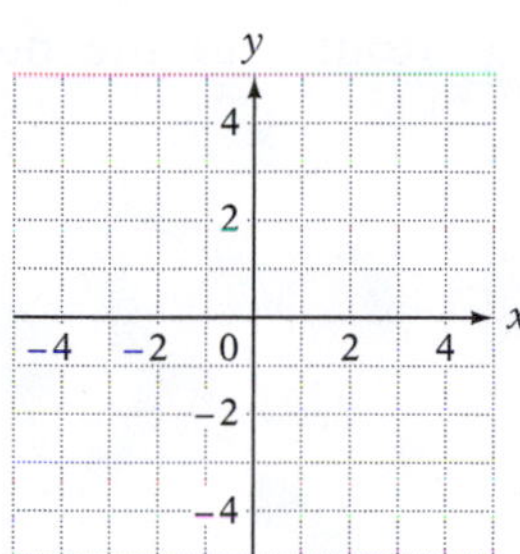

22. Graph: $f(x) = \log_2(2x - 1)$

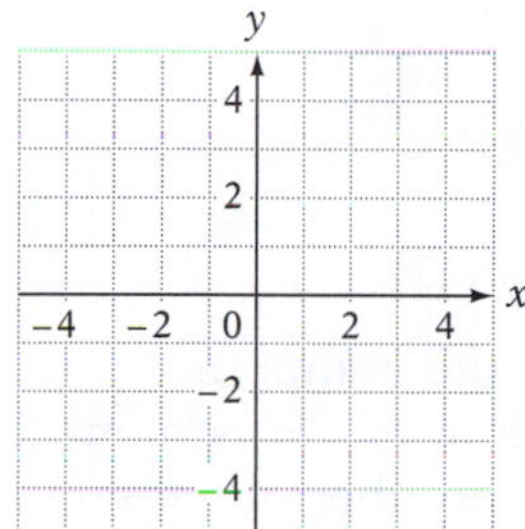

23. Evaluate: $\log_6 36$

24. Write $\frac{1}{3}(\log_2 x - \log_2 y)$ as a single logarithm with a coefficient of 1.

25. Solve for x: $9^{2x} = 3^{x+3}$

26. Evaluate $f(x) = \left(\frac{3}{5}\right)^x$ at $x = 0$.

27. Solve for x: $\log_5 x = -1$

28. Write $3^4 = 81$ in logarithmic form.

29. Solve for x: $\log x + \log(x - 2) = \log 15$

30. Write $\log_5 \sqrt[3]{x^2 y}$ in expanded form.

31. Solve for x: $9^{2x-5} = 9^{x-3}$

32. Evaluate $f(x) = 7^{x+2}$ at $x = -3$.

Copyright © Houghton Mifflin Company. All rights reserved.

33. Evaluate: $\log_2 16$

34. Solve for x: $\log_6 x = \log_6 2 + \log_6(2x - 3)$

35. Evaluate $\log_2 5$. Round to the nearest ten-thousandth.

36. Solve for x: $4^x = 8^{x-1}$

37. Solve for x: $\log_5 x = 4$

38. Write $2 \log_b x - 7 \log_b y$ as a single logarithm with a coefficient of 1.

39. Evaluate $f(x) = 5^{-x-1}$ at $x = -2$.

40. Solve $5^{x-2} = 7$ for x. Round to the nearest ten-thousandth.

41. **Investments** Use the compound interest formula $P = A(1 + i)^n$, where A is the original value of an investment, i is the interest rate per compounding period, n is the number of compounding periods, and P is the value of the investment after n periods, to find the value of an investment after 2 years. The amount of the investment is \$4000, and it is invested at 8% compounded monthly. Round to the nearest dollar.

42. **Seismology** An earthquake in Japan in September, 2003, had an intensity of $I = 199{,}526{,}232I_0$. Find the Richter scale magnitude of the earthquake. Use the Richter scale equation $M = \log \frac{I}{I_0}$, where M is the magnitude of an earthquake, I is the intensity of the shock waves, and I_0 is the measure of the intensity of a zero-level earthquake. Round to the nearest tenth.

43. **Radioactivity** Use the exponential decay equation $A = A_0\left(\frac{1}{2}\right)^{t/k}$, where A is the amount of a radioactive material present after time t, k is the half-life of the radioactive material, and A_0 is the original amount of radioactive material, to find the half-life of a material that decays from 25 mg to 15 mg in 20 days. Round to the nearest whole number.

44. **Sound** The number of decibels, D, of a sound can be given by the equation $D = 10(\log I + 16)$, where I is the power of the sound measured in watts. Find the number of decibels of sound emitted from a busy street corner for which the power of the sound is 5×10^{-6} watts. Round to the nearest whole number.

Copyright © Houghton Mifflin Company. All rights reserved.

Chapter 12 Test

1. Evaluate $f(x) = \left(\frac{2}{3}\right)^x$ at $x = 0$.

2. Evaluate $f(x) = 3^{x+1}$ at $x = -2$.

3. Graph: $f(x) = 2^x - 3$

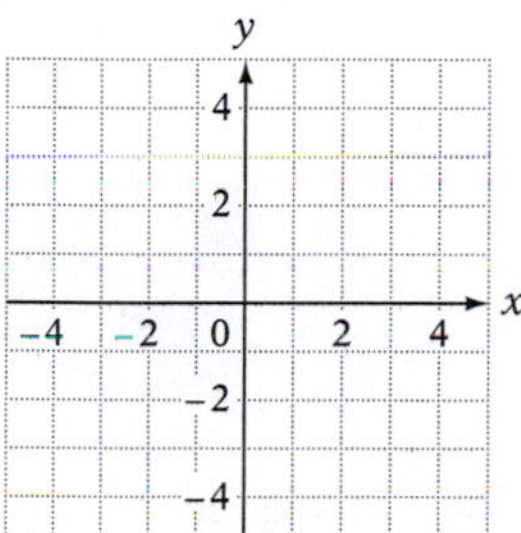

4. Graph: $f(x) = 2^x + 2$

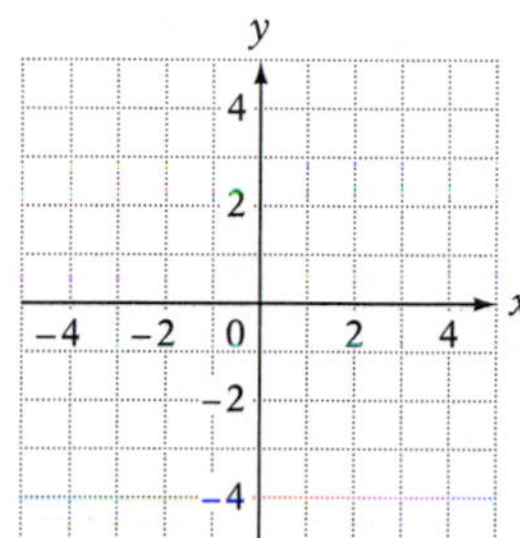

5. Evaluate: $\log_4 16$

6. Solve for x: $\log_3 x = -2$

7. Graph: $f(x) = \log_2(2x)$

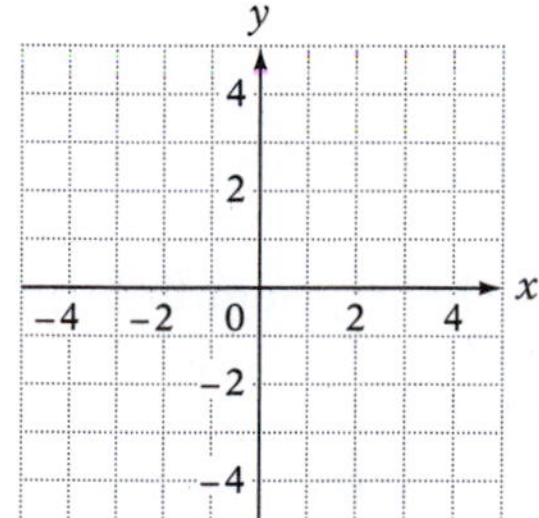

8. Graph: $f(x) = \log_3(x + 1)$

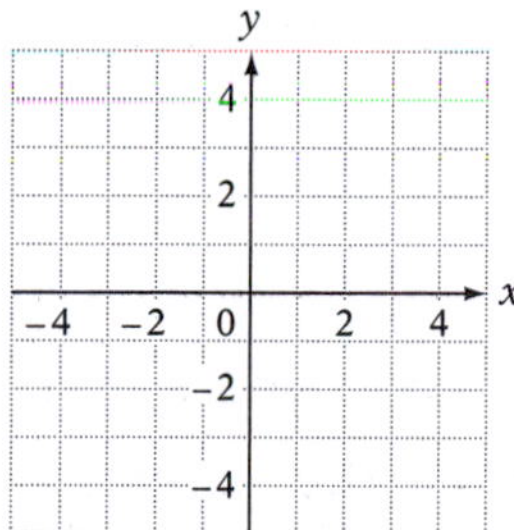

9. Write $\log_6 \sqrt{xy^3}$ in expanded form.

10. Write $\frac{1}{2}(\log_3 x - \log_3 y)$ as a single logarithm with a coefficient of 1.

Copyright © Houghton Mifflin Company. All rights reserved.

11. Write $\ln \frac{x}{\sqrt{z}}$ in expanded form.

12. Write $3 \ln x - \ln y - \frac{1}{2} \ln z$ as a single logarithm with a coefficient of 1.

13. Solve for x: $3^{7x+1} = 3^{4x-5}$

14. Solve for x: $8^x = 2^{x-6}$

15. Solve for x: $3^x = 17$

16. Solve for x: $\log x + \log(x - 4) = \log 12$

17. Solve for x: $\log_6 x + \log_6(x - 1) = 1$

18. Find $\log_5 9$.

19. Find $\log_3 19$.

20. **Radioactivity** Use the exponential decay equation $A = A_0 \left(\frac{1}{2}\right)^{t/k}$, where A is the amount of a radioactive material present after time t, k is the half-life of the material, and A_0 is the original amount of radioactive material, to find the half-life of a material that decays from 10 mg to 9 mg in 5 h. Round to the nearest whole number.

Copyright © Houghton Mifflin Company. All rights reserved.

Cumulative Review Exercises

1. Solve: $4 - 2[x - 3(2 - 3x) - 4x] = 2x$

2. Find the equation of the line that contains the point $(2, -2)$ and is parallel to the line $2x - y = 5$.

3. Factor: $4x^{2n} + 7x^n + 3$

4. Simplify: $\dfrac{1 - \dfrac{5}{x} + \dfrac{6}{x^2}}{1 + \dfrac{1}{x} - \dfrac{6}{x^2}}$

5. Simplify: $\dfrac{\sqrt{xy}}{\sqrt{x} - \sqrt{y}}$

6. Solve by completing the square: $x^2 - 4x - 6 = 0$

7. Write a quadratic equation that has integer coefficients and has solutions $\frac{1}{3}$ and -3.

8. Graph the solution set: $2x - y < 3$
$x + y < 1$

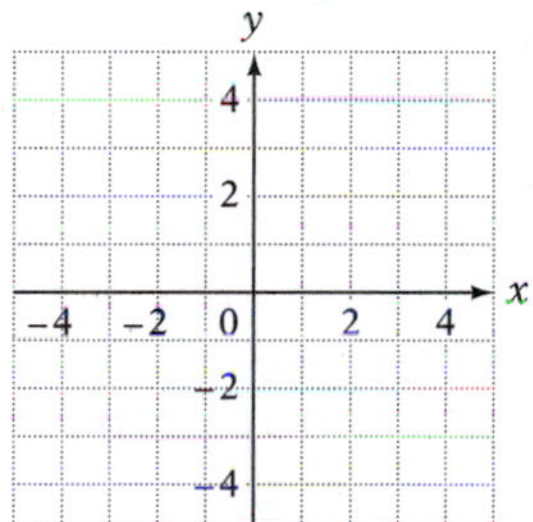

9. Solve by the addition method:
$$\begin{aligned} 3x - y + z &= 3 \\ x + y + 4z &= 7 \\ 3x - 2y + 3z &= 8 \end{aligned}$$

10. Simplify: $\dfrac{x - 4}{2 - x} - \dfrac{1 - 6x}{2x^2 - 7x + 6}$

11. Solve: $x^2 + 4x - 5 \le 0$
Write the solution set in set-builder notation.

12. Solve: $|2x - 5| \le 3$

13. Graph: $f(x) = \left(\dfrac{1}{2}\right)^x + 1$

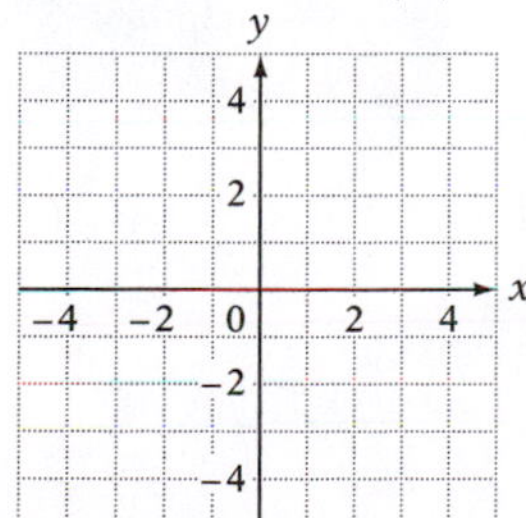

14. Graph: $f(x) = \log_2 x - 1$

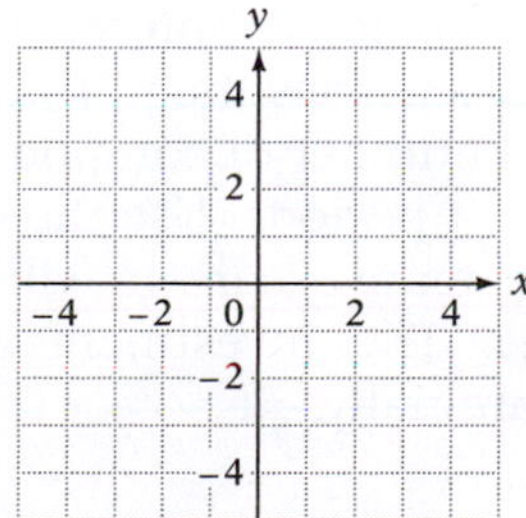

Copyright © Houghton Mifflin Company. All rights reserved.

15. Evaluate the function $f(x) = 2^{-x-1}$ at $x = -3$.

16. Solve for x: $\log_5 x = 3$

17. Write $3 \log_b x - 5 \log_b y$ as a single logarithm with a coefficient of 1.

18. Find $\log_3 7$. Round to the nearest ten-thousandth.

19. Solve for x: $4^{5x-2} = 4^{3x+2}$

20. Solve for x: $\log x + \log(2x + 3) = \log 2$

21. **Banking** A bank offers two types of business checking accounts. One account has a charge of \$5 per month plus 2 cents per check. The second account has a charge of \$2 per month plus 8 cents per check. How many checks can a customer who has the second type of account write if it is to cost the customer less than the first type of checking account?

22. **Mixtures** Find the cost per pound of a mixture made from 16 lb of chocolate that costs \$4.00 per pound and 24 lb of chocolate that costs \$2.50 per pound.

23. **Uniform Motion** A plane can fly at a rate of 225 mph in calm air. Traveling with the wind, the plane flew 1000 mi in the same amount of time that it took to fly 800 mi against the wind. Find the rate of the wind.

24. **Physics** The distance, d, that a spring stretches varies directly as the force, f, used to stretch the spring. If a force of 20 lb stretches a spring 6 in., how far will a force of 34 lb stretch the spring?

25. **Carpentry** A carpenter purchased 80 ft of redwood and 140 ft of fir for a total cost of \$67. A second purchase, at the same prices, included 140 ft of redwood and 100 ft of fir for a total cost of \$81. Find the cost of redwood and of fir.

26. **Investments** The compound interest formula is $A = P(1 + i)^n$, where P is the original value of an investment, i is the interest rate per compounding period, n is the total number of compounding periods, and A is the value of the investment after n periods. Use the compound interest formula to find approximately how many years it will take for an investment of \$5000 to double in value. The investment earns 7% annual interest and is compounded semiannually.

Copyright © Houghton Mifflin Company. All rights reserved.

Final Exam

1. Simplify: $12 - 8[3 - (-2)]^2 \div 5 - 3$

2. Evaluate $\dfrac{a^2 - b^2}{a - b}$ when $a = 3$ and $b = -4$.

3. Simplify: $5 - 2[3x - 7(2 - x) - 5x]$

4. Solve: $\dfrac{3}{4}x - 2 = 4$

5. Solve: $8 - |5 - 3x| = 1$

6. Find the volume of a sphere with a diameter of 8 ft. Round to the nearest tenth.

7. Graph $2x - 3y = 9$ using the x- and y-intercepts.

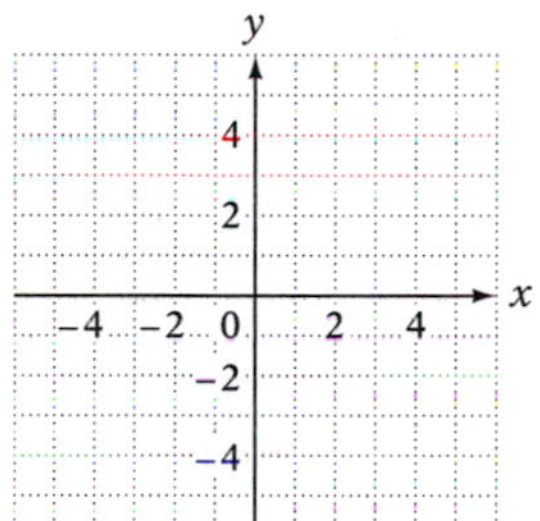

8. Find the equation of the line containing the points $(3, -2)$ and $(1, 4)$.

9. Find the equation of the line that contains the point $(-2, 1)$ and is perpendicular to the line $3x - 2y = 6$.

10. Simplify: $2a[5 - a(2 - 3a) - 2a] + 3a^2$

11. Factor: $8 - x^3y^3$

12. Factor: $x - y - x^3 + x^2y$

13. Divide: $(2x^3 - 7x^2 + 4) \div (2x - 3)$

14. Divide: $\dfrac{x^2 - 3x}{2x^2 - 3x - 5} \div \dfrac{4x - 12}{4x^2 - 4}$

15. Subtract: $\dfrac{x - 2}{x + 2} - \dfrac{x + 3}{x - 3}$

16. Simplify: $\dfrac{\dfrac{3}{x} + \dfrac{1}{x + 4}}{\dfrac{1}{x} + \dfrac{3}{x + 4}}$

Copyright © Houghton Mifflin Company. All rights reserved.

17. Solve: $\dfrac{5}{x-2} - \dfrac{5}{x^2-4} = \dfrac{1}{x+2}$

18. Solve $a_n = a_1 + (n-1)d$ for d.

19. Simplify: $\left(\dfrac{4x^2y^{-1}}{3x^{-1}y}\right)^{-2}\left(\dfrac{2x^{-1}y^2}{9x^{-2}y^2}\right)^3$

20. Simplify: $\left(\dfrac{3x^{2/3}y^{1/2}}{6x^2y^{4/3}}\right)^6$

21. Subtract: $x\sqrt{18x^2y^3} - y\sqrt{50x^4y}$

22. Simplify: $\dfrac{\sqrt{16x^5y^4}}{\sqrt{32xy^7}}$

23. Simplify: $\dfrac{3}{2+i}$

24. Write a quadratic equation that has integer coefficients and has solutions $-\frac{1}{2}$ and 2.

25. Solve by using the quadratic formula:
$2x^2 - 3x - 1 = 0$

26. Solve: $x^{2/3} - x^{1/3} - 6 = 0$

27. Graph: $f(x) = -x^2 + 4$

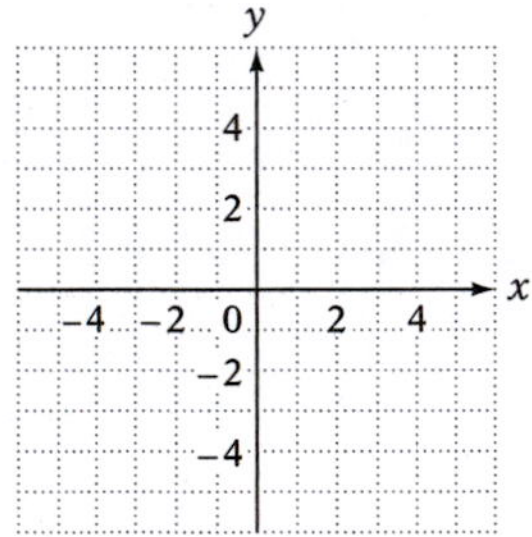

28. Graph: $f(x) = -\dfrac{1}{2}x - 3$

29. Solve: $\dfrac{2}{x} - \dfrac{2}{2x+3} = 1$

30. Find the inverse of the function
$f(x) = \dfrac{2}{3}x - 4$.

31. Solve by the addition method:
$3x - 2y = 1$
$5x - 3y = 3$

32. Simplify: $\sqrt{49x^6}$

Copyright © Houghton Mifflin Company. All rights reserved.

33. Solve: $2 - 3x < 6$ and $2x + 1 > 4$

34. Solve: $|2x + 5| < 3$

35. Graph the solution set: $3x + 2y > 6$

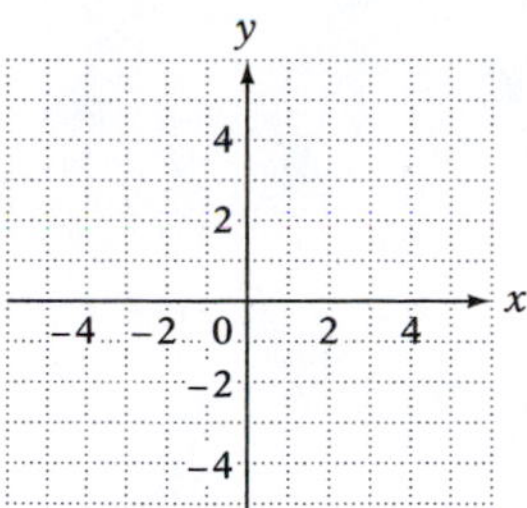

36. Graph: $f(x) = 3^{-x} - 2$

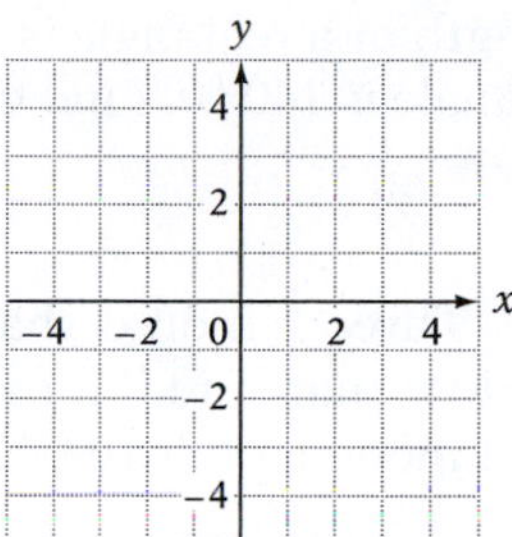

37. Graph: $f(x) = \log_2 (x + 1)$

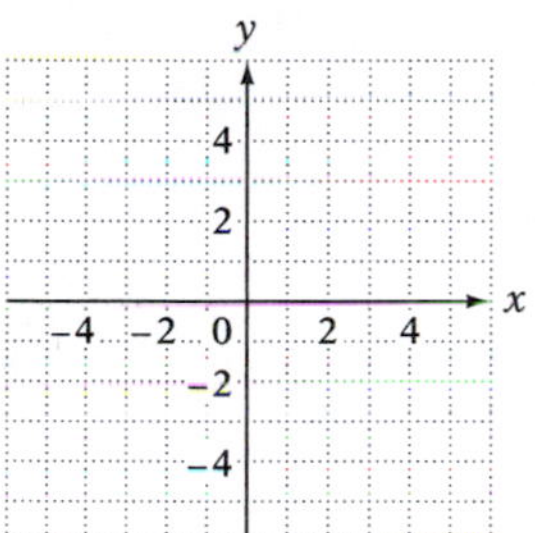

38. Write $2(\log_2 a - \log_2 b)$ as a single logarithm with a coefficient of 1.

39. Solve for x: $\log_3 x - \log_3 (x - 3) = \log_3 2$

40. **Education** An average score of 70–79 in a history class receives a C grade. A student has grades of 64, 58, 82, and 77 on four history tests. Find the range of scores on the fifth test that will give the student a C grade for the course.

41. **Uniform Motion** A jogger and a cyclist set out at 8 A.M. from the same point headed in the same direction. The average speed of the cyclist is two and a half times the average speed of the jogger. In 2 h, the cyclist is 24 mi ahead of the jogger. How far did the cyclist ride in that time?

Copyright © Houghton Mifflin Company. All rights reserved.

42. **Investments** You have a total of \$12,000 invested in two simple interest accounts. On one account, a money market fund, the annual simple interest rate is 8.5%. On the other account, a tax-free bond fund, the annual simple interest rate is 6.4%. The total annual interest earned by the two accounts is \$936. How much do you have invested in each account?

43. **Geometry** The length of a rectangle is 1 ft less than three times the width. The area of the rectangle is 140 ft^2. Find the length and width of the rectangle.

44. **The Stock Market** Three hundred shares of a utility stock earn a yearly dividend of \$486. How many additional shares of the utility stock would give a total dividend income of \$810?

45. **Travel** An account executive traveled 45 mi by car and then an additional 1050 mi by plane. The rate of the plane was seven times the rate of the car. The total time for the trip was $3\frac{1}{4}$ h. Find the rate of the plane.

46. **Physics** An object is dropped from the top of a building. Find the distance the object has fallen when the speed reaches 75 ft/s. Use the equation $v = \sqrt{64d}$, where v is the speed of the object and d is the distance. Round to the nearest whole number.

47. **Travel** A small plane made a trip of 660 mi in 5 h. The plane traveled the first 360 mi at a constant rate before increasing its speed by 30 mph. Then it traveled another 300 mi at the increased speed. Find the rate of the plane for the first 360 mi.

48. **Light** The intensity (L) of a light source is inversely proportional to the square of the distance (d) from the source. If the intensity is 8 foot-candles at a distance of 20 ft, what is the intensity when the distance is 4 ft?

49. **Travel** A motorboat traveling with the current can go 30 mi in 2 h. Against the current, it takes 3 h to go the same distance. Find the rate of the motorboat in calm water and the rate of the current.

50. **Investments** An investor deposits \$4000 into an account that earns 9% annual interest compounded monthly. Use the compound interest formula $P = A(1 + i)^n$, where A is the original value of the investment, i is the interest rate per compounding period, n is the total number of compounding periods, and P is the value of the investment after n periods, to find the value of the investment after 2 years. Round to the nearest cent.

Copyright © Houghton Mifflin Company. All rights reserved.

chapter R

Review of Introductory Algebra Topics

OBJECTIVES

Section R.1

A To evaluate a variable expression
B To simplify a variable expression

Section R.2

A To solve a first-degree equation in one variable
B To solve an inequality in one variable

Section R.3

A To graph points in a rectangular coordinate system
B To graph a linear equation in two variables
C To evaluate a function
D To find the equation of a line

Section R.4

A To multiply and divide monomials
B To add and subtract polynomials
C To multiply polynomials
D To divide polynomials
E To factor polynomials of the form $ax^2 + bx + c$

Copyright © Houghton Mifflin Company. All rights reserved.

Need help? For online student resources, such as section quizzes, visit this textbook's website at **math.college.hmco.com/students.**

R.1 Variable Expressions

Objective A To evaluate a variable expression

Whenever an expression contains more than one operation, the operations must be performed in a specified order, as listed below in the Order of Operations Agreement.

1.3A* Order of Operations Agreement

The Order of Operations Agreement

Step 1 Perform operations inside grouping symbols. Grouping symbols include parentheses ( ), brackets [ ], braces { }, the absolute value symbol | |, and fraction bars.

Step 2 Simplify exponential expressions.

Step 3 Do multiplication and division as they occur from left to right.

Step 4 Do addition and subtraction as they occur from left to right.

Example 1

Evaluate: $-2(7-3)^2 + 4 - 2(5-2)$

Solution

$-2(7-3)^2 + 4 - 2(5-2)$

$= -2(4)^2 + 4 - 2(3)$ • Perform operations inside parentheses.

$= -2(16) + 4 - 2(3)$ • Simplify the exponential expression.

$= -32 + 4 - 2(3)$ • Do the multiplication and division from left to right.

$= -32 + 4 - 6$

$= -28 - 6$ • Do the addition and subtraction from left to right.

$= -28 + (-6) = -34$

You Try It 1

Evaluate: $(-4)(6-8)^2 - (-12 \div 4)$

Your solution

Solution on p. S37

1.4A Evaluate variable expressions

A **variable** is a letter that represents a quantity that is unknown or that can change or vary. An expression that contains one or more variables is a **variable expression.** $3x - 4y + 7z$ is a variable expressions. It contains the variables x, y, and z.

Replacing a variable in a variable expression by a number and then simplifying the resulting numerical expression is called **evaluating the variable expression.** The number substituted for the variable is called the **value of the variable.** The result is called the **value of the variable expression.**

*Review this objective for more detailed coverage of this topic.

Copyright © Houghton Mifflin Company. All rights reserved.

Example 2

Evaluate $5ab^3 + 2a^2b^2 - 4$ when $a = 3$ and $b = -2$.

Solution

$5ab^3 + 2a^2b^2 - 4$

$5(3)(-2)^3 + 2(3)^2(-2)^2 - 4$ • Replace *a* by 3 and *b* by −2.

$= 5(3)(-8) + 2(9)(4) - 4$ • Use the Order of Operations Agreement to simplify the numerical expression.

$= -120 + 72 - 4$

$= -48 - 4$

$= -48 + (-4)$

$= -52$

You Try It 2

Evaluate $3xy^2 - 3x^2y$ when $x = -2$ and $y = 5$.

Your solution

Solution on p. S37

Objective B To simplify a variable expression

1.4B 1.4C Simplify variable expressions using the Properties of Addition and Multiplication

The Properties of Real Numbers are used to simplify variable expressions.

Example 3

Simplify: $-\frac{1}{3}(-3y)$

Solution

$-\frac{1}{3}(-3y)$

$= \left[-\frac{1}{3}(-3)\right]y$ • Use the Associative Property of Multiplication to regroup factors.

$= 1y$ • Use the Inverse Property of Multiplication.

$= y$ • Use the Multiplication Property of One.

You Try It 3

Simplify: $-5(-3a)$

Your solution

Solution on p. S37

A variable expression is shown at the right. The expression can be rewritten by writing subtraction as addition of the opposite. Note that the expression has four addends. The **terms** of a variable expression are the addends of the expression. The expression has four terms.

$3x^2 - 4xy + 5z - 2$

$3x^2 + (-4xy) + 5z + (-2)$

Four terms

$\underbrace{3x^2 - 4xy + 5z}_{\text{Variable terms}} \underbrace{- 2}_{\text{Constant term}}$

The terms $3x^2$, $-4xy$, and $5z$ are **variable terms.** The term -2 is a **constant term,** or simply a **constant.**

Copyright © Houghton Mifflin Company. All rights reserved.

Like terms of a variable expression are terms that have the same variable part. The terms $3x$ and $-7x$ are like terms. Constant terms are also like terms. Thus -6 and 9 are like terms.

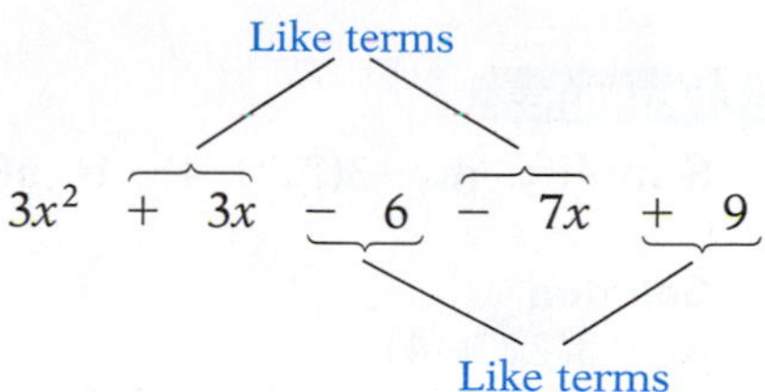

By using the Commutative Property of Multiplication, we can rewrite the Distributive Property as $ba + ca = (b + c)a$. This form of the Distributive Property is used to **combine like terms** of a variable expression by adding their coefficients. For instance,

$$7x + 9x = (7 + 9)x$$
$$= 16x$$

• Use the Distributive Property: $ba + ca = (b + c)a$.

Example 4

Simplify: $4x^2 + 5x - 6x^2 - 7x$

Solution

$4x^2 + 5x - 6x^2 - 7x$

$= 4x^2 - 6x^2 + 5x - 7x$

$= (4x^2 - 6x^2) + (5x - 7x)$

• Use the Associative and Commutative Properties of Addition to rearrange and group like terms.

$= -2x^2 + (-2x)$

$= -2x^2 - 2x$

• Use the Distributive Property to combine like terms.

You Try It 4

Simplify: $2z^2 - 5z - 3z^2 + 6z$

Your solution

Solution on p. S37

1.4D Simplify variable expressions using the Distributive Property

The Distributive Property also is used to remove parentheses from a variable expression. Here is an example.

$4(2x + 5z) = 4(2x) + 4(5z)$ • Use the Distributive Property: $a(b + c) = ab + ac$.

$= (4 \cdot 2)x + (4 \cdot 5)z$ • Use the Associative Property of Multiplication to regroup factors.

$= 8x + 20z$ • Multiply $4 \cdot 2$ and $4 \cdot 5$.

The Distributive Property can be extended to expressions containing more than two terms. For instance,

$$4(2x + 3y + 5z) = 4(2x) + 4(3y) + 4(5z)$$
$$= 8x + 12y + 20z$$

Copyright © Houghton Mifflin Company. All rights reserved.

Example 5

Simplify. **a.** $-3(2x + 4)$ **b.** $6(3x - 4y + z)$

Solution

a. $-3(2x + 4)$

$= -3(2x) + (-3)(4)$ • Use the Distributive Property.

$= -6x - 12$

b. $6(3x - 4y + z)$

$= 6(3x) - 6(4y) + 6(z)$ • Use the Distributive Property.

$= 18x - 24y + 6z$

You Try It 5

Simplify. **a.** $-3(5y - 2)$ **b.** $-2(4x + 2y - 6z)$

Your solution

Solution on p. S37

TAKE NOTE

Recall that the Distributive Property states that if a, b, and c are real numbers, then

$a(b + c) = ab + ac$

To simplify the expression $5 + 3(4x - 2)$, use the Distributive Property to remove the parentheses.

$5 + 3(4x - 2) = 5 + 3(4x) - 3(2)$ • Use the Distributive Property.

$= 5 + 12x - 6$

$= 12x - 1$ • Add the like terms 5 and −6.

Example 6

Simplify.

a. $3(2x - 4) - 5(3x + 2)$

b. $3a - 2[7a - 2(2a + 1)]$

Solution

a. $3(2x - 4) - 5(3x + 2)$

$= 6x - 12 - 15x - 10$ • Use the Distributive Property.

$= -9x - 22$ • Combine like terms.

b. $3a - 2[7a - 2(2a + 1)]$

$= 3a - 2[7a - 4a - 2]$ • Use the Distributive Property.

$= 3a - 2[3a - 2]$ • Combine like terms.

$= 3a - 6a + 4$ • Use the Distributive Property.

$= -3a + 4$ • Combine like terms.

You Try It 6

Simplify.

a. $7(-3x - 4y) - 3(3x + y)$

b. $2y - 3[5 - 3(3 + 2y)]$

Your solution

Solution on p. S37

Copyright © Houghton Mifflin Company. All rights reserved.

R.1 Exercises

Objective A To evaluate a variable expression

Evaluate the variable expression when $a = 2$, $b = 3$, and $c = -4$.

1. $a - 2c$

2. $-3a + 4b$

3. $3b - 3c$

4. $-3c + 4$

5. $16 \div (2c)$

6. $6b \div (-a)$

7. $3b - (a + c)^2$

8. $(a - b)^2 + 2c$

9. $(b - 3a)^2 + bc$

Evaluate the variable expression when $a = -1$, $b = 3$, $c = -2$, and $d = 4$.

10. $\dfrac{b - a}{d}$

11. $\dfrac{d - b}{a}$

12. $\dfrac{2d + b}{-a}$

13. $\dfrac{b - d}{c - a}$

14. $2(b + c) - 2a$

15. $3(b - a) - bc$

16. $\dfrac{-4bc}{2a + c}$

17. $\dfrac{abc}{b - d}$

18. $(d - a)^2 - (b - c)^2$

19. $(-b + d)^2 + (-a + c)^2$

20. $4ab + (2c)^2$

21. $3cd - (4a)^2$

Evaluate the variable expression when $a = 2.7$, $b = -1.6$, and $c = -0.8$.

22. $c^2 - ab$

23. $(a + b)^2 - c$

24. $\dfrac{b^3}{c} - 4a$

Objective B To simplify a variable expression

Simplify each of the following.

25. $x + 7x$

26. $12y + 9y$

27. $8b - 5b$

28. $4y - 11y$

29. $-12a + 17a$

30. $-15xy + 7xy$

Copyright © Houghton Mifflin Company. All rights reserved.

31. $4x + 5x + 2x$

32. $-5x^2 - 10x^2 + x^2$

33. $6x - 2y + 9x$

34. $3x - 7y - 6x + 4x$

35. $5a + 6a - 2a$

36. $2a - 5a + 3a$

37. $12y^2 + 10y^2$

38. $2z^2 - 9z^2$

39. $\frac{3}{4}x - \frac{1}{4}x$

40. $\frac{2}{5}y - \frac{3}{5}y$

41. $-4(5x)$

42. $-2(-8y)$

43. $(6a)(-4)$

44. $-5(7x^2)$

45. $\frac{1}{4}(4x)$

46. $\frac{12x}{5}\left(\frac{5}{12}\right)$

47. $\frac{1}{3}(21x)$

48. $-\frac{5}{8}(24a^2)$

49. $(36y)\left(\frac{1}{12}\right)$

50. $-(z + 4)$

51. $-3(a + 5)$

52. $(4 - 3b)9$

53. $(-2x - 6)8$

54. $3(5x^2 + 2x)$

55. $-5(2y^2 - 1)$

56. $4(x^2 - 3x + 5)$

57. $6(3x^2 - 2xy - y^2)$

58. $5a - (4a + 6)$

59. $3 - (10 + 8y)$

60. $12(y - 2) + 3(7 - 4y)$

61. $-5[2x + 3(5 - x)]$

62. $-3[2x - (x + 7)]$

63. $-5a - 2[2a - 4(a + 7)]$

Copyright © Houghton Mifflin Company. All rights reserved.

R.2 Equations and Inequalities

Objective A To solve a first-degree equation in one variable

An **equation** expresses the equality of two mathematical expressions. Each of the equations below is a **first-degree equation in one variable.** *First degree* means that the variable has an exponent of 1.

$$x + 11 = 14$$
$$3a + 5 = 8a$$
$$2(6y - 1) = 3$$
$$4 - 3(2n - 1) = 6n - 5$$

A **solution** of an equation is a number that, when substituted for the variable, results in a true equation.

3 is a solution of the equation $x + 4 = 7$ because $3 + 4 = 7$.
9 is not a solution of the equation $x + 4 = 7$ because $9 + 4 \neq 7$.

To **solve an equation** means to find a solution of the equation. In solving an equation, the goal is to rewrite the given equation with the variable alone on one side of the equation and a constant term on the other side of the equation.

variable = *constant* or *constant* = *variable*

The constant is the solution of the equation.

The following properties of equations are used to rewrite equations in this form.

2.1B 2.1C Solving equations using the Addition and Multiplication Properties of Equations

Properties of Equations

Addition Property of Equations

The same number can be added to each side of an equation without changing the solution of the equation. In symbols, the equation $a = b$ has the same solution as the equation $a + c = b + c$.

Multiplication Property of Equations

Each side of an equation can be multiplied by the same nonzero number without changing the solution of the equation. In symbols, if $c \neq 0$, then the equation $a = b$ has the same solution as the equation $ac = bc$.

Copyright © Houghton Mifflin Company. All rights reserved.

TAKE NOTE
Subtraction is defined as addition of the opposite.
$a - b = a + (-b)$

The Addition Property of Equations is used to remove a term from one side of the equation by adding the opposite of that term to each side of the equation. Because subtraction is defined in terms of addition, the Addition Property of Equations also makes it possible to subtract the same number from each side of an equation without changing the solution of the equation.

For example, to solve the equation $t + 9 = -4$, subtract the constant term (9) from each side of the equation.

$$t + 9 = -4$$
$$t + 9 - 9 = -4 - 9$$
$$t = -13$$

Now the variable is alone on one side of the equation and a constant term (-13) is on the other side. The solution is the constant. The solution is -13.

To solve $7 = y - 8$, add 8 to each side of the equation.

$$7 = y - 8$$
$$7 + 8 = y - 8 + 8$$
$$15 = y$$

TAKE NOTE
Division is defined as multiplication by the reciprocal.

$$a \div b = a \cdot \frac{1}{b}$$

The equation is in the form *constant* = *variable*. The solution is the constant. The solution is 15.

The Multiplication Property of Equations is used to remove a coefficient by multiplying each side of the equation by the reciprocal of the coefficient. Because division is defined in terms of multiplication, each side of an equation can be divided by the same nonzero number without changing the solution of the equation.

TAKE NOTE
When using the Multiplication Property of Equations, multiply each side of the equation by the reciprocal of the coefficient when the coefficient is a fraction. Divide each side of the equation by the coefficient when the coefficient is an integer or a decimal.

For example, to solve the equation $-5q = 120$, divide each side of the equation by the coefficient -5.

$$-5q = 120$$
$$\frac{-5q}{-5} = \frac{120}{-5}$$
$$q = -24$$

Now the variable is alone on one side of the equation and a constant (-24) is on the other side. The solution is the constant. The solution is -24.

2.2B 2.2C Solve general equations

In solving more complicated first-degree equations in one variable, use the following sequence of steps.

Steps for Solving a First-Degree Equation in One Variable

1. Use the Distributive Property to remove parentheses.
2. Combine any like terms on the right side of the equation and any like terms on the left side of the equation.
3. Use the Addition Property to rewrite the equation with only one variable term.
4. Use the Addition Property to rewrite the equation with only one constant term.
5. Use the Multiplication Property to rewrite the equation with the variable alone on one side of the equation and a constant on the other side of the equation.

If one of these steps is not needed to solve a given equation, proceed to the next step.

Copyright © Houghton Mifflin Company. All rights reserved.

Example 1

Solve.

a. $5x + 9 = 23 - 2x$

b. $8x - 3(4x - 5) = -2x + 6$

Solution

a.

$$5x + 9 = 23 - 2x$$
$$5x + 2x + 9 = 23 - 2x + 2x \quad \text{• Step 3}$$
$$7x + 9 = 23$$
$$7x + 9 - 9 = 23 - 9 \quad \text{• Step 4}$$
$$7x = 14$$
$$\frac{7x}{7} = \frac{14}{7} \quad \text{• Step 5}$$
$$x = 2$$

The solution is 2.

b.

$$8x - 3(4x - 5) = -2x + 6$$
$$8x - 12x + 15 = -2x + 6 \quad \text{• Step 1}$$
$$-4x + 15 = -2x + 6 \quad \text{• Step 2}$$
$$-4x + 2x + 15 = -2x + 2x + 6 \quad \text{• Step 3}$$
$$-2x + 15 = 6$$
$$-2x + 15 - 15 = 6 - 15 \quad \text{• Step 4}$$
$$-2x = -9$$
$$\frac{-2x}{-2} = \frac{-9}{-2} \quad \text{• Step 5}$$
$$x = \frac{9}{2}$$

The solution is $\frac{9}{2}$.

You Try It 1

Solve.

a. $4x + 3 = 7x + 9$

b. $4 - (5x - 8) = 4x + 3$

Your solution

Solution on pp. S37–S38

Copyright © Houghton Mifflin Company. All rights reserved.

Objective B To solve an inequality in one variable

An **inequality** contains the symbol >, <, ≥, or ≤. An inequality expresses the relative order of two mathematical expressions. Here are some examples of inequalities in one variable.

$$\left.\begin{array}{l} 4x \ge 12 \\ 2x + 7 \le 9 \\ x^2 + 1 > 3x \end{array}\right\} \text{Inequalities in one variable}$$

2.5A Solve an inequality in one variable

A **solution of an inequality in one variable** is a number that, when substituted for the variable, results in a true inequality. For the inequality $x < 4$ shown below, 3, 0, and -5 are solutions of the inequality because replacing the variable by these numbers results in a true inequality.

$x < 4$	$x < 4$	$x < 4$
$3 < 4$ True	$0 < 4$ True	$-5 < 4$ True

The number 7 is not a solution of the inequality $x < 4$ because $7 < 4$ is a false inequality.

Besides the numbers 3, 0, and -5, there are an infinite number of other solutions of the inequality $x < 4$. Any number less than 4 is a solution; for instance, -5.2, $\frac{5}{2}$, π, and 1 are also solutions of the inequality. The set of all the solutions of an inequality is called the **solution set of the inequality.** The solution set of the inequality $x < 4$ is written in set-builder notation as $\{x|x < 4\}$. This is read, "the set of all x such that x is less than 4."

The graph of the solution set of $x < 4$ is shown at the right.

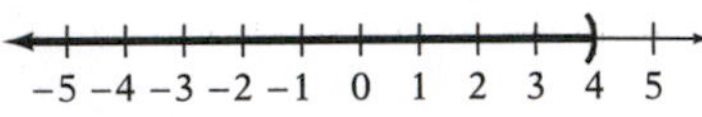

In solving an inequality, the goal is to rewrite the given inequality in the form

variable < *constant* or *variable* > *constant*

The Addition Property of Inequalities is used to rewrite an inequality in this form.

TAKE NOTE
The Addition Property of Inequalities states that the same number can be added to each side of an inequality without changing the solution set of the inequality.

The Addition Property of Inequalities

If $a > b$ and c is a real number, then the inequalities $a > b$ and $a + c > b + c$ have the same solution set.

If $a < b$ and c is a real number, then the inequalities $a < b$ and $a + c < b + c$ have the same solution set.

The Addition Property of Inequalities is also true for the symbols $\leq$ and $\geq$.

The Addition Property of Inequalities is used to remove a term from one side of an inequality by adding the additive inverse of that term to each side of the inequality. Because subtraction is defined in terms of addition, the same number can be subtracted from each side of an inequality without changing the solution set of the inequality.

As shown in the example below, the Addition Property of Inequalities applies to variable terms as well as to constants.

HOW TO Solve $4x - 5 \leq 3x - 2$. Write the solution set in set-builder notation.

$4x - 5 \leq 3x - 2$

$4x - 3x - 5 \leq 3x - 3x - 2$ • Subtract $3x$ from each side of the inequality.

$x - 5 \leq -2$ • Simplify.

$x - 5 + 5 \leq -2 + 5$ • Add 5 to each side of the inequality.

$x \leq 3$ • Simplify.

The solution set is $\{x|x \leq 3\}$.

Copyright © Houghton Mifflin Company. All rights reserved.

When multiplying or dividing an inequality by a number, the inequality symbol may be reversed, depending on whether the number is positive or negative. Look at the following two examples.

$3 < 5$		$3 < 5$	• Multiply by negative 2. The inequality symbol is reversed in order to make the inequality a true statement.
$2(3) < 2(5)$	• Multiply by positive 2. The inequality symbol remains the same.	$-2(3) > -2(5)$	
$6 < 10$	• $6 < 10$ is a true statement.	$-6 > -10$	

This is summarized in the Multiplication Property of Inequalities.

The Multiplication Property of Inequalities

Rule 1

If $a > b$ and $c > 0$, then $ac > bc$.
If $a < b$ and $c > 0$, then $ac < bc$.

Rule 2

If $a > b$ and $c < 0$, then $ac < bc$.
If $a < b$ and $c < 0$, then $ac > bc$.

Rule 1 states that when each side of an inequality is multiplied by a positive number, the inequality symbol remains the same. Rule 2 states that when each side of an inequality is multiplied by a negative number, the inequality symbol must be reversed.

Here are a few more examples of this property.

Rule 1		**Rule 2**	
$-4 < -2$	$5 > -3$	$7 < 9$	$-2 > -6$
$-4(2) < -2(2)$	$5(3) > -3(3)$	$7(-2) > 9(-2)$	$-2(-3) < -6(-3)$
$-8 < -4$	$15 > -9$	$-14 > -18$	$6 < 18$

The Multiplication Property of Inequalities is also true for the symbols $\leq$ and $\geq$.

Use the Multiplication Property of Inequalities to remove a coefficient other than 1 from one side of an inequality so that the inequality can be rewritten with the variable alone on one side of the inequality and a constant term on the other side.

Because division is defined in terms of multiplication, when each side of an inequality is divided by a positive number, the inequality symbol remains the same. When each side of an inequality is divided by a negative number, the inequality symbol must be reversed.

TAKE NOTE

Solving inequalities in one variable is similar to solving equations in one variable *except* that when you multiply or divide by a negative number, you must reverse the inequality symbol.

HOW TO Solve $-3x < 9$. Write the solution set in set-builder notation.

$-3x < 9$

$\frac{-3x}{-3} > \frac{9}{-3}$ • Divide each side of the inequality by the coefficient -3 and reverse the inequality symbol.

$x > -3$ • Simplify.

$\{x|x > -3\}$ • Write the answer in set-builder notation.

Copyright © Houghton Mifflin Company. All rights reserved.

Example 2

Solve $x + 3 > 4x + 6$. Write the solution set in set-builder notation.

Solution

$x + 3 > 4x + 6$

$x - 4x + 3 > 4x - 4x + 6$ • Subtract $4x$ from each side.

$-3x + 3 > 6$

$-3x + 3 - 3 > 6 - 3$ • Subtract 3 from each side.

$-3x > 3$

$\frac{-3x}{-3} < \frac{3}{-3}$ • Divide each side by -3. Reverse the inequality symbol.

$x < -1$

$\{x|x < -1\}$

You Try It 2

Solve $3x - 1 \leq 5x - 7$. Write the solution set in set-builder notation.

Your solution

Solution on p. S38

When an inequality contains parentheses, often the first step in solving the inequality is to use the Distributive Property to remove the parentheses.

Example 3

Solve $-2(x - 7) > 3 - 4(2x - 3)$. Write the solution set in set-builder notation.

Solution

$-2(x - 7) > 3 - 4(2x - 3)$

$-2x + 14 > 3 - 8x + 12$ • Use the Distributive Property.

$-2x + 14 > 15 - 8x$ • Combine like terms.

$-2x + 8x + 14 > 15 - 8x + 8x$ • Add $8x$ to each side.

$6x + 14 > 15$

$6x + 14 - 14 > 15 - 14$ • Subtract 14 from each side.

$6x > 1$

$\frac{6x}{6} > \frac{1}{6}$

$x > \frac{1}{6}$

$\left\{x|x > \frac{1}{6}\right\}$

You Try It 3

Solve $3 - 2(3x + 1) < 7 - 2x$. Write the solution set in set-builder notation.

Your solution

Solution on p. S38

Copyright © Houghton Mifflin Company. All rights reserved.

R.2 Exercises

Objective A To solve a first-degree equation in one variable

Solve.

1. $x + 7 = -5$
2. $9 + b = 21$
3. $-9 = z - 8$
4. $b - 11 = 11$
5. $-48 = 6z$
6. $-9a = -108$
7. $-\frac{3}{4}x = 15$
8. $\frac{5}{2}x = -10$
9. $-\frac{x}{4} = -2$
10. $\frac{2x}{5} = -8$
11. $4 - 2b = 2 - 4b$
12. $4y - 10 = 6 + 2y$
13. $5x - 3 = 9x - 7$
14. $3m + 5 = 2 - 6m$
15. $6a - 1 = 2 + 2a$
16. $5x + 7 = 8x + 5$
17. $2 - 6y = 5 - 7y$
18. $4b + 15 = 3 - 2b$
19. $2(x + 1) + 5x = 23$
20. $9n - 15 = 3(2n - 1)$
21. $7a - (3a - 4) = 12$
22. $5(3 - 2y) = 3 - 4y$
23. $9 - 7x = 4(1 - 3x)$
24. $2(3b + 5) - 1 = 10b + 1$
25. $2z - 2 = 5 - (9 - 6z)$
26. $4a + 3 = 7 - (5 - 8a)$
27. $5(6 - 2x) = 2(5 - 3x)$
28. $4(3y + 1) = 2(y - 8)$
29. $2(3b - 5) = 4(6b - 2)$
30. $3(x - 4) = 1 - (2x - 7)$

Objective B To solve an inequality in one variable

Solve. Write the answer in set-builder notation.

31. $x - 5 > -2$
32. $5 + n \geq 4$
33. $-2 + n \geq 0$

Copyright © Houghton Mifflin Company. All rights reserved.

34. $x - 3 < 2$

35. $8x \le -24$

36. $-4x < 8$

37. $3n > 0$

38. $-2n \le -8$

39. $2x - 1 > 7$

40. $5x - 2 \le 8$

41. $4 - 3x < 10$

42. $7 - 2x \ge 1$

43. $3x - 1 > 2x + 2$

44. $6x + 4 \le 8 + 5x$

45. $8x + 1 \ge 2x + 13$

46. $6x + 3 > 4x - 1$

47. $-3 - 4x > -11$

48. $4x - 2 < x - 11$

49. $4x - 2 > 3x + 1$

50. $7x + 5 \le 9 + 6x$

51. $9x + 2 \ge 3x + 14$

52. $8x + 1 > 6x - 3$

53. $-5 - 2x > -13$

54. $5x - 3 < x - 11$

55. $4(2x - 1) > 3x - 2(3x - 5)$

56. $2 - 5(x + 1) \ge 3(x - 1) - 8$

57. $3(4x + 3) \le 7 - 4(x - 2)$

58. $3 + 2(x + 5) \ge x + 5(x + 1) + 1$

59. $3 - 4(x + 2) \le 6 + 4(2x + 1)$

60. $12 - 2(3x - 2) \ge 5x - 2(5 - x)$

Copyright © Houghton Mifflin Company. All rights reserved.

R.3 Linear Equations in Two Variables

Objective A

To graph points in a rectangular coordinate system

A **rectangular coordinate system** is formed by two number lines, one horizontal and one vertical. The point of intersection is called the **origin.** The two axes are called the **coordinate axes** or simply the **axes.** Generally, the horizontal axis is labeled the x-axis, and the vertical axis is labeled the y-axis. In this case, the axes form what is called the ***xy*-plane**.

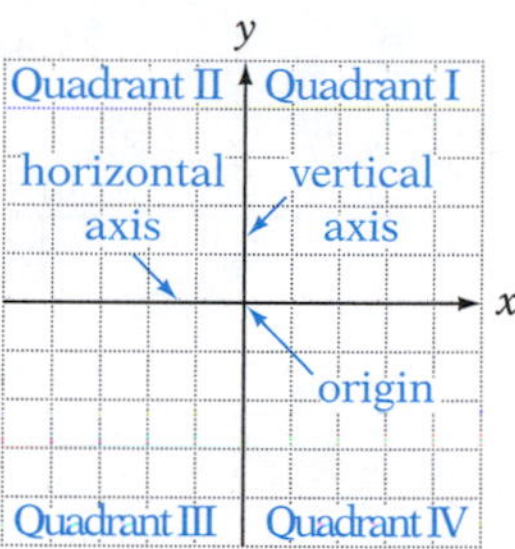

4.1A Points in the plane

Each point in the plane can be identified by a pair of numbers called an **ordered pair.** The first number of the ordered pair measures a horizontal change from the y-axis and is called the **abscissa** or ***x*-coordinate.** The second number of the ordered pair measures a vertical change from the x-axis and is called the **ordinate** or ***y*-coordinate.** The ordered pair (x, y) associated with a point is also called the **coordinates** of the point.

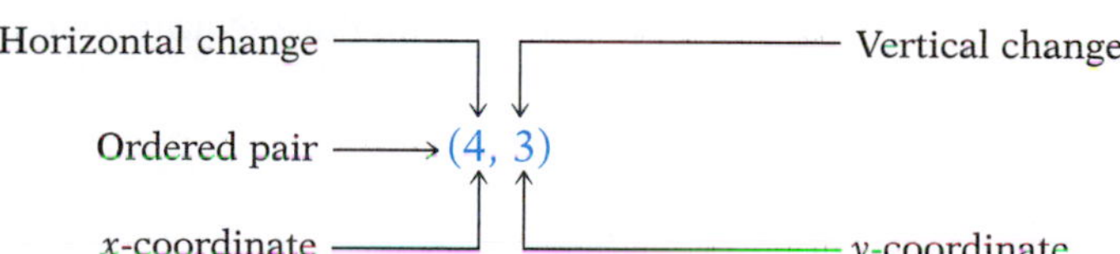

To **graph,** or **plot,** a point means to place a dot at the coordinates of the point. For example, to graph the ordered pair (4, 3), start at the origin. Move 4 units to the right and then 3 units up. Draw a dot. To graph $(-3, -4)$, start at the origin. Move 3 units to the left and then 4 units down. Draw a dot.

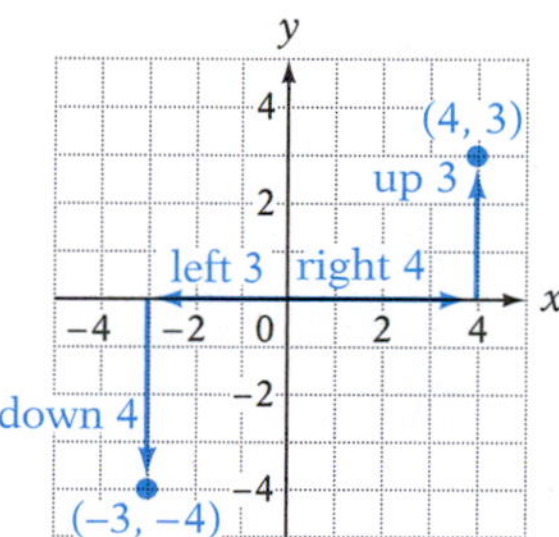

The **graph of an ordered pair** is the dot drawn at the coordinates of the point in the plane. The graphs of the ordered pairs (4, 3) and $(-3, -4)$ are shown above.

The graphs of the points (2, 3) and (3, 2) are different points. The *order* in which the numbers in an *ordered* pair are listed is important.

Copyright © Houghton Mifflin Company. All rights reserved.

Example 1

Graph the ordered pairs (−4, 2), (3, 4), (0, −1), (2, 0), and (−1, −3).

Solution

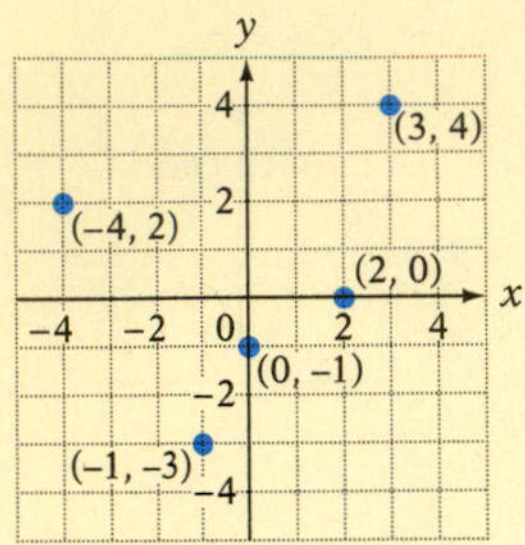

You Try It 1

Graph the ordered pairs (−2, 4), (4, 0), (0, 3), (−4, −3), and (5, −1).

Your solution

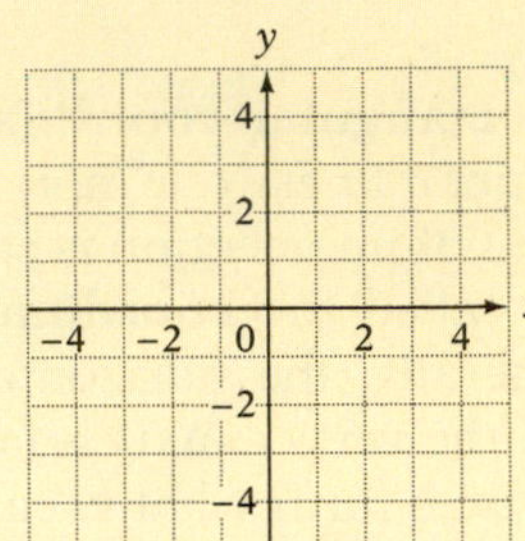

Solution on p. S38

Objective B To graph a linear equation in two variables

The equations below are examples of equations in two variables.

$$y = 3x - 4$$
$$2x - y = 7$$
$$y = x^2 + 1$$

4.1B Solutions of equations in two variables

A **solution of an equation in two variables** is an ordered pair (x, y) whose coordinates make the equation a true statement.

HOW TO Is the ordered pair (−4, 9) a solution of the equation $y = -2x + 1$?

$y = -2x + 1$	
9	$-2(-4) + 1$
9	$8 + 1$

$9 = 9$

- Replace x by −4 and y by 9.
- Simplify the right side.
- Compare the results. If the resulting equation is true, the ordered pair is a solution of the equation. If it is not true, the ordered pair is not a solution of the equation.

Yes, the ordered pair (−4, 9) is a solution of the equation $y = -2x + 1$.

Besides (−4, 9), there are many other ordered pairs that are solutions of the equation $y = -2x + 1$. For example, (0, 1), (3, −5), and $\left(-\frac{3}{2}, 4\right)$ are also solutions. In general, an equation in two variables has an infinite number of solutions. By choosing any value of x and substituting that value into the equation, we can calculate a corresponding value of y.

Copyright © Houghton Mifflin Company. All rights reserved.

HOW TO Find the ordered-pair solution of $y = \frac{2}{5}x - 4$ that corresponds to $x = 5$.

$y = \frac{2}{5}x - 4$

$y = \frac{2}{5}(5) - 4$ • We are given that $x = 5$. Replace x by 5 in the equation.

$y = 2 - 4$ • Simplify the right side.

$y = -2$ • When $x = 5$, $y = -2$.

The ordered-pair solution of $y = \frac{2}{5}x - 4$ for $x = 5$ is $(5, -2)$.

4.3A Graph an equation of the form $y = mx + b$

Solutions of an equation in two variables can be graphed in the rectangular coordinate system.

HOW TO Graph the ordered-pair solutions of $y = -2x + 1$ for $x = -2, -1, 0, 1$, and 2.

x	$y = -2x + 1$	y	(x, y)
−2	−2(−2) + 1	5	(−2, 5)
−1	−2(−1) + 1	3	(−1, 3)
0	−2(0) + 1	1	(0, 1)
1	−2(1) + 1	−1	(1, −1)
2	−2(2) + 1	−3	(2, −3)

• Use the given values of x to determine ordered-pair solutions of the equation. It is convenient to record these in a table.

The ordered-pair solutions of $y = -2x + 1$ for $x = -2, -1, 0, 1$, and 2 are $(-2, 5)$, $(-1, 3)$, $(0, 1)$, $(1, -1)$, and $(2, -3)$. These are graphed at the right.

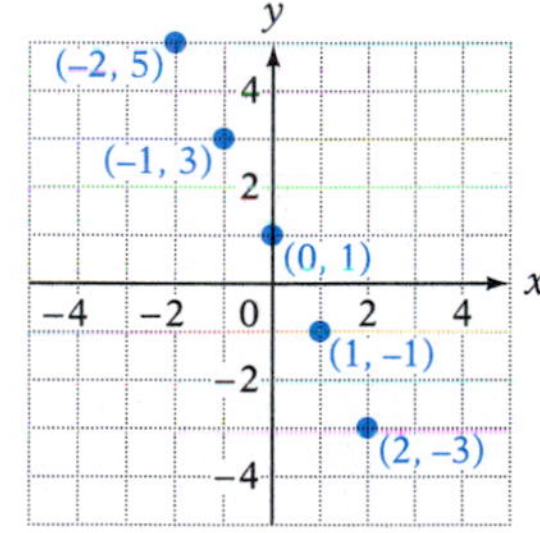

The **graph of an equation in two variables** is a graph of all the ordered-pair solutions of the equation. Consider the equation $y = -2x + 1$ above. The ordered-pair solutions $(-2, 5)$, $(-1, 3)$, $(0, 1)$, $(1, -1)$, and $(2, -3)$ are graphed in the figure above. We can choose values of x that are not integers to produce more ordered pairs to graph, such as $\left(\frac{5}{2}, -4\right)$ and $\left(-\frac{1}{2}, 2\right)$. Choosing still other values of x would result in more and more ordered pairs being graphed. The result would be so many dots that the graph would appear as a straight line, as shown below. This is the graph of $y = -2x + 1$.

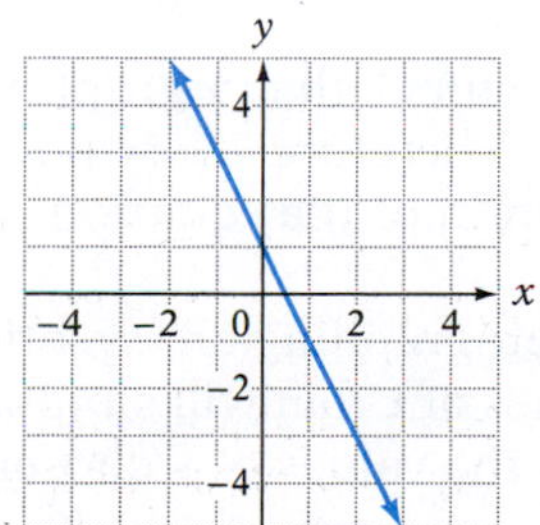

Copyright © Houghton Mifflin Company. All rights reserved.

The equation $y = -2x + 1$ is an example of a linear equation because its graph is a straight line. It is also called a first-degree equation in two variables because the exponent on each variable is 1.

TAKE NOTE

The equation $y = x^2 + 2x - 4$ is not a linear equation in two variables because there is a term with a variable squared. The equation $y = \frac{5}{x-3}$ is not a linear equation because a variable occurs in the denominator of a fraction.

Linear Equation in Two Variables

An equation of the form $y = mx + b$, where m is the coefficient of x and b is a constant, is a linear equation in two variables. The graph of a linear equation in two variables is a straight line.

Examples of linear equations are shown at the right.

$y = 5x + 3 \quad (m = 5, b = 3)$

$y = x - 4 \quad (m = 1, b = -4)$

$y = -\frac{3}{4}x \quad \left(m = -\frac{3}{4}, b = 0\right)$

To graph a linear equation, find ordered-pair solutions of the equation. Do this by choosing any value of x and finding the corresponding value of y. Repeat this procedure, choosing different values for x, until you have found the number of solutions desired. Because the graph of a linear equation in two variables is a straight line, and a straight line is determined by two points, it is necessary to find only two solutions. However, it is recommended that at least three points be used to ensure accuracy.

HOW TO Graph: $y = 2x - 3$

x	$y = 2x - 3$	y
0	$2(0) - 3$	-3
2	$2(2) - 3$	1
-1	$2(-1) - 3$	-5

- Choose any values of x. Then find the corresponding values of y. The numbers 0, 2, and -1 were chosen arbitrarily for x.

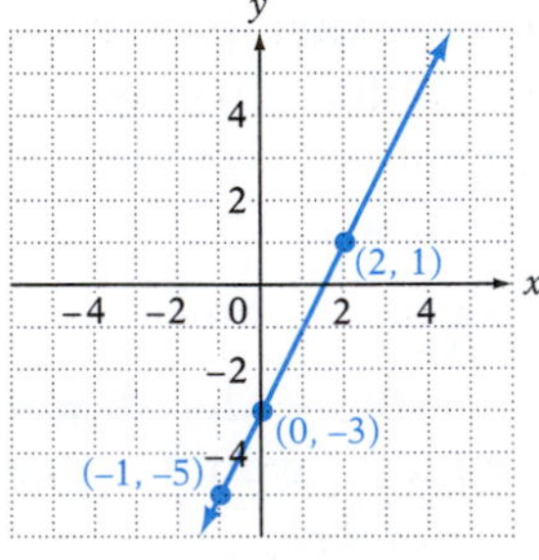

- Graph the ordered-pair solutions $(0, -3)$, $(2, 1)$, and $(-1, -5)$. Draw a straight line through the points.

Remember that a graph is a drawing of the ordered-pair solutions of an equation. Therefore, every point on the graph is a solution of the equation, and every solution of the equation is a point on the graph.

When graphing an equation of the form $y = mx + b$, if m is a fraction, choose values of x that will simplify the evaluation. This is illustrated in Example 2. Note that the values of x chosen are multiples of the denominator, 2.

Copyright © Houghton Mifflin Company. All rights reserved.

Example 2

Graph $y = -\frac{3}{2}x - 3$.

Solution

x	y
0	−3
−2	0
−4	3

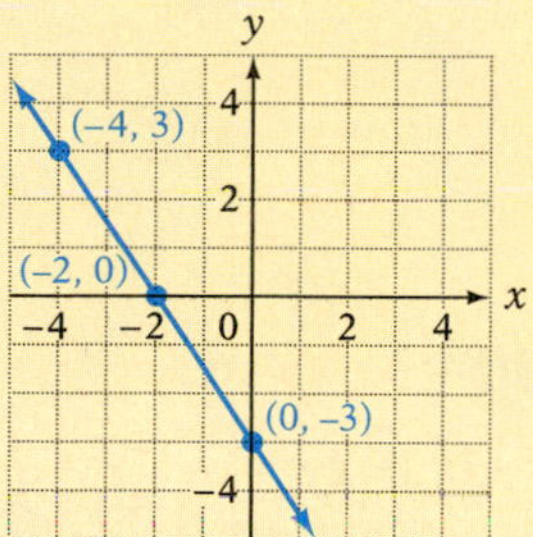

You Try It 2

Graph $y = \frac{3}{5}x - 4$.

Your solution

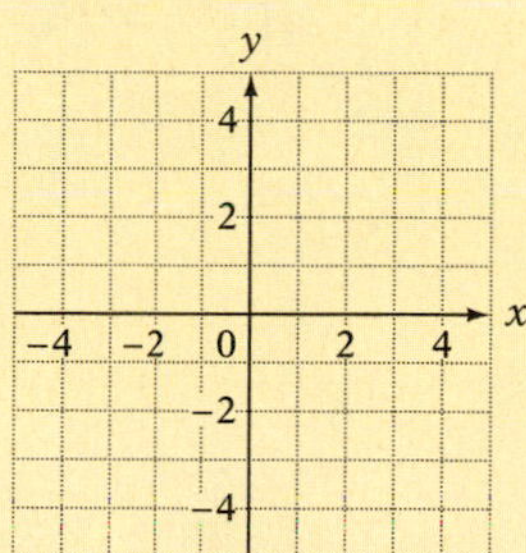

Solution on p. S38

4.3B Graph an equation of the form $Ax + By = C$

An equation of the form $Ax + By = C$ is also a linear equation in two variables. Examples of these equations are shown below.

$$3x + 4y = 12 \qquad (A = 3, B = 4, C = 12)$$
$$x - 5y = -10 \qquad (A = 1, B = -5, C = -10)$$
$$2x - y = 0 \qquad (A = 2, B = -1, C = 0)$$

One method of graphing an equation of the form $Ax + By = C$ involves first solving the equation for y and then following the same procedure used for graphing an equation of the form $y = mx + b$. To solve the equation for y means to rewrite the equation so that y is alone on one side of the equation and the term containing x and the constant are on the other side of the equation. The Addition and Multiplication Properties of Equations are used to rewrite an equation of the form $Ax + By = C$ in the form $y = mx + b$.

HOW TO Graph $3x + 2y = 6$.

$3x + 2y = 6$
- The equation is in the form $Ax + By = C$.

$2y = -3x + 6$
- Solve the equation for y. Subtract $3x$ from each side of the equation.

$y = -\frac{3}{2}x + 3$
- Divide each side of the equation by 2. Note that each term on the right side is divided by 2.
- Find at least three solutions.

x	y
0	3
2	0
4	−3

Copyright © Houghton Mifflin Company. All rights reserved.

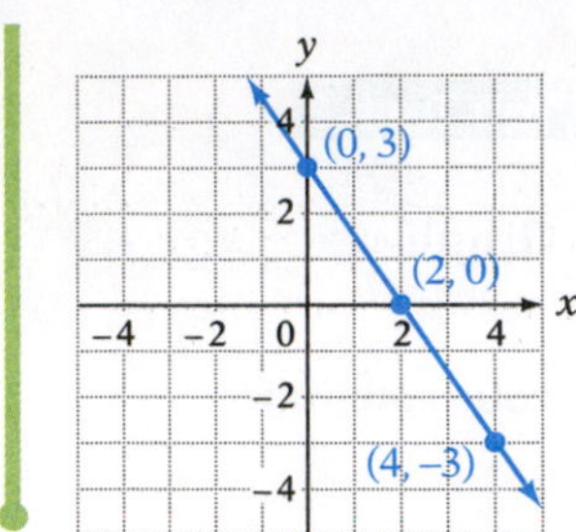

- Graph the ordered pairs (0, 3), (2, 0), and (4, −3). Draw a straight line through the points.

Example 3

Graph $2x + 3y = 9$.

Solution

$$2x + 3y = 9$$
$$3y = -2x + 9$$
$$y = -\frac{2}{3}x + 3$$

x	y
−3	5
0	3
3	1

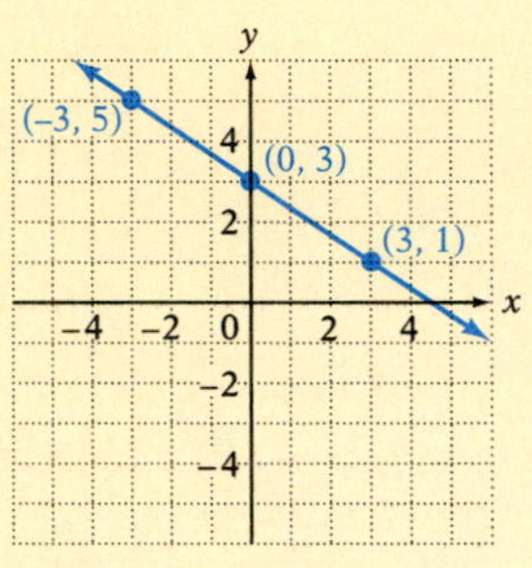

You Try It 3

Graph $-3x + 2y = 4$.

Your solution

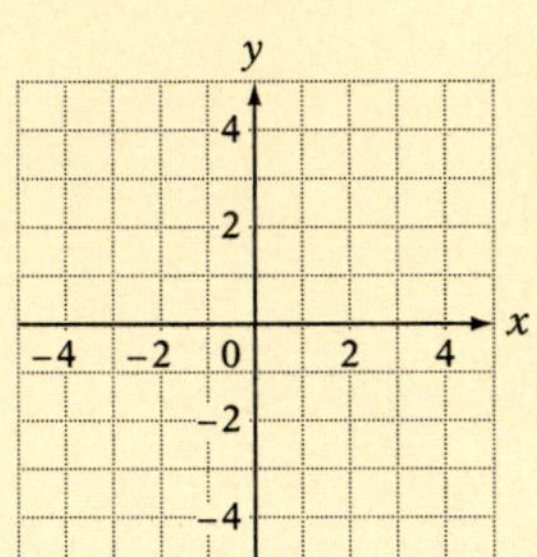

Solution on p. S38

4.4A Slope

The **slope** of a line is a measure of the slant of the line. The symbol for slope is m.

For an equation of the form $y = mx + b$, m is the slope. Here are a few examples:

The slope of the line $y = -2x + 5$ is -2.

The slope of the line $y = 8x$ is 8.

The slope of the line $y = \frac{3}{4}x - 1$ is $\frac{3}{4}$.

The slope of a line containing two points is the ratio of the change in the y values between the two points to the change in the x values.

TAKE NOTE

$$\text{Slope} = m = \frac{\text{change in } y}{\text{change in } x}$$

Slope Formula

The slope of the line containing the two points $P_1(x_1, y_1)$ and $P_2(x_2, y_2)$ is given by

$$m = \frac{y_2 - y_1}{x_2 - x_1}, x_1 \neq x_2$$

Copyright © Houghton Mifflin Company. All rights reserved.

Example 4

Find the slope of the line containing the points $(-4, -3)$ and $(-1, 1)$.

Solution

Let $(x_1, y_1) = (-4, -3)$ and $(x_2, y_2) = (-1, 1)$.

$$m = \frac{y_2 - y_1}{x_2 - x_1} = \frac{1 - (-3)}{-1 - (-4)} = \frac{4}{3}$$

The slope is $\frac{4}{3}$.

You Try It 4

Find the slope of the line containing the points $(-2, 3)$ and $(1, -3)$.

Your solution

Solution on p. S38

4.4B Slope-intercept form of a straight line

One important characteristic of the graph of a linear equation is its *intercepts*. An ***x*-intercept** is a point at which the graph crosses the x-axis. A ***y*-intercept** is a point at which the graph crosses the y-axis.

The graph of the equation $y = \frac{1}{2}x - 2$ is shown at the right. The x-intercept of the graph is $(4, 0)$. The y-intercept of the graph is $(0, -2)$.

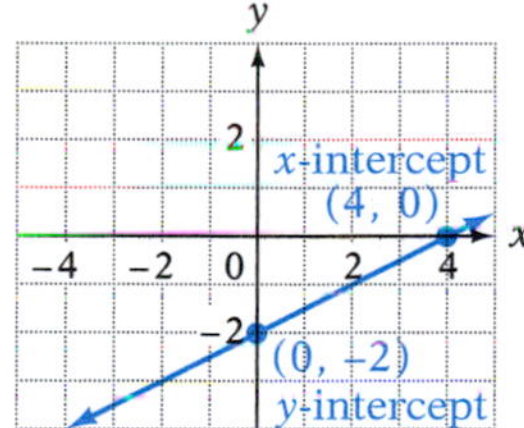

Note that at an x-intercept, the y-coordinate is 0. At a y-intercept, the x-coordinate is 0.

We can find the y-intercept of a linear equation by letting $x = 0$.

HOW TO Find the y-intercept of the graph of the equation $y = 3x + 4$.

$y = 3x + 4$

$y = 3(0) + 4$ • **To find the *y*-intercept, let *x* = 0.**

$y = 0 + 4$

$y = 4$

The y-intercept is $(0, 4)$.

Note that the constant term of $y = mx + b$ is b, and the y-intercept is $(0, b)$.

In general, **for any equation of the form $y = mx + b$, the y-intercept is $(0, b)$.**

Because the slope and the y-intercept can be determined directly from the equation $y = mx + b$, this equation is called the slope-intercept form of a straight line.

Copyright © Houghton Mifflin Company. All rights reserved.

Slope-Intercept Form of a Straight Line

The equation $y = mx + b$ is called the **slope-intercept form of a straight line.** The slope of the line is m, the coefficient of x. The y-intercept is $(0, b)$.

The following equations are written in slope-intercept form.

$y = -4x + 3$ Slope $= -4$, y-intercept $= (0, 3)$

$y = \frac{2}{5}x - 1$ Slope $= \frac{2}{5}$, y-intercept $= (0, -1)$

$y = -x$ Slope $= -1$, y-intercept $= (0, 0)$

When an equation is in slope-intercept form, it is possible to quickly draw a graph of the function.

HOW TO Graph $x + 2y = 4$ by using the slope and y-intercept.

Solve the equation for y.

$$x + 2y = 4$$
$$2y = -x + 4$$
$$y = -\frac{1}{2}x + 2$$

From the equation $y = -\frac{1}{2}x + 2$, the slope is $-\frac{1}{2}$ and the y-intercept is $(0, 2)$.

Rewrite the slope $-\frac{1}{2}$ as $\frac{-1}{2}$.

Beginning at the y-intercept, move right 2 units (change in x) and then down 1 unit (change in y).

The point whose coordinates are $(2, 1)$ is a second point on the graph. Draw a straight line through the points $(0, 2)$ and $(2, 1)$.

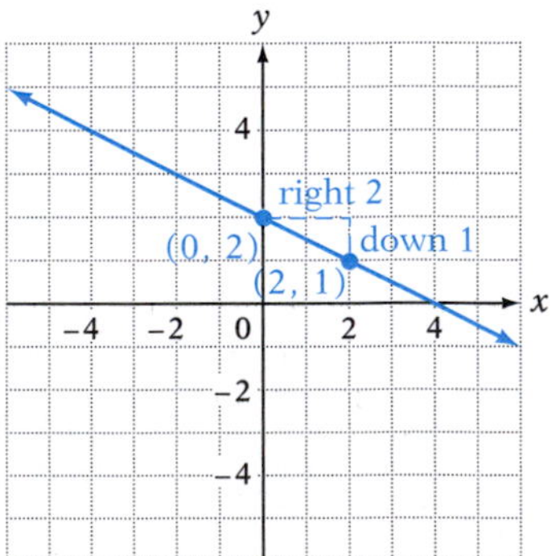

TAKE NOTE

Recall that slope $= m = \frac{\text{change in } y}{\text{change in } x}$. For the example at the right, $m = -\frac{1}{2} = \frac{-1}{2} = \frac{\text{change in } y}{\text{change in } x}$. Therefore, the change in y is -1 and the change in x is 2.

Example 5

Graph $y = -\frac{3}{2}x + 4$ by using the slope and the y-intercept.

You Try It 5

Graph $y = -\frac{2}{3}x + 2$ by using the slope and the y-intercept.

Copyright © Houghton Mifflin Company. All rights reserved.

Solution

From the equation, the slope is $-\frac{3}{2}$ and the y-intercept is (0, 4).

Rewrite the slope $-\frac{3}{2}$ as $\frac{-3}{2}$.

Place a dot at the y-intercept.

Starting at the y-intercept, move right 2 units (the change in x) and down 3 units (the change in y). Place a dot at that location.

Draw a line through the two points.

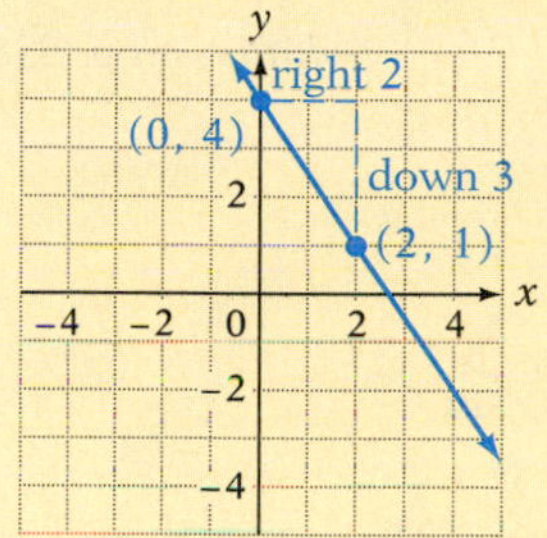

Your solution

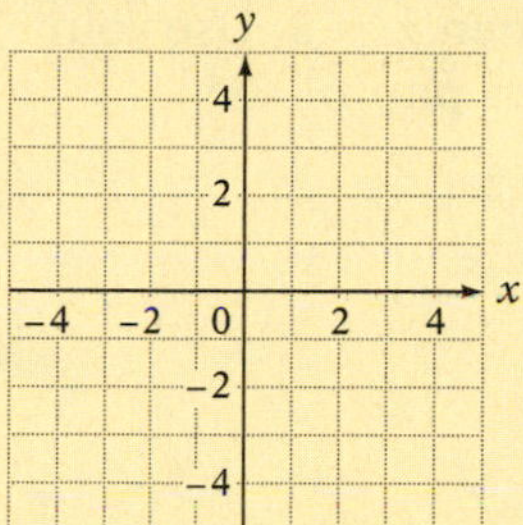

Solution on p. S38

Objective C **To evaluate a function**

4.2A Evaluate a function

A **relation** is a set of ordered pairs. A **function** is a relation in which no two ordered pairs have the same first coordinate and different second coordinates.

The relation {(0, 4), (1, 5), (1, 6), (2, 7)} is not a function because the ordered pairs (1, 5), and (1, 6) have the same first coordinate and different second coordinates.

The relation {(0, 4), (1, 5), (2, 6), (3, 7)} is a function.

The phrase "y is a function of x," or a similar phrase with different variables, is used to describe those equations in two variables that define functions. To emphasize that the equation represents a function, functional notation is used. For example, the square function is written in functional notation as follows:

TAKE NOTE
The symbol $f(x)$ is read "the value of f at x" or "f of x."

$$f(x) = x^2$$

The process of determining $f(x)$ for a given value of x is called **evaluating the function.** For instance, to evaluate $f(x) = x^2$ when $x = 3$, replace x by 3 and simplify.

$$f(x) = x^2$$
$$f(3) = 3^2 = 9$$

The value of the function is 9 when $x = 3$. An ordered pair of the function is (3, 9).

Copyright © Houghton Mifflin Company. All rights reserved.

Example 6

Evaluate $f(x) = 2x - 4$ when $x = 3$. Use your answer to write an ordered pair of the function.

Solution

$f(x) = 2x - 4$

$f(3) = 2(3) - 4$ • $x = 3$

$f(3) = 6 - 4$

$f(3) = 2$

An ordered pair of the function is (3, 2).

You Try It 6

Evaluate the function $f(x) = 4 - 2x$ at $x = -3$. Use your answer to write an ordered pair of the function.

Your solution

Solution on p. S38

Objective D To find the equation of a line

4.5A Find the equation of a line given a point and the slope

When the slope of a line and a point on the line are known, the equation of the line can be determined by using the point-slope formula.

Point-Slope Formula

Let m be the slope of a line, and let (x_1, y_1) be the coordinates of a point on the line. The equation of the line can be found using the point-slope formula:

$$y - y_1 = m(x - x_1)$$

Example 7

Find the equation of the line that contains the point $(1, -3)$ and has slope -2.

Solution

$y - y_1 = m(x - x_1)$

$y - (-3) = -2(x - 1)$ • $x_1 = 1, y_1 = -3, m = -2$

$y + 3 = -2x + 2$

$y = -2x - 1$

The equation of the line is $y = -2x - 1$.

You Try It 7

Find the equation of the line that contains the point $(-2, 2)$ and has slope $-\frac{1}{2}$.

Your solution

Solution on p. S38

Copyright © Houghton Mifflin Company. All rights reserved.

R.3 Exercises

Objective A To graph points in a rectangular coordinate system

Graph the ordered pairs.

1. (2, 3), (4, 0), (−4, 1), (−2, −2)

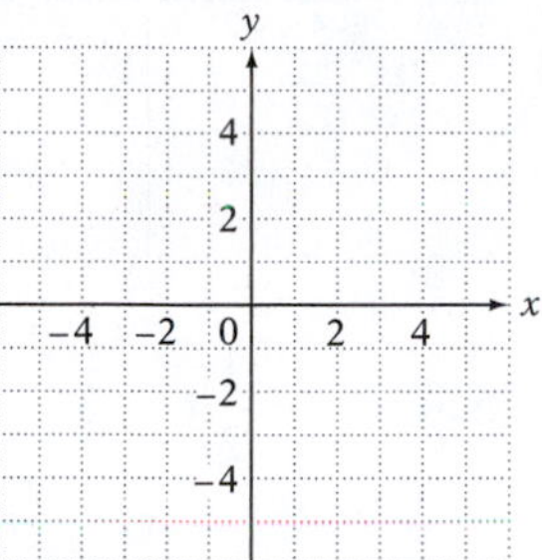

2. (0, 2), (−4, −1), (2, 0), (1, −3)

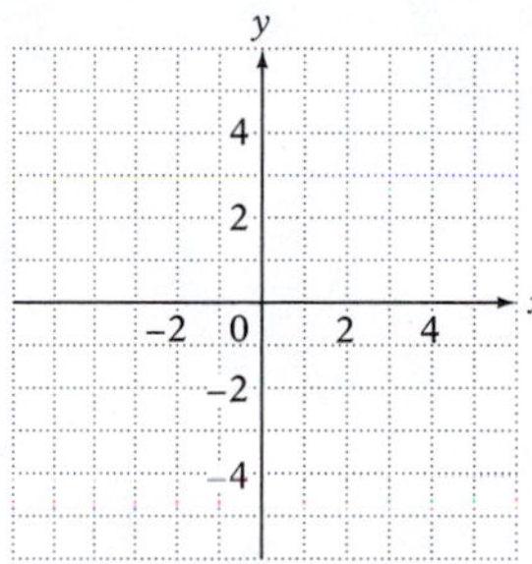

3. (−2, 5), (3, 4), (0, 0), (−3, −2)

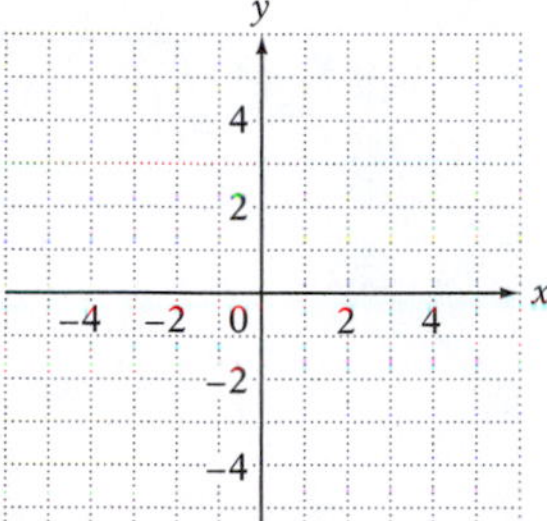

4. (1, −4), (−2, 0), (−1, −5), (0, 4)

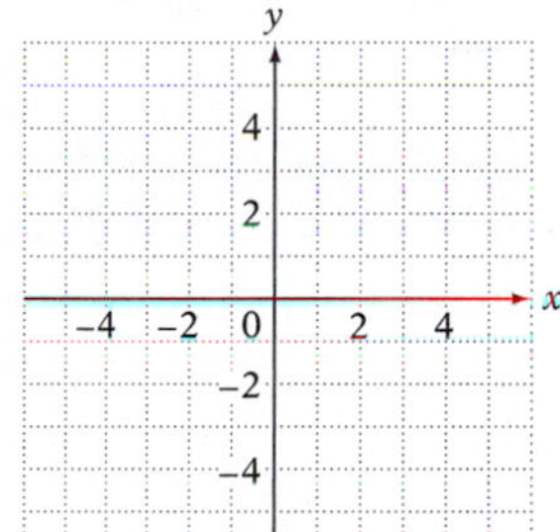

Objective B To graph a linear equation in two variables

Graph.

5. $y = 2x + 1$

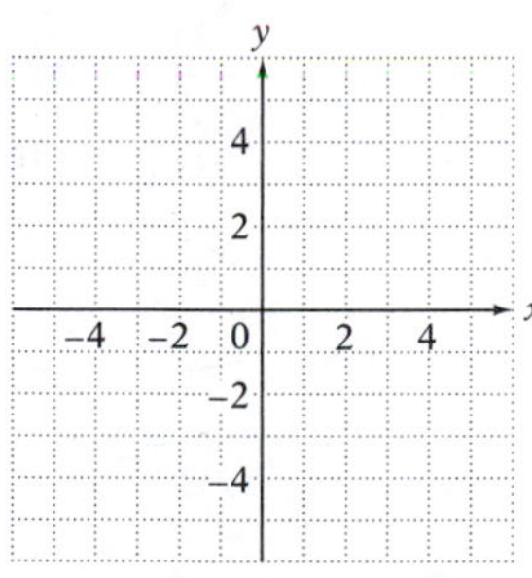

6. $y = 2x - 4$

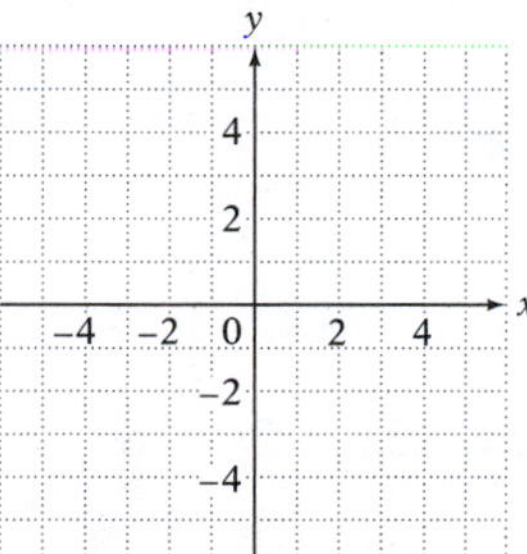

7. $y = -3x + 4$

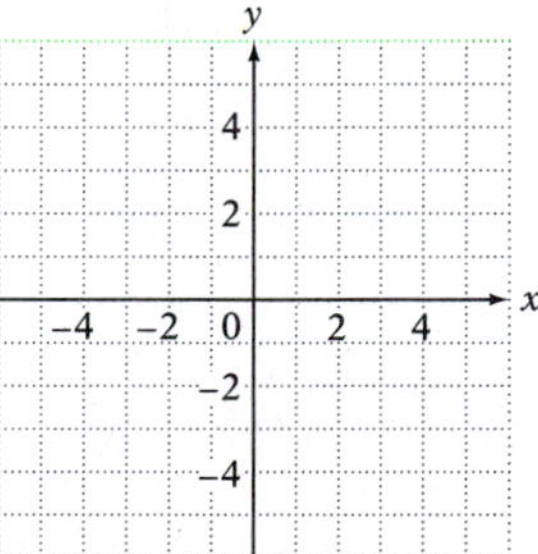

8. $y = -4x + 1$

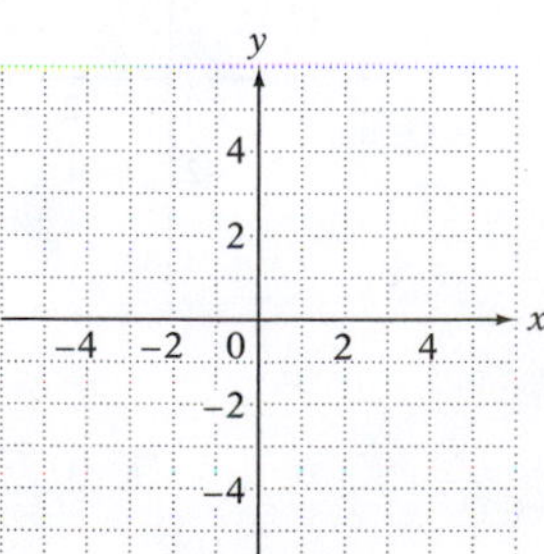

9. $y = 3x$

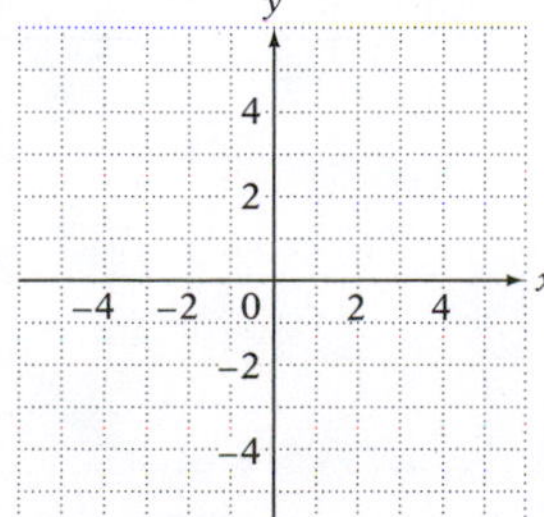

10. $y = 2x$

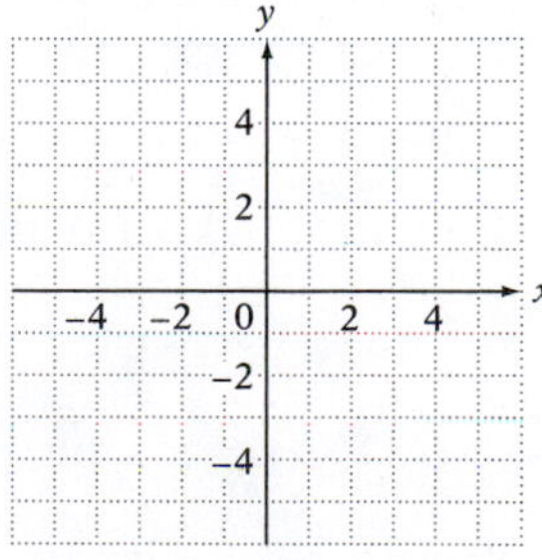

Copyright © Houghton Mifflin Company. All rights reserved.

11. $y = -\frac{4}{3}x$

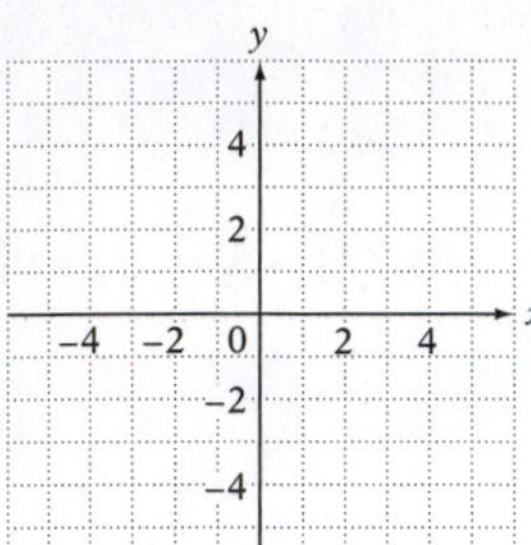

12. $y = -\frac{5}{2}x$

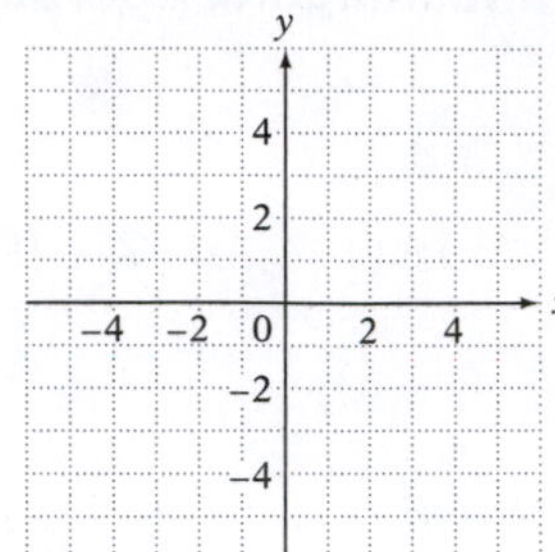

13. $y = \frac{3}{2}x - 1$

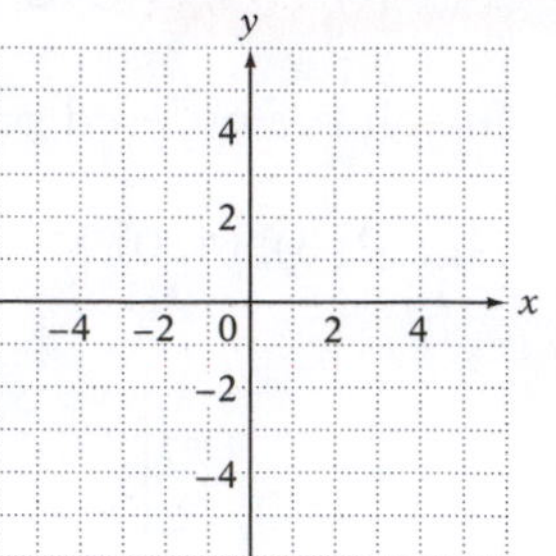

14. $y = \frac{2}{3}x + 1$

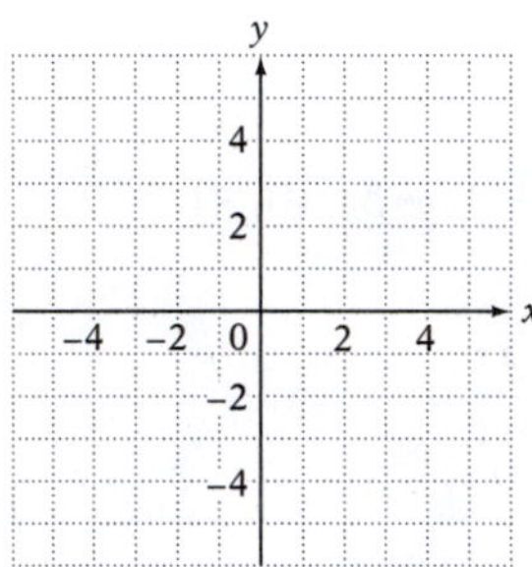

15. $y = -\frac{2}{3}x + 1$

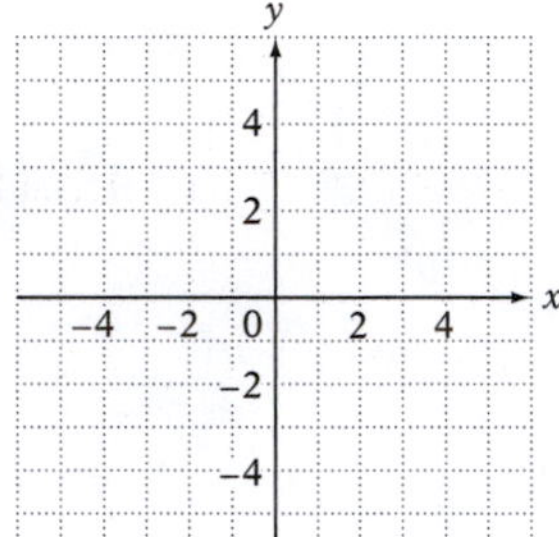

16. $y = -\frac{1}{2}x + 3$

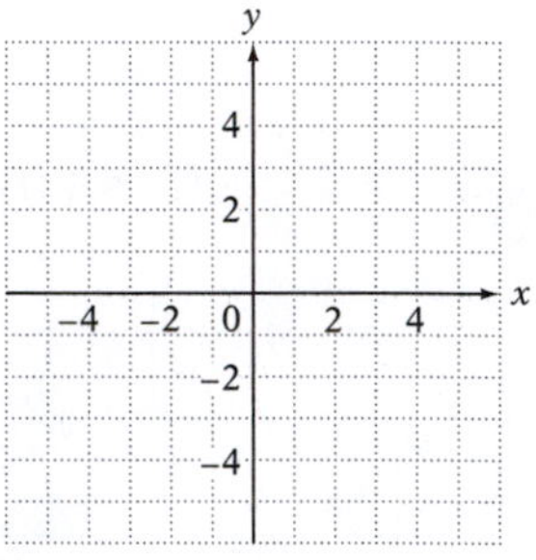

17. $2x + y = -3$

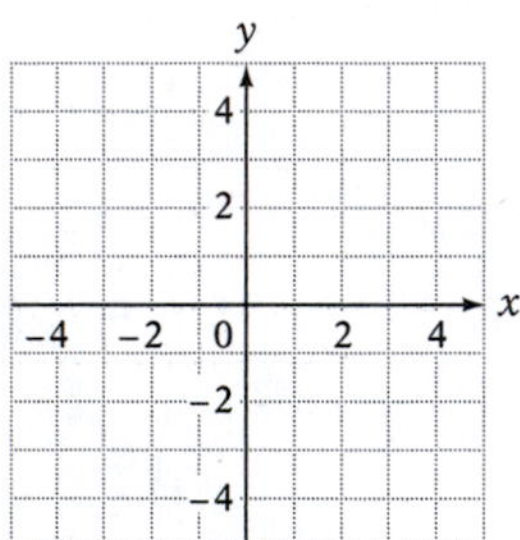

18. $2x - y = 3$

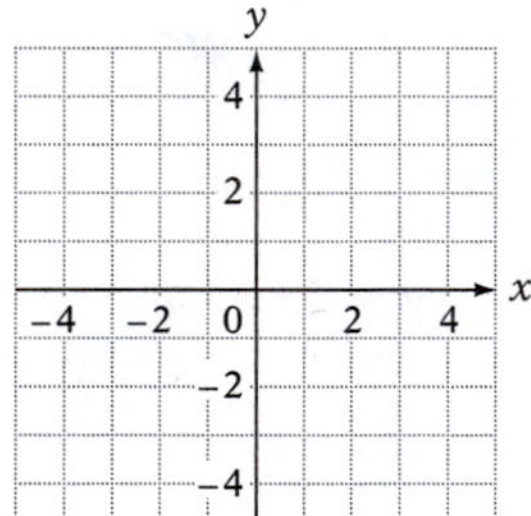

19. $x - 4y = 8$

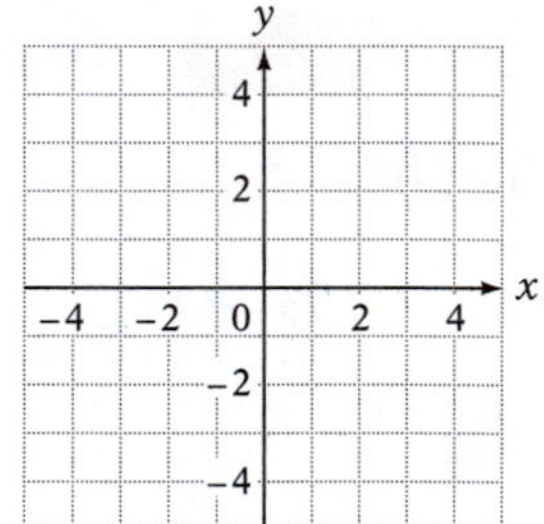

20. $2x + 5y = 10$

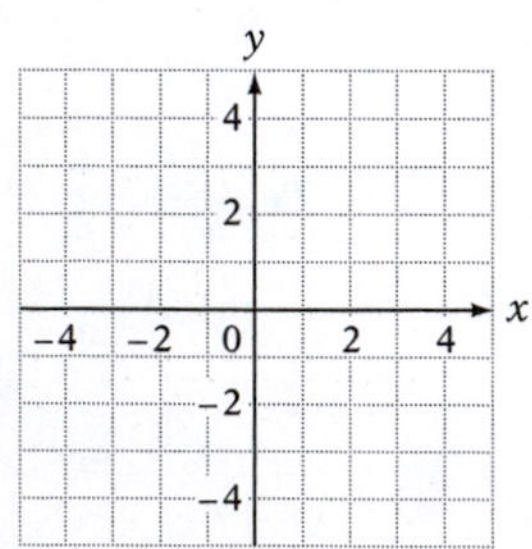

21. $3x - 2y = 8$

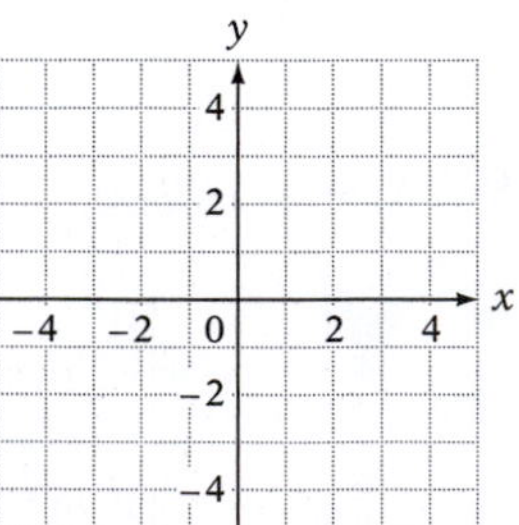

22. $3x - y = -2$

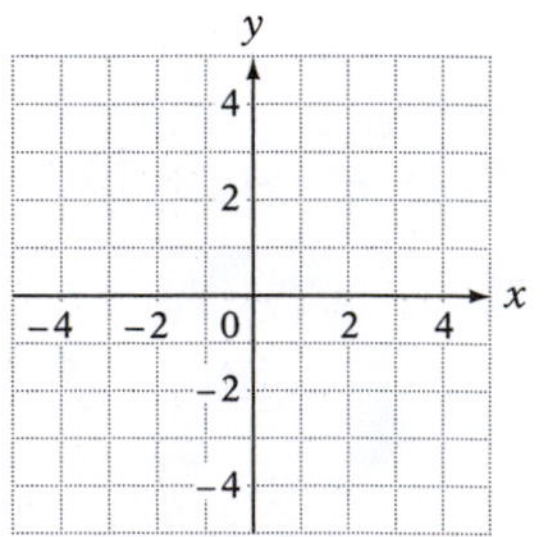

Copyright © Houghton Mifflin Company. All rights reserved.

Find the slope of the line containing the given points.

23. $P_1(2, 1), P_2(3, 4)$

24. $P_1(4, 2), P_2(3, 4)$

25. $P_1(-2, 1), P_2(2, 2)$

26. $P_1(-1, 3), P_2(2, 4)$

27. $P_1(1, 3), P_2(5, -3)$

28. $P_1(2, 4), P_2(4, -1)$

29. $P_1(-1, 2), P_2(-1, 3)$

30. $P_1(3, -4), P_2(3, 5)$

31. $P_1(5, 1), P_2(-2, 1)$

32. $P_1(4, -2), P_2(3, -2)$

33. $P_1(3, 0), P_2(2, -1)$

34. $P_1(0, -1), P_2(3, -2)$

Give the slope and the y-intercept of the graph of the equation.

35. $y = \frac{5}{2}x - 4$

36. $y = -3x + 7$

37. $y = x$

Graph by using the slope and the y-intercept.

38. $y = \frac{2}{3}x - 3$

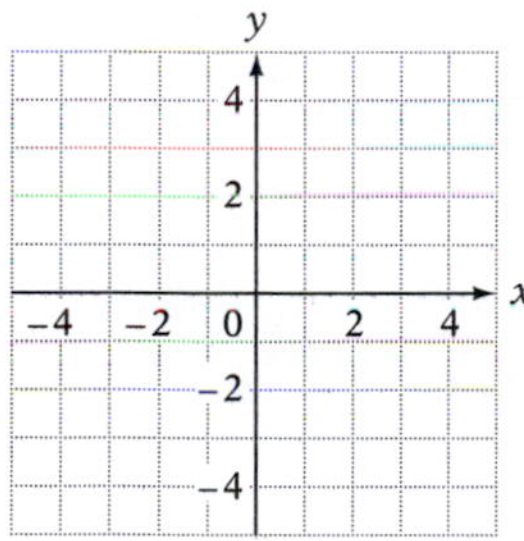

39. $y = \frac{1}{2}x + 2$

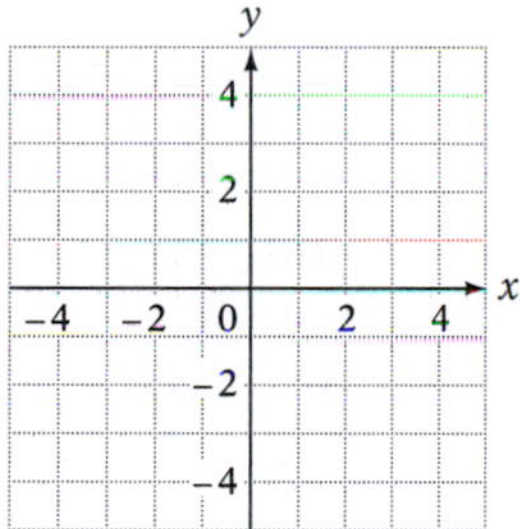

40. $y = \frac{3}{4}x$

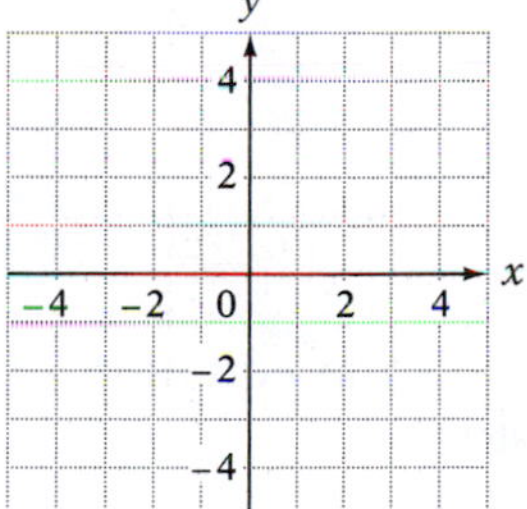

41. $y = -\frac{3}{2}x$

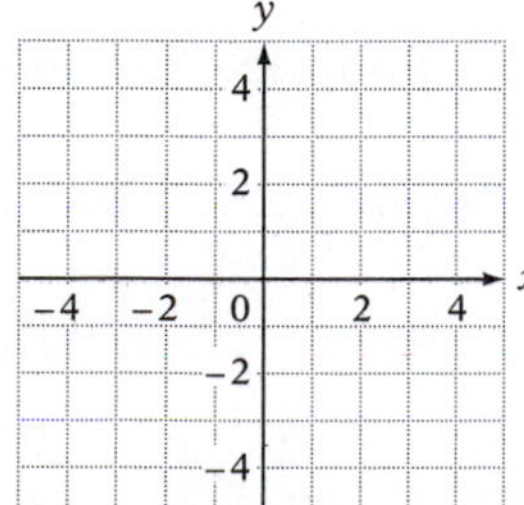

42. $y = \frac{2}{3}x - 1$

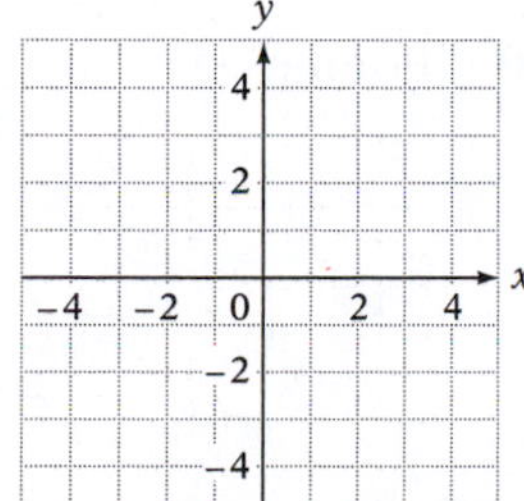

43. $y = -\frac{1}{2}x + 2$

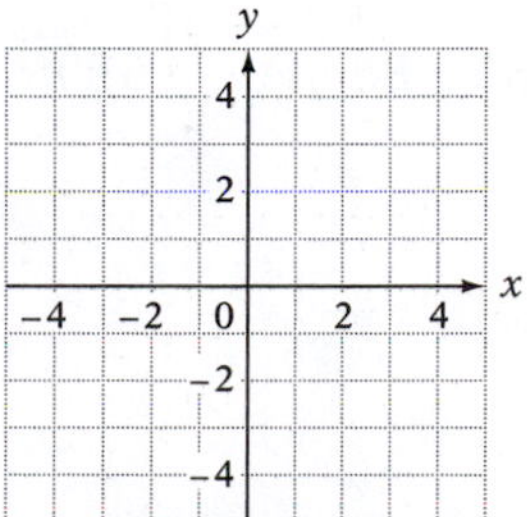

Copyright © Houghton Mifflin Company. All rights reserved.

44. $3x + 2y = 8$

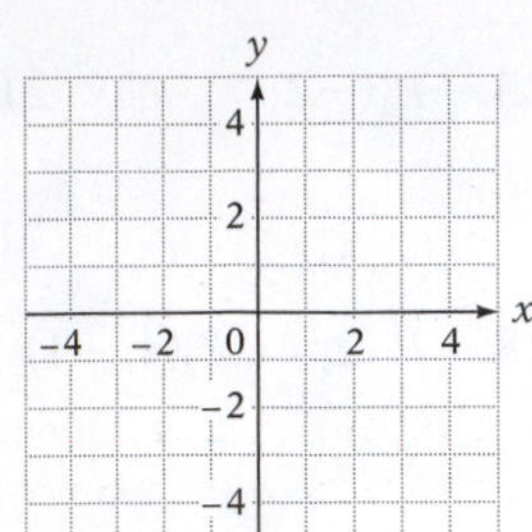

45. $x - 3y = 3$

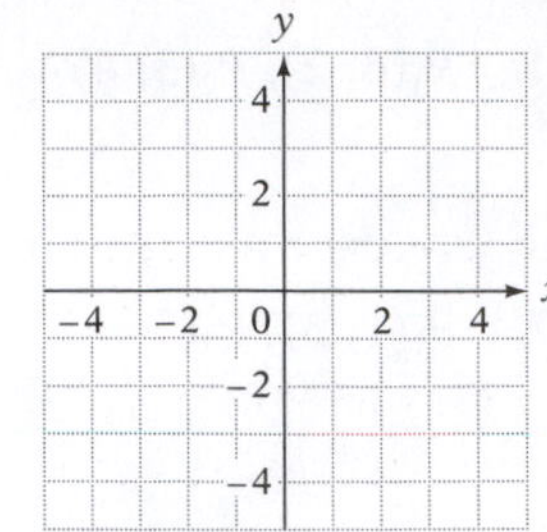

46. $4x + y = 2$

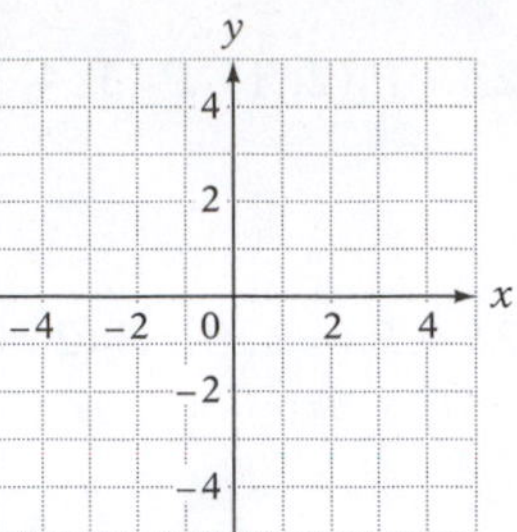

Objective C To evaluate a function

Evaluate the function for the given value of the variable. Use your answer to write an ordered pair of the function.

47. $g(x) = -3x + 1; x = -4$

48. $f(x) = 2x + 7; x = -2$

49. $p(x) = 6 - 8x; x = -1$

50. $h(x) = 4x - 2; x = 3$

51. $f(t) = t^2 - t - 3; t = 2$

52. $t(x) = 5 - 7x; x = 0$

53. $h(x) = -3x^2 + x - 1; x = -2$

54. $p(n) = n^2 - 4n - 7; n = -3$

55. $g(t) = 4t^3 - 2t; t = -1$

Objective D To find the equation of a line

Find the equation of the line that contains the given point and has the given slope.

56. Point $(2, -3)$; $m = 3$

57. Point $(-1, 2)$; $m = -3$

58. Point $(-3, 5)$; $m = 3$

59. Point $(4, -5)$; $m = -2$

60. Point $(3, 1)$; $m = \frac{1}{3}$

61. Point $(5, -3)$; $m = -\frac{3}{5}$

62. Point $(4, -2)$; $m = \frac{3}{4}$

63. Point $(-3, -2)$; $m = -\frac{2}{3}$

64. Point $(2, 3)$; $m = -\frac{1}{2}$

Copyright © Houghton Mifflin Company. All rights reserved.

R.4 Polynomials

Objective A To multiply and divide monomials

6.1A Multiply and
6.1B divide monomials

A **monomial** is a number, a variable, or a product of a number and variables. The following rules and definitions are used to multiply and divide monomials and to write monomials in simplest form.

Rule for Multiplying Exponential Expressions

If m and n are integers, then $x^m \cdot x^n = x^{m+n}$.

Rule for Simplifying Powers of Exponential Expressions

If m and n are integers, then $(x^m)^n = x^{mn}$.

Rule for Simplifying Powers of Products

If m, n, and p are integers, then $(x^m y^n)^p = x^{mp}y^{np}$.

Rule for Dividing Exponential Expressions

If m and n are integers and $x \neq 0$, then $\dfrac{x^m}{x^n} = x^{m-n}$.

Rule for Simplifying Powers of Quotients

If m, n, and p are integers and $y \neq 0$, then $\left(\dfrac{x^m}{y^n}\right)^p = \dfrac{x^{mp}}{y^{np}}$.

Definition of Zero as an Exponent

If $x \neq 0$, then $x^0 = 1$. The expression 0^0 is undefined.

Definition of Negative Exponents

If n is a positive integer and $x \neq 0$, then $x^{-n} = \dfrac{1}{x^n}$ and $\dfrac{1}{x^{-n}} = x^n$.

HOW TO Simplify: $(3x^4)^2(4x^3)$

$$\begin{aligned}(3x^4)^2(4x^3) &= (3^{1\cdot 2}x^{4\cdot 2})(4x^3)\\ &= (3^2x^8)(4x^3)\\ &= (9x^8)(4x^3)\\ &= (9\cdot 4)(x^8\cdot x^3)\\ &= 36x^{8+3}\\ &= 36x^{11}\end{aligned}$$

- Use the Rule for Simplifying Powers of Products to simplify $(3x^4)^2$.
- Use the Rule for Multiplying Exponential Expressions.

Copyright © Houghton Mifflin Company. All rights reserved.

An exponential expression is in simplest form when there are no negative exponents in the expression. For example, the expression y^{-7} is not in simplest form; use the Definition of Negative Exponents to rewrite the expression with a positive exponent: $y^{-7} = \frac{1}{y^7}$. The expression $\frac{1}{c^{-4}}$ is not in simplest form; use the Definition of Negative Exponents to rewrite the expression with a positive exponent: $\frac{1}{c^{-4}} = c^4$.

HOW TO Simplify: $\frac{6x^2}{8x^9}$

$$\frac{6x^2}{8x^9} = \frac{3x^2}{4x^9} = \frac{3x^{2-9}}{4}$$

- Divide the coefficients by their common factors. Use the Rule for Dividing Exponential Expressions.

$$= \frac{3x^{-7}}{4} = \frac{3}{4} \cdot \frac{x^{-7}}{1} = \frac{3}{4} \cdot \frac{1}{x^7}$$

- Rewrite the expression with only positive exponents.

$$= \frac{3}{4x^7}$$

HOW TO Simplify: $\left(\frac{a^4}{b^3}\right)^{-2}$

$$\left(\frac{a^4}{b^3}\right)^{-2} = \frac{a^{4(-2)}}{b^{3(-2)}} = \frac{a^{-8}}{b^{-6}}$$

- Use the Rule for Simplifying Powers of Quotients.

$$= \frac{b^6}{a^8}$$

- Rewrite the expression with positive exponents.

Example 1

Simplify.

a. $(-2x)(3x^{-2})^{-3}$

b. $\left(\frac{3a^2b^{-1}}{27a^{-3}b^{-4}}\right)^{-2}$

Solution

a. $(-2x)(3x^{-2})^{-3}$

$= (-2x)(3^{-3}x^6)$

- Use the Rule for Simplifying Powers of Products.

$= \frac{-2x \cdot x^6}{3^3}$

- Write the expression with positive exponents.

$= -\frac{2x^7}{27}$

- Use the Rule for Multiplying Exponential Expressions. Simplify 3^3.

You Try It 1

Simplify.

a. $(-2ab)(2a^3b^{-2})^{-3}$

b. $\left(\frac{2x^2y^{-4}}{4x^{-2}y^{-5}}\right)^{-3}$

Your solution

Copyright © Houghton Mifflin Company. All rights reserved.

b. Use the Rule for Simplifying Powers of Quotients. Then simplify the expression and write it with positive exponents.

$$\left(\frac{3a^2b^{-1}}{27a^{-3}b^{-4}}\right)^{-2}$$

$$= \left(\frac{a^2b^{-1}}{9a^{-3}b^{-4}}\right)^{-2}$$

$$= \frac{a^{2(-2)}b^{(-1)(-2)}}{9^{1(-2)}a^{(-3)(-2)}b^{(-4)(-2)}}$$

$$= \frac{a^{-4}b^2}{9^{-2}a^6b^8}$$

$$= 9^2a^{-4-6}b^{2-8}$$

$$= 81a^{-10}b^{-6} = \frac{81}{a^{10}b^6}$$

Solution on p. S39

Objective B **To add and subtract polynomials**    

A **polynomial** is a variable expression in which the terms are monomials. The polynomial $15t^2 - 2t + 3$ has three terms: $15t^2$, $-2t$, and 3. Note that each of these three terms is a monomial.

A polynomial of *one* term is a **monomial.** $-7x^2$ is a monomial.
A polynomial of *two* terms is a **binomial.** $4y + 3$ is a binomial.
A polynomial of *three* terms is a **trinomial.** $6b^2 + 5b - 8$ is a trinomial.

6.2B Add polynomials

Polynomials can be added by combining like terms. This is illustrated in Example 2 below.

Example 2

Add: $(8x^2 - 4x - 9) + (2x^2 + 9x - 9)$

Solution

$(8x^2 - 4x - 9) + (2x^2 + 9x - 9)$
$= (8x^2 + 2x^2) + (-4x + 9x) + (-9 - 9)$
$= 10x^2 + 5x - 18$

You Try It 2

Add: $(-4x^3 + 2x^2 - 8) + (4x^3 + 6x^2 - 7x + 5)$

Your solution

Solution on p. S39

Copyright © Houghton Mifflin Company. All rights reserved.

6.2B Subtract polynomials

The additive inverse of the polynomial $(3x^2 - 7x + 8)$ is $-(3x^2 - 7x + 8)$.

To find the additive inverse of a polynomial, change the sign of each term inside the parentheses. $-(3x^2 - 7x + 8) = -3x^2 + 7x - 8$

To subtract two polynomials, add the additive inverse of the second polynomial to the first.

HOW TO Simplify: $(5a^2 - a + 2) - (-2a^3 + 3a - 3)$

$(5a^2 - a + 2) - (-2a^3 + 3a - 3)$

$= (5a^2 - a + 2) + (2a^3 - 3a + 3)$
- Rewrite subtraction as addition of the additive inverse.

$= 2a^3 + 5a^2 - 4a + 5$
- Combine like terms.

Example 3

Subtract: $(7c^2 - 9c - 12) - (9c^2 + 5c - 8)$

Solution

$(7c^2 - 9c - 12) - (9c^2 + 5c - 8)$
$= (7c^2 - 9c - 12) + (-9c^2 - 5c + 8)$
$= -2c^2 - 14c - 4$

You Try It 3

Subtract:
$(-4w^3 + 8w - 8) - (3w^3 - 4w^2 + 2w - 1)$

Your solution

Solution on p. S39

Objective C To multiply polynomials

CD TUTOR

WEB

SSM

6.3A Multiply a polynomial by a monomial

The Distributive Property is used to multiply a polynomial by a monomial. Each term of the polynomial is multiplied by the monomial.

HOW TO Multiply: $3x^3(4x^4 - 2x + 5)$

$3x^3(4x^4 - 2x + 5)$

$= 3x^3(4x^4) - 3x^3(2x) + 3x^3(5)$
- Use the Distributive Property. Multiply each term of the polynomial by $3x^3$.

$= 12x^7 - 6x^4 + 15x^3$
- Use the Rule for Multiplying Exponential Expressions.

Example 4

Multiply: $2xy(3x^2 - xy + 2y^2)$

Solution

$2xy(3x^2 - xy + 2y^2)$
$= 2xy(3x^2) - 2xy(xy) + 2xy(2y^2)$
$= 6x^3y - 2x^2y^2 + 4xy^3$

You Try It 4

Multiply: $3mn^2(2m^2 - 3mn - 1)$

Your solution

Solution on p. S39

Copyright © Houghton Mifflin Company. All rights reserved.

6.3B Multiply two polynomials

A vertical format similar to that used for multiplication of whole numbers is used to multiply two polynomials. The product $(2y - 3)(y^2 + 2y + 5)$ is shown below.

$$\begin{array}{rl} y^2 + 2y + 5 & \\ 2y - 3 & \\ \hline -3y^2 - 6y - 15 & \text{This is } -3(y^2 + 2y + 5). \\ 2y^3 + 4y^2 + 10y \quad & \text{This is } 2y(y^2 + 2y + 5). \text{ Like terms are placed in the same columns.} \\ \hline 2y^3 + y^2 + 4y - 15 & \text{Add the terms in each column.} \end{array}$$

Example 5

Multiply: $(2a^2 + 4a - 5)(3a + 5)$

Solution

$$\begin{array}{r} 2a^2 + 4a - 5 \\ 3a + 5 \\ \hline 10a^2 + 20a - 25 \\ 6a^3 + 12a^2 - 15a \quad \\ \hline 6a^3 + 22a^2 + 5a - 25 \end{array}$$

• **Align like terms in the same column.**

You Try It 5

Multiply: $(3c^2 - 4c + 5)(2c - 3)$

Your solution

Solution on p. S39

6.3B Multiply two binomials

It is frequently necessary to multiply two binomials. The product is computed by using a method called FOIL. The letters of FOIL stand for **F**irst, **O**uter, **I**nner, and **L**ast. The FOIL method is based on the Distributive Property and involves adding the products of the first terms, the outer terms, the inner terms, and the last terms.

The product $(2x + 3)(3x + 4)$ is shown below using FOIL.

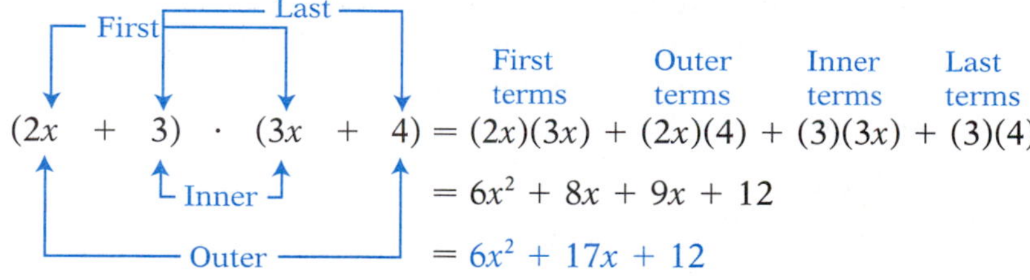

Example 6

Multiply: $(4x - 3)(2x + 5)$

Solution

$(4x - 3)(2x + 5)$

$= (4x)(2x) + (4x)(5) + (-3)(2x) + (-3)(5)$

$= 8x^2 + 20x + (-6x) + (-15)$

$= 8x^2 + 14x - 15$

You Try It 6

Multiply: $(4y - 7)(3y - 5)$

Your solution

Solution on p. S39

Copyright © Houghton Mifflin Company. All rights reserved.

Objective D To divide polynomials

6.4B Divide polynomials

To divide two polynomials, use a method similar to that used for division of whole numbers.

To divide $(x^2 - 5x + 8) \div (x - 3)$:

Step 1

$$\begin{array}{r} x \phantom{{}-5x+8} \\ x-3\overline{)x^2 - 5x + 8} \\ \underline{x^2 - 3x} \phantom{{}+8} \\ -2x + 8 \end{array}$$

Think: $x\overline{)x^2} = \dfrac{x^2}{x} = x$

Multiply: $x(x - 3) = x^2 - 3x$

Subtract: $(x^2 - 5x) - (x^2 - 3x) = -2x$

Bring down the +8.

Step 2

$$\begin{array}{r} x - 2 \\ x-3\overline{)x^2 - 5x + 8} \\ \underline{x^2 - 3x} \phantom{{}+8} \\ -2x + 8 \\ \underline{-2x + 6} \\ 2 \end{array}$$

Think: $x\overline{)-2x} = \dfrac{-2x}{x} = -2$

Multiply: $-2(x - 3) = -2x + 6$

Subtract: $(-2x + 8) - (-2x + 6) = 2$

The remainder is 2.

Check: (Quotient × Divisor) + Remainder = Dividend

$(x - 2)(x - 3) + 2 = x^2 - 3x - 2x + 6 + 2 = x^2 - 5x + 8$

$(x^2 - 5x + 8) \div (x - 3) = x - 2 + \dfrac{2}{x - 3}$

Example 7

Divide: $(6x + 2x^3 + 26) \div (x + 2)$

Solution

Arrange the terms of the dividend in descending order. There is no x^2 term in $2x^3 + 6x + 26$. Insert $0x^2$ for the missing term so that like terms will be in the same columns.

$$\begin{array}{r} 2x^2 - 4x + 14 \\ x+2\overline{)2x^3 + 0x^2 + 6x + 26} \\ \underline{2x^3 + 4x^2} \phantom{{}+6x+26} \\ -4x^2 + 6x \phantom{{}+26} \\ \underline{-4x^2 - 8x} \phantom{{}+26} \\ 14x + 26 \\ \underline{14x + 28} \\ -2 \end{array}$$

Check: $(x + 2)(2x^2 - 4x + 14) - 2$
$= 2x^3 + 6x + 28 - 2 = 2x^3 + 6x + 26$

$(6x + 2x^3 + 26) \div (x + 2) = 2x^2 - 4x + 14 - \dfrac{2}{x + 2}$

You Try It 7

Divide: $(x^3 - 7 - 2x) \div (x - 2)$

Your solution

Solution on p. S39

Copyright © Houghton Mifflin Company. All rights reserved.

Objective E To factor polynomials of the form $ax^2 + bx + c$

7.1A Factor a monomial from a polynomial

A polynomial is in **factored form** when it is written as a product of other polynomials. Factoring can be thought of as the reverse of multiplication.

Polynomial	**Factored Form**
$2x^3 + 6x^2 - 10x$	$= 2x(x^2 + 3x - 5)$
$x^2 - 3x - 28$	$= (x + 4)(x - 7)$

To factor out a common monomial from the terms of a polynomial, first find the greatest common factor (GCF) of the terms.

The GCF of two or more monomials is the product of the GCF of the coefficients and the common variable factors.

$$10a^3b = \mathbf{2} \cdot 5 \cdot \boldsymbol{a} \cdot \boldsymbol{a} \cdot a \cdot \boldsymbol{b}$$
$$4a^2b^2 = \mathbf{2} \cdot 2 \cdot \boldsymbol{a} \cdot \boldsymbol{a} \cdot \boldsymbol{b} \cdot b$$
$$\text{GCF} = \mathbf{2} \cdot \boldsymbol{a} \cdot \boldsymbol{a} \cdot \boldsymbol{b} = 2a^2b$$

Note that the exponent of each variable in the GCF is the same as the smallest exponent of that variable in either of the monomials.

The GCF of $10a^3b$ and $4a^2b^2$ is $2a^2b$.

HOW TO Factor: $5x^3 - 35x^2 + 10x$

The GCF is $5x$.
- Find the GCF of the terms $5x^3$, $-35x^2$, and $10x$.

$$\frac{5x^3 - 35x^2 + 10x}{5x} = x^2 - 7x + 2$$
- Divide each term of the polynomial by the GCF.

$$5x^3 - 35x^2 + 10x = 5x(x^2 - 7x + 2)$$
- Write the polynomial as the product of the GCF and the quotient found above.

$$5x(x^2 - 7x + 2) = 5x^3 - 35x^2 + 10x$$
- Check the factorization by multiplying.

Example 8

Factor: $16x^4y^5 + 8x^4y^2 - 12x^3y$

Solution

The GCF is $4x^3y$.

$$16x^4y^5 + 8x^4y^2 - 12x^3y = 4x^3y(4xy^4 + 2xy - 3)$$

You Try It 8

Factor: $6x^4y^2 - 9x^3y^2 + 12x^2y^4$

Your solution

Solution on p. S39

Copyright © Houghton Mifflin Company. All rights reserved.

A **quadratic trinomial** is a trinomial of the form $ax^2 + bx + c$, where a and b are coefficients and c is a constant. Examples of quadratic trinomials are shown below.

$$x^2 + 9x + 14 \qquad a = 1, b = 9, c = 14$$
$$x^2 - 2x - 15 \qquad a = 1, b = -2, c = -15$$
$$3x^2 - x + 4 \qquad a = 3, b = -1, c = 4$$

To **factor a quadratic trinomial** means to express the trinomial as the product of two binomials. For example,

Trinomial		**Factored Form**
$2x^2 - x - 1$	=	$(2x + 1)(x - 1)$
$y^2 - 3y + 2$	=	$(y - 1)(y - 2)$

7.2A Factor a trinomial of the form $x^2 + bx + c$

We will begin by factoring trinomials of the form $x^2 + bx + c$, where the coefficient of x^2 is 1.

The method by which factors of a trinomial are found is based on FOIL. Consider the following binomial products, noting the relationship between the constant term of the binomials and the terms of the trinomial.

Sum of the binomial constants

Product of the binomial constants

$$(x + 4)(x + 5) = x \cdot x + 5x + 4x + 4 \cdot 5 \qquad = x^2 + 9x + 20$$
$$(x - 6)(x + 8) = x \cdot x + 8x - 6x + (-6)(8) \qquad = x^2 + 2x - 48$$
$$(x - 3)(x - 2) = x \cdot x - 2x - 3x + (-3)(-2) \qquad = x^2 - 5x + 6$$

TAKE NOTE
Once the correct factors are found, it is not necessary to try the remaining factors.

TAKE NOTE
Always check the proposed factorization by multiplying the factors.

HOW TO Factor: $x^2 - 7x + 10$

Find two integers whose product is 10 and whose sum is -7.

Negative Factors of 10	**Sum**	
$-1, -10$	-11	
$-2, -5$	**-7**	• These are the correct factors.

$x^2 - 7x + 10 = (x - 2)(x - 5)$ • Write the trinomial as a product of factors.

Check:
$$(x - 2)(x - 5) = x^2 - 5x - 2x + 10$$
$$= x^2 - 7x + 10$$

Example 9

Factor: $x^2 + 6x - 27$

Solution
Two factors of -27 whose sum is 6 are -3 and 9.

$x^2 + 6x - 27 = (x - 3)(x + 9)$

You Try It 9

Factor: $x^2 - 8x + 15$

Your solution

Solution on p. S39

Copyright © Houghton Mifflin Company. All rights reserved.

7.3A Factor a trinomial of the form $ax^2 + bx + c$ by using trial factors

To use the trial factor method to factor a trinomial of the form $ax^2 + bx + c$, where $a \neq 1$, use the factors of a and the factors of c to write all of the possible binomial factors of the trinomial. Then use FOIL to determine the correct factorization. To reduce the number of trial factors that must be considered, remember the following guidelines.

Use the signs of the constant term and the coefficient of x in the trinomial to determine the signs of the binomial factors. If the constant term is positive, the signs of the binomial factors will be the same as the sign of the coefficient of x in the trinomial. If the sign of the constant term is negative, the constant terms in the binomials will have different signs.

HOW TO Factor: $2x^2 - 7x + 3$

Because the constant term is positive (+3) and the coefficient of x is negative (−7), the binomial constants will be negative.

Positive Factors of 2 ($a = 2$)	**Negative Factors of 3 ($c = 3$)**
1, 2	−1, −3

Write trial factors. Use the **O**uter and **I**nner products of FOIL to determine the middle term, $-7x$, of the trinomial.

Trial Factors	**Middle Term**	
$(x - 1)(2x - 3)$	$-3x - 2x = -5x$	
$\mathbf{(x - 3)(2x - 1)}$	$\mathbf{-x - 6x = -7x}$	• **$-7x$ is the middle term.**

$2x^2 - 7x + 3 = (x - 3)(2x - 1)$

Check: $(x - 3)(2x - 1) = 2x^2 - x - 6x + 3 = 2x^2 - 7x + 3$

HOW TO Factor: $5x^2 + 22x - 15$

The constant term is negative (−15). The binomial constants will have different signs.

Positive Factors of 5 ($a = 5$)	**Factors of −15 ($c = -15$)**
1, 5	−1, 15
	1, −15
	−3, 5
	3, −5

Write trial factors. Use the **O**uter and **I**nner products of FOIL to determine the middle term, $22x$, of the trinomial.

Trial Factors	**Middle Term**	
$(x - 1)(5x + 15)$	common factor	
$(x + 15)(5x - 1)$	$-x + 75x = 74x$	
$(x + 1)(5x - 15)$	common factor	
$(x - 15)(5x + 1)$	$x - 75x = -74x$	
$(x - 3)(5x + 5)$	common factor	
$\mathbf{(x + 5)(5x - 3)}$	$\mathbf{-3x + 25x = 22x}$	• **$22x$ is the middle term.**
$(x + 3)(5x - 5)$	common factor	
$(x - 5)(5x + 3)$	$3x - 25x = -22x$	

$5x^2 + 22x - 15 = (x + 5)(5x - 3)$

Check: $(x + 5)(5x - 3) = 5x^2 - 3x + 25x - 15 = 5x^2 + 22x - 15$

TAKE NOTE

It is not necessary to test trial factors that have a common factor. If the trinomial does not have a common factor, then its factors cannot have a common factor.

Copyright © Houghton Mifflin Company. All rights reserved.

Example 10

Factor: $2x^2 + 3x - 5$

Solution

Positive Factors of 2	Factors of −5
1, 2	1, −5
	−1, 5

Trial Factors	Middle Term
$(x + 1)(2x - 5)$	$-5x + 2x = -3x$
$(x - 5)(2x + 1)$	$x - 10x = -9x$
$(x - 1)(2x + 5)$	$5x - 2x = 3x$
$(x + 5)(2x - 1)$	$-x + 10x = 9x$

$2x^2 + 3x - 5 = (x - 1)(2x + 5)$

Check: $(x - 1)(2x + 5) = 2x^2 + 5x - 2x - 5$
$= 2x^2 + 3x - 5$

You Try It 10

Factor: $3x^2 - x - 2$

Your solution

Solution on p. S39

7.4D Factor completely

A polynomial is factored completely when it is written as a product of factors that are nonfactorable over the integers.

The first step in any factoring problem is to determine whether the terms of the polynomial have a common factor. If they do, factor it out first.

Example 11

Factor: $5x^2y + 60xy + 100y$

Solution

There is a common factor, $5y$.
Factor out the GCF.

$5x^2y + 60xy + 100y = 5y(x^2 + 12x + 20)$

Factor $x^2 + 12x + 20$. The two factors of 20 whose sum is 12 are 2 and 10.

$5y(x^2 + 12x + 20) = 5y(x + 2)(x + 10)$

$5x^2y + 60xy + 100y = 5y(x + 2)(x + 10)$

Check:
$5y(x + 2)(x + 10) = (5xy + 10y)(x + 10)$
$= 5x^2y + 50xy + 10xy + 100y$
$= 5x^2y + 60xy + 100y$

You Try It 11

Factor: $4a^3 - 4a^2 - 24a$

Your solution

Solution on p. S40

Copyright © Houghton Mifflin Company. All rights reserved.

R.4 Exercises

Objective A To multiply and divide monomials

Simplify.

1. $z^3 \cdot z \cdot z^4$
2. $b \cdot b^2 \cdot b^6$
3. $(x^3)^5$
4. $(b^2)^4$
5. $(x^2y^3)^6$
6. $(m^4n^2)^3$
7. $\dfrac{a^8}{a^2}$
8. $\dfrac{c^{12}}{c^5}$
9. $(-m^3n)(m^6n^2)$
10. $(-r^4t^3)(r^2t^9)$
11. $(-2a^3bc^2)^3$
12. $(-4xy^3z^2)^2$
13. $\dfrac{m^4n^7}{m^3n^5}$
14. $\dfrac{a^5b^6}{a^3b^2}$
15. $\dfrac{-16a^7}{24a^6}$
16. $\dfrac{18b^5}{-45b^4}$
17. $(9mn^4p)(-3mp^2)$
18. $(-3v^2wz)(-4vz^4)$
19. $(-2n^2)(-3n^4)^3$
20. $(-3m^3n)(-2m^2n^3)^3$
21. $\dfrac{14x^4y^6z^2}{16x^3y^9z}$
22. $\dfrac{25x^4y^7z^2}{20x^5y^9z^{11}}$
23. $(-2x^3y^2)^3(-xy^2)^4$
24. $(-m^4n^2)^5(-2m^3n^3)^3$
25. $4x^{-7}$
26. $-6y^{-1}$
27. $d^{-4}d^{-6}$
28. $x^{-3}x^{-5}$
29. $\dfrac{x^{-3}}{x^2}$
30. $\dfrac{x^4}{x^{-5}}$
31. $\dfrac{1}{3x^{-2}}$
32. $\dfrac{2}{5c^{-6}}$
33. $(x^2y^{-4})^3$
34. $(x^3y^5)^{-4}$
35. $(3x^{-1}y^{-2})^2$
36. $(5xy^{-3})^{-2}$

Copyright © Houghton Mifflin Company. All rights reserved.

37. $(2x^{-1})(x^{-3})$

38. $(-2x^{-5})(x^{7})$

39. $\dfrac{3x^{-2}y^{2}}{6xy^{2}}$

40. $\dfrac{2x^{-2}y}{8xy}$

41. $\dfrac{2x^{-1}y^{-4}}{4xy^{2}}$

42. $\dfrac{3a^{-2}b}{ab}$

43. $(x^{-2}y)^{2}(xy)^{-2}$

44. $(x^{-1}y^{2})^{-3}(x^{2}y^{-4})^{-3}$

45. $\left(\dfrac{x^{2}y^{-1}}{xy}\right)^{-4}$

46. $\left(\dfrac{x^{-2}y^{-4}}{x^{-2}y}\right)^{-2}$

47. $\left(\dfrac{4a^{-2}b}{8a^{3}b^{-4}}\right)^{2}$

48. $\left(\dfrac{6ab^{-2}}{3a^{-2}b}\right)^{-2}$

Objective B To add and subtract polynomials

Add or subtract.

49. $(4b^2 - 5b) + (3b^2 + 6b - 4)$

50. $(2c^2 - 4) + (6c^2 - 2c + 4)$

51. $(2a^2 - 7a + 10) + (a^2 + 4a + 7)$

52. $(-6x^2 + 7x + 3) + (3x^2 + x + 3)$

53. $(x^2 - 2x + 1) - (x^2 + 5x + 8)$

54. $(3x^2 + 2x - 2) - (5x^2 - 5x + 6)$

55. $(-2x^3 + x - 1) - (-x^2 + x - 3)$

56. $(2x^2 + 5x - 3) - (3x^3 + 2x - 5)$

57. $(x^3 - 7x + 4) + (2x^2 + x - 10)$

58. $(3y^3 + y^2 + 1) + (-4y^3 - 6y - 3)$

59. $(5x^3 + 7x - 7) + (10x^2 - 8x + 3)$

60. $(3y^3 + 4y + 9) + (2y^2 + 4y - 21)$

61. $(2y^3 + 6y - 2) - (y^3 + y^2 + 4)$

62. $(-2x^2 - x + 4) - (-x^3 + 3x - 2)$

63. $(4y^3 - y - 1) - (2y^2 - 3y + 3)$

64. $(3x^2 - 2x - 3) - (2x^3 - 2x^2 + 4)$

Copyright © Houghton Mifflin Company. All rights reserved.

Objective C To multiply polynomials

Multiply.

65. $4b(3b^3 - 12b^2 - 6)$

66. $-2a^2(3a^2 - 2a + 3)$

67. $3b(3b^4 - 3b^2 + 8)$

68. $-2x^2(2x^2 - 3x - 7)$

69. $-2x^2y(x^2 - 3xy + 2y^2)$

70. $3ab^2(3a^2 - 2ab + 4b^2)$

71. $(x^2 + 3x + 2)(x + 1)$

72. $(x^2 - 2x + 7)(x - 2)$

73. $(a - 3)(a^2 - 3a + 4)$

74. $(2x - 3)(x^2 - 3x + 5)$

75. $(-2b^2 - 3b + 4)(b - 5)$

76. $(-a^2 + 3a - 2)(2a - 1)$

77. $(x^3 - 3x + 2)(x - 4)$

78. $(y^3 + 4y^2 - 8)(2y - 1)$

79. $(y + 2)(y^3 + 2y^2 - 3y + 1)$

80. $(2a - 3)(2a^3 - 3a^2 + 2a - 1)$

81. $(a - 3)(a + 4)$

82. $(b - 6)(b + 3)$

83. $(y - 7)(y - 3)$

84. $(a - 8)(a - 9)$

85. $(2x + 1)(x + 7)$

86. $(y + 2)(5y + 1)$

87. $(3x - 1)(x + 4)$

88. $(7x - 2)(x + 4)$

89. $(4x - 3)(x - 7)$

90. $(2x - 3)(4x - 7)$

91. $(3y - 8)(y + 2)$

92. $(5y - 9)(y + 5)$

93. $(7a - 16)(3a - 5)$

94. $(5a - 12)(3a - 7)$

95. $(x + y)(2x + y)$

96. $(2a + b)(a + 3b)$

97. $(3x - 4y)(x - 2y)$

98. $(2a - b)(3a + 2b)$

Copyright © Houghton Mifflin Company. All rights reserved.

99. $(5a - 3b)(2a + 4b)$

100. $(2x + 3)(2x - 3)$

101. $(4x - 7)(4x + 7)$

Objective D To divide polynomials

Divide.

102. $(b^2 - 14 + 49) \div (b - 7)$

103. $(x^2 - x - 6) \div (x - 3)$

104. $(2x^2 + 5x + 2) \div (x + 2)$

105. $(2y^2 - 13y + 21) \div (y - 3)$

106. $(x^2 + 1) \div (x - 1)$

107. $(x^2 + 4) \div (x + 2)$

108. $(6x^2 - 7x) \div (3x - 2)$

109. $(6y^2 + 2y) \div (2y + 4)$

110. $(a^2 + 5a + 10) \div (a + 2)$

111. $(b^2 - 8b - 9) \div (b - 3)$

112. $(2y^2 - 9y + 8) \div (2y + 3)$

113. $(3x^2 + 5x - 4) \div (x - 4)$

114. $(8x + 3 + 4x^2) \div (2x - 1)$

115. $(10 + 21y + 10y^2) \div (2y + 3)$

116. $(x^3 + 3x^2 + 5x + 3) \div (x + 1)$

117. $(x^3 - 6x^2 + 7x - 2) \div (x - 1)$

118. $(x^4 - x^2 - 6) \div (x^2 + 2)$

119. $(x^4 + 3x^2 - 10) \div (x^2 - 2)$

Objective E To factor polynomials of the form $ax^2 + bx + c$

Factor.

120. $8x + 12$

121. $12y^2 - 5y$

122. $10x^4 - 12x^2$

123. $10x^2yz^2 + 15xy^3z$

124. $x^3 - 3x^2 - x$

125. $5x^2 - 15x + 35$

126. $3x^3 + 6x^2 + 9x$

127. $3y^4 - 9y^3 - 6y^2$

128. $2x^3 + 6x^2 - 14x$

129. $x^4y^4 - 3x^3y^3 + 6x^2y^2$

130. $4x^5y^5 - 8x^4y^4 + x^3y^3$

131. $16x^2y - 8x^3y^4 - 48x^2y^2$

132. $x^2 + 5x + 6$

133. $x^2 + x - 2$

134. $x^2 + x - 6$

Copyright © Houghton Mifflin Company. All rights reserved.

135. $a^2 + a - 12$ **136.** $a^2 - 2a - 35$ **137.** $a^2 - 3a + 2$

138. $a^2 - 5a + 4$ **139.** $b^2 + 7b - 8$ **140.** $y^2 + 6y - 55$

141. $z^2 - 4z - 45$ **142.** $y^2 - 8y + 15$ **143.** $z^2 - 14z + 45$

144. $p^2 + 12p + 27$ **145.** $b^2 + 9b + 20$ **146.** $y^2 - 8y + 32$

147. $y^2 - 9y + 81$ **148.** $p^2 + 24p + 63$ **149.** $x^2 - 15x + 56$

150. $5x^2 + 6x + 1$ **151.** $2y^2 + 7y + 3$ **152.** $2a^2 - 3a + 1$

153. $3a^2 - 4a + 1$ **154.** $4x^2 - 3x - 1$ **155.** $2x^2 - 5x - 3$

156. $6t^2 - 11t + 4$ **157.** $10t^2 + 11t + 3$ **158.** $8x^2 + 33x + 4$

159. $10z^2 + 3z - 4$ **160.** $3x^2 + 14x - 5$ **161.** $3z^2 + 95z + 10$

162. $8z^2 - 36z + 1$ **163.** $2t^2 - t - 10$ **164.** $2t^2 + 5t - 12$

165. $12y^2 + 19y + 5$ **166.** $5y^2 - 22y + 8$ **167.** $11a^2 - 54a - 5$

168. $4z^2 + 11z + 6$ **169.** $6b^2 - 13b + 6$ **170.** $6x^2 + 35x - 6$

171. $3x^2 + 15x + 18$ **172.** $3a^2 + 3a - 18$ **173.** $ab^2 + 7ab - 8a$

174. $3y^3 - 15y^2 + 18y$ **175.** $2y^4 - 26y^3 - 96y^2$ **176.** $3y^4 + 54y^3 + 135y^2$

177. $2x^3 - 11x^2 + 5x$ **178.** $2x^3 + 3x^2 - 5x$ **179.** $10t^2 - 5t - 50$

180. $16t^2 + 40t - 96$ **181.** $6p^3 + 5p^2 + p$ **182.** $12x^2y - 36xy + 27y$

Copyright © Houghton Mifflin Company. All rights reserved.

Appendix A

Keystroke Guide for the TI-83 and TI-83 Plus/TI-84 Plus

Basic Operations

Numerical calculations are performed on the **home screen.** You can always return to the home screen by pressing 2nd QUIT. Pressing CLEAR erases the home screen.

To evaluate the expression $-2(3 + 5) - 8 \div 4$, use the following keystrokes.

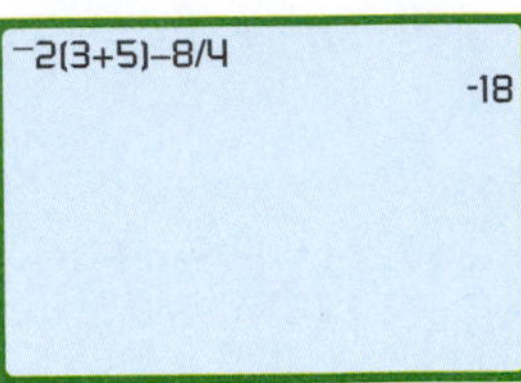

Note: There is a difference between the key to enter a negative number, (-), and the key for subtraction, −. You cannot use these keys interchangeably.

> **TAKE NOTE**
> The descriptions in the margins (for example, Basic Operations and Complex Numbers) are the same as those used in the text and are arranged alphabetically.

The 2nd key is used to access the commands in gold writing above a key. For instance, to evaluate the $\sqrt{49}$, press 2nd √ 49) ENTER.

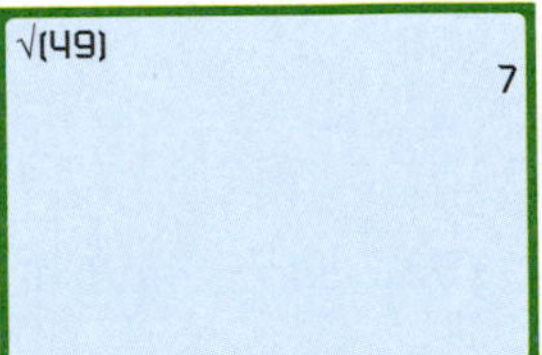

The ALPHA key is used to place a letter on the screen. One reason to do this is to store a value of a variable. The following keystrokes give A the value of 5.

5 STO▸ ALPHA A ENTER

This value is now available in calculations. For instance, we can find the value of $3a^2$ by using the following keystrokes: 3 ALPHA A x^2. To display the value of the variable on the screen, press 2nd RCL ALPHA A.

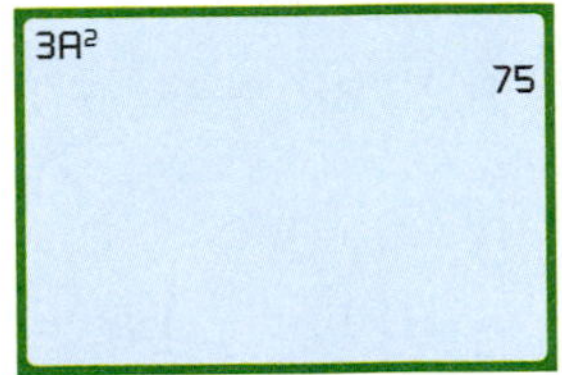

Note: When you use the ALPHA key, only capital letters are available on the TI-83 calculator.

Complex Numbers

To perform operations on complex numbers, first press MODE and then use the arrow keys to select a+bi. Then press ENTER 2nd QUIT.

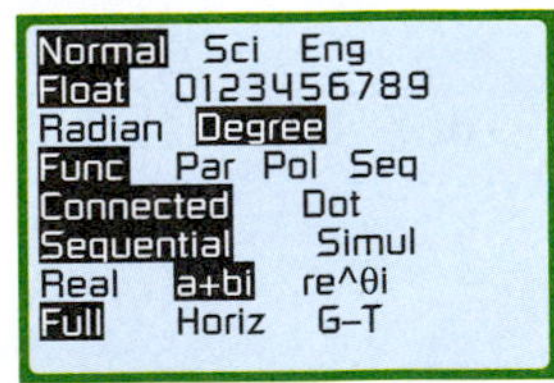

Addition of complex numbers To add $(3 + 4i) + (2 - 7i)$, use the keystrokes

(3 + 4 2nd *i*) +
(2 − 7 2nd *i*) ENTER.

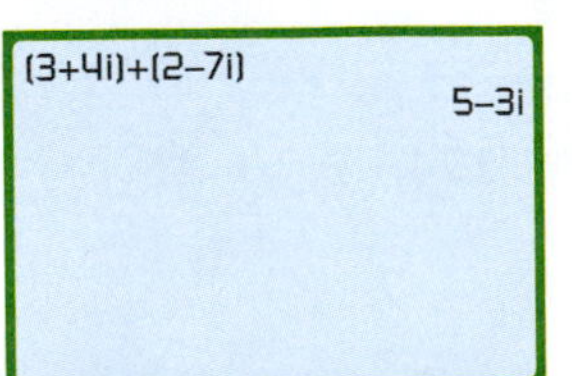

Division of complex numbers. To divide $\frac{26 + 2i}{2 + 4i}$, use the keystrokes (26 + 2 2nd *i*) ÷ (2 + 4 2nd *i*) ENTER.

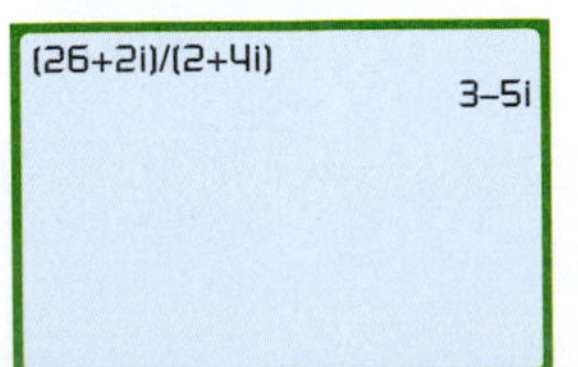

Note: Operations for subtraction and multiplication are similar.

Copyright © Houghton Mifflin Company. All rights reserved.

Additional operations on complex numbers can be found by selecting CPX under the MATH key.

To find the absolute value of $2 - 5i$, press MATH (scroll to CPX) (scroll to abs) ENTER (2 − 5 2nd i) ENTER.

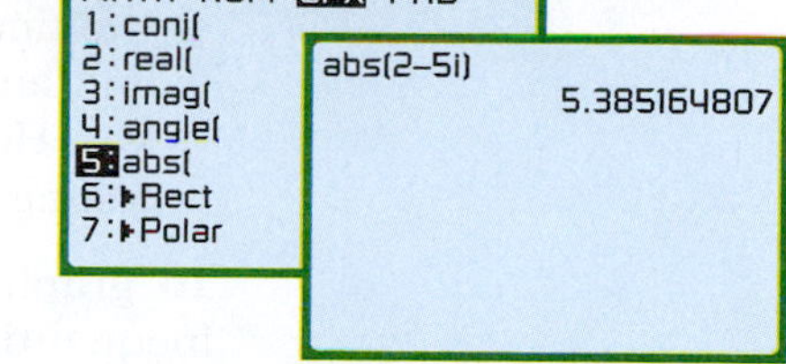
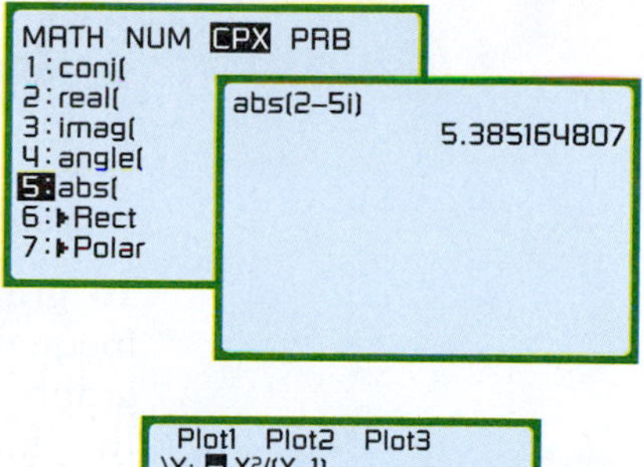

Evaluating Functions

There are various methods of evaluating a function but all methods require that the expression be entered as one of the ten functions Y_1 to Y_0. To evaluate $f(x) = \frac{x^2}{x - 1}$ when $x = -3$, enter the expression into, for instance, Y_1, and then press VARS ▸ 1 1 ((−) 3) ENTER.

> **TAKE NOTE**
> Use the down arrow key to scroll past Y_7 to see Y_8, Y_9, and Y_0.

Note: If you try to evaluate a function at a number that is not in the domain of the function, you will get an error message. For instance, 1 is not in the domain of $f(x) = \frac{x^2}{x - 1}$. If we try to evaluate the function at 1, the error screen at the right appears.

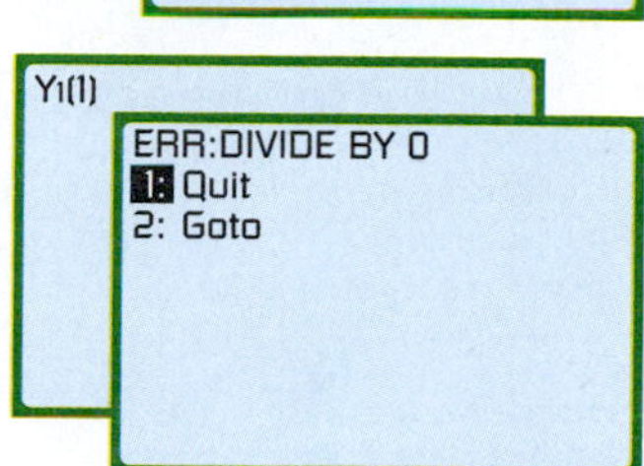

Evaluating Variable Expressions

To evaluate a variable expression, first store the values of each variable. Then enter the variable expression. For instance, to evaluate $s^2 + 2sl$ when $s = 4$ and $l = 5$, use the following keystrokes.

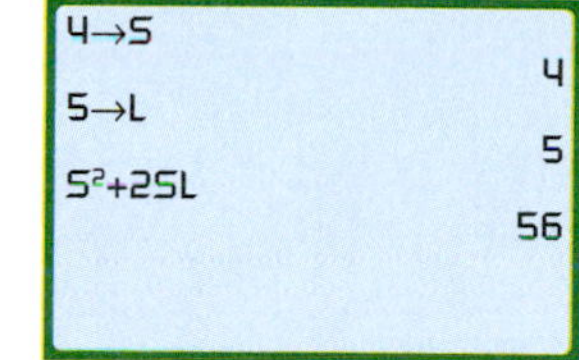

Graph

To graph a function, use the Y= key to enter the expression for the function, select a suitable viewing window, and then press GRAPH. For instance, to graph $f(x) = 0.1x^3 - 2x - 1$ in the standard viewing window, use the following keystrokes.

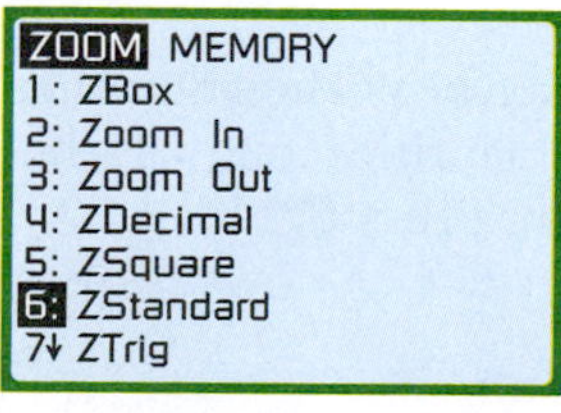
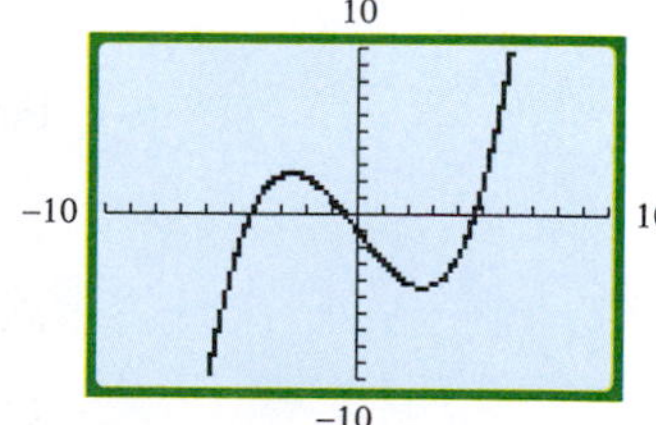

Note: For the keystrokes above, you do not have to scroll to 6. Alternatively, use ZOOM 6. This will select the standard viewing window and automatically start the graph. Use the WINDOW key to create a custom window for a graph.

Graphing Inequalities

To illustrate this feature, we will graph $y \leq 2x - 1$. Enter $2x - 1$ into Y_1. Because $y \leq 2x - 1$, we want to shade below the graph. Move the cursor to the left of Y_1 and press ENTER three times. Press GRAPH.

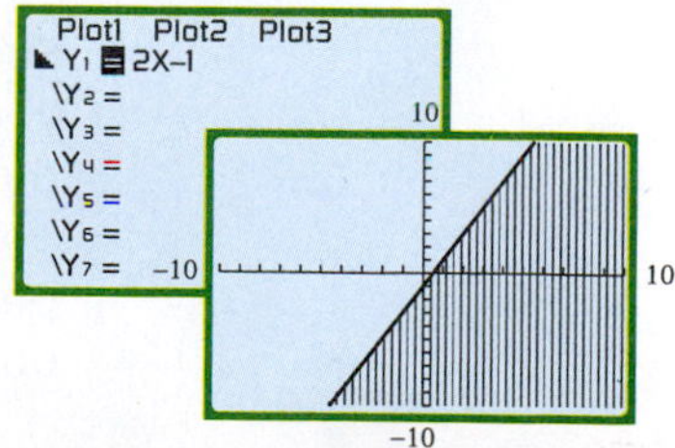

Copyright © Houghton Mifflin Company. All rights reserved.

Note: To shade above the graph, move the cursor to the left of Y_1 and press ENTER two times. An inequality with the symbol $\leq$ or $\geq$ should be graphed with a solid line, and an inequality with the symbol $<$ or $>$ should be graphed with a dashed line. However, the graph of a linear inequality on a graphing calculator does not distinguish between a solid line and a dashed line.

To graph the solution set of a system of inequalities, solve each inequality for y and graph each inequality. The solution set is the intersection of the two inequalities. The solution set of $\begin{aligned} 3x + 2y &> 10 \\ 4x - 3y &\leq 5 \end{aligned}$ is shown at the right.

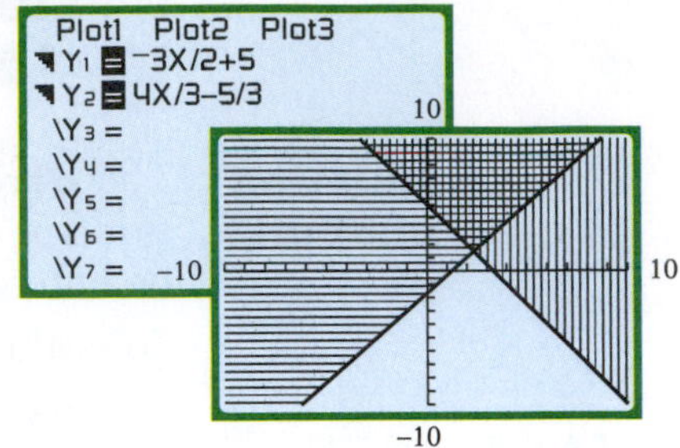

Intersect The INTERSECT feature is used to solve a system of equations. To illustrate this feature, we will use the system of equations $\begin{aligned} 2x - 3y &= 13 \\ 3x + 4y &= -6 \end{aligned}$.

Note: Some equations can be solved by this method. See the section "Solve an equation" below. Also, this method is used to find a number in the domain of a function for a given number in the range. See the section "Find a domain element."

Solve each of the equations in the system of equations for y. In this case, we have $y = \frac{2}{3}x - \frac{13}{3}$ and $y = -\frac{3}{4}x - \frac{3}{2}$.

Use the Y-editor to enter $\frac{2}{3}x - \frac{13}{3}$ into Y_1 and $-\frac{3}{4}x - \frac{3}{2}$ into Y_2. Graph the two functions in the standard viewing window. (If the window does not show the point of intersection of the two graphs, adjust the window until you can see the point of intersection.)

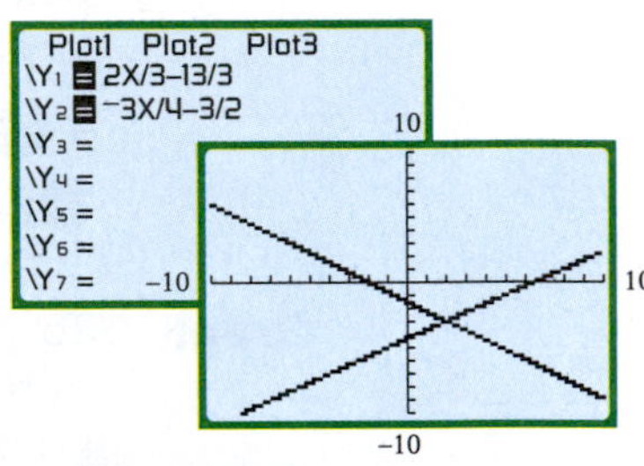

Press 2nd CALC (scroll to 5, intersect) ENTER.

Alternatively, you can just press 2nd CALC 5.

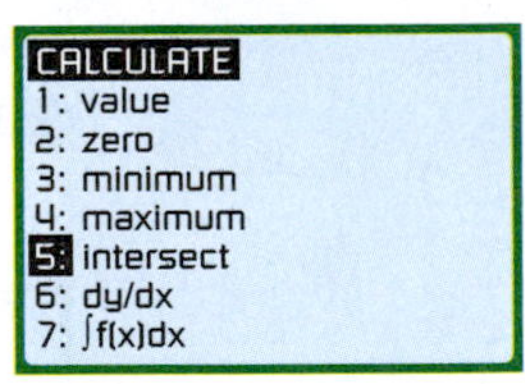

First curve? is shown at the bottom of the screen and identifies one of the two graphs on the screen. Press ENTER.

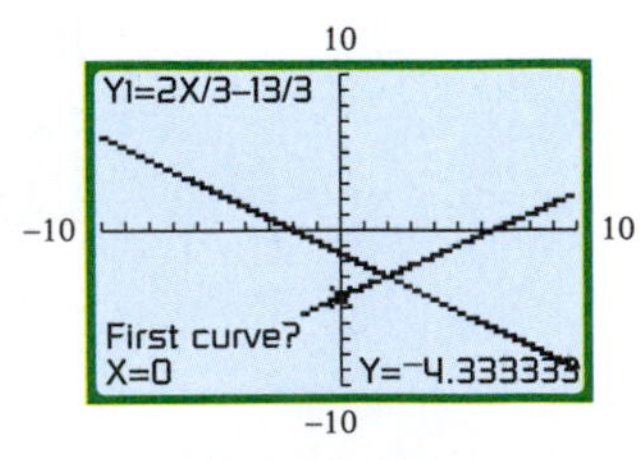

Second curve? is shown at the bottom of the screen and identifies the second of the two graphs on the screen. Press ENTER.

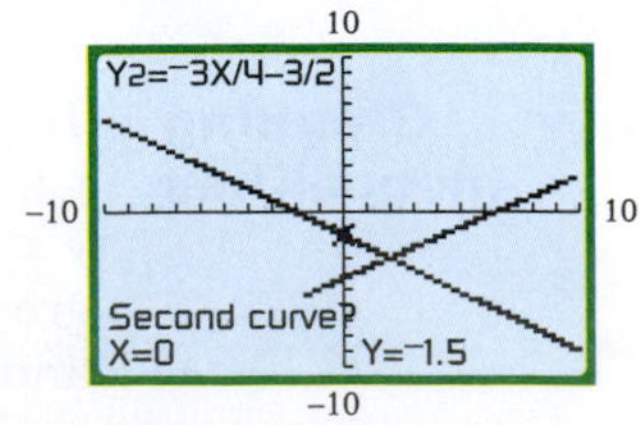

Guess? shown at the bottom of the screen asks you to use the left or right arrow key to move the cursor to the *approximate* location of the point of intersection. (If there are two or more points of intersection, it does not matter which one you choose first.) Press ENTER.

Copyright © Houghton Mifflin Company. All rights reserved.

The solution of the system of equations is $(2, -3)$.

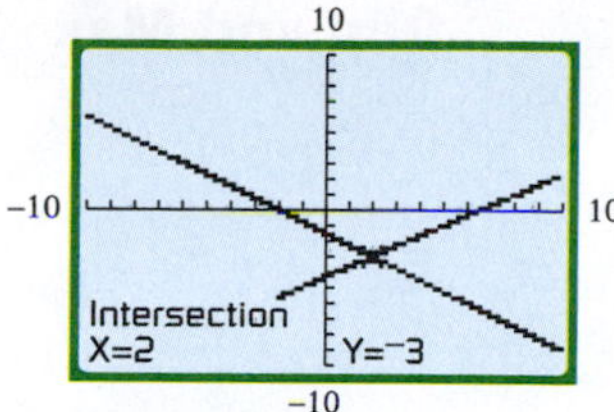

Solve an equation To illustrate the steps involved, we will solve the equation $2x + 4 = -3x - 1$. The idea is to write the equation as the system of equations $\begin{aligned} y &= 2x + 4 \\ y &= -3x - 1 \end{aligned}$ and then use the steps for solving a system of equations.

Use the Y-editor to enter the left and right sides of the equation into Y_1 and Y_2. Graph the two functions and then follow the steps for Intersect.

The solution is -1, the x-coordinate of the point of intersection.

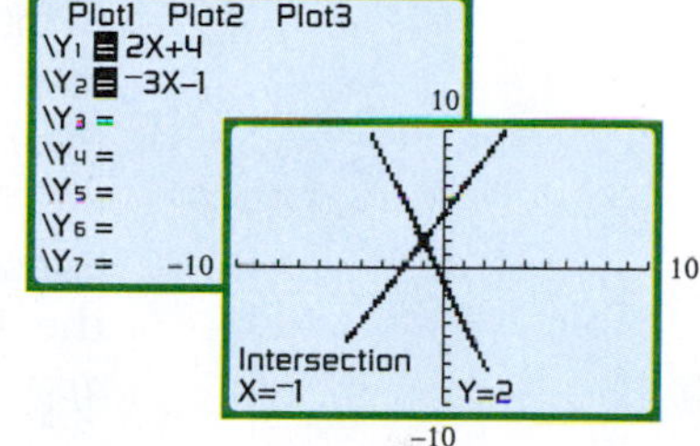

Find a domain element For this example, we will find a number in the domain of $f(x) = -\frac{2}{3}x + 2$ that corresponds to 4 in the range of the function. This is like solving the system of equations $y = -\frac{2}{3}x + 2$ and $y = 4$.

Use the Y= editor to enter the expression for the function in Y_1 and the desired output, 4, in Y_2. Graph the two functions and then follow the steps for Intersect.

The point of intersection is $(-3, 4)$. The number -3 in the domain of f produces an output of 4 in the range of f.

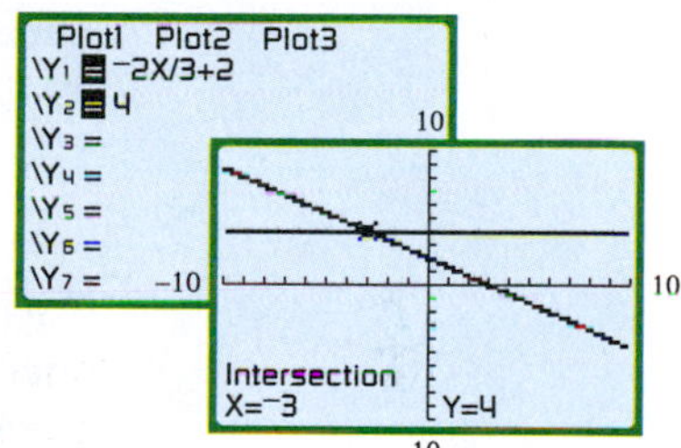

Math

Pressing MATH gives you access to many built-in functions. The following keystrokes will convert 0.125 to a fraction: .125 MATH 1 ENTER.

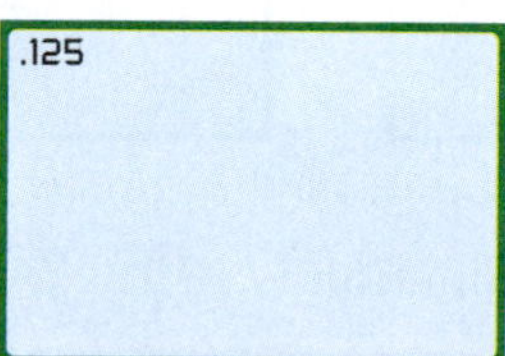

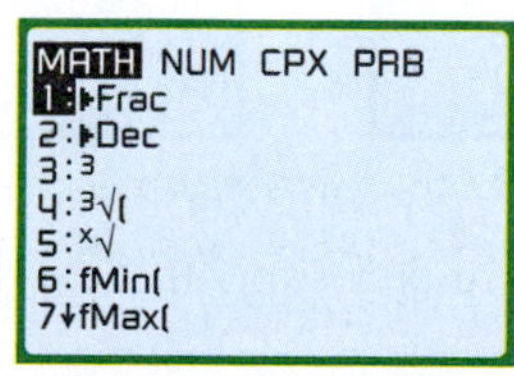

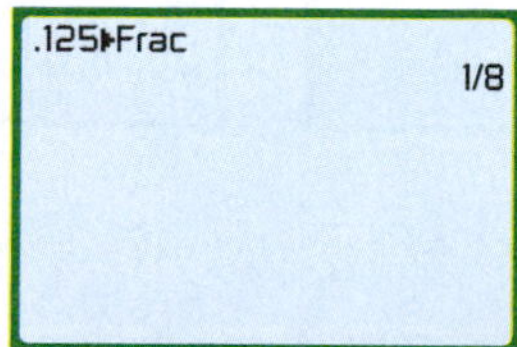

Additional built-in functions under MATH can be found by pressing MATH ▸. For instance, to evaluate $-|-25|$, press (−) MATH ▸ 1 (−) 25) ENTER.

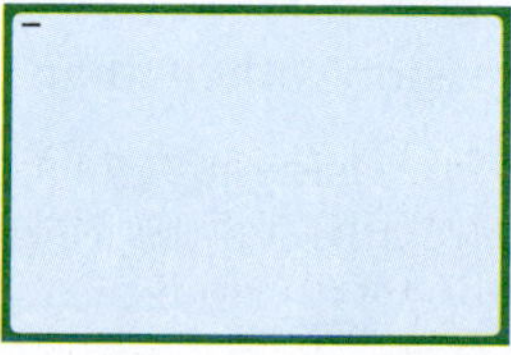

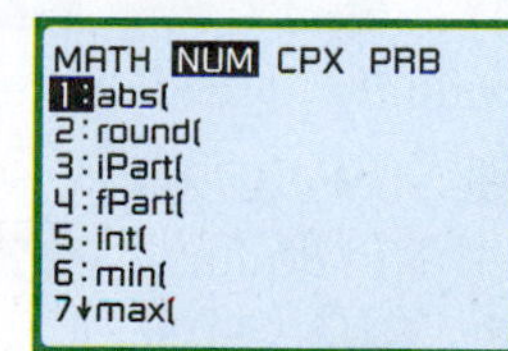

See your owner's manual for assistance with other functions under the MATH key.

Copyright © Houghton Mifflin Company. All rights reserved.

Min and Max

The local minimum and the local maximum values of a function are calculated by accessing the CALC menu. For this demonstration, we will find the minimum value and the maximum value of $f(x) = 0.2x^3 + 0.3x^2 - 3.6x + 2$.

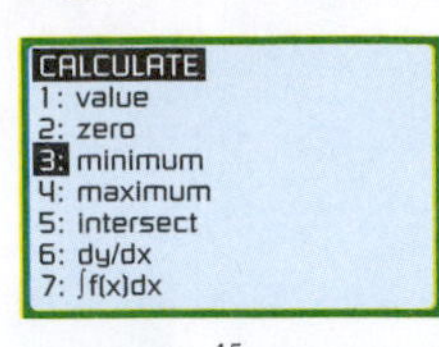

Enter the function into Y_1. Press [2nd] CALC (scroll to 3 for minimum of the function) [ENTER].

Alternatively, you can just press [2nd] CALC 3.

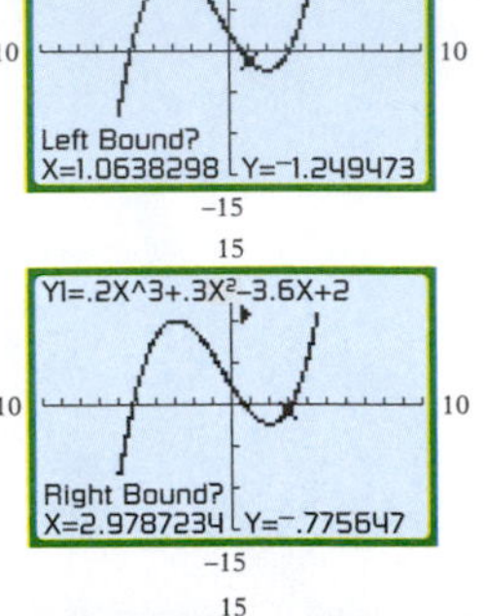

Left Bound? shown at the bottom of the screen asks you to use the left or right arrow key to move the cursor to the *left* of the minimum. Press [ENTER].

Right Bound? shown at the bottom of the screen asks you to use the left or right arrow key to move the cursor to the *right* of the minimum. Press [ENTER].

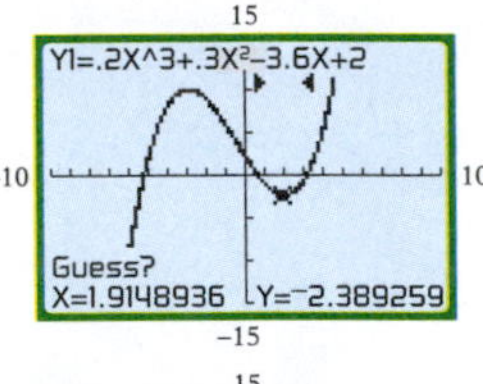

Guess? shown at the bottom of the screen asks you to use the left or right arrow key to move the cursor to the *approximate* location of the minimum. Press [ENTER].

The minimum value of the function is the y-coordinate. For this example, the minimum value of the function is -2.4.

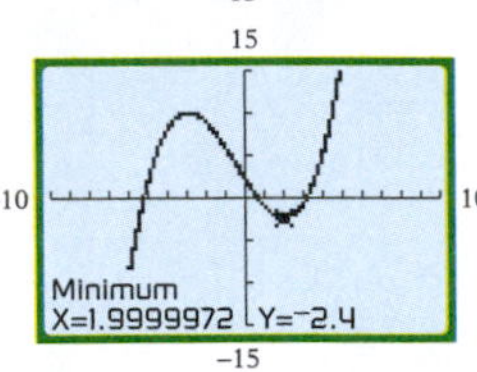

The x-coordinate for the minimum is 2. However, because of rounding errors in the calculation, it is shown as a number close to 2.

To find the maximum value of the function, follow the same steps as above except select maximum under the CALC menu. The screens for this calculation are shown below.

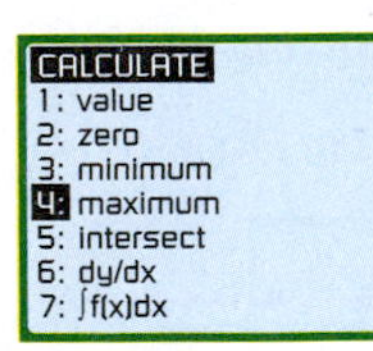

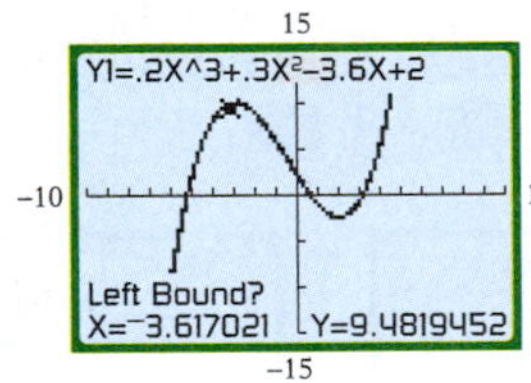

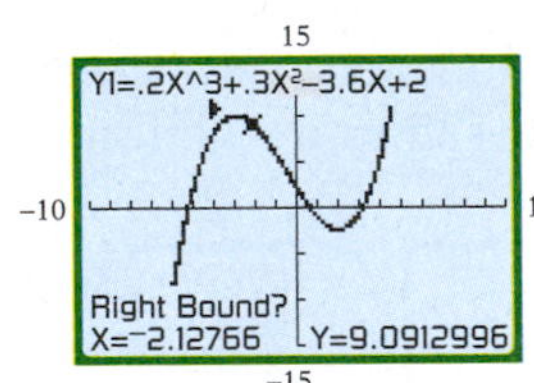

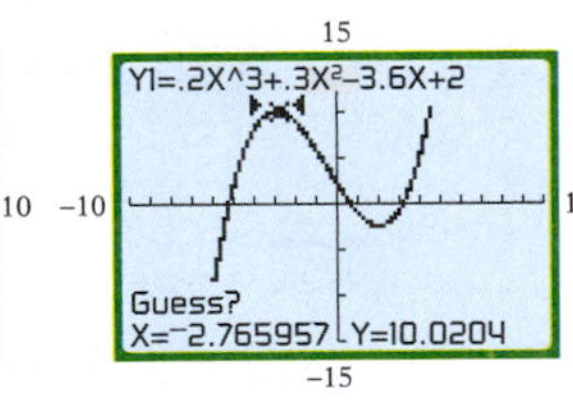

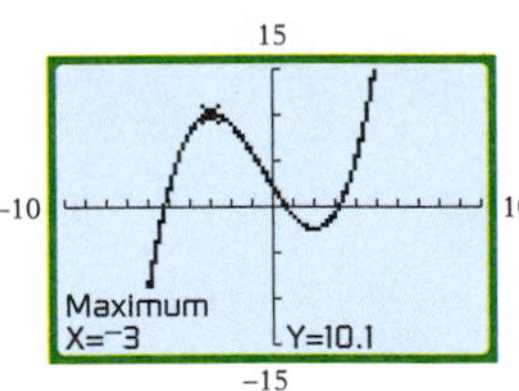

The maximum value of the function is 10.1.

Radical Expressions

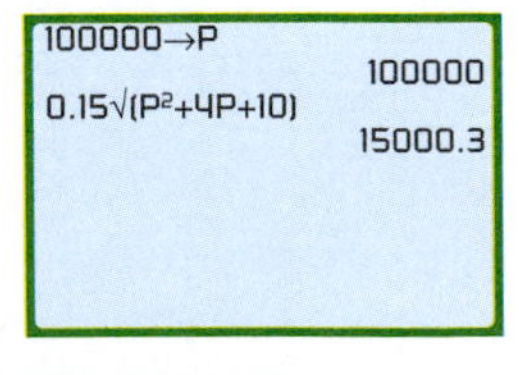

To evaluate a square-root expression, press [2nd] $\sqrt{\ }$.

For instance, to evaluate $0.15\sqrt{p^2 + 4p + 10}$ when $p = 100{,}000$, first store 100,000 in P. Then press 0.15 [2nd] $\sqrt{\ }$ [ALPHA] P [x^2] [+] 4 [ALPHA] P [+] 10 [ )] [ENTER].

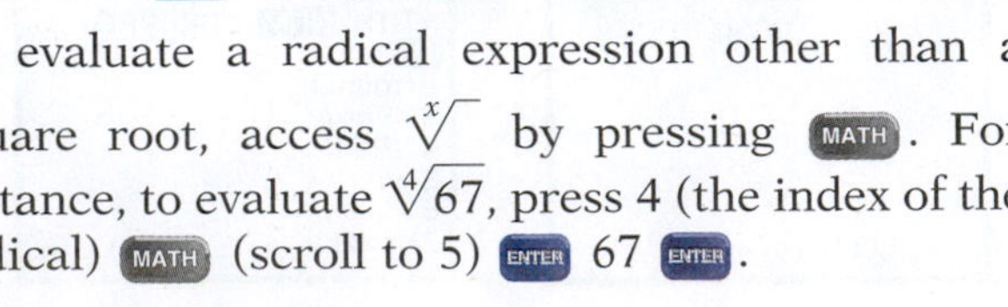

To evaluate a radical expression other than a square root, access $\sqrt[x]{\ }$ by pressing [MATH]. For instance, to evaluate $\sqrt[4]{67}$, press 4 (the index of the radical) [MATH] (scroll to 5) [ENTER] 67 [ENTER].

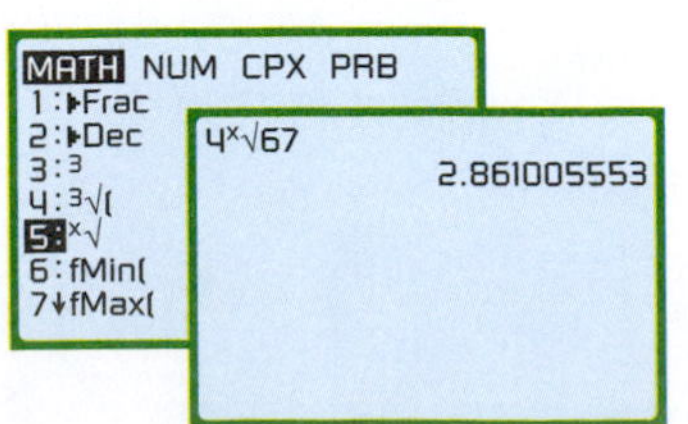

Copyright © Houghton Mifflin Company. All rights reserved.

Scientific Notation To enter a number in scientific notation, use 2nd EE. For instance, to find $\frac{3.45 \times 10^{-12}}{1.5 \times 10^{25}}$, press 3.45 2nd EE (-) 12 ÷ 1.5 2nd EE 25 ENTER. The answer is 2.3×10^{-37}.

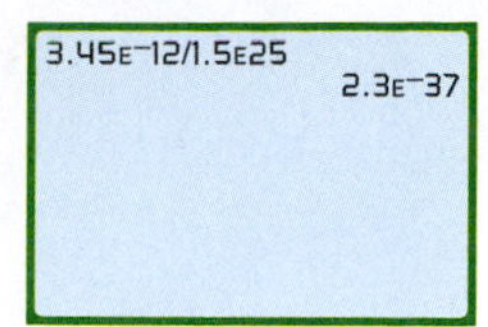

Table There are three steps in creating an input/output table for a function. First use the Y= editor to input the function. The second step is setting up the table, and the third step is displaying the table.

To set up the table, press 2nd TBLSET. TblStart is the first value of the independent variable in the input/output table. ΔTbl is the difference between successive values. Setting this to 1 means that, for this table, the input values are −2, −1, 0, 1, 2. . . . If ΔTbl = 0.5, then the input values are −2, −1.5, −1, −0.5, 0, 0.5, . . .

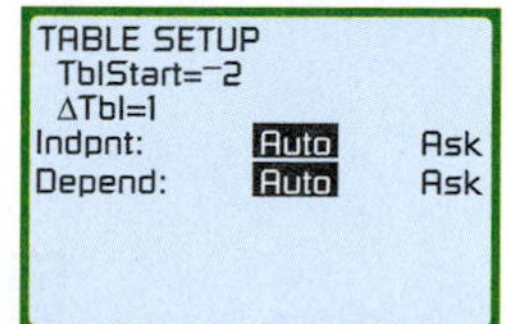

Indpnt is the independent variable. When this is set to Auto, values of the independent variable are automatically entered into the table. Depend is the dependent variable. When this is set to Auto, values of the dependent variable are automatically entered into the table.

To display the table, press 2nd TABLE. An input/output table for $f(x) = x^2 - 1$ is shown at the right.

Once the table is on the screen, the up and down arrow keys can be used to display more values in the table. For the table at the right, we used the up arrow key to move to $x = -7$.

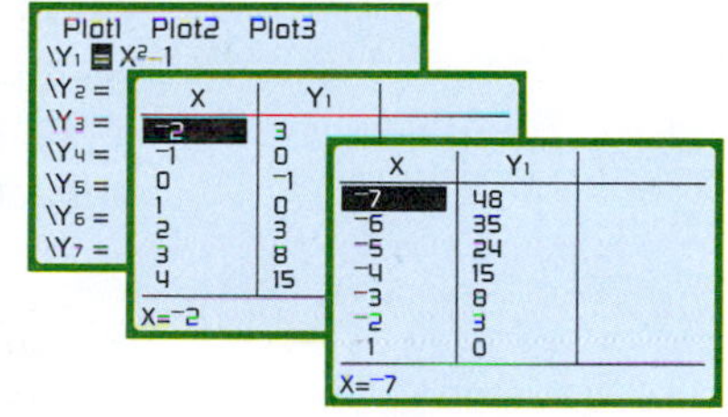

An input/output table for any given input can be created by selecting Ask for the independent variable. The table at the right shows an input/output table for $f(x) = \frac{4x}{x-2}$ for selected values of x. Note the word ERROR or ERR: when 2 is entered. This occurred because f is not defined when $x = 2$.

Note: Using the table feature in Ask mode is the same as evaluating a function for given values of the independent variable. For instance, from the table at the right, we have $f(4) = 8$.

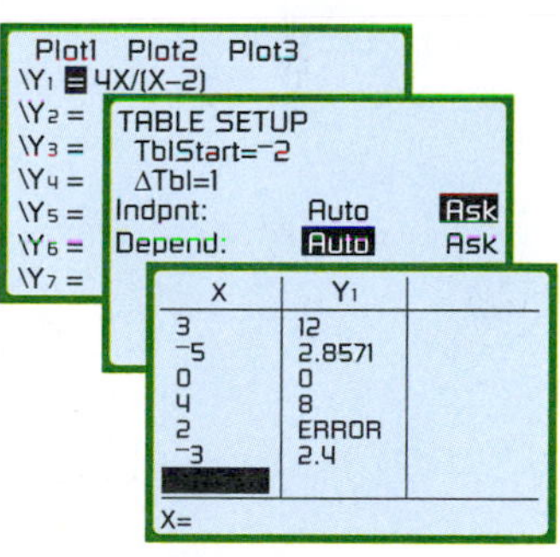

Test The TEST feature has many uses, one of which is to graph the solution set of a linear inequality in one variable. To illustrate this feature, we will graph the solution set of $x - 1 < 4$. Press Y= X,T,θ,n − 1 2nd TEST (scroll to 5) ENTER 4 GRAPH.

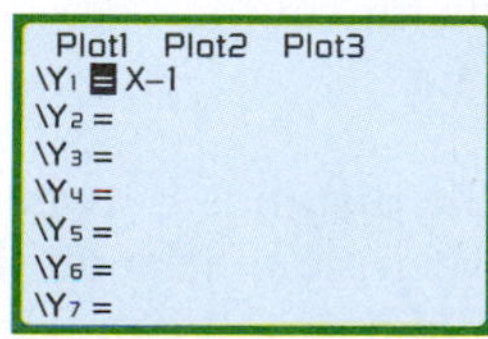

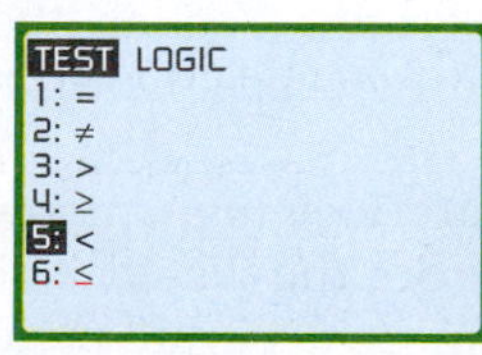

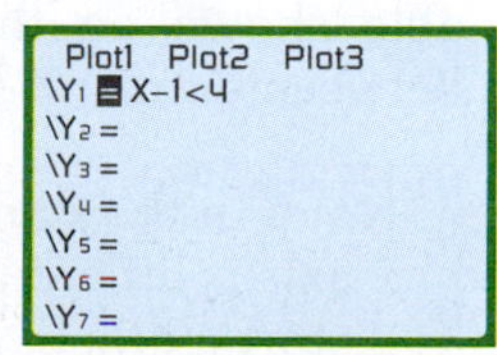

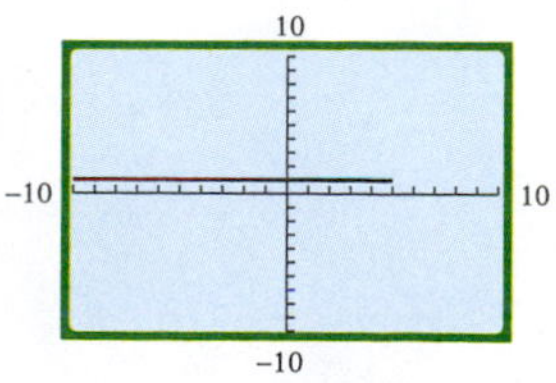

Copyright © Houghton Mifflin Company. All rights reserved.

Trace Once a graph is drawn, pressing TRACE will place a cursor on the screen, and the coordinates of the point below the cursor are shown at the bottom of the screen. Use the left and right arrow keys to move the cursor along the graph. For the graph at the right, we have $f(4.8) = 3.4592$, where $f(x) = 0.1x^3 - 2x + 2$ is shown at the top left of the screen.

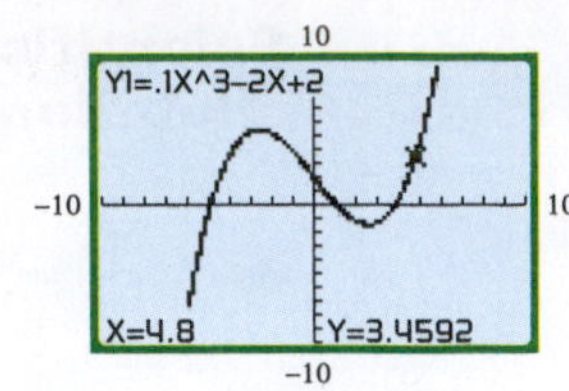

In TRACE mode, you can evaluate a function at any value of the independent variable that is within Xmin and Xmax. To do this, first graph the function. Now press TRACE (the value of x) ENTER. For the graph at the left below, we used $x = -3.5$. If a value of x is chosen outside the window, an error message is displayed.

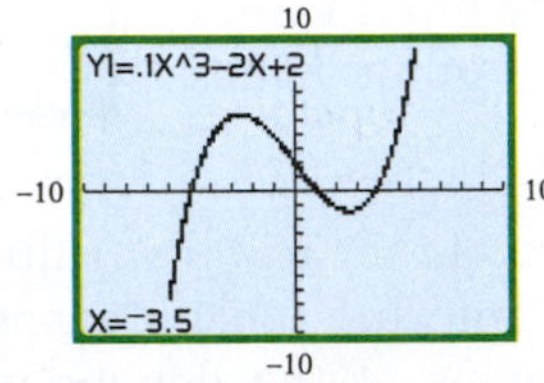

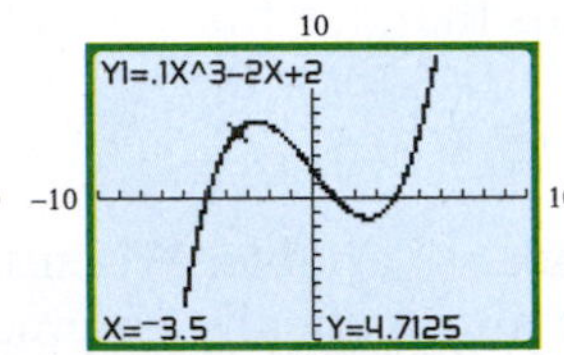

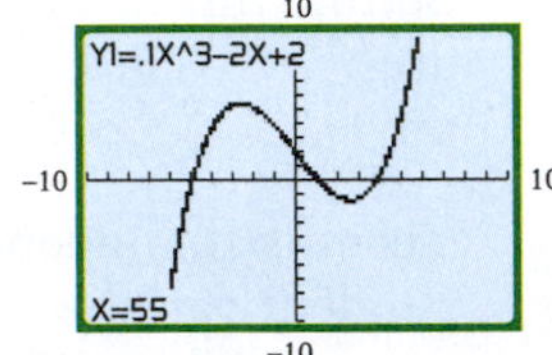

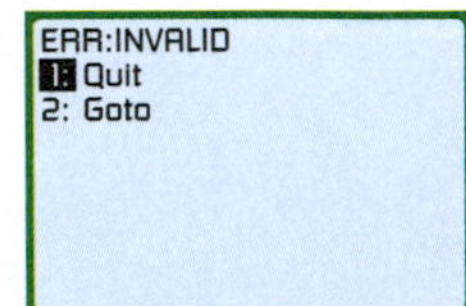

In the example above where we entered -3.5 for x, the value of the function was calculated as 4.7125. This means that $f(-3.5) = 4.7125$. The keystrokes 2nd QUIT VARS ◊ 11 MATH 1 ENTER will convert the decimal value to a fraction.

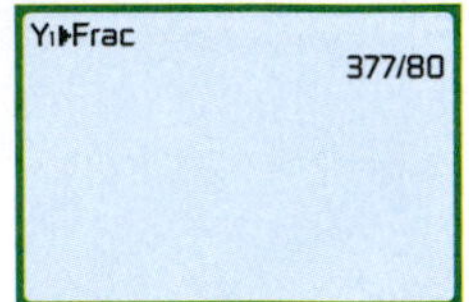

When the TRACE feature is used with two or more graphs, the up and down arrow keys are used to move between the graphs. The graphs below are for the functions $f(x) = 0.1x^3 - 2x + 2$ and $g(x) = 2x - 3$. By using the up and down arrows, we can place the cursor on either graph. The right and left arrows are used to move along the graph.

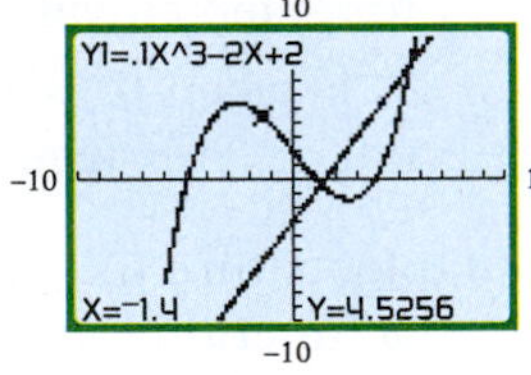

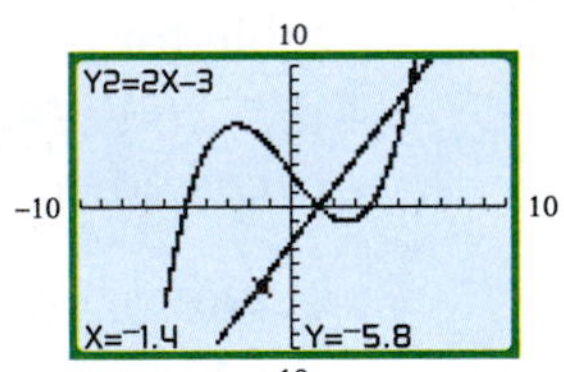

Window The viewing window for a graph is controlled by pressing WINDOW. Xmin and Xmax are the minimum value and maximum value, respectively, of the independent variable shown on the graph. Xscl is the distance between tic marks on the x-axis. Ymin and Ymax are the minimum value and maximum value, respectively, of the dependent variable shown on the graph. Yscl is the distance between tic marks on the y-axis. Leave Xres as 1.

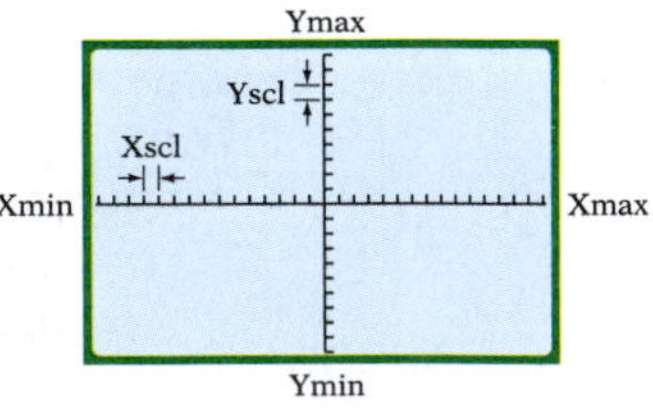

Note: In the standard viewing window, the distance between tic marks on the x-axis is different from the distance between tic marks on the y-axis. This will distort a graph. A more accurate picture of a graph can be created by using a square viewing window. See ZOOM.

Copyright © Houghton Mifflin Company. All rights reserved.

Y= The Y= editor is used to enter the expression for a function. There are ten possible functions, labeled Y_1 to Y_0, that can be active at any one time. For instance, to enter $f(x) = x^2 + 3x - 2$ as Y_1, use the following keystrokes.

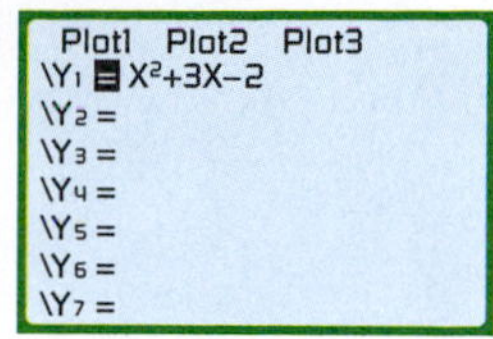

Y= X,T,θ,n x^2 + 3 X,T,θ,n − 2

Note: If an expression is already entered for Y_1, place the cursor anywhere on that expression and press CLEAR.

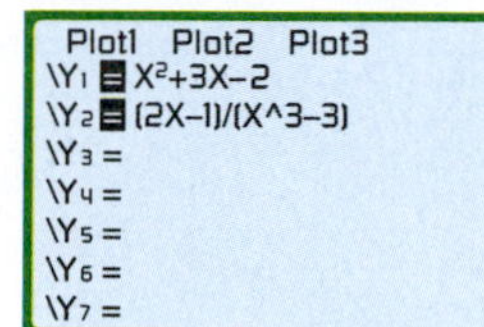

To enter $s = \frac{2v - 1}{v^3 - 3}$ into Y_2, place the cursor to the right of the equals sign for Y_2. Then press (2 X,T,θ,n − 1) ÷ (X,T,θ,n ^ 3 − 3).

Note: When we enter an equation, the independent variable, v in the expression above, is entered using X,T,θ,n. The dependent variable, s in the expression above, is one of Y_1 to Y_0. Also note the use of parentheses to ensure the correct order of operations.

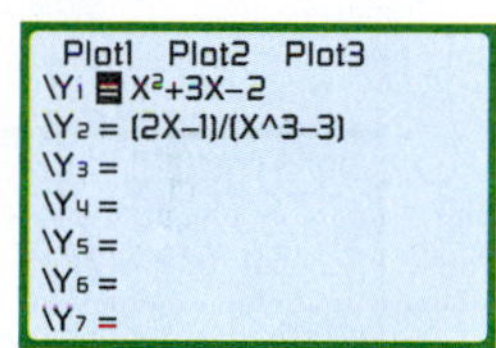

Observe the black rectangle that covers the equals sign for the two examples we have shown. This rectangle means that the function is "active." If we were to press GRAPH, then the graph of both functions would appear. You can make a function inactive by using the arrow keys to move the cursor over the equals sign of that function and then pressing ENTER. This will remove the black rectangle. We have done that for Y_2, as shown at the right. Now if GRAPH is pressed, only Y_1 will be graphed.

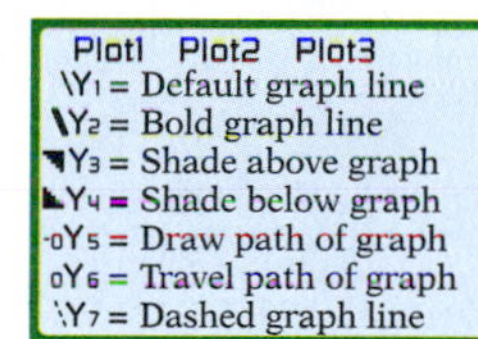

It is also possible to control the appearance of the graph by moving the cursor on the Y= screen to the left of any Y. With the cursor in this position, pressing ENTER will change the appearance of the graph. The options are shown at the right.

Zero

The ZERO feature of a graphing calculator is used for various calculations: to find the x-intercepts of a function, to solve some equations, and to find the zero of a function.

***x*-intercepts** To illustrate the procedure for finding x-intercepts, we will use $f(x) = x^2 + x - 2$.

First, use the Y-editor to enter the expression for the function and then graph the function in the standard viewing window. (It may be necessary to adjust this window so that the intercepts are visible.) Once the graph is displayed, use the keystrokes below to find the x-intercepts of the graph of the function.

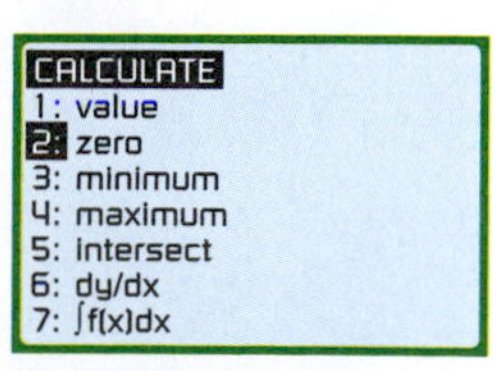

Press 2nd CALC (scroll to 2 for zero of the function) ENTER.

Alternatively, you can just press 2nd CALC 2.

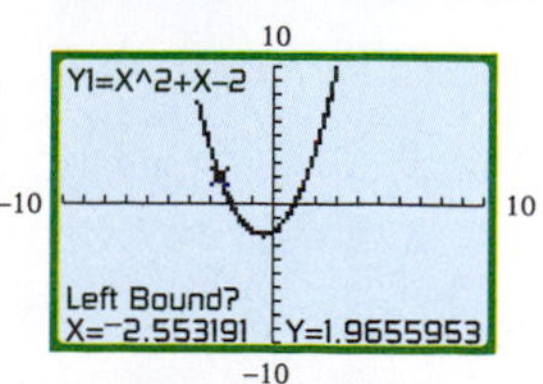

Left Bound? shown at the bottom of the screen asks you to use the left or right arrow key to move the cursor to the *left* of the desired x-intercept. Press ENTER.

Copyright © Houghton Mifflin Company. All rights reserved.

Right Bound? shown at the bottom of the screen asks you to use the left or right arrow key to move the cursor to the *right* of the desired x-intercept. Press ENTER.

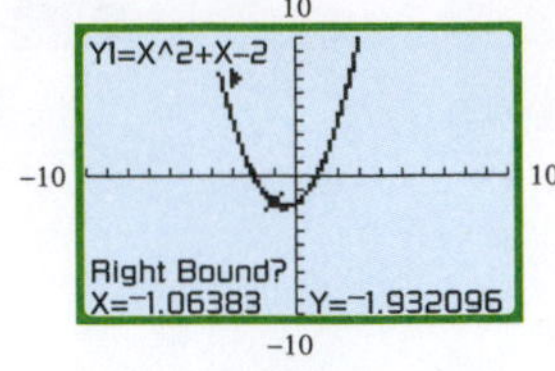

Guess? shown at the bottom of the screen asks you to use the left or right arrow key to move the cursor to the *approximate* location of the desired x-intercept. Press ENTER.

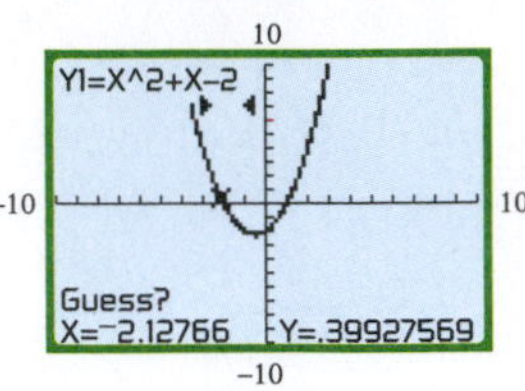

The x-coordinate of an x-intercept is -2. Therefore, an x-intercept is $(-2, 0)$.

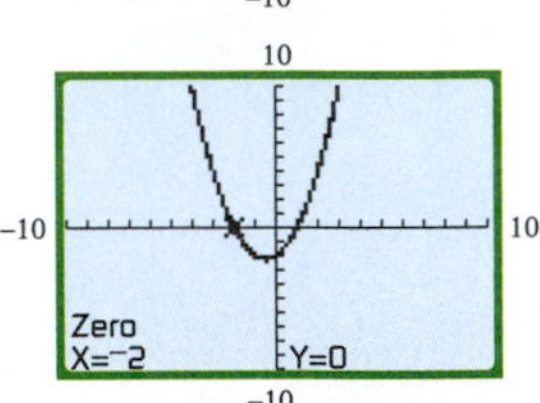

To find the other x-intercept, follow the same steps as above. The screens for this calculation are shown below.

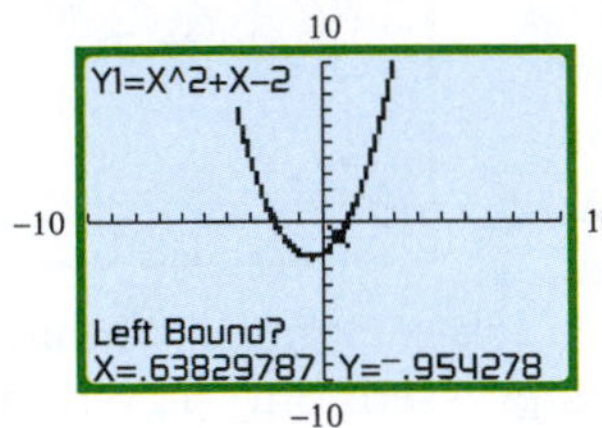

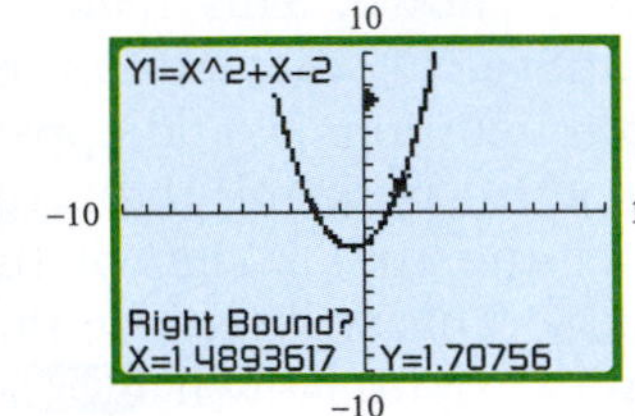

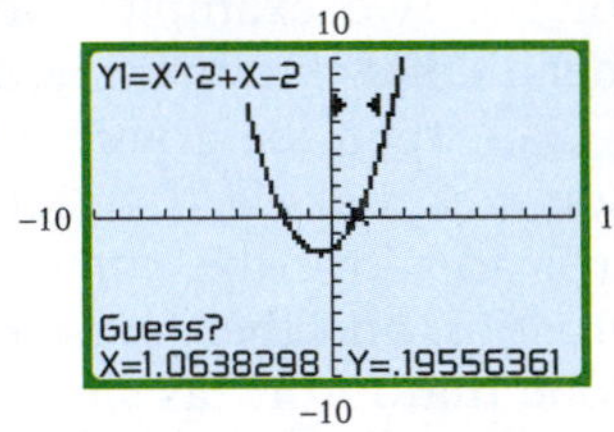

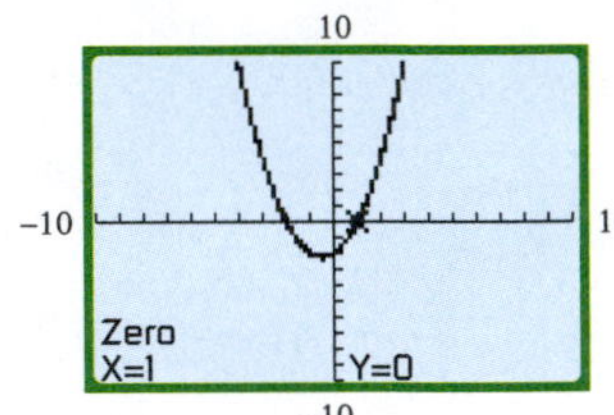

A second x-intercept is $(1, 0)$.

Solve an equation To use the ZERO feature to solve an equation, first rewrite the equation with all terms on one side. For instance, one way to solve the equation $x^3 - x + 1 = -2x + 3$ is first to rewrite it as $x^3 + x - 2 = 0$. Enter $x^3 + x - 2$ into Y1 and then follow the steps for finding x-intercepts.

Find the real zeros of a function To find the real zeros of a function, follow the steps for finding x-intercepts.

Zoom

Pressing ZOOM allows you to select some preset viewing windows. This key also gives you access to ZBox, Zoom In, and Zoom Out. These functions enable you to redraw a selected portion of a graph in a new window. Some windows used frequently in this text are shown below.

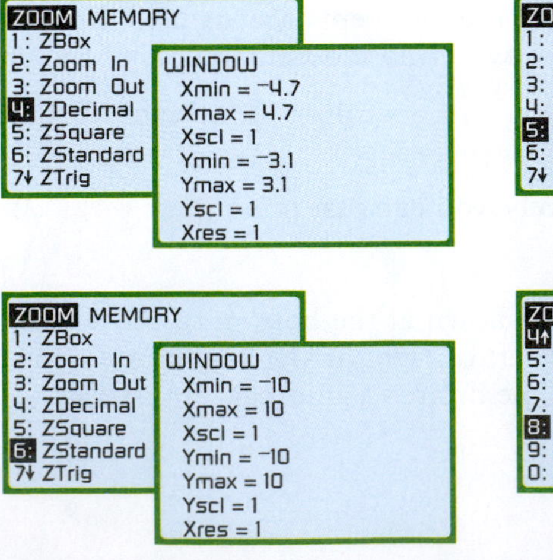

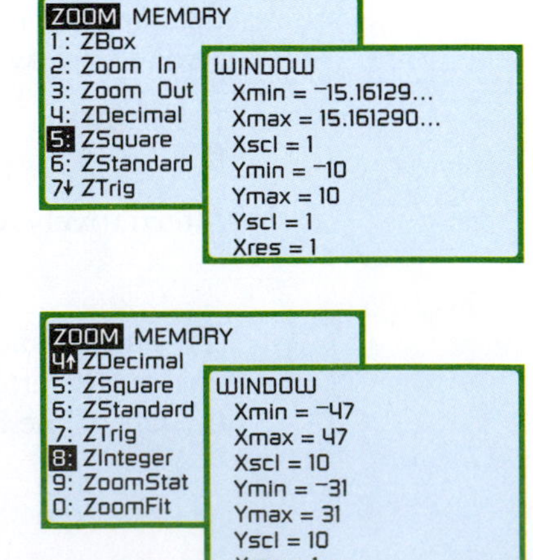

Copyright © Houghton Mifflin Company. All rights reserved.

Appendix B

Proofs and Tables

Proofs of Logarithmic Properties

In each of the following proofs of logarithmic properties, it is assumed that the Properties of Exponents are true for all real number exponents.

The Logarithm Property of the Product of Two Numbers

For any positive real numbers x, y, and b, $b \neq 1$, $\log_b xy = \log_b x + \log_b y$.

Proof: Let $\log_b x = m$ and $\log_b y = n$.

Write each equation in its equivalent exponential form. $\quad x = b^m \quad y = b^n$

Use substitution and the Properties of Exponents. $\quad xy = b^m b^n$

$xy = b^{m+n}$

Write the equation in its equivalent logarithmic form. $\quad \log_b xy = m + n$

Substitute $\log_b x$ for m and $\log_b y$ for n. $\quad \log_b xy = \log_b x + \log_b y$

The Logarithm Property of the Quotient of Two Numbers

For any positive real numbers x, y, and b, $b \neq 1$, $\log_b \frac{x}{y} = \log_b x - \log_b y$.

Proof: Let $\log_b x = m$ and $\log_b y = n$.

Write each equation in its equivalent exponential form. $\quad x = b^m \quad y = b^n$

Use substitution and the Properties of Exponents. $\quad \frac{x}{y} = \frac{b^m}{b^n}$

$\frac{x}{y} = b^{m-n}$

Write the equation in its equivalent logarithmic form. $\quad \log_b \frac{x}{y} = m - n$

Substitute $\log_b x$ for m and $\log_b y$ for n. $\quad \log_b \frac{x}{y} = \log_b x - \log_b y$

The Logarithm Property of the Power of a Number

For any real numbers x, r, and b, $b \neq 1$, $\log_b x^r = r \log_b x$.

Proof: Let $\log_b x = m$.

Write the equation in its equivalent exponential form. $\quad x = b^m$

Raise both sides to the r power. $\quad x^r = (b^m)^r$

$x^r = b^{mr}$

Write the equation in its equivalent logarithmic form. $\quad \log_b x^r = mr$

Substitute $\log_b x$ for m. $\quad \log_b x^r = r \log_b x$

Copyright © Houghton Mifflin Company. All rights reserved.

Table of Symbols

$+$	add
$-$	subtract
$\cdot, \times, (a)(b)$	multiply
$\frac{a}{b}, \div$	divide
$(\;)$	parentheses, a grouping symbol
$[\;]$	brackets, a grouping symbol
π	pi, a number approximately equal to $\frac{22}{7}$ or 3.14
$-a$	the opposite, or additive inverse, of a
$\frac{1}{a}$	the reciprocal, or multiplicative inverse, of a
$=$	is equal to
$\approx$	is approximately equal to
$\neq$	is not equal to
$<$	is less than
$\leq$	is less than or equal to
$>$	is greater than
$\geq$	is greater than or equal to
(a, b)	an ordered pair whose first component is a and whose second component is b
$^\circ$	degree (for angles)
$\sqrt{a}$	the principal square root of a
$\varnothing, \{\;\}$	the empty set
$\lvert a \rvert$	the absolute value of a
$\cup$	union of two sets
$\cap$	intersection of two sets
$\in$	is an element of (for sets)
$\notin$	is not an element of (for sets)

Table of Measurement Abbreviations

U.S. Customary System

Length		Capacity		Weight		Area	
in.	inches	oz	fluid ounces	oz	ounces	in^2	square inches
ft	feet	c	cups	lb	pounds	ft^2	square feet
yd	yards	qt	quarts				
mi	miles	gal	gallons				

Metric System

Length		Capacity		Weight/Mass		Area	
mm	millimeter (0.001 m)	ml	milliliter (0.001 L)	mg	milligram (0.001 g)	cm^2	square centimeters
cm	centimeter (0.01 m)	cl	centiliter (0.01 L)	cg	centigram (0.01 g)	m^2	square meters
dm	decimeter (0.1 m)	dl	deciliter (0.1 L)	dg	decigram (0.1 g)		
m	meter	L	liter	g	gram		
dam	decameter (10 m)	dal	decaliter (10 L)	dag	decagram (10 g)		
hm	hectometer (100 m)	hl	hectoliter (100 L)	hg	hectogram (100 g)		
km	kilometer (1000 m)	kl	kiloliter (1000 L)	kg	kilogram (1000 g)		

Time

h	hours	min	minutes	s	seconds

Copyright © Houghton Mifflin Company. All rights reserved.

Table of Properties

Properties of Real Numbers

The Associative Property of Addition
If a, b, and c are real numbers, then
$(a + b) + c = a + (b + c)$.

The Associative Property of Multiplication
If a, b, and c are real numbers, then
$(a \cdot b) \cdot c = a \cdot (b \cdot c)$.

The Commutative Property of Addition
If a and b are real numbers, then
$a + b = b + a$.

The Commutative Property of Multiplication
If a and b are real numbers, then
$a \cdot b = b \cdot a$.

The Addition Property of Zero
If a is a real number, then
$a + 0 = 0 + a = a$.

The Multiplication Property of One
If a is a real number, then
$a \cdot 1 = 1 \cdot a = a$.

The Multiplication Property of Zero
If a is a real number, then
$a \cdot 0 = 0 \cdot a = 0$.

The Inverse Property of Multiplication
If a is a real number and $a \neq 0$, then
$a \cdot \frac{1}{a} = \frac{1}{a} \cdot a = 1$.

The Inverse Property of Addition
If a is a real number, then
$a + (-a) = (-a) + a = 0$.

Distributive Property
If a, b, and c are real numbers, then
$a(b + c) = ab + ac$.

Properties of Equations

Addition Property of Equations
If $a = b$, then $a + c = b + c$.

Multiplication Property of Equations
If $a = b$ and $c \neq 0$, then $a \cdot c = b \cdot c$.

Properties of Inequalities

Addition Property of Inequalities
If $a > b$, then $a + c > b + c$.
If $a < b$, then $a + c < b + c$.

Multiplication Property of Inequalities
If $a > b$ and $c > 0$, then $ac > bc$.
If $a < b$ and $c > 0$, then $ac < bc$.
If $a > b$ and $c < 0$, then $ac < bc$.
If $a < b$ and $c < 0$, then $ac > bc$.

Properties of Exponents

If m and n are integers, then $x^m \cdot x^n = x^{m+n}$.
If m and n are integers, then $(x^m)^n = x^{mn}$.
If $x \neq 0$, then $x^0 = 1$.
If m and n are integers and $x \neq 0$, then $\frac{x^m}{x^n} = x^{m-n}$.

If m, n, and p are integers, then $(x^m \cdot y^n)^p = x^{mp}y^{np}$.
If n is a positive integer and $x \neq 0$, then
$x^{-n} = \frac{1}{x^n}$ and $\frac{1}{x^{-n}} = x^n$.
If m, n, and p are integers and $y \neq 0$, then $\left(\frac{x^m}{y^n}\right)^p = \frac{x^{mp}}{y^{np}}$.

Principle of Zero Products

If $a \cdot b = 0$, then $a = 0$ or $b = 0$.

Properties of Radical Expressions

If a and b are positive real numbers, then $\sqrt{ab} = \sqrt{a}\sqrt{b}$.

If a and b are positive real numbers, then $\sqrt{\frac{a}{b}} = \frac{\sqrt{a}}{\sqrt{b}}$.

Property of Squaring Both Sides of an Equation

If a and b are real numbers and $a = b$, then $a^2 = b^2$.

Properties of Logarithms

If x, y, and b are positive real numbers and $b \neq 1$, then
$\log_b(xy) = \log_b x + \log_b y$.
If x, y, and b are positive real numbers and $b \neq 1$, then
$\log_b \frac{x}{y} = \log_b x - \log_b y$.

If x and b are positive real numbers, $b \neq 1$, and r is any real number, then $\log_b x^r = r \log_b x$.
If x and b are positive real numbers and $b \neq 1$, then $\log_b b^x = x$.

Copyright © Houghton Mifflin Company. All rights reserved.

Table of Algebraic and Geometric Formulas

Slope of a Line

$m = \frac{y_2 - y_1}{x_2 - x_1}, x_1 \neq x_2$

Point-Slope Formula for a Line

$y - y_1 = m(x - x_1)$

Quadratic Formula

$x = \frac{-b \pm \sqrt{b^2 - 4ac}}{2a}$

discriminant $= b^2 - 4ac$

Perimeter and Area of a Triangle, and Sum of the Measures of the Angles

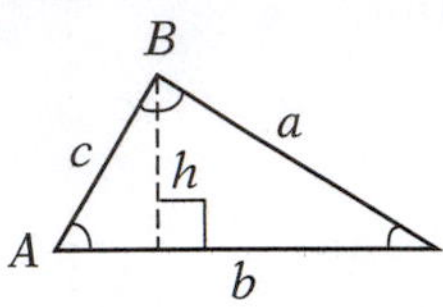

$P = a + b + c$

$A = \frac{1}{2}bh$

$A + B + C = 180°$

Pythagorean Theorem

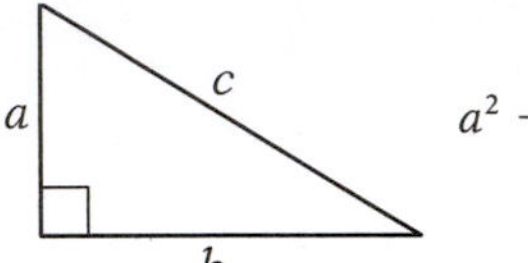

$a^2 + b^2 = c^2$

Perimeter and Area of a Rectangle

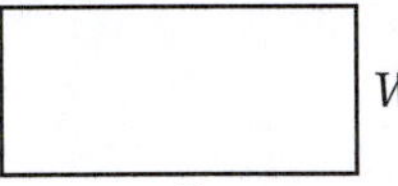

$P = 2L + 2W$
$A = LW$

Perimeter and Area of a Square

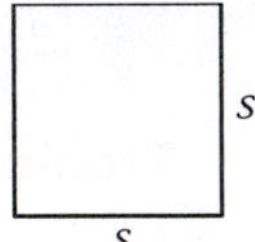

$P = 4s$
$A = s^2$

Area of a Trapezoid

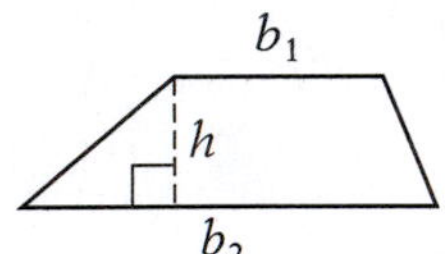

$A = \frac{1}{2}h(b_1 + b_2)$

Circumference and Area of a Circle

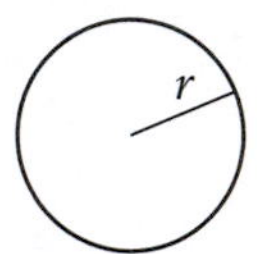

$C = 2\pi r$
$A = \pi r^2$

Volume and Surface Area of a Rectangular Solid

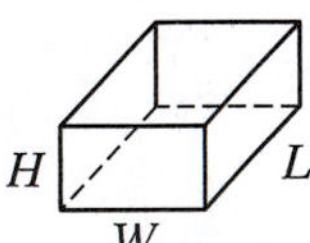

$V = LWH$
$SA = 2LW + 2LH + 2WH$

Volume and Surface Area of a Sphere

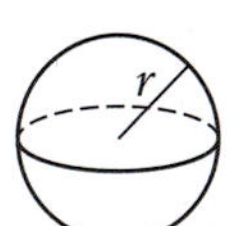

$V = \frac{4}{3}\pi r^3$

$SA = 4\pi r^2$

Volume and Surface Area of a Right Circular Cylinder

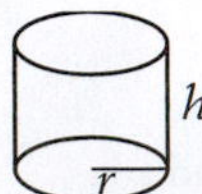

$V = \pi r^2 h$
$SA = 2\pi r^2 + 2\pi rh$

Volume and Surface Area of a Right Circular Cone

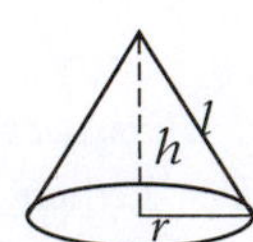

$V = \frac{1}{3}\pi r^2 h$

$SA = \pi r^2 + \pi rl$

Copyright © Houghton Mifflin Company. All rights reserved.

Solutions to Chapter 1 "You Try It"

SECTION 1.1

You Try It 1 Replace y by each of the elements of the set and determine whether the inequality is true.

$$\begin{aligned} y &> -1 \\ -5 &> -1 \text{ False} \\ -1 &> -1 \text{ False} \\ 5 &> -1 \text{ True} \end{aligned}$$

The inequality is true for 5.

You Try It 2 Replace z by each element of the set and determine the value of the expression.

$-z$	$\lvert z \rvert$
$-(-11) = 11$	$\lvert -11 \rvert = 11$
$-(0) = 0$	$\lvert 0 \rvert = 0$
$-(8) = -8$	$\lvert 8 \rvert = 8$

You Try It 3 $100 + (-43) = 57$

You Try It 4

$$\begin{aligned} &(-51) + 42 + 17 + (-102) \\ &= -9 + 17 + (-102) \\ &= 8 + (-102) \\ &= -94 \end{aligned}$$

You Try It 5

$$\begin{aligned} 19 - (-32) &= 19 + 32 \\ &= 51 \end{aligned}$$

You Try It 6

$$\begin{aligned} &-9 - (-12) - 17 - 4 \\ &= -9 + 12 + (-17) + (-4) \\ &= 3 + (-17) + (-4) \\ &= -14 + (-4) \\ &= -18 \end{aligned}$$

You Try It 7

$$\begin{aligned} 8(-9)10 &= -72(10) \\ &= -720 \end{aligned}$$

You Try It 8

$$\begin{aligned} (-2)3(-8)7 &= -6(-8)7 \\ &= 48(7) \\ &= 336 \end{aligned}$$

You Try It 9 $(-135) \div (-9) = 15$

You Try It 10 $\dfrac{-72}{4} = -18$

You Try It 11

$$\begin{aligned} -\frac{36}{-12} &= -(-3) \\ &= 3 \end{aligned}$$

You Try It 12

Strategy To find the average daily low temperature:

- Add the seven temperature readings.
- Divide the sum by 7.

Solution

$$\begin{aligned} &-6 + (-7) + 0 + (-5) + (-8) + (-1) + (-1) \\ &= -28 \\ &-28 \div 7 = -4 \end{aligned}$$

The average daily low temperature was -4°C.

SECTION 1.2

You Try It 1

$$\begin{array}{r} 0.444 \\ 9\overline{)4.000} \\ \underline{-3\,6} \\ 40 \\ \underline{-36} \\ 40 \\ \underline{-36} \\ 4 \end{array}$$

$\dfrac{4}{9} = 0.\overline{4}$

You Try It 2

$$125\% = 125\left(\frac{1}{100}\right) = \frac{125}{100} = \frac{5}{4}$$

$$125\% = 125(0.01) = 1.25$$

You Try It 3

$$\begin{aligned} \frac{1}{3} &= \frac{1}{3}(100\%) \\ &= \frac{100}{3}\% = 33\frac{1}{3}\% \end{aligned}$$

You Try It 4 $0.043 = 0.043(100\%) = 4.3\%$

You Try It 5 The LCM of 8, 6, and 4 is 24.

$$\begin{aligned} -\frac{7}{8} - \frac{5}{6} + \frac{3}{4} &= -\frac{21}{24} - \frac{20}{24} + \frac{18}{24} \\ &= \frac{-21}{24} + \frac{-20}{24} + \frac{18}{24} \\ &= \frac{-21 - 20 + 18}{24} \\ &= \frac{-23}{24} = -\frac{23}{24} \end{aligned}$$

You Try It 6

$$\begin{aligned} &16.127 - 67.91 \\ &= 16.127 + (-67.91) \\ &= -51.783 \end{aligned}$$

Copyright © Houghton Mifflin Company. All rights reserved.

You Try It 7
The quotient is positive.

$$-\frac{3}{8} \div \left(-\frac{5}{12}\right) = \frac{3}{8} \div \frac{5}{12} = \frac{3}{8} \cdot \frac{12}{5}$$
$$= \frac{3 \cdot 12}{8 \cdot 5}$$
$$= \frac{3 \cdot \overset{1}{\cancel{2}} \cdot \overset{1}{\cancel{2}} \cdot 3}{2 \cdot \underset{1}{\cancel{2}} \cdot \underset{1}{\cancel{2}} \cdot 5} = \frac{9}{10}$$

You Try It 8 The product is negative.

$$\begin{array}{r} 5.44 \\ \times\ 3.8 \\ \hline 4352 \\ 1632\ \ \\ \hline 20.672 \end{array}$$

$-5.44(3.8) = -20.672$

You Try It 9
$-6^3 = -(6 \cdot 6 \cdot 6) = -216$

You Try It 10
$(-3)^4 = (-3)(-3)(-3)(-3) = 81$

You Try It 11
$$(3^3)(-2)^3 = (3)(3)(3) \cdot (-2)(-2)(-2)$$
$$= 27(-8) = -216$$

You Try It 12
$$\left(-\frac{2}{5}\right)^2 = \left(-\frac{2}{5}\right)\left(-\frac{2}{5}\right) = \frac{4}{25}$$

You Try It 13
$$-3(0.3)^3 = -3(0.3)(0.3)(0.3)$$
$$= -0.9(0.3)(0.3)$$
$$= -0.27(0.3) = -0.081$$

You Try It 14
$$-5\sqrt{32} = -5\sqrt{16 \cdot 2} = -5\sqrt{16}\sqrt{2}$$
$$= -5 \cdot 4\sqrt{2} = -20\sqrt{2}$$

You Try It 15 $\sqrt{216} = \sqrt{36 \cdot 6} = \sqrt{36}\sqrt{6} = 6\sqrt{6}$

You Try It 16

Strategy To determine the annual net income for 2005, multiply the net income for the third quarter of 2004 (−2.1) by the number of quarters in one year (4).

Solution $4(-2.1) = -8.4$

The annual net income for Frontier Airlines for 2005 would be −\$8.4 million.

SECTION 1.3

You Try It 1
$$18 - 5[8 - 2(2 - 5)] \div 10$$
$$= 18 - 5[8 - 2(-3)] \div 10$$
$$= 18 - 5[8 + 6] \div 10$$
$$= 18 - 5[14] \div 10$$
$$= 18 - 70 \div 10$$
$$= 18 - 7$$
$$= 11$$

You Try It 2
$$36 \div (8 - 5)^2 - (-3)^2 \cdot 2$$
$$= 36 \div (3)^2 - (-3)^2 \cdot 2$$
$$= 36 \div 9 - 9 \cdot 2$$
$$= 4 - 9 \cdot 2$$
$$= 4 - 18$$
$$= -14$$

You Try It 3
$$(6.97 - 4.72)^2 \cdot 4.5 \div 0.05$$
$$= (2.25)^2 \cdot 4.5 \div 0.05$$
$$= 5.0625 \cdot 4.5 \div 0.05$$
$$= 22.78125 \div 0.05$$
$$= 455.625$$

SECTION 1.4

You Try It 1
$$\frac{a^2 + b^2}{a + b}$$
$$\frac{5^2 + (-3)^2}{5 + (-3)} = \frac{25 + 9}{5 + (-3)}$$
$$= \frac{34}{2}$$
$$= 17$$

You Try It 2
$$x^3 - 2(x + y) + z^2$$
$$(2)^3 - 2[2 + (-4)] + (-3)^2$$
$$= (2)^3 - 2(-2) + (-3)^2$$
$$= 8 - 2(-2) + 9$$
$$= 8 + 4 + 9$$
$$= 12 + 9$$
$$= 21$$

You Try It 3 $3a - 2b - 5a + 6b = -2a + 4b$

You Try It 4 $-3y^2 + 7 + 8y^2 - 14 = 5y^2 - 7$

You Try It 5 $-5(4y^2) = -20y^2$

You Try It 6 $-7(-2a) = 14a$

You Try It 7 $(-5x)(-2) = 10x$

You Try It 8 $-8(-2a + 7b) = 16a - 56b$

Copyright © Houghton Mifflin Company. All rights reserved.

You Try It 9 $(3a - 1)5 = 15a - 5$

You Try It 10 $2(x^2 - x + 7) = 2x^2 - 2x + 14$

You Try It 11
$$\begin{aligned} 3y - 2(y - 7x) &= 3y - 2y + 14x \\ &= 14x + y \end{aligned}$$

You Try It 12
$$\begin{aligned} -2(x - 2y) - (-x + 3y) &= -2x + 4y + x - 3y \\ &= -x + y \end{aligned}$$

You Try It 13
$$\begin{aligned} 3y - 2[x - 4(2 - 3y)] &= 3y - 2[x - 8 + 12y] \\ &= 3y - 2x + 16 - 24y \\ &= -2x - 21y + 16 \end{aligned}$$

You Try It 14 the unknown number: x
the difference between the number and sixty: $x - 60$

$5(x - 60)$; $5x - 300$

You Try It 15 the speed of the older model: s
the speed of the new model: $2s$

You Try It 16 the length of the longer piece: L
the length of the shorter piece: $6 - L$

SECTION 1.5

You Try It 1 $A = \{-9, -7, -5, -3, -1\}$

You Try It 2 $A = \{1, 3, 5, \ldots\}$

You Try It 3 $A \cup B = \{-2, -1, 0, 1, 2, 3, 4\}$

You Try It 4 $C \cap D = \{10, 16\}$

You Try It 5 $A \cap B = \varnothing$

You Try It 6 $\{x | x < 59, x \in$ positive even integers$\}$

You Try It 7 $\{x | x > -3, x \in$ real numbers$\}$

You Try It 8 The graph is the numbers greater than -2.

−5 −4 −3 −2 −1 0 1 2 3 4 5

You Try It 9 The graph is the numbers greater than -1 and the numbers less than -3.

−5 −4 −3 −2 −1 0 1 2 3 4 5

You Try It 10 The graph is the numbers less than or equal to 4 and greater than or equal to -4.

−5 −4 −3 −2 −1 0 1 2 3 4 5

You Try It 11 The graph is the real numbers.

−5 −4 −3 −2 −1 0 1 2 3 4 5

Solutions to Chapter 2 "You Try It"

SECTION 2.1

You Try It 1

$5 - 4x$	$= 8x + 2$
$5 - 4\left(\frac{1}{4}\right)$	$8\left(\frac{1}{4}\right) + 2$
$5 - 1$	$2 + 2$
4	$= 4$

Yes, $\frac{1}{4}$ is a solution.

You Try It 2

$10x - x^2$	$= 3x - 10$
$10(5) - (5)^2$	$3(5) - 10$
$50 - 25$	$15 - 10$
25	$\neq 5$

No, 5 is not a solution.

You Try It 3
$$\begin{aligned} \frac{5}{6} &= y - \frac{3}{8} \\ \frac{5}{6} + \frac{3}{8} &= y - \frac{3}{8} + \frac{3}{8} \\ \frac{29}{24} &= y \end{aligned}$$

The solution is $\frac{29}{24}$.

You Try It 4
$$\begin{aligned} -\frac{2x}{5} &= 6 \\ \left(-\frac{5}{2}\right)\left(-\frac{2}{5}x\right) &= \left(-\frac{5}{2}\right)(6) \\ x &= -15 \end{aligned}$$

• $-\frac{2x}{5} = -\frac{2}{5}x$

The solution is -15.

Copyright © Houghton Mifflin Company. All rights reserved.

You Try It 5

$$\begin{aligned} 4x - 8x &= 16 \\ -4x &= 16 \\ \frac{-4x}{-4} &= \frac{16}{-4} \\ x &= -4 \end{aligned}$$

The solution is −4.

You Try It 6

$$\begin{aligned} P \cdot B &= A \\ \frac{1}{6}B &= 18 \qquad \bullet\ 16\frac{2}{3}\% = \frac{1}{6} \\ 6 \cdot \frac{1}{6}B &= 6 \cdot 18 \\ B &= 108 \end{aligned}$$

18 is $16\frac{2}{3}\%$ of 108.

You Try It 7

Strategy Use the percent equation. $B = 83.3$, the total revenue received by the BCS; $A = 3.1$, the amount received by the college representing the Pac-10 conference; P is the unknown percent.

Solution

$$\begin{aligned} P \cdot B &= A \\ P(83.3) &= 3.1 \qquad \bullet\ B = 83.3, A = 3.1 \\ P &= \frac{3.1}{83.3} \approx 0.037 \end{aligned}$$

The college representing the Pac-10 conference received approximately 3.7% of the BCS revenue.

You Try It 8

Strategy

To find how much she must deposit into the bank account:

- Find the amount of interest earned on the municipal bond by solving $I = Prt$ for I using $P = 1000$, $r = 6.4\% = 0.064$, and $t = 1$.
- Solve $I = Prt$ for P using the amount of interest earned on the municipal bond as I. $r = 8\% = 0.08$, and $t = 1$.

Solution

$$\begin{aligned} I &= Prt \\ &= 1000(0.064)(1) = 64 \end{aligned}$$

The interest earned on the municipal bond was \$64.

$$\begin{aligned} I &= Prt \\ 64 &= P(0.08)(1) \qquad \bullet\ I = 64, r = 0.08, t = 1 \\ 64 &= 0.08P \\ \frac{64}{0.08} &= \frac{0.08P}{0.08} \\ 800 &= P \end{aligned}$$

Clarissa must invest \$800 in the bank account.

You Try It 9

Strategy To find the number of ounces of cereal in the bowl, solve $Q = Ar$ for A using $Q = 2$ and $r = 25\% = 0.25$.

Solution

$$\begin{aligned} Q &= Ar \\ 2 &= A(0.25) \qquad \bullet\ Q = 2, r = 0.25 \\ \frac{2}{0.25} &= \frac{A(0.25)}{0.25} \\ 8 &= A \end{aligned}$$

The cereal bowl contains 8 oz of cereal.

You Try It 10

Strategy To find the distance, solve the equation $d = rt$ for d. The time is 3 h. Therefore, $t = 3$. The plane is moving against the wind, which means the headwind is slowing the actual speed of the plane. 250 mph − 25 mph = 225 mph. Thus $r = 225$.

Solution

$$\begin{aligned} d &= rt \\ d &= 225(3) \qquad \bullet\ r = 225, t = 3 \\ &= 675 \end{aligned}$$

The plane travels 675 mi in 3 h.

SECTION 2.2

You Try It 1

$$\begin{aligned} 5x + 7 &= 10 \\ 5x + 7 - 7 &= 10 - 7 \qquad \bullet\ \text{Subtract 7.} \\ 5x &= 3 \\ \frac{5x}{5} &= \frac{3}{5} \qquad \bullet\ \text{Divide by 5.} \\ x &= \frac{3}{5} \end{aligned}$$

The solution is $\frac{3}{5}$.

You Try It 2

$$\begin{aligned} 2 &= 11 + 3x \\ 2 - 11 &= 11 - 11 + 3x \qquad \bullet\ \text{Subtract 11.} \\ -9 &= 3x \\ \frac{-9}{3} &= \frac{3x}{3} \qquad \bullet\ \text{Divide by 3.} \\ -3 &= x \end{aligned}$$

The solution is −3.

You Try It 3

$$\begin{aligned} \frac{5}{8} - \frac{2x}{3} &= \frac{5}{4} \\ \frac{5}{8} - \frac{5}{8} - \frac{2}{3}x &= \frac{5}{4} - \frac{5}{8} \qquad \bullet\ \text{Recall that } \frac{2x}{3} = \frac{2}{3}x. \\ -\frac{2}{3}x &= \frac{5}{8} \\ -\frac{3}{2}\left(-\frac{2}{3}x\right) &= -\frac{3}{2}\left(\frac{5}{8}\right) \qquad \bullet\ \text{Multiply by } -\frac{3}{2}. \end{aligned}$$

Copyright © Houghton Mifflin Company. All rights reserved.

$$x = -\frac{15}{16}$$

The solution is $-\frac{15}{16}$.

You Try It 4

$$\frac{2}{3}x + 3 = \frac{7}{2}$$
$$6\left(\frac{2}{3}x + 3\right) = 6\left(\frac{7}{2}\right)$$
$$6\left(\frac{2}{3}x\right) + 6(3) = 6\left(\frac{7}{2}\right)$$ • Distributive Property
$$4x + 18 = 21$$
$$4x + 18 - 18 = 21 - 18$$ • Subtract 18.
$$4x = 3$$
$$\frac{4x}{4} = \frac{3}{4}$$ • Divide by 4.
$$x = \frac{3}{4}$$

The solution is $\frac{3}{4}$.

You Try It 5

$$x - 5 + 4x = 25$$
$$5x - 5 = 25$$
$$5x - 5 + 5 = 25 + 5$$
$$5x = 30$$
$$\frac{5x}{5} = \frac{30}{5}$$
$$x = 6$$

The solution is 6.

You Try It 6

$$5x + 4 = 6 + 10x$$
$$5x - 10x + 4 = 6 + 10x - 10x$$ • Subtract 10*x*.
$$-5x + 4 = 6$$
$$-5x + 4 - 4 = 6 - 4$$ • Subtract 4.
$$-5x = 2$$
$$\frac{-5x}{-5} = \frac{2}{-5}$$ • Divide by −5.
$$x = -\frac{2}{5}$$

The solution is $-\frac{2}{5}$.

You Try It 7

$$5x - 10 - 3x = 6 - 4x$$
$$2x - 10 = 6 - 4x$$ • Combine like terms.
$$2x + 4x - 10 = 6 - 4x + 4x$$ • Add 4*x*.
$$6x - 10 = 6$$
$$6x - 10 + 10 = 6 + 10$$ • Add 10.
$$6x = 16$$
$$\frac{6x}{6} = \frac{16}{6}$$ • Divide by 6.
$$x = \frac{8}{3}$$

The solution is $\frac{8}{3}$.

You Try It 8

$$5x - 4(3 - 2x) = 2(3x - 2) + 6$$
$$5x - 12 + 8x = 6x - 4 + 6$$ • Distributive Property
$$13x - 12 = 6x + 2$$
$$13x - 6x - 12 = 6x - 6x + 2$$ • Subtract 6*x*.
$$7x - 12 = 2$$
$$7x - 12 + 12 = 2 + 12$$ • Add 12.
$$7x = 14$$
$$\frac{7x}{7} = \frac{14}{7}$$ • Divide by 7.
$$x = 2$$

The solution is 2.

You Try It 9

$$-2[3x - 5(2x - 3)] = 3x - 8$$
$$-2[3x - 10x + 15] = 3x - 8$$ • Distributive Property
$$-2[-7x + 15] = 3x - 8$$
$$14x - 30 = 3x - 8$$
$$14x - 3x - 30 = 3x - 3x - 8$$ • Subtract 3*x*.
$$11x - 30 = -8$$
$$11x - 30 + 30 = -8 + 30$$ • Add 30.
$$11x = 22$$
$$\frac{11x}{11} = \frac{22}{11}$$ • Divide by 11.
$$x = 2$$

The solution is 2.

You Try It 10

Strategy Given: $F_1 = 45$
$F_2 = 80$
$d = 25$
Unknown: x

Solution

$$F_1x = F_2(d - x)$$
$$45x = 80(25 - x)$$
$$45x = 2000 - 80x$$
$$45x + 80x = 2000 - 80x + 80x$$
$$125x = 2000$$
$$\frac{125x}{125} = \frac{2000}{125}$$
$$x = 16$$

The fulcrum is 16 ft from the 45-pound force.

Copyright © Houghton Mifflin Company. All rights reserved.

SECTION 2.3

You Try It 1

The smaller number: n
The larger number: $12 - n$

The total of three times the smaller number and six	amounts to	seven less than the product of four and the larger number

$$
\begin{aligned}
3n + 6 &= 4(12 - n) - 7 \\
3n + 6 &= 48 - 4n - 7 \\
3n + 6 &= 41 - 4n \\
3n + 4n + 6 &= 41 - 4n + 4n \\
7n + 6 &= 41 \\
7n + 6 - 6 &= 41 - 6 \\
7n &= 35 \\
\frac{7n}{7} &= \frac{35}{7} \\
n &= 5
\end{aligned}
$$

$12 - n = 12 - 5 = 7$

The smaller number is 5.
The larger number is 7.

You Try It 2

Strategy
- First integer: n
  Second integer: $n + 1$
  Third integer: $n + 2$
- The sum of the three integers is -6.

Solution

$$
\begin{aligned}
n + (n + 1) + (n + 2) &= -6 \\
3n + 3 &= -6 \\
3n &= -9 \\
n &= -3
\end{aligned}
$$

$n + 1 = -3 + 1 = -2$
$n + 2 = -3 + 2 = -1$

The three consecutive integers are -3, -2, and -1.

You Try It 3

Strategy
To find the number of tickets that you are purchasing, write and solve an equation using x to represent the number of tickets purchased.

Solution

\$3.50 plus \$17.50 for each ticket	is	\$161

$$
\begin{aligned}
3.50 + 17.50x &= 161 \\
3.50 - 3.50 + 17.50x &= 161 - 3.50 \\
17.50x &= 157.50 \\
\frac{17.50x}{17.50} &= \frac{157.50}{17.50} \\
x &= 9
\end{aligned}
$$

You are purchasing 9 tickets.

You Try It 4

Strategy
To find the length, write and solve an equation using x to represent the length of the shorter piece and $22 - x$ to represent the length of the longer piece.

Solution

The length of the longer piece	is	4 in. more than twice the length of the shorter piece

$$
\begin{aligned}
22 - x &= 2x + 4 \\
22 - x - 2x &= 2x - 2x + 4 \\
22 - 3x &= 4 \\
22 - 22 - 3x &= 4 - 22 \\
-3x &= -18 \\
\frac{-3x}{-3} &= \frac{-18}{-3} \\
x &= 6
\end{aligned}
$$

$22 - x = 22 - 6 = 16$

The length of the shorter piece is 6 in.
The length of the longer piece is 16 in.

SECTION 2.4

You Try It 1

Strategy
- Pounds of \$.55 fertilizer: x

	Amount	*Cost*	*Value*
\$.80 fertilizer	20	.80	0.80(20)
\$.55 fertilizer	x	.55	$0.55x$
\$.75 fertilizer	$20 + x$	.75	$0.75(20 + x)$

- The sum of the values before mixing equals the value after mixing.

Solution

$$
\begin{aligned}
0.80(20) + 0.55x &= 0.75(20 + x) \\
16 + 0.55x &= 15 + 0.75x \\
16 - 0.20x &= 15 \\
-0.20x &= -1 \\
x &= 5
\end{aligned}
$$

5 lb of the \$.55 fertilizer must be added.

Copyright © Houghton Mifflin Company. All rights reserved.

You Try It 2

Strategy • Liters of 6% solution: x

	Amount	Percent	Quantity
6% solution	x	0.06	$0.06x$
12% solution	5	0.12	5(0.12)
8% solution	$x + 5$	0.08	$0.08(x + 5)$

• The sum of the quantities before mixing equals the quantity after mixing.

Solution

$$0.06x + 5(0.12) = 0.08(x + 5)$$
$$0.06x + 0.60 = 0.08x + 0.40$$
$$-0.02x + 0.60 = 0.40$$
$$-0.02x = -0.20$$
$$x = 10$$

The pharmacist adds 10 L of the 6% solution to the 12% solution to get an 8% solution.

You Try It 3

Strategy • Rate of the first train: r
Rate of the second train: $2r$

	Rate	Time	Distance
1st train	r	3	$3r$
2nd train	$2r$	3	$3(2r)$

• The sum of the distances traveled by the two trains equals 288 mi.

Solution

$$3r + 3(2r) = 288$$
$$3r + 6r = 288$$
$$9r = 288$$
$$r = 32$$

$$2r = 2(32) = 64$$

The first train is traveling at 32 mph. The second train is traveling at 64 mph.

You Try It 4

Strategy • Time spent flying out: t
Time spent flying back: $5 - t$

	Rate	Time	Distance
Out	150	t	$150t$
Back	100	$5 - t$	$100(5 - t)$

• The distance out equals the distance back.

Solution

$$150t = 100(5 - t)$$
$$150t = 500 - 100t$$
$$250t = 500$$
$$t = 2 \quad \text{(The time out was 2 h.)}$$

The distance out $= 150t = 150(2)$
$= 300$ mi

The parcel of land was 300 mi away.

SECTION 2.5

You Try It 1

$2x - 1 < 6x + 7$
$-4x - 1 < 7$ • Subtract $6x$ from each side.
$-4x < 8$ • Add 1 to each side.
$\frac{-4x}{-4} > \frac{8}{-4}$ • Divide each side by -4.
$x > -2$
$\{x \mid x > -2\}$

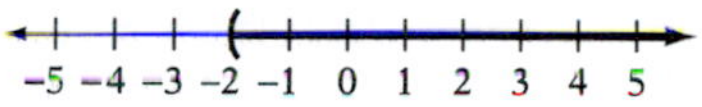

[Number line from −5 to 5 with open parenthesis at −2, shaded to the right]

You Try It 2

$$5x - 2 \le 4 - 3(x - 2)$$
$$5x - 2 \le 4 - 3x + 6$$
$$5x - 2 \le 10 - 3x$$
$$8x - 2 \le 10$$
$$8x \le 12$$
$$\frac{8x}{8} \le \frac{12}{8}$$
$$x \le \frac{3}{2}$$
$$\left\{x \,\middle|\, x \le \frac{3}{2}\right\}$$

You Try It 3

$-2 \le 5x + 3 \le 13$
$-2 - 3 \le 5x + 3 - 3 \le 13 - 3$ • Subtract 3 from each of the three parts.
$-5 \le 5x \le 10$
$\frac{-5}{5} \le \frac{5x}{5} \le \frac{10}{5}$ • Divide each of the three parts by 5.
$-1 \le x \le 2$
$\{x \mid -1 \le x \le 2\}$

You Try It 4

$2 - 3x > 11$ or $5 + 2x > 7$
$-3x > 9$ $\quad$ $2x > 2$
$x < -3$ $\quad$ $x > 1$
$\{x \mid x < -3\}$ $\quad$ $\{x \mid x > 1\}$

$\{x \mid x < -3\} \cup \{x \mid x > 1\}$
$= \{x \mid x < -3 \text{ or } x > 1\}$

Copyright © Houghton Mifflin Company. All rights reserved.

You Try It 5

Strategy To find the maximum height, substitute the given values in the inequality $\frac{1}{2}bh < A$ and solve.

Solution

$$\frac{1}{2}bh < A$$
$$\frac{1}{2}(12)(x + 2) < 50$$
$$6(x + 2) < 50$$
$$6x + 12 < 50$$
$$6x < 38$$
$$x < \frac{19}{3}$$

The largest integer less than $\frac{19}{3}$ is 6.

$x + 2 = 6 + 2 = 8$

The maximum height of the triangle is 8 in.

You Try It 6

Strategy To find the range of scores, write and solve an inequality using N to represent the score on the last test.

Solution

$$80 \le \frac{72 + 94 + 83 + 70 + N}{5} \le 89$$
$$80 \le \frac{319 + N}{5} \le 89$$
$$5 \cdot 80 \le 5\left(\frac{319 + N}{5}\right) \le 5 \cdot 89$$
$$400 \le 319 + N \le 445$$
$$400 - 319 \le 319 + N - 319 \le 445 - 319$$
$$81 \le N \le 126$$

Because 100 is the maximum score, the range of scores to receive a B grade is $81 \le N \le 100$.

SECTION 2.6

You Try It 1

$|2x - 3| = 5$

$$2x - 3 = 5 \qquad 2x - 3 = -5$$
$$2x = 8 \qquad 2x = -2 \qquad \text{• Add 3.}$$
$$x = 4 \qquad x = -1 \qquad \text{• Divide by 2.}$$

The solutions are 4 and -1.

You Try It 2 $|x - 3| = -2$

There is no solution to this equation because the absolute value of a number must be nonnegative.

You Try It 3

$$5 - |3x + 5| = 3$$
$$-|3x + 5| = -2 \qquad \text{• Subtract 5.}$$
$$|3x + 5| = 2 \qquad \text{• Multiply by } -1.$$
$$3x + 5 = 2 \qquad 3x + 5 = -2$$
$$3x = -3 \qquad 3x = -7$$
$$x = -1 \qquad x = -\frac{7}{3}$$

The solutions are -1 and $-\frac{7}{3}$.

You Try It 4

$$|3x + 2| < 8$$
$$-8 < 3x + 2 < 8$$
$$-8 - 2 < 3x + 2 - 2 < 8 - 2$$
$$-10 < 3x < 6$$
$$\frac{-10}{3} < \frac{3x}{3} < \frac{6}{3}$$
$$-\frac{10}{3} < x < 2$$
$$\left\{x \middle| -\frac{10}{3} < x < 2\right\}$$

You Try It 5 $|3x - 7| < 0$

The absolute value of a number must be nonnegative.

The solution set is the empty set.

$\varnothing$

You Try It 6 $|2x + 7| \ge -1$

The absolute value of a number is nonnegative.

The solution set is the set of real numbers.

You Try It 7

$|5x + 3| > 8$

$$5x + 3 < -8 \quad \text{or} \quad 5x + 3 > 8$$
$$5x < -11 \qquad 5x > 5$$
$$x < -\frac{11}{5} \qquad x > 1$$
$$\left\{x \middle| x < -\frac{11}{5}\right\} \qquad \{x | x > 1\}$$
$$\left\{x \middle| x < -\frac{11}{5}\right\} \cup \{x | x > 1\}$$
$$= \left\{x \middle| x < -\frac{11}{5} \text{ or } x > 1\right\}$$

Copyright © Houghton Mifflin Company. All rights reserved.

You Try It 8

Strategy

Let b represent the diameter of the bushing, T the tolerance, and d the lower and upper limits of the diameter. Solve the absolute value inequality $|d - b| \le T$ for d.

Solution

$$|d - b| \le T$$
$$|d - 2.55| \le 0.003$$
$$-0.003 \le d - 2.55 \le 0.003$$
$$-0.003 + 2.55 \le d - 2.55 + 2.55 \le 0.003 + 2.55$$
$$2.547 \le d \le 2.553$$

The lower and upper limits of the diameter of the bushing are 2.547 in. and 2.553 in.

Solutions to Chapter 3 "You Try It"

SECTION 3.1

You Try It 1

$$QR + RS + ST = QT$$
$$24 + RS + 17 = 62$$
$$41 + RS = 62$$
$$RS = 21$$

$RS = 21$ cm

You Try It 2

$$AC = AB + BC$$
$$AC = \frac{1}{4}(BC) + BC$$
$$AC = \frac{1}{4}(16) + 16$$
$$AC = 4 + 16$$
$$AC = 20$$

$AC = 20$ ft

You Try It 3

Strategy Supplementary angles are two angles whose sum is 180°. To find the supplement, let x represent the supplement of a 129° angle. Write an equation and solve for x.

Solution

$$x + 129° = 180°$$
$$x = 51°$$

The supplement of a 129° angle is a 51° angle.

You Try It 4

Strategy To find the measure of $\angle a$, write an equation using the fact that the sum of the measure of $\angle a$ and 68° is 118°. Solve for $\angle a$.

Solution

$$\angle a + 68° = 118°$$
$$\angle a = 50°$$

The measure of $\angle a$ is 50°.

You Try It 5

Strategy The angles labeled are adjacent angles of intersecting lines and are, therefore, supplementary angles. To find x, write an equation and solve for x.

Solution

$$(x + 16°) + 3x = 180°$$
$$4x + 16° = 180°$$
$$4x = 164°$$
$$x = 41°$$

You Try It 6

Strategy $3x = y$ because corresponding angles have the same measure. $y + (x + 40°) = 180°$ because adjacent angles of intersecting lines are supplementary angles. Substitute $3x$ for y and solve for x.

Solution

$$3x + (x + 40°) = 180°$$
$$4x + 40° = 180°$$
$$4x = 140°$$
$$x = 35°$$

You Try It 7

Strategy

- To find the measure of angle b, use the fact that $\angle b$ and $\angle x$ are supplementary angles.
- To find the measure of angle c, use the fact that the sum of the measures of the interior angles of a triangle is 180°.
- To find the measure of angle y, use the fact that $\angle c$ and $\angle y$ are vertical angles.

Copyright © Houghton Mifflin Company. All rights reserved.

Solution

$$\angle b + \angle x = 180°$$
$$\angle b + 100° = 180°$$
$$\angle b = 80°$$

$$\angle a + \angle b + \angle c = 180°$$
$$45° + 80° + \angle c = 180°$$
$$125° + \angle c = 180°$$
$$\angle c = 55°$$

$$\angle y = \angle c = 55°$$

You Try It 8

Strategy To find the measure of the third angle, use the facts that the measure of a right angle is 90° and the sum of the measures of the interior angles of a triangle is 180°. Write an equation using x to represent the measure of the third angle. Solve the equation for x.

Solution

$$x + 90° + 34° = 180°$$
$$x + 124° = 180°$$
$$x = 56°$$

The measure of the third angle is 56°.

SECTION 3.2

You Try It 1

Strategy To find the perimeter, use the formula for the perimeter of a square. Substitute 60 for s and solve for P.

Solution

$$P = 4s$$
$$P = 4(60)$$
$$P = 240$$

The perimeter of the infield is 240 ft.

You Try It 2

Strategy To find the perimeter, use the formula for the perimeter of a rectangle. Substitute 11 for L and $8\frac{1}{2}$ for W and solve for P.

Solution

$$P = 2L + 2W$$
$$P = 2(11) + 2\left(8\frac{1}{2}\right)$$
$$P = 2(11) + 2\left(\frac{17}{2}\right)$$
$$P = 22 + 17$$
$$P = 39$$

The perimeter of a standard piece of typing paper is 39 in.

You Try It 3

Strategy To find the circumference, use the circumference formula that involves the diameter. Leave the answer in terms of π.

Solution

$$C = \pi d$$
$$C = \pi(9)$$
$$C = 9\pi$$

The circumference is 9π in.

You Try It 4

Strategy To find the number of rolls of wallpaper to be purchased:

- Use the formula for the area of a rectangle to find the area of one wall.
- Multiply the area of one wall by the number of walls to be covered (2).
- Divide the area of wall to be covered by the area that one roll of wallpaper will cover (30).

Solution

$A = LW$

$A = 12 \cdot 8 = 96$ The area of one wall is 96 ft^2.

$2(96) = 192$ The area of the two walls is 192 ft^2.

$192 \div 30 = 6.4$

Because a portion of a seventh roll is needed, 7 rolls of wallpaper should be purchased.

You Try It 5

Strategy To find the area, use the formula for the area of a circle. An approximation is asked for; use the π key on a calculator. $r = 11$

Solution

$$A = \pi r^2$$
$$A = \pi(11)^2$$
$$A = 121\pi$$
$$A \approx 380.13$$

The area is approximately 380.13 cm^2.

SECTION 3.3

You Try It 1

Strategy To find the volume, use the formula for the volume of a cube. $s = 2.5$

Solution

$$V = s^3$$
$$V = (2.5)^3 = 15.625$$

The volume of the cube is 15.625 m^3.

Copyright © Houghton Mifflin Company. All rights reserved.

You Try It 2

Strategy To find the volume:

- Find the radius of the base of the cylinder. $d = 8$
- Use the formula for the volume of a cylinder. Leave the answer in terms of π.

Solution $r = \frac{1}{2}d = \frac{1}{2}(8) = 4$

$V = \pi r^2 h = \pi(4)^2(22) = \pi(16)(22)$
$= 352\pi$

The volume of the cylinder is 352π ft^3.

You Try It 3

Strategy To find the surface area of the cylinder:

- Find the radius of the base of the cylinder. $d = 6$
- Use the formula for the surface area of a cylinder. An approximation is asked for; use the π key on a calculator.

Solution $r = \frac{1}{2}d = \frac{1}{2}(6) = 3$

$SA = 2\pi r^2 + 2\pi rh$
$SA = 2\pi(3)^2 + 2\pi(3)(8)$
$SA = 2\pi(9) + 2\pi(3)(8)$
$SA = 18\pi + 48\pi$
$SA = 66\pi$
$SA \approx 207.35$

The surface area of the cylinder is approximately 207.35 ft^2.

You Try It 4

Strategy To find which solid has the larger surface area:

- Use the formula for the surface area of a cube to find the surface area of the cube. $s = 10$
- Find the radius of the sphere. $d = 8$
- Use the formula for the surface area of a sphere to find the surface area of the sphere. Because this number is to be compared to another number, use the π key on a calculator to approximate the surface area.
- Compare the two numbers.

Solution $SA = 6s^2$

$SA = 6(10)^2 = 6(100) = 600$

The surface area of the cube is 600 cm^2.

$r = \frac{1}{2}d = \frac{1}{2}(8) = 4$

$SA = 4\pi r^2$

$SA = 4\pi(4)^2 = 4\pi(16) = 64\pi \approx 201.06$

The surface area of the sphere is approximately 201.06 cm^2.

$600 > 201.06$

The cube has a larger surface area than the sphere.

Solutions to Chapter 4 "You Try It"

Section 4.1

You Try It 1

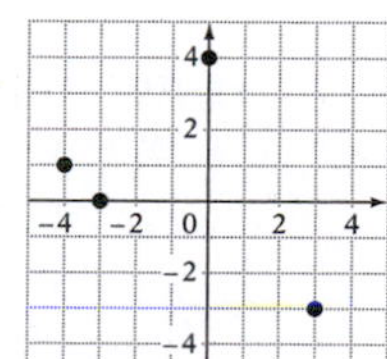

You Try It 2 The coordinates of A are $(4, -2)$.
The coordinates of B are $(-2, 4)$.
The abscissa of D is 0.
The ordinate of C is 0.

You Try It 3

$$\begin{array}{r|l} x - 3y = & -14 \\ \hline -2 - 3(4) & -14 \\ -2 - 12 & -14 \\ -14 = & -14 \end{array}$$

Yes, $(-2, 4)$ is a solution of $x - 3y = -14$.

You Try It 4 Replace x by -2 and solve for y.

$$y = \frac{3x}{x+1} = \frac{3(-2)}{-2+1} = \frac{-6}{-1} = 6$$

The ordered-pair solution is $(-2, 6)$.

Copyright © Houghton Mifflin Company. All rights reserved.

You Try It 5

x	$y = -\frac{1}{2}x + 2$	y	(x, y)
-4	$-\frac{1}{2}(-4) + 2$	4	$(-4, 4)$
-2	$-\frac{1}{2}(-2) + 2$	3	$(-2, 3)$
0	$-\frac{1}{2}(0) + 2$	2	$(0, 2)$
2	$-\frac{1}{2}(2) + 2$	1	$(2, 1)$

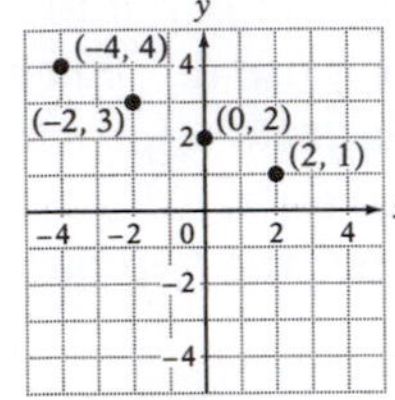

You Try It 6

Strategy To draw a scatter diagram:

- Draw a coordinate grid with the horizontal axis representing the year and the vertical axis representing the number of deaths.
- Graph the ordered pairs (1998, 14), (1999, 28), (2000, 17), (2001, 18), (2002, 23) and (2003, 29).

Solution

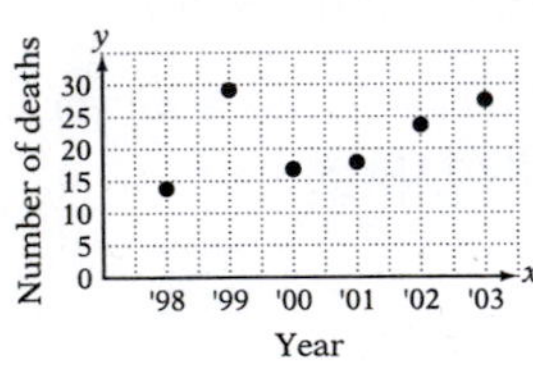

SECTION 4.2

You Try It 1 Domain: $\{-1, 3, 4, 6\}$
Range: $\{5\}$

- The domain is the set of first coordinates.

You Try It 2

$$G(x) = \frac{3x}{x + 2}$$

$$G(-4) = \frac{3(-4)}{-4 + 2} = \frac{-12}{-2} = 6$$

You Try It 3

$$f(x) = x^2 - 11$$
$$f(3h) = (3h)^2 - 11$$
$$= 9h^2 - 11$$

You Try It 4

$$h(z) = 3z + 1$$
$$h(0) = 3(0) + 1 = 1$$
$$h\left(\frac{1}{3}\right) = 3\left(\frac{1}{3}\right) + 1 = 2$$
$$h\left(\frac{2}{3}\right) = 3\left(\frac{2}{3}\right) + 1 = 3$$
$$h(1) = 3(1) + 1 = 4$$

The range is $\{1, 2, 3, 4\}$.

You Try It 5 $f(x) = \frac{2}{x - 5}$

For $x = 5$, $5 - 5 = 0$.

$f(5) = \frac{2}{5 - 5} = \frac{2}{0}$, which is not a real number.

5 is excluded from the domain of the function.

SECTION 4.3

You Try It 1

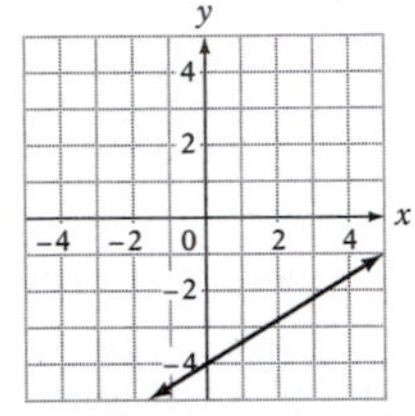

You Try It 2

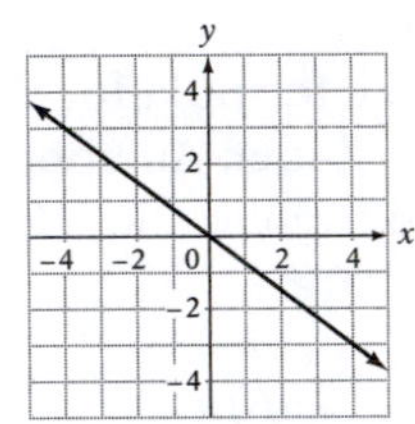

You Try It 3

$$-3x + 2y = 4$$
$$2y = 3x + 4$$
$$y = \frac{3}{2}x + 2$$

You Try It 4

$$y - 3 = 0$$
$$y = 3$$

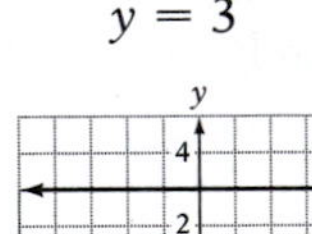

- The graph of $y = 3$ goes through the point (0, 3).

You Try It 5

x-intercept:

$$3x - y = 2$$
$$3x - 0 = 2$$
$$3x = 2$$
$$x = \frac{2}{3}$$

- Let $y = 0$.

x-intercept: $\left(\frac{2}{3}, 0\right)$

y-intercept:

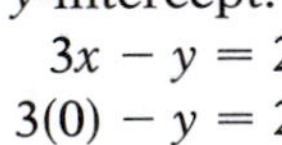

$$3x - y = 2$$
$$3(0) - y = 2$$
$$-y = 2$$
$$y = -2$$

- Let $x = 0$.

y-intercept: $(0, -2)$

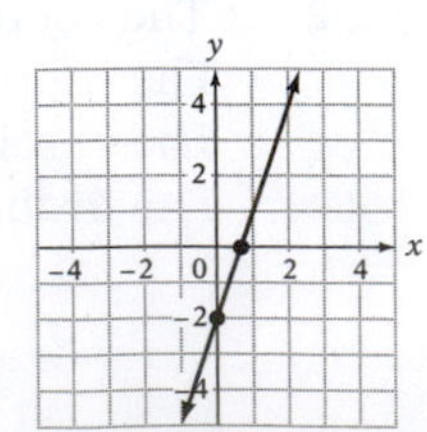

Copyright © Houghton Mifflin Company. All rights reserved.

You Try It 6

x-intercept:

$$y = \frac{1}{4}x + 1$$
$$0 = \frac{1}{4}x + 1$$
$$-\frac{1}{4}x = 1$$
$$x = -4$$

$(-4, 0)$

y-intercept:

$(0, b)$
$b = 1$
$(0, 1)$

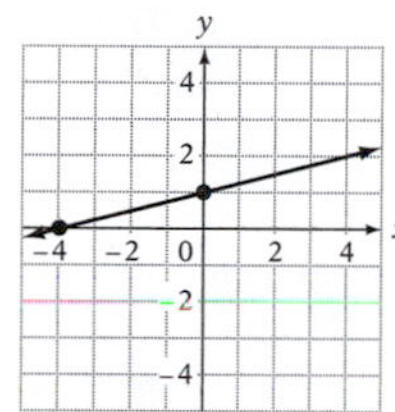

You Try It 7

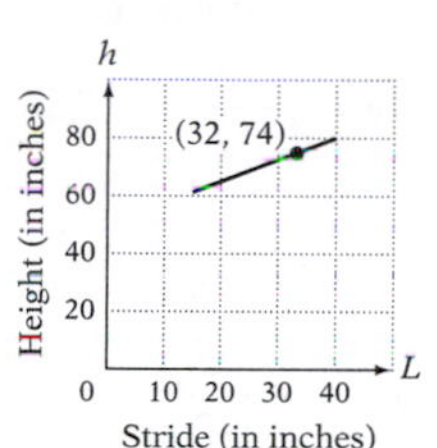

• Graph $h = \frac{3}{4}L + 50$. When $L = 20$, $h = 65$. When $L = 40$, $h = 80$.

The ordered pair (32, 74) means that a person with a stride of 32 in. is 74 in. tall.

SECTION 4.4

You Try It 1 Let $P_1 = (4, -3)$ and $P_2 = (2, 7)$.

$$m = \frac{y_2 - y_1}{x_2 - x_1} = \frac{7 - (-3)}{2 - 4} = \frac{10}{-2} = -5$$

The slope is -5.

You Try It 2 Let $P_1 = (6, -1)$ and $P_2 = (6, 7)$.

$$m = \frac{y_2 - y_1}{x_2 - x_1} = \frac{7 - (-1)}{6 - 6} = \frac{8}{0}$$

Division by zero is not defined.

The slope of the line is undefined.

You Try It 3 $P_1 = (5, 25{,}000)$, $P_2 = (2, 55{,}000)$

$$m = \frac{y_2 - y_1}{x_2 - x_1}$$

$$= \frac{55{,}000 - 25{,}000}{2 - 5}$$

• $(x_1, y_1) = (5, 25{,}000)$, $(x_2, y_2) = (2, 55{,}000)$

$$= \frac{30{,}000}{-3}$$
$$= -10{,}000$$

A slope of $-10{,}000$ means that the value of the recycling truck is decreasing by $10,000 per year.

You Try It 4

$$2x + 3y = 6$$
$$3y = -2x + 6$$
$$y = -\frac{2}{3}x + 2$$

$$m = -\frac{2}{3} = \frac{-2}{3}$$

y-intercept $= (0, 2)$

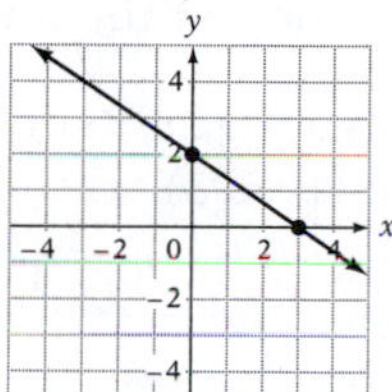

You Try It 5 $(x_1, y_1) = (-3, -2)$
$m = 3$

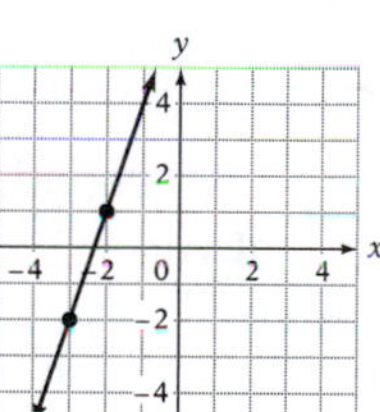

SECTION 4.5

You Try It 1 $m = -\frac{1}{3}$ $(x_1, y_1) = (-3, -2)$

$$y - y_1 = m(x - x_1)$$
$$y - (-2) = -\frac{1}{3}[x - (-3)]$$
$$y + 2 = -\frac{1}{3}(x + 3)$$
$$y + 2 = -\frac{1}{3}x - 1$$
$$y = -\frac{1}{3}x - 3$$

The equation of the line is $y = -\frac{1}{3}x - 3$.

Copyright © Houghton Mifflin Company. All rights reserved.

You Try It 2 $m = -3 \quad (x_1, y_1) = (4, -3)$

$$y - y_1 = m(x - x_1)$$
$$y - (-3) = -3(x - 4)$$
$$y + 3 = -3x + 12$$
$$y = -3x + 9$$

The equation of the line is $y = -3x + 9$.

You Try It 3 Let $(x_1, y_1) = (2, 0)$ and $(x_2, y_2) = (5, 3)$.

$$m = \frac{y_2 - y_1}{x_2 - x_1} = \frac{3 - 0}{5 - 2} = \frac{3}{3} = 1$$

$$y - y_1 = m(x - x_1)$$
$$y - 0 = 1(x - 2)$$
$$y = 1(x - 2)$$
$$y = x - 2$$

The equation of the line is $y = x - 2$.

You Try It 4 Let $(x_1, y_1) = (2, 3)$ and $(x_2, y_2) = (-5, 3)$.

$$m = \frac{y_2 - y_1}{x_2 - x_1} = \frac{3 - 3}{-5 - 2} = \frac{0}{-7} = 0$$

The line has zero slope.
The line is a horizontal line.
All points on the line have an ordinate of 3.
The equation of the line is $y = 3$.

You Try It 5

Strategy

- Select the independent and dependent variables. The function is to be used to predict the Celsius temperature, so that quantity is the dependent variable, y. The Fahrenheit temperature is the independent variable, x.
- From the given data, two ordered pairs are (212, 100) and (32, 0). Use these ordered pairs to determine the linear function.

Solution Let $(x_1, y_1) = (32, 0)$ and $(x_2, y_2) = (212, 100)$.

$$m = \frac{y_2 - y_1}{x_2 - x_1} = \frac{100 - 0}{212 - 32} = \frac{100}{180} = \frac{5}{9}$$
$$y - y_1 = m(x - x_1)$$
$$y - 0 = \frac{5}{9}(x - 32)$$
$$y = \frac{5}{9}(x - 32), \text{ or } C = \frac{5}{9}(F - 32).$$

The linear function is $f(F) = \frac{5}{9}(F - 32)$.

SECTION 4.6

You Try It 1

$$m_1 = \frac{1 - (-3)}{7 - (-2)} = \frac{4}{9}$$

• $(x_1, y_1) = (-2, -3)$, $(x_2, y_2) = (7, 1)$

$$m_2 = \frac{-5 - 1}{6 - 4} = \frac{-6}{2} = -3$$

• $(x_1, y_1) = (4, 1)$, $(x_2, y_2) = (6, -5)$

$$m_1 \cdot m_2 = \frac{4}{9}(-3) = -\frac{4}{3}$$

No, the lines are not perpendicular.

You Try It 2

$$5x + 2y = 2$$
$$2y = -5x + 2$$
$$y = -\frac{5}{2}x + 1$$
$$m_1 = -\frac{5}{2}$$
$$5x + 2y = -6$$
$$2y = -5x - 6$$
$$y = -\frac{5}{2}x - 3$$
$$m_2 = -\frac{5}{2}$$
$$m_1 = m_2 = -\frac{5}{2}$$

Yes, the lines are parallel.

You Try It 3

$$x - 4y = 3$$
$$-4y = -x + 3$$
$$y = \frac{1}{4}x - \frac{3}{4}$$
$$m_1 = \frac{1}{4}$$
$$m_1 \cdot m_2 = -1$$
$$\frac{1}{4} \cdot m_2 = -1$$
$$m_2 = -4$$
$$y - y_1 = m(x - x_1)$$
$$y - 2 = -4[x - (-2)]$$

• $(x_1, y_1) = (-2, 2)$

$$y - 2 = -4(x + 2)$$
$$y - 2 = -4x - 8$$
$$y = -4x - 6$$

The equation of the line is $y = -4x - 6$.

SECTION 4.7

You Try It 1

$$x + 3y > 6$$
$$3y > -x + 6$$
$$y > -\frac{1}{3}x + 2$$

Copyright © Houghton Mifflin Company. All rights reserved.

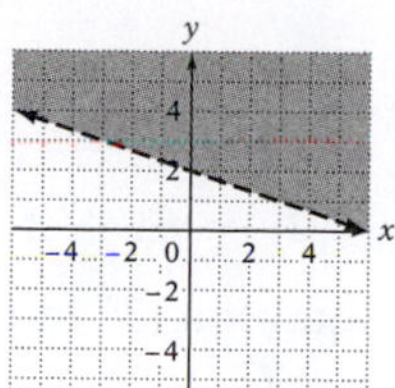

You Try It 2 $y < 2$

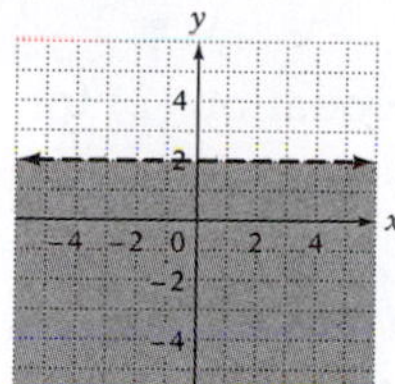

Solutions to Chapter 5 "You Try It"

SECTION 5.1

You Try It 1

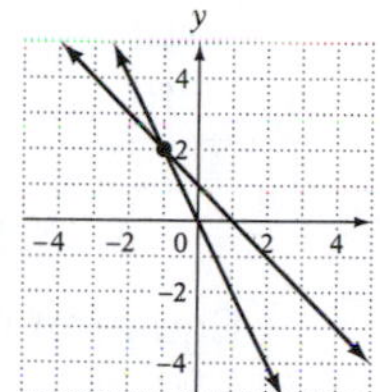

• **Find the point of intersection of the graphs of the equations.**

The solution is $(-1, 2)$.

You Try It 2

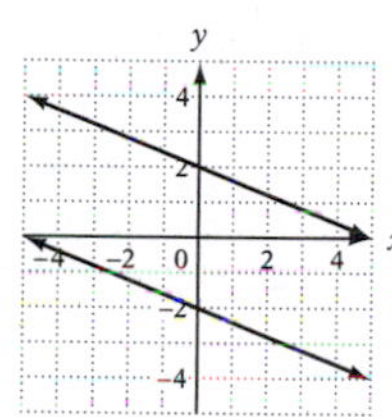

• **Graph the two equations.**

The lines are parallel and therefore do not intersect. The system of equations has no solution. The system of equations is inconsistent.

You Try It 3

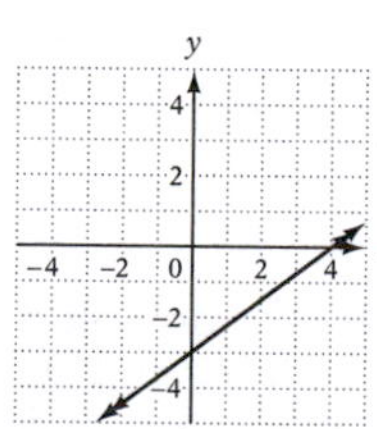

• **Graph the two equations.**

The two equations represent the same line. The system of equations is dependent. The solutions are the ordered pairs $\left(x, \frac{3}{4}x - 3\right)$.

You Try It 4

(1) $3x - y = 3$
(2) $6x + 3y = -4$

Solve Equation (1) for y.

$$\begin{aligned} 3x - y &= 3 \\ -y &= -3x + 3 \\ y &= 3x - 3 \end{aligned}$$

Substitute into Equation (2).

$$\begin{aligned} 6x + 3y &= -4 \\ 6x + 3(3x - 3) &= -4 \\ 6x + 9x - 9 &= -4 \\ 15x - 9 &= -4 \\ 15x &= 5 \\ x &= \frac{5}{15} = \frac{1}{3} \end{aligned}$$

Substitute the value of x into Equation (1).

$$\begin{aligned} 3x - y &= 3 \\ 3\left(\frac{1}{3}\right) - y &= 3 \\ 1 - y &= 3 \\ -y &= 2 \\ y &= -2 \end{aligned}$$

The solution is $\left(\frac{1}{3}, -2\right)$.

You Try It 5

(1) $y = 2x - 3$
(2) $3x - 2y = 6$

$$\begin{aligned} 3x - 2y &= 6 \\ 3x - 2(2x - 3) &= 6 \\ 3x - 4x + 6 &= 6 \\ -x + 6 &= 6 \\ -x &= 0 \\ x &= 0 \end{aligned}$$

Substitute the value of x into Equation (1).

$$\begin{aligned} y &= 2x - 3 \\ y &= 2(0) - 3 \\ y &= 0 - 3 \\ y &= -3 \end{aligned}$$

The solution is $(0, -3)$.

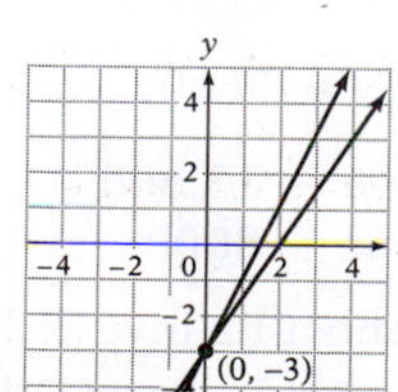

• **Gra**

Copyright © Houghton Mifflin Company. All rights reserved.

You Try It 6

(1) $6x - 3y = 6$
(2) $2x - y = 2$

Solve Equation (2) for y.

$$2x - y = 2$$
$$-y = -2x + 2$$
$$y = 2x - 2$$

Substitute into Equation (1).

$$6x - 3y = 6$$
$$6x - 3(2x - 2) = 6$$
$$6x - 6x + 6 = 6$$
$$6 = 6$$

The system of equations is dependent. The solutions are the ordered pairs $(x, 2x - 2)$.

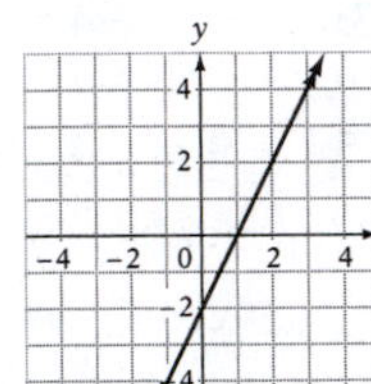

• Graph the two equations.

You Try It 7

Strategy

• Amount invested at 4.2%: x
Amount invested at 6%: y

	Principal	*Rate*	*Interest*
Amount at 4.2%	x	0.042	$0.042x$
Amount at 6%	y	0.06	$0.06y$

• The total investment is \$13,600. The two accounts earn the same interest.

Solution

$$x + y = 13{,}600$$
$$0.042x = 0.06y$$
$$x = \frac{10}{7}y$$

Substitute $\frac{10}{7}y$ for x in $x + y = 13{,}600$ and solve for y.

$$\frac{10}{7}y + y = 13{,}600$$
$$\frac{17}{7}y = 13{,}600$$
$$y = 5600$$
$$x + 5600 = 13{,}600$$
$$x = 8000$$

\$8000 must be invested at 4.2% and \$5600 must be invested at 6%.

SECTION 5.2

You Try It 1

(1) $2x + 5y = 6$
(2) $3x - 2y = 6x + 2$

Write Equation (2) in the form $Ax + By = C$.

$$3x - 2y = 6x + 2$$
$$-3x - 2y = 2$$

Solve the system: $2x + 5y = 6$
$-3x - 2y = 2$

Eliminate y.

$$2(2x + 5y) = 2(6)$$
$$5(-3x - 2y) = 5(2)$$

$$4x + 10y = 12$$
$$-15x - 10y = 10$$
$$-11x = 22 \quad \text{• Add the equations.}$$
$$x = -2 \quad \text{• Solve for } x.$$

Replace x in Equation (1).

$$2x + 5y = 6$$
$$2(-2) + 5y = 6$$
$$-4 + 5y = 6$$
$$5y = 10$$
$$y = 2$$

The solution is $(-2, 2)$.

You Try It 2

$$2x + y = 5$$
$$4x + 2y = 6$$

Eliminate y.

$$-2(2x + y) = -2(5)$$
$$4x + 2y = 6$$

$$-4x - 2y = -10$$
$$4x + 2y = 6$$
$$0x + 0y = -4 \quad \text{• Add the equations.}$$
$$0 = -4$$

This is not a true equation. The system is inconsistent and therefore has no solution.

You Try It 3

(1) $x - y + z = 6$
(2) $2x + 3y - z = 1$
(3) $x + 2y + 2z = 5$

Eliminate z. Add Equations (1) and (2).

$$x - y + z = 6$$
$$2x + 3y - z = 1$$
$$3x + 2y = 7 \quad \text{• Equation (4)}$$

Multiply Equation (2) by 2 and add to Equation (3).

$$4x + 6y - 2z = 2$$
$$x + 2y + 2z = 5$$
$$5x + 8y = 7 \quad \text{• Equation (5)}$$

Copyright © Houghton Mifflin Company. All rights reserved.

Solve the system of two equations.

(4) $3x + 2y = 7$
(5) $5x + 8y = 7$

Multiply Equation (4) by -4 and add to Equation (5).

$$\begin{aligned} -12x - 8y &= -28 \\ 5x + 8y &= 7 \\ -7x &= -21 \\ x &= 3 \end{aligned}$$

Replace x by 3 in Equation (4).

$$\begin{aligned} 3x + 2y &= 7 \\ 3(3) + 2y &= 7 \\ 9 + 2y &= 7 \\ 2y &= -2 \\ y &= -1 \end{aligned}$$

Replace x by 3 and y by -1 in Equation (1).

$$\begin{aligned} x - y + z &= 6 \\ 3 - (-1) + z &= 6 \\ 4 + z &= 6 \\ z &= 2 \end{aligned}$$

The solution is $(3, -1, 2)$.

SECTION 5.3

You Try It 1

Strategy

- Rate of the rowing team in calm water: t
  Rate of the current: c

	Rate	Time	Distance
With current	$t + c$	2	$2(t + c)$
Against current	$t - c$	2	$2(t - c)$

- The distance traveled with the current is 18 mi.
  The distance traveled against the current is 10 mi.

$$2(t + c) = 18$$
$$2(t - c) = 10$$

Solution

$2(t + c) = 18 \qquad \frac{1}{2} \cdot 2(t + c) = \frac{1}{2} \cdot 18$

$2(t - c) = 10 \qquad \frac{1}{2} \cdot 2(t - c) = \frac{1}{2} \cdot 10$

$$\begin{aligned} t + c &= 9 \\ t - c &= 5 \\ 2t &= 14 \\ t &= 7 \end{aligned}$$

$t + c = 9$
$7 + c = 9$ • **Substitute 7 for *t*.**
$c = 2$

The rate of the rowing team in calm water is 7 mph.
The rate of the current is 2 mph.

You Try It 2

Strategy • Cost of an orange tree: x
Cost of a grapefruit tree: y

First purchase:

	Amount	Unit Cost	Value
Orange trees	25	x	$25x$
Grapefruit trees	20	y	$20y$

Second purchase:

	Amount	Unit Cost	Value
Orange trees	20	x	$20x$
Grapefruit trees	30	y	$30y$

- The total of the first purchase was \$290.
  The total of the second purchase was \$330.

Solution

$25x + 20y = 290 \qquad 4(25x + 20y) = 4 \cdot 290$ • **Multiply by 4.**

$20x + 30y = 330 \qquad -5(20x + 30y) = -5 \cdot 330$ • **Multiply by −5.**

$$\begin{aligned} 100x + 80y &= 1160 \\ -100x - 150y &= -1650 \\ -70y &= -490 \\ y &= 7 \end{aligned}$$

$25x + 20y = 290$
$25x + 20(7) = 290$ • **$y = 7$**
$25x + 140 = 290$
$25x = 150$
$x = 6$

The cost of an orange tree is \$6.
The cost of a grapefruit tree is \$7.

SECTION 5.4

You Try It 1

Shade above the solid line $y = 2x - 3$.
Shade above the dotted line $y = -3x$.

The solution set of the system is the intersection of the solution sets of the individual inequalities.

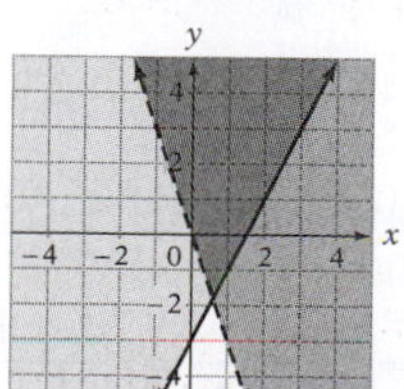

Copyright © Houghton Mifflin Company. All rights reserved.

You Try It 2

$$3x + 4y > 12$$
$$4y > -3x + 12$$
$$y > -\frac{3}{4}x + 3$$

Shade above the dotted line $y = -\frac{3}{4}x + 3$.

Shade below the dotted line $y = \frac{3}{4}x - 1$.

The solution set of the system is the intersection of the solution sets of the individual inequalities.

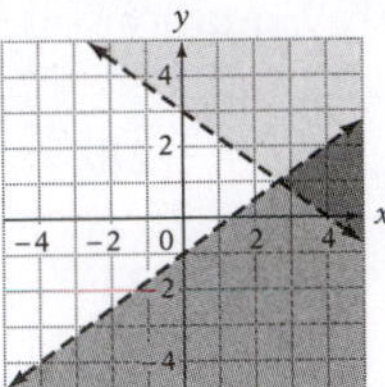

Solutions to Chapter 6 "You Try It"

SECTION 6.1

You Try It 1

$$(-3a^2b^4)(-2ab^3)^4 = (-3a^2b^4)[(-2)^4a^4b^{12}]$$
$$= (-3a^2b^4)(16a^4b^{12})$$
$$= -48a^6b^{16}$$

You Try It 2 $(y^{n-3})^2 = y^{(n-3)2}$ • **Multiply the exponents.**

$$= y^{2n-6}$$

You Try It 3

$$[(ab^3)^3]^4 = [a^3b^9]^4$$
$$= a^{12}b^{36}$$

You Try It 4

$$\frac{20r^{-2}t^{-5}}{-16r^{-3}s^{-2}} = -\frac{4 \cdot 5r^{-2-(-3)}s^2t^{-5}}{4 \cdot 4}$$
$$= -\frac{5rs^2}{4t^5}$$

You Try It 5

$$\frac{(9u^{-6}v^4)^{-1}}{(6u^{-3}v^{-2})^{-2}} = \frac{9^{-1}u^6v^{-4}}{6^{-2}u^6v^4}$$
$$= 9^{-1} \cdot 6^2u^0v^{-8}$$
$$= \frac{36}{9v^8}$$
$$= \frac{4}{v^8}$$

You Try It 6 $\frac{a^{2n+1}}{a^{n+3}} = a^{2n+1-(n+3)}$ • **Subtract the exponents.**

$$= a^{2n+1-n-3}$$
$$= a^{n-2}$$

You Try It 7 $942{,}000{,}000 = 9.42 \times 10^8$

You Try It 8 $2.7 \times 10^{-5} = 0.000027$

You Try It 9

$$\frac{5{,}600{,}000 \times 0.000000081}{900 \times 0.000000028}$$
$$= \frac{5.6 \times 10^6 \times 8.1 \times 10^{-8}}{9 \times 10^2 \times 2.8 \times 10^{-8}}$$
$$= \frac{(5.6)(8.1) \times 10^{6+(-8)-2-(-8)}}{(9)(2.8)}$$
$$= 1.8 \times 10^4 = 18{,}000$$

You Try It 10

Strategy To find the number of arithmetic operations:

- Find the reciprocal of 1×10^{-7}, which is the number of operations performed in 1 s.
- Write the number of seconds in 1 min (60) in scientific notation.
- Multiply the number of arithmetic operations per second by the number of seconds in 1 min.

Solution

$$\frac{1}{1 \times 10^{-7}} = 10^7$$
$$60 = 6 \times 10$$

$$6 \times 10 \times 10^7$$
$$6 \times 10^8$$

The computer can perform 6×10^8 operations in 1 min.

SECTION 6.2

You Try It 1

$$R(x) = -2x^4 - 5x^3 + 2x - 8$$
$$R(2) = -2(2)^4 - 5(2)^3 + 2(2) - 8$$ • **Replace *x* by 2. Simplify.**
$$= -2(16) - 5(8) + 4 - 8$$
$$= -32 - 40 + 4 - 8$$
$$= -76$$

You Try It 2 The leading coefficient is -3, the constant term is -12, and the degree is 4.

You Try It 3

a. Yes, this is a polynomial function.

b. No, this is not a polynomial function. A polynomial function does not have a variable expression raised to a negative power.

c. No, this is not a polynomial function. A polynomial function does not have a variable expression within a radical.

Copyright © Houghton Mifflin Company. All rights reserved.

You Try It 4

x	y
-4	5
-3	0
-2	-3
-1	-4
0	-3
1	0
2	5

You Try It 5

x	y
-3	28
-2	9
-1	2
0	1
1	0
2	-7
3	-26

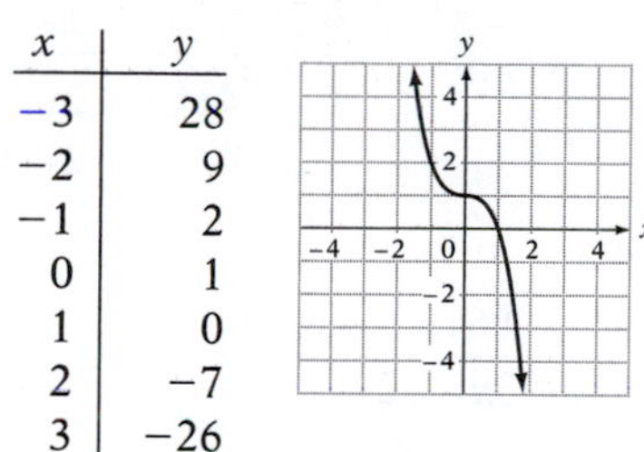

You Try It 6

$$\begin{array}{r} -3x^2 - 4x + 9 \\ \underline{-5x^2 - 7x + 1} \\ -8x^2 - 11x + 10 \end{array}$$

You Try It 7 Add the additive inverse of $6x^2 + 3x - 7$ to $-5x^2 + 2x - 3$.

$$\begin{array}{r} -5x^2 + 2x - 3 \\ \underline{-6x^2 - 3x + 7} \\ -11x^2 - x + 4 \end{array}$$

You Try It 8

$S(x) = (4x^3 - 3x^2 + 2) + (-2x^2 + 2x - 3)$
$= 4x^3 - 5x^2 + 2x - 1$

$S(-1) = 4(-1)^3 - 5(-1)^2 + 2(-1) - 1$
$= 4(-1) - 5(1) - 2 - 1$
$= -4 - 5 - 2 - 1$
$= -12$

You Try It 9

$D(x) = P(x) - R(x)$

$D(x) = (5x^{2n} - 3x^n - 7) - (-2x^{2n} - 5x^n + 8)$
$= (5x^{2n} - 3x^n - 7) + (2x^{2n} + 5x^n - 8)$
$= 7x^{2n} + 2x^n - 15$

SECTION 6.3

You Try It 1

$(2b^2 - 7b - 8)(-5b)$
$= 2b^2(-5b) - 7b(-5b) - 8(-5b)$ • Use the Distributive Property.
$= -10b^3 + 35b^2 + 40b$

You Try It 2 $x^2 - 2x[x - x(4x - 5) + x^2]$
$= x^2 - 2x[x - 4x^2 + 5x + x^2]$
$= x^2 - 2x[6x - 3x^2]$
$= x^2 - 12x^2 + 6x^3$
$= 6x^3 - 11x^2$

You Try It 3

$y^{n+3}(y^{n-2} - 3y^2 + 2)$
$= y^{n+3}(y^{n-2}) - (y^{n+3})(3y^2) + (y^{n+3})(2)$ • Use the Distributive Property.
$= y^{n+3+(n-2)} - 3y^{n+3+2} + 2y^{n+3}$
$= y^{2n+1} - 3y^{n+5} + 2y^{n+3}$

You Try It 4

$$\begin{array}{r} -2b^2 + 5b - 4 \\ \underline{- 3b + 2} \\ -4b^2 + 10b - 8 \\ \underline{6b^3 - 15b^2 + 12b} \\ 6b^3 - 19b^2 + 22b - 8 \end{array}$$

• $2(-2b^2 + 5b - 4)$
• $-3b(-2b^2 + 5b - 4)$

You Try It 5

$(3x - 4)(2x - 3) = 6x^2 - 9x - 8x + 12$ • FOIL
$= 6x^2 - 17x + 12$

You Try It 6

$(2x^n + y^n)(x^n - 4y^n)$
$= 2x^{2n} - 8x^ny^n + x^ny^n - 4y^{2n}$ • FOIL
$= 2x^{2n} - 7x^ny^n - 4y^{2n}$

You Try It 7 $(3x - 7)(3x + 7)$ • The sum and difference of two terms
$= 9x^2 - 49$

You Try It 8 $(2x^n + 3)(2x^n - 3)$ • The sum and difference of two terms
$= 4x^{2n} - 9$

You Try It 9 $(3x - 4y)^2$ • The square of a binomial
$= 9x^2 - 24xy + 16y^2$

You Try It 10 $(2x^n - 8)^2$ • The square of a binomial
$= 4x^{2n} - 32x^n + 64$

You Try It 11

Strategy To find the area, replace the variables b and h in the equation $A = \frac{1}{2}bh$ by the given values and solve for A.

Solution

$A = \dfrac{1}{2}bh$

$A = \dfrac{1}{2}(2x + 6)(x - 4)$

$A = (x + 3)(x - 4)$

$A = x^2 - 4x + 3x - 12$ • FOIL

$A = x^2 - x - 12$

The area is $(x^2 - x - 12)$ square feet.

You Try It 12

Strategy To find the volume, subtract the volume of the small rectangular solid from the volume of the large rectangular solid.

Large rectangular solid:
Length $= L_1 = 12x$
Width $= W_1 = 7x + 2$
Height $= H_1 = 5x - 4$

Copyright © Houghton Mifflin Company. All rights reserved.

Small rectangular solid:
Length $= L_2 = 12x$
Width $= W_2 = x$
Height $= H_2 = 2x$

Solution
$V =$ Volume of large rectangular solid $-$ volume of small rectangular solid

$$V = (L_1 \cdot W_1 \cdot H_1) - (L_2 \cdot W_2 \cdot H_2)$$
$$V = (12x)(7x + 2)(5x - 4) - (12x)(x)(2x)$$
$$= (84x^2 + 24x)(5x - 4) - (12x^2)(2x)$$
$$= (420x^3 - 336x^2 + 120x^2 - 96x) - 24x^3$$
$$= 396x^3 - 216x^2 - 96x$$

The volume is $(396x^3 - 216x^2 - 96x)$ cubic feet.

You Try It 13

Strategy To find the area, replace the variable r in the equation $A = \pi r^2$ by the given value and solve for A.

Solution
$$A = \pi r^2$$
$$A \approx 3.14(2x + 3)^2$$
$$= 3.14(4x^2 + 12x + 9)$$
$$= 12.56x^2 + 37.68x + 28.26$$

The area is $(12.56x^2 + 37.68x + 28.26)$ square centimeters.

SECTION 6.4

You Try It 1

$$\frac{4x^3y + 8x^2y^2 - 4xy^3}{2xy} = \frac{4x^3y}{2xy} + \frac{8x^2y^2}{2xy} - \frac{4xy^3}{2xy}$$
$$= 2x^2 + 4xy - 2y^2$$

Check: $2xy(2x^2 + 4xy - 2y^2) = 4x^3y + 8x^2y^2 - 4xy^3$

You Try It 2

$$\begin{array}{r} 5x - 1 \\ 3x + 4\overline{)15x^2 + 17x - 20} \\ \underline{15x^2 + 20x} \\ -3x - 20 \\ \underline{-3x - 4} \\ -16 \end{array}$$

$$\frac{15x^2 + 17x - 20}{3x + 4} = 5x - 1 - \frac{16}{3x + 4}$$

You Try It 3

$$\begin{array}{r} x^2 + 3x - 1 \\ 3x - 1\overline{)3x^3 + 8x^2 - 6x + 2} \\ \underline{3x^3 - x^2} \\ 9x^2 - 6x \\ \underline{9x^2 - 3x} \\ -3x + 2 \\ \underline{-3x + 1} \\ 1 \end{array}$$

$$\frac{3x^3 + 8x^2 - 6x + 2}{3x - 1} = x^2 + 3x - 1 + \frac{1}{3x - 1}$$

You Try It 4

$$\begin{array}{r} 3x^2 - 2x + 4 \\ x^2 - 3x + 2\overline{)3x^4 - 11x^3 + 16x^2 - 16x + 8} \\ \underline{3x^4 - 9x^3 + 6x^2} \\ -2x^3 + 10x^2 - 16x \\ \underline{-2x^3 + 6x^2 - 4x} \\ 4x^2 - 12x + 8 \\ \underline{4x^2 - 12x + 8} \\ 0 \end{array}$$

$$\frac{3x^4 - 11x^3 + 16x^2 - 16x + 8}{x^2 - 3x + 2} = 3x^2 - 2x + 4$$

You Try It 5

$$\begin{array}{r|rrr} -2 & 6 & 8 & -5 \\ & & -12 & 8 \\ \hline & 6 & -4 & 3 \end{array}$$

$$(6x^2 + 8x - 5) \div (x + 2)$$
$$= 6x - 4 + \frac{3}{x + 2}$$

You Try It 6

$$\begin{array}{r|rrrr} 2 & 5 & -12 & -8 & 16 \\ & & 10 & -4 & -24 \\ \hline & 5 & -2 & -12 & -8 \end{array}$$

$$(5x^3 - 12x^2 - 8x + 16) \div (x - 2)$$
$$= 5x^2 - 2x - 12 - \frac{8}{x - 2}$$

You Try It 7

$$\begin{array}{r|rrrrr} 3 & 2 & -3 & -8 & 0 & -2 \\ & & 6 & 9 & 3 & 9 \\ \hline & 2 & 3 & 1 & 3 & 7 \end{array}$$

$$(2x^4 - 3x^3 - 8x^2 - 2) \div (x - 3)$$
$$= 2x^3 + 3x^2 + x + 3 + \frac{7}{x - 3}$$

You Try It 8

$$\begin{array}{r|rrr} 2 & 2 & -3 & -5 \\ & & 4 & 2 \\ \hline & 2 & 1 & -3 \end{array}$$

$P(2) = -3$

You Try It 9

$$\begin{array}{r|rrrr} -3 & 2 & -5 & 0 & 7 \\ & & -6 & 33 & -99 \\ \hline & 2 & -11 & 33 & -92 \end{array}$$

$P(-3) = -92$

Copyright © Houghton Mifflin Company. All rights reserved.

Solutions to Chapter 7 "You Try It"

SECTION 7.1

You Try It 1 The GCF is $7a^2$.

$$14a^2 - 21a^4b = 7a^2(2) + 7a^2(-3a^2b)$$
$$= 7a^2(2 - 3a^2b)$$

You Try It 2 The GCF is 9.

$$27b^2 + 18b + 9$$
$$= 9(3b^2) + 9(2b) + 9(1)$$
$$= 9(3b^2 + 2b + 1)$$

You Try It 3
The GCF is $3x^2y^2$.

$$6x^4y^2 - 9x^3y^2 + 12x^2y^4$$
$$= 3x^2y^2(2x^2) + 3x^2y^2(-3x) + 3x^2y^2(4y^2)$$
$$= 3x^2y^2(2x^2 - 3x + 4y^2)$$

You Try It 4
$2y(5x - 2) - 3(2 - 5x)$
$= 2y(5x - 2) + 3(5x - 2)$ • $5x - 2$ is the common factor.
$= (5x - 2)(2y + 3)$

You Try It 5
$a^2 - 3a + 2ab - 6b$
$= (a^2 - 3a) + (2ab - 6b)$
$= a(a - 3) + 2b(a - 3)$ • $a - 3$ is the common factor.
$= (a - 3)(a + 2b)$

You Try It 6
$2mn^2 - n + 8mn - 4$
$= (2mn^2 - n) + (8mn - 4)$
$= n(2mn - 1) + 4(2mn - 1)$ • $2mn - 1$ is the common factor.
$= (2mn - 1)(n + 4)$

You Try It 7
$3xy - 9y - 12 + 4x$
$= (3xy - 9y) - (12 - 4x)$ • $-12 + 4x = -(12 - 4x)$
$= 3y(x - 3) - 4(3 - x)$ • $-(3 - x) = (x - 3)$
$= 3y(x - 3) + 4(x - 3)$ • $x - 3$ is the common factor.
$= (x - 3)(3y + 4)$

SECTION 7.2

You Try It 1
Find the positive factors of 20 whose sum is 9.

Factors	*Sum*
1, 20	21
2, 10	12
4, 5	9

$$x^2 + 9x + 20 = (x + 4)(x + 5)$$

You Try It 2
Find the factors of -18 whose sum is 7.

Factors	*Sum*
+1, −18	−17
−1, +18	17
+2, −9	−7
−2, +9	7
+3, −6	−3
−3, +6	3

$$x^2 + 7x - 18 = (x + 9)(x - 2)$$

You Try It 3
The GCF is $-2x$.

$$-2x^3 + 14x^2 - 12x = -2x(x^2 - 7x + 6)$$

Factor the trinomial $x^2 - 7x + 6$. Find two negative factors of 6 whose sum is -7.

Factors	*Sum*
−1, −6	−7
−2, −3	−5

$$-2x^3 + 14x^2 - 12x = -2x(x - 6)(x - 1)$$

You Try It 4
The GCF is 3.

$$3x^2 - 9xy - 12y^2 = 3(x^2 - 3xy - 4y^2)$$

Factor the trinomial.

Find the factors of -4 whose sum is -3.

Factors	*Sum*
+1, −4	−3
−1, +4	3
+2, −2	0

$$3x^2 - 9xy - 12y^2 = 3(x + y)(x - 4y)$$

SECTION 7.3

You Try It 1
Factor the trinomial $2x^2 - x - 3$.

Positive factors of 2: 1, 2

Factors of -3: $+1, -3$
$-1, +3$

Trial Factors	*Middle Term*
$(x + 1)(2x - 3)$	$-3x + 2x = -x$
$(x - 3)(2x + 1)$	$x - 6x = -5x$
$(x - 1)(2x + 3)$	$3x - 2x = x$
$(x + 3)(2x - 1)$	$-x + 6x = 5x$

$$2x^2 - x - 3 = (x + 1)(2x - 3)$$

Copyright © Houghton Mifflin Company. All rights reserved.

You Try It 2

The GCF is $-3y$.

$$-45y^3 + 12y^2 + 12y = -3y(15y^2 - 4y - 4)$$

Factor the trinomial $15y^2 - 4y - 4$.

Positive factors of 15: 1, 15; 3, 5

Factors of -4: 1, -4; -1, 4; 2, -2

Trial Factors	*Middle Term*
$(y + 1)(15y - 4)$	$-4y + 15y = 11y$
$(y - 4)(15y + 1)$	$y - 60y = -59y$
$(y - 1)(15y + 4)$	$4y - 15y = -11y$
$(y + 4)(15y - 1)$	$-y + 60y = 59y$
$(y + 2)(15y - 2)$	$-2y + 30y = 28y$
$(y - 2)(15y + 2)$	$2y - 30y = -28y$
$(3y + 1)(5y - 4)$	$-12y + 5y = -7y$
$(3y - 4)(5y + 1)$	$3y - 20y = -17y$
$(3y - 1)(5y + 4)$	$12y - 5y = 7y$
$(3y + 4)(5y - 1)$	$-3y + 20y = 17y$
$(3y + 2)(5y - 2)$	$-6y + 10y = 4y$
$(3y - 2)(5y + 2)$	$6y - 10y = -4y$

$$-45y^3 + 12y^2 + 12y = -3y(3y - 2)(5y + 2)$$

You Try It 3

Factors of -14 [$2(-7)$]	*Sum*
$+1, -14$	-13
$-1, +14$	13
$+2, -7$	-5
$-2, +7$	5

$$\begin{aligned}2a^2 + 13a - 7 &= 2a^2 - a + 14a - 7\\ &= (2a^2 - a) + (14a - 7)\\ &= a(2a - 1) + 7(2a - 1)\\ &= (2a - 1)(a + 7)\end{aligned}$$

$$2a^2 + 13a - 7 = (2a - 1)(a + 7)$$

You Try It 4

The GCF is $5x$.

$$15x^3 + 40x^2 - 80x = 5x(3x^2 + 8x - 16)$$

Factors of -48 [$3(-16)$]	*Sum*
$+1, -48$	-47
$-1, +48$	47
$+2, -24$	-22
$-2, +24$	22
$+3, -16$	-13
$-3, +16$	13
$+4, -12$	-8
$-4, +12$	8

$$\begin{aligned}3x^2 + 8x - 16 &= 3x^2 - 4x + 12x - 16\\ &= (3x^2 - 4x) + (12x - 16)\\ &= x(3x - 4) + 4(3x - 4)\\ &= (3x - 4)(x + 4)\end{aligned}$$

$$\begin{aligned}15x^3 + 40x^2 - 80x &= 5x(3x^2 + 8x - 16)\\ &= 5x(3x - 4)(x + 4)\end{aligned}$$

SECTION 7.4

You Try It 1

$$\begin{aligned}x^2 - 36y^4 &= x^2 - (6y^2)^2\\ &= (x + 6y^2)(x - 6y^2)\end{aligned}$$

• Difference of two squares

You Try It 2

$$9x^2 + 12x + 4 = (3x + 2)^2$$

• Perfect-square trinomial

You Try It 3

$$\begin{aligned}&(a + b)^2 - (a - b)^2\\ &= [(a + b) + (a - b)][(a + b) - (a - b)]\\ &= (a + b + a - b)(a + b - a + b)\\ &= (2a)(2b) = 4ab\end{aligned}$$

• Difference of two squares

You Try It 4

$$\begin{aligned}a^3b^3 - 27 &= (ab)^3 - 3^3\\ &= (ab - 3)(a^2b^2 + 3ab + 9)\end{aligned}$$

• Difference of two cubes

You Try It 5

$$\begin{aligned}8x^3 + y^3z^3 &= (2x)^3 + (yz)^3\\ &= (2x + yz)(4x^2 - 2xyz + y^2z^2)\end{aligned}$$

• Sum of two cubes

You Try It 6

$$\begin{aligned}&(x - y)^3 + (x + y)^3\\ &= [(x - y) + (x + y)]\\ &\quad \times [(x - y)^2 - (x - y)(x + y) + (x + y)^2]\\ &= 2x[x^2 - 2xy + y^2 - (x^2 - y^2)\\ &\quad + x^2 + 2xy + y^2]\\ &= 2x(x^2 - 2xy + y^2 - x^2 + y^2 + x^2\\ &\quad + 2xy + y^2)\\ &= 2x(x^2 + 3y^2)\end{aligned}$$

• Sum of two cubes

You Try It 7 Let $u = x^2$.

$$\begin{aligned}3x^4 + 4x^2 - 4 &= 3u^2 + 4u - 4\\ &= (u + 2)(3u - 2)\\ &= (x^2 + 2)(3x^2 - 2)\end{aligned}$$

You Try It 8

$$\begin{aligned}18x^3 - 6x^2 - 60x &= 6x(3x^2 - x - 10)\\ &= 6x(3x + 5)(x - 2)\end{aligned}$$

• GCF

You Try It 9

$$\begin{aligned}&4x - 4y - x^3 + x^2y\\ &= (4x - 4y) - (x^3 - x^2y)\\ &= 4(x - y) - x^2(x - y)\\ &= (x - y)(4 - x^2)\\ &= (x - y)(2 + x)(2 - x)\end{aligned}$$

• Factor by grouping.

Copyright © Houghton Mifflin Company. All rights reserved.

You Try It 10

$$\begin{aligned} x^{4n} - x^{2n}y^{2n} &= x^{2n+2n} - x^{2n}y^{2n} \\ &= x^{2n}(x^{2n} - y^{2n}) \\ &= x^{2n}[(x^n)^2 - (y^n)^2] \\ &= x^{2n}(x^n + y^n)(x^n - y^n) \end{aligned}$$

- GCF
- Difference of two squares

You Try It 11

$$\begin{aligned} ax^5 - ax^2y^6 &= ax^2(x^3 - y^6) \\ &= ax^2(x - y^2)(x^2 + xy^2 + y^4) \end{aligned}$$

- GCF
- Difference of two cubes

SECTION 7.5

You Try It 1

$2x(x + 7) = 0$

$2x = 0 \qquad x + 7 = 0$

$x = 0 \qquad x = -7$

- Principle of Zero Products

The solutions are 0 and -7.

You Try It 2

$$\begin{aligned} 4x^2 - 9 &= 0 \\ (2x - 3)(2x + 3) &= 0 \end{aligned}$$

- Difference of two squares

$2x - 3 = 0 \qquad 2x + 3 = 0$

$2x = 3 \qquad 2x = -3$

$x = \frac{3}{2} \qquad x = -\frac{3}{2}$

- Principle of Zero Products

The solutions are $\frac{3}{2}$ and $-\frac{3}{2}$.

You Try It 3

$$\begin{aligned} (x + 2)(x - 7) &= 52 \\ x^2 - 5x - 14 &= 52 \\ x^2 - 5x - 66 &= 0 \\ (x + 6)(x - 11) &= 0 \end{aligned}$$

$x + 6 = 0 \qquad x - 11 = 0$

$x = -6 \qquad x = 11$

- Principle of Zero Products

The solutions are -6 and 11.

You Try It 4

Strategy First consecutive positive integer: n
Second consecutive positive integer: $n + 1$

The sum of the squares of the two consecutive positive integers is 61.

Solution

$$\begin{aligned} n^2 + (n + 1)^2 &= 61 \\ n^2 + n^2 + 2n + 1 &= 61 \\ 2n^2 + 2n + 1 &= 61 \\ 2n^2 + 2n - 60 &= 0 \\ 2(n^2 + n - 30) &= 0 \\ 2(n - 5)(n + 6) &= 0 \end{aligned}$$

$n - 5 = 0 \qquad n + 6 = 0$

$n = 5 \qquad n = -6$

- Principle of Zero Products

Because -6 is not a positive integer, it is not a solution.

$n = 5$

$n + 1 = 5 + 1 = 6$

The two integers are 5 and 6.

You Try It 5

Strategy Width $= x$
Length $= 2x + 4$

The area of the rectangle is 96 in². Use the equation $A = L \cdot W$.

Solution

$$\begin{aligned} A &= L \cdot W \\ 96 &= (2x + 4)x \\ 96 &= 2x^2 + 4x \\ 0 &= 2x^2 + 4x - 96 \\ 0 &= 2(x^2 + 2x - 48) \\ 0 &= 2(x + 8)(x - 6) \end{aligned}$$

$x + 8 = 0 \qquad x - 6 = 0$

$x = -8 \qquad x = 6$

- Principle of Zero Products

Because the width cannot be a negative number, -8 is not a solution.

$x = 6$

$2x + 4 = 2(6) + 4 = 12 + 4 = 16$

The length is 16 in. The width is 6 in.

Solutions to Chapter 8 "You Try It"

SECTION 8.1

You Try It 1

$$\frac{6x^5y}{12x^2y^3} = \frac{\overset{1}{\cancel{2}} \cdot \overset{1}{\cancel{3}} \cdot x^5y}{\underset{1}{\cancel{2}} \cdot 2 \cdot \underset{1}{\cancel{3}} \cdot x^2y^3} = \frac{x^3}{2y^2}$$

You Try It 2

$$\begin{aligned} \frac{x^2 + 2x - 24}{16 - x^2} &= \frac{\overset{-1}{\cancel{(x - 4)}}(x + 6)}{\underset{1}{\cancel{(4 - x)}}(4 + x)} \\ &= -\frac{x + 6}{x + 4} \end{aligned}$$

- $(4 - x) = -1(x - 4)$

Copyright © Houghton Mifflin Company. All rights reserved.

You Try It 3

$$\frac{x^2+4x-12}{x^2-3x+2}=\frac{\overset{1}{\cancel{(x-2)}}(x+6)}{(x-1)\underset{1}{\cancel{(x-2)}}}=\frac{x+6}{x-1}$$

You Try It 4

$$\frac{12x^2+3x}{10x-15}\cdot\frac{8x-12}{9x+18}=\frac{3x(4x+1)}{5(2x-3)}\cdot\frac{4(2x-3)}{9(x+2)}$$

• **Factor.**

$$=\frac{\overset{1}{\cancel{3}}x(4x+1)\cdot 2\cdot 2\overset{1}{\cancel{(2x-3)}}}{5\underset{1}{\cancel{(2x-3)}}\cdot\underset{1}{\cancel{3}}\cdot 3(x+2)}$$

$$=\frac{4x(4x+1)}{15(x+2)}$$

You Try It 5

$$\frac{x^2+2x-15}{9-x^2}\cdot\frac{x^2-3x-18}{x^2-7x+6}$$

$$=\frac{(x-3)(x+5)}{(3-x)(3+x)}\cdot\frac{(x+3)(x-6)}{(x-1)(x-6)}$$

• **Factor.**

$$=\frac{\overset{-1}{\cancel{(x-3)}}(x+5)\cdot\overset{1}{\cancel{(x+3)}}\overset{1}{\cancel{(x-6)}}}{\underset{1}{\cancel{(3-x)}}\underset{1}{\cancel{(3+x)}}\cdot(x-1)\underset{1}{\cancel{(x-6)}}}=-\frac{x+5}{x-1}$$

You Try It 6

$$\frac{a^2}{4bc^2-2b^2c}\div\frac{a}{6bc-3b^2}$$

$$=\frac{a^2}{4bc^2-2b^2c}\cdot\frac{6bc-3b^2}{a}$$

• **Multiply by the reciprocal.**

$$=\frac{a^2\cdot 3\overset{1}{\cancel{b}}\overset{1}{\cancel{(2c-b)}}}{2\underset{1}{\cancel{b}}c\underset{1}{\cancel{(2c-b)}}\cdot a}=\frac{3a}{2c}$$

You Try It 7

$$\frac{3x^2+26x+16}{3x^2-7x-6}\div\frac{2x^2+9x-5}{x^2+2x-15}$$

$$=\frac{3x^2+26x+16}{3x^2-7x-6}\cdot\frac{x^2+2x-15}{2x^2+9x-5}$$

• **Multiply by the reciprocal.**

$$=\frac{\overset{1}{\cancel{(3x+2)}}(x+8)\cdot\overset{1}{\cancel{(x+5)}}\overset{1}{\cancel{(x-3)}}}{\underset{1}{\cancel{(3x+2)}}\underset{1}{\cancel{(x-3)}}\cdot(2x-1)\underset{1}{\cancel{(x+5)}}}=\frac{x+8}{2x-1}$$

SECTION 8.2

You Try It 1

The LCM is $(2x-5)(x+4)$.

$$\frac{2x}{2x-5}=\frac{2x}{2x-5}\cdot\frac{x+4}{x+4}=\frac{2x^2+8x}{(2x-5)(x+4)}$$

$$\frac{3}{x+4}=\frac{3}{x+4}\cdot\frac{2x-5}{2x-5}=\frac{6x-15}{(2x-5)(x+4)}$$

You Try It 2

$2x^2-11x+15=(x-3)(2x-5)$; $x^2-3x=x(x-3)$

The LCM is $x(x-3)(2x-5)$.

$$\frac{3x}{2x^2-11x+15}=\frac{3x}{(x-3)(2x-5)}\cdot\frac{x}{x}=\frac{3x^2}{x(x-3)(2x-5)}$$

$$\frac{x-2}{x^2-3x}=\frac{x-2}{x(x-3)}\cdot\frac{2x-5}{2x-5}=\frac{2x^2-9x+10}{x(x-3)(2x-5)}$$

You Try It 3

$2x-x^2=x(2-x)=-x(x-2)$;
$3x^2-5x-2=(x-2)(3x+1)$

The LCM is $x(x-2)(3x+1)$.

$$\frac{2x-7}{2x-x^2}=-\frac{2x-7}{x(x-2)}\cdot\frac{3x+1}{3x+1}=-\frac{6x^2-19x-7}{x(x-2)(3x+1)}$$

$$\frac{3x-2}{3x^2-5x-2}=\frac{3x-2}{(x-2)(3x+1)}\cdot\frac{x}{x}=\frac{3x^2-2x}{x(x-2)(3x+1)}$$

You Try It 4

The LCM is ab.

$$\frac{2}{b}-\frac{1}{a}+\frac{4}{ab}=\frac{2}{b}\cdot\frac{a}{a}-\frac{1}{a}\cdot\frac{b}{b}+\frac{4}{ab}$$

$$=\frac{2a}{ab}-\frac{b}{ab}+\frac{4}{ab}=\frac{2a-b+4}{ab}$$

You Try It 5

The LCM is $a(a-5)(a+5)$.

$$\frac{a-3}{a^2-5a}+\frac{a-9}{a^2-25}$$

$$=\frac{a-3}{a(a-5)}\cdot\frac{a+5}{a+5}+\frac{a-9}{(a-5)(a+5)}\cdot\frac{a}{a}$$

$$=\frac{(a-3)(a+5)+a(a-9)}{a(a-5)(a+5)}$$

$$=\frac{(a^2+2a-15)+(a^2-9a)}{a(a-5)(a+5)}$$

$$=\frac{a^2+2a-15+a^2-9a}{a(a-5)(a+5)}$$

$$=\frac{2a^2-7a-15}{a(a-5)(a+5)}=\frac{(2a+3)(a-5)}{a(a-5)(a+5)}$$

$$=\frac{(2a+3)\overset{1}{\cancel{(a-5)}}}{a\underset{1}{\cancel{(a-5)}}(a+5)}=\frac{2a+3}{a(a+5)}$$

You Try It 6

The LCM is $(x-4)(x+1)$.

$$\frac{2x}{x-4}-\frac{x-1}{x+1}+\frac{2}{x^2-3x-4}$$

$$=\frac{2x}{x-4}\cdot\frac{x+1}{x+1}-\frac{x-1}{x+1}\cdot\frac{x-4}{x-4}+\frac{2}{(x-4)(x+1)}$$

$$=\frac{2x(x+1)-(x-1)(x-4)+2}{(x-4)(x+1)}$$

$$=\frac{(2x^2+2x)-(x^2-5x+4)+2}{(x-4)(x+1)}$$

$$=\frac{x^2+7x-2}{(x-4)(x+1)}$$

Copyright © Houghton Mifflin Company. All rights reserved.

SECTION 8.3

You Try It 1

The LCM is $x - 3$.

$$\frac{2x + 5 + \frac{14}{x-3}}{4x + 16 + \frac{49}{x-3}} = \frac{2x + 5 + \frac{14}{x-3}}{4x + 16 + \frac{49}{x-3}} \cdot \frac{x-3}{x-3}$$

$$= \frac{(2x+5)(x-3) + \frac{14}{x-3}(x-3)}{(4x+16)(x-3) + \frac{49}{x-3}(x-3)}$$

$$= \frac{2x^2 - x - 15 + 14}{4x^2 + 4x - 48 + 49} = \frac{2x^2 - x - 1}{4x^2 + 4x + 1}$$

$$= \frac{(2x+1)(x-1)}{(2x+1)(2x+1)} = \frac{\overset{1}{\cancel{(2x+1)}}(x-1)}{\underset{1}{\cancel{(2x+1)}}(2x+1)} = \frac{x-1}{2x+1}$$

You Try It 2

The LCM of the denominators is x.

$$2 - \frac{1}{2 - \frac{1}{x}} = 2 - \frac{1}{2 - \frac{1}{x}} \cdot \frac{x}{x}$$

$$= 2 - \frac{1 \cdot x}{2 \cdot x - \frac{1}{x} \cdot x} = 2 - \frac{x}{2x - 1}$$

The LCM of the denominators is $2x - 1$.

$$2 - \frac{x}{2x-1} = 2 \cdot \frac{2x-1}{2x-1} - \frac{x}{2x-1}$$

$$= \frac{4x-2}{2x-1} - \frac{x}{2x-1}$$

$$= \frac{4x - 2 - x}{2x-1} = \frac{3x-2}{2x-1}$$

SECTION 8.4

You Try It 1

$$\frac{x}{x+6} = \frac{3}{x}$$
• **The LCM is $x(x + 6)$.**

$$\frac{x\overset{1}{\cancel{(x+6)}}}{1} \cdot \frac{x}{\underset{1}{\cancel{x+6}}} = \frac{x(x+6)}{1} \cdot \frac{3}{x}$$
• **Multiply by the LCM.**

$$x^2 = (x+6)3$$
• **Simplify.**

$$x^2 = 3x + 18$$
$$x^2 - 3x - 18 = 0$$
$$(x+3)(x-6) = 0$$
• **Factor.**

$$x + 3 = 0 \qquad x - 6 = 0$$
$$x = -3 \qquad x = 6$$
• **Principle of Zero Products**

Both -3 and 6 check as solutions.
The solutions are -3 and 6.

You Try It 2

$$\frac{5x}{x+2} = 3 - \frac{10}{x+2}$$
• **The LCM is $x + 2$.**

$$\frac{x+2}{1} \cdot \frac{5x}{x+2} = \frac{x+2}{1}\left(3 - \frac{10}{x+2}\right)$$
• **Clear denominators.**

$$\frac{\overset{1}{\cancel{x+2}}}{1} \cdot \frac{5x}{\underset{1}{\cancel{x+2}}} = \frac{x+2}{1} \cdot 3 - \frac{\overset{1}{\cancel{x+2}}}{1} \cdot \frac{10}{\underset{1}{\cancel{x+2}}}$$

$$5x = (x+2)3 - 10$$
• **Solve for x.**

$$5x = 3x + 6 - 10$$
$$5x = 3x - 4$$
$$2x = -4$$
$$x = -2$$

-2 does not check as a solution.
The equation has no solution.

SECTION 8.5

You Try It 1

$$\frac{2}{x+3} = \frac{6}{5x+5}$$

$$\frac{(x+3)(5x+5)}{1} \cdot \frac{2}{x+3} = \frac{(x+3)(5x+5)}{1} \cdot \frac{6}{5x+5}$$

$$\frac{\overset{1}{\cancel{(x+3)}}(5x+5)}{1} \cdot \frac{2}{\underset{1}{\cancel{x+3}}} = \frac{(x+3)\overset{1}{\cancel{(5x+5)}}}{1} \cdot \frac{6}{\underset{1}{\cancel{5x+5}}}$$

$$(5x+5)2 = (x+3)6$$
• **Solve for x.**

$$10x + 10 = 6x + 18$$
$$4x + 10 = 18$$
$$4x = 8$$
$$x = 2$$

The solution is 2.

You Try It 2

Strategy To find the total area that 256 ceramic tiles will cover, write and solve a proportion using x to represent the number of square feet that 256 tiles will cover.

Solution

$$\frac{9}{16} = \frac{x}{256}$$
• **Write a proportion.**

$$256\left(\frac{9}{16}\right) = 256\left(\frac{x}{256}\right)$$
• **Clear denominators.**

$$144 = x$$

An area of 144 ft^2 can be tiled using 256 ceramic tiles.

Copyright © Houghton Mifflin Company. All rights reserved.

You Try It 3

Strategy To find the area of triangle AOB:

- Solve a proportion to find the length of AO (the height of triangle AOB).
- Use the formula for the area of a triangle. AB is the base and AO is the height.

Solution

$$\frac{CD}{AB} = \frac{DO}{AO}$$ • Write a proportion.

$$\frac{4}{10} = \frac{3}{AO}$$ • Substitute.

$$10 \cdot AO \cdot \frac{4}{10} = 10 \cdot AO \cdot \frac{3}{AO}$$

$$4(AO) = 30$$

$$AO = 7.5$$

$$A = \frac{1}{2}bh$$ • Area of a triangle

$$= \frac{1}{2}(10)(7.5)$$ • Substitute.

$$= 37.5$$

The area of triangle AOB is 37.5 cm^2.

SECTION 8.6

You Try It 1

$$5x - 2y = 10$$

$$5x - 5x - 2y = -5x + 10$$ • Subtract $5x$.

$$-2y = -5x + 10$$

$$\frac{-2y}{-2} = \frac{-5x + 10}{-2}$$ • Divide by -2.

$$y = \frac{5}{2}x - 5$$

You Try It 2

$$s = \frac{A + L}{2}$$

$$2 \cdot s = 2\left(\frac{A + L}{2}\right)$$ • Multiply by 2.

$$2s = A + L$$

$$2s - A = A - A + L$$ • Subtract A.

$$2s - A = L$$

You Try It 3

$$S = a + (n - 1)d$$

$$S = a + nd - d$$

$$S - a = a - a + nd - d$$ • Subtract a.

$$S - a = nd - d$$

$$S - a + d = nd - d + d$$ • Add d.

$$S - a + d = nd$$

$$\frac{S - a + d}{d} = \frac{nd}{d}$$ • Divide by d.

$$\frac{S - a + d}{d} = n$$

You Try It 4

$$S = rS + C$$

$$S - rS = rS - rS + C$$ • Subtract rS.

$$S - rS = C$$

$$(1 - r)S = C$$ • Factor.

$$\frac{(1 - r)S}{1 - r} = \frac{C}{1 - r}$$ • Divide by $1 - r$.

$$S = \frac{C}{1 - r}$$

SECTION 8.7

You Try It 1

Strategy

- Time for one printer to complete the job: t

	Rate	*Time*	*Part*
1st printer	$\frac{1}{t}$	2	$\frac{2}{t}$
2nd printer	$\frac{1}{t}$	5	$\frac{5}{t}$

- The sum of the parts of the task completed must equal 1.

Solution

$$\frac{2}{t} + \frac{5}{t} = 1$$

$$t\left(\frac{2}{t} + \frac{5}{t}\right) = t \cdot 1$$

$$2 + 5 = t$$

$$7 = t$$

Working alone, one printer takes 7 h to print the payroll.

You Try It 2

Strategy

- Rate sailing across the lake: r
  Rate sailing back: $3r$

	Distance	*Rate*	*Time*
Across	6	r	$\frac{6}{r}$
Back	6	$3r$	$\frac{6}{3r}$

- The total time for the trip was 2 h.

Solution

$$\frac{6}{r} + \frac{6}{3r} = 2$$

$$3r\left(\frac{6}{r} + \frac{6}{3r}\right) = 3r(2)$$ • Multiply by the LCM, $3r$.

$$3r \cdot \frac{6}{r} + 3r \cdot \frac{6}{3r} = 6r$$

$$18 + 6 = 6r$$ • Solve for r.

$$24 = 6r$$

$$4 = r$$

The rate across the lake was 4 km/h.

Copyright © Houghton Mifflin Company. All rights reserved.

SECTION 8.8

You Try It 1

Strategy To find the distance:

- Write the basic direct variation equation, replace the variables by the given values, and solve for k.
- Write the direct variation equation, replacing k by its value. Substitute 5 for t and solve for s.

Solution

$$s = kt^2 \quad \text{• Direct variation equation}$$
$$64 = k(2)^2 \quad \text{• Replace } s \text{ by 64 and } t \text{ by 2.}$$
$$64 = k \cdot 4$$
$$16 = k$$

$$s = 16t^2 = 16(5)^2 = 400 \quad \text{• } k = 16, t = 5$$

The object will fall 400 ft in 5 s.

You Try It 2

Strategy To find the resistance:

- Write the basic inverse variation equation, replace the variables by the given values, and solve for k.
- Write the inverse variation equation, replacing k by its value. Substitute 0.02 for d and solve for R.

Solution

$$R = \frac{k}{d^2} \quad \text{• Inverse variation equation}$$
$$0.5 = \frac{k}{(0.01)^2} \quad \text{• Replace } R \text{ by 0.5 and } d \text{ by 0.01.}$$
$$0.5 = \frac{k}{0.0001}$$
$$0.00005 = k$$

$$R = \frac{0.00005}{d^2} = \frac{0.00005}{(0.02)^2} = 0.125 \quad \text{• } k = 0.00005,\ d = 0.02$$

The resistance is 0.125 ohm.

You Try It 3

Strategy To find the strength of the beam:

- Write the basic combined variation equation, replace the variables by the given values, and solve for k.
- Write the combined variation equation, replacing k by its value and substituting 4 for W, 8 for d, and 16 for L. Solve for s.

Solution

$$s = \frac{kWd^2}{L} \quad \text{• Combined variation equation}$$
$$1200 = \frac{k(2)(12)^2}{12} \quad \text{• Replace } s \text{ by 1200, } W \text{ by 2, } d \text{ by 12, and } L \text{ by 12.}$$
$$1200 = 24k$$
$$50 = k$$
$$s = \frac{50Wd^2}{L} \quad \text{• Replace } k \text{ by 50 in the combined variation equation.}$$
$$= \frac{50(4)8^2}{16} \quad \text{• Replace } W \text{ by 4, } d \text{ by 8, and } L \text{ by 16.}$$
$$= 800$$

The strength of the beam is 800 lb.

Solutions to Chapter 9 "You Try It"

SECTION 9.1

You Try It 1

$$16^{-3/4} = (2^4)^{-3/4} = 2^{-3} = \frac{1}{2^3} = \frac{1}{8}$$

You Try It 2 $(-81)^{3/4}$

The base of the exponential expression is negative, and the denominator of the exponent is a positive even number.

Therefore, $(-81)^{3/4}$ is not a real number.

You Try It 3

$$(x^{3/4}y^{1/2}z^{-2/3})^{-4/3} = x^{-1}y^{-2/3}z^{8/9} = \frac{z^{8/9}}{xy^{2/3}}$$

Copyright © Houghton Mifflin Company. All rights reserved.

You Try It 4

$$\left(\frac{16a^{-2}b^{4/3}}{9a^4b^{-2/3}}\right)^{-1/2} = \left(\frac{2^4a^{-6}b^2}{3^2}\right)^{-1/2}$$

- Use the Rule for Dividing Exponential Expressions.

$$= \frac{2^{-2}a^3b^{-1}}{3^{-1}}$$

- Use the Rule for Simplifying Powers of Products.

$$= \frac{3a^3}{2^2b} = \frac{3a^3}{4b}$$

You Try It 5 $(2x^3)^{3/4} = \sqrt[4]{(2x^3)^3}$
$= \sqrt[4]{8x^9}$

You Try It 6 $-5a^{5/6} = -5(a^5)^{1/6}$
$= -5\sqrt[6]{a^5}$

You Try It 7 $\sqrt[3]{3ab} = (3ab)^{1/3}$

You Try It 8 $\sqrt[4]{x^4 + y^4} = (x^4 + y^4)^{1/4}$

You Try It 9 $\sqrt[3]{-8x^{12}y^3} = -2x^4y$

You Try It 10 $-\sqrt[4]{81x^{12}y^8} = -3x^3y^2$

SECTION 9.2

You Try It 1 $\sqrt[5]{x^7} = \sqrt[5]{x^5 \cdot x^2}$
$= \sqrt[5]{x^5}\sqrt[5]{x^2}$
$= x\sqrt[5]{x^2}$

- x^5 is a perfect fifth power.

You Try It 2
$\sqrt[3]{-64x^8y^{18}} = \sqrt[3]{-64x^6y^{18}(x^2)}$
$= \sqrt[3]{-64x^6y^{18}}\sqrt[3]{x^2}$
$= -4x^2y^6\sqrt[3]{x^2}$

- $-64x^6y^{18}$ is a perfect third power.

You Try It 3
$3xy\sqrt[3]{81x^5y} - \sqrt[3]{192x^8y^4}$
$= 3xy\sqrt[3]{27x^3 \cdot 3x^2y} - \sqrt[3]{64x^6y^3 \cdot 3x^2y}$
$= 3xy\sqrt[3]{27x^3}\sqrt[3]{3x^2y} - \sqrt[3]{64x^6y^3}\sqrt[3]{3x^2y}$
$= 3xy \cdot 3x\sqrt[3]{3x^2y} - 4x^2y\sqrt[3]{3x^2y}$
$= 9x^2y\sqrt[3]{3x^2y} - 4x^2y\sqrt[3]{3x^2y} = 5x^2y\sqrt[3]{3x^2y}$

You Try It 4
$\sqrt{5b}(\sqrt{3b} - \sqrt{10})$
$= \sqrt{15b^2} - \sqrt{50b}$
$= \sqrt{b^2 \cdot 15} - \sqrt{25 \cdot 2b}$
$= \sqrt{b^2}\sqrt{15} - \sqrt{25}\sqrt{2b}$
$= b\sqrt{15} - 5\sqrt{2b}$

- The Distributive Property
- Simplify each radical expression.

You Try It 5
$(2\sqrt[3]{2x} - 3)(\sqrt[3]{2x} - 5)$
$= 2\sqrt[3]{4x^2} - 10\sqrt[3]{2x} - 3\sqrt[3]{2x} + 15$
$= 2\sqrt[3]{4x^2} - 13\sqrt[3]{2x} + 15$

- The FOIL method

You Try It 6
$(\sqrt{a} - 3\sqrt{y})(\sqrt{a} + 3\sqrt{y})$
$= (\sqrt{a})^2 - (3\sqrt{y})^2$
$= a - 9y$

- The expressions are conjugates.

You Try It 7

$$\frac{y}{\sqrt{3y}} = \frac{y}{\sqrt{3y}} \cdot \frac{\sqrt{3y}}{\sqrt{3y}} = \frac{y\sqrt{3y}}{\sqrt{9y^2}}$$

$$= \frac{y\sqrt{3y}}{3y} = \frac{\sqrt{3y}}{3}$$

- Rationalize the denominator.

You Try It 8

$$\frac{3x}{\sqrt[3]{3x^2}} = \frac{3x}{\sqrt[3]{3x^2}} \cdot \frac{\sqrt[3]{9x}}{\sqrt[3]{9x}} = \frac{3x\sqrt[3]{9x}}{\sqrt[3]{27x^3}}$$

$$= \frac{3x\sqrt[3]{9x}}{3x} = \sqrt[3]{9x}$$

- Rationalize the denominator.

You Try It 9

$$\frac{3 + \sqrt{6}}{2 - \sqrt{6}} = \frac{3 + \sqrt{6}}{2 - \sqrt{6}} \cdot \frac{2 + \sqrt{6}}{2 + \sqrt{6}}$$

$$= \frac{6 + 3\sqrt{6} + 2\sqrt{6} + (\sqrt{6})^2}{2^2 - (\sqrt{6})^2}$$

$$= \frac{6 + 5\sqrt{6} + 6}{4 - 6} = \frac{12 + 5\sqrt{6}}{-2}$$

$$= -\frac{12 + 5\sqrt{6}}{2}$$

- Rationalize the denominator.

SECTION 9.3

You Try It 1
$\sqrt{x} - \sqrt{x + 5} = 1$
$\sqrt{x} = 1 + \sqrt{x + 5}$
$(\sqrt{x})^2 = (1 + \sqrt{x + 5})^2$
$x = 1 + 2\sqrt{x + 5} + x + 5$
$-6 = 2\sqrt{x + 5}$
$-3 = \sqrt{x + 5}$
$(-3)^2 = (\sqrt{x + 5})^2$
$9 = x + 5$
$4 = x$

- Add $\sqrt{x + 5}$ to each side.
- Square each side.
- Square each side.

4 does not check as a solution. The equation has no solution.

You Try It 2
$\sqrt[4]{x - 8} = 3$
$(\sqrt[4]{x - 8})^4 = 3^4$
$x - 8 = 81$
$x = 89$

- Raise each side to the fourth power.

Copyright © Houghton Mifflin Company. All rights reserved.

Check:

$$\begin{array}{c|c} \sqrt[4]{x-8} = 3 & \\ \sqrt[4]{89-8} & 3 \\ \sqrt[4]{81} & 3 \\ 3 = 3 & \end{array}$$

The solution is 89.

You Try It 3

Strategy To find the diagonal, use the Pythagorean Theorem. One leg is the length of the rectangle. The second leg is the width of the rectangle. The hypotenuse is the diagonal of the rectangle.

Solution

$c^2 = a^2 + b^2$ • Pythagorean Theorem

$c^2 = (6)^2 + (3)^2$ • Replace a by 6 and b by 3.

$c^2 = 36 + 9$ • Solve for c.

$c^2 = 45$

$(c^2)^{1/2} = (45)^{1/2}$ • Raise each side to the $\frac{1}{2}$ power.

$c = \sqrt{45}$ • $a^{1/2} = \sqrt{a}$

$c \approx 6.7$

The diagonal is approximately 6.7 cm.

You Try It 4

Strategy To find the height, replace d in the equation with the given value and solve for h.

Solution

$d = \sqrt{1.5h}$

$5.5 = \sqrt{1.5h}$ • Replace d by 5.5.

$(5.5)^2 = (\sqrt{1.5h})^2$ • Square each side.

$30.25 = 1.5h$

$20.17 \approx h$

The periscope must be approximately 20.17 ft above the water.

You Try It 5

Strategy To find the distance, replace the variables v and a in the equation by the given values and solve for s.

Solution

$v = \sqrt{2as}$

$88 = \sqrt{2 \cdot 22s}$ • Replace v by 88 and a by 22.

$88 = \sqrt{44s}$

$(88)^2 = (\sqrt{44s})^2$ • Square each side.

$7744 = 44s$

$176 = s$

The distance required is 176 ft.

SECTION 9.4

You Try It 1 $\sqrt{-45} = i\sqrt{45} = i\sqrt{9 \cdot 5} = 3i\sqrt{5}$

You Try It 2
$$\begin{aligned} \sqrt{98} - \sqrt{-60} &= \sqrt{98} - i\sqrt{60} \\ &= \sqrt{49 \cdot 2} - i\sqrt{4 \cdot 15} \\ &= 7\sqrt{2} - 2i\sqrt{15} \end{aligned}$$

You Try It 3 $(-4 + 2i) - (6 - 8i) = -10 + 10i$

You Try It 4
$$\begin{aligned} &(16 - \sqrt{-45}) - (3 + \sqrt{-20}) \\ &= (16 - i\sqrt{45}) - (3 + i\sqrt{20}) \\ &= (16 - i\sqrt{9 \cdot 5}) - (3 + i\sqrt{4 \cdot 5}) \\ &= (16 - 3i\sqrt{5}) - (3 + 2i\sqrt{5}) \\ &= 13 - 5i\sqrt{5} \end{aligned}$$

You Try It 5 $(3 - 2i) + (-3 + 2i) = 0 + 0i = 0$

You Try It 6 $(-3i)(-10i) = 30i^2 = 30(-1) = -30$

You Try It 7
$$\begin{aligned} -\sqrt{-8} \cdot \sqrt{-5} &= -i\sqrt{8} \cdot i\sqrt{5} = -i^2\sqrt{40} \\ &= -(-1)\sqrt{40} = \sqrt{4 \cdot 10} = 2\sqrt{10} \end{aligned}$$

You Try It 8
$$\begin{aligned} -6i(3 + 4i) &= -18i - 24i^2 \\ &= -18i - 24(-1) \\ &= 24 - 18i \end{aligned}$$
• The Distributive Property

You Try It 9
$$\begin{aligned} \sqrt{-3}(\sqrt{27} - \sqrt{-6}) &= i\sqrt{3}(\sqrt{27} - i\sqrt{6}) \\ &= i\sqrt{81} - i^2\sqrt{18} \\ &= i\sqrt{81} - (-1)\sqrt{9 \cdot 2} \\ &= 9i + 3\sqrt{2} \\ &= 3\sqrt{2} + 9i \end{aligned}$$

You Try It 10
$$\begin{aligned} (4 - 3i)(2 - i) &= 8 - 4i - 6i + 3i^2 \\ &= 8 - 10i + 3i^2 \\ &= 8 - 10i + 3(-1) \\ &= 5 - 10i \end{aligned}$$
• FOIL

You Try It 11
$$\begin{aligned} (3 + 6i)(3 - 6i) &= 3^2 + 6^2 \\ &= 9 + 36 \\ &= 45 \end{aligned}$$
• Conjugates

Copyright © Houghton Mifflin Company. All rights reserved.

You Try It 12

$$\frac{2-3i}{4i} = \frac{2-3i}{4i}\cdot\frac{i}{i} = \frac{2i-3i^2}{4i^2} = \frac{2i-3(-1)}{4(-1)} = \frac{3+2i}{-4} = -\frac{3}{4} - \frac{1}{2}i$$

You Try It 13

$$\frac{2+5i}{3-2i} = \frac{2+5i}{3-2i}\cdot\frac{3+2i}{3+2i} = \frac{6+4i+15i+10i^2}{3^2+2^2} = \frac{6+19i+10(-1)}{9+4} = \frac{-4+19i}{13} = -\frac{4}{13} + \frac{19}{13}i$$

Solutions to Chapter 10 "You Try It"

SECTION 10.1

You Try It 1

$$2x^2 = 7x - 3$$
$$2x^2 - 7x + 3 = 0 \quad \text{• Write in standard form.}$$
$$(2x-1)(x-3) = 0 \quad \text{• Solve by factoring.}$$

$$2x - 1 = 0 \qquad x - 3 = 0$$
$$2x = 1 \qquad x = 3$$
$$x = \frac{1}{2}$$

The solutions are $\frac{1}{2}$ and 3.

You Try It 2

$$x^2 - 3ax - 4a^2 = 0$$
$$(x+a)(x-4a) = 0$$

$$x + a = 0 \qquad x - 4a = 0$$
$$x = -a \qquad x = 4a$$

The solutions are $-a$ and $4a$.

You Try It 3

$$(x - r_1)(x - r_2) = 0$$
$$(x-3)\left[x - \left(-\frac{1}{2}\right)\right] = 0 \quad \text{• } r_1 = 3, r_2 = -\frac{1}{2}$$
$$(x-3)\left(x + \frac{1}{2}\right) = 0$$
$$x^2 - \frac{5}{2}x - \frac{3}{2} = 0$$
$$2\left(x^2 - \frac{5}{2}x - \frac{3}{2}\right) = 2\cdot 0$$
$$2x^2 - 5x - 3 = 0$$

You Try It 4

$$2(x+1)^2 - 24 = 0$$
$$2(x+1)^2 = 24 \quad \text{• Solve for } (x+1)^2.$$
$$(x+1)^2 = 12$$
$$\sqrt{(x+1)^2} = \sqrt{12} \quad \text{• Take the square root of each side of the equation.}$$
$$x + 1 = \pm\sqrt{12}$$
$$x + 1 = \pm 2\sqrt{3}$$

$$x + 1 = 2\sqrt{3} \qquad x + 1 = -2\sqrt{3} \quad \text{• Solve for } x.$$
$$x = -1 + 2\sqrt{3} \qquad x = -1 - 2\sqrt{3}$$

The solutions are $-1 + 2\sqrt{3}$ and $-1 - 2\sqrt{3}$.

SECTION 10.2

You Try It 1

$$4x^2 - 4x - 1 = 0 \quad \text{• Write in the form } ax^2 + bx = -c.$$
$$4x^2 - 4x = 1$$
$$\frac{1}{4}(4x^2 - 4x) = \frac{1}{4}\cdot 1 \quad \text{• Multiply both sides by } \frac{1}{a}.$$
$$x^2 - x = \frac{1}{4}$$
$$x^2 - x + \frac{1}{4} = \frac{1}{4} + \frac{1}{4} \quad \text{• Complete the square.}$$
$$\left(x - \frac{1}{2}\right)^2 = \frac{2}{4} \quad \text{• Factor.}$$
$$\sqrt{\left(x - \frac{1}{2}\right)^2} = \sqrt{\frac{2}{4}} \quad \text{• Take square roots.}$$
$$x - \frac{1}{2} = \pm\frac{\sqrt{2}}{2}$$

$$x - \frac{1}{2} = \frac{\sqrt{2}}{2} \qquad x - \frac{1}{2} = -\frac{\sqrt{2}}{2} \quad \text{• Solve for } x.$$
$$x = \frac{1}{2} + \frac{\sqrt{2}}{2} \qquad x = \frac{1}{2} - \frac{\sqrt{2}}{2}$$

The solutions are $\frac{1+\sqrt{2}}{2}$ and $\frac{1-\sqrt{2}}{2}$.

Copyright © Houghton Mifflin Company. All rights reserved.

You Try It 2

$$x^2 + 4x + 8 = 0$$
$$x^2 + 4x = -8$$
$$x^2 + 4x + 4 = -8 + 4$$ • Complete the square.
$$(x + 2)^2 = -4$$ • Factor.
$$\sqrt{(x + 2)^2} = \sqrt{-4}$$ • Take square roots.
$$x + 2 = \pm 2i$$

$$x + 2 = 2i \qquad x + 2 = -2i$$ • Solve for x.
$$x = -2 + 2i \qquad x = -2 - 2i$$

The solutions are $-2 + 2i$ and $-2 - 2i$.

SECTION 10.3

You Try It 1

$$x^2 - 2x + 10 = 0$$
$$a = 1, b = -2, c = 10$$
$$x = \frac{-b \pm \sqrt{b^2 - 4ac}}{2a}$$
$$= \frac{-(-2) \pm \sqrt{(-2)^2 - 4(1)(10)}}{2 \cdot 1}$$
$$= \frac{2 \pm \sqrt{4 - 40}}{2} = \frac{2 \pm \sqrt{-36}}{2}$$
$$= \frac{2 \pm 6i}{2} = 1 \pm 3i$$

The solutions are $1 + 3i$ and $1 - 3i$.

You Try It 2

$$4x^2 = 4x - 1$$
$$4x^2 - 4x + 1 = 0$$ • Write in standard form.
$$a = 4, b = -4, c = 1$$
$$x = \frac{-b \pm \sqrt{b^2 - 4ac}}{2a}$$
$$= \frac{-(-4) \pm \sqrt{(-4)^2 - 4(4)(1)}}{2 \cdot 4}$$
$$= \frac{4 \pm \sqrt{16 - 16}}{8} = \frac{4 \pm \sqrt{0}}{8}$$
$$= \frac{4}{8} = \frac{1}{2}$$

The solution is $\frac{1}{2}$.

You Try It 3

$$3x^2 - x - 1 = 0$$
$$a = 3, b = -1, c = -1$$
$$b^2 - 4ac =$$
$$(-1)^2 - 4(3)(-1) = 1 + 12 = 13$$
$$13 > 0$$

Because the discriminant is greater than zero, the equation has two real number solutions.

SECTION 10.4

You Try It 1

$$x - 5x^{1/2} + 6 = 0$$
$$(x^{1/2})^2 - 5(x^{1/2}) + 6 = 0$$
$$u^2 - 5u + 6 = 0$$
$$(u - 2)(u - 3) = 0$$

$$u - 2 = 0 \qquad u - 3 = 0$$
$$u = 2 \qquad u = 3$$

Replace u by $x^{1/2}$.

$$x^{1/2} = 2 \qquad x^{1/2} = 3$$
$$\sqrt{x} = 2 \qquad \sqrt{x} = 3$$
$$(\sqrt{x})^2 = 2^2 \qquad (\sqrt{x})^2 = 3^2$$
$$x = 4 \qquad x = 9$$

4 and 9 check as solutions.
The solutions are 4 and 9.

You Try It 2

$$\sqrt{2x + 1} + x = 7$$
$$\sqrt{2x + 1} = 7 - x$$ • Solve for the radical.
$$(\sqrt{2x + 1})^2 = (7 - x)^2$$ • Square each side.
$$2x + 1 = 49 - 14x + x^2$$ • Write in standard form.
$$0 = x^2 - 16x + 48$$
$$0 = (x - 4)(x - 12)$$ • Solve by factoring.

$$x - 4 = 0 \qquad x - 12 = 0$$
$$x = 4 \qquad x = 12$$

4 checks as a solution.
12 does not check as a solution.

The solution is 4.

You Try It 3

$$\sqrt{2x - 1} + \sqrt{x} = 2$$

Solve for one of the radical expressions.

$$\sqrt{2x - 1} = 2 - \sqrt{x}$$
$$(\sqrt{2x - 1})^2 = (2 - \sqrt{x})^2$$ • Square each side.
$$2x - 1 = 4 - 4\sqrt{x} + x$$
$$x - 5 = -4\sqrt{x}$$
$$(x - 5)^2 = (-4\sqrt{x})^2$$ • Square each side.
$$x^2 - 10x + 25 = 16x$$
$$x^2 - 26x + 25 = 0$$
$$(x - 1)(x - 25) = 0$$

$$x - 1 = 0 \qquad x - 25 = 0$$
$$x = 1 \qquad x = 25$$

1 checks as a solution.
25 does not check as a solution.

The solution is 1.

Copyright © Houghton Mifflin Company. All rights reserved.

You Try It 4

$$3y + \frac{25}{3y-2} = -8$$

$$(3y-2)\left(3y + \frac{25}{3y-2}\right) = (3y-2)(-8)$$

• Clear denominators.

$$(3y-2)(3y) + (3y-2)\left(\frac{25}{3y-2}\right) = (3y-2)(-8)$$

$$9y^2 - 6y + 25 = -24y + 16$$

• Write in standard form.

$$9y^2 + 18y + 9 = 0$$
$$9(y^2 + 2y + 1) = 0$$

• Solve by factoring.

$$9(y+1)(y+1) = 0$$

$y + 1 = 0 \qquad y + 1 = 0$
$y = -1 \qquad y = -1$

The solution is -1.

SECTION 10.5

You Try It 1 $2x^2 - x - 10 \le 0$

$(2x - 5)(x + 2) \le 0$

$2x - 5$	$- - -$	$- - - - - - - - - -$	$+ + +$
$x + 2$	$- - -$	$+ + + + + + + + + +$	$+ + +$

$-3 \quad -2 \quad -1 \quad 0 \quad 1 \quad 2 \quad 3$

$-5\ -4\ -3\ -2\ -1\ 0\ 1\ 2\ 3\ 4\ 5$

$$\left\{x \mid -2 \le x \le \frac{5}{2}\right\}$$

SECTION 10.6

You Try It 1

Strategy
• This is a geometry problem.
• Width of the rectangle: W
Length of the rectangle: $W + 3$
• Use the equation $A = L \cdot W$.

Solution

$$A = L \cdot W$$
$$54 = (W+3)(W)$$
$$54 = W^2 + 3W$$
$$0 = W^2 + 3W - 54$$
$$0 = (W+9)(W-6)$$

$W + 9 = 0 \qquad W - 6 = 0$
$W = -9 \qquad W = 6$

The solution -9 is not possible.

$W + 3 = 6 + 3 = 9$

The length is 9 m.

Solutions to Chapter 11 "You Try It"

SECTION 11.1

You Try It 1

x-coordinate of vertex:

$$-\frac{b}{2a} = -\frac{4}{2(4)} = -\frac{1}{2}$$

y-coordinate of vertex:

$$y = 4x^2 + 4x + 1$$
$$= 4\left(-\frac{1}{2}\right)^2 + 4\left(-\frac{1}{2}\right) + 1$$
$$= 1 - 2 + 1$$
$$= 0$$

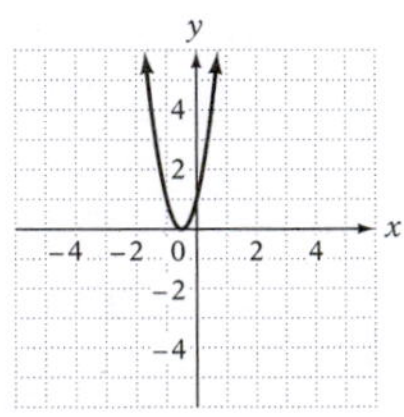

Vertex: $\left(-\frac{1}{2}, 0\right)$

Axis of symmetry: $x = -\frac{1}{2}$

You Try It 2

x-coordinate of vertex:

$$-\frac{b}{2a} = -\frac{-2}{2(-1)} = -1$$

y-coordinate of vertex:

$$f(x) = -x^2 - 2x - 1$$
$$f(-1) = -(-1)^2 - 2(-1) - 1$$
$$= -1 + 2 - 1$$
$$= 0$$

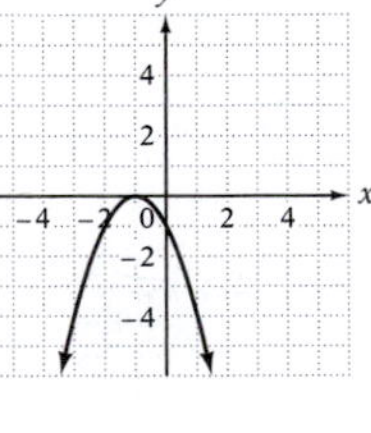

Vertex: $(-1, 0)$

The domain is $\{x|x \in \text{real numbers}\}$. The range is $\{y \mid y \le 0\}$.

You Try It 3

$y = x^2 + 3x + 4$
$0 = x^2 + 3x + 4$

Copyright © Houghton Mifflin Company. All rights reserved.

$$x = \frac{-b \pm \sqrt{b^2 - 4ac}}{2a}$$
$$= \frac{-3 \pm \sqrt{3^2 - 4(1)(4)}}{2 \cdot 1} \quad \bullet\ a = 1, b = 3, c = 4$$
$$= \frac{-3 \pm \sqrt{-7}}{2}$$
$$= \frac{-3 \pm i\sqrt{7}}{2}$$

The equation has no real number solutions. There are no x-intercepts.

You Try It 4

$g(x) = x^2 - x + 6$
$0 = x^2 - x + 6$

$$x = \frac{-b \pm \sqrt{b^2 - 4ac}}{2a}$$
$$= \frac{-(-1) \pm \sqrt{(-1)^2 - 4(1)(6)}}{2(1)} \quad \bullet\ a = 1, b = -1, c = 6$$
$$= \frac{1 \pm \sqrt{1 - 24}}{2}$$
$$= \frac{1 \pm \sqrt{-23}}{2} = \frac{1 \pm i\sqrt{23}}{2}$$

The zeros of the function are $\frac{1}{2} + \frac{\sqrt{23}}{2}i$ and $\frac{1}{2} - \frac{\sqrt{23}}{2}i$.

You Try It 5

$y = x^2 - x - 6$
$a = 1, b = -1, c = -6$

$b^2 - 4ac$
$(-1)^2 - 4(1)(-6) = 1 + 24 = 25$

Because the discriminant is greater than zero, the parabola has two x-intercepts.

You Try It 6

$f(x) = -3x^2 + 4x - 1$

$$x = -\frac{b}{2a} = -\frac{4}{2(-3)} = \frac{2}{3} \quad \bullet\ \text{The } x\text{-coordinate of the vertex}$$
$$f(x) = -3x^2 + 4x - 1$$
$$f\left(\frac{2}{3}\right) = -3\left(\frac{2}{3}\right)^2 + 4\left(\frac{2}{3}\right) - 1 \quad \bullet\ x = \frac{2}{3}$$
$$= -\frac{4}{3} + \frac{8}{3} - 1 = \frac{1}{3}$$

Because a is negative, the function has a maximum value.

The maximum value of the function is $\frac{1}{3}$.

You Try It 7

Strategy

- To find the time it takes the ball to reach its maximum height, find the t-coordinate of the vertex.
- To find the maximum height, evaluate the function at the t-coordinate of the vertex.

Solution

$$t = -\frac{b}{2a} = -\frac{64}{2(-16)} = 2 \quad \bullet\ \text{The } t\text{-coordinate of the vertex}$$

The ball reaches its maximum height in 2 s.

$s(t) = -16t^2 + 64t$
$s(2) = -16(2)^2 + 64(2) = -64 + 128 = 64 \quad \bullet\ t = 2$

The maximum height is 64 ft.

You Try It 8

Strategy

$$P = 2x + y$$
$$100 = 2x + y \quad \bullet\ P = 100$$
$$100 - 2x = y \quad \bullet\ \text{Solve for } y.$$

Express the area of the rectangle in terms of x.

$A = xy$
$A = x(100 - 2x) \quad \bullet\ y = 100 - 2x$
$A = -2x^2 + 100x$

- To find the width, find the x-coordinate of the vertex of $f(x) = -2x^2 + 100x$.
- To find the length, replace x in $y = 100 - 2x$ by the x-coordinate of the vertex.

Solution

$$x = -\frac{b}{2a} = -\frac{100}{2(-2)} = 25$$

The width is 25 ft.

$$100 - 2x = 100 - 2(25)$$
$$= 100 - 50 = 50$$

The length is 50 ft.

SECTION 11.2

You Try It 1 Any vertical line intersects the graph at most once. The graph is the graph of a function.

You Try It 2

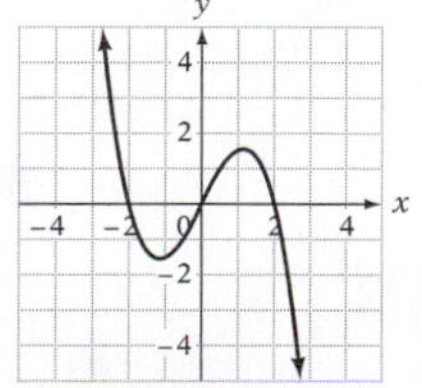

Domain: $\{x \mid x \in \text{real numbers}\}$
Range: $\{y \mid y \in \text{real numbers}\}$

Copyright © Houghton Mifflin Company. All rights reserved.

You Try It 3

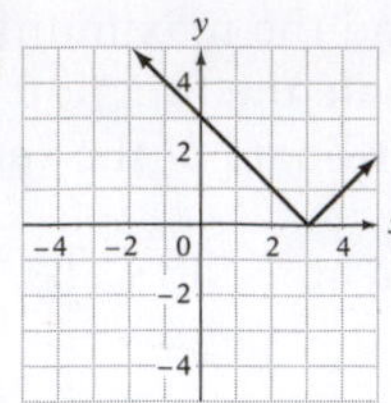

Domain: $\{x \mid x \in \text{real numbers}\}$
Range: $\{y \mid y \geq 0\}$

You Try It 4

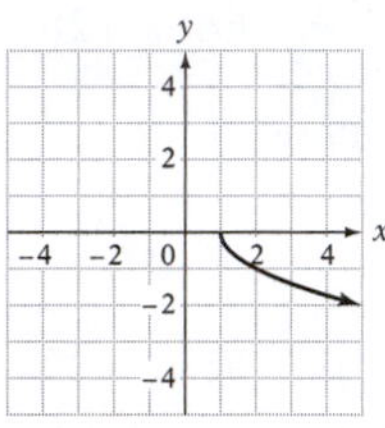

Domain: $\{x \mid x \geq 1\}$
Range: $\{y \mid y \leq 0\}$

SECTION 11.3

You Try It 1

$$
\begin{aligned}
(f+g)(-2) &= f(-2) + g(-2) \\
&= [(-2)^2 + 2(-2)] + [5(-2) - 2] \\
&= (4 - 4) + (-10 - 2) \\
&= -12
\end{aligned}
$$

$(f+g)(-2) = -12$

You Try It 2

$$
\begin{aligned}
(f \cdot g)(3) &= f(3) \cdot g(3) \\
&= (4 - 3^2) \cdot [3(3) - 4] \\
&= (4 - 9) \cdot (9 - 4) \\
&= (-5)(5) \\
&= -25
\end{aligned}
$$

$(f \cdot g)(3) = -25$

You Try It 3

$$
\begin{aligned}
\left(\frac{f}{g}\right)(4) &= \frac{f(4)}{g(4)} \\
&= \frac{4^2 - 4}{4^2 + 2 \cdot 4 + 1} \\
&= \frac{16 - 4}{16 + 8 + 1} \\
&= \frac{12}{25}
\end{aligned}
$$

$\left(\frac{f}{g}\right)(4) = \frac{12}{25}$

You Try It 4

$g(x) = x^2$
$g(-1) = (-1)^2 = 1$ • Evaluate g at -1.

$f(x) = 1 - 2x$ • Evaluate f at $g(-1) = 1$.
$f[g(-1)] = f(1) = 1 - 2(1) = -1$
$f[g(-1)] = -1$

You Try It 5

$M(s) = s^3 + 1$ • Evaluate M at $L(s)$.

$$
\begin{aligned}
M[L(s)] &= (s+1)^3 + 1 \\
&= s^3 + 3s^2 + 3s + 1 + 1 \\
&= s^3 + 3s^2 + 3s + 2
\end{aligned}
$$

$M[L(s)] = s^3 + 3s^2 + 3s + 2$

SECTION 11.4

You Try It 1 Because any horizontal line intersects the graph at most once, the graph is the graph of a 1–1 function.

You Try It 2

$$
\begin{aligned}
f(x) &= \frac{1}{2}x + 4 \\
y &= \frac{1}{2}x + 4 \\
x &= \frac{1}{2}y + 4 \\
x - 4 &= \frac{1}{2}y \\
2x - 8 &= y \\
f^{-1}(x) &= 2x - 8
\end{aligned}
$$

The inverse of the function is given by $f^{-1}(x) = 2x - 8$.

You Try It 3

$$
\begin{aligned}
f[g(x)] &= 2\left(\frac{1}{2}x - 3\right) - 6 \\
&= x - 6 - 6 = x - 12
\end{aligned}
$$

No, $g(x)$ is not the inverse of $f(x)$.

Copyright © Houghton Mifflin Company. All rights reserved.

Solutions to Chapter 12 "You Try It"

SECTION 12.1

You Try It 1

$f(x) = \left(\frac{2}{3}\right)^x$

$f(3) = \left(\frac{2}{3}\right)^3 = \frac{8}{27}$ • $x = 3$

$f(-2) = \left(\frac{2}{3}\right)^{-2} = \left(\frac{3}{2}\right)^2 = \frac{9}{4}$ • $x = -2$

You Try It 2

$f(x) = 2^{2x+1}$

$f(0) = 2^{2(0)+1} = 2^1 = 2$ • $x = 0$

$f(-2) = 2^{2(-2)+1} = 2^{-3} = \frac{1}{2^3} = \frac{1}{8}$ • $x = -2$

You Try It 3

$f(x) = e^{2x-1}$

$f(2) = e^{2\cdot 2-1} = e^3 \approx 20.0855$ • $x = 2$

$f(-2) = e^{2(-2)-1} = e^{-5} \approx 0.0067$ • $x = -2$

You Try It 4

x	y
-4	4
-2	2
0	1
2	$\frac{1}{2}$
4	$\frac{1}{4}$

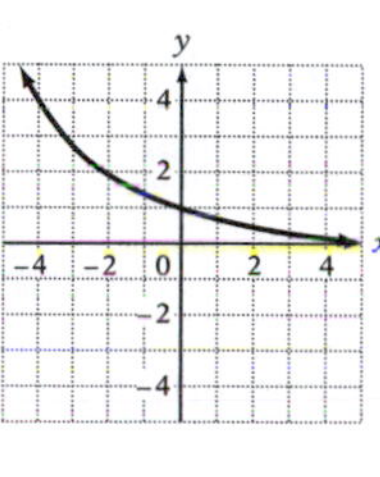

You Try It 5

x	y
-2	$\frac{5}{4}$
-1	$\frac{3}{2}$
0	2
1	3
2	5

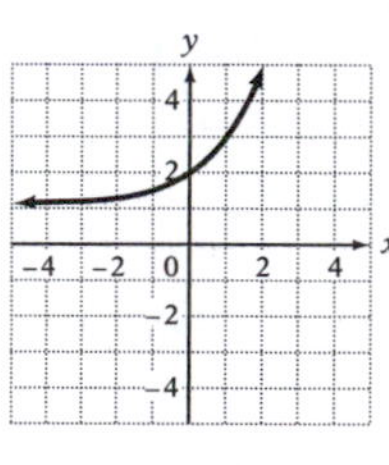

You Try It 6

x	y
-2	6
-1	4
0	3
1	$\frac{5}{2}$
2	$\frac{9}{4}$

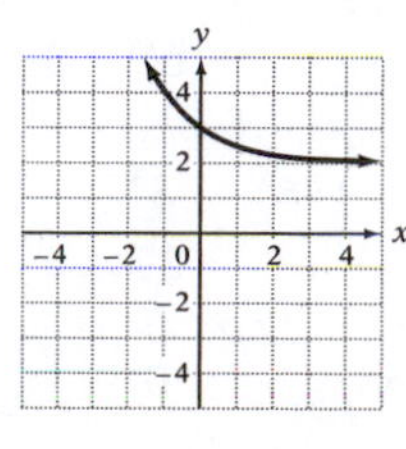

You Try It 7

x	y
-4	$\frac{9}{4}$
-2	$\frac{5}{2}$
0	3
2	4
4	6

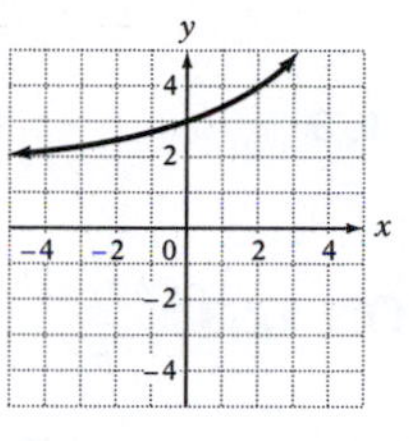

SECTION 12.2

You Try It 1

$\log_4 64 = x$ • Write an equation.

$64 = 4^x$ • Write the equivalent exponential form.

$4^3 = 4^x$ • $64 = 4^3$

$3 = x$ • The bases are the same. The exponents are equal.

$\log_4 64 = 3$

You Try It 2

$\log_2 x = -4$

$2^{-4} = x$ • Write the equivalent exponential form.

$\frac{1}{2^4} = x$

$\frac{1}{16} = x$

The solution is $\frac{1}{16}$.

You Try It 3

$\ln x = 3$

$e^3 = x$ • Write the equivalent exponential form.

$20.0855 \approx x$

You Try It 4

$\log_8 \sqrt[3]{xy^2} = \log_8 (xy^2)^{1/3} = \frac{1}{3}\log_8(xy^2)$ • Power Property

$= \frac{1}{3}(\log_8 x + \log_8 y^2)$ • Product Property

$= \frac{1}{3}(\log_8 x + 2\log_8 y)$ • Power Property

$= \frac{1}{3}\log_8 x + \frac{2}{3}\log_8 y$ • Distributive Property

You Try It 5

$\frac{1}{3}(\log_4 x - 2\log_4 y + \log_4 z)$

$= \frac{1}{3}(\log_4 x - \log_4 y^2 + \log_4 z)$

$= \frac{1}{3}\left(\log_4 \frac{x}{y^2} + \log_4 z\right) = \frac{1}{3}\left(\log_4 \frac{xz}{y^2}\right)$

$= \log_4\left(\frac{xz}{y^2}\right)^{1/3} = \log_4 \sqrt[3]{\frac{xz}{y^2}}$

Copyright © Houghton Mifflin Company. All rights reserved.

You Try It 6 Because $\log_b 1 = 0$, $\log_9 1 = 0$.

You Try It 7 $\log_3 0.834 = \dfrac{\ln 0.834}{\ln 3} \approx -0.16523$

You Try It 8 $\log_7 6.45 = \dfrac{\log 6.45}{\log 7} \approx 0.95795$

SECTION 12.3

You Try It 1

$f(x) = \log_2(x - 1)$

$y = \log_2(x - 1)$ • $f(x) = y$

$2^y = x - 1$ • Write the equivalent exponential equation.

$2^y + 1 = x$

x	y
$\frac{5}{4}$	-2
$\frac{3}{2}$	-1
2	0
3	1
5	2

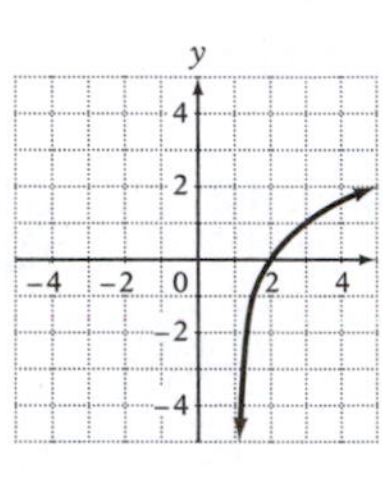

You Try It 2

$f(x) = \log_3(2x)$

$y = \log_3(2x)$ • $f(x) = y$

$3^y = 2x$ • Write the equivalent exponential equation.

$\dfrac{3^y}{2} = x$

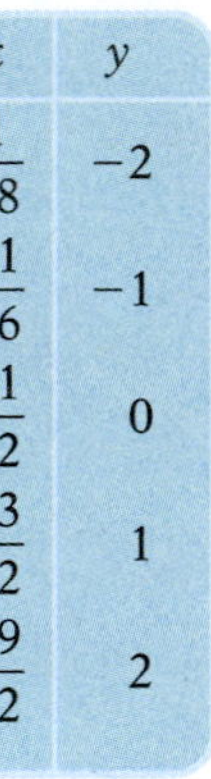

x	y
$\frac{1}{18}$	-2
$\frac{1}{6}$	-1
$\frac{1}{2}$	0
$\frac{3}{2}$	1
$\frac{9}{2}$	2

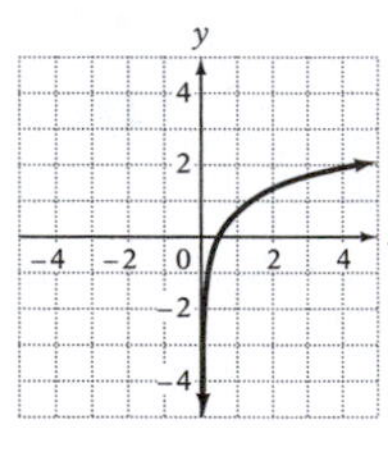

You Try It 3

$f(x) = -\log_3(x + 1)$

$y = -\log_3(x + 1)$ • $f(x) = y$

$-y = \log_3(x + 1)$ • Multiply both sides by -1.

$3^{-y} = x + 1$ • Write the equivalent exponential equation.

$3^{-y} - 1 = x$

x	y
8	-2
2	-1
0	0
$-\frac{2}{3}$	1
$-\frac{8}{9}$	2

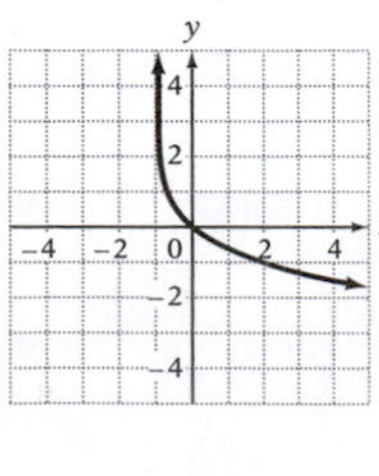

SECTION 12.4

You Try It 1

$(1.06)^n = 1.5$

$\log(1.06)^n = \log 1.5$ • Take the log of each side.

$n \log 1.06 = \log 1.5$ • Power Property

$n = \dfrac{\log 1.5}{\log 1.06}$ • Divide both sides by log 1.06.

$n \approx 6.9585$

The solution is approximately 6.9585.

You Try It 2

$4^{3x} = 25$

$\log 4^{3x} = \log 25$ • Take the log of each side.

$3x \log 4 = \log 25$ • Power Property

$3x = \dfrac{\log 25}{\log 4}$ • Divide both sides by log 4.

$3x \approx 2.3219$

$x \approx 0.7740$ • Divide both sides by 3.

The solution is approximately 0.7740.

You Try It 3

$\log_4(x^2 - 3x) = 1$

$4^1 = x^2 - 3x$ • Write in exponential form.

$4 = x^2 - 3x$

$0 = x^2 - 3x - 4$

$0 = (x + 1)(x - 4)$

$x + 1 = 0 \qquad x - 4 = 0$

$x = -1 \qquad x = 4$

The solutions are -1 and 4.

You Try It 4

$\log_3 x + \log_3(x + 3) = \log_3 4$

$\log_3[x(x + 3)] = \log_3 4$ • Product Property

$x(x + 3) = 4$ • 1–1 Property of Logarithms

$x^2 + 3x = 4$

$x^2 + 3x - 4 = 0$

$(x + 4)(x - 1) = 0$ • Solve by factoring.

$x + 4 = 0 \qquad x - 1 = 0$

$x = -4 \qquad x = 1$

-4 does not check as a solution.
The solution is 1.

You Try It 5

$\log_3 x + \log_3(x + 6) = 3$

$\log_3[x(x + 6)] = 3$ • Product Property

$x(x + 6) = 3^3$ • Write in exponential form.

$x^2 + 6x = 27$

$x^2 + 6x - 27 = 0$

$(x + 9)(x - 3) = 0$ • Solve by factoring.

$x + 9 = 0 \qquad x - 3 = 0$

$x = -9 \qquad x = 3$

-9 does not check as a solution.
The solution is 3.

Copyright © Houghton Mifflin Company. All rights reserved.

SECTION 12.5

You Try It 1

Strategy To find the hydrogen ion concentration, replace pH by 2.9 in the equation $pH = -\log(H^+)$ and solve for H^+.

Solution

$$\begin{aligned} pH &= -\log(H^+) \\ 2.9 &= -\log(H^+) \\ -2.9 &= \log(H^+) \\ 10^{-2.9} &= H^+ \\ 0.00126 &\approx H^+ \end{aligned}$$

• Multiply by −1.
• Write the equivalent exponential equation.

The hydrogen ion concentration is approximately 0.00126.

You Try It 2

Strategy To find the intensity, use the equation for the Richter scale magnitude of an earthquake, $M = \log \frac{I}{I_0}$. Replace M by 4.6 and solve for I.

Solution

$$\begin{aligned} M &= \log \frac{I}{I_0} \\ 4.6 &= \log \frac{I}{I_0} \\ 10^{4.6} &= \frac{I}{I_0} \\ 10^{4.6} I_0 &= I \\ 39{,}811 I_0 &\approx I \end{aligned}$$

• Replace M by 4.6.
• Write in exponential form.

The earthquake had an intensity that was approximately 39,811 times the intensity of a zero-level earthquake.

Solutions to Chapter R "You Try It"

SECTION R.1

You Try It 1

$$\begin{aligned} (-4)(6-8)^2 - (-12 \div 4) &= -4(-2)^2 - (-3) \\ &= -4(4) - (-3) \\ &= -16 - (-3) \\ &= -16 + 3 \\ &= -13 \end{aligned}$$

You Try It 2

$3xy^2 - 3x^2y$

$$\begin{aligned} 3(-2)(5)^2 - 3(-2)^2(5) &= 3(-2)(25) - 3(4)(5) \\ &= -6(25) - 3(4)(5) \\ &= -150 - 3(4)(5) \\ &= -150 - 12(5) \\ &= -150 - 60 \\ &= -150 + (-60) \\ &= -210 \end{aligned}$$

You Try It 3

$$\begin{aligned} -5(-3a) &= [-5(-3)]a \\ &= 15a \end{aligned}$$

You Try It 4

$$\begin{aligned} 2z^2 - 5z - 3z^2 + 6z &= 2z^2 - 3z^2 - 5z + 6z \\ &= (2z^2 - 3z^2) + (-5z + 6z) \\ &= -1z^2 + z \\ &= -z^2 + z \end{aligned}$$

You Try It 5

a. $$\begin{aligned} -3(5y - 2) &= -3(5y) - (-3)(2) \\ &= -15y + 6 \end{aligned}$$

b.
$$\begin{aligned} -2(4x + 2y - 6z) &= -2(4x) + (-2)(2y) - (-2)(6z) \\ &= -8x - 4y + 12z \end{aligned}$$

You Try It 6

a.
$$\begin{aligned} 7(-3x - 4y) - 3(3x + y) &= -21x - 28y - 9x - 3y \\ &= -30x - 31y \end{aligned}$$

b. $$\begin{aligned} 2y - 3[5 - 3(3 + 2y)] &= 2y - 3[5 - 9 - 6y] \\ &= 2y - 3[-4 - 6y] \\ &= 2y + 12 + 18y \\ &= 20y + 12 \end{aligned}$$

SECTION R.2

You Try It 1

a.
$$\begin{aligned} 4x + 3 &= 7x + 9 \\ 4x - 7x + 3 &= 7x - 7x + 9 \\ -3x + 3 &= 9 \\ -3x + 3 - 3 &= 9 - 3 \\ -3x &= 6 \\ \frac{-3x}{-3} &= \frac{6}{-3} \\ x &= -2 \end{aligned}$$

The solution is −2.

Copyright © Houghton Mifflin Company. All rights reserved.

b.
$$\begin{aligned}
4 - (5x - 8) &= 4x + 3\\
4 - 5x + 8 &= 4x + 3\\
-5x + 12 &= 4x + 3\\
-5x - 4x + 12 &= 4x - 4x + 3\\
-9x + 12 &= 3\\
-9x + 12 - 12 &= 3 - 12\\
-9x &= -9\\
\frac{-9x}{-9} &= \frac{-9}{-9}\\
x &= 1
\end{aligned}$$

The solution is 1.

You Try It 2
$$\begin{aligned}
3x - 1 &\le 5x - 7\\
3x - 5x - 1 &\le 5x - 5x - 7\\
-2x - 1 &\le -7\\
-2x - 1 + 1 &\le -7 + 1\\
-2x &\le -6\\
\frac{-2x}{-2} &\ge \frac{-6}{-2}\\
x &\ge 3
\end{aligned}$$

$\{x|x \ge 3\}$

You Try It 3
$$\begin{aligned}
3 - 2(3x + 1) &< 7 - 2x\\
3 - 6x - 2 &< 7 - 2x\\
1 - 6x &< 7 - 2x\\
1 - 6x + 2x &< 7 - 2x + 2x\\
1 - 4x &< 7\\
1 - 1 - 4x &< 7 - 1\\
-4x &< 6\\
\frac{-4x}{-4} &> \frac{6}{-4}\\
x &> -\frac{3}{2}
\end{aligned}$$

$\left\{x|x > -\frac{3}{2}\right\}$

SECTION R.3

You Try It 1

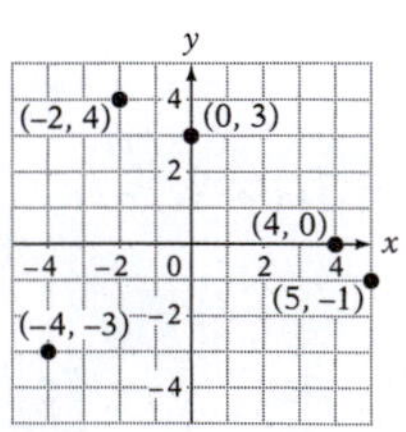

You Try It 2 $y = \frac{3}{5}x - 4$

x	y
5	−1
0	−4
−5	−7

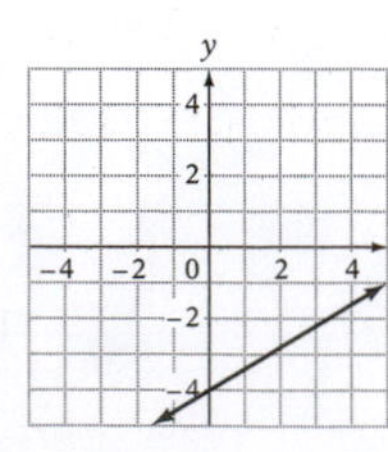

You Try It 3
$$\begin{aligned}
-3x + 2y &= 4\\
2y &= 3x + 4\\
y &= \frac{3}{2}x + 2
\end{aligned}$$

x	y
2	5
0	2
−2	−1

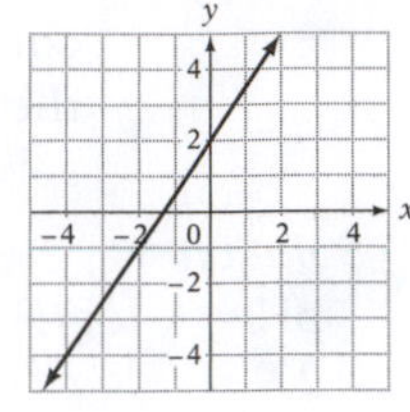

You Try It 4 Let $(x_1, y_1) = (-2, 3)$ and $(x_2, y_2) = (1, -3)$.

$$m = \frac{y_2 - y_1}{x_2 - x_1} = \frac{-3 - 3}{1 - (-2)} = \frac{-6}{3} = -2$$

The slope is −2.

You Try It 5 $y = -\frac{2}{3}x + 2$

$$m = -\frac{2}{3} = \frac{-2}{3}$$

y-intercept = (0, 2)

Place a dot at the y-intercept.
Starting at the y-intercept, move to the right 3 units (the change in x) and down 2 units (the change in y).
Place a dot at that location.
Draw a line through the two points.

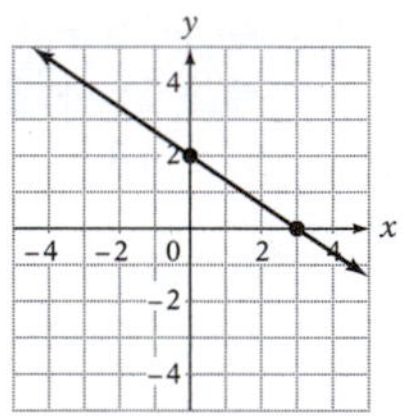

You Try It 6
$$\begin{aligned}
f(x) &= 4 - 2x\\
f(-3) &= 4 - 2(-3)\\
f(-3) &= 4 + 6\\
f(-3) &= 10
\end{aligned}$$

An ordered pair of the function is (−3, 10).

You Try It 7 $(x_1, y_1) = (-2, 2), m = -\frac{1}{2}$

$$\begin{aligned}
y - y_1 &= m(x - x_1)\\
y - 2 &= -\frac{1}{2}[x - (-2)]\\
y - 2 &= -\frac{1}{2}(x + 2)\\
y - 2 &= -\frac{1}{2}x - 1\\
y &= -\frac{1}{2}x + 1
\end{aligned}$$

Copyright © Houghton Mifflin Company. All rights reserved.

SECTION R.4

You Try It 1

a. $(-2ab)(2a^3b^{-2})^{-3} = (-2ab)(2^{-3}a^{-9}b^6)$
$= (-2^{-2})a^{-8}b^7$
$= -\dfrac{b^7}{2^2a^8}$
$= -\dfrac{b^7}{4a^8}$

b. $\left(\dfrac{2x^2y^{-4}}{4x^{-2}y^{-5}}\right)^{-3} = \left(\dfrac{x^2y^{-4}}{2x^{-2}y^{-5}}\right)^{-3}$
$= \dfrac{x^{2(-3)}y^{(-4)(-3)}}{2^{1(-3)}x^{(-2)(-3)}y^{(-5)(-3)}}$
$= \dfrac{x^{-6}y^{12}}{2^{-3}x^6y^{15}}$
$= 2^3x^{-6-6}y^{12-15}$
$= 2^3x^{-12}y^{-3}$
$= \dfrac{8}{x^{12}y^3}$

You Try It 2

$(-4x^3 + 2x^2 - 8) + (4x^3 + 6x^2 - 7x + 5)$
$= (-4x^3 + 4x^3) + (2x^2 + 6x^2) + (-7x) + (-8 + 5)$
$= 8x^2 - 7x - 3$

You Try It 3

$(-4w^3 + 8w - 8) - (3w^3 - 4w^2 + 2w - 1)$
$= (-4w^3 + 8w - 8) + (-3w^3 + 4w^2 - 2w + 1)$
$= (-4w^3 - 3w^3) + 4w^2 + (8w - 2w) + (-8 + 1)$
$= -7w^3 + 4w^2 + 6w - 7$

You Try It 4

$3mn^2(2m^2 - 3mn - 1)$
$= 3mn^2(2m^2) - 3mn^2(3mn) - 3mn^2(1)$
$= 6m^3n^2 - 9m^2n^3 - 3mn^2$

You Try It 5

$$\begin{array}{r} 3c^2 - 4c + 5 \\ 2c - 3 \\ \hline -9c^2 + 12c - 15 \\ 6c^3 - 8c^2 + 10c \\ \hline 6c^3 - 17c^2 + 22c - 15 \end{array}$$

You Try It 6

$(4y - 7)(3y - 5)$
$= (4y)(3y) + (4y)(-5) + (-7)(3y) + (-7)(-5)$
$= 12y^2 - 20y - 21y + 35$
$= 12y^2 - 41y + 35$

You Try It 7

$(x^3 - 7 - 2x) \div (x - 2)$

$$\begin{array}{r} x^2 + 2x + 2 \\ x - 2\overline{)x^3 + 0x^2 - 2x - 7} \\ \underline{x^3 - 2x^2 } \\ 2x^2 - 2x \\ \underline{2x^2 - 4x } \\ 2x - 7 \\ \underline{2x - 4} \\ -3 \end{array}$$

Check: $(x - 2)(x^2 + 2x + 2) - 3$
$= x^3 - 2x - 4 - 3$
$= x^3 - 2x - 7$

$(x^3 - 7 - 2x) \div (x - 2)$
$= x^2 + 2x + 2 - \dfrac{3}{x - 2}$

You Try It 8 The GCF is $3x^2y^2$.

$6x^4y^2 - 9x^3y^2 + 12x^2y^4$
$= 3x^2y^2(2x^2 - 3x + 4y^2)$

You Try It 9 Two factors of 15 whose sum is -8 are -3 and -5.

$x^2 - 8x + 15 = (x - 3)(x - 5)$

You Try It 10

$3x^2 - x - 2$

Positive Factors of 3	*Factors of −2*
1, 3	1, −2 −1, 2

Trial Factors	*Middle Term*
$(x + 1)(3x - 2)$	$-2x + 3x = x$
$(x - 2)(3x + 1)$	$x - 6x = -5x$
$(x - 1)(3x + 2)$	$2x - 3x = -x$
$(x + 2)(3x - 1)$	$-x + 6x = 5x$

$3x^2 - x - 2 = (x - 1)(3x + 2)$

Check: $(x - 1)(3x + 2) = 3x^2 + 2x - 3x - 2$
$= 3x^2 - x - 2$

Copyright © Houghton Mifflin Company. All rights reserved.

You Try It 11

$4a^3 - 4a^2 - 24a$

There is a common factor, $4a$. Factor out the GCF.

$4a^3 - 4a^2 - 24a = 4a(a^2 - a - 6)$

Factor $a^2 - a - 6$. The two factors of -6 whose sum is -1 are 2 and -3.

$4a(a^2 - a - 6) = 4a(a + 2)(a - 3)$

$4a^3 - 4a^2 - 24a = 4a(a + 2)(a - 3)$

Check:

$$\begin{aligned} 4a(a + 2)(a - 3) &= (4a^2 + 8a)(a - 3) \\ &= 4a^3 - 12a^2 + 8a^2 - 24a \\ &= 4a^3 - 4a^2 - 24a \end{aligned}$$

Copyright © Houghton Mifflin Company. All rights reserved.

Answers to Chapter 1 Selected Exercises

PREP TEST

1. 127.16 **2.** 55.107 **3.** 4517 **4.** 11,396 **5.** 24 **6.** 24 **7.** 4 **8.** $3 \cdot 7$ **9.** $\frac{2}{5}$

SECTION 1.1

1. $8 > -6$ **3.** $-12 < 1$ **5.** $42 > 19$ **7.** $0 > -31$ **9.** $53 > -46$ **11.** False **13.** True **15.** False **17.** True **19.** False **21.** $-23, -18$ **23.** 21, 37 **25.** -23 **27.** -4 **29.** 9 **31.** 28 **33.** 14 **35.** -77 **37.** 0 **39.** 74 **41.** -82 **43.** -81 **45.** $|-83| > |58|$ **47.** $|43| < |-52|$ **49.** $|-68| > |-42|$ **51.** $|-45| < |-61|$ **53.** $19, 0, -28$ **55.** $-45, 0, -17$ **59.** -11 **61.** -5 **63.** -83 **65.** -46 **67.** 0 **69.** -5 **71.** 9 **73.** 1 **75.** 8 **77.** -7 **79.** -9 **81.** 9 **83.** -3 **85.** 18 **87.** -10 **89.** -41 **91.** -12 **93.** 0 **95.** -9 **97.** 11 **99.** -18 **101.** 0 **103.** 2 **105.** -138 **107.** -8 **109.** -12 **111.** -20 **115.** 42 **117.** -28 **119.** 60 **121.** -253 **123.** -238 **125.** -114 **127.** -2 **129.** 8 **131.** -7 **133.** -12 **135.** -6 **137.** -7 **139.** 11 **141.** -14 **143.** 15 **145.** -16 **147.** 0 **149.** -29 **151.** Undefined **153.** -11 **155.** Undefined **157.** -105 **159.** 252 **161.** -240 **163.** 96 **165.** -216 **167.** -315 **169.** 420 **171.** 2880 **173.** -2772 **175.** 0 **177.** The difference in elevation is 7046 m. **179.** The difference between the highest and lowest elevations is greatest in Asia. **181.** The difference is 5°C. **183.** The student's score is 93 points. **185.** The difference between the average temperatures is 0°F. **187.** The difference is 9°F.

SECTION 1.2

1. 0.125 **3.** $0.\overline{2}$ **5.** $0.1\overline{6}$ **7.** 0.5625 **9.** $0.58\overline{3}$ **11.** 0.24 **13.** 0.225 **15.** $0.\overline{45}$ **19.** $\frac{2}{5}$, 0.40 **21.** $\frac{22}{25}$, 0.88 **23.** $\frac{8}{5}$, 1.6 **25.** $\frac{87}{100}$, 0.87 **27.** $\frac{9}{2}$, 4.50 **29.** $\frac{3}{70}$ **31.** $\frac{3}{8}$ **33.** $\frac{1}{400}$ **35.** $\frac{1}{16}$ **37.** $\frac{23}{400}$ **39.** 0.091 **41.** 0.167 **43.** 0.009 **45.** 0.0915 **47.** 0.1823 **49.** 37% **51.** 2% **53.** 12.5% **55.** 136% **57.** 0.4% **59.** 83% **61.** $37\frac{1}{2}\%$ **63.** $44\frac{4}{9}\%$ **65.** 45% **67.** 250% **69.** $\frac{5}{26}$ **71.** $\frac{11}{8}$ **73.** $\frac{1}{12}$ **75.** $\frac{7}{24}$ **77.** 0 **79.** $\frac{3}{8}$ **81.** $-\frac{7}{60}$ **83.** $-\frac{1}{16}$ **85.** -1.06 **87.** -23.845 **89.** -10.7893 **91.** -37.19 **93.** -17.5 **95.** 19.61 **97.** $-\frac{3}{8}$ **99.** $\frac{1}{10}$ **101.** $-\frac{4}{9}$ **103.** $-\frac{7}{30}$ **105.** $\frac{15}{64}$ **107.** $-\frac{10}{9}$ **109.** $-\frac{147}{32}$ **111.** $\frac{25}{8}$ **113.** $\frac{2}{3}$ **115.** 4.164 **117.** 4.347 **119.** -4.028 **121.** -2.22 **123.** -1.104 **125.** -2.59 **127.** -5.11 **129.** -2060.55 **131.** 2401 **133.** -64 **135.** -8 **137.** -125 **139.** $-\frac{27}{64}$ **141.** 3.375 **143.** -1 **145.** -8 **147.** -6750 **149.** -144 **151.** -18 **153.** 4 **155.** 7 **157.** $4\sqrt{2}$ **159.** $2\sqrt{2}$ **161.** $18\sqrt{2}$ **163.** $10\sqrt{10}$ **165.** $\sqrt{15}$ **167.** $\sqrt{29}$ **169.** $-54\sqrt{2}$ **171.** $3\sqrt{5}$ **173.** 0 **175.** $48\sqrt{2}$ **177.** 15.492 **179.** 16.971 **181.** 16 **183.** 16.583 **185.** 15.652 **187.** 18.762 **189a.** The average monthly net income was $-\$2.668$ million. **b.** The difference is \$3325.286 million. **191.** The amount spent on decorations is 25% of the total. **193.** 3, 4, 5, 6, 7, 8, 9

SECTION 1.3

3. 0 **5.** -11 **7.** 20 **9.** -10 **11.** 20 **13.** 29 **15.** 11 **17.** 7 **19.** -11 **21.** 6 **23.** 15 **25.** 4 **27.** 5 **29.** -1 **31.** 4 **33.** 0.51 **35.** 1.7 **37.** Row 1: $-\frac{1}{6}$, 0; Row 2: $-\frac{1}{2}$; Row 3: $\frac{1}{3}$, $\frac{1}{2}$ **39.** Your savings on gasoline would pay for the increased cost of the car in 39 months.

SECTION 1.4

1. -9 **3.** 41 **5.** -7 **7.** 13 **9.** -15 **11.** 41 **13.** 1 **15.** 5 **17.** 1 **19.** 57 **21.** 5 **23.** 8 **25.** -3 **27.** -2 **29.** -4 **31.** 10 **33.** -25 **35.** $25x$ **37.** $9a$ **39.** $2y$ **41.** $-12y - 3$ **43.** $9a$ **45.** $6ab$ **47.** $-12xy$ **49.** 0 **51.** $-\frac{1}{10}y$ **53.** $\frac{2}{9}y^2$ **55.** $20x$ **57.** $-4a$ **59.** $-2y^2$ **61.** $-2x + 8y$ **63.** $8x$

Copyright © Houghton Mifflin Company. All rights reserved.

65. $19a - 12b$ **67.** $-12x - 2y$ **69.** $-7x^2 - 5x$ **71.** $60x$ **73.** $-10a$ **75.** $30y$ **77.** $72x$ **79.** $-28a$
81. $108b$ **83.** $-56x^2$ **85.** x^2 **87.** x **89.** a **91.** b **93.** x **95.** n **97.** $2x$ **99.** $-2x$ **101.** $-15a^2$
103. $6y$ **105.** $3y$ **107.** $-2x$ **109.** $-9y$ **111.** $-x - 7$ **113.** $10x - 35$ **115.** $-5a - 80$ **117.** $-15y + 35$
119. $20 - 14b$ **121.** $-4x + 2y$ **123.** $18x^2 + 12x$ **125.** $10x - 35$ **127.** $-14x + 49$ **129.** $-30x^2 - 15$
131. $-24y^2 + 96$ **133.** $5x^2 + 5y^2$ **135.** $-\frac{1}{2}x + 2y$ **137.** $3x^2 + 6x - 18$ **139.** $-2y^2 + 4y - 8$
141. $-2x + 3y - \frac{1}{3}$ **143.** $10x^2 + 15x - 35$ **145.** $6x^2 + 3xy - 9y^2$ **147.** $-3a^2 - 5a + 4$ **149.** $-2x - 16$
151. $-12y - 9$ **153.** $7n - 7$ **155.** $-2x + 41$ **157.** $3y - 3$ **159.** $2a - 4b$ **161.** $-4x + 24$ **163.** $-2x - 16$
165. $-3x + 21$ **167.** $-4x + 12$ **169.** $-x + 50$ **171.** $\frac{x}{18}$ **173.** $x + 20$ **175.** $10(x - 50); 10x - 500$
177. $\frac{5}{8}x + 6$ **179.** $x - (x + 3); -3$ **181.** $4(x + 19); 4x + 76$ **183.** $\frac{15}{x + 12}$ **185.** $\frac{2}{3}(x + 7); \frac{2}{3}x + \frac{14}{3}$
187. $40 - \frac{x}{20}$ **189.** $x^2 + 2x$ **191.** $(x + 8) + \frac{1}{3}x; \frac{4}{3}x + 8$ **193.** $x + (x + 9); 2x + 9$ **195.** $x - (8 - x); 2x - 8$
197. $\frac{1}{3}x - \frac{5}{8}x; -\frac{7}{24}x$ **199.** $(x + 5) + 2; x + 7$ **201.** $2(6x + 7); 12x + 14$ **203.** Let n be the number of nations participating in 1896; $n + 1990$ **205.** Let g be the amount of oil in one container; $g, 20 - g$ **207.** Let p be the pounds of pecans produced in Texas; $\frac{1}{2}p$ **209.** Let d be the diameter of a baseball; $4d$ **211.** Yes **213.** $2x$ **215.** $\frac{1}{4}x$
217. $\frac{3}{5}x$

SECTION 1.5

1. $A = \{16, 17, 18, 19, 20, 21\}$ **3.** $A = \{9, 11, 13, 15, 17\}$ **5.** $A = \{b, c\}$ **9.** $A \cup B = \{3, 4, 5, 6\}$
11. $A \cup B = \{-10, -9, -8, 8, 9, 10\}$ **13.** $A \cup B = \{a, b, c, d, e, f\}$ **15.** $A \cup B = \{1, 3, 7, 9, 11, 13\}$
17. $A \cap B = \{4, 5\}$ **19.** $A \cap B = \varnothing$ **21.** $A \cap B = \{c, d, e\}$ **23.** $\{x | x > -5, x \in \text{negative integers}\}$
25. $\{x | x > 30, x \in \text{integers}\}$ **27.** $\{x | x > 5, x \in \text{even integers}\}$ **29.** $\{x | x > 8, x \in \text{real numbers}\}$

31. −5 −4 −3 −2 −1 0 1 2 3 4 5

33. −5 −4 −3 −2 −1 0 1 2 3 4 5

35. −5 −4 −3 −2 −1 0 1 2 3 4 5

37. −5 −4 −3 −2 −1 0 1 2 3 4 5

39. −5 −4 −3 −2 −1 0 1 2 3 4 5

41a. Never true **b.** Always true **c.** Always true **43a.** Yes **b.** Yes

CHAPTER 1 REVIEW EXERCISES*

1. $-4, 0$ [1.1A] **2.** 4 [1.1B] **3.** -5 [1.1B] **4.** -13 [1.1C] **5.** 1 [1.1C] **6.** -42 [1.1D]
7. -20 [1.1D] **8.** 0.28 [1.2A] **9.** 0.062 [1.2B] **10.** 62.5% [1.2B] **11.** $\frac{7}{12}$ [1.2C] **12.** -1.068 [1.2C]
13. $-\frac{72}{85}$ [1.2D] **14.** -4.6224 [1.2D] **15.** $\frac{16}{81}$ [1.2E] **16.** 12 [1.2F] **17.** $-6\sqrt{30}$ [1.2F]
18. 31 [1.3A] **19.** 29 [1.4A] **20.** $8a - 4b$ [1.4B] **21.** $36y$ [1.4C] **22.** $10x - 35$ [1.4D]
23. $7x + 46$ [1.4D] **24.** $-90x + 25$ [1.4D] **25.** $\{1, 3, 5, 7\}$ [1.5A] **26.** $A \cap B = \{1, 5, 9\}$ [1.5A]
27. −5 −4 −3 −2 −1 0 1 2 3 4 5 [1.5C] **28.** −5 −4 −3 −2 −1 0 1 2 3 4 5 [1.5C] **29.** The student's score is 98. [1.1E] **30.** 50.8% of the candy consumed was chocolate. [1.2G] **31.** $2x - \frac{1}{2}x; \frac{3}{2}x$ [1.4E]
32. Let A be the number of American League players' cards; $5A$ [1.4E] **33.** Let T be the number of ten-dollar bills; $35 - T$ [1.4E]

*Note: The numbers in brackets following the answers to the Chapter Review Exercises are a reference to the objective that corresponds to that problem. For example, the reference [1.2A] stands for Section 1.2, Objective A. This notation will be used for all Prep Tests, Chapter Review Exercises, Chapter Tests, and Cumulative Review Exercises throughout the text.

Copyright © Houghton Mifflin Company. All rights reserved.

CHAPTER 1 TEST

1. $-2 > -40$ [1.1A] **2.** 4 [1.1B] **3.** -4 [1.1B] **4.** -14 [1.1C] **5.** -16 [1.1C] **6.** 4 [1.1C]
7. 17 [1.1D] **8.** $0.\overline{7}$ [1.2A] **9.** $\frac{9}{20}$, 0.45 [1.2B] **10.** $\frac{1}{15}$ [1.2C] **11.** -5.3578 [1.2D] **12.** $-\frac{1}{2}$ [1.2D]
13. 12 [1.2E] **14.** $-6\sqrt{5}$ [1.2F] **15.** 17 [1.3A] **16.** 22 [1.4A] **17.** $5x$ [1.4B] **18.** $2x$ [1.4C]
19. $-6x^2 + 21y^2$ [1.4D] **20.** $-x + 6$ [1.4D] **21.** $-7x + 33$ [1.4D] **22.** $\{-2, -1, 0, 1, 2, 3\}$ [1.5A]
23. $\{x \mid x < -3, x \in \text{real numbers}\}$ [1.5B] **24.** $A \cup B = \{1, 2, 3, 4, 5, 6, 7, 8\}$ [1.5A]
25. −5 −4 −3 −2 −1 0 1 2 3 4 5 [1.5C] **26.** −5 −4 −3 −2 −1 0 1 2 3 4 5 [1.5C]
27. $10(x - 3)$; $10x - 30$ [1.4E] **28.** Let s be the speed of the catcher's return throw; $2s$ [1.4E]
29a. The balance of trade increased from the previous year in 1981, 1988, 1989, 1990, 1991, and 1995. [1.2G]
b. The difference was \$288.6 billion. [1.2G] **c.** The difference was greatest between 1999 and 2000. [1.2G]
d. The trade balance was approximately 4 times greater in 1990 than in 1980. [1.2G] **e.** The average trade balance per quarter for the year 2000 was −\$92.425 billion. [1.2G] **30.** The difference is 215.4°F. [1.2G]

Answers to Chapter 2 Selected Exercises

PREP TEST

1. 0.09 [1.2B] **2.** 75% [1.2B] **3.** 63 [1.4A] **4.** $0.65R$ [1.4B] **5.** $\frac{7}{6}x$ [1.4B] **6.** $9x - 18$ [1.4D]
7. $1.66x + 1.32$ [1.4D] **8.** $5 - 2n$ [1.4E] **9.** Speed of the old card: s; speed of the new card: $5s$ [1.4E]
10. $5 - x$ [1.4E]

SECTION 2.1

3. Yes **5.** No **7.** No **9.** Yes **11.** No **13.** Yes **15.** No **17.** Yes **19.** No **23.** 2 **25.** 15
27. 6 **29.** 3 **31.** 0 **33.** -7 **35.** -7 **37.** -12 **39.** -5 **41.** 15 **43.** 9 **45.** 14 **47.** -1
49. 1 **51.** $-\frac{1}{2}$ **53.** $-\frac{3}{4}$ **55.** $\frac{1}{12}$ **57.** $-\frac{7}{12}$ **59.** 0.6529 **61.** -0.283 **63.** 9.257 **67.** -3 **69.** 0
71. -2 **73.** 9 **75.** 80 **77.** 0 **79.** -7 **81.** 12 **83.** -18 **85.** 15 **87.** -20 **89.** 0 **91.** $\frac{8}{3}$
93. $\frac{1}{3}$ **95.** $-\frac{1}{2}$ **97.** $-\frac{3}{2}$ **99.** $\frac{15}{7}$ **101.** 4 **103.** 3 **105.** 4.745 **107.** 2.06 **109.** -2.13
111. Equal to **113.** 28 **115.** 0.72 **117.** 64 **119.** 24% **121.** 7.2 **123.** 400 **125.** 9 **127.** 25%
129. 200% **131.** 400 **133.** 7.7 **135.** 200 **137.** 400 **139.** 30 **141.** There are 4536 L of oxygen in the room. **143.** The median income was \$42,428. **145.** 47.9% of the U.S. population watched Super Bowl XXXVIII.
147. You need to know the number of people 3 years old and older in the U.S. **149.** Andrea must invest \$1875.
151. Octavia will earn the greater amount of interest. **153.** \$1500 was invested at 8%. **155.** There are 1.8 g of platinum in the necklace. **157.** There are 131.25 lb of wool in the carpet. **159.** The percent concentration of sugar is 50%. **161.** The percent concentration of the resulting mixture is 6%. **163.** The runner will travel 3 mi.
165. Marcella's average rate of speed is 36 mph. **167.** It would take Palmer 2.5 h to walk the course.
169. The two joggers will meet 40 min after they start. **171.** It will take them 0.5 h. **173.** $x = \frac{b}{a}, a \neq 0$
175. a. Answers will vary. **b.** Answers will vary.

SECTION 2.2

1. 3 **3.** 6 **5.** -1 **7.** -3 **9.** 2 **11.** 2 **13.** 5 **15.** -3 **17.** 6 **19.** 3 **21.** 1 **23.** 6 **25.** -7
27. 0 **29.** $\frac{3}{4}$ **31.** $\frac{4}{9}$ **33.** $\frac{1}{3}$ **35.** $-\frac{1}{2}$ **37.** $-\frac{3}{4}$ **39.** $\frac{1}{3}$ **41.** $-\frac{1}{6}$ **43.** 1 **45.** 1 **47.** 0 **49.** $\frac{13}{10}$

Copyright © Houghton Mifflin Company. All rights reserved.

51. $\frac{2}{5}$ **53.** $-\frac{4}{3}$ **55.** $-\frac{3}{2}$ **57.** 18 **59.** 8 **61.** -16 **63.** 25 **65.** $\frac{3}{4}$ **67.** $\frac{3}{8}$ **69.** $\frac{16}{9}$ **71.** $\frac{1}{18}$ **73.** $\frac{15}{2}$ **75.** $-\frac{18}{5}$ **77.** 2 **79.** 3 **81.** $x = 7$ **83.** $y = 3$ **85.** 2 **87.** 3 **89.** -1 **91.** 2 **93.** -2 **95.** -3 **97.** 0 **99.** -1 **101.** -3 **103.** -1 **105.** 4 **107.** $\frac{2}{3}$ **109.** $\frac{5}{6}$ **111.** $\frac{3}{4}$ **113.** -17 **115.** 41 **117.** 8 **119.** 1 **121.** 4 **123.** -1 **125.** -1 **127.** 24 **129.** 495 **131.** $\frac{1}{2}$ **133.** $-\frac{1}{3}$ **135.** $\frac{10}{3}$ **137.** $-\frac{1}{4}$ **139.** 0 **141.** -1 **143.** A force of 25 lb must be applied to the other end. **145.** The fulcrum is 6 ft from the 180-pound person. **147.** The fulcrum is 10 ft from the 128-pound acrobat. **149.** The minimum force to move the rock is 34.6 lb. **151.** The break-even point is 260 barbecues. **153.** The break-even point is 520 desk lamps. **155.** The break-even point is 3000 softball bats. **157.** No solution **159.** 0

SECTION 2.3

1. $x - 15 = 7$; 22 **3.** $7x = -21$; -3 **5.** $9 - x = 7$; 2 **7.** $5 - 2x = 1$; 2 **9.** $2x + 5 = 15$; 5 **11.** $4x - 6 = 22$; 7 **13.** $3(4x - 7) = 15$; 3 **15.** $3x = 2(20 - x)$; 8, 12 **17.** $2x - (14 - x) = 1$; 5, 9 **19.** 15, 17, 19 **21.** -1, 1, 3 **23.** 4, 6 **25.** 5, 7 **27.** The processor speed of the newer personal computer is $4.2\overline{6}$ GHz. **29.** The lengths of the sides are 6 ft, 6 ft, and 11 ft. **31.** The union member worked 168 h during March. **33.** 37 h of labor was required to paint the house. **35.** There are 1024 vertical pixels. **37.** The length is 13 m; the width is 8 m. **39.** The shorter piece is 3 ft; the longer piece is 9 ft. **41.** The larger scholarship is $5000.

SECTION 2.4

1. The amount of $1 herbs is 20 oz. **3.** The mixture will cost $1.84 per pound. **5.** The amount of caramel is 3 lb. **7.** The amount of olive oil is 2 c; the amount of vinegar is 8 c. **9.** The cost of the mixture is $3.00 per ounce. **11.** To make the mixture, 16 oz of the alloy are needed. **13.** The amount of almonds is 37 lb; the amount of walnuts is 63 lb. **15.** There were 228 adult tickets sold. **17.** The cost per pound of the sugar-coated cereal is $.70. **19.** The percent concentration of gold in the mixture is 24%. **21.** The amount of the 15% acid is 20 gal. **23.** The amount of the 25% wool yarn is 30 lb. **25.** The amount of 9% nitrogen plant food is 6.25 gal. **27.** The percent concentration of sugar in the mixture is 19%. **29.** 20 lb of 40% java bean coffee must be used. **31.** The amount of the 7% solution is 100 ml; the amount of the 4% solution is 200 ml. **33.** 150 oz of pure chocolate must be added. **35.** The percent concentration of the resulting alloy is 50%. **37.** The first plane is traveling at a rate of 105 mph; the second plane is traveling at a rate of 130 mph. **39.** The planes will be 3000 km apart at 11 A.M. **41.** In 2 h the cabin cruiser will be alongside the motorboat. **43.** The corporate offices are 120 mi from the airport. **45.** The rate of the car is 68 mph. **47.** The distance between the airports is 300 mi. **49.** The planes will pass each other 2.5 h after the plane leaves Seattle. **51.** The cyclists will meet after 1.5 h. **53.** The bus overtakes the car at 180 mi. **55.** 75 g of pure water must be added. **57.** 3.75 gal of 20% antifreeze must be drained. **59.** The bicyclist's average speed is $13\frac{1}{3}$ mph.

SECTION 2.5

3. a, c **5.** $\{x \mid x < 5\}$ −5 −4 −3 −2 −1 0 1 2 3 4 5 **7.** $\{x \mid x \le 2\}$ −5 −4 −3 −2 −1 0 1 2 3 4 5 **9.** $\{x \mid x < -4\}$ −5 −4 −3 −2 −1 0 1 2 3 4 5 **11.** $\{x \mid x > 3\}$ **13.** $\{x \mid x > 4\}$ **15.** $\{x \mid x \le 2\}$ **17.** $\{x \mid x > -2\}$ **19.** $\{x \mid x \ge 2\}$ **21.** $\{x \mid x > -2\}$ **23.** $\{x \mid x \le 3\}$ **25.** $\{x \mid x < 2\}$ **27.** $\{x \mid x < -3\}$ **29.** $\{x \mid x \le 5\}$ **31.** $\left\{x \mid x \ge -\frac{1}{2}\right\}$ **33.** $\left\{x \mid x < \frac{23}{16}\right\}$ **35.** $\left\{x \mid x < \frac{8}{3}\right\}$ **37.** $\{x \mid x > 1\}$ **39.** $\left\{x \mid x > \frac{14}{11}\right\}$ **41.** $\{x \mid x \le 1\}$ **43.** $\left\{x \mid x \le \frac{7}{4}\right\}$ **45.** $\left\{x \mid x \ge -\frac{5}{4}\right\}$ **47.** $\{x \mid x \le 2\}$ **51.** $\{x \mid -2 \le x \le 4\}$ **53.** $\{x \mid x < 3 \text{ or } x > 5\}$ **55.** $\{x \mid -4 < x < 2\}$ **57.** $\{x \mid x > 6 \text{ or } x < -4\}$ **59.** $\{x \mid x < -3\}$ **61.** $\varnothing$ **63.** $\{x \mid -2 < x < 1\}$ **65.** $\{x \mid x < -2 \text{ or } x > 2\}$ **67.** $\{x \mid 2 < x < 6\}$ **69.** $\{x \mid -3 < x < -2\}$ **71.** $\left\{x \mid x > 5 \text{ or } x < -\frac{5}{3}\right\}$ **73.** $\varnothing$ **75.** The set of real numbers

Copyright © Houghton Mifflin Company. All rights reserved.

77. $\left\{x \middle| \frac{17}{7} \le x \le \frac{45}{7}\right\}$ **79.** $\left\{x \middle| -5 < x < \frac{17}{3}\right\}$ **81.** The set of real numbers **83.** $\varnothing$ **85.** The smallest number is -12. **87.** The maximum width of the rectangle is 11 cm. **89.** The TopPage plan is less expensive when service is for more than 460 pages. **91.** Paying with cash is less expensive when the call is 7 min or less. **93.** $32° < F < 86°$ **95.** George's amount of sales must be \$44,000 or more. **97.** The first account is less expensive if more than 200 checks are written. **99.** $58 \le x \le 100$ **101.** The three even integers are 10, 12, and 14; or 12, 14, and 16; or 14, 16, and 18. **103. a.** Always true **b.** Sometimes true **c.** Sometimes true **d.** Sometimes true **e.** Always true

SECTION 2.6

1. Yes **3.** Yes **5.** 7 and -7 **7.** 4 and -4 **9.** 6 and -6 **11.** 7 and -7 **13.** No solution **15.** No solution **17.** 1 and -5 **19.** 8 and 2 **21.** 2 **23.** No solution **25.** $-\frac{3}{2}$ and 3 **27.** $\frac{3}{2}$ **29.** No solution **31.** 7 and -3 **33.** 2 and $-\frac{10}{3}$ **35.** 1 and 3 **37.** $\frac{3}{2}$ **39.** No solution **41.** $\frac{11}{6}$ and $-\frac{1}{6}$ **43.** $-\frac{1}{3}$ and -1 **45.** No solution **47.** 3 and 0 **49.** No solution **51.** 1 and $\frac{13}{3}$ **53.** No solution **55.** $\frac{7}{3}$ and $\frac{1}{3}$ **57.** $-\frac{1}{2}$ **59.** $-\frac{1}{2}$ and $-\frac{7}{2}$ **61.** $-\frac{8}{3}$ and $\frac{10}{3}$ **63.** No solution **65.** $\{x \mid x > 3 \text{ or } x < -3\}$ **67.** $\{x \mid x > 1 \text{ or } x < -3\}$ **69.** $\{x \mid 4 \le x \le 6\}$ **71.** $\{x \mid x \ge 5 \text{ or } x \le -1\}$ **73.** $\{x \mid -3 < x < 2\}$ **75.** $\left\{x \middle| x > 2 \text{ or } x < -\frac{14}{5}\right\}$ **77.** $\varnothing$ **79.** The set of real numbers **81.** $\left\{x \middle| x \le -\frac{1}{3} \text{ or } x \ge 3\right\}$ **83.** $\left\{x \middle| -2 \le x \le \frac{9}{2}\right\}$ **85.** $\{x \mid x = 2\}$ **87.** $\left\{x \middle| x < -2 \text{ or } x > \frac{22}{9}\right\}$ **89.** $\left\{x \middle| -\frac{3}{2} < x < \frac{9}{2}\right\}$ **91.** $\left\{x \middle| x < 0 \text{ or } x > \frac{4}{5}\right\}$ **93.** $\{x \mid x > 5 \text{ or } x < 0\}$ **95.** The lower and upper limits of the diameter of the bushing are 1.742 in. and 1.758 in. **97.** The lower and upper limits of the voltage of the electric motor are 195 volts and 245 volts. **99.** The lower and upper limits of the length of the piston rod are $9\frac{19}{32}$ in. and $9\frac{21}{32}$ in. **101.** The lower and upper limits of the resistor are 28,420 ohms and 29,580 ohms. **103.** The lower and upper limits of the resistor are 23,750 ohms and 26,250 ohms. **105. a.** $\{x \mid x \ge -3\}$ **b.** $\{a \mid a \le 4\}$ **107. a.** $\le$ **b.** $\ge$ **c.** $\ge$ **d.** $=$ **e.** $=$

CHAPTER 2 REVIEW EXERCISES

1. 6 [2.2B] **2.** $\left\{x \middle| x > \frac{5}{3}\right\}$ [2.5A] **3.** No [2.1A] **4.** -9 [2.1B] **5.** $\left\{x \middle| -3 < x < \frac{4}{3}\right\}$ [2.5B] **6.** $-\frac{40}{7}$ [2.2B] **7.** $-\frac{2}{3}$ [2.1C] **8.** -1 and 9 [2.6A] **9.** $\{x \mid 1 < x < 4\}$ [2.6B] **10.** $-\frac{9}{19}$ [2.2C] **11.** $\frac{26}{17}$ [2.2C] **12.** $\{x \mid x > 2 \text{ or } x < -2\}$ [2.5B] **13.** $\left\{x \middle| x \ge 2 \text{ or } x \le \frac{1}{2}\right\}$ [2.6B] **14.** 250% [2.1D] **15.** $-\frac{17}{2}$ [2.2C] **16.** No solution [2.6A] **17.** The set of real numbers [2.5B] **18.** $\frac{5}{2}$ [2.2C] **19.** -15 [2.1B] **20.** $-\frac{1}{12}$ [2.1B] **21.** 7 [2.1C] **22.** $\frac{2}{3}$ [2.1C] **23.** $\frac{8}{5}$ [2.2B] **24.** 3 [2.2B] **25.** 9 [2.2C] **26.** 8 [2.2C] **27.** $\{x \mid x < 1\}$ [2.5A] **28.** $\{x \mid x \ge -4\}$ [2.5A] **29.** $\left\{x \middle| x \le -\frac{39}{2}\right\}$ [2.5A] **30.** $\{x \mid x \ge 1\}$ [2.5A] **31.** $\{x \mid -1 < x < 2\}$ [2.5B] **32.** $-\frac{5}{2}, \frac{11}{2}$ [2.6A] **33.** $-\frac{8}{5}$ [2.6A] **34.** $\varnothing$ [2.6B] **35.** The island is 16 mi from the dock. [2.4C] **36.** The mixture must contain 52 gal of apple juice. [2.4A] **37.** The executive's amount of sales must be \$55,000 or more. [2.5C] **38.** $5x - 4 = 16; x = 4$ [2.3A] **39.** The lower and upper limits of the diameter of the bushing are 2.747 in. and 2.753 in. [2.6C] **40.** The integers are 6 and 14. [2.3A] **41.** $82 \le x \le 100$ [2.5C] **42.** The speed of the first plane is 520 mph. The speed of the second plane is 440 mph. [2.4C] **43.** 375 lb of the 30% tin alloy and 125 lb of the 70% tin alloy were used. [2.4B] **44.** The lower and upper limits of the length of the piston are $10\frac{11}{32}$ in. and $10\frac{13}{32}$ in. [2.6C]

Copyright © Houghton Mifflin Company. All rights reserved.

CHAPTER 2 TEST

1. -2 [2.1B] **2.** $-\frac{1}{8}$ [2.1B] **3.** $\frac{5}{6}$ [2.1C] **4.** 4 [2.2A] **5.** $\frac{32}{3}$ [2.2A] **6.** $-\frac{1}{5}$ [2.2B] **7.** 1 [2.2C] **8.** No [2.1A] **9.** $\frac{12}{7}$ [2.2C] **10.** $\{x \mid x \le -3\}$ [2.5A] **11.** 0.04 [2.1D] **12.** $\{x \mid x > -2\}$ [2.5B] **13.** $\varnothing$ [2.5B] **14.** $-\frac{9}{5}$ and 3 [2.6A] **15.** 7 and -2 [2.6A] **16.** $\left\{x \,\middle|\, \frac{1}{3} \le x \le 3\right\}$ [2.6B] **17.** $\left\{x \,\middle|\, x < -\frac{1}{2} \text{ or } x > 2\right\}$ [2.6B] **18.** It costs less to rent from Gambelli Agency if the car is driven less than 120 mi. [2.5C] **19.** The lower and upper limits of the diameter of the bushing are 2.648 in. and 2.652 in. [2.6C] **20.** The integers are 4 and 11. [2.3A] **21.** 1.25 gal are needed. [2.4B] **22.** The price of the hamburger mixture is $2.70 per pound. [2.4A] **23.** The jogger ran a total distance of 12 mi. [2.4C] **24.** The rate of the slower train is 60 mph. The rate of the faster train is 65 mph. [2.4C] **25.** It is necessary to add 100 oz of pure water. [2.4B]

CUMULATIVE REVIEW EXERCISES

1. 6 [1.1C] **2.** -48 [1.1D] **3.** $-\frac{19}{48}$ [1.2C] **4.** 54 [1.2E] **5.** $\frac{49}{40}$ [1.3A] **6.** 6 [1.4A] **7.** $-17x$ [1.4B] **8.** $-5a - 4b$ [1.4B] **9.** $2x$ [1.4C] **10.** $36y$ [1.4C] **11.** $2x^2 + 6x - 4$ [1.4D] **12.** $-4x + 14$ [1.4D] **13.** $6x - 34$ [1.4D] **14.** $A \cap B = \{-4, 0\}$ [1.5A] **15.** 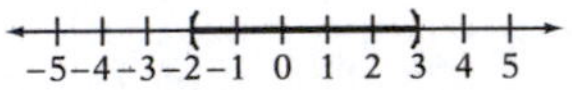

[1.5C] **16.** Yes [2.1A] **17.** -25 [2.1C] **18.** -3 [2.2A] **19.** 3 [2.2A] **20.** -3 [2.2B] **21.** $\frac{1}{2}$ [2.2B] **22.** 13 [2.2C] **23.** $\{x \mid x \le 1\}$ [2.5A] **24.** $\{x \mid -4 \le x \le 1\}$ [2.5B] **25.** -1 and 4 [2.6A] **26.** $\left\{x \,\middle|\, x > 2 \text{ or } x < -\frac{4}{3}\right\}$ [2.6B] **27.** $\frac{11}{20}$ [1.2B] **28.** 103% [1.2B] **29.** 25% of 120 is 30. [2.1D] **30.** $6x + 13 = 3x - 5$; $x = -6$ [2.3A] **31.** 20 lb of oat flour must be used. [2.4A] **32.** 25 g of pure gold must be added. [2.4B] **33.** The length of the track is 120 m. [2.4C]

Answers to Chapter 3 Selected Exercises

PREP TEST

1. 43 [2.1B] **2.** 51 [2.1B] **3.** 56 [1.3A] **4.** 56.52 [1.4A] **5.** 113.04 [1.4A] **6.** 120 [1.4A]

SECTION 3.1

1. 40°; acute **3.** 115°; obtuse **5.** 90°; right **7.** The complement is 28°. **9.** The supplement is 18°. **11.** The length of BC is 14 cm. **13.** The length of QS is 28 ft. **15.** The length of EG is 30 m. **17.** The measure of $\angle MON$ is 86°. **19.** 71° **21.** 30° **23.** 36° **25.** 127° **27.** 116° **29.** 20° **31.** 20° **33.** 20° **35.** 141° **37.** 106° **39.** 11° **41.** $\angle a = 38°$, $\angle b = 142°$ **43.** $\angle a = 47°$, $\angle b = 133°$ **45.** 20° **47.** 47° **49.** $\angle x = 155°$, $\angle y = 70°$ **51.** $\angle a = 45°$, $\angle b = 135°$ **53.** $90° - x$ **55.** The measure of the third angle is 60°. **57.** The measure of the third angle is 35°. **59.** The measure of the third angle is 102°. **61.** **a.** 1° **b.** 179° **65.** 360°

SECTION 3.2

1. Hexagon **3.** Pentagon **5.** Scalene **7.** Equilateral **9.** Obtuse **11.** Acute **13.** 56 in. **15.** 14 ft **17.** 47 mi **19.** 8π cm or approximately 25.13 cm **21.** 11π mi or approximately 34.56 mi **23.** 17π ft or approximately 53.41 ft **25.** The perimeter is 17.4 cm. **27.** The perimeter is 8 cm. **29.** The perimeter is 24 m. **31.** The perimeter is 48.8 cm. **33.** The perimeter is 17.5 in. **35.** The length of a diameter is 8.4 cm. **37.** The circumference is 1.5π in. **39.** The circumference is 226.19 cm. **41.** 60 ft of fencing should be purchased.

Copyright © Houghton Mifflin Company. All rights reserved.

43. The carpet must be nailed down along 44 ft. **45.** The length is 120 ft. **47.** The length of the third side is 10 in. **49.** The length of each side is 12 in. **51.** The length of a diameter is 2.55 cm. **53.** The length is 13.19 ft. **55.** The bicycle travels 50.27 ft. **57.** The circumference is 39,935.93 km. **59.** 60 ft^2 **61.** 20.25 in^2 **63.** 546 ft^2 **65.** 16π cm^2 or approximately 50.27 cm^2 **67.** 30.25π mi^2 or approximately 95.03 mi^2 **69.** 72.25π ft^2 or approximately 226.98 ft^2 **71.** The area is 156.25 cm^2. **73.** The area is 570 in^2. **75.** The area is 192 in^2. **77.** The area is 13.5 ft^2. **79.** The area is 330 cm^2. **81.** The area is 25π in^2. **83.** The area is 9.08 ft^2. **85.** The area is $10{,}000\pi$ in^2. **87.** The area is 126 ft^2. **89.** 7500 yd^2 must be purchased. **91.** The width is 10 in. **93.** The length of the base is 20 m. **95.** You should buy 2 qt. **97.** It will cost \$98. **99.** The increase in area is 113.10 in^2. **101.** The cost will be \$878. **103.** The area is 216 m^2. **105.** The cost is \$1600. **109.** $A = \frac{\pi d^2}{4}$

SECTION 3.3

1. 840 in^3 **3.** 15 ft^3 **5.** 4.5π cm^3 or approximately 14.14 cm^3 **7.** The volume is 34 m^3. **9.** The volume is 42.875 in^3. **11.** The volume is 36π ft^3. **13.** The volume is 8143.01 cm^3. **15.** The volume is 75π in^3. **17.** The volume is 120 in^3. **19.** The width is 2.5 ft. **21.** The radius of the base is 4.00 in. **23.** The length is 5 in. The width is 5 in. **25.** There are 75.40 m^3 in the tank. **27.** 94 m^2 **29.** 56 m^2 **31.** 96π in^2 or approximately 301.59 in^2 **33.** The surface area is 184 ft^2. **35.** The surface area is 69.36 m^2. **37.** The surface area is 225π cm^2. **39.** The surface area is 402.12 in^2. **41.** The surface area is 6π ft^2. **43.** The surface area is 297 in^2. **45.** The width is 3 cm. **47.** 11 cans of paint should be purchased. **49.** 456 in^2 of glass are needed. **51.** The surface area of the pyramid is 22.53 cm^2 larger. **53.** **a.** Always true **b.** Never true **c.** Sometimes true

CHAPTER 3 REVIEW EXERCISES

1. $\angle x = 22°$, $\angle y = 158°$ [3.1C] **2.** $\angle x = 68°$ [3.1B] **3.** The length of AC is 44 cm. [3.1A] **4.** 19° [3.1A] **5.** The volume is 96 cm^3. [3.3A] **6.** $\angle a = 138°$, $\angle b = 42°$ [3.1B] **7.** The surface area is 220 ft^2. [3.3B] **8.** The supplement is 148°. [3.1A] **9.** The area is 78 cm^2. [3.2B] **10.** The area is 63 m^2. [3.2B] **11.** The volume is 39 ft^3. [3.3A] **12.** The measure of the third angle is 95°. [3.1C] **13.** The length of the base is 8 cm. [3.2B] **14.** The volume is 288π mm^3. [3.3A] **15.** The volume is $\frac{784\pi}{3}$ cm^3. [3.3A] **16.** Each side measures 21.5 cm. [3.2A] **17.** 4 cans of paint should be purchased. [3.3B] **18.** 208 yd of fencing are needed. [3.2A] **19.** The area is 90.25 m^2. [3.2B] **20.** The area is 276 m^2. [3.2B]

CHAPTER 3 TEST

1. The radius is 0.75 m. [3.3A] **2.** The circumference is 31.42 cm. [3.2A] **3.** The perimeter is 26 ft. [3.2A] **4.** $BC = 3$ [3.1A] **5.** The volume is 268.08 ft^3. [3.3A] **6.** The area is 63.62 cm^2. [3.2B] **7.** $a = 100°$, $b = 80°$ [3.1B] **8.** 75° [3.1A] **9.** $a = 135°$, $b = 45°$ [3.1B] **10.** The area is 55 m^2. [3.2B] **11.** The volume is 169.65 m^3. [3.3A] **12.** The perimeter is 6.8 m. [3.2A] **13.** 58° [3.1A] **14.** The surface area is 164.93 ft^2. [3.3B] **15.** There are 113.10 in^2 more in the larger pizza. [3.2B] **16.** The measures of the other two angles are 58° and 90°. [3.1C] **17.** The bicycle travels 73.3 ft. [3.2A] **18.** The area of the room is 28 yd^2. [3.2B] **19.** The volume of the silo is 1145.11 ft^3. [3.3A] **20.** The area is 11 m^2. [3.2B]

CUMULATIVE REVIEW EXERCISES

1. -3, 0, and 1 [1.1A] **2.** 0.089 [1.2B] **3.** 35% [1.2B] **4.** $-\frac{2}{3}$ [1.2D] **5.** -24.51 [1.2D] **6.** $-5\sqrt{5}$ [1.2F] **7.** -28 [1.3A] **8.** -8 [1.4A] **9.** $-3m + 3n$ [1.4B] **10.** $21y$ [1.4C] **11.** $7x + 9$ [1.4D] **12.** $\{-2, -1\}$ [1.5A] **13.** $\{-10, 0, 10, 20, 30\}$ [1.5A] **14.** 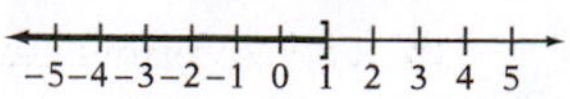

[1.5C] **15.** 5 [2.2B]

Copyright © Houghton Mifflin Company. All rights reserved.

16. $\frac{1}{2}$ [2.2C] **17.** $\{y \mid y \le -4\}$ [2.5A] **18.** $\{x \mid x \ge 2\}$ [2.5A] **19.** $\{x \mid x < -3 \text{ or } x > 4\}$ [2.5B] **20.** $\{x \mid 2 \le x \le 6\}$ [2.5B] **21.** $1, -\frac{1}{3}$ [2.6A] **22.** $\{x \mid 6 \le x \le 10\}$ [2.6B] **23.** $\angle x = 131°$ [3.1B] **24.** $4x - 10 = 2; x = 3$ [2.3B] **25.** The third angle measures 122°. [3.1C] **26.** Michael will earn \$312.50 from the two accounts in one year. [2.1D] **27.** The third side measures 4.5 m. [3.2A] **28.** The women's median annual earnings are 70.8% of the men's median annual earnings. [1.2G] **29.** The area is 20.25π cm². [3.2B] **30.** The height of the box is 3 ft. [3.3A]

Answers to Chapter 4 Selected Exercises

PREP TEST

1. $-4x + 12$ [1.4D] **2.** 10 [1.2F] **3.** -2 [1.3A] **4.** 11 [1.4A] **5.** 2.5 [1.4A] **6.** 5 [1.4A] **7.** 1 [1.4A] **8.** 4 [2.2A]

SECTION 4.1

1.

3. 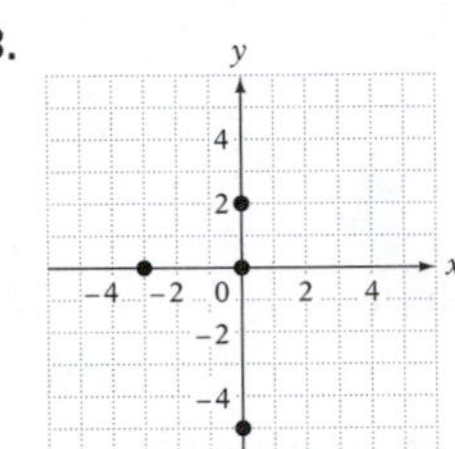

5. 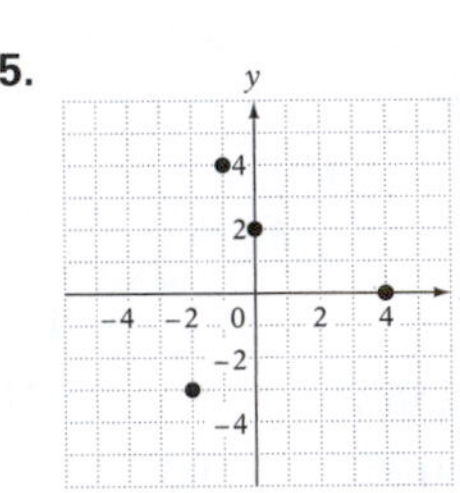

7. $A(2, 3), B(4, 0), C(-4, 1), D(-2, -2)$

9. $A(-2, 5), B(3, 4), C(0, 0), D(-3, -2)$ **11. a.** $2, -4$ **b.** $1, -3$ **15.** Yes **17.** No **19.** No **21.** No **23.** $(3, 7)$ **25.** $(6, 3)$ **27.** $(0, 1)$ **29.** $(-5, 0)$

31.

33. 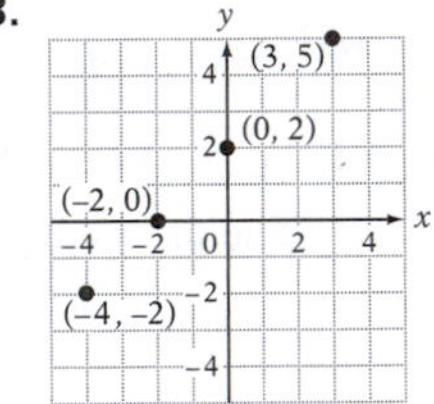

35. 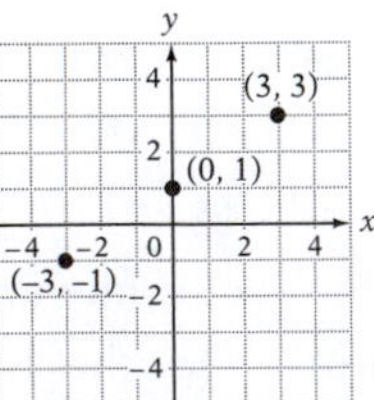

37. 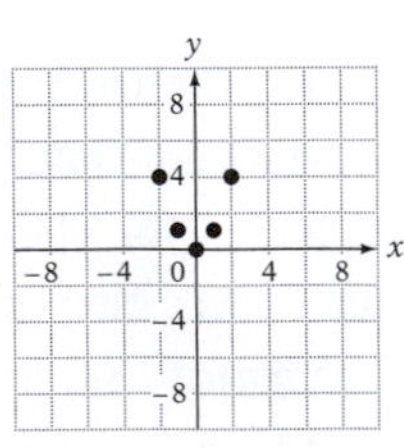

39. a. After 20 min., the temperature is 280°F. **b.** After 50 min., the temperature is 160°F.

41. 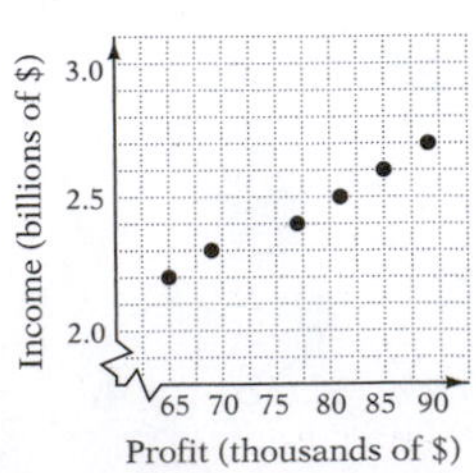

43. 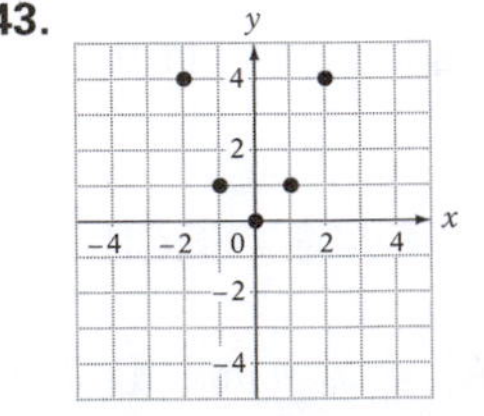

Copyright © Houghton Mifflin Company. All rights reserved.

SECTION 4.2

3. Yes **5.** Yes **7.** No **9.** Function **11.** Function **13.** Function **15.** Not a function **17. a.** Yes **b.** $y = \$34.75$ **19.** $f(3) = 11$ **21.** $f(0) = -4$ **23.** $G(0) = 4$ **25.** $G(-2) = 10$ **27.** $q(3) = 5$ **29.** $q(-2) = 0$ **31.** $F(4) = 24$ **33.** $F(-3) = -4$ **35.** $H(1) = 1$ **37.** $H(t) = \frac{3t}{t+2}$ **39.** $s(-1) = 6$ **41.** $s(a) = a^3 - 3a + 4$ **43.** $4h$ **45. a.** \$4.75 per game **b.** \$4.00 per game **47. a.** \$3000 **b.** \$950 **49.** Domain = {1, 2, 3, 4, 5} Range = {1, 4, 7, 10, 13}

51. Domain = {0, 2, 4, 6} Range = {1, 2, 3, 4} **53.** Domain = {1, 3, 5, 7, 9} Range = {0} **55.** Domain = {−2, −1, 0, 1, 2} Range = {0, 1, 2}

57. Domain = {−2, −1, 0, 1, 2} Range = {−3, 3, 6, 7, 9} **59.** 1 **61.** −8 **63.** None **65.** None **67.** 0 **69.** None **71.** None

73. −2 **75.** None **77.** Range = {−3, 1, 5, 9} **79.** Range = {−23, −13, −8, −3, 7} **81.** Range = {0, 1, 4}

83. Range = {2, 14, 26, 42} **85.** Range = $\left\{-5, \frac{5}{3}, 5\right\}$ **87.** Range = $\left\{-1, -\frac{1}{2}, -\frac{1}{3}, 1\right\}$ **89.** Range = {−38, −8, 2}

93. a. {(−2, −8), (−1, −1), (0, 0), (1, 1), (2, 8)} **b.** Yes, the set of ordered pairs defines a function because each member of the domain is assigned to exactly one member of the range. **95.** The power produced will be 50.625 watts.

97. a. 22.5 ft/s **b.** 30 ft/s **99. a.** 68°F **b.** 52°F

SECTION 4.3

1.

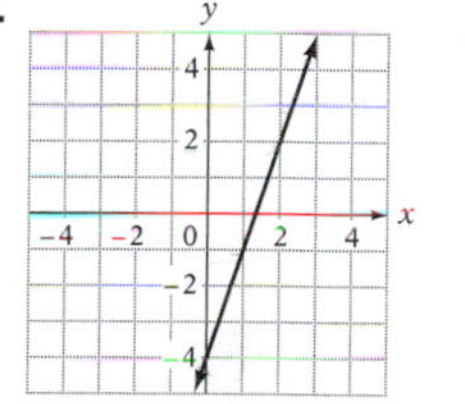

3.

5.

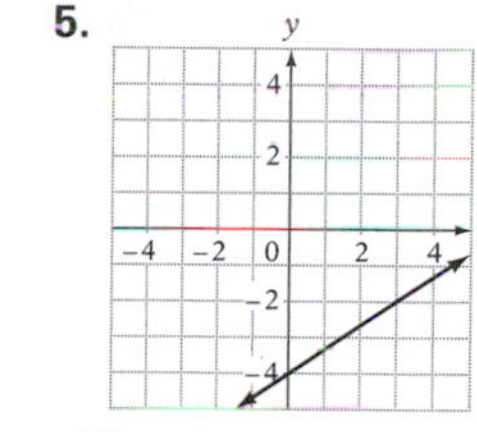

7.

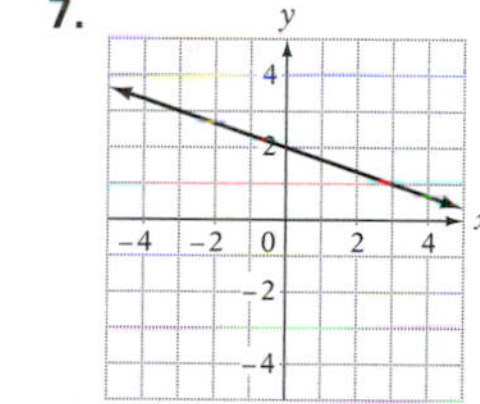

9.

11.

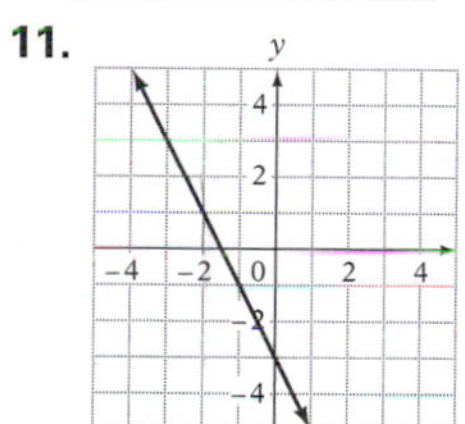

13.

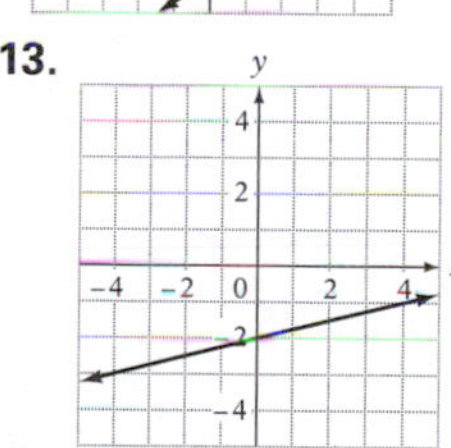

15.

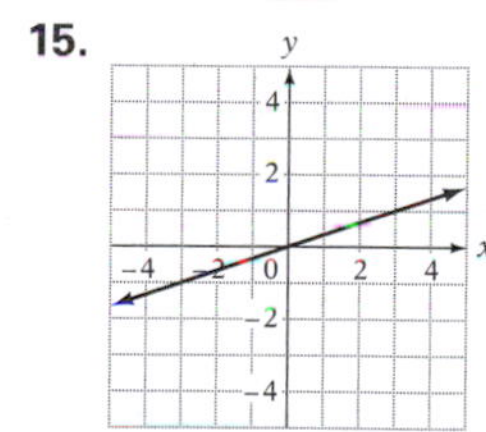

17.

19. x-intercept: (−4, 0); y-intercept: (0, 2)

21. x-intercept: $\left(\frac{9}{2}, 0\right)$; y-intercept: (0, −3)

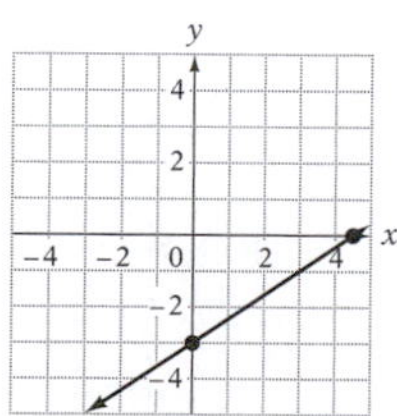

23. x-intercept: $\left(\frac{3}{2}, 0\right)$; y-intercept: (0, 3)

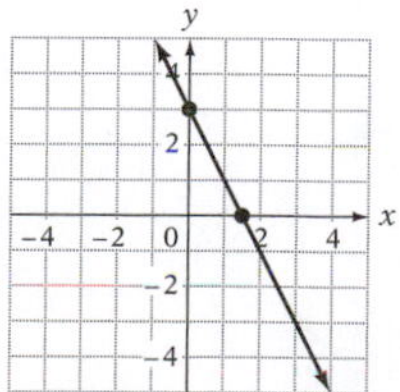

25. x-intercept: $\left(\frac{4}{3}, 0\right)$; y-intercept: (0, 2)

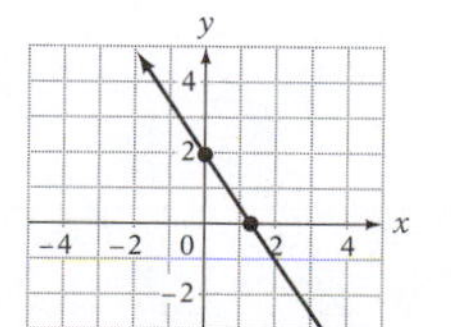

Copyright © Houghton Mifflin Company. All rights reserved.

27. x-intercept: $\left(\frac{3}{2}, 0\right)$; y-intercept: $(0, -2)$

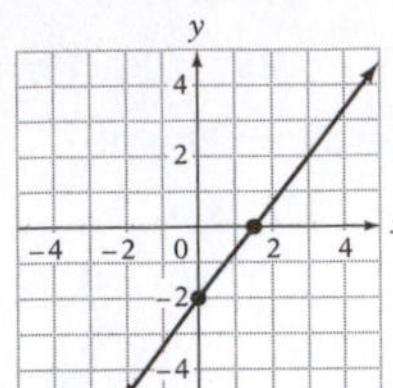

29. Marlys receives \$165 for tutoring 15 h.

31. The cost of receiving 32 messages is \$14.40.

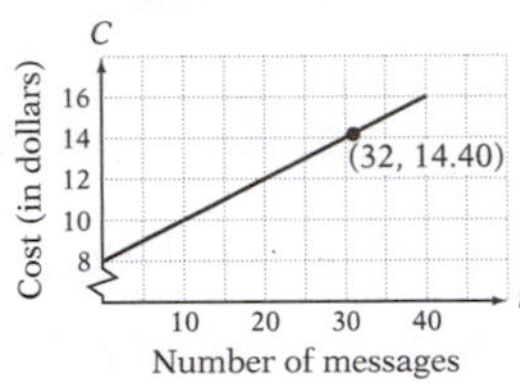

33. The cost of manufacturing 6000 compact discs is \$33,000.

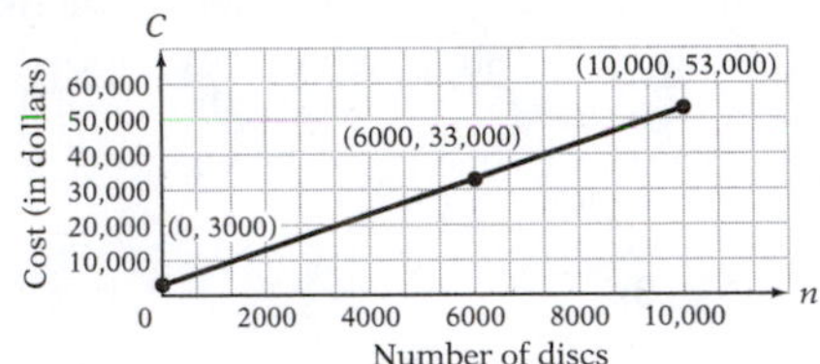

39.

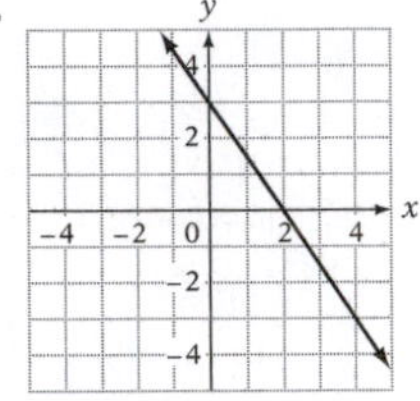

SECTION 4.4

1. -1 **3.** $\frac{1}{3}$ **5.** $-\frac{2}{3}$ **7.** $-\frac{3}{4}$ **9.** Undefined **11.** $\frac{7}{5}$ **13.** 0 **15.** $-\frac{1}{2}$ **17.** Undefined **19.** $m = 40$. The average speed of the motorist is 40 mph. **21.** $m = -5$. The temperature of the oven decreases 5°/min. **23.** $m = -0.05$. For each mile the car is driven, approximately 0.05 gallon of fuel is used. **25.** $m \approx 343.9$. The average speed of the runner was 343.9 m/min. **27.** No **29.** $-3, 5, -3, (0, 5)$ **31.** $4, 0, 4, (0, 0)$

33.

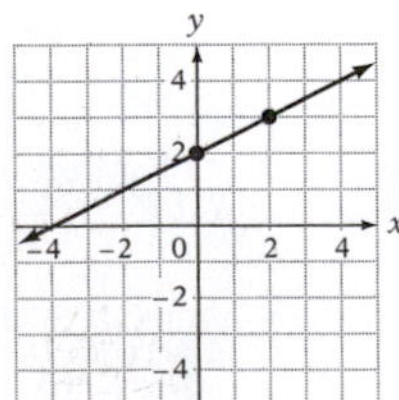

35.

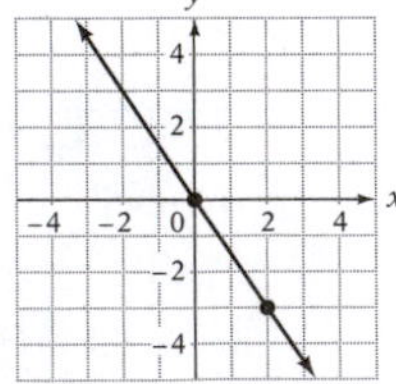

37.

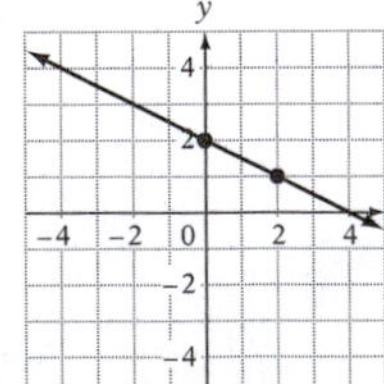

39.

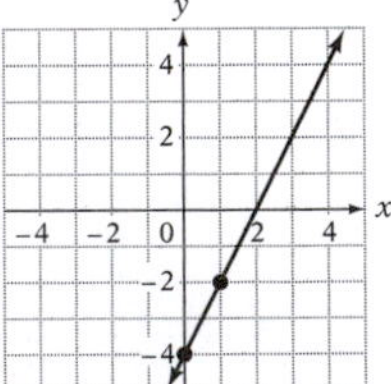

41.

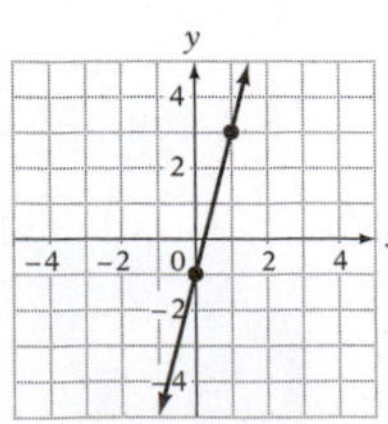

43.

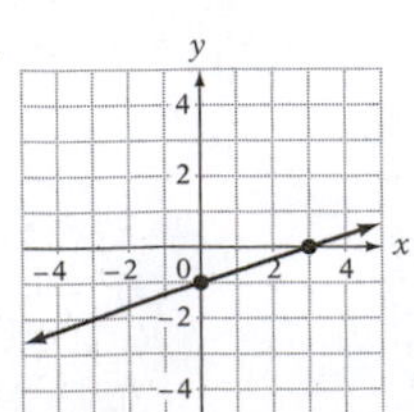

45.

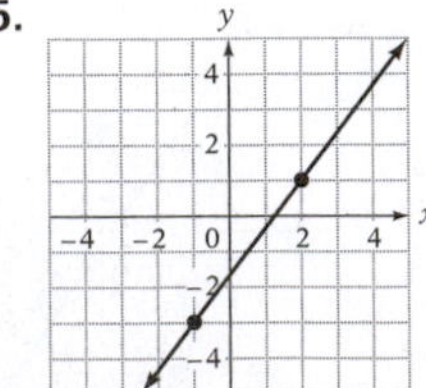

47.

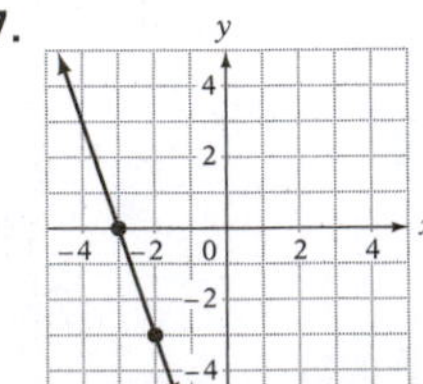

Copyright © Houghton Mifflin Company. All rights reserved.

49. **51.** Increases by 2 **53.** Increases by 2 **55.** Decreases by $\frac{2}{3}$

57. i. D; ii. C; iii. B; iv. F; v. E; vi. A **61.** $k = -1$ **63.** $k = \frac{13}{2}$

SECTION 4.5

3. $y = 2x + 5$ **5.** $y = \frac{1}{2}x + 2$ **7.** $y = \frac{5}{4}x + \frac{21}{4}$ **9.** $y = -\frac{5}{3}x + 5$ **11.** $y = -3x + 9$ **13.** $y = -3x + 4$

15. $y = \frac{2}{3}x - \frac{7}{3}$ **17.** $y = \frac{1}{2}x$ **19.** $y = 3x - 9$ **21.** $y = -\frac{2}{3}x + 7$ **23.** $y = -x - 3$ **25.** $y = \frac{7}{5}x - \frac{27}{5}$

27. $y = -\frac{2}{5}x + \frac{3}{5}$ **29.** $x = 3$ **31.** $y = -\frac{5}{4}x - \frac{15}{2}$ **33.** $y = -3$ **35.** $y = -2x + 3$ **37.** $x = -5$

39. $y = x + 2$ **41.** $y = -2x - 3$ **43.** $y = \frac{2}{3}x + \frac{5}{3}$ **45.** $y = \frac{1}{3}x + \frac{10}{3}$ **47.** $y = \frac{3}{2}x - \frac{1}{2}$ **49.** $y = -\frac{3}{2}x + 3$

51. $y = -1$ **53.** $y = x - 1$ **55.** $y = -x + 1$ **57.** $y = -\frac{8}{3}x + \frac{25}{3}$ **59.** $y = \frac{1}{2}x - 1$ **61.** $y = -4$

63. $y = \frac{3}{4}x$ **65.** $y = -\frac{4}{3}x + \frac{5}{3}$ **67.** $x = -2$ **69.** $y = x - 1$ **71.** $y = \frac{4}{3}x + \frac{7}{3}$ **73.** $y = -x + 3$

75. a. $y = 1200x,\ 0 \le x \le 26\frac{2}{3}$ **b.** The height of the plane 11 min after takeoff is 13,200 ft. **77. a.** $y = 85x + 30{,}000$ **b.** It will cost \$183,000 to build an 1800-ft^2 house. **79. a.** $y = -0.032x + 16,\ 0 \le x \le 500$ **b.** After 150 mi are driven, 11.2 gal are left in the tank. **81. a.** $y = -20x + 230{,}000$ **b.** 60,000 cars would be sold at \$8500 each. **83. a.** $y = 63x$ **b.** There are 315 Calories in a 5-ounce serving. **85.** $f(x) = x + 3$ **87.** 0 **89. a.** -10 **b.** 6 **95.** Answers will vary. The possible answers include (0, 4), (3, 2), and (9, −2).

SECTION 4.6

3. -5 **5.** $-\frac{1}{4}$ **7.** Yes **9.** No **11.** No **13.** Yes **15.** Yes **17.** Yes **19.** No **21.** Yes

23. $y = \frac{2}{3}x - \frac{8}{3}$ **25.** $y = \frac{1}{3}x - \frac{1}{3}$ **27.** $y = -\frac{5}{3}x - \frac{14}{3}$ **29.** $y = -2x + 15$ **31.** $\frac{A_1}{B_1} = \frac{A_2}{B_2}$ **33.** Any equation of the form $y = 2x + b$, where $b \neq -13$, or of the form $y = -\frac{3}{2}x + c$, where $c \neq 8$.

SECTION 4.7

3. Yes **5.** No **7.**

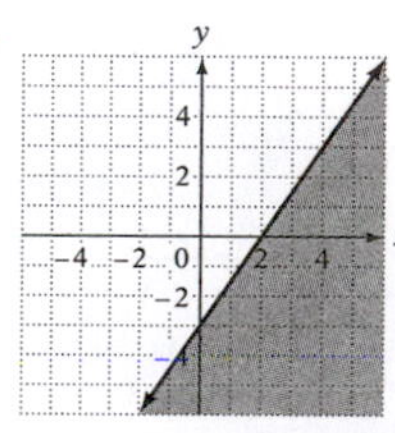

9.

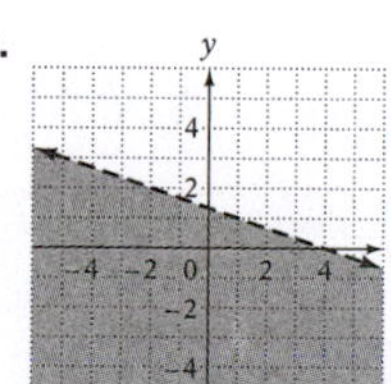

11.

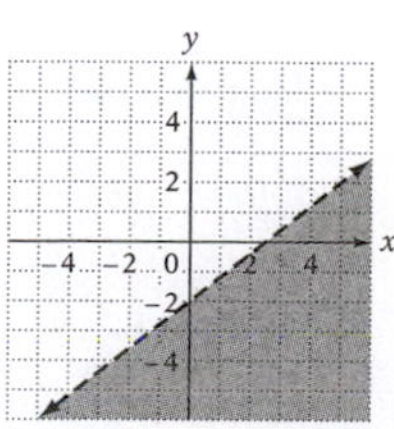

13.

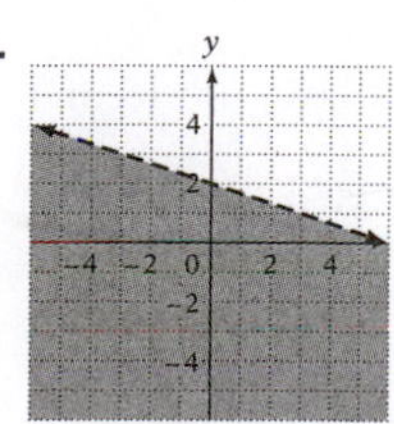

15.

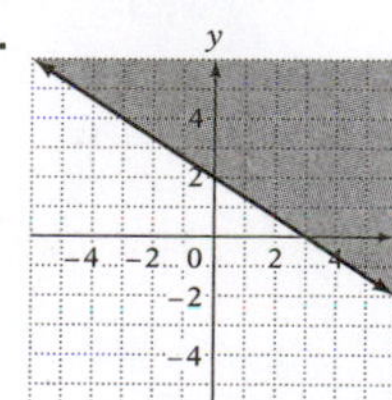

17.

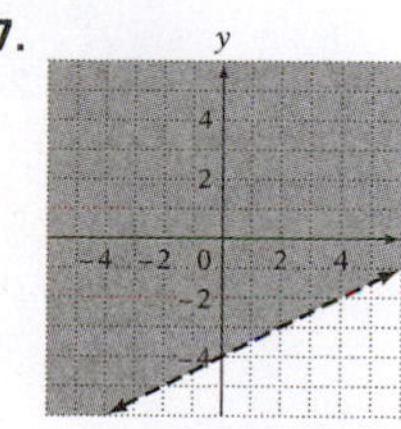

19.

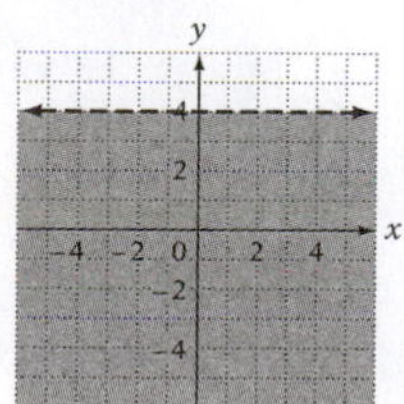

Copyright © Houghton Mifflin Company. All rights reserved.

21.

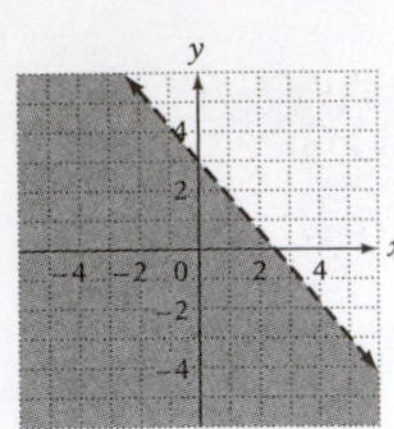

23.

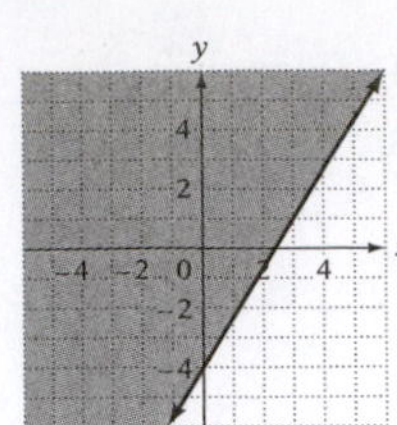

CHAPTER 4 REVIEW EXERCISES

1. (4, 2) [4.1B] **2.** $P(-2) = -2$ [4.2A]
$P(a) = 3a + 4$

3. [4.1B]

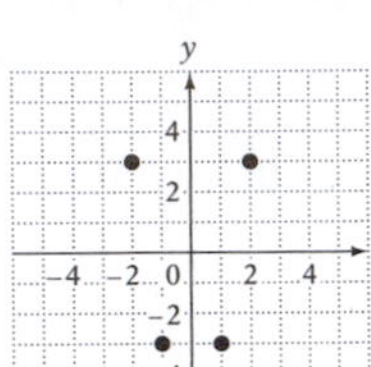

4. [4.1A]

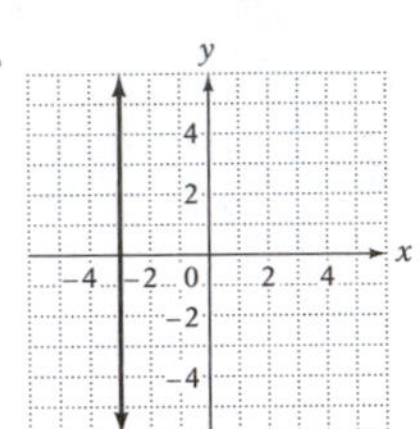

5. Range = $\{-1, 1, 5\}$ [4.2A] **6.** Domain = $\{-1, 0, 1, 5\}$ [4.2A]
Range = $\{0, 2, 4\}$ **7.** Yes [4.6A]

8. -4 [4.2A] **9.** x-intercept: $(-3, 0)$
y-intercept: $(0, -2)$

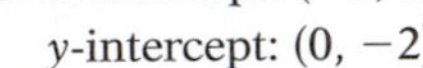

[4.3C]

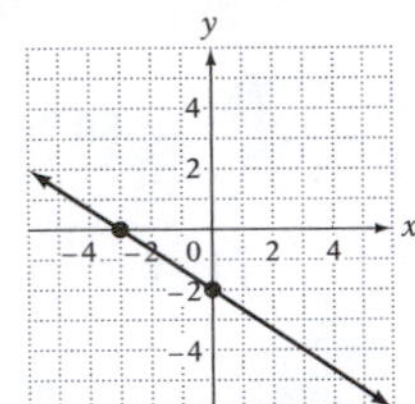

10. x-intercept: $(-2, 0)$
y-intercept: $(0, -3)$ [4.3C]

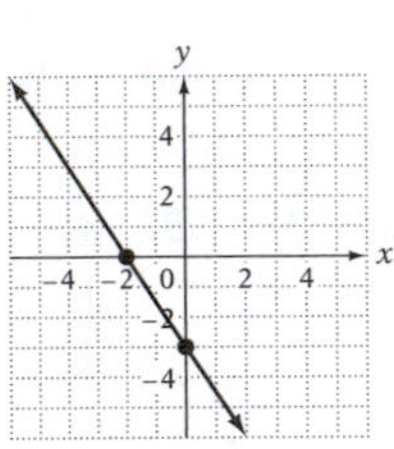

11. [4.3A]

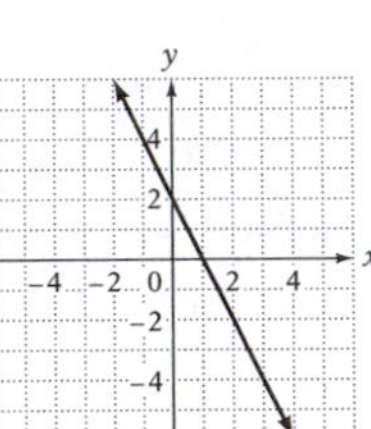

12. [4.3B]

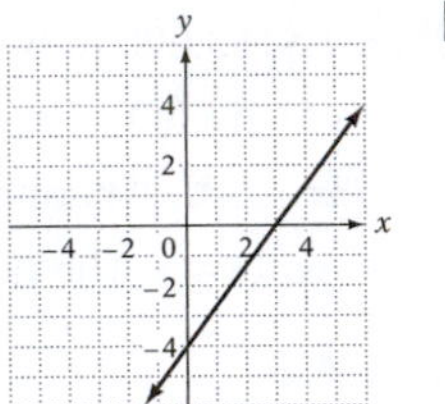

13. -1 [4.4A] **14.** $y = \frac{5}{2}x + \frac{23}{2}$ [4.5A]

15. [4.1A]

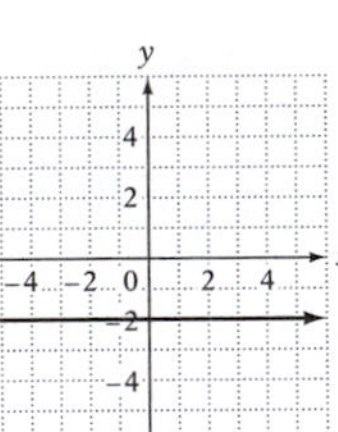

16. [4.4B]

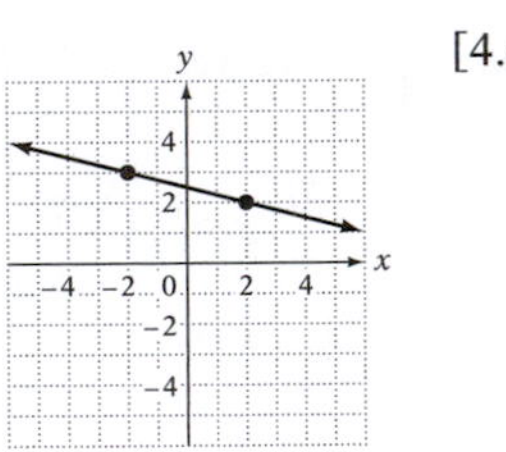

17. $\{-2, -1, 2\}$ [4.2A] **18. a.** $y = -x + 295; 0 \le x \le 295$ **b.** When the rate is \$120, 175 rooms will be occupied. [4.5C] **19.** $y = -4x - 5$ [4.6A] **20.** $y = \frac{5}{2}x + 8$ [4.6A]

21. [4.3B]

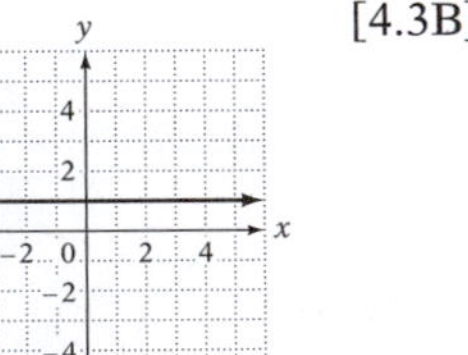

22. [4.3B]

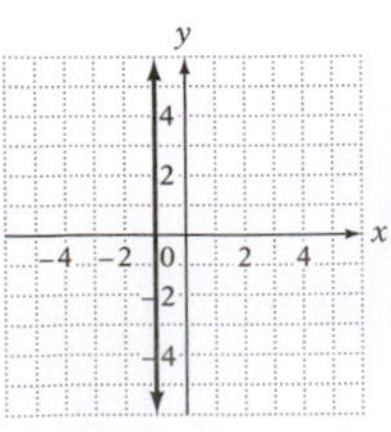

23. $y = -\frac{2}{3}x + 1$ [4.5A] **24.** $y = \frac{1}{4}x + 4$ [4.5B] **25.** $(-4, 0), (0, 6)$ [4.3C]

Copyright © Houghton Mifflin Company. All rights reserved.

26. $(-1, -7)$ [4.1B] **27.** 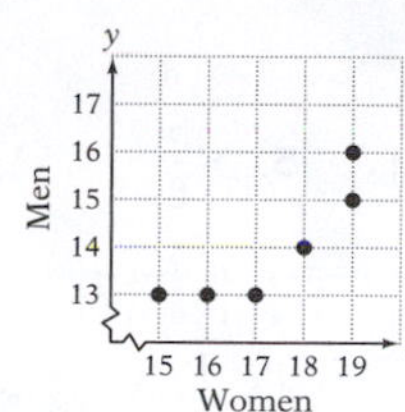

[4.1C] **28.** 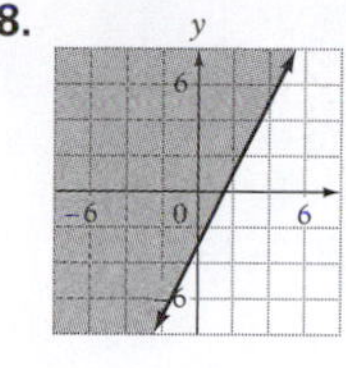 [4.7A] **29.** 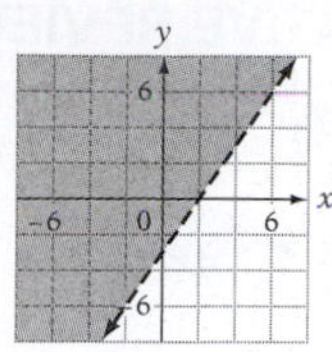 [4.7A]

30. $y = -\frac{7}{6}x + \frac{5}{3}$ [4.5B] **31.** $y = 2x$ [4.6A] **32.** $y = -3x + 7$ [4.6A] **33.** $y = \frac{3}{2}x + 2$ [4.6A]

34. 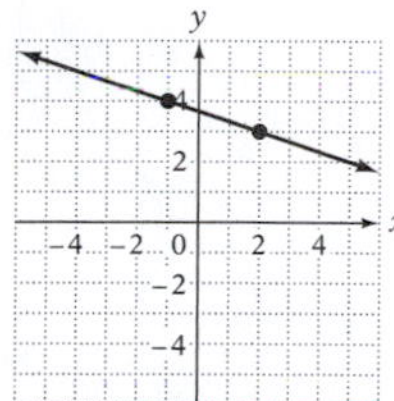 [4.4B] **35.** After 4 h, the car has traveled 220 mi. 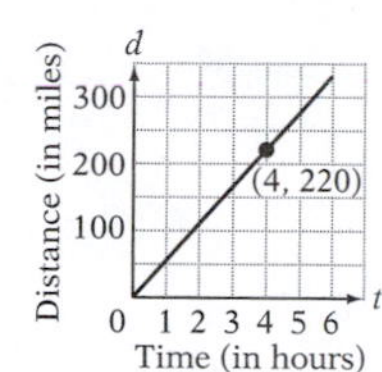

[4.3D]

36. The slope is 20. The manufacturing cost is $20 per calculator. [4.4A] **37. a.** $y = 80x + 25{,}000$
b. The house will cost $185,000 to build. [4.5C]

CHAPTER 4 TEST

1. 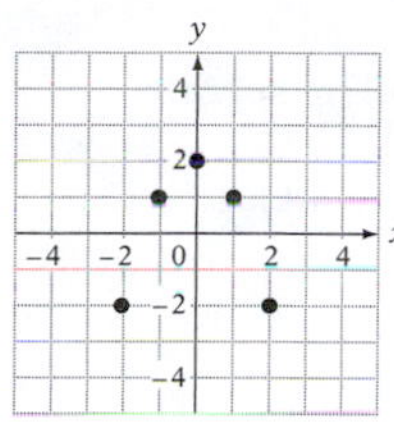 [4.1B] **2.** $(-3, 0)$ [4.1B] **3.** 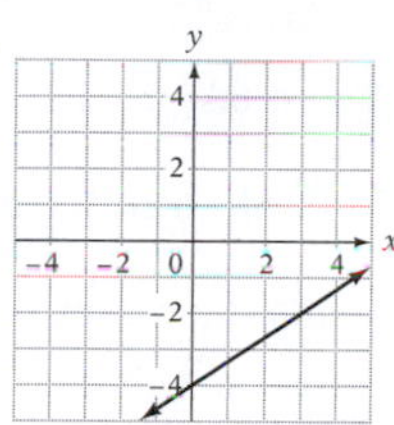 [4.3A] **4.** 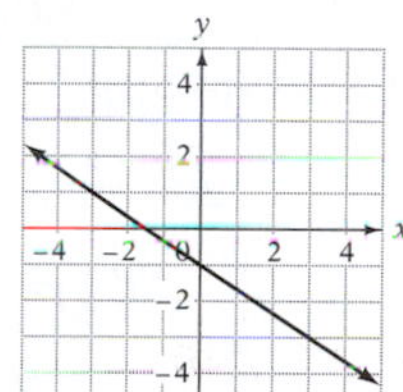 [4.3B]

5. $x = -2$ [4.5A] **6.** No [4.6A] **7.** $-\frac{1}{6}$ [4.4A] **8.** $P(2) = 9$ [4.2A] **9.** 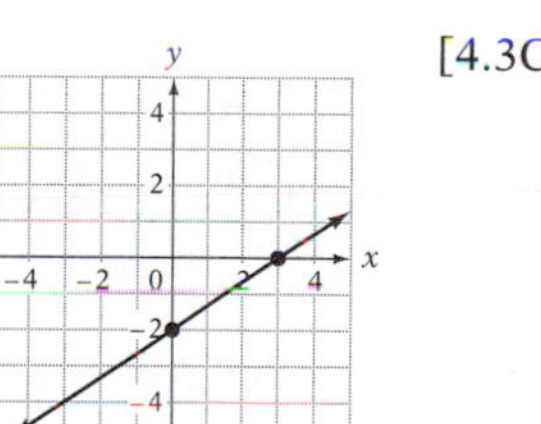 [4.3C]

10. 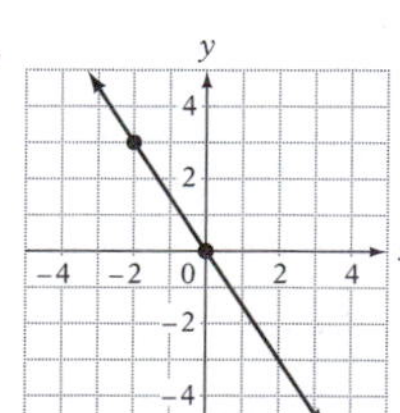 [4.4B] **11.** $y = \frac{2}{5}x + 4$ [4.5A] **12.** 0 [4.2A] **13.** $y = -\frac{7}{5}x + \frac{1}{5}$ [4.5B]

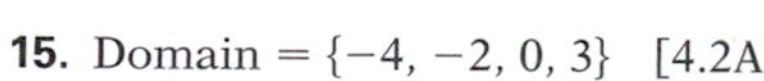

14. $y = -3$ [4.5A] **15.** Domain = $\{-4, -2, 0, 3\}$ [4.2A] **16.** $y = -\frac{3}{2}x + \frac{7}{2}$ [4.6A]
Range = $\{0, 2, 5\}$

17. $y = 2x + 1$ [4.6A] **18.** 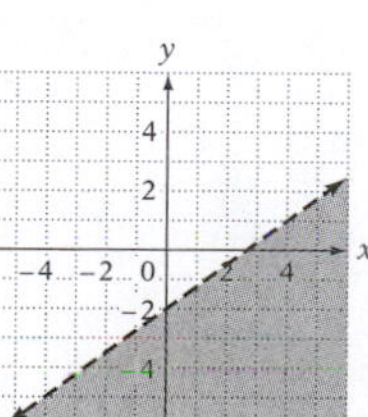 [4.7A] **19.** The slope is $-10{,}000$. The value of the house decreases by $10,000 each year. [4.4A] **20. a.** $y = -\frac{3}{10}x + 175$ **b.** When the tuition is $300, 85 students will enroll. [4.5C]

Copyright © Houghton Mifflin Company. All rights reserved.

CUMULATIVE REVIEW EXERCISES

1. $-5, -3$ [1.1A] **2.** 0.85 [1.2B] **3.** $9\sqrt{5}$ [1.2F] **4.** -12 [1.3A] **5.** $-\frac{5}{8}$ [1.4A]

6. $-4d - 9$ [1.4B] **7.** $-32z$ [1.4C] **8.** $-13x + 7y$ [1.4D] **9.**

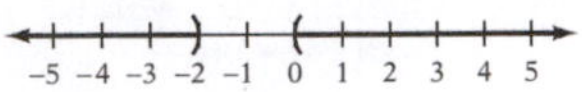

[1.5C]

10. $\frac{3}{2}$ [2.2A] **11.** 1 [2.2C] **12.** $\{x|x > -1\}$ [2.5A] **13.** $\varnothing$ [2.5B] **14.** $\left\{x|0 < x < \frac{10}{3}\right\}$ [2.6B]

15. $f(2) = 6$ [4.2A] **16.** The slope is 2. [4.4A] **17.**

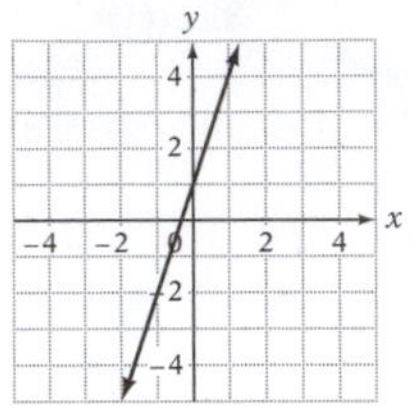

[4.3A]

18.

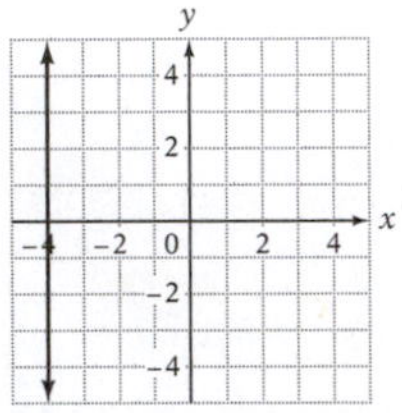

[4.3B]

19.

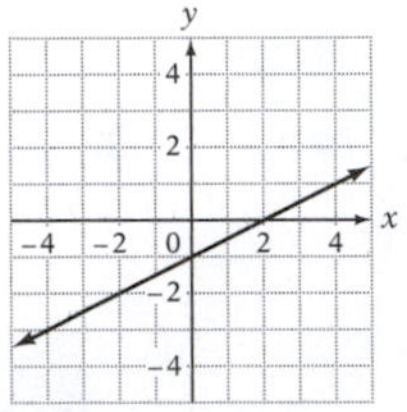

[4.4B]

20.

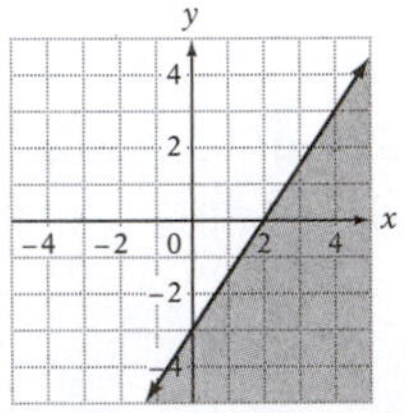

[4.7A]

21. $y = -\frac{1}{3}x - 2$ [4.5B] **22.** $y = -\frac{3}{2}x + 7$ [4.6A] **23.** The first plane is traveling at 400 mph. The second plane is traveling at 200 mph. [2.4C] **24.** 20 lb of $6 coffee and 40 lb of $9 coffee should be used. [2.4A]

25. The slope is $-\frac{10{,}000}{3}$. The value of the backhoe decreases by $3333.33 each year. [4.4A]

Answers to Chapter 5 Selected Exercises

PREP TEST

1. $6x + 5y$ [1.4D] **2.** 7 [1.4A] **3.** 0 [2.2A] **4.** -3 [2.2C] **5.** 1000 [2.2C]

6.

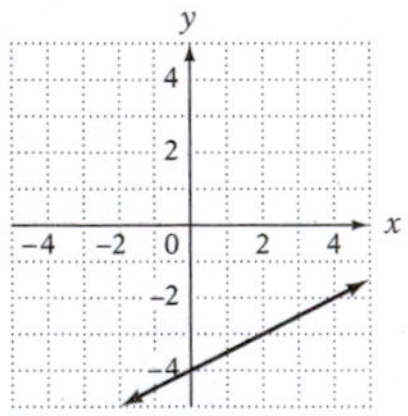

[4.3A]

7.

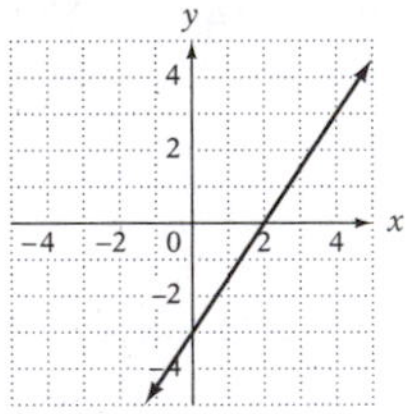

[4.3B]

8.

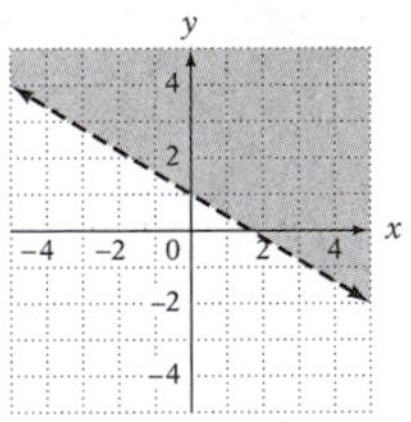

[4.7A]

SECTION 5.1

1. No **3.** Yes **5.** Independent **7.** Inconsistent **9.**

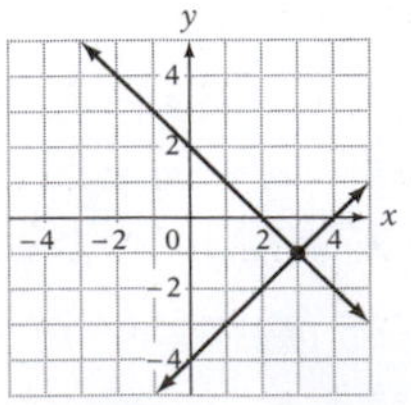

(3, −1)

11.

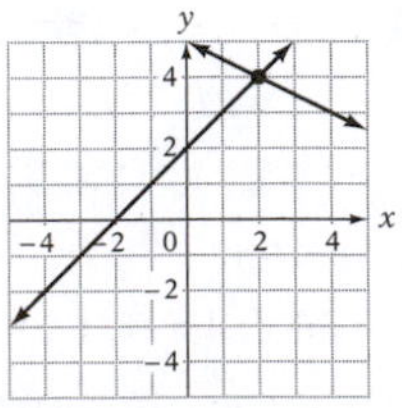

(2, 4)

Copyright © Houghton Mifflin Company. All rights reserved.

13. 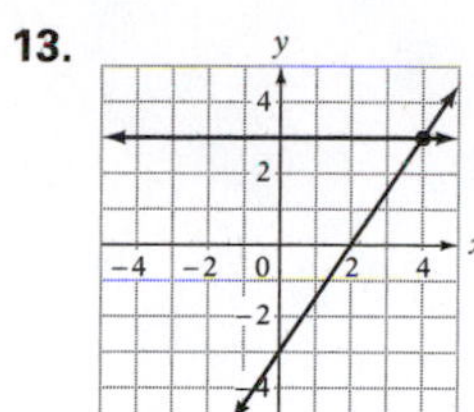
(4, 3)

15. 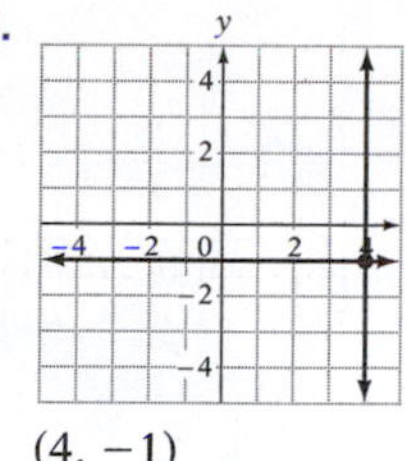
(4, −1)

17. 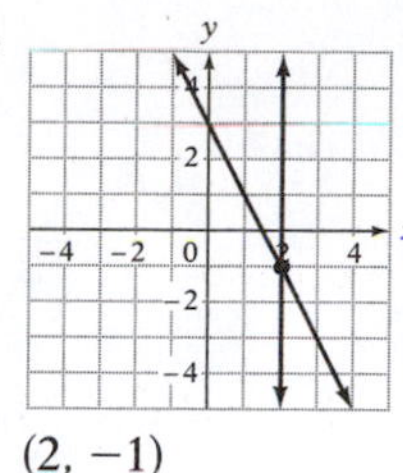
(2, −1)

19. 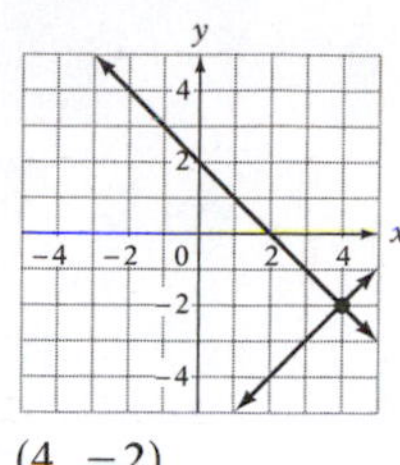
(4, −2)

21. 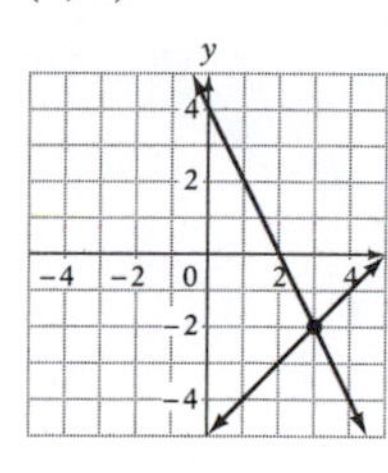

23. 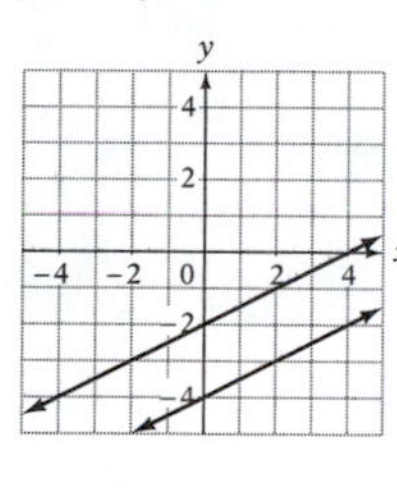

25. 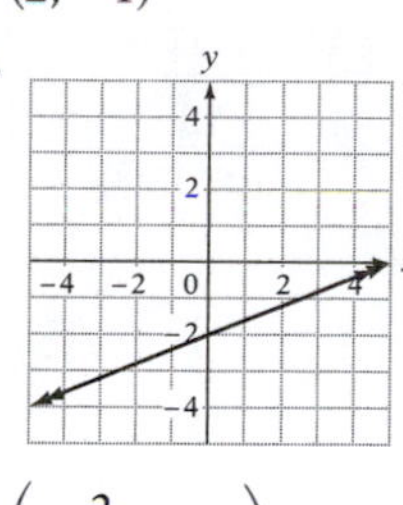

(3, −2) No solution $\left(x, \frac{2}{5}x - 2\right)$

27. (2, −1) **29.** (1, 2) **31.** (3, −2) **33.** $\left(-\frac{5}{2}, -4\right)$ **35.** (−1, 4) **37.** (−2, 5) **39.** (1, 5) **41.** (2, 0) **43.** $\left(\frac{2}{3}, -1\right)$ **45.** No solution **47.** (−2, 5) **49.** (0, 0) **51.** $\left(x, \frac{1}{2}x - 4\right)$ **53.** (1, 5) **55.** $\left(\frac{2}{3}, 3\right)$ **57.** The amount invested at 4.2% is \$5500. **59.** The amount invested at 6.5% must be \$4000. **61.** The amount invested at 3.5% is \$23,625. The amount invested at 4.5% is \$18,375. **63.** There is \$6000 invested in the mutual bond fund. **65.** (1.20, 1.40) **67.** (0.54, 1.03)

SECTION 5.2

1. (6, 1) **3.** (1, 1) **5.** (2, 1) **7.** (−2, 1) **9.** $(x, 3x - 4)$ **11.** $\left(-\frac{1}{2}, 2\right)$ **13.** No solution **15.** (−1, −2) **17.** (−5, 4) **19.** (2, 5) **21.** $\left(\frac{1}{2}, \frac{3}{4}\right)$ **23.** (0, 0) **25.** (−1, 3) **27.** $\left(\frac{2}{3}, -\frac{2}{3}\right)$ **29.** (−2, 3) **31.** (2, −1) **33.** (10, −5) **35.** $\left(-\frac{1}{2}, \frac{2}{3}\right)$ **37.** $\left(\frac{5}{3}, \frac{1}{3}\right)$ **39.** No solution **41.** (1, −1) **43.** (2, 1, 3) **45.** (1, −1, 2) **47.** (1, 2, 4) **49.** (2, −1, −2) **51.** (−2, −1, 3) **53.** No solution **55.** (1, 4, 1) **57.** (1, 3, 2) **59.** (1, −1, 3) **61.** (0, 2, 0) **63.** (1, 5, 2) **65.** (−2, 1, 1) **69.** (1, −1) **71.** (1, −1) **73.** $y = \frac{2x}{1 - 3x}, y = \frac{-3x}{x + 2}$

SECTION 5.3

1. The rate of the motorboat in calm water is 15 mph. The rate of the current is 3 mph. **3.** The rate of the plane in calm air is 502.5 mph. The rate of the wind is 47.5 mph. **5.** The rate of the team in calm water is 8 km/h. The rate of the current is 2 km/h. **7.** The rate of the plane in calm air is 180 mph. The rate of the wind is 20 mph. **9.** The rate of the plane in calm air is 110 mph. The rate of the wind is 10 mph. **11.** The cost per pound of the wheat flour is \$.65. The cost per pound of the rye flour is \$.70. **13.** The cost of the pine is \$1.10/ft. The cost of the redwood is \$3.30/ft. **15.** The cost of the wool carpet is \$52/yd. **17.** The company plans to manufacture 25 mountain bikes during the week. **19.** The chemist should use 480 g of the first alloy and 40 g of the second alloy. **21.** The Model V computer costs \$4000. **23.** There were 190 regular tickets, 350 member tickets, and 210 student tickets sold for the Saturday performance. **25.** The investor placed \$10,400 in the 8% account, \$5200 in the 6% account, and \$9400 in the 4% account. **27.** The measures of the two angles are 35° and 145°. **29.** The age of the oil painting is 85 years, and the age of the watercolor is 50 years.

Copyright © Houghton Mifflin Company. All rights reserved.

SECTION 5.4

1. 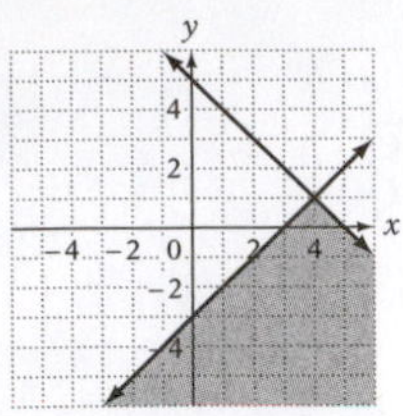
3. 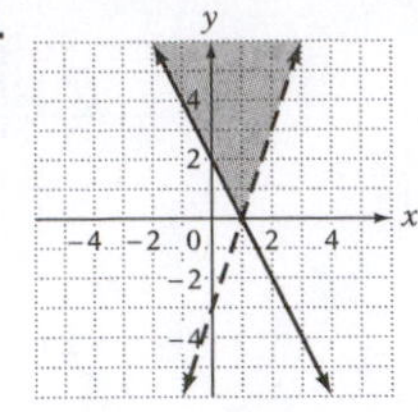
5. 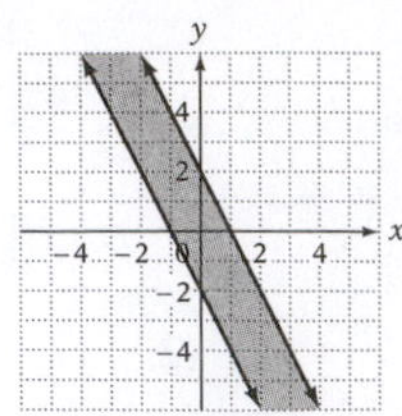
7. 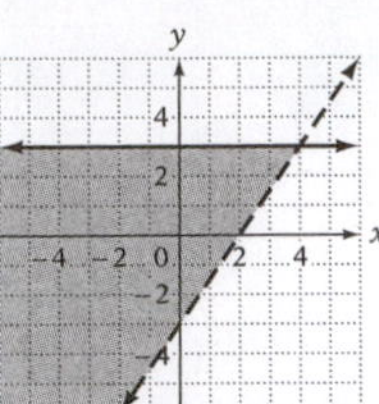
9. 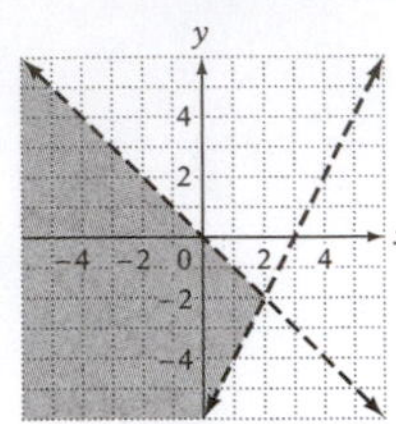
11. 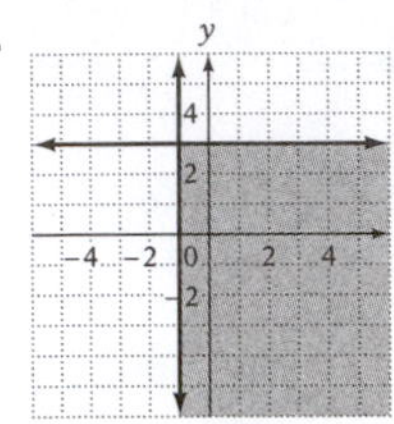
13. 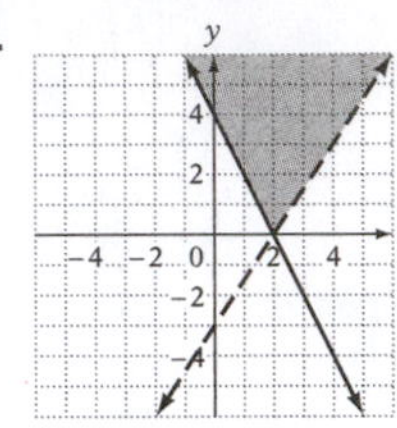
15. 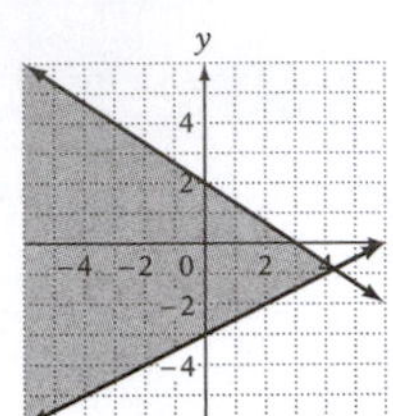
17. 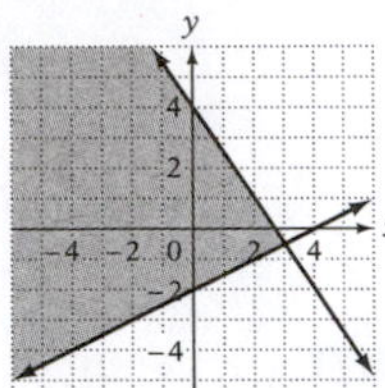
19. 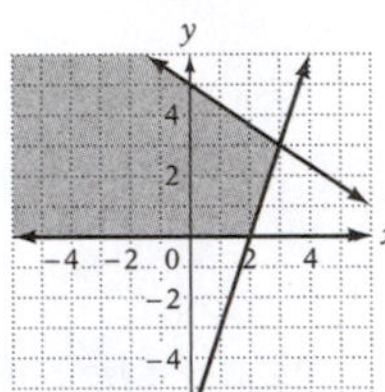
21. 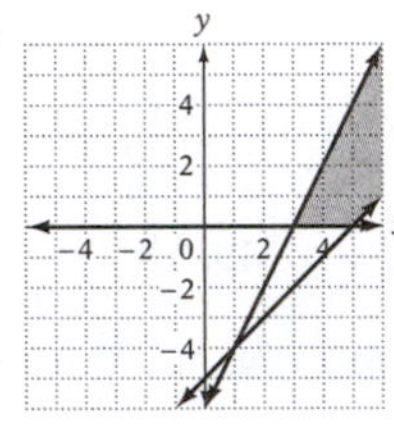
23. 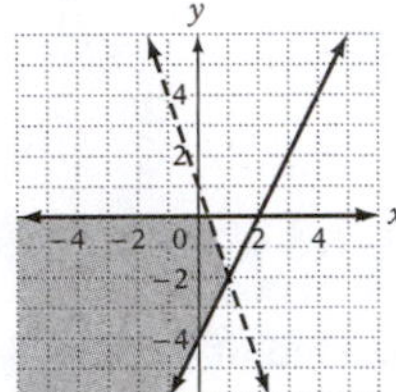

CHAPTER 5 REVIEW EXERCISES

1. $\left(6, -\frac{1}{2}\right)$ [5.1B] **2.** $(-4, 7)$ [5.2A] **3.** 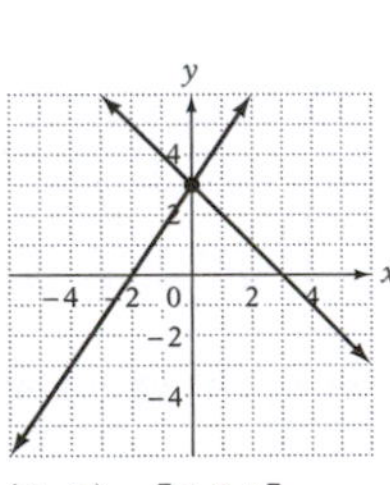 $(0, 3)$ [5.1A] **4.** 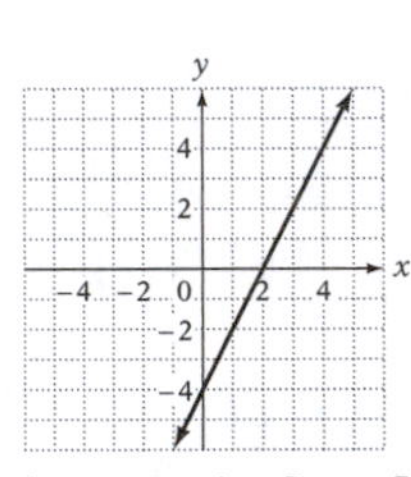 $(x, 2x - 4)$ [5.1A]

5. $\left(x, -\frac{1}{4}x + \frac{3}{2}\right)$ [5.1B] **6.** $\left(x, \frac{1}{3}x - 2\right)$ [5.2A] **7.** $(3, -1, -2)$ [5.2B] **8.** $(5, -2, 3)$ [5.2B] **9.** Yes [5.1A]

10. $\left(\frac{1}{2}, -\frac{5}{2}\right)$ [5.1B] **11.** $(3, -1)$ [5.1B] **12.** $(3, 2)$ [5.2A] **13.** $(-1, -3, 4)$ [5.2B] **14.** $(2, 3, -5)$ [5.2B]

15. 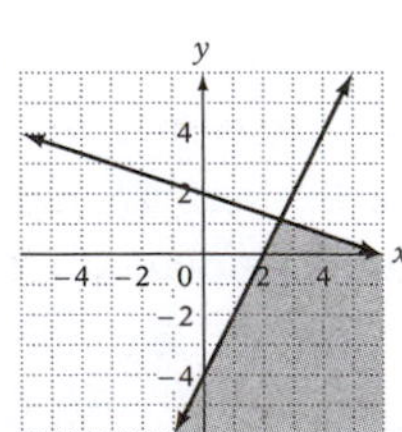 [5.4A] **16.** 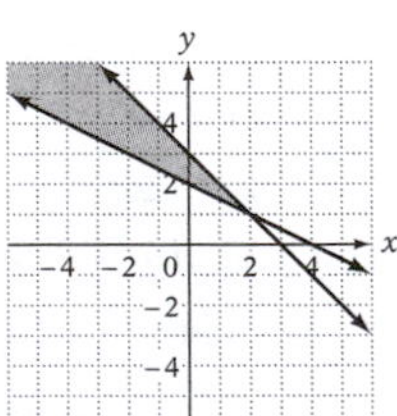 [5.4A]

17. The rate of the cabin cruiser in calm water is 16 mph. The rate of the current is 4 mph. [5.3A]
18. The rate of the plane in calm air is 175 mph. The rate of the wind is 25 mph. [5.3A] **19.** On Friday, 100 children attended. [5.3B] **20.** The amount invested at 3% is $5000. The amount invested at 7% is $15,000. [5.1C]

CHAPTER 5 TEST

1. 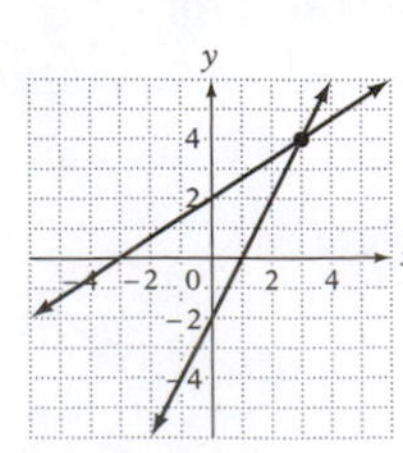 $(3, 4)$ [5.1A] **2.** 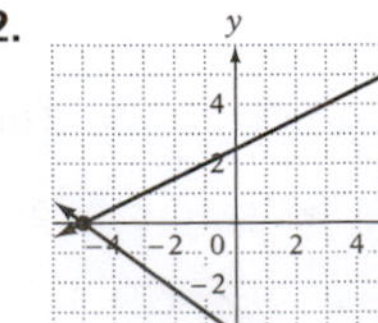 $(-5, 0)$ [5.1A]

Copyright © Houghton Mifflin Company. All rights reserved.

3. 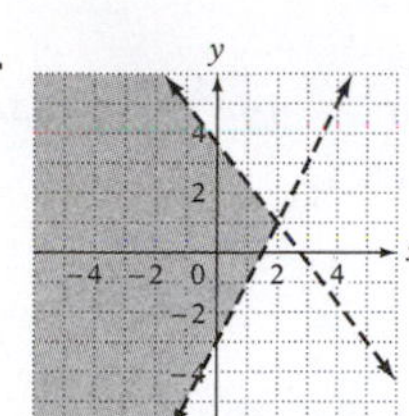 [5.4A] **4.** 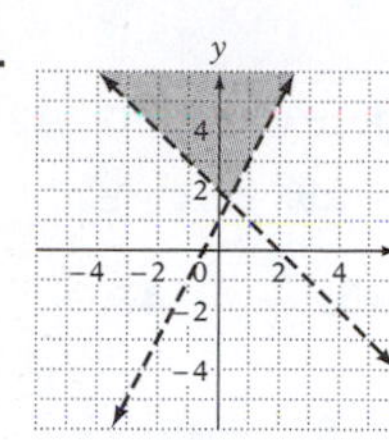 [5.4A] **5.** $\left(\frac{3}{4}, \frac{7}{8}\right)$ [5.1B]

6. $(-3, -4)$ [5.1B] **7.** $(2, -1)$ [5.1B] **8.** $(-2, 1)$ [5.2A] **9.** No solution [5.2A] **10.** $(1, 1)$ [5.2A]

11. No solution [5.2B] **12.** $(2, -1, -2)$ [5.2B] **13.** $\left(-\frac{1}{3}, -\frac{10}{3}\right)$ [5.1B] **14.** Yes [5.1A]

15. $\left(\frac{1}{5}, -\frac{6}{5}, \frac{3}{5}\right)$ [5.2B] **16.** The rate of the plane in calm air is 150 mph. The rate of the wind is 25 mph. [5.3A]

17. The cost of cotton is \$9/yd. The cost of wool is \$14/yd. [5.3B] **18.** The amount invested at 2.7% is \$9000. The amount invested at 5.1% is \$6000. [5.1C]

CUMULATIVE REVIEW EXERCISES

1. $-6\sqrt{10}$ [1.2F] **2.** 22 [2.2C] **3.** $3x - 24$ [1.4D] **4.** -4 [1.4A] **5.** $\{x|x < 6\}$ [2.5B]
6. $\{x| -4 < x < 8\}$ [2.6B] **7.** $\{x|x > 4 \text{ or } x < -1\}$ [2.6B] **8.** $F(2) = 1$ [4.2A]

9. –5 –4 –3 –2 –1 0 1 2 3 4 5 [1.5C] **10.** $y = -\frac{2}{3}x + \frac{5}{3}$ [4.5A] **11.** $y = 5x - 11$ [4.5B]

12. $y = -\frac{3}{2}x - 1$ [4.6A] **13.** 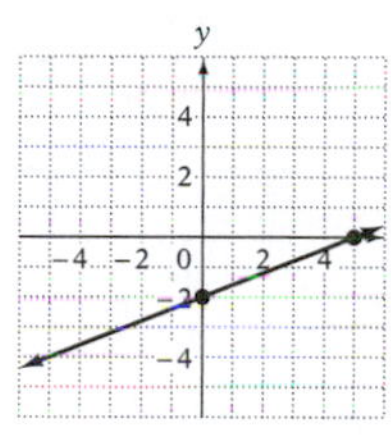 [4.3B] **14.** 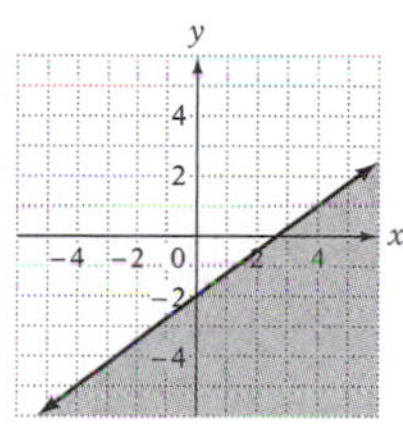 [4.7A]

15. 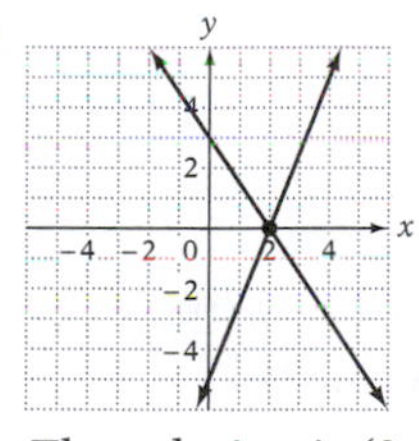 [5.1A]
The solution is (2, 0).

16. 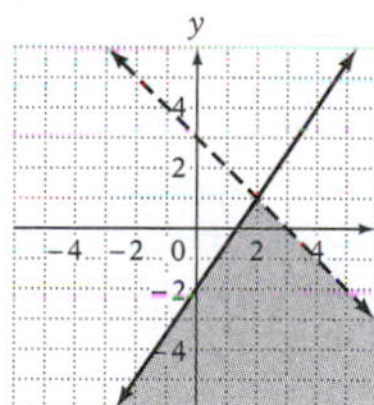 [5.4A] **17.** $(1, 0, -1)$ [5.2B] **18.** $(2, 1, -1)$ [5.2B]

19. $(2, -3)$ [5.2A] **20.** $(-5, -11)$ [5.1B] **21.** 60 ml of pure water must be used. [2.4A] **22.** The rate of the wind is 12.5 mph. [5.3A] **23.** One pound of steak costs \$10. [5.3B] **24.** The lower limit is 10,200 ohms. The upper limit is 13,800 ohms. [2.6C] **25.** The slope is 40. The slope represents the monthly income (in dollars) per thousand dollars in sales. [4.4A]

Answers to Chapter 6 Selected Exercises

PREP TEST

1. 1 [1.1C] **2.** -18 [1.1D] **3.** $\frac{2}{3}$ [1.1D] **4.** $-12y$ [1.4C] **5.** -8 [1.2E] **6.** $3a - 8b$ [1.4B]
7. $11x - 2y - 2$ [1.4D] **8.** 0 [1.4B] **9.** No [1.4B]

Copyright © Houghton Mifflin Company. All rights reserved.

SECTION 6.1

1. a^4b^4 **3.** $-18x^3y^4$ **5.** x^8y^{16} **7.** $81x^8y^{12}$ **9.** $729a^{10}b^6$ **11.** x^5y^{11} **13.** $729x^6$ **15.** $a^{18}b^{18}$ **17.** $4096x^{12}y^{12}$ **19.** $64a^{24}b^{18}$ **21.** x^{2n+1} **23.** y^{6n-2} **25.** a^{2n^2-6n} **27.** x^{15n+10} **29.** $-6x^5y^5z^4$ **31.** $-12a^2b^9c^2$ **33.** $-6x^4y^4z^5$ **35.** $-432a^7b^{11}$ **37.** $54a^{13}b^{17}$ **39.** 243 **41.** y^3 **43.** $\frac{a^3b^2}{4}$ **45.** $\frac{x}{y^4}$ **47.** $\frac{1}{2}$ **49.** $-\frac{1}{9}$ **51.** $\frac{1}{y^8}$ **53.** $\frac{1}{x^6y^{10}}$ **55.** x^5y^5 **57.** $\frac{a^8}{b^9}$ **59.** $\frac{1}{2187a}$ **61.** $\frac{y^6}{x^3}$ **63.** $\frac{y^4}{x^3}$ **65.** $-\frac{1}{2x^2}$ **67.** $-\frac{1}{243a^5b^{10}}$ **69.** $\frac{16x^8}{81y^4z^{12}}$ **71.** $\frac{3a^4}{4b^3}$ **73.** $\frac{-9a}{8b^6}$ **75.** $\frac{16x^2}{y^6}$ **77.** $\frac{1}{b^{4n}}$ **79.** $-\frac{1}{y^{6n}}$ **81.** y^{n-2} **83.** $\frac{1}{y^{2n}}$ **85.** $\frac{x^{n-5}}{y^6}$ **87.** $\frac{8b^{15}}{3a^{18}}$ **89.** 4.67×10^{-6} **91.** 1.7×10^{-10} **93.** 2×10^{11} **95.** 0.000000123 **97.** 8,200,000,000,000,000 **99.** 0.039 **101.** 150,000 **103.** 20,800,000 **105.** 0.000000015 **107.** 0.0000000000178 **109.** 140,000,000 **111.** 11,456,790 **113.** 0.000008 **115.** It would take a spaceship 2.24×10^{15} h to cross the galaxy. **117.** It takes the satellite 8.86×10^3 h to reach Saturn. **119.** A proton is 1.83664508×10^3 times heavier than an electron. **121.** Light travels 8.64×10^{12} m in 8 h. **123.** The signals from Earth to Mars traveled $1.0\overline{81} \times 10^7$ mi/min. **125.** The centrifuge makes one revolution in 1.5×10^{-7} s. **127. a.** $\frac{8}{5}$ **b.** $\frac{5}{4}$

SECTION 6.2

1. $P(3) = 13$ **3.** $R(2) = 10$ **5.** $f(-1) = -11$ **7.** Polynomial: **a.** -1 **b.** 8 **c.** 2 **9.** Not a polynomial **11.** Not a polynomial **13.** Polynomial: **a.** 3 **b.** π **c.** 5 **15.** Polynomial: **a.** -5 **b.** 2 **c.** 3 **17.** Polynomial: **a.** 14 **b.** 14 **c.** 0 **19.** **21.** **23.**

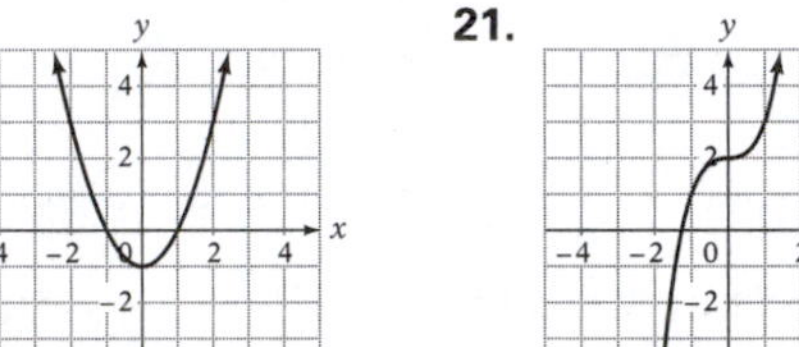

25. $6x^2 - 6x + 5$ **27.** $-x^2 + 1$ **29.** $5y^2 - 15y + 2$ **31.** $7a^2 - a + 2$ **33.** $3x^2 - 3xy - 2y^2$ **35.** $8x^2 - 2xy + 5y^2$ **37.** $S(x) = 3x^4 - 8x^2 + 2x$; $S(2) = 20$ **39. a.** $k = 8$ **b.** $k = -4$

SECTION 6.3

1. $2x^2 - 6x$ **3.** $6x^4 - 3x^3$ **5.** $6x^2y - 9xy^2$ **7.** $x^{n+1} + x^n$ **9.** $x^{2n} + x^ny^n$ **11.** $-4b^2 + 10b$ **13.** $-6a^4 + 4a^3 - 6a^2$ **15.** $-3y^4 - 4y^3 + 2y^2$ **17.** $-20x^2 + 15x^3 - 15x^4 - 20x^5$ **19.** $-2x^4y + 6x^3y^2 - 4x^2y^3$ **21.** $x^{3n} + x^{2n} + x^{n+1}$ **23.** $a^{2n+1} - 3a^{n+2} + 2a^{n+1}$ **25.** $5y^2 - 11y$ **27.** $6y^2 - 31y$ **29.** $9b^5 - 9b^3 + 24b$ **31.** $x^2 + 5x - 14$ **33.** $8y^2 + 2y - 21$ **35.** $8x^2 + 8xy - 30y^2$ **37.** $x^2y^2 + xy - 12$ **39.** $2x^4 - 15x^2 + 25$ **41.** $10x^4 - 15x^2y + 5y^2$ **43.** $x^{2n} - x^n - 6$ **45.** $6a^{2n} + a^n - 15$ **47.** $6a^{2n} + a^nb^n - 2b^{2n}$ **49.** $x^3 + 2x^2 - 11x + 20$ **51.** $10a^3 - 27a^2b + 26ab^2 - 12b^3$ **53.** $4x^5 - 16x^4 + 15x^3 - 4x^2 + 28x - 45$ **55.** $x^4 - 3x^3 - 6x^2 + 29x - 21$ **57.** $2a^3 + 7a^2 - 43a + 42$ **59.** $x^{3n} + 2x^{2n} + 2x^n + 1$ **61.** $x^{2n} - 2x^{2n}y^n + 4x^ny^n - 2x^ny^{2n} + 3y^{2n}$ **63.** $2y^5 - 10y^4 - y^3 - y^2 + 3$ **65.** $9x^2 - 4$ **67.** $36 - x^2$ **69.** $4a^2 - 9b^2$ **71.** $x^4 - 1$ **73.** $x^{2n} - 9$ **75.** $x^2 - 10x + 25$ **77.** $9a^2 + 30ab + 25b^2$ **79.** $x^4 - 6x^2 + 9$ **81.** $4x^4 - 12x^2y^2 + 9y^4$ **83.** $a^{2n} - 2a^nb^n + b^{2n}$ **85.** $-x^2 + 2xy$ **87.** $-4xy$ **89.** $(3x^2 + 10x - 8)$ square feet **91.** $(x^2 + 3x)$ square meters **93.** $(x^3 + 9x^2 + 27x + 27)$ cubic centimeters **95.** $(2x^3)$ cubic inches **97.** $(78.5x^2 + 125.6x + 50.24)$ square inches **99. a.** $a^3 - b^3$ **b.** $x^3 + y^3$ **101. a.** $k = 5$ **b.** $k = 1$ **103.** $a^2 + b^2$

SECTION 6.4

1. $x - 2$ **3.** $-x + 2$ **5.** $xy + 2$ **7.** $x^2 + 3x - 5$ **9.** $3b^3 + 4b^2 + 2b$ **11.** $a^4 - 6a^2 + 1$ **13.** $x + 8$ **15.** $x^2 + 3x + 6 + \frac{20}{x - 3}$ **17.** $3x + 5 + \frac{3}{2x + 1}$ **19.** $5x + 7 + \frac{2}{2x - 1}$ **21.** $4x^2 + 6x + 9 + \frac{18}{2x - 3}$ **23.** $3x^2 + 1 + \frac{1}{2x^2 - 5}$ **25.** $x^2 - 3x - 10$ **27.** $x^2 - 2x + 1 - \frac{1}{x - 3}$ **29.** $2x^3 - 3x^2 + x - 4$ **31.** $2x + \frac{x + 2}{x^2 + 2x - 1}$

Copyright © Houghton Mifflin Company. All rights reserved.

33. $x^2 + 4x + 6 + \dfrac{10x + 8}{x^2 - 2x - 1}$ **35.** $x^2 + 4 + \dfrac{3}{2x + 1}$ **37.** $2x - 8$ **39.** $3x - 8$ **41.** $3x + 3 - \dfrac{1}{x - 1}$
43. $2x^2 - 3x + 9$ **45.** $4x^2 + 8x + 15 + \dfrac{12}{x - 2}$ **47.** $2x^2 - 3x + 7 - \dfrac{8}{x + 4}$ **49.** $3x^3 + 2x^2 + 12x + 19 + \dfrac{33}{x - 2}$
51. $3x^3 - x + 4 - \dfrac{2}{x + 1}$ **53.** $3x + 1 + \dfrac{8}{x - 2}$ **55.** $P(3) = 8$ **57.** $R(4) = 43$ **59.** $P(-2) = -39$
61. $Z(-3) = -60$ **63.** $Q(2) = 31$ **65.** $F(-3) = 178$ **67.** $P(5) = 122$ **69.** $R(-3) = 302$ **71.** $Q(2) = 0$
73. a. $a^2 - ab + b^2$ **b.** $x^4 - x^3y + x^2y^2 - xy^3 + y^4$ **c.** $x^5 - x^4y + x^3y^2 - x^2y^3 + xy^4 - y^5$

CHAPTER 6 REVIEW EXERCISES

1. $21y^2 + 4y - 1$ [6.2B] **2.** $5x + 4 + \dfrac{6}{3x - 2}$ [6.4B] **3.** $144x^2y^{10}z^{14}$ [6.1B] **4.** $25y^2 - 70y + 49$ [6.3C]
5. $\dfrac{b^6}{a^4}$ [6.1B] **6.** 1 [6.4D] **7.** $4x^2 - 8xy + 5y^2$ [6.2B] **8.** $4b^4 + 12b^2 - 1$ [6.4A] **9.** $-\dfrac{1}{2a}$ [6.1B]
10. $-6a^3b^6$ [6.1A] **11.** $\dfrac{2x^3}{3}$ [6.1B] **12.** $4x^2 + 3x - 8 + \dfrac{50}{x + 6}$ [6.4B/6.4C] **13.** -7 [6.2A]
14. $13y^3 - 12y^2 - 5y - 1$ [6.2B] **15.** $b^2 + 5b + 2 + \dfrac{7}{b - 7}$ [6.4B/6.4C] **16.** $12x^5y^3 + 8x^3y^2 - 28x^2y^4$ [6.3A]
17. $2ax - 4ay - bx + 2by$ [6.3B] **18.** $8b^2 - 2b - 15$ [6.3B] **19.** $33x^2 + 24x$ [6.3A] **20.** $x^4y^8z^4$ [6.1A]
21. $16x^2 - 24xy + 9y^2$ [6.3C] **22.** $x^3 + 4x^2 + 16x + 64 + \dfrac{252}{x - 4}$ [6.4B/6.4C] **23.** $2x^2 - 5x - 2$ [6.2B]
24. $70xy^2z^6$ [6.1B] **25.** $-\dfrac{x^3}{4y^2z^3}$ [6.1B] **26.** 9.48×10^8 [6.1C] **27.** 2×10^{-6} [6.1C] **28.** 68 [6.4D]
29. $4x^4 - 2x^2 + 5$ [6.4A] **30.** $2x - 3 - \dfrac{4}{6x + 1}$ [6.4B] **31.** $a^{3n+3} - 5a^{2n+4} + 2a^{2n+3}$ [6.3A]
32. $x^4 + 3x^3 - 23x^2 - 29x + 6$ [6.3B] **33.** $-8x^3 - 14x^2 + 18x$ [6.3A] **34.** $6y^3 + 17y^2 - 2y - 21$ [6.3B]
35. $16u^{12}v^{16}$ [6.1A] **36.** $2x^3 + 9x^2 - 3x - 12$ [6.2B] **37.** $2x^2 + 3x - 8$ [6.2B] **38.** $a^2 - 49$ [6.3C]
39. $100a^{15}b^{13}$ [6.1A] **40.** 14,600,000 [6.1C] **41.** $-108x^{18}$ [6.1A] **42.** $2y - 9$ [6.4B] **43.** $-\dfrac{1}{16}$ [6.1B]
44. $10a^2 + 31a - 63$ [6.3B] **45.** $-x + 2 + \dfrac{1}{x + 3}$ [6.4B/6.4C] **46.** 1.27×10^{-7} [6.1C] **47.** $-4y + 8$ [6.4A]
48. $\dfrac{c^{10}}{2b^{17}}$ [6.1B] **49.** $6x^3 - 29x^2 + 14x + 24$ [6.3B] **50.** $\dfrac{x^4y^6}{9}$ [6.1B] **51.** $-54a^{13}b^5c^7$ [6.1A]
52. $25a^2 - 4b^2$ [6.3C] **53.** 0.00254 [6.1C] **54.** $8a^3b^3 - 4a^2b^4 + 6ab^5$ [6.3A] **55.**

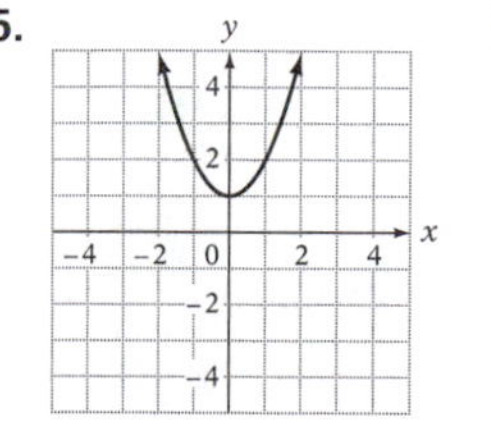

[6.2A]

56. a. 3 **b.** 8 **c.** 5 [6.2A] **57.** The mass of the moon is 8.103×10^{19} tons. [6.1D] **58.** The area is $(9x^2 - 12x + 4)$ square inches. [6.3D] **59.** The Great Galaxy of Andromeda is 1.291224×10^{19} mi from Earth. [6.1D]
60. The area is $(10x^2 - 29x - 21)$ square centimeters. [6.3D]

CHAPTER 6 TEST

1. $4x^3 - 6x^2$ [6.3A] **2.** -8 [6.4D] **3.** $-\dfrac{4}{x^6}$ [6.1B] **4.** $-6x^3y^6$ [6.1A] **5.** $x - 1 + \dfrac{2}{x + 1}$ [6.4B/6.4C]

6. $x^3 - 7x^2 + 17x - 15$ [6.3B] **7.** $-8a^6b^3$ [6.1A] **8.** $\dfrac{9y^{10}}{x^{10}}$ [6.1B] **9.** $a^2 + 3ab - 10b^2$ [6.3B] **10.** -3 [6.2A]
11. $x + 7$ [6.4B/6.4C] **12.** $6y^4 - 9y^3 + 18y^2$ [6.3A] **13.** $-4x^4 + 8x^3 - 3x^2 - 14x + 21$ [6.3B]
14. $16y^2 - 9$ [6.3C] **15.** $6x^3 + 3x^2 - 2x$ [6.4A] **16.** $8ab^4$ [6.1B] **17.** $\dfrac{2b^7}{a^{10}}$ [6.1B]

Copyright © Houghton Mifflin Company. All rights reserved.

18. $-5a^3 + 3a^2 - 4a + 3$ [6.2B] **19.** $4x^2 - 20x + 25$ [6.3C] **20.** $x^2 - 5x + 10 - \frac{23}{x+3}$ [6.4B/6.4C] **21.** $10x^2 - 43xy + 28y^2$ [6.3B] **22.** $3x^3 + 6x^2 - 8x + 3$ [6.2B] **23.** 3.02×10^{-9} [6.1C] **24.** There are 6.048×10^6 s in 10 weeks. [6.1D] **25.** The area of the circle is $(\pi x^2 - 10\pi x + 25\pi)$ square meters. [6.3D]

CUMULATIVE REVIEW EXERCISES

1. -3 and 3 [1.1A] **2.** -83 [1.1B] **3.** 6 [1.3A] **4.** $-\frac{5}{4}$ [1.4A] **5.** $-50\sqrt{3}$ [1.2F] **6.** The Inverse Property of Addition [1.4B] **7.** $-186x + 8$ [1.4D] **8.** $-\frac{1}{6}$ [2.2A] **9.** $-\frac{11}{4}$ [2.2B] **10.** -1 and $\frac{7}{3}$ [2.6A] **11.** 18 [4.2A] **12.** Yes [4.2A] **13.** $-\frac{1}{6}$ [4.4A] **14.** $y = -\frac{3}{2}x + \frac{1}{2}$ [4.5A] **15.** $y = \frac{2}{3}x + \frac{16}{3}$ [4.6A] **16.** $(-1, -2)$ [5.1B] **17.** $\left(-\frac{9}{7}, \frac{2}{7}, \frac{11}{7}\right)$ [5.2B] **18.** $5x - 3xy$ [1.4B] **19.** $4x^3 - 7x + 3$ [6.3B] **20.** 5.01×10^{-6} [6.1C] **21.**

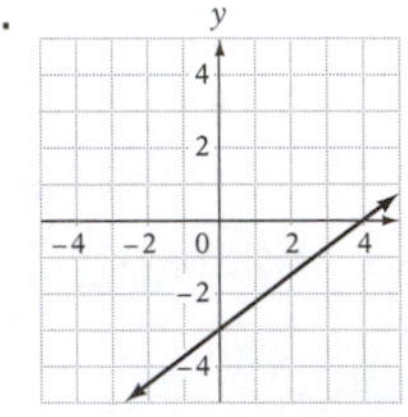

[4.3B] **22.** [4.7A]

23.

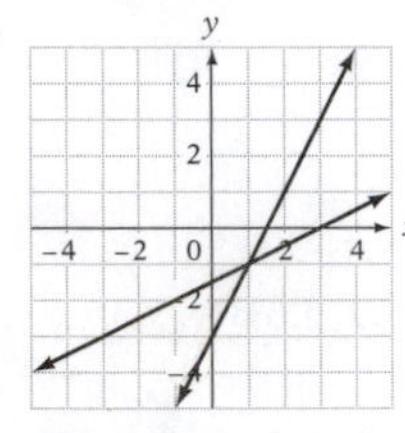

[5.1A]

The solution is $(1, -1)$.

24.

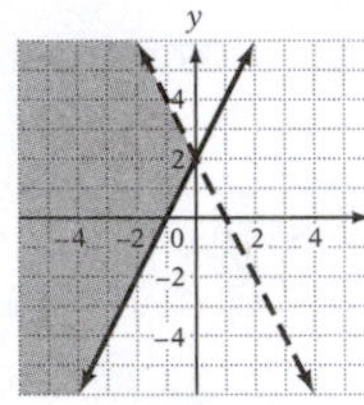

[5.4A] **25.** $\frac{b^5}{a^8}$ [6.1B] **26.** $\frac{y^2}{25x^6}$ [6.1B]

27. The two integers are 9 and 15. [2.3A] **28.** 40 oz of pure gold must be used. [2.4A] **29.** The cyclists are traveling at 5 mph and 7.5 mph. [2.4C] **30.** \$2000 is invested in the 4% account. [5.1C] **31.** The slope is 50. The slope represents the average speed in miles per hour. [4.4A] **32.** The length is 15 m. The width is 6 m. [3.2A] **33.** The area is $(4x^2 + 12x + 9)$ square meters. [6.3D]

Answers to Chapter 7 Selected Exercises

PREP TEST

1. $2 \cdot 3 \cdot 5$ [1.2D] **2.** $-12y + 15$ [1.4D] **3.** $-a + b$ [1.4D] **4.** $-3a + 3b$ [1.4D] **5.** 0 [2.1C] **6.** $-\frac{1}{2}$ [2.2A] **7.** $x^2 - 2x - 24$ [6.3B] **8.** $6x^2 - 11x - 10$ [6.3B] **9.** x^3 [6.1B] **10.** $3x^3y$ [6.1B]

SECTION 7.1

3. $5(a + 1)$ **5.** $8(2 - a^2)$ **7.** $4(2x + 3)$ **9.** $6(5a - 1)$ **11.** $x(7x - 3)$ **13.** $a^2(3 + 5a^3)$ **15.** $y(14y + 11)$ **17.** $2x(x^3 - 2)$ **19.** $2x^2(5x^2 - 6)$ **21.** $4a^5(2a^3 - 1)$ **23.** $xy(xy - 1)$ **25.** $3xy(xy^3 - 2)$ **27.** $xy(x - y^2)$ **29.** $5y(y^2 - 4y + 1)$ **31.** $3y^2(y^2 - 3y - 2)$ **33.** $3y(y^2 - 3y + 8)$ **35.** $a^2(6a^3 - 3a - 2)$ **37.** $ab(2a - 5ab + 7b)$ **39.** $2b(2b^4 + 3b^2 - 6)$ **41.** $x^2(8y^2 - 4y + 1)$ **43.** $(a + z)(y + 7)$ **45.** $(a - b)(3r + s)$ **47.** $(m - 7)(t - 7)$ **49.** $(4a - b)(2y + 1)$ **51.** $(x + 2)(x + 2y)$ **53.** $(p - 2)(p - 3r)$ **55.** $(a + 6)(b - 4)$ **57.** $(2z - 1)(z + y)$

Copyright © Houghton Mifflin Company. All rights reserved.

59. $(2v - 3y)(4v + 7)$ **61.** $(2x - 5)(x - 3y)$ **63.** $(y - 2)(3y - a)$ **65.** $(3x - y)(y + 1)$ **67.** $(3s + t)(t - 2)$ **69. a.** 28 **b.** 496 **71. a.** $r^2(\pi - 2)$ **b.** $2r^2(4 - \pi)$ **c.** $r^2(4 - \pi)$

SECTION 7.2

1. The same **3.** $(x + 1)(x + 2)$ **5.** $(x + 1)(x - 2)$ **7.** $(a + 4)(a - 3)$ **9.** $(a - 1)(a - 2)$ **11.** $(a + 2)(a - 1)$ **13.** $(b - 3)(b - 3)$ **15.** $(b + 8)(b - 1)$ **17.** $(y + 11)(y - 5)$ **19.** $(y - 2)(y - 3)$ **21.** $(z - 5)(z - 9)$ **23.** $(z + 8)(z - 20)$ **25.** $(p + 3)(p + 9)$ **27.** $(x + 10)(x + 10)$ **29.** $(b + 4)(b + 5)$ **31.** $(x + 3)(x - 14)$ **33.** $(b + 4)(b - 5)$ **35.** $(y + 3)(y - 17)$ **37.** $(p + 3)(p - 7)$ **39.** Nonfactorable over the integers **41.** $(x - 5)(x - 15)$ **43.** $(p + 3)(p + 21)$ **45.** $(x + 2)(x + 19)$ **47.** $(x + 9)(x - 4)$ **49.** $(a + 4)(a - 11)$ **51.** $(a - 3)(a - 18)$ **53.** $(z + 21)(z - 7)$ **55.** $(c + 12)(c - 15)$ **57.** $(p + 9)(p + 15)$ **59.** $(c + 2)(c + 9)$ **61.** $(x + 15)(x - 5)$ **63.** $(x + 25)(x - 4)$ **65.** $(b - 4)(b - 18)$ **67.** $(a + 45)(a - 3)$ **69.** $(b - 7)(b - 18)$ **71.** $(z + 12)(z + 12)$ **73.** $(x - 4)(x - 25)$ **75.** $(x + 16)(x - 7)$ **77.** $3(x + 2)(x + 3)$ **79.** $-(x - 2)(x + 6)$ **81.** $a(b + 8)(b - 1)$ **83.** $x(y + 3)(y + 5)$ **85.** $-2a(a + 1)(a + 2)$ **87.** $4y(y + 6)(y - 3)$ **89.** $2x(x^2 - x + 2)$ **91.** $6(z + 5)(z - 3)$ **93.** $3a(a + 3)(a - 6)$ **95.** $(x + 7y)(x - 3y)$ **97.** $(a - 5b)(a - 10b)$ **99.** $(s + 8t)(s - 6t)$ **101.** Nonfactorable over the integers **103.** $z^2(z + 10)(z - 8)$ **105.** $b^2(b + 2)(b - 5)$ **107.** $3y^2(y + 3)(y + 15)$ **109.** $-x^2(x + 1)(x - 12)$ **111.** $3y(x + 3)(x - 5)$ **113.** $-3x(x - 3)(x - 9)$ **115.** $(x - 3y)(x - 5y)$ **117.** $(a - 6b)(a - 7b)$ **119.** $(y + z)(y + 7z)$ **121.** $3y(x + 21)(x - 1)$ **123.** $3x(x + 4)(x - 3)$ **125.** $4z(z + 11)(z - 3)$ **127.** $4x(x + 3)(x - 1)$ **129.** $5(p + 12)(p - 7)$ **131.** $p^2(p + 12)(p - 3)$ **133.** $(t - 5s)(t - 7s)$ **135.** $(a + 3b)(a - 11b)$ **137.** $y(x + 6)(x - 9)$ **139.** $-36, 36, -12, 12$ **141.** $22, -22, 10, -10$ **143.** 6, 10, 12 **145.** 6, 10, 12 **147.** 4, 6

SECTION 7.3

1. $(x + 1)(2x + 1)$ **3.** $(y + 3)(2y + 1)$ **5.** $(a - 1)(2a - 1)$ **7.** $(b - 5)(2b - 1)$ **9.** $(x + 1)(2x - 1)$ **11.** $(x - 3)(2x + 1)$ **13.** $(t + 2)(2t - 5)$ **15.** $(p - 5)(3p - 1)$ **17.** $(3y - 1)(4y - 1)$ **19.** Nonfactorable over the integers **21.** $(2t - 1)(3t - 4)$ **23.** $(x + 4)(8x + 1)$ **25.** Nonfactorable over the integers **27.** $(3y + 1)(4y + 5)$ **29.** $(a + 7)(7a - 2)$ **31.** $(b - 4)(3b - 4)$ **33.** $(z - 14)(2z + 1)$ **35.** $(p + 8)(3p - 2)$ **37.** $2(x + 1)(2x + 1)$ **39.** $5(y - 1)(3y - 7)$ **41.** $x(x - 5)(2x - 1)$ **43.** $b(a - 4)(3a - 4)$ **45.** Nonfactorable over the integers **47.** $-3x(x + 4)(x - 3)$ **49.** $4(4y - 1)(5y - 1)$ **51.** $z(2z + 3)(4z + 1)$ **53.** $y(2x - 5)(3x + 2)$ **55.** $5(t + 2)(2t - 5)$ **57.** $p(p - 5)(3p - 1)$ **59.** $2(z + 4)(13z - 3)$ **61.** $2y(y - 4)(5y - 2)$ **63.** $yz(z + 2)(4z - 3)$ **65.** $3a(2a + 3)(7a - 3)$ **67.** $y(3x - 5y)(3x - 5y)$ **69.** $xy(3x - 4y)(3x - 4y)$ **71.** $(2x - 3)(3x - 4)$ **73.** $(b + 7)(5b - 2)$ **75.** $(3a + 8)(2a - 3)$ **77.** $(z + 2)(4z + 3)$ **79.** $(2p + 5)(11p - 2)$ **81.** $(y + 1)(8y + 9)$ **83.** $(6t - 5)(3t + 1)$ **85.** $(b + 12)(6b - 1)$ **87.** $(3x + 2)(3x + 2)$ **89.** $(2b - 3)(3b - 2)$ **91.** $(3b + 5)(11b - 7)$ **93.** $(3y - 4)(6y - 5)$ **95.** $(3a + 7)(5a - 3)$ **97.** $(2y - 5)(4y - 3)$ **99.** $(2z + 3)(4z - 5)$ **101.** Nonfactorable over the integers **103.** $(2z - 5)(5z - 2)$ **105.** $(6z + 5)(6z + 7)$ **107.** $(x + y)(3x - 2y)$ **109.** $(a + 2b)(3a - b)$ **111.** $(y - 2z)(4y - 3z)$ **113.** $-(z - 7)(z + 4)$ **115.** $-(x - 1)(x + 8)$ **117.** $3(x + 5)(3x - 4)$ **119.** $4(2x - 3)(3x - 2)$ **121.** $a^2(5a + 2)(7a - 1)$ **123.** $5(b - 7)(3b - 2)$ **125.** $(x - 7y)(3x - 5y)$ **127.** $3(8y - 1)(9y + 1)$ **129.** $-(x - 1)(x + 21)$ **133.** $x(x - 1)$ **135.** $(2y + 1)(y + 3)$ **137.** $(4y - 3)(y - 3)$ **139.** $-5, 5, -1, 1$ **141.** $-5, 5, -1, 1$ **143.** $-9, 9, -3, 3$

SECTION 7.4

1. 4; $25x^6$; $100x^4y^4$ **3.** $4z^4$ **5.** $9a^2b^3$ **7.** $(x + 4)(x - 4)$ **9.** $(2x + 1)(2x - 1)$ **11.** $(4x + 11)(4x - 11)$ **13.** $(1 + 3a)(1 - 3a)$ **15.** $(xy + 10)(xy - 10)$ **17.** Nonfactorable over the integers **19.** $(5 + ab)(5 - ab)$ **21.** $(a^n + 1)(a^n - 1)$ **23.** $(x - 6)^2$ **25.** $(b - 1)^2$ **27.** $(4x - 5)^2$ **29.** Nonfactorable over the integers **31.** Nonfactorable over the integers **33.** $(x + 3y)^2$ **35.** $(5a - 4b)^2$ **37.** $(x^n + 3)^2$ **39.** $(x - 7)(x - 1)$ **41.** $(x - y - a - b)(x - y + a + b)$ **43.** 8; x^9; $27c^{15}d^{18}$ **45.** $2x^3$ **47.** $4a^2b^6$ **49.** $(x - 3)(x^2 + 3x + 9)$ **51.** $(2x - 1)(4x^2 + 2x + 1)$ **53.** $(x - y)(x^2 + xy + y^2)$ **55.** $(m + n)(m^2 - mn + n^2)$ **57.** $(4x + 1)(16x^2 - 4x + 1)$ **59.** $(3x - 2y)(9x^2 + 6xy + 4y^2)$ **61.** $(xy + 4)(x^2y^2 - 4xy + 16)$ **63.** Nonfactorable over the integers **65.** Nonfactorable over the integers **67.** $(a - 2b)(a^2 - ab + b^2)$ **69.** $(x^{2n} + y^n)(x^{4n} - x^{2n}y^n + y^{2n})$

Copyright © Houghton Mifflin Company. All rights reserved.

71. $(x^n + 2)(x^{2n} - 2x^n + 4)$ **73.** $(xy - 5)(xy - 3)$ **75.** $(xy - 12)(xy - 5)$ **77.** $(x^2 - 3)(x^2 - 6)$ **79.** $(b^2 + 5)(b^2 - 18)$ **81.** $(x^2y^2 - 2)(x^2y^2 - 6)$ **83.** $(x^n + 1)(x^n + 2)$ **85.** $(3xy - 5)(xy - 3)$ **87.** $(2ab - 3)(3ab - 7)$ **89.** $(x^2 + 1)(2x^2 - 15)$ **91.** $(2x^n - 1)(x^n - 3)$ **93.** $(2a^n + 5)(3a^n + 2)$ **95.** $3(2x - 3)^2$ **97.** $a(3a - 1)(9a^2 + 3a + 1)$ **99.** $5(2x + 1)(2x - 1)$ **101.** $y^3(y + 11)(y - 5)$ **103.** $(4x^2 + 9)(2x + 3)(2x - 3)$ **105.** $2a(2 - a)(4 + 2a + a^2)$ **107.** $b^3(ab - 1)(a^2b^2 + ab + 1)$ **109.** $2x^2(2x - 5)^2$ **111.** $(x^2 + y^2)(x + y)(x - y)$ **113.** $(x^2 + y^2)(x^4 - x^2y^2 + y^4)$ **115.** Nonfactorable over the integers **117.** $2a(2a - 1)(4a^2 + 2a + 1)$ **119.** $a^2b^2(a + 4b)(a - 12b)$ **121.** $2b^2(3a + 5b)(4a - 9b)$ **123.** $(x - 2)^2(x + 2)$ **125.** $(2x + 1)(2x - 1)(x + y)(x - y)$ **127.** $x(x^n + 1)^2$ **129.** $b^n(b + 2)(3b - 2)$

SECTION 7.5

3. −3, −2 **5.** 7, 3 **7.** 0, 5 **9.** 0, 9 **11.** $0, -\frac{3}{2}$ **13.** $0, \frac{2}{3}$ **15.** −2, 5 **17.** −9, 9 **19.** $-\frac{7}{2}, \frac{7}{2}$ **21.** $-\frac{1}{3}, \frac{1}{3}$ **23.** −2, −4 **25.** −7, 2 **27.** $-\frac{1}{2}, 5$ **29.** $-\frac{1}{3}, -\frac{1}{2}$ **31.** 0, 3 **33.** 0, 7 **35.** −1, −4 **37.** 2, 3 **39.** $\frac{1}{2}, -4$ **41.** $\frac{1}{3}, 4$ **43.** 3, 9 **45.** −2, 9 **47.** −1, −2 **49.** −9, 5 **51.** −7, 4 **53.** −2, −3 **55.** −8, 9 **57.** 1, 4 **59.** −5, 2 **61.** The number is 6. **63.** The numbers are 2 and 4. **65.** The numbers are 4 and 5. **67.** The numbers are 3 and 7. **69.** There will be 12 consecutive numbers. **71.** There are six teams in the league. **73.** The object will hit the ground 3 s later. **75.** The golf ball will return to the ground 3.75 s later. **77.** The length is 15 in. The width is 5 in. **79.** The height of the triangle is 14 m. **81.** The dimensions of the type area are 4 in. by 7 in. **83.** The radius of the original circular lawn was approximately 3.81 ft. **85.** 3, 48 **87.** $-\frac{3}{2}, -5$ **89.** 0, 9

CHAPTER 7 REVIEW EXERCISES

1. $(b - 10)(b - 3)$ [7.2A] **2.** $(x - 3)(4x + 5)$ [7.1B] **3.** Nonfactorable over the integers [7.3A] **4.** $(7x^2y^2 + 3)(3x^2y^2 + 2)$ [7.4C] **5.** $7y^3(2y^6 - 7y^3 + 1)$ [7.1A] **6.** $(y + 9)(y - 4)$ [7.2A] **7.** $(2x - 7)(3x - 4)$ [7.3A] **8.** $3ab(4a + b)$ [7.1A] **9.** $(a^3 + 10)(a^3 - 10)$ [7.4A] **10.** $n^2(n - 3)(n + 1)$ [7.2B] **11.** $(6y - 1)(2y + 3)$ [7.3A] **12.** $2b(2b - 7)(3b - 4)$ [7.4D] **13.** $(3y^2 + 5z)(3y^2 - 5z)$ [7.4A] **14.** $(c + 6)(c + 2)$ [7.2A] **15.** $(6a - 5)(3a + 2)$ [7.3B] **16.** $\frac{1}{4}, -7$ [7.5A] **17.** $4x(x - 6)(x + 1)$ [7.2B] **18.** $(4a - 3b)(16a^2 + 12ab + 9b^2)$ [7.4B] **19.** $(a - 12)(2a + 5)$ [7.3B] **20.** 7, −3 [7.5A] **21.** $(3a - 5b)(7x + 2y)$ [7.1B] **22.** $(6x^4 - 1)(6x^4 - 5)$ [7.4C] **23.** $(2x + 5)(5x + 2y)$ [7.1B] **24.** $5(x - 3)(x + 2)$ [7.2B] **25.** $3(x + 6)^2$ [7.4D] **26.** $(x - 5)(3x - 2)$ [7.3B] **27.** The length is 100 yd. The width is 60 yd. [7.5B] **28.** The distance is 20 ft. [7.5B] **29.** The width of the frame is 1.5 in. [7.5B] **30.** The side of the original garden plot was 20 ft. [7.5B]

CHAPTER 7 TEST

1. $(b + 6)(a - 3)$ [7.1B] **2.** $2y^2(y - 8)(y + 1)$ [7.2B] **3.** $4(x + 4)(2x - 3)$ [7.3B] **4.** $(2x + 1)(3x + 8)$ [7.3A] **5.** $(a - 16)(a - 3)$ [7.2A] **6.** $2x(3x^2 - 4x + 5)$ [7.1A] **7.** $(x + 5)(x - 3)$ [7.2A] **8.** $-\frac{1}{2}, \frac{1}{2}$ [7.5A] **9.** $5(x^2 - 9x - 3)$ [7.1A] **10.** $(p + 6)^2$ [7.4A] **11.** 3, 5 [7.5A] **12.** $3(x + 2y)^2$ [7.4D] **13.** $(b + 4)(b - 4)$ [7.4A] **14.** $3y^2(2x + 1)(x + 1)$ [7.3B] **15.** $(3x - 2)(9x^2 + 6x + 4)$ [7.4B] **16.** $(2a^2 - 5)(3a^2 + 1)$ [7.4C] **17.** $(p + 1)(x - 1)$ [7.1B] **18.** $3(a + 5)(a - 5)$ [7.4D] **19.** Nonfactorable over the integers [7.3A] **20.** $(x - 12)(x + 3)$ [7.2A] **21.** $(2a - 3b)^2$ [7.4A] **22.** $(2x + 7y)(2x - 7y)$ [7.4A] **23.** $\frac{3}{2}, -7$ [7.5A] **24.** The two numbers are 7 and 3. [7.5B] **25.** The length is 15 cm. The width is 6 cm. [7.5B]

CUMULATIVE REVIEW EXERCISES

1. 7 [1.1C] **2.** 4 [1.3A] **3.** −7 [1.4A] **4.** $15x^2$ [1.4C] **5.** 12 [1.4D] **6.** $\frac{2}{3}$ [2.1C] **7.** $\frac{7}{4}$ [2.2B]

Copyright © Houghton Mifflin Company. All rights reserved.

8. 3 [2.2C] **9.** 45 [2.1D] **10.** 1 [4.2A] **11.**

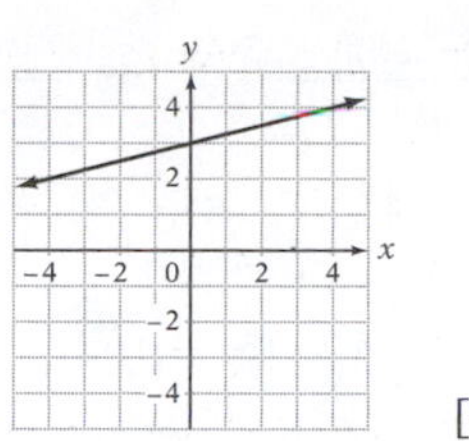

[4.3A] **12.**

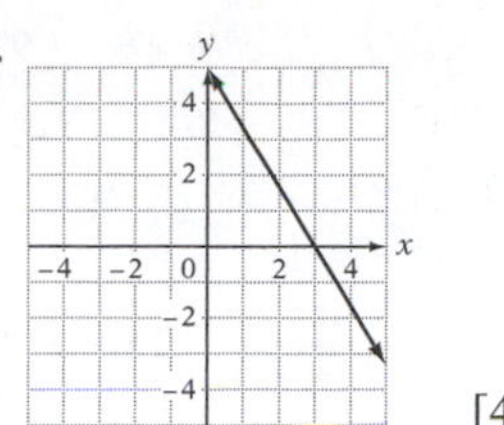

[4.3B]

13. $y = \frac{2}{3}x + 6$ [4.5A] **14.** (1, 6) [5.1B] **15.** (−1, −2) [5.2A] **16.** $9a^6b^4$ [6.1A] **17.** $x^3 - 3x^2 - 6x + 8$ [6.3B] **18.** $4x + 8 + \frac{21}{2x - 3}$ [6.4B] **19.** $\frac{y^6}{x^8}$ [6.1B] **20.** $(a - b)(3 - x)$ [7.1B] **21.** $5xy^2(3 - 4y^2)$ [7.1A] **22.** $(x - 7y)(x + 2y)$ [7.2A] **23.** $\frac{2}{3}$ and −7 [7.5A] **24.** $\frac{5}{2}$ and 4 [7.5A] **25.** The third angle measures 59°. [3.1C] **26.** The width is 15 ft. [3.2A] **27.** The pieces are 4 ft long and 6 ft long. [2.3B] **28.** $8000 is invested at 4%. [5.1C] **29.** The distance to the resort is 168 mi. [2.4C] **30.** The length of the base of the triangle is 12 in. [7.5B]

Answers to Chapter 8 Selected Exercises

PREP TEST

1. 50 [1.2C] **2.** $-\frac{1}{6}$ [1.2D] **3.** $-\frac{3}{2}$ [1.2D] **4.** $\frac{1}{24}$ [1.2C] **5.** $\frac{5}{24}$ [1.2C] **6.** $\frac{1}{3}$ [1.4A] **7.** −2 [2.2C] **8.** $\frac{10}{7}$ [2.2C] **9.** The rate of the first plane is 110 mph, and the rate of the second plane is 130 mph. [2.4C]

SECTION 8.1

3. $\frac{3}{4x}$ **5.** $\frac{1}{x + 3}$ **7.** −1 **9.** $\frac{2}{3y}$ **11.** $-\frac{3}{4x}$ **13.** $\frac{a}{b}$ **15.** $-\frac{2}{x}$ **17.** $\frac{y - 2}{y - 3}$ **19.** $\frac{x + 5}{x + 4}$ **21.** $\frac{x + 4}{x - 3}$ **23.** $-\frac{x + 2}{x + 5}$ **25.** $\frac{2(x + 2)}{x + 3}$ **27.** $\frac{2x - 1}{2x + 3}$ **29.** $-\frac{x + 7}{x + 6}$ **31.** $\frac{35ab^2}{24x^2y}$ **33.** $\frac{4x^3y^3}{3a^2}$ **35.** $\frac{3}{4}$ **37.** ab^2 **39.** $\frac{x^2(x - 1)}{y(x + 3)}$ **41.** $\frac{y(x - 1)}{x^2(x + 10)}$ **43.** $-ab^2$ **45.** $\frac{x + 5}{x + 4}$ **47.** 1 **49.** $-\frac{n - 10}{n - 7}$ **51.** $\frac{x(x + 2)}{2(x - 1)}$ **53.** $-\frac{x + 2}{x - 6}$ **55.** $\frac{x + 5}{x - 12}$ **59.** $\frac{7a^3y^2}{40bx}$ **61.** $\frac{4}{3}$ **63.** $\frac{3a}{2}$ **65.** $\frac{x^2(x + 4)}{y^2(x + 2)}$ **67.** $\frac{x(x - 2)}{y(x - 6)}$ **69.** $-\frac{3by}{ax}$ **71.** $\frac{(x + 6)(x - 3)}{(x + 7)(x - 6)}$ **73.** 1 **75.** $-\frac{x + 8}{x - 4}$ **77.** $\frac{2n + 1}{2n - 3}$ **81.** $\frac{x}{x + 8}$ **83.** $\frac{n - 2}{n + 3}$

SECTION 8.2

1. $\frac{9y^3}{12x^2y^4}, \frac{17x}{12x^2y^4}$ **3.** $\frac{2x^2 - 4x}{6x^2(x - 2)}, \frac{3x - 6}{6x^2(x - 2)}$ **5.** $\frac{3x - 1}{2x(x - 5)}, -\frac{6x^3 - 30x^2}{2x(x - 5)}$ **7.** $\frac{6x^2 + 9x}{(2x + 3)(2x - 3)}, \frac{10x^2 - 15x}{(2x + 3)(2x - 3)}$ **9.** $\frac{2x}{(x + 3)(x - 3)}, \frac{x^2 + 4x + 3}{(x + 3)(x - 3)}$ **11.** $\frac{6}{6(x + 2y)(x - 2y)}, \frac{5x + 10y}{6(x + 2y)(x - 2y)}$ **13.** $\frac{3x^2 - 3x}{(x + 1)(x - 1)^2}, \frac{5x^2 + 5x}{(x + 1)(x - 1)^2}$ **15.** $-\frac{x - 3}{(x - 2)(x^2 + 2x + 4)}, \frac{2x - 4}{(x - 2)(x^2 + 2x + 4)}$ **17.** $\frac{2x^2 + 6x}{(x - 1)(x + 3)^2}, -\frac{x^2 - x}{(x - 1)(x + 3)^2}$ **19.** $-\frac{12x^2 - 8x}{(2x - 3)(2x - 5)(3x - 2)}, \frac{6x^2 - 9x}{(2x - 3)(2x - 5)(3x - 2)}$ **21.** $\frac{5}{(3x - 4)(2x - 3)}, -\frac{4x^2 - 6x}{(3x - 4)(2x - 3)}, \frac{3x^2 - x - 4}{(3x - 4)(2x - 3)}$ **23.** $\frac{2x^2 + 10x}{(x + 5)(x - 3)}, -\frac{2x - 6}{(x + 5)(x - 3)}, -\frac{x - 1}{(x + 5)(x - 3)}$ **25.** $\frac{x - 5}{(x^n + 1)(x^n + 2)}, \frac{2x^{n+1} + 2x}{(x^n + 1)(x^n + 2)}$ **27.** $\frac{1}{2x^2}$ **29.** $\frac{1}{x + 2}$ **31.** $\frac{12ab - 9b + 8a}{30a^2b^2}$ **33.** $\frac{5 - 16b + 12a}{40ab}$ **35.** $\frac{7}{12x}$ **37.** $\frac{2xy - 8x + 3y}{10x^2y^2}$ **39.** $-\frac{a(2a - 13)}{(a + 1)(a - 2)}$ **41.** $\frac{5x^2 - 6x + 10}{(5x - 2)(2x - 5)}$

Copyright © Houghton Mifflin Company. All rights reserved.

$\frac{a}{b(a-b)}$ **45.** $\frac{a^2+18a-9}{a(a-3)}$ **47.** $\frac{17x^2+20x-25}{x(6x-5)}$ **49.** $\frac{6}{(x-3)^2(x+3)}$ **51.** $-\frac{2(x-1)}{(x+2)^2}$ **53.** $-\frac{5x^2-17x+8}{(x+4)(x-2)}$

55. $\frac{3x^n+2}{(x^n+1)(x^n-1)}$ **57.** 1 **59.** $\frac{x^2-52x+160}{4(x+3)(x-3)}$ **61.** $\frac{3x-1}{4x+1}$ **63.** $\frac{2(5x-3)}{(x+4)(x-3)(x+3)}$ **65.** $\frac{x-2}{x+3}$

67. 1 **69.** $\frac{x+1}{2x-1}$ **71.** $\frac{1}{2x-1}$ **73.** $\frac{1}{x^2+4}$ **75.** $\frac{3-a}{3a}$ **77.** $-\frac{2x^2+5x-2}{(x+2)(x+1)}$ **79.** $\frac{b-a}{b+2a}$ **81.** $\frac{2}{x+2}$

83. a. $\frac{b+6}{3(b-2)}$ **b.** $\frac{6(x-1)}{x(2x+1)}$ **85. a.** $f(4)=\frac{2}{3}$; $g(4)=4$; $S(4)=4\frac{2}{3}$; Yes, $f(4)+g(4)=S(4)$ **b.** $S(a)=f(a)+g(a)$

SECTION 8.3

3. $\frac{5}{23}$ **5.** $\frac{2}{5}$ **7.** $\frac{x}{x-1}$ **9.** $-\frac{a}{a+2}$ **11.** $-\frac{a-1}{a+1}$ **13.** $\frac{2}{5}$ **15.** $-\frac{1}{2}$ **17.** $\frac{x^2-x-1}{x^2+x+1}$ **19.** $\frac{x+2}{x-1}$

21. $-\frac{x+4}{2x+3}$ **23.** $\frac{a(a^2+a+1)}{a^2+1}$ **25.** $\frac{x+1}{x-4}$ **27.** $\frac{(x-1)(x+1)}{x^2+1}$ **29.** $-\frac{1}{x(x+h)}$ **31.** $\frac{x-2}{x-3}$ **33.** $\frac{x-3}{x+4}$

35. $\frac{3x+1}{x-5}$ **37.** $-\frac{2a+2}{7a-4}$ **39.** $\frac{x+y}{x-y}$ **41.** $-\frac{2x}{x^2+1}$ **43.** $-\frac{a^2}{1-2a}$ **45.** $-\frac{3x+2}{x-2}$ **47.** $\frac{3n^2+12n+8}{n(n+2)(n+4)}$

49. a. $\frac{2}{2+x}$ **b.** -1 **c.** $\frac{x-3}{x-2}$

SECTION 8.4

3. 3 **5.** 1 **7.** 9 **9.** 1 **11.** $\frac{1}{4}$ **13.** 1 **15.** -3 **17.** $\frac{1}{2}$ **19.** 8 **21.** 5 **23.** -1 **25.** 5

27. No solution **29.** 4, 2 **31.** $-\frac{3}{2}$, 4 **33.** 3 **35.** 4 **37.** 0 **39.** $-\frac{2}{5}$ **41.** 0, $-\frac{2}{3}$

SECTION 8.5

3. 9 **5.** 12 **7.** 7 **9.** 6 **11.** 1 **13.** -6 **15.** 4 **17.** $-\frac{2}{3}$ **19.** 20,000 voters voted in favor of the amendment. **21.** The distance between the two cities is 175 mi. **23.** The sales tax will be \$97.50 higher. **25.** The person is 67.5 in. tall. **27.** There are approximately 75 elk in the preserve. **29.** The length of side *AC* is 6.7 cm. **31.** The height is 2.9 m. **33.** The perimeter is 22.5 ft. **35.** The area is 48 m^2. **37.** The length of *BC* is 6.25 cm. **39.** The length of *DA* is 6 in. **41.** The length of *OP* is 13 cm. **43.** The distance across the river is 35 m. **45.** The first person won \$1.25 million. **47.** The player made 210 foul shots.

SECTION 8.6

1. $y=-3x+10$ **3.** $y=4x-3$ **5.** $y=-\frac{3}{2}x+3$ **7.** $y=\frac{2}{5}x-2$ **9.** $y=-\frac{2}{7}x+2$ **11.** $y=-\frac{1}{3}x+2$

13. $y=3x+8$ **15.** $y=-\frac{2}{3}x-3$ **17.** $x=-6y+10$ **19.** $x=\frac{1}{2}y+3$ **21.** $x=-\frac{3}{4}y+3$ **23.** $x=4y+3$

25. $t=\frac{d}{r}$ **27.** $T=\frac{PV}{nR}$ **29.** $l=\frac{P-2w}{2}$ **31.** $b_1=\frac{2A-hb_2}{h}$ **33.** $h=\frac{3V}{A}$ **35.** $S=C-Rt$ **37.** $P=\frac{A}{1+rt}$

39. $w=\frac{A}{S+1}$ **41. a.** $S=\frac{F+BV}{B}$ **b.** The required selling price is \$180. **c.** The required selling price is \$75.

SECTION 8.7

3. It will take 2 h to fill the fountain with both sprinklers working. **5.** With both skiploaders working together, it would take 3 h to remove the earth. **7.** With both computers working, it would take 30 h to solve the problem. **9.** It would take 30 min to cool the room with both air conditioners working. **11.** It would take the second pipeline 90 min to fill the tank. **13.** It would take the apprentice 15 h to construct the wall. **15.** It will take the second technician 3 h to complete the wiring. **17.** It would have taken one of the welders 40 h to complete the welds. **19.** It would have taken one machine 28 h to fill the boxes. **21.** The jogger ran 16 mi in 2 h. **23.** The rate of travel in the congested area was 20 mph. **25.** The rate of the jogger was 8 mph. The rate of the cyclist was 20 mph. **27.** The rate of the jet was 360 mph. **29.** Camille's walking rate is 4 mph. **31.** The rate of the car is 48 mph. **33.** The rate of the wind is

Copyright © Houghton Mifflin Company. All rights reserved.

20 mph. **35.** The rate of the gulf current is 6 mph. **37.** The rate of the trucker for the first 330 mi was 55 mph. **39.** The bus usually travels 60 mph.

SECTION 8.8

1. The profit is $80,000. **3.** The pressure is 6.75 lb/in^2. **5.** In 10 s, the object will fall 1600 ft. **7.** In 3 s, the ball will roll 54 ft. **9.** When the width is 4 ft, the length is 10 ft. **11.** The gear that has 36 teeth will make 30 rpm. **13.** The current is 7.5 amps. **15.** The intensity is 48 foot-candles when the distance is 5 ft. **17.** y is doubled. **19.** Inversely **21.** Inversely

CHAPTER 8 REVIEW EXERCISES

1. $\frac{b^3y}{10ax}$ [8.1C] **2.** $\frac{7x+22}{60x}$ [8.2B] **3.** $\frac{x-2}{3x-10}$ [8.3A] **4.** $-\frac{x+6}{x+3}$ [8.1A] **5.** $\frac{2x^4}{3y^7}$ [8.1A] **6.** 62 [8.5A] **7.** $\frac{(3y-2)^2}{(y-1)(y-2)}$ [8.1C] **8.** $\frac{by^3}{6ax^2}$ [8.1B] **9.** $\frac{x}{x-7}$ [8.3A] **10.** $\frac{3x^2-x}{(2x+3)(6x-1)(3x-1)}, \frac{24x^3-4x^2}{(2x+3)(6x-1)(3x-1)}$ [8.2A] **11.** $a=\frac{T-2bc}{2b+2c}$ [8.6A] **12.** 2 [8.4A] **13.** $c=\frac{100m}{i}$ [8.6A] **14.** The equation has no solution. [8.4A] **15.** $\frac{1}{x^2}$ [8.1C] **16.** $\frac{2y-3}{5y-7}$ [8.2B] **17.** $\frac{1}{x+3}$ [8.2B] **18.** $(5x-3)(2x-1)(4x-1)$ [8.2A] **19.** $y=-\frac{4}{9}x+2$ [8.6A] **20.** $\frac{8x+5}{3x-4}$ [8.1B] **21.** 5 [8.4A] **22.** $\frac{3x-1}{x-5}$ [8.2B] **23.** 10 [8.5A] **24.** 12 [8.5A] **25.** The perimeter of triangle ABC is 24 in. [8.5C] **26.** It would take 6 h to fill the pool using both hoses. [8.7A] **27.** The rate of the car is 45 mph. [8.7B] **28.** The rate of the wind is 20 mph. [8.7B] **29.** The pitcher's ERA is 1.35. [8.5B] **30.** The current is 2 amps. [8.8A]

CHAPTER 8 TEST

1. $\frac{2x^3}{3y^3}$ [8.1A] **2.** $-\frac{x+5}{x+1}$ [8.1A] **3.** $\frac{x+1}{x^3(x-2)}$ [8.1B] **4.** $\frac{(x-5)(2x-1)}{(x+3)(2x+5)}$ [8.1B] **5.** $\frac{x+5}{x+4}$ [8.1C] **6.** $3(2x-1)(x+1)$ [8.2A] **7.** $\frac{3(x+2)}{x(x+2)(x-2)}; \frac{x^2}{x(x+2)(x-2)}$ [8.2A] **8.** $\frac{2}{x+5}$ [8.2B] **9.** $\frac{5}{(2x-1)(3x+1)}$ [8.2B] **10.** $\frac{x^2-4x+5}{(x-2)(x+3)}$ [8.2B] **11.** $\frac{x-3}{x-2}$ [8.3A] **12.** 2 [8.4A] **13.** The equation has no solution. [8.4A] **14.** −1 [8.5A] **15.** The area is 64.8 m^2. [8.5C] **16.** $t=\frac{d-s}{r}$ [8.6A] **17.** 14 rolls of wallpaper are needed. [8.5B] **18.** It would take 10 min with both landscapers working. [8.7A] **19.** The rate of the cyclist is 10 mph. [8.7B] **20.** The resistance is 0.4 ohm. [8.8A]

CUMULATIVE REVIEW EXERCISES

1. $\frac{31}{30}$ [1.3A] **2.** 21 [1.4A] **3.** $5x-2y$ [1.4B] **4.** $-8x+26$ [1.4D] **5.** $-\frac{9}{2}$ [2.2A] **6.** −12 [2.2C] **7.** 10 [2.1D] **8.** $\{x \mid x \le 8\}$ [2.5A] **9.** The volume is 200 ft^3. [3.3A] **10.**

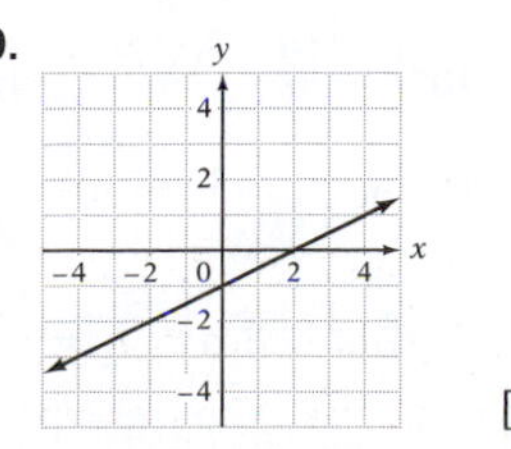

[4.3B] **11.** $\frac{3}{7}$ [4.2A] **12.** $y=\frac{3}{2}x+2$ [4.6A] **13.** (2, 1, −1) [5.2B] **14.** a^3b^7 [6.1A] **15.** $\frac{a^5}{b^6}$ [6.1B] **16.** 3.5×10^{-8} [6.1C] **17.** $-4a^4+6a^3-2a^2$ [6.3A] **18.** $a^2+ab-12b^2$ [6.3B] **19.** x^2+2x+4 [6.4B] **20.** $(y-6)(y-1)$ [7.2A] **21.** $(4x+1)(3x-1)$ [7.3A/7.3B] **22.** $a(2a-3)(a+5)$ [7.3A/7.3B]

Copyright © Houghton Mifflin Company. All rights reserved.

.4D] **24.** -3 and $\frac{5}{2}$ [7.5A] **25.** $\frac{2x^3}{3y^5}$ [8.1A] **26.** $-\frac{x-2}{x+5}$ [8.1A] **27.** 1 [8.1C]

1) [8.2B] **29.** $\frac{x+3}{x+5}$ [8.3A] **30.** 4 [8.4A] **31.** The alloy contains 70% silver. [2.4B]

cost \$160. [8.5B] **33.** It would take 6 min to fill the tank. [8.7A]

Answers to Chapter 9 Selected Exercises

PREP TEST

1. 16 [1.1D] **2.** 32 [1.2E] **3.** 9 [1.2D] **4.** $\frac{1}{12}$ [1.2C] **5.** $-5x-1$ [1.4D] **6.** $\frac{xy^5}{4}$ [6.1B] **7.** $9x^2-12x+4$ [6.3C] **8.** $-12x^2+14x+10$ [6.3B] **9.** $36x^2-1$ [6.3C] **10.** -1 and 15 [7.5A]

SECTION 9.1

1. 2 **3.** 27 **5.** $\frac{1}{9}$ **7.** 4 **9.** Not a real number **11.** $\frac{343}{125}$ **13.** x **15.** $y^{1/2}$ **17.** $x^{1/12}$ **19.** $a^{7/12}$ **21.** $\frac{1}{a}$ **23.** $\frac{1}{y}$ **25.** $y^{3/2}$ **27.** $\frac{1}{x}$ **29.** $\frac{1}{x^4}$ **31.** a **33.** $x^{3/10}$ **35.** a^3 **37.** $\frac{1}{x^{1/2}}$ **39.** $y^{1/9}$ **41.** x^4y **43.** $x^6y^3z^9$ **45.** $\frac{x}{y^2}$ **47.** $\frac{x^{3/2}}{y^{1/4}}$ **49.** $\frac{x^2}{y^8}$ **51.** $\frac{1}{x^{11/12}}$ **53.** $\frac{1}{y^{5/2}}$ **55.** $\frac{1}{b^{7/8}}$ **57.** a^5b^{13} **59.** $\frac{m^2}{4n^{3/2}}$ **61.** $\frac{y^{17/2}}{x^3}$ **63.** $\frac{16b^2}{a^{1/3}}$ **65.** y^2-y **67.** $a-a^2$ **69.** x^{4n} **71.** $x^{3n/2}$ **73.** $y^{3n/2}$ **75.** x^{2n^2} **77.** $x^{2n}y^n$ **79.** $x^{4n}y^{2n}$ **81.** $\sqrt[4]{3}$ **83.** $\sqrt{a^3}$ **85.** $\sqrt{32t^5}$ **87.** $-2\sqrt[3]{x^2}$ **89.** $\sqrt[3]{a^4b^2}$ **91.** $\sqrt[5]{a^6b^{12}}$ **93.** $\sqrt[4]{(4x-3)^3}$ **95.** $\frac{1}{\sqrt[3]{x^2}}$ **97.** $14^{1/2}$ **99.** $x^{1/3}$ **101.** $x^{4/3}$ **103.** $b^{3/5}$ **105.** $(2x^2)^{1/3}$ **107.** $-(3x^5)^{1/2}$ **109.** $3xy^{2/3}$ **111.** $(a^2-2)^{1/2}$ **113.** x^8 **115.** $-x^4$ **117.** xy^3 **119.** $-x^5y$ **121.** $4a^2b^6$ **123.** Not a real number **125.** $3x^3$ **127.** $-4x^3y^4$ **129.** $-x^2y^3$ **131.** x^4y^2 **133.** $3xy^5$ **135.** $2ab^2$ **137.** **a.** False; 2 **b.** True **c.** True **d.** False; $(a^n+b^n)^{1/n}$ **e.** False; $a+2a^{1/2}b^{1/2}+b$ **f.** False; $a^{n/m}$

SECTION 9.2

1. $x^2yz^2\sqrt{yz}$ **3.** $2ab^4\sqrt{2a}$ **5.** $3xyz^2\sqrt{5yz}$ **7.** Not a real number **9.** $a^5b^2\sqrt[3]{ab^2}$ **11.** $-5y\sqrt[3]{x^2y}$ **13.** $abc^2\sqrt[3]{ab^2}$ **15.** $2x^2y\sqrt[4]{xy}$ **17.** $-6\sqrt{x}$ **19.** $-2\sqrt{2}$ **21.** $3\sqrt{2b}+5\sqrt{3b}$ **23.** $-2xy\sqrt{2y}$ **25.** $6ab^2\sqrt{3ab}+3ab\sqrt{3ab}$ **27.** $-\sqrt[3]{2}$ **29.** $8b\sqrt[3]{2b^2}$ **31.** $3a\sqrt[4]{2a}$ **33.** $17\sqrt{2}-15\sqrt{5}$ **35.** $5b\sqrt{b}$ **37.** $-8xy\sqrt{2x}+2xy\sqrt{xy}$ **39.** $2y\sqrt[3]{2x}$ **41.** $-4ab\sqrt[4]{2b}$ **43.** 16 **45.** $2\sqrt[3]{4}$ **47.** $xy^3\sqrt{x}$ **49.** $8xy\sqrt{x}$ **51.** $2x^2y\sqrt[3]{2}$ **53.** $2ab\sqrt[4]{3a^2b}$ **55.** 6 **57.** $x-\sqrt{2x}$ **59.** $4x-8\sqrt{x}$ **61.** $x-6\sqrt{x}+9$ **63.** $84+16\sqrt{5}$ **65.** $672x^2y^2$ **67.** $4a^3b^3\sqrt[3]{a}$ **69.** $-8\sqrt{5}$ **71.** $x-y^2$ **73.** $12x-y$ **77.** $y\sqrt{5y}$ **79.** $b\sqrt{13b}$ **81.** $\frac{\sqrt{2}}{2}$ **83.** $\frac{2\sqrt{3y}}{3y}$ **85.** $\frac{3\sqrt{3a}}{a}$ **87.** $\frac{\sqrt{2y}}{2}$ **89.** $\frac{5\sqrt[3]{3}}{3}$ **91.** $\frac{5\sqrt[3]{9y^2}}{3y}$ **93.** $\frac{b\sqrt{2a}}{2a^2}$ **95.** $\frac{\sqrt{15x}}{5x}$ **97.** $2+2\sqrt{2}$ **99.** $-\frac{12+4\sqrt{2}}{7}$ **101.** $-\frac{10+5\sqrt{7}}{3}$ **103.** $-\frac{7\sqrt{x}+21}{x-9}$ **105.** $-\sqrt{6}+3-2\sqrt{2}+2\sqrt{3}$ **107.** $-\frac{17+5\sqrt{5}}{4}$ **109.** $\frac{8a-10\sqrt{ab}+3b}{16a-9b}$ **111.** $\frac{3-7\sqrt{y}+2y}{1-4y}$ **113.** **a.** False; $\sqrt[6]{432}$ **b.** True **c.** False; $x^{2/3}$ **d.** False; $\sqrt{x}+\sqrt{y}$ is in simplest form. **e.** False; $\sqrt[2]{2}+\sqrt[3]{3}$ is in simplest form. **f.** True **115.** $\sqrt[4]{a+b}$

SECTION 9.3

1. -2 **3.** 9 **5.** 7 **7.** $\frac{13}{3}$ **9.** 35 **11.** -7 **13.** -12 **15.** 9 **17.** 2 **19.** 2 **21.** 1 **23.** On the moon, an object will fall 24.75 ft in 3 s. **25.** The HDTV screen is approximately 7.15 in. wider. **27.** The length of the pendulum is approximately 7.30 ft. **29.** 2 **31.** $r=\sqrt[3]{\frac{3V}{4\pi}}$

Copyright © Houghton Mifflin Company. All rights reserved.

SECTION 9.4

3. $2i$ **5.** $7i\sqrt{2}$ **7.** $3i\sqrt{3}$ **9.** $4 + 2i$ **11.** $2\sqrt{3} - 3i\sqrt{2}$ **13.** $4\sqrt{10} - 7i\sqrt{3}$ **15.** $8 - i$ **17.** $-8 + 4i$ **19.** $6 - 6i$ **21.** $19 - 7i\sqrt{2}$ **23.** $6\sqrt{2} - 3i\sqrt{2}$ **25.** 63 **27.** -4 **29.** $-3\sqrt{2}$ **31.** $-4 + 12i$ **33.** $-2 + 4i$ **35.** $17 - i$ **37.** $8 + 27i$ **39.** 1 **41.** 1 **43.** $-3i$ **45.** $\frac{3}{4} + \frac{1}{2}i$ **47.** $\frac{10}{13} - \frac{2}{13}i$ **49.** $\frac{4}{5} + \frac{2}{5}i$ **51.** $-i$ **53.** $-\frac{\sqrt{5}}{5} + \frac{2\sqrt{5}}{5}i$ **55.** $\frac{3}{10} - \frac{11}{10}i$ **57.** $\frac{6}{5} + \frac{7}{5}i$ **59. a.** Yes **b.** Yes

CHAPTER 9 REVIEW EXERCISES

1. $20x^2y^2$ [9.1A] **2.** 7 [9.3A] **3.** $39 - 2i$ [9.4C] **4.** $7x^{2/3}y$ [9.1B] **5.** $6\sqrt{3} - 13$ [9.2C] **6.** -2 [9.3A] **7.** $\frac{1}{x^5}$ [9.1A] **8.** $\frac{8\sqrt{3y}}{3y}$ [9.2D] **9.** $-2a^2b^4$ [9.1C] **10.** $2a^2b\sqrt{2b}$ [9.2B] **11.** $\frac{x\sqrt{x} - x\sqrt{2} + 2\sqrt{x} - 2\sqrt{2}}{x - 2}$ [9.2D] **12.** $\frac{2}{3} - \frac{5}{3}i$ [9.4D] **13.** $3ab^3\sqrt{2a}$ [9.2A] **14.** $-4\sqrt{2} + 8i\sqrt{2}$ [9.4B] **15.** $5x^3y^3\sqrt[3]{2x^2y}$ [9.2B] **16.** $4xy^2\sqrt[3]{x^2}$ [9.2C] **17.** $7 + 3i$ [9.4C] **18.** $3\sqrt[4]{x^3}$ [9.1B] **19.** $-2ab^2\sqrt[5]{2a^3b^2}$ [9.2A] **20.** $-2 + 7i$ [9.4D] **21.** $-6\sqrt{2}$ [9.4C] **22.** 30 [9.3A] **23.** $3a^2b^3$ [9.1C] **24.** $5i\sqrt{2}$ [9.4A] **25.** $-12 + 10i$ [9.4B] **26.** $31 - 10\sqrt{6}$ [9.2C] **27.** $6x^2\sqrt{3y}$ [9.2B] **28.** $\frac{1}{3}$ [9.1A] **29.** $\frac{1}{a^{10}}$ [9.1A] **30.** $-7x^3y^8$ [9.1C] **31.** The amount of power is approximately 120 watts. [9.3B] **32.** The distance required is 242 ft. [9.3B] **33.** The distance is approximately 6.63 ft. [9.3B]

CHAPTER 9 TEST

1. $\frac{1}{2}x^{3/4}$ [9.1B] **2.** $-2x^2y\sqrt[3]{2x}$ [9.2B] **3.** $3\sqrt[5]{y^2}$ [9.1B] **4.** $18 + 16i$ [9.4C] **5.** $4x + 4\sqrt{xy} + y$ [9.2C] **6.** $r^{1/6}$ [9.1A] **7.** 4 [9.3A] **8.** $2xy^2$ [9.1C] **9.** $-4x\sqrt{3}$ [9.2C] **10.** $-3 + 2i$ [9.4B] **11.** $4x^2y^3\sqrt{2y}$ [9.2A] **12.** $14 + 10\sqrt{3}$ [9.2C] **13.** $10 + 2i$ [9.4C] **14.** 2 [9.2D] **15.** $8a\sqrt{2a}$ [9.2B] **16.** $2a - \sqrt{ab} - 15b$ [9.2C] **17.** $\frac{64x^3}{y^6}$ [9.1A] **18.** $\frac{x + \sqrt{xy}}{x - y}$ [9.2D] **19.** $-\frac{4}{5} + \frac{7}{5}i$ [9.4D] **20.** -3 [9.3A] **21.** $\frac{b^3}{8a^6}$ [9.1A] **22.** $3abc^2\sqrt[3]{ac}$ [9.2A] **23.** $\frac{4x^2}{y}$ [9.2D] **24.** -4 [9.4C] **25.** The distance is 576 ft. [9.3B]

CUMULATIVE REVIEW EXERCISES

1. 92 [1.3A] **2.** 56 [1.4A] **3.** $-10x + 1$ [1.4D] **4.** $\frac{3}{2}$ [2.2A] **5.** $\frac{2}{3}$ [2.2C] **6.** $\{x \mid x > 1\}$ [2.5A] **7.** $\frac{1}{3}, \frac{7}{3}$ [2.6A] **8.** $\{x \mid -6 \le x \le 3\}$ [2.6B] **9.** The area is 187.5 cm^2. [3.2B] **10.** The volume is 14 ft^3. [3.3A]

11. $m = \frac{3}{2}$ $b = 3$ [4.3B] **12.** [4.7A] **13.** $y = \frac{1}{3}x + \frac{7}{3}$ [4.5B]

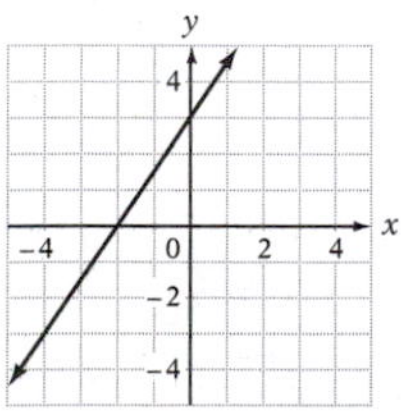

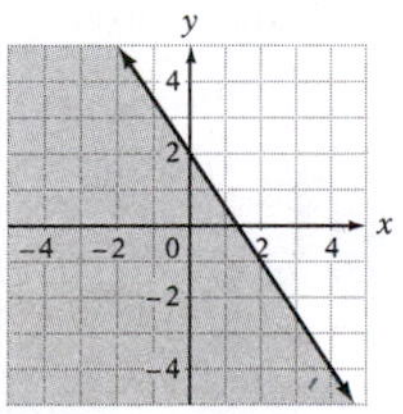

14. (3, 2) [5.2A] **15.** $2x^2y^2$ [6.1B] **16.** $(9x + y)(9x - y)$ [7.4A] **17.** $x(x^2 + 3)(x + 1)(x - 1)$ [7.4D] **18.** $C = R - nP$ [8.6A] **19.** $\frac{y^5}{x^4}$ [9.1A] **20.** $-x\sqrt{10x}$ [9.2B] **21.** $13 - 7\sqrt{3}$ [9.2C] **22.** $\sqrt{6} + \sqrt{2}$ [9.2D] **23.** $-\frac{1}{5} + \frac{3}{5}i$ [9.4D] **24.** -20 [9.3A] **25.** The length of side DE is 27 m. [8.5C] **26.** \$2500 is invested at 3.5%. [5.1C] **27.** The rate of the plane was 250 mph. [8.7B] **28.** It takes 1.25 s for light to travel to Earth from the moon. [6.1D] **29.** The slope is 0.08. The slope represents the simple interest rate on the investment. The interest rate is 8%. [4.4A] **30.** The periscope must be approximately 32.7 ft above the water. [9.3B]

Copyright © Houghton Mifflin Company. All rights reserved.

Answers to Chapter 10 Selected Exercises

PREP TEST

1. $3\sqrt{2}$ [1.2F] **2.** $3i$ [9.4A] **3.** $\frac{2x-1}{x-1}$ [8.2B] **4.** 8 [1.4A] **5.** Yes [7.4A] **6.** $(2x-1)^2$ [7.4A]
7. $(3x+2)(3x-2)$ [7.4A] **8.** 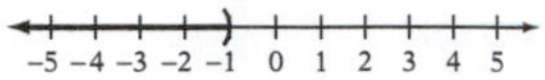

[1.5C] **9.** −3 and 5 [7.5A] **10.** 4 [8.5A]

SECTION 10.1

3. $2x^2-4x-5=0$; $a=2, b=-4, c=-5$ **5.** $4x^2-5x+6=0$; $a=4, b=-5, c=6$ **7.** 0 and 4 **9.** −5 and 5
11. −2 and 3 **13.** 3 **15.** 0 and 2 **17.** −2 and 5 **19.** 2 and 5 **21.** $-\frac{3}{2}$ and 6 **23.** $\frac{1}{4}$ and 2 **25.** −4 and $\frac{1}{3}$
27. $-\frac{2}{3}$ and $\frac{9}{2}$ **29.** −4 and $\frac{1}{4}$ **31.** −2 and 9 **33.** −2 and $-\frac{3}{4}$ **35.** −5 and 2 **37.** −4 and $-\frac{3}{2}$ **39.** $2b$ and $7b$
41. $-c$ and $7c$ **43.** $-b$ and $-\frac{b}{2}$ **45.** $\frac{2a}{3}$ and $4a$ **47.** $-\frac{a}{3}$ and $3a$ **49.** $-\frac{3y}{2}$ and $-\frac{y}{2}$ **51.** $-\frac{4a}{3}$ and $-\frac{a}{2}$
53. $x^2-7x+10=0$ **55.** $x^2+6x+8=0$ **57.** $x^2-5x-6=0$ **59.** $x^2-9=0$ **61.** $x^2-8x+16=0$
63. $x^2-5x=0$ **65.** $x^2-3x=0$ **67.** $2x^2-7x+3=0$ **69.** $4x^2-5x-6=0$ **71.** $3x^2+11x+10=0$
73. $9x^2-4=0$ **75.** $6x^2-5x+1=0$ **77.** $10x^2-7x-6=0$ **79.** $8x^2+6x+1=0$ **81.** $50x^2-25x-3=0$
83. −7 and 7 **85.** $-2i$ and $2i$ **87.** −2 and 2 **89.** $-\frac{9}{2}$ and $\frac{9}{2}$ **91.** $-7i$ and $7i$ **93.** $-4\sqrt{3}$ and $4\sqrt{3}$
95. $-5\sqrt{3}$ and $5\sqrt{3}$ **97.** $-3i\sqrt{2}$ and $3i\sqrt{2}$ **99.** −5 and 7 **101.** −6 and 0 **103.** −7 and 3 **105.** 0 and 1
107. $-5-\sqrt{6}$ and $-5+\sqrt{6}$ **109.** $3-3i\sqrt{5}$ and $3+3i\sqrt{5}$ **111.** $-\frac{2-9\sqrt{2}}{3}$ and $-\frac{2+9\sqrt{2}}{3}$ **113.** $x^2-2=0$
115. $x^2+1=0$ **117.** $x^2-8=0$ **119.** $x^2+2=0$ **121.** $-\frac{3b}{a}$ and $\frac{3b}{a}$ **123.** $-a-2$ and $-a+2$ **125.** $-\frac{1}{2}$

SECTION 10.2

1. −1 and 5 **3.** −9 and 1 **5.** 3 **7.** $-2-\sqrt{11}$ and $-2+\sqrt{11}$ **9.** $3-\sqrt{2}$ and $3+\sqrt{2}$ **11.** $1-i$ and $1+i$
13. −3 and 8 **15.** −9 and 4 **17.** $\frac{3-\sqrt{5}}{2}$ and $\frac{3+\sqrt{5}}{2}$ **19.** $\frac{1-\sqrt{5}}{2}$ and $\frac{1+\sqrt{5}}{2}$ **21.** $3-\sqrt{13}$ and $3+\sqrt{13}$
23. 3 and 5 **25.** $2-3i$ and $2+3i$ **27.** $-3-2i$ and $-3+2i$ **29.** $1-3\sqrt{2}$ and $1+3\sqrt{2}$
31. $\frac{1-\sqrt{17}}{2}$ and $\frac{1+\sqrt{17}}{2}$ **33.** $1-2i\sqrt{3}$ and $1+2i\sqrt{3}$ **35.** $\frac{1}{2}-i$ and $\frac{1}{2}+i$ **37.** $\frac{1}{3}-\frac{1}{3}i$ and $\frac{1}{3}+\frac{1}{3}i$
39. $\frac{2-\sqrt{14}}{2}$ and $\frac{2+\sqrt{14}}{2}$ **41.** $-\frac{3}{2}$ and 1 **43.** $1-\sqrt{5}$ and $1+\sqrt{5}$ **45.** $\frac{1}{2}$ and 5 **47.** $2-\sqrt{5}$ and $2+\sqrt{5}$
49. −3.236 and 1.236 **51.** 0.293 and 1.707 **53.** −0.809 and 0.309 **55.** $-a$ and $2a$ **57.** $-5a$ and $2a$
59. No; the ball will have gone only 197.2 ft when it hits the ground.

SECTION 10.3

3. −2 and 5 **5.** −9 and 4 **7.** $4-2\sqrt{22}$ and $4+2\sqrt{22}$ **9.** −8 and 3 **11.** $\frac{-5-\sqrt{33}}{4}$ and $\frac{-5+\sqrt{33}}{4}$
13. $-\frac{1}{4}$ and $\frac{3}{2}$ **15.** $7-3\sqrt{5}$ and $7+3\sqrt{5}$ **17.** $\frac{1-\sqrt{3}}{2}$ and $\frac{1+\sqrt{3}}{2}$ **19.** $-1-i$ and $-1+i$
21. $1-2i$ and $1+2i$ **23.** $2-3i$ and $2+3i$ **25.** $\frac{1-\sqrt{11}}{2}$ and $\frac{1+\sqrt{11}}{2}$ **27.** $-\frac{3}{2}-\frac{1}{2}i$ and $-\frac{3}{2}+\frac{1}{2}i$
29. $\frac{3}{4}-\frac{3\sqrt{3}}{4}i$ and $\frac{3}{4}+\frac{3\sqrt{3}}{4}i$ **31.** 0.394 and 7.606 **33.** −4.236 and 0.236 **35.** −1.351 and 1.851
37. Two complex number solutions **39.** One real number solution **41.** Two real number solutions
43. No. The arrow does not reach a height of 275 ft. (The discriminant is less than zero.) **45.** $\{p \mid p < 9\}$
47. $\{p \mid p > 1\}$ **49.** $-2i$ and i

Copyright © Houghton Mifflin Company. All rights reserved.

SECTION 10.4

1. $-2, 2, -3,$ and 3 **3.** $-2, 2, -\sqrt{2},$ and $\sqrt{2}$ **5.** 1 and 4 **7.** 16 **9.** $-2i, 2i, -1,$ and 1 **11.** $-4i, 4i, -2,$ and 2 **13.** 16 **15.** 1 and 512 **17.** $-\frac{2}{3}, \frac{2}{3}, -1,$ and 1 **19.** 3 **21.** 9 **23.** -1 and 2 **25.** 0 and 2 **27.** $-\frac{1}{2}$ and 2 **29.** -2 **31.** 1 **33.** 1 **35.** -3 **37.** -1 and 10 **39.** $-\frac{1}{2} - \frac{\sqrt{7}}{2}i$ and $-\frac{1}{2} + \frac{\sqrt{7}}{2}i$ **41.** -3 and 1 **43.** -1 and 0 **45.** $-\frac{1}{3}$ and $\frac{1}{2}$ **47.** $-\frac{2}{3}$ and 6 **49.** $\frac{4}{3}$ and 3 **51.** $-\frac{1}{4}$ and 3 **53.** 9 and 36 **55.** $-\sqrt{5}$ or $\sqrt{5}$

SECTION 10.5

3. −5 −4 −3 −2 −1 0 1 2 3 4 5 $\{x \mid x < -2 \text{ or } x > 4\}$ **5.** −5 −4 −3 −2 −1 0 1 2 3 4 5 $\{x \mid x \le 1 \text{ or } x \ge 2\}$

7. −5 −4 −3 −2 −1 0 1 2 3 4 5 $\{x \mid -3 < x < 4\}$ **9.** −5 −4 −3 −2 −1 0 1 2 3 4 5 $\{x \mid x < -2 \text{ or } 1 < x < 3\}$

11. −5 −4 −3 −2 −1 0 1 2 3 4 5 $\{x \mid -4 \le x \le 1 \text{ or } x \ge 2\}$ **13.** −5 −4 −3 −2 −1 0 1 2 3 4 5 $\{x \mid x < -2 \text{ or } x > 4\}$

15. −5 −4 −3 −2 −1 0 1 2 3 4 5 $\{x \mid -1 < x \le 3\}$ **17.** −5 −4 −3 −2 −1 0 1 2 3 4 5 $\{x \mid x \le -2 \text{ or } 1 \le x < 3\}$

19. $\{x \mid x > 4 \text{ or } x < -4\}$ **21.** $\{x \mid -3 \le x \le 12\}$ **23.** $\left\{x \,\middle|\, \frac{1}{2} < x < \frac{3}{2}\right\}$ **25.** $\left\{x \,\middle|\, x < 1 \text{ or } x > \frac{5}{2}\right\}$

27. $\{x \mid x < -1 \text{ or } 1 < x \le 2\}$ **29.** $\left\{x \,\middle|\, \frac{1}{2} < x \le 1\right\}$ **31.** $\{x \mid 2 < x \le 3\}$ **33.** $\{x \mid x > 5 \text{ or } -4 < x < -1\}$

35. −4 −2 0 2 4 **37.** −4 −2 0 2 4 **39.** −4 −2 0 2 4

SECTION 10.6

1. The height is 3 cm. The base is 14 cm. **3.** The dimensions of Colorado are approximately 272 mi by 383 mi. **5.** The maximum speed is 33 mph. **7.** The rocket takes 12.5 s to return to Earth. **9.** A driver can be going approximately 72.5 km/h and still be able to stop within 150 m. **11.** It would take the larger pipe 6 min to fill the tank. It would take the smaller pipe 12 min to fill the tank. **13.** The rate of the wind was approximately 108 mph. **15.** The rowing rate of the guide is 6 mph. **17.** The radius of the cone is 1.5 in.

CHAPTER 10 REVIEW EXERCISES

1. 0 and $\frac{3}{2}$ [10.1A] **2.** $-2c$ and $\frac{c}{2}$ [10.1A] **3.** $-4\sqrt{3}$ and $4\sqrt{3}$ [10.1C] **4.** $-\frac{1}{2} - 2i$ and $-\frac{1}{2} + 2i$ [10.1C] **5.** -3 and -1 [10.2A] **6.** $\frac{7 - 2\sqrt{7}}{7}$ and $\frac{7 + 2\sqrt{7}}{7}$ [10.2A] **7.** $\frac{3}{4}$ and $\frac{4}{3}$ [10.3A] **8.** $\frac{1}{2} - \frac{\sqrt{31}}{2}i$ and $\frac{1}{2} + \frac{\sqrt{31}}{2}i$ [10.3A] **9.** $x^2 + 3x = 0$ [10.1B] **10.** $12x^2 - x - 6 = 0$ [10.1B] **11.** $1 - i\sqrt{7}$ and $1 + i\sqrt{7}$ [10.2A] **12.** $-2i$ and $2i$ [10.2A] **13.** $\frac{11 - \sqrt{73}}{6}$ and $\frac{11 + \sqrt{73}}{6}$ [10.3A] **14.** Two real number solutions [10.3A] **15.** $\left\{x \,\middle|\, -3 < x < \frac{5}{2}\right\}$ [10.5A] **16.** $\left\{x \,\middle|\, x \le -4 \text{ or } -\frac{3}{2} \le x \le 2\right\}$ [10.5A] **17.** -64 and 27 [10.4A] **18.** $\frac{5}{4}$ [10.4B] **19.** -1 and 3 [10.4C] **20.** -1 [10.4C] **21.** −5 −4 −3 −2 −1 0 1 2 3 4 5 $\left\{x \,\middle|\, x < \frac{3}{2} \text{ or } x \ge 2\right\}$ [10.5A] **22.** −5 −4 −3 −2 −1 0 1 2 3 4 5 $\left\{x \,\middle|\, x \le -3 \text{ or } \frac{1}{2} \le x < 4\right\}$ [10.5A] **23.** 4 [10.4B] **24.** 5 [10.4B] **25.** $\frac{-3 - \sqrt{249}}{10}$ and $\frac{-3 + \sqrt{249}}{10}$ [10.4C] **26.** $\frac{-11 - \sqrt{129}}{2}$ and $\frac{-11 + \sqrt{129}}{2}$ [10.4C] **27.** The width of the rectangle is 5 cm. The length of the rectangle is 12 cm. [10.6A] **28.** The integers are 2, 4, and 6 or $-6, -4,$ and -2. [10.6A] **29.** Working alone, the new computer can print the payroll in 12 min. [10.6A] **30.** The sculling crew's rate of rowing in calm water is 6 mph. [10.6A]

Copyright © Houghton Mifflin Company. All rights reserved.

CHAPTER 10 TEST

1. -4 and $\frac{2}{3}$ [10.1A] **2.** $-\frac{2}{3}$ and $\frac{3}{2}$ [10.1A] **3.** $x^2 - 9 = 0$ [10.1B] **4.** $2x^2 + 7x - 4 = 0$ [10.1B] **5.** $2 - 2\sqrt{2}$ and $2 + 2\sqrt{2}$ [10.1C] **6.** $3 - \sqrt{11}$ and $3 + \sqrt{11}$ [10.2A] **7.** $\frac{3 - \sqrt{15}}{3}$ and $\frac{3 + \sqrt{15}}{3}$ [10.2A] **8.** $\frac{1 - \sqrt{3}}{2}$ and $\frac{1 + \sqrt{3}}{2}$ [10.3A] **9.** $-2 - 2i\sqrt{2}$ and $-2 + 2i\sqrt{2}$ [10.3A] **10.** Two real number solutions [10.3A] **11.** Two complex number solutions [10.3A] **12.** $\frac{1}{4}$ [10.4A] **13.** $\sqrt{3}, -\sqrt{3}, 1,$ and -1 [10.4A] **14.** 4 [10.4B] **15.** No solution [10.4B] **16.** -9 and 2 [10.4C] **17.** −5 −4 −3 −2 −1 0 1 2 3 4 5 $\{x \mid x < -4 \text{ or } 2 < x < 4\}$ [10.5A] **18.** −5 −4 −3 −2 −1 0 1 2 3 4 5 $\left\{x \mid -4 < x \le -\frac{3}{2}\right\}$ [10.5A] **19.** The ball hits the basket approximately 1.88 s after it is released. [10.6A] **20.** The rate of the canoe in calm water is 4 mph. [10.6A]

CUMULATIVE REVIEW EXERCISES

1. 14 [1.4A] **2.** $\left\{x \mid -2 < x < \frac{10}{3}\right\}$ [2.6B] **3.** The volume is 54π m³. [3.3A] **4.** $-\frac{7}{3}$ [4.2A] **5.** $-\frac{3}{2}$ [4.4A] **6.** $\left(\frac{5}{2}, 0\right), (0, -3)$ [4.3C] **7.** $y = x + 1$ [4.6A] **8.** $(1, -1, 2)$ [5.2B] **9.**

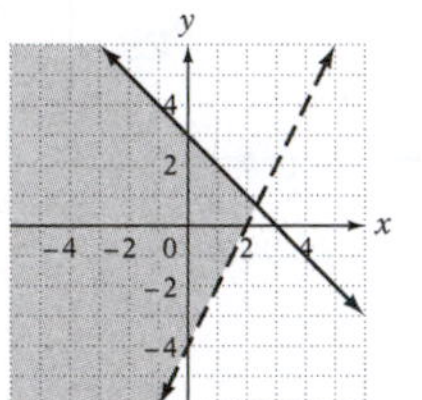

[5.4A]

10. The height of triangle *DEF* is 16 cm. [8.5C] **11.** $x^2 - 3x - 4 - \frac{6}{3x - 4}$ [6.4B] **12.** $-3xy(x^2 - 2xy + 3y^2)$ [7.1A] **13.** $(2x - 5)(3x + 4)$ [7.3A/7.3B] **14.** $\frac{x}{2}$ [8.1B] **15.** $-\frac{3}{2}$ and -1 [8.4A] **16.** $b = \frac{2S - an}{n}$ [8.6A] **17.** $1 - a$ [9.1A] **18.** $-8 - 14i$ [9.4C] **19.** 0, 1 [9.3A] **20.** $2, -2, \sqrt{2}, -\sqrt{2}$ [10.4A] **21.** The lower limit is $9\frac{23}{64}$ in. The upper limit is $9\frac{25}{64}$ in. [2.6C] **22.** The area is $(x^2 + 6x - 16)$ square feet. [6.3D] **23.** The slope is $-\frac{25{,}000}{3}$. The building decreases \$8333.33 in value each year. [4.4A] **24.** The ladder will reach 15 feet up on the house. [9.3B] **25.** There are two complex number solutions. [10.3A]

Answers to Chapter 11 Selected Exercises

PREP TEST

1. 1 [1.4A] **2.** -7 [1.4A] **3.** 30 [4.2A] **4.** $h^2 + 4h - 1$ [4.2A] **5.** $-\frac{2}{3}$ and 3 [10.1A] **6.** $2 - \sqrt{3}$ and $2 + \sqrt{3}$ [10.3A] **7.** $y = \frac{1}{2}x - 2$ [8.6A] **8.** Domain: $\{-2, 3, 4, 6\}$; Range: $\{4, 5, 6\}$. The relation is a function. [4.2A] **9.** [4.3B]

y 4 2 x −4 −2 0 2 4 −2 −4

Copyright © Houghton Mifflin Company. All rights reserved.

SECTION 11.1

5. -5 **7.** $x = 7$ **9.**

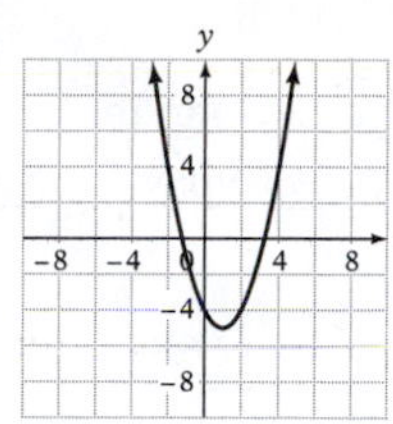

Vertex: $(1, -5)$

Axis of symmetry: $x = 1$

11.

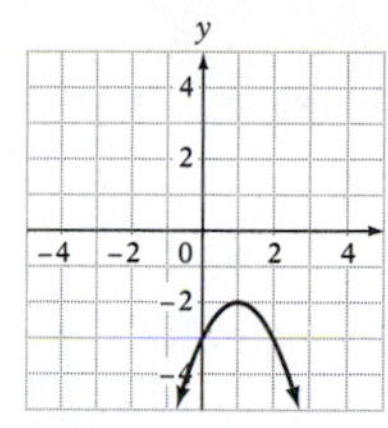

Vertex: $(1, -2)$

Axis of symmetry: $x = 1$

13.

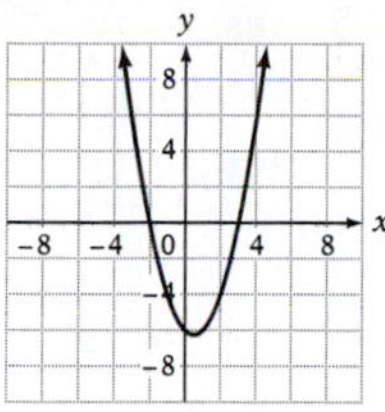

Vertex: $\left(\frac{1}{2}, -\frac{25}{4}\right)$

Axis of symmetry: $x = \frac{1}{2}$

15.

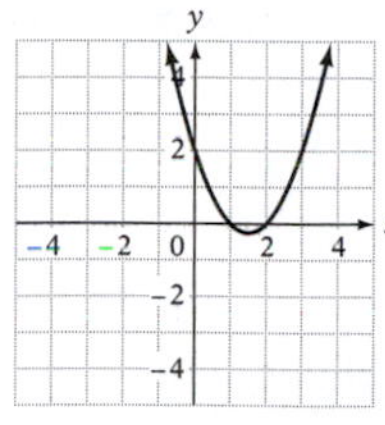

Vertex: $\left(\frac{3}{2}, -\frac{1}{4}\right)$

Axis of symmetry: $x = \frac{3}{2}$

17.

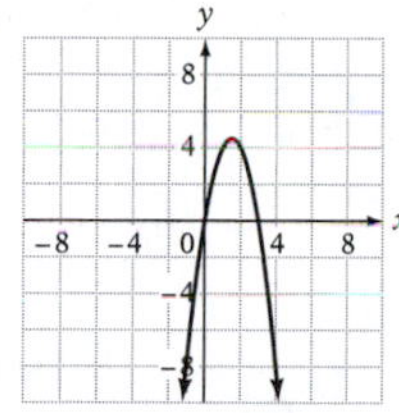

Vertex: $\left(\frac{3}{2}, \frac{9}{2}\right)$

Axis of symmetry: $x = \frac{3}{2}$

19.

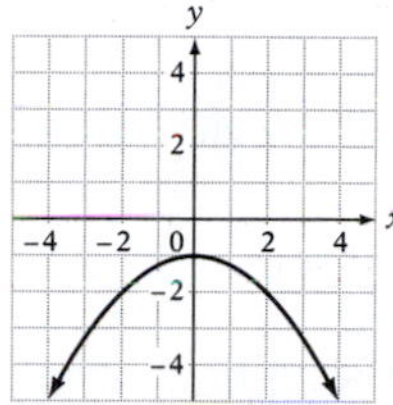

Vertex: $(0, -1)$

Axis of symmetry: $x = 0$

21.

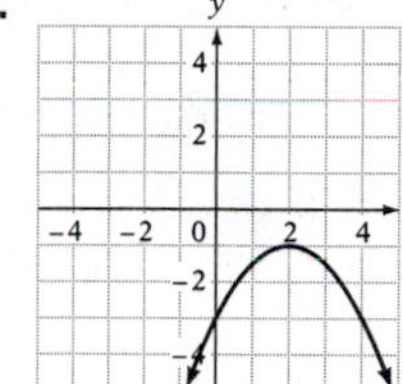

Vertex: $(2, -1)$

Axis of symmetry: $x = 2$

23.

Vertex: $\left(1, -\frac{5}{2}\right)$

Axis of symmetry: $x = 1$

25. Domain: $\{x \mid x \in \text{real numbers}\}$ Range: $\{y \mid y \geq -5\}$ **27.** Domain: $\{x \mid x \in \text{real numbers}\}$ Range: $\{y \mid y \leq 0\}$

29. Domain: $\{x \mid x \in \text{real numbers}\}$ Range: $\{y \mid y \geq -7\}$ **33.** $(3, 0)$ and $(-3, 0)$ **35.** $(0, 0)$ and $(-2, 0)$ **37.** $(4, 0)$ and $(-2, 0)$

39. $\left(-\frac{1}{2}, 0\right)$ and $(3, 0)$ **41.** $(-2 + \sqrt{7}, 0)$ and $(-2 - \sqrt{7}, 0)$ **43.** No x-intercepts **45.** 3 **47.** $\frac{3 + \sqrt{41}}{2}$ and $\frac{3 - \sqrt{41}}{2}$ **49.** 0 and $\frac{4}{3}$ **51.** $i\sqrt{2}$ and $-i\sqrt{2}$ **53.** $\frac{1}{6} + \frac{\sqrt{47}}{6}i$ and $\frac{1}{6} - \frac{\sqrt{47}}{6}i$ **55.** $\frac{1 + \sqrt{41}}{4}$ and $\frac{1 - \sqrt{41}}{4}$

57. Two **59.** One **61.** No x-intercepts **63.** Two **65.** No x-intercepts **67.** Two **69.** $(-4, 0)$ and $(5, 0)$

73. a. Minimum **b.** Maximum **c.** Minimum **75.** Minimum value: -2 **77.** Maximum value: -3

79. Maximum value: $\frac{9}{8}$ **81.** Minimum value: $-\frac{11}{4}$ **83.** Maximum value: $\frac{9}{4}$ **85.** Minimum value: $-\frac{1}{12}$

87. Parabola **b** has the greatest maximum value. **89.** The maximum value of the function is -5. **91.** A price of \$250 will give the maximum revenue. **93.** The diver reaches a height of 13.1 m above the water. **95.** The minimum height is 24.36 ft. **97.** The water will land at a height of $31\frac{2}{3}$ ft. **99.** The car can be traveling 20 mph and still stop at a stop sign 44 ft away. **101.** The two numbers are 10 and 10. **103.** The length is 100 ft and the width is 50 ft.

105. $k = 16$

Copyright © Houghton Mifflin Company. All rights reserved.

SECTION 11.2

1. Yes **3.** No **5.** Yes

7. Domain: $\{x \mid x \in \text{real numbers}\}$ Range: $\{y \mid y \geq 0\}$

9. 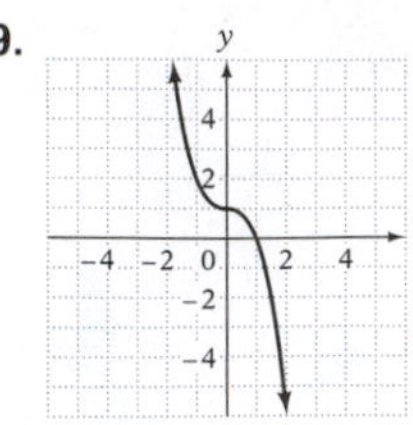 Domain: $\{x \mid x \in \text{real numbers}\}$ Range: $\{y \mid y \in \text{real numbers}\}$

11. 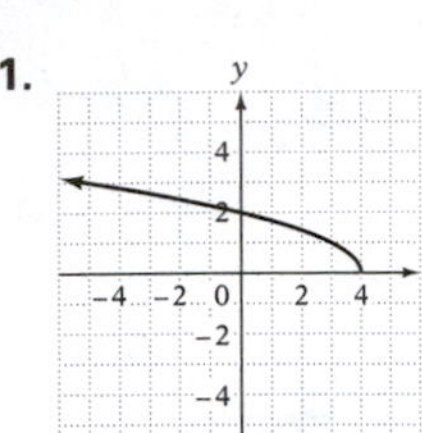 Domain: $\{x \mid x \leq 4\}$ Range: $\{y \mid y \geq 0\}$

13. Domain: $\{x \mid x \in \text{real numbers}\}$ Range: $\{y \mid y \in \text{real numbers}\}$

15. 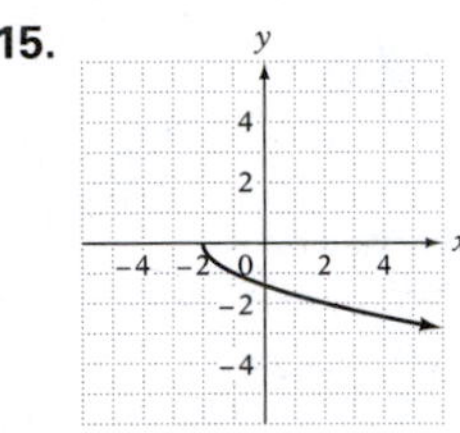 Domain: $\{x \mid x \geq -2\}$ Range: $\{y \mid y \leq 0\}$

17. 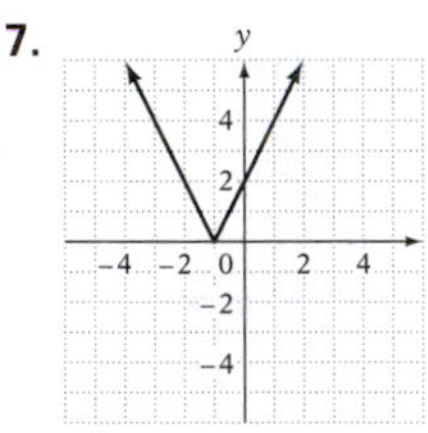 Domain: $\{x \mid x \in \text{real numbers}\}$ Range: $\{y \mid y \geq 0\}$

19. $a = 18$ **21.** 17 **23.** $f(14) = 8$ **25.** $\{x \mid -2 < x < 2\}$ **27.** $x = 1$

SECTION 11.3

1. 5 **3.** 1 **5.** 0 **7.** $-\frac{29}{4}$ **9.** $\frac{2}{3}$ **11.** 2 **13.** 39 **15.** -8 **17.** $-\frac{4}{5}$ **19.** -2 **21.** 7
25. -13 **27.** -29 **29.** $8x - 13$ **31.** 4 **33.** 3 **35.** $x + 4$ **37.** 1 **39.** 5 **41.** $x^2 + 1$
43. 5 **45.** 11 **47.** $3x^2 + 3x + 5$ **49.** -3 **51.** -27 **53.** $x^3 - 6x^2 + 12x - 8$
55. a. $S(M(x)) = 80 + \frac{16000}{x}$ **b.** \$83.20 **c.** When 5000 digital cameras are manufactured, the camera store sells each camera for \$83.20. **57. a.** $d[r(p)] = 0.90p - 1350$ **b.** $r[d(p)] = 0.90p - 1500$
c. $r[d(p)]$ (The cost is less.) **59.** 0 **61.** -2 **63.** 7 **65.** $h^2 + 6h$ **67.** $2 + h$ **69.** $2a + h$ **71.** -1
73. -6 **75.** $6x - 13$

SECTION 11.4

3. Yes **5.** No **7.** Yes **9.** No **11.** No **13.** No **17.** $\{(0, 1), (3, 2), (8, 3), (15, 4)\}$ **19.** No inverse
21. $\{(-2, 0), (5, -1), (3, 3), (6, -4)\}$ **23.** No inverse **25.** $f^{-1}(x) = \frac{1}{4}x + 2$ **27.** $f^{-1}(x) = \frac{1}{2}x - 2$
29. $f^{-1}(x) = 2x + 2$ **31.** $f^{-1}(x) = -\frac{1}{2}x + 1$ **33.** $f^{-1}(x) = \frac{3}{2}x - 6$ **35.** $f^{-1}(x) = -3x + 3$ **37.** $f^{-1}(x) = \frac{1}{2}x + \frac{5}{2}$
39. $f^{-1}(x) = \frac{1}{5}x + \frac{2}{5}$ **41.** $f^{-1}(x) = \frac{1}{6}x + \frac{1}{2}$ **43.** $\frac{5}{3}$ **45.** 3 **47.** Yes; Yes **49.** Yes **51.** No
53. Yes **55.** Yes **57.**

59. 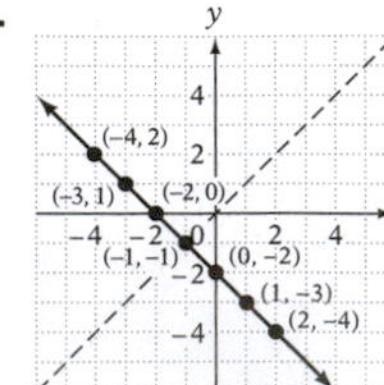

61. 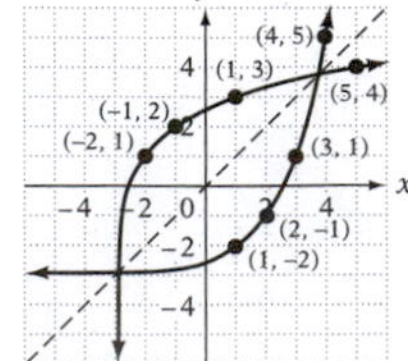

65. 5 **67.** 0 **69.** -4

Copyright © Houghton Mifflin Company. All rights reserved.

CHAPTER 11 REVIEW EXERCISES

1. Yes [11.2A] **2.** Yes [11.4A] **3.**

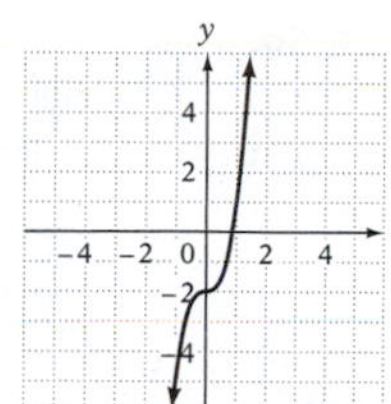

Domain: $\{x|x \in \text{real numbers}\}$
Range: $\{y|y \in \text{real numbers}\}$ [11.2A]

4.

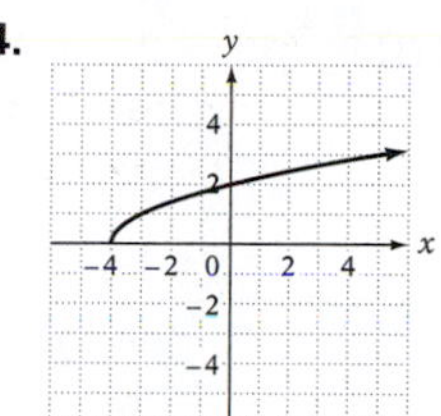

Domain: $\{x|x \geq -4\}$
Range: $\{y|y \geq 0\}$ [11.2A]

5. Two [11.1B] **6.** (0, 0) and (−3, 0) [11.1B] **7.** $-\frac{1}{3} + \frac{\sqrt{5}}{3}i$ and $-\frac{1}{3} - \frac{\sqrt{5}}{3}i$ [11.1B] **8.** 3 [11.1C]

9. $-\frac{17}{4}$ [11.1C] **10.** 5 [11.3B] **11.** 10 [11.3B] **12.** $12x^2 + 12x - 1$ [11.3B] **13.** Yes [11.4B]

14.

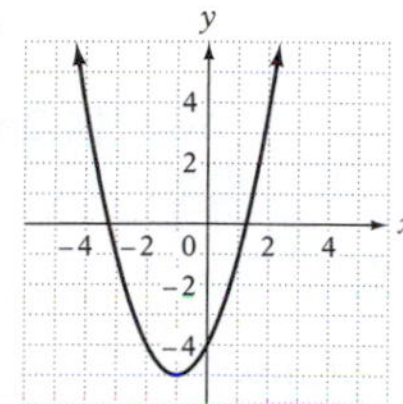

Domain: $\{x|x \in \text{real numbers}\}$
Range: $\{y|y \geq -5\}$ [11.1A]

15.

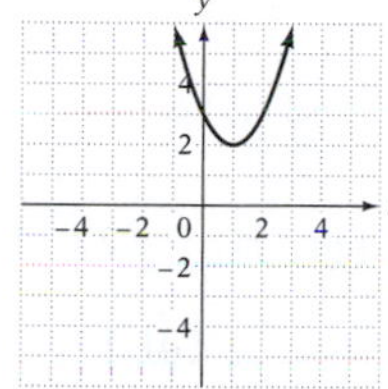

Vertex: (1, 2)
Axis of symmetry: $x = 1$ [11.1A]

16.

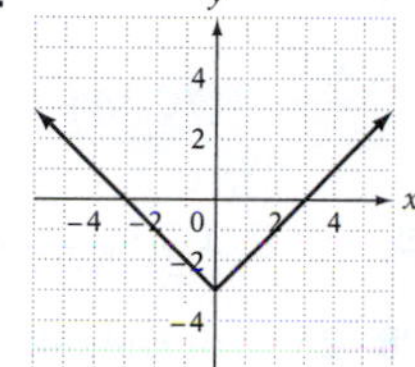

Domain: $\{x|x \in \text{real numbers}\}$
Range: $\{y|y \geq -3\}$ [11.2A]

17. No [11.4A] **18.** 7 [11.3A] **19.** −9 [11.3A] **20.** 70 [11.3A] **21.** $\frac{12}{7}$ [11.3A] **22.** $6x^2 + 3x - 16$ [11.3B] **23.** $f^{-1}(x) = -\frac{1}{6}x + \frac{2}{3}$ [11.4B] **24.** $f^{-1}(x) = \frac{3}{2}x + 18$ [11.4B] **25.** $f^{-1}(x) = 2x - 16$ [11.4B]

26. The dimensions are 7 ft by 7 ft. [11.1D]

CHAPTER 11 TEST

1. $\frac{3}{4} + \frac{\sqrt{23}}{4}i$ and $\frac{3}{4} - \frac{\sqrt{23}}{4}i$ [11.1B] **2.** $\frac{-3 + \sqrt{41}}{2}$ and $\frac{-3 - \sqrt{41}}{2}$ [11.1B] **3.** Two real zeros [11.1B]

4. −2 [11.3A] **5.** 234 [11.3A] **6.** $-\frac{13}{2}$ [11.3A] **7.** −5 [11.3A] **8.** 5 [11.3B]

9. $f[g(x)] = 2x^2 - 4x - 5$ [11.3B] **10.** 9 [11.1C] **11.** $f^{-1}(x) = \frac{1}{4}x + \frac{1}{2}$ [11.4B] **12.** $f^{-1}(x) = 4x + 16$ [11.4B]

13. {(6, 2), (5, 3), (4, 4), (3, 5)} [11.4B] **14.** Yes [11.4B] **15.** 100 [11.1D]

16.

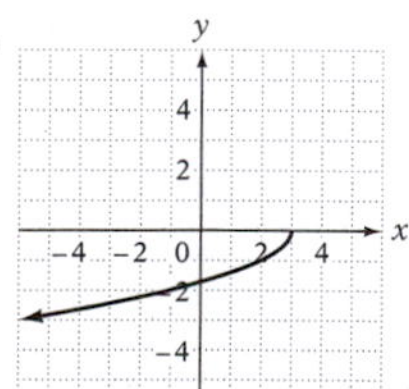

Domain: $\{x|x \leq 3\}$
Range: $\{y|y \leq 0\}$ [11.2A]

17.

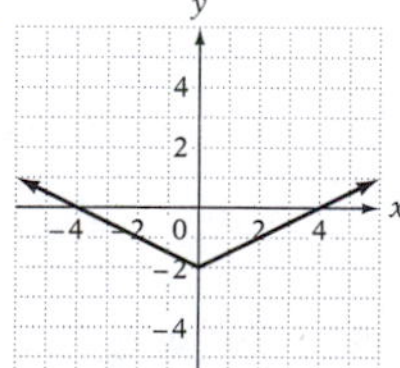

Domain: $\{x|x \in \text{real numbers}\}$
Range: $\{y|y \geq -2\}$ [11.2A]

18.

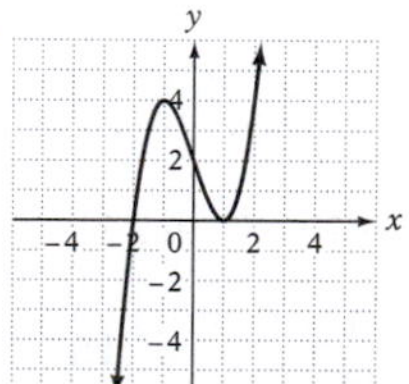

Domain: $\{x|x \in \text{real numbers}\}$
Range: $\{y|y \in \text{real numbers}\}$ [11.2A]

19. No [11.4A] **20.** Dimensions of 50 cm by 50 cm would give a maximum area of 2500 cm². [11.1D]

CUMULATIVE REVIEW EXERCISES

1. $-\frac{23}{4}$ [1.4A] **2.** −5 −4 −3 −2 −1 0 1 2 3 4 5 [1.5C] **3.** 3 [2.2C] **4.** $\{x\,|\,x < -2 \text{ or } x > 3\}$ [2.5B]

5. The set of all real numbers [2.6B] **6.** $-\frac{a^{10}}{12b^4}$ [6.1B] **7.** $2x^3 - 4x^2 - 17x + 4$ [6.3B] **8.** $\left(\frac{1}{2}, 3\right)$ [5.2A]

Copyright © Houghton Mifflin Company. All rights reserved.

9. $xy(x + 3y)(x - 2y)$ [7.2B] **10.** -3 and 8 [7.5A] **11.** $\{x \mid x < -3 \text{ or } x > 5\}$ [10.5A]

12. $\frac{5}{2x - 1}$ [8.2B] **13.** -2 [8.4A] **14.** $-3 - 2i$ [9.4D] **15.**

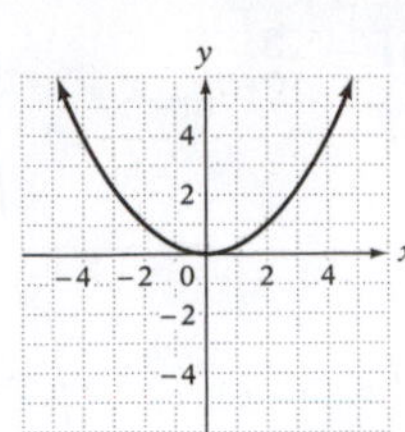

Vertex: (0, 0) [11.1A]
Axis of symmetry: $x = 0$

16.

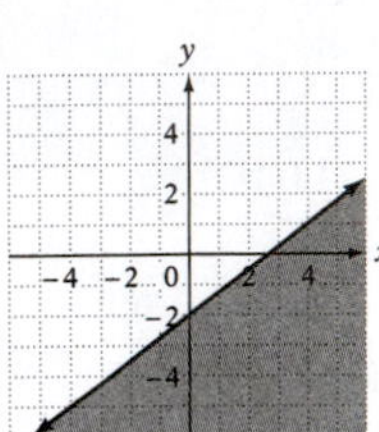

[4.7A] **17.** $y = -2x - 2$ [4.5B] **18.** $y = -\frac{3}{2}x - \frac{7}{2}$ [4.6A]

19. $\frac{1}{2} + \frac{\sqrt{3}}{6}i$ and $\frac{1}{2} - \frac{\sqrt{3}}{6}i$ [10.3A] **20.** 3 [10.4B] **21.** -3 [11.1C] **22.** $\{1, 2, 4, 5\}$ [4.2A] **23.** Yes [4.2A]

24. 2 [9.3A] **25.** 10 [11.3B] **26.** $f^{-1}(x) = -\frac{1}{3}x + 3$ [11.4B] **27.** The cost per pound of the mixture is $3.96. [2.4A] **28.** 25 lb of the 80% copper alloy must be used. [2.4B] **29.** An additional 4.5 oz of insecticide are required. [8.5B] **30.** It would take the larger pipe 4 min to fill the tank. [10.6A] **31.** A force of 40 lb will stretch the string 24 in. [8.8A] **32.** The frequency is 80 vibrations/min. [8.8A]

Answers to Chapter 12 Selected Exercises

PREP TEST

1. $\frac{1}{9}$ [6.1B] **2.** 16 [6.1B] **3.** -3 [6.1B] **4.** 0; 108 [4.2A/6.1B] **5.** -6 [2.2B] **6.** -2 and 8 [10.1A]

7. 6326.60 [1.4A] **8.**

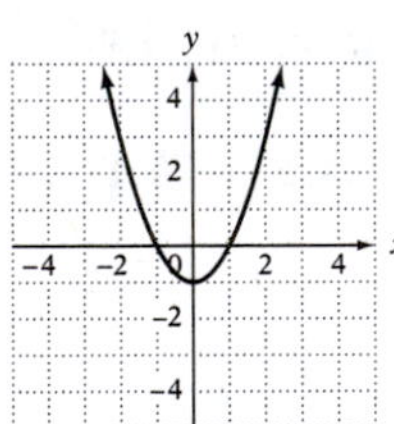

[11.1A]

SECTION 12.1

3. c **5. a.** 9 **b.** 1 **c.** $\frac{1}{9}$ **7. a.** 16 **b.** 4 **c.** $\frac{1}{4}$ **9. a.** 1 **b.** $\frac{1}{8}$ **c.** 16 **11. a.** 7.3891 **b.** 0.3679 **c.** 1.2840 **13. a.** 54.5982 **b.** 1 **c.** 0.1353 **15. a.** 16 **b.** 16 **c.** 1 **17. a.** 0.1353 **b.** 0.1353 **c.** 0.0111 **19.**

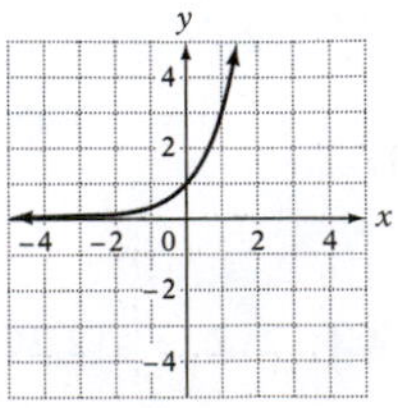

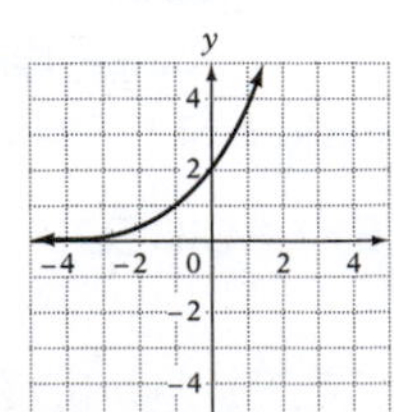

21.

Copyright © Houghton Mifflin Company. All rights reserved.

23. 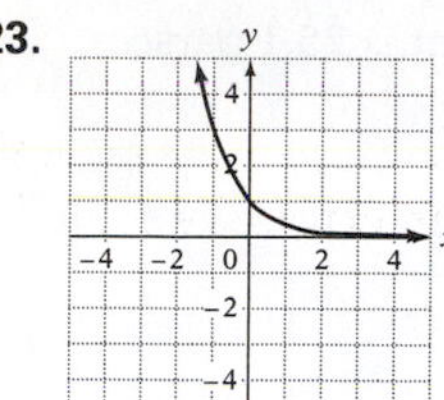

25. 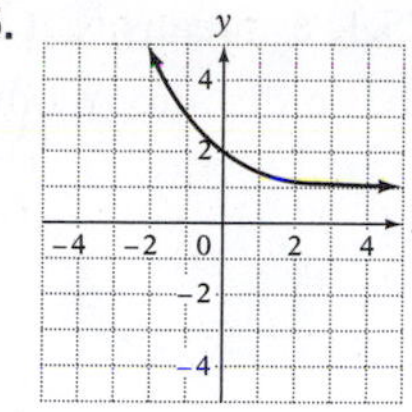

27. 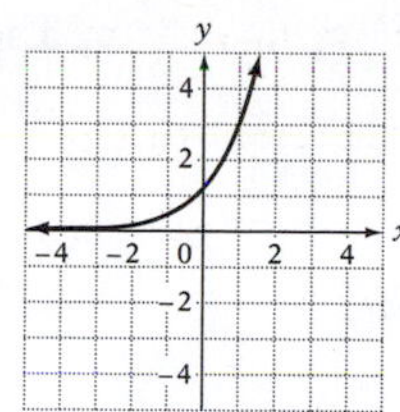

29. 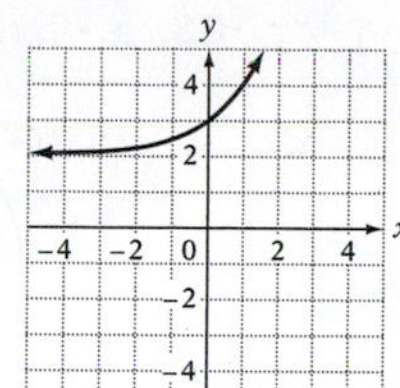

31. **b** and **d** 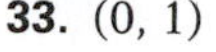 **33.** (0, 1) **35.** No x-intercept; (0, 1)

37. 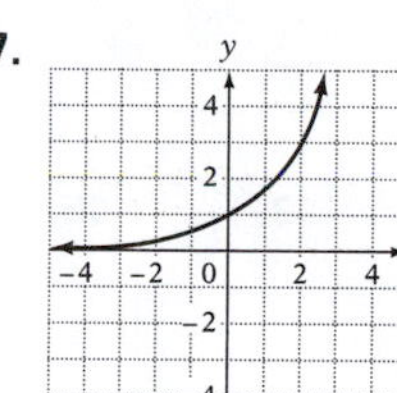

39. 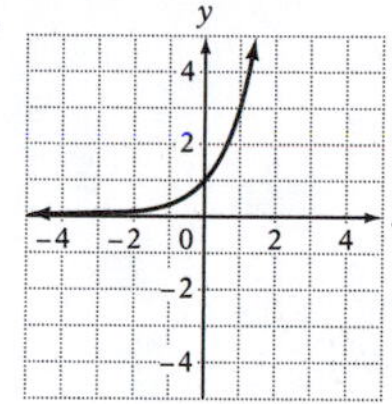

41. a. 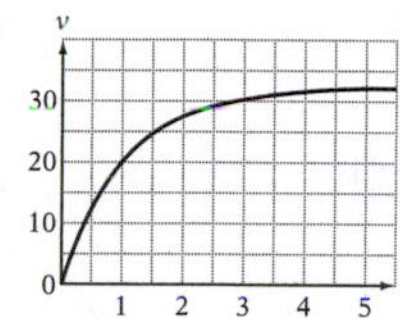

b. The point (2, 27.7) means that after 2 s, the object is falling at a speed of 27.7 ft/s.

SECTION 12.2

3. $\log_5 25 = 2$ **5.** $\log_4 \frac{1}{16} = -2$ **7.** $\log_{10} x = y$ **9.** $\log_a w = x$ **11.** $3^2 = 9$ **13.** $10^{-2} = 0.01$ **15.** $e^y = x$

17. $b^v = u$ **19.** 4 **21.** 7 **23.** 2 **25.** 3 **27.** 0 **29.** 4 **31.** 9 **33.** 64 **35.** $\frac{1}{7}$ **37.** 1

39. 316.23 **41.** 0.02 **43.** 7.39 **45.** 0.61 **49.** $\log_3(x^3y^2)$ **51.** $\ln \frac{x^4}{y^2}$ **53.** $\log_7 x^3$ **55.** $\ln(x^3y^4)$

57. $\log_4(x^2y^2)$ **59.** $\log_3 \frac{x^2z^2}{y}$ **61.** $\ln \frac{x}{y^2z}$ **63.** $\log_6 \sqrt{\frac{x}{y}}$ **65.** $\log_4 \frac{s^2r^2}{t^4}$ **67.** $\ln \frac{x}{y^2z^2}$ **69.** $\log_2 \frac{t^3v^2}{r^2}$

71. $\log_4 \sqrt{\frac{x^3z}{y^2}}$ **73.** $\ln \sqrt{\frac{x}{y^3}}$ **75.** $\log_2 \frac{\sqrt{xz}}{\sqrt[3]{y^2}}$ **77.** $\log_8 x + \log_8 z$ **79.** $5 \log_3 x$ **81.** $\log_b r - \log_b s$

83. $2 \log_3 x + 6 \log_3 y$ **85.** $3 \log_7 u - 4 \log_7 v$ **87.** $2 \log_2 r + 2 \log_2 s$ **89.** $2 \ln x + \ln y + \ln z$

91. $\log_5 x + 2 \log_5 y - 4 \log_5 z$ **93.** $2 \log_8 x - \log_8 y - 2 \log_8 z$ **95.** $\frac{3}{2} \log_4 x + \frac{1}{2} \log_4 y$ **97.** $\frac{3}{2} \log_7 x - \frac{1}{2} \log_7 y$

99. $\log_3 t - \frac{1}{2} \log_3 x$ **101.** 0.8451 **103.** −0.2218 **105.** 1.3863 **107.** 1.0415 **109.** 0.8617 **111.** 2.1133

113. −0.6309 **115.** 0.2727 **117.** 1.6826 **119.** 1.9266 **121.** 0.6752 **123.** 2.6125 **125. a.** False **b.** True **c.** False **d.** True **e.** False **f.** True

SECTION 12.3

5. 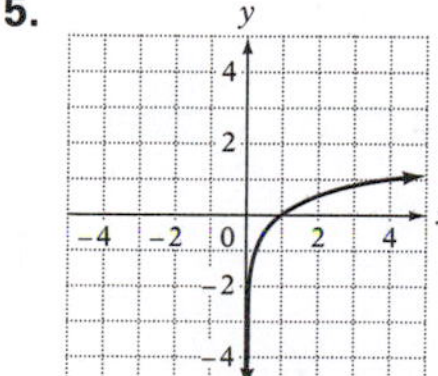

7. 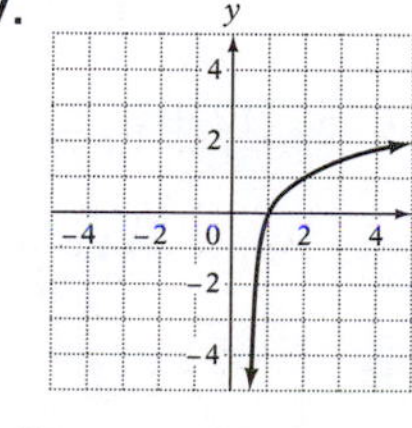

9. 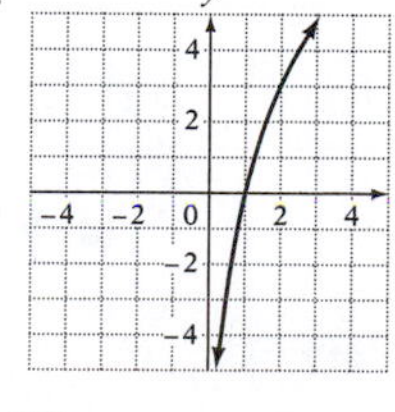

11. 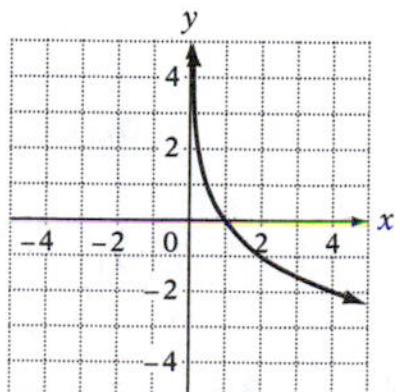

13. 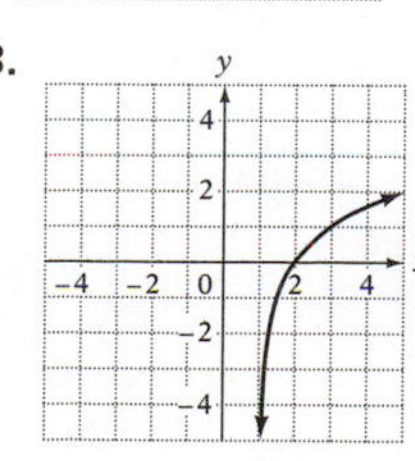

15. 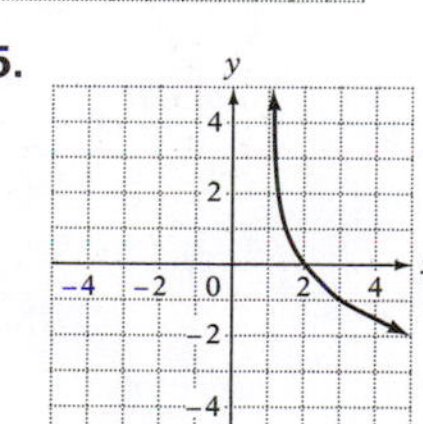

17. 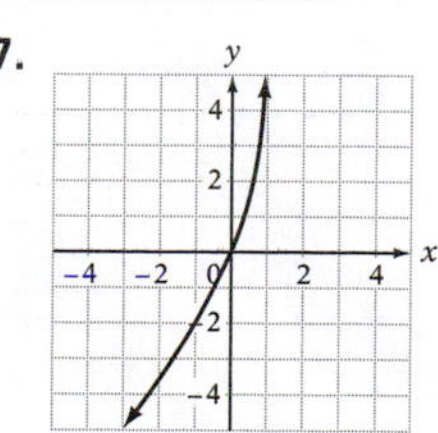

19. 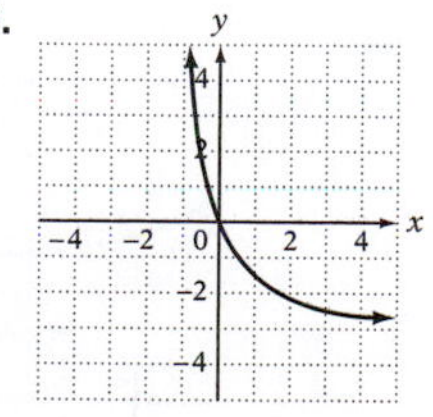

Copyright © Houghton Mifflin Company. All rights reserved.

21.

23. a. 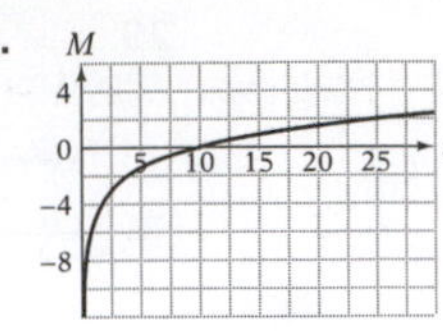

b. The point (25.1, 2) means that a star that is 25.1 parsecs from Earth has a distance modulus of 2.

SECTION 12.4

3. $-\frac{1}{3}$ **5.** −3 **7.** 1 **9.** 6 **11.** $\frac{4}{7}$ **13.** $\frac{9}{8}$ **15.** 1.1133 **17.** 1.0986 **19.** 1.3222 **21.** −2.8074 **23.** 3.5850 **25.** 1.1309 **29.** $\frac{11}{2}$ **31.** −4 and 2 **33.** $\frac{5}{3}$ **35.** $\frac{1}{2}$ **37.** $\frac{9}{2}$ **39.** 17.5327 **41.** 3 **43.** 2 **45.** No solution **47.** 1.7233 **49.** 0.8060 **51.** 5.6910 **53. a.**

b. It will take 2.64 s for the object to travel 100 ft.

SECTION 12.5

1. The value of the investment after 2 years is \$1172. **3.** The investment will be worth \$15,000 in approximately 18 years. **5. a.** After 3 h, the technetium level will be 21.2 mg. **b.** The technetium level will reach 20 mg after 3.5 h. **7.** The half-life is 2.5 years. **9. a.** At 40 km above Earth's surface, the atmospheric pressure is approximately 0.098 newton/cm^2. **b.** On Earth's surface, the atmospheric pressure is approximately 10.13 newtons/cm^2. **c.** The atmospheric pressure decreases as you rise above Earth's surface. **11.** The pH of milk is 6.4. **13.** The depth is 2.5 m. **15.** Normal conversation emits 65 decibels. **17.** The average time of a Major League baseball game increased by 22.5 min. **19.** The thickness of the copper is 0.4 cm. **21.** The Richter scale magnitude of the earthquake was 6.8. **23.** The intensity of the earthquake was 794,328,235I_0. **25.** The magnitude of the earthquake for the seismogram given is 5.3. **27.** The magnitude of the earthquake for the seismogram given is 5.6. **29.** The investment has a value of \$3210.06 after 5 years.

CHAPTER 12 REVIEW EXERCISES

1. 1 [12.1A] **2.** $5^2 = 25$ [12.2A] **3.** [12.1B] **4.** [12.3A]

5. $\frac{2}{5}\log_3 x + \frac{4}{5}\log_3 y$ [12.2B] **6.** $\log_3\frac{x^2}{y^5}$ [12.2B] **7.** −12 [12.4A] **8.** $\frac{1}{4}$ [12.4B] **9.** 1.7251 [12.2C] **10.** 32 [12.2A] **11.** 79 [12.4B] **12.** 1000 [12.2A] **13.** $\log_7\sqrt[3]{xy^4}$ [12.2B] **14.** $\frac{1}{2}(5\log_8 x - 3\log_8 y)$ [12.2B] **15.** $\log_2 32 = 5$ [12.2A] **16.** 0.4278 [12.2C] **17.** −0.535 [12.4A] **18.** $\frac{3}{2}$ [12.1A] **19.** $\frac{2}{3}$ [12.4B] **20.** 2 [12.4B] **21.** [12.1B] **22.** [12.3A] **23.** 2 [12.2A]

Copyright © Houghton Mifflin Company. All rights reserved.

24. $\log_2 \sqrt[3]{\frac{x}{y}}$ [12.2B] **25.** 1 [12.4A] **26.** 1 [12.1A] **27.** $\frac{1}{5}$ [12.2A] **28.** $\log_3 81 = 4$ [12.2A]
29. 5 [12.4B] **30.** $\frac{2}{3}\log_5 x + \frac{1}{3}\log_5 y$ [12.2B] **31.** 2 [12.4A] **32.** $\frac{1}{7}$ [12.1A] **33.** 4 [12.2A]
34. 2 [12.4B] **35.** 2.3219 [12.2C] **36.** 3 [12.4A] **37.** 625 [12.2A] **38.** $\log_b \frac{x^2}{y^7}$ [12.2B]
39. 5 [12.1A] **40.** 3.2091 [12.4A] **41.** The value of the investment in 2 years is $4692. [12.5A]
42. The Richter scale magnitude of the earthquake is 8.3. [12.5A] **43.** The half-life is 27 days. [12.5A]
44. The sound emitted from a busy street corner is 107 decibels. [12.5A]

CHAPTER 12 TEST

1. $f(0) = 1$ [12.1A] **2.** $f(-2) = \frac{1}{3}$ [12.1A] **3.** 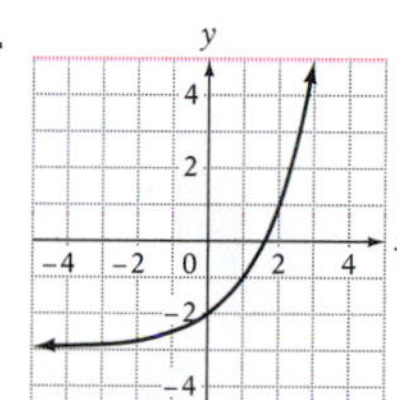 [12.1B] **4.** 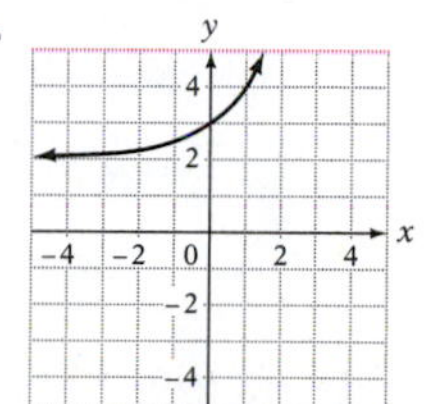 [12.1B]

5. 2 [12.2A] **6.** $\frac{1}{9}$ [12.2A] **7.** 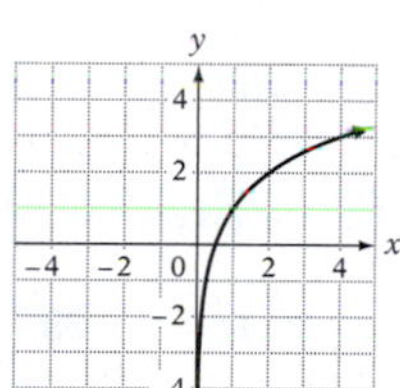 [12.3A] **8.** 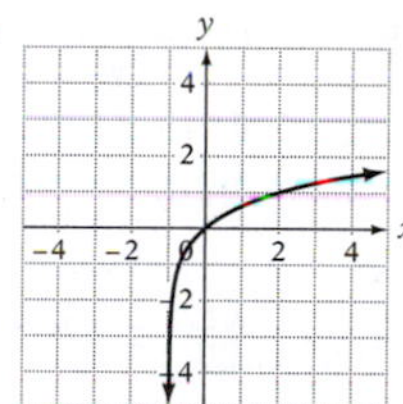 [12.3A]

9. $\frac{1}{2}(\log_6 x + 3\log_6 y)$ [12.2B] **10.** $\log_3 \sqrt{\frac{x}{y}}$ [12.2B] **11.** $\ln x - \frac{1}{2}\ln z$ [12.2B] **12.** $\ln \frac{x^3}{y\sqrt{z}}$ [12.2B]
13. -2 [12.4A] **14.** -3 [12.4A] **15.** 2.5789 [12.4A] **16.** 6 [12.4B] **17.** 3 [12.4B]
18. 1.3652 [12.2C] **19.** 2.6801 [12.2C] **20.** The half-life is 33 h. [12.5A]

CUMULATIVE REVIEW EXERCISES

1. $\frac{8}{7}$ [2.2C] **2.** $y = 2x - 6$ [4.6A] **3.** $(4x^n + 3)(x^n + 1)$ [7.3A/7.3B] **4.** $\frac{x-3}{x+3}$ [8.3A] **5.** $\frac{x\sqrt{y} + y\sqrt{x}}{x - y}$ [9.2D]
6. $2 + \sqrt{10}$ and $2 - \sqrt{10}$ [10.2A] **7.** $3x^2 + 8x - 3 = 0$ [10.1B] **8.** 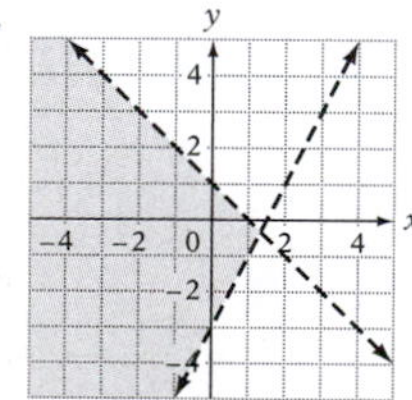 [5.4A]

9. $(0, -1, 2)$ [5.2B] **10.** $-\frac{2x^2 - 17x + 13}{(x-2)(2x-3)}$ [8.2B] **11.** $\{x \mid -5 \le x \le 1\}$ [10.5A] **12.** $\{x \mid 1 \le x \le 4\}$ [2.6B]
13. 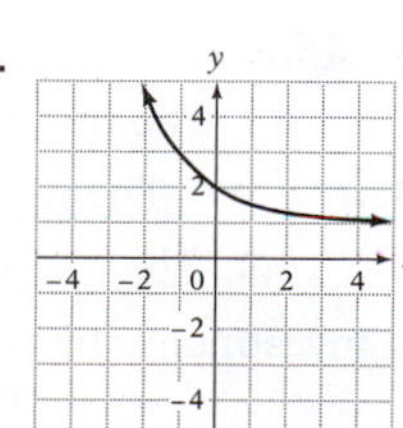 [12.1B] **14.** 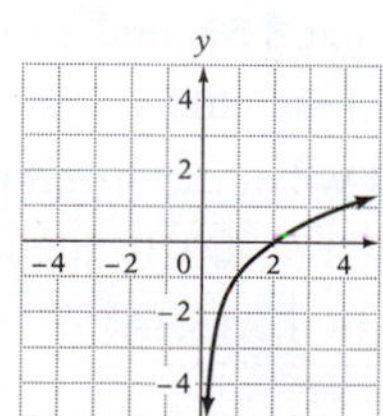 [12.3A] **15.** 4 [12.1A]

16. 125 [12.2A] **17.** $\log_b \frac{x^3}{y^5}$ [12.2B] **18.** 1.7712 [12.2C] **19.** 2 [12.4A] **20.** $\frac{1}{2}$ [12.4B]

Copyright © Houghton Mifflin Company. All rights reserved.

21. The customer can write at most 49 checks. [2.5C] **22.** The cost per pound of the mixture is \$3.10. [2.4A] **23.** The rate of the wind is 25 mph. [8.7B] **24.** The spring will stretch 10.2 in. [8.8A] **25.** The cost of redwood is \$.40 per foot. The cost of fir is \$.25 per foot. [5.3B] **26.** In approximately 10 years, the investment will be worth \$10,000. [12.5A]

Answers to Final Exam

1. -31 [1.3A] **2.** -1 [1.4A] **3.** $-10x + 33$ [1.4D] **4.** 8 [2.2A] **5.** $4, -\frac{2}{3}$ [2.6A] **6.** The volume is 268.1 ft³. [3.3A] **7.**

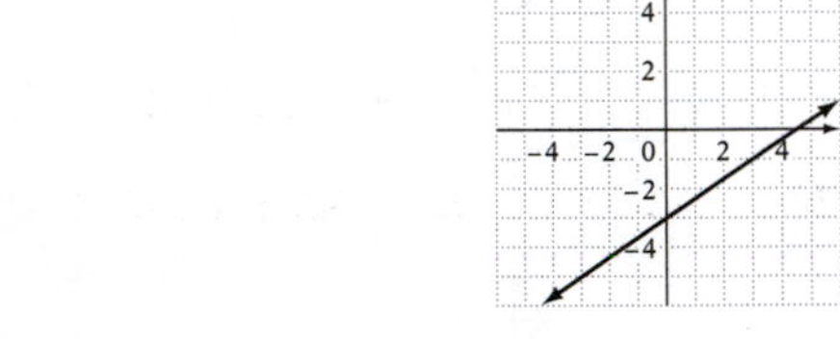

[4.3C] **8.** $y = -3x + 7$ [4.5B] **9.** $y = -\frac{2}{3}x - \frac{1}{3}$ [4.6A]

10. $6a^3 - 5a^2 + 10a$ [6.3A] **11.** $(2 - xy)(4 + 2xy + x^2y^2)$ [7.4B] **12.** $(x - y)(1 + x)(1 - x)$ [7.4D]

13. $x^2 - 2x - 3 - \frac{5}{2x - 3}$ [6.4B] **14.** $\frac{x(x - 1)}{2x - 5}$ [8.1C] **15.** $\frac{-10x}{(x + 2)(x - 3)}$ [8.2B] **16.** $\frac{x + 3}{x + 1}$ [8.3A]

17. $-\frac{7}{4}$ [8.4A] **18.** $d = \frac{a_n - a_1}{n - 1}$ [8.6A] **19.** $\frac{y^4}{162x^3}$ [6.1B] **20.** $\frac{1}{64x^8y^5}$ [9.1A] **21.** $-2x^2y\sqrt{2y}$ [9.2B]

22. $\frac{x^2\sqrt{2y}}{2y^2}$ [9.2D] **23.** $\frac{6}{5} - \frac{3}{5}i$ [9.4D] **24.** $2x^2 - 3x - 2 = 0$ [10.1B] **25.** $\frac{3 + \sqrt{17}}{4}, \frac{3 - \sqrt{17}}{4}$ [10.3A]

26. $-8, 27$ [10.4A] **27.** [11.2A] **28.** [4.3A]

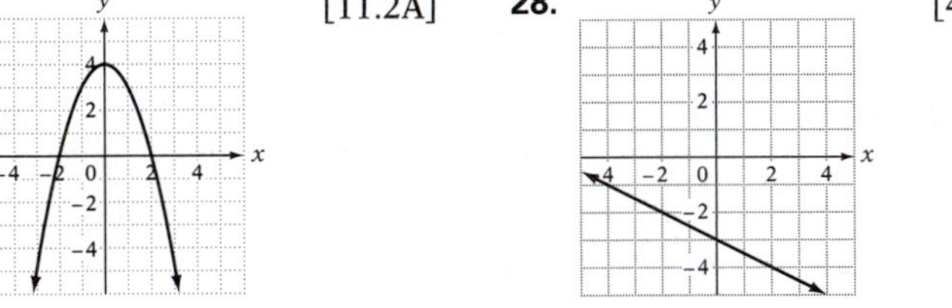

29. $-2, \frac{3}{2}$ [10.4C] **30.** $f^{-1}(x) = \frac{3}{2}x + 6$ [11.4B] **31.** $(3, 4)$ [5.2A] **32.** $7x^3$ [9.1C]

33. $\left\{x \mid x > \frac{3}{2}\right\}$ [2.5B] **34.** $\{x \mid -4 < x < -1\}$ [2.6B]

35. [4.7A] **36.** [12.1B] **37.** [12.3A]

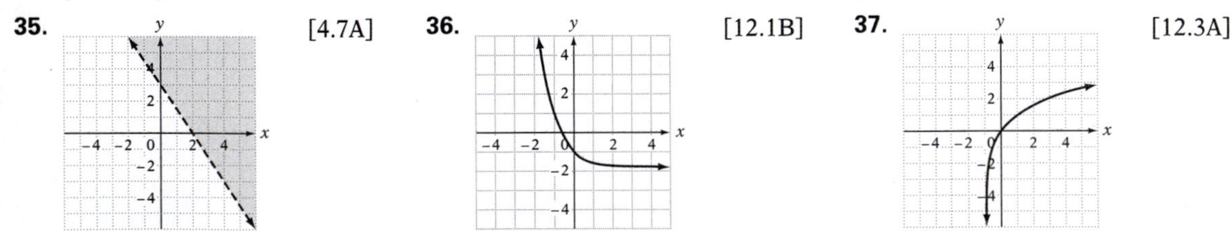

38. $\log_2 \frac{a^2}{b^2}$ [12.2B] **39.** 6 [12.4B] **40.** The range of scores is $69 \le x \le 100$. [2.5C]

41. The cyclist rode 40 mi. [2.4C] **42.** There is \$8000 invested at 8.5% and \$4000 invested at 6.4%. [5.1C] **43.** The length is 20 ft. The width is 7 ft. [10.6A] **44.** An additional 200 shares are needed. [8.5B] **45.** The rate of the plane was 420 mph. [8.7B] **46.** The object has fallen 88 ft when the speed reaches 75 ft/s. [9.3B] **47.** The rate of the plane for the first 360 mi was 120 mph. [8.7B] **48.** The intensity is 200 foot-candles. [8.8A] **49.** The rate of the boat in calm water is 12.5 mph. The rate of the current is 2.5 mph. [5.3A] **50.** The value of the investment after 2 years is \$4785.65. [12.5A]

Copyright © Houghton Mifflin Company. All rights reserved.

Answers to Chapter R Selected Exercises

SECTION R.1

1. 10 **3.** 21 **5.** -2 **7.** 5 **9.** -3 **11.** -1 **13.** 1 **15.** 18 **17.** -6 **19.** 2 **21.** -40 **23.** 2.01
25. $8x$ **27.** $3b$ **29.** $5a$ **31.** $11x$ **33.** $15x - 2y$ **35.** $9a$ **37.** $22y^2$ **39.** $\frac{1}{2}x$ **41.** $-20x$ **43.** $-24a$
45. x **47.** $7x$ **49.** $3y$ **51.** $-3a - 15$ **53.** $-16x - 48$ **55.** $-10y^2 + 5$ **57.** $18x^2 - 12xy - 6y^2$
59. $-8y - 7$ **61.** $5x - 75$ **63.** $-a + 56$

SECTION R.2

1. -12 **3.** -1 **5.** -8 **7.** -20 **9.** 8 **11.** -1 **13.** 1 **15.** $\frac{3}{4}$ **17.** 3 **19.** 3 **21.** 2 **23.** -1
25. $\frac{1}{2}$ **27.** 5 **29.** $-\frac{1}{9}$ **31.** $\{x \mid x > 3\}$ **33.** $\{n \mid n \geq 2\}$ **35.** $\{x \mid x \leq -3\}$ **37.** $\{n \mid n > 0\}$ **39.** $\{x \mid x > 4\}$
41. $\{x \mid x > -2\}$ **43.** $\{x \mid x > 3\}$ **45.** $\{x \mid x \geq 2\}$ **47.** $\{x \mid x < 2\}$ **49.** $\{x \mid x > 3\}$ **51.** $\{x \mid x \geq 2\}$ **53.** $\{x \mid x < 4\}$
55. $\left\{x \mid x > \frac{14}{11}\right\}$ **57.** $\left\{x \mid x \leq \frac{3}{8}\right\}$ **59.** $\left\{x \mid x \geq -\frac{5}{4}\right\}$

SECTION R.3

1.
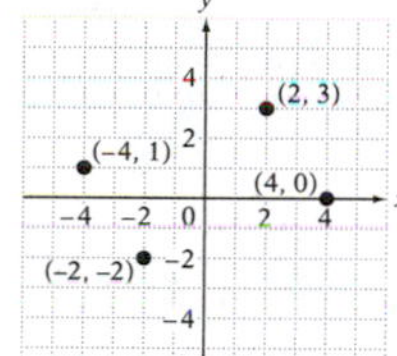

3.
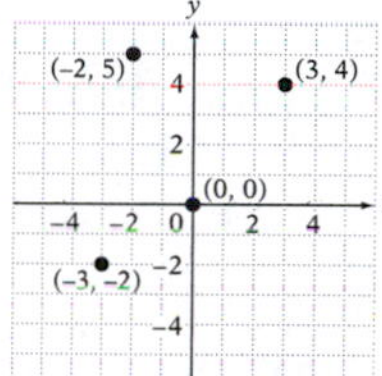

5.
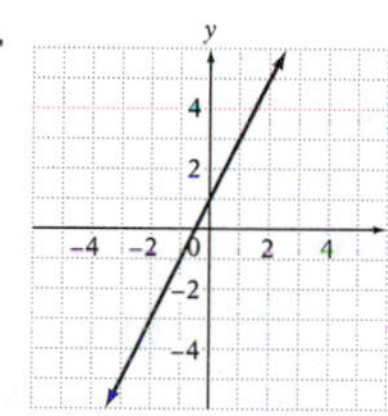

7.
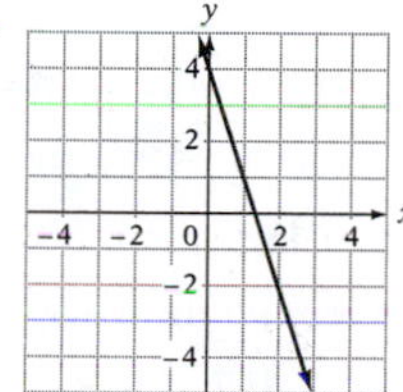

9.
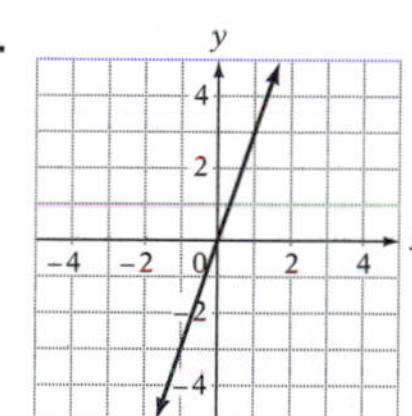

11.
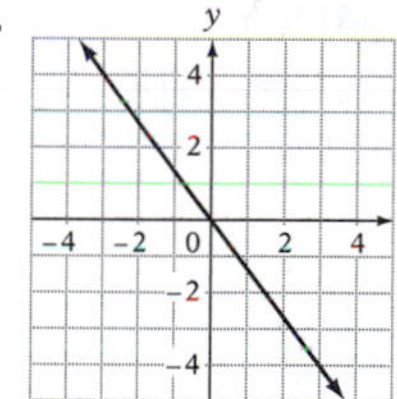

13.
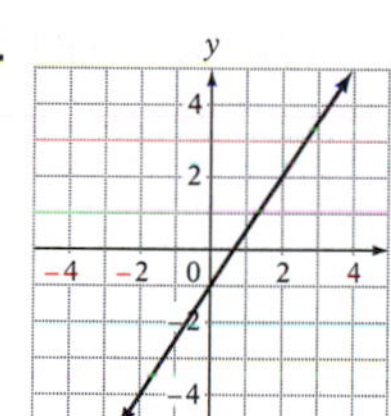

15.
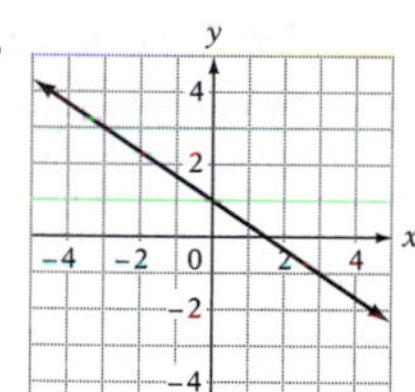

17.
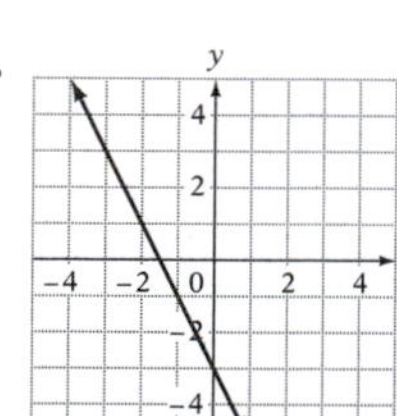

19.
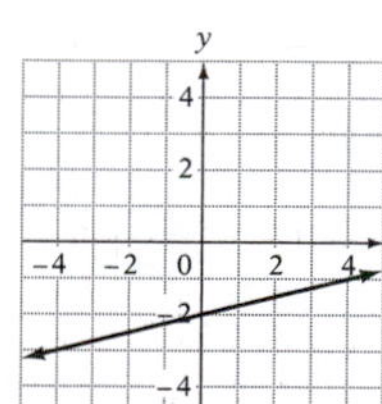

21.
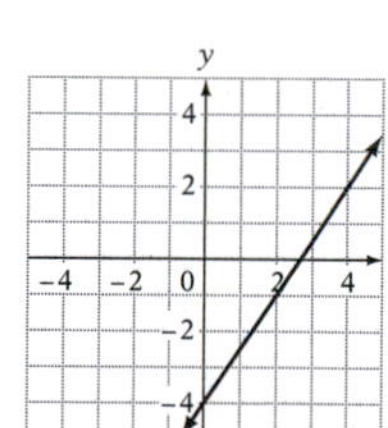

23. 3 **25.** $\frac{1}{4}$ **27.** $-\frac{3}{2}$

29. Undefined **31.** 0 **33.** 1 **35.** $m = \frac{5}{2}, b = (0, -4)$ **37.** $m = 1, b = (0, 0)$

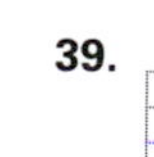

39.
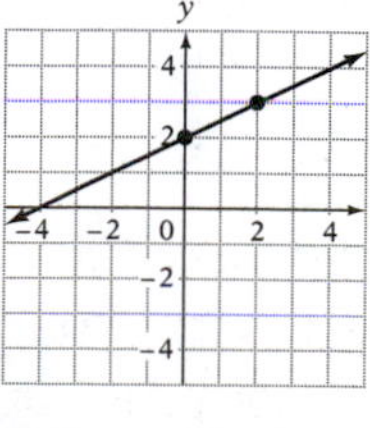

41.
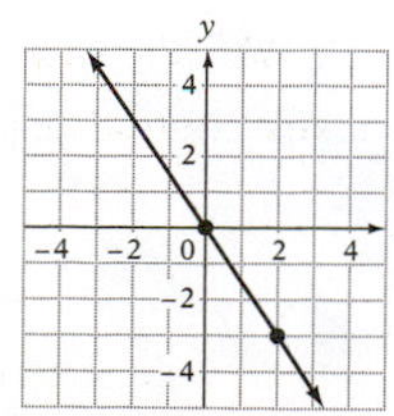

43.
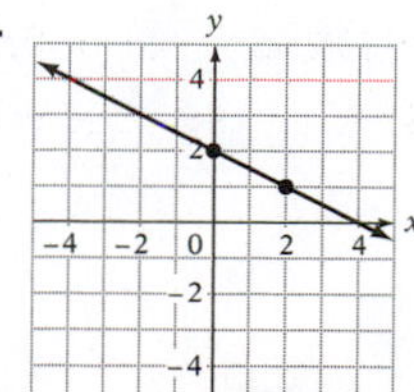

45.
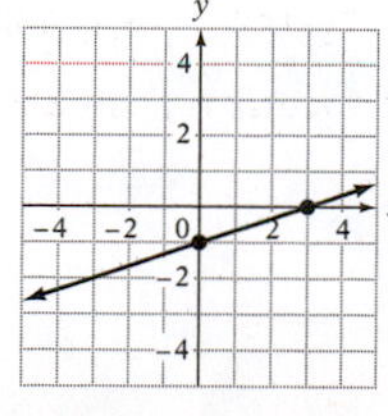

47. 13; $(-4, 13)$ **49.** 14; $(-1$

Copyright © Houghton Mifflin Company. All rights reserved.

51. $-1; (2, -1)$ **53.** $-15; (-2, -15)$ **55.** $-2; (-1, -2)$ **57.** $y = -3x - 1$ **59.** $y = -2x + 3$ **61.** $y = -\frac{3}{5}x$

63. $y = -\frac{2}{3}x - 4$

SECTION R.4

1. z^8 **3.** x^{15} **5.** $x^{12}y^{18}$ **7.** a^6 **9.** $-m^9n^3$ **11.** $-8a^9b^3c^6$ **13.** mn^2 **15.** $-\frac{2a}{3}$ **17.** $-27m^2n^4p^3$

19. $54n^{14}$ **21.** $\frac{7xz}{8y^3}$ **23.** $-8x^{13}y^{14}$ **25.** $\frac{4}{x^7}$ **27.** $\frac{1}{d^{10}}$ **29.** $\frac{1}{x^5}$ **31.** $\frac{x^2}{3}$ **33.** $\frac{x^6}{y^{12}}$ **35.** $\frac{9}{x^2y^4}$ **37.** $\frac{2}{x^4}$

39. $\frac{1}{2x^3}$ **41.** $\frac{1}{2x^2y^6}$ **43.** $\frac{1}{x^6}$ **45.** $\frac{y^8}{x^4}$ **47.** $\frac{b^{10}}{4a^{10}}$ **49.** $7b^2 + b - 4$ **51.** $3a^2 - 3a + 17$ **53.** $-7x - 7$

55. $-2x^3 + x^2 + 2$ **57.** $x^3 + 2x^2 - 6x - 6$ **59.** $5x^3 + 10x^2 - x - 4$ **61.** $y^3 - y^2 + 6y - 6$

63. $4y^3 - 2y^2 + 2y - 4$ **65.** $12b^4 - 48b^3 - 24b$ **67.** $9b^5 - 9b^3 + 24b$ **69.** $-2x^4y + 6x^3y^2 - 4x^2y^3$

71. $x^3 + 4x^2 + 5x + 2$ **73.** $a^3 - 6a^2 + 13a - 12$ **75.** $-2b^3 + 7b^2 + 19b - 20$ **77.** $x^4 - 4x^3 - 3x^2 + 14x - 8$

79. $y^4 + 4y^3 + y^2 - 5y + 2$ **81.** $a^2 + a - 12$ **83.** $y^2 - 10y + 21$ **85.** $2x^2 + 15x + 7$ **87.** $3x^2 + 11x - 4$

89. $4x^2 - 31x + 21$ **91.** $3y^2 - 2y - 16$ **93.** $21a^2 - 83a + 80$ **95.** $2x^2 + 3xy + y^2$ **97.** $3x^2 - 10xy + 8y^2$

99. $10a^2 + 14ab - 12b^2$ **101.** $16x^2 - 49$ **103.** $x + 2$ **105.** $2y - 7$ **107.** $x - 2 + \frac{8}{x + 2}$

109. $3y - 5 + \frac{20}{2y + 4}$ **111.** $b - 5 - \frac{24}{b - 3}$ **113.** $3x + 17 + \frac{64}{x - 4}$ **115.** $5y + 3 + \frac{1}{2y + 3}$ **117.** $x^2 - 5x + 2$

119. $x^2 + 5$ **121.** $y(12y - 5)$ **123.** $5xyz(2xz + 3y^2)$ **125.** $5(x^2 - 3x + 7)$ **127.** $3y^2(y^2 - 3y - 2)$

129. $x^2y^2(x^2y^2 - 3xy + 6)$ **131.** $8x^2y(2 - xy^3 - 6y)$ **133.** $(x + 2)(x - 1)$ **135.** $(a + 4)(a - 3)$

137. $(a - 1)(a - 2)$ **139.** $(b + 8)(b - 1)$ **141.** $(z + 5)(z - 9)$ **143.** $(z - 5)(z - 9)$ **145.** $(b + 4)(b + 5)$

147. Nonfactorable over the integers **149.** $(x - 7)(x - 8)$ **151.** $(y + 3)(2y + 1)$ **153.** $(a - 1)(3a - 1)$

155. $(x - 3)(2x + 1)$ **157.** $(2t + 1)(5t + 3)$ **159.** $(2z - 1)(5z + 4)$ **161.** Nonfactorable over the integers

163. $(t + 2)(2t - 5)$ **165.** $(3y + 1)(4y + 5)$ **167.** $(a - 5)(11a + 1)$ **169.** $(2b - 3)(3b - 2)$

171. $3(x + 2)(x + 3)$ **173.** $a(b + 8)(b - 1)$ **175.** $2y^2(y + 3)(y - 16)$ **177.** $x(x - 5)(2x - 1)$

179. $5(t + 2)(2t - 5)$ **181.** $p(2p + 1)(3p + 1)$

Copyright © Houghton Mifflin Company. All rights reserved.

Glossary

abscissa The first number of an ordered pair; it measures a horizontal distance and is also called the first coordinate of an ordered pair. [4.1]

absolute value of a number The distance of the number from zero on the number line. [1.1, 2.6]

absolute value inequality An inequality that contains the absolute-value symbol. [2.6]

absolute value equation An equation that contains the absolute-value symbol. [2.6]

acute angle An angle whose measure is between 0° and 90°. [3.1]

acute triangle A triangle that has three acute angles. [3.2]

addend In addition, a number being added. [1.1]

addition method An algebraic method of finding an exact solution of a system of linear equations. [5.2]

additive inverse of a polynomial The polynomial with the sign of every term changed. [6.2]

additive inverses Numbers that are the same distance from zero on the number line but lie on different sides of zero; also called opposites. [1.1, 1.4]

adjacent angles Two angles that share a common side. [3.1]

alternate exterior angles Two nonadjacent angles that are on opposite sides of the transversal and lie outside the parallel lines. [3.1]

alternate interior angles Two nonadjacent angles that are on opposite sides of the transversal and between the parallel lines. [3.1]

analytic geometry Geometry in which a coordinate system is used to study relationships between variables. [4.1]

angle Figure formed when two rays start from the same point. [3.1]

area A measure of the amount of surface in a region. [3.2]

asymptotes The two straight lines that a hyperbola "approaches." [11.5 – online]

axes The two number lines that form a rectangular coordinate system; also called coordinate axes. [4.1]

axis of symmetry of a parabola A line of symmetry that passes through the vertex of the parabola and is parallel to the y-axis for an equation of the form $y = ax^2 + bx + c$ or parallel to the x-axis for an equation of the form $x = ay^2 + by + c$. [11.2, 11.5 – online]

base In an exponential expression, the number that is taken as a factor as many times as indicated by the exponent. [1.2]

basic percent equation Percent times base equals amount. [2.1]

binomial A polynomial of two terms. [6.2]

center of a circle The central point that is equidistant from all the points that make up a circle. [3.2, 11.5 – online]

center of an ellipse The intersection of the two axes of symmetry of the ellipse. [11.5 – online]

circle Plane figure in which all points are the same distance from its center. [3.2, 11.5 – online]

circumference The perimeter of a circle. [3.2]

clearing denominators Removing denominators from an equation that contains fractions by multiplying each side of the equation by the LCM of the denominators. [2.2]

coefficient The number part of a variable term. [1.4]

cofactor of an element of a matrix $(-1)^{i+j}$ times the minor of that element, where i is the row number of the element and j is its column number. [5.5 – online]

combined variation A variation in which two or more types of variation occur at the same time. [8.8]

combining like terms Using the Distributive Property to add the coefficients of like variable terms; adding like terms of a variable expression. [1.4]

common logarithms Logarithms to the base 10. [12.2]

common monomial factor A monomial factor that is a factor of the terms in a polynomial. [7.1]

complementary angles Two angles whose measures have the sum 90°. [3.1]

completing the square Adding to a binomial the constant term that makes it a perfect-square trinomial. [10.2]

complex fraction A fraction whose numerator or denominator contains one or more fractions. [8.3]

complex number A number of the form $a + bi$, where a and b are real numbers and $i = \sqrt{-1}$. [9.4]

composition of two functions The operation on two functions f and g denoted by $f \circ g$. The value of the composition of f and g is given by $(f \circ g)(x) = f[g(x)]$. [11.3]

compound inequality Two inequalities joined with a connective word such as *and* or *or*. [2.5]

compound interest Interest that is computed not only on the original principal but also on the interest already earned. [12.5]

conic section A curve that can be constructed from the intersection of a plane and a right circular cone. The four conic sections are the parabola, hyperbola, ellipse, and circle. [11.5 – online]

conjugates Binomial expressions that differ only in the sign of a term. The expressions $a + b$ and $a - b$ are conjugates. [9.2]

consecutive even integers Even integers that follow one another in order. [2.3]

consecutive integers Integers that follow one another in order. [2.3]

consecutive odd integers Odd integers that follow one another in order. [2.3]

constant function A function given by $f(x) = b$, where b is a constant. Its graph is a horizontal line passing through $(0, b)$. [4.3]

constant of proportionality k in a variation equation; also called the constant of variation. [8.8]

constant of variation k in a variation equation; also called the constant of proportionality. [8.8]

Copyright © Houghton Mifflin Company. All rights reserved.

constant term A term that includes no variable part; also called a constant. [1.4]

coordinate axes The two number lines that form a rectangular coordinate system; also called axes. [4.1]

coordinates of a point The numbers in the ordered pair that is associated with the point. [4.1]

corresponding angles Two angles that are on the same side of the transversal and are both acute angles or are both obtuse angles. [3.1]

cube A rectangular solid in which all six faces are squares. [3.3]

cube root of a perfect cube One of the three equal factors of the perfect cube. [7.4]

cubic function A third-degree polynomial function. [6.2]

decimal notation Notation in which a number consists of a whole-number part, a decimal point, and a decimal part. [1.2]

degree A unit used to measure angles. [3.1]

degree of a monomial The sum of the exponents of the variables. [6.1]

degree of a polynomial The greatest of the degrees of any of the polynomial's terms. [6.2]

dependent system of equations A system of equations whose graphs coincide. [5.1]

dependent variable In a function, the variable whose value depends on the value of another variable known as the independent variable. [4.2]

descending order The terms of a polynomial in one variable are arranged in descending order when the exponents of the variable decrease from left to right. [6.2]

determinant A number associated with a square matrix. [5.5 – online]

diameter of a circle A line segment with endpoints on the circle and going through the center. [3.2]

diameter of a sphere A line segment with endpoints on the sphere and going through the center. [3.3]

difference of two perfect squares A polynomial in the form of $a^2 - b^2$. [7.4]

difference of two perfect cubes A polynomial in the form $a^3 - b^3$. [7.4]

direct variation A special function that can be expressed as the equation $y = kx$, where k is a constant called the constant of variation or the constant of proportionality. [8.8]

discriminant For an equation of the form $ax^2 + bx + c = 0$, the quantity $b^2 - 4ac$ is called the discriminant. [10.3]

domain The set of the first coordinates of all the ordered pairs of a relation. [4.2]

double root When a quadratic equation has two solutions that are the same number, the solution is called a double root of the equation. [10.1]

element of a matrix A number in a matrix. [5.5 – online]

elements of a set The objects in the set. [1.1, 1.5]

ellipse An oval shape that is one of the conic sections. [11.5 – online]

empty set The set that contains no elements; also called the null set. [1.5]

equation A statement of the equality of two mathematical expressions. [2.1]

equilateral triangle A triangle in which all three sides are of equal length. [3.2]

equivalent equations Equations that have the same solution. [2.1]

evaluating a function Replacing x in $f(x)$ with some value and then simplifying the numerical expression that results. [4.2]

evaluating a variable expression Replacing each variable by its value and then simplifying the resulting numerical expression. [1.4]

even integer An integer that is divisible by 2. [2.3]

expanding by cofactors A technique for finding the value of a 3×3 or larger determinant. [5.5 – online]

exponent In an exponential expression, the raised number that indicates how many times the factor, or base, occurs in the multiplication. [1.2]

exponential equation An equation in which the variable occurs in the exponent. [12.4]

exponential form The expression 2^6 is in exponential form. Compare *factored form.* [1.2]

exponential function The exponential function with base b is defined by $f(x) = b^x$, where b is a positive real number not equal to one. [12.1]

exterior angle An angle adjacent to an interior angle of a triangle. [3.1]

extraneous solution When each side of an equation is raised to an even power, the resulting equation may have a solution that is not a solution of the original equation. Such a solution is called an extraneous solution. [9.3]

factor In multiplication, a number being multiplied. [1.1]

factored form The multiplication $2 \cdot 2 \cdot 2 \cdot 2 \cdot 2 \cdot 2$ is in factored form. Compare *exponential form.* [1.2]

factoring a polynomial Writing the polynomial as a product of other polynomials. [7.1]

factoring a quadratic trinomial Expressing the trinomial as the product of two binomials. [7.2]

FOIL A method of finding the product of two binomials. The letters stand for First, Outer, Inner, and Last. [6.3]

formula A literal equation that states a rule about measurement. [8.6]

function A relation in which no two ordered pairs that have the same first coordinate have different second coordinates. [4.2]

functional notation A function designated by $f(x)$, which is the value of the function at x. [4.2]

graph of an integer A heavy dot directly above the number on the number line. [1.1]

graph of a function A graph of the ordered pairs that belong to the function. [4.3]

graph of an ordered pair The dot drawn at the coordinates of the point in the plane. [4.1]

Copyright © Houghton Mifflin Company. All rights reserved.

graphing a point in the plane Placing a dot at the location given by the ordered pair; also called plotting a point in the plane. [4.1]

greater than A number that lies to the right of another number on the number line is said to be greater than that number. [1.1]

greatest common factor (GCF) The greatest common factor of two or more integers is the greatest integer that is a factor of all the integers. [7.1]

greatest common factor (GCF) of two or more monomials The greatest common factor of two or more monomials is the product of the GCF of the coefficients and the common variable factors. [7.1]

grouping symbols Parentheses ( ), brackets [ ], braces { }, the absolute value symbol, and the fraction bar. [1.3]

half-plane The solution set of a linear inequality in two variables. [4.7]

horizontal-line test A graph of a function represents the graph of a one-to-one function if any horizontal line intersects the graph at no more than one point. [11.4]

hyperbola A conic section formed by the intersection of a cone and a plane perpendicular to the base of the cone. [11.5 – online]

hypotenuse In a right triangle, the side opposite the 90° angle. [9.3]

imaginary number A number of the form ai, where a is a real number and $i = \sqrt{-1}$. [9.4]

imaginary part of a complex number For the complex number $a + bi$, b is the imaginary part. [9.4]

inconsistent system of equations A system of equations that has no solution. [5.1]

independent system of equations A system of equations whose graphs intersect at only one point. [5.1]

independent variable In a function, the variable that varies independently and whose value determines the value of the dependent variable. [4.2]

index In the expression $\sqrt[n]{a}$, n is the index of the radical. [9.1]

inequality An expression that contains the symbol $>$, $<$, $\geq$ (is greater than or equal to), or $\leq$ (is less than or equal to). [1.5]

integers The numbers . . . , −3, −2, −1, 0, 1, 2, 3, ... [1.1]

interior angle of a triangle Angle within the region enclosed by a triangle. [3.1]

intersecting lines Lines that cross at a point in the plane. [3.1]

intersection of two sets The set that contains all elements that are common to both of the sets. [1.5]

inverse of a function The set of ordered pairs formed by reversing the coordinates of each ordered pair of the function. [11.4]

inverse variation A function that can be expressed as the equation $y = \frac{k}{x}$, where k is a constant. [8.8]

irrational number The decimal representation of an irrational number never terminates or repeats and can only be approximated. [1.2, 9.2]

isosceles triangle A triangle that has two sides of equal length; the angles opposite the equal sides are of equal measure. [3.2]

joint variation A variation in which a variable varies directly as the product of two or more variables. A joint variation can be expressed as the equation $z = kxy$, where k is a constant. [8.8]

leading coefficient In a polynomial, the coefficient of the variable with the largest exponent. [6.2]

least common denominator The smallest number that is a multiple of each denominator in question. [1.2]

least common multiple (LCM) The LCM of two or more numbers is the smallest number that is a multiple of each of those numbers. [1.2]

least common multiple (LCM) of two or more polynomials The simplest polynomial of least degree that contains the factors of each polynomial. [8.2]

leg In a right triangle, one of the two sides that are not opposite the 90° angle. [9.3]

less than A number that lies to the left of another number on the number line is said to be less than that number. [1.1]

like terms Terms of a variable expression that have the same variable part. Having no variable part, constant terms are like terms. [1.4]

line Having no width, it extends indefinitely in two directions in a plane. [3.1]

linear equation in three variables An equation of the form $Ax + By + Cz = D$ where A, B, and C are coefficients of the variables and D is a constant. [5.2]

linear equation in two variables An equation of the form $y = mx + b$, where m is the coefficient of x and b is a constant; also called a linear function. [4.3]

linear function A function that can be expressed in the form $y = mx + b$. Its graph is a straight line. [4.3, 6.2]

linear inequality in two variables An inequality of the form $y > mx + b$ or $Ax + By > C$. The symbol $>$ could be replaced by $\geq$, $<$, or $\leq$. [4.7]

line segment Part of a line that has two endpoints. [3.1]

literal equation An equation that contains more than one variable. [8.6]

logarithm For b greater than zero and not equal to 1, the statement $y = \log_b x$ (the logarithm of x to the base b) is equivalent to $x = b^y$. [12.2]

lower limit In a tolerance, the lowest acceptable value. [2.6]

main fraction bar The fraction bar that is placed between the numerator and denominator of a complex fraction. [8.3]

matrix A rectangular array of numbers. [5.5 – online]

maximum value of a function The greatest value that the function can take on. [11.1]

minimum value of a function The least value that the function can take on. [11.1]

minor of an element The minor of an element in a 3×3 determinant is the 2×2 determinant obtained by eliminating the row and column that contain that element. [5.5 – online]

Copyright © Houghton Mifflin Company. All rights reserved.

monomial A number, a variable, or a product of a number and variables; a polynomial of one term. [6.1, 6.2]

multiplication The process of finding the product of two numbers. [1.1]

multiplicative inverse The multiplicative inverse of a nonzero real number a is $\frac{1}{a}$; also called the reciprocal. [1.4]

natural exponential function The function defined by $f(x) = e^x$, where $e \approx 2.71828$. [12.1]

natural logarithm When e (the base of the natural exponential function) is used as the base of a logarithm, the logarithm is referred to as the natural logarithm and is abbreviated $\ln x$. [12.2]

natural numbers The numbers 1, 2, 3. . . . ; also called the positive integers. [1.1]

negative integers The numbers. . . , −3, −2, −1. [1.1]

negative reciprocal The negative reciprocal of a nonzero real number a is $-\frac{1}{a}$. [11.1]

negative slope The slope of a line that slants downward to the right. [4.4]

nonfactorable over the integers A polynomial is nonfactorable over the integers if it does not factor using only integers. [7.2]

n*th root of *a A number b such that $b^n = a$. The nth root of a can be written $a^{1/n}$ or $\sqrt[n]{a}$. [9.1]

null set The set that contains no elements; also called the empty set. [1.5]

numerical coefficient The number part of a variable term. When the numerical coefficient is 1 or −1, the 1 is usually not written. [1.4]

obtuse angle An angle whose measure is between 90° and 180°. [3.1]

obtuse triangle A triangle that has one obtuse angle. [3.2]

odd integer An integer that is not divisible by 2. [2.3]

one-to-one function In a one-to-one function, given any y, there is only one x that can be paired with the given y. [11.4]

opposites Numbers that are the same distance from zero on the number line but lie on different sides of zero; also called additive inverses. [1.1]

order $m \times n$ A matrix of m rows and n columns is of order $m \times n$. [5.5 – online]

Order of Operations Agreement A set of rules that tells us in what order to perform the operations that occur in a numerical expression. [1.3]

ordered pair A pair of numbers expressed in the form (a, b) and used to locate a point in the plane determined by a rectangular coordinate system. [4.1]

ordered triple Three numbers expressed in the form (x, y, z) and used to locate a point in the xyz-coordinate system. [5.2]

ordinate The second number of an ordered pair; it measures a vertical distance and is also called the second coordinate of an ordered pair. [4.1]

origin The point of intersection of the two number lines that form a rectangular coordinate system. [4.1]

parabola The graph of a quadratic function is called a parabola. [11.1]

parallel lines Lines that never meet; the distance between them is always the same. In a rectangular coordinate system, parallel lines have the same slope and thus do not intersect. [3.1, 4.6]

parallelogram Four-sided plane figure with opposite sides parallel. [3.2]

percent Parts of 100. [1.2]

perfect cube The product of the same three factors. [7.4]

perfect square The product of a term and itself. [1.2, 7.4]

perfect-square trinomial The square of a binomial. [7.4]

perimeter The distance around a plane geometric figure. [3.2]

perpendicular lines Intersecting lines that form right angles. The slopes of perpendicular lines are negative reciprocals of each other. [3.1, 4.6]

plane A flat surface that extends indefinitely. [3.1]

plane figure A figure that lies entirely in a plane. [3.1]

plotting a point in the plane Placing a dot at the location given by the ordered pair; also called graphing a point in the plane. [4.1]

point-slope formula The equation $y - y_1 = m(x - x_1)$, where m is the slope of a line and (x_1, y_1) is a point on the line. [4.5]

polygon A closed figure determined by three or more line segments that lie in a plane. [3.2]

polynomial A variable expression in which the terms are monomials. [6.2]

positive integers The numbers 1, 2, 3. . . . ; also called the natural numbers. [1.1]

positive slope The slope of a line that slants upward to the right. [4.4]

prime polynomial A polynomial that is nonfactorable over the integers. [7.2]

principal square root The positive square root of a number. [1.2, 9.1]

product In multiplication, the result of multiplying two numbers. [1.1]

product of the sum and difference of two terms A polynomial that can be expressed in the form $(a + b)(a - b)$. [7.4]

proportion An equation that states the equality of two ratios or rates. [8.4]

Pythagorean Theorem The square of the hypotenuse of a right triangle is equal to the sum of the squares of the two legs. [9.3]

quadrant One of the four regions into which a rectangular coordinate system divides the plane. [4.1]

quadratic equation An equation of the form $ax^2 + bx + c = 0$, where a and b are coefficients, c is a constant, and $a \neq 0$; also called a second-degree equation. [7.5, 10.1]

© Houghton Mifflin Company. All rights reserved.

quadratic equation in standard form A quadratic equation written in descending order and set equal to zero. [7.5]

quadratic formula A general formula, derived by applying the method of completing the square to the standard form of a quadratic equation, used to solve quadratic equations. [10.3]

quadratic function A function that can be expressed by the equation $f(x) = ax^2 + bx + c$, where a is not equal to zero. [6.2, 11.1]

quadratic inequality An inequality that can be written in the form $ax^2 + bx + c < 0$ or $ax^2 + bx + c > 0$, where a is not equal to zero. The symbols $\leq$ and $\geq$ can also be used. [10.5]

quadratic trinomial A trinomial of the form $ax^2 + bx + c$, where a and b are nonzero coefficients and c is a nonzero constant. [7.4]

quadrilateral A four-sided closed figure. [3.2]

quotient In division, the result of dividing the divisor into the dividend. [1.1]

radical equation An equation that contains a variable expression in a radicand. [9.3]

radical sign The symbol $\sqrt{}$, which is used to indicate the positive, or principal, square root of a number. [1.2, 9.1]

radicand In a radical expression, the expression under the radical sign. [1.2, 9.1]

radius of a circle Line segment from the center of the circle to a point on the circle. [3.2, 11.5 – online]

radius of a sphere A line segment going from the center to a point on the sphere. [3.3]

range The set of the second coordinates of all the ordered pairs of a relation. [4.2]

rate The quotient of two quantities that have different units. [8.5]

rate of work That part of a task that is completed in one unit of time. [8.7]

ratio The quotient of two quantities that have the same unit. [8.5]

rational expression A fraction in which the numerator or denominator is a polynomial. [8.1]

rational number A number of the form $\frac{a}{b}$, where a and b are integers and b is not equal to zero. [1.2]

rationalizing the denominator The procedure used to remove a radical from the denominator of a fraction. [9.2]

ray Line that starts at a point and extends indefinitely in one direction. [3.1]

real numbers The rational numbers and the irrational numbers taken together. [1.2]

real part of a complex number For the complex number $a + bi$, a is the real part. [9.4]

reciprocal The reciprocal of a nonzero real number a is $\frac{1}{a}$; also called the multiplicative inverse. [1.2, 1.4]

reciprocal of a rational expression The rational expression with the numerator and denominator interchanged. [8.1]

rectangle A parallelogram that has four right angles. [3.2]

rectangular coordinate system A coordinate system formed by two number lines, one horizontal and one vertical, that intersect at the zero point of each line. [4.1]

rectangular solid A solid in which all six faces are rectangles. [3.3]

regular polygon A polygon in which each side has the same length and each angle has the same measure. [3.2]

relation A set of ordered pairs. [4.2]

repeating decimal A decimal that is formed when dividing the numerator of its fractional counterpart by the denominator results in a decimal part wherein a block of digits repeats infinitely. [1.2]

right angle An angle whose measure is 90°. [3.1]

right triangle A triangle that contains a 90° angle. [3.1]

roster method A method of designating a set by enclosing a list of its elements in braces. [1.5]

scalene triangle A triangle that has no sides of equal length; no two of its angles are of equal measure. [3.2]

scatter diagram A graph of collected data as points in a coordinate system. [4.1]

scientific notation Notation in which a number is expressed as the product of a number between 1 and 10 and a power of 10. [6.1]

second-degree equation An equation of the form $ax^2 + bx + c = 0$, where a and b are coefficients, c is a constant, and $a \neq 0$; also called a quadratic equation. [10.1]

set A collection of objects. [1.1, 1.5]

set-builder notation A method of designating a set that makes use of a variable and a certain property that only elements of that set possess. [1.5]

similar objects Similar objects have the same shape but not necessarily the same size. [8.5]

simplest form of a rational expression A rational expression is in simplest form when the numerator and denominator have no common factors. [8.1]

slope A measure of the slant, or tilt, of a line. The symbol for slope is m. [4.4]

slope-intercept form of a straight line The equation $y = mx + b$, where m is the slope of the line and $(0, b)$ is the y-intercept. [4.4]

solution of a system of equations in three variables An ordered triple that is a solution of each equation of the system. [5.2]

solution of a system of equations in two variables An ordered pair that is a solution of each equation of the system. [5.1]

solution of an equation A number that, when substituted for the variable, results in a true equation. [2.1]

solution of an equation in three variables An ordered triple (x, y, z) whose coordinates make the equation a true statement. [5.2]

Copyright © Houghton Mifflin Company. All rights reserved.

solution of an equation in two variables An ordered pair whose coordinates make the equation a true statement. [4.1]

solution set of a system of inequalities The intersection of the solution sets of the individual inequalities. [5.4]

solution set of an inequality A set of numbers, each element of which, when substituted for the variable, results in a true inequality. [2.5]

solving an equation Finding a solution of the equation. [2.1]

sphere A solid in which all points are the same distance from point O, which is called the center of the sphere. [3.3]

square A rectangle with four equal sides. [3.2]

square of a binomial A polynomial that can be expressed in the form $(a + b)^2$. [6.3]

square root A square root of a positive number x is a number a for which $a^2 = x$. [1.2]

square matrix A matrix that has the same number of rows as columns. [5.5 – online]

square root of a perfect square One of the two equal factors of the perfect square. [7.4]

standard form of a quadratic equation A quadratic equation is in standard form when the polynomial is in descending order and equal to zero. [7.5, 10.1]

substitution method An algebraic method of finding an exact solution of a system of linear equations. [5.1]

straight angle An angle whose measure is 180°. [3.1]

sum In addition, the total of two or more numbers. [1.1]

supplementary angles Two angles whose measures have the sum 180°. [3.1]

synthetic division A shorter method of dividing a polynomial by a binomial of the form $x - a$. This method uses only the coefficients of the variable terms. [6.4]

system of equations Two or more equations considered together. [5.1]

system of inequalities Two or more inequalities considered together. [5.4]

terminating decimal A decimal that is formed when dividing the numerator of its fractional counterpart by the denominator results in a remainder of zero. [1.2]

terms of a variable expression The addends of the expression. [1.4]

tolerance of a component The amount by which it is acceptable for the component to vary from a given measurement. [2.6]

transversal A line that intersects two other lines at two different points. [3.1]

triangle A three-sided closed figure. [3.1]

trinomial A polynomial of three terms. [6.2]

undefined slope The slope of a vertical line is undefined. [4.4]

uniform motion The motion of an object whose speed and direction do not change. [2.1]

union of two sets The set that contains all elements that belong to either of the sets. [1.5]

upper limit In a tolerance, the greatest acceptable value. [2.6]

value of a function The value of the dependent variable for a given value of the independent variable. [4.2]

variable A letter of the alphabet used to stand for a number that is unknown or that can change. [1.1]

variable expression An expression that contains one or more variables. [1.4]

variable part In a variable term, the variable or variables and their exponents. [1.4]

variable term A term composed of a numerical coefficient and a variable part. When the numerical coefficient is 1 or −1, the 1 is usually not written. [1.4]

vertex Point at which the rays that form an angle meet. [3.1]

vertex of a parabola The point on the parabola with the smallest y-coordinate or the largest y-coordinate. [11.1]

vertical angles Two angles that are on opposite sides of the intersection of two lines. [3.1]

vertical-line test A graph defines a function if any vertical line intersects the graph at no more than one point. [11.2]

volume A measure of the amount of space inside a closed surface. [3.3]

***x*-coordinate** The abscissa in an xy-coordinate system. [4.1]

***x*-intercept** The point at which a graph crosses the x-axis. [4.3]

***y*-coordinate** The ordinate in an xy-coordinate system. [4.1]

***y*-intercept** The point at which a graph crosses the y-axis. [4.3]

zero of a function A value of x for which $f(x) = 0$. [11.1]

zero slope The slope of a horizontal line. [4.4]

Copyright © Houghton Mifflin Company. All rights reserved.

Index

Copyright © Houghton Mifflin Company. All rights reserved.

D

E

Copyright © Houghton Mifflin Company. All rights reserved.

F

Copyright © Houghton Mifflin Company. All rights reserved.

Copyright © Houghton Mifflin Company. All rights reserved.

Q

R

© ... Houghton Mifflin Company. All rights reserved.

S

Copyright © Houghton Mifflin Company. All rights reserved.

T

U

V

Index of Applications

(Continued from inside front cover)

Copyright © Houghton Mifflin Company. All rights reserved.

TI-30X IIS

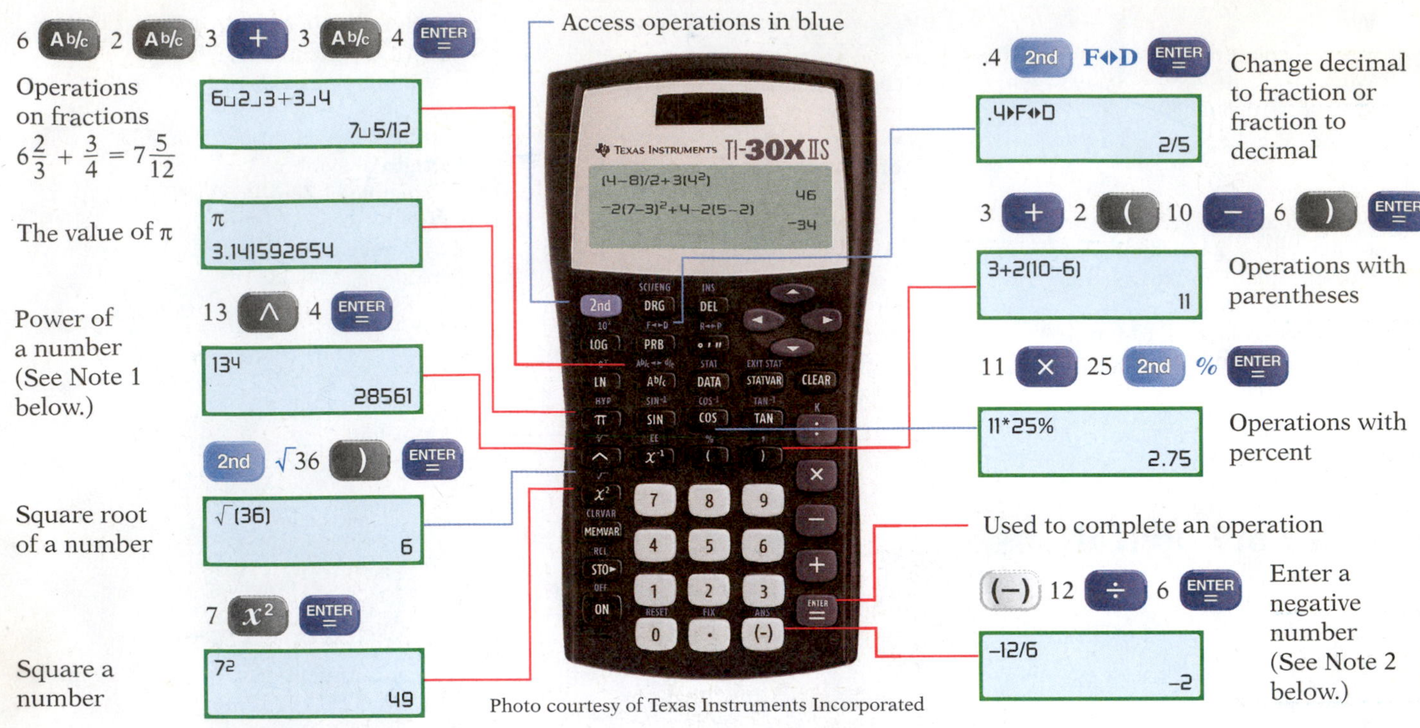

fx-300MS

Access operations in gold

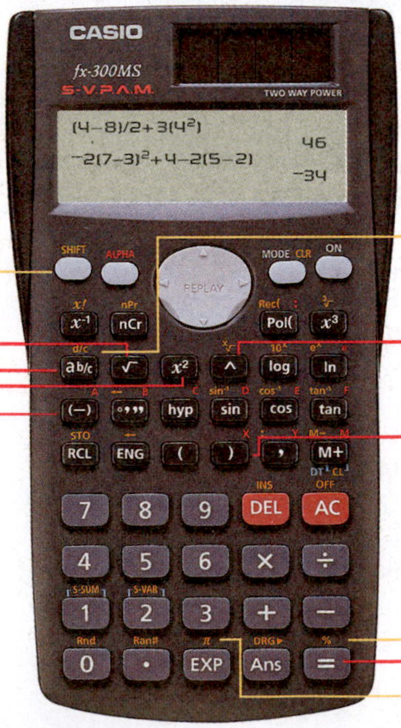

√ 36 =

Square root of a number

√36
6

6 a b/c 2 a b/c 3 + 3 a b/c 4 =

Operations on fractions

$6\frac{2}{3} + \frac{3}{4} = 7\frac{5}{12}$

6⌟2⌟3+3⌟4
7⌟5⌟12

7 x^2 =

Square a number

7²
49

Enter a negative number (See Note 2 below.)

(−) 12 ÷ 6 =

−12÷6
−2

.4 = SHIFT d/c

.4
2⌟5

Change decimal to fraction

13 ∧ 4 =

13^4
28561

Power of a number (See Note 1 below.)

3 + 2 (10 − 6) =

3+2(10−6)
11

Operations with parentheses

11 × 25 SHIFT % =

11x25%
2.75

Operations with percent

Used to complete an operation

SHIFT π =

π
3.141592654

The value of π

Photo courtesy of Casio, Inc.

NOTE 1: Some calculators use the y^x key to calculate a power. For those calculators, enter 13 y^x 4 = to evaluate 13^4.

NOTE 2: Some calculators use the +/− key to enter a negative number. For those calculators, enter 12 +/− ÷ 6 = to calculate $-12 \div 6$.